# HANDBOOK OF

# Media for Clinical and Public Health Microbiology

# HANDBOOK OF

# Media for Clinical and Public Health Microbiology

**Ronald M. Atlas • James W. Snyder**

**CRC Press**
Taylor & Francis Group
Boca Raton  London  New York

CRC Press is an imprint of the
Taylor & Francis Group, an **informa** business

CRC Press
Taylor & Francis Group
6000 Broken Sound Parkway NW, Suite 300
Boca Raton, FL 33487-2742

First issued in paperback 2019

© 2014 by Taylor & Francis Group, LLC
CRC Press is an imprint of Taylor & Francis Group, an Informa business

No claim to original U.S. Government works

ISBN-13: 978-1-4665-8292-7 (hbk)
ISBN-13: 978-0-367-37931-5 (pbk)

---

**Library of Congress Cataloging-in-Publication Data**

---

Atlas, Ronald M., 1946- author.
   Handbook of media for clinical and public health microbiology / Ronald M. Atlas, James W. Snyder.
   p. ; cm.
   Parallel title: Media for clinical and public health microbiology
   Includes bibliographical references and index.
   ISBN 978-1-4665-8292-7 (hardcover : alk. paper)
   I. Snyder, James W., author. II. Title. III. Title: Media for clinical and public health microbiology.
   [DNLM: 1.  Culture Media--Handbooks. 2.  Communicable Diseases--etiology--Handbooks. 3.  Microbiological Techniques--Handbooks.  QW 39]

RA643
616.9--dc23                                                                                                     2013034924

---

**Visit the Taylor & Francis Web site at**
**http://www.taylorandfrancis.com**

**and the CRC Press Web site at**
**http://www.crcpress.com**

# TABLE OF CONTENTS

xi

# Preface

Almost 1,800 media are described in the *Handbook of Media for Clinical and Public Health Microbiology*, including newly described media for the cultivation of emerging pathogens. Diseases caused by emerging pathogens that are responsible for increased rates of morbidity and mortality rates, such as *Escherichia coli* O157:H7, methicillin-resistant *Staphylococcus aureus* (MRSA), vancomycin-resistant enterococci (VRE), and carbapenem-resistant enterococci (CRE) have raised special concerns and various media included in the *Handbook* have been designed for the specific cultivation and identification of these pathogens.

Many of the new media included in the *Handbook of Media for Clinical and Public Health Microbiology* permit the cultivation of bacteria, fungi, and viruses that are currently causing major medical problems around the world. These media are very important for the rapid detection of pathogenic microorganisms and the diagnosis of individuals with specific infectious diseases and for preventing the spread of pathogens via food, water, and environmental sources. Several of the new media described in the *Handbook of Media for Clinical and Public Health Microbiology* include chromogenic or fluorogenic substrates that permit the rapid detection of specific pathogens.

An important function of the *Handbook of Media for Clinical and Public Health Microbiology* is to provide descriptions of the media that are used to cultivate and identify microorganisms from clinical specimens and those of public health significance. The *Handbook* provides a compilation of the formulations, methods of preparation, and applications for media used in clinical and public health microbiology laboratories. Each listing is alphabetical and includes medium composition, instruction for preparation, commercial sources, storage/shelf life, and intended uses.

The format of the *Handbook* allows easy reference to information needed to prepare media for the cultivation of microorganisms relevant to clinical diagnostics. The *Handbook of Media for Clinical and Public Health Microbiologyy* includes descriptions of expected results as they apply to microorganisms of importance for the examination of clinical specimens, foods, water, and other specimens of public health significance.

# About the Authors

**Ronald M. Atlas** is Professor of Biology at the University of Louisville. He received his B.S. from the State University of New York at Stony Brook in 1968 and his M.S. and Ph.D. from Rutgers in 1970 and 1972, respectively. Dr. Atlas has received a number of honors including: The University of Louisville Excellence in Research Award, the Johnson and Johnson Fellowship for Biology, and the American Society for Microbiology Award in Applied and Environmental Microbiology. He has an honorary doctorate from the University of Guelf. He is a fellow of the American Academy of Microbiology. He has authored several microbiology textbooks and handbooks of microbiological media. His research has included development of diagnostic systems for pathogenic microorganisms. He has served on the editorial boards of *Applied and Environmental Microbiology, Bio-Science, Biotechniques, Environmental Microbiology, Biosecurity, and Bioterrorism,* and *Journal of Industrial Microbiology.* He is the previous editor of *Critical Reviews in Microbiology.* He also has served as President of the American Society for Microbiology.

**James W. Snyder** is Professor of Pathology in the Department of Pathology and Laboratory Medicine at the University of Louisville School of Medicine and serves as the Director of Microbiology for the University of Louisville Hospital. He received his B.S. in 1969 and M.S. in 1970 from Eastern Kentucky University and Ph.D. in 1974 from the University of Dayton. Dr. Snyder is a diplomate of the American Board of Medical Microbiology (ABMM) and a Fellow in the American Academy of Microbiology. He has received a number of honors including: The South Central Association for Clinical Microbiology Outstanding Contribution to Microbiology Award, the Spirit of Louisville Foundation Award, and election to the Alpha Omega Alpha Honor Medical Society. He is the former Editor of *The Yearbook of Clinical Microbiology* and the *Certified Clinical Microbiology Data Base.* He is currently a member of the *Critical Reviews in Microbiology* editorial board and serves on the Professional Affairs Committee of the American Society for Microbiology. He maintains an active research program in both applied and basic clinical microbiology, including molecular techniques for the detection of pathogens and the assessment and comparison of automated and nonautomated diagnostic systems and tracking of microbial incidence and *in vitro* activity of antibiotics in support of local and national surveillance programs.

## Diagnostic Microbiology: Isolation and Identification of Pathogens

The definitive laboratory diagnosis of an infectious disease is dependant on the detection and/or the isolation and identification of the pathogenic microorganism or the detection of antigens or antibodies specifically associated with the pathogen. Cultivation of microbes has three major benefits: 1) growth and isolation of flora present in the specimen; 2) facilitates the microbiologist's efforts to determine which organism(s) is/are most likely responsible for the infection or are contaminants or colonizers; 3) obtain sufficient growth for use in assessing cultural characteristics and definitive identification. Traditional methods for the identification of pathogens depend on microscopic observations, phenotypic characteristics observed in culture, and metabolic changes following growth in a variety of substrates. Many differential media and selective media utilize color changes to highlight specific cultural features or substrate utilization. These growth-dependent methods provide a reliable and relatively accurate means for identifying the pathogen in promoting the laboratory diagnosis for many infectious diseases.

A wide range of biochemical, serological, and nucleic acid-based procedures are available for the definitive identification of microbial isolates. Accuracy, reliability, and speed are important factors that influence the selection of clinical identification protocols. The selection of the specific procedures to be employed for the identification of pathogenic isolates is guided by the need for presumptive or definitive identification of the organism at the family or genus and species level, based on the observation of and characterization of colonial morphology, phenotypic characteristics, and other key growth characteristics on primary isolation medium.

Identification of pathogenic filamentous fungi and protozoa is generally dependant on the morphological characteristics of the organism. Additionally, the identification of the former relies primarily on cultural and microscopic characteristics, and a limited number of biochemical tests. Conventional identification schemes for bacteria and yeasts rely on the determination of a variety of biochemical features. In general, fewer than 20 biochemical tests are required to identify most clinical bacterial isolates to the species level. The primary objective is to differentiate the isolates present in the specimen using a minimal number of tests to accurately define distinct taxa.

## Isolation and Culture Procedures

A variety of procedures are employed for the collection, isolation, detection, and identification of pathogenic microorganisms from different anatomical sites including tissue, body fluids, pulmonary secretions, and blood. Laboratory procedures are designed to screen and facilitate the recovery of etiologic agents of disease that predominate in particular clinical syndromes. When the manifestation of a disease suggests that the disease may be caused by a rare pathogen and/or routine screening fails to detect a probable causative microorganism, additional specialized isolation procedures may be required.

As the clinical microbiology laboratory slowly enters into the new era of "molecular diagnostics," the dependency on traditional culture methods involving the production and preparation of growth supporting media will continue to play an integral role in our efforts to detect and characterize both known and "new" microbial pathogens. Furthermore, the major goal of the clinical microbiology laboratory, as stated by Dr. Raymond Bartlett over twenty years ago, has not and will not change, "to provide information of maximal and epidemiological usefulness as rapidly as is consistent with acceptable accuracy and minimal cost." The latter two criteria serve as the driving forces that

account for the slow introduction and acceptance by the clinical microbiology community of methods that have been developed for either the noncultural detection or cultural confirmation of pathogenic microorganisms. Many procedures, including the more recent nucleic acid probes and gene amplification methods, remain cost prohibitive for many laboratories, sometimes lack acceptable sensitivity and/or specificity, and are not practical or adaptable for use in the service-oriented, clinical laboratory. Until such problems are resolved, microbiologists will continue to rely on the availability and utilization of a variety of media for the detection, isolation, characterization, and identification of primary and opportunistic microbial pathogens. The selection of a specific procedure(s) to be employed for the definitive identification of a pathogen at the family or genus and species level is influenced by the information gleaned from observing colonial morphology, pigment production, and other characteristics that are observed following growth of the microorganism on either selective, differential, or general purpose media.

These concepts also apply to fungi and viruses.(many virology laboratories have discontinued conventional tissue culture and replaced it with molecular techniques) The limited availability of practical and cost-effective detection methods requires that these microorganisms be recovered in culture before confirmation of identity can be accomplished using key morphological and biochemical characteristics, the type of cytopathic effect (CPE), or the application of culture confirmation methods such as serological, direct fluorescent antibiody, nucleic acid probes, or amplification technologies. Culture media will continue to be developed and used for the cultivation of microorganisms despite the inevitable impact that noncultural methods will have on the rapid detection of microbial pathogens and ultimately affect, in a beneficial manner, patient outcome.

## Media for the Isolation and Identification of Microorganisms of Public Health Concern

Increased emphasis on the microbiological safety of foods has led to greater testing of foods for potentially pathogenic microorganisms. Targeted quality control in the food industry has become essential for preventing outbreaks of foodborne disease and for controlling food spoilage. Many foods are routinely examined for the presence of disease-causing and food spoilage microorganisms. Many of these quality control procedures involve the cultivation of microorganisms using a variety of media that are described in the *Handbook of Media for Clinical and Public Health Microbiology*.

Given that an estimated 14 million cases of foodborne disease with known microbial pathogens occur every year in the United States alone, the Food and Drug Administration and U.S. Department of Agriculture have been implementing new regulations and guidance for the microbial testing of a variety of foods. Similarly, Europe and other regions of the developed world are implementing stricter oversight of food safety to ensure that foods are free of potentially dangerous pathogens, e.g., *Salmonella* spp, *Vibrio* spp., *Listeria monocytogenes*, *Campylobacter jejuni*, Shiga-producing *E. coli*, etc.

Emerging diseases caused by foodborne pathogens, such as *Escherichia coli* O157:H7, that have high mortality rates have raised special concerns. This has encouraged the development of new media for the cultivation of bacteria, fungi, and viruses that can aid in the rapid detection of microorganisms in foods and the protection of the food supply. Many of these new media include chromogenic or fluorogenic substrates that permit the specific rapid detection of pathogens of concern.

The ingestion of foods and water containing toxins and human pathogens can cause a variety of diseases. It is essential to prevent contamination of food products with human pathogens and to control the potential proliferation of toxin-producing microorganisms, which can result in food poisoning and the transmission of foodborne pathogens. In some cases, the growth of pathogenic microorganisms in a food is accompanied by obvious signs of spoilage, such as the production of gas or foul-smelling compounds, which give a clear indication that the product should not be eaten. In other cases, there are no obvious signs of spoilage to indicate the presence of pathogens or toxins. It is therefore necessary to carry out inspection programs aimed at detecting the possible contamination of food products with pathogenic microorganisms. As a consequence of the high quality control procedures employed in the food industry, few occurrences of foodborne disease are traced back to the food processor. Rather, most outbreaks of food poisoning are caused by improper postprocess handling of the food, often as a result of improper preparation and/or handling at home or in a food service setting.

Quality control laboratories in the food industry routinely perform tests to detect pathogens in food products and to ensure that the numbers and types of microorganisms associated with a food are not likely to cause serious food spoilage and/or health problems. In some cases, quality control procedures are aimed at detecting the presence of specific microorganisms, but in most cases the tests consist of examining the food for indicator organisms. For example, coliform counts are routinely performed on representative samples of many food products as an indication of possible fecal contamination, since food contaminated with fecal material has a relatively high probability of containing human pathogens. Foods such as hamburger often have 100,000 total bacteria per gram, but as long as they do not contain any *Salmonella* or other pathogens, they are considered safe for human consumption.

In the United States, agencies involved in the surveillance and regulation of foods are the Department of Agriculture (USDA), the Food and Drug Administration (FDA), and state boards of health. Standards based on the presence or absence of pathogens, such as those set by Congress and regulated by the USDA and FDA, do provide the safeguards needed for assurance of the safety of food products. On the international front, standards for the microbial testing of foods are established by the International Organization for Standardization (ISO). Both the FDA and ISO standards specify the media and procedures that are to be employed for the microbial analyses of specific foods.

## Media for the Isolation and Identification of Microorganisms from Clinical and Public Health Specimens

The *Handbook of Media for Clinical and Public Health Microbiology* includes both classic and modern media used for the identification and maintenance of diverse bacteria described in the *Manual* for medically important microorganisms.

## References

Atlas of the Clinical Microbiology of Infectious Diseases, Volume 1: Bacterial Agents. 2003. Bottone, E.J. Parthenon Pub. Group, New York.

Atlas of the Clinical Microbiology of Infectious Diseases, Volume 2: Viral, Fungal, and Parasitic Agents, 2006. Bottone, E.J. Parthenon Pub. Group, New York.

BD Diagnostic. Difco & BBL Manual: Dehydrated Culture Media and Reagents for Microbiology. 2003. Becton, Dickinson and Co., Sparks, MD. http://www.bd.com/ds/technicalCenter/inserts/difcoBblManual.asp

Biohazardous Bloodborne and Airborne Pathogens: A Guide for Clinical and Laboratory Safety. 2004. Langerman, N. Lewis Pub., Boca Raton, FL.

Bridson, E.Y., ed. *The Oxoid Manual.* 1998. Unipath Ltd. Basingstoke, Hampshire, England. http://www.oxoid.com/UK/blue/catbrowse/catbrowse.asp

Clinical and Laboratory Standards Institute. 2004. Quality Assurance for Commercially Prepared Microbiological Culture Media. Standard M22-A3. Clinical and Laboratory Standards Institute, Wayne, PA.

Clinical Bacteriology. 2003. Struthers, J.K. and R.P. Westran. ASM Press, Washington DC.

Clinical Microbiology Procedures Handbook. 2004. Isenberg, H.D., L. Clarke, P. Della-Latta, G.A. Denys, S.D. Douglas, L.S. Garcia, K.C. Hazen, J.F. Hindler, and S.G. Jenkins. ASM Press, Washington DC.

Clinical Mycology. 2003. Anaissie, E.J., M.R. McGinnis, and M.A. Pfaller. Churchill Livingston, New York.

Clinical Mycology. 2003. Dismukes, W.E., P.G. Pappas, and J.D. Sobel. Oxford University Press, New York.

Color Atlas and Textbook of Diagnostic Microbiology. 1992. Koneman, E.W., S.D. Allen, W.M. Janda, P.C. Schreckenberger, and W.C. Winn, Jr., eds. J.B. Lippincott Co., Philadelphia, PA.

Corry, J. E. L., Curtis, G D. W., and Baird, R. M. 2003. *Handbook of Culture Media for Food Microbiology,* 2nd ed. Elsevier, Amsterdam.

Curtis, G. D. and Lee, W. H. 1995. Culture media and methods for the isolation of *Listeria monocytogenes. International Journal of Food Microbiology* 26(1):1–13.

de Boer, E. 1992. Isolation of *Yersinia enterocolitica* from foods. *International Journal of Food Microbiology* 17(2):75–84.

Diagnostic Microbiology. 1990. Finegold, S.M. and W.J. Martin. C.V. Mosby Co., St. Louis, MO.

Domig, K. J., Mayer, H. K., and Kneifel, W. 2003. Methods used for the isolation, enumeration, characterisation and identification of *Enterococcus* spp. 1. Media for isolation and enumeration. *International Journal of Food Microbiology* 88(2-3):147–164.

Donovan, T. J. and van Netten, P. 1995. Culture media for the isolation and enumeration of pathogenic *Vibrio* species in foods and environmental samples. *International Journal of Food Microbiology* 26(1):77–91.

Downes, F. and Ito, K. 2001. *Compendium of Methods for the Microbiological Examination of Foods.* American Public Health Association, Washington, D.C.

HiMedia. 2006. *The HiVeg Manual.* HiMedia Laboratories Pvt. Limited. Mumbai, India.

HiMedia. 2009. *The HiMedia Manual.* HiMedia Laboratories Pvt. Limited. Mumbai, India.

Manual of Clinical Microbiology, tenth edition. 2011. Versalovic, J., K. C. Carroll, G. Funke, J. H. Jorgensen, M. L. Landry, and D. W. Warnock, eds. ASM Press, Washington, DC.

Media for Isolation–Cultivation–Identification– Maintenance of Medical Bacteria. 1985. McFaddin, J.F. Williams and Wilkins, Baltimore, MD.

Pocket Guide to Clinical Microbiology. 2004. Murray, P. R. ASM Press, Washington DC.

Practical Guide to Clinical Virology. 2002. Haaheim L.R., J.R. Pattison, and R.J. Whitley. John Wiley & Sons, Chichester, UK.

Principles and Practice of Clinical Virology. 2004. Zuckerman, A.J., J.E. Banatvala, P. Griffiths, J.R. Pattison, and B. Schoub. John Wiley & Sons, Chichester, UK.

Textbook of Diagnostic Microbiology. 2000. Mahon, C. Saunders, Philadelphia.

## Web Resources

Below is a list of Web sites that provide information about media and microbial cultures.

Bacteria/Culture Media Protocols http://www.protocol-online.org/prot/Microbiology/Bacteria/Culture_Media___Plates/index.html

BD (Becton, Dickinson and Company) http://www.bd.com/

Gibco Invitrogen Cell Culture Products

http://www.invitrogen.com/site/us/en/home/Applications/Cell-Cuture.html?cid=invggl123000000000095s&

Hardy Diagnostics
http://www.hardydiagnostics.com/?gclid=CMifuc62tJsCFR7yDAodZlWRQg

HiMedia. http://www.himedialabs.com/

Oxoid Ltd. http://www.oxoid.com/uk/blue/index.asp

U. S. Food and Drug Administration FDA Bacteriological Analytical Manual Online (BAM)
http://www.fda.gov/Food/ScienceResearch/LaboratoryMethods/BacteriologicalAnalyticalManualBAM/default.htm

U. S. Environmental Protection Agency. Microbiological Methods. http://www.epa.gov/nerlcwww/online.htm

## A 3 Agar

**Composition** per 202.4mL:

Agar base ................................................................. 140.0mL
Supplement solution ................................................ 62.4mL

pH 6.0 ± 0.2 at 25°C

**Agar Base:**
**Composition** per liter:

Pancreatic digest of casein ...................................... 17.0g
Ionagar No. 2 .............................................................. 7.5g
NaCl ............................................................................ 5.0g
Papaic digest of soybean meal .................................. 3.0g
$K_2HPO_4$ .................................................................... 2.5g
Glucose ...................................................................... 2.5g

**Source:** Ionagar No. 2 is available from Thermo Scientific.

**Preparation of Agar Base:** Add components, except agar, to distilled/deionized water and bring volume to 1.0L. Adjust pH to 5.5. Add agar. Mix thoroughly. Gently heat and bring to boiling. Distribute into screw-capped bottles in 140.0mL volumes. Autoclave for 15 min at 15 psi pressure–121°C. Cool to 45°–50°C.

**Supplement Solution:**
**Composition** per 62.4mL:

Horse serum-urea solution ...................................... 40.0mL
Fresh yeast extract solution ..................................... 20.0mL
Penicillin solution ..................................................... 2.0mL
Phenol Red solution .................................................. 0.4mL

**Preparation of Supplement Solution:** Aseptically combine components. Mix thoroughly.

**Horse Serum-Urea Solution:**
**Composition** per 40.0mL:

Urea ............................................................................ 0.2g
Horse serum, unheated ........................................... 40.0mL

**Preparation of Horse Serum-Urea Solution:** Add urea to 40.0mL of horse serum. Mix thoroughly. Filter sterilize.

**Fresh Yeast Extract Solution:**
**Composition:**

Baker's yeast, live, pressed, starch-free .................. 25.0g

**Preparation of Fresh Yeast Extract Solution:** Add the live Baker's yeast to 100.0mL of distilled/deionized water. Autoclave for 90 min at 15 psi pressure–121°C. Allow to stand. Remove supernatant solution. Adjust pH to 6.6–6.8. Filter sterilize.

**Penicillin Solution:**
**Composition** per 10.0mL:

Penicillin G ...................................................... 1,000,000U

**Preparation of Penicillin Solution:** Add penicillin to distilled/deionized water and bring volume to 10.0mL. Mix thoroughly. Filter sterilize.

**Phenol Red Solution:**
**Composition** per 10.0mL:

Phenol Red ................................................................. 0.1g

**Preparation of Phenol Red Solution:** Add Phenol Red to distilled/deionized water and bring volume to 10.0mL. Mix thoroughly. Filter sterilize.

**Preparation of Medium:** Aseptically combine 140.0mL of cooled, sterile agar base and 62.4mL of sterile supplement solution. Mix thoroughly. Pour into sterile Petri dishes or distribute into sterile tubes.

**Storage/Shelf Life:** Store dehydrated media in the dark in a sealed container below 30°C. Prepared media should be stored under refrigeration (2-8°C). Media should be used within 60 days of preparation. Media should not be used if there are any signs of deterioration (shrinking, cracking, or discoloration) or contamination, or if the expiration date supplied by the manufacturer has passed.

**Use:** For the cultivation of *Ureaplasma urealyticum* from urine.

## A 3B Agar

**Composition** per 101.5mL:

Agar base ................................................................. 80.0mL
Supplement solution ................................................ 21.5mL

pH 6.0 ± 0.2 at 25°C

**Agar Base:**
**Composition** per liter:

Pancreatic digest of casein ...................................... 17.0g
Ionagar No. 2 .............................................................. 7.5g
NaCl ............................................................................ 5.0g
Papaic digest of soybean meal .................................. 3.0g
$K_2HPO_4$ .................................................................... 2.5g
Glucose ...................................................................... 2.5g

**Source:** Ionagar No. 2 is available from Thermo Scientific.

**Preparation of Agar Base:** Add components, except agar, to distilled/deionized water and bring volume to 1.0L. Adjust pH to 5.5. Add agar. Mix thoroughly. Gently heat and bring to boiling. Distribute into screw-capped bottles in 80.0mL volumes. Autoclave for 15 min at 15 psi pressure–121°C. Cool to 45°–50°C.

**Supplement Solution:**
**Composition** per 21.5mL:

Horse serum-urea solution ...................................... 20.0mL
Penicillin solution ..................................................... 1.0mL
L-Cysteine·HCl·$H_2O$ solution ................................ 0.5mL

**Preparation of Supplement Solution:** Aseptically combine components. Mix thoroughly.

**Horse Serum-Urea Solution:**
**Composition** per 40.0mL:

Urea ............................................................................ 0.2g
Horse serum, unheated ........................................... 40.0mL

**Preparation of Horse Serum-Urea Solution:** Add urea to 40.0mL of horse serum. Mix thoroughly. Filter sterilize.

**Penicillin Solution:**
**Composition** per 10.0mL:

Penicillin G ...................................................... 1,000,000U

**Preparation of Penicillin Solution:** Add penicillin to distilled/deionized water and bring volume to 10.0mL. Mix thoroughly. Filter sterilize.

**L-Cysteine·HCl·$H_2O$ Solution:**
**Composition** per 10.0mL:

L-Cysteine·HCl·$H_2O$ ................................................ 0.2g

**Preparation of L-Cysteine·HCl·$H_2O$ Solution:** Add L-cysteine·HCl·$H_2O$ to distilled/deionized water and bring volume to 10.0mL. Mix thoroughly. Filter sterilize.

**Preparation of Medium:** Aseptically combine 80.0mL of cooled, sterile agar base and 21.5mL of sterile supplement solution. Mix thoroughly.

**Storage/Shelf Life:** Store dehydrated media in the dark in a sealed container below 30°C. Prepared media should be stored under refrigeration (2-8°C). Media should be used within 60 days of preparation. Media should not be used if there are any signs of deterioration (shrinking, cracking, or discoloration) or contamination, or if the expiration date supplied by the manufacturer has passed.

**Use:** For the cultivation of *Ureaplasma urealyticum* from urine.

## A 7 Agar
### (Shepard's Differential Agar)

**Composition** per 205.7mL:

Agar base ............................................................. 160.0mL
Supplement solution ............................................... 45.7mL

pH 6.0 ± 0.2 at 25°C

**Agar Base:**

**Composition** per 165.0mL:

Pancreatic digest of casein ........................................ 2.72g
Agar ...................................................................... 2.1g
NaCl ...................................................................... 0.8g
Papaic digest of soybean meal ................................... 0.48g
$K_2HPO_4$ ................................................................ 0.4g
Glucose .................................................................. 0.4g
$MnSO_4 \cdot H_2O$ ....................................................... 0.15g

**Preparation of Agar Base:** Add components, except agar, to distilled/deionized water and bring volume to 165.0mL. Adjust pH to 5.5. Add agar. Mix thoroughly. Autoclave for 15 min at 15 psi pressure–121°C. Cool to 45°–50°C.

**Supplement Solution:**

**Composition** per 45.72mL:

Horse serum, unheated ............................................ 40.0mL
Fresh yeast extract solution ....................................... 2.0mL
Penicillin solution .................................................... 2.0mL
CVA enrichment ...................................................... 1.0mL
L-Cysteine·HCl·$H_2O$ solution .................................. 0.5mL
Urea solution ......................................................... 0.22mL

**Preparation of Supplement Solution:** Aseptically combine components. Mix thoroughly.

**Fresh Yeast Extract Solution:**

**Composition** per 100.0mL:

Baker's yeast, live, pressed, starch-free ....................... 25.0g

**Preparation of Fresh Yeast Extract Solution:** Add the live Baker's yeast to 100.0mL of distilled/deionized water. Autoclave for 90 min at 15 psi pressure–121°C. Allow to stand. Remove supernatant solution. Adjust pH to 6.6–6.8. Filter sterilize.

**Penicillin Solution:**

**Composition** per 10.0mL:

Penicillin G ...................................................... 1,000,000U

**Preparation of Penicillin Solution:** Add penicillin to distilled/deionized water and bring volume to 10.0mL. Mix thoroughly. Filter sterilize.

**CVA Enrichment:**

**Composition** per liter:

Glucose ................................................................ 100.0g
L-Cysteine·HCl·$H_2O$ ............................................... 25.9g
L-Glutamine ............................................................ 10.0g
L-Cystine·2HCl ........................................................ 1.0g
Adenine ................................................................. 1.0g

Nicotinamide adenine dinucleotide .............................. 0.25g
Cocarboxylase .......................................................... 0.1g
Guanine·HCl ............................................................ 0.03g
$Fe(NO_3)_3$ .............................................................. 0.02g
Vitamin $B_{12}$ .......................................................... 0.01g
*p*-Aminobenzoic acid ............................................ 0.013g
Thiamine·HCl .......................................................... 3.0mg

**Preparation of CVA Enrichment:** Add components to distilled/deionized water and bring volume to 1.0L. Mix thoroughly. Filter sterilize.

**L-Cysteine·HCl·$H_2O$ Solution:**

**Composition** per 10.0mL:

L-Cysteine·HCl·$H_2O$ ............................................... 0.4g

**Preparation of L-Cysteine·HCl·$H_2O$ Solution:** Add L-cysteine·HCl·$H_2O$ solution to distilled/deionized water and bring volume to 10.0mL. Mix thoroughly. Filter sterilize.

**Urea Solution:**

**Composition** per 10.0mL:

Urea, ultrapure ........................................................ 1.0g

**Preparation of Urea Solution:** Add urea to distilled/deionized water and bring volume to 10.0mL. Mix thoroughly. Filter sterilize.

**Preparation of Medium:** Aseptically combine 160.0mL of cooled, sterile agar base and 45.9mL of sterile supplement solution. Mix thoroughly. Pour into sterile Petri dishes or distribute into sterile tubes.

**Storage/Shelf Life:** Store dehydrated media in the dark in a sealed container below 30°C. Prepared media should be stored under refrigeration (2-8°C). Media should be used within 60 days of preparation. Media should not be used if there are any signs of deterioration (shrinking, cracking, or discoloration) or contamination, or if the expiration date supplied by the manufacturer has passed.

**Use:** For the cultivation and differentiation of *Ureaplasma urealyticum* from urine based on its ability to produce ammonia from urea. Bacteria that produce ammonia appear as golden to dark brown colonies. Also used for the cultivation of other *Ureaplasma* species.

## A 7 Agar, Modified

**Composition** per 205.7mL:

Agar base ............................................................. 160.0mL
Supplement solution ............................................... 45.7mL

pH 6.0 ± 0.2 at 25°C

**Agar Base:**

**Composition** per 165.0mL:

Agar ..................................................................... 10.0g
Pancreatic digest of casein ........................................ 2.72g
NaCl ...................................................................... 0.8g
Papaic digest of soybean meal ................................... 0.48g
$K_2HPO_4$ ................................................................ 0.4g
Glucose .................................................................. 0.4g
$MnSO_4 \cdot H_2O$ ....................................................... 0.15g

**Preparation of Agar Base:** Add components, except agar, to distilled/deionized water and bring volume to 165.0mL. Adjust pH to 5.5. Add agar. Mix thoroughly. Autoclave for 15 min at 15 psi pressure–121°C. Cool to 45°–50°C.

**Supplement Solution:**

**Composition** per 45.72mL:

Horse serum, unheated ............................................ 40.0mL
Fresh yeast extract solution ....................................... 2.0mL

Penicillin solution ...................................................... 2.0mL
CVA enrichment............................................................ 1.0mL
L-Cysteine·HCl·H₂O solution........................................ 0.5mL
Urea solution................................................................ 0.22mL

**Preparation of Supplement Solution:** Aseptically combine components. Mix thoroughly.

**Fresh Yeast Extract Solution:**
**Composition** per 100.0mL:
Baker's yeast, live, pressed, starch-free.......................25.0g

**Preparation of Fresh Yeast Extract Solution:** Add the live Baker's yeast to 100.0mL of distilled/deionized water. Autoclave for 90 min at 15 psi pressure–121°C. Allow to stand. Remove supernatant solution. Adjust pH to 6.6–6.8. Filter sterilize.

**Penicillin Solution:**
**Composition** per 10.0mL:
Penicillin G ..................................................... 1,000,000U

**Preparation of Penicillin Solution:** Add penicillin to distilled/deionized water and bring volume to 10.0mL. Mix thoroughly. Filter sterilize.

**CVA Enrichment:**
**Composition** per liter:
Glucose ..........................................................100.0g
L-Cysteine·HCl·H₂O........................................................25.9g
L-Glutamine..............................................................10.0g
L-Cystine·2HCl.............................................................1.0g
Adenine......................................................................1.0g
Nicotinamide adenine dinucleotide ................................0.25g
Cocarboxylase.............................................................0.1g
Guanine·HCl...............................................................0.03g
Fe(NO₃)₃ ...................................................................0.02g
*p*-Aminobenzoic acid..................................................0.013g
Vitamin B₁₂ ...............................................................0.01g
Thiamine·HCl ...............................................................3.0mg

**Preparation of CVA Enrichment:** Add components to distilled/deionized water and bring volume to 1.0L. Mix thoroughly. Filter sterilize.

**L-Cysteine·HCl·H₂O Solution:**
**Composition** per 10.0mL:
L-Cysteine·HCl·H₂O........................................................0.4g

**Preparation of L-Cysteine·HCl·H₂O Solution:** Add L-cysteine·HCl·H₂O solution to distilled/deionized water and bring volume to 10.0mL. Mix thoroughly. Filter sterilize.

**Urea Solution:**
**Composition** per 10.0mL:
Urea, ultrapure ..........................................................1.0g

**Preparation of Urea Solution:** Add urea to distilled/deionized water and bring volume to 10.0mL. Mix thoroughly. Filter sterilize.

**Preparation of Medium:** Aseptically combine 160.0mL of cooled, sterile agar base and 45.9mL of sterile supplement solution. Mix thoroughly. Pour into sterile Petri dishes or distribute into sterile tubes.

**Storage/Shelf Life:** Store dehydrated media in the dark in a sealed container below 30°C. Prepared media should be stored under refrigeration (2-8°C). Media should be used within 60 days of preparation. Media should not be used if there are any signs of deterioration (shrinking, cracking, or discoloration) or contamination, or if the expiration date supplied by the manufacturer has passed.

**Use:** For the cultivation and differentiation of *Ureaplasma urealyticum* from urine based on its ability to produce ammonia from urea. Bacteria that produce ammonia appear as golden to dark brown colonies. Also used for the cultivation of other *Ureaplasma* species.

## A 7B Agar
**Composition** per 205.7mL:
Agar base ................................................................ 160.0mL
Supplement solution ................................................... 45.7mL
pH 6.0 ± 0.2 at 25°C

**Agar Base:**
**Composition** per 165.0mL:
Pancreatic digest of casein............................................2.72g
Agar .......................................................................2.1g
NaCl.......................................................................0.8g
Papaic digest of soybean meal.....................................0.48g
K₂HPO₄...................................................................0.4g
Glucose....................................................................0.4g
Putrescine·2HCl.........................................................0.33g
MnSO₄·H₂O...............................................................0.15g

**Preparation of Agar Base:** Add components, except agar, to distilled/deionized water and bring volume to 165.0mL. Adjust pH to 5.5. Add agar. Mix thoroughly. Autoclave for 15 min at 15 psi pressure–121°C. Cool to 45°–50°C.

**Supplement Solution:**
**Composition** per 45.72mL:
Horse serum, unheated................................................. 40.0mL
Fresh yeast extract solution ........................................ 2.0mL
Penicillin solution ..................................................... 2.0mL
CVA enrichment ........................................................ 1.0mL
L-Cysteine·HCl·H₂O solution........................................ 0.5mL
Urea solution............................................................ 0.22mL

**Preparation of Supplement Solution:** Aseptically combine components. Mix thoroughly.

**Fresh Yeast Extract Solution:**
**Composition** per 100.0mL:
Baker's yeast, live, pressed, starch-free.......................25.0g

**Preparation of Fresh Yeast Extract Solution:** Add the live Baker's yeast to 100.0mL of distilled/deionized water. Autoclave for 90 min at 15 psi pressure–121°C. Allow to stand. Remove supernatant solution. Adjust pH to 6.6–6.8. Filter sterilize.

**Penicillin Solution:**
**Composition** per 10.0mL:
Penicillin G ..................................................... 1,000,000U

**Preparation of Penicillin Solution:** Add penicillin to distilled/deionized water and bring volume to 10.0mL. Mix thoroughly. Filter sterilize.

**CVA Enrichment:**
**Composition** per liter:
Glucose ..........................................................100.0g
L-Cysteine·HCl·H₂O ....................................................25.9g
L-Glutamine .............................................................10.0g
L-Cystine·2HCl.............................................................1.0g
Adenine....................................................................1.0g
Nicotinamide adenine dinucleotide ...............................0.25g
Cocarboxylase.............................................................0.1g
Guanine·HCl...............................................................0.03g
Fe(NO₃)₃ ...................................................................0.02g

p-Aminobenzoic acid ................................................0.013g
Vitamin B$_{12}$ ...........................................................0.01g
Thiamine·HCl ........................................................3.0mg

**Preparation of CVA Enrichment:** Add components to distilled/deionized water and bring volume to 1.0L. Mix thoroughly. Filter sterilize.

**L-Cysteine·HCl·H$_2$O Solution:**
**Composition** per 10.0mL:
L-Cysteine·HCl·H$_2$O......................................................0.4g

**Preparation of L-Cysteine·HCl·H$_2$O Solution:** Add L-cysteine·HCl·H$_2$O solution to distilled/deionized water and bring volume to 10.0mL. Mix thoroughly. Filter sterilize.

**Urea Solution:**
**Composition** per 10.0mL:
Urea, ultrapure ...........................................................1.0g

**Preparation of Urea Solution:** Add urea to distilled/deionized water and bring volume to 10.0mL. Mix thoroughly. Filter sterilize.

**Preparation of Medium:** Aseptically combine 160.0mL of cooled, sterile agar base and 45.9mL of sterile supplement solution. Mix thoroughly. Pour into sterile Petri dishes or distribute into sterile tubes.

**Storage/Shelf Life:** Store dehydrated media in the dark in a sealed container below 30°C. Prepared media should be stored under refrigeration (2-8°C). Media should be used within 60 days of preparation. Media should not be used if there are any signs of deterioration (shrinking, cracking, or discoloration) or contamination, or if the expiration date supplied by the manufacturer has passed.

**Use:** For the cultivation and differentiation of *Ureaplasma urealyticum* from urine based on its ability to produce ammonia from urea. Bacteria that produce ammonia appear as golden to dark brown colonies. Also used for the cultivation of other *Ureaplasma* species.

# A 8B Agar
**Composition** per 84.6mL:
Agar base ....................................................... 80.0mL
Supplement solution ....................................... 4.6mL
pH 6.0 ± 0.2 at 25°C

**Agar Base:**
**Composition** per 165.0mL:
Pancreatic digest of casein.................................2.72g
Agar ...........................................................................2.1g
NaCl.........................................................................0.8g
Papaic digest of soybean meal ..........................0.48g
K$_2$HPO$_4$...................................................................0.4g
Glucose ....................................................................0.4g
MnSO$_4$·H$_2$O ........................................................0.15g
CaCl$_2$·2H$_2$O...........................................................0.03g
Putrescine·2HCl .....................................................34.0mg

**Preparation of Agar Base:** Add components, except agar, to distilled/deionized water and bring volume to 165.0mL. Adjust pH to 5.5. Add agar. Mix thoroughly. Autoclave for 15 min at 15 psi pressure–121°C. Cool to 45°–50°C.

**Supplement Solution:**
**Composition** per 4.6mL:
Horse serum, unheated................................... 1.0mL
Fresh yeast extract solution............................ 1.0mL
Penicillin solution ......................................... 1.0mL
Urea solution.................................................. 1.0mL

L-Cysteine·HCl·H$_2$O solution .......................... 0.5mL
GHL tripeptide solution.................................... 0.1mL

**Preparation of Supplement Solution:** Aseptically combine components. Mix thoroughly.

**Fresh Yeast Extract Solution:**
**Composition** per 100.0mL:
Baker's yeast, live, pressed, starch-free.........................25.0g

**Preparation of Fresh Yeast Extract Solution:** Add the live Baker's yeast to 100.0mL of distilled/deionized water. Autoclave for 90 min at 15 psi pressure–121°C. Allow to stand. Remove supernatant solution. Adjust pH to 6.6–6.8. Filter sterilize.

**Penicillin Solution:**
**Composition** per 10.0mL:
Penicillin G ............................................. 1,000,000U

**Preparation of Penicillin Solution:** Add penicillin to distilled/deionized water and bring volume to 10.0mL. Mix thoroughly. Filter sterilize.

**GHL Tripeptide Solution:**
**Composition** per 10.0mL:
GHL (Glycyl-L-histidyl-L-lysineacetate) tripeptide ........................0.2g

**Preparation of GHL Tripeptide Solution:** Add GHL (Glycyl-L-histidyl-L-lysine acetate) tripeptide to distilled/deionized water and bring volume to 10.0mL. Mix thoroughly. Filter sterilize.

**L-Cysteine·HCl·H$_2$O Solution:**
**Composition** per 10.0mL:
L-Cysteine·HCl·H$_2$O......................................................0.4g

**Preparation of L-Cysteine·HCl·H$_2$O Solution:** Add L-cysteine·HCl·H$_2$O solution to distilled/deionized water and bring volume to 10.0mL. Mix thoroughly. Filter sterilize.

**Urea Solution:**
**Composition** per 10.0mL:
Urea, ultrapure ...........................................................1.0g

**Preparation of Urea Solution:** Add urea to distilled/deionized water and bring volume to 10.0mL. Mix thoroughly. Filter sterilize.

**Preparation of Medium:** Aseptically combine 80.0mL of cooled, sterile agar base and 4.6mL of sterile supplement solution. Mix thoroughly. Pour into sterile Petri dishes or distribute into sterile tubes.

**Storage/Shelf Life:** Store dehydrated media in the dark in a sealed container below 30°C. Prepared media should be stored under refrigeration (2-8°C). Media should be used within 60 days of preparation. Media should not be used if there are any signs of deterioration (shrinking, cracking, or discoloration) or contamination, or if the expiration date supplied by the manufacturer has passed.

**Use:** For the cultivation of *Ureaplasma urealyticum* from urine. Also used for the cultivation of other *Ureaplasma* species.

# Acetamide Agar
**Composition** per liter:
Agar ...........................................................................15.0g
Acetamide..................................................................10.0g
NaCl.............................................................................5.0g
K$_2$HPO$_4$.....................................................................1.0g
NH$_4$H$_2$PO$_4$...............................................................1.0g
MgSO$_4$·7H$_2$O...........................................................0.2g
Bromthymol Blue......................................................0.08g
pH 6.9 ± 0.2 at 25°C

**Preparation of Medium:** Add components to distilled/deionized water and bring volume to 1.0L. Mix thoroughly. Gently heat and bring to boiling. Adjust pH. Distribute into tubes or flasks. Autoclave for 15 min at 15 psi pressure–121°C. Cool tubes in a slanted position to produce a long slant.

**Use:** For the differentiation of nonfermentative Gram-negative bacteria, especially *Pseudomonas aeruginosa*. Can be used as a confirmatory test for water analysis for public health surveillance. Bacteria that deamidate acetamide turn the medium blue.

## Acetamide Agar

**Composition** per liter:

| | |
|---|---|
| Agar | 15.0g |
| Acetamide | 10.0g |
| NaCl | 5.0g |
| K$_2$HPO$_4$ | 1.39g |
| KH$_2$PO$_4$ | 0.73g |
| MgSO$_4$·7H$_2$O | 0.5g |
| Phenol Red | 0.012g |

pH 6.9 ± 0.2 at 25°C

**Source:** This medium is available as a premixed powder from BD Diagnostic Systems.

**Preparation of Medium:** Add components to distilled/deionized water and bring volume to 1.0L. Mix thoroughly. Gently heat and bring to boiling. Adjust pH. Distribute into tubes or flasks. Autoclave for 15 min at 15 psi pressure–121°C. Cool tubes in a slanted position to produce a long slant.

**Storage/Shelf Life:** Store dehydrated media in the dark in a sealed container below 30°C. Prepared media should be stored under refrigeration (2-8°C). Media should be used within 60 days of preparation. Media should not be used if there are any signs of deterioration (shrinking, cracking, or discoloration) or contamination, or if the expiration date supplied by the manufacturer has passed.

**Use:** For the differentiation of nonfermentative Gram-negative bacteria, especially *Pseudomonas aeruginosa*. Can be used as a confirmatory test for water analysis. Bacteria that deamidate acetamide turn the medium red.

## Acetamide Broth

**Composition** per liter:

| | |
|---|---|
| Acetamide | 10.0g |
| NaCl | 5.0g |
| K$_2$HPO$_4$ | 1.39g |
| KH$_2$PO$_4$ | 0.73g |
| MgSO$_4$·7H$_2$O | 0.5g |
| Phenol Red | 0.012g |

pH 6.9 ± 0.2 at 25°C

**Preparation of Medium:** Add components to distilled/deionized water and bring volume to 1.0L. Mix thoroughly. Adjust pH. Autoclave for 15 min at 15 psi pressure–121°C.

**Storage/Shelf Life:** Store dehydrated media in the dark in a sealed container below 30°C. Prepared media should be stored under refrigeration (2-8°C). Media should be used within 60 days of preparation. Media should not be used if there are any signs of deterioration (discoloration) or contamination, or if the expiration date supplied by the manufacturer has passed.

**Use:** For the differentiation of nonfermentative Gram-negative bacteria, especially *Pseudomonas aeruginosa*. Can be used as a confirmatory test for water analysis. Bacteria that deamidate acetamide turn the broth purplish red.

## Acetamide Broth

**Composition** per liter:

| | |
|---|---|
| Acetamide | 2.0g |
| KH$_2$PO$_4$ | 1.0g |
| NaCl | 0.2g |
| MgSO$_4$, anhydrous | 0.2g |
| Na$_2$MoO$_4$·2H$_2$O | 5.0mg |
| FeSO$_4$ | 0.5mg |

pH 7.0 ± 0.2 at 25°C

**Source:** This medium is available from HiMedia.

**Preparation of Medium:** Add components, except acetamide, to distilled/deionized water and bring volume to 1.0L. Mix thoroughly. Add acetamide. Adjust pH to 7.0. Autoclave for 15 min at 15 psi pressure–121°C.

**Storage/Shelf Life:** Store dehydrated media in the dark in a sealed container below 30°C. Prepared media should be stored under refrigeration (2-8°C). Media should be used within 60 days of preparation. Media should not be used if there are any signs of deterioration (discoloration) or contamination, or if the expiration date supplied by the manufacturer has passed.

**Use:** For the differentiation of nonfermentative Gram-negative bacteria, especially *Pseudomonas aeruginosa*.

## Acetate Differential Agar
### (Sodium Acetate Agar)
### (Simmons' Citrate Agar, Modified)

**Composition** per liter:

| | |
|---|---|
| Agar | 20.0g |
| NaCl | 5.0g |
| Sodium acetate | 2.0g |
| (NH$_4$)H$_2$PO$_4$ | 1.0g |
| K$_2$HPO$_4$ | 1.0g |
| MgSO$_4$·7H$_2$O | 0.2g |
| Bromthymol Blue | 0.08g |

pH 6.8 ± 0.2 at 25°C

**Source:** This medium is available as a premixed powder from BD Diagnostic Systems.

**Preparation of Medium:** Add components to cold distilled/deionized water and bring volume to 1.0L. Mix thoroughly. Gently heat and bring to boiling. Distribute into tubes to produce a 1 cm butt and 30 cm slant. Autoclave for 15 min at 15 psi pressure–121°C. Cool tubes in a slanted position.

**Storage/Shelf Life:** Store dehydrated media in the dark in a sealed container below 30°C. Prepared media should be stored under refrigeration (2-8°C). Media should be used within 60 days of preparation. Media should not be used if there are any signs of deterioration (shrinking, cracking, or discoloration) or contamination, or if the expiration date supplied by the manufacturer has passed.

**Use:** For the differentiation of *Shigella* species from *Escherichia coli* and also for the differentiation of nonfermenting Gram-negative bacteria. Bacteria that can utilize acetate as the sole carbon source turn the medium blue.

## Acid Broth

**Composition** per liter:

Glucose ...................................................................5.0g
Proteose peptone ....................................................5.0g
Yeast extract...........................................................5.0g
$K_2HPO_4$ ...................................................................4.0g

pH $5.0 \pm 0.2$ at 25°C

**Preparation of Medium:** Add components to distilled/deionized water and bring volume to 1.0L. Mix thoroughly. Distribute into tubes or flasks. Autoclave for 15 min at 15 psi pressure–121°C.

**Storage/Shelf Life:** Store dehydrated media in the dark in a sealed container below 30°C. Prepared media should be stored under refrigeration (2-8°C). Media should be used within 60 days of preparation. Media should not be used if there are any signs of deterioration (discoloration) or contamination, or if the expiration date supplied by the manufacturer has passed.

**Use:** For the isolation of bacteria from canned foods.

## Acid Broth

**Composition** per liter:

Invert sugar ..........................................................10.0g
Peptic digest of animal tissue...........................10.0g
Yeast extract...........................................................7.5g

pH $4.0 \pm 0.2$ at 25°C

**Source:** This medium is available as a premixed powder from Hi-Media.

**Preparation of Medium:** Add components to distilled/deionized water and bring volume to 1.0L. Mix thoroughly. Distribute into tubes or flasks. Autoclave for 15 min at 15 psi pressure–121°C.

**Use:** For the isolation of bacteria from canned foods.

## Acid Egg Medium

**Composition** per 1640.0mL:

Potato starch..........................................................30.0g
$KH_2PO_4$....................................................................12.3g
Malachite Green......................................................0.4g
$MgSO_4 \cdot 7H_2O$ .......................................................0.3g
Penicillin G ..................................................... 100,000IU
Fresh egg mixture ................................................. 1.0L
Glycerol ...................................................................12.0mL

**Source:** This medium is available as a prepared medium from Thermo Scientific.

**Preparation of Medium:** Add components to 1.0L of fresh egg mixture. Mix thoroughly. Gently heat and bring to boiling. Bring volume to 1640.0mL with distilled/deionized water. Distribute into tubes or flasks. Autoclave for 15 min at 15 psi pressure–121°C with tubes in an upright position.

**Storage/Shelf Life:** Store dehydrated media in the dark in a sealed container below 30°C. Prepared media should be stored under refrigeration (2-8°C). Media should be used within 60 days of preparation. Media should not be used if there are any signs of deterioration (discoloration) or contamination, or if the expiration date supplied by the manufacturer has passed.

**Use:** For the cultivation and maintenance of *Mycobacterium tuberculosis*.

## Acid HiVeg Broth

Sucrose...................................................................10.0g
Plant peptone ........................................................10.0g
Yeast extract...........................................................7.5g

pH $4.0 \pm 0.2$ at 25°C

**Source:** This medium is available as a premixed powder from Hi-Media.

**Preparation of Medium:** Add components to distilled/deionized water and bring volume to 1.0L. Mix thoroughly. Distribute into tubes or flasks. Autoclave for 15 min at 15 psi pressure–121°C. Aseptically adjust pH to 4.0.

**Storage/Shelf Life:** Store dehydrated media in the dark in a sealed container below 30°C. Prepared media should be stored under refrigeration (2-8°C). Media should be used within 60 days of preparation. Media should not be used if there are any signs of deterioration (discoloration) or contamination, or if the expiration date supplied by the manufacturer has passed.

**Use:** For the isolation of acid tolerant bacteria from canned foods.

## Acid Products Test Broth

**Composition** per liter:

Invert sugar ..........................................................10.0g
Peptone ..................................................................10.0g
Yeast extract...........................................................7.5g

pH $4.0 \pm 0.2$ at 25°C

**Preparation of Medium:** Add components to distilled/deionized water and bring volume to 1.0L. Mix thoroughly. Gently heat while stirring and bring to boiling. Cool to 25°C. Adjust pH to 4.0 with 25% tartaric acid solution. Distribute into screw-capped flasks in 300.0mL volumes. Autoclave for 15 min at 15 psi pressure–121°C.

**Storage/Shelf Life:** Store dehydrated media in the dark in a sealed container below 30°C. Prepared media should be stored under refrigeration (2-8°C). Media should be used within 60 days of preparation. Media should not be used if there are any signs of deterioration (discoloration) or contamination, or if the expiration date supplied by the manufacturer has passed.

**Use:** For the cultivation of acid tolerant microorganisms from foods. For the sterility testing of canned foods.

## Actidione® Agar
## (Cycloheximide Agar)

**Composition** per liter:

Glucose ...................................................................50.0g
Agar .........................................................................15.0g
Pancreatic digest of casein..................................5.0g
Yeast extract...........................................................4.0g
$KH_2PO_4$....................................................................0.55g
KCl.............................................................................0.425g
$CaCl_2 \cdot 2H_2O$ .....................................................0.125g
$MgSO_4 \cdot 7H_2O$ ...................................................0.125g
Bromocresol Green...............................................22.0mg
Actidione® (cycloheximide)................................10.0mg
$FeCl_3$........................................................................2.5mg

pH $5.5 \pm 0.2$ at 25°C

**Source:** Actidione® Agar is available as a prepared medium from Thermo Scientific.

**Caution:** Cycloheximide is toxic. Avoid skin contact or aerosol formation and inhalation.

**Preparation of Medium:** Add components to distilled/deionized water and bring volume to 1.0L. Mix thoroughly. Gently heat and bring to boiling. Distribute into tubes or flasks. Autoclave for 15 min at 15 psi pressure–121°C. Pour into sterile Petri dishes or leave in tubes.

**Storage/Shelf Life:** Store dehydrated media in the dark in a sealed container below 30°C. Prepared media should be stored under refrigeration (2-8°C). Media should be used within 60 days of preparation. Media should not be used if there are any signs of deterioration (shrinking, cracking, or discoloration) or contamination, or if the expiration date supplied by the manufacturer has passed.

**Use:** For the enumeration and detection of bacteria in specimens containing large numbers of yeasts and molds.

## Actidione HiVeg Agar with Actidione®
## (Actidione HiVeg Agar with Cycloheximide)
**Composition** per liter:

| | |
|---|---|
| Glucose | 50.0g |
| Agar | 15.0g |
| Plant hydrolysate | 5.0g |
| Yeast extract | 4.0g |
| $KH_2PO_4$ | 0.55g |
| KCl | 0.425g |
| $CaCl_2 \cdot 2H_2O$ | 0.125g |
| $MgSO_4 \cdot 7H_2O$ | 0.125g |
| Bromocresol Green | 22.0mg |
| Actidione® (cycloheximide) | 10.0mg |
| $MnSO_4 \cdot 4H_2O$ | 2.5mg |
| $FeCl_3$ | 2.5mg |

pH 5.5 ± 0.2 at 25°C

**Source:** This medium is available as a premixed powder from HiMedia.

**Caution:** Cycloheximide is toxic. Avoid skin contact or aerosol formation and inhalation.

**Preparation of Medium:** Add components to distilled/deionized water and bring volume to 1.0L. Mix thoroughly. Gently heat and bring to boiling. Distribute into tubes or flasks. Autoclave for 15 min at 15 psi pressure–121°C. Pour into sterile Petri dishes or leave in tubes.

**Storage/Shelf Life:** Store dehydrated media in the dark in a sealed container below 30°C. Prepared media should be stored under refrigeration (2-8°C). Media should be used within 60 days of preparation. Media should not be used if there are any signs of deterioration (shrinking, cracking, or discoloration) or contamination, or if the expiration date supplied by the manufacturer has passed.

**Use:** For the enumeration and detection of bacteria in specimens containing large numbers of yeasts and molds.

## Actidione HiVeg Agar Base with Actidione®
**Composition** per liter:

| | |
|---|---|
| Glucose | 50.0g |
| Agar | 15.0g |
| Plant hydrolysate | 5.0g |
| Yeast extract | 4.0g |
| $KH_2PO_4$ | 0.55g |
| KCl | 0.425g |
| $CaCl_2 \cdot 2H_2O$ | 0.125g |

| | |
|---|---|
| $MgSO_4 \cdot 7H_2O$ | 0.125g |
| Bromocresol Green | 22.0mg |
| $MnSO_4 \cdot 4H_2O$ | 2.5mg |
| $FeCl_3$ | 2.5mg |
| Cycloheximide solution | 10.0mL |

pH 5.5 ± 0.2 at 25°C

**Source:** This medium, without actidione (cycloheximide), is available as a premixed powder from HiMedia.

**Cycloheximide Solution:**
**Composition** per 10.0mL:
Cycloheximide ... 0.025g

**Preparation of Cycloheximide Solution:** Add cycloheximide to distilled/deionized water and bring volume to 10.0mL. Mix thoroughly. Filter sterilize.

**Caution:** Cycloheximide is toxic. Avoid skin contact or aerosol formation and inhalation.

**Preparation of Medium:** Add components, except cycloheximide solution, to distilled/deionized water and bring volume to 990.0mL. Mix thoroughly. Gently heat and bring to boiling. Distribute into tubes or flasks. Autoclave for 15 min at 15 psi pressure–121°C. Cool to 50°C. Aseptically add 10.0mL cycloheximide solution. Pour into sterile Petri dishes or leave in tubes.

**Storage/Shelf Life:** Store dehydrated media in the dark in a sealed container below 30°C. Prepared media should be stored under refrigeration (2-8°C). Media should be used within 60 days of preparation. Media should not be used if there are any signs of deterioration (shrinking, cracking, or discoloration) or contamination, or if the expiration date supplied by the manufacturer has passed.

**Use:** For the enumeration and detection of bacteria in specimens containing large numbers of yeasts and molds.

## AE Sporulation Medium, Modified
**Composition** per 1079.2mL:

| | |
|---|---|
| Polypeptone™ | 10.0g |
| Yeast extract | 10.0g |
| $Na_2HPO_4$ | 4.36g |
| Ammonium acetate | 1.5g |
| $KH_2PO_4$ | 0.25g |
| $MgSO_4 \cdot 7H_2O$ | 0.2g |
| Raffinose solution | 39.6mL |
| $Na_2CO_3$ solution | 13.2mL |
| $CoCl_2 \cdot 6H_2O$ solution | 13.2mL |
| Sodium ascorbate solution | 13.2mL |

pH 7.8 ± 0.1 at 25°C

**Raffinose Solution:**
**Composition** per 100.0mL:
Raffinose ... 10.0g

**Preparation of Raffinose Solution:** Add raffinose to distilled/deionized water and bring volume to 100.0mL. Mix thoroughly. Filter sterilize.

**$Na_2CO_3$ Solution:**
**Composition** per 100.0mL:
$Na_2CO_3$ ... 7.0g

**Preparation of $Na_2CO_3$ Solution:** Add $Na_2CO_3$ to distilled/deionized water and bring volume to 100.0mL. Mix thoroughly. Filter sterilize.

**CoCl₂·6H₂O Solution:**
**Composition** per 100.0mL:
CoCl₂·6H₂O ..................................................................0.32g

**Preparation of CoCl₂·6H₂O Solution:** Add CoCl₂·6H₂O to distilled/deionized water and bring volume to 100.0mL. Mix thoroughly. Filter sterilize.

**Sodium Ascorbate Solution:**
**Composition** per 100.0mL:
Sodium ascorbate ..........................................................1.5g

**Preparation of Sodium Ascorbate Solution:** Add sodium ascorbate to distilled/deionized water and bring volume to 100.0mL. Mix thoroughly. Filter sterilize. Use freshly prepared solution.

**Preparation of Medium:** Add components—except raffinose solution, Na₂CO₃ solution, CoCl₂·6H₂O solution, and sodium ascorbate solution—to distilled/deionized water and bring volume to 1.0L. Mix thoroughly. Adjust pH to 7.5 using 2*M* sodium carbonate solution. Distribute into tubes in 15.0mL volumes. Autoclave for 15 min at 15 psi pressure–121°C. Aseptically add 0.6mL of sterile raffinose solution, 0.2mL of sterile Na₂CO₃ solution, and 0.2mL of sterile CoCl₂·6H₂O solution to each tube. Mix thoroughly. Prior to inoculation, steam medium for 10 min. Cool to 25°C. Aseptically add 0.2mL of sterile sodium ascorbate solution to each tube.

**Storage/Shelf Life:** Store dehydrated media in the dark in a sealed container below 30°C. Prepared media should be stored under refrigeration (2-8°C). Media should be used within 60 days of preparation. Media should not be used if there are any signs of deterioration (discoloration) or contamination, or if the expiration date supplied by the manufacturer has passed.

**Use:** For the cultivation and sporulation of *Clostridium perfringens*.

## Aero Pseudo Selective Agar

**Composition** per liter:
Agar .................................................................................12.0g
Starch, soluble ..............................................................20.0g
Sodium glutamate ...........................................................2.0g
KH₂PO₄ .............................................................................2.0g
MgSO₄·7H₂O .....................................................................0.5g
Phenol Red .....................................................................0.36g

pH 7.2 ± 0.2 at 25°C

**Source:** This medium is available from HiMedia.

**Selective Supplement Solution:**
**Composition** per 10.0mL:
Pimaricin .......................................................................0.01g
Penicillin G ....................................................100,000 units

**Preparation of Selective Supplement Solution:** Add components to distilled/deionized water and bring volume to 10.0mL. Mix thoroughly. Filter sterilize.

**Preparation of Medium:** Add components, except selective supplement solution, to distilled/deionized water and bring volume to 990.0mL. Mix thoroughly. Autoclave for 15 min at 15 psi pressure–121°C. Cool to 50°C. Aseptically add selective supplement solution. Mix thoroughly. Pour into Petri dishes or aseptically distribute into sterile tubes.

**Storage/Shelf Life:** Store dehydrated media in the dark in a sealed container below 30°C. Prepared media should be stored under refrigeration (2-8°C). Media should be used within 60 days of preparation. Media should not be used if there are any signs of deterioration (shrink-

ing, cracking, or discoloration) or contamination, or if the expiration date supplied by the manufacturer has passed.

**Use:** For the selective cultivation of *Pseudomonas* spp. and *Aeromonas* spp. For the detection of *Aeromonas* and *Pseudomonas* in foods, water, and food processing equipment.

## *Aeromonas* Differential Agar
## (Dextrin Fuchsin Sulfite Agar)

**Composition** per liter:
Dextrin ...........................................................................15.0g
Agar .................................................................................13.0g
Pancreatic digest of casein ..........................................10.0g
Na₂HPO₄ ..........................................................................7.75g
NaCl ..................................................................................5.0g
Beef extract .....................................................................3.0g
Na₂SO₃ ..............................................................................1.6g
Acid Fuchsin solution ................................................50.0mL

pH 7.5 ± 0.2 at 25°C

**Acid Fuchsin Solution:**
**Composition** per 50.0mL:
Acid Fuchsin ..................................................................0.25g
Aqueous dioxan, 5% ...................................................50.0mL

**Preparation of Acid Fuchsin Solution:** Add Acid Fuchsin to 50.0mL of 5% aqueous dioxan. Mix well to dissolve.

**Caution:** Acid Fuchsin is a potential carcinogen and care must be taken to avoid inhalation of the powdered dye and contamination of the skin.

**Preparation of Medium:** Add components to distilled/deionized water and bring volume to 1.0L. Mix thoroughly. Gently heat while stirring and bring to boiling. Distribute into tubes or flasks. Autoclave for 15 min at 15 psi pressure–121°C. Pour into sterile Petri dishes or leave in tubes.

**Storage/Shelf Life:** Store dehydrated media in the dark in a sealed container below 30°C. Prepared media should be stored under refrigeration (2-8°C). Media should be used within 60 days of preparation. Media should not be used if there are any signs of deterioration (shrinking, cracking, or discoloration) or contamination, or if the expiration date supplied by the manufacturer has passed.

**Use:** For the isolation and differentiation of *Aeromonas* species from other Gram-negative rods such as *Pseudomonas* and Enterobacteriaceae. Specimens with low numbers of *Aeromonas* may first be enriched by growth in starch broth for 4–9 days. After 24 hrs of growth on this agar, colonies are sprayed with Nadi reagent (1% solution of *N,N,N′,N′*-tetramethyl-*p*-phenylene-diammonium dichloride). A positive Nadi reaction (dextrin degradation) is indicated by a purple color at the periphery of the colony. Dextrin fermentation is also indicated by red colonies. *Aeromonas* species appear as large, convex, dark red colonies with a purple periphery.

## *Aeromonas hydrophila* Medium

**Composition** per liter:
Inositol ...........................................................................10.0g
Pancreatic digest of casein ..........................................10.0g
L-Ornithine·HCl ..............................................................5.0g
Proteose peptone .............................................................5.0g
Agar ...................................................................................3.0g
Yeast extract .....................................................................3.0g
Mannitol ...........................................................................1.0g

Ferric ammonium citrate................................................0.5g
Na₂S₂O₃·5H₂O.........................................................0.4g

$Na_2S_2O_3 \cdot 5H_2O$ ................................................0.4g

Bromcresol Purple ..................................................0.02g

pH 6.7 ± 0.2 at 25°C

**Preparation of Medium:** Add components to distilled/deionized water and bring volume to 1.0L. Mix thoroughly. Gently heat until dissolved. Adjust pH to 6.7. Distribute into tubes in 5.0mL volumes. Autoclave for 12 min at 15 psi pressure–121°C.

**Storage/Shelf Life:** Store dehydrated media in the dark in a sealed container below 30°C. Prepared media should be stored under refrigeration (2-8°C). Media should be used within 60 days of preparation. Media should not be used if there are any signs of deterioration (discoloration) or contamination, or if the expiration date supplied by the manufacturer has passed.

**Use:** For the isolation and cultivation of *Aeromonas hydrophila*.

## *Aeromonas* Isolation Medium

**Composition** per liter:

Agar ...............................................................12.5g
$Na_2S_2O_3$ .........................................................10.67g
Special peptone ....................................................5.0g
NaCl...............................................................5.0g
Xylose ............................................................3.75g
L-Lysine·HCl.......................................................3.5g
Yeast extract......................................................3.0g
Sorbitol...........................................................3.0g
Bile salts ........................................................3.0g
Inositol ..........................................................2.5g
L-Arginine·HCl.....................................................2.0g
Lactose ...........................................................1.5g
Ferric ammonium citrate............................................0.8g
Bromthymol Blue ...................................................0.04g
Thymol Blue .......................................................0.04g
Ampicillin solution ...............................................2.5mL

pH 8.0 ± 0.1 at 25°C

**Source:** This medium without ampicillin is available from Sigma Aldrich.

**Ampicillin Solution:**
**Composition** per 5.0mL:
Ampicillin.........................................................10.0mg

**Preparation of Ampicillin Solution:** Add ampicillin to distilled/deionized water and bring volume to 5.0mL. Mix thoroughly. Filter sterilize.

**Preparation of Medium:** Add components, except ampicillin solution, to distilled/deionized water and bring volume to 1.0L. Mix thoroughly. Gently heat and bring to boiling. Do not autoclave. Cool to 50°C. Aseptically add 2.5mL of ampicillin solution. Pour into sterile Petri dishes.

**Storage/Shelf Life:** Store dehydrated media in the dark in a sealed container below 30°C. Prepared media should be stored under refrigeration (2-8°C). Media should be used within 60 days of preparation. Media should not be used if there are any signs of deterioration (shrinking, cracking, or discoloration) or contamination, or if the expiration date supplied by the manufacturer has passed.

**Use:** For the isolation and selective differentiation of *Aeromonas hydrophila* and other *Aeromonas* species from clinical specimens and foods. *Aeromonas* species appear as small (0.5–1.5mm), dark green colonies with darker centers.

## *Aeromonas* Isolation HiVeg Medium

**Composition** per liter:

Agar ..............................................................12.5g
$Na_2S_2O_3$ .........................................................10.67g
Plant special peptone .............................................5.0g
NaCl...............................................................5.0g
Xylose ............................................................3.75g
L-Lysine·HCl.......................................................3.5g
Yeast extract......................................................3.0g
Sorbitol...........................................................3.0g
Synthetic detergent ...............................................3.0g
Inositol ..........................................................2.5g
L-Arginine·HCl.....................................................2.0g
Lactose ...........................................................1.5g
Ferric ammonium citrate............................................0.8g
Bromthymol Blue ...................................................0.04g
Thymol Blue .......................................................0.04g
Ampicillin solution ...............................................2.5mL

pH 8.0 ± 0.1 at 25°C

**Source:** This medium without ampicillin is available from HiMedia.

**Ampicillin Solution:**
**Composition** per 5.0mL:
Ampicillin.........................................................10.0mg

**Preparation of Ampicillin Solution:** Add ampicillin to distilled/deionized water and bring volume to 5.0mL. Mix thoroughly. Filter sterilize.

**Source:** Ampicillin supplement solution is also available from HiMedia.

**Preparation of Medium:** Add components, except ampicillin solution, to distilled/deionized water and bring volume to 1.0L. Mix thoroughly. Gently heat and bring to boiling. Do not autoclave. Cool to 50°C. Aseptically add 2.5mL of ampicillin solution. Pour into sterile Petri dishes.

**Storage/Shelf Life:** Store dehydrated media in the dark in a sealed container below 30°C. Prepared media should be stored under refrigeration (2-8°C). Media should be used within 60 days of preparation. Media should not be used if there are any signs of deterioration (shrinking, cracking, or discoloration) or contamination, or if the expiration date supplied by the manufacturer has passed.

**Use:** For the isolation and selective differentiation of *Aeromonas hydrophila* and other *Aeromonas* species from clinical specimens and foods. *Aeromonas* species appear as small (0.5–1.5mm), dark green colonies with darker centers.

## *Aeromonas* Medium
## (Ryan's *Aeromonas* Medium)

**Composition** per liter:

Agar ..............................................................12.5g
$Na_2S_2O_3$ .........................................................10.67g
Proteose peptone ..................................................5.0g
NaCl...............................................................5.0g
Xylose ............................................................3.75g
L-Lysine·HCl.......................................................3.5g
Yeast extract......................................................3.0g
Sorbitol ..........................................................3.0g
Bile salts No. 3...................................................3.0g
Inositol ..........................................................2.5g
L-Arginine·HCl.....................................................2.0g
Lactose ...........................................................1.5g

Ferric ammonium citrate ...............................................0.8g
Bromthymol Blue .......................................................0.04g
Thymol Blue ...............................................................0.04g

pH 8.0 ± 0.1 at 25°C

**Source:** This medium is available as a dehydrated powder from Thermo Scientific.

**Preparation of Medium:** Add components to distilled/deionized water and bring volume to 1.0L. Mix thoroughly. Gently heat and bring to boiling. Do not autoclave. Cool to 50°C and aseptically add 5.0mg of ampicillin. Pour into sterile Petri dishes.

**Storage/Shelf Life:** Store dehydrated media in the dark in a sealed container below 30°C. Prepared media should be stored under refrigeration (2-8°C). Media should be used within 60 days of preparation. Media should not be used if there are any signs of deterioration (shrinking, cracking, or discoloration) or contamination, or if the expiration date supplied by the manufacturer has passed.

**Use:** For the isolation and selective differentiation of *Aeromonas hydrophila* and other *Aeromonas* species from clinical speciments and nonclinical specimens of public health importance. *Aeromonas* species appear as small (0.5–1.5mm), dark green colonies with darker centers.

## AH5 Medium

**Composition** per 205.9mL:

Agar base .............................................................. 160.0mL
Supplement solution ................................................ 45.9mL

pH 6.0 ± 0.2 at 25°C

**Agar Base:**

**Composition** per 165.0mL:

Pancreatic digest of casein ..........................................2.72g
Agar .............................................................................2.1g
NaCl ............................................................................0.8g
Papaic digest of soybean meal .....................................0.48g
$K_2HPO_4$ ....................................................................0.4g
Glucose ........................................................................0.4g

**Preparation of Agar Base:** Add components, except agar, to distilled/deionized water and bring volume to 165.0mL. Adjust pH to 5.5. Add agar. Mix thoroughly. Autoclave for 15 min at 15 psi pressure–121°C. Cool to 45°–50°C.

**Supplement Solution:**

**Composition** per 45.9mL:

Horse serum, unheated ............................................... 40.0mL
Fresh yeast extract solution............................................. 2.0mL
Penicillin solution .......................................................... 2.0mL
CVA enrichment............................................................. 1.0mL
L-Cysteine·HCl·H$_2$O solution ....................................... 0.5mL
Urea solution.................................................................. 0.4mL

**Preparation of Supplement Solution:** Aseptically combine components. Mix thoroughly.

**Fresh Yeast Extract Solution:**

**Composition** per 100.0mL:

Baker's yeast, live, pressed, starch-free ...........................25.0g

**Preparation of Fresh Yeast Extract Solution:** Add the live Baker's yeast to 100.0mL of distilled/deionized water. Autoclave for 90 min at 15 psi pressure–121°C. Allow to stand. Remove supernatant solution. Adjust pH to 6.6–6.8.

**Penicillin Solution:**

**Composition** per 10.0mL:

Penicillin G ..................................................... 1,000,000U

**Preparation of Penicillin Solution:** Add penicillin to distilled/deionized water and bring volume to 10.0mL. Mix thoroughly. Filter sterilize.

**CVA Enrichment:**

**Composition** per liter:

Glucose ....................................................................100.0g
L-Cysteine·HCl·H$_2$O...................................................25.9g
L-Glutamine ................................................................ 10.0g
L-Cystine·2HCl ............................................................. 1.0g
Adenine........................................................................ 1.0g
Nicotinamide adenine dinucleotide .................................0.25g
Cocarboxylase ..............................................................0.1g
Guanine·HCl ...............................................................0.03g
$Fe(NO_3)_3$ ..................................................................0.02g
*p*-Aminobenzoic acid...................................................0.013g
Vitamin $B_{12}$ ..............................................................0.01g
Thiamine·HCl ..............................................................3.0mg

**Preparation of CVA Enrichment:** Add components to distilled/deionized water and bring volume to 1.0L. Mix thoroughly. Filter sterilize.

**L-Cysteine·HCl·H$_2$O Solution:**

**Composition** per 10.0mL:

L-Cysteine·HCl·H$_2$O.......................................................0.4g

**Preparation of L-Cysteine·HCl·H$_2$O Solution:** Add L-cysteine·HCl·H$_2$O solution to distilled/deionized water and bring volume to 10.0mL. Mix thoroughly. Filter sterilize.

**Urea Solution:**

**Composition** per 10.0mL:

Urea.............................................................................1.0g

**Preparation of Urea Solution:** Add urea to distilled/deionized water and bring volume to 10.0mL. Mix thoroughly. Filter sterilize.

**Preparation of Medium:** Aseptically combine cooled, sterile components. Mix thoroughly. Pour into sterile Petri dishes or distribute into sterile tubes.

**Storage/Shelf Life:** Store dehydrated media in the dark in a sealed container below 30°C. Prepared media should be stored under refrigeration (2-8°C). Media should be used within 60 days of preparation. Media should not be used if there are any signs of deterioration (discoloration) or contamination, or if the expiration date supplied by the manufacturer has passed.

**Use:** For the cultivation of *Ureaplasma urealyticum* from urine and exudates and for the cultivation of other *Ureaplasma* species.

## AKI Medium

**Composition** per liter:

Peptone ......................................................................15.0g
NaCl............................................................................5.0g
Yeast extract.................................................................4.0g
Sodium bicarbonate solution ....................................... 30.0mL

pH 7.2 ± 0.2 at 25°C

**Sodium Bicarbonate Solution:**

**Composition** per 100.0mL:

$NaHCO_3$ ....................................................................10.0g

**Preparation of Sodium Bicarbonate Solution:** Add sodium bicarbonate to distilled/deionized water and bring volume to 100.0mL. Mix thoroughly. Filter sterilize. Use freshly prepared solution.

**Preparation of Medium:** Add components, except sodium bicarbonate solution, to distilled/deionized water and bring volume to 970.0mL. Mix thoroughly. Autoclave for 15 min at 15 psi pressure–121°C. Cool to 45°–50°C. Aseptically add sterile sodium bicarbonate solution. Mix thoroughly. Aseptically distribute into sterile tubes or flasks. Prepare medium freshly.

**Storage/Shelf Life:** Store dehydrated media in the dark in a sealed container below 30°C. Prepared media should be stored under refrigeration (2-8°C). Media should be used within 60 days of preparation. Media should not be used if there are any signs of deterioration (discoloration) or contamination, or if the expiration date supplied by the manufacturer has passed.

**Use:** For the cultivation of *Vibrio cholerae* and other *Vibrio* species.

## Alginate Utilization Medium

**Composition** per liter:
| | |
|---|---|
| Solution B | 500.0mL |
| Solution A | 400.0mL |
| Solution C | 100.0mL |

**Solution A:**
**Composition** per 400.0mL:
| | |
|---|---|
| Marine salts | 38.0g |

**Preparation of Solution A:** Add marine salts to distilled/deionized water and bring volume to 400.0mL. Mix thoroughly. Autoclave for 15 min at 15 psi pressure–121°C.

**Solution B:**
**Composition** per 500.0mL:
| | |
|---|---|
| Agar | 20.0g |
| Sodium alginate | 10.0g |

**Preparation of Solution B:** Add components to distilled/deionized water and bring volume to 500.0mL. Mix thoroughly. Autoclave for 15 min at 15 psi pressure–121°C.

**Solution C:**
**Composition** per 100.0mL:
| | |
|---|---|
| Tris·HCl buffer | 0.067g |
| $NaNO_3$ | 0.047g |
| Ferric EDTA | 66.5mg |
| Sodium glycerophosphate | 6.67mg |
| Thiamine·HCl | 67.0µg |
| Vitamin $B_{12}$ | 1.3µg |
| Biotin | 0.67µg |

**Preparation of Solution C:** Add components to distilled/deionized water and bring volume to 100.0mL. Mix thoroughly. Filter sterilize.

**Preparation of Medium:** Aseptically combine solutions A, B, and C. For liquid medium, omit agar from solution B.

**Storage/Shelf Life:** Store dehydrated media in the dark in a sealed container below 30°C. Prepared media should be stored under refrigeration (2-8°C). Media should be used within 60 days of preparation. Media should not be used if there are any signs of deterioration (shrinking, cracking, or discoloration) or contamination, or if the expiration date supplied by the manufacturer has passed.

**Use:** For the cultivation of microorganisms that can utilize alginate as a carbon source. Growth on alginate (production of alginase) is a diagnostic test used in the differentiation of *Vibrio* species.

## Alkaline HiVeg Peptone Water

**Composition** per liter:
| | |
|---|---|
| NaCl | 10.0g |
| Plant peptone | 10.0g |

pH 8.4 ± 0.2 at 25°C

**Source:** This medium is available as a premixed powder from Hi-Media.

**Preparation of Medium:** Add components to distilled/deionized water and bring volume to 1.0L. Mix thoroughly. Adjust pH to 8.5. Distribute into tubes or flasks. Autoclave for 10 min at 15 psi pressure–121°C.

**Storage/Shelf Life:** Store dehydrated media in the dark in a sealed container below 30°C. Prepared media should be stored under refrigeration (2-8°C). Media should be used within 60 days of preparation. Media should not be used if there are any signs of deterioration (discoloration) or contamination, or if the expiration date supplied by the manufacturer has passed.

**Use:** For the enrichment of *Vibrio* species from foods.

## Alkaline Peptone Agar

**Composition** per liter:
| | |
|---|---|
| NaCl | 20.0g |
| Agar | 15.0g |
| Peptone | 10.0g |

pH 8.5 ± 0.2 at 25°C

**Preparation of Medium:** Add components to distilled/deionized water and bring volume to 1.0L. Mix thoroughly. Gently heat and bring to boiling. Adjust pH to 8.5. Distribute into tubes. Autoclave for 15 min at 15 psi pressure–121°C. Allow tubes to cool in a slanted position.

**Storage/Shelf Life:** Store dehydrated media in the dark in a sealed container below 30°C. Prepared media should be stored under refrigeration (2-8°C). Media should be used within 60 days of preparation. Media should not be used if there are any signs of deterioration (shrinking, cracking, or discoloration) or contamination, or if the expiration date supplied by the manufacturer has passed.

**Use:** For the cultivation of *Vibrio cholerae* and other *Vibrio* species.

## Alkaline Peptone Salt Broth
## (APS Broth)

**Composition** per liter:
| | |
|---|---|
| NaCl | 30.0g |
| Peptone | 10.0g |

**Preparation of Medium:** Add components to distilled/deionized water and bring volume to 1.0L. Mix thoroughly. Adjust pH to 8.5. Distribute into tubes in 10.0mL volumes. Autoclave for 10 min at 15 psi pressure–121°C.

**Storage/Shelf Life:** Store dehydrated media in the dark in a sealed container below 30°C. Prepared media should be stored under refrigeration (2-8°C). Media should be used within 60 days of preparation. Media should not be used if there are any signs of deterioration (discoloration) or contamination, or if the expiration date supplied by the manufacturer has passed.

**Use:** For the cultivation of *Vibrio cholerae* and other *Vibrio* species from foods.

## Alkaline Peptone Water

**Composition** per liter:

NaCl .............................................................................. 10.0g
Peptone ......................................................................... 10.0g

pH $8.5 \pm 0.2$ at 25°C

**Preparation of Medium:** Add components to distilled/deionized water and bring volume to 1.0L. Mix thoroughly. Adjust pH to 8.5. Distribute into tubes or flasks. Autoclave for 10 min at 15 psi pressure–121°C.

**Storage/Shelf Life:** Store dehydrated media in the dark in a sealed container below 30°C. Prepared media should be stored under refrigeration (2-8°C). Media should be used within 60 days of preparation. Media should not be used if there are any signs of deterioration (discoloration) or contamination.

**Use:** For the cultivation and transport of *Vibrio cholerae* and other *Vibrio* species from foods.

## ALOA Medium
## (Agar *Listeria* Ottavani & Agosti)
## (BAM M10a)

**Composition** per liter:

Agar ............................................................................... 18.0g
Peptone ......................................................................... 18.0g
LiCl ............................................................................... 10.0g
Yeast extract ................................................................. 10.0g
Tryptone .......................................................................... 6.0g
NaCl ................................................................................ 5.0g
$Na_2HPO_4$ ....................................................................... 2.5g
Na-pyruvate ..................................................................... 2.0g
Glucose ........................................................................... 2.0g
Mg-glycerophosphate ...................................................... 1.0g
$MgSO_4$ ............................................................................ 0.5g
5-Bromo4-chloro-indolyl-β-D-glucopyranoside ............. 0.05g
Phosphatidylinositol solution ..................................... 50.0mL
Nalidixic acid solution ................................................. 5.0mL
Ceftazidime solution ...................................................... 5.0mL
Cycloheximide solution ................................................. 5.0mL
Polymyxin B solution ..................................................... 5.0mL

pH $7.2 \pm 0.2$ at 25°C

**Caution:** Lithium chloride is harmful. Avoid bodily contact and inhalation of vapors. On contact with skin wash with plenty of wate.

**Nalidixic Acid Solution:**
**Composition** per 5.0mL:

Nalidixic acid ................................................................ 0.02g

**Preparation of Nalidixic Acid Solution:** Add nalidixic acid to distilled/deionized water and bring volume to 5.0mL. Mix thoroughly. Filter sterilize.

**Ceftazidime Solution:**
**Composition** per 5.0mL:

Ceftazidime .................................................................... 0.02g

**Preparation of Ceftazidime Solution:** Add ceftazidime to distilled/deionized water and bring volume to 5.0mL. Mix thoroughly. Filter sterilize.

**Cycloheximide Solution:**
**Composition** per 5.0mL:

Cycloheximide ................................................................ 0.05g
Ethanol ......................................................................... 2.5mL

**Preparation of Cycloheximide Solution:** Add cycloheximide to 2.5mL of ethanol. Mix thoroughly. Bring volume to 5.0mL with distilled/deionized water. Filter sterilize.

**Caution:** Cycloheximide is toxic. Avoid skin contact or aerosol formation and inhalation.

**Polymyxin B Solution:**
**Composition** per 5.0mL:

Polymyxin B ............................................................... 76700U

**Preparation of Polymyxin B Solution:** Add polymyxin B to distilled/deionized water and bring volume to 5.0mL. Mix thoroughly. Filter sterilize.

**Phosphatidylinositol Solution:**
**Composition** per 50.0mL:

L-α-phosphatidylinositol .................................................. 2.0g

**Preparation of Phosphatidylinositol Solution:** Add L-α-phosphotidylinositol to cold distilled/deionized water and bring volume to 50.0mL. Stir for 30 min so a homogeneous suspension is obtained. Autoclave for 15 min at 15 psi pressure–121°C. Cool to 48–50°C.

**Preparation of Medium:** Add components, except phosphatidylinositol solution, nalidixic acid solution, cetazidime solution, cycloheximide solution, and polymyxin B solution, to distilled/deionized water and bring volume to 930.0mL. Mix thoroughly. Adjust the pH to 7.2. Gently heat and bring to boiling. Autoclave for 15 min at 15 psi pressure–121°C. Cool to 45°–50°C. Aseptically add 50.0mL sterile phosphatidylinositol solution, 5.0mL sterile nalidixic acid solution, 5.0mL sterile cetazidime solution, 5.0mL sterile cycloheximide solution, and 5.0mL sterile polymyxin B solution. Mix thoroughly. Pour into Petri dishes or distribute into sterile tubes.

**Storage/Shelf Life:** Store dehydrated media in the dark in a sealed container below 30°C. Prepared media should be stored under refrigeration (2-8°C). Media should be used within 60 days of preparation. Media should not be used if there are any signs of deterioration (shrinking, cracking, or discoloration) or contamination, or if the expiration date supplied by the manufacturer has passed.

**Use:** For the isolaltion and cultivation of *Listeria* spp.

## ALOA Medium
## (Agar *Listeria* Ottavani & Agosti)
## (BAM M10a)

**Composition** per liter:

Agar ............................................................................... 18.0g
Peptone ......................................................................... 18.0g
LiCl ............................................................................... 10.0g
Yeast extract ................................................................. 10.0g
Tryptone .......................................................................... 6.0g
NaCl ................................................................................ 5.0g
$Na_2HPO_4$ ....................................................................... 2.5g
Na-pyruvate ..................................................................... 2.0g
Glucose ........................................................................... 2.0g
Mg-glycerophosphate ...................................................... 1.0g
$MgSO_4$ ............................................................................ 0.5g
5-Bromo4-chloro-indolyl-β-D-glucopyranoside ............. 0.05g
Phosphatidylinositol solution ..................................... 50.0mL
Amphotericin B solution .............................................. 10.0mL
Nalidixic acid solution ................................................. 5.0mL

Ceftazidime solution ................................................................ 5.0mL
Polymyxin B solution ............................................................. 5.0mL

pH 7.2 ± 0.2 at 25°C

**Caution:** Lithium chloride is harmful. Avoid bodily contact and inhalation of vapors. On contact with skin wash with plenty of water immediately.

**Nalidixic Acid Solution:**
**Composition** per 5.0mL:
Nalidixic acid ......................................................................0.02g

**Preparation of Nalidixic Acid Solution:** Add nalidixic acid to distilled/deionized water and bring volume to 5.0mL. Mix thoroughly. Filter sterilize.

**Ceftazidime Solution:**
**Composition** per 5.0mL:
Ceftazidime ..........................................................................0.02g

**Preparation of Ceftazidime Solution:** Add ceftazidime to distilled/deionized water and bring volume to 5.0mL. Mix thoroughly. Filter sterilize.

**Amphotericin B Solution:**
**Composition** per 10.0mL:
Amphotericin B......................................................................0.01g
Dimethylforamide................................................................ 7.5mL
HCL, 1$M$..................................................................................2.5mL

**Preparation of Amphotericin B Solution:** Add amphotericin B to 2.5mL of 1$M$ HCl. Mix thoroughly. Add 7.5 mL of dimethlyforamide. Mix thoroughly. Filter sterilize.

**Polymyxin B Solution:**
**Composition** per 5.0mL:
Polymyxin B .........................................................................76700U

**Preparation of Polymyxin B Solution:** Add polymyxin B to distilled/deionized water and bring volume to 5.0mL. Mix thoroughly. Filter sterilize.

**Phosphatidylinositol Solution:**
**Composition** per 50.0mL:
L-α-phosphatidylinositol...............................................................2.0g

**Preparation of Phosphatidylinositol Solution:** Add L-α-phosphotidylinositol to cold distilled/deionized water and bring volume to 50.0mL. Stir for 30 min so a homogeneous suspension is obtained. Autoclave for 15 min at 15 psi pressure–121°C. Cool to 48–50°C.

**Preparation of Medium:** Add components, except phosphatidylinositol solution, nalidixic acid solution, cetazidime solution, amphotericin B solution, and polymyxin B solution, to distilled/deionized water and bring volume to 920.0mL. Mix thoroughly. Adjust the pH to 7.2. Gently heat and bring to boiling. Autoclave for 15 min at 15 psi pressure–121°C. Cool to 45°–50°C. Aseptically add 50.0mL sterile phosphatidylinositol solution, 5.0mL sterile nalidixic acid solution, 5.0mL sterile cetazidime solution, 10.0mL sterile amphotericin B solution, and 5.0mL sterile polymyxin B solution. Mix thoroughly. Pour into Petri dishes or distribute into sterile tubes.

**Storage/Shelf Life:** Store dehydrated media in the dark in a sealed container below 30°C. Prepared media should be stored under refrigeration (2-8°C). Media should be used within 60 days of preparation. Media should not be used if there are any signs of deterioration (shrinking, cracking, or discoloration) or contamination, or if the expiration date supplied by the manufacturer has passed.

**Use:** For the isolaltion and cultivation of *Listeria* spp. For the isolation and cultivation of *Literia* spp. according to ISO standard 11290.

## Amies Modified Transport Medium with Charcoal
**Composition** per liter:
Charcoal.................................................................................10.0g
Agar .......................................................................................4.0g
NaCl........................................................................................3.0g
Na$_2$HPO$_4$...............................................................................1.15g
Sodium thioglycolate ............................................................1.0g
KCl..........................................................................................0.2g
KH$_2$PO$_4$....................................................................................0.2g
CaCl$_2$·2H$_2$O ...........................................................................0.1g
MgCl$_2$·6H$_2$O ..........................................................................0.1g

pH 7.2 ± 0.2 at 25°C

**Source:** This medium is available as a premixed powder from BD Diagnostic Systems.

**Preparation of Medium:** Add components to distilled/deionized water and bring volume to 1.0L. Mix thoroughly. Gently heat and bring to boiling. Distribute into flasks or tubes. Autoclave for 20 min at 15 psi pressure–121°C. While cooling, turn tubes to uniformly suspend charcoal.

**Storage/Shelf Life:** Store dehydrated media in the dark in a sealed container below 30°C. Prepared media should be stored under refrigeration (2-8°C). Media should be used within 60 days of preparation. Media should not be used if there are any signs of deterioration (discoloration) or contamination, or if the expiration date supplied by the manufacturer has passed.

**Use:** For the transport of swab specimens to prolong the survival of microorganisms, especially *Neisseria gonorrhoeae,* between collection and culturing. Addition of charcoal to this medium neutralizes metabolic products that may be toxic to *Neisseria gonorrhoeae.*

## Amies Modified Transport Medium with Charcoal
**Composition** per liter:
Charcoal.................................................................................10.0g
NaCl........................................................................................8.0g
Agar .......................................................................................3.6g
Na$_2$HPO$_4$...............................................................................1.15g
Sodium thioglycolate ............................................................1.0g
KCl..........................................................................................0.2g
CaCl$_2$·2H$_2$O ...........................................................................0.1g
MgCl$_2$·6H$_2$O ..........................................................................0.1g
KH$_2$PO$_4$....................................................................................0.2g

pH 7.2 ± 0.2 at 25°C

**Source:** This medium is available as a premixed powder from BD Diagnostic Systems and Thermo Scientific.

**Preparation of Medium:** Add components to distilled/deionized water and bring volume to 1.0L. Mix thoroughly. Gently heat and bring to boiling. Distribute into flasks or tubes. Autoclave for 20 min at 15 psi pressure–121°C. While cooling, turn tubes to uniformly suspend charcoal.

**Storage/Shelf Life:** Store dehydrated media in the dark in a sealed container below 30°C. Prepared media should be stored under refrigeration (2-8°C). Media should be used within 60 days of preparation. Media should not be used if there are any signs of deterioration (discoloration) or contamination, or if the expiration date supplied by the manufacturer has passed.

**Use:** For the transport of swab specimens to prolong the survival of microorganisms, especially *Neisseria gonorrhoeae,* between collection and culturing. Addition of charcoal to this medium neutralizes metabolic products that may be toxic to *Neisseria gonorrhoeae.*

### Amies Transport Medium without Charcoal
**Composition** per liter:

| | |
|---|---|
| Agar | 4.0g |
| NaCl | 3.0g |
| Na$_2$HPO$_4$ | 1.15g |
| Sodium thioglycolate | 1.0g |
| KCl | 0.2g |
| CaCl$_2$·2H$_2$O | 0.1g |
| MgCl$_2$·6H$_2$O | 0.1g |
| KH$_2$PO$_4$ | 0.2g |

pH 7.2 ± 0.2 at 25°C

**Source:** This medium is available as a premixed powder from BD Diagnostic Systems.

**Preparation of Medium:** Add components to distilled/deionized water and bring volume to 1.0L. Mix thoroughly. Gently heat and bring to boiling. Distribute into flasks or tubes. Autoclave for 20 min at 15 psi pressure–121°C.

**Use:** For the transport of swab specimens to prolong the survival of microorganisms, especially *Neisseria gonorrhoeae,* between collection and culturing.

### Amies Transport Medium without Charcoal
**Composition** per liter:

| | |
|---|---|
| NaCl | 8.0g |
| Agar | 3.6g |
| Na$_2$HPO$_4$ | 1.15g |
| Sodium thioglycolate | 1.0g |
| KCl | 0.2g |
| CaCl$_2$·2H$_2$O | 0.1g |
| MgCl$_2$·6H$_2$O | 0.1g |
| KH$_2$PO$_4$ | 0.2g |

pH 7.2 ± 0.2 at 25°C

**Source:** This medium is available as a premixed powder from BD Diagnostic Systems.

**Preparation of Medium:** Add components to distilled/deionized water and bring volume to 1.0L. Mix thoroughly. Gently heat and bring to boiling. Distribute into flasks or tubes. Autoclave for 20 min at 15 psi pressure–121°C.

**Storage/Shelf Life:** Store dehydrated media in the dark in a sealed container below 30°C. Prepared media should be stored under refrigeration (2-8°C). Media should be used within 60 days of preparation. Media should not be used if there are any signs of deterioration (discoloration) or contamination, or if the expiration date supplied by the manufacturer has passed.

**Use:** For the transport of swab specimens to prolong the survival of microorganisms, especially *Neisseria gonorrhoeae,* between collection and culturing.

### Ampicillin Dextrin Agar
**Composition** per liter:

| | |
|---|---|
| Agar | 15.0g |
| Dextrin | 10.0g |
| NaCl | 3.0g |
| Yeast extract | 2.0g |

| | |
|---|---|
| KCl | 2.0g |
| MgSO$_4$·7H$_2$O | 0.2g |
| FeCl$_3$·4H$_2$O | 0.1g |
| Selective supplement solution | 10.0mL |

pH 8.0 ± 0.2 at 25°C

**Source:** This medium is available from HiMedia.

**Selective Supplement Solution:**
**Composition** per 10.0mL:

| | |
|---|---|
| Sodium deoxycholate | 100.0mg |
| Ampicillin | 10.0mg |

**Preparation of Selective Supplement Solution:** Add components to distilled/deionized water and bring volume to 10.0mL. Mix thoroughly. Filter sterilize.

**Preparation of Medium:** Add components, except selective supplement solution and kanamycin solution, to distilled/deionized water and bring volume to 990.0mL. Mix thoroughly. Autoclave for 15 min at 15 psi pressure–121°C. Aseptically add 10.0mL of sterile selective supplement solution. Mix thoroughly. Pour into Petri dishes or aseptically distribute into sterile tubes.

**Storage/Shelf Life:** Store dehydrated media in the dark in a sealed container below 30°C. Prepared media should be stored under refrigeration (2-8°C). Media should be used within 60 days of preparation. Media should not be used if there are any signs of deterioration (shrinking, cracking, or discoloration) or contamination, or if the expiration date supplied by the manufacturer has passed.

**Use:** For the selective isolation and differentiation of *Aeromonas* spp. from water samples.

### Ampicillin Dextrin Agar with Vancomycin (ADA-V)
**Composition** per liter:

| | |
|---|---|
| Agar | 13.0g |
| Dextrin | 11.4g |
| Tryptose | 5.0g |
| NaCl | 3.0g |
| KCl | 2.0g |
| Yeast extract | 2.0g |
| MgSO$_4$·7H$_2$O | 1.0g |
| Bromothymol Blue | 0.08g |
| FeCl$_3$·6H$_2$O | 0.06g |
| Sodium deoxycholate | 1.0mg |
| Ampicillin solution | 10.0mL |
| Vancomycin solution | 10.0mL |

pH 8.0 ± 0.2 at 25°C

**Ampicillin Solution:**
**Composition** per 10.0mL:

| | |
|---|---|
| Ampicillin | 10.0mg |

**Preparation of Ampicillin Solution:** Add components to distilled/deionized water and bring volume to 10.0mL. Mix thoroughly. Filter sterilize.

**Vancomycin Solution:**
**Composition** per 10.0mL:

| | |
|---|---|
| Vancomycin | 2.0mg |

**Preparation of Vancomycin Solution:** Add components to distilled/deionized water and bring volume to 10.0mL. Mix thoroughly. Filter sterilize.

**Preparation of Medium:** Add components, except sodium deoxycholate, ampicillin solution, and vancomycin solution, to distilled/deionized water and bring volume to 980.0mL. Mix thoroughly. Adjust pH to 8.0. Add sodium deoxycholate. Autoclave for 15 min at 15 psi pressure–121°C. Cool to 50°C. Aseptically add ampicillin and vancomycin solutions. Mix thoroughly. Pour into Petri dishes or aseptically distribute into sterile tubes.

**Storage/Shelf Life:** Store dehydrated media in the dark in a sealed container below 30°C. Prepared media should be stored under refrigeration (2-8°C). Media should be used within 60 days of preparation. Media should not be used if there are any signs of deterioration (shrinking, cracking, or discoloration) or contamination, or if the expiration date supplied by the manufacturer has passed.

**Use:** For the selective isolation and differentiation of *Aeromonas* spp. from water samples.

## Anaerobic Basal Agar with Blood
**Composition** per liter:
| | |
|---|---|
| Peptic digest of animal tissue | 16.0g |
| Agar | 12.0g |
| Yeast extract | 7.0g |
| NaCl | 5.0g |
| Starch | 1.0g |
| Glucose | 1.0g |
| Sodium pyruvate | 1.0g |
| L-Arginine | 1.0g |
| Sodium succinate | 0.5g |
| $Fe_4(P_2O_7)\cdot H_2O$ | 0.5g |
| $NaHCO_3$ | 0.4g |
| L-Cysteine HCl | 0.25g |
| Dithiothreitol | 0.25g |
| Hemin | 5.0mg |
| Vitamin K | 5.0mg |
| Horse blood, defibrinated | 100.0mL |

pH 7.0 ± 0.2 at 25°C

**Source:** This medium is available from HiMedia.

**Preparation of Medium:** Add components, except horse blood, to distilled/deionized water and bring volume to 900.0mL. Mix thoroughly. Adjust pH to 7.0. Autoclave for 15 min at 15 psi pressure–121°C. Cool to 50°–55°C. Aseptically add 100.0mL of sterile horse blood. Mix thoroughly. Pour into sterile Petri dishes or leave in tubes.

**Storage/Shelf Life:** Store dehydrated media in the dark in a sealed container below 30°C. Prepared media should be stored under refrigeration (2-8°C). Media should be used within 60 days of preparation. Media should not be used if there are any signs of deterioration (shrinking, cracking, or discoloration) or contamination, or if the expiration date supplied by the manufacturer has passed.

**Use:** For the cultivation of anaerobic microorganisms, especially *Bacteroides* species and other fastidious anaerobes.

## Anaerobic Blood Agar Base with Blood and Neomycin (Anaerobe Neomycin 5% Sheep Blood Agar)
**Composition** per liter:
| | |
|---|---|
| Casein enzymic hydrolysate | 14.5g |
| Agar | 14.0g |
| Papaic digest of soybean meal | 5.0g |
| NaCl | 5.0g |
| Growth factors | 1.5g |

| | |
|---|---|
| Sheep blood, sterile defibrinated | 50.0mL |
| Selective supplement solution | 10.0mL |

pH 7.3 ± 0.2 at 25°C

**Source:** This medium is available from HiMedia.

**Selective Supplement Solution:**
**Composition** per 10.0mL:
| | |
|---|---|
| Neomycin sulfate | 30.0mg |

**Preparation of Selective Supplement Solution:** Add neomycin sulfate to distilled/deionized water and bring volume to 10.0mL. Mix thoroughly. Filter sterilize.

**Preparation of Medium:** Add components, except sheep blood and selective supplement, to distilled/deionized water and bring volume to 940.0mL. Mix thoroughly. Adjust pH to 7.3. Autoclave for 15 min at 15 psi pressure–121°C. Cool to 50°–55°C. Aseptically add 50.0mL of sterile sheep blood and 10.0mL of selective supplement solution. Mix thoroughly. Pour into sterile Petri dishes or leave in tubes.

**Storage/Shelf Life:** Store dehydrated media in the dark in a sealed container below 30°C. Prepared media should be stored under refrigeration (2-8°C). Media should be used within 60 days of preparation. Media should not be used if there are any signs of deterioration (shrinking, cracking, or discoloration) or contamination, or if the expiration date supplied by the manufacturer has passed.

**Use:** For the isolation and cultivation of Group A and Group B streptococci from throat cultures and other clinical samples.

## Anaerobic CNA Agar (Anaerobic Colistin Nalidixic Acid Agar)
**Composition** per liter:
| | |
|---|---|
| Agar | 13.0g |
| Pancreatic digest of casein | 12.0g |
| Peptic digest of animal tissue | 5.0g |
| NaCl | 5.0g |
| Yeast extract | 3.0g |
| Beef extract | 3.0g |
| Cornstarch | 1.0g |
| Glucose | 1.0g |
| L-Cysteine·HCl·$H_2O$ | 0.5g |
| Vitamin $K_1$ | 10.0mg |
| Hemin | 10.0mg |
| Colistin | 10.0mg |
| Nalidixic acid | 10.0mg |
| Sheep blood, defibrinated | 50.0mL |

**Source:** This medium is available as a premixed powder from BD Diagnostic Systems.

**Preparation of Medium:** Add components, except sheep blood, to distilled/deionized water and bring volume to 950.0mL. Mix thoroughly. Gently heat and bring to boiling. Autoclave for 15 min at 15 psi pressure–121°C. Cool to 45°–50°C. Aseptically add 50.0mL of sterile, defibrinated sheep blood. Mix thoroughly. Pour into sterile Petri dishes.

**Storage/Shelf Life:** Store dehydrated media in the dark in a sealed container below 30°C. Prepared media should be stored under refrigeration (2-8°C). Media should be used within 60 days of preparation. Media should not be used if there are any signs of deterioration (shrinking, cracking, or discoloration) or contamination, or if the expiration date supplied by the manufacturer has passed.

**Use:** For the selective isolation of anaerobic streptococci.

## Anaerobic CNA Agar Base with Blood

**Composition** per liter:

| | |
|---|---|
| Agar | 13.5g |
| Casein enzymic hydrolysate | 12.0g |
| Peptic digest of animal tissue | 5.0g |
| NaCl | 5.0g |
| Yeast extract | 3.0g |
| Beef extract | 3.0g |
| Corn starch | 1.0g |
| Glucose | 1.0g |
| L-Cystine hydrochloride | 0.5g |
| Dithiothreitol (DTE) | 0.1g |
| Vitamin $K_1$ | 0.01g |
| Hemin | 0.01g |
| Colistin | 0.01g |
| Nalidixic acid | 0.01g |
| Sheep blood, sterile defibrinated | 50.0mL |

pH 7.0 ± 0.2 at 25°C

**Source:** This medium is available from HiMedia.

**Preparation of Medium:** Add components, except sheep blood, to distilled/deionized water and bring volume to 950.0mL. Mix thoroughly. Adjust pH to 7.0. Autoclave for 15 min at 15 psi pressure–121°C. Cool to 50°–55°C. Aseptically add 50.0mL of sterile sheep blood. Mix thoroughly. Pour into sterile Petri dishes or leave in tubes.

**Storage/Shelf Life:** Store dehydrated media in the dark in a sealed container below 30°C. Prepared media should be stored under refrigeration (2-8°C). Media should be used within 60 days of preparation. Media should not be used if there are any signs of deterioration (shrinking, cracking, or discoloration) or contamination, or if the expiration date supplied by the manufacturer has passed.

**Use:** For the selective isolation of anaerobic streptococci.

## Anaerobic Egg Yolk Agar

**Composition** per 1080.0mL:

| | |
|---|---|
| Agar | 20.0g |
| Proteose peptone | 20.0g |
| NaCl | 5.0g |
| Pancreatic digest of casein | 5.0g |
| Yeast extract | 5.0g |
| Egg yolk emulsion, 50% | 80.0mL |

pH 7.0 ± 0.2 at 25°C

**Egg Yolk Emulsion, 50%:**
**Composition** per 100.0mL:

| | |
|---|---|
| Chicken egg yolks | 11 |
| Whole chicken egg | 1 |
| NaCl (0.9% solution) | 50.0mL |

**Preparation of Egg Yolk Emulsion, 50%:** Soak whole eggs with 1:100 dilution of saturated mercuric chloride solution for 1 min. Crack eggs and separate yolks from whites. Mix egg yolks with 1 chicken egg. Beat to form emulsion. Measure 50.0mL of egg yolk emulsion and add to 50.0mL of 0.9% NaCl solution. Mix thoroughly. Filter sterilize. Warm to 45°–50°C.

**Preparation of Medium:** Add components, except egg yolk emulsion, 50%, to distilled/deionized water and bring volume to 1.0L. Mix thoroughly. Gently heat and bring to boiling. Autoclave for 15 min at 15 psi pressure–121°C. Cool to 45°–50°C. Aseptically add 80.0mL of sterile egg yolk emulsion, 50%. Mix thoroughly. Pour into sterile Petri dishes or distribute into sterile tubes. Allow plates to dry at 35°C for 24 h.

**Storage/Shelf Life:** Store dehydrated media in the dark in a sealed container below 30°C. Prepared media should be stored under refrigeration (2-8°C). Media should be used within 60 days of preparation. Media should not be used if there are any signs of deterioration (shrinking, cracking, or discoloration) or contamination, or if the expiration date supplied by the manufacturer has passed.

**Use:** For the cultivation of *Clostridium* species. For the cultivation of *Yersinia enterocolitica*.

## Anaerobic Egg Yolk Agar

**Composition** per liter:

| | |
|---|---|
| Agar | 20.0g |
| Proteose peptone | 20.0g |
| Pancreatic digest of casein | 5.0g |
| NaCl | 5.0g |
| Yeast extract | 5.0g |
| Egg yolk emulsion, 50% | 20.0mL |

pH 7.0 ± 0.2 at 25°C

**Egg Yolk Emulsion, 50%:**
**Composition** per 100.0mL:

| | |
|---|---|
| Chicken egg yolks | 2 |
| NaCl (0.9% solution) | 10.0mL |

**Preparation of Egg Yolk Emulsion, 50%:** Soak eggs with 1:100 dilution of saturated mercuric chloride solution for 1 min. Crack eggs and separate yolks from whites. Beat to form emulsion. Measure 10.0mL of egg yolk emulsion and add to 10.0mL of 0.9% NaCl solution. Mix thoroughly. Filter sterilize. Warm to 45°–50°C.

**Preparation of Medium:** Add components, except egg yolk emulsion, to distilled/deionized water and bring volume to 980.0mL. Mix thoroughly. Gently heat and bring to boiling. Autoclave for 15 min at 15 psi pressure–121°C. Cool to 45°–50°C. Aseptically add sterile egg yolk emulsion. Mix thoroughly. Pour into sterile Petri dishes. Allow plates to dry at 35°C for 24 h.

**Storage/Shelf Life:** Store dehydrated media in the dark in a sealed container below 30°C. Prepared media should be stored under refrigeration (2-8°C). Media should be used within 60 days of preparation. Media should not be used if there are any signs of deterioration (shrinking, cracking, or discoloration) or contamination, or if the expiration date supplied by the manufacturer has passed.

**Use:** For the cultivation of *Yersinia enterocolitica*.

## Anaerobic Egg Yolk Agar (BAM M12)

**Composition** per liter:

| | |
|---|---|
| Agar | 20.0g |
| Proteose peptone | 20.0g |
| Pancreatic digest of casein | 5.0g |
| NaCl | 5.0g |
| Yeast extract | 5.0g |
| Egg yolk emulsion, 50% | 80.0mL |

pH 7.0 ± 0.2 at 25°C

**Egg Yolk Emulsion, 50%:**
**Composition** per 80.0mL:

| | |
|---|---|
| Chicken egg yolks | 2 or more |
| NaCl (0.85% solution) | 40.0mL |

**Preparation of Egg Yolk Emulsion, 50%:** Wash fresh eggs with stiff brush and drain. Soak eggs in 70% ethanol for 1 hour. Crack eggs aseptically and separate yolks from whites. Drain contents of yolk sacs into sterile stoppered graduate cylinder and discard sacs. Measure 40.0mL of egg yolk emulsion and add 40.0mL of 0.85% NaCl solution. Mix thoroughly by inverting graduate cylinder. Warm to 45°–50°C.

**Preparation of Medium:** Add components, except egg yolk emulsion, to distilled/deionized water and bring volume to 1.0L. Mix thoroughly. Gently heat and bring to boiling. Autoclave for 15 min at 15 psi pressure–121°C. Cool to 45°–50°C. Aseptically add 80.0mL sterile egg yolk emulsion. Mix thoroughly. Pour into sterile Petri dishes. Allow plates to dry at ambient temperature for 2–3 days or at 35°C for 24 h. Check plates for contamination before use.

**Storage/Shelf Life:** Store dehydrated media in the dark in a sealed container below 30°C. Prepared media should be stored under refrigeration (2-8°C). Media should be used within 60 days of preparation. Media should not be used if there are any signs of deterioration (shrinking, cracking, or discoloration) or contamination, or if the expiration date supplied by the manufacturer has passed.

**Use:** For the cultivation of *Yersinia enterocolitica.*

## Anaerobic Egg Yolk Base with Egg Yolk Emulsion
**Composition** per liter:

| | |
|---|---|
| Agar | 20.0g |
| Proteose peptone | 20.0g |
| Casein enzymatic hydrolysate | 5.0g |
| Yeast extract | 5.0g |
| NaCl | 5.0g |
| Egg yolk emulsion | 80.0mL |

pH 7.0 ± 0.2 at 25°C

**Source:** This medium is available from HiMedia.

**Egg Yolk Emulsion:**
**Composition** per liter:

| | |
|---|---|
| Egg yolks | 30.0mL |
| NaCl, 0.9% solution | 70.0mL |

**Preparation of Egg Yolk Emulsion:** Soak eggs with 1:100 dilution of saturated mercuric chloride solution for 1 min. Crack 11 eggs and separate yolks from whites. Mix egg yolks. Measure 30.0mL of egg yolk emulsion and add to 70.0mL of 0.9% sterile NaCl solution. Mix thoroughly. Warm to 45°–50°C.

**Preparation of Medium:** Add components, except egg yolk emulsion, to distilled/deionized water and bring volume to 920.0mL. Mix thoroughly. Gently heat and bring to boiling. Autoclave for 15 min at 15 psi pressure–121°C. Cool to 45°–50°C. Aseptically add sterile egg yolk emulsion. Mix thoroughly. Pour into sterile Petri dishes.

**Storage/Shelf Life:** Store dehydrated media in the dark in a sealed container below 30°C. Prepared media should be stored under refrigeration (2-8°C). Media should be used within 60 days of preparation. Media should not be used if there are any signs of deterioration (shrinking, cracking, or discoloration) or contamination, or if the expiration date supplied by the manufacturer has passed.

**Use:** For the cultivation of *Clostridium perfringens* from foods.

## Anaerobic HiVeg Agar Base with Egg Yolk Emulsion
**Composition** per liter:

| | |
|---|---|
| Agar | 20.0g |
| Plant petone No. 3 | 20.0g |
| Plant hydrolysate | 5.0g |
| NaCl | 5.0g |

| | |
|---|---|
| Yeast extract | 5.0g |
| Egg yolk emulsion | 100.0mL |

pH 7.2 ± 0.2 at 25°C

**Source:** This medium, without egg yolk emulsion, is available as a premixed powder from HiMedia.

**Egg Yolk Emulsion:**
**Composition** per liter:

| | |
|---|---|
| Egg yolks | 30.0mL |
| NaCl, 0.9% solution | 70.0mL |

**Preparation of Egg Yolk Emulsion:** Soak eggs with 1:100 dilution of saturated mercuric chloride solution for 1 min. Crack 11 eggs and separate yolks from whites. Mix egg yolks. Measure 30.0mL of egg yolk emulsion and add to 70.0mL of 0.9% sterile NaCl solution. Mix thoroughly. Warm to 45°–50°C.

**Preparation of Medium:** Add components, except egg yolk emulsion, to distilled/deionized water and bring volume to 900.0mL. Mix thoroughly. Gently heat and bring to boiling. Autoclave for 15 min at 15 psi pressure–121°C. Cool to 45°–50°C. Aseptically add sterile egg yolk emulsion. Mix thoroughly. Pour into sterile Petri dishes.

**Storage/Shelf Life:** Store dehydrated media in the dark in a sealed container below 30°C. Prepared media should be stored under refrigeration (2-8°C). Media should be used within 60 days of preparation. Media should not be used if there are any signs of deterioration (shrinking, cracking, or discoloration) or contamination, or if the expiration date supplied by the manufacturer has passed.

**Use:** For the cultivation of *Clostridium perfringens* from foods.

## Anaerobic HiVeg Agar without Dextrose
**Composition** per liter:

| | |
|---|---|
| Plant hydrolysate | 17.5g |
| Agar | 15.0g |
| NaCl | 2.5g |
| Sodium thioglycolate | 2.0g |
| Sodium formaldehyde sulfoxylate | 1.0g |
| Methylene Blue | 2.0mg |

pH 7.2 ± 0.2 at 25°C

**Source:** This medium is available as a premixed powder from HiMedia.

**Preparation of Medium:** Add components to distilled/deionized water and bring volume to 1.0L. Mix thoroughly. Gently heat and bring to boiling. Distribute into tubes or flasks. Autoclave for 15 min at 15 psi pressure–121°C. Pour into sterile Petri dishes or leave in tubes.

**Storage/Shelf Life:** Store dehydrated media in the dark in a sealed container below 30°C. Prepared media should be stored under refrigeration (2-8°C). Media should be used within 60 days of preparation. Media should not be used if there are any signs of deterioration (shrinking, cracking, or discoloration) or contamination, or if the expiration date supplied by the manufacturer has passed.

**Use:** For the cultivation of a variety of anaerobic microorganisms. With added blood for the detection of hemolytic activity of clostridia, streptococci, and other anaerobic bacteria. With added carbohydrate for fermentation studies.

## Anaerobic HiVeg Agar without Dextrose and Eh Indicator
**Composition** per liter:

| | |
|---|---|
| Plant hydrolysate | 20.0g |
| Agar | 15.0g |

NaCl ................................................................................5.0g
Sodium thioglycolate ...................................................2.0g
Sodium formaldehyde sulfoxylate ...............................1.0g

pH 7.2 ± 0.2 at 25°C

**Source:** This medium is available as a premixed powder from Hi-Media.

**Preparation of Medium:** Add components to distilled/deionized water and bring volume to 1.0L. Mix thoroughly. Gently heat and bring to boiling. Distribute into tubes or flasks. Autoclave for 15 min at 15 psi pressure–121°C. Pour into sterile Petri dishes or leave in tubes.

**Storage/Shelf Life:** Store dehydrated media in the dark in a sealed container below 30°C. Prepared media should be stored under refrigeration (2-8°C). Media should be used within 60 days of preparation. Media should not be used if there are any signs of deterioration (shrinking, cracking, or discoloration) or contamination, or if the expiration date supplied by the manufacturer has passed.

**Use:** For the cultivation of a variety of anaerobic microorganisms. With added blood for the detection of hemolytic activity of clostridia, streptococci, and other anaerobic bacteria.

## Anaerobic LKV Blood Agar

**Composition** per liter:

Agar ................................................................................15.0g
Pancreatic digest of casein ..........................................13.0g
Peptic digest of animal tissue .....................................10.0g
NaCl ................................................................................5.0g
Yeast extract ...................................................................2.0g
Glucose ...........................................................................1.0g
NaHSO$_3$ ...........................................................................0.1g
Sheep blood, laked .................................................. 50.0mL
Antibiotic solution .................................................. 10.0mL
Hemin solution ........................................................... 1.0mL
Vitamin K$_1$ solution ................................................... 1.0mL

pH 7.1–7.8 at 25°C

**Source:** This medium is available as a premixed powder from BD Diagnostic Systems.

**Antibiotic Solution:**
**Composition** per 10.0mL:

Kanamycin .....................................................................0.075g
Vancomycin ...................................................................7.5mg

**Preparation of Antibiotic Solution:** Add components to distilled/deionized water and bring volume to 10.0mL. Mix thoroughly. Filter sterilize.

**Vitamin K$_1$ Solution:**
**Composition** per 100.0mL:

Vitamin K$_1$ ......................................................................0.1g
Ethanol ...................................................................... 99.0mL

**Preparation of Vitamin K$_1$ Solution:** Add vitamin K$_1$ to 99.0mL of absolute ethanol. Mix thoroughly.

**Hemin Solution:**
**Composition** per 100.0mL:

Hemin ..............................................................................0.01g
NaOH (1$N$ solution) ................................................. 20.0mL

**Preparation of Hemin Solution:** Add hemin to 20.0mL of 1$N$ NaOH solution. Mix thoroughly. Bring volume to 100.0mL with distilled/deionized water.

**Preparation of Medium:** Add components—except sheep blood, antibiotic solution, and vitamin K$_1$ solution—to distilled/deionized water and bring volume to 939.0mL. Mix thoroughly. Gently heat and bring to boiling. Autoclave for 15 min at 15 psi pressure–121°C. Cool to 45°–50°C. Aseptically add 50.0mL of sterile sheep blood, 10.0mL of sterile antibiotic solution, and 1.0mL of sterile vitamin K$_1$ solution. Mix thoroughly. Pour into sterile Petri dishes or distribute into sterile tubes.

**Storage/Shelf Life:** Store dehydrated media in the dark in a sealed container below 30°C. Prepared media should be stored under refrigeration (2-8°C). Media should be used within 60 days of preparation. Media should not be used if there are any signs of deterioration (shrinking, cracking, or discoloration) or contamination, or if the expiration date supplied by the manufacturer has passed.

**Use:** For the isolation and cultivation of anaerobic Gram-negative microorganisms, especially *Bacteroides* species.

## Anaerobic Thioglycollate Medium Base with Serum

**Composition** per liter:

Casein enzymic hydrolysate ........................................17.0g
Meat extract ....................................................................7.5g
D-Glucose ........................................................................6.0g
Liver hydrolysate ...........................................................3.0g
Papaic digest of soybean meal .....................................3.0g
NaCl ................................................................................2.5g
Agar ................................................................................0.7g
Sodium thioglycollate ...................................................0.5g
L-Cysteine .......................................................................0.25g
Na$_2$SO$_3$ ...........................................................................0.1g
Serum, sterile ......................................................... 100.0mL

pH 7.3 ± 0.2 at 25°C

**Source:** This medium is available from HiMedia.

**Preparation of Medium:** Add components, except serum, to distilled/deionized water and bring volume to 900.0mL. Mix thoroughly. Adjust pH to 7.0. Autoclave for 15 min at 15 psi pressure–121°C. Cool to 50°–55°C. Aseptically add 100.0mL of sterile serum. Mix thoroughly. Aseptically distribute into sterile tubes.

**Storage/Shelf Life:** Store dehydrated media in the dark in a sealed container below 30°C. Prepared media should be stored under refrigeration (2-8°C). Media should be used within 60 days of preparation. Media should not be used if there are any signs of deterioration (shrinking, cracking, or discoloration) or contamination, or if the expiration date supplied by the manufacturer has passed.

**Use:** For the selective isolation of anaerobic bacteria.

## Anaerobic Trypticase™ Soy Agar with Calf Blood

**Composition** per liter:

Pancreatic digest of casein ..........................................15.0g
Agar ................................................................................15.0g
Papaic digest of soybean meal .....................................5.0g
NaCl ................................................................................5.0g
Calf blood, defibrinated ........................................ 100.0mL

pH 7.3 ± 0.2 at 25°C

**Preparation of Medium:** Add components, except calf blood, to distilled/deionized water and bring volume to 900.0mL. Mix thoroughly. Prepare medium anaerobically with 80% N$_2$ + 10% CO$_2$ + 10% H$_2$. Gently heat while stirring and bring to boiling for 1 min. Autoclave for 15 min at 15 psi pressure–121°C. Do not overheat. Cool to 45°–50°C.

Aseptically add 100.0mL sterile, defibrinated calf blood. Pour into sterile Petri dishes.

**Storage/Shelf Life:** Store dehydrated media in the dark in a sealed container below 30°C. Prepared media should be stored under refrigeration (2-8°C). Media should be used within 60 days of preparation. Media should not be used if there are any signs of deterioration (shrinking, cracking, or discoloration) or contamination, or if the expiration date supplied by the manufacturer has passed.

**Use:** For the isolation and cultivation of fastidious as well as nonfastidious microorganisms. For the differentiation of *Haemophilus* species.

## Andrade Peptone Water
**Composition** per liter:
Peptic digest of animal tissue.........................................................10.0g
NaCl...................................................................................................5.0g
Andrade indicator ............................................................................0.1g
pH 7.4 ± 0.2 at 25°C

**Source:** This medium is available as a premixed powder from Hi-Media.

**Preparation of Medium:** Add components to distilled/deionized water and bring volume to 1.0L. Mix thoroughly. Gently heat and bring to boiling. Distribute into tubes or flasks. Autoclave for 15 min at 15 psi pressure–121°C. Pour into sterile Petri dishes or leave in tubes.

**Caution:** Acid Fuchsin in Andrade indicator is a potential carcinogen and care must be taken to avoid inhalation of the powdered dye and contact with the skin.

**Storage/Shelf Life:** Store dehydrated media in the dark in a sealed container below 30°C. Prepared media should be stored under refrigeration (2-8°C). Media should be used within 60 days of preparation. Media should not be used if there are any signs of deterioration (discoloration) or contamination, or if the expiration date supplied by the manufacturer has passed.

**Use:** For the determination of carbohydrate fermentation reactions of microorganisms, particularly members of the Enterobacteriaceae. A specific carbohydrate is added to the medium to test the fermentation of that carbohydrate. A Durham tube is used to collect gas produced during the fermentation reaction. Acid production is indicated by a pink color.

## Andrade Peptone Water with HiVeg Extract No. 1
**Composition** per liter:
Plant peptone...................................................................................10.0g
NaCl...................................................................................................5.0g
Plant extract No. 1 ...........................................................................3.0g
Andrade indicator ............................................................................0.1g
pH 7.4 ± 0.2 at 25°C

**Source:** This medium is available as a premixed powder from Hi-Media.

**Preparation of Medium:** Add components to distilled/deionized water and bring volume to 1.0L. Mix thoroughly. Gently heat and bring to boiling. Distribute into tubes or flasks. Autoclave for 15 min at 15 psi pressure–121°C. Pour into sterile Petri dishes or leave in tubes.

**Caution:** Acid Fuchsin in Andrade indicator is a potential carcinogen and care must be taken to avoid inhalation of the powdered dye and contact with the skin.

**Storage/Shelf Life:** Store dehydrated media in the dark in a sealed container below 30°C. Prepared media should be stored under refrigeration (2-8°C). Media should be used within 60 days of preparation. Media should not be used if there are any signs of deterioration (discoloration) or contamination, or if the expiration date supplied by the manufacturer has passed.

**Use:** With added carbohydrates, for the determination of carbohydrate fermentation reactions of microorganisms, particularly members of the Enterobacteriaceae. A specific carbohydrate is added to the medium to test the fermentation of that carbohydrate. A Durham tube is used to collect gas produced during the fermentation reaction. Acid production is indicated by a pink color.

## Andrade Peptone Water with Meat Extract
**Composition** per liter:
Peptic digest of animal tissue ........................................................10.0g
NaCl...................................................................................................5.0g
Meat extract .....................................................................................3.0g
Andrade indicator ............................................................................0.1g
pH 7.4 ± 0.2 at 25°C

**Source:** This medium is available as a premixed powder from Hi-Media.

**Preparation of Medium:** Add components to distilled/deionized water and bring volume to 1.0L. Mix thoroughly. Gently heat and bring to boiling. Distribute into tubes or flasks. Autoclave for 15 min at 15 psi pressure–121°C. Pour into sterile Petri dishes or leave in tubes.

**Caution:** Acid Fuchsin in Andrade indicator is a potential carcinogen and care must be taken to avoid inhalation of the powdered dye and contact with the skin.

**Storage/Shelf Life:** Store dehydrated media in the dark in a sealed container below 30°C. Prepared media should be stored under refrigeration (2-8°C). Media should be used within 60 days of preparation. Media should not be used if there are any signs of deterioration (discoloration) or contamination, or if the expiration date supplied by the manufacturer has passed.

**Use:** With added carbohydrates, for the determination of carbohydrate fermentation reactions of microorganisms, particularly members of the Enterobacteriaceae. A specific carbohydrate is added to the medium to test the fermentation of that carbohydrate. A Durham tube is used to collect gas produced during the fermentation reaction. Acid production is indicated by a pink color.

## Andrade's Broth
**Composition** per liter:
Pancreatic digest of gelatin............................................................10.0g
NaCl...................................................................................................5.0g
Beef extract......................................................................................3.0g
Andrade's indicator..........................................................................10.0mL
Carbohydrate solution......................................................................50.0mL
pH 7.4 ± 0.2 at 25°C

**Source:** This medium is available as a prepared medium from BD Diagnostic Systems, in tubes containing adonitol, arabinose, cellobiose, dulcitol, fructose, galactose, glucose, inositol, lactose, maltose, mannitol, raffinose, rhamnose, salicin, sorbitol, sucrose, trehalose, or xylose.

### Andrade's Indicator
**Composition** per 100.0mL:
NaOH (1*N* solution)........................................................................16.0mL
Acid Fuchsin....................................................................................0.1g

**Preparation of Andrade's Indicator:** Add Acid Fuchsin to NaOH solution and bring volume to 100.0mL with distilled/deionized water.

**Carbohydrate Solution:**
**Composition** per 100.0mL:
Carbohydrate.............................................................................10.0g

**Preparation of Carbohydrate Solution:** Add carbohydrate to distilled/deionized water and bring volume to 100.0mL. Adonitol, arabinose, cellobiose, dulcitol, fructose, galactose, glucose, inositol, lactose, maltose, mannitol, raffinose, rhamnose, salicin, sorbitol, sucrose, trehalose, xylose, or other carbohydrates may be used. Mix thoroughly. Filter sterilize.

**Preparation of Medium:** Add components, except carbohydrate solution, to distilled/deionized water and bring volume to 1.0L. Mix thoroughly. Gently heat and bring to boiling. Distribute in 10.0mL volumes into test tubes containing inverted Durham tubes. Autoclave for 15 min at 15 psi pressure–121°C. Cool to 25°C. Add 0.5mL of sterile carbohydrate solution to each tube.

**Caution:** Acid Fuchsin is a potential carcinogen and care must be taken to avoid inhalation of the powdered dye and contact with the skin.

**Storage/Shelf Life:** Store dehydrated media in the dark in a sealed container below 30°C. Prepared media should be stored under refrigeration (2-8°C). Media should be used within 60 days of preparation. Media should not be used if there are any signs of deterioration (discoloration) or contamination, or if the expiration date supplied by the manufacturer has passed.

**Use:** For the determination of carbohydrate fermentation reactions of microorganisms, particularly members of the Enterobacteriaceae. A Durham tube is used to collect gas produced during the fermentation reaction. Acid production is indicated by a pink color.

### Andrade's Carbohydrate Broth and Indicator (BAM M13)

**Composition** per liter:
Pancreatic digest of gelatin...........................................10.0g
NaCl.................................................................................10.0g
Beef extract.......................................................................3.0g
Carbohydrate solution.................................................. 100.0mL
Andrade's indicator...................................................... 10.0mL
pH 7.2 ± 0.2 at 25°C

**Source:** This medium is available as a prepared medium from BBL Microbiology Systems, in tubes containing adonitol, arabinose, cellobiose, glucose, dulcitol, fructose, galactose, inositol, lactose, maltose, mannitol, raffinose, rhamnose, salicin, sorbitol, sucrose, trehalose, or xylose.

**Andrade's Indicator**
**Composition** per 26.0mL:
NaOH (1*N* solution)..................................................... 16.0mL
Acid Fuchsin..................................................................0.21g

**Preparation of Andrade's Indicator:** Add Acid Fuchsin to NaOH solution and bring volume to 26.0mL with distilled/deionized water.

**Carbohydrate Solution:**
**Composition** per 100.0mL:
Carbohydrate.............................................................5.0–10.0g

**Preparation of Carbohydrate Solution:** Add carbohydrate to distilled/deionized water and bring volume to 100.0mL. For glucose,

lactose, sucrose, and mannitol, add 10.0g to distilled/deionized water and bring volume to 100.0mL. For dulcitol, salicin, and other carbohydrates, add 5.0g to distilled/deionized water and bring volume to 100.0mL. Mix thoroughly. Filter sterilize.

**Preparation of Medium:** Add components, except carbohydrate solution, to distilled/deionized water and bring volume to 1.0L. Mix thoroughly. Gently heat and bring to boiling. Cool. Aseptically add 100mL of sterile carbohydrate solution to 900mL of sterile medium. Mix thoroughly. Aseptically distribute into tubes or flasks. Alternately, prior to autoclaving, distribute 9.0mL volumes into test tubes containing inverted Durham tubes. Autoclave for 15 min at 15 psi pressure–121°C. Cool to 25°C. Add 1.0mL of sterile carbohydrate solution to each tube.

**Caution:** Acid Fuchsin is a potential carcinogen and care must be taken to avoid inhalation of the powdered dye and contact with the skin.

**Storage/Shelf Life:** Store dehydrated media in the dark in a sealed container below 30°C. Prepared media should be stored under refrigeration (2-8°C). Media should be used within 60 days of preparation. Media should not be used if there are any signs of deterioration (discoloration) or contamination, or if the expiration date supplied by the manufacturer has passed.

**Use:** For the determination of carbohydrate fermentation reactions of microorganisms, particularly members of the Enterobacteriaceae. A Durham tube is used to collect gas produced during the fermentation reaction. Acid production is indicated by a pink color.

### Anthracis Chromogenic Agar

**Composition** per liter:
Proprietary.

**Source:** This medium is available as a premixed powder from BIO-SYNTH International, Inc.

**Preparation of Medium:** Per manufacturer's directions.

**Storage/Shelf Life:** Store in the dark. Chromogenic agars are especially light and temperature sensitive; protect from light, excessive heat, moisture, and freezing. Prepared media plates can be kept for one day at ambient temperature. Plates can be stored for 45 days under refrigeration (2-8°C) if properly prepared and protected from light and dehydration. Do not use after the expiration date supplied by the manufacturer.

**Use:** For the rapid identification and isolation of *Bacillus anthracis* based on the detection of phosphatidylcholine-specific phospholipase C activity by 5-bromo–4-chloro–3-indoxyl-cholinphosphate hydrolysis. The medium incorporates chromogenic substrates for detecting specific enzyme activities in *Bacillus anthracis, B. cereus,* and *B. thuringiensis.* The enzymes targeted by the chromogenic medium are not present in other *Bacillus* species, allowing for specific isolation of these three *Bacillus* species. Inclusion of inhibitory compounds into the medium prevents the growth of environmental contaminants. The use of proprietary chromogenic substrates, X-IP and X-CP, allows for the differentiation of *Bacillus anthracis* from near-neighbors *B. cereus* and *B. thuringiensis.* Cream to pale teal-blue colonies of *Bacillus anthracis* after 20–24h, teal-blue colonies of *Bacillus anthracis* after 36–48 h at 35–37°C. Dark teal-blue colonies of *Bacillus cereus/Bacillus thuringiensis* after 20–24h at 35–37°C.

### Antibiotic Assay Medium No. 1 (Seed Agar)

**Composition** per liter:
Agar...............................................................................15.0g
Peptone............................................................................6.0g

Casein enzymatic hydrolysate ........................................................4.0g
Yeast extract...............................................................................3.0g
Beef extract ..............................................................................1.5g
Glucose .....................................................................................1.0g

pH 6.6 ± 0.2 at 25°C

**Source:** This medium is available as a premixed powder from Hi-Media.

**Preparation of Medium:** Add components to distilled/deionized water and bring volume to 1.0L. Mix thoroughly. Gently heat and bring to boiling. Distribute into tubes or flasks. Autoclave for 15 min at 15 psi pressure–121°C. Pour into sterile Petri dishes or leave in tubes.

**Storage/Shelf Life:** Store dehydrated media in the dark in a sealed container below 30°C. Prepared media should be stored under refrigeration (2-8°C). Media should be used within 60 days of preparation. Media should not be used if there are any signs of deterioration (shrinking, cracking, or discoloration) or contamination, or if the expiration date supplied by the manufacturer has passed.

**Use:** For antibiotic assay testing. Widely employed as seed agar in the preparation of plates for microbiological agar diffusion antibiotic assays.

## Antibiotic Assay Medium No. 2
### (Base Agar)

**Composition** per liter:

Agar .......................................................................................15.0g
Peptone......................................................................................6.0g
Yeast extract...............................................................................3.0g
Beef extract ..............................................................................1.5g

pH 6.6 ± 0.2 at 25°C

**Source:** This medium is available as a premixed powder from Hi-Media.

**Preparation of Medium:** Add components to distilled/deionized water and bring volume to 1.0L. Mix thoroughly. Gently heat and bring to boiling. Distribute into tubes or flasks. Autoclave for 15 min at 15 psi pressure–121°C. Pour into sterile Petri dishes or leave in tubes.

**Storage/Shelf Life:** Store dehydrated media in the dark in a sealed container below 30°C. Prepared media should be stored under refrigeration (2-8°C). Media should be used within 60 days of preparation. Media should not be used if there are any signs of deterioration (shrinking, cracking, or discoloration) or contamination, or if the expiration date supplied by the manufacturer has passed.

**Use:** For antibiotic assay testing. For use as a base layer in antibiotic assay testing. Especially useful for the plate assay of bacitracin and penicillin G.

## Antibiotic Assay Medium No. 3
### (Assay Broth)

**Composition** per liter:

Peptone......................................................................................5.0g
$K_2HPO_4$.................................................................................3.68g
NaCl ..........................................................................................3.5g
Beef extract ..............................................................................1.5g
Yeast extract...............................................................................1.5g
$KH_2PO_4$................................................................................1.32g
Glucose .....................................................................................1.0g

pH 7.0 ± 0.2 at 25°C

**Source:** This medium is available as a premixed powder from Hi-Media.

**Preparation of Medium:** Add components to distilled/deionized water and bring volume to 1.0L. Mix thoroughly. Gently heat and bring to boiling. Distribute into tubes or flasks. Autoclave for 15 min at 15 psi pressure–121°C.

**Storage/Shelf Life:** Store dehydrated media in the dark in a sealed container below 30°C. Prepared media should be stored under refrigeration (2-8°C). Media should be used within 60 days of preparation. Media should not be used if there are any signs of deterioration (discoloration) or contamination, or if the expiration date supplied by the manufacturer has passed.

**Use:** For antibiotic assay testing. Used for the serial dilution assay of penicillins and other antibiotics. Used in the turbidimetric assay of penicillin and tetracycline with *Staphylococcus aureus*.

## Antibiotic Assay Medium No. 4
### (Yeast Beef Agar)

**Composition** per liter:

Agar .......................................................................................15.0g
Peptone......................................................................................6.0g
Yeast extract...............................................................................3.0g
Beef extract ..............................................................................1.5g
Glucose .....................................................................................1.0g

pH 6.6 ± 0.2 at 25°C

**Source:** This medium is available as a premixed powder from Hi-Media.

**Preparation of Medium:** Add components to distilled/deionized water and bring volume to 1.0L. Mix thoroughly. Gently heat and bring to boiling. Distribute into tubes or flasks. Autoclave for 15 min at 15 psi pressure–121°C. Pour into sterile Petri dishes or leave in tubes.

**Storage/Shelf Life:** Store dehydrated media in the dark in a sealed container below 30°C. Prepared media should be stored under refrigeration (2-8°C). Media should be used within 60 days of preparation. Media should not be used if there are any signs of deterioration (shrinking, cracking, or discoloration) or contamination, or if the expiration date supplied by the manufacturer has passed.

**Use:** For antibiotic assay testing.

## Antibiotic Assay Medium No. 5
### (Streptomycin Assay Agar with Yeast Extract)

**Composition** per liter:

Agar .......................................................................................15.0g
Peptone......................................................................................6.0g
Yeast extract...............................................................................3.0g
Beef extract ..............................................................................1.5g

pH 7.9 ± 0.2 at 25°C

**Source:** This medium is available as a premixed powder from Hi-Media.

**Preparation of Medium:** Add components to distilled/deionized water and bring volume to 1.0L. Mix thoroughly. Gently heat and bring to boiling. Distribute into tubes or flasks. Autoclave for 15 min at 15 psi pressure–121°C. Pour into sterile Petri dishes or leave in tubes.

**Storage/Shelf Life:** Store dehydrated media in the dark in a sealed container below 30°C. Prepared media should be stored under refrigeration (2-8°C). Media should be used within 60 days of preparation. Media should not be used if there are any signs of deterioration (shrink-

ing, cracking, or discoloration) or contamination, or if the expiration date supplied by the manufacturer has passed.

**Use:** For antibiotic assay testing. For the streptomycin assay using the cylinder plate technique and *Bacillus subtilis* as test organism.

### Antibiotic Assay Medium No. 6

**Composition** per liter:

Casein enzymatic hydrolysate ....................................................17.0g
NaCl.............................................................................................5.0g
Papaic digest of soybean meal.....................................................3.0g
Glucose ........................................................................................2.5g
K$_2$HPO$_4$.................................................................................2.5g
MnSO$_4$·H$_2$O .........................................................................0.03g

pH 7.0 ± 0.1 at 25°C

**Source:** This medium is available as a premixed powder from Hi-Media.

**Preparation of Medium:** Add components to distilled/deionized water and bring volume to 1.0L. Mix thoroughly. Gently heat and bring to boiling. Distribute into tubes or flasks. Autoclave for 15 min at 15 psi pressure–121°C. Pour into sterile Petri dishes.

**Storage/Shelf Life:** Store dehydrated media in the dark in a sealed container below 30°C. Prepared media should be stored under refrigeration (2-8°C). Media should be used within 60 days of preparation. Media should not be used if there are any signs of deterioration (discoloration) or contamination, or if the expiration date supplied by the manufacturer has passed.

**Use:** For antibiotic assay testing. For inoculum development and spore induction of *Bacillus subtilis* for antibiotic assays.

### Antibiotic Assay Medium No. 8
### (Base Agar with low pH)

**Composition** per liter:

Agar ...........................................................................................15.0g
Peptone.........................................................................................6.0g
Yeast extract.................................................................................3.0g
Beef extract ..................................................................................1.5g

pH 5.9 ± 0.1 at 25°C

**Source:** This medium is available as a premixed powder from Hi-Media.

**Preparation of Medium:** Add components to distilled/deionized water and bring volume to 1.0L. Mix thoroughly. Gently heat and bring to boiling. Distribute into tubes or flasks. Autoclave for 15 min at 15 psi pressure–121°C. Pour into sterile Petri dishes.

**Storage/Shelf Life:** Store dehydrated media in the dark in a sealed container below 30°C. Prepared media should be stored under refrigeration (2-8°C). Media should be used within 60 days of preparation. Media should not be used if there are any signs of deterioration (shrinking, cracking, or discoloration) or contamination, or if the expiration date supplied by the manufacturer has passed.

**Use:** For antibiotic assay testing. For use as the base agar and the seed agar in the plate assay of tetracycline. For use as the seed agar in the plate assay of vancomycin, mitomycin, and mithramycin.

### Antibiotic Assay Medium No. 9
### (Polymyxin Base Agar)

**Composition** per liter:

Agar ...........................................................................................20.0g
Casein enzymatic hydrolysate ....................................................17.0g

NaCl.............................................................................................5.0g
Papaic digest of soybean meal.....................................................3.0g
K$_2$HPO$_4$.................................................................................2.5g
Glucose ........................................................................................2.5g

pH 7.2 ± 0.1 at 25°C

**Source:** This medium is available as a premixed powder from Hi-Media.

**Preparation of Medium:** Add components to distilled/deionized water and bring volume to 1.0L. Mix thoroughly. Gently heat and bring to boiling. Distribute into tubes or flasks. Autoclave for 15 min at 15 psi pressure–121°C. Pour into sterile Petri dishes.

**Storage/Shelf Life:** Store dehydrated media in the dark in a sealed container below 30°C. Prepared media should be stored under refrigeration (2-8°C). Media should be used within 60 days of preparation. Media should not be used if there are any signs of deterioration (shrinking, cracking, or discoloration) or contamination, or if the expiration date supplied by the manufacturer has passed.

**Use:** For antibiotic assay testing. For base agar for the plate assay of carbenicillin, colistimethate, and polymyxin B.

### Antibiotic Assay Medium No. 10
### (Polymyxin Seed Agar)

**Composition** per liter:

Casein enzymatic hydrolysate ....................................................17.0g
Agar ...........................................................................................12.0g
NaCl.............................................................................................5.0g
Papaic digest of soybean meal.....................................................3.0g
K$_2$HPO$_4$.................................................................................2.5g
Glucose ........................................................................................2.5g

pH 7.2 ± 0.2 at 25°C

**Source:** This medium, without polysorbate 80, is available as a premixed powder from HiMedia.

**Preparation of Medium:** Add components to distilled/deionized water and bring volume to 1.0L. Mix thoroughly. Gently heat and bring to boiling. Distribute into tubes or flasks. Autoclave for 15 min at 15 psi pressure–121°C. Pour into sterile Petri dishes.

**Storage/Shelf Life:** Store dehydrated media in the dark in a sealed container below 30°C. Prepared media should be stored under refrigeration (2-8°C). Media should be used within 60 days of preparation. Media should not be used if there are any signs of deterioration (shrinking, cracking, or discoloration) or contamination, or if the expiration date supplied by the manufacturer has passed.

**Use:** For antibiotic assay testing. For seed agar for the plate assay of carbenicillin, colistimethate, and polymyxin B.

### Antibiotic Assay Medium No. 11
### (Neomycin, Erythromycin Assay Agar)

**Composition** per liter:

Agar ...........................................................................................15.0g
Peptone.........................................................................................6.0g
Casein enzymatic hydrolysate ......................................................4.0g
Yeast extract.................................................................................3.0g
Beef extract ..................................................................................1.5g
Glucose ........................................................................................1.0g

pH 8.3 ± 0.2 at 25°C

**Source:** This medium, without polysorbate 80, is available as a premixed powder from HiMedia.

**Preparation of Medium:** Add components to distilled/deionized water and bring volume to 1.0L. Mix thoroughly. Gently heat and bring to boiling. Distribute into tubes or flasks. Autoclave for 15 min at 15 psi pressure–121°C. Pour into sterile Petri dishes.

**Storage/Shelf Life:** Store dehydrated media in the dark in a sealed container below 30°C. Prepared media should be stored under refrigeration (2-8°C). Media should be used within 60 days of preparation. Media should not be used if there are any signs of deterioration (shrinking, cracking, or discoloration) or contamination, or if the expiration date supplied by the manufacturer has passed.

**Use:** For antibiotic assay testing. For analyzing the neomycin content in pharmaceutical peparations.

## Antibiotic Assay Medium No. 20
### (Yeast Beef Broth)
**Composition** per liter:

| | |
|---|---|
| Peptone | 15.0g |
| Glucose | 11.0g |
| Yeast extract | 6.5g |
| $K_2HPO_4$ | 3.68g |
| NaCl | 3.5g |
| Beef extract | 1.5g |
| $KH_2PO_4$ | 1.32g |

pH 6.6 ± 0.2 at 25°C

**Source:** This medium is available as a premixed powder from Hi-Media.

**Preparation of Medium:** Add components to distilled/deionized water and bring volume to 1.0L. Mix thoroughly. Gently heat and bring to boiling. Distribute into tubes or flasks. Autoclave for 15 min at 15 psi pressure–121°C.

**Storage/Shelf Life:** Store dehydrated media in the dark in a sealed container below 30°C. Prepared media should be stored under refrigeration (2-8°C). Media should be used within 60 days of preparation. Media should not be used if there are any signs of deterioration (discoloration) or contamination, or if the expiration date supplied by the manufacturer has passed.

**Use:** For assaying the mycostatic activity of pharmaceutical preparations.

## Antibiotic Assay Medium No. 32
**Composition** per liter:

| | |
|---|---|
| Agar | 15.0g |
| Peptone | 6.0g |
| Casein enzymatic hydrolysate | 4.0g |
| Yeast extract | 3.0g |
| Beef extract | 1.5g |
| Glucose | 1.0g |
| $MnSO_4 \cdot 4H_2O$ | 0.3g |

pH 6.6 ± 0.2 at 25°C

**Source:** This medium is available as a premixed powder from Hi-Media.

**Preparation of Medium:** Add components to distilled/deionized water and bring volume to 1.0L. Mix thoroughly. Gently heat and bring to boiling. Distribute into tubes or flasks. Autoclave for 15 min at 15 psi pressure–121°C. Pour into sterile Petri dishes.

**Storage/Shelf Life:** Store dehydrated media in the dark in a sealed container below 30°C. Prepared media should be stored under refriger-

ation (2-8°C). Media should be used within 60 days of preparation. Media should not be used if there are any signs of deterioration (shrinking, cracking, or discoloration) or contamination, or if the expiration date supplied by the manufacturer has passed.

**Use:** For preparing inoculum of *Bacillus subtilis* ATCC 6633 during assay of dihydrostreptomycin and vancomycin.

## Antibiotic Assay Medium No. 34
**Composition** per liter:

| | |
|---|---|
| Peptone | 10.0g |
| Beef extract | 10.0g |
| Glycerol | 10.0g |
| NaCl | 3.0g |

pH 7.0 ± 0.2 at 25°C

**Source:** This medium is available as a premixed powder from Hi-Media.

**Preparation of Medium:** Add components to distilled/deionized water and bring volume to 1.0L. Mix thoroughly. Gently heat and bring to boiling. Distribute into tubes or flasks. Autoclave for 15 min at 15 psi pressure–121°C. Pour into sterile Petri dishes.

**Storage/Shelf Life:** Store dehydrated media in the dark in a sealed container below 30°C. Prepared media should be stored under refrigeration (2-8°C). Media should be used within 60 days of preparation. Media should not be used if there are any signs of deterioration (discoloration) or contamination, or if the expiration date supplied by the manufacturer has passed.

**Use:** For antibiotic assay effectiveness testing of bleomycin using *Mycobacterium smegmatis* ATCC 607.

## Antibiotic Assay Medium No. 35
**Composition** per liter:

| | |
|---|---|
| Agar | 17.0 |
| Peptone | 10.0g |
| Beef extract | 10.0g |
| NaCl | 3.0g |

pH 7.0 ± 0.2 at 25°C

**Source:** This medium is available as a premixed powder from Hi-Media.

**Preparation of Medium:** Add components to distilled/deionized water and bring volume to 1.0L. Mix thoroughly. Gently heat and bring to boiling. Distribute into tubes or flasks. Autoclave for 15 min at 15 psi pressure–121°C. Pour into sterile Petri dishes.

**Storage/Shelf Life:** Store dehydrated media in the dark in a sealed container below 30°C. Prepared media should be stored under refrigeration (2-8°C). Media should be used within 60 days of preparation. Media should not be used if there are any signs of deterioration (shrinking, cracking, or discoloration) or contamination, or if the expiration date supplied by the manufacturer has passed.

**Use:** For antibiotic assay effectiveness testing of bleomycin using *Mycobacterium smegmatis* ATCC 607.

## Antibiotic Assay Medium No. 36
**Composition** per liter:

| | |
|---|---|
| Agar | 15.0g |
| Casein enzymatic hydrolysate | 15.0g |

NaCl ..................................................................................5.0g
Papaic digest of soybean meal ........................................5.0g

pH 7.3 ± 0.2 at 25°C

**Source:** This medium is available as a premixed powder from Hi-Media.

**Preparation of Medium:** Add components to distilled/deionized water and bring volume to 1.0L. Mix thoroughly. Gently heat and bring to boiling. Distribute into tubes or flasks. Autoclave for 15 min at 15 psi pressure–121°C. Pour into sterile Petri dishes.

**Storage/Shelf Life:** Store dehydrated media in the dark in a sealed container below 30°C. Prepared media should be stored under refrigeration (2-8°C). Media should be used within 60 days of preparation. Media should not be used if there are any signs of deterioration (shrinking, cracking, or discoloration) or contamination, or if the expiration date supplied by the manufacturer has passed.

**Use:** A general purpose medium for cultivating a wide variety of fastidious microorganisms.

## Antibiotic Assay Medium No. 37

**Composition** per liter:

Casein enzymatic hydrolysate .......................................17.0g
NaCl ..................................................................................5.0g
Papaic digest of soybean meal ........................................3.0g
Glucose .............................................................................2.5g
K$_2$HPO$_4$ ..........................................................................2.5g

pH 7.3 ± 0.2 at 25°C

**Source:** This medium is available as a premixed powder from Hi-Media.

**Preparation of Medium:** Add components to distilled/deionized water and bring volume to 1.0L. Mix thoroughly. Distribute into tubes or flasks. Autoclave for 10 min at 15 psi pressure–121°C.

**Storage/Shelf Life:** Store dehydrated media in the dark in a sealed container below 30°C. Prepared media should be stored under refrigeration (2-8°C). Media should be used within 60 days of preparation. Media should not be used if there are any signs of deterioration (discoloration) or contamination, or if the expiration date supplied by the manufacturer has passed.

**Use:** A general purpose medium for cultivating a wide variety of fastidious microorganisms.

## Antibiotic Assay Medium No. 38

**Composition** per liter:

Agar ................................................................................15.0g
Peptone...........................................................................15.0g
Glucose .............................................................................5.5g
Papaic digest of soybean meal ........................................5.0g
NaCl ..................................................................................4.0g
L-Cysteine·HCl·H$_2$O......................................................0.7g
Na$_2$SO$_3$ .........................................................................0.2g

pH 7.0 ± 0.2 at 25°C

**Source:** This medium is available as a premixed powder from Hi-Media.

**Preparation of Medium:** Add components to distilled/deionized water and bring volume to 1.0L. Mix thoroughly. Gently heat and bring to boiling. Distribute into tubes or flasks. Autoclave for 15 min at 15 psi pressure–121°C. Pour into sterile Petri dishes.

**Storage/Shelf Life:** Store dehydrated media in the dark in a sealed container below 30°C. Prepared media should be stored under refrigeration (2-8°C). Media should be used within 60 days of preparation. Media should not be used if there are any signs of deterioration (shrinking, cracking, or discoloration) or contamination, or if the expiration date supplied by the manufacturer has passed.

**Use:** For microbiological assay of ticarcillin using *Pseudomonas aeruginosa* ATCC 29336.

## Antibiotic Assay Medium No. 39

**Composition** per liter:

Peptone .............................................................................5.0g
K$_2$HPO$_4$..........................................................................3.68g
NaCl ..................................................................................3.5g
Beef extract ......................................................................1.5g
Yeast extract .....................................................................1.5g
KH$_2$PO$_4$ ..........................................................................1.32g
Glucose .............................................................................1.0g

pH 7.9 ± 0.2 at 25°C

**Source:** This medium is available as a premixed powder from Hi-Media.

**Preparation of Medium:** Add components to distilled/deionized water and bring volume to 1.0L. Mix thoroughly. Gently heat and bring to boiling. Distribute into tubes or flasks. Autoclave for 15 min at 15 psi pressure–121°C.

**Storage/Shelf Life:** Store dehydrated media in the dark in a sealed container below 30°C. Prepared media should be stored under refrigeration (2-8°C). Media should be used within 60 days of preparation. Media should not be used if there are any signs of deterioration (discoloration) or contamination, or if the expiration date supplied by the manufacturer has passed.

**Use:** For the microbiological assay of neomycin and streptomycin using *Klebsiella pneumoniae* ATCC 10031 as the test organism.

## Antibiotic Assay Medium No. 40

**Composition** per liter:

Yeast extract...................................................................20.0g
Agar ................................................................................10.0g
Glucose ...........................................................................10.0g
Casein enzymatic hydrolysate .........................................2.5g
Peptone .............................................................................2.5g
KH$_2$PO$_4$...........................................................................2.0g
Tween™ 80........................................................................0.1g

pH 6.7 ± 0.2 at 25°C

**Source:** This medium is available as a premixed powder from Hi-Media.

**Preparation of Medium:** Add components to distilled/deionized water and bring volume to 1.0L. Mix thoroughly. Gently heat and bring to boiling. Distribute into tubes or flasks. Autoclave for 15 min at 15 psi pressure–121°C. Pour into sterile Petri dishes.

**Storage/Shelf Life:** Store dehydrated media in the dark in a sealed container below 30°C. Prepared media should be stored under refrigeration (2-8°C). Media should be used within 60 days of preparation. Media should not be used if there are any signs of deterioration (shrinking, cracking, or discoloration) or contamination, or if the expiration date supplied by the manufacturer has passed.

**Use:** For the microbiological assay of Thiostrepton using *Streptococcus faecium* ATCC 10541.

## Antibiotic Assay Medium No. 41

**Composition** per liter:

Glucose ................................................................20.0g
Sodium citrate ......................................................10.0g
Casein enzymatic hydrolysate ...............................9.0g
Yeast extract..........................................................5.0g
K$_2$HPO$_4$..............................................................1.0g
KH$_2$PO$_4$............................................................1.0g

pH 6.8 ± 0.2 at 25°C

**Source:** This medium is available as a premixed powder from Hi-Media.

**Preparation of Medium:** Add components to distilled/deionized water and bring volume to 1.0L. Mix thoroughly. Distribute into tubes or flasks. Autoclave for 10 min at 15 psi pressure–121°C.

**Storage/Shelf Life:** Store dehydrated media in the dark in a sealed container below 30°C. Prepared media should be stored under refrigeration (2-8°C). Media should be used within 60 days of preparation. Media should not be used if there are any signs of deterioration (discoloration) or contamination, or if the expiration date supplied by the manufacturer has passed.

**Use:** For the microbiological assay of thiostrepton using *Streptococcus faecium* ATCC 10541.

## Antibiotic Assay Medium B

**Composition** per liter:

Casein enzymatic hydrolysate .............................17.0g
Agar ....................................................................15.0g
Glucose .................................................................5.0g
NaCl.......................................................................5.0g
Papaic digest of soybean meal ..............................3.0g
K$_2$HPO$_4$............................................................2.5g

pH 7.3 ± 0.2 at 25°C

**Preparation of Medium:** Add components to distilled/deionized water and bring volume to 1.0L. Mix thoroughly. Gently heat and bring to boiling. Distribute into tubes or flasks. Autoclave for 15 min at 15 psi pressure–121°C. Pour into sterile Petri dishes.

**Storage/Shelf Life:** Store dehydrated media in the dark in a sealed container below 30°C. Prepared media should be stored under refrigeration (2-8°C). Media should be used within 60 days of preparation. Media should not be used if there are any signs of deterioration (shrinking, cracking, or discoloration) or contamination, or if the expiration date supplied by the manufacturer has passed.

**Use:** For the microbiological assay of colistimethate using *Bordetella bronchiseptica* or *Escherichia coli.*

## Antibiotic Assay Medium C

**Composition** per liter:

Peptone..................................................................6.0g
K$_2$HPO$_4$............................................................3.68g
NaCl.......................................................................3.5g
Yeast extract..........................................................3.0g
Beef extract...........................................................1.5g
KH$_2$PO$_4$............................................................1.32g
Glucose .................................................................1.0g

pH 7.0 ± 0.2 at 25°C

**Preparation of Medium:** Add components to distilled/deionized water and bring volume to 1.0L. Mix thoroughly. Distribute into tubes or flasks. Autoclave for 10 min at 15 psi pressure–121°C.

**Storage/Shelf Life:** Store dehydrated media in the dark in a sealed container below 30°C. Prepared media should be stored under refrigeration (2-8°C). Media should be used within 60 days of preparation. Media should not be used if there are any signs of deterioration (discoloration) or contamination, or if the expiration date supplied by the manufacturer has passed.

**Use:** For the microbiological assay of Rifampin using *Escherichia coli*, Colistimethate using *Escherichia coli*, erythromycin, framycetin, gentamicin, gramicidin, kanamycin, neomycin, and vancomycin using *Staphylococcus aureus*, and gramicin using *Enterococcus hirae.*

## Antibiotic Assay Medium D

**Composition** per liter:

Casein peptone......................................................5.0g
K$_2$HPO$_4$............................................................3.68g
NaCl.......................................................................3.5g
KNO$_3$ ...................................................................2.0g
Heart extract..........................................................1.5g
Yeast extract..........................................................1.5g
KH$_2$PO$_4$............................................................1.32g

pH 7.0 ± 0.2 at 25°C

**Preparation of Medium:** Add components to distilled/deionized water and bring volume to 1.0L. Mix thoroughly. Distribute into tubes or flasks. Autoclave for 10 min at 15 psi pressure–121°C.

**Storage/Shelf Life:** Store dehydrated media in the dark in a sealed container below 30°C. Prepared media should be stored under refrigeration (2-8°C). Media should be used within 60 days of preparation. Media should not be used if there are any signs of deterioration (discoloration) or contamination, or if the expiration date supplied by the manufacturer has passed.

**Use:** For the microbiological assay of erythromycin using *Klebsiella pneumoniae.*

## Antibiotic Assay Medium E

**Composition** per liter:

Na$_2$HPO$_4$·12H$_2$O........................................26.9g
Agar ....................................................................10.0g
Peptone .................................................................5.0g
Meat extract ..........................................................3.0g

pH 7.9 ± 0.2 at 25°C

**Preparation of Medium:** Add components to distilled/deionized water and bring volume to 1.0L. Mix thoroughly. Gently heat and bring to boiling. Distribute into tubes or flasks. Autoclave for 15 min at 15 psi pressure–121°C. Pour into sterile Petri dishes.

**Storage/Shelf Life:** Store dehydrated media in the dark in a sealed container below 30°C. Prepared media should be stored under refrigeration (2-8°C). Media should be used within 60 days of preparation. Media should not be used if there are any signs of deterioration (shrinking, cracking, or discoloration) or contamination, or if the expiration date supplied by the manufacturer has passed.

**Use:** For the microbiological assay of framycetin using *Bacillus subtilus.*

## Antibiotic Assay Medium G

**Composition** per liter:

Agar ....................................................................15.0g
Meat extract ........................................................10.0g

Peptone.................................................................................10.0g
NaCl.......................................................................................3.0g

pH 7.0 ± 0.2 at 25°C

**Preparation of Medium:** Add components to distilled/deionized water and bring volume to 1.0L. Mix thoroughly. Gently heat and bring to boiling. Distribute into tubes or flasks. Autoclave for 15 min at 15 psi pressure–121°C. Pour into sterile Petri dishes.

**Storage/Shelf Life:** Store dehydrated media in the dark in a sealed container below 30°C. Prepared media should be stored under refrigeration (2-8°C). Media should be used within 60 days of preparation. Media should not be used if there are any signs of deterioration (shrinking, cracking, or discoloration) or contamination, or if the expiration date supplied by the manufacturer has passed.

**Use:** For the microbiological assay of bleomycin using *Mycobacterium smegmatis*.

## Antibiotic Assay Medium H

**Composition** per liter:

D-Glucose......................................................................10.0g
Casein enzymatic hydrolysate .......................................6.0g
Yeast extract..................................................................2.0g

pH 8.0 ± 0.2 at 25°C

**Preparation of Medium:** Add components to distilled/deionized water and bring volume to 1.0L. Mix thoroughly. Distribute into tubes or flasks. Autoclave for 10 min at 15 psi pressure–121°C.

**Storage/Shelf Life:** Store dehydrated media in the dark in a sealed container below 30°C. Prepared media should be stored under refrigeration (2-8°C). Media should be used within 60 days of preparation. Media should not be used if there are any signs of deterioration (discoloration) or contamination, or if the expiration date supplied by the manufacturer has passed.

**Use:** For the microbiological assay of apramycin using *Salmonella cholerasuis*.

## Antibiotic Assay Medium L-AODC

**Composition** per liter:

Agar ...............................................................................15.0g
Glucose, anhydrous.......................................................10.0g
Yeast extract..................................................................2.5g
K$_2$HPO$_4$.................................................................0.69g
KH$_2$PO$_4$................................................................0.45g

pH 6.0 ± 0.2 at 25°C

**Preparation of Medium:** Add components to distilled/deionized water and bring volume to 1.0L. Mix thoroughly. Gently heat and bring to boiling. Distribute into tubes or flasks. Autoclave for 15 min at 15 psi pressure–121°C. Pour into sterile Petri dishes.

**Storage/Shelf Life:** Store dehydrated media in the dark in a sealed container below 30°C. Prepared media should be stored under refrigeration (2-8°C). Media should be used within 60 days of preparation. Media should not be used if there are any signs of deterioration (shrinking, cracking, or discoloration) or contamination, or if the expiration date supplied by the manufacturer has passed.

**Use:** For the microbiological assay of monensin using *Bacillus subtilus*.

## Antibiotic Assay Medium M-AODC

**Composition** per liter:

Agar ...............................................................................20.0g
Glucose ..........................................................................10.0g
Yeast extract..................................................................2.5g
K$_2$HPO$_4$.................................................................0.69g
KH$_2$PO$_4$................................................................0.45g

pH 6.0 ± 0.2 at 25°C

**Preparation of Medium:** Add components to distilled/deionized water and bring volume to 1.0L. Mix thoroughly. Gently heat and bring to boiling. Distribute into tubes or flasks. Autoclave for 15 min at 15 psi pressure–121°C. Pour into sterile Petri dishes.

**Storage/Shelf Life:** Store dehydrated media in the dark in a sealed container below 30°C. Prepared media should be stored under refrigeration (2-8°C). Media should be used within 60 days of preparation. Media should not be used if there are any signs of deterioration (shrinking, cracking, or discoloration) or contamination, or if the expiration date supplied by the manufacturer has passed.

**Use:** For the microbiological assay of monensin using *Bacillus subtilus*.

## Antibiotic HiVeg Assay Medium No. 1
### (Antibiotic HiVeg Assay Medium - A)
### (Seed HiVeg Agar)

**Composition** per liter:

Agar ...............................................................................15.0g
Plant peptone ................................................................6.0g
Plant hydrolysate ..........................................................4.0g
Yeast extract..................................................................3.0g
Plant extract ..................................................................1.5g
Glucose ..........................................................................1.0g

pH 6.6 ± 0.2 at 25°C

**Source:** This medium is available as a premixed powder from Hi-Media.

**Preparation of Medium:** Add components to distilled/deionized water and bring volume to 1.0L. Mix thoroughly. Gently heat and bring to boiling. Distribute into tubes or flasks. Autoclave for 15 min at 15 psi pressure–121°C. Pour into sterile Petri dishes or leave in tubes.

**Storage/Shelf Life:** Store dehydrated media in the dark in a sealed container below 30°C. Prepared media should be stored under refrigeration (2-8°C). Media should be used within 60 days of preparation. Media should not be used if there are any signs of deterioration (shrinking, cracking, or discoloration) or contamination, or if the expiration date supplied by the manufacturer has passed.

**Use:** For antibiotic assay testing. Widely employed as seed agar in the preparation of plates for microbiological agar diffusion antibiotic assays.

## Antibiotic HiVeg Assay Medium No. 2
### (Antibiotic HiVeg Assay Medium - B)
### (Seed HiVeg Agar)

**Composition** per liter:

Agar ...............................................................................15.0g
Plant peptone ................................................................6.0g
Yeast extract..................................................................3.0g
Plant extract ..................................................................1.5g

pH 6.6 ± 0.2 at 25°C

**Source:** This medium is available as a premixed powder from Hi-Media.

**Preparation of Medium:** Add components to distilled/deionized water and bring volume to 1.0L. Mix thoroughly. Gently heat and bring to boiling. Distribute into tubes or flasks. Autoclave for 15 min at 15 psi pressure–121°C. Pour into sterile Petri dishes or leave in tubes.

**Storage/Shelf Life:** Store dehydrated media in the dark in a sealed container below 30°C. Prepared media should be stored under refrigeration (2-8°C). Media should be used within 60 days of preparation. Media should not be used if there are any signs of deterioration (shrinking, cracking, or discoloration) or contamination, or if the expiration date supplied by the manufacturer has passed.

**Use:** For antibiotic assay testing. For use as a base layer in antibiotic assay testing. Especially useful for the plate assay of bacitracin and penicillin G.

## Antibiotic HiVeg Assay Medium No. 3
### (Antibiotic HiVeg Assay Medium - C)
**Composition** per liter:

| | |
|---|---|
| Plant peptone | 5.0g |
| $K_2HPO_4$ | 3.68g |
| NaCl | 3.5g |
| Yeast extract | 1.5g |
| Plant extract | 1.5g |
| $KH_2PO_4$ | 1.32g |
| Glucose | 1.0g |

pH 7.0 ± 0.2 at 25°C

**Source:** This medium is available as a premixed powder from Hi-Media.

**Preparation of Medium:** Add components to distilled/deionized water and bring volume to 1.0L. Mix thoroughly. Gently heat and bring to boiling. Distribute into tubes or flasks. Autoclave for 15 min at 15 psi pressure–121°C.

**Storage/Shelf Life:** Store dehydrated media in the dark in a sealed container below 30°C. Prepared media should be stored under refrigeration (2-8°C). Media should be used within 60 days of preparation. Media should not be used if there are any signs of deterioration (shrinking, cracking, or discoloration) or contamination, or if the expiration date supplied by the manufacturer has passed.

**Use:** For antibiotic assay testing. Used for the serial dilution assay of penicillins and other antibiotics. Used in the turbidimetric assay of penicillin and tetracycline with *Staphylococcus aureus*.

## Antibiotic HiVeg Assay Medium No. 4
### (Yeast Beef HiVeg Agar)
**Composition** per liter:

| | |
|---|---|
| Agar | 15.0g |
| Plant peptone | 6.0g |
| Yeast extract | 3.0g |
| Plant extract | 1.5g |
| Glucose | 1.0g |

pH 6.6 ± 0.2 at 25°C

**Source:** This medium is available as a premixed powder from Hi-Media.

**Preparation of Medium:** Add components to distilled/deionized water and bring volume to 1.0L. Mix thoroughly. Gently heat and bring

to boiling. Distribute into tubes or flasks. Autoclave for 15 min at 15 psi pressure–121°C. Pour into sterile Petri dishes or leave in tubes.

**Storage/Shelf Life:** Store dehydrated media in the dark in a sealed container below 30°C. Prepared media should be stored under refrigeration (2-8°C). Media should be used within 60 days of preparation. Media should not be used if there are any signs of deterioration (shrinking, cracking, or discoloration) or contamination, or if the expiration date supplied by the manufacturer has passed.

**Use:** For antibiotic assay testing.

## Antibiotic HiVeg Assay Medium No. 5
### (Streptomycin HiVeg Agar with Yeast Extract)
### (Antibiotic HiVeg Assay Medium - E)
**Composition** per liter:

| | |
|---|---|
| Agar | 15.0g |
| Plant peptone | 6.0g |
| Yeast extract | 3.0g |
| Plant extract | 1.5g |

pH 7.9 ± 0.2 at 25°C

**Source:** This medium is available as a premixed powder from Hi-Media.

**Preparation of Medium:** Add components to distilled/deionized water and bring volume to 1.0L. Mix thoroughly. Gently heat and bring to boiling. Distribute into tubes or flasks. Autoclave for 15 min at 15 psi pressure–121°C. Pour into sterile Petri dishes or leave in tubes.

**Storage/Shelf Life:** Store dehydrated media in the dark in a sealed container below 30°C. Prepared media should be stored under refrigeration (2-8°C). Media should be used within 60 days of preparation. Media should not be used if there are any signs of deterioration (shrinking, cracking, or discoloration) or contamination, or if the expiration date supplied by the manufacturer has passed.

**Use:** For antibiotic assay testing. For the streptomycin assay using the cylinder plate technique and *Bacillus subtilis* as test organism.

## Antibiotic HiVeg Assay Medium No. 6
**Composition** per liter:

| | |
|---|---|
| Plant hydrolysate | 17.0g |
| NaCl | 5.0g |
| Papaic digest of soybean meal | 3.0g |
| Glucose | 2.5g |
| $K_2HPO_4$ | 2.5g |
| $MnSO_4 \cdot H_2O$ | 0.03g |

pH 7.0 ± 0.1 at 25°C

**Source:** This medium is available as a premixed powder from Hi-Media.

**Preparation of Medium:** Add components to distilled/deionized water and bring volume to 1.0L. Mix thoroughly. Gently heat and bring to boiling. Distribute into tubes or flasks. Autoclave for 15 min at 15 psi pressure–121°C. Pour into sterile Petri dishes.

**Storage/Shelf Life:** Store dehydrated media in the dark in a sealed container below 30°C. Prepared media should be stored under refrigeration (2-8°C). Media should be used within 60 days of preparation. Media should not be used if there are any signs of deterioration (shrinking, cracking, or discoloration) or contamination, or if the expiration date supplied by the manufacturer has passed.

**Use:** For antibiotic assay testing. For inoculum development and spore induction of *Bacillus subtilis* for antibiotic assays.

## Antibiotic HiVeg Assay Medium No. 8
### (Base HiVeg Agar w/ low pH)
### (Antibiotic HiVeg Assay Medium F)

**Composition** per liter:

| | |
|---|---|
| Agar | 15.0g |
| Plant peptone | 6.0g |
| Yeast extract | 3.0g |
| Plant extract | 1.5g |

pH 5.9 ± 0.1 at 25°C

**Source:** This medium is available as a premixed powder from HiMedia.

**Preparation of Medium:** Add components to distilled/deionized water and bring volume to 1.0L. Mix thoroughly. Gently heat and bring to boiling. Distribute into tubes or flasks. Autoclave for 15 min at 15 psi pressure–121°C. Pour into sterile Petri dishes.

**Storage/Shelf Life:** Store dehydrated media in the dark in a sealed container below 30°C. Prepared media should be stored under refrigeration (2-8°C). Media should be used within 60 days of preparation. Media should not be used if there are any signs of deterioration (shrinking, cracking, or discoloration) or contamination, or if the expiration date supplied by the manufacturer has passed.

**Use:** For antibiotic assay testing. For use as the base agar and the seed agar in the plate assay of tetracycline. For use as the seed agar in the plate assay of vancomycin, mitomycin, and mithramycin.

## Antibiotic HiVeg Assay Medium No. 9
### (Polymyxin HiVeg Base Agar)

**Composition** per liter:

| | |
|---|---|
| Agar | 20.0g |
| Plant hydrolysate | 17.0g |
| NaCl | 5.0g |
| Papaic digest of soybean meal | 3.0g |
| $K_2HPO_4$ | 2.5g |
| Glucose | 2.5g |

pH 7.2 ± 0.1 at 25°C

**Source:** This medium is available as a premixed powder from HiMedia.

**Preparation of Medium:** Add components to distilled/deionized water and bring volume to 1.0L. Mix thoroughly. Gently heat and bring to boiling. Distribute into tubes or flasks. Autoclave for 15 min at 15 psi pressure–121°C. Pour into sterile Petri dishes.

**Storage/Shelf Life:** Store dehydrated media in the dark in a sealed container below 30°C. Prepared media should be stored under refrigeration (2-8°C). Media should be used within 60 days of preparation. Media should not be used if there are any signs of deterioration (shrinking, cracking, or discoloration) or contamination, or if the expiration date supplied by the manufacturer has passed.

**Use:** For antibiotic assay testing. For base agar for the plate assay of carbenicillin, colistimethate, and polymyxin B.

## Antibiotic HiVeg Assay Medium No. 10
### (Polymyxin Seed HiVeg Agar)
### (Antibiotic HiVeg Assay Medium H)

**Composition** per liter:

| | |
|---|---|
| Plant hydrolysate | 17.0g |
| Agar | 12.0g |
| Polysorbate 80 | 10.0g |
| NaCl | 5.0g |
| Papaic digest of soybean meal | 3.0g |
| $K_2HPO_4$ | 2.5g |
| Glucose | 2.5g |

pH 7.2 ± 0.2 at 25°C

**Source:** This medium, without polysorbate 80, is available as a premixed powder from HiMedia.

**Preparation of Medium:** Add components to distilled/deionized water and bring volume to 1.0L. Mix thoroughly. Gently heat and bring to boiling. Distribute into tubes or flasks. Autoclave for 15 min at 15 psi pressure–121°C. Pour into sterile Petri dishes.

**Storage/Shelf Life:** Store dehydrated media in the dark in a sealed container below 30°C. Prepared media should be stored under refrigeration (2-8°C). Media should be used within 60 days of preparation. Media should not be used if there are any signs of deterioration (shrinking, cracking, or discoloration) or contamination, or if the expiration date supplied by the manufacturer has passed.

**Use:** For antibiotic assay testing. For seed agar for the plate assay of carbenicillin, colistimethate, and polymyxin B.

## Antibiotic HiVeg Assay Medium No. 11
### (Neomycin, Erythromycin HiVeg Assay Agar)

**Composition** per liter:

| | |
|---|---|
| Agar | 15.0g |
| Plant peptone | 6.0g |
| Plant hydrolysate | 4.0g |
| Yeast extract | 3.0g |
| Plant extract | 1.5g |
| Glucose | 1.0g |

pH 8.3 ± 0.2 at 25°C

**Source:** This medium, without polysorbate 80, is available as a premixed powder from HiMedia.

**Preparation of Medium:** Add components to distilled/deionized water and bring volume to 1.0L. Mix thoroughly. Gently heat and bring to boiling. Distribute into tubes or flasks. Autoclave for 15 min at 15 psi pressure–121°C. Pour into sterile Petri dishes.

**Storage/Shelf Life:** Store dehydrated media in the dark in a sealed container below 30°C. Prepared media should be stored under refrigeration (2-8°C). Media should be used within 60 days of preparation. Media should not be used if there are any signs of deterioration (shrinking, cracking, or discoloration) or contamination, or if the expiration date supplied by the manufacturer has passed.

**Use:** For antibiotic assay testing. For analyzing the neomycin content in pharmaceutical preparations.

## Antibiotic HiVeg Assay Medium No. 20
### (Yeast Beef HiVeg Broth)

**Composition** per liter:

| | |
|---|---|
| Plant peptone | 15.0g |
| Glucose | 11.0g |
| Yeast extract | 6.5g |
| $K_2HPO_4$ | 3.68g |
| NaCl | 3.5g |
| Plant extract | 1.5g |
| $KH_2PO_4$ | 1.32g |

pH 6.6 ± 0.2 at 25°C

**Source:** This medium is available as a premixed powder from Hi-Media.

**Preparation of Medium:** Add components to distilled/deionized water and bring volume to 1.0L. Mix thoroughly. Gently heat and bring to boiling. Distribute into tubes or flasks. Autoclave for 15 min at 15 psi pressure–121°C.

**Storage/Shelf Life:** Store dehydrated media in the dark in a sealed container below 30°C. Prepared media should be stored under refrigeration (2-8°C). Media should be used within 60 days of preparation. Media should not be used if there are any signs of deterioration (discoloration) or contamination, or if the expiration date supplied by the manufacturer has passed.

**Use:** For assaying the mycostatic activity of pharmaceutical preparations.

### Antibiotic HiVeg Assay Medium No. 32

**Composition** per liter:
Agar ...................................................................................15.0g
Plant peptone........................................................................6.0g
Plant hydrolysate...................................................................4.0g
Yeast extract........................................................................3.0g
Plant extract .........................................................................1.5g
Glucose ...............................................................................1.0g
MnSO$_4$·4H$_2$O .....................................................................0.3g
pH 6.6 ± 0.2 at 25°C

**Source:** This medium is available as a premixed powder from Hi-Media.

**Preparation of Medium:** Add components to distilled/deionized water and bring volume to 1.0L. Mix thoroughly. Gently heat and bring to boiling. Distribute into tubes or flasks. Autoclave for 15 min at 15 psi pressure–121°C. Pour into sterile Petri dishes.

**Storage/Shelf Life:** Store dehydrated media in the dark in a sealed container below 30°C. Prepared media should be stored under refrigeration (2-8°C). Media should be used within 60 days of preparation. Media should not be used if there are any signs of deterioration (shrinking, cracking, or discoloration) or contamination, or if the expiration date supplied by the manufacturer has passed.

**Use:** For preparing inoculum of *Bacillus subtilis* ATCC 6633 during assay of dihydrostreptomycin and vancomycin.

### Antibiotic HiVeg Assay Medium No. 35 (Antibiotic HiVeg Assay Medium - I)

**Composition** per liter:
Agar ...................................................................................17.0g
Plant peptone......................................................................10.0g
Plant extract .......................................................................10.0g
Glycerol .............................................................................10.0g
NaCl.....................................................................................3.0g
pH 7.0 ± 0.2 at 25°C

**Source:** This medium is available as a premixed powder from Hi-Media.

**Preparation of Medium:** Add components to distilled/deionized water and bring volume to 1.0L. Mix thoroughly. Gently heat and bring to boiling. Distribute into tubes or flasks. Autoclave for 15 min at 15 psi pressure–121°C. Pour into sterile Petri dishes.

**Storage/Shelf Life:** Store dehydrated media in the dark in a sealed container below 30°C. Prepared media should be stored under refriger-

ation (2-8°C). Media should be used within 60 days of preparation. Media should not be used if there are any signs of deterioration (shrinking, cracking, or discoloration) or contamination, or if the expiration date supplied by the manufacturer has passed.

**Use:** For antibiotic assay effectiveness testing of bleomycin using *Mycobacterium smegmatis* ATCC 607.

### Antibiotic HiVeg Assay Medium No. 36 (Antibiotic HiVeg Assay Medium - J)

**Composition** per liter:
Agar ...................................................................................15.0g
Plant hydrolysate ...............................................................15.0g
NaCl.....................................................................................5.0g
Papaic digest of soybean meal..............................................5.0g
pH 7.3 ± 0.2 at 25°C

**Source:** This medium is available as a premixed powder from Hi-Media.

**Preparation of Medium:** Add components to distilled/deionized water and bring volume to 1.0L. Mix thoroughly. Gently heat and bring to boiling. Distribute into tubes or flasks. Autoclave for 15 min at 15 psi pressure–121°C. Pour into sterile Petri dishes.

**Storage/Shelf Life:** Store dehydrated media in the dark in a sealed container below 30°C. Prepared media should be stored under refrigeration (2-8°C). Media should be used within 60 days of preparation. Media should not be used if there are any signs of deterioration (shrinking, cracking, or discoloration) or contamination, or if the expiration date supplied by the manufacturer has passed.

**Use:** A general purpose medium for cultivating a wide variety of fastidious microorganisms.

### Antibiotic HiVeg Assay Medium No. 37

**Composition** per liter:
Plant hydrolysate ...............................................................15.0g
NaCl.....................................................................................5.0g
Papaic digest of soybean meal..............................................5.0g
pH 7.3 ± 0.2 at 25°C

**Source:** This medium is available as a premixed powder from Hi-Media.

**Preparation of Medium:** Add components to distilled/deionized water and bring volume to 1.0L. Mix thoroughly. Distribute into tubes or flasks. Autoclave for 10 min at 15 psi pressure–121°C.

**Storage/Shelf Life:** Store dehydrated media in the dark in a sealed container below 30°C. Prepared media should be stored under refrigeration (2-8°C). Media should be used within 60 days of preparation. Media should not be used if there are any signs of deterioration (shrinking, cracking, or discoloration) or contamination, or if the expiration date supplied by the manufacturer has passed.

**Use:** A general purpose medium for cultivating a wide variety of fastidious microorganisms.

### Antibiotic HiVeg Assay Medium No. 38

**Composition** per liter:
Agar ...................................................................................15.0g
Plant peptone .....................................................................15.0g
Glucose ...............................................................................5.5g
Papaic digest of soybean meal..............................................5.0g
NaCl.....................................................................................4.0g

L-Cysteine·HCl·H₂O..............................................0.7g

Na₂SO₃.....................................................................0.2g

pH 7.2 ± 0.2 at 25°C

**Source:** This medium is available as a premixed powder from Hi-Media.

**Preparation of Medium:** Add components to distilled/deionized water and bring volume to 1.0L. Mix thoroughly. Gently heat and bring to boiling. Distribute into tubes or flasks. Autoclave for 15 min at 15 psi pressure–121°C. Pour into sterile Petri dishes.

**Storage/Shelf Life:** Store dehydrated media in the dark in a sealed container below 30°C. Prepared media should be stored under refrigeration (2-8°C). Media should be used within 60 days of preparation. Media should not be used if there are any signs of deterioration (shrinking, cracking, or discoloration) or contamination, or if the expiration date supplied by the manufacturer has passed.

**Use:** For microbiological assay of ticarcillin using *Pseudomonas aeruginosa* ATCC 29336.

## Antibiotic HiVeg Assay Medium No. 39

**Composition** per liter:

Plant peptone.............................................................5.0g

K₂HPO₄.....................................................................3.68g

NaCl...........................................................................3.5g

Plant extract ..............................................................1.5g

Yeast extract..............................................................1.5g

KH₂PO₄......................................................................1.32g

Glucose ......................................................................1.0g

pH 7.9 ± 0.2 at 25°C

**Source:** This medium is available as a premixed powder from Hi-Media.

**Preparation of Medium:** Add components to distilled/deionized water and bring volume to 1.0L. Mix thoroughly. Gently heat and bring to boiling. Distribute into tubes or flasks. Autoclave for 15 min at 15 psi pressure–121°C.

**Use:** For the microbiological assay of neomycin and streptomycin using *Klebsiella pneumoniae* ATCC 10031 as the test organism.

## Antibiotic HiVeg Assay Medium No. 40

**Composition** per liter:

Yeast extract..............................................................20.0g

Agar ...........................................................................10.0g

Glucose ......................................................................10.0g

Plant hydrolysate.......................................................2.5g

Plant peptone.............................................................2.5g

KH₂PO₄......................................................................2.0g

Tween™ 80 ................................................................0.1g

pH 6.7 ± 0.2 at 25°C

**Source:** This medium is available as a premixed powder from Hi-Media.

**Preparation of Medium:** Add components to distilled/deionized water and bring volume to 1.0L. Mix thoroughly. Gently heat and bring to boiling. Distribute into tubes or flasks. Autoclave for 15 min at 15 psi pressure–121°C. Pour into sterile Petri dishes.

**Storage/Shelf Life:** Store dehydrated media in the dark in a sealed container below 30°C. Prepared media should be stored under refrigeration (2-8°C). Media should be used within 60 days of preparation. Media should not be used if there are any signs of deterioration (shrink-

ing, cracking, or discoloration) or contamination, or if the expiration date supplied by the manufacturer has passed.

**Use:** For the microbiological assay of thiostrepton using *Streptococcus faecium* ATCC 10541.

## Antibiotic HiVeg Assay Medium No. 41

**Composition** per liter:

Glucose ......................................................................20.0g

Sodium citrate ...........................................................10.0g

Plant hydrolysate.......................................................9.0g

Yeast extract..............................................................5.0g

K₂HPO₄......................................................................1.0g

KH₂PO₄......................................................................1.0g

pH 6.8 ± 0.2 at 25°C

**Source:** This medium is available as a premixed powder from Hi-Media.

**Preparation of Medium:** Add components to distilled/deionized water and bring volume to 1.0L. Mix thoroughly. Distribute into tubes or flasks. Autoclave for 10 min at 15 psi pressure–121°C.

**Storage/Shelf Life:** Store dehydrated media in the dark in a sealed container below 30°C. Prepared media should be stored under refrigeration (2-8°C). Media should be used within 60 days of preparation. Media should not be used if there are any signs of deterioration (discoloration) or contamination, or if the expiration date supplied by the manufacturer has passed.

**Use:** For the microbiological assay of thiostrepton using *Streptococcus faecium* ATCC 10541.

## Antibiotic Medium 1
## (Penassay Seed Agar)
## (Seed Agar)/(Agar Medium A)

**Composition** per liter:

Agar ...........................................................................15.0g

Pancreatic digest of gelatin.......................................6.0g

Pancreatic digest of casein........................................4.0g

Yeast extract..............................................................3.0g

Beef extract................................................................1.5g

Glucose ......................................................................1.0g

pH 6.6 ± 0.1 at 25°C

**Source:** This medium is available as a premixed powder from BD Diagnostic Systems.

**Preparation of Medium:** Add components to distilled/deionized water and bring volume to 1.0L. Mix thoroughly. Gently heat and bring to boiling. Distribute into tubes or flasks. Autoclave for 15 min at 15 psi pressure–121°C. Pour into sterile Petri dishes or leave in tubes.

**Storage/Shelf Life:** Store dehydrated media in the dark in a sealed container below 30°C. Prepared media should be stored under refrigeration (2-8°C). Media should be used within 60 days of preparation. Media should not be used if there are any signs of deterioration (shrinking, cracking, or discoloration) or contamination, or if the expiration date supplied by the manufacturer has passed.

**Use:** For antibiotic assay testing, detection of antibiotics in milk, and determination of the antimicrobial effectiveness of antibiotics.

## Antibiotic Medium 1 with Tetracycline

**Composition** per liter:

Agar ...........................................................................15.0g

Pancreatic digest of gelatin.......................................6.0g

Pancreatic digest of casein............................................................4.0g
Yeast extract...............................................................................3.0g
Beef extract ...............................................................................1.5g
Glucose .....................................................................................1.0g
Tetracycline solution................................................................ 10.0mL

pH 6.6 ± 0.1 at 25°C

**Tetracycline Solution:**
**Composition** per 10.0mL:
Tetracycline.....................................................................................0.02g

**Preparation of Tetracycline Solution:** Add tetracycline to distilled/deionized water and bring volume to 10.0mL. Mix thoroughly. Filter sterilize.

**Preparation of Medium:** Add components, except tetracycline solution, to distilled/deionized water and bring volume to 1.0L. Mix thoroughly. Gently heat and bring to boiling. Autoclave for 15 min at 15 psi pressure–121°C. Cool to 45°–50°C. Aseptically add sterile tetracycline solution. Mix thoroughly. Pour into sterile Petri dishes or distribute into sterile tubes.

**Storage/Shelf Life:** Store dehydrated media in the dark in a sealed container below 30°C. Prepared media should be stored under refrigeration (2-8°C). Media should be used within 60 days of preparation. Media should not be used if there are any signs of deterioration (shrinking, cracking, or discoloration) or contamination, or if the expiration date supplied by the manufacturer has passed.

**Use:** For the selective cultivation and maintenance of *Salmonella choleraesuis*.

## Antibiotic Medium 1 with Tetracycline, Streptomycin, and Chloramphenicol

**Composition** per liter:
Agar .........................................................................................15.0g
Pancreatic digest of gelatin ............................................................6.0g
Pancreatic digest of casein .............................................................4.0g
Yeast extract...............................................................................3.0g
Beef extract ...............................................................................1.5g
Glucose .....................................................................................1.0g
Antibiotic solution .................................................................. 10.0mL

pH 6.6 ± 0.1 at 25°C

**Antibiotic Solution:**
**Composition** per 10.0mL:
Tetracycline....................................................................................0.02g
Streptomycin...................................................................................0.02g
Chloramphenicol..............................................................................0.02g

**Preparation of Antibiotic Solution:** Add components to distilled/deionized water and bring volume to 10.0mL. Mix thoroughly. Filter sterilize.

**Preparation of Medium:** Add components, except antibiotic solution, to distilled/deionized water and bring volume to 1.0L. Mix thoroughly. Gently heat and bring to boiling. Autoclave for 15 min at 15 psi pressure–121°C. Cool to 45°–50°C. Aseptically add sterile antibiotic solution. Mix thoroughly. Pour into sterile Petri dishes or distribute into sterile tubes.

**Storage/Shelf Life:** Store dehydrated media in the dark in a sealed container below 30°C. Prepared media should be stored under refrigeration (2-8°C). Media should be used within 60 days of preparation. Media should not be used if there are any signs of deterioration (shrinking, cracking, or discoloration) or contamination, or if the expiration date supplied by the manufacturer has passed.

**Use:** For the selective cultivation and maintenance of *Salmonella choleraesuis*.

## Antibiotic Medium 2
## (Base Agar)
## (Penassay Base Agar)

**Composition** per liter:
Agar ..........................................................................................15.0g
Pancreatic digest of gelatin ............................................................6.0g
Yeast extract...............................................................................3.0g
Beef extract ...............................................................................1.5g

pH 6.6 ± 0.1 at 25°C

**Source:** This medium is available as a premixed powder from BD Diagnostic Systems and Thermo Scientific.

**Preparation of Medium:** Add components to distilled/deionized water and bring volume to 1.0L. Mix thoroughly. Gently heat and bring to boiling. Distribute into tubes or flasks. Autoclave for 15 min at 15 psi pressure–121°C. Pour into sterile Petri dishes.

**Storage/Shelf Life:** Store dehydrated media in the dark in a sealed container below 30°C. Prepared media should be stored under refrigeration (2-8°C). Media should be used within 60 days of preparation. Media should not be used if there are any signs of deterioration (shrinking, cracking, or discoloration) or contamination.

**Use:** For use as a base layer in antibiotic assay testing. Especially useful for the plate assay of bacitracin and penicillin G.

## Antibiotic Medium 3
## (Penassay Broth)

**Composition** per liter:
Pancreatic digest of gelatin ............................................................5.0g
NaCl..........................................................................................3.5g
Yeast extract...............................................................................1.5g
Beef extract ...............................................................................1.5g
Glucose .....................................................................................1.0g
$K_2HPO_4$...................................................................................3.68g
$KH_2PO_4$...................................................................................1.32g

pH 7.0 ± 0.05 at 25°C

**Source:** This medium is available as a premixed powder from BD Diagnostic Systems and Thermo Scientific.

**Preparation of Medium:** Add components to distilled/deionized water and bring volume to 1.0L. Mix thoroughly. Gently heat and bring to boiling. Distribute into tubes or flasks. Autoclave for 15 min at 15 psi pressure–121°C.

**Storage/Shelf Life:** Store dehydrated media in the dark in a sealed container below 30°C. Prepared media should be stored under refrigeration (2-8°C). Media should be used within 60 days of preparation. Media should not be used if there are any signs of deterioration (discoloration) or contamination, or if the expiration date supplied by the manufacturer has passed.

**Use:** For antibiotic assay testing. Used for the serial dilution assay of penicillins and other antibiotics. Used in the turbidimetric assay of penicillin and tetracycline with *Staphylococcus aureus*. For the cultivation and maintenance of *Bacillus subtilis, Salmonella choleraesuis,* and *Staphylococcus aureus*. For the cloning of plasmids in *Streptococcus mutans*.

## Antibiotic Medium 3 Plus
**Composition** per liter:
Agar ..........................................................................................15.0g
Peptone .......................................................................................5.0g

| | |
|---|---|
| $K_2HPO_4$ | 3.68g |
| NaCl | 3.5g |
| Yeast extract | 2.5g |
| Glucose | 1.75g |
| Beef extract | 1.5g |
| $KH_2PO_4$ | 1.32g |

pH 7.0 ± 0.05 at 25°C

**Preparation of Medium:** Add components to distilled/deionized water and bring volume to 1.0L. Mix thoroughly. Gently heat and bring to boiling. Distribute into tubes or flasks. Autoclave for 15 min at 15 psi pressure–121°C. Pour into sterile Petri dishes or leave in tubes.

**Storage/Shelf Life:** Store dehydrated media in the dark in a sealed container below 30°C. Prepared media should be stored under refrigeration (2-8°C). Media should be used within 60 days of preparation. Media should not be used if there are any signs of deterioration (shrinking, cracking, or discoloration) or contamination, or if the expiration date supplied by the manufacturer has passed.

**Use:** For antibiotic assay testing and for the cultivation of *Escherichia coli.*

## Antibiotic Medium 4 (Yeast Beef Agar) (Agar Medium C)

**Composition** per liter:

| | |
|---|---|
| Agar | 15.0g |
| Pancreatic digest of gelatin | 6.0g |
| Yeast extract | 3.0g |
| Beef extract | 1.5g |
| Glucose | 1.0g |

pH 6.6 ± 0.05 at 25°C

**Source:** This medium is available as a premixed powder from BD Diagnostic Systems.

**Preparation of Medium:** Add components to distilled/deionized water and bring volume to 1.0L. Mix thoroughly. Gently heat and bring to boiling. Distribute into tubes or flasks. Autoclave for 15 min at 15 psi pressure–121°C. Pour into sterile Petri dishes or leave in tubes.

**Storage/Shelf Life:** Store dehydrated media in the dark in a sealed container below 30°C. Prepared media should be stored under refrigeration (2-8°C). Media should be used within 60 days of preparation. Media should not be used if there are any signs of deterioration (shrinking, cracking, or discoloration) or contamination, or if the expiration date supplied by the manufacturer has passed.

**Use:** For antibiotic assay testing.

## Antibiotic Medium 5 (Streptomycin Assay Agar with Yeast Extract)

**Composition** per liter:

| | |
|---|---|
| Agar | 15.0g |
| Pancreatic digest of gelatin | 6.0g |
| Yeast extract | 3.0g |
| Beef extract | 1.5g |

pH 7.9 ± 0.1 at 25°C

**Source:** This medium is available as a premixed powder from BD Diagnostic Systems and Thermo Scientific.

**Preparation of Medium:** Add components to distilled/deionized water and bring volume to 1.0L. Mix thoroughly. Gently heat and bring to boiling. Distribute into tubes or flasks. Autoclave for 15 min at 15 psi pressure–121°C. Pour into sterile Petri dishes.

**Storage/Shelf Life:** Store dehydrated media in the dark in a sealed container below 30°C. Prepared media should be stored under refrigeration (2-8°C). Media should be used within 60 days of preparation. Media should not be used if there are any signs of deterioration (shrinking, cracking, or discoloration) or contamination, or if the expiration date supplied by the manufacturer has passed.

**Use:** For antibiotic assay testing. For the streptomycin assay using the cylinder plate technique and *Bacillus subtilis* as test organism.

## Antibiotic Medium 6

**Composition** per liter:

| | |
|---|---|
| Pancreatic digest of casein | 17.0g |
| NaCl | 5.0g |
| Papaic digest of soybean meal | 3.0g |
| Glucose | 2.5g |
| $K_2HPO_4$ | 2.5g |
| $MnSO_4 \cdot H_2O$ | 0.03g |

pH 7.0 ± 0.1 at 25°C

**Source:** This medium is available as a premixed powder from BD Diagnostic Systems.

**Preparation of Medium:** Add components to distilled/deionized water and bring volume to 1.0L. Mix thoroughly. Gently heat and bring to boiling. Distribute into tubes or flasks. Autoclave for 15 min at 15 psi pressure–121°C. Pour into sterile Petri dishes.

**Storage/Shelf Life:** Store dehydrated media in the dark in a sealed container below 30°C. Prepared media should be stored under refrigeration (2-8°C). Media should be used within 60 days of preparation. Media should not be used if there are any signs of deterioration (discoloration) or contamination, or if the expiration date supplied by the manufacturer has passed.

**Use:** For antibiotic assay testing.

## Antibiotic Medium 7

**Composition** per liter:

| | |
|---|---|
| Agar | 15.0g |
| Pancreatic digest of gelatin | 6.0g |
| Yeast extract | 3.0g |
| Beef extract | 1.5g |

pH 7.0 ± 0.1 at 25°C

**Preparation of Medium:** Add components to distilled/deionized water and bring volume to 1.0L. Mix thoroughly. Gently heat and bring to boiling. Adjust pH to 7.0. Distribute into tubes or flasks. Autoclave for 15 min at 15 psi pressure–121°C. Pour into sterile Petri dishes.

**Storage/Shelf Life:** Store dehydrated media in the dark in a sealed container below 30°C. Prepared media should be stored under refrigeration (2-8°C). Media should be used within 60 days of preparation. Media should not be used if there are any signs of deterioration (shrinking, cracking, or discoloration) or contamination, or if the expiration date supplied by the manufacturer has passed.

**Use:** For use as a base layer in antibiotic assay testing. Especially useful for the plate assay of bacitracin and penicillin G.

## Antibiotic Medium 8 (Base Agar with Low pH)

**Composition** per liter:

| | |
|---|---|
| Agar | 15.0g |
| Pancreatic digest of gelatin | 6.0g |

Yeast extract ...............................................................3.0g
Beef extract ...............................................................1.5g

pH 5.9 ± 0.1 at 25°C

**Source:** This medium is available as a premixed powder from BD Diagnostic Systems.

**Preparation of Medium:** Add components to distilled/deionized water and bring volume to 1.0L. Mix thoroughly. Gently heat and bring to boiling. Distribute into tubes or flasks. Autoclave for 15 min at 15 psi pressure–121°C. Pour into sterile Petri dishes.

**Storage/Shelf Life:** Store dehydrated media in the dark in a sealed container below 30°C. Prepared media should be stored under refrigeration (2-8°C). Media should be used within 60 days of preparation. Media should not be used if there are any signs of deterioration (shrinking, cracking, or discoloration) or contamination, or if the expiration date supplied by the manufacturer has passed.

**Use:** For antibiotic assay testing. For use as the base agar and the seed agar in the plate assay of tetracycline. For use as the seed agar in the plate assay of vancomycin, mitomycin, and mithramycin.

## Antibiotic Medium 9
### (Polymyxin Base Agar)

**Composition** per liter:

Agar ...............................................................20.0g
Pancreatic digest of casein ...............................................................17.0g
NaCl ...............................................................5.0g
Papaic digest of soybean meal ...............................................................3.0g
K$_2$HPO$_4$...............................................................2.5g
Glucose ...............................................................2.5g

pH 7.2 ± 0.1 at 25°C

**Source:** This medium is available as a premixed powder from BD Diagnostic Systems.

**Preparation of Medium:** Add components to distilled/deionized water and bring volume to 1.0L. Mix thoroughly. Gently heat and bring to boiling. Distribute into tubes or flasks. Autoclave for 15 min at 15 psi pressure–121°C. Pour into sterile Petri dishes.

**Storage/Shelf Life:** Store dehydrated media in the dark in a sealed container below 30°C. Prepared media should be stored under refrigeration (2-8°C). Media should be used within 60 days of preparation. Media should not be used if there are any signs of deterioration (shrinking, cracking, or discoloration) or contamination, or if the expiration date supplied by the manufacturer has passed.

**Use:** For antibiotic assay testing. For base agar for the plate assay of carbenicillin, colistimethate, and polymyxin B.

## Antibiotic Medium 10
### (Polymyxin Seed Agar)

**Composition** per liter:

Pancreatic digest of casein ...............................................................17.0g
Agar ...............................................................12.0g
Polysorbate 80 ...............................................................10.0g
NaCl ...............................................................5.0g
Papaic digest of soybean meal ...............................................................3.0g
K$_2$HPO$_4$...............................................................2.5g
Glucose ...............................................................2.5g

pH 7.3 ± 0.2 at 25°C

**Source:** This medium is available as a premixed powder from BD Diagnostic Systems.

**Preparation of Medium:** Add components to distilled/deionized water and bring volume to 1.0L. Mix thoroughly. Gently heat and bring to boiling. Distribute into tubes or flasks. Autoclave for 15 min at 15 psi pressure–121°C. Pour into sterile Petri dishes.

**Storage/Shelf Life:** Store dehydrated media in the dark in a sealed container below 30°C. Prepared media should be stored under refrigeration (2-8°C). Media should be used within 60 days of preparation. Media should not be used if there are any signs of deterioration (shrinking, cracking, or discoloration) or contamination, or if the expiration date supplied by the manufacturer has passed.

**Use:** For antibiotic assay testing. For seed agar for the plate assay of carbenicillin, colistimethate, and polymyxin B.

## Antibiotic Medium 11
### (Neomycin Assay Agar)

**Composition** per liter:

Agar ...............................................................15.0g
Pancreatic digest of gelatin ...............................................................6.0g
Pancreatic digest of casein ...............................................................4.0g
Yeast extract ...............................................................3.0g
Beef extract ...............................................................1.5g
Glucose ...............................................................1.0g

pH 8.0 ± 0.1 at 25°C

**Source:** This medium is available as a premixed powder from BD Diagnostic Systems and Thermo Scientific.

**Preparation of Medium:** Add components to distilled/deionized water and bring volume to 1.0L. Mix thoroughly. Gently heat and bring to boiling. Distribute into tubes or flasks. Autoclave for 15 min at 15 psi pressure–121°C. Pour into sterile Petri dishes.

**Storage/Shelf Life:** Store dehydrated media in the dark in a sealed container below 30°C. Prepared media should be stored under refrigeration (2-8°C). Media should be used within 60 days of preparation. Media should not be used if there are any signs of deterioration (shrinking, cracking, or discoloration) or contamination, or if the expiration date supplied by the manufacturer has passed.

**Use:** For antibiotic assay testing. For base agar and seed agar for the plate assay to test the effectiveness of neomycin sulfate, amoxicillin, ampicillin, clindamycin, cyclacillin, erythromycin, gentamycin, neomycin, oleandomycin, and sisomycin.

## Antibiotic Medium 12

**Composition** per liter:

Agar ...............................................................25.0g
Peptone ...............................................................10.0g
Glucose ...............................................................10.0g
NaCl ...............................................................10.0g
Yeast extract ...............................................................5.0g
Beef extract ...............................................................2.5g

pH 6.0 ± 0.1 at 25°C

**Source:** This medium is available as a premixed powder from BD Diagnostic Systems.

**Preparation of Medium:** Add components to distilled/deionized water and bring volume to 1.0L. Mix thoroughly. Gently heat and bring to boiling. Distribute into tubes or flasks. Autoclave for 15 min at 15 psi pressure–121°C. Pour into sterile Petri dishes.

**Storage/Shelf Life:** Store dehydrated media in the dark in a sealed container below 30°C. Prepared media should be stored under refrigeration (2-8°C). Media should be used within 60 days of preparation.

Media should not be used if there are any signs of deterioration (shrinking, cracking, or discoloration) or contamination, or if the expiration date supplied by the manufacturer has passed.

**Use:** For antibiotic assay effectiveness testing.

## Antibiotic Medium 21

**Composition** per liter:

| | |
|---|---|
| Glucose | 11.0g |
| Pancreatic digest of gelatin | 5.0g |
| $K_2HPO_4$ | 3.68g |
| NaCl | 3.5g |
| Yeast extract | 1.5g |
| Beef extract | 1.5g |
| $KH_2PO_4$ | 1.32g |

pH 6.6 ± 0.2 at 25°C

**Preparation of Medium:** Add components to distilled/deionized water and bring volume to 1.0L. Mix thoroughly. Gently heat and bring to boiling. Distribute into tubes or flasks. Autoclave for 15 min at 15 psi pressure–121°C. Pour into sterile Petri dishes.

**Storage/Shelf Life:** Store dehydrated media in the dark in a sealed container below 30°C. Prepared media should be stored under refrigeration (2-8°C). Media should be used within 60 days of preparation. Media should not be used if there are any signs of deterioration (discoloration) or contamination, or if the expiration date supplied by the manufacturer has passed.

**Use:** For assaying the mycostatic activity of pharmaceutical preparations.

## Antibiotic Sulfonamide Sensitivity Test Agar (ASS Agar)

**Composition** per liter:

| | |
|---|---|
| Agar | 12.0g |
| Proteose peptone | 10.0g |
| Beef extract | 10.0g |
| NaCl | 3.0g |
| Glucose | 2.0g |
| $Na_2HPO_4$ | 2.0g |
| Sodium acetate | 1.0g |
| Adenine | 0.01g |
| Guanine | 0.01g |
| Uracil | 0.01g |
| Xanthine | 0.01g |

pH 7.4 ± 0.2 at 25°C

**Preparation of Medium:** Add components to distilled/deionized water and bring volume to 1.0L. Mix thoroughly. Gently heat and bring to boiling. Distribute into tubes or flasks. Autoclave for 15 min at 15 psi pressure–121°C. Pour into sterile Petri dishes or leave in tubes.

**Storage/Shelf Life:** Store dehydrated media in the dark in a sealed container below 30°C. Prepared media should be stored under refrigeration (2-8°C). Media should be used within 60 days of preparation. Media should not be used if there are any signs of deterioration (shrinking, cracking, or discoloration) or contamination, or if the expiration date supplied by the manufacturer has passed.

**Use:** For testing the antimicrobial effectiveness of antibiotics and sulfonamides. For detecting the presence of antimicrobial substances in milk, urine, and other fluids.

## Antimicrobial Inhibitor Test Agar pH 6.0

**Composition** per liter:

| | |
|---|---|
| Agar | 13.0g |
| NaCl | 5.0g |
| Tryptone | 3.5g |
| Meat extract | 3.5g |
| *Bacillus subtilis* spore suspension | 1.0mL |

pH 6.0 ± 0.2 at 25°C

**Preparation of Medium:** Add components, except *Bacillus subtilis* spore suspension, to distilled/deionized water and bring volume to 990.0mL. Mix thoroughly. Adjust pH to 6.0. Gently heat and bring to boiling. Distribute into tubes or flasks. Autoclave for 15 min at 15 psi pressure–121°C. Cool to 50°C. Add *Bacillus subtilis* spore suspension. Mix thoroughly. Pour into sterile Petri dishes.

**Storage/Shelf Life:** Store dehydrated media in the dark in a sealed container below 30°C. Prepared media should be stored under refrigeration (2-8°C). Media should be used within 60 days of preparation. Media should not be used if there are any signs of deterioration (shrinking, cracking, or discoloration) or contamination, or if the expiration date supplied by the manufacturer has passed.

**Use:** For residual analysis of antimicrobial components in meat and organ samples, using *Bacillus subtilis* ATCC 6633 as test organism.

## Antimicrobial Inhibitor Test Agar pH 7.2

**Composition** per liter:

| | |
|---|---|
| Agar | 13.0g |
| Peptone | 7.0g |
| NaCl | 5.0g |
| $Na_3PO_4 \cdot 12H_2O$ | 0.8g |
| Selective supplement solution | 10.0mL |
| *Bacillus subtilis* spore suspension | 1.0mL |

pH 7.2 ± 0.2 at 25°C

**Selective Supplement Solution:**
**Composition** per 10.0mL:

| | |
|---|---|
| Trimethoprim | 5.0mg |

**Preparation of Selective Supplement Solution:** Add trimethoprim to distilled/deionized water and bring volume to 10.0mL. Mix thoroughly. Filter sterilize.

**Preparation of Medium:** Add components, except selective supplement solution and *Bacillus subtilis* spore suspension, to distilled/deionized water and bring volume to 1.0L. Mix thoroughly. Adjust pH to 7.2. Gently heat and bring to boiling. Distribute into tubes or flasks. Autoclave for 15 min at 15 psi pressure–121°C. Cool to 50°C. Add selective supplement solution and *Bacillus subtilis* spore suspension. Mix thoroughly. Pour into sterile Petri dishes.

**Storage/Shelf Life:** Store dehydrated media in the dark in a sealed container below 30°C. Prepared media should be stored under refrigeration (2-8°C). Media should be used within 60 days of preparation. Media should not be used if there are any signs of deterioration (shrinking, cracking, or discoloration) or contamination, or if the expiration date supplied by the manufacturer has passed.

**Use:** For residual analysis of antimicrobial components in meat and organ samples, using *Bacillus subtilis* ATCC 6633 as test organism.

## Antimicrobial Inhibitor Test Agar pH 8.0

**Composition** per liter:

| | |
|---|---|
| Tryptone | 3.5g |
| Meat extract | 3.5g |

NaCl.........................................................................5.0g
Na$_3$PO$_4$·12H$_2$O.........................................................2.5g
Agar .......................................................................13.0g
*Micrococcus luteus* suspension................................ 10.0mL
*Bacillus subtilis* spore suspension............................... 1.0mL

pH 8.0 ± 0.2 at 25°C

**Preparation of Medium:** Add components, except *Bacillus subtilis* spore suspension, to distilled/deionized water and bring volume to 990.0mL. Mix thoroughly. Adjust pH to 8.0. Gently heat and bring to boiling. Distribute into tubes or flasks. Autoclave for 15 min at 15 psi pressure–121°C. Cool to 50°C. Add *Bacillus subtilis* spore suspension and *Micrococcus luteus* (10$^4$ CFU per mL) suspension. Mix thoroughly. Pour into sterile Petri dishes.

**Storage/Shelf Life:** Store dehydrated media in the dark in a sealed container below 30°C. Prepared media should be stored under refrigeration (2-8°C). Media should be used within 60 days of preparation. Media should not be used if there are any signs of deterioration (shrinking, cracking, or discoloration) or contamination, or if the expiration date supplied by the manufacturer has passed.

**Use:** For residual analysis of antimicrobial components in meat and organ samples, using *Bacillus subtilis* ATCC 6633 and *Micrococcus luteus* ATCC 9341 as test organisms.

### AquaCHROM™

**Composition** per liter:
Peptones and growth regulators ...................................20.0g
Chromogenic and selective mix .......................................2.3g

pH 7.1± 0.2 at 25°C

**Source:** This medium is available from CHROMagar, Paris, France.

**Preparation of Medium:** Add 2.23g of the medium to 100mL water samples.

**Storage/Shelf Life:** Store in the dark. Chromogenic agars are especially light and temperature sensitive; protect from light, excessive heat, moisture, and freezing. Store at 15-30°C. Do not use after the expiration date supplied by the manufacturer.

**Use:** For the detection of *E. coli* and coliforms in water samples. Green to blue-green color after 24 hour incubation indicates the presence of *E. coli*. A yellow color indicates presence of coliforms other than *E. coli*.

### Arabinose Agar Base with Selective Supplement

**Composition** per liter:
Peptone, special ...........................................................23.0g
Agar .......................................................................15.0g
Arabinose......................................................................10.0g
NaCl.............................................................................5.0g
Corn starch....................................................................1.0g
Phenol Red....................................................................0.1g
Selective supplement solution ..................................... 10.0mL

pH 7.8 ± 0.2 at 25°C

**Source:** This medium is available from HiMedia.

**Selective Supplement Solution:**
**Composition** per 10.0mL:
Thallium acetate..............................................................0.2g
Nalidixic acid ..............................................................25.0mg

**Preparation of Selective Supplement Solution:** Add components to distilled/deionized water and bring volume to 10.0mL. Mix thoroughly. Filter sterilize.

**Preparation of Medium:** Add components, except selective supplement solution, to distilled/deionized water and bring volume to 990.0mL. Mix thoroughly. Adjust pH to 7.8. Gently heat and bring to boiling. Distribute into tubes or flasks. Gently heat and bring to boil. Do not autoclave. Cool to 50°C. Add selective supplement solution. Mix thoroughly. Pour into sterile Petri dishes.

**Storage/Shelf Life:** Store dehydrated media in the dark in a sealed container below 30°C. Prepared media should be stored under refrigeration (2-8°C). Media should be used within 60 days of preparation. Media should not be used if there are any signs of deterioration (shrinking, cracking, or discoloration) or contamination, or if the expiration date supplied by the manufacturer has passed.

**Use:** For selective isolation of *Enterococcus faecium* from feces, sewage, and water supplies.

### Arylsulfatase Agar
### (Wayne Sulfatase Agar)

**Composition** per liter:
Agar .......................................................................15.0g
Na$_2$HPO$_4$....................................................................2.5g
L-Asparagine.................................................................1.0g
KH$_2$PO$_4$.......................................................................1.0g
K$_2$HPO$_4$.......................................................................1.0g
Trisodium phenolphthalein sulfate .............................0.65g
Pancreatic digest of casein...........................................0.5g
Ferric ammonium citrate..............................................0.05g
MgSO$_4$·7H$_2$O..............................................................0.01g
CaCl$_2$·2H$_2$O ...............................................................0.5mg
ZnSO$_4$·7H$_2$O ..............................................................0.1mg
CuSO$_4$ ........................................................................0.1mg
Glycerol ..................................................................... 10.0mL

pH 7.0 ± 0.2 at 25°C

**Source:** This medium is available as a premixed powder from BD Diagnostic Systems.

**Preparation of Medium:** Add glycerol to approximately 800.0mL of distilled/deionized water. Mix thoroughly. Add remaining components and bring volume to 1.0L with distilled/deionized water. Mix thoroughly. Gently heat and bring to boiling. Distribute into tubes. Autoclave for 15 min at 15 psi pressure–121°C. Cool tubes in an upright position.

**Storage/Shelf Life:** Store dehydrated media in the dark in a sealed container below 30°C. Prepared media should be stored under refrigeration (2-8°C). Media should be used within 60 days of preparation. Media should not be used if there are any signs of deterioration (shrinking, cracking, or discoloration) or contamination, or if the expiration date supplied by the manufacturer has passed.

**Use:** For the biochemical differentiation of species of *Mycobacterium*. Inoculate tubes with *Mycobacterium* cultures and incubate aerobically at 35°C for 3–14 days. Add 0.5–1.0mL of 2$N$ Na$_2$CO$_3$ to each tube and observe color change within 30 min. Development of a pink color is indicative of *Mycobacterium fortuitum* or *Mycobacterium chelonae*. *Mycobacterium tuberculosis* gives a negative reaction.

### Ashdown's Medium

**Composition** per liter:
Casein hydrolysate.......................................................15.0g
Agar .......................................................................12.0g
NaCl.............................................................................5.0g
Soy peptone ..................................................................5.0g

Glycerol .........................................................................40.0g
Crystal Violet ................................................................5.0mg
Neutral Red ..................................................................50.0mg
Gentamicin solution ...................................................... 10.0mL

## Gentamicin Solution:
**Composition** per 10.0mL:
Gentamicin......................................................................4.0mg

**Preparation of Gentamicin Solution:** Add gentamicin to distilled/deionized water and bring volume to 10.0mL. Mix thoroughly. Filter sterilize.

**Preparation of Medium:** Add components—except gentamicin solution—to distilled/deionized water and bring volume to 990.0mL. Mix thoroughly. Gently heat and bring to boiling. Autoclave for 15 min at 15 psi pressure–121°C. Cool to 45°–50°C. Aseptically add 10.0mL of freshly prepared sterile gentamicin solution. Mix thoroughly. Pour into sterile Petri dishes or distribute into sterile tubes.

**Storage/Shelf Life:** Store dehydrated media in the dark in a sealed container below 30°C. Prepared media should be stored under refrigeration (2-8°C). Media should be used within 60 days of preparation. Media should not be used if there are any signs of deterioration (shrinking, cracking, or discoloration) or contamination, or if the expiration date supplied by the manufacturer has passed.

**Use:** For the selective isolation and differentiation of *Burkholderia pseudomallei* from clinical specimens such as sputum. Use within a week of preparation.

## Ashdown's Medium
**Composition** per liter:
Casein hydrolysate .........................................................17.0g
Agar ..............................................................................15.0g
NaCl................................................................................5.0g
Glucose ...........................................................................5.0g
Soy peptone....................................................................3.0g
K$_2$HPO$_4$.............................................................................2.5g
Neutral Red ..................................................................50.0mg
Glycerol .........................................................................40.0g
Crystal Violet ................................................................5.0mg
Gentamicin solution ...................................................... 10.0mL

## Gentamicin Solution:
**Composition** per 10.0mL:
Gentamicin......................................................................4.0mg

**Preparation of Gentamicin Solution:** Add gentamicin to distilled/deionized water and bring volume to 10.0mL. Mix thoroughly. Filter sterilize.

**Preparation of Medium:** Add components—except gentamicin solution—to distilled/deionized water and bring volume to 990.0mL. Mix thoroughly. Gently heat and bring to boiling. Autoclave for 15 min at 15 psi pressure–121°C. Cool to 45°–50°C. Aseptically add 10.0mL of freshly prepared sterile gentamicin solution. Mix thoroughly. Pour into sterile Petri dishes or distribute into sterile tubes.

**Storage/Shelf Life:** Store dehydrated media in the dark in a sealed container below 30°C. Prepared media should be stored under refrigeration (2-8°C). Media should be used within 60 days of preparation. Media should not be used if there are any signs of deterioration (shrinking, cracking, or discoloration) or contamination, or if the expiration date supplied by the manufacturer has passed.

**Use:** For the selective isolation and differentiation of *Burkholderia pseudomallei* from clinical specimens such as sputum. Use within a week of preparation.

## Asparagine Broth
**Composition** per liter:
DL-Asparagine.................................................................30.0g
K$_2$HPO$_4$.............................................................................1.0g
MgSO$_4$·7H$_2$O .................................................................0.5g
pH 6.9–7.2 at 25°C

**Preparation of Medium:** Add components to distilled/deionized water and bring volume to 1.0L. Mix well until dissolved. Adjust pH to between 6.9 and 7.2. Distribute into tubes or flasks. Autoclave for 15 min at 15 psi pressure–121°C.

**Storage/Shelf Life:** Store dehydrated media in the dark in a sealed container below 30°C. Prepared media should be stored under refrigeration (2-8°C). Media should be used within 60 days of preparation. Media should not be used if there are any signs of deterioration (discoloration) or contamination, or if the expiration date supplied by the manufacturer has passed.

**Use:** For a presumptive test medium in the differentiation of nonfermentative Gram-negative bacteria, especially *Pseudomonas aeruginosa*. For use in the multiple tube technique in the microbiological analysis of recreational waters.

## ATS Medium
## (American Trudeau Society Medium)
**Composition** per liter:
Potato ............................................................................20.0g
Malachite Green...............................................................0.2g
Egg yolk emulsion ........................................................ 500.0mL
Glycerol ........................................................................ 10.0mL
pH 6.5–7.0 at 25°C

**Source:** This medium is available as a prepared medium from BD Diagnostic Systems.

## Egg Yolk Emulsion:
**Composition**:
Chicken egg yolks............................................................ 11
Whole chicken egg .......................................................... 1

**Preparation of Egg Yolk Emulsion:** Soak eggs with 1:100 dilution of saturated mercuric chloride solution for 1 min. Crack eggs and separate yolks from whites. Mix egg yolks with 1 chicken egg.

**Preparation of Medium:** Add components to distilled/deionized water and bring volume to 1.0L. Distribute into tubes. Autoclave for 15 min at 15 psi pressure–121°C in a slanted position.

**Storage/Shelf Life:** Store dehydrated media in the dark in a sealed container below 30°C. Prepared media should be stored under refrigeration (2-8°C). Media should be used within 60 days of preparation. Media should not be used if there are any signs of deterioration (discoloration) or contamination, or if the expiration date supplied by the manufacturer has passed.

**Use:** For the isolation and cultivation of *Mycobacterium* species other than *Mycobacterium leprae*. Especially useful for the detection of *Mycobacterium tuberculosis* from clinical specimens such as cerebrospinal fluid, pleural fluid, and tissues.

## Azide Blood Agar

**Composition** per liter:

Agar ........................................................................15.0g
Pancreatic digest of casein ......................................5.0g
Peptic digest of animal tissue..................................5.0g
NaCl........................................................................5.0g
Beef extract ............................................................3.0g
NaN$_3$ ......................................................................0.2g
Sheep blood, defibrinated ................................ 50.0mL

pH 7.2 ± 0.2 at 25°C

**Source:** This medium is available as a premixed powder from BD Diagnostic Systems and Thermo Scientific.

**Caution:** Sodium azide is toxic. Azides also react with metals and disposal must be highly diluted.

**Preparation of Medium:** Add components, except sheep blood, to distilled/deionized water and bring volume to 950.0mL. Mix thoroughly. Gently heat and bring to boiling. Autoclave for 15 min at 15 psi pressure–121°C. Cool to 45–50°C. Aseptically add 50.0mL of sterile defibrinated sheep blood. Pour into sterile Petri dishes or distribute into sterile tubes. Allow tubes to cool in a slanted position.

**Storage/Shelf Life:** Store dehydrated media in the dark in a sealed container below 30°C. Prepared media should be stored under refrigeration (2-8°C). Media should be used within 60 days of preparation. Media should not be used if there are any signs of deterioration (shrinking, cracking, or discoloration) or contamination, or if the expiration date supplied by the manufacturer has passed.

**Use:** For the isolation and differentiation of streptococci and staphylococci from specimens containing mixed flora and from nonclinical specimens such as water and sewage.

## Azide Blood Agar Base with Blood

**Composition** per liter:

Agar ........................................................................15.0g
Peptone, special ...................................................10.0g
NaCl........................................................................5.0g
Beef........................................................................3.0g
NaN$_3$ ......................................................................0.2g
Sheep blood, defibrinated ................................ 50.0mL

pH 7.2 ± 0.2 at 25°C

**Source:** This medium without sheep blood is available as a premixed powder from HiMedia.

**Caution:** Sodium azide is toxic. Azides also react with metals and disposal must be highly diluted.

**Preparation of Medium:** Add components, except sheep blood, to distilled/deionized water and bring volume to 950.0mL. Mix thoroughly. Gently heat and bring to boiling. Autoclave for 15 min at 15 psi pressure–121°C. Pour into sterile Petri dishes or leave in tubes.

**Storage/Shelf Life:** Store dehydrated media in the dark in a sealed container below 30°C. Prepared media should be stored under refrigeration (2-8°C). Media should be used within 60 days of preparation. Media should not be used if there are any signs of deterioration (shrinking, cracking, or discoloration) or contamination, or if the expiration date supplied by the manufacturer has passed.

**Use:** For the isolation and differentiation of streptococci and staphylococci from specimens containing mixed flora and from nonclinical specimens such as water and sewage.

## Azide Blood Agar Base, HiVeg with Blood

**Composition** per liter:

Agar ........................................................................15.0g
Plant special peptone ...........................................10.0g
NaCl........................................................................5.0g
Plant extract ..........................................................3.0g
NaN$_3$ ......................................................................0.2g
Sheep blood, defibrinated ................................ 50.0mL

pH 7.2 ± 0.2 at 25°C

**Source:** This medium without sheep blood is available as a premixed powder from HiMedia.

**Caution:** Sodium azide is toxic. Azides also react with metals and disposal must be highly diluted.

**Preparation of Medium:** Add components, except sheep blood, to distilled/deionized water and bring volume to 950.0mL. Mix thoroughly. Gently heat and bring to boiling. Autoclave for 15 min at 15 psi pressure–121°C. Pour into sterile Petri dishes or leave in tubes.

**Storage/Shelf Life:** Store dehydrated media in the dark in a sealed container below 30°C. Prepared media should be stored under refrigeration (2-8°C). Media should be used within 60 days of preparation. Media should not be used if there are any signs of deterioration (shrinking, cracking, or discoloration) or contamination, or if the expiration date supplied by the manufacturer has passed.

**Use:** For the isolation and differentiation of streptococci and staphylococci from specimens containing mixed flora and from nonclinical specimens such as water and sewage.

## Azide Dextrose HiVeg Broth

**Composition** per liter:

Plant special peptone ...........................................15.0g
Glucose ...................................................................7.5g
NaCl........................................................................7.5g
Plant extract ..........................................................4.5g
NaN$_3$ ......................................................................0.2g

pH 7.2 ± 0.2 at 25°C

**Source:** This medium is available as a premixed powder from HiMedia.

**Caution:** Sodium azide is toxic. Azides also react with metals and disposal must be highly diluted.

**Preparation of Medium:** Add components to distilled/deionized water and bring volume to 1.0L. Mix thoroughly. Gently heat and bring to boiling. Autoclave for 15 min at 12 psi pressure–118°C.

**Storage/Shelf Life:** Store dehydrated media in the dark in a sealed container below 30°C. Prepared media should be stored under refrigeration (2-8°C). Media should be used within 60 days of preparation. Media should not be used if there are any signs of deterioration (discoloration) or contamination, or if the expiration date supplied by the manufacturer has passed.

**Use:** For the detection and enrichment of fecal streptococci in water and sewage. Also used in the multiple-tube technique as a presumptive test for the presence of fecal streptococci in water, sewage, food, and other materials suspected of sewage contamination.

## Azide Medium

**Composition** per liter:

Peptone ................................................................10.0g
K$_2$HPO$_4$ .................................................................5.0g

| | |
|---|---|
| Glucose | 5.0g |
| NaCl | 5.0g |
| Yeast extract | 3.0g |
| $KH_2PO_4$ | 2.0g |
| $NaN_3$ | 0.25g |
| Bromcresol Purple solution | 2.0mL |

pH 7.2 ± 0.2 at 25°C

**Bromcresol Purple Solution:**
**Composition** per 10.0mL:

| | |
|---|---|
| Bromcresol Purple | 0.16g |
| Ethanol | 10.0mL |

**Preparation of Bromcresol Purple Solution:** Add Bromcresol Purple to ethanol and bring volume to 10.0mL. Mix thoroughly.

**Caution:** Sodium azide is toxic. Azides also react with metals and disposal must be highly diluted.

**Preparation of Medium:** Add components to distilled/deionized water and bring volume to 1.0L. Mix thoroughly. Distribute into tubes or flasks. Autoclave for 15 min at 15 psi pressure–121°C.

**Storage/Shelf Life:** Store dehydrated media in the dark in a sealed container below 30°C. Prepared media should be stored under refrigeration (2-8°C). Media should be used within 60 days of preparation. Media should not be used if there are any signs of deterioration (discoloration) or contamination, or if the expiration date supplied by the manufacturer has passed.

**Use:** For the cultivation of *Streptococcus* species and *Staphylococcus* species from clinical speciments and nonclinical specimens of public health importance.

## BACARA™ Agar
### (*Bacillus cereus* Chromogenic Medium)
**Composition** per liter:
Proprietary

**Source:** This medium is available from AES CHEMUNEX, a bio-Mérieux Company Rue Maryse Bastié - Ker Lann - CS17219 - F-35172 BRUZ Cedex - France.

**Storage/Shelf Life:** Store in the dark under refrigeration (2-8°C). Chromogenic agars are especially light and temperature sensitive; protect from light, excessive heat, moisture, and freezing. Do not use after the expiration date supplied by the manufacturer.

**Use:** For enumeration of *Bacillus cereus* in food. The medium inhibits background flora.

## *Bacillus cereus* Agar Base
## with Egg Yolk Emulsion and Polymyxin
**Composition** per liter:

| | |
|---|---|
| Agar | 15.0g |
| Sodium pyruvate | 10.0g |
| Mannitol | 10.0g |
| $Na_2HPO_4$ | 2.5g |
| NaCl | 2.0g |
| Peptone | 1.0g |
| $KH_2PO_4$ | 0.25g |
| Bromthymol Blue | 0.12g |
| $MgSO_4 \cdot 7H_2O$ | 0.1g |
| Egg yolk emulsion | 100.0mL |
| Selective supplement solution | 10.0mL |

pH 7.2 ± 0.2 at 25°C

**Source:** This medium, without egg yolk emulsion, is available as a premixed powder from HiMedia.

**Selective Supplement Solution:**
**Composition** per 10.0mL:

| | |
|---|---|
| Polymyxin B | 100,000 U |

**Preparation of Selective Supplement Solution:** Add components to distilled/deionized water and bring volume to 10.0mL. Mix thoroughly. Filter sterilize.

**Egg Yolk Emulsion:**
**Composition** per liter:

| | |
|---|---|
| Egg yolks | 30.0mL |
| NaCl, 0.9% solution | 70.0mL |

**Preparation of Egg Yolk Emulsion:** Soak eggs with 1:100 dilution of saturated mercuric chloride solution for 1 min. Crack 11 eggs and separate yolks from whites. Mix egg yolks. Measure 30.0mL of egg yolk emulsion and add to 70.0mL of 0.9% sterile NaCl solution. Mix thoroughly. Warm to 45°–50°C.

**Preparation of Medium:** Add components, except egg yolk emulsion, and selective supplement solution, to distilled/deionized water and bring volume to 890.0mL. Mix thoroughly. Gently heat and bring to boiling. Autoclave for 15 min at 15 psi pressure–121°C. Cool to 45°–50°C. Aseptically add 100.0mL egg yolk emulsion and 10.0mL sterile selective supplement solution. Mix well. Pour into sterile Petri dishes or sterile tubes.

**Storage/Shelf Life:** Store dehydrated media in the dark in a sealed container below 30°C. Prepared media should be stored under refrigeration (2-8°C). Media should be used within 60 days of preparation. Media should not be used if there are any signs of deterioration (shrinking, cracking, or discoloration) or contamination, or if the expiration date supplied by the manufacturer has passed.

**Use:** For the isolation, detection, and enumeration of *Bacillus cereus*.

## *Bacillus cereus* HiVeg Agar Base
## with Egg Yolk Emulsion
**Composition** per liter:

| | |
|---|---|
| Agar | 15.0g |
| Sodium pyruvate | 10.0g |
| Mannitol | 10.0g |
| $Na_2HPO_4$ | 2.5g |
| NaCl | 2.0g |
| Plant peptone | 1.0g |
| $KH_2PO_4$ | 0.25g |
| Bromthymol Blue | 0.12g |
| $MgSO_4 \cdot 7H_2O$ | 0.1g |
| Egg yolk emulsion | 100.0mL |

pH 7.2 ± 0.2 at 25°C

**Source:** This medium, without egg yolk emulsion, is available as a premixed powder from HiMedia.

**Egg Yolk Emulsion:**
**Composition** per liter:

| | |
|---|---|
| Egg yolks | 30.0mL |
| NaCl, 0.9% solution | 70.0mL |

**Preparation of Egg Yolk Emulsion:** Soak eggs with 1:100 dilution of saturated mercuric chloride solution for 1 min. Crack 11 eggs and separate yolks from whites. Mix egg yolks. Measure 30.0mL of egg yolk emulsion and add to 70.0mL of 0.9% sterile NaCl solution. Mix thoroughly. Warm to 45°–50°C.

**Preparation of Medium:** Add components, except egg yolk emulsion, to distilled/deionized water and bring volume to 900.0mL. Mix thoroughly. Gently heat and bring to boiling. Autoclave for 15 min at 15 psi pressure–121°C. Cool to 45°–50°C. Aseptically add 100.0mL egg yolk emulsion. Mix well. Pour into sterile Petri dishes or sterile tubes.

**Storage/Shelf Life:** Store dehydrated media in the dark in a sealed container below 30°C. Prepared media should be stored under refrigeration (2-8°C). Media should be used within 60 days of preparation. Media should not be used if there are any signs of deterioration (shrinking, cracking, or discoloration) or contamination, or if the expiration date supplied by the manufacturer has passed.

**Use:** For the isolation, detection, and enumeration of *Bacillus cereus.*

## *Bacillus cereus* Medium
### (BCM)

**Composition** per 110.0mL:

| | |
|---|---|
| Agar | 2.0g |
| D-Mannitol | 1.0g |
| $(NH_4)_2PO_4$ | 0.1g |
| KCl | 0.02g |
| $MgSO_4 \cdot 7H_2O$ | 0.02g |
| Yeast extract | 0.02g |
| Bromcresol Purple | 4.0mg |
| Egg yolk emulsion, 20% | 10.0mL |

pH 7.0 ± 0.2 at 25°C

**Egg Yolk Emulsion, 20%:**
**Composition** per 100.0mL:

| | |
|---|---|
| Chicken egg yolks | 11 |
| Whole chicken egg | 1 |
| NaCl (0.9% solution) | 80.0mL |

**Preparation of Egg Yolk Emulsion, 20%:** Soak eggs with 1:100 dilution of saturated mercuric chloride solution for 1 min. Crack eggs and separate yolks from whites. Mix egg yolks with 1 chicken egg. Measure 20.0mL of egg yolk emulsion and add to 80.0mL of 0.9% NaCl solution. Mix thoroughly. Filter sterilize. Warm to 45°–50°C.

**Preparation of Medium:** Add components—except egg yolk emulsion, 20%—to distilled/deionized water and bring volume to 100.0mL. Mix thoroughly. Gently heat and bring to boiling. Autoclave for 15 min at 15 psi pressure–121°C. Cool to 45°–50°C. Aseptically add 10.0mL of sterile egg yolk emulsion, 20%. Mix thoroughly. Pour into sterile Petri dishes or distribute into sterile tubes.

**Storage/Shelf Life:** Store dehydrated media in the dark in a sealed container below 30°C. Prepared media should be stored under refrigeration (2-8°C). Media should be used within 60 days of preparation. Media should not be used if there are any signs of deterioration (shrinking, cracking, or discoloration) or contamination, or if the expiration date supplied by the manufacturer has passed.

**Use:** For the cultivation of *Bacillus cereus.*

## *Bacillus cereus* Selective Agar Base

**Composition** per liter:

| | |
|---|---|
| Agar | 15.0g |
| Sodium pyruvate | 10.0g |

| | |
|---|---|
| Mannitol | 10.0g |
| $Na_2HPO_4$ | 2.5g |
| NaCl | 2.0g |
| Peptone | 1.0g |
| $KH_2PO_4$ | 0.25g |
| Bromthymol Blue | 0.12g |
| $MgSO_4 \cdot 7H_2O$ | 0.1g |
| Egg yolk emulsion | 25.0mL |
| Polymyxin B solution | 10.0mL |

pH 7.2 ± 0.2 at 25°C

**Source:** This medium is available as a premixed powder from Thermo Scientific.

**Egg Yolk Emulsion:**
**Composition**:

| | |
|---|---|
| Chicken egg yolks | 11 |
| Whole chicken egg | 1 |

**Preparation of Egg Yolk Emulsion:** Soak eggs with 1:100 dilution of saturated mercuric chloride solution for 1 min. Crack eggs and separate yolks from whites. Mix egg yolks with 1 chicken egg.

**Polymyxin B Solution:**
**Composition** per 10.0mL:

| | |
|---|---|
| Polymyxin B | 100,000U |

**Preparation of Polymyxin B Solution:** Add polymyxin B to distilled/deionized water and bring volume to 10.0mL. Mix thoroughly. Filter sterilize.

**Preparation of Medium:** Add components, except egg yolk emulsion and polymyxin B solution, to distilled/deionized water and bring volume to 965.0mL. Gently heat and bring to boiling. Distribute into tubes or flasks. Autoclave for 15 min at 15 psi pressure–121°C. Cool to 50°C. Aseptically add sterile polymyxin B and 25.0mL of sterile egg yolk emulsion. Mix thoroughly. Pour into sterile Petri dishes or leave in tubes.

**Storage/Shelf Life:** Store dehydrated media in the dark in a sealed container below 30°C. Prepared media should be stored under refrigeration (2-8°C). Media should be used within 60 days of preparation. Media should not be used if there are any signs of deterioration (shrinking, cracking, or discoloration) or contamination, or if the expiration date supplied by the manufacturer has passed.

**Use:** For the selection and presumptive identification of *Bacillus cereus.* Also for the isolation and enumeration of these bacteria. *Bacillus cereus* grows as moderate-sized (5mm) crenated colonies, which are turquoise, surrounded by a precipitate of egg yolk, which is also turquoise.

## *Bacteroides* Bile Esculin Agar
### (BBE Agar)

**Composition** per liter:

| | |
|---|---|
| Oxgall | 20.0g |
| Pancreatic digest of casein | 15.0g |
| Agar | 15.0g |
| Papaic digest of soybean meal | 5.0g |
| NaCl | 5.0g |
| Esculin | 1.0g |
| Ferric ammonium citrate | 0.5g |
| Gentamicin solution | 2.5mL |
| Hemin solution | 2.5mL |
| Vitamin $K_1$ solution | 1.0mL |

pH 7.0 ± 0.2 at 25°C

**Source:** This medium is available as a premixed powder from BD Diagnostic Systems.

**Gentamicin Solution:**
**Composition** per 10.0mL:
Gentamicin ............................................................................0.4mg

**Preparation of Gentamicin Solution:** Add gentamicin to 10.0mL of distilled/deionized water. Mix thoroughly. Filter sterilize.

**Hemin Solution:**
**Composition** per 100.0mL:
Hemin.......................................................................................0.5g
NaOH (1*N* solution)..........................................................10.0mL

**Preparation of Hemin Solution:** Add components to 100.0mL of distilled/deionized water. Mix thoroughly. Autoclave for 15 min at 15 psi pressure–121°C. Cool to 45°–50°C.

**Vitamin K₁ Solution:**
**Composition** per 100.0mL:
Vitamin K₁ ..............................................................................1.0g
Ethanol ................................................................................99.0mL

**Preparation of Vitamin K₁ Solution:** Add vitamin K₁ to 99.0mL of absolute ethanol. Mix thoroughly. Filter sterilize.

**Preparation of Medium:** Add components, except hemin solution, gentamicin solution, and vitamin K₁ solution, to distilled/deionized water and bring volume to 994.0mL. Mix thoroughly. Gently heat and bring to boiling. Autoclave for 15 min at 15 psi pressure–121°C. Cool to 45°–50°C. Aseptically add 2.5mL of sterile hemin solution, 2.5mL of sterile gentamicin solution, and 1.0mL of sterile vitamin K₁ solution.

**Storage/Shelf Life:** Store dehydrated media in the dark in a sealed container below 30°C. Prepared media should be stored under refrigeration (2-8°C). Media should be used within 60 days of preparation. Media should not be used if there are any signs of deterioration (shrinking, cracking, or discoloration) or contamination, or if the expiration date supplied by the manufacturer has passed.

**Use:** For the selection and presumptive identification of the *Bacteriodes fragilis* group. For the differentiation of *Bacteroides* species based on the hydrolysis of esculin and presence of catalase. After incubation for 48 hr, bacteria of the *Bacteroides fragilis* group appear as gray, circular, raised colonies larger than 1.0mm. Esculin hydrolysis is indicated by the presence of a blackened zone around the colonies.

## *Bacteroides* HiVeg Agar Base
## with Selective Supplement
## (BBE)

**Composition** per liter:
Plant hydrolysate.....................................................................25.0g
Agar ........................................................................................15.0g
Papaic digest of soybean meal ...............................................10.0g
NaCl..........................................................................................5.0g
Synthetic detergent No. II........................................................2.0g
Esculin ......................................................................................1.0g
Ferric ammonium citrate...........................................................0.5g
Fe₄(P₂O₇)₃·H₂O.........................................................................0.01g
Vitamin K₁ ..............................................................................0.01g
Selective supplement solution .............................................10.0mL
pH 7.2 ± 0.2 at 25°C

**Source:** This medium, without selective supplement, is available as a premixed powder from HiMedia.

**Selective Supplement Solution:**
**Composition** per 10.0mL:
Gentamicin............................................................................0.1mg

**Preparation of Selective Supplement Solution:** Add gentamicin to distilled/deionized water and bring volume to 10.0mL. Mix thoroughly. Filter sterilize.

**Preparation of Medium:** Add components, except selective supplement, to distilled/deionized water and bring volume to 990.0mL. Mix thoroughly. Heat with frequent agitation and boil for 1 min to completely dissolve. Autoclave for 15 min at 15 psi pressure–121°C. Cool to 50°–55°C. Add 10.0mL of sterile selective supplement. Mix thoroughly. Pour into sterile Petri dishes or leave in tubes.

**Storage/Shelf Life:** Store dehydrated media in the dark in a sealed container below 30°C. Prepared media should be stored under refrigeration (2-8°C). Media should be used within 60 days of preparation. Media should not be used if there are any signs of deterioration (shrinking, cracking, or discoloration) or contamination, or if the expiration date supplied by the manufacturer has passed.

**Use:** For the selection and presumptive identification of the *Bacteriodes fragilis* group. For the differentiation of *Bacteroides* species based on the hydrolysis of esculin and presence of catalase. After incubation for 48 hr, bacteria of the *Bacteroides fragilis* group appear as gray, circular, raised colonies larger than 1.0mm. Esculin hydrolysis is indicated by the presence of a blackened zone around the colonies.

## BAGG Broth
## (Buffered Azide Glucose Glycerol Broth)

**Composition** per liter:
Pancreatic digest of casein...........................................10.0g
Peptic digest of animal tissue .....................................10.0g
Glucose ..........................................................................5.0g
NaCl................................................................................5.0g
K₂HPO₄............................................................................4.0g
KH₂PO₄............................................................................1.5g
NaN₃ ...............................................................................0.5g
Bromcresol Purple .......................................................0.015g
Glycerol ...................................................................... 5.0mL
pH 6.9 ± 0.2 at 25°C

**Source:** This medium is available as a premixed powder from BD Diagnostic Systems.

**Caution:** Sodium azide is toxic. Azides also react with metals and disposal must be highly diluted.

**Preparation of Medium:** Add 5.0mL of glycerol to 900.0mL of distilled/deionized water. Add remaining components and bring volume to 1.0L. Mix thoroughly. Gently heat and bring to boiling. Distribute into tubes in 10.0mL volumes. Autoclave for 15 min at 10 psi pressure–116°C.

**Storage/Shelf Life:** Store dehydrated media in the dark in a sealed container below 30°C. Prepared media should be stored under refrigeration (2-8°C). Media should be used within 60 days of preparation. Media should not be used if there are any signs of deterioration (discoloration) or contamination, or if the expiration date supplied by the manufacturer has passed.

**Use:** For the cultivation of fecal streptococci from a variety of clinical speciments and nonclinical specimens of public health importance. It is recommended for qualitative presumptive and confirmatory tests for fecal streptococci.

## BAGG Broth Base with Glycerol
### (Buffered Azide Glucose Glycerol Broth Base)

**Composition** per liter:

Tryptose ...................................................................20.0g
Glucose .....................................................................5.0g
NaCl...........................................................................5.0g
$K_2HPO_4$.................................................................4.0g
$KH_2PO_4$................................................................1.5g
$NaN_3$ .....................................................................0.5g
Bromcresol Purple ................................................0.015g
Glycerol ................................................................. 5.0mL

pH 6.9 ± 0.2 at 25°C

**Source:** This medium without glycerol is available as a premixed powder from HiMedia.

**Caution:** Sodium azide is toxic. Azides also react with metals and disposal must be highly diluted.

**Preparation of Medium:** Add 5.0mL of glycerol to 900.0mL of distilled/deionized water. Add remaining components and bring volume to 1.0L. Mix thoroughly. Gently heat and bring to boiling. Distribute into tubes in 10.0mL volumes. Autoclave for 15 min at 10 psi pressure–115°C.

**Storage/Shelf Life:** Store dehydrated media in the dark in a sealed container below 30°C. Prepared media should be stored under refrigeration (2-8°C). Media should be used within 60 days of preparation. Media should not be used if there are any signs of deterioration (discoloration) or contamination, or if the expiration date supplied by the manufacturer has passed.

**Use:** For the cultivation of fecal streptococci from a variety of clinical speciments and nonclinical specimens of public health importance. It is recommended for qualitative presumptive and confirmatory tests for fecal streptococci.

## BAGG HiVeg Broth Base with Glycerol
### (Buffered Azide Glucose Glycerol HiVeg Broth Base)

**Composition** per liter:

Plant hydrolysate No. 1.........................................20.0g
Glucose .....................................................................5.0g
NaCl...........................................................................5.0g
$K_2HPO_4$.................................................................4.0g
$KH_2PO_4$................................................................1.5g
$NaN_3$ .....................................................................0.5g
Bromcresol Purple ................................................0.015g
Glycerol ................................................................. 5.0mL

pH 6.9 ± 0.2 at 25°C

**Source:** This medium without glycerol is available as a premixed powder from HiMedia.

**Caution:** Sodium azide is toxic. Azides also react with metals and disposal must be highly diluted.

**Preparation of Medium:** Add 5.0mL of glycerol to 900.0mL of distilled/deionized water. Add remaining components and bring volume to 1.0L. Mix thoroughly. Gently heat and bring to boiling. Distribute into tubes in 10.0mL volumes. Autoclave for 15 min at 10 psi pressure–115°C.

**Storage/Shelf Life:** Store dehydrated media in the dark in a sealed container below 30°C. Prepared media should be stored under refrigeration (2-8°C). Media should be used within 60 days of preparation. Media should not be used if there are any signs of deterioration (discol-

oration) or contamination, or if the expiration date supplied by the manufacturer has passed.

**Use:** For the cultivation of fecal streptococci from a variety of clinical speciments and nonclinical specimens of public health importance. It is recommended for qualitative presumptive and confirmatory tests for fecal streptococci.

## Baird-Parker Agar

**Composition** per liter:

Agar .........................................................................17.0g
Glycine......................................................................12.0g
Sodium pyruvate ......................................................10.0g
Pancreatic digest of casein......................................10.0g
Beef extract................................................................5.0g
LiCl.............................................................................5.0g
Yeast extract...............................................................1.0g

pH 7.0 ± 0.2 at 25°C

**Source:** This medium is available as a premixed powder from Thermo Scientific and BD Diagnostic Systems.

**Caution:** Lithium chloride is harmful. Avoid bodily contact and inhalation of vapors. On contact with skin wash with plenty of water immediately.

**Preparation of Medium:** Add components to distilled/deionized water and bring volume to 1.0L. Mix thoroughly. Gently heat and bring to boiling. Autoclave for 15 min at 15 psi pressure–121°C. Cool to 45°–50°C. Pour into sterile Petri dishes.

**Storage/Shelf Life:** Store dehydrated media in the dark in a sealed container below 30°C. Prepared media should be stored under refrigeration (2-8°C). Media should be used within 60 days of preparation. Media should not be used if there are any signs of deterioration (shrinking, cracking, or discoloration) or contamination, or if the expiration date supplied by the manufacturer has passed.

**Use:** Used as a base for the preparation of egg-tellurite-glycine-pyruvate agar for the selective isolation and enumeration of coagulase-positive staphylococci from food, skin, and other specimens.

## Baird-Parker Agar

**Composition** per liter:

Agar .........................................................................17.0g
Glycine......................................................................12.0g
Sodium pyruvate ......................................................10.0g
Pancreatic digest of casein......................................10.0g
Beef extract................................................................5.0g
LiCl.............................................................................5.0g
Yeast extract...............................................................1.0g
Sulfamethazine solution........................................ 10.0mL

pH 7.0 ± 0.2 at 25°C

**Caution:** Lithium chloride is harmful. Avoid bodily contact and inhalation of vapors. On contact with skin wash with plenty of water immediately.

**Sulfamethazine Solution:**
**Composition** per 10.0mL:

Sulfamethazine .........................................................0.05g

**Preparation of Sulfamethazine Solution:** Add sulfamethazine to distilled/deionized water and bring volume to 10.0mL. Mix thoroughly. Filter sterilize.

**Preparation of Medium:** Add components, except sulfamethazine solution, to distilled/deionized water and bring volume to 990.0mL. Mix thoroughly. Gently heat and bring to boiling. Autoclave for 15 min at 15 psi pressure–121°C. Cool to 45°–50°C. Aseptically add sterile sulfamethazine solution. Mix thoroughly. Pour into sterile Petri dishes or distribute into sterile tubes.

**Storage/Shelf Life:** Store dehydrated media in the dark in a sealed container below 30°C. Prepared media should be stored under refrigeration (2-8°C). Media should be used within 60 days of preparation. Media should not be used if there are any signs of deterioration (shrinking, cracking, or discoloration) or contamination, or if the expiration date supplied by the manufacturer has passed.

**Use:** Used as a base for the preparation of egg-tellurite-glycine-pyruvate agar for the selective isolation and enumeration of coagulase-positive staphylococci from food, skin, and other specimens.

## Baird-Parker Agar Base
## with Egg Yolk Tellurite Enrichment
**Composition** per liter:

| | |
|---|---|
| Agar | 20.0g |
| Glycine | 12.0g |
| Casein enzymatic hydrolysate | 10.0g |
| Sodium pyruvate | 10.0g |
| Plant extract | 5.0g |
| LiCl | 5.0g |
| Yeast extract | 1.0g |
| Egg yolk tellurite enrichment | 50.0mL |

pH 7.0 ± 0.2 at 25°C

**Source:** This medium is available as a premixed powder from Hi-Media.

**Caution:** Lithium chloride is harmful. Avoid bodily contact and inhalation of vapors. On contact with skin wash with plenty of water immediately.

**Egg Yolk Tellurite Enrichment:**
**Composition** per 100.0mL:

| | |
|---|---|
| Chicken egg yolks | 10 |
| $K_2TeO_3$ | 0.15g |
| NaCl (0.9% solution) | 50.0mL |

**Preparation of Egg Yolk Tellurite Enrichment:** Soak eggs with 1:100 dilution of saturated mercuric chloride solution for 1 min. Crack 11 eggs and separate yolks from whites. Mix egg yolks. Measure 30.0mL of egg yolk emulsion and add to 70.0mL of 0.9% NaCl solution. Mix thoroughly. Add 0.15g $K_2TeO_3$. Filter sterilize. Warm to 45°–50°C.

**Caution:** Potassium tellurite is toxic.

**Preparation of Medium:** Add components, except egg yolk tellurite enrichment, to distilled/deionized water and bring volume to 950.0mL. Mix thoroughly. Gently heat and bring to boiling. Autoclave for 15 min at 15 psi pressure–121°C. Cool to 45°–50°C. Aseptically add 50 mL of egg yolk tellurite enrichment. Mix well. Pour into sterile Petri dishes or sterile tubes.

**Storage/Shelf Life:** Store dehydrated media in the dark in a sealed container below 30°C. Prepared media should be stored under refrigeration (2-8°C). Media should be used within 60 days of preparation. Media should not be used if there are any signs of deterioration (shrinking, cracking, or discoloration) or contamination, or if the expiration date supplied by the manufacturer has passed.

**Use:** For the selective isolation and enumeration of coagulase-positive staphylococci.

## Baird-Parker Agar Base, HiVeg
## with Egg Yolk Tellurite Enrichment
**Composition** per liter:

| | |
|---|---|
| Agar | 20.0g |
| Glycine | 12.0g |
| Plant hydrolysate | 10.0g |
| Sodium pyruvate | 10.0g |
| Plant extract | 5.0g |
| LiCl | 5.0g |
| Yeast extract | 1.0g |
| Egg yolk tellurite enrichment | 50.0mL |

pH 7.0 ± 0.2 at 25°C

**Source:** This medium is available as a premixed powder from Hi-Media.

**Caution:** Lithium chloride is harmful. Avoid bodily contact and inhalation of vapors. On contact with skin wash with plenty of water immediately.

**Egg Yolk Tellurite Enrichment:**
**Composition** per 100.0mL:

| | |
|---|---|
| Chicken egg yolks | 10 |
| $K_2TeO_3$ | 0.15g |
| NaCl (0.9% solution) | 50.0mL |

**Preparation of Egg Yolk Tellurite Enrichment:** Soak eggs with 1:100 dilution of saturated mercuric chloride solution for 1 min. Crack 11 eggs and separate yolks from whites. Mix egg yolks. Measure 30.0mL of egg yolk emulsion and add to 70.0mL of 0.9% NaCl solution. Mix thoroughly. Add 0.15g $K_2TeO_3$. Filter sterilize. Warm to 45°–50°C.

**Caution:** Potassium tellurite is toxic.

**Preparation of Medium:** Add components, except egg yolk tellurite enrichment, to distilled/deionized water and bring volume to 950.0mL. Mix thoroughly. Gently heat and bring to boiling. Autoclave for 15 min at 15 psi pressure–121°C. Cool to 45°–50°C. Aseptically add 50 mL of egg yolk tellurite enrichment. Mix well. Pour into sterile Petri dishes or sterile tubes.

**Storage/Shelf Life:** Store dehydrated media in the dark in a sealed container below 30°C. Prepared media should be stored under refrigeration (2-8°C). Media should be used within 60 days of preparation. Media should not be used if there are any signs of deterioration (shrinking, cracking, or discoloration) or contamination, or if the expiration date supplied by the manufacturer has passed.

**Use:** For the selective isolation and enumeration of coagulase-positive staphylococci.

## Baird-Parker Agar, Supplemented
**Composition** per liter:

| | |
|---|---|
| Agar | 17.0g |
| Glycine | 12.0g |
| Sodium pyruvate | 10.0g |
| Pancreatic digest of casein | 10.0g |
| Beef extract | 5.0g |
| LiCl | 5.0g |

Yeast extract.............................................................1.0g
RPF supplement................................................ 100.0mL
pH 7.0 ± 0.2 at 25°C

**Caution:** Lithium chloride is harmful. Avoid bodily contact and inhalation of vapors. On contact with skin wash with plenty of water immediately.

### RPF Supplement:
**Composition** per 100.0mL:
Bovine fibrinogen ........................................................3.75g
Trypsin inhibitor .......................................................25.0mg
K$_2$TeO$_3$...............................................................25.0mg
Rabbit plasma ...........................................................25.0mL

**Caution:** Potassium tellurite is toxic.

**Preparation of RPF Supplement:** Add components to distilled/deionized water and bring volume to 100.0mL. Mix thoroughly. Filter sterilize.

**Preparation of Medium:** Add components, except RPF supplement, to distilled/deionized water and bring volume to 900.0mL. Mix thoroughly. Gently heat and bring to boiling. Autoclave for 15 min at 15 psi pressure–121°C. Cool to 45°–50°C. Aseptically add 100.0mL of filter-sterilized RPF supplement. Mix thoroughly but gently. Pour into sterile Petri dishes.

**Storage/Shelf Life:** Store dehydrated media in the dark in a sealed container below 30°C. Prepared media should be stored under refrigeration (2-8°C). Media should be used within 60 days of preparation. Media should not be used if there are any signs of deterioration (shrinking, cracking, or discoloration) or contamination, or if the expiration date supplied by the manufacturer has passed.

**Use:** For the selective isolation and enumeration of coagulase-positive staphylococci from food, skin, and other specimens. For the differentiation and identification of staphylococci on the basis of their ability to coagulate plasma. Colonies surrounded by an opaque zone of coagulated plasma are diagnostic for *Staphylococcus aureus*.

## Baird-Parker Egg Yolk Agar (ISO)
**Composition** per 1050.0mL:
Agar ........................................................................20.0g
L-Glycine.................................................................12.0g
Pancreatic digest of casein.....................................10.0g
Sodium pyruvate .....................................................10.0g
Meat extract .............................................................5.0g
LiCl...........................................................................5.0g
Yeast extract.............................................................1.0g
Egg yolk tellurite enrichment ............................... 50.0mL
pH 7.2 ± 0.2 at 25°C

**Source:** This medium is available as a premixed powder from Thermo Scientific.

**Caution:** Lithium chloride is harmful. Avoid bodily contact and inhalation of vapors. On contact with skin wash with plenty of water immediately.

### Egg Yolk Tellurite Enrichment:
**Composition** per 100.0mL:
Chicken egg yolks.........................................................10
K$_2$TeO$_3$................................................................0.15g
NaCl (0.9% solution) ............................................. 50.0mL

**Preparation of Egg Yolk Tellurite Enrichment:** Soak eggs with 1:100 dilution of saturated mercuric chloride solution for 1 min. Crack 11 eggs and separate yolks from whites. Mix egg yolks. Measure

30.0mL of egg yolk emulsion and add to 70.0mL of 0.9% NaCl solution. Mix thoroughly. Add 0.15g K$_2$TeO$_3$. Filter sterilize. Warm to 45°–50°C.

**Caution:** Potassium tellurite is toxic.

**Preparation of Medium:** Add components, except egg yolk tellurite enrichment, to distilled/deionized water and bring volume to 950.0mL. Mix thoroughly. Gently heat and bring to boiling. Autoclave for 15 min at 15 psi pressure–121°C. Cool to 45°–50°C. Aseptically add 50 mL of egg yolk tellurite enrichment. Mix well. Pour into sterile Petri dishes or sterile tubes.

**Storage/Shelf Life:** Store dehydrated media in the dark in a sealed container below 30°C. Prepared media should be stored under refrigeration (2-8°C). Media should be used within 60 days of preparation. Media should not be used if there are any signs of deterioration (shrinking, cracking, or discoloration) or contamination, or if the expiration date supplied by the manufacturer has passed.

**Use:** For the selective isolation and enumeration of coagulase-positive staphylococci. A selective medium for the isolation and enumeration of coagulase-positive staphylococci from food, with formulation conforming to that recommended in ISO 6888-1:1999.

## Baird-Parker Medium (BAM M17)
**Composition** per liter:
Agar ........................................................................20.0g
Glycine....................................................................12.0g
Sodium pyruvate .....................................................10.0g
Pancreatic digest of casein.....................................10.0g
Beef extract..............................................................5.0g
LiCl·6H$_2$O...............................................................5.0g
Yeast extract.............................................................1.0g
Egg yolk tellurite enrichment ............................... 50.0mL
pH 7.0 ± 0.2 at 25°C

**Caution:** Lithium chloride is harmful. Avoid bodily contact and inhalation of vapors. On contact with skin wash with plenty of water immediately.

### Egg Yolk Tellurite Enrichment:
**Composition** per 100.0mL:
Chicken egg yolks.........................................................10
K$_2$TeO$_3$................................................................0.15g
NaCl (0.9% solution) ............................................. 50.0mL

**Preparation of Egg Yolk Tellurite Enrichment:** Soak eggs with 1:100 dilution of saturated mercuric chloride solution for 1 min. Crack 11 eggs and separate yolks from whites. Mix egg yolks. Measure 30.0mL of egg yolk emulsion and add to 70.0mL of 0.9% NaCl solution. Mix thoroughly. Add 0.15g K$_2$TeO$_3$. Filter sterilize. Warm to 45°–50°C.

**Caution:** Potassium tellurite is toxic.

**Source:** This medium is available as a premixed powder from BD Diagnostic Systems.

**Preparation of Medium:** Add components, except EY tellurite enrichment, to distilled/deionized water and bring volume to 950.0mL. Mix thoroughly. Gently heat and bring to boiling. Autoclave for 15 min at 15 psi pressure–121°C. Cool to 48°–50°C. Aseptically add 50.0mL of sterile EY tellurite enrichment. Mix thoroughly. Pour into sterile Petri dishes. The medium must be densely opaque. Dry plates before use. Plates can be stored for up to 5 days at 20–25°C before use.

**Storage/Shelf Life:** Store dehydrated media in the dark in a sealed container below 30°C. Prepared media should be stored under refrigeration (2-8°C). Media should be used within 60 days of preparation. Media should not be used if there are any signs of deterioration (shrinking, cracking, or discoloration) or contamination, or if the expiration date supplied by the manufacturer has passed.

**Use:** For the selective isolation and enumeration of coagulase-positive staphylococci from foods.

## Baird-Parker Medium
### (BAM M17)

**Composition** per liter:

| | |
|---|---|
| Agar | 20.0g |
| Glycine | 12.0g |
| Sodium pyruvate | 10.0g |
| Pancreatic digest of casein | 10.0g |
| Beef extract | 5.0g |
| LiCl·6H$_2$O | 5.0g |
| Yeast extract | 1.0g |
| Egg yolk tellurite enrichment | 50.0mL |

pH 7.0 ± 0.2 at 25°C

**Caution:** Lithium chloride is harmful. Avoid bodily contact and inhalation of vapors. On contact with skin wash with plenty of water immediately.

**Egg Yolk Tellurite Enrichment:**

**Composition** per 100.0mL:

| | |
|---|---|
| Chicken egg yolks | 10 |
| K$_2$TeO$_3$ | 0.15g |
| NaCl (0.9% solution) | 50.0mL |

**Preparation of Egg Yolk Tellurite Enrichment:** Soak eggs with 1:100 dilution of saturated mercuric chloride solution for 1 min. Crack 11 eggs and separate yolks from whites. Mix egg yolks. Measure 30.0mL of egg yolk emulsion and add to 70.0mL of 0.9% NaCl solution. Mix thoroughly. Add 0.15g K$_2$TeO$_3$. Filter sterilize. Warm to 45°–50°C.

**Caution:** Potassium tellurite is toxic.

**Source:** This medium is available as a premixed powder from BD Diagnostic Systems.

**Preparation of Medium:** Add components, except egg yolk tellurite enrichment, to distilled/deionized water and bring volume to 950.0mL. Mix thoroughly. Gently heat and bring to boiling. Autoclave for 15 min at 15 psi pressure–121°C. Cool to 48°–50°C. Aseptically add 50.0mL of sterile egg yolk tellurite enrichment. Mix thoroughly. Pour into sterile Petri dishes. The medium must be densely opqaue. Dry plates before use. Plates can be stored for up to 5 days at 20–25°C before use.

**Storage/Shelf Life:** Store dehydrated media in the dark in a sealed container below 30°C. Prepared media should be stored under refrigeration (2-8°C). Media should be used within 60 days of preparation. Media should not be used if there are any signs of deterioration (shrinking, cracking, or discoloration) or contamination, or if the expiration date supplied by the manufacturer has passed.

**Use:** For the selective isolation and enumeration of coagulase-positive staphylococci from foods.

## BCG Glucose Agar
### (Snyder Test Agar)

**Composition** per liter:

| | |
|---|---|
| Agar | 20.0g |
| Glucose | 20.0g |
| Peptic digest of animal tissue | 20.0g |
| NaCl | 5.0g |
| Bromcresol Green | 0.02 |

pH 4.8 ± 0.2 at 25°C

**Source:** This medium is available as a premixed powder from Hi-Media.

**Preparation of Medium:** Add components to distilled/deionized water and bring volume to 1.0L. Mix thoroughly. Gently heat and bring to boiling. Distribute into tubes in 10.0mL volumes. Autoclave for 15 min at 15 psi pressure–121°C. Do not overheat. Pour into sterile Petri dishes or leave in tubes.

**Storage/Shelf Life:** Store dehydrated media in the dark in a sealed container below 30°C. Prepared media should be stored under refrigeration (2-8°C). Media should be used within 60 days of preparation. Media should not be used if there are any signs of deterioration (shrinking, cracking, or discoloration) or contamination, or if the expiration date supplied by the manufacturer has passed.

**Use:** For the cultivation and enumeration of lactobacilli in saliva and indication of dental caries activity.

## BCG Glucose HiVeg Agar
### (Snyder Test HiVeg Agar)

**Composition** per liter:

| | |
|---|---|
| Agar | 20.0g |
| Glucose | 20.0g |
| Plant peptone | 20.0g |
| NaCl | 5.0g |
| Bromcresol Green | 0.02 |

pH 4.8 ± 0.2 at 25°C

**Source:** This medium is available as a premixed powder from Hi-Media.

**Preparation of Medium:** Add components to distilled/deionized water and bring volume to 1.0L. Mix thoroughly. Gently heat and bring to boiling. Distribute into tubes in 10.0mL volumes. Autoclave for 15 min at 15 psi pressure–121°C. Do not overheat. Pour into sterile Petri dishes or leave in tubes.

**Storage/Shelf Life:** Store dehydrated media in the dark in a sealed container below 30°C. Prepared media should be stored under refrigeration (2-8°C). Media should be used within 60 days of preparation. Media should not be used if there are any signs of deterioration (shrinking, cracking, or discoloration) or contamination, or if the expiration date supplied by the manufacturer has passed.

**Use:** For the cultivation and enumeration of lactobacilli in saliva and indication of dental caries activity.

## BCM *Bacillus cereus* Group Plating Medium

**Composition** per liter:

Proprietary

**Source:** This medium is available from Biosynth International, Inc.

**Storage/Shelf Life:** Store in the dark under refrigeration (2-8°C). Chromogenic agars are especially light and temperature sensitive; protect from light, excessive heat, moisture, and freezing. Do not use after the expiration date supplied by the manufacturer.

**Use:** For detection of *Bacillus cereus* in food. The medium contains 5-bromo-4-chloro-3-indoxyl-myoinositol-1-phosphate as a chromogenic substrate, which changes from colorless to turquoise upon enzymatic cleavage. *B. cereus*, *B. mycoides*, *B. thuringiensis*, and *B. weihenstephanensis* secrete phosphatidylinositol phospholipase C and so grow as turquoise colonies with species-specific morphologies.

## BCM O157:H7(+) Plating Medium

**Composition** per liter:

Proprietary

**Source:** This medium is available from Biosynth International, Inc.

**Storage/Shelf Life:** Store in the dark under refrigeration (2-8°C). Chromogenic agars are especially light and temperature sensitive; protect from light, excessive heat, moisture, and freezing. Do not use after the expiration date supplied by the manufacturer.

**Use:** For detection of this highly pathogenic EHEC serovar BCM *O157:H7(+)*.

## BCP Azide Broth
## (Bromcresol Purple Azide Broth)

**Composition** per liter:

| | |
|---|---|
| Casein peptone | 10.0g |
| Yeast extract | 10.0g |
| D-Glucose | 5.0g |
| NaCl | 5.0g |
| $K_2HPO_4$ | 2.7g |
| $KH_2PO_4$ | 2.7g |
| $NaN_3$ | 0.5g |
| Bromcresol Purple | 0.032g |

pH 6.9 ± 0.2 at 25°C

**Caution:** Sodium azide is toxic. Azides also react with metals and disposal must be highly diluted.

**Preparation of Medium:** Add components to distilled/deionized water to 1.0L. Mix thoroughly. Gently heat to boiling. Distribute into tubes or flasks. Autoclave for 15 min at 15 psi pressure–121°C.

**Storage/Shelf Life:** Store dehydrated media in the dark in a sealed container below 30°C. Prepared media should be stored under refrigeration (2-8°C). Media should be used within 60 days of preparation. Media should not be used if there are any signs of deterioration (discoloration) or contamination, or if the expiration date supplied by the manufacturer has passed.

**Use:** For use in the confirmation test for the presence of fecal streptococci in water and wastewater.

## BCP D Agar
## (Bromcresol Purple Deoxycholate Agar)

**Composition** per liter:

| | |
|---|---|
| Agar | 25.0g |
| Lactose | 10.0g |
| Sucrose | 10.0g |
| Pancreatic digest of casein | 7.5g |
| Thiopeptone | 7.5g |
| NaCl | 5.0g |
| Yeast extract | 2.0g |

| | |
|---|---|
| Sodium citrate | 2.0g |
| Sodium deoxycholate | 1.0g |
| Bromcresol Purple | 0.02g |

pH 7.2 ± 0.2 at 25°C

**Preparation of Medium:** Add components to distilled/deionized water and bring volume to 1.0L. Mix thoroughly. Gently heat and bring to boiling. Pour into sterile Petri dishes without sterilization. Do not autoclave. Use the same day.

**Storage/Shelf Life:** Store dehydrated media in the dark in a sealed container below 30°C. Prepared media should be stored under refrigeration (2-8°C). Media should be used within 60 days of preparation. Media should not be used if there are any signs of deterioration (shrinking, cracking, or discoloration) or contamination, or if the expiration date supplied by the manufacturer has passed.

**Use:** For the isolation, cultivation, and differentiation of Gram-negative enteric bacilli from clinical speciments and nonclinical specimens of public health importance. For the isolation, cultivation, and identification of microorganisms from fecal specimens. For the isolation and cultivation of *Salmonella*, *Shigella*, and other nonlactose- and nonsucrose-fermenting microorganisms. Nonlactose/nonsucrose fermenting microorganisms appear as colorless or blue colonies. Lactose/sucrose-fermenting microorganisms, such as coliform bacteria, appear as yellow-opaque white colonies surrounded by a zone of precipitated deoxycholate.

## BCP DCLS Agar
## (Bromcresol Purple Deoxycholate
## Citrate Lactose Sucrose Agar)

**Composition** per liter:

| | |
|---|---|
| Agar | 14.0g |
| Sodium citrate | 10.0g |
| Lactose | 7.5g |
| Sucrose | 7.5g |
| Pancreatic digest of casein | 7.5g |
| Peptone | 7.5g |
| NaCl | 5.0g |
| $Na_2S_2O_3 \cdot 5H_2O$ | 5.0g |
| Yeast extract | 3.0g |
| Meat extract | 3.0g |
| Sodium deoxycholate | 2.5g |
| Bromcresol Purple | 0.02g |

pH 7.2 ± 0.2 at 25°C

**Preparation of Medium:** Add components to distilled/deionized water and bring volume to 1.0L. Mix thoroughly. Gently heat and bring to boiling. Pour into sterile Petri dishes without sterilization. Do not autoclave. Use the same day.

**Storage/Shelf Life:** Store dehydrated media in the dark in a sealed container below 30°C. Prepared media should be used on the same day as preparation.

**Use:** For the differential isolation of Gram-negative enteric bacilli from clinical speciments and nonclinical specimens of public health importance. For the isolation and identification of microorganisms from fecal specimens. For the isolation of *Salmonella*, *Shigella*, and other nonlactose- and nonsucrose-fermenting microorganisms. Nonlactose/nonsucrose-fermenting microorganisms appear as colorless or blue colonies. Lactose/sucrose-fermenting microorganisms, such as coliform bacteria, appear as yellow-opaque white colonies surrounded by a zone of precipitated deoxycholate.

## BCYE Agar with Cysteine
## (BCYE Alpha Base)
## (Buffered Charcoal Yeast Extract Agar)

**Composition** per liter:

| | |
|---|---|
| Agar | 15.0g |
| Yeast extract | 10.0g |
| ACES buffer (2-[(2-amino-2-oxoethyl)-amino]-ethane sulfonic acid) | 10.0g |
| Charcoal, activated | 2.0g |
| α-Ketoglutarate | 1.0g |
| L-Cysteine·HCl·H$_2$O | 0.4g |
| Fe$_4$(P$_2$O$_7$)$_3$·9H$_2$O | 0.25g |
| L-Cysteine solution | 4.0mL |

pH 6.9 ± 0.2 at 25°C

**Source:** This medium is available as a premixed powder from BD Diagnostic Systems.

**L-Cysteine Solution:**

**Composition** per 10.0mL:

| | |
|---|---|
| L-cysteine·HCl·H$_2$O | 0.4g |

**Preparation of L-Cysteine Solution:** Add L-cysteine·HCl·H$_2$O to distilled/deionized water and bring volume to 10.0mL. Mix thoroughly. Filter sterilize.

**Preparation of Medium:** Add components, except L-cysteine solution, to distilled/deionized water and bring volume to 1.0L. Mix thoroughly. Adjust medium to pH 6.9 with 1$N$ KOH. Heat gently and bring to boil for 1 min. Autoclave for 15 min at 15 psi pressure–121°C. Cool to 50°–55°C. Aseptically add 4.0mL of L-cysteine solution. Mix thoroughly. Pour into sterile Petri dishes with constant agitation to keep charcoal in suspension.

**Storage/Shelf Life:** Store dehydrated media in the dark in a sealed container below 30°C. Prepared media should be stored under refrigeration (2-8°C). Media should be used within 60 days of preparation. Media should not be used if there are any signs of deterioration (shrinking, cracking, or discoloration) or contamination, or if the expiration date supplied by the manufacturer has passed.

**Use:** For the isolation, cultivation, and maintenance of *Legionella pneumophila* and other *Legionella* species from environmental and clinical specimens.

## BCYE Differential Agar
## (Buffered Charcoal Yeast Extract Differential Agar)

**Composition** per liter:

| | |
|---|---|
| Agar | 15.0g |
| Yeast extract | 10.0g |
| ACES buffer (2-[(2-amino-2-oxoethyl)-amino]-ethane sulfonic acid) | 10.0g |
| Charcoal, activated | 2.0g |
| α-Ketoglutarate | 1.0g |
| L-Cysteine·HCl·H$_2$O | 0.4g |
| Fe$_4$(P$_2$O$_7$)$_3$·9H$_2$O | 0.25g |
| Bromcresol Purple | 0.01g |
| Bromthymol Blue | 0.01g |

pH 6.9 ± 0.2 at 25°C

**Source:** This medium is available as a premixed powder from BD Diagnostic Systems.

**Preparation of Medium:** Add components, except L-cysteine·HCl·H$_2$O, to distilled/deionized water and bring volume to

1.0L. Mix thoroughly. Adjust medium to pH 6.9 with 1$N$ KOH. Heat gently and bring to boiling for 1 min. Autoclave for 15 min at 15 psi pressure–121°C. Cool to 50°–55°C. Add 4.0mL of a 10% solution of L-cysteine·HCl·H$_2$O that has been filter sterilized. Mix thoroughly. Pour into sterile Petri dishes with constant agitation to keep charcoal in suspension.

**Storage/Shelf Life:** Store dehydrated media in the dark in a sealed container below 30°C. Prepared media should be stored under refrigeration (2-8°C). Media should be used within 60 days of preparation. Media should not be used if there are any signs of deterioration (shrinking, cracking, or discoloration) or contamination, or if the expiration date supplied by the manufacturer has passed.

**Use:** For the isolation, cultivation, and maintenance of *Legionella pneumophila* and other *Legionella* species from environmental and clinical specimens. For the presumptive differential identification of *Legionella* species based on colony color and morphology. *Legionella pneumophila* appears as light blue/green colonies. *Legionella micdadei* appears as blue/gray or dark blue colonies.

## BCYE Medium, Diphasic Blood Culture
## (Buffered Charcoal Yeast Extract Medium, Diphasic Blood Culture)

**Composition** per liter:

| | |
|---|---|
| Agar phase | 1.0L |
| Broth phase | 1.0L |

pH 6.9 ± 0.2 at 25°C

**Agar Phase:**

**Composition** per liter:

| | |
|---|---|
| Agar | 20.0g |
| ACES buffer (2-[(2-amino-2-oxoethyl)-amino]-ethane sulfonic acid) | 10.0g |
| Yeast extract | 10.0g |
| Charcoal, activated, acid washed | 4.0g |
| KOH | 2.8g |
| α-Ketoglutarate | 1.0g |
| L-Cysteine·HCl·H$_2$O solution | 10.0mL |
| Fe$_4$(P$_2$O$_7$)$_3$·9H$_2$O solution | 10.0mL |

**L-Cysteine·HCl·H$_2$O Solution:**

**Composition** per 10.0mL:

| | |
|---|---|
| L-Cysteine·HCl·H$_2$O | 0.4g |

**Preparation of L-Cysteine·HCl·H$_2$O Solution:** Add L-cysteine·HCl·H$_2$O to distilled/deionized water and bring volume to 10.0mL. Mix thoroughly. Filter sterilize.

**Fe$_4$(P$_2$O$_7$)$_3$·9H$_2$O Solution:**

**Composition** per 10.0mL:

| | |
|---|---|
| Fe$_4$(P$_2$O$_7$)$_3$·9H$_2$O | 0.25g |

**Preparation of Fe$_4$(P$_2$O$_7$)$_3$·9H$_2$O Solution:** Add Fe$_4$(P$_2$O$_7$)$_3$·9H$_2$O to distilled/deionized water and bring volume to 10.0mL. Mix thoroughly. Filter sterilize.

**Preparation of Agar Phase:** Add components, except L-cysteine·HCl·H$_2$O solution and Fe$_4$(P$_2$O$_7$)$_3$ solution, to distilled/deionized water and bring volume to 980.0mL. Mix thoroughly. Adjust medium to pH 6.9 with 1$N$ KOH. Heat gently and bring to boiling for 1 min. Autoclave for 15 min at 15 psi pressure–121°C. Cool to 50°–55°C. Aseptically add the L-cysteine·HCl·H$_2$O solution and Fe$_4$(P$_2$O$_7$)$_3$·9H$_2$O solution. Mix thoroughly.

**Broth Phase:**
**Composition** per liter:
ACES buffer (2-[(2-amino-2-oxoethyl)-
amino]-ethane sulfonic acid)..............................10.0g
Yeast extract...........................................................10.0g
Charcoal, activated, acid washed...........................4.0g
KOH.........................................................................2.4g
α-Ketoglutarate.......................................................1.0g
Sodium polyaneolsulfonate.....................................0.3g
L-Cysteine·HCl·H₂O solution ...........................10.0mL
Fe₄(P₂O₇)₃·9H₂O solution ..................................10.0mL

**L-Cysteine·HCl·H₂O Solution:**
**Composition** per 10.0mL:
L-Cysteine·HCl·H₂O................................................0.4g

**Preparation of L-Cysteine·HCl·H₂O Solution:** Add L-cysteine·HCl·H₂O to distilled/deionized water and bring volume to 10.0mL. Mix thoroughly. Filter sterilize.

**Fe₄(P₂O₇)₃·9H₂O Solution:**
**Composition** per 10.0mL:
Fe₄(P₂O₇)₃·9H₂O....................................................0.25g

**Preparation of Fe₄(P₂O₇)₃·9H₂O Solution:** Add Fe₄(P₂O₇)₃·9H₂O to distilled/deionized water and bring volume to 10.0mL. Mix thoroughly. Filter sterilize.

**Preparation of Broth Phase:** Add components, except L-cysteine·HCl·H₂O solution and Fe₄(P₂O₇)₃ solution, to distilled/deionized water and bring volume to 980.0mL. Mix thoroughly. Adjust medium to pH 6.9 with 1N KOH. Heat gently and bring to boiling for 1 min. Autoclave for 15 min at 15 psi pressure–121°C. Cool to 50–55°C. Aseptically add the cysteine·HCl·H₂O solution and Fe₄(P₂O₇)₃·9H₂O solution. Mix thoroughly.

**Preparation of Medium:** Aseptically distribute cooled sterile agar phase into sterile blood culture bottles in 100.0mL volumes. Allow bottles to cool in a slanted position. Aseptically add 50.0mL of sterile broth phase to each blood culture bottle.

**Storage/Shelf Life:** Store dehydrated media in the dark in a sealed container below 30°C. Prepared media should be stored under refrigeration (2-8°C). Media should be used within 60 days of preparation. Media should not be used if there are any signs of deterioration (shrinking, cracking, or discoloration) or contamination, or if the expiration date supplied by the manufacturer has passed.

**Use:** For the isolation and cultivation of *Legionella pneumophila* and other *Legionella* species from blood samples.

## BCYE Selective Agar with CCVC
### (Buffered Charcoal Yeast Extract Selective Agar with Cephalothin, Colistin, Vancomycin, and Cycloheximide)
**Composition** per 1014.0mL:
Agar ......................................................................15.0g
Yeast extract...........................................................10.0g
ACES buffer (2-[(2-amino-2-oxoethyl)-
amino]-ethane sulfonic acid)..............................10.0g
Charcoal, activated..................................................2.0g
α-Ketoglutarate.......................................................1.0g
Fe₄(P₂O₇)₃·9H₂O....................................................0.25g
Antibiotic solution ..............................................10.0mL
Cysteine·HCl·H₂O solution...................................4.0mL
pH 6.9 ± 0.2 at 25°C

**Source:** This medium is available as a premixed powder from BD Diagnostic Systems.

**L-Cysteine·HCl·H₂O Solution:**
**Composition** per 10.0mL:
L-Cysteine·HCl·H₂O.................................................1.0g

**Preparation of L-Cysteine·HCl·H₂O Solution:** Add L-cysteine·HCl·H₂O to distilled/deionized water and bring volume to 10.0mL. Mix thoroughly. Filter sterilize.

**Antibiotic Solution:**
**Composition** per 10.0mL:
Cycloheximide.....................................................80.0mg
Colistin.................................................................16.0mg
Cephalothin...........................................................4.0mg
Vancomycin...........................................................0.5mg

**Preparation of Antibiotic Solution:** Add components to distilled/deionized water and bring volume to 10.0mL. Mix thoroughly. Filter sterilize.

**Caution:** Cycloheximide is toxic. Avoid skin contact or aerosol formation and inhalation.

**Preparation of Medium:** Add components, except L-cysteine and antibiotic solutions, to distilled/deionized water and bring volume to 1.0L. Mix thoroughly. Adjust medium to pH 6.9 with 1N KOH. Heat gently and bring to boil for 1 min. Autoclave for 15 min at 15 psi pressure–121°C. Cool to 50°–55°C. Add 4.0mL of L-cysteine·HCl·H₂O solution and 10.0mL of sterile antibiotic solution. Mix thoroughly. Pour into sterile Petri dishes with constant agitation to keep charcoal in suspension.

**Storage/Shelf Life:** Store dehydrated media in the dark in a sealed container below 30°C. Prepared media should be stored under refrigeration (2-8°C). Media should be used within 60 days of preparation. Media should not be used if there are any signs of deterioration (shrinking, cracking, or discoloration) or contamination, or if the expiration date supplied by the manufacturer has passed.

**Use:** For the isolation, cultivation, and maintenance of *Legionella pneumophila* and other *Legionella* species from environmental and clinical specimens. For the selective recovery of *Legionella pneumophila* while reducing contaminating microorganisms from environmental water samples.

## BCYE Selective Agar with GPVA
### (Buffered Charcoal Yeast Extract Selective Agar with Glycine, Polymyxin B, Vancomycin, and Anisomycin)
**Composition** per 1014.0mL:
Agar ......................................................................15.0g
Yeast extract...........................................................10.0g
ACES buffer (2-[(2-amino-2-oxoethyl)-
amino]-ethane sulfonic acid)..............................10.0g
Charcoal, activated..................................................2.0g
α-Ketoglutarate.......................................................1.0g
Fe₄(P₂O₇)₃·9H₂O....................................................0.25g
Antibiotic solution ..............................................10.0mL
L-Cysteine·HCl·H₂O solution ..............................4.0mL
pH 6.9 ± 0.2 at 25°C

**L-Cysteine·HCl·H₂O Solution:**
**Composition** per 10.0mL:
L-Cysteine·HCl·H₂O.................................................1.0g

**Preparation of L-Cysteine·HCl·H₂O Solution:** Add L-cysteine·HCl·H₂O to distilled/deionized water and bring volume to 10.0mL. Mix thoroughly. Filter sterilize.

**Antibiotic Solution:**
**Composition** per 10.0mL:

Glycine ................................................................3.0g
Anisomycin .......................................................0.08g
Vancomycin ........................................................5.0mg
Polymyxin B ..............................................100,000U

**Preparation of Antibiotic Solution:** Add components to distilled/deionized water and bring volume to 10.0mL. Mix thoroughly. Filter sterilize.

**Preparation of Medium:** Add components, except L-cysteine·HCl·H₂O solution and antibiotic solution, to distilled/deionized water and bring volume to 1.0L. Mix thoroughly. Adjust medium to pH 6.9 with 1N KOH. Heat gently and bring to boil for 1 min. Autoclave for 15 min at 15 psi pressure–121°C. Cool to 50°–55°C. Add 4.0mL of L-cysteine·HCl·H₂O solution and 10.0mL of sterile antibiotic solution. Mix thoroughly. Pour into sterile Petri dishes with constant agitation to keep charcoal in suspension.

**Storage/Shelf Life:** Store dehydrated media in the dark in a sealed container below 30°C. Prepared media should be stored under refrigeration (2-8°C). Media should be used within 60 days of preparation. Media should not be used if there are any signs of deterioration (shrinking, cracking, or discoloration) or contamination, or if the expiration date supplied by the manufacturer has passed.

**Use:** For the isolation, cultivation, and maintenance of *Legionella pneumophila* and other *Legionella* species from environmental and clinical specimens. For the selective recovery of *Legionella pneumophila* while reducing contaminating microorganisms from potable water samples.

## BCYE Selective Agar with GVPC
### (Buffered Charcoal Yeast Extract Selective Agar with Glycine, Vancomycin, Polymyxin B, and Cycloheximide)
**Composition** per 1014.0mL:

Agar ....................................................................15.0g
Yeast extract .......................................................10.0g
ACES buffer (2-[(2-amino-2-oxoethyl)-
    amino]-ethane sulfonic acid) ...........................10.0g
Charcoal, activated...............................................2.0g
α-Ketoglutarate....................................................1.0g
Fe₄(P₂O₇)₃·9H₂O................................................0.25g
Antibiotic solution ...........................................10.0mL
L-Cysteine·HCl·H₂O solution ...........................4.0mL
pH 6.9 ± 0.2 at 25°C

**Source:** This medium is available as a premixed powder from Thermo Scientific.

**L-Cysteine·HCl·H₂O Solution:**
**Composition** per 10.0mL:

L-Cysteine·HCl·H₂O.............................................1.0g

**Preparation of L-Cysteine·HCl·H₂O Solution:** Add L-cysteine·HCl·H₂O to distilled/deionized water and bring volume to 10.0mL. Mix thoroughly. Filter sterilize.

**Antibiotic Solution:**
**Composition** per 10.0mL:

Glycine................................................................3.0g
Cycloheximide....................................................0.08g

Vancomycin ...........................................................1.0mg
Polymyxin B .....................................................79,200U

**Preparation of Antibiotic Solution:** Add components to distilled/deionized water and bring volume to 10.0mL. Mix thoroughly. Filter sterilize.

**Caution:** Cycloheximide is toxic. Avoid skin contact or aerosol formation and inhalation.

**Preparation of Medium:** Add components, except L-cysteine·HCl·H₂O solution and antibiotic solution, to distilled/deionized water and bring volume to 1.0L. Mix thoroughly. Adjust medium to pH 6.9 with 1N KOH. Heat gently and bring to boil for 1 min. Autoclave for 15 min at 15 psi pressure–121°C. Cool to 50°–55°C. Add 4.0mL of L-cysteine·HCl·H₂O solution and 10.0mL of sterile antibiotic solution. Mix thoroughly. Pour into sterile Petri dishes with constant agitation to keep charcoal in suspension.

**Storage/Shelf Life:** Store dehydrated media in the dark in a sealed container below 30°C. Prepared media should be stored under refrigeration (2-8°C). Media should be used within 60 days of preparation. Media should not be used if there are any signs of deterioration (shrinking, cracking, or discoloration) or contamination, or if the expiration date supplied by the manufacturer has passed.

**Use:** For the isolation, cultivation, and maintenance of *Legionella pneumophila* and other *Legionella* species from environmental and clinical specimens. For the selective recovery of *Legionella pneumophila* while reducing contaminating microorganisms from potable water samples.

## BCYE Selective Agar with PAC
### (Buffered Charcoal Yeast Extract Selective Agar with Polymyxin B, Anisomycin, and Cefamandole)
**Composition** per 1014.0mL:

Agar ....................................................................15.0g
Yeast extract .......................................................10.0g
ACES buffer (2-[(2-amino-2-oxoethyl)-
    amino]-ethane sulfonic acid) ...........................10.0g
Charcoal, activated ...............................................2.0g
α-Ketoglutarate.....................................................1.0g
Fe₄(P₂O₇)₃·9H₂O................................................0.25g
Antibiotic solution ...........................................10.0mL
L-Cysteine·HCl·H₂O solution ...........................4.0mL
pH 6.9 ± 0.2 at 25°C

**Source:** This medium is available as a premixed powder from BD Diagnostic Systems.

**L-Cysteine·HCl·H₂O Solution:**
**Composition** per 10.0mL:

L-Cysteine·HCl·H₂O.............................................1.0g

**Preparation of L-Cysteine·HCl·H₂O Solution:** Add L-cysteine·HCl·H₂O to distilled/deionized water and bring volume to 10.0mL. Mix thoroughly. Filter sterilize.

**Antibiotic Solution:**
**Composition** per 10.0mL:

Polymyxin B ......................................................80,000 U
Anisomycin.........................................................80.0mg
Cefamandole.........................................................2.0mg

**Preparation of Antibiotic Solution:** Add components to distilled/deionized water and bring volume to 10.0mL. Mix thoroughly. Filter sterilize.

**Preparation of Medium:** Add components, except L-cysteine·HCl·H$_2$O solution and antibiotic solution, to distilled/deionized water and bring volume to 1.0L. Mix thoroughly. Adjust medium to pH 6.9 with 1$N$ KOH. Heat gently and bring to boiling for 1 min. Autoclave for 15 min at 15 psi pressure–121°C. Cool to 50°–55°C. Add 4.0mL of L-cysteine·HCl·H$_2$O solution and 10.0mL of sterile antibiotic solution. Mix thoroughly. Pour into sterile Petri dishes with constant agitation to keep charcoal in suspension.

**Storage/Shelf Life:** Store dehydrated media in the dark in a sealed container below 30°C. Prepared media should be stored under refrigeration (2-8°C). Media should be used within 60 days of preparation. Media should not be used if there are any signs of deterioration (shrinking, cracking, or discoloration) or contamination, or if the expiration date supplied by the manufacturer has passed.

**Use:** For the isolation, cultivation, and maintenance of *Legionella pneumophila* and other *Legionella* species from environmental and clinical specimens. For the selective recovery of *Legionella pneumophila* while reducing contaminating microorganisms from potable water samples.

## BCYE Selective Agar with PAV
### (Buffered Charcoal Yeast Extract Selective Agar with Polymyxin B, Anisomicin, and Vancomycin) (Wadowsky–Yee Medium)

**Composition** per 1014.0mL:

| | |
|---|---|
| Agar | 15.0g |
| Yeast extract | 10.0g |
| ACES buffer (2-[(2-amino-2-oxoethyl)-amino]-ethane sulfonic acid) | 10.0g |
| Charcoal, activated | 2.0g |
| α-Ketoglutarate | 1.0g |
| Fe$_4$(P$_2$O$_7$)$_3$·9H$_2$O | 0.25g |
| Antibiotic solution | 10.0mL |
| L-Cysteine·HCl·H$_2$O solution | 4.0mL |

pH 6.9 ± 0.2 at 25°C

**Source:** This medium is available as a premixed powder from BD Diagnostic Systems.

**L-Cysteine·HCl·H$_2$O Solution:**
**Composition** per 10.0mL:

| | |
|---|---|
| L-Cysteine·HCl·H$_2$O | 1.0g |

**Preparation of L-Cysteine·HCl·H$_2$O Solution:** Add L-cysteine·HCl·H$_2$O to distilled/deionized water and bring volume to 10.0mL. Mix thoroughly. Filter sterilize.

**Antibiotic Solution:**
**Composition** per 10.0mL:

| | |
|---|---|
| Anisomycin | 80.0mg |
| Vancomycin | 0.5mg |
| Polymyxin B | 40,000 U |

**Preparation of Antibiotic Solution:** Add components to distilled/deionized water and bring volume to 10.0mL. Mix thoroughly. Filter sterilize.

**Preparation of Medium:** Add components, except L-cysteine and antibiotic solution, to distilled/deionized water and bring volume to 1.0L. Mix thoroughly. Adjust medium to pH 6.9 with 1$N$ KOH. Heat gently and bring to boil for 1 min. Autoclave for 15 min at 15 psi pressure–121°C. Cool to 50°–55°C. Add 4.0mL of L-cysteine·HCl·H$_2$O solution and 10.0mL of sterile antibiotic solution. Mix thoroughly. Pour into sterile Petri dishes with constant agitation to keep charcoal in suspension.

**Storage/Shelf Life:** Store dehydrated media in the dark in a sealed container below 30°C. Prepared media should be stored under refrigeration (2-8°C). Media should be used within 60 days of preparation. Media should not be used if there are any signs of deterioration (shrinking, cracking, or discoloration) or contamination, or if the expiration date supplied by the manufacturer has passed.

**Use:** For the isolation, cultivation, and maintenance of *Legionella pneumophila* and other *Legionella* species from environmental and clinical specimens. For the selective recovery of *Legionella pneumophila* while reducing contaminating microorganisms from potable water samples.

## BCYEα with Alb
### (Buffered Charcoal Yeast Extract Agar with Albumin)

**Composition** per liter:

| | |
|---|---|
| Agar | 15.0g |
| Yeast extract | 10.0g |
| ACES buffer (2-[(2-amino-2-oxoethyl)-amino]-ethane sulfonic acid) | 10.0g |
| Charcoal, activated | 2.0g |
| α-Ketoglutarate | 1.0g |
| Bovine serum albumin solution | 10.0mL |
| L-Cysteine·HCl·H$_2$O solution | 10.0mL |
| Fe$_4$(P$_2$O$_7$)$_3$·9H$_2$O solution | 10.0mL |

pH 6.9 ± 0.2 at 25°C

**Bovine Serum Albumin Solution:**
**Composition** per 10.0mL:

| | |
|---|---|
| Bovine serum albumin | 0.1g |

**Preparation of Bovine Serum Albumin Solution:** Add bovine serum albumin to distilled/deionized water and bring volume to 10.0mL. Mix thoroughly. Filter sterilize.

**L-Cysteine·HCl·H$_2$O Solution:**
**Composition** per 10.0mL:

| | |
|---|---|
| L-Cysteine·HCl·H$_2$O | 0.4g |

**Preparation of L-Cysteine·HCl·H$_2$O Solution:** Add L-cysteine·HCl·H$_2$O to distilled/deionized water and bring volume to 10.0mL. Mix thoroughly. Filter sterilize.

**Fe$_4$(P$_2$O$_7$)$_3$·9H$_2$O Solution:**
**Composition** per 10.0mL:

| | |
|---|---|
| Fe$_4$(P$_2$O$_7$)$_3$·9H$_2$O | 0.25g |

**Preparation of Fe$_4$(P$_2$O$_7$)$_3$·9H$_2$O Solution:** Add Fe$_4$(P$_2$O$_7$)$_3$·9H$_2$O to distilled/deionized water and bring volume to 10.0mL. Mix thoroughly. Filter sterilize.

**Preparation of Medium:** Add components—except Fe$_4$(P$_2$O$_7$)$_3$·9H$_2$O solution, L-cysteine·HCl·H$_2$O solution, and bovine serum albumin solution—to distilled/deionized water and bring volume to 970.0mL. Mix thoroughly. Adjust medium to pH 6.9 with 1$N$ KOH. Heat gently and bring to boiling for 1 min. Autoclave for 15 min at 15 psi pressure–121°C. Cool to 50°–55°C. Aseptically add 10.0mL of sterile bovine serum albumin solution, the Fe$_4$(P$_2$O$_7$)$_3$·9H$_2$O solution, and the L-cysteine·HCl·H$_2$O solution. Mix thoroughly. Pour into sterile Petri dishes with constant agitation to keep charcoal in suspension.

**Storage/Shelf Life:** Store dehydrated media in the dark in a sealed container below 30°C. Prepared media should be stored under refrigeration (2-8°C). Media should be used within 60 days of preparation. Media should not be used if there are any signs of deterioration (shrink-

ing, cracking, or discoloration) or contamination, or if the expiration date supplied by the manufacturer has passed.

**Use:** For the isolation, cultivation, and maintenance of *Legionella pneumophila* and other *Legionella* species from environmental and clinical specimens.

## BCYEα without L-Cysteine
### (Buffered Charcoal Yeast Extract Agar without L-Cysteine)

**Composition** per liter:

| | |
|---|---|
| Agar | 15.0g |
| Yeast extract | 10.0g |
| ACES buffer (2-[(2-amino-2-oxoethyl)-amino]-ethane sulfonic acid) | 10.0g |
| Charcoal, activated | 2.0g |
| α-Ketoglutarate | 1.0g |
| $Fe_4(P_2O_7)_3 \cdot 9H_2O$ solution | 10.0mL |

pH $6.9 \pm 0.2$ at 25°C

**$Fe_4(P_2O_7)_3 \cdot 9H_2O$ Solution:**
**Composition** per 10.0mL:

| | |
|---|---|
| $Fe_4(P_2O_7)_3 \cdot 9H_2O$ | 0.25g |

**Preparation of $Fe_4(P_2O_7)_3 \cdot 9H_2O$ Solution:** Add $Fe_4(P_2O_7)_3 \cdot 9H_2O$ to distilled/deionized water and bring volume to 10.0mL. Mix thoroughly. Filter sterilize.

**Preparation of Medium:** Add components, except $Fe_4(P_2O_7)_3 \cdot 9H_2O$ solution, to distilled/deionized water and bring volume to 990.0mL. Mix thoroughly. Adjust medium to pH 6.9 with 1*N* KOH. Heat gently and bring to boiling for 1 min. Autoclave for 15 min at 15 psi pressure–121°C. Cool to 50°–55°C. Aseptically add 10.0mL of sterile $Fe_4(P_2O_7)_3 \cdot 9H_2O$ solution. Mix thoroughly. Pour into sterile Petri dishes with constant agitation to keep charcoal in suspension.

**Storage/Shelf Life:** Store dehydrated media in the dark in a sealed container below 30°C. Prepared media should be stored under refrigeration (2-8°C). Media should be used within 60 days of preparation. Media should not be used if there are any signs of deterioration (shrinking, cracking, or discoloration) or contamination, or if the expiration date supplied by the manufacturer has passed.

**Use:** For the isolation, cultivation, and maintenance of *Legionella pneumophila* and other *Legionella* species from environmental and clinical specimens.

## B.D.G. Broth, Hajna

**Composition** per liter:

| | |
|---|---|
| Tryptose | 20.0g |
| Glucose | 5.0g |
| NaCl | 5.0g |
| $K_2HPO_4$ | 4.0g |
| $KH_2PO_4$ | 1.5g |
| Sodium deoxycholate | 0.1g |

pH $7.0 \pm 0.2$ at 25°C

**Source:** This medium is available from HiMedia.

**Preparation of Medium:** Add components to distilled/deionized water and bring volume to 1.0L. Mix thoroughly. Distribute into tubes with inverted Durham tubes. Autoclave for 15 min at 15 psi pressure–121°C.

**Storage/Shelf Life:** Store dehydrated media in the dark in a sealed container below 30°C. Prepared media should be stored under refriger-

ation (2-8°C). Media should be used within 60 days of preparation. Media should not be used if there are any signs of deterioration (discoloration) or contamination, or if the expiration date supplied by the manufacturer has passed.

**Use:** For the selective enrichment and cultivation of enteric bacilli from food and in treated drinking water.

## BG Sulfa Agar
### (Brilliant Green Sulfapyridine Agar)

**Composition** per liter:

| | |
|---|---|
| Agar | 20.0g |
| Proteose peptone No. 3 | 10.0g |
| Lactose | 10.0g |
| Sucrose | 10.0g |
| NaCl | 5.0g |
| Yeast extract | 3.0g |
| Sodium sulfapyridine | 1.0g |
| Brilliant Green | 0.125g |

pH $6.9 \pm 0.2$ at 25°C

**Source:** This medium is available as a premixed powder from BD Diagnostic Systems.

**Preparation of Medium:** Add components to distilled/deionized water and bring volume to 1.0L. Mix thoroughly. Heat gently to boiling. Distribute into tubes or flasks. Autoclave for no longer than 15 min at 15 psi pressure–121°C. Pour into sterile Petri dishes if desired.

**Storage/Shelf Life:** Store dehydrated media in the dark in a sealed container below 30°C. Prepared media should be stored under refrigeration (2-8°C). Media should be used within 60 days of preparation. Media should not be used if there are any signs of deterioration (shrinking, cracking, or discoloration) or contamination, or if the expiration date supplied by the manufacturer has passed.

**Use:** For the selective isolation of *Salmonella* species other than *Salmonella typhi* from food, dairy products, eggs and egg products, and feed. *Salmonella* appear as red, pink, or white colonies surrounded by zones of bright red.

## BG Sulfa HiVeg Agar
### (Brilliant Green Sulfa HiVeg Agar)

**Composition** per liter:

| | |
|---|---|
| Agar | 20.0g |
| Plant peptone No. 3 | 10.0g |
| Lactose | 10.0g |
| Sucrose | 10.0g |
| NaCl | 5.0g |
| Yeast extract | 3.0g |
| Sodium sulphapyridine | 1.0g |
| Phenol Red | 0.08g |
| Brilliant Green | 12.5mg |

pH $6.9 \pm 0.2$ at 25°C

**Source:** This medium is available as a premixed powder from HiMedia.

**Preparation of Medium:** Add components to distilled/deionized water and bring volume to 1.0L. Mix thoroughly. Heat gently to boiling. Distribute into tubes or flasks. Autoclave for no longer than 15 min at 15 psi pressure–121°C. Pour into sterile Petri dishes if desired.

**Storage/Shelf Life:** Store dehydrated media in the dark in a sealed container below 30°C. Prepared media should be stored under refrigeration (2-8°C). Media should be used within 60 days of preparation.

Media should not be used if there are any signs of deterioration (shrinking, cracking, or discoloration) or contamination, or if the expiration date supplied by the manufacturer has passed.

**Use:** For the selective isolation of *Salmonella* species other than *Salmonella typhi* from food, dairy products, eggs and egg products, and feed. *Salmonella* appear as red, pink, or white colonies surrounded by zones of bright red.

## Bile Broth Base, HiVeg with Streptokinase

**Composition** per liter:
Plant peptone.............................................................20.0g
NaCl.........................................................................5.0g
Synthetic detergent No. V............................................5.0g
Streptokinase solution............................................. 1.0mL
pH 7.1 ± 0.2 at 25°C

**Source:** This medium, without streptokinase solution, is available as a premixed powder from HiMedia.

**Streptokinase Solution:**
**Composition** per 1.0mL:
Streptokinase.................................................. 100,000 units

**Streptokinase Solution:** Add streptokinase to distilled/deionized water and bring volume to 1.0mL. Mix thoroughly. Filter sterilize.

**Preparation of Medium:** Add components to distilled/deionized water and bring volume to 1.0L. Mix thoroughly. Heat gently and bring to boiling. Distribute into tubes or flasks. Autoclave for 15 min at 15 psi pressure–121°C. Cool to 40°C. Aseptically add 1.0mL of streptokinase solution. Mix thoroughly. If desired, carbohydrate may also be added to this medium prior to sterilization.

**Storage/Shelf Life:** Store dehydrated media in the dark in a sealed container below 30°C. Prepared media should be stored under refrigeration (2-8°C). Media should be used within 60 days of preparation. Media should not be used if there are any signs of deterioration (discoloration) or contamination, or if the expiration date supplied by the manufacturer has passed.

**Use:** For the culture of bacteria from blood clots from patients with suspected enteric fever.

## Bile Broth Base with Streptokinase

**Composition** per liter:
Peptone.....................................................................20.0g
NaCl.........................................................................5.0g
Synthetic detergent No. V............................................5.0g
Streptokinase solution............................................. 1.0mL
pH 7.1 ± 0.2 at 25°C

**Source:** This medium, without streptokinase solution, is available as a premixed powder from HiMedia.

**Streptokinase Solution:**
**Composition** per 1.0mL:
Streptokinase.................................................. 100,000 units

**Streptokinase Solution:** Add streptokinase to distilled/deionized water and bring volume to 1.0mL. Mix thoroughly. Filter sterilize.

**Preparation of Medium:** Add components to distilled/deionized water and bring volume to 1.0L. Mix thoroughly. Heat gently and bring to boiling. Distribute into tubes or flasks. Autoclave for 15 min at 15 psi pressure–121°C. Cool to 40°C. Aseptically add 1.0mL of streptokinase solution. Mix thoroughly. If desired, carbohydrate may also be added to this medium prior to sterilization.

**Storage/Shelf Life:** Store dehydrated media in the dark in a sealed container below 30°C. Prepared media should be stored under refrigeration (2-8°C). Media should be used within 60 days of preparation. Media should not be used if there are any signs of deterioration (discoloration) or contamination, or if the expiration date supplied by the manufacturer has passed.

**Use:** For the culture of bacteria from blood clots from patients with suspected enteric fever.

## Bile Esculin Agar

**Composition** per liter:
Oxgall .....................................................................20.0g
Agar ........................................................................15.0g
Pancreatic digest of gelatin.........................................5.0g
Beef extract...............................................................3.0g
Esculin .....................................................................1.0g
Ferric citrate.............................................................0.5g
Horse serum ........................................................ 50.0mL
pH 6.8 ± 0.2 at 25°C

**Source:** This medium is available as a premixed powder from Thermo Scientific and BD Diagnostic Systems.

**Preparation of Medium:** Add components, except horse serum, to distilled/deionized water and bring volume to 950.0L. Mix thoroughly and heat with frequent agitation until boiling. Autoclave for 15 min at 15 psi pressure–121°C. Cool to 45°–50°C. Aseptically add 50.0mL of filter sterilized horse serum. Distribute into sterile Petri dishes or test tubes. Cool tubes in a slanted position.

**Storage/Shelf Life:** Store dehydrated media in the dark in a sealed container below 30°C. Prepared media should be stored under refrigeration (2-8°C). Media should be used within 60 days of preparation. Media should not be used if there are any signs of deterioration (shrinking, cracking, or discoloration) or contamination, or if the expiration date supplied by the manufacturer has passed.

**Use:** For differentiation between group D streptococci and nongroup D streptococci. To differentiate members of the Enterobacteriaceae, particularly *Klebsiella, Enterobacter*, and *Serratia,* from other enteric bacteria. To differentiate *Listeria monocytogenes.* Bile tolerance and esculin hydrolysis (seen as a dark brown to black complex) are presumptive for enterococci (group D streptococci).

## Bile Esculin Agar

**Composition** per liter:
Esculin ....................................................................1.0g
Bile esculin agar base ............................................... 1.0L
pH 6.6 ± 0.2 at 25°C

**Bile Esculin Agar Base:**
**Composition** per liter:
Oxgall ....................................................................40.0g
Agar ........................................................................15.0g
Peptone ....................................................................5.0g
Beef extract...............................................................3.0g
Ferric citrate.............................................................0.5g

**Source:** This medium is available as a premixed powder from BD Diagnostic Systems.

**Preparation of Bile Esculin Agar Base:** Add components to distilled/deionized water and bring volume to 1.0L. Mix thoroughly.

**Preparation of Medium:** Add desired amount of esculin—typically 1.0g—to bile esculin agar base. Mix thoroughly and heat with frequent agitation until boiling. Autoclave for 15 min at 15 psi pressure–121°C. Cool to 45°–50°C. Distribute into sterile Petri dishes or test tubes. Cool tubes in a slanted position.

**Storage/Shelf Life:** Store dehydrated media in the dark in a sealed container below 30°C. Prepared media should be stored under refrigeration (2-8°C). Media should be used within 60 days of preparation. Media should not be used if there are any signs of deterioration (shrinking, cracking, or discoloration) or contamination, or if the expiration date supplied by the manufacturer has passed.

**Use:** For the isolation and presumptive identification of group D streptococci.

## Bile Esculin Agar
## (BAM M18)

**Composition** per liter:

| | |
|---|---|
| Oxgall | 40.0g |
| Agar | 15.0g |
| Pancreatic digest of gelatin | 5.0g |
| Beef extract | 3.0g |
| Esculin | 1.0g |
| Ferric citrate | 0.5g |

pH 6.6 ± 0.2 at 25°C

**Preparation of Medium:** Add components to distilled/deionized water and bring volume to 1.0L. Mix thoroughly and heat with frequent agitation until boiling. Autoclave for 15 min at 15 psi pressure–121°C. Distribute into sterile Petri dishes or test tubes. Cool tubes in a slanted position.

**Storage/Shelf Life:** Store dehydrated media in the dark in a sealed container below 30°C. Prepared media should be stored under refrigeration (2-8°C). Media should be used within 60 days of preparation. Media should not be used if there are any signs of deterioration (shrinking, cracking, or discoloration) or contamination, or if the expiration date supplied by the manufacturer has passed.

**Use:** For differentiation between group D streptococci and nongroup D streptococci. To differentiate members of the Enterobacteriaceae, particularly *Klebsiella, Enterobacter,* and *Serratia,* from other enteric bacteria. To differentiate *Listeria monocytogenes.* Bile tolerance and esculin hydrolysis (seen as a dark brown to black complex) are presumptive for enterococci (group D streptococci).

## Bile Esculin Agar

**Composition** per liter:

| | |
|---|---|
| Bile salts | 40.0g |
| Agar | 15.0g |
| Pancreatic digest of animal tissue | 5.0g |
| Beef extract | 3.0g |
| Esculin | 1.0g |
| Ferric citrate | 0.5g |

pH 6.6 ± 0.2 at 25°C

**Preparation of Medium:** Add components to distilled/deionized water and bring volume to 1.0L. Mix thoroughly and heat with frequent agitation until boiling. Autoclave for 15 min at 15 psi pressure–121°C. Distribute into sterile Petri dishes or test tubes.

**Storage/Shelf Life:** Store dehydrated media in the dark in a sealed container below 30°C. Prepared media should be stored under refrigeration (2-8°C). Media should be used within 60 days of preparation. Media should not be used if there are any signs of deterioration (shrinking, cracking, or discoloration) or contamination, or if the expiration date supplied by the manufacturer has passed.

**Use:** For the isolation and identification of *Yersinia enterocolitica.*

## Bile Esculin Agar, HiVeg

**Composition** per liter:

| | |
|---|---|
| Plant peptone | 25.0g |
| Agar | 15.0g |
| Plant hydrolysate | 15.0g |
| Plant extract | 6.0g |
| Synthetic detergent No. II | 2.0g |
| Esculin | 1.0g |
| Ferric citrate | 0.5g |

pH 6.6 ± 0.2 at 25°C

**Source:** This medium is available as a premixed powder from Hi-Media.

**Preparation of Medium:** Add components to distilled/deionized water and bring volume to 1.0L. Mix thoroughly and heat with frequent agitation until boiling. Autoclave for 15 min at 15 psi pressure–121°C. Distribute into sterile Petri dishes or test tubes. Cool tubes in a slanted position.

**Storage/Shelf Life:** Store dehydrated media in the dark in a sealed container below 30°C. Prepared media should be stored under refrigeration (2-8°C). Media should be used within 60 days of preparation. Media should not be used if there are any signs of deterioration (shrinking, cracking, or discoloration) or contamination, or if the expiration date supplied by the manufacturer has passed.

**Use:** For the isolation and presumptive identification of group D streptococci.

## Bile Esculin Agar with Kanamycin

**Composition** per liter:

| | |
|---|---|
| Oxgall | 20.0g |
| Agar | 15.0g |
| Beef extract | 3.0g |
| Esculin | 1.0g |
| Ferric citrate | 0.5g |
| Hemin | 10.0mg |
| Vitamin K$_1$ | 10.0mg |
| Horse serum | 50.0mL |
| Kanamycin solution | 10.0mL |

pH 7.1 ± 0.2 at 25°C

**Source:** This medium is available as a premixed powder from BD Diagnostic Systems.

**Kanamycin Solution:**
**Composition** per 10.0mL:

| | |
|---|---|
| Kanamycin | 1.0g |

**Preparation of Kanamycin Solution:** Add kanamycin to distilled/deionized water and bring volume to 10.0mL. Mix thoroughly. Filter sterilize.

**Preparation of Medium:** Add components to distilled/deionized water and bring volume to 1.0L. Mix thoroughly and heat with frequent agitation until boiling. Autoclave for 15 min at 15 psi pressure–121°C. Cool to 45°–50°C. Aseptically add 50.0mL of 5% filter-sterilized horse serum and 10.0mL of sterile kanamycin solution. Distribute into test tubes or flasks. Cool tubes in a slanted position.

**Storage/Shelf Life:** Store dehydrated media in the dark in a sealed container below 30°C. Prepared media should be stored under refrigeration (2-8°C). Media should be used within 60 days of preparation. Media should not be used if there are any signs of deterioration (shrink-

ing, cracking, or discoloration) or contamination, or if the expiration date supplied by the manufacturer has passed.

**Use:** For the selective isolation and/or presumptive identification of bacteria of the *Bacteroides fragilis* group from specimens containing mixed flora. Examine colonies with a long-wavelength UV light. Pigmented colonies of the *Bacteroides* group will fluoresce red-orange. Growth on this medium with blackening of the medium is presumptive for *Bacteroides fragilis*.

## Bile Esculin Azide Agar

**Composition** per liter:
| | |
|---|---|
| Pancreatic digest of casein | 17.0g |
| Agar | 15.0g |
| Oxgall | 10.0g |
| NaCl | 5.0g |
| Yeast extract | 5.0g |
| Proteose peptone No. 3 | 3.0g |
| Esculin | 1.0g |
| Ferric ammonium citrate | 0.5g |
| NaN$_3$ | 0.15g |

pH 7.1 ± 0.2 at 25°C

**Source:** This medium is available as a premixed powder from BD Diagnostic Systems.

**Caution:** Sodium azide is toxic. Azides also react with metals and disposal must be highly diluted.

**Preparation**: Add components to distilled/deionized water and bring volume to 1.0L. Mix thoroughly and heat with frequent agitation until boiling. Distribute into tubes or flasks. Autoclave for 15 min at 15 psi pressure–121°C. Cool to 45°–50°C. Pour into sterile Petri dishes or leave in tubes. Cool tubes in a slanted position.

**Storage/Shelf Life:** Store dehydrated media in the dark in a sealed container below 30°C. Prepared media should be stored under refrigeration (2-8°C). Media should be used within 60 days of preparation. Media should not be used if there are any signs of deterioration (shrinking, cracking, or discoloration) or contamination, or if the expiration date supplied by the manufacturer has passed.

**Use:** For the isolation and presumptive identification of group D streptococci.

## Bile Esculin Azide HiVeg Agar

**Composition** per liter:
| | |
|---|---|
| Plant hydrolysate | 20.0g |
| Agar | 15.0g |
| Plant extract | 5.0g |
| Plant peptone No. 3 | 5.0g |
| NaCl | 5.0g |
| Synthetic detergent No. II | 5.0g |
| Esculin | 1.0g |
| Ferric ammonium citrate | 0.5g |
| NaN$_3$ | 0.15g |

pH 7.1 ± 0.2 at 25°C

**Source:** This medium is available as a premixed powder from HiMedia.

**Caution:** Sodium azide is toxic. Azides also react with metals and disposal must be highly diluted.

**Preparation**: Add components to distilled/deionized water and bring volume to 1.0L. Mix thoroughly and heat with frequent agitation until boiling. Distribute into tubes or flasks. Autoclave for 15 min at 15 psi pressure–121°C. Cool to 45°–50°C. Pour into sterile Petri dishes or leave in tubes. Cool tubes in a slanted position.

**Storage/Shelf Life:** Store dehydrated media in the dark in a sealed container below 30°C. Prepared media should be stored under refrigeration (2-8°C). Media should be used within 60 days of preparation. Media should not be used if there are any signs of deterioration (shrinking, cracking, or discoloration) or contamination, or if the expiration date supplied by the manufacturer has passed.

**Use:** For the isolation and presumptive identification of group D streptococci.

## Bile Esculin HiVeg Agar Base with Esculin

**Composition** per liter:
| | |
|---|---|
| Plant peptone | 22.0g |
| Agar | 15.0g |
| Plant hydrolysate | 15.0g |
| Plant extract | 6.0g |
| Synthetic detergent No. II | 5.0g |
| Ferric citrate | 0.5g |
| Esculin solution | 4.0mL |

pH 6.8 ± 0.2 at 25°C

**Source:** This medium, without esculin, is available as a premixed powder from HiMedia.

**Esculin Solution:**
**Composition** per 4.0mL:
| | |
|---|---|
| Esculin | 1.0g |

**Esculin Solution:** Add esculin to distilled/deionized water and bring volume to 4.0mL. Mix thoroughly. Filter sterilize.

**Preparation of Medium:** Add components, except esculin solution, to distilled/deionized water and bring volume to 1.0L. Mix thoroughly. Gently heat and bring to boiling. Autoclave for 15 min at 15 psi pressure–121°C. Cool to 50°C. Aseptically add 4.0mL of sterile esculin solution. Mix thoroughly. Pour into sterile Petri dishes or distribute into sterile test tubes.

**Storage/Shelf Life:** Store dehydrated media in the dark in a sealed container below 30°C. Prepared media should be stored under refrigeration (2-8°C). Media should be used within 60 days of preparation. Media should not be used if there are any signs of deterioration (shrinking, cracking, or discoloration) or contamination, or if the expiration date supplied by the manufacturer has passed.

**Use:** For the isolation and presumptive identification of group D streptococci.

## Bile Esculin HiVeg Agar with Kanamycin

**Composition** per liter:
| | |
|---|---|
| Plant peptone no. 2 | 17.0g |
| Agar | 15.0g |
| Plant extract | 6.0g |
| Synthetic detergent | 5.0g |
| Esculin | 1.0g |
| Ferric citrate | 0.5g |
| Kanamycin | 0.1g |
| Fe$_4$(P$_2$O$_7$)$_3$·H$_2$O | 0.01g |
| Vitamin K$_1$ | 0.01g |

pH 7.1 ± 0.2 at 25°C

**Source:** This medium is available as a premixed powder from HiMedia.

**Preparation**: Add components to distilled/deionized water and bring volume to 1.0L. Mix thoroughly and heat with frequent agitation until boiling. Distribute into tubes or flasks. Autoclave for 15 min at 15 psi pressure–121°C. Cool to 45°–50°C. Pour into sterile Petri dishes or leave in tubes. Cool tubes in a slanted position.

**Storage/Shelf Life:** Store dehydrated media in the dark in a sealed container below 30°C. Prepared media should be stored under refrigeration (2-8°C). Media should be used within 60 days of preparation. Media should not be used if there are any signs of deterioration (shrinking, cracking, or discoloration) or contamination, or if the expiration date supplied by the manufacturer has passed.

**Use:** For the selective isolation and/or presumptive identification of bacteria of the *Bacteroides fragilis* group from specimens containing mixed flora. Examine colonies with a long-wavelength UV light. Pigmented colonies of the *Bacteroides* group will fluoresce red-orange. Growth on this medium with blackening of the medium is presumptive for *Bacteroides fragilis*.

## Bile Oxalate Sorbose Broth
## (BOS Broth)

**Composition** per liter:
| | |
|---|---|
| $Na_2HPO_4$ | 9.14g |
| Sodium oxalate | 5.0g |
| Bile salts | 2.0g |
| NaCl | 1.0g |
| $CaCl_2 \cdot 2H_2O$ | 0.01g |
| $MgSO_4 \cdot 7H_2O$ | 0.01g |
| Asparagine solution | 100.0mL |
| Methionine solution | 100.0mL |
| Sorbose solution | 100.0mL |
| Yeast extract solution | 10.0mL |
| Sodium pyruvate solution | 10.0mL |
| Metanil Yellow solution | 10.0mL |
| Sodium nitrofurantoin solution | 10.0mL |
| Irgasan® solution | 1.0mL |

pH 7.6 ± 0.2 at 25°C

**Asparagine Solution:**
**Composition** per 100.0mL:
| | |
|---|---|
| Asparagine | 1.0g |

**Preparation of Asparagine Solution:** Add asparagine to distilled/deionized water and bring volume to 100.0mL. Mix thoroughly. Filter sterilize.

**Methionine Solution:**
**Composition** per 100.0mL:
| | |
|---|---|
| Methionine | 1.0g |

**Preparation of Methionine Solution:** Add methionine to distilled/deionized water and bring volume to 100.0mL. Mix thoroughly. Filter sterilize.

**Sorbose Solution:**
**Composition** per 100.0mL:
| | |
|---|---|
| Sorbose | 10.0g |

**Preparation of Sorbose Solution:** Add sorbose to distilled/deionized water and bring volume to 100.0mL. Mix thoroughly. Filter sterilize.

**Yeast Extract Solution:**
**Composition** per 10.0mL:
| | |
|---|---|
| Yeast extract | 0.025g |

**Preparation of Yeast Extract Solution:** Add yeast extract to distilled/deionized water and bring volume to 10.0mL. Mix thoroughly. Filter sterilize.

**Sodium Pyruvate Solution:**
**Composition** per 10.0mL:
| | |
|---|---|
| Sodium pyruvate | 0.05g |

**Preparation of Sodium Pyruvate Solution:** Add sodium pyruvate to distilled/deionized water and bring volume to 10.0mL. Mix thoroughly. Filter sterilize.

**Metanil Yellow Solution:**
**Composition** per 10.0mL:
| | |
|---|---|
| Metanil Yellow | 0.025g |

**Preparation of Metanil Yellow Solution:** Add Metanil Yellow to distilled/deionized water and bring volume to 10.0mL. Mix thoroughly. Filter sterilize.

**Sodium Nitrofurantoin Solution:**
**Composition** per 10.0mL:
| | |
|---|---|
| Sodium nitrofurantoin | 0.01g |

**Preparation of Sodium Nitrofurantoin Solution:** Add sodium nitrofurantoin to distilled/deionized water and bring volume to 10.0mL. Mix thoroughly. Filter sterilize.

**Irgasan® Solution:**
**Composition** per 10.0mL:
| | |
|---|---|
| Irgasan | 0.04g |
| Ethanol (95% solution) | 10.0mL |

**Preparation of Irgasan Solution:** Add Irgasan to 10.0mL of ethanol. Mix thoroughly. Filter sterilize.

**Preparation of Medium:** Add components, except asparagine solution, methionine solution, sorbose solution, yeast extract solution, sodium pyruvate solution, Metanil Yellow solution, sodium nitrofurantoin solution, and Irgasan solution, to distilled/deionized water and bring volume to 659.0mL. Mix thoroughly. Gently heat and bring to boiling. Autoclave for 15 min at 15 psi pressure–121°C. Cool to 45°–50°C. Aseptically add 100.0mL of sterile asparagine solution, 100.0mL of sterile methionine solution, 100.0mL of sterile sorbose solution, 10.0mL of sterile yeast extract solution, 10.0mL of sterile sodium pyruvate solution, 10.0mL of sterile Metanil Yellow solution, 10.0mL of sterile sodium nitrofurantoin solution, and 1.0mL of sterile Irgasan solution. Mix thoroughly. Pour into sterile Petri dishes or distribute into sterile tubes.

**Storage/Shelf Life:** Store dehydrated media in the dark in a sealed container below 30°C. Prepared media should be stored under refrigeration (2-8°C). Media should be used within 60 days of preparation. Media should not be used if there are any signs of deterioration (discoloration) or contamination, or if the expiration date supplied by the manufacturer has passed.

**Use:** For the isolation and cultivation of *Yersinia enterocolitica* from foods.

## Bile Peptone Transport Medium

**Composition** per liter:
| | |
|---|---|
| Casein enzymatic hydrolysate | 10.0g |
| NaCl | 10.0g |
| Sodium taurocholate | 5.0g |

pH 8.5 ± 0.2 at 25°C

**Source:** This medium is available from HiMedia.

**Preparation of Medium:** Add components to distilled/deionized water and bring volume to 1.0L. Mix thoroughly. Distribute into tubes. Autoclave for 15 min at 15 psi pressure–121°C.

**Storage/Shelf Life:** Store dehydrated media in the dark in a sealed container below 30°C. Prepared media should be stored under refrigeration (2-8°C). Media should be used within 60 days of preparation. Media should not be used if there are any signs of deterioration (discoloration) or contamination, or if the expiration date supplied by the manufacturer has passed.

**Use:** For the transport and preservation of *Vibrio cholerae.*

## Bile Salt Agar with Streptokinase
**Composition** per liter:
| | |
|---|---|
| Agar | 18.0g |
| Peptone | 10.0g |
| Meat extract | 10.0g |
| NaCl | 5.0g |
| Sodium taurocholate | 5.0g |
| Streptokinase solution | 1.0mL |

pH 8.2 ± 0.2 at 25°C

**Source:** This medium, without streptokinase solution, is available as a premixed powder from HiMedia.

**Streptokinase Solution:**
**Composition** per 1.0mL:
Streptokinase ................ 100,000 U

**Streptokinase Solution:** Add streptokinase to distilled/deionized water and bring volume to 1.0mL. Mix thoroughly. Filter sterilize.

**Preparation of Medium:** Add components to distilled/deionized water and bring volume to 1.0L. Mix thoroughly. Heat gently and bring to boiling. Distribute into tubes or flasks. Autoclave for 15 min at 15 psi pressure–121°C. Cool to 40°C. Aseptically add 1.0mL of streptokinase solution. Mix thoroughly. If desired, carbohydrate may also be added to this medium prior to sterilization.

**Storage/Shelf Life:** Store dehydrated media in the dark in a sealed container below 30°C. Prepared media should be stored under refrigeration (2-8°C). Media should be used within 60 days of preparation. Media should not be used if there are any signs of deterioration (shrinking, cracking, or discoloration) or contamination, or if the expiration date supplied by the manufacturer has passed.

**Use:** For the isolation and cultivation of bile tolerant enteric bacilli.

## Bile Salts Brilliant Green Starch Agar (BBGS Agar)
**Composition** per liter:
| | |
|---|---|
| Agar | 15.0g |
| Soluble starch | 10.0g |
| Proteose peptone | 10.0g |
| Beef extract | 5.0g |
| Bile salts | 5.0g |
| Brilliant Green (0.05% solution) | 1.0mL |

pH 7.2 ± 0.2 at 25°C

**Preparation of Medium:** Add components to distilled/deionized water and bring volume to 1.0L. Mix thoroughly. Gently heat while stirring and bring to boiling. Distribute into tubes or flasks. Autoclave for 15 min at 15 psi pressure–121°C. Pour into sterile Petri dishes or leave in tubes.

**Storage/Shelf Life:** Store dehydrated media in the dark in a sealed container below 30°C. Prepared media should be stored under refriger-

ation (2-8°C). Media should be used within 60 days of preparation. Media should not be used if there are any signs of deterioration (shrinking, cracking, or discoloration) or contamination, or if the expiration date supplied by the manufacturer has passed.

**Use:** For the isolation and cultivation of *Aeromonas hydrophila* from foods.

## BIN Medium
**Composition** per liter:
| | |
|---|---|
| Beef heart, infusion from | 250.0g |
| Calf brains, infusion from | 200.0g |
| Agar | 15.0g |
| Proteose peptone | 10.0g |
| NaCl | 5.0g |
| Na$_2$HPO$_4$ | 2.5g |
| Glucose | 2.0g |
| Irgasan solution | 4.0mL |
| Crystal Violet solution | 1.0mL |
| Sodium cholate solution | 1.0mL |
| Sodium deoxycholate solution | 1.0mL |
| Nystatin solution | 1.0mL |

pH 7.4 ± 0.2 at 25°C

**Sodium Cholate Solution:**
**Composition** per 100.0mL:
Sodium cholate .................. 5.0g

**Preparation of Sodium Cholate Solution:** Add sodium cholate to distilled/deionized water and bring volume to 100.0mL. Mix thoroughly. Gently heat while stirring and bring to boiling. Autoclave for 15 min at 15 psi pressure–121°C. Cool to 25°C.

**Sodium Deoxycholate Solution:**
**Composition** per 100.0mL:
Sodium deoxycholate .................. 5.0g

**Preparation of Sodium Deoxycholate Solution:** Add sodium deoxycholate to distilled/deionized water and bring volume to 100.0mL. Mix thoroughly. Gently heat while stirring and bring to boiling. Autoclave for 15 min at 15 psi pressure–121°C. Cool 25°C.

**Irgasan Solution:**
**Composition** per 50.0mL:
| | |
|---|---|
| Irgasan DP300 | 10.0mg |
| Ethanol, 90% | 50.0mL |

**Preparation of Irgasan Solution:** Add irgasan to 90% ethanol and bring volume to 50.0mL. Mix thoroughly.

**Crystal Violet Solution:**
**Composition** per 10.0mL:
Crystal Violet .................. 10.0mg

**Preparation of Crystal Violet Solution:** Add Crystal Violet to distilled/deionized water and bring volume to 10.0mL. Mix thoroughly. Gently heat while stirring and bring to boiling. Autoclave for 15 min at 15 psi pressure–121°C. Cool to 25°C.

**Nystatin Solution:**
**Composition** per 10.0mL:
Nystatin .................. 2.5g

**Preparation of Nystatin Solution:** Add nystatin to distilled/deionized water and bring volume to 10.0mL. Mix thoroughly. Filter sterilize.

**Preparation of Medium:** Add components, except irgasan solution, Crystal Violet solution, sodium cholate solution, sodium deoxycholate solution, and nystatin solution, to distilled/deionized water and bring volume to 992.0mL. Mix thoroughly. Gently heat while stirring and bring to boiling. Autoclave for 15 min at 15 psi pressure–121°C. Cool to 85°C. Aseptically add 4.0mL irgasam solution. Mix thoroughly to volatilize the ethanol. Cool to 50°C. Aseptically add 1.0mL each of Crystal Violet solution, sodium cholate solution, sodium deoxycholate solution, and nystatin solution. Mix thoroughly. Pour into sterile Petri dishes.

**Storage/Shelf Life:** Store dehydrated media in the dark in a sealed container below 30°C. Prepared media should be stored under refrigeration (2-8°C). Media should be used within 60 days of preparation. Media should not be used if there are any signs of deterioration (shrinking, cracking, or discoloration) or contamination, or if the expiration date supplied by the manufacturer has passed.

**Use:** For the efficient detection of *Yersinia pestis* from clinical and other specimens. The formulation of this medium is based on brain heart infusion agar, to which the selective agents irgasan, cholate salts, Crystal Violet, and nystatin are introduced to enhance efficiency of recovery of *Y. pestis*.

### Biosynth Chromogenic Medium for *Listeria monocytogenes* (BCM for *Listeria monocytogenes*) (BAM M17a)

**Composition** per liter:
Proprietary

**Source:** This medium is available from Biosynth International, Inc.

**Storage/Shelf Life:** Store in the dark under refrigeration (2-8°C). Chromogenic agars are especially light and temperature sensitive; protect from light, excessive heat, moisture, and freezing. Do not use after the expiration date supplied by the manufacturer.

**Use:** To differentiate *Listeria monocytogenes* and *L. ivanovii* from other *Listeria* spp. Supplements render the medium selective. Differential activity for all *Listeria* species is based upon a chromogenic substrate included in the medium. This is a complete test system with a fluorogenic selective enrichment broth and a chromogenic plating medium, both detecting the virulence factor phosphatidylinositol specific phospholipase C (PI-PLC). The medium contains a substrate for phosphatidylinositol-specific phospholipase C (PlcA) enzymes. The selective enrichment broth is fluorogenic. The plating medium for rapid detection and enumeration of pathogenic *Listeria* combines cleavage of the chromogenic PI-PLC substrate with the additional production of a white precipitate surrounding the target colonies.

### Bismuth Sulfite Agar

**Composition** per liter:

| | |
|---|---|
| Agar | 20.0g |
| $Bi_2(SO_3)_3$ | 8.0g |
| Pancreatic digest of casein | 5.0g |
| Peptic digest of animal tissue | 5.0g |
| Beef extract | 5.0g |
| Glucose | 5.0g |
| $Na_2HPO_4$ | 4.0g |
| $FeSO_4 \cdot 7H_2O$ | 0.3g |

pH 7.5 ± 0.2 at 25°C

**Source:** This medium is available as a premixed powder from Thermo Scientific and BD Diagnostic Systems.

**Preparation of Medium:** Add components to distilled/deionized water and bring volume to 1.0L. Mix thoroughly and heat with frequent agitation until boiling. Boil for 1 min. Do not autoclave. Cool to 45°–50°C. Pour into sterile Petri dishes while gently shaking flask to disperse precipitate. Use plates the same day as prepared.

**Storage/Shelf Life:** Store dehydrated media in the dark in a sealed container below 30°C. Prepared media should be stored under refrigeration (2-8°C). Media should be used within 60 days of preparation. Media should not be used if there are any signs of deterioration (shrinking, cracking, or discoloration) or contamination, or if the expiration date supplied by the manufacturer has passed.

**Use:** For the selective isolation and identification of *Salmonella typhi* and other enteric bacilli. *Salmonella typhi* appears as flat, black, "rabbit-eye" colonies surrounded by a zone of black with a metallic sheen.

### Bismuth Sulfite Agar

**Composition** per liter:

| | |
|---|---|
| Agar | 20.0g |
| Peptic digest of animal tissue | 10.0g |
| Bismuth sulfite indicator | 8.0g |
| Glucose | 5.0g |
| Beef extract | 5.0g |
| $Na_2HPO_4$ | 4.0g |
| $FeSO_4$ | 0.3g |
| Brilliant Green | 0.025g |

pH 7.7 ± 0.2 at 25°C

**Source:** This medium is available from HiMedia.

**Preparation of Medium:** Add components to distilled/deionized water and bring volume to 1.0L. Mix thoroughly and heat with frequent agitation until boiling. Boil for 1 min. Do not autoclave. Cool to 45°–50°C. Mix thoroughly. The sensitivity of the medium depends largely upon uniform dispersion of precipitated bismuth sulfite in the final gel, which should be dispersed before pouring the plates. Pour into sterile Petri dishes while gently shaking flask to disperse precipitate. Use plates the same day as prepared.

**Storage/Shelf Life:** Store dehydrated media in the dark in a sealed container below 30°C. Prepared media should be stored under refrigeration (2-8°C). Media should be used within 60 days of preparation. Media should not be used if there are any signs of deterioration (shrinking, cracking, or discoloration) or contamination, or if the expiration date supplied by the manufacturer has passed.

**Use:** For the selective isolation and identification of *Salmonella typhi* and other enteric bacilli. *Salmonella typhi* appears as flat, black, "rabbit-eye" colonies surrounded by a zone of black with a metallic sheen.

### Bismuth Sulfite Agar, HiVeg

**Composition** per liter:

| | |
|---|---|
| Agar | 20.0g |
| Plant peptone | 10.0g |
| Bismuth sulfite indicator | 8.0g |
| Glucose | 5.0g |
| Plant extract | 5.0g |
| $Na_2HPO_4$ | 4.0g |
| $FeSO_4$ | 0.3g |
| Brilliant Green | 0.025g |

pH 7.7 ± 0.2 at 25°C

**Preparation of Medium:** Add components to distilled/deionized water and bring volume to 1.0L. Mix thoroughly and heat with frequent agitation until boiling. Boil for 1 min. Do not autoclave. Cool to 45°–50°C. Mix thoroughly. The sensitivity of the medium depends largely upon uniform dispersion of precipitated bismuth sulfite in the final gel, which should be dispersed before pouring the plates. Pour into sterile Petri dishes while gently shaking flask to disperse precipitate. Use plates the same day as prepared.

**Storage/Shelf Life:** Store dehydrated media in the dark in a sealed container below 30°C. Prepared media should be stored under refrigeration (2-8°C). Media should be used within 60 days of preparation. Media should not be used if there are any signs of deterioration (shrinking, cracking, or discoloration) or contamination, or if the expiration date supplied by the manufacturer has passed.

**Use:** For the selective isolation and identification of *Salmonella typhi* and other enteric bacilli. *Salmonella typhi* appears as flat, black, "rabbit-eye" colonies surrounded by a zone of black with a metallic sheen.

### Bismuth Sulfite Agar, Modified

**Composition** per liter:

| | |
|---|---|
| Agar | 12.7g |
| Bismuth sulfite indicator | 8.0g |
| Glucose | 5.0g |
| Beef extract | 5.0g |
| Peptic digest of animal tissue | 5.0g |
| $Na_2HPO_4$ | 4.0g |
| $FeSO_4$ | 0.3g |
| Brilliant Green | 0.016 |

pH 7.5 ± 0.2 at 25°C

**Source:** This medium is available as a premixed powder from Hi-Media.

**Preparation of Medium:** Add components to distilled/deionized water and bring volume to 1.0L. Mix thoroughly and heat with frequent agitation until boiling. Boil for 1 min. Do not autoclave. Cool to 45°–50°C. Mix thoroughly. The sensitivity of the medium depends largely upon uniform dispersion of precipitated bismuth sulfite in the final gel, which should be dispersed before pouring the plates. Pour into sterile Petri dishes while gently shaking flask to disperse precipitate. Use plates the same day as prepared.

**Storage/Shelf Life:** Store dehydrated media in the dark in a sealed container below 30°C. Prepared media should be stored under refrigeration (2-8°C). Media should be used within 60 days of preparation. Media should not be used if there are any signs of deterioration (shrinking, cracking, or discoloration) or contamination, or if the expiration date supplied by the manufacturer has passed.

**Use:** For the selective isolation and identification of *Salmonella typhi* and other enteric bacilli. *Salmonella typhi* appears as flat, black, "rabbit-eye" colonies surrounded by a zone of black with a metallic sheen.

### Bismuth Sulfite Agar, Modified, HiVeg

**Composition** per liter:

| | |
|---|---|
| Agar | 12.7g |
| Bismuth sulfite indicator | 8.0g |
| Glucose | 5.0g |
| Plant extract | 5.0g |
| Plant peptone | 5.0g |
| $Na_2HPO_4$ | 4.0g |
| $FeSO_4$ | 0.3g |
| Brilliant Green | 0.016 |

pH 7.5 ± 0.2 at 25°C

**Source:** This medium is available as a premixed powder from Hi-Media.

**Preparation of Medium:** Add components to distilled/deionized water and bring volume to 1.0L. Mix thoroughly and heat with frequent agitation until boiling. Boil for 1 min. Do not autoclave. Cool to 45°–50°C. Mix thoroughly. The sensitivity of the medium depends largely upon uniform dispersion of precipitated bismuth sulfite in the final gel, which should be dispersed before pouring the plates. Pour into sterile Petri dishes while gently shaking flask to disperse precipitate. Use plates the same day as prepared.

**Storage/Shelf Life:** Store dehydrated media in the dark in a sealed container below 30°C. Prepared media should be stored under refrigeration (2-8°C). Media should be used within 60 days of preparation. Media should not be used if there are any signs of deterioration (shrinking, cracking, or discoloration) or contamination, or if the expiration date supplied by the manufacturer has passed.

**Use:** For the selective isolation and identification of *Salmonella typhi* and other enteric bacilli. *Salmonella typhi* appears as flat, black, "rabbit-eye" colonies surrounded by a zone of black with a metallic sheen.

### Bismuth Sulfite Agar Wilson and Blair (BAM 19)

**Composition** per liter:

| | |
|---|---|
| Agar | 20.0g |
| Pancreatic digest of casein | 10.0g |
| $Bi_2(SO_3)_3$ | 8.0g |
| Beef extract | 5.0g |
| Glucose | 5.0g |
| $Na_2HPO_4$ | 4.0g |
| $FeSO_4 \cdot 7H_2O$ | 0.3g |
| Brilliant Green | 0.025g |

pH 7.7 ± 0.2 at 25°C

**Preparation of Medium:** Add components to distilled/deionized water and bring volume to 1.0L. Mix thoroughly and heat with frequent agitation until boiling. Boil for 1 min. Do not autoclave. Cool to 45°–50°C. Pour into sterile Petri dishes while gently shaking flask to disperse precipitate. Let plates dry for about 2h with lids partially removed. Use plates within one day of preparation; medium loses selectivity after 48h.

**Storage/Shelf Life:** Store dehydrated media in the dark in a sealed container below 30°C. Prepared media should be stored under refrigeration (2-8°C). Media should be used within 60 days of preparation. Media should not be used if there are any signs of deterioration (shrinking, cracking, or discoloration) or contamination, or if the expiration date supplied by the manufacturer has passed.

**Use:** For the selective isolation and identification of *Salmonella typhi* and other enteric bacilli. *Salmonella typhi* appears as flat, black, "rabbit-eye" colonies surrounded by a zone of black with a metallic sheen.

### Bismuth Sulfite Broth (m-Bismuth Sulfite Broth)

**Composition** per liter:

| | |
|---|---|
| $Bi_2(SO_3)_3$ | 16.0g |
| Pancreatic digest of casein | 10.0g |
| Peptic digest of animal tissue | 10.0g |
| Beef extract | 10.0g |
| Glucose | 10.0g |

Na$_2$HPO$_4$ ..................................................................8.0g
FeSO$_4$·7H$_2$O ...........................................................0.6g
pH 7.7 ± 0.2 at 25°C

**Preparation of Medium:** Add components to distilled/deionized water and bring volume to 1.0L. Mix thoroughly and heat with frequent agitation until boiling. Boil for 1 min. Do not autoclave. Cool to 45°–50°C. Mix to disperse the precipitate and aseptically distribute into sterile tubes or flasks. Use 2.0–2.2mL of medium for each membrane filter.

**Storage/Shelf Life:** Store dehydrated media in the dark in a sealed container below 30°C. Prepared media should be stored under refrigeration (2-8°C). Media should be used within 60 days of preparation. Media should not be used if there are any signs of deterioration (discoloration) or contamination, or if the expiration date supplied by the manufacturer has passed.

**Use:** For the selective isolation of *Salmonella typhi* and other enteric bacilli and for the detection of *Salmonella* by the membrane filter method.

## BLE HiVeg Broth Base
## with *Listeria* Selective Supplement
## (Buffered *Listeria* Enrichment HiVeg Broth Base with *Listeria* Selective Supplement)

**Composition** per liter:
Plant hydrolysate .......................................................17.0g
Na$_2$HPO$_4$, anhydrous ...............................................9.6g
Yeast extract .................................................................6.0g
NaCl ..............................................................................5.0g
Papaic digest of soybean meal ....................................3.0g
KH$_2$PO$_4$ ....................................................................2.5g
Glucose .........................................................................2.5g
Sodium pyruvate ..........................................................1.0g
*Listeria* selective supplement ..................................5.0mL
pH 7.3 ± 0.2 at 25°C

***Listeria* Selective Supplement:**
**Composition** per 5.0mL:
Cycloheximide ...........................................................50.0mg
Nalidixic acid ............................................................40.0mg
Acriflavin hydrochloride ...........................................15.0mg

**Preparation of *Listeria* Selective Supplement:** Add components to distilled/deionized water and bring volume to 5.0mL. Mix thoroughly. Filter sterilize.

**Caution:** Cycloheximide is toxic. Avoid skin contact or aerosol formation and inhalation.

**Source:** This medium, without *Listeria* selective supplement, is available as a premixed powder from HiMedia.

**Preparation of Medium:** Add components, except *Listeria* selective supplement, to distilled/deionized water and bring volume to 1.0L. Mix thoroughly. Gently heat and bring to boiling. Autoclave for 15 min at 15 psi pressure–121°C. Cool to 45°–50°C. Aseptically add 50.0mL of sterile *Listeria* selective supplement. Mix thoroughly.

**Storage/Shelf Life:** Store dehydrated media in the dark in a sealed container below 30°C. Prepared media should be stored under refrigeration (2-8°C). Media should be used within 60 days of preparation. Media should not be used if there are any signs of deterioration (discoloration) or contamination, or if the expiration date supplied by the manufacturer has passed.

**Use:** For the enrichment and isolation of *Listeria monocytogenes*.

## Blood Agar

**Composition** per liter:
Agar .............................................................................15.0g
Pancreatic digest of casein .........................................15.0g
Papaic digest of soybean meal .....................................5.0g
NaCl ..............................................................................5.0g
Sheep blood, defibrinated ..........................................50.0mL
pH 7.6 ± 0.2 at 25°C

**Preparation of Medium:** Add components, except sheep blood, to distilled/deionized water and bring volume to 950.0mL. Mix thoroughly. Gently heat and bring to boiling. Autoclave for 15 min at 15 psi pressure–121°C. Cool to 45°–50°C. Aseptically add 50.0mL of sterile sheep blood. Mix thoroughly. Pour into sterile Petri dishes in 20.0mL volumes.

**Storage/Shelf Life:** Store dehydrated media in the dark in a sealed container below 30°C. Prepared media should be stored under refrigeration (2-8°C). Media should be used within 60 days of preparation. Media should not be used if there are any signs of deterioration (shrinking, cracking, or discoloration) or contamination, or if the expiration date supplied by the manufacturer has passed.

**Use:** For the cultivation of fastidious microorganisms.

## Blood Agar Base

**Composition** per liter:
Agar .............................................................................15.0g
Beef extract ................................................................10.0g
Peptone ........................................................................10.0g
NaCl ..............................................................................5.0g
Sheep blood, defibrinated ..........................................50.0mL
pH 7.3 ± 0.2 at 25°C

**Source:** This medium is available as a premixed powder from Thermo Scientific.

**Preparation of Medium:** Add components, except sheep blood, to distilled/deionized water and bring volume to 950.0mL. Mix thoroughly. Heat with frequent agitation and boil for 1 min to completely dissolve. Autoclave for 15 min at 15 psi pressure–121°C. Cool to 45°–50°C. Aseptically add 50.0mL of sterile, defibrinated sheep blood. Mix thoroughly and pour into sterile Petri dishes.

**Storage/Shelf Life:** Store dehydrated media in the dark in a sealed container below 30°C. Prepared media should be stored under refrigeration (2-8°C). Media should be used within 60 days of preparation. Media should not be used if there are any signs of deterioration (shrinking, cracking, or discoloration) or contamination, or if the expiration date supplied by the manufacturer has passed.

**Use:** For the isolation, cultivation, and detection of hemolytic activity of streptococci and other fastidious microorganisms.

## Blood Agar Base

**Composition** per liter:
Beef heart, infusion from ..........................................500.0g
Agar .............................................................................15.0g
Tryptose ......................................................................10.0g
NaCl ..............................................................................5.0g
pH 6.8 ± 0.2 at 25°C

**Source:** This medium is available as a premixed powder from BD Diagnostic Systems.

**Preparation of Medium:** Add components to distilled/deionized water and bring volume to 1.0L. Mix thoroughly. Heat with frequent agitation and boil for 1 min to completely dissolve. Autoclave for 15 min at 15 psi pressure–121°C. Cool the basal medium to 45°–50°C. Aseptically add sterile, defibrinated blood to a final concentration of 5%. Mix thoroughly and pour into sterile Petri dishes.

**Storage/Shelf Life:** Store dehydrated media in the dark in a sealed container below 30°C. Prepared media should be stored under refrigeration (2-8°C). Media should be used within 60 days of preparation. Media should not be used if there are any signs of deterioration (shrinking, cracking, or discoloration) or contamination, or if the expiration date supplied by the manufacturer has passed.

**Use:** For the isolation, cultivation, and detection of hemolytic activity of staphylococci, streptococci, and other fastidious microorganisms.

## Blood Agar Base
### (BAM M20a)

**Composition** per liter:
| | |
|---|---|
| Beef heart, infusion from | 500.0g |
| Agar | 15.0g |
| Tryptose | 10.0g |
| NaCl | 5.0g |
| Sheep blood, defibrinated | 50.0mL |

pH 6.8 ± 0.2 at 25°C

**Preparation of Medium:** Add components, except sheep blood, to distilled/deionized water and bring volume to 950.0mL. Mix thoroughly. Heat with frequent agitation and boil for 1 min to completely dissolve. Autoclave for 15 min at 15 psi pressure–121°C. Cool to 45°–50°C. Aseptically add 50.0mL of sterile, defibrinated sheep blood. Mix thoroughly and pour into sterile Petri dishes.

**Storage/Shelf Life:** Store dehydrated media in the dark in a sealed container below 30°C. Prepared media should be stored under refrigeration (2-8°C). Media should be used within 60 days of preparation. Media should not be used if there are any signs of deterioration (shrinking, cracking, or discoloration) or contamination, or if the expiration date supplied by the manufacturer has passed.

**Use:** For the isolation, cultivation, and detection of hemolytic activity of staphylococci, streptococci, and other fastidious microorganisms.

## Blood Agar Base
### (Infusion Agar)

**Composition** per liter:
| | |
|---|---|
| Agar | 15.0g |
| Pancreatic digest of casein | 13.0g |
| NaCl | 5.0g |
| Yeast extract | 5.0g |
| Heart muscle, solids from infusion | 2.0g |
| Sheep blood, defibrinated | 50.0mL |

pH 7.3 ± 0.2 at 25°C

**Source:** This medium is available as a premixed powder from BD Diagnostic Systems.

**Preparation of Medium:** Add components, except sheep blood, to distilled/deionized water and bring volume to 950.0mL. Mix thoroughly. Heat with frequent agitation and boil for 1 min to completely dissolve. Autoclave for 15 min at 15 psi pressure–121°C. Cool to 45°–

50°C. Aseptically add 50.0mL of sterile, defibrinated sheep blood. Mix thoroughly and pour into sterile Petri dishes.

**Storage/Shelf Life:** Store dehydrated media in the dark in a sealed container below 30°C. Prepared media should be stored under refrigeration (2-8°C). Media should be used within 60 days of preparation. Media should not be used if there are any signs of deterioration (shrinking, cracking, or discoloration) or contamination, or if the expiration date supplied by the manufacturer has passed.

**Use:** For the isolation, cultivation, and detection of hemolytic activity of streptococci and other fastidious microorganisms.

## Blood Agar Base
### (Infusion Agar)
### (FDA Medium M21)

**Composition** per liter:
| | |
|---|---|
| Heart muscle, infusion from | 375.0g |
| Agar | 15.0g |
| Thiotone | 10.0g |
| NaCl | 5.0g |

pH 7.3 ± 0.2 at 25°C

**Preparation of Medium:** Add components to distilled/deionized water and bring volume to 1.0L. Mix thoroughly. Gently heat and bring to boiling. Distribute into tubes or flasks. Autoclave for 20 min at 15 psi pressure–121°C. Pour into sterile Petri dishes or leave in tubes.

**Storage/Shelf Life:** Store dehydrated media in the dark in a sealed container below 30°C. Prepared media should be stored under refrigeration (2-8°C). Media should be used within 60 days of preparation. Media should not be used if there are any signs of deterioration (shrinking, cracking, or discoloration) or contamination, or if the expiration date supplied by the manufacturer has passed.

**Use:** For the cultivation of a variety of microorganisms. For the preparation of blood agar by the addition of sterile blood.

## Blood Agar Base with Blood

**Composition** per liter:
| | |
|---|---|
| Agar | 15.0g |
| Beef extract | 10.0g |
| Tryptose | 10.0g |
| NaCl | 5.0g |
| Sheep blood, defibrinated | 50.0mL |

pH 7.3 ± 0.2 at 25°C

**Source:** This medium without blood is available as a premixed powder from HiMedia.

**Preparation of Medium:** Add components, except sheep blood, to distilled/deionized water and bring volume to 950.0mL. Mix thoroughly. Heat with frequent agitation and boil for 1 min to completely dissolve. Autoclave for 15 min at 15 psi pressure–121°C. Cool to 45°–50°C. Aseptically add 50.0mL of sterile, defibrinated sheep blood. Mix thoroughly and pour into sterile Petri dishes.

**Storage/Shelf Life:** Store dehydrated media in the dark in a sealed container below 30°C. Prepared media should be stored under refrigeration (2-8°C). Media should be used within 60 days of preparation. Media should not be used if there are any signs of deterioration (shrinking, cracking, or discoloration) or contamination, or if the expiration date supplied by the manufacturer has passed.

**Use:** For the isolation, cultivation, and detection of hemolytic activity of streptococci and other fastidious microorganisms.

## Blood Agar Base, HiVeg with Blood

**Composition** per liter:

Agar .................................................................. 15.0g
Plant hydrolysate No. 1 .................................. 10.0g
Plant infusion ................................................ 10.0g
NaCl ................................................................. 5.0g
Sheep blood, defibrinated .......................... 50.0mL

pH 7.3 ± 0.2 at 25°C

**Source:** This medium without blood is available as a premixed powder from HiMedia.

**Preparation of Medium:** Add components, except sheep blood, to distilled/deionized water and bring volume to 950.0mL. Mix thoroughly. Heat with frequent agitation and boil for 1 min to completely dissolve. Autoclave for 15 min at 15 psi pressure–121°C. Cool to 45°–50°C. Aseptically add 50.0mL of sterile, defibrinated sheep blood. Mix thoroughly and pour into sterile Petri dishes.

**Storage/Shelf Life:** Store dehydrated media in the dark in a sealed container below 30°C. Prepared media should be stored under refrigeration (2-8°C). Media should be used within 60 days of preparation. Media should not be used if there are any signs of deterioration (shrinking, cracking, or discoloration) or contamination, or if the expiration date supplied by the manufacturer has passed.

**Use:** For the isolation, cultivation, and detection of hemolytic activity of streptococci and other fastidious microorganisms.

## Blood Agar Base, Improved

**Composition** per liter:

Enzymatic digest of casein ......................... 15.0g
Agar .................................................................. 14.0g
NaCl ................................................................... 5.0g
Enzymatic digest of animal tissue ................ 4.0g
Yeast extract .................................................... 2.0g
Corn starch ....................................................... 1.0g
Blood, defibrinated (sheep, rabbit, or horse) ........... 50-100mL

pH 7.0 ± 0.2 at 25°C

**Source:** This medium is available from Acumedia, Neogen Corporation.

**Preparation of Medium:** Add components to distilled/deionized water and bring volume to 1.0L. Mix thoroughly. Heat with frequent agitation and boil for 1 min to completely dissolve. Autoclave for 15 min at 15 psi pressure–121°C. Cool the basal medium to 45°–50°C. Aseptically add sterile, defibrinated blood to a final concentration of 5-10%. Mix thoroughly and pour into sterile Petri dishes.

**Storage/Shelf Life:** Store dehydrated media in the dark in a sealed container below 30°C. Prepared media should be stored under refrigeration (2-8°C). Media should be used within 60 days of preparation. Media should not be used if there are any signs of deterioration (shrinking, cracking, or discoloration) or contamination, or if the expiration date supplied by the manufacturer has passed.

**Use:** For isolating, cultivating, and determining hemolytic reactions of fastidious pathogenic microorganisms.

## Blood Agar Base, Sheep

**Composition** per liter:

Pancreatic digest of casein ......................... 14.0g
Agar ................................................................ 12.5g
NaCl ................................................................... 5.0g
Peptone ............................................................ 4.5g

Yeast extract ..................................................... 4.5g
Sheep blood, defibrinated .......................... 70.0mL

ph 7.3 ± 0.2 at 25°C

**Source:** This medium is available as a premixed powder from Thermo Scientific.

**Preparation:** Add components to distilled/deionized water and bring volume to 1.0L. Mix thoroughly. Autoclave for 15 min at 15 psi pressure–121°C. Cool the basal medium to 45°–50°C. Aseptically add 70.0mL of sterile, defibrinated sheep blood. Pour into sterile Petri dishes.

**Storage/Shelf Life:** Store dehydrated media in the dark in a sealed container below 30°C. Prepared media should be stored under refrigeration (2-8°C). Media should be used within 60 days of preparation. Media should not be used if there are any signs of deterioration (shrinking, cracking, or discoloration) or contamination, or if the expiration date supplied by the manufacturer has passed.

**Use:** For giving improved hemolytic reactions with sheep blood.

## Blood Agar Base with Low pH, HiVeg with Blood

**Composition** per liter:

Agar .................................................................. 15.0g
Plant hydrolysate No. 1 .................................. 10.0g
Plant infusion ................................................ 10.0g
NaCl ................................................................... 5.0g
Sheep blood, defibrinated .......................... 50.0mL

pH 6.8 ± 0.2 at 25°C

**Source:** This medium without blood is available as a premixed powder from HiMedia.

**Preparation of Medium:** Add components, except sheep blood, to distilled/deionized water and bring volume to 950.0mL. Mix thoroughly. Heat with frequent agitation and boil for 1 min to completely dissolve. Autoclave for 15 min at 15 psi pressure–121°C. Cool to 45°–50°C. Aseptically add 50.0mL of sterile, defibrinated sheep blood. Mix thoroughly and pour into sterile Petri dishes.

**Storage/Shelf Life:** Store dehydrated media in the dark in a sealed container below 30°C. Prepared media should be stored under refrigeration (2-8°C). Media should be used within 60 days of preparation. Media should not be used if there are any signs of deterioration (shrinking, cracking, or discoloration) or contamination, or if the expiration date supplied by the manufacturer has passed.

**Use:** For the detection of the hemolytic reactions of streptococci and other fastidious microorganisms. The slightly acid pH of this medium enhances distinct hemolytic reactions.

## Blood Agar Base with Peptone

**Composition** per liter:

Agar .................................................................. 15.0g
Beef extract .................................................... 10.0g
Peptone ........................................................... 10.0g
NaCl ................................................................... 5.0g

pH 7.3 ± 0.2 at 25°C

**Preparation of Medium:** Add components to distilled/deionized water and bring volume to 1.0L. Mix thoroughly. Gently heat and bring to boiling. Distribute into tubes or flasks. Autoclave for 15 min at 15 psi pressure–121°C. Pour into sterile Petri dishes or leave in tubes.

**Storage/Shelf Life:** Store dehydrated media in the dark in a sealed container below 30°C. Prepared media should be stored under refriger-

ation (2-8°C). Media should be used within 60 days of preparation. Media should not be used if there are any signs of deterioration (shrinking, cracking, or discoloration) or contamination, or if the expiration date supplied by the manufacturer has passed.

**Use:** For use as a base to which blood can be added; for the isolation, cultivation, and detection of hemolytic activity of streptococci and other fastidious microorganisms.

## Blood Agar with Low pH

**Composition** per liter:

| | |
|---|---|
| Beef heart, solids from infusion | 500.0g |
| Agar | 15.0g |
| Tryptose | 10.0g |
| NaCl | 5.0g |
| Sheep blood, defibrinated | 50.0mL |

pH 6. 8 ± 0.2 at 25°C

**Source:** This medium is available as a premixed powder from BD Diagnostic Systems.

**Preparation of Medium:** Add components, except sheep blood, to distilled/deionized water and bring volume to 950.0mL. Mix thoroughly. Heat with frequent agitation and boil for 1 min to completely dissolve. Autoclave for 15 min at 15 psi pressure–121°C. Cool to 45°–50°C. Aseptically add 50.0mL of sterile, defibrinated sheep blood. Mix thoroughly and pour into sterile Petri dishes.

**Storage/Shelf Life:** Store dehydrated media in the dark in a sealed container below 30°C. Prepared media should be stored under refrigeration (2-8°C). Media should be used within 60 days of preparation. Media should not be used if there are any signs of deterioration (shrinking, cracking, or discoloration) or contamination, or if the expiration date supplied by the manufacturer has passed.

**Use:** For the detection of the hemolytic reactions of streptococci and other fastidious microorganisms. The slightly acid pH of this medium enhances distinct hemolytic reactions.

## Blood Agar No. 2

**Composition** per liter:

| | |
|---|---|
| Proteose peptone | 15.0g |
| Agar | 12.0g |
| NaCl | 5.0g |
| Yeast extract | 5.0g |
| Liver digest | 2.5g |

pH 7.4 ± 0.2 at 25°C

**Source:** This medium is available as a premixed powder from BD Diagnostic Systems and Thermo Scientific.

**Preparation of Medium:** Add components to distilled/deionized water and bring volume to 1.0L. Mix thoroughly. Heat with frequent agitation and boil for 1 min to completely dissolve. Autoclave for 15 min at 15 psi pressure–121°C. Cool the basal medium to 45°–50°C. Aseptically add sterile, defibrinated blood to a final concentration of 7%. Pour into sterile Petri dishes.

**Storage/Shelf Life:** Store dehydrated media in the dark in a sealed container below 30°C. Prepared media should be stored under refrigeration (2-8°C). Media should be used within 60 days of preparation. Media should not be used if there are any signs of deterioration (shrinking, cracking, or discoloration) or contamination, or if the expiration date supplied by the manufacturer has passed.

**Use:** For the isolation, cultivation, and detection of hemolytic activity of streptococci, pneumococci, and other particularly fastidious microorganisms.

## Blood Free *Campylobacter* Selectivity HiVeg Agar Base

**Composition** per liter:

| | |
|---|---|
| Agar | 12.0g |
| Plant extract | 10.0g |
| Plant peptone | 10.0g |
| NaCl | 5.0g |
| Charcoal, bacteriological | 4.0g |
| Plant hydrolysate | 3.0g |
| Synthetic detergent No. III | 1.0g |
| FeSO$_4$ | 0.25g |
| Sodium pyruvate | 0.25g |
| Sodium deoxycholate solution | 10.0mL |
| Cefazolin solution | 1.0mL |

pH 7.4 ± 0.2 at 25°C

**Source:** This medium, without deoxycholate and cefazolin solutions, is available as a premixed powder from HiMedia.

**Sodium Deoxycholate Solution:**

**Composition** per 100.0mL:

| | |
|---|---|
| Sodium deoxycholate | 10.0g |

**Preparation of Sodium Deoxycholate Solution:** Add sodium deoxycholate to distilled/deionized water and bring volume to 100.0mL. Mix thoroughly. Gently heat while stirring and bring to boiling. Autoclave for 15 min at 15 psi pressure–121°C. Cool to 25°C.

**Cefazolin Solution:**

**Composition** per 10.0mL:

| | |
|---|---|
| Cefazolin | 0.1g |

**Preparation of Cefazolin Solution:** Add cefazolin to distilled/deionized water and bring volume to 10.0mL. Mix thoroughly. Filter sterilize.

**Preparation of Medium:** Add components, except cefazolin solution and sodium deoxycholate solution, to distilled/deionized water and bring volume to 990.0mL. Mix thoroughly. Heat with frequent agitation and boil for 1 min to completely dissolve. Autoclave for 15 min at 15 psi pressure–121°C. Cool to 50°–55°C. Add 10.0mL of sterile sodium deoxycholate solution and 1.0mL of sterile cefazolin solution. Mix thoroughly. Pour into sterile Petri dishes.

**Storage/Shelf Life:** Store dehydrated media in the dark in a sealed container below 30°C. Prepared media should be stored under refrigeration (2-8°C). Media should be used within 60 days of preparation. Media should not be used if there are any signs of deterioration (shrinking, cracking, or discoloration) or contamination, or if the expiration date supplied by the manufacturer has passed.

**Use:** For the selective isolation of *Campylobacter* species, especially *Campylobacter jejuni* from human feces.

## Blood Glucose Cystine Agar

**Composition** per 100.0mL:

| | |
|---|---|
| Nutrient agar | 85.0mL |
| Glucose cystine solution | 10.0mL |
| Human blood, fresh | 5.0mL |

pH 6.8 ± 0.2 at 25°C

**Nutrient Agar:**
**Composition** per liter:

Agar ........................................................................15.0g
Pancreatic digest of gelatin ......................................5.0g
Beef extract ..............................................................3.0g

**Source:** Nutrient agar is available as a premixed powder from BD Diagnostic Systems.

**Preparation of Nutrient Agar:** Add components to distilled/deionized water and bring volume to 1.0L. Mix thoroughly. Gently heat while stirring and bring to boiling. Distribute into tubes or flasks. Autoclave for 15 min at 15 psi pressure–121°C. Cool to 45°–50°C.

**Glucose Cystine Solution:**
**Composition** per 50.0mL:

Glucose ...................................................................12.5g
L-Cystine·HCl.............................................................0.5g

**Preparation of Glucose Cystine Solution:** Add components to distilled/deionized water and bring volume to 50.0mL. Mix thoroughly. Filter sterilize.

**Preparation of Medium:** To 85.0mL of cooled, sterile agar solution, aseptically add 10.0mL of sterile glucose cystine solution and 5.0mL of human blood. Mix thoroughly. Pour into sterile Petri dishes or distribute into sterile tubes.

**Storage/Shelf Life:** Store dehydrated media in the dark in a sealed container below 30°C. Prepared media should be stored under refrigeration (2-8°C). Media should be used within 60 days of preparation. Media should not be used if there are any signs of deterioration (shrinking, cracking, or discoloration) or contamination, or if the expiration date supplied by the manufacturer has passed.

**Use:** For the cultivation of *Francisella tularensis*.

## BMPA-α Medium
## (Edelstein BMPA-α Medium)

**Composition** per liter:

Agar .........................................................................13.0g
Yeast extract ............................................................10.0g
ACES buffer (2-[(2-amino-2-oxoethyl)-
    amino]-ethane sulfonic acid) .................................2.0g
Charcoal, activated.....................................................2.0g
α-Ketoglutarate..........................................................0.2g
$Fe_4(P_2O_7)_3 \cdot 9H_2O$.........................................................0.05g
Antibiotic inhibitor ................................................ 10.0mL
L-Cysteine·HCl·H$_2$O solution.............................. 10.0mL
pH 6.9 ± 0.2 at 25°C

**Source:** This medium is available as premixed vials from Thermo Scientific.

**Antibiotic Inhibitor:**
**Composition** per 10.0mL:

Anisomycin...............................................................0.08g
Cefamandole ............................................................4.0mg
Polymyxin B .........................................................80,000U

**Preparation of Antibiotic Inhibitor:** Add components to distilled/deionized water and bring volume to 10.0mL. Mix thoroughly. Filter sterilize.

**L-Cysteine·HCl·H$_2$O Solution:**
**Composition** per 10.0mL:

L-Cysteine·HCl·H$_2$O.................................................0.08g

**Preparation of L-Cysteine·HCl·H$_2$O Solution:** Add L-cysteine·HCl·H$_2$O to distilled/deionized water and bring volume to 10.0mL. Mix thoroughly. Filter sterilize.

**Preparation of Medium:** Add components, except antibiotic inhibitor and L-cysteine·HCl·H$_2$O solution, to distilled/deionized water and bring volume to 980.0mL. Mix thoroughly. Adjust medium to pH 6.9 with 1$N$ KOH. Heat gently and bring to boiling for 1 min. Autoclave for 15 min at 15 psi pressure–121°C. Cool to 50°–55°C. Add 10.0mL of the sterile L-cysteine·HCl·H$_2$O solution and 10.0mL of the sterile antibiotic solution. Mix thoroughly. Pour into sterile Petri dishes with constant agitation to keep charcoal in suspension.

**Storage/Shelf Life:** Store dehydrated media in the dark in a sealed container below 30°C. Prepared media should be stored under refrigeration (2-8°C). Media should be used within 60 days of preparation. Media should not be used if there are any signs of deterioration (shrinking, cracking, or discoloration) or contamination, or if the expiration date supplied by the manufacturer has passed.

**Use:** For the selective isolation and cultivation of *Legionella pneumophila* and other *Legionella* species.

## BMPA-α Medium
## (Semiselective Medium for *Legionella pneumophila*)

**Composition** per liter:

Agar ........................................................................15.0g
Yeast extract ............................................................10.0g
ACES buffer (2-[(2-amino-2-oxoethyl)-
    amino]-ethane sulfonic acid) ...............................10.0g
Charcoal, activated ....................................................2.0g
α-Ketoglutarate..........................................................1.0g
$Fe_4(P_2O_7)_3 \cdot 9H_2O$.........................................................0.25g
Antibiotic inhibitor ................................................ 10.0mL
L-Cysteine·HCl·H$_2$O solution.............................. 10.0mL
pH 6.9 ± 0.2 at 25°C

**Antibiotic Inhibitor:**
**Composition** per 10.0mL:

Anisomycin...............................................................0.08g
Cefamandole ............................................................4.0mg
Polymyxin B .........................................................80,000U

**Preparation of Antibiotic Inhibitor:** Add components to distilled/deionized water and bring volume to 10.0mL. Mix thoroughly. Filter sterilize.

**L-Cysteine·HCl·H$_2$O Solution:**
**Composition** per 10.0mL:

L-Cysteine·HCl·H$_2$O.................................................0.4g

**Preparation of L-Cysteine·HCl·H$_2$O Solution:** Add L-cysteine·HCl·H$_2$O to distilled/deionized water and bring volume to 10.0mL. Mix thoroughly. Filter sterilize.

**Preparation of Medium:** Add components, except antibiotic inhibitor and L-cysteine·HCl·H$_2$O solution, to distilled/deionized water and bring volume to 980.0mL. Mix thoroughly. Adjust medium to pH 6.9 with 1$N$ KOH. Heat gently and bring to boiling for 1 min. Autoclave for 15 min at 15 psi pressure–121°C. Cool to 50°–55°C. Add 10.0mL of the sterile L-cysteine·HCl·H$_2$O solution and 10.0mL of the sterile antibiotic solution. Mix thoroughly. Pour into sterile Petri dishes with constant agitation to keep charcoal in suspension.

**Storage/Shelf Life:** Store dehydrated media in the dark in a sealed container below 30°C. Prepared media should be stored under refrigeration (2-8°C). Media should be used within 60 days of preparation.

Media should not be used if there are any signs of deterioration (shrinking, cracking, or discoloration) or contamination, or if the expiration date supplied by the manufacturer has passed.

**Use:** For the selective isolation and cultivation of *Legionella pneumophila* and other *Legionella* species.

## Bolton Broth

**Composition** per 505mL:

| | |
|---|---|
| Bolton selective enrichment broth base | 500.0mL |
| Horse blood, lysed | 25.0mL |
| Bolton selective supplement soution | 5.0mL |

pH 7.4 ± 0.2 at 25°C

**Bolton Selective Enrichment Broth Base:**

**Composition** per liter:

| | |
|---|---|
| Peptone | 10.0g |
| Lactalbumin hydrolysate | 5.0g |
| Yeast extract | 5.0g |
| NaCl | 5.0g |
| α-Ketoglutarate | 1.0g |
| Na-pyruvate | 0.5g |
| Na-metabisulfite | 0.5g |
| $Na_2CO_3$ | 0.6g |
| Hemin | 0.01g |

**Preparation of Bolton Selective Enrichment Broth Base:** Add components to distilled/deionized water and bring volume to 1.0L. Mix thoroughly. Autoclave for 15 min at 15 psi pressure–121°C. Cool to 50°C.

**Bolton Broth Supplement Solution:**

**Composition** per 5.0mL

| | |
|---|---|
| Vancomycin | 10.0mg |
| Cefoperazone | 10.0mg |
| Trimethoprim | 10.0mg |
| Cycloheximide | 10.0mg |
| Ethanol | 2.5mL |

**Preparation of Bolton Supplement Solution:** Add antibiotics to 2.5mL ethanol. Mix thoroughly. Bring volume to 5.0mL with distilled/deionized water. Mix thoroughly. Filter sterilize.

**Caution:** Cycloheximide is toxic. Avoid skin contact or aerosol formation and inhalation.

**Preparation of Medium:** Aseptically combine 500.0mL warm Bolton selective enrichment broth base, 25.0mL lysed horse blood, and 5.0mL Bolton selective supplement soution.

**Storage/Shelf Life:** Store dehydrated media in the dark in a sealed container below 30°C. Prepared media should be stored under refrigeration (2-8°C). Media should be used within 60 days of preparation. Media should not be used if there are any signs of deterioration (discoloration) or contamination, or if the expiration date supplied by the manufacturer has passed.

**Use:** For the enrichment of *Campylobacter* spp. from foods.

## *Bordetella pertussis* Selective Medium with Bordet-Gengou Agar Base

**Composition** per 1210.0mL:

| | |
|---|---|
| Bordet-Gengou agar base | 1.0L |
| Horse blood, defibrinated | 200.0mL |
| Cephalexin solution | 10.0mL |

pH 6.7± 0.2 at 25°C

**Source:** This medium is available as a premixed powder from Thermo Scientific.

**Bordet-Gengou Agar Base:**

**Composition** per liter:

| | |
|---|---|
| Agar | 20.0g |
| NaCl | 5.5g |
| Pancreatic digest of casein | 5.0g |
| Peptic digest of animal tissue | 5.0g |

**Preparation of Bordet-Gengou Agar Base:** Add components to 1.0L of 1% glycerol solution. Autoclave for 15 min at 15 psi pressure–121°C. Cool to 50°C.

**Cephalexin Solution:**

**Composition** per 10.0mL:

| | |
|---|---|
| Cephalexin | 0.04g |

**Preparation of Cephalexin Solution:** Add cephalexin to distilled/deionized water and bring volume to 10.0mL. Mix thoroughly. Filter sterilize.

**Preparation of Medium:** Aseptically add 10.0mL of sterile cephalexin solution and 200.0mL of defibrinated horse blood to 1.0L Bordet-Gengou agar base. Mix thoroughly and pour into sterile Petri dishes.

**Storage/Shelf Life:** Store dehydrated media in the dark in a sealed container below 30°C. Prepared media should be stored under refrigeration (2-8°C). Media should be used within 60 days of preparation. Media should not be used if there are any signs of deterioration (shrinking, cracking, or discoloration) or contamination, or if the expiration date supplied by the manufacturer has passed.

**Use:** For the selective isolation and presumptive identification of *Bordetella pertussis* and *Bordetella parapertussis*. *Bordetella pertussis* appears as small, nearly transparent, "bisected pearl-like" colonies.

## *Bordetella pertussis* Selective Medium with Charcoal Agar Base

**Composition** per 1110.0mL:

| | |
|---|---|
| Charcoal agar base | 1.0L |
| Horse blood, defibrinated | 100.0mL |
| Cephalexin solution | 10.0mL |

pH 6.7± 0.2 at 25°C

**Source:** This medium is available as a premixed powder from Thermo Scientific.

**Charcoal Agar Base:**

**Composition** per liter:

| | |
|---|---|
| Agar | 12.0g |
| Beef extract | 10.0g |
| Starch | 10.0g |
| NaCl | 5.0g |
| Pancreatic digest of casein | 5.0g |
| Peptic digest of animal tissue | 5.0g |
| Charcoal | 4.0g |
| Nicotinic acid | 1.0mg |

**Preparation of Charcoal Agar Base:** Add components of charcoal agar base to distilled/deionized water and bring volume to 1.0L. Autoclave for 15 min at 15 psi pressure–121°C. Cool to 50°C.

**Cephalexin Solution:**

**Composition** per 10.0mL:

| | |
|---|---|
| Cephalexin | 0.04g |

**Preparation of Cephalexin Solution:** Add cephalexin to distilled/deionized water and bring volume to 10.0mL. Mix thoroughly. Filter sterilize.

**Preparation of Medium:** Aseptically add 10.0mL of sterile cephalexin solution and 100.0mL of defibrinated horse blood to charcoal agar base. Mix thoroughly and pour into sterile Petri dishes.

**Storage/Shelf Life:** Store dehydrated media in the dark in a sealed container below 30°C. Prepared media should be stored under refrigeration (2-8°C). Media should be used within 60 days of preparation. Media should not be used if there are any signs of deterioration (shrinking, cracking, or discoloration) or contamination, or if the expiration date supplied by the manufacturer has passed.

**Use:** For the selective isolation and presumptive identification of *Bordetella pertussis* and *Bordetella parapertussis. Bordetella pertussis* appears as small, pale, shiny colonies.

## Bordet-Gengou Agar

**Composition** per liter:

| | |
|---|---|
| Agar | 20.0g |
| Glycerol | 10.0g |
| NaCl | 5.5g |
| Pancreatic digest of casein | 5.0g |
| Peptic digest of animal tissue | 5.0g |
| Potato, solids from infusion | 4.5g |
| Rabbit blood | 200.0mL |

pH 6.7± 0.2 at 25°C

**Source:** This medium is available as a premixed powder from Thermo Scientific and BD Diagnostic Systems.

**Preparation of Medium:** Add 10.0g of glycerol to 980.0mL of distilled/deionized water. Add other components, except rabbit blood, to the glycerol solution. Mix thoroughly. Heat with occasional agitation of the medium. Boil for 1 min. Autoclave for 15 min at 15 psi pressure–121°C. Cool medium to 50°C. Aseptically add 200.0mL of rabbit blood (prewarmed to 35°C) to a concentration of 15–30%. 150.0–200.0mL of sterile, defibrinated horse blood may be used in place of rabbit blood. Mix thoroughly and pour plates or prepare slants.

**Storage/Shelf Life:** Store dehydrated media in the dark in a sealed container below 30°C. Prepared media should be stored under refrigeration (2-8°C). Media should be used within 60 days of preparation. Media should not be used if there are any signs of deterioration (shrinking, cracking, or discoloration) or contamination, or if the expiration date supplied by the manufacturer has passed.

**Use:** For the detection and isolation of *Bordetella pertussis* and *Bordetella parapertussis* from clinical specimens. The medium is rendered selective by the addition of methicillin. *Bordetella pertussis* appears as small (<1mm), smooth, pearl-like colonies surrounded by a narrow zone of hemolysis. *Bordetella parapertussis* appears as brown, nonshiny colonies with a green-black coloration on the reverse side. *Bordetella bronchiseptica* appears as brown, nonshiny, moderately sized colonies with a roughly pitted surface.

## Bordet-Gengou Agar Base with Rabbit Blood and Glycerol

**Composition** per liter:

| | |
|---|---|
| Potatoes, infusion from | 125.0g |
| Agar | 20.0g |
| Peptic digest of animal tissue | 10.0g |
| NaCl | 5.5g |

| | |
|---|---|
| Rabbit blood | 200.0mL |
| Glycerol | 10.0mL |

pH 6.7± 0.2 at 25°C

**Source:** This medium without glycerol and blood is available as a premixed powder from HiMedia.

**Preparation of Medium:** Add 10.0g of glycerol to 790.0mL of distilled/deionized water. Add other components, except rabbit blood, to the glycerol solution. Mix thoroughly. Heat with occasional agitation of the medium. Boil for 1 min. Autoclave for 15 min at 15 psi pressure–121°C. Cool medium to 50°C. Aseptically add 200.0mL of rabbit blood (prewarmed to 35°C). Note: 150.0–200.0mL of sterile, defibrinated horse blood may be used in place of rabbit blood. Mix thoroughly and pour plates or prepare slants.

**Storage/Shelf Life:** Store dehydrated media in the dark in a sealed container below 30°C. Prepared media should be stored under refrigeration (2-8°C). Media should be used within 60 days of preparation. Media should not be used if there are any signs of deterioration (shrinking, cracking, or discoloration) or contamination, or if the expiration date supplied by the manufacturer has passed.

**Use:** For the detection and isolation of *Bordetella pertussis* and *Bordetella parapertussis* from clinical specimens. *Bordetella pertussis* appears as small (<1mm), smooth, pearl-like colonies surrounded by a narrow zone of hemolysis. *Bordetella parapertussis* appears as brown, nonshiny colonies with a green-black coloration on the reverse side. *Bordetella bronchiseptica* appears as brown, nonshiny, moderately sized colonies with a roughly pitted surface.

## Bordet-Gengou Agar Base with 1.6% Agar

**Composition** per liter:

| | |
|---|---|
| Potatoes, infusion from | 125.0g |
| Agar | 16.0g |
| Peptic digest of animal tissue | 10.0g |
| NaCl | 5.5g |
| Rabbit blood | 200.0mL |
| Glycerol | 10.0mL |

pH 6.7± 0.2 at 25°C

**Source:** This medium without glycerol and blood is available as a premixed powder from HiMedia.

**Preparation of Medium:** Add 10.0g of glycerol to 790.0mL of distilled/deionized water. Add other components, except rabbit blood, to the glycerol solution. Mix thoroughly. Heat with occasional agitation of the medium. Boil for 1 min. Autoclave for 15 min at 15 psi pressure–121°C. Cool medium to 50°C. Aseptically add 200.0mL of rabbit blood (prewarmed to 35°C). 150.0–200.0mL of sterile, defibrinated horse blood may be used in place of rabbit blood. Mix thoroughly and pour plates or prepare slants.

**Storage/Shelf Life:** Store dehydrated media in the dark in a sealed container below 30°C. Prepared media should be stored under refrigeration (2-8°C). Media should be used within 60 days of preparation. Media should not be used if there are any signs of deterioration (shrinking, cracking, or discoloration) or contamination, or if the expiration date supplied by the manufacturer has passed.

**Use:** For the detection and isolation of *Bordetella pertussis* and *Bordetella parapertussis* from clinical specimens. *Bordetella pertussis* appears as small (<1mm), smooth, pearl-like colonies surrounded by a narrow zone of hemolysis. *Bordetella parapertussis* appears as brown, nonshiny colonies with a green-black coloration on the reverse side. *Bordetella bronchiseptica* appears as brown, nonshiny, moderately sized colonies with a roughly pitted surface.

## Bordet-Gengou HiVeg Agar Base with 1.6% Agar

**Composition** per liter:

| | |
|---|---|
| Potatoes, infusion from | 125.0g |
| Agar | 16.0g |
| Plant peptone | 10.0g |
| NaCl | 5.5g |
| Rabbit blood | 200.0mL |
| Glycerol | 10.0mL |

pH 6.7± 0.2 at 25°C

**Source:** This medium without glycerol and blood is available as a pre-mixed powder from HiMedia.

**Preparation of Medium:** Add 10.0g of glycerol to 790.0mL of distilled/deionized water. Add other components, except rabbit blood, to the glycerol solution. Mix thoroughly. Heat with occasional agitation of the medium. Boil for 1 min. Autoclave for 15 min at 15 psi pressure–121°C. Cool medium to 50°C. Aseptically add 200.0mL of rabbit blood (prewarmed to 35°C). 150.0–200.0mL of sterile, defibrinated horse blood may be used in place of rabbit blood. Mix thoroughly and pour plates or prepare slants.

**Storage/Shelf Life:** Store dehydrated media in the dark in a sealed container below 30°C. Prepared media should be stored under refrigeration (2-8°C). Media should be used within 60 days of preparation. Media should not be used if there are any signs of deterioration (shrinking, cracking, or discoloration) or contamination, or if the expiration date supplied by the manufacturer has passed.

**Use:** For the detection and isolation of *Bordetella pertussis* and *Bordetella parapertussis* from clinical specimens. *Bordetella pertussis* appears as small (<1mm), smooth, pearl-like colonies surrounded by a narrow zone of hemolysis. *Bordetella parapertussis* appears as brown, nonshiny colonies with a green-black coloration on the reverse side. *Bordetella bronchiseptica* appears as brown, nonshiny, moderately sized colonies with a roughly pitted surface.

## Bordet-Gengou HiVeg Agar Base with Rabbit Blood and Glycerol

**Composition** per liter:

| | |
|---|---|
| Potatoes, infusion from | 125.0g |
| Agar | 20.0g |
| Plant peptone | 10.0g |
| NaCl | 5.5g |
| Rabbit blood | 200.0mL |
| Glycerol | 10.0mL |

pH 6.7± 0.2 at 25°C

**Source:** This medium without glycerol and blood is available as a pre-mixed powder from HiMedia.

**Preparation of Medium:** Add 10.0g of glycerol to 790.0mL of distilled/deionized water. Add other components, except rabbit blood, to the glycerol solution. Mix thoroughly. Heat with occasional agitation of the medium. Boil for 1 min. Autoclave for 15 min at 15 psi pressure–121°C. Cool medium to 50°C. Aseptically add 200.0mL of rabbit blood (prewarmed to 35°C). Note: 150.0–200.0mL of sterile, defibrinated horse blood may be used in place of rabbit blood. Mix thoroughly and pour plates or prepare slants.

**Storage/Shelf Life:** Store dehydrated media in the dark in a sealed container below 30°C. Prepared media should be stored under refrigeration (2-8°C). Media should be used within 60 days of preparation. Media should not be used if there are any signs of deterioration (shrinking, cracking, or discoloration) or contamination, or if the expiration date supplied by the manufacturer has passed.

**Use:** For the detection and isolation of *Bordetella pertussis* and *Bordetella parapertussis* from clinical specimens. *Bordetella pertussis* appears as small (<1mm), smooth, pearl-like colonies surrounded by a narrow zone of hemolysis. *Bordetella parapertussis* appears as brown, nonshiny colonies with a green-black coloration on the reverse side. *Bordetella bronchiseptica* appears as brown, nonshiny, moderately sized colonies with a roughly pitted surface.

## Bordet-Gengou Medium

**Composition** per liter:

| | |
|---|---|
| Agar | 20.0g |
| Glycerol | 10.0g |
| Proteose peptone | 10.0g |
| NaCl | 5.5g |
| Pancreatic digest of casein | 5.0g |
| Peptic digest of animal tissue | 5.0g |
| Potato, solids from infusion | 4.5g |
| Rabbit blood | 150.0mL |

pH 6.7± 0.2 at 25°C

**Source:** This medium is available as a premixed powder from Thermo Scientific and BD Diagnostic Systems.

**Preparation of Medium:** Add 10.0g of glycerol to 980.0mL of distilled/deionized water. Add other components, except rabbit blood, to the glycerol solution. Mix thoroughly. Heat with occasional agitation of the medium. Boil for 1 min. Autoclave for 15 min at 15 psi pressure–121°C. Cool medium to 50°C. Aseptically add 150.0mL of rabbit blood (prewarmed to 35°C). Mix thoroughly. Pour into sterile Petri dishes or distribute into sterile tubes. Allow tubes to cool in a slanted position.

**Storage/Shelf Life:** Store dehydrated media in the dark in a sealed container below 30°C. Prepared media should be stored under refrigeration (2-8°C). Media should be used within 60 days of preparation. Media should not be used if there are any signs of deterioration (shrinking, cracking, or discoloration) or contamination, or if the expiration date supplied by the manufacturer has passed.

**Use:** For the detection and isolation of *Bordetella pertussis* and *Bordetella parapertussis* from clinical specimens. The medium is rendered selective by the addition of methicillin. *Bordetella pertussis* appears as small (<1mm), smooth, pearl-like colonies surrounded by a narrow zone of hemolysis. *Bordetella parapertussis* appears as brown, nonshiny colonies with a green-black coloration on the reverse side. *Bordetella bronchiseptica* appears as brown, nonshiny, moderately sized colonies with a roughly pitted surface.

## Bovine Serum Albumin Tween™ 80 Broth (BSA Tween™ 80 Broth)

**Composition** per liter:

| | |
|---|---|
| Basal medium | 900.0mL |
| Albumin supplement | 100.0mL |

pH 7.4 ± 0.2 at 25°C

**Basal Medium:**

**Composition** per liter:

| | |
|---|---|
| $Na_2HPO_4$ | 1.0g |
| NaCl | 1.0g |
| $KH_2PO_4$ | 0.3g |
| Glycerol (10% solution) | 1.0mL |
| $NH_4Cl$ (25% solution) | 1.0mL |
| Sodium pyruvate (10% solution) | 1.0mL |
| Thiamine (0.5% solution | 1.0mL |

**Preparation of Basal Medium:** Add components to distilled/deionized water and bring volume to 1.0L. Mix thoroughly. Adjust pH to 7.4. Autoclave for 15 min at 15 psi pressure–121°C. Cool to 25°C.

**Albumin Supplement:**
**Composition** per 100.0mL:

| | |
|---|---|
| Bovine albumin | 10.0g |
| Tween™ 80 (10% solution) | 12.5mL |
| $FeSO_4$ (0.5% solution) | 10.0mL |
| $MgCl_2$–$CaCl_2$ solution | 1.0mL |
| Cyanocobalamin (0.02% solution) | 1.0mL |
| $ZnSO_4$ (0.4% solution) | 1.0mL |

**Preparation of Albumin Supplement:** Add components to distilled/deionized water and bring volume to 100.0mL. Mix thoroughly. Adjust pH to 7.4. Filter sterilize.

**$MgCl_2$–$CaCl_2$ Solution:**
**Composition** per 100.0mL:

| | |
|---|---|
| $CaCl_2$·$2H_2O$ | 1.5g |
| $MgCl_2$·$6H_2O$ | 1.5g |

**Preparation of $MgCl_2$–$CaCl_2$ Solution:** Add components to distilled/deionized water and bring volume to 100.0mL. Mix thoroughly.

**Preparation of Medium:** To 900.0mL of cooled, sterile basal medium, aseptically add 100.0mL of sterile albumin supplement. Mix thoroughly. Aseptically distribute into sterile tubes or flasks.

**Storage/Shelf Life:** Store dehydrated media in the dark in a sealed container below 30°C. Prepared media should be stored under refrigeration (2-8°C). Media should be used within 60 days of preparation. Media should not be used if there are any signs of deterioration (discoloration) or contamination, or if the expiration date supplied by the manufacturer has passed.

**Use:** For the isolation and cultivation of *Leptospira* species.

**Bovine Albumin Tween™ 80 Medium,**
**Ellinghausen and McCullough, Modified**
**(Albumin Fatty Acid Broth, *Leptospira* Medium)**

**Composition** per liter:

| | |
|---|---|
| Basal medium | 900.0mL |
| Albumin fatty acid supplement | 100.0mL |

**Basal Medium:**
**Composition** per liter:

| | |
|---|---|
| $Na_2HPO_4$, anhydrous | 1.0g |
| NaCl | 1.0g |
| $KH_2PO_4$, anhydrous | 0.3g |
| $NH_4Cl$ (25% solution) | 1.0mL |
| Glycerol (10% solution) | 1.0mL |
| Sodium pyruvate (10% solution) | 1.0mL |
| Thiamine·HCl (0.5% solution) | 1.0mL |

pH 7.4 ± 0.2 at 25°C

**Preparation of Basal Medium:** Add components to distilled/deionized water and bring volume to 1.0L. Mix thoroughly. Adjust pH to 7.4. Gently heat and bring to boiling. Autoclave for 15 min at 15 psi pressure–121°C. Cool to 25°C.

**Albumin Fatty Acid Supplement:**
**Composition** per 200.0mL:

| | |
|---|---|
| Bovine albumin fraction V | 20.0g |
| Polysorbate (Tween™) 80 (10% solution) | 25.0mL |
| $FeSO_4$·$7H_2O$ (0.5% solution) | 20.0mL |
| $CaCl_2$·$2H_2O$ (1.5% solution) | 2.0mL |
| $MgCl_2$·$2H_2O$ (1.5% solution) | 2.0mL |
| Vitamin $B_{12}$ (0.2% solution) | 2.0mL |
| $ZnSO_4$·$7H_2O$ (0.4% solution) | 2.0mL |
| $CuSO_4$·$5H_2O$ (0.3% solution) | 0.2mL |

**Preparation of Albumin Fatty Acid Supplement:** Add bovine albumin to 100.0mL of distilled/deionized water. Mix thoroughly. Add remaining components while stirring. Adjust pH to 7.4. Bring volume to 200.0mL with distilled/deionized water. Filter sterilize. Store this supplement at –20°C.

**Preparation of Medium:** Aseptically combine 100.0mL of sterile albumin fatty acid supplement and 900.0mL of sterile basal medium. Mix thoroughly. Aseptically distribute into sterile tubes or flasks.

**Storage/Shelf Life:** Store dehydrated media in the dark in a sealed container below 30°C. Prepared media should be stored under refrigeration (2-8°C). Media should be used within 60 days of preparation. Media should not be used if there are any signs of deterioration (discoloration) or contamination, or if the expiration date supplied by the manufacturer has passed.

**Use:** For the cultivation of *Leptospira* species.

**Bovine Albumin Tween™ 80 Semisolid Medium,**
**Ellinghausen and McCullough, Modified**
**(Albumin Fatty Acid Semisolid Medium, Modified)**

**Composition** per liter:

| | |
|---|---|
| Basal medium | 900.0mL |
| Albumin fatty acid supplement | 100.0mL |

**Basal Medium:**
**Composition** per liter:

| | |
|---|---|
| Agar | 2.2g |
| $Na_2HPO_4$, anhydrous | 1.0g |
| NaCl | 1.0g |
| $KH_2PO_4$, anhydrous | 0.3g |
| $NH_4Cl$ (25% solution) | 1.0mL |
| Glycerol (10% solution) | 1.0mL |
| Sodium pyruvate (10% solution) | 1.0mL |
| Thiamine·HCl (0.5% solution) | 1.0mL |

pH 7.4 ± 0.2 at 25°C

**Preparation of Basal Medium:** Add components to distilled/deionized water and bring volume to 1.0L. Mix thoroughly. Adjust pH to 7.4. Gently heat and bring to boiling. Autoclave for 15 min at 15 psi pressure–121°C. Cool to 25°C.

**Albumin Fatty Acid Supplement:**
**Composition** per 200.0mL:

| | |
|---|---|
| Bovine albumin fraction V | 20.0g |
| Polysorbate (Tween™) 80 (10% solution) | 25.0mL |
| $FeSO_4$·$7H_2O$ (0.5% solution) | 20.0mL |
| $CaCl_2$·$2H_2O$ (1.5% solution) | 2.0mL |
| $MgCl_2$·$2H_2O$ (1.5% solution) | 2.0mL |
| Vitamin $B_{12}$ (0.2% solution) | 2.0mL |
| $ZnSO_4$·$7H_2O$ (0.4% solution) | 2.0mL |
| $CuSO_4$·$5H_2O$ (0.3% solution) | 0.2mL |

**Preparation of Albumin Fatty Acid Supplement:** Add bovine albumin to 100.0mL of distilled/deionized water. Mix thoroughly. Add remaining components while stirring. Adjust pH to 7.4. Bring volume to 200.0mL with distilled/deionized water. Filter sterilize. Store this supplement at –20°C.

**Preparation of Medium:** Aseptically combine 100.0mL of sterile albumin fatty acid supplement and 900.0mL of sterile basal medium. Mix thoroughly. Aseptically distribute into sterile tubes or flasks.

**Storage/Shelf Life:** Store dehydrated media in the dark in a sealed container below 30°C. Prepared media should be stored under refrigeration (2-8°C). Media should be used within 60 days of preparation. Media should not be used if there are any signs of deterioration (discoloration) or contamination, or if the expiration date supplied by the manufacturer has passed.

**Use:** For the cultivation of *Leptospira* species.

## BPL Agar
### (Brilliant Green Phenol Red Agar)

**Composition** per liter:

Lactose ...............................................................15.0g
Agar .....................................................................13.0g
NaCl.......................................................................5.0g
Meat peptone.........................................................7.0g
Phenol Red...........................................................0.04g
Brilliant Green .....................................................5.0mg

pH 6.5 ± 0.2 at 25°C

**Source:** This medium is available as a premixed powder from Hi-Media.

**Preparation of Medium:** Add components to distilled/deionized water and bring volume to 1.0L. Mix thoroughly. Gently heat and bring to boiling. Distribute into tubes or flasks. Autoclave for 15 min at 15 psi pressure–121°C. Pour into sterile Petri dishes or leave in tubes.

**Storage/Shelf Life:** Store dehydrated media in the dark in a sealed container below 30°C. Prepared media should be stored under refrigeration (2-8°C). Media should be used within 60 days of preparation. Media should not be used if there are any signs of deterioration (shrinking, cracking, or discoloration) or contamination, or if the expiration date supplied by the manufacturer has passed.

**Use:** For the cultivation of *Salmonella* species, with the exception of *S. typhi*, from feces, urine, meat, milk, and other materials.

## BPL HiVeg Agar
### (Brilliant Green Phenol Red HiVeg Agar)

**Composition** per liter:

Lactose ................................................................15.0g
Agar ......................................................................13.0g
Plant peptone No. 1...............................................7.0g
NaCl.......................................................................5.0g
Phenol Red...........................................................0.04g
Brilliant Green .....................................................5.0mg

pH 6.5 ± 0.2 at 25°C

**Source:** This medium is available as a premixed powder from Hi-Media.

**Preparation of Medium:** Add components to distilled/deionized water and bring volume to 1.0L. Mix thoroughly. Gently heat and bring to boiling. Distribute into tubes or flasks. Autoclave for 15 min at 15 psi pressure–121°C. Pour into sterile Petri dishes or leave in tubes.

**Storage/Shelf Life:** Store dehydrated media in the dark in a sealed container below 30°C. Prepared media should be stored under refrigeration (2-8°C). Media should be used within 60 days of preparation. Media should not be used if there are any signs of deterioration (shrink-

ing, cracking, or discoloration) or contamination, or if the expiration date supplied by the manufacturer has passed.

**Use:** For the cultivation of *Salmonella* species, with the exception of *S. typhi*, from feces, urine, meat, milk and other materials.

## Brain Heart Infusion
### (BHI)

**Composition** per liter:

Pancreatic digest of gelatin.................................14.5g
Brain heart, solids from infusion ..........................6.0g
Peptic digest of animal tissue ...............................6.0g
NaCl.......................................................................5.0g
Glucose ..................................................................3.0g
Na₂HPO₄................................................................2.5g

pH 7.4 ± 0.2 at 25°C

**Source:** This medium is available as a premixed powder from Thermo Scientific and BD Diagnostic Systems.

**Preparation of Medium:** Add components to distilled/deionized water and bring volume to 1.0L. Mix thoroughly. Distribute into tubes or flasks. Autoclave for 15 min at 15 psi pressure–121°C.

**Storage/Shelf Life:** Store dehydrated media in the dark in a sealed container below 30°C. Prepared media should be stored under refrigeration (2-8°C). Media should be used within 60 days of preparation. Media should not be used if there are any signs of deterioration (discoloration) or contamination, or if the expiration date supplied by the manufacturer has passed.

**Use:** For the cultivation of fastidious and nonfastidious microorganisms, including aerobic and anaerobic bacteria, from a variety of clinical speciments and nonclinical specimens of public health importance. It is particularly useful for culturing streptococci, pneumococci, and meningococci. It is also used for the preparation of inocula for use in antimicrobial susceptibility tests and as a base for blood culture.

## Brain Heart Infusion Agar
### (BAM M24 Medium 2)

**Composition** per liter:

Agar .....................................................................15.0g
Pancreatic digest of gelatin.................................14.5g
Brain heart, solids from infusion ..........................6.0g
Peptic digest of animal tissue ...............................6.0g
NaCl.......................................................................5.0g
Glucose ..................................................................3.0g
Na₂HPO₄................................................................2.5g

pH 7.4 ± 0.2 at 25°C

**Source:** This medium is available as a premixed powder from BD Diagnostic Systems.

**Preparation of Medium:** Add components to distilled/deionized water and bring volume to 1.0L. Mix thoroughly. Distribute into tubes or flasks while shaking to distribute precipitate. Autoclave for 15 min at 15 psi pressure–121°C. Mix thoroughly. Pour into sterile Petri dishes.

**Storage/Shelf Life:** Store dehydrated media in the dark in a sealed container below 30°C. Prepared media should be stored under refrigeration (2-8°C). Media should be used within 60 days of preparation. Media should not be used if there are any signs of deterioration (shrinking, cracking, or discoloration) or contamination, or if the expiration date supplied by the manufacturer has passed.

**Use:** For the cultivation of a wide variety of fastidious microorganisms, including bacteria, yeasts, and molds.

## Brain Heart Infusion Agar
### (BHI Agar)

**Composition** per liter:

Beef heart infusion ................................................................250.0g
Calf brain infusion ..............................................................200.0g
Agar ........................................................................................13.5g
Proteose peptone ..................................................................10.0g
NaCl ..........................................................................................5.0g
$Na_2HPO_4 \cdot 12H_2O$ ..........................................................................2.5g
Glucose ....................................................................................2.0g

pH 7.4 ± 0.2 at 25°C

**Preparation of Medium:** Add components to distilled/deionized water and bring volume to 1.0L. Mix thoroughly. Gently heat and bring to boiling. Distribute into tubes or flasks. Autoclave for 15 min at 15 psi pressure–121°C. Pour into sterile Petri dishes or leave in tubes.

**Storage/Shelf Life:** Store dehydrated media in the dark in a sealed container below 30°C. Prepared media should be stored under refrigeration (2-8°C). Media should be used within 60 days of preparation. Media should not be used if there are any signs of deterioration (shrinking, cracking, or discoloration) or contamination, or if the expiration date supplied by the manufacturer has passed.

**Use:** For the cultivation of a variety of fastidious and nonfastidious aerobic and anaerobic microorganisms.

## Brain Heart Infusion Agar
### (BAM M24 Medium 2)

**Composition** per liter:

Agar ........................................................................................15.0g
Pancreatic digest of gelatin ................................................14.5g
Brain heart, solids from infusion ..........................................6.0g
Peptic digest of animal tissue................................................6.0g
NaCl..........................................................................................5.0g
Glucose ....................................................................................3.0g
$Na_2HPO_4$..................................................................................2.5g

pH 7.4 ± 0.2 at 25°C

**Source:** This medium is available as a premixed powder from BD Diagnostic Systems.

**Preparation of Medium:** Add components to distilled/deionized water and bring volume to 1.0L. Mix thoroughly. Distribute into tubes or flasks while shaking to distribute precipitate. Autoclave for 15 min at 15 psi pressure–121°C. Mix thoroughly. Pour into sterile Petri dishes.

**Storage/Shelf Life:** Store dehydrated media in the dark in a sealed container below 30°C. Prepared media should be stored under refrigeration (2-8°C). Media should be used within 60 days of preparation. Media should not be used if there are any signs of deterioration (shrinking, cracking, or discoloration) or contamination, or if the expiration date supplied by the manufacturer has passed.

**Use:** For the cultivation of a wide variety of fastidious microorganisms, including bacteria.

## Brain Heart Infusion Agar 0.7%
### (BAM M23)

**Composition** per liter:

Pancreatic digest of gelatin ................................................14.5g
Agar ..........................................................................................7.0g

Brain heart, solids from infusion ..........................................6.0g
Peptic digest of animal tissue ................................................6.0g
NaCl..........................................................................................5.0g
Glucose ....................................................................................3.0g
$Na_2HPO_4$..................................................................................2.5g

pH 5.3 ± 0.2 at 25°C

**Source:** This medium without agar is available as a premixed powder from BD Diagnostic Systems.

**Preparation of Medium:** Add components, except agar, to distilled/deionized water and bring volume to 1.0L. Mix thoroughly. Adjust pH to 5.3 with 1*N* HCl. Mix thoroughly. Add agar. Gently heat and bring to boiling. Distribute into tubes. Autoclave for 10 min at 15 psi pressure–121°C.

**Storage/Shelf Life:** Store dehydrated media in the dark in a sealed container below 30°C. Prepared media should be stored under refrigeration (2-8°C). Media should be used within 60 days of preparation. Media should not be used if there are any signs of deterioration (shrinking, cracking, or discoloration) or contamination, or if the expiration date supplied by the manufacturer has passed.

**Use:** For the detection of staphylococcal enterotoxin.

## Brain Heart Infusion Agar 0.7%
### (BHI Agar 0.7%)
### (BAM M23)

**Composition** per liter:

Beef heart infusion ................................................................250.0g
Calf brain infusion ..............................................................200.0g
Proteose peptone ..................................................................10.0g
Agar ..........................................................................................7.0g
NaCl ..........................................................................................5.0g
$Na_2HPO_4 \cdot 12H_2O$ ..........................................................................2.5g
Glucose ....................................................................................2.0g

pH 5.3 ± 0.2 at 25°C

**Source:** This medium without agar is available as a premixed powder from BD Diagnostic Systems.

**Preparation of Medium:** Add components, except agar, to distilled/deionized water and bring volume to 1.0L. Mix thoroughly. Adjust pH to 5.3 with 1*N* HCl. Mix thoroughly. Add agar. Gently heat and bring to boiling. Distribute into tubes. Autoclave for 10 min at 15 psi pressure–121°C.

**Storage/Shelf Life:** Store dehydrated media in the dark in a sealed container below 30°C. Prepared media should be stored under refrigeration (2-8°C). Media should be used within 60 days of preparation. Media should not be used if there are any signs of deterioration (shrinking, cracking, or discoloration) or contamination, or if the expiration date supplied by the manufacturer has passed.

**Use:** For the detection of staphylococcal enterotoxin.

## Brain Heart Infusion Agar 0.7%
### (BHI Agar 0.7%)
### (BAM M23)

**Composition** per liter:

Pancreatic digest of gelatin ................................................14.5g
Agar ..........................................................................................7.0g
Brain heart, solids from infusion ..........................................6.0g
Peptic digest of animal tissue ................................................6.0g

NaCl .................................................................................5.0g
Glucose ...........................................................................3.0g
Na$_2$HPO$_4$ .......................................................................2.5g

pH 5.3 ± 0.2 at 25°C

**Source:** This medium without agar is available as a premixed powder from BD Diagnostic Systems.

**Preparation of Medium:** Add components, except agar, to distilled/deionized water and bring volume to 1.0L. Mix thoroughly. Adjust pH to 5.3 with 1*N* HCl. Mix thoroughly. Add agar. Gently heat and bring to boiling. Distribute into tubes. Autoclave for 10 min at 15 psi pressure–121°C.

**Storage/Shelf Life:** Store dehydrated media in the dark in a sealed container below 30°C. Prepared media should be stored under refrigeration (2-8°C). Media should be used within 60 days of preparation. Media should not be used if there are any signs of deterioration (shrinking, cracking, or discoloration) or contamination, or if the expiration date supplied by the manufacturer has passed.

**Use:** For the detection of staphylococcal enterotoxin.

### Brain Heart Infusion Agar with 1% Agar, HiVeg

**Composition** per liter:
Agar ................................................................................15.0g
Plant infusion ................................................................10.0g
Plant peptone No. 3 .......................................................10.0g
Plant special infusion .......................................................7.5g
NaCl ................................................................................5.0g
Na$_2$HPO$_4$ .......................................................................2.5g
Glucose ...........................................................................2.0g

pH 7.4 ± 0.2 at 25°C

**Source:** This medium is available as a premixed powder from Hi-Media.

**Preparation of Medium:** Add components to distilled/deionized water and bring volume to 1.0L. Mix thoroughly. Gently heat and bring to boiling. Distribute into tubes or flasks. Autoclave for 15 min at 15 psi pressure–121°C. Pour into sterile Petri dishes or leave in tubes.

**Storage/Shelf Life:** Store dehydrated media in the dark in a sealed container below 30°C. Prepared media should be stored under refrigeration (2-8°C). Media should be used within 60 days of preparation. Media should not be used if there are any signs of deterioration (shrinking, cracking, or discoloration) or contamination, or if the expiration date supplied by the manufacturer has passed.

**Use:** For the cultivation of a variety of fastidious pathogenic bacterias.

### Brain Heart Infusion Agar with 1% Agar, HiVeg

**Composition** per liter:
Agar ................................................................................10.0g
Plant infusion ................................................................10.0g
Plant peptone No. 3 .......................................................10.0g
Plant special infusion .......................................................7.5g
NaCl ................................................................................5.0g
Na$_2$HPO$_4$ .......................................................................2.5g
Glucose ...........................................................................2.0g

pH 7.4 ± 0.2 at 25°C

**Source:** This medium is available as a premixed powder from Hi-Media.

**Preparation of Medium:** Add components to distilled/deionized water and bring volume to 1.0L. Mix thoroughly. Gently heat and bring

to boiling. Distribute into tubes or flasks. Autoclave for 15 min at 15 psi pressure–121°C. Pour into sterile Petri dishes or leave in tubes.

**Storage/Shelf Life:** Store dehydrated media in the dark in a sealed container below 30°C. Prepared media should be stored under refrigeration (2-8°C). Media should be used within 60 days of preparation. Media should not be used if there are any signs of deterioration (shrinking, cracking, or discoloration) or contamination, or if the expiration date supplied by the manufacturer has passed.

**Use:** For the cultivation of a variety of fastidious pathogenic bacterias.

### Brain Heart Infusion Agar with 1% Agar, HiVeg with Penicillin

**Composition** per liter:
Agar ................................................................................15.0g
Plant infusion ................................................................10.0g
Plant peptone No. 3 .......................................................10.0g
Plant special infusion .......................................................7.5g
NaCl ................................................................................5.0g
Na$_2$HPO$_4$ .......................................................................2.5g
Glucose ...........................................................................2.0g
Penicillin solution ........................................................2.0mL

pH 7.4 ± 0.2 at 25°C

**Source:** This medium is available as a premixed powder from Hi-Media.

**Penicillin Solution:**
**Composition** per 2.0mL:
Penicillin .........................................................................0.1g

**Preparation of Penicillin Solution:** Add penicillin to ethanol and bring volume to 2.0mL. Mix thoroughly. Filter sterilize.

**Preparation of Medium:** Add components to distilled/deionized water and bring volume to 1.0L. Mix thoroughly. Gently heat and bring to boiling. Autoclave for 15 min at 15 psi pressure–121°C. Cool to 45–50°C. Aseptically add 2.0mL penicillin solution. Mix thoroughly. Pour into sterile Petri dishes or leave in tubes.

**Storage/Shelf Life:** Store dehydrated media in the dark in a sealed container below 30°C. Prepared media should be stored under refrigeration (2-8°C). Media should be used within 60 days of preparation. Media should not be used if there are any signs of deterioration (shrinking, cracking, or discoloration) or contamination, or if the expiration date supplied by the manufacturer has passed.

**Use:** For the cultivation of a variety of fastidious pathogenic bacterias.

### Brain Heart Infusion with PABA (Brain Heart Infusion with *p*-Aminobenzoic Acid)

**Composition** per liter:
Pancreatic digest of gelatin ...........................................14.5g
Brain heart, solids from infusion .....................................6.0g
Peptic digest of animal tissue ..........................................6.0g
NaCl ................................................................................5.0g
Glucose ...........................................................................3.0g
Na$_2$HPO$_4$ .......................................................................2.5g
*p*-Aminobenzoic acid ...................................................0.05g

pH 7.4 ± 0.2 at 25°C

**Source:** This medium is available as a premixed powder from BD Diagnostic Systems.

**Preparation of Medium:** Add components to distilled/deionized water and bring volume to 1.0L. Mix thoroughly. The addition of 1.0g agar to the medium enhances the growth of anaerobic and microaerophilic microorganisms. Heat with frequent agitation and boil for 1 min to dissolve. Distribute into tubes or flasks. Autoclave for 15 min at 15 psi pressure–121°C.

**Storage/Shelf Life:** Store dehydrated media in the dark in a sealed container below 30°C. Prepared media should be stored under refrigeration (2-8°C). Media should be used within 60 days of preparation. Media should not be used if there are any signs of deterioration (discoloration) or contamination, or if the expiration date supplied by the manufacturer has passed.

**Use:** For the detection of microorganisms in the blood of patients who have received sulfonamide therapy.

## Brain Heart Infusion with PABA and Agar (Brain Heart Infusion with *p*-Aminobenzoic Acid and Agar)

**Composition** per liter:
Pancreatic digest of gelatin ................................... 14.5g
Brain heart, solids from infusion ............................. 6.0g
Peptic digest of animal tissue ................................ 6.0g
NaCl ......................................................... 5.0g
Glucose ...................................................... 3.0g
Na$_2$HPO$_4$ ................................................ 2.5g
Agar ......................................................... 1.0g
*p*-Aminobenzoic acid ........................................ 0.05g
pH 7.4 ± 0.2 at 25°C

**Source:** This medium is available as a premixed powder from BD Diagnostic Systems.

**Preparation of Medium:** Add components to distilled/deionized water and bring volume to 1.0L. Mix thoroughly. The addition of 1.0g of agar to the medium enhances the growth of anaerobic and microaerophilic microorganisms. Heat with frequent agitation and boil for 1 min to dissolve. Distribute into tubes or flasks. Autoclave for 15 min at 15 psi pressure–121°C.

**Storage/Shelf Life:** Store dehydrated media in the dark in a sealed container below 30°C. Prepared media should be stored under refrigeration (2-8°C). Media should be used within 60 days of preparation. Media should not be used if there are any signs of deterioration (discoloration) or contamination, or if the expiration date supplied by the manufacturer has passed.

**Use:** For the detection of microorganisms in the blood of patients who have received sulfonamide therapy.

## Brain Heart Infusion with PABA and Agar, HiVeg

**Composition** per liter:
Plant infusion ............................................... 10.0g
Plant peptone ................................................ 10.0g
Plant special infusion ....................................... 7.5g
Na$_2$HPO$_4$ ................................................ 2.5g
NaCl ......................................................... 5.0g
Glucose ...................................................... 2.0g
Agar ......................................................... 1.0g
*p*-Aminobenzoic acid (PABA) ................................. 0.05g
pH 7.4 ± 0.2 at 25°C

**Source:** This medium is available as a premixed powder from Hi-Media.

**Preparation of Medium:** Add components to distilled/deionized water and bring volume to 1.0L. Mix thoroughly. The addition of 1.0g

agar to the medium enhances the growth of anaerobic and microaerophilic microorganisms. Heat with frequent agitation and boil for 1 min to dissolve. Distribute into tubes or flasks. Autoclave for 15 min at 15 psi pressure–121°C.

**Storage/Shelf Life:** Store dehydrated media in the dark in a sealed container below 30°C. Prepared media should be stored under refrigeration (2-8°C). Media should be used within 60 days of preparation. Media should not be used if there are any signs of deterioration (discoloration) or contamination, or if the expiration date supplied by the manufacturer has passed.

**Use:** For the detection of microorganisms in the blood of patients who have received sulfonamide therapy.

## Brain Heart Infusion with Para-amino Benzoic Acid, HiVeg

**Composition** per liter:
Plant infusion ............................................... 10.0g
Plant peptone No. 3 .......................................... 10.0g
Plant special infusion ....................................... 7.5g
NaCl ......................................................... 5.0g
Na$_2$HPO$_4$ ................................................ 2.5g
Glucose ...................................................... 2.0g
*p*-Aminobenzoic acid (PABA) ................................. 0.05g
pH 7.4 ± 0.2 at 25°C

**Source:** This medium is available as a premixed powder from Hi-Media.

**Preparation of Medium:** Add components to distilled/deionized water and bring volume to 1.0L. Mix thoroughly. The addition of 1.0g agar to the medium enhances the growth of anaerobic and microaerophilic microorganisms. Heat with frequent agitation and boil for 1 min to dissolve. Distribute into tubes or flasks. Autoclave for 15 min at 15 psi pressure–121°C.

**Storage/Shelf Life:** Store dehydrated media in the dark in a sealed container below 30°C. Prepared media should be stored under refrigeration (2-8°C). Media should be used within 60 days of preparation. Media should not be used if there are any signs of deterioration (discoloration) or contamination, or if the expiration date supplied by the manufacturer has passed.

**Use:** For the detection of microorganisms in the blood of patients who have received sulfonamide therapy.

## Brain Heart Infusion with 3% Sodium Chloride

**Composition** per liter:
NaCl ......................................................... 30.0g
Pancreatic digest of gelatin ................................. 14.5g
Brain heart, solids from infusion ............................ 6.0g
Peptic digest of animal tissue ............................... 6.0g
Glucose ...................................................... 3.0g
Na$_2$HPO$_4$ ................................................ 2.5g
pH 7.4 ± 0.2 at 25°C

**Preparation of Medium:** Add components to distilled/deionized water and bring volume to 1.0L. Mix thoroughly. Distribute into tubes or flasks. Autoclave for 15 min at 15 psi pressure–121°C.

**Storage/Shelf Life:** Store dehydrated media in the dark in a sealed container below 30°C. Prepared media should be stored under refrigeration (2-8°C). Media should be used within 60 days of preparation. Media should not be used if there are any signs of deterioration (discol-

oration) or contamination, or if the expiration date supplied by the manufacturer has passed.

**Use:** For the cultivation of *Vibrio parahaemolyticus*.

## BRILA MUG Broth
### (Brilliant Green 2%-Bile MUG Broth)

**Composition** per liter:

| | |
|---|---|
| Ox-bile (dried) | 20.0g |
| Peptone | 10.0g |
| Lactose | 10.0g |
| L-Tryptophan | 1.0g |
| Brillant Green | 0.133g |
| 4-Methylumbelliferyl-ß-D-glucuronide | 0.1g |

pH 7.2 ± 0.2 at 37°C

**Source:** This medium is available from Fluka, Sigma-Aldrich.

**Preparation of Medium:** Add components to distilled/deionized water and bring volume to 1.0L. Mix thoroughly. Distribute into test tubes that contain an inverted Durham tube in 10.0mL volumes. Autoclave for 15 min at 15 psi pressure–121°C.

**Storage/Shelf Life:** Store dehydrated media in the dark in a sealed container below 30°C. Prepared media should be stored under refrigeration (2-8°C). Media should not be used if there are any signs of deterioration (discoloration) or contamination, or if the expiration date supplied by the manufacturer has passed.

**Use:** For the detection of *E. coli* and coliforms. Bile and Brilliant Green extensively inhibit the growth of accompanying flora, in particular Gram-positive microorganisms. The presence of *E. coli* results in fluorescence in the UV. A positive indole test and possibly gas formation from lactose fermentation provide confirmation. β-D-glucoronidase, which is produced by *E. coli*, cleaves 4-methylumbelliferyl-β-D-glucuronide to 4-methylumbelliferone and glucuronide. The fluorogen 4-methylumbelliferone can be detected under a long wavelength UV lamp.The broth can be used in conjunction with the MPN method for *E. coli* and coliform enumeration in the water of bathing areas.

## Brilliance™ *Bacillus cereus* Agar
### (Chromogenic *Bacillus cereus* Agar)

**Composition** per liter:

| | |
|---|---|
| Agar | 13.0g |
| Peptone | 10.0g |
| Sodium pyruvate | 10.0g |
| Yeast extract | 4.0g |
| Na$_2$HPO$_4$ | 2.52g |
| Chromogenic mix | 1.2g |
| KH$_2$PO$_4$ | 0.28g |
| *Bacillus cereus* selective supplement | 10.0mL |

pH 7.2 ± 0.2 at 25°C

**Source:** This medium is available as a premixed powder from Thermo Scientific.

**Bacillus cereus Selective Supplement:**

**Composition** per 10.0mL:

| | |
|---|---|
| Trimethoprim | 10.0mg |
| Polymyxin B | 106,000 U |

**Preparation of *Bacillus cereus* Selective Supplement Solution:** Add components to distilled/deionized water and bring volume to 10.0mL. Mix thoroughly. Filter sterilize.

**Preparation of Medium:** Add components, except *Bacillus cereus* selective supplement. to distilled/deionized water and bring volume to 1.0L. Mix thoroughly. Gently heat while stirring and bring to boiling. Autoclave for 15 min at 15 psi pressure–121°C. Cool to 50°C. Aseptically add 5.0mL *Bacillus cereus* selective supplement. Mix thouroughly. Pour into sterile Petri dishes.

**Storage/Shelf Life:** Store in the dark under refrigeration (2-8°C). Chromogenic agars are especially light and temperature sensitive; protect from light, excessive heat, moisture, and freezing. Do not use after the expiration date supplied by the manufacturer.

**Use:** For the isolation and differentiation of *Bacillus cereus* from food samples.

## Brilliance CampyCount Agar

**Composition** per liter:

| | |
|---|---|
| Defined salt mix | 18.9g |
| Agar | 12.0g |
| Amino acid mix | 3.7g |

pH 7.63± 0.2 at 25°C

**Source:** This medium is available as a prepared medium from Thermo Scientific.

**Preparation of Medium:** Available as prepared plates.

**Storage/Shelf Life:** Prepared media should be stored under refrigeration (2-8°C). Media should not be used if there are any signs of deterioration (shrinking, cracking, or discoloration) or contamination, or if the expiration date supplied by the manufacturer has passed.

**Use:** For the enumeration of *Campylobacter jejuni* and *Campylobacter coli* from poultry and related samples. *Campylobacter* produce distinct dark red colonies.

## Brilliance CRE Agar

**Composition** per liter:

| | |
|---|---|
| Agar | 15.0g |
| Peptones | 15.0g |
| TiO$_2$ | 5.1g |
| Carbohydrates | 2.0g |
| Chromogenic mix | 1.0g |
| Antibiotic mix | 19.0mL |

**Source:** This medium is available as a prepared medium from Thermo Scientific.

**Preparation of Medium:** Available as prepared plates.

**Storage/Shelf Life:** Prepared media should be stored under refrigeration (2-8°C). Media should not be used if there are any signs of deterioration (shrinking, cracking, or discoloration) or contamination, or if the expiration date supplied by the manufacturer has passed.

**Use:** For the presumptive identification of carbapenem-resistant *E. coli* and the *Klebsiella*, *Enterobacter*, *Serratia,* and *Citrobacter* (KESC) group, direct from clinical samples, in 18 hours.

## Brilliance™ *E. coli*/Coliform Agar
### (Chromogenic *E. coli*/Coliform Agar)

**Composition** per liter:

| | |
|---|---|
| Chromogenic mix | 20.3g |
| Agar | 15.0g |
| Peptone | 5.0g |
| NaCl | 5.0g |
| Na$_2$HPO$_4$ | 3.5g |

Yeast extract......................................................................3.0g
Lactose.............................................................................2.5g
KH$_2$PO$_4$.....................................................................1.5g
Neutral Red.....................................................................0.03g

pH 7.0 ± 0.2 at 25°C

**Source:** This medium is available as a premixed powder from Thermo Scientific.

**Preparation of Medium:** Add components to distilled/deionized water and bring volume to 1.0L. Mix thoroughly. Gently heat and bring to boiling. Distribute into tubes or flasks. Autoclave for 15 min at 15 psi pressure–121°C. Pour into sterile Petri dishes or leave in tubes.

**Storage/Shelf Life:** Store in the dark under refrigeration (2-8°C). Chromogenic agars are especially light and temperature sensitive; protect from light, excessive heat, moisture, and freezing. Do not use after the expiration date supplied by the manufacturer.

**Use:** For the presumptive identification of *Escherichia coli* and coliforms from food and environmental samples. *E. coli* forms pink colonies.

## Brilliance™ *E. coli*/Coliform Selective Agar

**Composition** per liter:

Agar ..................................................................................10.6g
Peptone...............................................................................8.0g
NaCl....................................................................................5.0g
Na$_2$HPO$_4$....................................................................2.2g
KH$_2$PO$_4$.....................................................................1.8g
Chromogenic mix ............................................................0.35g
Sodium lauryl sulfate .......................................................0.1g

pH 6.7 ± 0.2 at 25°C

**Source:** This medium is available as a premixed powder from Thermo Scientific.

**Preparation of Medium:** Add components to distilled/deionized water and bring volume to 1.0L. Mix thoroughly. Gently heat and bring to boiling. Do not autoclave. Cool to 45°C. Pour into sterile Petri dishes or leave in tubes.

**Storage/Shelf Life:** Store in the dark under refrigeration (2-8°C). Chromogenic agars are especially light and temperature sensitive; protect from light, excessive heat, moisture, and freezing. Do not use after the expiration date supplied by the manufacturer.

**Use:** For the presumptive identification of *Escherichia coli* and coliforms from food and environmental samples. *E. coli* forms pink-purple colonies.

## Brilliance™ *Enterobacter sakazakii* Agar
## (DFI Agar)
## (Druggan, Forsythe, and Iverson Agar)

**Composition** per liter:

Agar ..................................................................................15.0g
Tryptone.............................................................................15.0g
Soya peptone.......................................................................5.0g
NaCl....................................................................................5.0g
Ferric ammonium citrate....................................................1.0g
Sodium desoxycholate.........................................................1.0g
Sodium thiosulphate............................................................1.0g
Chromogen.........................................................................0.1g

pH 7.3 ± 0.2 at 25°C

**Source:** This medium is available as a premixed powder from Thermo Scientific.

**Preparation of Medium:** Add components to distilled/deionized water and bring volume to 1.0L. Mix thoroughly. Gently heat and bring to boiling. Distribute into tubes or flasks. Autoclave for 15 min at 15 psi pressure–121°C. Pour into sterile Petri dishes or leave in tubes.

**Storage/Shelf Life:** Store in the dark under refrigeration (2-8°C). Chromogenic agars are especially light and temperature sensitive; protect from light, excessive heat, moisture, and freezing. Do not use after the expiration date supplied by the manufacturer.

**Use:** For the differentiation and enumeration of *Enterobacter sakazakii* from infant formula and other food samples. A chromogenic medium for the isolation and differentiation of *Enterobacter sakazakii* (now *Cronobacter sakazakii*) from food and dairy samples, according to the formulation by Druggan, Forsythe, and Iverson.

## Brilliance™ ESBL Agar

**Composition** per liter:

Agar ..................................................................................15.0g
Peptones.............................................................................12.0g
NaCl....................................................................................5.0g
Phosphate buffers...............................................................4.0g
Chromogenic mix ..............................................................4.0g
Antibiotic mix...................................................................0.28g

pH 7.2 ± 0.2 at 25°C

**Source:** This medium is available as a premixed powder from Thermo Scientific.

**Preparation of Medium:** Add components to distilled/deionized water and bring volume to 1.0L. Mix thoroughly. Gently heat while stirring and bring to boiling. Do not autoclave. Cool to 45°C. Mix thoroughly. Pour into sterile Petri dishes.

**Storage/Shelf Life:** Store in the dark under refrigeration (2-8°C). Chromogenic agars are especially light and temperature sensitive; protect from light, excessive heat, moisture, and freezing. Do not use after the expiration date supplied by the manufacturer.

**Use:** For the detection of Extended Spectrum β-Lactamase-producing organisms. The medium provides presumptive identification of ESBL-producing *E. coli* and the *Klebsiella, Enterobacter, Serratia,* and *Citrobacter* group (KESC), direct from clinical samples, in 24 h.

## Brilliance™ GBS Agar

**Composition** per liter:
Proprietary

**Source:** This medium is available as a premixed powder from Thermo Scientific.

**Preparation of Medium:** Add components, except antibiotic cocktail, to distilled/deionized water and bring volume to 1.0L. Mix thoroughly. Gently heat while stirring and bring to boiling. Autoclave for 15 min at 15 psi pressure–121°C. Cool to 50°C. Mix thoroughly. Pour into sterile Petri dishes.

**Storage/Shelf Life:** Store in the dark under refrigeration (2-8°C). Chromogenic agars are especially light and temperature sensitive; protect from light, excessive heat, moisture, and freezing. Do not use after the expiration date supplied by the manufacturer.

**Use:** For detection of group B streptococci, which produce bright pink colonies. Other bacteria are either inhibited or produce dark blue-purple colonies.

## Brilliance™ *Listeria* Agar
## (Oxoid Chromogenic *Listeria* Agar)
## (OCLA)

**Composition** per liter:

| | |
|---|---|
| Peptone | 18.5g |
| LiCl | 15.0g |
| Agar | 14.0g |
| NaCl | 9.5g |
| Yeast extract | 4.0g |
| Maltose | 4.0g |
| Sodium pyruvate | 2.0g |
| X-glucoside chromogenic mix | 0.2g |
| Differential lecithin solution | 40.0mL |
| Selective supplement solution | 20.0mL |

pH 7.2 ± 0.2 at 25°C

**Source:** This medium is available as a premixed powder from Thermo Scientific.

**Caution:** Lithium chloride is harmful. Avoid bodily contact and inhalation of vapors. On contact with skin wash with plenty of water immediately.

**Differential Lecithin Solution:**
**Composition** per 40.0mL:

| | |
|---|---|
| Lecithin | Proprietary |

**Preparation of Differential Lecithin Solution:** Available as premixed solution.

**Selective Supplement Solution:**
**Composition** per 20.0mL:

| | |
|---|---|
| Nalidixic acid | 26.0mg |
| Polymyxin B | 10.0mg |
| Ceftazidime | 6.0mg |
| Amphotericin | 10.0mg |

**Preparation of Selective Supplement Solution:** Add components to distilled/deionized water and bring volume to 20.0mL. Mix thoroughly. Filter sterilize.

**Preparation of Medium:** Add components, except differential lecithin solution and selective supplement solution, to distilled/deionized water and bring volume to 940.0mL. Mix thoroughly. Gently heat while stirring and bring to boiling. Autoclave for 15 min at 15 psi pressure–121°C. Cool to 46°C. Aseptically add differential lecithin solution and selective supplement solution. Mix thoroughly. Pour into sterile Petri dishes.

**Storage/Shelf Life:** Store in the dark under refrigeration (2-8°C). Chromogenic agars are especially light and temperature sensitive; protect from light, excessive heat, moisture, and freezing. Do not use after the expiration date supplied by the manufacturer.

**Use:** For the isolation, enumeration, and presumptive identification of *Listeria* species and *Listeria monocytogenes* from food samples. A chromogenic agar for the selective growth and differentiation of *Listeria monocytogenes* and *Listeria* spp.

## Brilliance™ MRSA Agar

**Composition** per liter:

| | |
|---|---|
| Peptone mix | 25.0g |
| Salt mix | 25.0g |
| Agar | 15.0g |
| Kaolin | 13.0g |
| Chromogenic mix | 2.0g |
| Antibiotic cocktail | 4.0mL |

pH 7.2 ± 0.2 at 25°C

**Source:** This medium is available as a premixed powder from Thermo Scientific.

**Preparation of Medium:** Add components, except antibiotic cocktail, to distilled/deionized water and bring volume to 1.0L. Mix thoroughly. Gently heat while stirring and bring to boiling. Autoclave for 15 min at 15 psi pressure–121°C. Cool to 50°C. Aseptically add antibiotic cocktail. Mix thoroughly. Pour into sterile Petri dishes.

**Storage/Shelf Life:** Store in the dark under refrigeration (2-8°C). Chromogenic agars are especially light and temperature sensitive; protect from light, excessive heat, moisture, and freezing. Do not use after the expiration date supplied by the manufacturer.

**Use:** For universal MRSA screening. Oxoid Brilliance MRSA Agar incorporates a novel chromogen that yields a blue color as a result of phosphatase activity, indicative of many staphylococci including *Staphylococcus aureus*. To allow the medium to differentiate MRSA accurately, it contains a combination of antibacterial compounds designed to inhibit the growth of a wide variety of competitor organisms and MSSA. Also included are compounds to suppress the expression of phosphatase activity in other staphylococci, thus ensuring a high level of sensitivity and specificity.

## Brilliance™ MRSA 2 Agar

**Composition** per liter:

Proprietary

**Source:** This medium is available as a premixed powder from Thermo Scientific.

**Preparation of Medium:** Add components, except antibiotic cocktail, to distilled/deionized water and bring volume to 1.0L. Mix thoroughly. Gently heat while stirring and bring to boiling. Autoclave for 15 min at 15 psi pressure–121°C. Cool to 50°C. Aseptically add antibiotic cocktail. Mix thoroughly. Pour into sterile Petri dishes.

**Storage/Shelf Life:** Store in the dark under refrigeration (2-8°C). Chromogenic agars are especially light and temperature sensitive; protect from light, excessive heat, moisture, and freezing. Do not use after the expiration date supplied by the manufacturer.

**Use:** For universal MRSA screening. Oxoid Brilliance MRSA Agar incorporates a novel chromogen that yields a blue color as a result of phosphatase activity, indicative of many staphylococci including *Staphylococcus aureus*. A novel pink counterstain improves ease of interpretation. To allow the medium to differentiate MRSA accurately, it contains a combination of antibacterial compounds designed to inhibit the growth of a wide variety of competitor organisms and MSSA. Also included are compounds to suppress the expression of phosphatase activity in other staphylococci, thus ensuring a high level of sensitivity and specificity.

## Brilliance™ *Salmonella* Agar

**Composition** per liter:

| | |
|---|---|
| Chromogenic mix | 25.0g |
| Agar | 15.0g |
| Inhibigen™ mix | 14.0g |
| *Salmonella* selective supplement solution | 10.0mL |

pH 7.3 ± 0.2 at 25°C

**Source:** This medium is available as a premixed powder from Thermo Scientific.

*Salmonella* **Selective Supplement Solution:**
**Composition** per 10.0mL:
Novobiocin.................................................................10.0mg
Cefsulodin.................................................................24.0mg

**Preparation of Selective Supplement Solution:** Add components to distilled/deionized water and bring volume to 10.0mL. Mix thoroughly. Filter sterilize.

**Preparation of Medium:** Add components, except *Salmonella* selective supplement solution, to distilled/deionized water and bring volume to 1.0L. Mix thoroughly. Add 10.0mL *Salmonella* selective supplement. It is critical that the selective supplement is added prior to heating. Gently heat while stirring and bring to boiling. Do not autoclave. Cool to 50°C. Mix thoroughly. Pour into sterile Petri dishes.

**Storage/Shelf Life:** Store in the dark under refrigeration (2-8°C). Chromogenic agars are especially light and temperature sensitive; protect from light, excessive heat, moisture, and freezing. Do not use after the expiration date supplied by the manufacturer.

**Use:** For the presumptive detection and identification of *Salmonella* spp. from foods and clinical specimens. The Inhibigen contained in this medium specifically targets *E.coli*, a particular benefit when testing fecal samples. Additional compounds are added to suppress growth of other competing flora. Differentiation of *Salmonella* from the other organisms that grow on Brilliance *Salmonella* Agar is achieved through the inclusion of two chromogens that target specific enzymes: caprylate esterase and β-glucosidase. The action of the enzymes on the chromogens results in a build-up of color within the colony. The color produced depends on which enzymes the organisms possess. The action of caprylate esterase present in all salmonellae results in a purple colony. Some Enterobacteriaceae species also produce caprylate esterase, but these are differentiated from *Salmonella* by a β-glucosidase substrate. This results in blue colonies, which are easy to distinguish from the purple *Salmonella* colonies.

## Brilliance Staph 24 Agar

**Composition** per liter:
Peptones.....................................................................21.0g
Agar ............................................................................14.0g
Chromogenic mix .........................................................5.0g
LiCl.............................................................................5.0g
Sodium pyruvate ...........................................................4.0g

pH 7.2± 0.2 at 25°C

**Source:** This medium is available as a prepared medium from Thermo Scientific.

**Caution:** Lithium chloride is harmful. Avoid bodily contact and inhalation of vapors. On contact with skin wash with plenty of water immediately.

**Preparation of Medium:** Available as prepared plates.

**Storage/Shelf Life:** Prepared media should be stored under refrigeration (2-8°C). Media should not be used if there are any signs of deterioration (shrinking, cracking, or discoloration) or contamination, or if the expiration date supplied by the manufacturer has passed.

**Use:** For the isolation and enumeration of coagulase-positive staphylococci in foods within 24 hours. Coagulase-positive staphylococci (CPS) grow as dark blue colonies on a clear background.

## Brilliance™ UTI Agar

**Composition** per liter:
Chromogenic mix ........................................................26.3g

Agar ............................................................................15.0g
Peptone .......................................................................15.0g

pH 7.0 ± 0.2 at 25°C

**Source:** This medium is available as a premixed powder from Thermo Scientific.

**Preparation of Medium:** Add components to distilled/deionized water and bring volume to 1.0L. Mix thoroughly. Gently heat and bring to boiling. Distribute into tubes or flasks. Autoclave for 15 min at 15 psi pressure–121°C. Cool to 50°C. Mix thoroughly. Pour into sterile Petri dishes or leave in tubes.

**Storage/Shelf Life:** Store in the dark under refrigeration (2-8°C). Chromogenic agars are especially light and temperature sensitive; protect from light, excessive heat, moisture, and freezing. Do not use after the expiration date supplied by the manufacturer.

**Use:** For the presumptive identification and differentiation of all the main microorganisms that cause urinary tract infections (UTIs). Brilliance UTI Agar contains two specific chromogenic substrates that are cleaved by enzymes produced by *Enterococcus* spp., *Escherichia coli*, and coliforms. In addition, it contains phenylalanine and tryptophan, which provide an indication of tryptophan deaminase activity, indicating the presence of *Proteus* spp., *Morganella* spp., and *Providencia* spp.

## Brilliance™ UTI Clarity Agar

**Composition** per liter:
Chromogenic mix .......................................................17.0g
Agar ............................................................................10.0g
Peptone .........................................................................9.0g
Tryptophan....................................................................1.0g

pH 7.0 ± 0.2 at 25°C

**Source:** This medium is available as a premixed powder from Thermo Scientific.

**Preparation of Medium:** Add components to distilled/deionized water and bring volume to 1.0L. Mix thoroughly. Gently heat and bring to boiling. Distribute into tubes or flasks. Autoclave for 15 min at 15 psi pressure–121°C. Pour into sterile Petri dishes or leave in tubes.

**Storage/Shelf Life:** Store in the dark under refrigeration (2-8°C). Chromogenic agars are especially light and temperature sensitive; protect from light, excessive heat, moisture, and freezing. Do not use after the expiration date supplied by the manufacturer.

**Use:** For the detection and differentiation of coliform bacteria. For the presumptive identification of the main pathogens that cause infection of the urinary tract. Brilliance UTI Clarity Agar contains two chromogenic substrates that are cleaved by enzymes produced by *E. coli*, *Enterococcus* spp., and coliforms. Of the two chromogens included in the medium, one is metabolized by β-galactosidase, an enzyme produced by *E. coli*, which grow as pink colonies. The other is cleaved by β-glucosidase enzyme activity, allowing the specific detection of enterococci, which form blue or turquoise colonies. Cleavage of both the chromogens gives dark blue or purple colonies, and indicates the organism is a coliform. The tryptophan in the medium is an indicator of tryptophan deaminase activity, resulting in colonies of *Proteus*, *Morganella*, and *Providencia* spp. with brown halos.

## Brilliance VRE Agar

**Composition** per liter:
Peptones.....................................................................25.0g
Agar ............................................................................12.5g

| | |
|---|---|
| Salt mix | 12.0g |
| TiO$_2$ | 1.0g |
| Chromogenic mix | 0.45g |
| Antibiotic cocktail including vancomycin | 5.0mL |

pH 6.5± 0.2 at 25°C

**Source:** This medium is available as a prepared medium from Thermo Scientific.

**Preparation of Medium:** Available as prepared plates.

**Storage/Shelf Life:** Prepared media should be stored under refrigeration (2-8°C). Media should not be used if there are any signs of deterioration (shrinking, cracking, or discoloration) or contamination, or if the expiration date supplied by the manufacturer has passed.

**Use:** For the detection of vancomycin-resistant enterococci (VRE). For the presumptive identification of *Enterococcus faecium* and *Enterococcus faecalis*, direct from clinical samples in 24 hours. VRE grow as either light blue colonies (*E. faecalis)* or as indigo-purple colonies (*E. faecium),* both of which are very easy to read against the new, semi-opaque background.

## Brilliant Green Agar

**Composition** per liter:

| | |
|---|---|
| Agar | 20.0g |
| Lactose | 10.0g |
| Sucrose | 10.0g |
| Peptic digest of animal tissue | 5.0g |
| Pancreatic digest of casein | 5.0g |
| NaCl | 5.0g |
| Phenol Red | 0.08g |
| Brilliant Green | 0.0125g |

pH 6.9 ± 0.2 at 25°C

**Source:** This medium is available as a premixed powder from Thermo Scientific and BD Diagnostic Systems.

**Preparation of Medium:** Add components to distilled/deionized water and bring volume to 1.0L. Mix thoroughly. Gently heat and bring to boiling. Distribute into tubes or flasks. Autoclave for 15 min at 15 psi pressure–121°C. Pour into sterile Petri dishes.

**Storage/Shelf Life:** Store dehydrated media in the dark in a sealed container below 30°C. Prepared media should be stored under refrigeration (2-8°C). Media should be used within 60 days of preparation. Media should not be used if there are any signs of deterioration (shrinking, cracking, or discoloration) or contamination, or if the expiration date supplied by the manufacturer has passed.

**Use:** For the selective isolation of *Salmonella* other than *Salmonella typhi* from feces and other specimens, and food and dairy products. *Salmonella* other than *Salmonella typhi* appear as red/pink/white colonies surrounded by a zone of red in the agar, indicating nonlactose/sucrose fermentation. *Proteus* or *Pseudomonas* species may appear as small red colonies. Lactose- or sucrose-fermenting bacteria appear as yellow-green colonies surrounded by a zone of yellow-green in the agar.

## Brilliant Green Agar Base
## with Phosphates and Sulfa Supplement

**Composition** per liter:

| | |
|---|---|
| Agar | 12.0g |
| Sucrose | 10.0g |
| Plant peptone | 10.0g |
| Lactose | 10.0g |
| Plant extract | 5.0g |
| Yeast extract | 3.0g |
| Na$_2$HPO$_4$ | 1.0g |
| NaH$_2$PO$_4$ | 0.6g |
| Phenol Red | 0.09g |
| Brilliant Green | 4.7mg |

pH 6.9 ± 0.2 at 25°C

**Source:** This medium, without sulfa supplement, is available as a premixed powder from HiMedia.

**Sulfa Supplement Solution:**
**Composition** per 10.0mL:

| | |
|---|---|
| Sodium sulfacetamide | 1.0g |
| Sodium mandelate | 0.25g |

**Preparation of Sulfa Supplement Solution:** Add components to distilled/deionized water and bring volume to 10.0mL. Mix thoroughly. Filter sterilize.

**Preparation of Medium:** Add components, except sulfa supplement solution, to distilled/deionized water and bring volume to 990.0mL. Mix thoroughly. Autoclave for 15 min at 15 psi pressure–121°C. Aseptically add 10.0mL of sterile sulfa supplement solution. Mix thoroughly. Pour into sterile Petri dishes.

**Storage/Shelf Life:** Store dehydrated media in the dark in a sealed container below 30°C. Prepared media should be stored under refrigeration (2-8°C). Media should be used within 60 days of preparation. Media should not be used if there are any signs of deterioration (shrinking, cracking, or discoloration) or contamination, or if the expiration date supplied by the manufacturer has passed.

**Use:** For the selective isolation of *Salmonella* other than *Salmonella typhi* from feces and other specimens while inhibiting *Escherichia coli, Proteus,* and *Pseudomonas* species.

## Brilliant Green Agar, Modified

**Composition** per liter:

| | |
|---|---|
| Agar | 12.0g |
| Lactose | 10.0g |
| Sucrose | 10.0g |
| Beef extract | 5.0g |
| Peptone | 5.0g |
| NaCl | 5.0g |
| Yeast extract | 3.0g |
| Na$_2$HPO$_4$ | 1.0g |
| NaH$_2$PO$_4$ | 0.6g |
| Phenol Red | 0.09g |
| Brilliant Green | 4.7mg |

pH 6.9 ± 0.2 at 25°C

**Source:** This medium is available as a premixed powder from Thermo Scientific.

**Preparation of Medium:** Add components to distilled/deionized water and bring volume to 1.0L. Mix thoroughly. Gently heat and bring to boiling. Do not autoclave. Cool to 45°–50°C. Addition of 1.0g of sodium sulfacetamide and 250.0mg of sodium mandelate enhances inhibition of contaminating microorganisms. Pour into sterile Petri dishes.

**Storage/Shelf Life:** Store dehydrated media in the dark in a sealed container below 30°C. Prepared media should be stored under refrigeration (2-8°C). Media should be used within 60 days of preparation. Media should not be used if there are any signs of deterioration (shrinking, cracking, or discoloration) or contamination, or if the expiration date supplied by the manufacturer has passed.

**Use:** For the selective isolation of *Salmonella* other than *Salmonella typhi* from feces and other specimens, and food and dairy products. *Salmonella* other than *Salmonella typhi* appear as red/pink/white colonies surrounded by a zone of red in the agar, indicating nonlactose/sucrose fermentation. *Proteus* or *Pseudomonas* species may appear as small red colonies. Lactose- or sucrose-fermenting bacteria appear as yellow-green colonies surrounded by a zone of yellow-green in the agar.

## Brilliant Green Agar with Sulfadiazine

**Composition** per liter:

| | |
|---|---|
| Agar | 20.0g |
| Lactose | 10.0g |
| Sucrose | 10.0g |
| Pancreatic digest of casein | 5.0g |
| Peptic digest of animal tissue | 5.0g |
| NaCl | 5.0g |
| Yeast extract | 3.0g |
| Phenol Red | 0.08g |
| Sulfadiazine | 0.08g |
| Brilliant Green | 0.0125g |

pH 6.9 ± 0.2 at 25°C

**Source:** This medium is available as a premixed powder from BD Diagnostic Systems.

**Preparation of Medium:** Add components to distilled/deionized water and bring volume to 1.0L. Mix thoroughly. Gently heat and bring to boiling. Distribute into tubes or flasks. Autoclave for 15 min at 15 psi pressure–121°C. Pour into sterile Petri dishes.

**Storage/Shelf Life:** Store dehydrated media in the dark in a sealed container below 30°C. Prepared media should be stored under refrigeration (2-8°C). Media should be used within 60 days of preparation. Media should not be used if there are any signs of deterioration (shrinking, cracking, or discoloration) or contamination, or if the expiration date supplied by the manufacturer has passed.

**Use:** For the selective detection of *Salmonella* in foods, especially from egg products. *Salmonella* other than *Salmonella typhi* appear as red/pink colonies surrounded by a zone of red in the agar indicating nonlactose/sucrose fermentation. *Proteus* or *Pseudomonas* species may appear as small red colonies. Lactose- or sucrose-fermenting bacteria appear as yellow-green colonies surrounded by a zone of yellow-green in the agar.

## Brilliant Green Bile Agar

**Composition** per liter:

| | |
|---|---|
| Noble agar | 10.15g |
| Pancreatic digest of gelatin | 8.25g |
| Lactose | 1.9g |
| Na$_2$SO$_3$ | 0.205g |
| Basic Fuchsin | 0.078g |
| Erioglaucine | 0.065g |
| FeCl$_3$ | 0.0295g |
| KH$_2$PO$_4$ | 0.015g |
| Oxgall, dehydrated | 2.95mg |
| Brilliant Green | 0.03mg |

pH 6.9 ± 0.2 at 25°C

**Source:** This medium is available as a premixed powder from BD Diagnostic Systems.

**Caution:** Basic Fuchsin is a potential carcinogen and care must be taken to avoid inhalation of the powdered dye and contamination of the skin.

**Preparation of Medium:** Add components to distilled/deionized water and bring volume to 1.0L. For plating 10.0mL samples, prepare the medium double strength. Mix thoroughly. Gently heat and bring to boiling. Distribute into tubes or flasks. Autoclave for 15 min at 15 psi pressure–121°C. Pour into sterile Petri dishes. Care should be taken to avoid exposure of the prepared medium to light.

**Storage/Shelf Life:** Store dehydrated media in the dark in a sealed container below 30°C. Prepared media should be stored under refrigeration (2-8°C). Media should be used within 60 days of preparation. Media should not be used if there are any signs of deterioration (shrinking, cracking, or discoloration) or contamination, or if the expiration date supplied by the manufacturer has passed.

**Use:** For the detection and enumeration of coliform bacteria in materials of sanitary importance such as water, sewage, and foods. *Escherichia coli* appears as dark red colonies with a pink halo. *Enterobacter* species appear as pink colonies.

## Brilliant Green Bile Broth
## (Brilliant Green Lactose Bile Broth)

**Composition** per liter:

| | |
|---|---|
| Oxgall, dehydrated | 20.0g |
| Lactose | 10.0g |
| Pancreatic digest of gelatin | 10.0g |
| Brilliant Green | 0.013g |

pH 7.2 ± 0.2 at 25°C

**Source:** This medium is available as a premixed powder from BD Diagnostic Systems and Thermo Scientific.

**Preparation of Medium:** Add components to distilled/deionized water and bring volume to 1.0L. Mix thoroughly. Distribute into tubes containing inverted Durham tubes, in 10.0mL amounts for testing 1.0mL or less of sample. Autoclave for 12 min (not longer than 15 min) at 15 psi pressure–121°C. After sterilization, cool the broth rapidly. Medium is sensitive to light.

**Storage/Shelf Life:** Store dehydrated media in the dark in a sealed container below 30°C. Prepared media should be stored under refrigeration (2-8°C). Media should be used within 60 days of preparation. Media should not be used if there are any signs of deterioration (discoloration) or contamination, or if the expiration date supplied by the manufacturer has passed.

**Use:** For the detection of coliform microorganisms in foods, dairy products, water, and wastewater, as well as in other materials of sanitary importance. Turbidity in the broth and gas in the Durham tube are positive indications of *Escherichia coli*.

## Brilliant Green Bile Broth with MUG

**Composition** per liter:

| | |
|---|---|
| Oxgall, dehydrated | 20.0g |
| Lactose | 10.0g |
| Pancreatic digest of gelatin | 10.0g |
| MUG (4-Methyl umbelliferyl-β-D-glucuronide) | 0.05g |
| Brilliant Green | 0.013g |

pH 7.2 ± 0.2 at 25°C

**Source:** This medium is available as a premixed powder from BD Diagnostic Systems.

**Preparation of Medium:** Add components to distilled/deionized water and bring volume to 1.0L. Mix thoroughly. Distribute into tubes containing inverted Durham tubes, in 10.0mL amounts for testing

1.0mL or less of sample. Autoclave for 12 min (not longer than 15 min) at 15 psi pressure–121°C. After sterilization, cool the broth rapidly.

**Storage/Shelf Life:** Store dehydrated media in the dark in a sealed container below 30°C. Prepared media should be stored under refrigeration (2-8°C). Media should be used within 60 days of preparation. Media should not be used if there are any signs of deterioration (discoloration) or contamination, or if the expiration date supplied by the manufacturer has passed.

**Use:** For the detection of coliform microorganisms in foods, dairy products, water, and wastewater, as well as in other materials of sanitary importance. The presence of *Escherichia coli* and other coliforms is determined by the presence of fluorescence in the tube.

### Brilliant Green 2%-Bile Broth, Fluorocult®
### (Fluorocult Brilliant Green 2%-Bile Broth)
### (BRILA)

**Composition** per liter:

| | |
|---|---|
| Ox bile, dried | 20.0g |
| Peptone | 10.0g |
| Lactose | 10.0g |
| L-Tryptophan | 1.0g |
| 4-Methylumbelliferyl-β-D-glucuronide | 0.1g |
| Brilliant Green | 0.0133g |

pH 7.2 ± 0.2 at 25°C

**Source:** This medium is available from Merck.

**Preparation of Medium:** Add components to distilled/deionized water and bring volume to 1.0L. Mix thoroughly. Gently heat and bring to boiling. Cool. Distribute into test tubes containing inverted Durham tubes. Autoclave for 15 min at 15 psi pressure–121°C. Do not autoclave longer. The prepared broth is clear and green.

**Storage/Shelf Life:** Store dehydrated media in the dark in a sealed container below 30°C. Prepared media should be stored under refrigeration (2-8°C). Media should be used within 60 days of preparation. Media should not be used if there are any signs of deterioration (discoloration) or contamination, or if the expiration date supplied by the manufacturer has passed.

**Use:** For the cultivation of *Escherichia coli*. Bile and Brilliant Green almost completely inhibit the growth of undesired microbial flora, in particular Gram-positive microorganisms. *E. coli* shows a positive fluorescence under UV light (366 nm). A positive indole reaction and, if necessary, gas formation due to fermenting lactose, confirm the findings.

### Brilliant Green Broth
### (m-Brilliant Green Broth)

**Composition** per liter:

| | |
|---|---|
| Proteose peptone No. 3 | 20.0g |
| Lactose | 20.0g |
| Sucrose | 20.0g |
| NaCl | 10.0g |
| Yeast extract | 6.0g |
| Phenol Red | 0.16g |
| Brilliant Green | 0.025g |

pH 6.9 ± 0.2 at 25°C

**Source:** This medium is available as a premixed powder from BD Diagnostic Systems.

**Preparation of Medium:** Add components to distilled/deionized water and bring volume to 1.0L. Mix thoroughly. Gently heat with fre-

quent mixing. Boil for 1 min. Cool to 25°C. Add 2.0mL to each sterile absorbent filter used.

**Storage/Shelf Life:** Store dehydrated media in the dark in a sealed container below 30°C. Prepared media should be stored under refrigeration (2-8°C). Media should be used within 60 days of preparation. Media should not be used if there are any signs of deterioration (discoloration) or contamination, or if the expiration date supplied by the manufacturer has passed.

**Use:** For the selective isolation and differentiation of *Salmonella* from polluted water by the membrane filter method.

### Brilliant Green HiVeg Agar

**Composition** per liter:

| | |
|---|---|
| Agar | 10.15g |
| Plant peptone | 8.25g |
| Lactose | 1.9g |
| Basic Fuchsin | 776.0mg |
| Erioglaucine | 649.0mg |
| FeCl₃ | 295.0mg |
| Na₂SO₃ | 205.0mg |
| KH₂PO₄ | 15.3mg |
| Synthetic detergent No. II | 2.95mg |
| Brilliant Green | 29.5μg |

pH 6.9 ± 0.2 at 25°C

**Source:** This medium is available as a premixed powder from Hi-Media.

**Preparation of Medium:** Add components to distilled/deionized water and bring volume to 1.0L. Mix thoroughly. Gently heat and bring to boiling. Distribute into tubes or flasks. Autoclave for 15 min at 15 psi pressure–121°C. Pour into sterile Petri dishes.

**Storage/Shelf Life:** Store dehydrated media in the dark in a sealed container below 30°C. Prepared media should be stored under refrigeration (2-8°C). Media should be used within 60 days of preparation. Media should not be used if there are any signs of deterioration (shrinking, cracking, or discoloration) or contamination, or if the expiration date supplied by the manufacturer has passed.

**Use:** For the selective isolation of *Salmonella* other than *Salmonella typhi* from feces and other specimens, and food and dairy products. *Salmonella* other than *Salmonella typhi* appear as red/pink/white colonies surrounded by a zone of red in the agar, indicating nonlactose/sucrose fermentation. *Proteus* or *Pseudomonas* species may appear as small red colonies. Lactose- or sucrose-fermenting bacteria appear as yellow-green colonies surrounded by a zone of yellow-green in the agar.

### Brilliant Green HiVeg Agar Base
### with Sulfa Supplement

**Composition** per liter:

| | |
|---|---|
| Agar | 20.0g |
| Plant peptone No. 3 | 10.0g |
| Lactose | 10.0g |
| Sucrose | 10.0g |
| NaCl | 5.0g |
| Yeast extract | 3.0g |
| Phenol Red | 0.08g |
| Brilliant Green | 125.0mg |
| Sulfa supplement solution | 10.0mL |

pH 6.9 ± 0.2 at 25°C

**Source:** This medium, without sulfa supplement, is available as a premixed powder from HiMedia.

**Sulfa Supplement Solution:**
**Composition** per 10.0mL:
Sodium sulfacetamide........................................................1.0g
Sodium mandelate............................................................0.25g

**Preparation of Sulfa Supplement Solution:** Add components to distilled/deionized water and bring volume to 10.0mL. Mix thoroughly. Filter sterilize.

**Preparation of Medium:** Add components, except sulfa supplement solution, to distilled/deionized water and bring volume to 990.0mL. Mix thoroughly. Autoclave for 15 min at 15 psi pressure–121°C. Aseptically add 10.0mL of sterile sulfa supplement solution. Mix thoroughly. Aseptically distribute into sterile tubes or flasks.

**Storage/Shelf Life:** Store dehydrated media in the dark in a sealed container below 30°C. Prepared media should be stored under refrigeration (2-8°C). Media should be used within 60 days of preparation. Media should not be used if there are any signs of deterioration (shrinking, cracking, or discoloration) or contamination, or if the expiration date supplied by the manufacturer has passed.

**Use:** For the selective isolation of *Salmonella* other than *Salmonella typhi* from feces and other specimens.

## Brilliant Green HiVeg Agar Base
## Modified with Sulfa Supplement
**Composition** per liter:
Agar ...............................................................................12.0g
Plant peptone No. 3.........................................................10.0g
Lactose ...........................................................................10.0g
Sucrose ...........................................................................10.0g
NaCl..................................................................................5.0g
Yeast extract.....................................................................3.0g
Phenol Red.......................................................................0.08g
Brilliant Green ..............................................................125.0mg
Sulfa supplement solution............................................. 10.0mL
pH 6.9 ± 0.2 at 25°C

**Source:** This medium, without sulfa supplement, is available as a premixed powder from HiMedia.

**Sulfa Supplement Solution:**
**Composition** per 10.0mL:
Sodium sulfacetamide........................................................1.0g
Sodium mandelate............................................................0.25g

**Preparation of Sulfa Supplement Solution:** Add components to distilled/deionized water and bring volume to 10.0mL. Mix thoroughly. Filter sterilize.

**Preparation of Medium:** Add components, except sulfa supplement solution, to distilled/deionized water and bring volume to 990.0mL. Mix thoroughly. Autoclave for 15 min at 15 psi pressure–121°C. Aseptically add 10.0mL of sterile sulfa supplement solution. Mix thoroughly. Pour into sterile Petri dishes.

**Storage/Shelf Life:** Store dehydrated media in the dark in a sealed container below 30°C. Prepared media should be stored under refrigeration (2-8°C). Media should be used within 60 days of preparation. Media should not be used if there are any signs of deterioration (shrinking, cracking, or discoloration) or contamination, or if the expiration date supplied by the manufacturer has passed.

**Use:** For the selective isolation of *Salmonella* other than *Salmonella typhi* from feces and other specimens.

## Brilliant Green HiVeg Broth 2%
**Composition** per liter:
Plant peptone ..................................................................25.0g
Lactose............................................................................10.0g
Synthetic detergent No. II..................................................5.0g
Brilliant Green ...............................................................13.3mg
pH 7.2 ± 0.2 at 25°C

**Source:** This medium is available as a premixed powder from HiMedia.

**Preparation of Medium:** Add components to distilled/deionized water and bring volume to 1.0L. Mix thoroughly. Gently heat with frequent mixing. Autoclave for 15 min at 15 psi pressure–121°C.

**Storage/Shelf Life:** Store dehydrated media in the dark in a sealed container below 30°C. Prepared media should be stored under refrigeration (2-8°C). Media should be used within 60 days of preparation. Media should not be used if there are any signs of deterioration (discoloration) or contamination, or if the expiration date supplied by the manufacturer has passed.

**Use:** For the selective isolation and differentiation of *Salmonella* from polluted water by the membrane filter method.

## Brilliant Green Lactose Bile Broth
## (BAM M25)
**Composition** per liter:
Oxgall, dehydrated............................................................20.0g
Lactose............................................................................10.0g
Pancreatic digest of gelatin..............................................10.0g
Brilliant Green ...............................................................0.0133g
pH 7.2 ± 0.1 at 25°C

**Source:** This medium is available as a premixed powder from BD Diagnostic Systems and Thermo Scientific.

**Preparation of Medium:** Add lactose and pancreatic digest of gelatin to distilled/deionized water and bring volume to 500.0mL. Mix thoroughly. Add 20.0g oxgall dissolved in 200.0mL distilled/deionized water. The pH of this solution should be 7.0–7.5. Mix thoroughly. Bring volume to 975.0mL with distilled/deionized water. Adjust pH to 7.4. Add 13.3mL of 0.1% aqueous Brilliant Green in distilled/deionized water. Adjust volume to 1.0L with distilled/deionized water. Distribute into tubes containing inverted Durham tubes in 10.0mL amounts for testing 1.0mL or less of sample. Make sure that the fluid level covers the inverted vials. Autoclave for 15 min at 15 psi pressure–121°C. After sterilization, cool the broth rapidly. Medium is sensitive to light.The final pH should be 7.2 ± 0.1 at 25°C.

**Storage/Shelf Life:** Store dehydrated media in the dark in a sealed container below 30°C. Prepared media should be stored under refrigeration (2-8°C). Media should be used within 60 days of preparation. Media should not be used if there are any signs of deterioration (discoloration) or contamination, or if the expiration date supplied by the manufacturer has passed.

**Use:** For the detection of coliform microorganisms in foods, dairy products, water, and wastewater as well as in other materials of sanitary importance. Turbidity in the broth and gas in the Durham tube are positive indications of *Escherichia coli*.

## BROLACIN MUG Agar
## (Bromothymol Blue Lactose Cystine MUG Agar)
## (C.L.E.D. MUG Agar)

**Composition** per liter:

Agar ................................................................12.0g
Lactose ............................................................10.0g
Universal peptone ...............................................4.0g
Casein peptone ...................................................4.0g
Meat extract ......................................................3.0g
L-cystine ........................................................0.128g
4-Methylumbelliferyl-β-D-glucuronide .................0.1g
Bromthymol Blue .............................................0.02g

pH 7.3 ± 0.2 at 37°C

**Source:** This medium is available from Fluka, Sigma-Aldrich.

**Preparation of Medium:** Add components to distilled/deionized water and bring volume to 1.0L. Mix thoroughly. Gently heat while stirring and bring to boiling. Autoclave for 15 min at 15 psi pressure–121°C. Cool to 50°C. Pour into sterile Petri dishes.

**Storage/Shelf Life:** Store dehydrated media in the dark in a sealed container below 30°C. Prepared media should be stored under refrigeration (2-8°C). Media should not be used if there are any signs of deterioration (shrinking, cracking, or discoloration) or contamination, or if the expiration date supplied by the manufacturer has passed.

**Use:** For the enumeration, isolation, and identification of microorganisms in urine. Growth of all urinary microorganisms is favored. Lactose catabolism produces a color change of Bromothymol Blue to yellow. Alkalization gives a color change to deep-blue. β-D-Glucoronidase, which is produced by *E. coli*, cleaves 4-methylumbelliferyl-β-D-glucuronide to 4-methylumbelliferone and glucuronide. The fluorogen 4-methylumbelliferone can be detected under a long wavelength UV lamp, permitting differentiation of *E. coli* colonies.

## Bromcresol Purple Broth

**Composition** per liter:

Peptone.............................................................10.0g
NaCl..................................................................5.0g
Beef extract .......................................................3.0g
Bromcresol Purple ............................................0.04g
Carbohydrate solution......................................10.0mL

pH 7.0 ± 0.2 at 25°C

**Carbohydrate Solution:**
**Composition** per 10.0mL:
Carbohydrate.......................................................5.0g

**Preparation of Carbohydrate Solution:** Add carbohydrate to distilled/deionized water and bring volume to 10.0mL. Mix thoroughly. Filter sterilize.

**Preparation of Medium:** Add components to distilled/deionized water and bring volume to 1.0L. Mix thoroughly. Gently heat and bring to boiling. Distribute into test tubes that contain an inverted Durham tube. Autoclave for 10 min at 15 psi pressure–121°C.

**Storage/Shelf Life:** Store dehydrated media in the dark in a sealed container below 30°C. Prepared media should be stored under refrigeration (2-8°C). Media should be used within 60 days of preparation. Media should not be used if there are any signs of deterioration (discoloration) or contamination, or if the expiration date supplied by the manufacturer has passed.

**Use:** For the differentiation of a variety of microorganisms based on their fermentation of specific carbohydrates. Bacteria that ferment the specific carbohydrate turn the medium yellow. When bacteria produce gas, the gas is trapped in the Durham tube.

## Bromo Cresol Purple Azide HiVeg Broth

**Composition** per liter:

Plant hydrolysate .............................................10.0g
Yeast extract....................................................10.0g
NaCl..................................................................5.0g
D-Glucose..........................................................5.0g
K$_2$HPO$_4$ .............................................................2.7g
KH$_2$PO$_4$ .............................................................2.7g
NaN$_3$ .................................................................0.5g
Bromo Cresol Purple .....................................32.0mg

pH 7.0 ± 0.2 at 25°C

**Source:** This medium is available as a premixed powder from HiMedia.

**Caution:** Sodium azide is toxic. Azides also react with metals and disposal must be highly diluted.

**Preparation of Medium:** Add components to distilled/deionized water to 1.0L. Mix thoroughly. Gently heat to boiling. Distribute into tubes or flasks. Autoclave for 15 min at 10 psi pressure–115°C.

**Storage/Shelf Life:** Store dehydrated media in the dark in a sealed container below 30°C. Prepared media should be stored under refrigeration (2-8°C). Media should be used within 60 days of preparation. Media should not be used if there are any signs of deterioration (discoloration) or contamination, or if the expiration date supplied by the manufacturer has passed.

**Use:** For use in the confirmation test for the presence of fecal streptococci in water and wastewater.

## Bromo Cresol Purple HiVeg Broth Base

**Composition** per liter:

Plant peptone ...................................................10.0g
Carbohydrate (test compound) .........................10.0g
NaCl..................................................................5.0g
Plant extract ......................................................3.0g
Bromo Cresol Purple ........................................0.04g

pH 7.0 ± 0.2 at 25°C

**Source:** This medium without carbohydrate is available as a premixed powder from HiMedia.

**Preparation of Medium:** Add components to distilled/deionized water and bring volume to 1.0L. Mix thoroughly. Distribute into tubes in 12–15mL volumes. Autoclave for 10 min at 15 psi pressure–121°C. Carbohydrate solutions are added to test bacterial fermentative abilities.

**Storage/Shelf Life:** Store dehydrated media in the dark in a sealed container below 30°C. Prepared media should be stored under refrigeration (2-8°C). Media should be used within 60 days of preparation. Media should not be used if there are any signs of deterioration (discoloration) or contamination, or if the expiration date supplied by the manufacturer has passed.

**Use:** For the cultivation and differentiation of bacteria based on their ability to ferment various carbohydrates. Bacteria that ferment the carbohydrate turn the medium yellow.

## Bromthymol Blue Agar

**Composition** per liter:

| | |
|---|---|
| Agar | 11.0g |
| Peptone | 10.0g |
| NaCl | 5.0g |
| Yeast extract | 5.0g |
| Lactose (33% solution) | 27.0mL |
| Bromthymol Blue (1% solution) | 10.0mL |
| Sodium thiosulfate (50% solution) | 2.0mL |
| Glucose (33% solution) | 1.2mL |
| Maranil solution (5% solution) | 1.0mL |

pH 7.7–7.8 at 25°C

**Preparation of Medium:** Add agar, peptone, NaCl, and yeast extract to distilled/deionized water and bring volume to 1.0L. Mix thoroughly. Adjust pH to 8.0. Autoclave for 20 min at 15 psi pressure–121°C. Cool to 45°–50°C. Filter sterilize the lactose solution, Bromthymol Blue solution, sodium thiosulfate solution, glucose solution, and maranil solution separately. To the cooled, sterile agar solution aseptically add 27.0mL of sterile lactose solution, 10.0mL of sterile Bromthymol Blue solution, 2.0mL of sterile sodium thiosulfate solution, 1.2mL of sterile glucose solution, and 1.0mL of sterile maranil solution. Mix thoroughly. Adjust pH to 7.7–7.8. Pour into sterile Petri dishes or distribute into sterile tubes.

**Storage/Shelf Life:** Store dehydrated media in the dark in a sealed container below 30°C. Prepared media should be stored under refrigeration (2-8°C). Media should be used within 60 days of preparation. Media should not be used if there are any signs of deterioration (shrinking, cracking, or discoloration) or contamination, or if the expiration date supplied by the manufacturer has passed.

**Use:** For the selective isolation and cultivation of members of the Enterobacteriaceae.

## *Brucella* Agar

**Composition** per liter:

| | |
|---|---|
| Agar | 15.0g |
| Pancreatic digest of casein | 10.0g |
| Peptic digest of animal tissue | 10.0g |
| NaCl | 5.0g |
| Yeast extract | 2.0g |
| Glucose | 1.0g |
| NaHSO₃ | 0.1g |
| Horse blood, defibrinated | 100.0mL |

pH 7.0 ± 0.2 at 25°C

**Source:** This medium is available as a premixed powder from BD Diagnostic Systems and Thermo Scientific.

**Preparation of Medium:** Add components to distilled/deionized water and bring volume to 900.0mL. Mix thoroughly. Heat gently with frequent mixing. Boil for 1 min. Autoclave for 15 min at 15 psi pressure–121°C. Cool to 45°–50°C. Add 100.0mL of sterile defibrinated horse blood. Mix gently and pour into sterile Petri dishes.

**Storage/Shelf Life:** Store dehydrated media in the dark in a sealed container below 30°C. Prepared media should be stored under refrigeration (2-8°C). Media should be used within 60 days of preparation. Media should not be used if there are any signs of deterioration (shrinking, cracking, or discoloration) or contamination, or if the expiration date supplied by the manufacturer has passed.

**Use:** For the isolation and cultivation of *Brucella* species and a variety of other nonfastidious and fastidious microorganisms from various clinical speciments and nonclinical specimens of public health importance.

## *Brucella* Agar Base *Campylobacter* Medium

**Composition** per 1100.0mL:

| | |
|---|---|
| Cycloheximide (actidione) | 0.05g |
| Sodium cephazolin | 0.015g |
| Novobiocin | 5.0mg |
| Bacitracin | 25,000U |
| Colistin sulfate | 10,000U |
| *Brucella* agar base | 1.0L |
| Horse blood, defibrinated | 100.0mL |

### *Brucella* Agar Base
**Composition** per liter:

| | |
|---|---|
| Agar | 15.0g |
| Pancreatic digest of casein | 10.0g |
| Peptic digest of animal tissue | 10.0g |
| NaCl | 5.0g |
| Yeast extract | 2.0g |
| Glucose | 1.0g |
| NaHSO₃ | 0.1g |

**Preparation of *Brucella* Agar Base:** Add components to distilled/deionized water and bring volume to 1.0L. Mix thoroughly.

### Optional Supplement:
**Composition** per 10.0mL:

| | |
|---|---|
| Sodium pyruvate | 0.25g |
| NaHSO₃ | 0.25g |
| FeSO₄·7H₂O | 0.25g |

**Preparation of Optional Supplement:** Add components to distilled/deionized water and bring volume to 10.0mL. Filter sterilize.

**Caution:** Cycloheximide is toxic. Avoid skin contact or aerosol formation and inhalation.

**Preparation of Medium:** Add components to 1.0L of prepared *Brucella* agar base. Mix thoroughly. Autoclave for 15 min at 15 psi pressure–121°C. Cool to 45°–50°C. Add 100.0mL of sterile, defibrinated horse blood. Addition of 10.0mL of optional supplement will improve growth. Mix thoroughly. Pour into sterile Petri dishes.

**Storage/Shelf Life:** Store dehydrated media in the dark in a sealed container below 30°C. Prepared media should be stored under refrigeration (2-8°C). Media should be used within 60 days of preparation. Media should not be used if there are any signs of deterioration (shrinking, cracking, or discoloration) or contamination, or if the expiration date supplied by the manufacturer has passed.

**Use:** For the selective isolation and cultivation of *Campylobacter jejuni* from fecal specimens or rectal swabs.

## *Brucella* Agar

**Composition** per liter:

| | |
|---|---|
| Agar | 15.0g |
| Pancreatic digest of casein | 10.0g |
| Peptic digest of animal tissue | 10.0g |
| NaCl | 5.0g |
| Yeast extract | 2.0g |
| Glucose | 1.0g |
| NaHSO₃ | 0.1g |
| Horse blood, defibrinated | 100.0mL |

pH 7.0 ± 0.2 at 25°C

**Source:** This medium is available as a premixed powder from BD Diagnostic Systems and Thermo Scientific.

**Preparation of Medium:** Add components to distilled/deionized water and bring volume to 900.0mL. Mix thoroughly. Heat gently with frequent mixing. Boil for 1 min. Autoclave for 15 min at 15 psi pressure–121°C. Cool to 45°–50°C. Add 100.0mL of sterile defibrinated horse blood. Mix gently and pour into sterile Petri dishes.

**Storage/Shelf Life:** Store dehydrated media in the dark in a sealed container below 30°C. Prepared media should be stored under refrigeration (2-8°C). Media should be used within 60 days of preparation. Media should not be used if there are any signs of deterioration (shrinking, cracking, or discoloration) or contamination, or if the expiration date supplied by the manufacturer has passed.

**Use:** For the cultivation and maintenance of *Brucella* species. For the isolation and cultivation of nonfastidious and fastidious microorganisms from a variety of clinical speciments and nonclinical specimens of public health importance.

## *Brucella* Albimi Broth with Sheep Blood
**Composition** per liter:
| | |
|---|---|
| Pancreatic digest of casein | 10.0g |
| Peptic digest of animal tissue | 10.0g |
| NaCl | 5.0g |
| Yeast extract | 2.0g |
| Glucose | 1.0g |
| NaHSO₃ | 0.1g |
| Sheep blood, defibrinated | 100.0mL |

pH 7.0 ± 0.2 at 25°C

**Preparation of Medium:** Add components, except sheep blood, to distilled/deionized water and bring volume to 900.0mL. Mix thoroughly. Heat gently with frequent mixing. Boil for 1 min. Autoclave for 15 min at 15 psi pressure–121°C. Cool to 45°–50°C. Aseptically add 100.0mL of sterile defibrinated sheep blood. Mix thoroughly. Aseptically distribute into sterile tubes or flasks.

**Storage/Shelf Life:** Store dehydrated media in the dark in a sealed container below 30°C. Prepared media should be stored under refrigeration (2-8°C). Media should be used within 60 days of preparation. Media should not be used if there are any signs of deterioration (shrinking, cracking, or discoloration) or contamination, or if the expiration date supplied by the manufacturer has passed.

**Use:** For the cultivation and maintenance of *Helicobacter nemestrinae* and *Helicobacter pylori*.

## *Brucella* Anaerobic Blood Agar
**Composition** per liter:
| | |
|---|---|
| Vitamin K₁ | 0.01g |
| Anaerobic agar base | 1000.0mL |
| Sheep blood, sterile, defibrinated | 50.0mL |

**Anaerobic Agar Base**
**Composition** per liter:
| | |
|---|---|
| Pancreatic digest of casein | 17.5g |
| Agar | 15.0g |
| Glucose | 10.0g |
| Papaic digest of soybean meal | 2.5g |
| NaCl | 2.5g |
| Sodium thioglycolate | 2.0g |
| Sodium formaldehyde sulfoxylate | 1.0g |
| L-Cystine·HCl·H₂O | 0.4g |
| Methylene Blue | 0.002g |

pH 7.0 ± 0.2 at 25°C

**Preparation of Anaerobic Agar Base:** Add components to distilled/deionized water and bring volume to 1.0L. Mix thoroughly. Autoclave for 15 min at 15 psi pressure–121°C. Cool to 45°–50°C.

**Preparation of Medium:** To 950.0mL of cooled, sterile anaerobic agar base, aseptically add 10.0mg of vitamin K₁ and 50.0mL of sterile, defibrinated sheep blood.

**Storage/Shelf Life:** Store dehydrated media in the dark in a sealed container below 30°C. Prepared media should be stored under refrigeration (2-8°C). Media should be used within 60 days of preparation. Media should not be used if there are any signs of deterioration (shrinking, cracking, or discoloration) or contamination, or if the expiration date supplied by the manufacturer has passed.

**Use:** For the isolation of anaerobes.

## *Brucella* Blood Agar with Hemin and Vitamin K₁
**Composition** per liter:
| | |
|---|---|
| Agar | 15.0g |
| Pancreatic digest of casein | 10.0g |
| Peptic digest of animal tissue | 10.0g |
| NaCl | 5.0g |
| Yeast extract | 2.0g |
| Glucose | 1.0g |
| NaHSO₃ | 0.1g |
| Vitamin K₁ | 1.0mL |
| Hemin | 1.0mL |
| Sheep blood, defibrinated | 50.0mL |

**Source:** This medium is available as a prepared medium from BD Diagnostic Systems.

**Vitamin K₁ Solution:**
**Composition** per 100.0mL:
| | |
|---|---|
| Vitamin K₁ | 1.0g |
| Ethanol, absolute | 99.0mL |

**Preparation of Vitamin K₁ Solution:** Add vitamin K₁ to 99.0mL of absolute ethanol. Mix thoroughly. Filter sterilize.

**Hemin Solution:**
**Composition** per 100.0mL:
| | |
|---|---|
| Hemin | 1.0g |
| NaOH (1*N* solution) | 20.0mL |

**Preparation of Hemin Solution:** Add hemin to 20.0mL of 1*N* NaOH solution. Mix thoroughly. Bring volume to 100.0mL with distilled/deionized water.

**Preparation of Medium:** Add components, except vitamin K₁ solution and sheep blood, to distilled/deionized water and bring volume to 949.0mL. Mix thoroughly. Gently heat and bring to boiling. Autoclave for 15 min at 15 psi pressure–121°C. Cool to 45°–50°C. Aseptically add 1.0mL of sterile vitamin K₁ solution and 50.0mL of sterile defibrinated sheep blood. Mix gently and pour into sterile Petri dishes.

**Storage/Shelf Life:** Store dehydrated media in the dark in a sealed container below 30°C. Prepared media should be stored under refrigeration (2-8°C). Media should be used within 60 days of preparation. Media should not be used if there are any signs of deterioration (shrinking, cracking, or discoloration) or contamination, or if the expiration date supplied by the manufacturer has passed.

**Use:** For the isolation and cultivation of anaerobic microorganisms from clinical speciments and nonclinical specimens of public health importance. After growth on agar plates, colonies should be examined under a dissecting microscope under long-wave UV light. Members of the pigmented *Bacteroides* group appear as red/orange fluorescent colonies.

## *Brucella* Blood Culture Broth

**Composition** per liter:

| | |
|---|---|
| Sucrose | 100.0g |
| Hemin | 0.5g |
| Sodium polyanetholsulfonate (SPS) | 0.25g |
| *Brucella* broth base | 1000.0mL |
| Vitamin $K_1$ solution | 1.0mL |

pH $7.0 \pm 0.2$ at 25°C

### *Brucella* Broth Base:
**Composition** per liter:

| | |
|---|---|
| Pancreatic digest of casein | 10.0g |
| Peptic digest of animal tissue | 10.0g |
| NaCl | 5.0g |
| Yeast extract | 2.0g |
| Glucose | 1.0g |
| $NaHSO_3$ | 0.1g |

**Preparation of *Brucella* Broth Base:** Add components to distilled/deionized water and bring volume to 1.0L. Mix thoroughly.

### Vitamin $K_1$ Solution:
**Composition** per 100.0mL:

| | |
|---|---|
| Vitamin $K_1$ | 1.09g |
| Ethanol, absolute | 99.0mL |

**Preparation of Vitamin $K_1$ Solution:** Add vitamin $K_1$ to 99.0mL of absolute ethanol. Store in the dark at 4°C.

**Preparation of Medium:** Add components, except vitamin $K_1$ solution, to prepared *Brucella* broth base. Autoclave for 15 min at 15 psi pressure–121°C. Cool to 45°–50°C. Aseptically add 1.0mL of vitamin $K_1$ solution. Distribute into sterile tubes or flasks.

**Use:** For the isolation and cultivation of microorganisms from blood. Especially useful for the cultivation of anaerobes.

## *Brucella* Broth Base *Campylobacter* Medium

**Composition** per liter:

| | |
|---|---|
| Cycloheximide (Actidione®) | 50.0mg |
| Sodium cephazolin | 15.0mg |
| Novobiocin | 5.0mg |
| Bacitracin | 25,000U |
| Colistin sulfate | 10,000U |
| *Brucella* broth base | 900.0mL |
| Horse blood, defibrinated | 100.0mL |

pH $7.0 \pm 0.2$ at 25°C

**Caution:** Cycloheximide is toxic. Avoid skin contact or aerosol formation and inhalation.

### *Brucella* Broth Base:
**Composition** per liter:

| | |
|---|---|
| Pancreatic digest of casein | 10.0g |
| Peptic digest of animal tissue | 10.0g |
| NaCl | 5.0g |
| Yeast extract | 2.0g |
| Glucose | 1.0g |
| $NaHSO_3$ | 0.1g |

**Preparation of *Brucella* Broth Base:** Add components to distilled/deionized water and bring volume to 1.0L. Mix thoroughly.

### Optional Supplement:
**Composition** per 10.0mL:

| | |
|---|---|
| Sodium pyruvate | 0.25g |
| $NaHSO_3$ | 0.25g |
| $FeSO_4·7H_2O$ | 0.25g |

**Preparation of Optional Supplement:** Add components to distilled/deionized water and bring volume to 10.0mL. Filter sterilize.

**Preparation of Medium:** Add components, except horse blood, to 900.0mL of prepared *Brucella* broth base. Mix thoroughly. Autoclave for 15 min at 15 psi pressure–121°C. Cool to 45°–50°C. Aseptically add 100.0mL of sterile, defibrinated horse blood. Addition of 10.0mL of optional supplement will improve growth. Mix thoroughly. Pour into sterile Petri dishes.

**Storage/Shelf Life:** Store dehydrated media in the dark in a sealed container below 30°C. Prepared media should be stored under refrigeration (2-8°C). Media should be used within 60 days of preparation. Media should not be used if there are any signs of deterioration (discoloration) or contamination, or if the expiration date supplied by the manufacturer has passed.

**Use:** For the selective isolation and cultivation of *Campylobacter jejuni* from fecal specimens or rectal swabs. Addition of the optional supplement improves growth.

## *Brucella* Broth
## (*Brucella* Albimi Broth)

**Composition** per liter:

| | |
|---|---|
| Pancreatic digest of casein | 10.0g |
| Peptic digest of animal tissue | 10.0g |
| NaCl | 5.0g |
| Yeast extract | 2.0g |
| Glucose | 1.0g |
| $NaHSO_3$ | 0.1g |
| Horse blood, defibrinated | 50.0mL |

pH $7.0 \pm 0.2$ at 25°C

**Source:** This medium is available as a premixed powder from BD Diagnostic Systems.

**Preparation of Medium:** Add components, except horse blood, to distilled/deionized water and bring volume to 950.0mL. Mix thoroughly. Heat gently with frequent mixing. Boil for 1 min. Autoclave for 15 min at 15 psi pressure–121°C. Cool to 45°–50°C. Aseptically add 50.0mL of sterile horse blood. Mix thoroughly. Aseptically distribute into sterile tubes or flasks.

**Storage/Shelf Life:** Store dehydrated media in the dark in a sealed container below 30°C. Prepared media should be stored under refrigeration (2-8°C). Media should be used within 60 days of preparation. Media should not be used if there are any signs of deterioration (discoloration) or contamination, or if the expiration date supplied by the manufacturer has passed.

**Use:** For the cultivation and maintenance of *Campylobacter coli*, *Campylobacter fecalis*, and *Brucella* species. Also used for the isolation and cultivation of a wide variety of fastidious and nonfastidious microorganisms.

## *Brucella* HiVeg Agar Base
## with Blood and Selective Supplement

**Composition** per liter:

| | |
|---|---|
| Agar | 15.0g |
| Plant hydrolysate | 10.0g |

Plant peptone.................................................................10.0g
NaCl.................................................................................5.0g
Yeast extract..................................................................2.0g
Glucose..........................................................................1.0g
NaHSO$_3$.........................................................................0.1g
Horse blood, defibrinated .......................................100.0mL
Selective supplement ................................................10.0mL
pH 7.0 ± 0.2 at 25°C

**Source:** This medium, without horse blood or selective supplement, is available as a premixed powder from HiMedia.

**Selective Supplement:**
**Composition** per 10.0mL:
Cycloheximide...............................................................0.1g
Vancomycin.................................................................20.0mg
Nalidixic acid...............................................................5.0mg
Nystatin.............................................................1,000,000 U
Bacitracin...........................................................250,000 U
Polymyxin B sulfate.............................................50,000 U

**Preparation of Selective Supplement:** Add components to distilled/deionized water and bring volume to 10.0mL. Mix thoroughly. Filter sterilize.

**Caution:** Cycloheximide is toxic. Avoid skin contact or aerosol formation and inhalation.

**Preparation of Medium:** Add components, except blood and selective supplement, to distilled/deionized water and bring volume to 900.0mL. Mix thoroughly. Heat gently with frequent mixing. Boil for 1 min. Autoclave for 15 min at 15 psi pressure–121°C. Cool to 45°–50°C. Add 100.0mL of sterile defibrinated horse blood and 10.0mL sterile selective supplement. Mix thoroughly. Pour into sterile Petri dishes.

**Storage/Shelf Life:** Store dehydrated media in the dark in a sealed container below 30°C. Prepared media should be stored under refrigeration (2-8°C). Media should be used within 60 days of preparation. Media should not be used if there are any signs of deterioration (discoloration) or contamination, or if the expiration date supplied by the manufacturer has passed.

**Use:** For the cultivation and maintenance of *Brucella* species. For the isolation and cultivation of nonfastidious and fastidious microorganisms from a variety of clinical speciments and nonclinical specimens of public health importance.

### *Brucella* HiVeg Agar Base, Modified with Blood and Selective Supplement
**Composition** per liter:
Agar ..............................................................................15.0g
Plant hydrolysate...........................................................15.0g
Plant peptone..................................................................5.0g
NaCl.................................................................................5.0g
Yeast extract..................................................................2.0g
Glucose..........................................................................1.0g
Sodium citrate...............................................................1.0g
NaHSO$_3$.........................................................................0.1g
Horse blood, defibrinated .......................................100.0mL
Selective supplement ................................................10.0mL
pH 7.0 ± 0.2 at 25°C

**Source:** This medium, without horse blood or selective supplement, is available as a premixed powder from HiMedia.

**Selective Supplement:**
**Composition** per 10.0mL:
Cycloheximide...............................................................0.1g
Vancomycin.................................................................20.0mg
Nalidixic acid...............................................................5.0mg
Nystatin.............................................................1,000,000 U
Bacitracin...........................................................250,000 U
Polymyxin B sulfate.............................................50,000 U

**Preparation of Selective Supplement:** Add components to distilled/deionized water and bring volume to 10.0mL. Mix thoroughly. Filter sterilize.

**Caution:** Cycloheximide is toxic. Avoid skin contact or aerosol formation and inhalation.

**Preparation of Medium:** Add components, except blood and selective supplement, to distilled/deionized water and bring volume to 900.0mL. Mix thoroughly. Heat gently with frequent mixing. Boil for 1 min. Autoclave for 15 min at 15 psi pressure–121°C. Cool to 45°–50°C. Add 100.0mL of sterile defibrinated horse blood and 10.0mL sterile selective supplement. Pour into sterile Petri dishes.

**Storage/Shelf Life:** Store dehydrated media in the dark in a sealed container below 30°C. Prepared media should be stored under refrigeration (2-8°C). Media should be used within 60 days of preparation. Media should not be used if there are any signs of deterioration (shrinking, cracking, or discoloration) or contamination, or if the expiration date supplied by the manufacturer has passed.

**Use:** For the cultivation and maintenance of *Brucella* species. For the isolation and cultivation of nonfastidious and fastidious microorganisms from a variety of clinical speciments and nonclinical specimens of public health importance.

### *Brucella* Laked Sheep Blood Agar with Kanamycin and Vancomycin
**Composition** per liter:
Agar ..............................................................................15.0g
Pancreatic digest of casein............................................10.0g
Peptic digest of animal tissue .......................................10.0g
NaCl.................................................................................5.0g
Dextrose..........................................................................1.0g
Yeast extract..................................................................1.0g
NaCl.................................................................................5.0g
NaHSO$_3$ .......................................................................0.1g
Sheep blood, defibrinated laked .................................50.0mL
Antibiotic solution .....................................................10.0mL
Hemin solution..............................................................1.0mL
Vitamin K$_1$ solution....................................................1.0mL

**Source:** This medium is available as a premixed powder from BD Diagnostic Systems.

**Antibiotic Solution:**
**Composition** per 10.0mL:
Kanamycin.......................................................................0.1g
Vancomycin.................................................................7.5mg

**Preparation of Antibiotic Solution:** Add components to distilled/deionized water and bring volume to 10.0mL. Mix thoroughly. Filter sterilize.

**Vitamin K$_1$ Solution:**
**Composition** per 100.0mL:
Vitamin K$_1$ ...................................................................0.1g
Ethanol........................................................................99.0mL

**Preparation of Vitamin K₁ Solution:** Add vitamin $K_1$ to 99.0mL of absolute ethanol. Mix thoroughly.

**Hemin Solution:**
**Composition** per 100.0mL:
Hemin..............................................................................0.01g
NaOH (1*N* solution)......................................................20.0mL

**Preparation of Hemin Solution:** Add hemin to 20.0mL of 1*N* NaOH solution. Mix thoroughly. Bring volume to 100.0mL with distilled/deionized water.

**Preparation of Medium:** Add components—except sheep blood, antibiotic solution, and vitamin $K_1$ solution—to distilled/deionized water and bring volume to 939.0mL. Mix thoroughly. Gently heat and bring to boiling. Autoclave for 15 min at 15 psi pressure–121°C. Cool to 45°–50°C. Aseptically add 50.0mL of sterile sheep blood, 10.0mL of sterile antibiotic solution, and 1.0mL of sterile vitamin $K_1$ solution. Mix thoroughly. Pour into sterile Petri dishes or distribute into sterile tubes.

**Storage/Shelf Life:** Store dehydrated media in the dark in a sealed container below 30°C. Prepared media should be stored under refrigeration (2-8°C). Media should be used within 60 days of preparation. Media should not be used if there are any signs of deterioration (shrinking, cracking, or discoloration) or contamination, or if the expiration date supplied by the manufacturer has passed.

**Use:** For the selective isolation of fastidious and slow growing obligately anaerobic bacteria from clinical specimens. The laked blood improves pigmentation of the *Prevotella melaniogenica - P. asaccharolytica* group.

### *Brucella* Medium Base

**Composition** per liter:
Agar.........................................................................15.0g
Glucose...................................................................10.0g
Peptone...................................................................10.0g
Beef extract..............................................................5.0g
NaCl.........................................................................5.0g

pH 7.5 ± 0.2 at 25°C

**Preparation:** Add components to distilled/deionized water and bring volume to 1.0L. Mix thoroughly. Heat gently and bring to boiling. Distribute into tubes or flasks. Autoclave for 15 min at 15 psi pressure–121°C. Pour into sterile Petri dishes or leave in tubes.

**Storage/Shelf Life:** Store dehydrated media in the dark in a sealed container below 30°C. Prepared media should be stored under refrigeration (2-8°C). Media should be used within 60 days of preparation. Media should not be used if there are any signs of deterioration (shrinking, cracking, or discoloration) or contamination, or if the expiration date supplied by the manufacturer has passed.

**Use:** For the isolation of *Campylobacter* species.

### *Brucella* Selective Medium

**Composition** per liter:
Beef heart, infusion from.......................................500.0g
Agar.........................................................................15.0g
Tryptose..................................................................10.0g
NaCl.........................................................................5.0g
Glucose....................................................................2.5g
Gelatin......................................................................1.0g

Sheep blood..........................................................100.0mL
Antibiotic solution..................................................10.0mL

pH 7.4 ± 0.2 at 25°C

**Antibiotic Solution:**
**Composition** per 10.0mL:
Cycloheximide..........................................................1.0g
Bacitracin.........................................................250,000U
Circulin............................................................250,000U
Polymyxin B......................................................100,000U

**Preparation of Antibiotic Solution:** Add components to distilled/deionized water and bring volume to 10.0mL. Mix thoroughly. Filter sterilize.

**Caution:** Cycloheximide is toxic. Avoid skin contact or aerosol formation and inhalation.

**Preparation of Medium:** Add components, except sheep blood and antibiotic solution, to distilled/deionized water and bring volume to 890.0mL. Mix thoroughly. Gently heat and bring to boiling. Autoclave for 15 min at 15 psi pressure–121°C. Cool to 45°–50°C. Aseptically add 100.0mL of sterile sheep blood and 10.0mL of sterile antibiotic solution. Mix thoroughly. Pour into sterile Petri dishes or distribute into sterile tubes.

**Storage/Shelf Life:** Store dehydrated media in the dark in a sealed container below 30°C. Prepared media should be stored under refrigeration (2-8°C). Media should be used within 60 days of preparation. Media should not be used if there are any signs of deterioration (shrinking, cracking, or discoloration) or contamination, or if the expiration date supplied by the manufacturer has passed.

**Use:** For the selective isolation and cultivation of *Brucella* species.

### BSK Medium
### (Barbour-Stoenner-Kelly Medium)

**Composition** per 1260.0mL:
Bovine albumin fraction V ........................................50.0g
HEPES (*N*-[2-hydroxyethyl]piperazine-*N'*-2-
    ethanesulfonic acid) buffer......................................6.0g
Neopeptone..............................................................5.0g
Glucose....................................................................5.0g
NaHCO₃.....................................................................2.2g
Sodium pyruvate.......................................................0.8g
Sodium citrate..........................................................0.7g
*N*-Acetylglucosamine..............................................0.4g
Gelatin solution.....................................................200.0mL
CMRL 1066, without glutamine,
    without bicarbonate, 10X.....................................100.0mL
Rabbit serum.........................................................72.0mL

pH 7.6–7.65 at 25°C

**Gelatin Solution:**
**Composition** per 200.0mL:
Gelatin....................................................................14.0g

**Preparation of Gelatin Solution:** Add gelatin to distilled/deionized water and bring volume to 200.0mL. Heat gently to boiling. Mix thoroughly. Filter sterilize.

**CMRL 1066 Medium without Glutamine, without Bicarbonate, 10X:**
**Composition** per liter:
NaCl.........................................................................6.8g
D-Glucose.................................................................1.0g

| | |
|---|---|
| KCl | 0.4g |
| L-Cysteine·HCl·H$_2$O | 0.26g |
| CaCl$_2$, anhydrous | 0.2g |
| MgSO$_4$·7H$_2$O | 0.2g |
| NaH$_2$PO$_4$·H$_2$O | 0.14g |
| Sodium acetate·3H$_2$O | 0.083g |
| L-Glutamic acid | 0.075g |
| L-Arginine·HCl | 0.07g |
| L-Lysine·HCl | 0.07g |
| L-Leucine | 0.06g |
| Glycine | 0.05g |
| Ascorbic acid | 0.05g |
| L-Proline | 0.04g |
| L-Tyrosine | 0.04g |
| L-Aspartic acid | 0.03g |
| L-Threonine | 0.03g |
| L-Alanine | 0.025g |
| L-Phenylalanine | 0.025g |
| L-Serine | 0.025g |
| L-Valine | 0.025g |
| L-Cystine | 0.02g |
| L-Histidine·HCl·H$_2$O | 0.02g |
| L-Isoleucine | 0.02g |
| Phenol red | 0.02g |
| L-Methionine | 0.015g |
| Deoxyadenosine | 0.01g |
| Deoxycytidine | 0.01g |
| Deoxyguanosine | 0.01g |
| Glutathione, reduced | 0.01g |
| Thymidine | 0.01g |
| Hydroxy-L-proline | 0.01g |
| L-Tryptophan | 0.01g |
| Nicotinamide adenine dinucleotide | 7.0mg |
| Tween™ 80 | 5.0mg |
| Sodium glucoronate·H$_2$O | 4.2mg |
| Coenzyme A | 2.5mg |
| Cocarboxylase | 1.0mg |
| Flavin adenine dinucleotide | 1.0mg |
| Nicotinamide adenine dinucleotide phosphate | 1.0mg |
| Uridine triphosphate | 1.0mg |
| Choline chloride | 0.5mg |
| Cholesterol | 0.2mg |
| 5-Methyldeoxycytidine | 0.1mg |
| Inositol | 0.05mg |
| p-Aminobenzoic acid | 0.05mg |
| Niacin | 0.025mg |
| Niacinamide | 0.025mg |
| Pyridoxine | 0.025mg |
| Pyridoxal·HCl | 0.025mg |
| Biotin | 0.01mg |
| D-Calcium pantothenate | 0.01mg |
| Folic acid | 0.01mg |
| Riboflavin | 0.01mg |
| Thiamine·HCl | 0.01mg |

pH 7.2 ± 0.2 at 25°C

**Preparation of CMRL 1066 Medium without Glutamine, without Bicarbonate, 10X:** Add components to distilled/deionized water and bring volume to 1.0L. Mix thoroughly. Adjust pH to 7.2. Filter sterilize.

**Preparation of Medium:** Add components, except gelatin solution and rabbit serum, to 628.0mL of glass-distilled water. Mix thoroughly.

Adjust pH to 7.6–7.65. Add 200.0mL of 7% aqueous gelatin solution. Filter sterilize entire medium. Aseptically add 72.0mL of sterile rabbit serum.

**Storage/Shelf Life:** Store dehydrated media in the dark in a sealed container below 30°C. Prepared media should be stored under refrigeration (2-8°C). Media should be used within 60 days of preparation. Media should not be used if there are any signs of deterioration (discoloration) or contamination, or if the expiration date supplied by the manufacturer has passed.

**Use:** For the cultivation of a wide variety of microorganisms in a chemically defined medium. For the cultivation of *Borrelia* and *Spirochaeta* species.

## BSK Medium, Modified

**Composition** per 1264.0mL:

| | |
|---|---|
| Bovine serum albumin, fraction V | 50.0g |
| HEPES (N-[2-hydroxymethyl]piperazine-N′ [ethane sulfonate]) buffer | 6.0g |
| Neopeptone | 5.0g |
| Glucose | 5.0g |
| Yeastolate | 2.54g |
| NaHCO$_3$ | 2.2g |
| Sodium pyruvate | 0.8g |
| Sodium citrate | 0.7g |
| MgSO$_4$·7H$_2$O | 0.6g |
| N-Acetylglucosamine | 0.4g |
| CaCl$_2$·2H$_2$O | 0.07g |
| CMRL 1066, 10X without glutamine or NaHCO$_3$ | 100.0mL |
| Rabbit serum, heat inactivated | 64.0mL |

pH 7.5 ± 0.2 at 25°C

**CMRL 1066, 10X without Glutamine or NaHCO$_3$:**
**Composition** per liter:

| | |
|---|---|
| NaCl | 6.8g |
| D-Glucose | 1.0g |
| KCl | 0.4g |
| L-Cysteine·HCl·H$_2$O | 0.26g |
| CaCl$_2$, anhydrous | 0.2g |
| MgSO$_4$·7H$_2$O | 0.2g |
| NaH$_2$PO$_4$·H$_2$O | 0.14g |
| Sodium acetate·3H$_2$O | 0.083g |
| L-Glutamic acid | 0.075g |
| L-Arginine·HCl | 0.070g |
| L-Lysine·HCl | 0.070g |
| L-Leucine | 0.060g |
| Glycine | 0.050g |
| Ascorbic acid | 0.050g |
| L-Proline | 0.040g |
| L-Tyrosine | 0.040g |
| L-Aspartic acid | 0.030g |
| L-Threonine | 0.030g |
| L-Alanine | 0.025g |
| L-Phenylalanine | 0.025g |
| L-Serine | 0.025g |
| L-Valine | 0.025g |
| L-Cystine | 0.020g |
| L-Histidine·HCl·H$_2$O | 0.020g |
| L-Isoleucine | 0.020g |
| Phenol Red | 0.020g |
| L-Methionine | 0.015g |
| Deoxyadenosine | 0.010g |

| | |
|---|---|
| Deoxycytidine | 0.010g |
| Deoxyguanosine | 0.010g |
| Glutathione, reduced | 0.010g |
| Thymidine | 0.010g |
| Hydroxy-L-proline | 0.010g |
| L-Tryptophan | 0.010g |
| Nicotinamide adenine dinucleotide | 7.0mg |
| Tween™ 80 | 5.0mg |
| Sodium glucoronate·$H_2O$ | 4.2mg |
| Coenzyme A | 2.5mg |
| Cocarboxylase | 1.0mg |
| Flavin adenine dinucleotide | 1.0mg |
| Nicotinamide adenine dinucleotide phosphate | 1.0mg |
| Uridine triphosphate | 1.0mg |
| Choline chloride | 0.50mg |
| Cholesterol | 0.20mg |
| 5-Methyldeoxycytidine | 0.10mg |
| Inositol | 0.05mg |
| p-Aminobenzoic acid | 0.05mg |
| Niacin | 0.025mg |
| Niacinamide | 0.025mg |
| Pyridoxine | 0.025mg |
| Pyridoxal·HCl | 0.025mg |
| Biotin | 0.01mg |
| D-Calcium pantothenate | 0.01mg |
| Folic acid | 0.01mg |
| Riboflavin | 0.01mg |
| Thiamine·HCl | 0.01mg |

**Preparation of CMRL 1066, 10X Without Glutamine or NaHCO$_3$:** Add components to distilled/deionized water and bring volume to 1.0L. Mix thoroughly. Adjust pH to 7.2. Filter sterilize.

**Preparation of Medium:** Add components, except CMRL 1066, 10X without glutamine or NaHCO$_3$ and rabbit serum, to distilled/deionized water and bring volume to 1100.0mL. Mix thoroughly. Adjust pH to 7.5 with NaOH. Filter sterilize. Aseptically add 100.0mL of sterile CMRL 1066, 10X without glutamine or NaHCO$_3$ and 64.0mL of sterile rabbit serum. Mix thoroughly. Aseptically distribute 10.0mL volumes into sterile 16 × 125.0mm test tubes.

**Storage/Shelf Life:** Store dehydrated media in the dark in a sealed container below 30°C. Prepared media should be stored under refrigeration (2-8°C). Media should be used within 60 days of preparation. Media should not be used if there are any signs of deterioration (discoloration) or contamination, or if the expiration date supplied by the manufacturer has passed.

**Use:** For the cultivation of *Borrelia afzelii, Borrelia burgdorferi,* and *Borrelia gorinii.*

## BSK Medium, Revised

**Composition** per 1164.0mL:

| | |
|---|---|
| Bovine serum albumin fraction V | 50.0g |
| HEPES (*N*-[2-hydroxyethyl]piperazine-*N*′-2-ethanesulfonic acid) buffer | 6.0g |
| Neopeptone | 5.0g |
| Glucose | 5.0g |
| TC-Yeastolate | 2.54g |
| NaHCO$_3$ | 2.2g |
| Sodium pyruvate | 0.8g |
| Sodium citrate | 0.7g |

| | |
|---|---|
| N-Acetylglucosamine | 0.4g |
| CMRL 1066, without glutamine, without bicarbonate, 10X | 100.0mL |
| Rabbit serum | 64.0mL |

pH 7.6–7.65 at 25°C

### CMRL 1066 Medium without Glutamine, without Bicarbonate, 10X:

**Composition** per liter:

| | |
|---|---|
| NaCl | 6.8g |
| D-Glucose | 1.0g |
| KCl | 0.4g |
| L-Cysteine·HCl·$H_2O$ | 0.26g |
| CaCl$_2$, anhydrous | 0.2g |
| MgSO$_4$·7$H_2O$ | 0.2g |
| NaH$_2$PO$_4$·$H_2O$ | 0.14g |
| Sodium acetate·3$H_2O$ | 0.083g |
| L-Glutamic acid | 0.075g |
| L-Arginine·HCl | 0.070g |
| L-Lysine·HCl | 0.070g |
| L-Leucine | 0.060g |
| Glycine | 0.050g |
| Ascorbic acid | 0.050g |
| L-Proline | 0.040g |
| L-Tyrosine | 0.040g |
| L-Aspartic acid | 0.030g |
| L-Threonine | 0.030g |
| L-Alanine | 0.025g |
| L-Phenylalanine | 0.025g |
| L-Serine | 0.025g |
| L-Valine | 0.025g |
| L-Cystine | 0.020g |
| L-Histidine·HCl·$H_2O$ | 0.020g |
| L-Isoleucine | 0.020g |
| Phenol red | 0.020g |
| L-Methionine | 0.015g |
| Deoxyadenosine | 0.010g |
| Deoxycytidine | 0.010g |
| Deoxyguanosine | 0.010g |
| Glutathione, reduced | 0.010g |
| Thymidine | 0.010g |
| Hydroxy-L-proline | 0.010g |
| L-Tryptophan | 0.010g |
| Nicotinamide adenine dinucleotide | 7.0mg |
| Tween™ 80 | 5.0mg |
| Sodium glucoronate·$H_2O$ | 4.2mg |
| Coenzyme A | 2.5mg |
| Cocarboxylase | 1.0mg |
| Flavin adenine dinucleotide | 1.0mg |
| Nicotinamide adenine dinucleotide phosphate | 1.0mg |
| Uridine triphosphate | 1.0mg |
| Choline chloride | 0.50mg |
| Cholesterol | 0.20mg |
| 5-Methyldeoxycytidine | 0.10mg |
| Inositol | 0.05mg |
| p-Aminobenzoic acid | 0.05mg |
| Niacin | 0.025mg |
| Niacinamide | 0.025mg |
| Pyridoxine | 0.025mg |
| Pyridoxal·HCl | 0.025mg |
| Biotin | 0.01mg |
| Calcium DL-pantothenate | 0.01mg |
| Folic acid | 0.01mg |

Riboflavin ....................................................................0.01mg
Thiamine·HCl ..............................................................0.01mg
pH 7.2 ± 0.2 at 25°C

**Preparation of CMRL 1066 Medium without Glutamine, without Bicarbonate, 10X:** Add components to distilled/deionized water and bring volume to 1.0L. Mix thoroughly. Adjust pH to 7.2. Filter sterilize.

**Preparation of Medium:** Add components, except rabbit serum and CMRL 1066, to 1.0L of glass-distilled/deionized water. Mix thoroughly. Adjust pH to 7.5 with NaOH. Filter sterilize. Aseptically add 100.0mL of sterile CMRL 1066 and 64.0mL of sterile rabbit serum. Adjust final pH to 7.5–7.6. Aseptically distribute into sterile tubes or flasks.

**Storage/Shelf Life:** Store dehydrated media in the dark in a sealed container below 30°C. Prepared media should be stored under refrigeration (2-8°C). Media should be used within 60 days of preparation. Media should not be used if there are any signs of deterioration (discoloration) or contamination, or if the expiration date supplied by the manufacturer has passed.

**Use:** For the cultivation of *Borrelia burgdorferi, Borrelia afzelii, Borrelia garinii, Borrelia anserina,* and *Borrelia japonica.*

### BTB Lactose Agar
### (Bromthymol Blue Lactose Agar)
**Composition** per liter:
Agar ...........................................................................15.0g
Lactose.......................................................................10.0g
Proteose peptone .........................................................5.0g
Beef extract.................................................................3.0g
Bromthymol Blue .......................................................0.17g
pH 8.7–7.2 at 25°C

**Preparation of Medium:** Add components to distilled/deionized water and bring volume to 1.0L. Mix thoroughly. Heat gently with frequent mixing. Bring to boiling. Distribute into tubes or flasks. Autoclave for 15 min at 15 psi pressure–121°C. Pour into sterile Petri dishes if desired.

**Storage/Shelf Life:** Store dehydrated media in the dark in a sealed container below 30°C. Prepared media should be stored under refrigeration (2-8°C). Media should be used within 60 days of preparation. Media should not be used if there are any signs of deterioration (shrinking, cracking, or discoloration) or contamination, or if the expiration date supplied by the manufacturer has passed.

**Use:** For the isolation and cultivation of pathogenic staphylococci.

### BTB Lactose HiVeg Agar
### (Bromthymol Blue Lactose HiVeg Agar)
**Composition** per liter:
Agar ...........................................................................15.0g
Lactose.......................................................................10.0g
Plant peptone No. 3......................................................5.0g
Plant extract ...............................................................3.0g
Bromthymol Blue .......................................................0.17g
pH 8.6 ± 0.2 at 25°C

**Source:** This medium is available as a premixed powder from Hi-Media.

**Preparation of Medium:** Add components to distilled/deionized water and bring volume to 1.0L. Mix thoroughly. Heat gently with frequent mixing. Bring to boiling. Distribute into tubes or flasks. Auto-

clave for 15 min at 15 psi pressure–121°C. Pour into sterile Petri dishes if desired.

**Storage/Shelf Life:** Store dehydrated media in the dark in a sealed container below 30°C. Prepared media should be stored under refrigeration (2-8°C). Media should be used within 60 days of preparation. Media should not be used if there are any signs of deterioration (shrinking, cracking, or discoloration) or contamination, or if the expiration date supplied by the manufacturer has passed.

**Use:** For the isolation and cultivation of pathogenic staphylococci.

### B.T.B. Lactose Agar, Modified
### (Lactose Blue HiVeg Agar)
**Composition** per liter:
Lactose.......................................................................15.5g
Agar ...........................................................................13.0g
NaCl.............................................................................5.0g
Casein enzymatic hydrolysate ......................................3.5g
Peptone .......................................................................3.5g
Bromthymol Blue .......................................................0.04g
pH 7.0 ± 0.2 at 25°C

**Source:** This medium is available as a premixed powder from Hi-Media.

**Preparation of Medium:** Add components to distilled/deionized water and bring volume to 1.0L. Mix thoroughly. Heat gently with frequent mixing. Bring to boiling. Distribute into tubes or flasks. Autoclave for 15 min at 15 psi pressure–121°C. Pour into sterile Petri dishes if desired.

**Storage/Shelf Life:** Store dehydrated media in the dark in a sealed container below 30°C. Prepared media should be stored under refrigeration (2-8°C). Media should be used within 60 days of preparation. Media should not be used if there are any signs of deterioration (shrinking, cracking, or discoloration) or contamination, or if the expiration date supplied by the manufacturer has passed.

**Use:** For the isolation and cultivation of pathogenic staphylococci.

### B.T.B. Lactose HiVeg Agar, Modified
### (Lactose Blue HiVeg Agar)
**Composition** per liter:
Lactose.......................................................................15.5g
Agar ...........................................................................13.0g
NaCl.............................................................................5.0g
Plant hydrolysate .........................................................3.5g
Plant peptone ..............................................................3.5g
Bromthymol Blue .......................................................0.04g
pH 7.0 ± 0.2 at 25°C

**Source:** This medium is available as a premixed powder from Hi-Media.

**Preparation of Medium:** Add components to distilled/deionized water and bring volume to 1.0L. Mix thoroughly. Heat gently with frequent mixing. Bring to boiling. Distribute into tubes or flasks. Autoclave for 15 min at 15 psi pressure–121°C. Pour into sterile Petri dishes if desired.

**Storage/Shelf Life:** Store dehydrated media in the dark in a sealed container below 30°C. Prepared media should be stored under refrigeration (2-8°C). Media should not be used if there are any signs of deterioration (shrink-

ing, cracking, or discoloration) or contamination, or if the expiration date supplied by the manufacturer has passed.

**Use:** For the isolation and cultivation of pathogenic staphylococci.

# Buffered Charcoal Yeast Extract Differential Agar (DIFF/BCYE)

**Composition** per 1014.0mL:

| | |
|---|---|
| Agar | 17.0g |
| ACES (2-[(2-amino-2-oxoethyl)-amino]-ethane sulfonic acid) buffer | 10.0g |
| Yeast extract | 10.0g |
| Charcoal, activated | 1.5g |
| $Fe_4(P_2O_7)_3 \cdot 9H_2O$ | 0.25g |
| Bromcresol Purple | 0.01g |
| Bromthymol Blue | 0.01g |
| Antibiotic solution | 10.0mL |
| L-Cysteine·HCl·H$_2$O solution | 4.0mL |

pH 6.9 ± 0.2 at 25°C

**Antibiotic Solution:**
**Composition** per 10.0mL:

| | |
|---|---|
| Vancomycin | 1.0mg |
| Polymyxin B | 50,000U |

**Preparation of Antibiotic Solution:** Add components to distilled/deionized water and bring volume to 10.0mL. Mix thoroughly. Filter sterilize.

**L-Cysteine·HCl·H$_2$O Solution:**
**Composition** per 10.0mL:

| | |
|---|---|
| L-Cysteine·HCl·H$_2$O | 1.0g |

**Preparation of L-Cysteine·HCl·H$_2$O Solution:** Add 1.0g of L-cysteine·HCl·H$_2$O to distilled/deionized water and bring volume to 10.0mL. Mix thoroughly. Filter sterilize.

**Preparation of Medium:** Add components, except L-cysteine·HCl·H$_2$O solution and antibiotic solution, to distilled/deionized water and bring volume to 1.0L. Mix thoroughly. Adjust medium to pH 6.9 with 1$N$ KOH. Heat gently and bring to boil for 1 min. Autoclave for 15 min at 15 psi pressure–121°C. Cool to 50°–55°C. Add 4.0mL of sterile L-cysteine·HCl·H$_2$O solution and 10.0mL of sterile antibiotic solution. Mix thoroughly. Pour into sterile Petri dishes with constant agitation to keep charcoal in suspension.

**Storage/Shelf Life:** Store dehydrated media in the dark in a sealed container below 30°C. Prepared media should be stored under refrigeration (2-8°C). Media should be used within 60 days of preparation. Media should not be used if there are any signs of deterioration (shrinking, cracking, or discoloration) or contamination, or if the expiration date supplied by the manufacturer has passed.

**Use:** For the isolation, cultivation, and maintenance of *Legionella pneumophila* and other *Legionella* species from environmental and clinical specimens. For the selective recovery of *Legionella pneumophila* while reducing contaminating microorganisms from environmental water samples.

# *Burkholderia cepacia* Agar

**Composition** per liter:

| | |
|---|---|
| Agar | 12.0g |
| Sodium pyruvate | 7.0g |
| Peptone | 5.0g |
| KH$_2$PO$_4$ | 4.4g |
| Yeast extract | 4.0g |

| | |
|---|---|
| Bile salts | 1.5g |
| Na$_2$HPO$_4$ | 1.4g |
| (NH$_4$)$_2$SO$_4$ | 1.0g |
| MgSO$_4$ | 0.2 |
| Phenol Red | 0.02g |
| Fe(NH$_4$)$_2$(SO$_4$)$_2$·6H$_2$O | 0.01g |
| Crystal Violet | 0.001g |
| Selective supplement solution | 10.0mL |

pH 6.2 ± 0.2 at 25°C

**Source:** This medium is available as a premixed powder from Thermo Scientific.

**Selective Supplement Solution:**
**Composition** per 10.0mL:

| | |
|---|---|
| Polymyxin B | 150,000IU |
| Ticarcillin | 100.0mg |
| Gentamicin | 5.0mg |

**Preparation of Selective Supplement Solution:** Add components to distilled/deionized water and bring volume to 10.0mL. Mix thoroughly. Filter sterilize.

**Preparation of Medium:** Add components, except selective supplement solution, to distilled/deionized water and bring volume to 990.0mL. Mix thoroughly. Gently heat while stirring and bring to boiling. Autoclave for 15 min at 15 psi pressure–121°C. Cool to 50°C. Aseptically add 10.0mL selective supplement solution. Mix thoroughly. Pour into sterile Petri dishes.

**Storage/Shelf Life:** Store dehydrated media in the dark in a sealed container below 30°C. Prepared media should be stored under refrigeration (2-8°C). Media should be used within 60 days of preparation. Media should not be used if there are any signs of deterioration (shrinking, cracking, or discoloration) or contamination, or if the expiration date supplied by the manufacturer has passed.

**Use:** For the selective isolation of *Burkholderia cepacia* from the respiratory secretions of patients with cystic fibrosis and for routine testing of non-sterile inorganic salt solutions containing preservative. Slow growing *B. cepacia* can be missed on conventional media such as blood or MacConkey agar due to overgrowth caused by other faster growing organisms found in the respiratory tract of CF patients such as mucoid *Klebsiella* species, *Pseudomonas aeruginosa*, and *Staphylococcus* species. This may lead to the infection being missed or wrongly diagnosed.

# *Burkholderia pseudomallei* Selective Agar (BPSA)

**Composition** per liter:

| | |
|---|---|
| Agar | 15.0g |
| Pancreatic Digest of Casein | 5.0g |
| Maltose | 4.0g |
| Yeast Extract | 2.5g |
| Glucose | 1.0g |
| Neutral Red | 0.1g |
| Gentamicin solution | 10.0mL |
| Glycerol | 10.0mL |
| Nile Blue solution | 1.0mL |

pH 7.0 ± 0.2 at 25°C

**Gentamicin Solution:**
**Composition** per 10.0mL:

| | |
|---|---|
| Gentamicin | 20.0mg |

**Preparation of Gentamicin Solution:** Add gentamicin to distilled/deionized water and bring volume to 10.0mL. Mix thoroughly. Filter sterilize.

**Nile Blue Solution:**
**Composition** per 10.0mL:
Nile Blue .......................................................................0.2g

**Preparation of Nile Blue Solution:** Add Nile blue to 10.0mL of a 1% solution of dimethyl sulfoxide. Mix thoroughly. Filter sterilize.

**Preparation of Medium:** Add components, except gentamcin solution, Nile Blue solution, and glycerol, to distilled/deionized water and bring volume to 979.0mL. Mix thoroughly. Gently heat while stirring and bring to boiling. Autoclave for 15 min at 15 psi pressure–121°C. Cool to 45°C. Aseptically add 10.0mL sterile gentamicin solution, 1.0mL sterile Nile blue solution, and 10.0mL filter-sterilized glycerol. Mix thoroughly for 5 min on a heated magnetic stirrer at 40°C. Pour into sterile Petri dishes.

**Storage/Shelf Life:** Store dehydrated media in the dark in a sealed container below 30°C. Prepared media should be stored under refrigeration (2-8°C). Media should be used within 60 days of preparation. Media should not be used if there are any signs of deterioration (shrinking, cracking, or discoloration) or contamination, or if the expiration date supplied by the manufacturer has passed.

**Use:** For the cultivation of *Burkholderia pseudomallei* from clinical specimens collected from non-sterile sites with improved recovery of the more easily inhibited strains of *B. pseudomallei*.

## CAE Agar Base with Triphenyltetrazolium Chloride
### (Citrate Azide *Enterococcus* HiVeg Agar Base)
**Composition** per liter:
Agar .................................................................................15.0g
Casein enzymatic hydrolysate ......................................15.0g
Sodium citrate ................................................................15.0g
KH$_2$PO$_4$........................................................................5.0g
Yeast extract.....................................................................5.0g
Na$_2$CO$_3$ ........................................................................2.0g
Polysorbate 80..................................................................1.0g
NaN$_3$ ...............................................................................0.4g
2,3,5-Triphenyltetrazolium chloride solution .......... 10.0mL
pH 7.0 ± 0.2 at 25°C

**Source:** This medium, without triphenyltetrazolium chloride solution, is available as a premixed powder from HiMedia.

**Caution:** Sodium azide is toxic. Azides also react with metals and disposal must be highly diluted.

**2,3,5-Triphenyltetrazolium Chloride Solution:**
**Composition** per 10.0mL:
2,3,5-Triphenyltetrazolium chloride ...............................0.1g

**Preparation of 2,3,5-Triphenyltetrazolium Chloride Solution:** Add 2,3,5-triphenyltetrazolium chloride to distilled/deionized water and bring volume to 10.0mL. Mix thoroughly. Filter sterilize.

**Preparation of Medium:** Add components, except 2,3,5-triphenyltetrazolium chloride solution, to distilled/deionized water and bring volume to 990.0mL. Mix thoroughly. Gently heat and bring to boiling. Autoclave for 10 min at 15 psi pressure–121°C. Cool to 45°–50°C. Aseptically add 10.0mL 2,3,5-triphenyltetrazolium chloride solution.

Mix thoroughly. Pour into sterile Petri dishes or distribute into sterile tubes.

**Storage/Shelf Life:** Store dehydrated media in the dark in a sealed container below 30°C. Prepared media should be stored under refrigeration (2-8°C). Media should be used within 60 days of preparation. Media should not be used if there are any signs of deterioration (shrinking, cracking, or discoloration) or contamination, or if the expiration date supplied by the manufacturer has passed.

**Use:** For the isolation, cultivation, and enumeration of entercocci in water, sewage, and feces by the membrane filter method. For the direct plating of specimens for the detection and enumeration of fecal streptococci.

## CAE HiVeg Agar Base
## with Triphenyltetrazolium Chloride
### (Citrate Azide *Enterococcus* HiVeg Agar Base)
**Composition** per liter:
Agar .................................................................................15.0g
Plant hydrolysate ...........................................................15.0g
Sodium citrate ................................................................15.0g
KH$_2$PO$_4$........................................................................5.0g
Yeast extract.....................................................................5.0g
Na$_2$CO$_3$ ........................................................................2.0g
Polysorbate 80 .................................................................1.0g
NaN$_3$ ...............................................................................0.4g
2,3,5-Triphenyltetrazolium chloride solution .......... 10.0mL
pH 7.0 ± 0.2 at 25°C

**Source:** This medium, without triphenyltetrazolium chloride solution, is available as a premixed powder from HiMedia.

**Caution:** Sodium azide is toxic. Azides also react with metals and disposal must be highly diluted.

**2,3,5-Triphenyltetrazolium Chloride Solution:**
**Composition** per 10.0mL:
2,3,5-Triphenyltetrazolium chloride ...............................0.1g

**Preparation of 2,3,5-Triphenyltetrazolium Chloride Solution:** Add 2,3,5-triphenyltetrazolium chloride to distilled/deionized water and bring volume to 10.0mL. Mix thoroughly. Filter sterilize.

**Preparation of Medium:** Add components, except 2,3,5-triphenyltetrazolium chloride solution, to distilled/deionized water and bring volume to 990.0mL. Mix thoroughly. Gently heat and bring to boiling. Autoclave for 10 min at 15 psi pressure–121°C. Cool to 45°–50°C. Aseptically add 10.0mL 2,3,5-triphenyltetrazolium chloride solution. Mix thoroughly. Pour into sterile Petri dishes or distribute into sterile tubes.

**Storage/Shelf Life:** Store dehydrated media in the dark in a sealed container below 30°C. Prepared media should be stored under refrigeration (2-8°C). Media should be used within 60 days of preparation. Media should not be used if there are any signs of deterioration (shrinking, cracking, or discoloration) or contamination, or if the expiration date supplied by the manufacturer has passed.

**Use:** For the isolation, cultivation, and enumeration of entercocci in water, sewage, and feces by the membrane filter method. For the direct plating of specimens for the detection and enumeration of fecal streptococci.

## CAL Agar
### (Cellobiose Arginine Lysine Agar)
### (*Yersinia* Isolation Agar)

**Composition** per liter:

| | |
|---|---|
| Agar | 20.0g |
| L-Arginine·HCl | 6.5g |
| L-Lysine·HCl | 6.5g |
| NaCl | 5.0g |
| Cellobiose | 3.5g |
| Yeast extract | 3.0g |
| Sodium deoxycholate | 1.5g |
| Neutral Red | 0.03g |

pH 7.3 ± 0.2 at 25°C

**Preparation of Medium:** Add components to distilled/deionized water and bring volume to 1.0L. Mix thoroughly. Heat to boiling. Do not autoclave. Pour into sterile Petri dishes.

**Storage/Shelf Life:** Store dehydrated media in the dark in a sealed container below 30°C. Prepared media should be stored under refrigeration (2-8°C). Media should be used within 60 days of preparation. Media should not be used if there are any signs of deterioration (shrinking, cracking, or discoloration) or contamination, or if the expiration date supplied by the manufacturer has passed.

**Use:** For the isolation and characterization of *Yersinia enterocolitica* from fecal specimens and enumeration of *Yersinia enterocolitica* from water and other liquid specimens.

## CAL HiVeg Agar
### (Cellobiose Arginine Lysine HiVeg Agar)

**Composition** per liter:

| | |
|---|---|
| Agar | 20.0g |
| L-Arginine | 6.5g |
| L-Lysine hydrochloride | 6.5g |
| NaCl | 5.0g |
| Cellobiose | 3.5g |
| Yeast extract | 3.0g |
| Synthetic detergent No. III | 1.5g |
| Neutral Red | 0.03g |

pH 7.1 ± 0.2 at 25°C

**Source:** This medium is available as a premixed powder from Hi-Media.

**Preparation of Medium:** Add components to distilled/deionized water and bring volume to 1.0L. Mix thoroughly. Heat to boiling. Do not autoclave. Pour into sterile Petri dishes.

**Storage/Shelf Life:** Store dehydrated media in the dark in a sealed container below 30°C. Prepared media should be stored under refrigeration (2-8°C). Media should be used within 60 days of preparation. Media should not be used if there are any signs of deterioration (shrinking, cracking, or discoloration) or contamination, or if the expiration date supplied by the manufacturer has passed.

**Use:** For the isolation and characterization of *Yersinia enterocolitica* from fecal specimens and enumeration of *Y. enterocolitica* from water.

## Campy THIO Medium

**Composition** per liter:

| | |
|---|---|
| Pancreatic digest of casein | 20.0g |
| Agar | 15.0g |
| NaCl | 2.5g |
| K$_2$HPO$_4$ | 1.5g |
| Sodium thioglycolate | 0.6g |
| L-Cystine | 0.4g |
| Na$_2$SO$_3$ | 0.2g |
| Antibiotic supplement | 10.0mL |

**Antibiotic Supplement:**
**Composition** per 10.0mL:

| | |
|---|---|
| Cephalothin | 15.0mg |
| Vancomycin | 10.0mg |
| Trimethoprim | 5.0mg |
| Amphotericin B | 2.0mg |
| Polymyxin B | 2500U |

**Preparation of Antibiotic Supplement:** Add components to 10.0mL of distilled/deionized water. Filter sterilize.

**Preparation of Medium:** Add components, except antibiotic solution, to distilled/deionized water and bring volume to 990.0mL. Mix thoroughly. Gently heat and bring to boiling. Autoclave for 15 min at 15 psi pressure–121°C. Cool to 45°–50°C. Aseptically add 10.0mL of sterile antibiotic solution. Mix thoroughly. Aseptically distribute into sterile screw-capped tubes in 3.0mL volumes for 1.5cm swabs or 5.0mL volumes for 3.0cm swabs.

**Storage/Shelf Life:** Store dehydrated media in the dark in a sealed container below 30°C. Prepared media should be stored under refrigeration (2-8°C). Media should be used within 60 days of preparation. Media should not be used if there are any signs of deterioration (shrinking, cracking, or discoloration) or contamination, or if the expiration date supplied by the manufacturer has passed.

**Use:** For the maintenance—as a holding or transport medium—of *Campylobacter* species isolated from clinical specimens on swabs.

## *Campylobacter* Agar with 5 Antimicrobics and 10% Sheep Blood

**Composition** per liter:

| | |
|---|---|
| Agar | 15.0g |
| Pancreatic digest of casein | 10.0g |
| Peptic digest of animal tissue | 10.0g |
| NaCl | 5.0g |
| Yeast extract | 2.0g |
| Glucose | 1.0g |
| NaHSO$_3$ | 0.1g |
| Sheep blood, defibrinated | 100.0mL |
| Antibiotic supplement | 10.0mL |

pH 7.2 ± 0.2 at 25°C

**Source:** This medium is available as a prepared medium from BD Diagnostic Systems.

**Antibiotic Supplement:**
**Composition** per 10.0mL:

| | |
|---|---|
| Cephalothin | 0.015g |
| Vancomycin | 0.01g |
| Trimethoprim | 5.0mg |
| Amphotericin B | 2.0mg |
| Polymyxin B | 2500U |

**Preparation of Antibiotic Supplement:** Add components to 10.0mL of distilled/deionized water. Filter sterilize.

**Preparation of Medium:** Add components, except sheep blood and antibiotic solution, to distilled/deionized water and bring volume to 890.0mL. Mix thoroughly. Gently heat and bring to boiling. Autoclave for 15 min at 15 psi pressure–121°C. Cool to 45°–50°C. Aseptically add 100.0mL of sterile sheep blood and 10.0mL of sterile antibiotic solution. Mix thoroughly. Pour into sterile Petri dishes or distribute into sterile tubes.

**Storage/Shelf Life:** Store dehydrated media in the dark in a sealed container below 30°C. Prepared media should be stored under refrigeration (2-8°C). Media should be used within 60 days of preparation. Media should not be used if there are any signs of deterioration (shrinking, cracking, or discoloration) or contamination, or if the expiration date supplied by the manufacturer has passed.

**Use:** For the primary selective isolation and cultivation of *Campylobacter jejuni* from human fecal specimens.

## *Campylobacter* Agar, Blaser's
### (Blaser's *Campylobacter* Agar)

**Composition** per liter:
| | |
|---|---|
| *Campylobacter* agar base | 990.0mL |
| Supplement B | 10.0mL |

pH 7.4 ± 0.2 at 25°C

**Campylobacter Agar Base:**
**Composition** per liter:
| | |
|---|---|
| Proteose peptone | 15.0g |
| Agar | 12.0g |
| NaCl | 5.0g |
| Yeast extract | 5.0g |
| Liver digest | 2.5g |

**Source:** *Campylobacter* agar base and *Campylobacter* antimicrobic supplement B are available as a premixed powder from BD Diagnostic Systems.

**Preparation of *Campylobacter* Agar Base:** Add components to distilled/deionized water and bring volume to 990.0mL. Mix thoroughly. Gently heat and bring to boiling. Autoclave for 15 min at 15 psi pressure–121°C. Cool to 45°–50°C.

**Supplement B:**
**Composition** per 10.0mL:
| | |
|---|---|
| Cephalothin | 15.0mg |
| Vancomycin | 10.0mg |
| Trimethoprim | 5.0mg |
| Amphotericin B | 2.0mg |
| Polymyxin B | 2500U |

**Preparation of Supplement B:** Add components to 10.0mL of distilled/deionized water. Filter sterilize.

**Preparation of Medium:** Prepare 990.0mL of *Campylobacter* agar base. Autoclave and cool to 45°–50°C. Aseptically add 10.0mL of sterile supplement B. Mix thoroughly. Pour into sterile Petri dishes.

**Storage/Shelf Life:** Store dehydrated media in the dark in a sealed container below 30°C. Prepared media should be stored under refrigeration (2-8°C). Media should be used within 60 days of preparation. Media should not be used if there are any signs of deterioration (shrinking, cracking, or discoloration) or contamination, or if the expiration date supplied by the manufacturer has passed.

**Use:** For the selective isolation of *Campylobacter jejuni* from fecal specimens, food, and environmental specimens.

## *Campylobacter* Agar, Skirrow's
### (Skirrow's *Campylobacter* Agar)

**Composition** per liter:
| | |
|---|---|
| *Campylobacter* agar base | 990.0mL |

| | |
|---|---|
| Supplement S | 10.0mL |

pH 7.4 ± 0.2 at 25°C

**Campylobacter Agar Base:**
**Composition** per liter:
| | |
|---|---|
| Proteose peptone | 15.0g |
| Agar | 12.0g |
| NaCl | 5.0g |
| Yeast extract | 5.0g |
| Liver digest | 2.5g |

**Source:** *Campylobacter* agar base and *Campylobacter* antimicrobic supplement S are available as a premixed powder from BD Diagnostic Systems.

**Preparation of *Campylobacter* Agar Base:** Add components to distilled/deionized water and bring volume to 990.0mL. Mix thoroughly. Gently heat and bring to boiling. Autoclave for 15 min at 15 psi pressure–121°C. Cool to 45°–50°C.

**Supplement S:**
**Composition** per 10.0mL:
| | |
|---|---|
| Vancomycin | 10.0mg |
| Trimethoprim | 5.0mg |
| Polymyxin B | 2500U |

**Preparation of Supplement S:** Add components to 10.0mL of distilled/deionized water. Filter sterilize.

**Preparation of Medium:** Prepare 990.0mL of *Campylobacter* agar base. Autoclave and cool to 45°–50°C. Aseptically add 10.0mL of sterile supplement S. Mix thoroughly. Pour into sterile Petri dishes.

**Storage/Shelf Life:** Store dehydrated media in the dark in a sealed container below 30°C. Prepared media should be stored under refrigeration (2-8°C). Media should be used within 60 days of preparation. Media should not be used if there are any signs of deterioration (shrinking, cracking, or discoloration) or contamination, or if the expiration date supplied by the manufacturer has passed.

**Use:** For the selective isolation of *Campylobacter jejuni* from fecal specimens, food, and environmental specimens.

## *Campylobacter* Blood-Free Agar Base, Modified
### (CCDA, Modified)

**Composition** per liter:
| | |
|---|---|
| Peptone | 20.0g |
| Agar | 12.0g |
| NaCl | 5.0g |
| Activated charcoal | 4.0g |
| Casein hydrolysate | 3.0g |
| Na-desoxycholate | 1.0g |
| Na-pyruvate | 0.25g |
| $FeSO_4$ | 0.25g |
| Cefoperazone-amphotericin B solution | 10.0mL |

pH 7.4 ± 0.2 at 25°C

**Cefoperazone-Amphotericin B Solution:**
**Composition** per 10.0mL:
| | |
|---|---|
| Cefoperazone | 0.016g |
| Amphotericin B | 0.005g |

**Preparation of Cefoperazone-Amphotericin B Solution:** Add cefoperazone and amphotericin B to distilled/deionized water and bring volume to 10.0mL. Mix thoroughly. Filter sterilize.

**Preparation of Medium:** Add components except cefoperazone-amphotericin B solution to distilled/deionized water and bring volume to 990.0mL. Mix thoroughly. Autoclave for 15 min at 15 psi pressure–121°C. Cool to 45°–50°C. Aseptically add 10.0mL of cefoperazone-amphotericin B solution. Mix thoroughly. Pour into sterile Petri dishes or aseptically distribute into tubes or flasks.

**Storage/Shelf Life:** Store dehydrated media in the dark in a sealed container below 30°C. Prepared media should be stored under refrigeration (2-8°C). Media should be used within 60 days of preparation. Media should not be used if there are any signs of deterioration (shrinking, cracking, or discoloration) or contamination, or if the expiration date supplied by the manufacturer has passed.

**Use:** For the isolation of *Campylobacter* spp. from foods. The use of *Campylobacter* Blood-Free Selective Agar is specified by the UK Ministry of Agriculture, Fisheries and Food (MAFF) in a validated method for isolation of *Campylobacter* from foods. Amphotericin largely reduces the growth of yeasts and molds. Cefoperazone especially inhibits Enterobacteriaceae.

### *Campylobacter* Charcoal Differential Agar (CCDA)
### (Preston Blood-Free Medium)
**Composition** per liter:
| | |
|---|---|
| Agar | 12.0g |
| Beef extract | 10.0g |
| Peptone | 10.0g |
| NaCl | 5.0g |
| Charcoal | 4.0g |
| Casein hydrolysate | 3.0g |
| Sodium deoxycholate | 1.0g |
| FeSO$_4$ | 0.25g |
| Sodium pyruvate | 0.25g |
| Cefoperazone solution | 10.0mL |

pH 7.5 ± 0.2 at 25°C

**Cefoperazone Solution:**
**Composition** per 10.0mL:
| | |
|---|---|
| Sodium cefoperazone | 0.032g |

**Preparation of Cefoperazone Solution:** Add sodium cefoperazone to distilled/deionized water and bring volume to 10.0mL. Mix thoroughly. Filter sterilize.

**Preparation of Medium:** Add components, except cefoperazone solution, to distilled/deionized water and bring volume to 990.0mL. Mix thoroughly. Gently heat and bring to boiling. Autoclave for 15 min at 15 psi pressure–121°C. Cool to 45°–50°C. Aseptically add 10.0mL of sterile cefoperazone solution. Mix thoroughly. Pour into sterile Petri dishes or distribute into sterile tubes.

**Storage/Shelf Life:** Store dehydrated media in the dark in a sealed container below 30°C. Prepared media should be stored under refrigeration (2-8°C). Media should be used within 60 days of preparation. Media should not be used if there are any signs of deterioration (shrinking, cracking, or discoloration) or contamination, or if the expiration date supplied by the manufacturer has passed.

**Use:** For the cultivation of *Campylobacter* species.

### *Campylobacter* Enrichment Broth
**Composition** per liter:
| | |
|---|---|
| Beef extract | 10.0g |
| Peptone | 10.0g |
| Yeast extract | 6.0g |
| NaCl | 5.0g |

| | |
|---|---|
| Horse blood, laked | 50.0mL |
| FBP solution | 4.0mL |
| Antibiotic solution | 4.0mL |

pH 7.5 ± 0.2 at 25°C

**Horse Blood, Laked:**
**Composition** per 50.0mL:
| | |
|---|---|
| Horse blood, fresh | 50.0mL |

**Preparation of Horse Blood, Laked:** Add blood to a sterile polypropylene bottle. Freeze overnight at –20°C. Thaw at 8°C. Refreeze at –20°C. Thaw again at 8°C.

**FBP Solution:**
**Composition** per 100.0mL:
| | |
|---|---|
| FeSO$_4$ | 6.25g |
| NaHSO$_3$ | 6.25g |
| Sodium pyruvate | 6.25g |

**Preparation of FBP Solution:** Add components to distilled/deionized water and bring volume to 100.0mL. Mix thoroughly. Filter sterilize.

**Antibiotic Solution:**
**Composition** per 10.0mL:
| | |
|---|---|
| Cycloheximide | 0.1g |
| Sodium cefoperazone | 0.03g |
| Trimethoprim lactate | 0.0125g |
| Rifampicin | 0.01g |

**Preparation of Antibiotic Solution:** Add components to distilled/deionized water and bring volume to 10.0mL. Mix thoroughly. Filter sterilize.

**Caution:** Cycloheximide is toxic. Avoid skin contact or aerosol formation and inhalation.

**Preparation of Medium:** Add components—except laked horse blood, FBP solution, and antibiotic solution—to distilled/deionized water and bring volume to 942.0mL. Mix thoroughly. Gently heat and bring to boiling. Autoclave for 15 min at 15 psi pressure–121°C. Cool to 45°–50°C. Aseptically add 50.0mL of sterile laked horse blood, 4.0mL of FBP solution, and 4.0mL of antibiotic solution. Mix thoroughly. Pour into sterile Petri dishes or distribute into sterile tubes.

**Storage/Shelf Life:** Store dehydrated media in the dark in a sealed container below 30°C. Prepared media should be stored under refrigeration (2-8°C). Media should be used within 60 days of preparation. Media should not be used if there are any signs of deterioration (discoloration) or contamination, or if the expiration date supplied by the manufacturer has passed.

**Use:** For the isolation and cultivation of *Campylobacter* species from dairy products.

### *Campylobacter* Enrichment Broth
### (FDA Medium M29)
**Composition** per 1024.0mL:
| | |
|---|---|
| Basal medium | 950.0mL |
| Horse blood, lysed | 50.0mL |
| Cefoperazone solution | 8.0mL |
| FBP solution | 4.0mL |
| Trimethoprim lactate solution | 4.0mL |
| Vancomycin solution | 4.0mL |
| Cycloheximide solution | 4.0mL |

pH 7.5 ± 0.2 at 25°C

## Basal Medium:

**Composition** per 950.0mL:

Beef extract............................................................10.0g
Peptone..................................................................10.0g
Yeast extract............................................................6.0g
NaCl........................................................................5.0g

**Preparation of Basal Medium:** Add components to distilled/deionized water and bring volume to 950.0mL. Mix thoroughly. Autoclave for 15 min at 15 psi pressure–121°C. Cool to 45°–50°C.

## FBP Solution:

**Composition** per 100.0mL:

$FeSO_4·7H_2O$..........................................................6.25g
$Na_2S_2O_5$ ................................................................6.25g
Sodium pyruvate.....................................................6.25g

**Preparation of FBP:** Add components to distilled/deionized water and bring volume to 100.0mL. Mix thoroughly. Filter sterilize.

## Cefoperazone Solution:

**Composition** per 10.0mL:

Cefoperazone ........................................................0.037g

**Preparation of Cefoperazone Solution:** Add cefoperazone to distilled/deionized water and bring volume to 10.0mL. Mix thoroughly. Filter sterilize.

## Trimethoprim Lactate Solution:

**Composition** per 10.0mL:

Trimethoprim lactate..............................................0.031g

**Preparation of Trimethoprim Lactate Solution:** Add trimethoprim lactate to distilled/deionized water and bring volume to 10.0mL. Mix thoroughly. Filter sterilize.

## Vancomycin Solution:

**Composition** per 10.0mL:

Vancomycin ..........................................................0.025g

**Preparation of Vancomycin Solution:** Add vancomycin to distilled/deionized water and bring volume to 10.0mL. Mix thoroughly. Filter sterilize.

## Cycloheximide Solution:

**Composition** per 10.0mL:

Cycloheximide .......................................................0.025g

**Preparation of Cycloheximide Solution:** Add cycloheximide to distilled/deionized water and bring volume to 10.0mL. Mix thoroughly. Filter sterilize.

**Caution:** Cycloheximide is toxic. Avoid skin contact or aerosol formation and inhalation.

**Preparation of Medium:** To 950.0mL of cooled sterile basal medium, aseptically add 50.0mL of lysed (fresh, frozen, and thawed) horse blood, 4.0mL of sterile FBP solution, 8.0mL of sterile cefoperazone solution, 4.0mL of sterile trimethoprim lactate solution, 4.0mL of sterile vancomycin solution, and 4.0mL of sterile cycloheximide solution. Mix thoroughly. Aseptically distribute into sterile screw-capped tubes or bottles. Close caps tightly to reduce $O_2$ absorption. Use within 2 weeks.

**Storage/Shelf Life:** Store dehydrated media in the dark in a sealed container below 30°C. Prepared media should be stored under refrigeration (2-8°C). Media should be used within 60 days of preparation. Media should not be used if there are any signs of deterioration (discoloration) or contamination, or if the expiration date supplied by the manufacturer has passed.

**Use:** For the selective isolation and cultivation of *Campylobacter* species.

## *Campylobacter* Enrichment Broth
## (FDA Medium M29)

**Composition** per 1020.0mL:

Basal medium .................................................... 950.0mL
Horse blood, lysed ............................................... 50.0mL
FBP solution ........................................................ 4.0mL
Cefoperazone solution .......................................... 4.0mL
Trimethoprim lactate solution............................... 4.0mL
Rifampicin solution .............................................. 4.0mL
Cycloheximide solution ........................................ 4.0mL

pH 7.5 ± 0.2 at 25°C

## Basal Medium:

**Composition** per 950.0mL:

Beef extract............................................................10.0g
Peptone .................................................................10.0g
Yeast extract............................................................6.0g
NaCl........................................................................5.0g

**Preparation of Basal Medium:** Add components to distilled/deionized water and bring volume to 950.0mL. Mix thoroughly. Autoclave for 15 min at 15 psi pressure–121°C. Cool to 45°–50°C.

## FBP Solution:

**Composition** per 100.0mL:

$FeSO_4·7H_2O$..........................................................6.25g
$Na_2S_2O_5$ ................................................................6.25g
Sodium pyruvate.....................................................6.25g

**Preparation of FBP Solution:** Add components to distilled/deionized water and bring volume to 100.0mL. Mix thoroughly. Filter sterilize.

## Cefoperazone Solution:

**Composition** per 10.0mL:

Cefoperazone ........................................................0.037g

**Preparation of Cefoperazone Solution:** Add cefoperazone to distilled/deionized water and bring volume to 10.0mL. Mix thoroughly. Filter sterilize.

## Trimethoprim Lactate Solution:

**Composition** per 10.0mL:

Trimethoprim lactate..............................................0.031g

**Preparation of Trimethoprim Lactate Solution:** Add trimethoprim lactate to distilled/deionized water and bring volume to 10.0mL. Mix thoroughly. Filter sterilize.

## Rifampicin Solution:

**Composition** per 100.0mL:

Rifampicin .............................................................0.25g
Ethanol, absolute.................................................50.0mL

**Preparation of Rifampicin Solution:** Add rifampicin to 50.0mL of ethanol. Mix thoroughly. Bring volume to 100.0mL with distilled/deionized water. Filter sterilize.

## Cycloheximide Solution:

**Composition** per 10.0mL:

Cycloheximide.......................................................0.025g

**Preparation of Cycloheximide Solution:** Add cycloheximide to distilled/deionized water and bring volume to 10.0mL. Mix thoroughly. Filter sterilize.

**Caution:** Cycloheximide is toxic. Avoid skin contact or aerosol formation and inhalation.

**Preparation of Medium:** To 950.0mL of cooled sterile basal medium, aseptically add 50.0mL of lysed (fresh, frozen, and thawed) horse blood, 4.0mL of sterile FBP solution, 4.0mL of sterile cefoperazone solution, 4.0mL of sterile trimethoprim lactate solution, 4.0mL of sterile rifampicin solution, and 4.0mL of sterile cycloheximide solution. Mix thoroughly. Aseptically distribute into sterile screw-capped tubes or bottles. Close caps tightly to reduce $O_2$ absorption. Use within 2 weeks.

**Storage/Shelf Life:** Store dehydrated media in the dark in a sealed container below 30°C. Prepared media should be stored under refrigeration (2-8°C). Media should be used within 60 days of preparation. Media should not be used if there are any signs of deterioration (discoloration) or contamination, or if the expiration date supplied by the manufacturer has passed.

**Use:** For the selective isolation and cultivation of *Campylobacter* species from dairy products.

### *Campylobacter* Enrichment Broth

**Composition** per liter:
| | |
|---|---|
| Beef extract | 10.0g |
| Peptone | 10.0g |
| Yeast extract | 6.0g |
| NaCl | 5.0g |
| Horse blood, laked | 50.0mL |
| FBP solution | 4.0mL |
| Antibiotic solution | 4.0mL |

pH 7.5 ± 0.2 at 25°C

**Horse Blood, Laked:**
**Composition** per 50.0mL:
| | |
|---|---|
| Horse blood, fresh | 50.0mL |

**Preparation of Horse Blood, Laked:** Add blood to a sterile polypropylene bottle. Freeze overnight at –20°C. Thaw at 8°C. Refreeze at –20°C. Thaw again at 8°C.

**FBP Solution:**
**Composition** per 100.0mL:
| | |
|---|---|
| FeSO₄ | 6.25g |
| NaHSO₃ | 6.25g |
| Sodium pyruvate | 6.25g |

**Preparation of FBP Solution:** Add components to distilled/deionized water and bring volume to 100.0mL. Mix thoroughly. Filter sterilize.

**Antibiotic Solution:**
**Composition** per 10.0mL:
| | |
|---|---|
| Cycloheximide | 0.1g |
| Sodium cefoperazone | 0.03g |
| Trimethoprim lactate | 0.0125g |
| Vancomycin | 0.01g |

**Preparation of Antibiotic Solution:** Add components to distilled/deionized water and bring volume to 10.0mL. Mix thoroughly. Filter sterilize.

**Caution:** Cycloheximide is toxic. Avoid skin contact or aerosol formation and inhalation.

**Preparation of Medium:** Add components—except horse blood, FBP solution, and antibiotic solution—to distilled/deionized water and bring volume to 942.0mL. Mix thoroughly. Gently heat and bring to boiling. Autoclave for 15 min at 15 psi pressure–121°C. Cool to 45°–50°C. Aseptically add 50.0mL of sterile horse blood, 4.0mL of FBP so-

lution, and 4.0mL of antibiotic solution. Mix thoroughly. Pour into sterile Petri dishes or distribute into sterile tubes.

**Storage/Shelf Life:** Store dehydrated media in the dark in a sealed container below 30°C. Prepared media should be stored under refrigeration (2-8°C). Media should be used within 60 days of preparation. Media should not be used if there are any signs of deterioration (discoloration) or contamination, or if the expiration date supplied by the manufacturer has passed.

**Use:** For the isolation and cultivation of *Campylobacter* species from foods.

### *Campylobacter* Enrichment Broth
### (BAM M28a)

**Composition** per 1016.0mL:
| | |
|---|---|
| Peptone | 10.0g |
| Yeast extract | 5.0g |
| Lactalbumin hydrolysate | 5.0g |
| NaCl | 5.0g |
| α-Ketoglutamic acid | 1.0g |
| Sodium pyruvate | 0.5g |
| Na₂CO₃ | 0.6g |
| Na₂S₂O₅ | 0.5g |
| Hemin | 0.01g |
| Cefoperazone solution | 4.0mL |
| Trimethoprim lactate solution | 4.0mL |
| Vancomycin solution | 4.0mL |
| Cycloheximide or amphotericin B solution | 4.0mL |

pH 7.4 ± 0.2 at 25°C

**Cefoperazone Solution:**
**Composition** per 10.0mL:
| | |
|---|---|
| Cefoperazone | 0.05g |

**Preparation of Cefoperazone Solution:** Add cefoperazone to distilled/deionized water and bring volume to 10.0mL. Mix thoroughly. Filter sterilize. Can be stored for 5 days at 4°C, 14 days at –20°C, and 5 months at –70°C.

**Trimethoprim Solution:**
**Composition** per 10.0mL:
| | |
|---|---|
| Trimethoprim lactate | 0.066g |

**Preparation of Trimethoprim Solution:** Add trimethoprim lactate to distilled/deionized water and bring volume to 10.0mL. Mix thoroughly. Filter sterilize. Alternately add 0.05g trimethoprim hydrochloride to 3.0mL 0.05*N* HCl. Heat to 50°C. Stir until dissolved. Add distilled/deionized water and bring volume to 10.0mL. Mix thoroughly. Filter sterilize. Can be stored for 1 year at 4°C.

**Vancomycin Solution:**
**Composition** per 10.0mL:
| | |
|---|---|
| Vancomycin | 0.05g |

**Preparation of Vancomycin Solution:** Add vancomycin to distilled/deionized water and bring volume to 10.0mL. Mix thoroughly. Filter sterilize. Can be stored for 2 months at 4°C.

**Cycloheximide Solution:**
**Composition** per 10.0mL:
| | |
|---|---|
| Cycloheximide | 0.025g |
| Ethanol | 2.0mL |

**Preparation of Cycloheximide Solution:** Add cycloheximide to 2.0mL ethanol to dissolve. Mix thoroughly. Add distilled/deionized water and bring volume to 10.0mL. Mix thoroughly. Filter sterilize. Can be stored for 1 year at 4°C.

**Caution:** Cycloheximide is toxic. Avoid skin contact or aerosol formation and inhalation.

**Amphotericin B Solution:**
**Composition** per 10.0mL:
Amphotericin B..................................................................0.005g

**Preparation of Amphotericin B Solution:** Add Amphotericin B to distilled/deionized water and bring volume to 10.0mL. Mix thoroughly. Filter sterilize. Can be stored for 1 year at –20°C.

**Preparation of Medium:** Add components, except antimicrobic solutions, to distilled/deionized water and bring volume to 1.0L. Mix thoroughly. Autoclave for 15 min at 15 psi pressure–121°C. Cool to 25°C. Aseptically add 4.0mL of sterile cefoperazone solution, 4.0mL of sterile trimethoprim lactate solution, 4.0mL of sterile vancomycin solution, and either 4.0mL of sterile cycloheximide solution or 4.0mL of sterile amphotericin B solution. Mix thoroughly. Aseptically distribute into sterile screw-capped tubes or bottles. Close caps tightly to reduce $O_2$ absorption. Use within 2 weeks.

**Storage/Shelf Life:** Store dehydrated media in the dark in a sealed container below 30°C. Prepared media should be stored under refrigeration (2-8°C). Media should be used within 60 days of preparation. Media should not be used if there are any signs of deterioration (discoloration) or contamination, or if the expiration date supplied by the manufacturer has passed.

**Use:** For the selective isolation and cultivation of *Campylobacter* species. For the pre-enrichment of *Campylobacter* spp. in food samples. This medium aids resuscitation of sublethally injured cells and overcomes the damaging effects of food processing.

## *Campylobacter* Enrichment HiVeg Broth Base with Blood and Antibiotics (Preston Enrichment HiVeg Broth Base)

**Composition** per liter:
Plant extract ......................................................................10.0g
Plant peptone.....................................................................10.0g
NaCl....................................................................................5.0g
Horse blood, lysed ..........................................................50.0mL
Antibiotic supplement .....................................................10.0mL
pH 7.5 ± 0.2 at 25°C

**Source:** This medium, without horse blood and antibiotic supplement, is available as a premixed powder from HiMedia.

**Antibiotic Supplement:**
**Composition** per 10.0mL:
Cycloheximide....................................................................0.1g
Rifampicin .........................................................................0.01g
Trimethoprim lactate.........................................................0.01g
Polmyxin B .....................................................................5000U

**Preparation of Antibiotic Supplement:** Add components to 10.0mL of 50:50 acetone:distilled/deionized water. Filter sterilize.

**Caution:** Cycloheximide is toxic. Avoid skin contact or aerosol formation and inhalation.

**Preparation of Medium:** Add components, except horse blood and antibiotic supplement, to distilled/deionized water and bring volume to 940.0mL. Mix thoroughly. Gently heat and bring to boiling. Autoclave for 15 min at 15 psi pressure–121°C. Cool to 45°–50°C. Aseptically add 50.0mL of lysed horse blood and 10.0mL of sterile antibiotic supplement. Mix thoroughly. Pour into sterile Petri dishes.

**Storage/Shelf Life:** Store dehydrated media in the dark in a sealed container below 30°C. Prepared media should be stored under refrigeration (2-8°C). Media should be used within 60 days of preparation. Media should not be used if there are any signs of deterioration (discoloration) or contamination, or if the expiration date supplied by the manufacturer has passed.

**Use:** For the selective isolation of *Campylobacter* species.

## *Campylobacter fetus* Medium

**Composition** per 1160.0mL:
Fluid thioglycolate agar ....................................................1.0L
Sheep blood, defibrinated ...............................................150.0mL
Antibiotic solution .........................................................10.0mL
pH 7.1 ± 0.2 at 25°C

**Fluid Thioglycolate Agar:**
**Composition** per liter:
Agar ..................................................................................15.0g
Pancreatic digest of casein...............................................15.0g
Glucose ..............................................................................5.5g
Yeast extract......................................................................5.0g
NaCl...................................................................................2.5g
Agar .................................................................................0.75g
L-Cystine ..........................................................................0.5g
Sodium thioglycolate ........................................................0.5g
Resazurin ........................................................................1.0mg

**Preparation of Fluid Thioglycolate Agar:** Add components to distilled/deionized water and bring volume to 1.0L. Mix thoroughly. Gently heat and bring to boiling. Autoclave for 15 min at 15 psi pressure–121°C. Cool to 25°C.

**Antibiotic Solution:**
**Composition** per 10.0mL:
Cycloheximide..................................................................0.05g
Novobiocin .......................................................................5.0mg
Bacitracin....................................................................25,000U
Polymyxin B sulfate .......................................................10,000U

**Preparation of Antibiotic Solution:** Add components to distilled/deionized water and bring volume to 10.0mL. Mix thoroughly. Filter sterilize.

**Caution:** Cycloheximide is toxic. Avoid skin contact or aerosol formation and inhalation.

**Preparation of Medium:** To 1.0L of cooled, sterile, fluid thioglycolate agar, aseptically add 150.0mL of sterile sheep blood and 10.0mL of sterile antibiotic solution.

**Storage/Shelf Life:** Store dehydrated media in the dark in a sealed container below 30°C. Prepared media should be stored under refrigeration (2-8°C). Media should be used within 60 days of preparation. Media should not be used if there are any signs of deterioration (discoloration) or contamination, or if the expiration date supplied by the manufacturer has passed.

**Use:** For the isolation and cultivation of *Campylobacter fetus* from human specimens.

## *Campylobacter fetus* Selective Medium

**Composition** per liter:
Fluid thioglycolate agar ....................................................1.0L
Sheep blood, defibrinated ...............................................150.0mL
Antibiotic solution .........................................................10.0mL

**Fluid Thioglycolate Agar:**

**Composition** per liter:

| | |
|---|---|
| Agar | 15.0g |
| Pancreatic digest of casein | 15.0g |
| Glucose | 5.5g |
| Yeast extract | 5.0g |
| NaCl | 2.5g |
| Agar | 0.75g |
| L-Cystine | 0.5g |
| Sodium thioglycolate | 0.5g |
| Resazurin | 1.0mg |

**Preparation of Fluid Thioglycolate Agar:** Add components to distilled/deionized water and bring volume to 1.0L. Mix thoroughly. Gently heat and bring to boiling. Autoclave for 15 min at 15 psi pressure–121°C. Cool to 25°C.

**Antibiotic Solution:**

**Composition** per 10.0mL:

| | |
|---|---|
| Cycloheximide | 0.05g |
| Cephalothin | 0.02g |
| Novobiocin | 5.0mg |
| Bacitracin | 25,000U |
| Colistin | 10,000U |

**Preparation of Antibiotic Solution:** Add components to distilled/deionized water and bring volume to 10.0mL. Mix thoroughly. Filter sterilize.

**Caution:** Cycloheximide is toxic. Avoid skin contact or aerosol formation and inhalation.

**Preparation of Medium:** To 1.0L of cooled, sterile, fluid thioglycolate agar, aseptically add 150.0mL of sterile sheep blood and 10.0mL of sterile antibiotic solution.

**Storage/Shelf Life:** Store dehydrated media in the dark in a sealed container below 30°C. Prepared media should be stored under refrigeration (2-8°C). Media should be used within 60 days of preparation. Media should not be used if there are any signs of deterioration (discoloration) or contamination, or if the expiration date supplied by the manufacturer has passed.

**Use:** For the isolation and cultivation of *Campylobacter fetus*.

## *Campylobacter* HiVeg Agar Base with Blood and Antibiotic Supplement

**Composition** per liter:

| | |
|---|---|
| Plant peptone No. 3 | 15.0g |
| Agar | 12.0g |
| NaCl | 5.0g |
| Yeast extract | 5.0g |
| Plant extract No. 2 | 2.5g |
| Horse blood, lysed | 50.0mL |
| Antibiotic supplement | 10.0mL |

pH 7.4 ± 0.2 at 25°C

**Source:** This medium, without horse blood and antibiotic supplement, is available as a premixed powder from HiMedia.

**Antibiotic Supplement:**

**Composition** per 10.0mL:

| | |
|---|---|
| Cycloheximide | 0.1g |
| Rifampicin | 0.01g |
| Trimethoprim lactate | 0.01g |
| Polmyxin B | 5000U |

**Preparation of Antibiotic Supplement:** Add components to 10.0mL of 50:50 acetone:distilled/deionized water. Filter sterilize.

**Caution:** Cycloheximide is toxic. Avoid skin contact or aerosol formation and inhalation.

**Preparation of Medium:** Add components, except horse blood and antibiotic supplement, to distilled/deionized water and bring volume to 940.0mL. Mix thoroughly. Gently heat and bring to boiling. Autoclave for 15 min at 15 psi pressure–121°C. Cool to 45°–50°C. Aseptically add 50.0mL of lysed horse blood and 10.0mL of sterile antibiotic supplement. Mix thoroughly. Pour into sterile Petri dishes.

**Storage/Shelf Life:** Store dehydrated media in the dark in a sealed container below 30°C. Prepared media should be stored under refrigeration (2-8°C). Media should be used within 60 days of preparation. Media should not be used if there are any signs of deterioration (shrinking, cracking, or discoloration) or contamination, or if the expiration date supplied by the manufacturer has passed.

**Use:** For the cultivation of *Campylobacter* species. Normally used with a selective supplement to suppress the growth of other bacterial species.

## *Campylobacter* HiVeg Agar Base with Blood and Selective Supplement

**Composition** per liter:

| | |
|---|---|
| Plant peptone No. 3 | 15.0g |
| Agar | 12.0g |
| NaCl | 5.0g |
| Yeast extract | 5.0g |
| Plant extract No. 2 | 2.5g |
| Horse blood, lysed | 50.0mL |
| Selectrive supplement solution | 10.0mL |

pH 7.4 ± 0.2 at 25°C

**Source:** This medium, without horse blood and antibiotic supplement, is available as a premixed powder from HiMedia.

**Selective Supplement Solution:**

**Composition** per 10.0mL:

| | |
|---|---|
| Cephalothin | 15.0mg |
| Vancomycin | 10.0mg |
| Trimethoprim | 5.0mg |
| Amphotericin B | 2.0mg |
| Polymyxin B sulfate | 2500 U |

**Preparation of Selective Supplement Solution:** Add components to distilled/deionized water and bring volume to 10.0mL. Mix thoroughly. Filter sterilize.

**Preparation of Medium:** Add components, except horse blood and selective supplement, to distilled/deionized water and bring volume to 940.0mL. Mix thoroughly. Gently heat and bring to boiling. Autoclave for 15 min at 15 psi pressure–121°C. Cool to 45°–50°C. Aseptically add 50.0mL of lysed horse blood and 10.0mL of sterile selective supplement. Mix thoroughly. Pour into sterile Petri dishes.

**Storage/Shelf Life:** Store dehydrated media in the dark in a sealed container below 30°C. Prepared media should be stored under refrigeration (2-8°C). Media should be used within 60 days of preparation. Media should not be used if there are any signs of deterioration (shrinking, cracking, or discoloration) or contamination, or if the expiration date supplied by the manufacturer has passed.

**Use:** For the cultivation of *Campylobacter* species. Normally used with a selective supplement to suppress the growth of other bacterial species.

## *Campylobacter* Isolation Agar A

**Composition** per liter:

| | |
|---|---|
| Agar | 12.0g |
| Beef extract | 10.0g |
| Peptone | 10.0g |
| NaCl | 5.0g |
| Charcoal | 4.0g |
| Casein hydrolysate | 3.0g |
| Yeast extract | 2.0g |
| Sodium deoxycholate | 1.0g |
| $FeSO_4$ | 0.25g |
| Sodium pyruvate | 0.25g |
| Antibiotic solution | 10.0mL |

pH $7.4 \pm 0.2$ at 25°C

**Antibiotic Solution:**
**Composition** per 10.0mL:

| | |
|---|---|
| Cycloheximide | 0.1g |
| Sodium cefoperazone | 0.03g |

**Preparation of Antibiotic Solution:** Add components to distilled/deionized water and bring volume to 10.0mL. Mix thoroughly. Filter sterilize.

**Caution:** Cycloheximide is toxic. Avoid skin contact or aerosol formation and inhalation.

**Preparation of Medium:** Add components, except antibiotic solution, to distilled/deionized water and bring volume to 990.0mL. Mix thoroughly. Gently heat and bring to boiling. Autoclave for 15 min at 15 psi pressure–121°C. Cool to 45°–50°C. Aseptically add sterile antibiotic solution. Mix thoroughly. Pour into sterile Petri dishes. Swirl flask while pouring to distribute charcoal.

**Storage/Shelf Life:** Store dehydrated media in the dark in a sealed container below 30°C. Prepared media should be stored under refrigeration (2-8°C). Media should be used within 60 days of preparation. Media should not be used if there are any signs of deterioration (shrinking, cracking, or discoloration) or contamination, or if the expiration date supplied by the manufacturer has passed.

**Use:** For the isolation and cultivation of *Campylobacter* species.

## *Campylobacter* Isolation Agar B
### (*Campy* Cefex Agar)

**Composition** per liter:

| | |
|---|---|
| Agar | 15.0g |
| Pancreatic digest of casein | 10.0g |
| Peptic digest of animal tissue | 10.0g |
| NaCl | 5.0g |
| Yeast extract | 2.0g |
| Glucose | 1.0g |
| $FeSO_4$ | 0.5g |
| Sodium pyruvate | 0.5g |
| $NaHSO_3$ | 0.35g |
| Horse blood, laked | 50.0mL |
| Antibiotic solution | 10.0mL |

pH $7.0 \pm 0.2$ at 25°C

**Horse Blood, Laked:**
**Composition** per 50.0mL:

| | |
|---|---|
| Horse blood, fresh | 50.0mL |

**Preparation of Horse Blood, Laked:** Add blood to a sterile polypropylene bottle. Freeze overnight at –20°C. Thaw at 8°C. Refreeze at –20°C. Thaw again at 8°C.

**Antibiotic Solution:**
**Composition** per 10.0mL:

| | |
|---|---|
| Cycloheximide | 0.1g |
| Sodium cefoperazone | 0.033g |

**Preparation of Antibiotic Solution:** Add components to distilled/deionized water and bring volume to 10.0mL. Mix thoroughly. Filter sterilize.

**Caution:** Cycloheximide is toxic. Avoid skin contact or aerosol formation and inhalation.

**Preparation of Medium:** Add components, except horse blood and antibiotic solution, to distilled/deionized water and bring volume to 940.0mL. Mix thoroughly. Gently heat and bring to boiling. Autoclave for 15 min at 15 psi pressure–121°C. Cool to 45°–50°C. Aseptically add sterile horse blood and antibiotic solution. Mix thoroughly. Pour into sterile Petri dishes or distribute into sterile tubes.

**Storage/Shelf Life:** Store dehydrated media in the dark in a sealed container below 30°C. Prepared media should be stored under refrigeration (2-8°C). Media should be used within 60 days of preparation. Media should not be used if there are any signs of deterioration (shrinking, cracking, or discoloration) or contamination, or if the expiration date supplied by the manufacturer has passed.

**Use:** For the isolation and cultivation of *Campylobacter* species.

## *Campylobacter* Selective Medium, Blaser-Wang
### (Blaser–Wang *Campylobacter* Medium)
### (Blaser's Agar)
### (Campy BAP Medium)

**Composition** per liter:

| | |
|---|---|
| *Brucella* agar base | 890.0mL |
| Sheep blood | 100.0mL |
| Antibiotic supplement | 10.0mL |

**Brucella Agar Base:**
**Composition** per 890.0mL:

| | |
|---|---|
| Agar | 15.0g |
| Glucose | 10.0g |
| Pancreatic digest of casein | 10.0g |
| NaCl | 5.0g |
| Peptic digest of animal tissue | 5.0g |

pH $7.5 \pm 0.2$ at 25°C

**Preparation of *Brucella* Agar Base:** Add components to distilled/deionized water and bring volume to 890.0mL. Mix thoroughly. Gently heat and bring to boiling. Autoclave for 15 min at 15 psi pressure–121°C. Cool to 45°–50°C.

**Antibiotic Supplement:**
**Composition** per 10.0mL:

| | |
|---|---|
| Cephalothin | 15.0mg |
| Vancomycin | 10.0mg |
| Trimethoprim | 5.0mg |
| Amphotericin B | 2.0mg |
| Polymyxin B | 2500U |

**Preparation of Antibiotic Supplement:** Add components to 10.0mL of distilled/deionized water. Filter sterilize.

**Preparation of Medium:** Prepare 890.0mL of *Brucella* agar base. Sterilize as directed. Cool to 50°–55°C and add 100.0mL of sheep blood or 50.0–70.0mL of laked horse blood. Laked blood is prepared by freezing whole blood overnight and thawing to room temperature.

Aseptically add 10.0mL of sterile antibiotic supplement. Mix thoroughly. Pour into sterile Petri dishes.

**Storage/Shelf Life:** Store dehydrated media in the dark in a sealed container below 30°C. Prepared media should be stored under refrigeration (2-8°C). Media should be used within 60 days of preparation. Media should not be used if there are any signs of deterioration (shrinking, cracking, or discoloration) or contamination, or if the expiration date supplied by the manufacturer has passed.

**Use:** For the selective isolation of *Campylobacter* species.

## *Campylobacter* Selective Medium, Blaser-Wang (Blaser–Wang *Campylobacter* Medium)

**Composition** per liter:
| | |
|---|---|
| Columbia agar base | 890.0mL |
| Sheep blood | 100.0mL |
| Antibiotic supplement | 10.0mL |

pH 7.3 ± 0.2 at 25°C

**Antibiotic Supplement:**
**Composition** per 10.0mL:
| | |
|---|---|
| Cephalothin | 15.0mg |
| Vancomycin | 10.0mg |
| Trimethoprim | 5.0mg |
| Amphotericin B | 2.0mg |
| Polymyxin B | 2,500U |

**Preparation of Antibiotic Supplement:** Add components to 10.0mL of distilled/deionized water. Filter sterilize.

**Columbia Agar Base:**
**Composition** per liter:
| | |
|---|---|
| Special peptone | 25.0g |
| Agar | 10.0g |
| NaCl | 5.0g |
| Starch | 1.0g |

**Preparation of Columbia Agar Base:** Add components to distilled/deionized water and bring volume to 890.0mL. Mix thoroughly. Gently heat and bring to boiling. Autoclave for 15 min at 15 psi pressure–121°C. Cool to 45°–50°C.

**Preparation of Medium:** To 890.0mL of cooled, sterile Columbia agar base, aseptically add 100.0mL of sheep blood or 50.0–70.0mL of laked horse blood. Laked blood is prepared by freezing whole blood overnight and thawing to room temperature. Aseptically add 10.0mL of sterile antibiotic supplement. Mix thoroughly. Pour into sterile Petri dishes.

**Storage/Shelf Life:** Store dehydrated media in the dark in a sealed container below 30°C. Prepared media should be stored under refrigeration (2-8°C). Media should be used within 60 days of preparation. Media should not be used if there are any signs of deterioration (shrinking, cracking, or discoloration) or contamination, or if the expiration date supplied by the manufacturer has passed.

**Use:** For the selective isolation of *Campylobacter* species.

## *Campylobacter* Selective Medium, Butzler's (Butzler's *Campylobacter* Medium)

**Composition** per liter:
| | |
|---|---|
| *Brucella* agar base | 940.0mL |
| Sheep or horse blood, defibrinated | 50.0mL |
| Antibiotic supplement | 10.0mL |

pH 7.5 ± 0.2 at 25°C

*Brucella* **Agar Base:**
**Composition** per liter:
| | |
|---|---|
| Agar | 15.0g |
| Glucose | 10.0g |
| Pancreatic digest of casein | 10.0g |
| NaCl | 5.0g |
| Peptic digest of animal tissue | 5.0g |

**Preparation of *Brucella* Agar Base:** Add components to distilled/deionized water and bring volume to 940.0mL. Mix thoroughly. Gently heat and bring to boiling. Autoclave for 15 min at 15 psi pressure–121°C. Cool to 45°–50°C.

**Antibiotic Supplement:**
**Composition** per 10.0mL:
| | |
|---|---|
| Cycloheximide | 50.0mg |
| Cephazolin | 15.0mg |
| Novobiocin | 5.0mg |
| Bacitracin | 25,000U |
| Colistin sulfate | 10,000U |

**Preparation of Antibiotic Supplement:** Add components to 10.0mL of distilled/deionized water. Filter sterilize.

**Caution:** Cycloheximide is toxic. Avoid skin contact or aerosol formation and inhalation.

**Preparation of Medium:** To 940.0mL of cooled, sterile *Brucella* agar base, aseptically add 50.0mL of defibrinated sheep or horse blood and 10.0mL of sterile antibiotic supplement. Mix thoroughly. For enhanced growth, medium may also be supplemented with 0.25g of $Fe_2SO_4 \cdot H_2O$, 0.25g of sodium metabisulfite, and 0.25g of sodium pyruvate. Pour into sterile Petri dishes.

**Storage/Shelf Life:** Store dehydrated media in the dark in a sealed container below 30°C. Prepared media should be stored under refrigeration (2-8°C). Media should be used within 60 days of preparation. Media should not be used if there are any signs of deterioration (shrinking, cracking, or discoloration) or contamination, or if the expiration date supplied by the manufacturer has passed.

**Use:** For the selective isolation of *Campylobacter* species.

## *Campylobacter* Selective Medium, Butzler's (Butzler's *Campylobacter* Medium)

**Composition** per liter:
| | |
|---|---|
| Columbia agar base | 940.0mL |
| Blood, horse or sheep | 50.0mL |
| Antibiotic supplement | 10.0mL |

pH 7.3 ± 0.2 at 25°C

**Columbia Agar Base:**
**Composition** per liter:
| | |
|---|---|
| Peptone | 25.0g |
| Agar | 10.0g |
| NaCl | 5.0g |
| Starch | 1.0g |

**Preparation of Columbia Agar Base:** Add components to distilled/deionized water and bring volume to 940.0mL. Mix thoroughly. Gently heat and bring to boiling. Autoclave for 15 min at 15 psi pressure–121°C. Cool to 45°–50°C.

**Antibiotic Supplement:**
**Composition** per 10.0mL:
| | |
|---|---|
| Cycloheximide | 50.0mg |
| Cephazolin | 15.0mg |
| Novobiocin | 5.0mg |

Bacitracin ..................................................................25,000U
Colistin sulfate ...........................................................10,000U

**Preparation of Antibiotic Supplement:** Add components to 10.0mL of distilled/deionized water. Filter sterilize.

**Caution:** Cycloheximide is toxic. Avoid skin contact or aerosol formation and inhalation.

**Preparation of Medium:** To 940.0mL of cooled, sterile Columbia agar base, aseptically add 50.0mL of defibrinated sheep or horse blood and 10.0mL of sterile antibiotic supplement. Mix thoroughly. The medium may also be supplemented with 0.25g of $Fe_2SO_4·H_2O$, 0.25g of sodium metabisulfite, and 0.25g of sodium pyruvate. Pour into sterile Petri dishes.

**Storage/Shelf Life:** Store dehydrated media in the dark in a sealed container below 30°C. Prepared media should be stored under refrigeration (2-8°C). Media should be used within 60 days of preparation. Media should not be used if there are any signs of deterioration (shrinking, cracking, or discoloration) or contamination, or if the expiration date supplied by the manufacturer has passed.

**Use:** For the selective isolation of *Campylobacter* species.

## *Campylobacter* Selective Medium, Karmali's (Karmali's *Campylobacter* Medium)

**Composition** per liter:
Activated charcoal .............................................................4.0g
Columbia agar base..........................................................990.0mL
Antibiotic supplement.......................................................10.0mL
pH 7.4 ± 0.2 at 25°C

**Source:** This medium is available as a premixed powder from Thermo Scientific.

**Columbia Agar Base:**
**Composition** per 990.0mL:
Peptone...........................................................................25.0g
Agar ................................................................................10.0g
NaCl..................................................................................5.0g
Starch ...............................................................................1.0g

**Preparation of Columbia Agar Base:** Add components to distilled/deionized water and bring volume to 990.0mL. Mix thoroughly. Gently heat and bring to boiling. Autoclave for 15 min at 15 psi pressure–121°C. Cool to 45°–50°C.

**Antibiotic Supplement:**
**Composition** per 10.0mL:
Sodium pyruvate .............................................................0.05g
Cycloheximide.................................................................0.05g
Cefoperazone.................................................................0.016g
Hemin.............................................................................0.016g
Vancomycin .....................................................................0.01g

**Preparation of Antibiotic Supplement:** Add components to 10.0mL of distilled/deionized water. Filter sterilize.

**Caution:** Cycloheximide is toxic. Avoid skin contact or aerosol formation and inhalation.

**Preparation of Medium:** Prepare 990.0mL of Columbia agar base. Sterilize as directed. Cool to 50°–55°C. Add defibrinated sheep or horse blood to a final concentration of 5–7%. Add 10.0mL of sterile antibiotic supplement. Mix thoroughly. For enhanced growth, medium may also be supplemented with 0.25g of $Fe_2SO_4·H_2O$, 0.25g of sodium metabisulfite, and 0.25g of sodium pyruvate. Pour into sterile Petri dishes. Swirl while pouring to keep charcoal in suspension.

**Storage/Shelf Life:** Store dehydrated media in the dark in a sealed container below 30°C. Prepared media should be stored under refrigeration (2-8°C). Media should be used within 60 days of preparation. Media should not be used if there are any signs of deterioration (shrinking, cracking, or discoloration) or contamination, or if the expiration date supplied by the manufacturer has passed.

**Use:** For the selective isolation of *Campylobacter* species.

## *Campylobacter* Selective Medium, Preston's (Preston's *Campylobacter* Medium)

**Composition** per liter:
*Campylobacter* agar base....................................................940.0mL
Horse blood, lysed ...........................................................50.0mL
Antibiotic supplement.......................................................10.0mL
pH 7.5 ± 0.2 at 25°C

**Campylobacter Agar Base:**
**Composition** per liter:
Agar ...............................................................................12.0g
Beef extract....................................................................10.0g
Peptone ..........................................................................10.0g
NaCl..................................................................................5.0g

**Preparation of *Campylobacter* Agar Base:** Add components to distilled/deionized water and bring volume to 940.0mL. Mix thoroughly. Gently heat and bring to boiling. Autoclave for 15 min at 15 psi pressure–121°C. Cool to 45°–50°C.

**Antibiotic Supplement:**
**Composition** per 10.0mL:
Cycloheximide..................................................................0.1g
Rifampicin ......................................................................0.01g
Trimethoprim lactate.......................................................0.01g
Polmyxin B ...................................................................5000U

**Preparation of Antibiotic Supplement:** Add components to 10.0mL of 50:50 acetone:distilled/deionized water. Filter sterilize.

**Caution:** Cycloheximide is toxic. Avoid skin contact or aerosol formation and inhalation.

**Preparation of Medium:** To 940.0mL of cooled, sterile *Campylobacter* agar base, aseptically add 50.0mL of lysed horse blood and 10.0mL of sterile antibiotic supplement. Mix thoroughly. Pour into sterile Petri dishes.

**Storage/Shelf Life:** Store dehydrated media in the dark in a sealed container below 30°C. Prepared media should be stored under refrigeration (2-8°C). Media should be used within 60 days of preparation. Media should not be used if there are any signs of deterioration (shrinking, cracking, or discoloration) or contamination, or if the expiration date supplied by the manufacturer has passed.

**Use:** For the selective isolation of *Campylobacter* species.

## *Campylobacter* Thioglycolate Medium with 5 Antimicrobics

**Composition** per liter:
Pancreatic digest of casein...............................................17.0g
Glucose ............................................................................6.0g
Papaic digest of soybean meal............................................3.0g
NaCl..................................................................................2.5g
Agar ..................................................................................1.6g
Sodium thioglycolate.........................................................0.5g

Na$_2$SO$_3$..............................................................................0.1g
Antibiotic supplement solution ................................. 10.0mL
pH 7.0 ± 0.2 at 25°C

## Antibiotic Supplement Solution:
**Composition** per 10.0mL:
Cephalothin.....................................................................0.015g
Vancomycin ......................................................................0.01g
Trimethoprim ..................................................................5.0mg
Amphotericin B................................................................2.0mg
Polymyxin B ....................................................................2500U

**Preparation of Antibiotic Supplement Solution:** Add components to 10.0mL of distilled/deionized water. Mix thoroughly. Filter sterilize.

**Preparation of Medium:** Add components, except cephalothin, vancomycin, trimethoprim, amphotericin, and polymyxin B, to distilled deionized water and bring volume to 990.0mL. Mix thoroughly. Gently heat and bring to boiling. Autoclave for 15 min at 15 psi pressure–121°C. Cool to 45°–50°C. Add 10.0mL of sterile antibiotic supplement. Mix thoroughly. Pour into sterile Petri dishes.

**Storage/Shelf Life:** Store dehydrated media in the dark in a sealed container below 30°C. Prepared media should be stored under refrigeration (2-8°C). Media should be used within 60 days of preparation. Media should not be used if there are any signs of deterioration (discoloration) or contamination, or if the expiration date supplied by the manufacturer has passed.

**Use:** For the maintenence—as a holding medium or transport medium—of fecal specimens or swabs suspected of containing *Campylobacter jejuni* or other *Campylobacter* species when immediate inoculation of *Campylobacter* growth medium is unavailable.

## Cary and Blair Transport Medium
**Composition** per liter:
Agar ...................................................................................5.0g
NaCl...................................................................................5.0g
Sodium thioglycolate .......................................................1.5g
Na$_2$HPO$_4$.............................................................................1.1g
CaCl$_2$ solution ............................................................. 9.0mL
pH 8.0 ± 0.5 at 25°C

**Source:** This medium is available as a premixed powder from BD Diagnostic Systems and Thermo Scientific.

## CaCl$_2$ Solution:
**Composition** per 10.0mL:
CaCl$_2$.................................................................................0.1g

**Preparation of CaCl$_2$ Solution:** Add CaCl$_2$ to distilled/deionized water and bring volume to 10.0mL. Mix thoroughly. Filter sterilize.

**Preparation of Medium:** Add components to distilled/deionized water and bring volume to 1.0L. Mix thoroughly and heat gently until boiling. Cool to 50°C. Add 9.0mL of a 1% CaCl$_2$ solution. Adjust the pH to 8.4. Distribute into screw-capped tubes in 7.0mL volumes. Sterilize under flowing steam for 15 min. After sterilization, tighten the screwcaps.

**Storage/Shelf Life:** Store dehydrated media in the dark in a sealed container below 30°C. Prepared media should be stored under refrigeration (2-8°C). Media should be used within 60 days of preparation. Media should not be used if there are any signs of deterioration (discoloration) or contamination, or if the expiration date supplied by the manufacturer has passed.

**Use:** For the maintenance—as a holding medium or transport medium—of clinical specimens during collection or shipment.

## Cary and Blair Transport Medium, Modified
**Composition** per liter:
Agar ...................................................................................5.0g
NaCl...................................................................................5.0g
Sodium thioglycolate .......................................................1.5g
L-Cysteine·HCl·H$_2$O.........................................................0.5g
CaCl$_2$·2H$_2$O.....................................................................0.1g
Na$_2$HPO$_4$.............................................................................0.1g
NaHSO$_3$ ...........................................................................0.1g
Resazurin solution ......................................................... 4.0mL
pH 8.4 ± 0.2 at 25°C

## Resazurin Solution:
**Composition** per 380.0mL:
Resazurin ........................................................................0.05g
Ethanol (95% solution).............................................. 200.0mL

**Preparation of Resazurin Solution:** Add resazurin to 200.0mL of ethanol. Mix thoroughly. Bring volume to 380.0mL with distilled/deionized water.

**Preparation of Medium:** Add components, except L-cysteine·HCl·H$_2$O, to distilled/deionized water and bring volume to 1.0L. Mix thoroughly. Gas the solution with 100% $CO_2$ for 10–15 min. Add the L-cysteine·HCl·H$_2$O. Mix thoroughly. Adjust pH to 8.4. Anaerobically distribute into tubes under 100% N$_2$. Cap tubes with butyl rubber stoppers. Autoclave for 15 min at 0 psi pressure–100°C on 3 consecutive days.

**Storage/Shelf Life:** Store dehydrated media in the dark in a sealed container below 30°C. Prepared media should be stored under refrigeration (2-8°C). Media should be used within 60 days of preparation. Media should not be used if there are any signs of deterioration (discoloration) or contamination, or if the expiration date supplied by the manufacturer has passed.

**Use:** For the maintenance—as a holding medium—of clinical specimens during collection or shipment.

## Casamino Peptone Czapek Medium
**Composition** per liter:
Sucrose............................................................................30.0g
Agar .................................................................................15.0g
Peptone .............................................................................2.0g
Casamino acids .................................................................1.0g
K$_2$HPO$_4$..............................................................................1.0g
KCl....................................................................................0.5g
MgSO$_4$·7H$_2$O....................................................................0.5g
FeSO$_4$·7H$_2$O....................................................................0.01g

**Preparation of Medium:** Add components to distilled/deionized water and bring volume to 1.0L. Mix thoroughly. Gently heat to boiling. Distribute into tubes or flasks. Autoclave for 15 min at 15 psi pressure–121°C. Pour into sterile Petri dishes or leave in tubes.

**Storage/Shelf Life:** Store dehydrated media in the dark in a sealed container below 30°C. Prepared media should be stored under refrigeration (2-8°C). Media should be used within 60 days of preparation. Media should not be used if there are any signs of deterioration (shrinking, cracking, or discoloration) or contamination, or if the expiration date supplied by the manufacturer has passed.

**Use:** For the cultivation and maintenance of *Actinoplanes* species, *Pseudonocardia compacta*, and *Streptomyces* species.

## Casein Agar

**Composition** per liter:

Agar .................................................................10.0g
Skim milk.......................................................50.0mL

**Preparation of Medium:** Add components to distilled/deionized water and bring volume to 1.0L. Mix thoroughly. Gently heat and bring to boiling. Distribute into tubes or flasks. Autoclave for 15 min at 15 psi pressure–121°C. Pour into sterile Petri dishes or leave in tubes.

**Storage/Shelf Life:** Store dehydrated media in the dark in a sealed container below 30°C. Prepared media should be stored under refrigeration (2-8°C). Media should be used within 60 days of preparation. Media should not be used if there are any signs of deterioration (shrinking, cracking, or discoloration) or contamination, or if the expiration date supplied by the manufacturer has passed.

**Use:** For the cultivation and differentiation of aerobic actinomycetes based on casein utilization. Bacteria that utilize casein, such as *Streptomyces* and *Actinomadura* species, appear as colonies surrounded by a clear zone. *Nocardia asteroides*, *Nocardia caviae*, and *Mycobacterium fortuitum* do not utilize casein.

## Casein Hydrolysate Yeast Extract HiVeg Broth
## (CAYE HiVeg Broth)

**Composition** per liter:

Plant acid hydrolysate ...................................30.0g
Yeast extract....................................................4.0g
Glucose ............................................................2.0g
K₂HPO₄.............................................................0.5g

$$pH\ 7.0 \pm 0.2\ at\ 25°C$$

**Source:** This medium is available as a premixed powder from Hi-Media.

**Preparation of Medium:** Add components to distilled/deionized water and bring volume to 1.0L. Mix thoroughly. Gently heat and bring to boiling. Distribute into tubes or flasks. Autoclave for 15 min at 15 psi pressure–121°C.

**Storage/Shelf Life:** Store dehydrated media in the dark in a sealed container below 30°C. Prepared media should be stored under refrigeration (2-8°C). Media should be used within 60 days of preparation. Media should not be used if there are any signs of deterioration (discoloration) or contamination, or if the expiration date supplied by the manufacturer has passed.

**Use:** For cultivation of *Vibrio cholerae* while testing enterotoxigenicity.

## Casman Agar Base

**Composition** per liter:

Noble agar.......................................................14.0g
Proteose peptone No. 3 ..................................10.0g
Tryptose .........................................................10.0g
NaCl..................................................................5.0g
Beef extract......................................................3.0g
Cornstarch .......................................................1.0g
Glucose ............................................................0.5g
*p*-Aminobenzoic acid.....................................0.05g
Nicotinamide...................................................0.05g

Blood................................................................50.0mL
Water-lysed blood solution ...........................1.5mL

$$pH\ 7.3 \pm 0.2\ at\ 25°C$$

**Water-Lysed Blood Solution:**
**Composition** per 8.0mL:
Blood..................................................................2.0mL

**Preparation of Water-Lysed Blood Solution:** Add blood to distilled/deionized water and bring volume to 8.0mL. Mix thoroughly. Filter sterilize.

**Preparation of Medium:** Add components, except blood and water-lysed blood solution, to distilled/deionized water and bring volume to 948.5mL. Mix thoroughly. Gently heat to boiling. Autoclave for 15 min at 15 psi pressure–121°C. Cool to 50°C. Aseptically add 50.0mL of sterile blood and 1.5mL of sterile water-lysed blood solution (one part blood to three parts water). Water-lysed blood may be omitted if sterile blood is partially lysed due to storage. Mix thoroughly. Pour into sterile Petri dishes or distribute into sterile tubes.

**Storage/Shelf Life:** Store dehydrated media in the dark in a sealed container below 30°C. Prepared media should be stored under refrigeration (2-8°C). Media should be used within 60 days of preparation. Media should not be used if there are any signs of deterioration (shrinking, cracking, or discoloration) or contamination, or if the expiration date supplied by the manufacturer has passed.

**Use:** For the isolation of fastidious bacteria from clinical specimens. For the cultivation under reduced oxygen tension of fastidious microorganisms such as *Haemophilus influenzae*, *Neisseria meningitidis*, and *Neisseria gonorrhoeae*.

## Casman Agar Base with Rabbit Blood
## (Casman-Medium)
## (DSMZ Medium 439)

**Composition** per liter:

Noble agar.......................................................14.0g
Proteose peptone No. 3 ..................................10.0g
Tryptose .........................................................10.0g
NaCl..................................................................5.0g
Beef extract......................................................3.0g
Cornstarch .......................................................1.0g
Glucose ............................................................0.5g
*p*-Aminobenzoic acid.....................................0.05g
Nicotinamide...................................................0.05g
Rabbit blood....................................................50.0mL
Water-lysed blood solution ...........................1.5mL

$$pH\ 7.3 \pm 0.2\ at\ 25°C$$

**Source:** Casman agar base is available as a premixed powder from BD Diagnostic Systems.

**Water-Lysed Blood Solution:**
**Composition** per 8.0mL:
Rabbit blood......................................................2.0mL

**Preparation of Water-Lysed Blood Solution:** Add blood to distilled/deionized water and bring volume to 8.0mL. Mix thoroughly. Filter sterilize.

**Preparation of Medium:** Add components, except rabbit blood and water-lysed blood solution, to distilled/deionized water and bring volume to 950.0L. Mix thoroughly. Gently heat to boiling. Autoclave for 15 min at 15 psi pressure–121°C. Cool to 50°C. Aseptically add 50.0mL of sterile rabbit blood and 1.5mL of sterile water-lysed blood solution. Water-lysed blood may be omitted if sterile blood is partially lysed due to storage. Mix thoroughly. Pour into sterile Petri dishes or distribute into sterile tubes.

**Storage/Shelf Life:** Store dehydrated media in the dark in a sealed container below 30°C. Prepared media should be stored under refrigeration (2-8°C). Media should be used within 60 days of preparation. Media should not be used if there are any signs of deterioration (shrinking, cracking, or discoloration) or contamination, or if the expiration date supplied by the manufacturer has passed.

**Use:** For the cultivation and maintenance of *Gardnerella vaginalis*.

### Casman HiVeg Agar Base with Blood
**Composition** per liter:
| | |
|---|---|
| Agar | 14.0g |
| Plant hydrolysate No. 1 | 10.0g |
| Plant peptone No. 3 | 10.0g |
| NaCl | 5.0g |
| Plant extract | 3.0g |
| Corn starch | 1.0g |
| Glucose | 0.5g |
| Nicotinamide | 0.05g |
| *p*-Amino benzoic acid (PABA) | 0.05g |
| Blood | 50.0mL |
| Water-lysed blood solution | 1.5mL |

pH 7.3 ± 0.2 at 25°C

**Source:** This medium, without blood and water-lysed blood solution, is available as a premixed powder from HiMedia.

**Water-Lysed Blood Solution:**
**Composition** per 8.0mL:
| | |
|---|---|
| Blood | 2.0mL |

**Preparation of Water-Lysed Blood Solution:** Add blood to distilled/deionized water and bring volume to 8.0mL. Mix thoroughly. Filter sterilize.

**Preparation of Medium:** Add components, except blood and water-lysed blood solution, to distilled/deionized water and bring volume to 948.5mL. Mix thoroughly. Gently heat to boiling. Autoclave for 15 min at 15 psi pressure–121°C. Cool to 50°C. Aseptically add 50.0mL of sterile blood and 1.5mL of sterile water-lysed blood solution (one part blood to three parts water). Water-lysed blood may be omitted if sterile blood is partially lysed due to storage. Mix thoroughly. Pour into sterile Petri dishes or distribute into sterile tubes.

**Storage/Shelf Life:** Store dehydrated media in the dark in a sealed container below 30°C. Prepared media should be stored under refrigeration (2-8°C). Media should be used within 60 days of preparation. Media should not be used if there are any signs of deterioration (shrinking, cracking, or discoloration) or contamination, or if the expiration date supplied by the manufacturer has passed.

**Use:** For the isolation of fastidious bacteria from clinical specimens. For the cultivation under reduced oxygen tension of fastidious microorganisms such as *Haemophilus influenzae*, *Neisseria meningitidis*, and *Neisseria gonorrhoeae*.

### Casman HiVeg Broth Base with Blood
**Composition** per liter:
| | |
|---|---|
| Plant hydrolysate No. 1 | 10.0g |
| Plant peptone No. 3 | 10.0g |
| NaCl | 5.0g |
| Plant extract | 3.0g |
| Corn starch | 1.0g |

| | |
|---|---|
| Glucose | 0.5g |
| Nicotinamide | 0.05g |
| *p*-Amino benzoic acid (PABA) | 0.05g |
| Blood | 50.0mL |
| Water-lysed blood solution | 1.5mL |

pH 7.3 ± 0.2 at 25°C

**Source:** This medium, without blood and water-lysed blood solution, is available as a premixed powder from HiMedia.

**Water-Lysed Blood Solution:**
**Composition** per 8.0mL:
| | |
|---|---|
| Blood | 2.0mL |

**Preparation of Water-Lysed Blood Solution:** Add blood to distilled/deionized water and bring volume to 8.0mL. Mix thoroughly. Filter sterilize.

**Preparation of Medium:** Add components, except blood and water-lysed blood solution, to distilled/deionized water and bring volume to 948.5mL. Mix thoroughly. Gently heat to boiling. Autoclave for 15 min at 15 psi pressure–121°C. Cool to 50°C. Aseptically add 50.0mL of sterile blood and 1.5mL of sterile water-lysed blood solution (one part blood to three parts water). Water-lysed blood may be omitted if sterile blood is partially lysed due to storage. Mix thoroughly.

**Storage/Shelf Life:** Store dehydrated media in the dark in a sealed container below 30°C. Prepared media should be stored under refrigeration (2-8°C). Media should be used within 60 days of preparation. Media should not be used if there are any signs of deterioration (shrinking, cracking, or discoloration) or contamination, or if the expiration date supplied by the manufacturer has passed.

**Use:** For the cultivation of fastidious bacteria from clinical specimens. For the cultivation under reduced oxygen tension of fastidious microorganisms such as *Haemophilus influenzae*, *Neisseria meningitidis*, and *Neisseria gonorrhoeae*.

### CASO MUG Agar
**Composition** per liter:
| | |
|---|---|
| Casein peptone | 16.0g |
| Agar | 13.0g |
| NaCl | 6.0g |
| Soy peptone | 5.0g |
| Tryptophan | 1.0g |
| 4-Methylumbelliferyl-β-D-glucuronide | 0.07g |

pH 7.3 ± 0.2 at 25°C

**Source:** This medium is available from Fluka, Sigma-Aldrich.

**Preparation of Medium:** Add components to distilled/deionized water and bring volume to 1.0L. Mix thoroughly. Gently heat while stirring and bring to boiling. Autoclave for 15 min at 15 psi pressure–121°C. Cool to 50°C. Pour into sterile Petri dishes.

**Storage/Shelf Life:** Store dehydrated media in the dark in a sealed container below 30°C. Prepared media should be stored under refrigeration (2-8°C). Media should be used within 60 days of preparation. Media should not be used if there are any signs of deterioration (shrinking, cracking, or discoloration) or contamination, or if the expiration date supplied by the manufacturer has passed.

**Storage/Shelf Life:** Store dehydrated media in the dark in a sealed container below 30°C. Prepared media should be stored under refrigeration (2-8°C). This medium has a limited shelf life. Media should not be used if there are any signs of deterioration (shrinking, cracking, or discoloration) or contamination, or if the expiration date supplied by the manufacturer has passed.

**Use:** This universal medium without indicator or inhibitor is intended for a broad range of application, including enumeration and cultivation of a wide variety of microorganisms. It is also suitable for the cultivation of more fastidious microorganisms. β-D-glucoronidase, which is produced by *E. coli*, cleaves 4-methylumbelliferyl-β-D-glucuronide to 4-methylumbelliferone and glucuronide. The fluorogen 4-methylumbelliferone can be detected under a long wavelength UV lamp. A positive indole reaction provides confirmation.

## CCAT Medium
### (*Campylobacter* Blood Free Preson Agar with Cefoperazone, Amphotericin, and Teicoplanin)
**Composition** per liter:

| | |
|---|---|
| Agar | 12.0g |
| Beef extract | 10.0g |
| Peptone | 10.0g |
| NaCl | 5.0g |
| Charcoal | 4.0g |
| Casein hydrolysate | 3.0g |
| Sodium deoxycholate | 1.0g |
| FeSO$_4$ | 0.25g |
| Sodium pyruvate | 0.25g |
| Selective supplement solution | 10.0mL |

pH 7.5 ± 0.2 at 25°C

**Source:** This medium is available as a premixed powder from Thermo Scientific.

**Selective Supplement Solution:**
**Composition** per 10.0mL:

| | |
|---|---|
| Amphotericin | 10.0mg |
| Sodium cefoperazone | 8.0mg |
| Teicoplanin | 4.0mg |

**Preparation of Selective Supplement Solution:** Add sodium cefoperazone to distilled/deionized water and bring volume to 10.0mL. Mix thoroughly. Filter sterilize.

**Preparation of Medium:** Add components, except selective supplement solution, to distilled/deionized water and bring volume to 990.0mL. Mix thoroughly. Gently heat and bring to boiling. Autoclave for 15 min at 15 psi pressure–121°C. Cool to 45°–50°C. Aseptically add 10.0mL of sterile selective supplement solution. Mix thoroughly. Pour into sterile Petri dishes or distribute into sterile tubes.

**Storage/Shelf Life:** Store dehydrated media in the dark in a sealed container below 30°C. Prepared media should be stored under refrigeration (2-8°C). Media should be used within 60 days of preparation. Media should not be used if there are any signs of deterioration (shrinking, cracking, or discoloration) or contamination, or if the expiration date supplied by the manufacturer has passed.

**Use:** For the cultivation of *Campylobacter* species. For the isolation of Campylobacter spp., especially *Campylobacter upsaliensis*.

## CCD Agar with Pyruvate and Cefazolin
### (Blood-free Selective Medium)
**Composition** per liter:

| | |
|---|---|
| Agar | 12.0g |
| Beef extract | 10.0g |
| Peptone | 10.0g |
| NaCl | 5.0g |
| Charcoal, bacteriological | 4.0g |
| Casein hydrolysate | 3.0g |
| Sodium deoxycholate solution | 10.0mL |

| | |
|---|---|
| FeSO$_4$ solution | 5.0mL |
| Sodium pyruvate solution | 5.0mL |
| Cefazolin solution | 1.0mL |

pH 7.4 ± 0.2 at 25°C

**Source:** This medium, without deoxycholate and cefazolin solutions, is available as a premixed powder from HiMedia.

**FeSO$_4$ Solution:**
**Composition** per 10.0mL:

| | |
|---|---|
| FeSO$_4$ | 0.5g |

**Preparation of FeSO$_4$ Solution:** Add FeSO$_4$ to distilled/deionized water and bring volume to 10.0mL. Mix thoroughly. Gently heat while stirring and bring to boiling. Autoclave for 15 min at 15 psi pressure–121°C. Cool 25°C.

**Sodium Pyruvate Solution:**
**Composition** per 10.0mL:

| | |
|---|---|
| Sodium pyruvate | 0.5g |

**Preparation of Sodium Pyruvate Solution:** Add sodium pyruvate to distilled/deionized water and bring volume to 10.0mL. Mix thoroughly. Gently heat while stirring and bring to boiling. Autoclave for 15 min at 15 psi pressure–121°C. Cool 25°C.

**Sodium Deoxycholate Solution:**
**Composition** per 100.0mL:

| | |
|---|---|
| Sodium deoxycholate | 10.0g |

**Preparation of Sodium Deoxycholate Solution:** Add sodium deoxycholate to distilled/deionized water and bring volume to 100.0mL. Mix thoroughly. Gently heat while stirring and bring to boiling. Autoclave for 15 min at 15 psi pressure–121°C. Cool 25°C.

**Cefazolin Solution:**
**Composition** per 10.0mL:

| | |
|---|---|
| Cefazolin | 0.1g |

**Preparation of Cefazolin Solution:** Add cefazolin to distilled/deionized water and bring volume to 10.0mL. Mix thoroughly. Filter sterilize.

**Preparation of Medium:** Add components, except cefazolin solution and sodium deoxycholate solution, to distilled/deionized water and bring volume to 990.0mL. Mix thoroughly. Heat with frequent agitation and boil for 1 min to completely dissolve. Autoclave for 15 min at 15 psi pressure–121°C. Cool to 50°–55°C. Add 10.0mL of sterile sodium deoxycholate solution and 1.0mL of sterile cefazolin solution. Mix thoroughly. Pour into sterile Petri dishes.

**Storage/Shelf Life:** Store dehydrated media in the dark in a sealed container below 30°C. Prepared media should be stored under refrigeration (2-8°C). Media should be used within 60 days of preparation. Media should not be used if there are any signs of deterioration (shrinking, cracking, or discoloration) or contamination, or if the expiration date supplied by the manufacturer has passed.

**Use:** For the selective isolation of *Campylobacter* species, especially *Campylobacter jejuni* from human feces.

## CCVC Medium
### (Cephalothin Cycloheximide Vancomycin Colistin Medium
**Composition** per liter:

| | |
|---|---|
| BCYE-alpha base | 990.0mL |
| Antibiotic supplement solution | 10.0mL |

pH 6.9 ± 0.2 at 25°C

**Source:** This medium is available as a premixed powder from BD Diagnostic Systems.

**BCYE-Alpha Base:**
**Composition** per liter:

| | |
|---|---|
| Agar | 15.0g |
| Yeast extract | 10.0g |
| ACES buffer (2-[(2-amino-2-oxoethyl)-amino]-ethane sulfonic acid) | 10.0g |
| Charcoal, activated | 2.0g |
| α-Ketoglutarate | 1.0g |
| $Fe_4(P_2O_7)_3 \cdot 9H_2O$ | 0.25g |
| L-Cysteine·HCl·H$_2$O solution | 10.0mL |

**L-Cysteine·HCl·H$_2$O Solution:**
**Composition** per 10.0mL:

| | |
|---|---|
| L-Cysteine·HCl·H$_2$O | 0.4g |

**Preparation of L-Cysteine·HCl·H$_2$O Solution:** Add L-cysteine·HCl·H$_2$O to distilled/deionized water and bring volume to 10.0mL. Mix thoroughly. Filter sterilize.

**Preparation of BCYE-Alpha Base:** Add components, except L-cysteine·HCl·H$_2$O solution, to distilled/deionized water and bring volume to 990.0mL. Mix thoroughly. Adjust medium to pH 6.9 with 1$N$ KOH. Heat gently and bring to boiling for 1 min. Autoclave for 15 min at 15 psi pressure–121°C. Cool to 50°–55°C. Add 4.0mL of L-cysteine·HCl·H$_2$O solution. Mix thoroughly.

**Antibiotic Supplement Solution:**
**Composition** per 10.0mL:

| | |
|---|---|
| Cycloheximide | 80.0mg |
| Colistin | 16.0mg |
| Cephalothin | 4.0mg |
| Vancomycin | 0.5mg |

**Preparation of Antibiotic Supplement Solution:** Add components to 10.0mL of distilled/deionized water. Filter sterilize.

**Caution:** Cycloheximide is toxic. Avoid skin contact or aerosol formation and inhalation.

**Preparation of Medium:** To cooled BCYE-alpha base, add 10.0mL sterile antibiotic supplement. Mix thoroughly. Adjust pH to 6.9 with sterile 1$N$ KOH. Pour into sterile Petri dishes with constant agitation to keep charcoal in suspension.

**Storage/Shelf Life:** Store dehydrated media in the dark in a sealed container below 30°C. Prepared media should be stored under refrigeration (2-8°C). Media should be used within 60 days of preparation. Media should not be used if there are any signs of deterioration (shrinking, cracking, or discoloration) or contamination, or if the expiration date supplied by the manufacturer has passed.

**Use:** For the selective isolation and cultivation of *Legionella* species from environmental samples.

## CDC Anaerobe Blood Agar

**Composition** per liter:

| | |
|---|---|
| Agar | 20.0g |
| Pancreatic digest of casein | 15.0g |
| Papaic digest of soybean meal | 5.0g |
| NaCl | 5.0g |
| Yeast extract | 5.0g |
| L-Cystine | 0.4g |
| Sheep blood, defibrinated | 50.0mL |
| Vitamin K$_1$ solution | 1.0mL |
| Hemin solution | 0.5mL |

pH 7.5 ± 0.2 at 25°C

**Source:** This medium is available as a prepared medium from BD Diagnostic Systems.

**Vitamin K$_1$ Solution:**
**Composition** per 100.0mL:

| | |
|---|---|
| Vitamin K$_1$ | 1.0g |
| Ethanol | 99.0mL |

**Preparation of Vitamin K$_1$ Solution:** Add vitamin K$_1$ to 99.0mL of absolute ethanol. Mix thoroughly. Filter sterilize.

**Hemin Solution:**
**Composition** per 100.0mL:

| | |
|---|---|
| Hemin | 1.0g |
| NaOH (1$N$ solution) | 20.0mL |

**Preparation of Hemin Solution:** Add hemin to 20.0mL of 1$N$ NaOH solution. Mix thoroughly. Bring volume to 100.0mL with distilled/deionized water.

**Preparation of Medium:** Add components, except vitamin K$_1$ and sheep blood, to distilled/deionized water and bring volume to 949.0mL. Mix thoroughly. Heat gently and bring to boiling for 1 min. Autoclave for 15 min at 15 psi pressure–121°C. Cool to 50°–55°C. Aseptically add 1.0mL of vitamin K$_1$ solution and 50.0mL of sterile, defibrinated sheep blood. Mix thoroughly. Pour into sterile Petri dishes.

**Storage/Shelf Life:** Store dehydrated media in the dark in a sealed container below 30°C. Prepared media should be stored under refrigeration (2-8°C). Media should be used within 60 days of preparation. Media should not be used if there are any signs of deterioration (shrinking, cracking, or discoloration) or contamination, or if the expiration date supplied by the manufacturer has passed.

**Use:** For the isolation and cultivation of fastidious and slow-growing, obligate anaerobic bacteria from a variety of clinical speciments and nonclinical specimens of public health importance. For the isolation and cultivation of *Actinomyces israelii, Bacteroides melaninogenicus, Bacteroides thetaiotaomicron, Clostridium haemolyticum,* and *Fusobacterium necrophorum.*

## CDC Anaerobe Blood Agar with Kanamycin and Vancomycin

**Composition** per liter:

| | |
|---|---|
| Agar | 20.0g |
| Pancreatic digest of casein | 15.0g |
| NaCl | 5.0g |
| Papaic digest of soybean meal | 5.0g |
| Yeast extract | 5.0g |
| L-Cystine | 0.4g |
| Sheep blood, defibrinated | 50.0mL |
| Antibiotic solution | 10.0mL |
| Vitamin K$_1$ solution | 1.0mL |
| Hemin solution | 0.5mL |

pH 7.5 ± 0.2 at 25°C

**Source:** This medium is available as a prepared medium from BD Diagnostic Systems.

**Antibiotic Solution:**

**Composition** per 10.0mL:

Kanamycin...............................................................0.1g
Vancomycin..........................................................7.5mg

**Preparation of Antibiotic Solution:** Add components to distilled/deionized water and bring volume to 10.0mL. Mix thoroughly. Filter sterilize.

**Vitamin K₁ Solution:**

**Composition** per 100.0mL:

Vitamin K₁ ..............................................................1.0g
Ethanol.................................................................99.0mL

**Preparation of Vitamin K₁ Solution:** Add vitamin K₁ to 99.0mL of absolute ethanol. Mix thoroughly. Filter sterilize.

**Hemin Solution:**

**Composition** per 100.0mL:

Hemin......................................................................1.0g
NaOH (1*N* solution)..........................................20.0mL

**Preparation of Hemin Solution:** Add hemin to 20.0mL of 1*N* NaOH solution. Mix thoroughly. Bring volume to 100.0mL with distilled/deionized water.

**Preparation of Medium:** Add components, except vitamin K₁ solution and sheep blood, to distilled/deionized water and bring volume to 949.0mL. Mix thoroughly. Heat gently and bring to boiling for 1 min. Autoclave for 15 min at 15 psi pressure–121°C. Cool to 50°–55°C. Aseptically add 1.0mL of sterile vitamin K₁ solution and 50.0mL of sterile, defibrinated sheep blood. Mix thoroughly. Pour into sterile Petri dishes.

**Storage/Shelf Life:** Store dehydrated media in the dark in a sealed container below 30°C. Prepared media should be stored under refrigeration (2-8°C). Media should be used within 60 days of preparation. Media should not be used if there are any signs of deterioration (shrinking, cracking, or discoloration) or contamination.

**Use:** For the selective isolation of fastidious and slow-growing, obligate anaerobic Gram-negative bacteria, especially *Bacteroides* species, from a variety of clinical speciments and nonclinical specimens of public health importance.

## CDC Anaerobe Blood Agar with Phenylethyl Alcohol (CDC Anaerobe Blood Agar with PEA)

**Composition** per liter:

Agar......................................................................20.0g
Pancreatic digest of casein...............................15.0g
NaCl.......................................................................5.0g
Papaic digest of soybean meal...........................5.0g
Yeast extract..........................................................5.0g
L-Cystine...............................................................0.4g
Sheep blood, defibrinated................................50.0mL
Vitamin K₁ solution .........................................10.0mL
Hemin solution...................................................0.5mL

pH 7.5 ± 0.2 at 25°C

**Source:** This medium is available as a prepared medium from BD Diagnostic Systems.

**Vitamin K₁ Solution:**

**Composition** per 100.0mL:

Vitamin K₁ ..............................................................0.1g
Phenylethyl alcohol............................................25.0g
Ethanol.................................................................74.0mL

**Preparation of Vitamin K₁ Solution:** Add components to 74.0mL of absolute ethanol. Mix thoroughly. Filter sterilize.

**Hemin Solution:**

**Composition** per 100.0mL:

Hemin......................................................................1.0g
NaOH (1*N* solution)..........................................20.0mL

**Preparation of Hemin Solution:** Add hemin to 20.0mL of 1*N* NaOH solution. Mix thoroughly. Bring volume to 100.0mL with distilled/deionized water.

**Preparation of Medium:** Add components, except vitamin K₁ solution and sheep blood, to distilled/deionized water and bring volume to 940.0mL. Mix thoroughly. Heat gently and bring to boiling for 1 min. Autoclave for 15 min at 15 psi pressure–121°C. Cool to 50°–55°C. Aseptically add 1.0mL of vitamin K₁ solution and 50.0mL of sterile, defibrinated sheep blood. Mix thoroughly. Pour into sterile Petri dishes.

**Storage/Shelf Life:** Store dehydrated media in the dark in a sealed container below 30°C. Prepared media should be stored under refrigeration (2-8°C). Media should be used within 60 days of preparation. Media should not be used if there are any signs of deterioration (shrinking, cracking, or discoloration) or contamination, or if the expiration date supplied by the manufacturer has passed.

**Use:** For the selective isolation of fastidious and slow-growing, obligate anaerobic bacteria from a variety of clinical speciments and nonclinical specimens of public health importance.

## CDC Anaerobe Laked Blood Agar with Kanamycin and Vancomycin (CDC Anaerobe Laked Blood Agar with KV)

**Composition** per liter:

Agar......................................................................20.0g
Pancreatic digest of casein...............................15.0g
Papaic digest of soybean meal...........................5.0g
Yeast extract..........................................................5.0g
L-Cystine...............................................................0.4g
Sheep blood, defibrinated, laked .....................50.0mL
Antibiotic solution ...........................................10.0mL
Vitamin K₁ solution ...........................................1.0mL
Hemin solution...................................................0.5mL

pH 7.5 ± 0.2 at 25°C

**Source:** This medium is available as a prepared medium from BD Diagnostic Systems.

**Antibiotic Solution:**

**Composition** per 10.0mL:

Kanamycin...............................................................0.1g
Vancomycin..........................................................7.5mg

**Preparation of Antibiotic Solution:** Add components to distilled/deionized water and bring volume to 10.0mL. Mix thoroughly. Filter sterilize.

**Vitamin K₁ Solution:**

**Composition** per 100.0mL:

Vitamin K₁ ..............................................................1.0g
Ethanol.................................................................99.0mL

**Preparation of Vitamin K₁ Solution:** Add vitamin K₁ to 99.0mL of absolute ethanol. Mix thoroughly. Filter sterilize.

**Hemin Solution:**
**Composition** per 100.0mL:
Hemin.................................................................1.0g
NaOH (1*N* solution)................................... 20.0mL

**Preparation of Hemin Solution:** Add hemin to 20.0mL of 1*N* NaOH solution. Mix thoroughly. Bring volume to 100.0mL with distilled/deionized water.

**Preparation of Medium:** Add components, except antibiotic solution, vitamin K₁, and laked sheep blood, to distilled/deionized water and bring volume to 939.0mL. Mix thoroughly. Heat gently and bring to boiling for 1 min. Autoclave for 15 min at 15 psi pressure–121°C. Cool to 50°–55°C. Aseptically add 1.0mL of sterile vitamin K₁ solution and 10.0mL of sterile antibiotic solution. Mix thoroughly. Aseptically add 50.0mL of sterile, defibrinated, laked sheep blood. Laked blood is prepared by freezing whole blood overnight and thawing to room temperature. Mix thoroughly. Pour into sterile Petri dishes.

**Storage/Shelf Life:** Store dehydrated media in the dark in a sealed container below 30°C. Prepared media should be stored under refrigeration (2-8°C). Media should be used within 60 days of preparation. Media should not be used if there are any signs of deterioration (shrinking, cracking, or discoloration) or contamination, or if the expiration date supplied by the manufacturer has passed.

**Use:** For the selective isolation of fastidious and slow-growing, obligate anaerobic bacteria from a variety of clinical speciments and nonclinical specimens of public health importance.

## Cefiximine Rhamnose Sorbitol MacConkey Agar (CR-SMAC Agar Base)

**Composition** per liter:
Peptone............................................................20.0g
Agar ................................................................15.0g
Sorbitol............................................................10.0g
NaCl..................................................................5.0g
Rhamnose...........................................................5.0g
Bile Salts No. 3 ..................................................1.5g
Neutral Red ......................................................0.03g
Crystal Violet ..................................................0.001g
Selective supplement solution .......................... 10.0mL
pH 7.1 ± 0.2 at 25°C

**Source:** This medium is available as a premixed powder from Thermo Scientific.

**Selective Supplement Solution:**
**Composition** per 10.0mL:
Cefiximine .......................................................0.05mg

**Preparation of Selective Supplement Solution:** Add cefiximine to distilled/deionized water and bring volume to 10.0mL. Mix thoroughly. Filter sterilize.

**Preparation of Medium:** Add components, except selective supplement solution, to distilled/deionized water and bring volume to 990.0mL. Mix thoroughly. Gently heat while stirring and bring to boiling. Autoclave for 15 min at 15 psi pressure–121°C. Cool to 50°C. Aseptially add selective supplement solution. Mix thoroughly. Pour into sterile Petri dishes.

**Storage/Shelf Life:** Store dehydrated media in the dark in a sealed container below 30°C. Prepared media should be stored under refrigeration (2-8°C). Media should be used within 60 days of preparation. Media should not be used if there are any signs of deterioration (shrink-

ing, cracking, or discoloration) or contamination, or if the expiration date supplied by the manufacturer has passed.

**Use:** For the detection of *Escherichia coli* O157:H7. This is a selective, differential medium based on Sorbitol MacConkey Agar with added rhamnose and cefixime. This medium provides a selective base with improved differentiation of *E. coli* O157. The addition of rhamnose aids in the differentiation of *Escherichia coli* O157 from background flora. Cefixime reduces the level of competing flora, particularly *Proteus* spp., that often account for large numbers of non-sorbitol fermenting colonies. *E. coli* O157 do not usually ferment sorbitol or rhamnose, so will appear as straw colored colonies. However, rhamnose is fermented by most sorbitol negative *E. coli* of other serogroups. These colonies will be pink/red and will not be counted as presumptive *E. coli* O157 colonies.

## Cefsulodin Irgasan® Novobiocin Agar (CIN Agar) (*Yersinia* Selective Agar) (BAM M35)

**Composition** per 1008.0mL:
Basal medium ................................................ 757.0mL
Desoxycholate solution.................................. 200.0mL
Cefsulodin solution ......................................... 10.0mL
Novobiocin solution ........................................ 10.0mL
Crystal Violet solution ..................................... 10.0mL
Strontium chloride solution .............................. 10.0mL
Neutral Red solution ........................................ 10.0mL
NaOH, 5*N* ....................................................... 1.0mL
Irgasan solution................................................. 1.0mL
pH 7.4 ± 0.2 at 25°C

**Basal Medium:**
**Composition** per 757.0mL:
Mannitol...........................................................20.0g
Special peptone .................................................20.0g
Agar ................................................................12.0g
Sodium pyruvate.................................................2.0g
Yeast extract.......................................................2.0g
NaCl..................................................................1.0g
Magnesium sulfate solution................................ 1.0mL

**Preparation of Basal Medium:** Add components to distilled/deionized water and bring volume to 757.0mL. Mix thoroughly. Gently heat and bring to boiling with stirring. Cool to about 80°C by placing in a 50°C water bath for about 10 min.

**Magnesium Sulfate Solution:**
**Composition** per 10mL:
MgSO₄·7H₂O .....................................................0.1g

**Preparation of Magnesium Sulfate Solution:** Add MgSO₄·7H₂O to distilled/deionized water and bring volume to 10.0mL. Mix thoroughly.

**Irgasan Solution:**
**Composition** per 10mL:
Irgasan (triclosan) ............................................0.04g

**Preparation of Irgasan Solution:** Add irgasan to 95% ethanol and bring volume to 10.0mL. Mix thoroughly. Can be stored for 4 weeks at –20°C.

**Desoxycholate Solution:**
**Composition** per 200.0mL:
Na-desoxycholate ..............................................0.5g

**Preparation of Desoxycholate Solution:** Add desoxycholate to distilled/deionized water and bring volume to 200.0mL. Mix thoroughly. Gently heat and bring to boiling with stirring. Cool to 50–55°C.

**Neutral Red Solution:**
**Composition** per 10.0mL:
Neutral Red .................................................................30.0mg

**Preparation of Neutral Red Solution:** Add Neutral Red to 10.0mL of distilled/deionized water. Mix thoroughly. Autoclave for 15 min at 15 psi pressure–121°C. Cool to 25°C.

**Crystal Violet Solution:**
**Composition** per 10.0mL:
Crystal Violet ..................................................................1.0mg

**Preparation of Crystal Violet Solution:** Add Crystal Violet to 10.0mL of distilled/deionized water. Mix thoroughly. Autoclave for 15 min at 15 psi pressure–121°C. Cool to 25°C.

**Cefsulodin Solution:**
**Composition** per 10.0mL:
Cefsulodin ....................................................................15.0mg

**Preparation of Cefsulodin Solution:** Add cefsulodin to 10.0mL of distilled/deionized water. Mix thoroughly. Filter sterilize.

**Novobiocin Solution:**
**Composition** per 10.0mL:
Novobiocin......................................................................2.5mg

**Preparation of Novobiocin Solution:** Add novobiocin to 10.0mL of distilled/deionized water. Mix thoroughly. Filter sterilize.

**Strontium Chloride Solution:**
**Composition** per 10.0mL:
SrCl$_2$·6H$_2$O.......................................................................1.0g

**Preparation of Strontium Chloride:** Add strontium chloride to 10.0mL of distilled/deionized water. Mix thoroughly. Filter sterilize.

**Preparation of Medium:** Add 1.0mL irgasan solution to 757.0mL basal medium. Mix thoroughly. Cool to 50–55°C. Add 200.0mL desoxychlolate solution. Mix thoroughly. Solution should remain clear. Aseptically add 1.0mL 5$N$ NaOH, 10.0mL Neutral Red solution, 10.0mL Crystal Violet solution, 10.0mL cefsulodin solution, and 10.0mL novobiocin solution. Mix thoroughly. Slowly add 10.0mL strontium chloride solution while continuously stirring. Adjust pH to 7.4 with 5$N$ NaOH. Pour into sterile Petri dishes or distribute into sterile tubes.

**Storage/Shelf Life:** Store dehydrated media in the dark in a sealed container below 30°C. Prepared media should be stored under refrigeration (2-8°C). Media should be used within 60 days of preparation. Media should not be used if there are any signs of deterioration (shrinking, cracking, or discoloration) or contamination, or if the expiration date supplied by the manufacturer has passed.

**Use**: For the selective isolation and differentiation of *Yersinia enterocolitica* based on mannitol fermentation. *Yersinia enterocolitica* appears as "bull's eye" colonies with deep red centers surrounded by a transparent periphery.

## Cellobiose Polymyxin B Colistin Agar, Modified

**Composition** per liter:
Solution 1 .................................................................. 900.0mL
Solution 2 .................................................................. 100.0mL

pH 7.6 ± 0.2 at 25°C

**Solution 1:**
**Composition** per 900.0mL:
NaCl.............................................................................20.0g
Agar .............................................................................15.0g
Peptone ........................................................................10.0g
Beef extract....................................................................5.0g
1000× dye stock solution .............................................. 1.0mL

**Preparation of Solution 1:** Add components to distilled/deionized water and bring volume to 900.0mL. Mix thoroughly. Adjust pH to 7.6. Gently heat and bring to boiling. Do not autoclave. Cool to 48°–55°C.

**1000X Dye Stock Solution:**
**Composition** per 100.0mL:
Bromthymol Blue ..........................................................4.0g
Cresol Red .....................................................................4.0g
Ethanol (95% solution)............................................ 100.0mL

**Preparation of 1000X Dye Stock Solution:** Add Bromthymol Blue and Cresol Red to 100.0mL of ethanol. Mix thoroughly.

**Solution 2:**
**Composition** per 100.0mL:
Cellobiose .................................................................... 10.0g
Colistin..................................................................400,000U
Polymyxin B ..........................................................100,000U

**Preparation of Solution 2:** Add cellobiose to distilled/deionized water and bring volume to 100.0mL. Mix thoroughly. Gently heat until dissolved. Cool to 25°C. Add colistin and polymyxin B. Mix thoroughly.

**Preparation of Medium:** Combine cooled solution 1 and solution 2. Mix thoroughly. Do not autoclave. Pour into sterile Petri dishes.

**Storage/Shelf Life:** Store dehydrated media in the dark in a sealed container below 30°C. Prepared media should be stored under refrigeration (2-8°C). Media should be used within 60 days of preparation. Media should not be used if there are any signs of deterioration (shrinking, cracking, or discoloration) or contamination,.

**Use:** For the cultivation of *Vibrio* species from foods.

## Cellobiose Polymyxin Colistin Agar (CPC Agar)

**Composition** per liter:
Solution A............................................................... 900.0mL
Solution B ............................................................... 100.0mL

pH 7.6 ± 0.2 at 25°C

**Solution A:**
**Composition** per 900.0mL:
NaCl.............................................................................20.0g
Agar .............................................................................15.0g
Peptone ........................................................................10.0g
Beef extract....................................................................5.0g
Bromthymol Blue .........................................................0.04g
Cresol Red ....................................................................0.04g

**Preparation of Solution A:** Add components to distilled/deionized water and bring volume to 900.0mL. Mix thoroughly. Adjust pH to 7.6. Gently heat and bring to boiling. Autoclave for 15 min at 15 psi pressure–121°C. Cool to 50°–55°C.

**Solution B:**
**Composition** per 100.0mL:
Cellobiose .................................................................... 15.0g
Colistin................................................................1,360,000U
Polymyxin B ..........................................................100,000U

**Preparation of Solution B:** Add components to distilled/deionized water and bring volume to 100.0mL. Mix thoroughly. Filter sterilize.

**Preparation of Medium:** Aseptically combine 900.0mL of cooled, sterile solution A and 100.0mL of sterile solution B. Mix thoroughly. Pour into sterile Petri dishes. Use within 7 days.

**Storage/Shelf Life:** Store dehydrated media in the dark in a sealed container below 30°C. Prepared media should be stored under refrigeration (2–8°C). Media should be used within 60 days of preparation. Media should not be used if there are any signs of deterioration (shrinking, cracking, or discoloration) or contamination, or if the expiration date supplied by the manufacturer has passed.

**Use:** For the cultivation and identification of *Vibrio* species from foods.

### Cetrimide Agar, Non-USP

**Composition** per liter:
Beef heart, solids from infusion................................................500.0g
Agar ............................................................................15.0g
Tryptose ......................................................................10.0g
NaCl...........................................................................5.0g
Cetrimide .....................................................................0.9g

pH 7.2 ± 0.2 at 25°C

**Preparation of Medium:** Add components to distilled/deionized water and bring volume to 1.0L. Mix thoroughly. Gently heat and bring to boiling. Distribute into tubes or flasks. Autoclave for 15 min at 13 psi pressure–118°C. Pour into sterile Petri dishes or leave in tubes.

**Storage/Shelf Life:** Store dehydrated media in the dark in a sealed container below 30°C. Prepared media should be stored under refrigeration (2–8°C). Media should be used within 60 days of preparation. Media should not be used if there are any signs of deterioration (shrinking, cracking, or discoloration) or contamination, or if the expiration date supplied by the manufacturer has passed.

**Use:** For the selective isolation, cultivation, and identification of *Pseudomonas aeruginosa* and other Gram-negative, nonfermentative bacteria.

### Cetrimide Agar, USP
### (Pseudosel® Agar)

**Composition** per liter:
Pancreatic digest of gelatin..............................................20.0g
Agar ..........................................................................13.6g
K$_2$SO$_4$......................................................................10.0g
MgCl$_2$.........................................................................1.4g
Cetrimide .....................................................................0.3g
Glycerol .......................................................................10.0mL

pH 7.2 ± 0.2 at 25°C

**Source:** This medium is available as a premixed powder from BD Diagnostic Systems.

**Preparation of Medium:** Add components to distilled/deionized water and bring volume to 1.0L. Mix thoroughly. Gently heat and bring to boiling. Distribute into tubes or flasks. Autoclave for 15 min at 13 psi pressure–118°C. Pour into sterile Petri dishes or leave in tubes.

**Storage/Shelf Life:** Store dehydrated media in the dark in a sealed container below 30°C. Prepared media should be stored under refrigeration (2–8°C). Media should be used within 60 days of preparation. Media should not be used if there are any signs of deterioration (shrinking, cracking, or discoloration) or contamination, or if the expiration date supplied by the manufacturer has passed.

**Use:** For the selective isolation, cultivation, and identification of *Pseudomonas aeruginosa* and other Gram-negative, nonfermentative bacteria.

### Cetrimide Agar Base with Glycerol

**Composition** per liter:
Pancreatic digest of gelatin..............................................20.0g
K$_2$SO$_4$......................................................................10.0g
MgCl$_2$.........................................................................1.4g
Cetrimide .....................................................................0.3g
Glycerol .......................................................................10.0mL

pH 7.2 ± 0.2 at 25°C

**Source:** This medium, without glycerol, is available as a premixed powder from HiMedia.

**Preparation of Medium:** Add components to distilled/deionized water and bring volume to 1.0L. Mix thoroughly. Gently heat and bring to boiling. Distribute into tubes or flasks. Autoclave for 15 min at 13 psi pressure–118°C. Pour into sterile Petri dishes or leave in tubes.

**Storage/Shelf Life:** Store dehydrated media in the dark in a sealed container below 30°C. Prepared media should be stored under refrigeration (2–8°C). Media should be used within 60 days of preparation. Media should not be used if there are any signs of deterioration (shrinking, cracking, or discoloration) or contamination, or if the expiration date supplied by the manufacturer has passed.

**Use:** For the selective isolation, cultivation, and identification of *Pseudomonas aeruginosa* and other Gram-negative, nonfermentative bacteria.

### Cetrimide Agar Base with Glycerol and Nalidixic Selective Supplement

**Composition** per liter:
Pancreatic digest of gelatin..............................................20.0g
K$_2$SO$_4$......................................................................10.0g
MgCl$_2$.........................................................................1.4g
Cetrimide .....................................................................0.3g
Glycerol .......................................................................10.0mL
Nalidixic selective supplement..........................................5.0mL

pH 7.2 ± 0.2 at 25°C

**Source:** This medium, without glycerol and nalidixic acid supplement, is available as a premixed powder from HiMedia.

**Nalidixic Selective Supplement:**
**Composition** per 5.0mL:
Nalidixic acid................................................................15.0mg

**Preparation of Nalidixic Selective Supplement:** Add nalidixic acid to distilled/deionized water and bring volume to 50.0mL. Mix thoroughly. Filter sterilize.

**Preparation of Medium:** Add components, except nalidixic selective supplement, to distilled/deionized water and bring volume to 1.0L. Mix thoroughly. Gently heat and bring to boiling. Distribute into tubes or flasks. Autoclave for 15 min at 15 psi pressure–121°C. Cool to 50°C. Aseptically add 5.0mL sterile nalidixic selective supplement. Mix thoroughly. Pour into sterile Petri dishes or leave in tubes.

**Storage/Shelf Life:** Store dehydrated media in the dark in a sealed container below 30°C. Prepared media should be stored under refrigeration (2–8°C). Media should be used within 60 days of preparation. Media should not be used if there are any signs of deterioration (shrink-

ing, cracking, or discoloration) or contamination, or if the expiration date supplied by the manufacturer has passed.

**Use:** For the selective isolation, cultivation, and identification of *Pseudomonas aeruginosa* and other Gram-negative, nonfermentative bacteria.

## Cetrimide HiVeg Agar Base with Glycerol
**Composition** per liter:
Plant peptone No. 2................................................20.0g
$K_2SO_4$ ........................................................10.0g
$MgCl_2$..........................................................1.4g
Cetrimide..........................................................0.3g
Glycerol ..................................................... 10.0mL
pH 7.2 ± 0.2 at 25°C

**Source:** This medium, without glycerol, is available as a premixed powder from HiMedia.

**Preparation of Medium:** Add components to distilled/deionized water and bring volume to 1.0L. Mix thoroughly. Gently heat and bring to boiling. Distribute into tubes or flasks. Autoclave for 15 min at 13 psi pressure–118°C. Pour into sterile Petri dishes or leave in tubes.

**Storage/Shelf Life:** Store dehydrated media in the dark in a sealed container below 30°C. Prepared media should be stored under refrigeration (2-8°C). Media should be used within 60 days of preparation. Media should not be used if there are any signs of deterioration (shrinking, cracking, or discoloration) or contamination, or if the expiration date supplied by the manufacturer has passed.

**Use:** For the selective isolation, cultivation, and identification of *Pseudomonas aeruginosa* and other Gram-negative, nonfermentative bacteria.

## Cetrimide HiVeg Agar Base with Glycerol and Nalidixic Selective Supplement
**Composition** per liter:
Plant peptone No. 2................................................20.0g
$K_2SO_4$ ........................................................10.0g
$MgCl_2$..........................................................1.4g
Cetrimide..........................................................0.3g
Glycerol ..................................................... 10.0mL
Nalidixic selective supplement .................................. 5.0mL
pH 7.2 ± 0.2 at 25°C

**Source:** This medium, without glycerol and nalidixic acid supplement, is available as a premixed powder from HiMedia.

**Nalidixic Selective Supplement:**
**Composition** per 5.0mL:
Nalidixic acid.....................................................15.0mg

**Preparation of Nalidixic Selective Supplement:** Add nalidixic to distilled/deionized water and bring volume to 50.0mL. Mix thoroughly. Filter sterilize.

**Preparation of Medium:** Add components, except nalidixic selective supplement, to distilled/deionized water and bring volume to 1.0L. Mix thoroughly. Gently heat and bring to boiling. Distribute into tubes or flasks. Autoclave for 15 min at 15 psi pressure–121°C. Cool to 50°C. Aseptically add 5.0mL sterile nalidixic selective supplement. Mix thoroughly. Pour into sterile Petri dishes or leave in tubes.

**Storage/Shelf Life:** Store dehydrated media in the dark in a sealed container below 30°C. Prepared media should be stored under refrigeration (2-8°C). Media should be used within 60 days of preparation.

Media should not be used if there are any signs of deterioration (shrinking, cracking, or discoloration) or contamination, or if the expiration date supplied by the manufacturer has passed.

**Use:** For the selective isolation, cultivation, and identification of *Pseudomonas aeruginosa* and other Gram-negative, nonfermentative bacteria.

## CFAT Medium
## (Cadmium Fluoride Acriflavin Tellurite Medium)
**Composition** per liter:
Pancreatic digest of casein....................................17.0g
Agar ............................................................15.0g
Glucose..........................................................7.5g
NaCl.............................................................5.0g
Papaic digest of soybean meal...................................3.0g
$K_2HPO_4$.......................................................2.5g
NaF..............................................................0.8g
$CdSO_4$.......................................................0.013g
$K_2TeO_3$......................................................2.5mg
Neutral acriflavin..............................................1.2mg
Basic Fuchsin..................................................0.25mg
Sheep blood, defibrinated .................................. 50.0mL

**Caution:** Potassium tellurite is toxic.

**Preparation of Medium:** Add components, except sheep blood, to distilled/deionized water and bring volume to 950.0mL. Mix thoroughly. Gently heat and bring to boiling. Autoclave for 15 min at 15 psi pressure–121°C. Cool to 45°–50°C. Add 50.0mL of sterile, defibrinated sheep blood. Mix thoroughly. Pour into sterile Petri dishes or leave in tubes.

**Storage/Shelf Life:** Store dehydrated media in the dark in a sealed container below 30°C. Prepared media should be stored under refrigeration (2-8°C). Media should be used within 60 days of preparation. Media should not be used if there are any signs of deterioration (shrinking, cracking, or discoloration) or contamination, or if the expiration date supplied by the manufacturer has passed.

**Use:** For the isolation, cultivation, and enumeration of *Actinomyces viscosus* and *Actinomyces naeslundii* from clinical specimens, especially dental plaque.

## Chapman Stone Agar
**Composition** per liter:
$(NH_4)_2SO_4$ ..................................................75.0g
NaCl............................................................55.0g
Gelatin.........................................................30.0g
Agar ...........................................................15.0g
D-Mannitol .....................................................10.0g
Pancreatic digest of casein....................................10.0g
$K_2HPO_4$......................................................5.0g
Yeast extract...................................................2.0g
pH 7.0 ± 0.2 at 25°C

**Source:** This medium is available as a premixed powder from BD Diagnostic Systems.

**Preparation of Medium:** Add components to distilled/deionized water and bring volume to 1.0L. Mix thoroughly. Autoclave for 10 min at 15 psi pressure–121°C. Pour into sterile Petri dishes while the medium is still hot. Add 25.0mL of medium per Petri dish.

**Storage/Shelf Life:** Store dehydrated media in the dark in a sealed container below 30°C. Prepared media should be stored under refriger-

ation (2-8°C). Media should be used within 60 days of preparation. Media should not be used if there are any signs of deterioration (shrinking, cracking, or discoloration) or contamination, or if the expiration date supplied by the manufacturer has passed.

**Use:** For the isolation of staphylococci from a variety of specimens.

### Charcoal Agar with Horse Blood
**Composition** per liter:

| | |
|---|---|
| Agar | 12.0g |
| Beef extract | 10.0g |
| Peptone | 10.0g |
| Starch | 10.0g |
| NaCl | 5.0g |
| Charcoal, bacteriological | 4.0g |
| Nicotinic acid | 1.0mg |
| Horse blood, defibrinated | 100.0mL |

pH 7.4 ± 0.2 at 25°C

**Preparation of Medium:** Add components to distilled/deionized water and bring volume to 900.0L. Mix thoroughly. Gently heat and bring to boiling with frequent stirring. Autoclave for 15 min at 15 psi pressure–121°C. Cool to 80°C. Aseptically add 100.0mL of sterile, defibrinated horse blood. Maintain at 80°C for 10 min to form chocolate agar. Pour into sterile Petri dishes or distribute into tubes. Shake flask while dispensing to keep charcoal in suspension.

**Storage/Shelf Life:** Store dehydrated media in the dark in a sealed container below 30°C. Prepared media should be stored under refrigeration (2-8°C). Media should be used within 60 days of preparation. Media should not be used if there are any signs of deterioration (shrinking, cracking, or discoloration) or contamination, or if the expiration date supplied by the manufacturer has passed.

**Use:** For the cultivation and isolation of *Haemophilus influenzae*.

### Charcoal Agar with Horse Blood and Cepahalexin
**Composition** per liter:

| | |
|---|---|
| Agar | 12.0g |
| Beef extract | 10.0g |
| Peptone | 10.0g |
| Starch | 10.0g |
| NaCl | 5.0g |
| Charcoal | 4.0g |
| Nicotinic acid | 1.0mg |
| Horse blood, defibrinated | 100.0mL |
| Cephalexin solution | 10.0mL |

pH 7.4 ± 0.2 at 25°C

**Cephalexin Solution:**
**Composition** per 10.0mL:

| | |
|---|---|
| Cephalexin | 0.04g |

**Preparation of Cephalexin Solution:** Add cephalexin to distilled/deionized water and bring volume to 10.0mL. Mix thoroughly. Filter sterilize.

**Preparation of Medium:** Add components, except cephalexin solution and horse blood, to distilled/deionized water and bring volume to 890.0L. Mix thoroughly. Gently heat and bring to boiling with frequent stirring. Autoclave for 15 min at 15 psi pressure–121°C. Cool to 45°–50°C. Aseptically add 100.0mL of sterile, defibrinated horse blood and 10.0mL of sterile cephalexin solution. Pour into sterile Petri dishes or distribute into tubes. Shake flask while dispensing to keep charcoal in suspension.

**Storage/Shelf Life:** Store dehydrated media in the dark in a sealed container below 30°C. Prepared media should be stored under refrigeration (2-8°C). Media should be used within 60 days of preparation. Media should not be used if there are any signs of deterioration (shrinking, cracking, or discoloration) or contamination, or if the expiration date supplied by the manufacturer has passed.

**Use:** For the cultivation and isolation of *Bordetella pertussis*.

### *Chlamydia* Isolation Medium
**Composition** per 500.0mL:

| | |
|---|---|
| Eagle minimum essential medium with Earle salts, 10X | 50.0mL |
| Fetal calf serum | 50.0mL |
| Selective supplement | 10.0mL |
| L-Glutamine solution | 5.0mL |

pH 7.4 ± 0.2 at 25°C

**Eagle Minimum Essential Medium with Earle Salts, 10X:**
**Composition** per liter:

| | |
|---|---|
| NaCl | 6.8g |
| Glucose | 1.0g |
| KCl | 0.4g |
| CaCl$_2$·2H$_2$O | 0.2g |
| MgCl$_2$·6H$_2$O | 0.2g |
| NaH$_2$PO$_4$ | 0.15g |
| L-Arginine | 0.1g |
| L-Lysine | 0.06g |
| L-Isoleucine | 0.05g |
| L-Leucine | 0.05g |
| L-Threonine | 0.05g |
| L-Valine | 0.05g |
| L-Tyrosine | 0.04g |
| L-Phenylalanine | 0.03g |
| L-Histidine | 0.03g |
| L-Cystine | 0.02g |
| L-Methionine | 0.02g |
| L-Tryptophan | 0.01g |
| *i*-Inositol | 2.0mg |
| Calcium pantothenate | 1.0mg |
| Choline chloride | 1.0mg |
| Folic acid | 1.0mg |
| Nicotinamide | 1.0mg |
| Pyridoxal | 1.0mg |
| Thiamine·HCl | 1.0mg |
| Riboflavin | 0.1mg |

**Preparation of Eagle Minimum Essential Medium with Earle Salts, 10X:** Add components to distilled/deionized water and bring volume to 1.0L. Mix thoroughly. Adjust pH to 7.4 with 7.5% Na$_2$CO$_3$ solution. Filter sterilize.

**Selective Supplement:**
**Composition** per 10.0mL:

| | |
|---|---|
| Glucose | 0.594g |
| Vancomycin | 0.05g |
| Gentamicin | 0.01g |
| Amphotericin B | 2.0mg |
| Cycloheximide | 2.0mg |

**Preparation of Selective Supplement:** Add components to distilled/deionized water and bring volume to 10.0mL. Mix thoroughly. Filter sterilize.

**Caution:** Cycloheximide is toxic. Avoid skin contact or aerosol formation and inhalation.

**Glutamine Solution:**
**Composition** per 100.0mL:
L-Glutamine..................................................................2.92g
NaCl (0.85% solution) ...........................................100.0mL

**Preparation of Glutamine Solution:** Add the glutamine to the 0.85% NaCl solution. Mix thoroughly. Filter sterilize.

**Preparation of Medium:** Aseptically combine 50.0mL of sterile Eagle minimum essential medium with Earle salts, 10X, 50.0mL of fetal calf serum, 10.0mL of selective supplement, and 5.0mL of sterile glutamine solution. Bring volume to 500.0mL with sterile distilled/deionized water. Mix thoroughly. Aseptically distribute into sterile tubes or flasks.

**Storage/Shelf Life:** Store dehydrated media in the dark in a sealed container below 30°C. Prepared media should be stored under refrigeration (2-8°C). Media should be used within 60 days of preparation. Media should not be used if there are any signs of deterioration (shrinking, cracking, or discoloration) or contamination, or if the expiration date supplied by the manufacturer has passed.

**Use:** For the isolation and cultivation of *Chlamydia* species.

## CHO Medium Base
## (Carbohydrate Medium Base)
**Composition** per liter:
Pancreatic digest of casein...........................................15.0g
Yeast extract.................................................................7.0g
NaCl.............................................................................2.5g
Agar ...........................................................................0.75g
Sodium thioglycolate....................................................0.5g
L-Cystine ...................................................................0.25g
Ascorbic acid...............................................................0.1g
Bromthymol Blue .......................................................0.01g
pH 7.0 ± 0.2 at 25°C

**Preparation of Medium:** Add components to distilled/deionized water and bring volume to 1.0L. Mix thoroughly. Gently heat and bring to boiling. Distribute into tubes or flasks. Autoclave for 15 min at 15 psi pressure–121°C. Cool to 45°–50°C.

**Storage/Shelf Life:** Store dehydrated media in the dark in a sealed container below 30°C. Prepared media should be stored under refrigeration (2-8°C). Media should be used within 60 days of preparation. Media should not be used if there are any signs of deterioration (discoloration) or contamination, or if the expiration date supplied by the manufacturer has passed.

**Use:** Used as a basal medium to which carbohydrates are added for fermentation studies of anaerobic bacteria. Generally, 6.25mL of a 10% filter-sterilized solution of carbohydrate is added to the sterile basal medium.

## Chocolate Agar
**Composition** per liter:
Agar ..........................................................................15.0g
Pantone.......................................................................10.0g
Bitone.........................................................................10.0g
NaCl.............................................................................5.0g
Tryptic digest of beef heart ...........................................3.0g
Cornstarch ...................................................................1.0g
Sheep blood, defibrinated ........................................100.0mL
Supplement B...........................................................10.0mL
pH 7.3 ± 0.2 at 25°C

**Supplement B:**
**Composition** per 10.0mL:
Cephalothin................................................................15.0mg
Vancomycin ..............................................................10.0mg
Trimethoprim...............................................................5.0mg
Amphotericin B ...........................................................2.0mg
Polymyxin B...............................................................2500U

**Preparation of Supplement B:** Add components to 10.0mL of distilled/deionized water. Mix thoroughly. Filter sterilize.

**Source:** Supplement B is available from BD Diagnostic Systems.

**Preparation of Medium:** Add components, except supplement B solution and sheep blood, to distilled/deionized water and bring volume to 890.0mL. Mix thoroughly. Gently heat until boiling. Autoclave for 15 min at 15 psi pressure–121°C. Cool to 45°–50°C. Aseptically add 100.0mL of sterile, defibrinated sheep blood. Gently heat while stirring and bring to 85°C for 5–10 min. Cool to 50°C. Aseptically add 10.0mL of sterile supplement B. Mix thoroughly. Pour into sterile Petri dishes or distribute into sterile tubes.

**Storage/Shelf Life:** Store dehydrated media in the dark in a sealed container below 30°C. Prepared media should be stored under refrigeration (2-8°C). Media should be used within 60 days of preparation. Media should not be used if there are any signs of deterioration (shrinking, cracking, or discoloration) or contamination, or if the expiration date supplied by the manufacturer has passed.

**Use:** For the isolation and cultivation of a variety of fastidious microorganisms.

## Chocolate Agar
**Composition** per liter:
Proteose peptone No. 3 ...............................................15.0g
Agar ..........................................................................10.0g
NaCl.............................................................................5.0g
K$_2$HPO$_4$.............................................................................4.0g
Cornstarch ...................................................................1.0g
KH$_2$PO$_4$.............................................................................1.0g
Hemoglobin solution ..............................................100.0mL
Supplement B...........................................................10.0mL
pH 7.0 ± 0.2 at 25°C

**Source:** This medium is available from BD Diagnostic Systems.

**Supplement B:**
**Composition** per 10.0mL:
Cephalothin................................................................15.0mg
Vancomycin ..............................................................10.0mg
Trimethoprim...............................................................5.0mg
Amphotericin B ...........................................................2.0mg
Polymyxin B...............................................................2500U

**Preparation of Supplement B:** Add components to distilled/deionized water and bring volume to 10.0mL. Mix thoroughly. Filter sterilize.

**Hemoglobin Solution:**
**Composition** per 100.0mL:
Hemoglobin ...............................................................10.0g

**Preparation of Hemoglobin Solution:** Add hemoglobin to distilled/deionized water and bring volume to 100.0mL. Mix thoroughly. Filter sterilize.

**Preparation of Medium:** Add components, except hemoglobin solution and supplement B, to distilled/deionized water and bring volume to 990.0mL. Mix thoroughly. Gently heat and bring to boiling. Autoclave for 15 min at 15 psi pressure–121°C. Cool to 45°–50°C. Aseptically add 100.0mL of sterile hemoglobin solution. Gently heat while stirring and bring to 85°C for 5–10 min. Cool to 50°C. Aseptically add 10.0mL of sterile supplement B. Mix thoroughly. Pour into sterile Petri dishes or distribute into sterile tubes.

**Storage/Shelf Life:** Store dehydrated media in the dark in a sealed container below 30°C. Prepared media should be stored under refrigeration (2-8°C). Media should be used within 60 days of preparation. Media should not be used if there are any signs of deterioration (shrinking, cracking, or discoloration) or contamination, or if the expiration date supplied by the manufacturer has passed.

**Use:** For the isolation and cultivation of fastidious microorganisms.

## Chocolate Agar, Enriched

**Composition** per liter:

GC medium base.................................................................. 740.0mL
Hemoglobin solution.............................................................. 250.0mL
Supplement B......................................................................... 10.0mL

pH 7.3 ± 0.2 at 25°C

**Source:** This medium is available from BD Diagnostic Systems.

### GC Medium Base:

**Composition** per 740.0mL:

Agar ...................................................................................20.0g
Proteose peptone No. 3 ........................................................15.0g
NaCl.......................................................................................5.0g
K$_2$HPO$_4$..............................................................................4.0g
Glucose ..................................................................................1.5g
Cornstarch .............................................................................1.0g
KH$_2$PO$_4$.............................................................................1.0g

pH 7.2 ± 0.2 at 25°C

**Preparation of GC Medium Base:** Add components to distilled/deionized water and bring volume to 740.0mL. Mix thoroughly. Gently heat until boiling. Autoclave for 15 min at 15 psi pressure–121°C. Cool to 45°–50°C.

### Hemoglobin Solution:

**Composition** per 250.0mL:

Hemoglobin ..........................................................................10.0g

**Preparation of Hemoglobin Solution:** Add hemoglobin to distilled/deionized water and bring volume to 250.0mL. Mix thoroughly. Autoclave for 15 min at 15 psi pressure–121°C. Cool to 45°–50°C.

### Supplement B:

**Composition** per 10.0mL:

Cephalothin..........................................................................15.0mg
Vancomycin ..........................................................................10.0mg
Trimethoprim ..........................................................................5.0mg
Amphotericin B........................................................................2.0mg
Polymyxin B ..........................................................................2500U

**Preparation of Supplement B:** Add components to distilled/deionized water and bring volume to 10.0mL. Mix thoroughly. Filter sterilize.

**Preparation of Medium:** To 740.0mL of cooled sterile GC medium base, aseptically add 250.0mL of sterile hemoglobin solution and 10.0mL of sterile supplement B. Mix thoroughly. Pour into sterile Petri dishes or distribute into sterile tubes.

**Storage/Shelf Life:** Store dehydrated media in the dark in a sealed container below 30°C. Prepared media should be stored under refrigeration (2-8°C). Media should be used within 60 days of preparation. Media should not be used if there are any signs of deterioration (shrinking, cracking, or discoloration) or contamination, or if the expiration date supplied by the manufacturer has passed.

**Use:** For the cultivation of fastidious microorganisms, especially *Neisseria* species.

## Chocolate Agar, Enriched

**Composition** per liter:

GC medium base.................................................................. 740.0mL
Hemoglobin solution.............................................................. 250.0mL
Supplement VX ..................................................................... 10.0mL

pH 7.3 ± 0.2 at 25°C

**Source:** This medium is available from BD Diagnostic Systems.

### GC Medium Base:

**Composition** per 740.0mL:

Proteose peptone No. 3 ........................................................15.0g
Agar ...................................................................................20.0g
NaCl.......................................................................................5.0g
K$_2$HPO$_4$..............................................................................4.0g
Glucose ..................................................................................1.5g
Cornstarch .............................................................................1.0g
KH$_2$PO$_4$.............................................................................1.0g

pH 7.2 ± 0.2 at 25°C

**Preparation of GC Medium Base:** Add components to distilled/deionized water and bring volume to 740.0mL. Mix thoroughly. Gently heat until boiling. Autoclave for 15 min at 15 psi pressure–121°C. Cool to 45°–50°C.

### Hemoglobin Solution:

**Composition** per 250.0mL:

Hemoglobin ..........................................................................10.0g

**Preparation of Hemoglobin Solution:** Add hemoglobin to distilled/deionized water and bring volume to 250.0mL. Mix thoroughly. Autoclave for 15 min at 15 psi pressure–121°C. Cool to 45°–50°C.

### Supplement VX:

**Composition** per 10.0mL:

Supplement VX contains essential growth factors.

**Preparation of Supplement VX:** Add components to distilled/deionized water and bring volume to 10.0mL. Mix thoroughly. Filter sterilize.

**Preparation of Medium:** To 740.0mL of cooled sterile GC medium base, aseptically add 250.0mL of sterile hemoglobin solution and 10.0mL of sterile supplement VX. Mix thoroughly. Pour into sterile Petri dishes or distribute into sterile tubes.

**Storage/Shelf Life:** Store dehydrated media in the dark in a sealed container below 30°C. Prepared media should be stored under refrigeration (2-8°C). Media should be used within 60 days of preparation. Media should not be used if there are any signs of deterioration (shrinking, cracking, or discoloration) or contamination, or if the expiration date supplied by the manufacturer has passed.

**Use:** For the cultivation of fastidious microorganisms, especially *Neisseria* species.

## Chocolate Agar-*Bartonella* C-29

**Composition** per 1010.0mL:

GC agar base solution............................................................500.0 ml
Hemoglobin solution ..............................................................500.0 ml
IsoVitaleX® enrichment............................................................ 10.0mL

## IsoVitaleX® Enrichment:
**Composition** per liter:

| | |
|---|---|
| Glucose | 100.0g |
| L-Cysteine·HCl | 25.9g |
| L-Glutamine | 10.0g |
| L-Cystine | 1.1g |
| Adenine | 1.0g |
| Nicotinamide adenine dinucleotide | 0.25g |
| Vitamin B$_{12}$ | 0.1g |
| Thiamine pyrophosphate | 0.1g |
| Guanine·HCl | 0.03g |
| Fe(NO$_3$)$_3$·6H$_2$O | 0.02g |
| *p*-Aminobenzoic acid | 0.013g |
| Thiamine·HCl | 3.0mg |

**Preparation of IsoVitaleX®:** Add components to distilled/deionized water and bring volume to 1.0L. Mix thoroughly. Filter sterilize.

## GC Agar Base Solution:
**Composition** per 500.0mL:

| | |
|---|---|
| Agar | 10.0g |
| Pancreatic digest of casein | 7.5g |
| Peptic digest of animal tissue | 7.5g |
| NaCl | 5.0g |
| K$_2$HPO$_4$ | 4.0g |
| Cornstarch | 1.0g |
| KH$_2$PO$_4$ | 1.0g |

**Preparation of GC Agar Base:** Add components to distilled/deionized water and bring volume to 500.0mL. Mix thoroughly. Gently heat until boiling. Autoclave for 15 min at 15 psi pressure–121°C. Cool to 45°–50°C.

## Hemoglobin Solution:
**Composition** per 500.0mL:

| | |
|---|---|
| Hemoglobin | 10.0g |

**Preparation of Hemoglobin Solution:** Add hemoglobin to distilled/deionized water and bring volume to 500.0mL. Mix thoroughly. Gently heat until boiling. Autoclave for 15 min at 15 psi pressure–121°C. Cool to 45°–50°C.

**Preparation of Medium:** Aseptically combine 500.0mL sterile, cooled GC agar base solution and 500.0mL cooled sterile hemoglobin solution. Aseptically add 10.0mL of sterile IsoVitaleX® enrichment. Mix thoroughly. Pour into sterile Petri dishes or distribute into sterile tubes.

**Storage/Shelf Life:** Store dehydrated media in the dark in a sealed container below 30°C. Prepared media should be stored under refrigeration (2-8°C). Media should be used within 60 days of preparation. Media should not be used if there are any signs of deterioration (shrinking, cracking, or discoloration) or contamination, or if the expiration date supplied by the manufacturer has passed.

**Use:** For the isolation and cultivation of fastidious microorganisms, especially *Neisseria* and *Haemophilus* species, from a variety of clinical specimens.

## Chocolate Agar Base
## with Hemoglobin and Yeast Autolysate
**Composition** per liter:

| | |
|---|---|
| Proteose peptone | 20.0g |
| Agar | 15.0g |
| Na$_2$HPO$_4$ | 5.0g |
| NaCl | 5.0g |
| Glucose | 0.5g |
| Hemoglobin solution | 500.0mL |
| Yeast autolysate solution | 20.0mL |

pH 7.3 ± 0.2 at 25°C

**Source:** This medium is available from HiMedia.

## Hemoglobin Solution:
**Composition** per 500.0mL:

| | |
|---|---|
| Bovine hemoglobin | 10.0g |

**Preparation of Hemoglobin Solution:** Add bovine hemoglobin to distilled/deionized water and bring volume to 500.0mL. Mix thoroughly. Autoclave for 15 min at 15 psi pressure–121°C. Cool to 45°–50°C.

## Yeast Autolysate Solution:
**Composition** per 20.0mL:

| | |
|---|---|
| Yeast autolysate | 10.0g |
| Glucose | 1.0g |
| NaHCO$_3$ | 0.15g |

**Preparation of Yeast Autolysate Solution:** Add components to distilled/deionized water and bring volume to 20.0mL. Mix thoroughly. Filter sterilize.

**Preparation of Medium:** Add components, except hemoglobin and yeast autolysate solutions, to distilled/deionized water and bring volume to 480.0mL. Mix thoroughly. Gently heat until boiling. Autoclave for 15 min at 15 psi pressure–121°C. Cool to 45°–50°C. Add 500.0mL sterile hemoglobin solution and 20.0mL sterile yeast autolysate solution. Mix thoroughly. Pour into sterile Petri dishes or distribute into sterile tubes.

**Storage/Shelf Life:** Store dehydrated media in the dark in a sealed container below 30°C. Prepared media should be stored under refrigeration (2-8°C). Media should be used within 60 days of preparation. Media should not be used if there are any signs of deterioration (shrinking, cracking, or discoloration) or contamination, or if the expiration date supplied by the manufacturer has passed.

**Use:** For the isolation of *Neisseria gonorrhoeae* from chronic and acute cases of gonococcal infections.

## Chocolate Agar Base
## with Hemoglobin and Vitamino Growth Supplement
**Composition** per liter:

| | |
|---|---|
| Proteose peptone | 20.0g |
| Agar | 15.0g |
| Na$_2$HPO$_4$ | 5.0g |
| NaCl | 5.0g |
| Glucose | 0.5g |
| Hemoglobin solution | 500.0mL |
| Vitamino growth supplement solution | 10.0mL |

pH 7.3 ± 0.2 at 25°C

**Source:** This medium is available from HiMedia.

## Hemoglobin Solution:
**Composition** per 500.0mL:

| | |
|---|---|
| Bovine hemoglobin | 10.0g |

**Preparation of Hemoglobin Solution:** Add bovine hemoglobin to distilled/deionized water and bring volume to 500.0mL. Mix thoroughly. Autoclave for 15 min at 15 psi pressure–121°C. Cool to 45°–50°C.

## Vitamino Growth Supplement Solution:
**Composition** per 10.0mL:

| | |
|---|---|
| L-Glutamine | 0.2g |
| Adenine sulfate | 20.0mg |
| Guanine hydrochlroide | 0.6mg |

p-Aminobenzoic acid (PABA) ..................................................0.26mg

Vitamin B$_{12}$ ..................................................................0.2mg

**Preparation of Vitamino Growth Supplement Solution:** Add components to distilled/deionized water and bring volume to 10.0mL. Mix thoroughly. Filter sterilize.

**Preparation of Medium:** Add components, except hemoglobin and Vitamino growth supplement solutions, to distilled/deionized water and bring volume to 480.0mL. Mix thoroughly. Gently heat until boiling. Autoclave for 15 min at 15 psi pressure–121°C. Cool to 45°–50°C. Add 500.0mL sterile hemoglobin solution and 10.0mL sterile Vitamino growth supplement solution. Mix thoroughly. Pour into sterile Petri dishes or distribute into sterile tubes.

**Storage/Shelf Life:** Store dehydrated media in the dark in a sealed container below 30°C. Prepared media should be stored under refrigeration (2-8°C). Media should be used within 60 days of preparation. Media should not be used if there are any signs of deterioration (shrinking, cracking, or discoloration) or contamination, or if the expiration date supplied by the manufacturer has passed.

**Use:** For the isolation of *Neisseria gonorrhoeae* from chronic and acute cases of gonococcal infections.

## Chocolate II Agar

**Composition** per liter:

Agar ............................................................................12.0g

Casein enzymic hydrolysate ........................................7.5g

Meat extract ................................................................7.5g

NaCl ..............................................................................5.0g

K$_2$HPO$_4$ ........................................................................4.0g

Corn starch ..................................................................1.0g

KH$_2$PO$_4$ ........................................................................1.0g

Vitamin B$_{12}$ ..................................................................0.2mg

Hemoglobin solution..............................................500.0mL

pH 7.3 ± 0.2 at 25°C

**Source:** This medium is available from HiMedia.

**Hemoglobin Solution:**

**Composition** per 500.0mL:

Bovine hemoglobin..................................................10.0g

**Preparation of Hemoglobin Solution:** Add bovine hemoglobin to distilled/deionized water and bring volume to 500.0mL. Mix thoroughly. Autoclave for 15 min at 15 psi pressure–121°C. Cool to 45°–50°C.

**Preparation of Medium:** Add components, except hemoglobin solution, to distilled/deionized water and bring volume to 500.0mL. Mix thoroughly. Gently heat until boiling. Autoclave for 15 min at 15 psi pressure–121°C. Cool to 45°–50°C. Add 500.0mL sterile hemoglobin solution. Mix thoroughly. Pour into sterile Petri dishes or distribute into sterile tubes.

**Storage/Shelf Life:** Store dehydrated media in the dark in a sealed container below 30°C. Prepared media should be stored under refrigeration (2-8°C). Media should be used within 60 days of preparation. Media should not be used if there are any signs of deterioration (shrinking, cracking, or discoloration) or contamination, or if the expiration date supplied by the manufacturer has passed.

**Use:** For the isolation of *Neisseria* and *Haemophilus* species from a variety of clinical specimens.

## Chocolate No. 2 Agar Base with Supplements

**Composition** per liter:

Agar ............................................................................12.0g

Casein enzymic hydrolysate ........................................7.5g

Meat extract ................................................................7.5g

NaCl ..............................................................................5.0g

K$_2$HPO$_4$ ........................................................................4.0g

Corn starch ..................................................................1.0g

KH$_2$PO$_4$ ........................................................................1.0g

Hemoglobin solution ............................................480.0mL

Supplement solution .............................................. 40.0mL

pH 7.3 ± 0.2 at 25°C

**Source:** This medium is available from HiMedia.

**Hemoglobin Solution:**

**Composition** per 500.0mL:

Hemoglobin ..............................................................10.0g

**Preparation of Hemoglobin Solution:** Add hemoglobin to distilled/deionized water and bring volume to 500.0mL. Mix thoroughly. Filter sterilize.

**Supplement Solution:**

**Composition** per 40.0mL:

p-Aminobenzoic acid..............................................259.0mg

L-Glutamine ...........................................................100.0mg

Adenine sulfate........................................................10.0mg

NAD............................................................................2.5mg

Vitamin B$_{12}$..............................................................1.0mg

Cocarboxylase............................................................1.0mg

Guanine·HCl..............................................................0.3mg

Fe(NO$_3$)$_3$..................................................................0.2mg

L-Cysteine·HCl........................................................0.13mg

Thiamine·HCl..........................................................0.03mg

**Preparation of Supplement Solution:** Add components to distilled/deionized water and bring volume to 40.0mL. Mix thoroughly. Filter sterilize.

**Preparation of Medium:** Add components, except hemoglobin solution and supplement solution, to distilled/deionized water and bring volume to 480.0mL. Mix thoroughly. Autoclave for 15 min at 15 psi pressure–121°C. Cool to 50°C. Aseptically add hemoglobin and supplement solutions. Mix thoroughly. Pour into Petri dishes or aseptically distribute into sterile tubes.

**Storage/Shelf Life:** Store dehydrated media in the dark in a sealed container below 30°C. Prepared media should be stored under refrigeration (2-8°C). Media should be used within 60 days of preparation. Media should not be used if there are any signs of deterioration (shrinking, cracking, or discoloration) or contamination, or if the expiration date supplied by the manufacturer has passed.

**Use:** For the isolation of *Neisseria* spp. and *Haemophilus* spp. from a variety of clinical specimens.

## Chocolate No. 2 Agar Base with Hemoglobin

**Composition** per liter:

Agar ............................................................................12.0g

Hemoglobin ..............................................................10.0g

Pancreatic digest of casein........................................7.5g

Selected meat peptone ..............................................7.5g

NaCl..............................................................................5.0g

K$_2$HPO$_4$ ........................................................................4.0g

Cornstarch ..................................................................................1.0g

KH$_2$PO$_4$ ..................................................................................1.0g

**Preparation of Medium:** Add components to distilled/deionized water and bring volume to 1.0L. Mix thoroughly. Gently heat to boiling. Autoclave for 15 min at 15 psi pressure–121°C. Pour into sterile Petri dishes or leave in tubes.

**Storage/Shelf Life:** Store dehydrated media in the dark in a sealed container below 30°C. Prepared media should be stored under refrigeration (2-8°C). Media should be used within 60 days of preparation. Media should not be used if there are any signs of deterioration (shrinking, cracking, or discoloration) or contamination, or if the expiration date supplied by the manufacturer has passed.

**Use:** For the isolation and cultivation of fastidious microorganisms.

## Chocolate II Agar with Hemoglobin and IsoVitaleX®
## (GCII Agar with Hemoglobin and IsoVitaleX®)

**Composition** per liter:

GCII agar base ........................................................ 990.0mL

IsoVitaleX® enrichment............................................. 10.0mL

pH 7.3 ± 0.2 at 25°C

**Source:** This medium is available as a prepared medium from BD Diagnostic Systems.

### GCII Agar Base:

**Composition** per liter:

Agar .......................................................................12.0g

Hemoglobin ...............................................................10.0g

Pancreatic digest of casein............................................7.5g

Selected meat peptone ..................................................7.5g

NaCl.........................................................................5.0g

K$_2$HPO$_4$ ...................................................................4.0g

Cornstarch .................................................................1.0g

KH$_2$PO$_4$ ...................................................................1.0g

**Preparation of GCII Agar Base:** Add components to distilled/deionized water and bring volume to 1.0L. Mix thoroughly. Gently heat to boiling. Autoclave for 15 min at 15 psi pressure–121°C. Cool to 45°–50°C.

### IsoVitaleX® Enrichment:

**Composition** per liter:

Glucose ..................................................................100.0g

L-Cysteine·HCl..........................................................25.9g

L-Glutamine.............................................................10.0g

L-Cystine ..................................................................1.1g

Adenine.....................................................................1.0g

Nicotinamide adenine dinucleotide ................................0.25g

Vitamin B$_{12}$ ...............................................................0.1g

Thiamine pyrophosphate................................................0.1g

Guanine·HCl .............................................................0.03g

Fe(NO$_3$)$_3$·6H$_2$O .......................................................0.02g

*p*-Aminobenzoic acid................................................0.013g

Thiamine·HCl .........................................................3.0mg

**Preparation of IsoVitaleX®:** Add components to distilled/deionized water and bring volume to 1.0L. Mix thoroughly. Filter sterilize.

**Preparation of Medium:** Aseptically add 10.0mL of sterile IsoVitaleX® enrichment to 990.0L of sterile, cooled GCII agar base. Mix thoroughly. Pour into sterile Petri dishes or distribute into sterile tubes.

**Storage/Shelf Life:** Store dehydrated media in the dark in a sealed container below 30°C. Prepared media should be stored under refriger-

ation (2-8°C). Media should be used within 60 days of preparation. Media should not be used if there are any signs of deterioration (shrinking, cracking, or discoloration) or contamination, or if the expiration date supplied by the manufacturer has passed.

**Use:** For the isolation and cultivation of fastidious microorganisms, especially *Neisseria* and *Haemophilus* species, from a variety of clinical specimens.

## Chocolate HiVeg Agar Base
## with Hemoglobin and Yeast Autolysate

**Composition** per liter:

Plant peptone No. 3........................................................20.0g

Agar .......................................................................15.0g

Na$_2$HPO$_4$ ...................................................................5.0g

NaCl.........................................................................5.0g

Glucose ......................................................................0.5g

Hemoglobin solution ............................................... 500.0mL

Yeast autolysate solution ............................................ 20.0mL

pH 7.3 ± 0.2 at 25°C

**Source:** This medium, without hemoglobin or yeast autolysate, is available as a premixed powder from HiMedia.

**Hemoglobin Solution:**

**Composition** per 500.0mL:

Bovine hemoglobin........................................................10.0g

**Preparation of Hemoglobin Solution:** Add bovine hemoglobin to distilled/deionized water and bring volume to 500.0mL. Mix thoroughly. Autoclave for 15 min at 15 psi pressure–121°C. Cool to 45°–50°C.

**Yeast Autolysate Solution:**

**Composition** per 20.0mL:

Yeast autolysate .........................................................10.0g

Glucose ......................................................................1.0g

NaHCO$_3$....................................................................0.15g

**Preparation of Yeast Autolysate Solution:** Add components to distilled/deionized water and bring volume to 20.0mL. Mix thoroughly. Filter sterilize.

**Preparation of Medium:** Add components, except hemoglobin and yeast autolysate solutions, to distilled/deionized water and bring volume to 480.0mL. Mix thoroughly. Gently heat until boiling. Autoclave for 15 min at 15 psi pressure–121°C. Cool to 45°–50°C. Add 500.0mL sterile hemoglobin solution and 20.0mL sterile yeast autolysate solution. Mix thoroughly. Pour into sterile Petri dishes or distribute into sterile tubes.

**Storage/Shelf Life:** Store dehydrated media in the dark in a sealed container below 30°C. Prepared media should be stored under refrigeration (2-8°C). Media should be used within 60 days of preparation. Media should not be used if there are any signs of deterioration (shrinking, cracking, or discoloration) or contamination, or if the expiration date supplied by the manufacturer has passed.

**Use:** For the isolation of *Neisseria gonorrhoeae* from chronic and acute cases of gonococcal infections.

## Chocolate HiVeg Agar Base with Hemoglobin
## and Vitamino Growth Supplement

**Composition** per liter:

Plant peptone No. 3........................................................20.0g

Agar .......................................................................15.0g

Na$_2$HPO$_4$ ...................................................................5.0g

NaCl......................................................................................5.0g
Glucose ................................................................................0.5g
Hemoglobin solution.................................................... 500.0mL
Vitamino growth supplement solution.................................... 10.0mL

pH 7.3 ± 0.2 at 25°C

**Source:** This medium, without hemoglobin or vitamino growth supplement, is available as a premixed powder from HiMedia.

**Hemoglobin Solution:**
**Composition** per 500.0mL:
Bovine hemoglobin....................................................................10.0g

**Preparation of Hemoglobin Solution:** Add bovine hemoglobin to distilled/deionized water and bring volume to 500.0mL. Mix thoroughly. Autoclave for 15 min at 15 psi pressure–121°C. Cool to 45°–50°C.

**Vitamino Growth Supplement Solution:**
**Composition** per 10.0mL:
L-Glutamine ............................................................................0.2g
Adenine sulfate .....................................................................20.0mg
Guanine hydrochlroide ...........................................................0.6mg
p-Aminobenzoic acid (PABA)................................................0.26mg
Vitamin B$_{12}$.........................................................................0.2mg

**Preparation of Vitamino Growth Supplement Solution:** Add components to distilled/deionized water and bring volume to 10.0mL. Mix thoroughly. Filter sterilize.

**Preparation of Medium:** Add components, except hemoglobin and Vitamino growth supplement solutions, to distilled/deionized water and bring volume to 480.0mL. Mix thoroughly. Gently heat until boiling. Autoclave for 15 min at 15 psi pressure–121°C. Cool to 45°–50°C. Add 500.0mL sterile hemoglobin solution and 10.0mL sterile Vitamino growth supplement solution. Mix thoroughly. Pour into sterile Petri dishes or distribute into sterile tubes.

**Storage/Shelf Life:** Store dehydrated media in the dark in a sealed container below 30°C. Prepared media should be stored under refrigeration (2-8°C). Media should be used within 60 days of preparation. Media should not be used if there are any signs of deterioration (shrinking, cracking, or discoloration) or contamination, or if the expiration date supplied by the manufacturer has passed.

**Use:** For the isolation of *Neisseria gonorrhoeae* from chronic and acute cases of gonococcal infections.

## Chocolate No. 2 HiVeg Agar Base with Hemoglobin
**Composition** per liter:
Agar ......................................................................................12.0g
Plant extract No.I.....................................................................7.5g
Plant hydrolysate.....................................................................7.5g
NaCl.......................................................................................5.0g
K$_2$HPO$_4$..................................................................................4.0g
Corn starch.............................................................................1.0g
KH$_2$PO$_4$..................................................................................1.0g
Vitamin B$_{12}$.........................................................................0.2mg
Hemoglobin solution.................................................... 500.0mL

pH 7.3 ± 0.2 at 25°C

**Source:** This medium, without hemoglobin, is available as a premixed powder from HiMedia.

**Hemoglobin Solution:**
**Composition** per 500.0mL:
Bovine hemoglobin....................................................................10.0g

**Preparation of Hemoglobin Solution:** Add bovine hemoglobin to distilled/deionized water and bring volume to 500.0mL. Mix thoroughly. Autoclave for 15 min at 15 psi pressure–121°C. Cool to 45°–50°C.

**Preparation of Medium:** Add components, except hemoglobin solution, to distilled/deionized water and bring volume to 500.0mL. Mix thoroughly. Gently heat until boiling. Autoclave for 15 min at 15 psi pressure–121°C. Cool to 45°–50°C. Add 500.0mL sterile hemoglobin solution. Mix thoroughly. Pour into sterile Petri dishes or distribute into sterile tubes.

**Storage/Shelf Life:** Store dehydrated media in the dark in a sealed container below 30°C. Prepared media should be stored under refrigeration (2-8°C). Media should be used within 60 days of preparation. Media should not be used if there are any signs of deterioration (shrinking, cracking, or discoloration) or contamination, or if the expiration date supplied by the manufacturer has passed.

**Use:** For the isolation of *Neisseria* and *Haemophilus* species from a variety of clinical specimens.

## Chocolate Tellurite Agar
## (Tellurite Blood Agar)
**Composition** per liter:
Agar ......................................................................................10.0g
Casein/meat (50/50) peptone ................................................10.0g
Hemoglobin ..........................................................................10.0g
NaCl.......................................................................................5.0g
K$_2$HPO$_4$..................................................................................4.0g
Cornstarch .............................................................................1.0g
KH$_2$PO$_4$..................................................................................1.0g
K$_2$TeO$_3$.................................................................................0.1g
Bio-X enrichment ............................................................. 10.0mL

**Bio-X Enrichment:**
**Composition** per liter:
Glucose .................................................................................100.0g
L-Cysteine·HCl ......................................................................25.9g
L-Glutamate ...........................................................................10.0g
L-Cystine ................................................................................1.1g
Adenine...................................................................................1.0g
Cocarboxylase........................................................................0.1g
Guanine·HCl ...........................................................................0.03g
FeNO$_3$ ...................................................................................0.02g
*p*-Aminobenzoic acid...........................................................0.013g
Vitamin B$_{12}$..........................................................................0.01g
NAD (nicotinamide adenine dinucleotide)............................250.0mg
Thiamine·HCl .........................................................................3.0mg

pH 7.2 ± 0.2 at 25°C

**Preparation of Bio-X Enrichment:** Add components to distilled/deionized water and bring volume to 1.0L. Mix thoroughly. Filter sterilize.

**Caution:** Potassium tellurite is toxic.

**Preparation of Medium:** Add components, except Bio-X enrichment, to distilled/deionized water and bring volume to 990.0mL. Mix thoroughly. Gently heat and bring to boiling. Autoclave for 15 min at 15 psi pressure–121°C. Cool to 45°–50°C. Aseptically add filter-sterilized Bio-X enrichment. Mix thoroughly. Pour into sterile Petri dishes or distribute into sterile tubes.

**Storage/Shelf Life:** Store dehydrated media in the dark in a sealed container below 30°C. Prepared media should be stored under refrigeration (2-8°C). Media should be used within 60 days of preparation.

Media should not be used if there are any signs of deterioration (shrinking, cracking, or discoloration) or contamination, or if the expiration date supplied by the manufacturer has passed.

**Use:** For the selective isolation and cultivation of *Corynebacterium* species. *Corynebacterium diphtheriae* appears as gray-black colonies.

## Cholera HiVeg Medium Base with Tellurite and Blood
**Composition** per liter:

| | |
|---|---|
| NaCl | 20.0g |
| Agar | 10.0g |
| Plant extract | 10.0g |
| Plant peptone | 10.0g |
| Sucrose | 10.0g |
| $Na_2CO_3$ | 5.0g |
| Sodium lauryl sulfate | 0.1g |
| Sheep blood, defibrinated | 50.0mL |
| Tellurite solution | 2.0mL |

pH $8.5 \pm 0.2$ at 25°C

**Source:** This medium, without tellurite or blood, is available as a premixed powder from HiMedia.

**Tellurite Solution:**
**Composition** per 10.0mL:

| | |
|---|---|
| $K_2TeO_3$ | 0.1g |

**Preparation of Tellurite Solution:** Add $K_2TeO_3$ to distilled/deionized water and bring volume to 100.0mL. Mix thoroughly. Filter sterilize.

**Caution:** Potassium tellurite is toxic.

**Preparation of Medium:** Add components to distilled/deionized water and bring volume to 1.0L. Mix thoroughly. Gently heat and bring to boiling. Do not autoclave. Cool to 70°C. Aseptically add 2.0mL of sterile tellurite solution and 50.0mL of sterile defibrinated blood. Maintain at 70°C for several minutes. Cool to 50°C. Pour into sterile Petri dishes or leave in tubes.

**Storage/Shelf Life:** Store dehydrated media in the dark in a sealed container below 30°C. Prepared media should be stored under refrigeration (2-8°C). Media should be used within 60 days of preparation. Media should not be used if there are any signs of deterioration (shrinking, cracking, or discoloration) or contamination, or if the expiration date supplied by the manufacturer has passed.

**Use:** For the isolation of pathogenic vibrios, especially *Vibrio cholerae*. For the selective isolation of *Vibrio* species from specimens grossly contaminated with Enterobacteriaceae.

## Cholera Medium Base with Tellurite and Blood
**Composition** per liter:

| | |
|---|---|
| NaCl | 20.0g |
| Agar | 10.0g |
| Peptic digest of animal tissue | 10.0g |
| Beef extract | 10.0g |
| Sucrose | 10.0g |
| $Na_2CO_3$ | 5.0g |
| Sodium lauryl sulfate | 0.1g |
| Sheep blood, defibrinated | 50.0mL |
| Tellurite solution | 2.0mL |

pH $8.5 \pm 0.2$ at 25°C

**Source:** This medium, without tellurite or blood, is available as a premixed powder from HiMedia.

**Tellurite Solution:**
**Composition** per 10.0mL:

| | |
|---|---|
| $K_2TeO_3$ | 0.1g |

**Preparation of Tellurite Solution:** Add $K_2TeO_3$ to distilled/deionized water and bring volume to 100.0mL. Mix thoroughly. Filter sterilize.

**Caution:** Potassium tellurite is toxic.

**Preparation of Medium:** Add components to distilled/deionized water and bring volume to 1.0L. Mix thoroughly. Gently heat and bring to boiling. Do not autoclave. Cool to 70°C. Aseptically add 2.0mL of sterile tellurite soltuion and 50.0mL of sterile defibrinated blood. Maintain at 70°C for several minutes. Cool to 50°C. Pour into sterile Petri dishes or leave in tubes.

**Storage/Shelf Life:** Store dehydrated media in the dark in a sealed container below 30°C. Prepared media should be stored under refrigeration (2-8°C). Media should be used within 60 days of preparation. Media should not be used if there are any signs of deterioration (shrinking, cracking, or discoloration) or contamination, or if the expiration date supplied by the manufacturer has passed.

**Use:** For the isolation of pathogenic vibrios, especially *Vibrio cholerae*. For the selective isolation of *Vibrio* species from specimens grossly contaminated with Enterobacteriaceae.

## Cholera Medium TCBS
**Composition** per liter:

| | |
|---|---|
| Sucrose | 20.0g |
| Agar | 14.0g |
| Peptone | 10.0g |
| $Na_2S_2O_3$ | 10.0g |
| Sodium citrate | 10.0g |
| NaCl | 10.0g |
| Ox bile | 8.0g |
| Yeast extract | 5.0g |
| Ferric citrate | 1.0g |
| Bromthymol Blue | 0.04g |
| Thymol Blue | 0.04g |

pH $8.6 \pm 0.2$ at 25°C

**Source:** This medium is available as a premixed powder from Thermo Scientific.

**Preparation of Medium:** Add components to distilled/deionized water and bring volume to 1.0mL. Mix thoroughly. Gently heat while stirring and bring to boiling. Do not autoclave. Cool to 45°C. Pour into sterile Petri dishes.

**Storage/Shelf Life:** Store dehydrated media in the dark in a sealed container below 30°C. Prepared media should be stored under refrigeration (2-8°C). Media should be used within 60 days of preparation. Media should not be used if there are any signs of deterioration (shrinking, cracking, or discoloration) or contamination, or if the expiration date supplied by the manufacturer has passed.

**Use:** For the isolation of pathogenic vibrios, especially *Vibrio cholerae*. This medium is suitable for the growth of *Vibrio cholerae*, *Vibrio parahaemolyticus*, and most other *Vibrios*. Most of the Enterobacteriaceae encountered in feces are totally suppressed for at least 24 hours. Slight growth of *Proteus* species and *Enterococcus faecalis* may occur but the colonies are easily distinguished from vibrio colonies. While inhibiting non-vibrios, it promotes rapid growth of pathogenic vibrios after overnight incubation at 35°C. *Vibrio cholerae* El Tor biotype forms yellow colonies, *Vibrio parahaemolyticus* forms blue-green colonies, *Vibrio alginolyticus* forms yellow colonies, *Vibrio metschnikovii* forms yellow colonies, *Vibrio fluvialis* forms yellow colonies, *Vibrio vulnificus* forms blue-green colonies, *Vibrio mimicus* forms blue-green colonies, *Enterococcus* species form yellow colonies, *Proteus* species form yellow-green colonies, *Pseudomonas* species form blue-green colonies and some strains of *Aeromonas hydrophila* produce yellow colonies, but *Plesimonas shigelloides* does not usually grow well on this medium.

## Chopped Liver Broth

**Composition** per liter:
Fresh beef liver ................................................................500.0g
Peptone..........................................................................10.0g
$K_2HPO_4$ ........................................................................1.0g
Soluble starch.................................................................1.0g

pH 7.0 ± 0.2 at 25°C

**Source:** This medium is available as a premixed powder from Hi-Media.

**Preparation of Medium:** Grind fresh beef liver. Add to 1.0L of distilled/deionized water. Gently heat and bring to boiling. Continue boiling for 60 min. Cool to 25°C. Adjust pH to 7.0. Gently heat and bring to boiling. Continue boiling for 10 min. Filter through cheesecloth. Save chopped liver particles. To filtrate, add remaining components. Bring volume to 1.0L with distilled/deionized water. Adjust pH to 7.0. Filter through Whatman #1 filter paper. Add chopped liver particles to test tubes to a depth of 1.2–2.5 cm. Add 10.0mL of broth to each tube. Autoclave for 15 min at 15 psi pressure–121°C.

**Storage/Shelf Life:** Store dehydrated media in the dark in a sealed container below 30°C. Prepared media should be stored under refrigeration (2-8°C). Media should be used within 60 days of preparation. Media should not be used if there are any signs of deterioration (discoloration) or contamination, or if the expiration date supplied by the manufacturer has passed.

**Use:** For the isolation and cultivation of *Clostridium botulinum*, *Clostridium perfringens*, and other anaerobic bacteria from foods.

## Chopped Liver HiVeg Broth

**Composition** per liter:
Plant infusion ...............................................................100.0g
Plant peptone................................................................10.0g
$K_2HPO_4$........................................................................1.0g
Starch, soluble................................................................1.0g

pH 7.0 ± 0.2 at 25°C

**Source:** This medium is available as a premixed powder from Hi-Media.

**Preparation of Medium:** Add components to distilled/deionized water and bring volume to 1.0L. Mix thoroughly. Gently heat and bring to boiling. Distribute into tubes or flasks. Autoclave for 20 min at 15 psi pressure–121°C.

**Storage/Shelf Life:** Store dehydrated media in the dark in a sealed container below 30°C. Prepared media should be stored under refrigeration (2-8°C). Media should be used within 60 days of preparation. Media should not be used if there are any signs of deterioration (discoloration) or contamination, or if the expiration date supplied by the manufacturer has passed.

**Use:** For the isolation and cultivation of *Clostridium botulinum*, *Clostridium perfringens*, and other anaerobic bacteria from foods.

## Chopped Meat Agar

**Composition** per liter:
Ground meat, fat free .....................................................500.0g
Pancreatic digest of casein.............................................30.0g
Agar ...............................................................................15.0g
$K_2HPO_4$.........................................................................5.0g
Yeast extract ...................................................................5.0g
L-Cysteine·HCl ...............................................................0.5g
Resazurin ........................................................................1.0mg
NaOH (1$N$ solution).......................................................25.0mL

pH 7.0 ± 0.2 at 25°C

**Preparation of Medium:** Use lean beef or horse meat. Remove fat and connective tissue. Grind finely. Add ground meat and 25.0mL of NaOH solution to distilled/deionized water and bring volume to 1025.0mL. Gently heat and bring to boiling. Continue boiling for 15 min without stirring. Cool to room temperature. Remove fat from surface. Filter and retain both meat particles and filtrate. Adjust volume of filtrate to 1.0L with distilled/deionized water. Add pancreatic digest of casein, agar, $K_2HPO_4$, yeast extract, and resazurin. Gently heat and bring to boiling. Boil for 1–2 min. Add L-cysteine·HCl. Mix thoroughly. Distribute 7.0mL into tubes that contain meat particles (1 part meat particles to 5 parts fluid). Autoclave for 30 min at 15 psi pressure–121°C.

**Storage/Shelf Life:** Store dehydrated media in the dark in a sealed container below 30°C. Prepared media should be stored under refrigeration (2-8°C). Media should be used within 60 days of preparation. Media should not be used if there are any signs of deterioration (shrinking, cracking, or discoloration) or contamination, or if the expiration date supplied by the manufacturer has passed.

**Use:** For the cultivation of various anaerobes.

## Chopped Meat Broth

**Composition** per liter:
Ground meat, fat free .....................................................500.0g
Pancreatic digest of casein.............................................30.0g
$K_2HPO_4$.........................................................................5.0g
Yeast extract ...................................................................5.0g
L-Cysteine·HCl ...............................................................0.5g
Resazurin ........................................................................1.0mg
NaOH (1$N$ solution).......................................................25.0mL

pH 7.0 ± 0.2 at 25°C

**Preparation of Medium:** Use lean beef or horse meat. Remove fat and connective tissue. Grind finely. Add ground meat and 25.0mL of NaOH solution to distilled/deionized water and bring volume to 1025.0mL. Gently heat and bring to boiling. Continue boiling for 15 min without stirring. Cool to room temperature. Remove fat from surface. Filter and retain both meat particles and filtrate. Adjust volume of filtrate to 1.0L with distilled/deionized water. Add pancreatic digest of casein, $K_2HPO_4$, yeast extract, and resazurin. Gently heat and bring to boiling. Boil for 1–2 min. Add L-cysteine·HCl. Mix thoroughly. Distribute 7.0mL into tubes that contain meat particles (1 part meat particles to 5 parts fluid). Autoclave for 30 min at 15 psi pressure–121°C.

**Storage/Shelf Life:** Store dehydrated media in the dark in a sealed container below 30°C. Prepared media should be stored under refrigeration (2-8°C). Media should be used within 60 days of preparation. Media should not be used if there are any signs of deterioration (discoloration) or contamination, or if the expiration date supplied by the manufacturer has passed.

**Use:** For the cultivation of various anaerobes.

## Christensen Agar

**Composition** per liter:

Agar ....................................................................15.0g
NaCl ......................................................................5.0g
Sodium citrate ......................................................3.0g
KH$_2$PO$_4$ ...................................................................1.0g
L-Cysteine·HCl·H$_2$O ...........................................0.1g
Phenol Red .......................................................12.0mg

pH 6.9 ± 0.2 at 25°C

**Preparation of Medium:** Add components to distilled/deionized water and bring volume to 1.0L. Mix thoroughly. Gently heat and bring to boiling. Dispense into tubes or flasks. Autoclave for 15 min at 15 psi pressure–121°C. Pour into sterile Petri dishes or leave in tubes. Allow tubes to cool in a slanted position.

**Storage/Shelf Life:** Store dehydrated media in the dark in a sealed container below 30°C. Prepared media should be stored under refrigeration (2–8°C). Media should be used within 60 days of preparation. Media should not be used if there are any signs of deterioration (shrinking, cracking, or discoloration) or contamination, or if the expiration date supplied by the manufacturer has passed.

**Use:** For the differentiation of enteric pathogens, especially members of the Enterobacteriaceae, and coliforms based on their ability to utilize citrate as a carbon source. Bacteria that can utilize citrate turn the medium pink-red.

## Christensen Agar

**Composition** per liter:

Agar ....................................................................15.0g
NaCl ......................................................................5.0g
Sodium citrate ......................................................3.0g
KH$_2$PO$_4$ ...................................................................1.0g
Yeast extract .........................................................0.5g
Glucose ................................................................0.2g
L-Cysteine·HCl·H$_2$O ...........................................0.1g
Phenol Red .......................................................12.0mg

pH 6.9 ± 0.2 at 25°C

**Source:** This medium is available as a premixed powder from BD Diagnostic Systems.

**Preparation of Medium:** Add components to distilled/deionized water and bring volume to 1.0L. Mix thoroughly. Gently heat and bring to boiling. Dispense into tubes or flasks. Autoclave for 15 min at 15 psi pressure–121°C. Pour into sterile Petri dishes or leave in tubes. Allow tubes to cool in a slanted position.

**Storage/Shelf Life:** Store dehydrated media in the dark in a sealed container below 30°C. Prepared media should be stored under refrigeration (2–8°C). Media should be used within 60 days of preparation. Media should not be used if there are any signs of deterioration (shrinking, cracking, or discoloration) or contamination, or if the expiration date supplied by the manufacturer has passed.

**Use:** For the differentiation of enteric pathogens, especially members of the Enterobacteriaceae, and coliforms based on their ability to utilize citrate as a carbon source. Bacteria that can utilize citrate turn the medium pink-red.

## Christensen Citrate Agar
### (BAM M39)

**Composition** per liter:

Agar ....................................................................15.0g
NaCl ......................................................................5.0g

Sodium citrate ......................................................3.0g
KH$_2$PO$_4$ ...................................................................1.0g
Yeast extract .........................................................0.5g
Ferric ammonium citrate ......................................0.4g
L-Cysteine·HCl·H$_2$O ...........................................0.1g
Na$_2$S$_2$O$_5$ ...............................................................0.08g
Phenol Red .......................................................12.0mg

pH 6.9 ± 0.2 at 25°C

**Preparation of Medium:** Add components to distilled/deionized water and bring volume to 1.0L. Mix thoroughly. Gently heat and bring to boiling. Dispense into tubes or flasks. Autoclave for 15 min at 15 psi pressure–121°C. Pour into sterile Petri dishes or leave in tubes. Allow tubes to cool in a slanted position.

**Storage/Shelf Life:** Store dehydrated media in the dark in a sealed container below 30°C. Prepared media should be stored under refrigeration (2–8°C). Media should be used within 60 days of preparation. Media should not be used if there are any signs of deterioration (shrinking, cracking, or discoloration) or contamination, or if the expiration date supplied by the manufacturer has passed.

**Use:** For the differentiation of enteric pathogens, especially members of the Enterobacteriaceae, and coliforms based on their ability to utilize citrate as a carbon source. Bacteria that can utilize citrate turn the medium pink-red.

## Christensen Citrate Agar, Modified
### (Citrate Agar)

**Composition** per liter:

Agar ....................................................................12.0g
NaCl ......................................................................5.0g
Sodium citrate ......................................................3.8g
KH$_2$PO$_4$ ...................................................................1.0g
Yeast extract .........................................................0.5g
Glucose ................................................................0.2g
L-Cysteine·HCl·H$_2$O ...........................................0.1g
Phenol Red ..........................................................0.02g

pH 6.7 ± 0.2 at 25°C

**Preparation of Medium:** Add components to distilled/deionized water and bring volume to 1.0L. Mix thoroughly. Gently heat and bring to boiling. Dispense into tubes or flasks. Autoclave for 15 min at 15 psi pressure–121°C. Pour into sterile Petri dishes or leave in tubes. Allow tubes to cool in a slanted position.

**Storage/Shelf Life:** Store dehydrated media in the dark in a sealed container below 30°C. Prepared media should be stored under refrigeration (2–8°C). Media should be used within 60 days of preparation. Media should not be used if there are any signs of deterioration (shrinking, cracking, or discoloration) or contamination, or if the expiration date supplied by the manufacturer has passed.

**Use:** For the differentiation of enteric pathogens, especially members of the Enterobacteriaceae, and coliforms based on their ability to utilize citrate as a carbon source. Bacteria that can utilize citrate turn the medium pink-red.

## Christensen Citrate Sulfide Medium

**Composition** per liter:

Agar ....................................................................15.0g
NaCl ......................................................................5.0g
Sodium citrate·2H$_2$O ...........................................3.0g
KH$_2$HPO$_4$ .................................................................1.0g
Yeast extract .........................................................0.5g

Ferric citrate....................................................................0.2g
Ammonium citrate ...........................................................0.2g
Glucose ............................................................................0.2g
L-Cysteine·HCl·H$_2$O .........................................................0.1g
Na$_2$S$_2$O$_3$·5H$_2$O .............................................................0.08g
Phenol Red....................................................................0.012g

pH 6.7± 0.2 at 25°C

**Preparation of Medium:** Add components to distilled/deionized water and bring volume to 1.0L. Mix thoroughly. Gently heat and bring to boiling. Dispense into tubes or flasks. Autoclave for 15 min at 15 psi pressure–121°C. Pour into sterile Petri dishes or leave in tubes. Allow tubes to cool in a slanted position.

**Storage/Shelf Life:** Store dehydrated media in the dark in a sealed container below 30°C. Prepared media should be stored under refrigeration (2-8°C). Media should be used within 60 days of preparation. Media should not be used if there are any signs of deterioration (shrinking, cracking, or discoloration) or contamination, or if the expiration date supplied by the manufacturer has passed.

**Use:** For the differentiation of enteric pathogens, especially members of the Enterobacteriaceae, and coliforms based on their ability to utilize citrate as a carbon source and production of H$_2$S. Bacteria that can utilize citrate turn the medium pink-red. H$_2$S production appears as a blackening of the butt of the tube.

## Christensen Citrate Sulfite Agar

**Composition** per liter:

Agar ...............................................................................14.0g
NaCl..................................................................................5.0g
Sodium citrate·2H$_2$O .......................................................3.0g
KH$_2$HPO$_4$...........................................................................1.0g
Yeast extract....................................................................0.5g
Ferric ammonium citrate...................................................0.4g
Ammonium citrate ...........................................................0.2g
Glucose ............................................................................0.2g
L-Cysteine·HCl·H$_2$O .........................................................0.1g
Na$_2$S$_2$O$_3$·5H$_2$O .............................................................0.08g
Phenol Red....................................................................0.012g

pH 6.7± 0.2 at 25°C

**Preparation of Medium:** Add components to distilled/deionized water and bring volume to 1.0L. Mix thoroughly. Gently heat and bring to boiling. Dispense into tubes or flasks. Autoclave for 15 min at 15 psi pressure–121°C. Pour into sterile Petri dishes or leave in tubes. Allow tubes to cool in a slanted position.

**Storage/Shelf Life:** Store dehydrated media in the dark in a sealed container below 30°C. Prepared media should be stored under refrigeration (2-8°C). Media should be used within 60 days of preparation. Media should not be used if there are any signs of deterioration (shrinking, cracking, or discoloration) or contamination, or if the expiration date supplied by the manufacturer has passed.

**Use:** For the differentiation of enteric pathogens, especially members of the Enterobacteriaceae, and coliforms based on their ability to utilize citrate as a carbon source and production of H$_2$S. Bacteria that can utilize citrate turn the medium pink-red. H$_2$S production appears as a blackening of the butt of the tube.

## Christensen's Urea Agar

**Composition** per liter:

Agar ...............................................................................15.0g
NaCl..................................................................................5.0g

KH$_2$PO$_4$...........................................................................2.0g
Peptone.............................................................................1.0g
Glucose ............................................................................1.0g
Phenol Red....................................................................0.012g
Urea solution............................................................100.0mL

pH 6.8 ± 0.1 at 25°C

**Urea Solution:**
**Composition** per 100.0mL:
Urea.................................................................................20.0g

**Preparation of Urea:** Add urea to 100.0mL of distilled/deionized water. Mix thoroughly. Filter sterilize.

**Preparation of Medium:** Add components, except urea solution, to distilled/deionized water and bring volume to 900.0mL. Mix thoroughly. Gently heat and bring to boiling. Autoclave for 15 min at 15 psi pressure–121°C. Cool to 50–55°C. Aseptically add 100.0mL of sterile urea solution. Mix thoroughly. Pour into Petri dishes or distribute into sterile tubes. Allow tubes to solidify in a slanted position.

**Storage/Shelf Life:** Store dehydrated media in the dark in a sealed container below 30°C. Prepared media should be stored under refrigeration (2-8°C). Media should be used within 60 days of preparation. Media should not be used if there are any signs of deterioration (shrinking, cracking, or discoloration) or contamination, or if the expiration date supplied by the manufacturer has passed.

**Use:** For the differentiation of a variety of microorganisms, especially members of the Enterobacteriaceae, aerobic actinomycetes, streptococci, and nonfermenting Gram-negative bacteria, on the basis of urease production.

## CHROMagar™ *Acinetobacter*

**Composition** per liter:

Agar ...............................................................................15.0g
Peptone and yeast extract.................................................12.0g
Salts..................................................................................4.0g
Chromogenic mix .............................................................1.8g
Growth and regulator factors ...........................................1.0g

pH 7.0± 0.2 at 25°C

**Source:** This medium is available from CHROMagar, Paris, France.

**Preparation of Medium:** Add components to distilled/deionized water and bring volume to 1.0L. Mix thoroughly. Gently heat and bring to boiling. Do not autoclave. Cool to 50°C. Pour into sterile Petri dishes.

**Storage/Shelf Life:** Store in the dark. Chromogenic agars are especially light and temperature sensitive; protect from light, excessive heat, moisture, and freezing. Prepared media plates can be kept for one day at ambient temperature. Plates can be stored at least one week under refrigeration (2-8°C) if properly prepared and protected from light and dehydration. Do not use after the expiration date supplied by the manufacturer.

**Use:** For the detection of *Acinetobacter* sp. which produce red colonies.

## CHROMagar™ *Acinetobacter* with MDR Supplement
**Composition** per liter:

Agar ...............................................................................15.0g
Peptone and yeast extract.................................................12.0g
Salts..................................................................................4.0g
Chromogenic mix .............................................................1.8g

Growth and regulator factors ..........................................................1.0g
MDR selective supplement solution .........................................5.0mL
pH 7.0± 0.2 at 25°C

**Source:** This medium is available from CHROMagar, Paris, France.

**MDR Selective Supplement Solution**
**Composition** per 5.0mL:
MDR Selective supplement .........................................................1 vial

**Preparation of MDR Selective Supplement Solution:** Add 1 vial of MDR selective supplement to distilled/deionized water and bring volume to 5.0mL. Mix thoroughly.

**Preparation of Medium:** Add components, except MDR selective supplement solution, to distilled/deionized water and bring volume to 995mL. Mix thoroughly. Gently heat and bring to boiling. Do not autoclave. Cool to 50°C. Aseptically add 5.0mL of MDR selective supplement solution. Mix thoroughly. Pour into sterile Petri dishes.

**Storage/Shelf Life:** Store in the dark. Chromogenic agars are especially light and temperature sensitive; protect from light, excessive heat, moisture, and freezing. Prepared media plates can be kept for one day at ambient temperature. Plates can be stored at least one week under refrigeration (2-8°C) if properly prepared and protected from light and dehydration. Do not use after the expiration date supplied by the manufacturer.

**Use:** For the detection of multi-drug resistant (MDR) *Acinetobacter* which produce red colonies.

## CHROMagar™ *B. cereus*

**Composition** per liter:
Agar ........................................................................................15.0g
NaCl .......................................................................................10.0g
Peptone and yeast extract.............................................................8.0g
Chromogenic mix ........................................................................0.3g
*B. cereus* supplement solution .................................................. 40.0mL
pH 6.8± 0.2 at 25°C

**Source:** This medium is available from CHROMagar, Paris, France.

***B. cereus* Supplement Solution:**
**Composition** per 40.0mL:
ChromAgar™ *B. cereus* supplement ..............................................3.0g

**Preparation of *B. cereus* Supplement Solution:** Slowly dispense 3.0g of CHROMagar *B. cereus* supplement into 40ml of purified water. Add a magnetic bar and homogenize during at least 30 minutes rotating the magnetic bar at high speed (~1200 rpm) until obtaining a creamy homogeneous suspension.

**Preparation of Medium:** Add components, except *B. cereus* supplement solution, to distilled/deionized water and bring volume to 960mL. Mix thoroughly. Distribute into tubes or flasks. Autoclave for 15 min at 15 psi pressure–121°C. Cool to 50°C. Aseptically add 40.0mL of *B. cereus* supplement solution. Mix thoroughly. Pour into sterile Petri dishes.

**Storage/Shelf Life:** Store in the dark. Chromogenic agars are especially light and temperature sensitive; protect from light, excessive heat, moisture, and freezing. Prepared media plates can be kept for one day at ambient temperature. Plates can be stored at least one week under refrigeration (2-8°C) if properly prepared and protected from light and dehydration. Do not use after the expiration date supplied by the manufacturer.

**Use:** For the detection and differentiation of *Bacillus cereus* group from foods. *B. cereus* colonies appear blue with a white halo.

## CHROMagar™ CTX

**Composition** per liter:
Agar .......................................................................................15.0g
Peptone and yeast extract...............................................................8.0g
NaCl.........................................................................................5.0g
Chromogenic mix .........................................................................4.8g
CTX supplement solution (freshly prepared) ...........................2.0mL
pH 7.2± 0.2 at 25°C

**Source:** This medium is available from CHROMagar, Paris, France.

**CTX Supplement Solution:**
**Composition** per 10.0mL:
Ethanol................................................................................... 5.0mL
CHROMagar™ CTX supplement .................................................1.25g

**Preparation of CTX Supplement Solution:** Add ethanol and CHROMagar™ CTX supplement to distilled/deionized water and bring volume to 10.0mL. Mix thoroughly. Prepare freshly.

**Preparation of Medium:** Add components, except CHROMagar™ CTX supplement solution, to distilled/deionized water and bring volume to 998mL. Mix thoroughly. Gently heat and bring to boiling. Do not autoclave. Cool to 50°C. Aseptically add 2.0mL of CHROMagar™ CTX supplement solution. Mix thoroughly. Pour into sterile Petri dishes.

**Storage/Shelf Life:** Store in the dark. Chromogenic agars are especially light and temperature sensitive; protect from light, excessive heat, moisture, and freezing. Prepared media plates can be kept for one day at ambient temperature. Plates can be stored at least one week under refrigeration (2-8°C) if properly prepared and protected from light and dehydration. Do not use after the expiration date supplied by the manufacturer.

**Use:** For the detection of bacteria with ESBL CTX-M-type resistance. *E.coli* that produce CTX-M–type â-lactamases (CTX-Ms) form blue colonies. Other enterobacteriaceae CTX-M form mauve colonies.

## CHROMagar™ *E. coli*

**Composition** per liter:
Agar .......................................................................................15.0g
Chromogenic mix .......................................................................9.0g
Peptone and yeast extract .............................................................8.3g
Sodium chloride .........................................................................5.0g
pH 6.0± 0.2 at 25°C

**Source:** This medium is available from CHROMagar, Paris, France.

**Preparation of Medium:** Add components to distilled/deionized water and bring volume to 1.0L. Mix thoroughly. Distribute into tubes or flasks. Autoclave for 15 min at 15 psi pressure–121°C. Pour into sterile Petri dishes.

**Storage/Shelf Life:** Store in the dark. Chromogenic agars are especially light and temperature sensitive; protect from light, excessive heat, moisture, and freezing. Prepared media plates can be kept for one day at ambient temperature. Plates can be stored at least one week under refrigeration (2-8°C) if properly prepared and protected from light and dehydration. Do not use after the expiration date supplied by the manufacturer.

**Use:** For the detection and enumeration of *E. coli.*

## CHROMagar™ ECC

**Composition** per liter:

Agar .......................................................................15.0g

Peptone and yeast extract.............................................8.0g

NaCl..........................................................................5.0g

Chromogenic mix ......................................................4.8g

pH 7.2± 0.2 at 25°C

**Source:** This medium is available from CHROMagar, Paris, France.

**Preparation of Medium:** Add components, to distilled/deionized water and bring volume to 1.0L. Mix thoroughly. Gently heat and bring to boiling. Do not autoclave. Heat until complete fusion of agar grains (large bubbles replacing foam: about 2 minutes). Cool to 50°C. Pour into sterile Petri dishes.

**Storage/Shelf Life:** Store in the dark. Chromogenic agars are especially light and temperature sensitive; protect from light, excessive heat, moisture, and freezing. Prepared media plates can be kept for one day at ambient temperature. Plates can be stored at least one week under refrigeration (2-8°C) if properly prepared and protected from light and dehydration. Do not use after the expiration date supplied by the manufacturer.

**Use:** For the detection and enumeration of *Escherichia coli* and coliforms. *E. coli* form blue colonies. Other fecal coliforms form mauve colonies.

## CHROMagar™ ECC with Cefsulodin

**Composition** per liter:

Agar .......................................................................15.0g

Peptone and yeast extract.............................................8.0g

NaCl..........................................................................5.0g

Chromogenic mix ......................................................4.8g

Cefsulodin solution ............................................... 10.0mL

pH 7.2± 0.2 at 25°C

**Source:** This medium is available from CHROMagar, Paris, France.

**Cefsulodin Solution:**
**Composition** per 10.0mL:

Cefsulodin.................................................................5.0mg

**Preparation of Cefsulodin Solution:** Add cefsulodin to distilled/deionized water and bring volume to 10.0mL. Mix thoroughly. Filter sterilize.

**Preparation of Medium:** Add components, except cefsulodin solution, to distilled/deionized water and bring volume to 990mL. Mix thoroughly. Gently heat and bring to boiling. Do not autoclave. Cool to 50°C. Aseptically add 10.0mL of cefsulodin solution. Mix thoroughly. Pour into sterile Petri dishes.

**Storage/Shelf Life:** Store in the dark. Chromogenic agars are especially light and temperature sensitive; protect from light, excessive heat, moisture, and freezing. Prepared media plates can be kept for one day at ambient temperature. Plates can be stored at least one week under refrigeration (2-8°C) if properly prepared and protected from light and dehydration. Do not use after the expiration date supplied by the manufacturer.

**Use:** For the selective detection and enumeration of *Escherichia coli* and coliforms with suppression of *Pseudomonas* and *Aeromonas* spp. *E. coli* form blue colonies. Other fecal coliforms form mauve colonies.

## CHROMagar™ ESBL

**Composition** per liter:

Peptone and yeast extract .........................................17.0g

Agar .......................................................................15.0g

Chromogenic mix ......................................................1.0g

ESBL selective supplement solution ...................... 10.0mL

pH 7.0± 0.2 at 25°C

**Source:** This medium is available from CHROMagar, Paris, France.

**ESBL Selective Supplement Solution**
**Composition** per 10.0mL:

EBSL selective supplement .......................................0.57g

**Preparation of ESBL Selective Supplement Solution:** Add ESBL selective supplement to distilled/deionized water and bring volume to 10.0mL. Mix thoroughly to obtain a homogenous suspension with an opaque yellowish appearance. Must be freshly prepared.

**Preparation of Medium:** Add components, except ESBL selective supplement solution, to distilled/deionized water and bring volume to 990mL. Mix thoroughly. Gently heat and bring to boiling. Autoclave for 15 min at 15 psi pressure–121°C. Cool to 50°C. Pour into sterile Petri dishes. Aseptically add 10.0mL of ESBL selective supplement solution. Mix thoroughly. Pour into sterile Petri dishes.

**Storage/Shelf Life:** Store in the dark. Chromogenic agars are especially light and temperature sensitive; protect from light, excessive heat, moisture, and freezing. Prepared media plates can be kept for one day at ambient temperature. Plates can be stored at least one week under refrigeration (2-8°C) if properly prepared and protected from light and dehydration. Do not use after the expiration date supplied by the manufacturer.

**Use:** For the detection of Gram-negative bacteria producing extended spectrum beta-lactamase. Extended spectrum β-lactam resistant strains of *E. coli* form dark pink to reddish colonies. Extended spectrum β-lactam resistant strains of *Klebsiella, Enterobacter*, and *Citrobacter* form metallic blue colonies. Extended spectrum β-lactam resistant strains of *Proteus* colonies with a brown halo.

## CHROMagar™ KPC

**Composition** per liter:

Peptone and yeast extract .........................................17.0g

Agar .......................................................................15.0g

Chromogenic mix ......................................................1.0g

KPC selective supplement solution ....................... 10.0mL

pH 7.0± 0.2 at 25°C

**Source:** This medium is available from CHROMagar, Paris, France.

**KPC Selective Supplement Solution**
**Composition** per 10.0mL:

KPC selective supplement ..........................................0.4g

**Preparation of KPC Selective Supplement Solution:** Add KPC selective supplement to distilled/deionized water and bring volume to 10.0mL. Mix thoroughly to obtain a homogenous suspension with an opaque yellowish appearance. Must be freshly prepared.

**Preparation of Medium:** Add components, except KPC selective supplement solution, to distilled/deionized water and bring volume to 990mL. Mix thoroughly. Gently heat and bring to boiling. Autoclave for 15 min at 15 psi pressure–121°C. Cool to 50°C. Pour into sterile Petri dishes. Aseptically add 10.0mL of KPC selective supplement solution. Mix thoroughly. Pour into sterile Petri dishes.

**Storage/Shelf Life:** Store in the dark. Chromogenic agars are especially light and temperature sensitive; protect from light, excessive heat, moisture, and freezing. Prepared media plates can be kept for one day at ambient temperature. Plates can be stored at least one week under refrigeration (2-8°C) if properly prepared and protected from light and dehydration. Do not use after the expiration date supplied by the manufacturer.

**Use:** For the detection of Gram-negative bacteria with a reduced susceptibility to most carbapenem antimicrobic agents. Carbapenem-resistant strains of *E. coli* form dark pink to reddish colonies. Carbapenem-resistant strains of *Klebsiella, Enterobacter*, and *Citrobacter* form blue colonies. Carbapenem-resistant strains of *Pseudomonas* form cream translucent colonies.

## CHROMagar™ *Listeria*
**Composition** per liter:
Proprietary

**Source:** CHROMagar *Listeria* is available from CHROMagar Microbiology.

**Preparation of Medium:** Add components to distilled/deionized water and bring volume to 1.0L. Mix thoroughly. Gently heat in a boiling water bath or steam bath. Shake periodically during heating to dissolve components. Heat long enough with shaking every 5 min to ensure complete dissolution. Do not overheat. Adding tellurite can increase specificity. Cool to 45–50°C. Pour into sterile Petri dishes.

**Storage/Shelf Life:** Store in the dark. Chromogenic agars are especially light and temperature sensitive; protect from light, excessive heat, moisture, and freezing. Prepared media plates can be kept for one day at ambient temperature. Plates can be stored at least one week under refrigeration (2-8°C) if properly prepared and protected from light and dehydration. Do not use after the expiration date supplied by the manufacturer.

**Use:** For the differentiation and presumptive identification of *Listeria monocytogenes,* which form blue colonies surrounded by white halos.

## CHROMagar™ MRSA
**Composition** per liter:
Chromopeptone................................................................40.0g
NaCl................................................................................25.0g
Agar ................................................................................14.0g
Chromogenic mix ..............................................................0.5g
Inhibitory agents ............................................................0.07g
Cefoxitin ........................................................................6.0mg

**Source:** CHROMagar MRSA is available from CHROMagar Microbiology. Prepared medium is also available from BD Diagnostic Systems.

**Preparation of Medium:** Add components to distilled/deionized water and bring volume to 1.0L. Mix thoroughly. Gently heat in a boiling water bath or steam bath. Shake periodically during heating to dissolve components. Heat long enough with shaking every 5 min to ensure complete dissolution. Do not overheat. Adding tellurite can increase specificity. Cool to 45–50°C. Pour into sterile Petri dishes.

**Storage/Shelf Life:** Store in the dark. Chromogenic agars are especially light and temperature sensitive; protect from light, excessive heat, moisture, and freezing. Prepared media plates can be kept for one

day at ambient temperature. Plates can be stored at least one week under refrigeration (2-8°C) if properly prepared and protected from light and dehydration. Do not use after the expiration date supplied by the manufacturer.

**Use:** For the qualitative direct detection of nasal colonization by methicillin-resistant *Staphylococcus aureus* (MRSA) to aid in the prevention and control of MRSA infections in healthcare settings.

## CHROMagar™ MRSA
**Composition** per liter:
Peptones and yeast extract............................................40.0g
Salts...............................................................................25.0g
Agar ...............................................................................15.0g
Chromogenic mix ..............................................................2.5g
MRSA solution ............................................................. 1.0mL
pH 6.9± 0.2 at 25°C

**Source:** This medium is available from CHROMagar, Paris, France.

**MRSA Solution**
**Composition** per 20.0mL:
CHROMagar™ MRSA supplement ..........................................1pkg

**Preparation of MRSA Solution:** Add 1 package of CHROMagar™ MRSA supplement to distilled/deionized water and bring volume to 20.0mL. Mix thoroughly.

**Preparation of Medium:** Add components, except MRSA solution, to distilled/deionized water and bring volume to 999mL. Mix thoroughly. Gently heat and bring to boiling. Do not autoclave. Cool to 50°C. Aseptically add 1.0mL of MRSA solution. Mix thoroughly. Pour into sterile Petri dishes.

**Storage/Shelf Life:** Store in the dark. Chromogenic agars are especially light and temperature sensitive; protect from light, excessive heat, moisture, and freezing. Prepared media plates can be kept for one day at ambient temperature. Plates can be stored at least one week under refrigeration (2-8°C) if properly prepared and protected from light and dehydration. Do not use after the expiration date supplied by the manufacturer.

**Use:** For the isolation and differentiation of methicillin-resistant *Staphylococcus aureus* (MRSA) including low level MRSA. Methicillin-resistant *Staphylococcus aureus* (MRSA) form rose to mauve colonies.

## CHROMagar™ O157
**Composition** per liter:
Agar ...............................................................................15.0g
Peptone and yeast extract............................................13.0g
Chromogenic mix ..............................................................1.2g
pH 6.9± 0.2 at 25°C

**Source:** This medium is available from CHROMagar, Paris, France.

**Preparation of Medium:** Add components to distilled/deionized water and bring volume to 1.0L. Mix thoroughly. Gently heat and bring to boiling. Do not autoclave. Cool to 50°C. Pour into sterile Petri dishes.

**Storage/Shelf Life:** Store in the dark. Chromogenic agars are especially light and temperature sensitive; protect from light, excessive heat, moisture, and freezing. Prepared media plates can be kept for one day at ambient temperature. Plates can be stored at least one week under refrigeration (2-8°C) if properly prepared and protected from light and dehydration. Do not use after the expiration date supplied by the manufacturer.

**Use:** For the selective isolation and differentiation of *E. coli* O157 in foods and clinical specimens. *E. coli* O157 produce mauve colonies.

## CHROMagar™ O157 with Cefsulodin

**Composition** per liter:

| | |
|---|---|
| Agar | 15.0g |
| Peptone and yeast extract | 13.0g |
| Chromogenic mix | 1.2g |
| Cefsulodin solution | 10.0mL |

pH 6.9± 0.2 at 25°C

**Source:** This medium is available from CHROMagar, Paris, France.

**Cefsulodin Solution:**
**Composition** per 10.0mL:

| | |
|---|---|
| Cefsulodin | 5.0mg |

**Preparation of Cefsulodin Solution:** Add cefsulodin to distilled/deionized water and bring volume to 10.0mL. Mix thoroughly. Filter sterilize.

**Preparation of Medium:** Add components, except cefsulodin solution, to distilled/deionized water and bring volume to 990mL. Mix thoroughly. Gently heat and bring to boiling. Do not autoclave. Cool to 50°C. Aseptically add 10.0mL of cefsulodin solution. Mix thoroughly. Pour into sterile Petri dishes.

**Storage/Shelf Life:** Store in the dark. Chromogenic agars are especially light and temperature sensitive; protect from light, excessive heat, moisture, and freezing. Prepared media plates can be kept for one day at ambient temperature. Plates can be stored at least one week under refrigeration (2-8°C) if properly prepared and protected from light and dehydration. Do not use after the expiration date supplied by the manufacturer.

**Use:** For the selective isolation and differentiation of *E. coli* O157 in foods and clinical specimens with high levels of *Aeromonas*. *E.coli* O157 produce mauve colonies.

## CHROMagar™ O157 with Potassium Tellurite

**Composition** per liter:

| | |
|---|---|
| Agar | 15.0g |
| Peptone and yeast extract | 13.0g |
| Chromogenic mix | 1.2g |
| Potassium tellurite solution | 10.0mL |

pH 6.9± 0.2 at 25°C

**Source:** This medium is available from CHROMagar, Paris, France..

**Potassium Tellurite Solution:**
**Composition** per 10.0mL:

| | |
|---|---|
| K$_2$TeO$_3$ | 2.5mg |

**Preparation of Potassium Tellurite Solution:** Add potassium tellurite to distilled/deionized water and bring volume to 10.0mL. Mix thoroughly. Filter sterilize.

**Preparation of Medium:** Add components, except potassium tellurite solution, to distilled/deionized water and bring volume to 990mL. Mix thoroughly. Gently heat and bring to boiling. Do not autoclave. Cool to 50°C. Aseptically add 10.0mL of potassium tellurite solution. Mix thoroughly. Pour into sterile Petri dishes.

**Storage/Shelf Life:** Store in the dark. Chromogenic agars are especially light and temperature sensitive; protect from light, excessive heat, moisture, and freezing. Prepared media plates can be kept for one day at ambient temperature. Plates can be stored at least one week under refrigeration (2-8°C) if properly prepared and protected from light

and dehydration. Do not use after the expiration date supplied by the manufacturer.

**Use:** For the selective isolation and differentiation of *E. coli* O157 in foods and clinical specimens. Potassium tellurite increases the selectivity of the medium and the specificity of *E. coli* O157 detection. *E.coli* O157 produce mauve colonies.

## CHROMagar™ O157 with Cefixime

**Composition** per liter:

| | |
|---|---|
| Agar | 15.0g |
| Peptone and yeast extract | 13.0g |
| Chromogenic mix | 1.2g |
| Cefixime solution | 10.0mL |

pH 6.9± 0.2 at 25°C

**Source:** This medium is available from CHROMagar, Paris, France..

**Cefixime Solution:**
**Composition** per 10.0mL:

| | |
|---|---|
| Cefixime | 0.025mg |

**Preparation of Cefixime Solution:** Add cefixime to distilled/deionized water and bring volume to 10.0mL. Mix thoroughly. Filter sterilize.

**Preparation of Medium:** Add components, except cefixime solution, to distilled/deionized water and bring volume to 990mL. Mix thoroughly. Gently heat and bring to boiling. Do not autoclave. Cool to 50°C. Aseptically add 10.0mL of cefixime solution. Mix thoroughly. Pour into sterile Petri dishes.

**Storage/Shelf Life:** Store in the dark. Chromogenic agars are especially light and temperature sensitive; protect from light, excessive heat, moisture, and freezing. Prepared media plates can be kept for one day at ambient temperature. Plates can be stored at least one week under refrigeration (2-8°C) if properly prepared and protected from light and dehydration. Do not use after the expiration date supplied by the manufacturer.

**Use:** For the selective isolation and differentiation of *E. coli* O157 in foods and clinical specimens with high levels of *Proteus*. *E.coli* O157 produce mauve colonies.

## CHROMagar™ Orientation

**Composition** per liter:

| | |
|---|---|
| Peptone and yeast extract | 17.0g |
| Agar | 15.0g |
| Chromogenic mix | 1.0g |

pH 7.0± 0.2 at 25°C

**Source:** This medium is available from CHROMagar, Paris, France.

**Preparation of Medium:** Add components to distilled/deionized water and bring volume to 1.0L. Mix thoroughly. Distribute into tubes or flasks. Autoclave for 15 min at 15 psi pressure–121°C. Pour into sterile Petri dishes.

**Storage/Shelf Life:** Store in the dark. Chromogenic agars are especially light and temperature sensitive; protect from light, excessive heat, moisture, and freezing. Prepared media plates can be kept for one day at ambient temperature. Plates can be stored at least one week under refrigeration (2-8°C) if properly prepared and protected from light and dehydration. Do not use after the expiration date supplied by the manufacturer.

**Use:** For the detection, enhanced isolation, and differentiation of urinary tract pathogens. *E. coli* form dark pink to reddish colonies. *En-*

*terococcus* form turquoise blue colonies. *Klebsiella*, *Enterobacter*, and *Citrobacter* form metallic blue colonies. *Proteus* form colonies with a brown halo. *Pseudomonas* form cream translucent colonies.

## CHROMagar™ Orientation with Tween™ 80

**Composition** per liter:

Peptone and yeast extract.............................................17.0g
Agar ...........................................................................15.0g
Chromogenic mix ........................................................1.0g
Tween™ 80..................................................................0.5g

pH 7.0± 0.2 at 25°C

**Source:** This medium is available from CHROMagar, Paris, France.

**Preparation of Medium:** Add components to distilled/deionized water and bring volume to 1.0L. Mix thoroughly. Distribute into tubes or flasks. Autoclave for 15 min at 15 psi pressure–121°C. Pour into sterile Petri dishes.

**Storage/Shelf Life:** Store in the dark. Chromogenic agars are especially light and temperature sensitive; protect from light, excessive heat, moisture, and freezing. Prepared media plates can be kept for one day at ambient temperature. Plates can be stored at least one week under refrigeration (2-8°C) if properly prepared and protected from light and dehydration. Do not use after the expiration date supplied by the manufacturer.

**Use:** For the enhanced isolation and differentiation of urinary tract pathogens. *E. coli* form dark pink to reddish colonies. *Enterococcus* form turquoise blue colonies. *Klebsiella*, *Enterobacter*, and *Citrobacter* form metallic blue colonies. *Proteus* form colonies with a brown halo. *Pseudomonas* form cream translucent colonies.

## CHROMagar™ *Pseudomonas*

**Composition** per liter:

Proprietary

**Source:** CHROMagar *Pseudomonas* is available from CHROMagar Microbiology.

**Preparation of Medium:** Add components to distilled/deionized water and bring volume to 1.0L. Mix thoroughly. Gently heat in a boiling water bath or steam bath. Shake periodically during heating to dissolve components. Heat long enough with shaking every 5 min to ensure complete dissolution. Do not overheat. Cool to 45–50°C. Pour into sterile Petri dishes.

**Storage/Shelf Life:** Store in the dark. Chromogenic agars are especially light and temperature sensitive; protect from light, excessive heat, moisture, and freezing. Prepared media plates can be kept for one day at ambient temperature. Plates can be stored at least one week under refrigeration (2-8°C) if properly prepared and protected from light and dehydration. Do not use after the expiration date supplied by the manufacturer.

**Use:** For the detection of *Pseudomonas*. For the simultaneous detection and enumeration of *Pseudomonas aeruginosa* with markedly different coloring (blue colonies).

## CHROMagar™ *Salmonella*

**Composition** per liter:

Agar ...........................................................................15.0g
Chromogenic and selective mix....................................12.9g
Peptone and yeast extract..............................................7.0g

pH 7.6± 0.2 at 25°C

**Source:** This medium is available from CHROMagar, Paris, France.

**Preparation of Medium:** Add components to distilled/deionized water and bring volume to 1.0L. Mix thoroughly. Gently heat and bring to boiling. Do not autoclave. Cool to 50°C. Pour into sterile Petri dishes.

**Storage/Shelf Life:** Store in the dark. Chromogenic agars are especially light and temperature sensitive; protect from light, excessive heat, moisture, and freezing. Prepared media plates can be kept for one day at ambient temperature. Plates can be stored at least one week under refrigeration (2-8°C) if properly prepared and protected from light and dehydration. Do not use after the expiration date supplied by the manufacturer.

**Use:** For the detection and isolation of *Salmonella* spp., including *S. typhi* and *S. paratyphi* in clinical specimens. *Salmonella* including *S. typhi* form mauve colonies.

## CHROMagar™ *Salmonella* with Cefsulodin

**Composition** per liter:

Agar ...........................................................................15.0g
Chromogenic and selective mix....................................12.9g
Peptone and yeast extract..............................................7.0g
Cefsulodin solution.................................................. 10.0mL

pH 7.6± 0.2 at 25°C

**Source:** This medium is available from CHROMagar, Paris, France.

**Cefsulodin Solution:**
**Composition** per 10.0mL:

Cefsulodin...................................................................5.0mg

**Preparation of Cefsulodin Solution:** Add cefsulodin to distilled/deionized water and bring volume to 10.0mL. Mix thoroughly. Filter sterilize.

**Preparation of Medium:** Add components, except cefsulodin solution, to distilled/deionized water and bring volume to 990mL. Mix thoroughly. Gently heat and bring to boiling. Do not autoclave. Cool to 50°C. Aseptically add 10.0mL of cefsulodin solution. Mix thoroughly. Pour into sterile Petri dishes.

**Storage/Shelf Life:** Store in the dark. Chromogenic agars are especially light and temperature sensitive; protect from light, excessive heat, moisture, and freezing. Prepared media plates can be kept for one day at ambient temperature. Plates can be stored at least one week under refrigeration (2-8°C) if properly prepared and protected from light and dehydration. Do not use after the expiration date supplied by the manufacturer.

**Use:** For the selective detection and isolation of *Salmonella* spp., including *S. typhi* and *S. paratyphi* in clinical specimens. *Salmonella* including *S. typhi* form mauve colonies.

## CHROMagar™ *Salmonella* Plus

**Composition** per liter:

Agar ...........................................................................15.0g
Salts............................................................................8.5g
Peptone and yeast extract..............................................8.0g
Supplement .................................................................6.0g
Chromogenic mix .........................................................1.3g

pH 7.5± 0.2 at 25°C

**Source:** This medium is available from CHROMagar, Paris, France.

**Preparation of Medium:** Add components to distilled/deionized water and bring volume to 1.0L. Mix thoroughly. Gently heat and bring to boiling. Do not autoclave. Cool to 50°C. Pour into sterile Petri dishes.

**Storage/Shelf Life:** Store in the dark. Chromogenic agars are especially light and temperature sensitive; protect from light, excessive heat, moisture, and freezing. Prepared media plates can be kept for one day at ambient temperature. Plates can be stored at least one week under refrigeration (2-8°C) if properly prepared and protected from light and dehydration. Do not use after the expiration date supplied by the manufacturer.

**Use:** For the detection and isolation of *Salmonella* species including lactose positive *Salmonella*. Detection of lactose positive *Salmonella* is now required in food microbiology by the ISO 6579:2002 standard. *Salmonella* (including *S. typhi*, *S. paratyphi* A and lactose positive *Salmonella*) produce mauve colonies.

## CHROMagar™ *Salmonella* Plus with Cefsulodin

**Composition** per liter:

| | |
|---|---|
| Agar | 15.0g |
| Salts | 8.5g |
| Peptone and yeast extract | 8.0g |
| Supplement | 6.0g |
| Chromogenic mix | 1.3g |
| Cefsulodin solution | 10.0mL |

pH 7.5± 0.2 at 25°C

**Source:** This medium is available from CHROMagar, Paris, France.

**Cefsulodin Solution:**
**Composition** per 10.0mL:

| | |
|---|---|
| Cefsulodin | 5.0mg |

**Preparation of Cefsulodin Solution:** Add cefsulodin to distilled/deionized water and bring volume to 10.0mL. Mix thoroughly. Filter sterilize.

**Preparation of Medium:** Add components, except cefsulodin solution, to distilled/deionized water and bring volume to 990mL. Mix thoroughly. Gently heat and bring to boiling. Do not autoclave. Cool to 50°C. Aseptically add 10.0mL of cefsulodin solution. Mix thoroughly. Pour into sterile Petri dishes.

**Storage/Shelf Life:** Store in the dark. Chromogenic agars are especially light and temperature sensitive; protect from light, excessive heat, moisture, and freezing. Prepared media plates can be kept for one day at ambient temperature. Plates can be stored at least one week under refrigeration (2-8°C) if properly prepared and protected from light and dehydration. Do not use after the expiration date supplied by the manufacturer.

**Use:** For the selective detection and isolation of *Salmonella* species including lactose positive *Salmonella* from samples with high *Aeromonas* concentrations. Detection of lactose positive *Salmonella* is now required in food microbiology by the ISO 6579:2002 standard. *Salmonella* (including *S. typhi*, *S. paratyphi* A and lactose positive *Salmonella*) produce mauve colonies.

## CHROMagar™ Staph aureus

**Composition** per liter:

| | |
|---|---|
| Peptone and yeast extract | 40.0g |
| Salts | 25.0g |
| Agar | 15.0g |
| Chromogenic mix | 2.5g |

pH 6.9± 0.2 at 25°C

**Source:** This medium is available from CHROMagar, Paris, France.

**Preparation of Medium:** Add components, to distilled/deionized water and bring volume to 1.0L. Mix thoroughly. Gently heat and bring to boiling. Do not autoclave. Cool to 50°C. Pour into sterile Petri dishes.

**Storage/Shelf Life:** Store in the dark. Chromogenic agars are especially light and temperature sensitive; protect from light, excessive heat, moisture, and freezing. Prepared media plates can be kept for one day at ambient temperature. Plates can be stored at least one week under refrigeration (2-8°C) if properly prepared and protected from light and dehydration. Do not use after the expiration date supplied by the manufacturer.

**Use:** For the isolation and direct differentiation of *Staphylococcus aureus* in clinical specimens. S. aureua forms pink to mauve colonies.

## CHROMagar™ STEC

**Composition** per liter:

| | |
|---|---|
| Agar | 15.0g |
| Peptones and yeast extract | 8.0g |
| Salts | 5.2g |
| Chromogenic mix | 2.6g |
| STEC supplement solution | 10.0mL |

pH 6.9± 0.2 at 25°C

**Source:** This medium is available from CHROMagar, Paris, France.

**STEC Supplement Solution:**
**Composition** per 10.0mL:

| | |
|---|---|
| ChromAgar STEC supplement mix | 2 vials |

**Preparation of STEC Supplement Solution:** Add ChromAgar STEC supplement mix to distilled/deionized water and bring volume to 10.0mL. Mix thoroughly. Filter sterilize.

**Preparation of Medium:** Add components, except STEC supplement solution, to distilled/deionized water and bring volume to 990mL. Mix thoroughly. Gently heat and bring to boiling. Do not autoclave. Cool to 50°C. Aseptically add 10.0mL of STEC supplement solution. Mix thoroughly. Pour into sterile Petri dishes.

**Storage/Shelf Life:** Store in the dark. Chromogenic agars are especially light and temperature sensitive; protect from light, excessive heat, moisture, and freezing. Prepared media plates can be kept for one day at ambient temperature. Plates can be stored at least one week under refrigeration (2-8°C) if properly prepared and protected from light and dehydration. Do not use after the expiration date supplied by the manufacturer.

**Use:** For the detection of Shiga-Toxin producing *E. coli* (STEC) in foods, environmental samples, and fecal specimens. Most common Shiga-Toxin *E. coli* serotypes form mauve colonies. Other Enterobacteriacae form colorless or blue colonies.

## CHROMagar™ *Streptb*

**Composition** per liter:
Proprietary

**Source:** CHROMagar *StrepB* is available from CHROMagar Microbiology.

**Preparation of Medium:** Add components to distilled/deionized water and bring volume to 1.0L. Mix thoroughly. Gently heat in a boiling water bath or steam bath. Shake periodically during heating to dissolve components. Heat long enough with shaking every 5 min to ensure complete dissolution. Do not overheat. Cool to 45–50°C. Pour into sterile Petri dishes.

**Storage/Shelf Life:** Store in the dark. Chromogenic agars are especially light and temperature sensitive; protect from light, excessive heat, moisture, and freezing. Prepared media plates can be kept for one day at ambient temperature. Plates can be stored at least one week under refrigeration (2–8°C) if properly prepared and protected from light and dehydration. Do not use after the expiration date supplied by the manufacturer.

**Use:** For the differentiation and presumptive identification of *Streptococcus* B (*Streptococcus agalactiae*) based upon color formation. *Streptococcus* B forms mauve to pink colonies.

## CHROMagar™ *Vibrio*

**Composition** per liter:

| | |
|---|---|
| Salts | 51.4g |
| Agar | 15.0g |
| Peptone and yeast extract | 8.0g |
| Chromogenic mix | 0.3g |

pH 9.0± 0.2 at 25°C

**Source:** This medium is available from CHROMagar, Paris, France.

**Preparation of Medium:** Add components, to distilled/deionized water and bring volume to 1.0L. Mix thoroughly. Gently heat and bring to boiling. Do not autoclave. Cool to 50°C. Pour into sterile Petri dishes.

**Storage/Shelf Life:** Store in the dark. Chromogenic agars are especially light and temperature sensitive; protect from light, excessive heat, moisture, and freezing. Prepared media plates can be kept for one day at ambient temperature. Plates can be stored at least one week under refrigeration (2–8°C) if properly prepared and protected from light and dehydration. Do not use after the expiration date supplied by the manufacturer.

**Use:** For the isolation and detection of *V. parahaemolyticus, V. vulnificus,* and *V. cholerae. V. parahaemolyticus* form mauve colonies. *V. vulnificus and V. cholera* form green blue to turquoise blue colonies. *V. alginolyticus* form colorless colonies.

## CHROMagar™ VRE

**Composition** per liter:

Proprietary

**Source:** CHROMagar VRE is available from CHROMagar Microbiology.

**Preparation of Medium:** Add components to distilled/deionized water and bring volume to 1.0L. Mix thoroughly. Gently heat in a boiling water bath or steam bath. Shake periodically during heating to dissolve components. Heat long enough with shaking every 5 min to ensure complete dissolution. Do not overheat. Cool to 45–50°C. Pour into sterile Petri dishes.

**Storage/Shelf Life:** Store in the dark. Chromogenic agars are especially light and temperature sensitive; protect from light, excessive heat, moisture, and freezing. Prepared media plates can be kept for one day at ambient temperature. Plates can be stored at least one week under refrigeration (2–8°C) if properly prepared and protected from light and dehydration. Do not use after the expiration date supplied by the manufacturer.

**Use:** For the differentiation and presumptive identification of vancomycin-resistant *Enterococcus* (*Enterococcus faecalis/E. facecium*). Vancomycin-resistant *Enterococcus* strains form rose to mauve colonies.

## CHROMagar™ *Y. enterocolitica*

**Composition** per liter:

| | |
|---|---|
| Agar | 15.0g |
| Peptones | 20.0g |
| Salts | 5.0g |
| Chromogenic and selective mix | 1.4g |
| *Y. enterocolitica* supplement solution | 1.0mL |

pH 7.0± 0.2 at 25°C

**Source:** This medium is available from CHROMagar, Paris, France.

### *Y. enterocolitica* Supplement Solution:

**Composition** per 10.0mL:

| | |
|---|---|
| CHROMagar *Y. enterocolitica* Supplement | 1.0g |

**Preparation of *Y. enterocolitica* Supplement Solution:** Add CHROMagar *Y. enterocolitica* supplement to distilled/deionized water and bring volume to 10.0mL. Mix thoroughly. Filter sterilize.

**Preparation of Medium:** Add components, except *Y. enterocolitica* supplement solution, to distilled/deionized water and bring volume to 990mL. Mix thoroughly. Gently heat and bring to boiling. Do not autoclave. Heat until complete fusion of the agar grains has taken place (large bubbles replacing foam). Cool to 50°C. Aseptically add 10.0mL of *Y. enterocolitica* supplement solution. Mix thoroughly. Pour into sterile Petri dishes.

**Storage/Shelf Life:** Store in the dark. Chromogenic agars are especially light and temperature sensitive; protect from light, excessive heat, moisture, and freezing. Prepared media plates can be kept for one day at ambient temperature. Plates can be stored at least one week under refrigeration (2–8°C) if properly prepared and protected from light and dehydration. Do not use after the expiration date supplied by the manufacturer.

**Use:** For the detection and direct differentiation of pathogenic *Yersinia enterocolitica*. Pathogenic *Y. enterocolitica* produce mauve colonies. Non pathogenic *Y. enterocolitica* and other microbes (*Citrobacter, Enterobacter, Aeromonas,* etc). produce metallic blue colonies.

## chromID™ *C. difficile*

**Composition** per liter:
Proprietary.

**Source:** This medium is available from bioMérieux.

**Preparation of Medium:** Available as a prepared medium.

**Storage/Shelf Life:** Prepared media should be stored in the dark under refrigeration (2–8°C). Chromogenic agars are especially light and temperature sensitive; protect from light, excessive heat, moisture, and freezing. Media should not be used if there are any signs of deterioration (shrinking, cracking, or discoloration) or contamination, or if the expiration date supplied by the manufacturer has passed.

**Use:** For the identification and isolation of *Clostridium difficile*. As part of a comprehensive infection prevention program, chromID *C. diff* can help institutions identify the reservoir and control the spread of this pathogenic organism.

## chromID™ CARBA Agar

**Composition** per liter:
Proprietary.

**Source:** This medium is available from bioMérieux.

**Preparation of Medium:** Available as a prepared medium.

**Storage/Shelf Life:** Prepared media should be stored in the dark under refrigeration (2-8°C). Chromogenic agars are especially light and temperature sensitive; protect from light, excessive heat, moisture, and freezing. Media should not be used if there are any signs of deterioration (shrinking, cracking, or discoloration) or contamination, or if the expiration date supplied by the manufacturer has passed.

**Use:** For the screening of carbapenemase producing enterobacteriaceae.

## chromID™ CPS

**Composition** per liter:
Proprietary.

**Source:** This medium is available from bioMérieux.

**Preparation of Medium:** Available as a prepared medium.

**Storage/Shelf Life:** Prepared media should be stored in the dark under refrigeration (2-8°C). Chromogenic agars are especially light and temperature sensitive; protect from light, excessive heat, moisture, and freezing. Media should not be used if there are any signs of deterioration (shrinking, cracking, or discoloration) or contamination, or if the expiration date supplied by the manufacturer has passed.

**Use:** For the isolation, enumeration and direct identification of *E. coli*, *Proteus,* Enterococci, and *Klebsiella–Enterobacter–Serratia–Citrobacter* (KESC) in a single step using urine specimens. *E. coli* form pink to burgundy colonies. *Proteus* form dark brown colonies. Enterococci form turquoise colonies.

## chromID™ ESBL

**Composition** per liter:
Proprietary.

**Source:** This medium is available from bioMérieux.

**Preparation of Medium:** Available as a prepared medium.

**Storage/Shelf Life:** Prepared media should be stored in the dark under refrigeration (2-8°C). Chromogenic agars are especially light and temperature sensitive; protect from light, excessive heat, moisture, and freezing. Media should not be used if there are any signs of deterioration (shrinking, cracking, or discoloration) or contamination, or if the expiration date supplied by the manufacturer has passed.

**Use:** For the screening of extended spectrum β-lactamase-producing enterobacteria (ESBL). *Escherichia coli* form pink to burgundy coloration colonies. *Klebsiella, Enterobacter, Serratia,* and *Citrobacter* (KESC) form green/blue to browny-green coloration colonies. Proteeae (*Proteus, Providencia,* and *Moraganella*) form dark to light brown coloration colonies of deaminase-expressing strains.

## chromID™ MRSA

**Composition** per liter:
Proprietary.

**Source:** This medium is available from bioMérieux.

**Preparation of Medium:** Available as a prepared medium.

**Storage/Shelf Life:** Prepared media should be stored in the dark under refrigeration (2-8°C). Chromogenic agars are especially light and temperature sensitive; protect from light, excessive heat, moisture, and freezing. Media should not be used if there are any signs of deterioration (shrinking, cracking, or discoloration) or contamination, or if the expiration date supplied by the manufacturer has passed.

**Use:** For the direct and definitive detection of methicillin-resistant *Staphylococcus aureus*. For use in healthcare units to actively reinforce MRSA surveillance culture and control in healthcare-associated infections. MRSA strains are inciated by green colored colonies resulting from alpha-glucuronidase producing colonies in the presence of the antibiotic cephoxitin.

## chromID™ *P. aeruginosa*

**Composition** per liter:
Proprietary.

**Source:** This medium is available from bioMérieux.

**Preparation of Medium:** Available as a prepared medium.

**Storage/Shelf Life:** Prepared media should be stored in the dark under refrigeration (2-8°C). Chromogenic agars are especially light and temperature sensitive; protect from light, excessive heat, moisture, and freezing. Media should not be used if there are any signs of deterioration (shrinking, cracking, or discoloration) or contamination, or if the expiration date supplied by the manufacturer has passed.

**Use:** For the direct identification of *Pseudomonas aeruginosa* based on the specific violet coloration of amino-peptidase-producing colonies.

## chromID™ *S. aureus*

**Composition** per liter:
Proprietary.

**Source:** This medium is available from bioMérieux.

**Preparation of Medium:** Available as a prepared medium.

**Storage/Shelf Life:** Prepared media should be stored in the dark under refrigeration (2-8°C). Chromogenic agars are especially light and temperature sensitive; protect from light, excessive heat, moisture, and freezing. Media should not be used if there are any signs of deterioration (shrinking, cracking, or discoloration) or contamination, or if the expiration date supplied by the manufacturer has passed.

**Use:** For the direct identification of *S. aureus* and the selective isolation of staphylococci. Direct identification of *S. aureus* is based on the spontaneous green coloration of α-glucosidase-producing colonies.

## chromID™ *Salmonella*

**Composition** per liter:
Proprietary.

**Source:** This medium is available from bioMérieux.

**Preparation of Medium:** Available as a prepared medium.

**Storage/Shelf Life:** Prepared media should be stored in the dark under refrigeration (2-8°C). Chromogenic agars are especially light and temperature sensitive; protect from light, excessive heat, moisture, and freezing. Media should not be used if there are any signs of deterioration (shrinking, cracking, or discoloration) or contamination, or if the expiration date supplied by the manufacturer has passed.

**Use:** For the selective isolation of all *Salmonella* serotypes. Three chromogenic substrates optimize the selective isolation and differentiation of *Salmonella*. Specific detection of the esterase enzymatic activity on a colorless background yields pale-pink to mauve colonies of *Salmonella* within 18-24h.

## chromID™ Strepto B

**Composition** per liter:
Proprietary.

**Source:** This medium is available from bioMérieux.

**Preparation of Medium:** Available as a prepared medium.

**Storage/Shelf Life:** Prepared media should be stored in the dark under refrigeration (2-8°C). Chromogenic agars are especially light and temperature sensitive; protect from light, excessive heat, moisture, and freezing. Media should not be used if there are any signs of deterioration (shrinking, cracking, or discoloration) or contamination, or if the expiration date supplied by the manufacturer has passed.

**Use:** For the screening of all *Streptococcus agalactiae*. All group B streptococci form pale pink to red colonies which are round and pearly after 18-24h.

## chromID™ *Vibrio*

**Composition** per liter:
Proprietary.

**Source:** This medium is available from bioMérieux.

**Preparation of Medium:** Available as a prepared medium.

**Storage/Shelf Life:** Prepared media should be stored in the dark under refrigeration (2-8°C). Chromogenic agars are especially light and temperature sensitive; protect from light, excessive heat, moisture, and freezing. Media should not be used if there are any signs of deterioration (shrinking, cracking, or discoloration) or contamination, or if the expiration date supplied by the manufacturer has passed.

**Use:** For the selective isolation of *Vibrio* and the presumptive identification of *V. cholerae* and *V. parahaemolyticus*. Presumptive identification of *V. cholerae* is through the blue-green coloration of beta-galactosidase producing colonies. *V. parahaemolyticus* form pink colonies through arabinose assimilation.

## chromID™ VRE

**Composition** per liter:
Proprietary.

**Source:** This medium is available from bioMérieux.

**Preparation of Medium:** Available as a prepared medium.

**Storage/Shelf Life:** Prepared media should be stored in the dark under refrigeration (2-8°C). Chromogenic agars are especially light and temperature sensitive; protect from light, excessive heat, moisture, and freezing. Media should not be used if there are any signs of deterioration (shrinking, cracking, or discoloration) or contamination, or if the expiration date supplied by the manufacturer has passed.

**Use:** For the rapid and reliable qualitative detection of *Enterococcus faecium* and *E. faecalis* showing acquired vancomycin resistance. As part of a comprehensive infection prevention program, chromID VRE can help institutions identify the reservoir and control the spread of these pathogenic organisms. *Enterococcus faecium* VRE form violet colonies. *E. faecalis* VRE form blue to green colonies.

## Chromogenic *E. coli*/Coliform Medium

**Composition** per liter:
Chromogenic mix ........................................................20.3g
Agar .............................................................................15.0g
Peptone.........................................................................5.0g
NaCl..............................................................................5.0g

Na$_2$HPO$_4$................................................................................3.5g
Yeast extract................................................................3.0g
Lactose.........................................................................2.5g
NaH$_2$PO$_4$..................................................................................1.5g
Neutral Red...............................................................0.03g

pH 6.8 ± 0.2 at 25°C

**Preparation of Medium:** Add components to distilled/deionized water and bring volume to 1.0L. Mix thoroughly. Gently heat and bring to boiling. Dispense into tubes or flasks. Autoclave for 15 min at 15 psi pressure–121°C. Pour into sterile Petri dishes or leave in tubes.

**Storage/Shelf Life:** Store dehydrated media in the dark in a sealed container below 30°C. Prepared media should be stored in the dark under refrigeration (2-8°C). Chromogenic agars are especially light and temperature sensitive; protect from light, excessive heat, moisture, and freezing. Media should not be used if there are any signs of deterioration (shrinking, cracking, or discoloration) or contamination, or if the expiration date supplied by the manufacturer has passed.

**Use:** For the differentiation between *Escherichia coli* and other coliforms in cultures produced from food samples. Agar base uses two enzyme substrates to differentiate between *E. coli* and other coliforms. One chromogenic substrate is cleaved by the enzyme glucuronidase which is specific for *E. coli* and produced by approximately 97% of strains. The second chromogenic substrate is cleaved by galactosidase, an enzyme produced by the majority of coliforms. This results in purple *E. coli* colonies, as they are able to cleave both chromogenic substrates and pink coliform colonies as they are only able to cleave the galactosidase chromogen.

## Chromogenic *Enterobacter sakazakii* Agar, DFI Formulation

**Composition** per liter:
Agar ............................................................................15.0g
Tryptone.....................................................................15.0g
Soya peptone................................................................5.0g
NaCl..............................................................................5.0g
Ferric ammonium citrate..............................................1.0g
Sodium deoxycholate....................................................1.0g
Na$_2$S$_2$O$_3$.......................................................................................1.0g
Chromogen ...................................................................0.1g

pH 7.3 ± 0.2 at 25°C

**Preparation of Medium:** Add components to distilled/deionized water and bring volume to 1.0L. Mix thoroughly. Gently heat and bring to boiling. Dispense into tubes or flasks. Autoclave for 15 min at 15 psi pressure–121°C. Pour into sterile Petri dishes or leave in tubes. Allow tubes to cool in a slanted position.

**Storage/Shelf Life:** Store dehydrated media in the dark in a sealed container below 30°C. Prepared media should be stored in the dark under refrigeration (2-8°C). Chromogenic agars are especially light and temperature sensitive; protect from light, excessive heat, moisture, and freezing. Media should not be used if there are any signs of deterioration (shrinking, cracking, or discoloration) or contamination, or if the expiration date supplied by the manufacturer has passed.

**Use:** For the differentiation and enumeration of *Enterobacter sakazakii* from infant formula and other food samples. The enzyme α-glucosidase, present in *E. sakazakii*, hydrolyzes the substrate 5-bromo-4-chloro-3-indolyl-α,D-glucopyranoside, thus producing blue-green colonies on this pale yellow medium. *Proteus vulgaris* is also weakly α-glucosidase positive and could grow to give colonies of a similar color to *E. sakazakii*. However, on this medium, *Proteus* spp. grow as grey colonies: they produce hydrogen sulphide in the presence of ferric ions forming ferrous sulphide. Deoxycholate inhibits the growth of most Gram-positive organisms.

## Chromogenic *Listeria* Agar

**Composition** per liter:

| | |
|---|---|
| Peptone | 18.5g |
| LiCl | 15.0g |
| Agar | 14.0g |
| NaCl | 9.5g |
| Yeast extract | 4.0g |
| Maltose | 4.0g |
| Sodium pyruvate | 2.0g |
| X-glucoside chromogenic mix | 0.2g |
| Differential lecithin solution | 40.0mL |
| Selective supplement solution | 20.0mL |

pH 7.2 ± 0.2 at 25°C

**Source:** This medium is available from Thermo Scientific.

**Caution:** Lithium chloride is harmful. Avoid bodily contact and inhalation of vapors. On contact with skin wash with plenty of water immediately.

**Differential Lecithin Solution:**

**Composition** per 40.0mL:

| | |
|---|---|
| Lecithin | Proprietary |

**Preparation of Differential Lecithin Solution:** Available as premixed solution.

**Selective Supplement Solution:**

**Composition** per 20.0mL:

| | |
|---|---|
| Nalidixic acid | 26.0mg |
| Polymyxin B | 10.0mg |
| Amphotericin | 10.0mg |
| Ceftazidime | 6.0mg |

**Preparation of Selective Supplement Solution:** Add components to distilled/deionized water and bring volume to 20.0mL. Mix thoroughly. Filter sterilize.

**Preparation of Medium:** Add components, except differential lecithin solution and selective supplement solution, to distilled/deionized water and bring volume to 940.0mL. Mix thoroughly. Gently heat while stirring and bring to boiling. Autoclave for 15 min at 15 psi pressure–121°C. Cool to 46°C. Aseptically add differential lecithin solution and selective supplement solution. Mix thoroughly. Pour into sterile Petri dishes.

**Storage/Shelf Life:** Store dehydrated media in the dark in a sealed container below 30°C. Prepared media should be stored in the dark under refrigeration (2-8°C). Chromogenic agars are especially light and temperature sensitive; protect from light, excessive heat, moisture, and freezing. Media should not be used if there are any signs of deterioration (shrinking, cracking, or discoloration) or contamination, or if the expiration date supplied by the manufacturer has passed.

**Use:** For the isolation, enumeration, and presumptive identification of *Listeria* spp. and *Listeria monocytogenes*. This selective medium contains the substrate lecithin, which permits differentiation of *L. monocytogenes* and *L. Ivanovii* from other *Listeria* species. Differential activity for all *Listeria* species is due to the addition of a chromogenic substrate.

## Chromogenic *Listeria* Agar (ISO)

**Composition** per liter:

| | |
|---|---|
| Enzymatic digest of animal tissue | 18.0g |
| Agar | 12.0g |
| LiCl | 10.0g |
| Yeast extract | 10.0g |
| Enzymatic digest of casein | 6.0g |
| NaCl | 5.0g |
| Na$_2$HPO$_4$, anhydrous | 2.5g |
| Glucose | 2.0g |
| Sodium pyruvate | 2.0g |
| Magnesium glycerophosphage | 1.0g |
| MgSO$_4$, anhydrous | 0.5g |
| X-glucoside chromogenic mix | 0.05g |
| L-α-phosphotidylinositol | 40.0mL |
| Selective supplement solution | 20.0mL |

pH 7.2 ± 0.2 at 25°C

**Source:** This medium is available from Thermo Scientific.

**Caution:** Lithium chloride is harmful. Avoid bodily contact and inhalation of vapors. On contact with skin wash with plenty of water immediately.

**Selective Supplement Solution:**

**Composition** per 20.0mL:

| | |
|---|---|
| Nalidixic acid | 20.0mg |
| Ceftazidime | 20.0mg |
| Amphotericin | 10.0mg |
| Polymyxin B | 76,700 U |

**Preparation of Selective Supplement Solution:** Add components to distilled/deionized water and bring volume to 20.0mL. Mix thoroughly. Filter sterilize.

**Preparation of Medium:** Add components, except L-α-phosphotidylinositol and selective supplement solution, to distilled/deionized water and bring volume to 940.0mL. Mix thoroughly. Gently heat while stirring and bring to boiling. Autoclave for 15 min at 15 psi pressure–121°C. Cool to 46°C. Aseptically add L-α-phosphotidylinositol and selective supplement solution. Mix thoroughly. Pour into sterile Petri dishes.

**Storage/Shelf Life:** Store dehydrated media in the dark in a sealed container below 30°C. Prepared media should be stored in the dark under refrigeration (2-8°C). Chromogenic agars are especially light and temperature sensitive; protect from light, excessive heat, moisture, and freezing. Media should not be used if there are any signs of deterioration (shrinking, cracking, or discoloration) or contamination, or if the expiration date supplied by the manufacturer has passed.

**Use:** For the isolation, enumeration, and presumptive identification of *Listeria* spp. and *Listeria monocytogenes*. This selective medium contains the substrate lecithin, which permits differentiation of *L. monocytogenes* and *L. Ivanovii* from other *Listeria* species.

## Chromogenic *Listeria* Agar (ISO) Modified

**Composition** per liter:

| | |
|---|---|
| Enzymatic digest of animal tissue | 18.0g |
| Agar | 12.0g |
| LiCl | 10.0g |
| Yeast extract | 10.0g |
| Enzymatic digest of casein | 6.0g |
| NaCl | 5.0g |
| Na$_2$HPO$_4$, anhydrous | 2.5g |
| Glucose | 2.0g |

Sodium pyruvate ..................................................................2.0g
Magnesium glycerophosphage ..............................................1.0g
MgSO$_4$, anhydrous.................................................................0.5g
X-glucoside chromogenic mix ............................................0.05g
Differential lecithin solution............................................40.0mL
Selective supplement solution ...........................................20.0mL
pH 7.2 ± 0.2 at 25°C

**Source:** This medium is available from Thermo Scientific.

**Caution:** Lithium chloride is harmful. Avoid bodily contact and inhalation of vapors. On contact with skin wash with plenty of water immediately.

**Differential Lecithin Solution:**
**Composition** per 40.0mL:
Lecithin ...............................................................Proprietary

**Preparation of Differential Lecithin Solution:** Available as premixed solution.

**Selective Supplement Solution:**
**Composition** per 20.0mL:
Nalidixic acid ..................................................................20.0mg
Ceftazidime......................................................................20.0mg
Amphotericin ...................................................................10.0mg
Polymyxin B ..................................................................76,700 U

**Preparation of Selective Supplement Solution:** Add components to distilled/deionized water and bring volume to 20.0mL. Mix thoroughly. Filter sterilize.

**Preparation of Medium:** Add components, except differential lecithin solution and selective supplement solution, to distilled/deionized water and bring volume to 940.0mL. Mix thoroughly. Gently heat while stirring and bring to boiling. Autoclave for 15 min at 15 psi pressure–121°C. Cool to 46°C. Aseptically add differential lecithin solution and selective supplement solution. Mix thoroughly. Pour into sterile Petri dishes.

**Storage/Shelf Life:** Store dehydrated media in the dark in a sealed container below 30°C. Prepared media should be stored in the dark under refrigeration (2-8°C). Chromogenic agars are especially light and temperature sensitive; protect from light, excessive heat, moisture, and freezing. Media should not be used if there are any signs of deterioration (shrinking, cracking, or discoloration) or contamination, or if the expiration date supplied by the manufacturer has passed.

**Use:** For the isolation, enumeration, and presumptive identification of *Listeria* spp. and *Listeria monocytogenes*. This selective medium contains the substrate lecithin, which permits differentiation of *L. monocytogenes* and *L. Ivanovii* from other *Listeria* species.

### Chromogenic *Salmonella* Esterase Agar (CSE Agar)
**Composition** per liter:
Agar .......................................................................12.0g
Lactose ...................................................................14.65
Peptone....................................................................4.0g
Tryptone ..................................................................4.0g
Tween™ 20 ..............................................................3.0g
Lab Lemco ...............................................................3.0g
Na$_3$-citrate dihydrate................................................0.5g
L-cysteine ...............................................................0.128g
Tris ..........................................................................0.06g
SLA-octonoate solution .......................................50.0mL

Novobiocin solution..............................................10.0mL
Ethyl 4-dimethylaminobenzoate solution ..............10.0mL
pH 7.0 ± 0.2 at 25°C

**Novobiocin Solution:**
**Composition** per 10.0mL:
Novobiocin .............................................................70.0mg

**Preparation of Novobiocin Solution:** Add novobiocin to distilled/deionized water and bring volume to 10.0mL. Mix thoroughly. Filter sterilize.

**Ethyl 4-dimethylaminobenzoate Solution:**
**Composition** per 10.0mL:
Ethyl 4-dimethylaminobenzoate.................................0.35g
Methanol ...................................................................8.0mL

**Preparation of Ethyl 4-dimethylaminobenzoate Solution:** Add ethyl 4-dimethylaminobenzoate to 8.0mL methanol. Mix thoroughly. Bring volume to 10.0mL with distilled/deionized water. Mix thoroughly. Filter sterilize.

**SLA-Octonoate Solution:**
**Composition** per 50.0mL:
4-[2-(4-octanoyloxy-3,5-dimethoxyphenyl)-vinyl]-quinolinium-1-(propan-3-yl carboxylic acid) bromide (SLPA-octanoate; bromide form) ..................................................0.3223g

**Preparation of SLA-Octonoate Solution:** Add SLA-octonoate to distilled/deionized water and bring volume to 50.0mL. Mix thoroughly. Filter sterilize.

**Preparation of Medium:** Add components, except novobiocin solution, SLA-octonoate solution, and ethyl 4-dimethylaminobenzoate solution, to distilled/deionized water and bring volume to 920.0mL. Mix thoroughly. Gently heat and bring to boiling. Autoclave for 15 min at 15 psi pressure–121°C. Cool to 50°C. Aseptically add 10.0mL novobiocin solution, 50.0mL SLA-octonoate solution, and 10.0mL ethyl 4-dimethylaminobenzoate solution. Mix thoroughly. Pour into sterile Petri dishes.

**Storage/Shelf Life:** Store dehydrated media in the dark in a sealed container below 30°C. Prepared media should be stored in the dark under refrigeration (2-8°C). Chromogenic agars are especially light and temperature sensitive; protect from light, excessive heat, moisture, and freezing. Media should not be used if there are any signs of deterioration (shrinking, cracking, or discoloration) or contamination, or if the expiration date supplied by the manufacturer has passed.

**Use:** For the detection of *Salmonella* spp. in clinical specimens. For the differentiation of *Salmonella* spp.

### Chromogenic Substrate Broth
**Composition** per liter:
NaCl.........................................................................10.0g
HEPES (*N*-[2-hydroxyethyl]piperazine-*N'*-[2-ethanesulfonic acid]) buffer ...............................................6.9g
(NH$_4$)$_2$SO$_4$ ..............................................................5.0g
*o*-Nitrophenyl-β-D-galactopyranoside .......................0.5g
Solanium ....................................................................0.5g
MgSO$_4$ .......................................................................0.1g
4-Methylumbelliferyl-β-D-glucuronide ...................0.075g
CaCl$_2$........................................................................0.05g
Na$_2$SO$_3$....................................................................0.04g
Amphotericin B .........................................................1.0mg
MnSO$_4$ .......................................................................0.5mg
ZnSO$_4$........................................................................0.5mg

**Preparation of Medium:** Add components to distilled/deionized water and bring volume to 1.0L. Mix thoroughly. Distribute into tubes or flasks. Autoclave for 15 min at 15 psi pressure–121°C.

**Storage/Shelf Life:** Store dehydrated media in the dark in a sealed container below 30°C. Prepared media should be stored in the dark under refrigeration (2-8°C). Chromogenic media are especially light and temperature sensitive; protect from light, excessive heat, moisture, and freezing. Media should not be used if there are any signs of deterioration (discoloration) or contamination, or if the expiration date supplied by the manufacturer has passed.

**Use:** For the detection of coliform bacteria based on their hydrolysis of chromogenic substrates by production of β-D-galactopyranosidase. Bacteria that produce β-D-galactopyranosidase turn the medium yellow.

## Chromogenic Urinary Tract Infection (UTI) Medium
**Composition** per liter:

| | |
|---|---|
| Chromogenic mix | 26.3g |
| Peptone | 15.0g |
| Agar | 15.0g |

pH 6.8 ± 0.2 at 25°C

**Source:** This medium is available as a premixed powder from Thermo Scientific.

**Preparation of Medium:** Add components to distilled/deionized water and bring volume to 1.0L. Mix thoroughly. Gently heat while stirring and bring to boiling. Autoclave for 15 min at 15 psi pressure–121°C. Cool to 50°C. Pour into sterile Petri dishes.

**Storage/Shelf Life:** Store dehydrated media in the dark in a sealed container below 30°C. Prepared media should be stored in the dark under refrigeration (2-8°C). Chromogenic agars are especially light and temperature sensitive; protect from light, excessive heat, moisture, and freezing. Media should not be used if there are any signs of deterioration (shrinking, cracking, or discoloration) or contamination, or if the expiration date supplied by the manufacturer has passed.

**Use:** For the presumptive identification and differentiation of all the main microorganisms that cause urinary tract infections (UTIs). The medium contains two specific chromogenic substrates that are cleaved by enzymes produced by *Enterococcus* spp., *Escherichia coli*, and coliforms. In addition, it contains phenylalanine and tryptophan which provide an indication of tryptophan deaminase activity, indicating the presence of *Proteus* spp., *Morganella* spp., and *Providencia* spp. It is based on electrolyte deficient CLED Medium, which provides a valuable non-inhibitory diagnostic agar for plate culture of other urinary organisms, while preventing the swarming of *Proteus* spp. One chromogen, X-Gluc, is targeted towards β-glucosidase, and allows the specific detection of enterococci through the formation of blue colonies. The other chromogen, Red-Gal, is cleaved by the enzyme β-galactosidase, which is produced by *Escherichia coli*, resulting in pink colonies. Cleavage of both chromogens occurs in the presence of coliforms, resulting in purple colonies. The medium also contains tryptophan, which acts as an indicator of tryptophan deaminase activity, resulting in colonies of *Proteus*, *Morganella*, and *Providencia* spp. appearing brown.

## Chromogenic UTI Medium, Clear
**Composition** per liter:

| | |
|---|---|
| Peptone | 15.0g |
| Agar | 15.0g |
| Chromogenic mix | 13.0g |

pH 7.0 ± 0.2 at 25°C

**Source:** This medium is available as a premixed powder from Thermo Scientific.

**Preparation of Medium:** Add components to distilled/deionized water and bring volume to 1.0L. Mix thoroughly. Gently heat while stirring and bring to boiling. Autoclave for 15 min at 15 psi pressure–121°C. Cool to 50°C. Pour into sterile Petri dishes.

**Storage/Shelf Life:** Store dehydrated media in the dark in a sealed container below 30°C. Prepared media should be stored in the dark under refrigeration (2-8°C). Chromogenic agars are especially light and temperature sensitive; protect from light, excessive heat, moisture, and freezing. Media should not be used if there are any signs of deterioration (shrinking, cracking, or discoloration) or contamination, or if the expiration date supplied by the manufacturer has passed.

**Use:** For the presumptive identification and differentiation of all the main microorganisms that cause urinary tract infections (UTIs). This medium uses the same chromogenic substrates as the existing opaque Chromogenc UTI Medium but has a clear background to make multiple sample testing easier. The medium contains two specific chromogenic substrates that are cleaved by enzymes produced by *Enterococcus* spp., *E. coli*, and coliforms. In addition, it contains tryptophan, which indicates tryptophan deaminase activity (TDA), indicating the presence of *Proteus* spp. It is based on Cystine Lactose Electrolyte Deficient (CLED) Medium, which provides a valuable non-inhibitory diagnostic agar for plate culture of other urinary organisms, while preventing the swarming of *Proteus* spp. The chromogen, X-glucoside, is targeted towards ß-glucosidase enzyme activity, and allows the specific detection of enterococci through the formation of blue colonies. The other chromogen, Red-Galactoside, is cleaved by the enzyme ß-galactosidase, which is produced by *E. coli*, resulting in pink colonies. Cleavage of both the chromogens by members of the coliform group results in purple colonies. The medium also contains tryptophan, which acts as an indicator of tryptophan deaminase activity (TDA), resulting in halos around the colonies of *Proteus*, *Morganella*, and *Providencia* spp.

## CIN Agar
## (*Yersinia* Selective Agar)
## (Cefsulodin Irgasan® Novobiocin Agar)
**Composition** per liter:

| | |
|---|---|
| Mannitol | 20.0g |
| Agar | 12.0g |
| Pancreatic digest of gelatin | 10.0g |
| Beef extract | 5.0g |
| Peptic digest of animal tissue | 5.0g |
| Sodium pyruvate | 2.0g |
| Yeast extract | 2.0g |
| NaCl | 1.0g |
| Sodium deoxycholate | 0.5g |
| Neutral Red | 0.03g |
| Cefsulodin | 0.015g |
| Irgasan®(triclosan) | 4.0mg |
| Novobiocin | 2.5mg |
| Crystal Violet | 1.0mg |

pH 7.4 ± 0.2 at 25°C

**Source:** This medium is available as a premixed powder from BD Diagnostic Systems.

**Preparation of Medium:** Add components, except cefsulodin and novobiocin, to distilled/deionized water and bring volume to 1.0L. Heat, mixing continuously, until boiling. Do not autoclave. Cool to

45°–50°C. Aseptically add cefsulodin and novobiocin. Mix thoroughly. Pour into sterile Petri dishes or distribute into sterile tubes.

**Storage/Shelf Life:** Store dehydrated media in the dark in a sealed container below 30°C. Prepared media should be stored under refrigeration (2-8°C). Media should be used within 60 days of preparation. Media should not be used if there are any signs of deterioration (shrinking, cracking, or discoloration) or contamination, or if the expiration date supplied by the manufacturer has passed.

**Use:** For the selective isolation and differentiation of *Yersinia enterocolitica* from a variety of clinical speciments and nonclinical specimens of public health importance based on mannitol fermentation. *Yersinia enterocolitica* appears as "bull's eye" colonies with deep red centers surrounded by a transparent periphery.

## Citrate Azide Tween Carbonate Base
**Composition** per liter:

| | |
|---|---|
| Agar | 15.0g |
| Casein enzymic hydrolysate | 15.0g |
| Sodium citrate | 15.0g |
| Yeast extract | 5.0g |
| KH$_2$PO$_4$ | 5.0g |
| Tween 80 | 1.0g |
| Selective supplement solution | 10.0mL |

pH 7.0 ± 0.2 at 25°C

**Source:** This medium is available from HiMedia.

**Selective Supplement Solution:**
**Composition** per 10.0mL:

| | |
|---|---|
| Na$_2$CO$_3$ | 1.0g |
| NaN$_3$ | 0.2g |
| 2,3,5, Triphenyltetrazolium chloride | 0.05g |

**Preparation of Selective Supplement Solution:** Add components to distilled/deionized water and bring volume to 10.0mL. Mix thoroughly. Filter sterilize.

**Caution:** Sodium azide is toxic. Azides also react with metals and disposal must be highly diluted.

**Preparation of Medium:** Add components, except selective supplement solution, to distilled/deionized water and bring volume to 990.0mL. Mix thoroughly. Autoclave for 15 min at 15 psi pressure–121°C. Cool to 50°C. Aseptically add selective supplement solution. Mix thoroughly. Pour into Petri dishes or aseptically distribute into sterile tubes.

**Storage/Shelf Life:** Store dehydrated media in the dark in a sealed container below 30°C. Prepared media should be stored under refrigeration (2-8°C). Media should be used within 60 days of preparation. Media should not be used if there are any signs of deterioration (shrinking, cracking, or discoloration) or contamination, or if the expiration date supplied by the manufacturer has passed.

**Use:** For the identification of enterococci in meat, meat products, dairy products, and other foodstuffs.

## CLED Agar
## (Cystine Lactose Electrolyte Deficient Agar)
## (Brolacin Agar)
**Composition** per liter:

| | |
|---|---|
| Agar | 15.0g |
| Lactose | 10.0g |
| Pancreatic digest of casein | 4.0g |

| | |
|---|---|
| Pancreatic digest of gelatin | 4.0g |
| Beef extract | 3.0g |
| L-Cystine | 0.128g |
| Bromthymol Blue | 0.02g |

pH 7.3 ± 0.2 at 25°C

**Preparation of Medium:** Add components to distilled/deionized water and bring volume to 1.0L. Mix thoroughly. Gently heat while stirring and bring to boiling. Autoclave for 15 min at 15 psi pressure–121°C. Cool to 50°–55°C. Pour into sterile Petri dishes or distribute into sterile tubes.

**Storage/Shelf Life:** Store dehydrated media in the dark in a sealed container below 30°C. Prepared media should be stored under refrigeration (2-8°C). Media should be used within 60 days of preparation. Media should not be used if there are any signs of deterioration (shrinking, cracking, or discoloration) or contamination, or if the expiration date supplied by the manufacturer has passed.

**Use:** For the isolation, enumeration, and presumptive identification of microorganisms from urine.

## CLED Agar with Andrade's Indicator
## (Cystine Lactose Electrolyte Deficient Agar with Andrade's Indicator)
**Composition** per liter:

| | |
|---|---|
| Agar | 15.0g |
| Pancreatic digest of casein | 10.0g |
| Peptone | 4.0g |
| Beef extract | 3.0g |
| L-Cystine | 0.128g |
| Bromthymol Blue | 0.02g |
| Andrade's indicator | 10.0mL |

pH 7.5 ± 0.2 at 25°C

**Source:** This medium is available as a premixed powder from Thermo Scientific.

**Caution:** Acid Fuchsin is a potential carcinogen and care must be taken to avoid inhalation of the powdered dye and contamination of the skin.

**Andrade's Indicator:**
**Composition** per 100.0mL:

| | |
|---|---|
| NaOH (1$N$ solution) | 16.0mL |
| Acid Fuchsin | 0.1g |

**Preparation of Andrade's Indicator:** Add Acid Fuchsin to NaOH solution and bring volume to 100.0mL with distilled/deionized water.

**Preparation of Medium:** Add components to distilled/deionized water and bring volume to 1.0L. Mix thoroughly. Gently heat while stirring and bring to boiling. Autoclave for 15 min at 15 psi pressure–121°C. Cool to 50°–55°C. Pour into sterile Petri dishes or distribute into sterile tubes.

**Storage/Shelf Life:** Store dehydrated media in the dark in a sealed container below 30°C. Prepared media should be stored under refrigeration (2-8°C). Media should be used within 60 days of preparation. Media should not be used if there are any signs of deterioration (shrinking, cracking, or discoloration) or contamination, or if the expiration date supplied by the manufacturer has passed.

**Use:** For the differentiation of microorganisms based on colony characteristics.

## CLED HiVeg Agar with Andrade's Indicator

**Composition** per liter:

Agar .......................................................................15.0g
Lactose ..................................................................10.0g
Plant hydrolysate......................................................4.0g
Plant peptone............................................................4.0g
Plant extract ............................................................3.0g
L-Cystine................................................................0.128g
Andrade's indicator....................................................0.1g
Bromo Thymol Blue .................................................0.02g

pH 7.5 ± 0.2 at 25°C

**Source:** This medium is available as a premixed powder from Hi-Media.

**Caution:** Acid Fuchsin in Andrade indicator is a potential carcinogen and care must be taken to avoid inhalation of the powdered dye and contamination of the skin.

**Preparation of Medium:** Add components to distilled/deionized water and bring volume to 1.0L. Mix thoroughly. Gently heat while stirring and bring to boiling. Autoclave for 15 min at 15 psi pressure–121°C. Cool to 50°–55°C. Pour into sterile Petri dishes or distribute into sterile tubes.

**Storage/Shelf Life:** Store dehydrated media in the dark in a sealed container below 30°C. Prepared media should be stored under refrigeration (2-8°C). Media should be used within 60 days of preparation. Media should not be used if there are any signs of deterioration (shrinking, cracking, or discoloration) or contamination, or if the expiration date supplied by the manufacturer has passed.

**Use:** For the isolation and differentiation of urinary pathogens on the basis of lactose fermentation.

## CLED HiVeg Agar with Bromthymol Blue

**Composition** per liter:

Agar .......................................................................15.0g
Lactose ..................................................................10.0g
Plant hydrolysate......................................................4.0g
Plant peptone............................................................4.0g
Plant extract ............................................................3.0g
L-Cystine................................................................0.128g
Bromthymol Blue .....................................................0.02g

pH 7.3 ± 0.2 at 25°C

**Source:** This medium is available as a premixed powder from HiMedia.

**Preparation of Medium:** Add components to distilled/deionized water and bring volume to 1.0L. Mix thoroughly. Gently heat while stirring and bring to boiling. Autoclave for 15 min at 15 psi pressure–121°C. Cool to 50°–55°C. Pour into sterile Petri dishes or distribute into sterile tubes.

**Storage/Shelf Life:** Store dehydrated media in the dark in a sealed container below 30°C. Prepared media should be stored under refrigeration (2-8°C). Media should be used within 60 days of preparation. Media should not be used if there are any signs of deterioration (shrinking, cracking, or discoloration) or contamination, or if the expiration date supplied by the manufacturer has passed.

**Use:** For the isolation and differentiation of urinary pathogens on the basis of lactose fermentation.

## C.L.E.D. HiVeg Agar Base without Indicator

**Composition** per liter:

Agar .......................................................................15.0g
Lactose ..................................................................10.0g

Plant hydrolysate......................................................4.0g
Plant peptone............................................................4.0g
Plant extract ............................................................3.0g
L-Cystine................................................................0.128g

pH 7.3 ± 0.2 at 25°C

**Source:** This medium is available as a premixed powder from Hi-Media.

**Preparation of Medium:** Add components to distilled/deionized water and bring volume to 1.0L. Mix thoroughly. Gently heat while stirring and bring to boiling. Autoclave for 15 min at 15 psi pressure–121°C. Cool to 50°–55°C. Pour into sterile Petri dishes or distribute into sterile tubes.

**Storage/Shelf Life:** Store dehydrated media in the dark in a sealed container below 30°C. Prepared media should be stored under refrigeration (2-8°C). Media should be used within 60 days of preparation. Media should not be used if there are any signs of deterioration (shrinking, cracking, or discoloration) or contamination, or if the expiration date supplied by the manufacturer has passed.

**Use:** For the isolation, enumeration, and presumptive identification of bacterial flora in the urinary tract.

## Clostridial Agar

**Composition** per liter:

Casein enzymatic hydrolysate ..................................17.0g
Agar .....................................................................14.5g
Glucose ...................................................................6.0g
Papaic digest of soybean meal....................................3.0g
NaCl........................................................................2.5g
Na-thioglycolate ......................................................1.8g
Sodium formaldehyde sulphoxylate ...........................1.0g
L-Cystine................................................................0.25g
NaN₃.......................................................................0.2g
Neomycin sulfate ...................................................0.15g

pH 7.0 ± 0.2 at 25°C

**Source:** This medium is available as a premixed powder from Hi-Media.

**Caution:** Sodium azide is toxic. Azides also react with metals and disposal must be highly diluted.

**Preparation of Medium:** Add components to distilled/deionized water and bring volume to 1.0L. Mix thoroughly. Gently heat while stirring and bring to boiling. Autoclave for 15 min at 15 psi pressure–121°C. Cool to 50°–55°C. Pour into sterile Petri dishes or distribute into sterile tubes.

**Storage/Shelf Life:** Store dehydrated media in the dark in a sealed container below 30°C. Prepared media should be stored under refrigeration (2-8°C). Media should be used within 60 days of preparation. Media should not be used if there are any signs of deterioration (shrinking, cracking, or discoloration) or contamination, or if the expiration date supplied by the manufacturer has passed.

**Use:** For the selective isolation of pathogenic Clostridia from mixed flora.

## Clostridial HiVeg Agar

**Composition** per liter:

Plant hydrolysate .....................................................17.0g
Agar .....................................................................14.5g
Glucose ...................................................................6.0g
Papaic digest of soybean meal....................................3.0g
NaCl........................................................................2.5g

Na-thioglycolate..............................................................1.8g
Sodium formaldehyde sulphoxylate ...............................1.0g
L-Cystine.......................................................................0.25g
NaN$_3$.............................................................................0.2g
Neomycin sulfate ..........................................................0.15g

pH 7.0 ± 0.2 at 25°C

**Source:** This medium is available as a premixed powder from Hi-Media.

**Caution:** Sodium azide is toxic. Azides also react with metals and disposal must be highly diluted.

**Preparation of Medium:** Add components to distilled/deionized water and bring volume to 1.0L. Mix thoroughly. Gently heat while stirring and bring to boiling. Autoclave for 15 min at 15 psi pressure–121°C. Cool to 50°–55°C. Pour into sterile Petri dishes or distribute into sterile tubes.

**Storage/Shelf Life:** Store dehydrated media in the dark in a sealed container below 30°C. Prepared media should be stored under refrigeration (2-8°C). Media should be used within 60 days of preparation. Media should not be used if there are any signs of deterioration (shrinking, cracking, or discoloration) or contamination, or if the expiration date supplied by the manufacturer has passed.

**Use:** For the selective isolation of pathogenic Clostridia from mixed flora.

## *Clostridium botulinum* Isolation Agar
## (CBI Agar)

**Composition** per 1033.0mL:
Egg yolk agar base....................................................900.0mL
Egg yolk emulsion, 50%............................................100.0mL
Cycloserine solution ...................................................25.0mL
Sulfamethoxazole solution............................................4.0mL
Trimethoprim solution ..................................................4.0mL

pH7.4 ± 0.2 at 25°C

**Egg Yolk Agar Base:**

**Composition** per 900.0mL:
Pancreatic digest of casein...........................................40.0g
Agar ..............................................................................20.0g
Na$_2$HPO$_4$.......................................................................5.0g
Yeast extract...................................................................5.0g
Glucose ..........................................................................2.0g
NaCl................................................................................2.0g
MgSO$_4$·7H$_2$O solution ...............................................0.2mL

**Preparation of Egg Yolk Agar Base:** Add components to distilled/deionized water and bring volume to 900.0mL. Mix thoroughly. Gently heat to boiling. Autoclave for 15 min at 15 psi pressure–121°C. Cool to 45°–50°C.

**MgSO$_4$·7H$_2$O Solution:**

**Composition** per 100.0mL:
MgSO$_4$·7H$_2$O ...............................................................5.0g

**Preparation of MgSO$_4$·7H$_2$O Solution:** Add MgSO$_4$·7H$_2$O to distilled/deionized water and bring volume to 100.0mL. Mix thoroughly.

**Cycloserine Solution:**

**Composition** per 100.0mL:
Cycloserine ....................................................................1.0g

**Preparation of Cycloserine Solution:** Add cycloserine to distilled/deionized water and bring volume to 100.0mL. Mix thoroughly. Filter sterilize.

**Sulfamethoxazole Solution:**

**Composition** per 100.0mL:
Sulfamethoxazole ...........................................................1.9g

**Preparation of Sulfamethoxazole Solution:** Add sulfamethoxazole to distilled/deionized water and bring volume to 50.0mL. Add sufficient 10% NaOH to dissolve. Bring volume to 100.0mL with distilled/deionized water. Mix thoroughly. Filter sterilize.

**Trimethoprim Solution:**

**Composition** per 100.0mL:
Trimethoprim ..................................................................0.1g

**Preparation of Trimethoprim Solution:** Add trimethoprim to distilled/deionized water and bring volume to 50.0mL. Gently heat to 55°C. Add sufficient 0.05$N$ HCl to dissolve. Bring volume to 100.0mL with distilled/deionized water. Mix thoroughly. Filter sterilize.

**Egg Yolk Emulsion, 50%:**

**Composition** per 100.0mL:
Chicken egg yolks............................................................11
Whole chicken egg .............................................................1
NaCl (0.9% solution)...................................................50.0mL

**Preparation of Egg Yolk Emulsion, 50%:** Soak eggs with 1:100 dilution of saturated mercuric chloride solution for 1 min. Crack eggs and separate yolks from whites. Mix egg yolks with 1 chicken egg. Beat to form emulsion. Measure 50.0mL of egg yolk emulsion and add to 50.0mL of 0.9% NaCl solution. Mix thoroughly. Filter sterilize. Warm to 45°–50°C.

**Preparation of Medium:** Aseptically add warmed, sterile egg yolk emulsion, 50%, and sterile cycloserine solution, sterile sulfamethoxazole solution, and sterile trimethoprim solution to cooled, sterile egg yolk agar base. Mix thoroughly. Pour into sterile Petri dishes.

**Storage/Shelf Life:** Store dehydrated media in the dark in a sealed container below 30°C. Prepared media should be stored under refrigeration (2-8°C). Media should be used within 60 days of preparation. Media should not be used if there are any signs of deterioration (shrinking, cracking, or discoloration) or contamination, or if the expiration date supplied by the manufacturer has passed.

**Use:** For isolation, cultivation, and differentiation based on lipase activity of *Clostridium botulinum* types A, B, and F from fecal specimens associated with foodborne and infant botulism. *Clostridium botulinum* types A, B, and F appear as raised colonies surrounded by an opaque zone. Other *Clostridium* species and *Clostridium botulinum* type G appear as pinpoint colonies with no opaque zone.

## *Clostridium botulinum* Isolation HiVeg Agar

**Composition** per liter:
Plant hydrolysate ..........................................................40.0g
Agar ..............................................................................20.0g
Na$_2$HPO$_4$.......................................................................5.0g
Yeast extract...................................................................5.0g
Glucose ..........................................................................2.0g
NaCl................................................................................2.0g
MgSO$_4$ .........................................................................0.01g
Egg yolk emulsion, 50%............................................100.0mL
Cycloserine solution ...................................................25.0mL
Sulfamethoxazole solution............................................4.0mL
Trimethoprim solution ..................................................4.0mL

pH 7.4 ± 0.2 at 25°C

**Source:** This medium, without egg yolk emulsion, cycloserine solution, sulfmethoxazole solution, and trimethoprim solution, is available as a premixed powder (*C. botulinum* Isolation HiVeg Agar Base) from HiMedia.

### Cycloserine Solution:
**Composition** per 100.0mL:

Cycloserine ..................................................................................1.0g

**Preparation of Cycloserine Solution:** Add cycloserine to distilled/deionized water and bring volume to 100.0mL. Mix thoroughly. Filter sterilize.

### Sulfamethoxazole Solution:
**Composition** per 100.0mL:

Sulfamethoxazole..........................................................................1.9g

**Preparation of Sulfamethoxazole Solution:** Add sulfamethoxazole to distilled/deionized water and bring volume to 50.0mL. Add sufficient 10% NaOH to dissolve. Bring volume to 100.0mL with distilled/deionized water. Mix thoroughly. Filter sterilize.

### Trimethoprim Solution:
**Composition** per 100.0mL:

Trimethoprim ...............................................................................0.1g

**Preparation of Trimethoprim Solution:** Add trimethoprim to distilled/deionized water and bring volume to 50.0mL. Gently heat to 55°C. Add sufficient 0.05$N$ HCl to dissolve. Bring volume to 100.0mL with distilled/deionized water. Mix thoroughly. Filter sterilize.

### Egg Yolk Emulsion, 50%:
**Composition** per 100.0mL:

Chicken egg yolks...........................................................................11
Whole chicken egg............................................................................1
NaCl (0.9% solution) ................................................................50.0mL

**Preparation of Egg Yolk Emulsion, 50%:** Soak eggs with 1:100 dilution of saturated mercuric chloride solution for 1 min. Crack eggs and separate yolks from whites. Mix egg yolks with 1 chicken egg. Beat to form emulsion. Measure 50.0mL of egg yolk emulsion and add to 50.0mL of 0.9% NaCl solution. Mix thoroughly. Filter sterilize. Warm to 45°–50°C.

**Preparation of Medium:** Add components, except egg yolk emulsion, cycloserine solution, sulfmethoxazole solution, and trimethoprim solution, to distilled/deionized water and bring volume to 900.0mL. Mix thoroughly. Gently heat to boiling. Autoclave for 15 min at 15 psi pressure–121°C. Cool to 45°–50°C. Aseptically add warmed, sterile egg yolk emulsion, 50%, and sterile cycloserine solution, sterile sulfamethoxazole solution, and sterile trimethoprim solution. Mix thoroughly. Pour into sterile Petri dishes.

**Storage/Shelf Life:** Store dehydrated media in the dark in a sealed container below 30°C. Prepared media should be stored under refrigeration (2-8°C). Media should be used within 60 days of preparation. Media should not be used if there are any signs of deterioration (shrinking, cracking, or discoloration) or contamination, or if the expiration date supplied by the manufacturer has passed.

**Use:** For isolation, cultivation, and differentiation based on lipase activity of *Clostridium botulinum* types A, B, and F from fecal specimens associated with foodborne and infant botulism. *Clostridium botulinum* types A, B, and F appear as raised colonies surrounded by an opaque zone. Other *Clostridium* species and *Clostridium botulinum* type G appear as pinpoint colonies with no opaque zone.

### *Clostridium difficile* Agar
**Composition** per liter:

*Clostridium difficile* agar base ...............................................920.0mL

Horse blood, defibrinated ........................................................70.0mL
*Clostridium difficile* selective supplement...............................10.0mL
pH 7.4 ± 0.2 at 25°C

**Source:** This medium is available as a premixed powder from Thermo Scientific.

### *Clostridium difficile* Agar Base:
**Composition** per 920.0mL:

Proteose peptone.........................................................................40.0g
Agar .............................................................................................15.0g
Fructose.........................................................................................6.0g
Na$_2$HPO$_4$..................................................................................5.0g
NaCl...............................................................................................2.0g
KH$_2$PO$_4$......................................................................................1.0g
MgSO$_4$·7H$_2$O.............................................................................0.1g

**Preparation of *Clostridium difficile* Agar Base:** Add components to distilled/deionized water and bring volume to 920.0mL. Mix thoroughly. Gently heat to boiling. Autoclave for 15 min at 15 psi pressure–121°C. Cool to 45°–50°C.

### *Clostridium difficile* Selective Supplement:
**Composition** per 10.0mL:

D-Cycloserine...........................................................................500.0mg
Cefoxitin ..................................................................................16.0mg

**Preparation of *Clostridium difficile* Selective Supplement:** Add components to distilled/deionized water and bring volume to 10.0mL. Mix thoroughly. Filter sterilize.

**Preparation of Medium:** Add 10.0mL of sterile *Clostridium difficile* selective supplement and 70.0mL of sterile, defibrinated horse blood to 920.0mL of cooled, sterile *Clostridium difficile* agar base. Mix thoroughly. Pour into sterile Petri dishes or distribute into sterile tubes.

**Storage/Shelf Life:** Store dehydrated media in the dark in a sealed container below 30°C. Prepared media should be stored under refrigeration (2-8°C). Media should be used within 60 days of preparation. Media should not be used if there are any signs of deterioration (shrinking, cracking, or discoloration) or contamination, or if the expiration date supplied by the manufacturer has passed.

**Use:** For the selective isolation and cultivation of *Clostridium difficile* from clinical speciments and nonclinical specimens of public health importance.

### *Clostridium difficile* Agar (Cycloserine Cefoxitin Fructose Agar) (CCFA)
**Composition** per liter:

Peptic digest of animal tissue .....................................................32.0g
Agar .............................................................................................20.0g
Fructose.........................................................................................6.0g
Na$_2$HPO$_4$..................................................................................5.0g
NaCl...............................................................................................2.0g
KH$_2$PO$_4$......................................................................................1.0g
Cycloserine .................................................................................0.25g
MgSO$_4$ .......................................................................................0.1g
Neutral Red ................................................................................0.03g
Cefoxitin solution .....................................................................10.0mL
pH 7.2 ± 0.2 at 25°C

**Source:** This medium is available as a premixed powder from BD Diagnostic Systems.

**Cefoxitin Solution:**
**Composition** per 10.0mL:
Cefoxitin ...............................................................16.0mg

**Preparation of Cefoxitin Solution:** Add cefoxitin to distilled/deionized water and bring volume to 10.0mL. Mix thoroughly. Filter sterilize.

**Preparation of Medium:** Add components, except cefoxitin solution, to distilled/deionized water and bring volume to 990.0mL. Mix thoroughly. Gently heat to boiling. Autoclave for 15 min at 15 psi pressure–121°C. Cool to 45°–50°C. Aseptically add 10.0mL of sterile cefoxitin solution. Mix thoroughly. Pour into sterile Petri dishes or distribute into sterile tubes.

**Storage/Shelf Life:** Store dehydrated media in the dark in a sealed container below 30°C. Prepared media should be stored under refrigeration (2-8°C). Media should be used within 60 days of preparation. Media should not be used if there are any signs of deterioration (shrinking, cracking, or discoloration) or contamination, or if the expiration date supplied by the manufacturer has passed.

**Use:** For the selective isolation and cultivation of *Clostridium difficile* from clinical speciments and nonclinical specimens of public health importance.

### *Clostridium difficile* HiVeg Agar Base
**Composition** per liter:
Plant peptone No. 3.......................................................40.0g
Agar ..............................................................................15.0g
Fructose...........................................................................6.0g
Na$_2$HPO$_4$...................................................................5.0g
NaCl.................................................................................2.0g
KH$_2$PO$_4$.....................................................................1.0g
MgSO$_4$...........................................................................0.1g
Horse blood, defibrinated ........................................ 70.0mL
*Clostridium difficile* selective supplement................. 10.0mL
pH 7.4 ± 0.2 at 25°C

**Source:** This medium, without blood or selective supplement, is available as a premixed powder from HiMedia.

#### *Clostridium difficile* Selective Supplement:
**Composition** per 10.0mL:
D-Cycloserine .............................................................500.0mg
Cefoxitin ......................................................................16.0mg

**Preparation of *Clostridium difficile* Selective Supplement:** Add components to distilled/deionized water and bring volume to 10.0mL. Mix thoroughly. Filter sterilize.

**Preparation of Medium:** Add components, except blood and selective supplement, to distilled/deionized water and bring volume to 920.0mL. Mix thoroughly. Gently heat to boiling. Autoclave for 15 min at 15 psi pressure–121°C. Cool to 45°–50°C. Add 10.0mL of sterile *Clostridium difficile* selective supplement and 70.0mL of sterile, defibrinated horse blood. Mix thoroughly. Pour into sterile Petri dishes or distribute into sterile tubes.

**Storage/Shelf Life:** Store dehydrated media in the dark in a sealed container below 30°C. Prepared media should be stored under refrigeration (2-8°C). Media should be used within 60 days of preparation. Media should not be used if there are any signs of deterioration (shrinking, cracking, or discoloration) or contamination, or if the expiration date supplied by the manufacturer has passed.

**Use:** For the selective isolation and cultivation of *Clostridium difficile* from fecal specimens.

### *Clostridium perfringens* Agar, OPSP (Perfringens Agar, OPSP)
**Composition** per liter:
Pancreatic digest of casein.........................................15.0g
Agar ..............................................................................10.0g
Liver extract...................................................................7.0g
Papaic digest of soybean meal......................................5.0g
Yeast extract...................................................................5.0g
Tris(hydroxymethyl)aminomethane buffer....................1.5g
Ferric ammonium citrate................................................1.0g
Na$_2$S$_2$O$_5$...................................................................1.0g
Antibiotic inhibitor .................................................. 10.0mL
pH 7.3 ± 0.2 at 25°C

**Source:** This medium is available as a premixed powder from Thermo Scientific.

**Antibiotic Inhibitor:**
**Composition** per 10.0mL:
Sodium sulfadiazine........................................................0.1g
Oleandomycin phosphate...............................................0.5mg
Polymyxin B ........................................................... 10,000U

**Preparation of Antibiotic Inhibitor:** Add components to distilled/deionized water and bring volume to 10.0mL. Mix thoroughly. Filter sterilize.

**Preparation of Medium:** Add components, except antibiotic inhibitor, to distilled/deionized water and bring volume to 990.0mL. Mix thoroughly. Gently heat and bring to boiling. Autoclave for 15 min at 15 psi pressure–121°C. Cool to 45°–50°C. Aseptically add sterile antibiotic inhibitor. Mix thoroughly. Pour into sterile Petri dishes or distribute into sterile tubes.

**Storage/Shelf Life:** Store dehydrated media in the dark in a sealed container below 30°C. Prepared media should be stored under refrigeration (2-8°C). Media should be used within 60 days of preparation. Media should not be used if there are any signs of deterioration (shrinking, cracking, or discoloration) or contamination, or if the expiration date supplied by the manufacturer has passed.

**Use:** For the presumptive identification and enumeration of *Clostridium perfringens* in foods.

### *Clostridium* Selective Agar (Clostrisel Agar)
**Composition** per liter:
Pancreatic digest of casein.........................................17.0g
Agar ..............................................................................14.0g
Glucose...........................................................................6.0g
Papaic digest of soybean meal......................................3.0g
NaCl.................................................................................2.5g
Sodium thioglycolate .....................................................1.8g
Sodium formaldehyde sulfoxylate.................................1.0g
L-Cystine ......................................................................0.25g
NaN$_3$ .............................................................................0.15g
Neomycin sulfate..........................................................0.15g
pH 7.0 ± 0.2 at 25°C

**Source:** This medium is available as a premixed powder from BD Diagnostic Systems.

**Preparation of Medium:** Add components to distilled/deionized water and bring volume to 1.0L. Mix thoroughly. Gently heat while stirring and bring to boiling. Distribute into tubes or flasks. Autoclave for 15 min at 15 psi pressure–118°C. Pour into sterile Petri dishes or leave in tubes.

**Caution:** Sodium azide is toxic. Azides also react with metals and disposal must be highly diluted.

**Storage/Shelf Life:** Store dehydrated media in the dark in a sealed container below 30°C. Prepared media should be stored under refrigeration (2-8°C). Media should be used within 60 days of preparation. Media should not be used if there are any signs of deterioration (shrinking, cracking, or discoloration) or contamination, or if the expiration date supplied by the manufacturer has passed.

**Use:** For the selective isolation of pathogenic *Clostridium* species from specimens containing mixed flora, e.g., from wounds, fecal specimens, soil, and other specimens.

## Coagulase Agar Base

**Composition** per liter:

| | |
|---|---|
| Agar | 25.0g |
| Brain heart infusion | 10.5g |
| Pancreatic digest of casein | 10.5g |
| D-Mannitol | 10.0g |
| Brain heart infusion | 5.0g |
| NaCl | 3.5g |
| Papaic digest of soybean meal | 3.5g |
| Bromcresol Purple | 0.02g |
| Rabbit plasma | 100.0mL |

pH 7.4 ± 0.2 at 25°C

**Preparation of Medium:** Add components, except rabbit plasma, to distilled/deionized water and bring volume to 1.0L. Mix thoroughly. Gently heat, while stirring, until boiling. Distribute into tubes or flasks. Autoclave for 15 min at 15 psi pressure–121°C. Cool to 45°–50°C. Add rabbit plasma to a final concentration of 7–15%. Mix thoroughly. Pour into sterile Petri dishes in 18.0mL volume per plate.

**Storage/Shelf Life:** Store dehydrated media in the dark in a sealed container below 30°C. Prepared media should be stored under refrigeration (2-8°C). Media should be used within 60 days of preparation. Media should not be used if there are any signs of deterioration (shrinking, cracking, or discoloration) or contamination, or if the expiration date supplied by the manufacturer has passed.

**Use:** For the cultivation and differentiation of *Staphylococcus aureus* from other *Staphylococcus* species based on coagulase production.

## Coagulase Mannitol Agar

**Composition** per liter:

| | |
|---|---|
| Agar | 14.5g |
| Pancreatic digest of casein | 10.5g |
| D-Mannitol | 10.0g |
| Brain heart infusion | 5.0g |
| NaCl | 3.5g |
| Papaic digest of soybean meal | 3.5g |
| Bromcresol Purple | 0.02g |
| Rabbit plasma with 0.15% EDTA | 100.0mL |

pH 7.3 ± 0.2 at 25°C

**Source:** This medium is available as a premixed powder from BD Diagnostic Systems.

**Preparation of Medium:** Add components, except rabbit plasma, to distilled/deionized water and bring volume to 1.0L. Mix thoroughly. Gently heat while stirring until boiling. Distribute into tubes or flasks. Autoclave for 15 min at 15 psi pressure–121°C. Cool to 45°–50°C. For

detection of coagulase activity add rabbit plasma with 0.15% EDTA to a final concentration of 7–15%. Mix thoroughly. Pour into sterile Petri dishes.

**Storage/Shelf Life:** Store dehydrated media in the dark in a sealed container below 30°C. Prepared media should be stored under refrigeration (2-8°C). Media should be used within 60 days of preparation. Media should not be used if there are any signs of deterioration (shrinking, cracking, or discoloration) or contamination, or if the expiration date supplied by the manufacturer has passed.

**Use:** For the cultivation and differentiation of *Staphylococcus aureus* from other *Staphylococcus* species based on coagulase production and mannitol fermentation.

## Coagulase Mannitol HiVeg Agar Base with Plasma

**Composition** per liter:

| | |
|---|---|
| Agar | 14.5g |
| Plant hydrolysate | 10.5g |
| Mannitol | 10.0g |
| Plant special infusion | 5.0g |
| Papaic digest of soybean meal | 3.5g |
| NaCl | 3.5g |
| Bromcresol Purple | 0.02 |
| Rabbit plasma | 100.0–150.0mL |

pH 7.3 ± 0.2 at 25°C

**Source:** This medium is available as a premixed powder from Hi-Media.

**Preparation of Medium:** Add components, except rabbit plasma, to distilled/deionized water and bring volume to 1.0L. Mix thoroughly. Gently heat while stirring until boiling. Distribute into tubes or flasks. Autoclave for 15 min at 15 psi pressure–121°C. Cool to 45°–50°C. For detection of coagulase activity add rabbit plasma with 0.15% EDTA to a final concentration of 7–15%. Mix thoroughly. Pour into sterile Petri dishes.

**Storage/Shelf Life:** Store dehydrated media in the dark in a sealed container below 30°C. Prepared media should be stored under refrigeration (2-8°C). Media should be used within 60 days of preparation. Media should not be used if there are any signs of deterioration (shrinking, cracking, or discoloration) or contamination, or if the expiration date supplied by the manufacturer has passed.

**Use:** For the cultivation and differentiation of *Staphylococcus aureus* from other *Staphylococcus* species based on coagulase production and mannitol fermentation. For the primary isolation and identification of pathogenic staphylococci from clinical specimens or for classifying pure cultures.

## Coagulase Mannitol Broth Base with Plasma

**Composition** per liter:

| | |
|---|---|
| Heart muscle, infusion from | 375.0g |
| D-Mannitol | 10.0g |
| Peptic digest of animal tissue | 10.0g |
| NaCl | 5.0g |
| Phenol Red | 0.025g |
| Rabbit plasma, sterile, pretested normal | 120.0–150.0mL |

pH 7.3 ± 0.2 at 25°C

**Source:** This medium is available as a premixed powder from Hi-Media.

**Preparation of Medium:** Add components, except rabbit plasma, to distilled/deionized water and bring volume to 1.0L. Mix thoroughly.

Gently heat while stirring until boiling. Distribute into tubes or flasks. Autoclave for 15 min at 15 psi pressure–121°C. Cool to 45°–50°C. For detection of coagulase activity add rabbit plasma with 0.15% EDTA to a final concentration of 12–15%. Mix thoroughly.

**Storage/Shelf Life:** Store dehydrated media in the dark in a sealed container below 30°C. Prepared media should be stored under refrigeration (2-8°C). Media should be used within 60 days of preparation. Media should not be used if there are any signs of deterioration (discoloration) or contamination, or if the expiration date supplied by the manufacturer has passed.

**Use:** For the cultivation and differentiation of *Staphylococcus aureus* from other *Staphylococcus* species based on coagulase production and mannitol fermentation. For the simultaneous detection of coagulase production and mannitol fermentation in the differentiation of staphylococci.

## Coagulase Mannitol HiVeg Broth Base with Plasma
**Composition** per liter:
| | |
|---|---|
| D-Mannitol | 10.0g |
| Plant infusion | 10.0g |
| Plant peptone | 10.0g |
| NaCl | 5.0g |
| Phenol Red | 0.025g |
| Rabbit plasma, strerile, pretested normal | 120.0-150.0mL |

pH 7.3 ± 0.2 at 25°C

**Source:** This medium is available as a premixed powder from Hi-Media.

**Preparation of Medium:** Add components, except rabbit plasma, to distilled/deionized water and bring volume to 1.0L. Mix thoroughly. Gently heat while stirring until boiling. Distribute into tubes or flasks. Autoclave for 15 min at 15 psi pressure–121°C. Cool to 45°–50°C. For detection of coagulase activity add rabbit plasma with 0.15% EDTA to a final concentration of 12–15%. Mix thoroughly.

**Storage/Shelf Life:** Store dehydrated media in the dark in a sealed container below 30°C. Prepared media should be stored under refrigeration (2-8°C). Media should be used within 60 days of preparation. Media should not be used if there are any signs of deterioration (discoloration) or contamination, or if the expiration date supplied by the manufacturer has passed.

**Use:** For the cultivation and differentiation of *Staphylococcus aureus* from other *Staphylococcus* species based on coagulase production and mannitol fermentation. For the simultaneous detection of coagulase production and mannitol fermentation in the differentiation of staphylococci.

## COBA
## (Colistin Oxolinic Acid Blood Agar)
**Composition** per liter:
| | |
|---|---|
| Columbia agar base | 930.0mL |
| Horse blood, defibrinated, sterile | 50.0mL |
| Colistin sulfate solution | 10.0mL |
| Oxolinic acid solution | 10.0mL |

pH 7.3 ± 0.2 at 25°C

**Columbia Agar Base:**
**Composition** per 930.0mL:
| | |
|---|---|
| Agar | 13.5g |
| Pancreatic digest of casein | 10.0g |
| Peptic digest of animal tissue | 10.0g |
| NaCl | 5.0g |
| Beef extract | 3.0g |
| Yeast extract | 3.0g |
| Cornstarch | 1.0g |

**Preparation of Columbia Agar Base:** Add components to distilled/deionized water and bring volume to 930.0mL. Mix thoroughly. Gently heat until boiling. Autoclave for 15 min at 15 psi pressure–121°C. Cool to 45°–50°C.

**Colistin Sulfate Solution:**
**Composition** per 10.0mL:
| | |
|---|---|
| Colistin sulfate | 10.0mg |

**Preparation of Colistin Sulfate Solution:** Add colistin sulfate to distilled/deionized water and bring volume to 10.0mL. Mix thoroughly. Filter sterilize.

**Oxolinic Acid Solution:**
**Composition** per 10.0mL:
| | |
|---|---|
| Oxolinic acid | 5.0–10.0mg |

**Preparation of Oxolinic Acid Solution:** Add oxolinic acid to distilled/deionized water and bring volume to 10.0mL. Mix thoroughly. Filter sterilize.

**Preparation of Medium:** To 930.0mL of sterile, cooled Columbia agar base, add sterile colistin sulfate, sterile oxolinic acid, and sterile, defibrinated horse blood. Mix thoroughly. Pour into sterile Petri dishes.

**Storage/Shelf Life:** Store dehydrated media in the dark in a sealed container below 30°C. Prepared media should be stored under refrigeration (2-8°C). Media should be used within 60 days of preparation. Media should not be used if there are any signs of deterioration (shrinking, cracking, or discoloration) or contamination, or if the expiration date supplied by the manufacturer has passed.

**Use:** For the isolation and cultivation of streptococci in pure culture from mixed flora in clinical specimens.

## Coletsos Medium
**Composition** per 1625mL:
| | |
|---|---|
| Potato starch | 10.0g |
| Gelatin | 4.0g |
| Asparagine | 2.25g |
| KH$_2$PO$_4$ | 1.5g |
| Na-glutamate | 1.0g |
| Na-pyruvate | 1.0g |
| Mg-citrate | 0.375g |
| Litmus | 0.25g |
| Malachite green | 0.25g |
| MgSO$_4$ | 0.15g |
| Activated carbon | 0.1g |
| Oligonucleotide mixture | 3.0mg |
| Egg mixture | 625.0mL |
| Glycerol | 7.5mL |

**Egg Mixture:**
**Composition** per liter:
| | |
|---|---|
| Whole eggs | 18–24 |

**Preparation of Egg Mixture:** Use fresh eggs, less than 1 week old. Scrub the shells with soap. Let stand in a soap solution for 30 min. Rinse in running water. Soak eggs in 70% ethanol for 15 min. Break the eggs into a sterile container. Separate egg whites from egg yolks. Combine 8 parts egg white with 2 parts egg yolk. Homogenize by shaking. Filter through four layers of sterile cheesecloth into a sterile graduated cylinder. Bring volume to 1.0L distilled/deionized water.

**Preparation of Medium:** Add glycerol to 600.0mL of distilled/deionized water. Mix thoroughly. Add remaining components, except egg mixture. Bring volume to 1.0L. Mix thoroughly. Gently heat while stirring and bring to boiling. Autoclave for 15 min at 15 psi pressure–121°C. Cool to 50°C. Aseptically add 625.0mL of egg mixture. Mix thoroughly. Distribute into sterile screw-capped tubes. Place tubes in a slanted position. Inspissate at 85°C (moist heat) for 45 min.

**Storage/Shelf Life:** Store dehydrated media in the dark in a sealed container below 30°C. Prepared media should be stored under refrigeration (2-8°C). Media should be used within 60 days of preparation. Media should not be used if there are any signs of deterioration (shrinking, cracking, or discoloration) or contamination, or if the expiration date supplied by the manufacturer has passed.

**Use:** For the cultivation of *Mycobacterium tuberculosis*.

## Coletsos Selective Medium

**Composition** per 1625mL:

| | |
|---|---|
| Potato starch | 10.0g |
| Gelatin | 4.0g |
| Asparagine | 2.25g |
| KH$_2$PO$_4$ | 1.5g |
| Na-glutamate | 1.0g |
| Na-pyruvate | 1.0g |
| Mg-citrate | 0.375g |
| Litmus | 0.25g |
| Malachite green | 0.25g |
| MgSO$_4$ | 0.15g |
| Activated carbon | 0.1g |
| Oligonucleotide mixture | 3.0mg |
| Egg mixture solution | 625.0mL |
| Glycerol | 7.5mL |
| Nalidixic acid solution | 1.0mL |
| Lincomycin solution | 1.0mL |
| Cycloheximide solution | 1.0mL |

**Nalidixic Acid Solution:**

**Composition** per 100.0mL:

| | |
|---|---|
| Nalidixic acid | 0.5g |

**Preparation of Nalidixic Acid Solution:** Add nalidixic acid to distilled/deionized water and bring volume to 100.0mL. Mix thoroughly. Filter sterilize.

**Cycloheximide Solution:**

**Composition** per 100.0mL:

| | |
|---|---|
| Cycloheximide | 1.5g |
| Ethanol | 40.0mL |

**Preparation of Cycloheximide Solution:** Add cycloheximide to 40.0mL of ethanol. Mix thoroughly. Bring volume to 100.0mL with distilled/deionized water. Filter sterilize.

**Caution:** Cycloheximide is toxic. Avoid skin contact or aerosol formation and inhalation.

**Lincomycin Solution:**

**Composition** per 100.0mL:

| | |
|---|---|
| Lincomycin | 0.5g |

**Preparation of Lioncomycin Solution:** Add lincomycin to distilled/deionized water and bring volume to 100.0mL. Mix thoroughly. Filter sterilize.

**Egg Mixture Solution:**

**Composition** per liter:

| | |
|---|---|
| Whole eggs | 18–24 |

**Preparation of Egg Mixture Solution:** Use fresh eggs, less than 1 week old. Scrub the shells with soap. Let stand in a soap solution for 30 min. Rinse in running water. Soak eggs in 70% ethanol for 15 min. Break the eggs into a sterile container. Separate egg whites from egg yolks. Combine 8 parts egg white with 2 parts egg yolk. Homogenize by shaking. Filter through four layers of sterile cheesecloth into a sterile graduated cylinder. Bring volume to 1.0L with distilled/deionized water.

**Preparation of Medium:** Add glycerol to 600.0mL of distilled/deionized water. Mix thoroughly. Add remaining components, except egg mixture, lincomycin solution, cycloheximide solution, and nalidixic acid solution. Mix thoroughly. Bring volume to 1.0L. Gently heat while stirring and bring to boiling. Autoclave for 15 min at 15 psi pressure–121°C. Cool to 50°C. Aseptically add 625.0mL of egg mixture. Mix thoroughly. Aseptically add 1.0mL cycloheximide solution, 1.0mL lincomycin solution, and 1.0mL nalidixic acid solution. Distribute into sterile screw-capped tubes. Place tubes in a slanted position. Inspissate at 85°C (moist heat) for 45 min.

**Storage/Shelf Life:** Store dehydrated media in the dark in a sealed container below 30°C. Prepared media should be stored under refrigeration (2-8°C). Media should be used within 60 days of preparation. Media should not be used if there are any signs of deterioration (shrinking, cracking, or discoloration) or contamination, or if the expiration date supplied by the manufacturer has passed.

**Use:** For the isolation and cultivation of *Mycobacterium tuberculosis*.

## Coli ID

**Composition** per liter:

Proprietary

**Source:** This medium is available from bioMérieux.

**Storage/Shelf Life:** Prepared media should be stored under refrigeration (2-8°C). Media should not be used if there are any signs of deterioration (shrinking, cracking, or discoloration) or contamination, or if the expiration date supplied by the manufacturer has passed.

**Use:** A selective chromogenic medium for the detection and enumeration of *E. coli* at 44°C, and simultaneous enumeration of *E. coli* and other coliforms at 37°C, from food products.

## Coliform Agar, Chromocult® (Chromocult Coliform Agar)

**Composition** per liter:

| | |
|---|---|
| Agar | 10.0g |
| NaCl | 5.0g |
| Peptone | 3.0g |
| Na$_2$HPO$_4$ | 2.7g |
| NaH$_2$PO$_4$ | 2.2g |
| Tryptophan | 1.0g |
| Na-pyruvate | 1.0g |
| Chromogenic mixture | 0.4g |
| Tergitol 7 | 0.15g |

pH 7.0 ± 0.2 at 25°C

**Source:** This medium is available from Merck.

**Preparation of Medium:** Add components to distilled/deionized water and bring volume to 1.0L. Mix well and warm gently until dissolved. Autoclave for 15 min at 15 psi pressure–121°C. Pour into sterile Petri dishes. Some turbidity may occur, but this does not effect the performance.

**Storage/Shelf Life:** Store dehydrated media in the dark in a sealed container below 30°C. Prepared media should be stored in the dark under refrigeration (2-8°C). Chromogenic agars are especially light and temperature sensitive; protect from light, excessive heat, moisture, and freezing. Media should not be used if there are any signs of deterioration (shrinking, cracking, or discoloration) or contamination, or if the expiration date supplied by the manufacturer has passed.

**Use:** For the detection of *E. coli* and coliform bacteria in foods. The interaction of selected peptones, pyruvate, sorbitol, and phosphate buffer guarantees rapid colony growth, even for sublethally injured coliforms. The growth of Gram-positive bacteria as well as some Gram-negative bacteria is largely inhibited by the content of Tergitol 7 which has no negative effect on the growth of the coliform bacteria. A combination of two chromogenic substrates allows for the simultaneous detection of total coliforms and *E. coli.* The characteristic enzyme for coliforms, β-D-galactosidase, cleaves the Salmon-GAL substrate and causes a salmon to red color of the coliform colonies. The substrate X-glucuronide is used for the identification of β-D-glucuronidase, which is characteristic for *E. coli. E. coli* cleaves both Salmon-GAL and X-glucuronide, so that positive colonies take on a dark-blue to violet color. These are easily distinguished from other coliform colonies, which have a salmon to red color. As part of an additional confirmation of *E. coli*, the inclusion of tryptophan improves the indole reaction, thereby increasing detection reliability when it is used in combination with the Salmon-GAL and X-glucuronide reaction.

## Coliform Agar ES, Chromocult®
## (Chromocult Coliform Agar ES)
## (Chromocult Enhanced Selectivity Agar)

**Composition** per liter:

| | |
|---|---|
| Agar | 10.0g |
| MOPS | 10.0g |
| KCl | 7.5g |
| Peptone | 5.0g |
| Bile salts | 1.15g |
| Na-propionate | 0.5g |
| 6-Chloro-3-indoxyl-β-D-galactopyranoside | 0.15g |
| 5-Bromo-4-chloro-3-indoxyl-β-D-glucuronic acid | 0.1g |
| Isopropyl-β-D-thiogalactopyranoside | 0.1g |

pH 7.0 ± 0.2 at 25°C

**Source:** This medium is available from Merck.

**Preparation of Medium:** Add components to distilled/deionized water and bring volume to 1.0L. Mix thoroughly and heat with frequent agitation until components are completely dissolved (approximately 45 min). Do not autoclave. Cool to 45°–50°C. Pour into sterile Petri dishes. The plates should be clear and colorless.

**Storage/Shelf Life:** Store dehydrated media in the dark in a sealed container below 30°C. Prepared media should be stored in the dark under refrigeration (2-8°C). Chromogenic agars are especially light and temperature sensitive; protect from light, excessive heat, moisture, and freezing. Media should not be used if there are any signs of deterioration (shrinking, cracking, or discoloration) or contamination, or if the expiration date supplied by the manufacturer has passed.

**Use:** For the detection of *E.coli* and total coliforms. The combination of suitable peptones and the buffering using MOPS allows rapid growth of coliforms and an optimal transformation of the chromogenic substrates. The amount of bile salts and propionate largely inhibit growth of Gram-positive and Gram-negative accompanying flora. The simultaneous detection of total coliforms and *E.coli* is achieved using

the combination of two chromogrenic substrates. The substrate Salmon™--β-D-GAL is split by β-D-galactosidase, characteristic for coliforms, resulting in a salmon to red coloration of coliform colonies. The detection of the β-D-glucuronidase, characteristic for *E. coli,* is cleaved via the substrate X-β-D-glucuronide, causing a blue coloration of positive colonies. As *E. coli* splits Salmon™-β-D-GAL as well as X-β-D-glucuronide, the colonies turn to a dark violet color and can be easily differentiated from the other coliforms being salmon-red.

## Colorex™ *Acinetobacter*

**Composition** per liter:

| | |
|---|---|
| Agar | 15.0g |
| Peptone and yeast extract | 12.0g |
| Salts | 4.0g |
| Chromogenic mix | 1.8g |
| Growth and regulator factors | 1.0g |

pH 7.0± 0.2 at 25°C

**Source:** Available as prepared plates from E&O Laboratories, Bonnybridge Scotland.

**Storage/Shelf Life:** Store in the dark. Chromogenic agars are especially light and temperature sensitive; protect from light, excessive heat, moisture, and freezing. Prepared media plates can be kept for one day at ambient temperature. Plates can be stored at least one week under refrigeration (2-8°C) if properly prepared and protected from light and dehydration. Do not use after the expiration date supplied by the manufacturer.

**Use:** For the detection of *Acinetobacter* sp., which produce red colonies.

## Colorex™ Extended Spectrum Beta Lactamase (ESBL)

**Composition** per liter:

| | |
|---|---|
| Peptone and yeast extract | 17.0g |
| Agar | 15.0g |
| Chromogenic mix | 1.0g |
| EBSL selective supplement | 0.57g |

pH 7.0± 0.2 at 25°C

**Source:** Available as prepared plates from E&O Laboratories, Bonnybridge Scotland.

**Storage/Shelf Life:** Store in the dark. Chromogenic agars are especially light and temperature sensitive; protect from light, excessive heat, moisture, and freezing. Prepared media plates can be kept for one day at ambient temperature. Plates can be stored at least one week under refrigeration (2-8°C) if properly prepared and protected from light and dehydration. Do not use after the expiration date supplied by the manufacturer.

**Use:** For the detection of Gram-negative bacteria producing extended spectrum beta-lactamase. Extended spectrum β-lactam resistant strains of *E. coli* form dark pink to reddish colonies. Extended spectrum β-lactam resistant strains of *Klebsiella, Enterobacter*, and *Citrobacter* form metallic blue colonies. Extended spectrum β-lactam resistant strains of *Proteus* colonies with a brown halo.

## Colorex™ *Klebsiella pneumoniae* Carbapenemase (KPC)

**Composition** per liter:

| | |
|---|---|
| Peptone and yeast extract | 17.0g |
| Agar | 15.0g |

Chromogenic mix ........................................................1.0g
KPC selective supplement .......................................0.4g

pH 7.0± 0.2 at 25°C

**Source:** Available as prepared plates from E&O Laboratories, Bonnybridge Scotland.

**Storage/Shelf Life:** Store in the dark. Chromogenic agars are especially light and temperature sensitive; protect from light, excessive heat, moisture, and freezing. Prepared media plates can be kept for one day at ambient temperature. Plates can be stored at least one week under refrigeration (2-8°C) if properly prepared and protected from light and dehydration. Do not use after the expiration date supplied by the manufacturer.

**Use:** For the detection of Gram-negative bacteria with a reduced susceptibility to most carbapenem antimicrobic agents. Carbapenem-resistant strains of *E. coli* form dark pink to reddish colonies. Carbapenem-resistant strains of *Klebsiella, Enterobacter,* and *Citrobacter* form blue colonies. Carbapenem-resistant strains of *Pseudomonas* form cream translucent colonies.

## Colorex™ KPC

**Composition** per liter:

Agar .......................................................................15.0g
Peptones................................................................20.0g
Salts.........................................................................5.0g
Chromogenic and selective mix.........................1.4g
*Y. enterocolitica* supplement solution ....................1.0mL

pH 7.0± 0.2 at 25°C

**Source:** This medium is available from E&O Laboratories, Bonnybridge, United Kingdom.

**Preparation of Medium:** Add components, except *Y. enterocolitica* supplement solution, to distilled/deionized water and bring volume to 990mL. Mix thoroughly. Gently heat and bring to boiling. Do not autoclave. Heat until complete fusion of the agar grains has taken place (large bubbles replacing foam). Cool to 50°C. Aseptically add 10.0mL of *Y. enterocolitica* supplement solution. Mix thoroughly. Pour into sterile Petri dishes.

**Storage/Shelf Life:** Store in the dark. Chromogenic agars are especially light and temperature sensitive; protect from light, excessive heat, moisture, and freezing. Prepared media plates can be kept for one day at ambient temperature. Plates can be stored at least one week under refrigeration (2-8°C) if properly prepared and protected from light and dehydration. Do not use after the expiration date supplied by the manufacturer.

**Use:** For the detection and direct differentiation of pathogenic *Yersinia enterocolitica.* Pathogenic *Y. enterocolitica* produce mauve colonies. Non pathogenic *Y. enterocolitica and* other microbes (*Citrobacter, Enterobacter, Aeromonas,* etc). produce metallic blue colonies.

## Colorex™ *Listeria*

**Composition** per liter:

Proprietary

**Source:** Available as prepared plates from E&O Laboratories, Bonnybridge Scotland.

**Use:** For the differentiation and presumptive identification of *Listeria monocytogenes,* which form blue colonies surrounded by white halos.

## Colorex™ MRSA

**Composition** per liter:

Chrompeptone................................................................40.0g
NaCl................................................................................25.0g
Agar.................................................................................14.0g
Chromogenic mix ...........................................................0.5g
Inhibitory agents ...........................................................0.07g
Cefoxitin ........................................................................6.0mg

**Source:** Available as prepared plates from E&O Laboratories, Bonnybridge Scotland.

**Storage/Shelf Life:** Store in the dark. Chromogenic agars are especially light and temperature sensitive; protect from light, excessive heat, moisture, and freezing. Prepared media plates can be kept for one day at ambient temperature. Plates can be stored at least one week under refrigeration (2-8°C) if properly prepared and protected from light and dehydration. Do not use after the expiration date supplied by the manufacturer.

**Use:** For the qualitative direct detection of nasal colonization by methicillin-resistant *Staphylococcus aureus* (MRSA) to aid in the prevention and control of MRSA infections in healthcare settings.

## Colorex™ O157

**Composition** per liter:

Agar .......................................................................15.0g
Peptone and yeast extract......................................13.0g
Chromogenic mix ...................................................1.2g

pH 6.9± 0.2 at 25°C

**Source:** Available as prepared plates from E&O Laboratories, Bonnybridge Scotland.

**Storage/Shelf Life:** Store in the dark. Chromogenic agars are especially light and temperature sensitive; protect from light, excessive heat, moisture, and freezing. Prepared media plates can be kept for one day at ambient temperature. Plates can be stored at least one week under refrigeration (2-8°C) if properly prepared and protected from light and dehydration. Do not use after the expiration date supplied by the manufacturer.

**Use:** For the selective isolation and differentiation of *E. coli* O157 in foods and clinical specimens. *E. coli* O157 produce mauve colonies.

## Colorex™ O157 with Cefixime andTellurite

**Composition** per liter:

Agar .......................................................................15.0g
Peptone and yeast extract......................................13.0g
Chromogenic mix ...................................................1.2g
$K_2TeO_3$.................................................................2.5mg
Cafixime ...............................................................0.025mg

pH 6.9± 0.2 at 25°C

**Source:** Available as prepared plates from E&O Laboratories, Bonnybridge Scotland.

**Storage/Shelf Life:** Store in the dark. Chromogenic agars are especially light and temperature sensitive; protect from light, excessive heat, moisture, and freezing. Prepared media plates can be kept for one day at ambient temperature. Plates can be stored at least one week under refrigeration (2-8°C) if properly prepared and protected from light and dehydration. Do not use after the expiration date supplied by the manufacturer.

**Use:** For the selective isolation and differentiation of *E. coli* O157 in foods and clinical specimens. Potassium tellurite increases the selectivity of the medium and the specificity of *E. coli* O157 detection. *E.coli* O157 produce mauve colonies.

## Colorex™ Orientation

**Composition** per liter:

Peptone and yeast extract .............................................17.0g
Agar ........................................................................15.0g
Chromogenic mix .........................................................1.0g

pH 7.0± 0.2 at 25°C

**Source:** Available as prepared plates from E&O Laboratories, Bonnybridge Scotland.

**Storage/Shelf Life:** Store in the dark. Chromogenic agars are especially light and temperature sensitive; protect from light, excessive heat, moisture, and freezing. Prepared media plates can be kept for one day at ambient temperature. Plates can be stored at least one week under refrigeration (2-8°C) if properly prepared and protected from light and dehydration. Do not use after the expiration date supplied by the manufacturer.

**Use:** For the detection enhanced isolation and differentiation of urinary tract pathogens. *E. coli* form dark pink to reddish colonies. *Enterococcus* form turquoise blue colonies. *Klebsiella, Enterobacter*, and *Citrobacter* form metallic blue colonies. *Proteus* form colonies with a brown halo. *Pseudomonas* form cream translucent colonies.

## Colorex™ *Salmonella* Plus

**Composition** per liter:

Agar ........................................................................15.0g
Salts........................................................................8.5g
Peptone and yeast extract.................................................8.0g
Supplement .................................................................6.0g
Chromogenic mix ...........................................................1.3g

pH 7.5± 0.2 at 25°C

**Source:** Available as prepared plates from E&O Laboratories, Bonnybridge Scotland.

**Preparation of Medium:** Add components, to distilled/deionized water and bring volume to 1.0L. Mix thoroughly. Gently heat and bring to boiling. Do not autoclave. Cool to 50°C. Pour into sterile Petri dishes.

**Storage/Shelf Life:** Store in the dark. Chromogenic agars are especially light and temperature sensitive; protect from light, excessive heat, moisture, and freezing. Prepared media plates can be kept for one day at ambient temperature. Plates can be stored at least one week under refrigeration (2-8°C) if properly prepared and protected from light and dehydration. Do not use after the expiration date supplied by the manufacturer.

**Use:** For the detection and isolation of *Salmonella* species including lactose positive *Salmonella*. Detection of lactose positive *Salmonella* is now required in food microbiology by the ISO 6579:2002 standard. *Salmonella* (including *S. typhi, S. paratyphi* A and lactose positive *Salmonella*) produce mauve colonies.

## Colorex™ *Staph aureus*

**Composition** per liter:

Peptone and yeast extract.................................................40.0g
Salts........................................................................25.0g
Agar ........................................................................15.0g
Chromogenic mix ...........................................................2.5g

pH 6.9± 0.2 at 25°C

**Source:** Available as prepared plates from E&O Laboratories, Bonnybridge Scotland.

**Storage/Shelf Life:** Store in the dark. Chromogenic agars are especially light and temperature sensitive; protect from light, excessive heat, moisture, and freezing. Prepared media plates can be kept for one day at ambient temperature. Plates can be stored at least one week under refrigeration (2-8°C) if properly prepared and protected from light and dehydration. Do not use after the expiration date supplied by the manufacturer.

**Use:** For the isolation and direct differentiation of *Staphylococcus aureus* in clinical specimens. S. aureua forms pink to mauve colonies.

## Colorex™ *Strep*B

**Composition** per liter:
Proprietary

**Source:** Available as prepared plates from E&O Laboratories, Bonnybridge Scotland.

**Storage/Shelf Life:** Store in the dark. Chromogenic agars are especially light and temperature sensitive; protect from light, excessive heat, moisture, and freezing. Prepared media plates can be kept for one day at ambient temperature. Plates can be stored at least one week under refrigeration (2-8°C) if properly prepared and protected from light and dehydration. Do not use after the expiration date supplied by the manufacturer.

**Use:** For the differentiation and presumptive identification of *Streptococcus* B (*Streptococcus agalactiae*) based upon color formation. *Streptococcus* B forms mauve to pink colonies.

## Colorex™ Vancomycin-Resistant *Enterococcus* (VRE)

**Composition** per liter:
Proprietary

**Source:** Available as prepared plates from E&O Laboratories, Bonnybridge Scotland.

**Storage/Shelf Life:** Store in the dark. Chromogenic agars are especially light and temperature sensitive; protect from light, excessive heat, moisture, and freezing. Prepared media plates can be kept for one day at ambient temperature. Plates can be stored at least one week under refrigeration (2-8°C) if properly prepared and protected from light and dehydration. Do not use after the expiration date supplied by the manufacturer.

**Use:** For the differentiation and presumptive identification of vancomycin-resistant *Enterococcus* (*Enterococcus faecalis/E. facecium*). Vancomycin-resistant *Enterococcus* strains form rose to mauve colonies.

## Colorex™ *Vibrio*

**Composition** per liter:

Salts........................................................................51.4g
Agar ........................................................................15.0g
Peptone and yeast extract.................................................8.0g
Chromogenic mix ...........................................................0.3g

pH 9.0± 0.2 at 25°C

**Source:** Available as prepared plates from E&O Laboratories, Bonnybridge Scotland.

**Storage/Shelf Life:** Store in the dark. Chromogenic agars are especially light and temperature sensitive; protect from light, excessive heat, moisture, and freezing. Prepared media plates can be kept for one day at ambient temperature. Plates can be stored at least one week under refrigeration (2-8°C) if properly prepared and protected from light and dehydration. Do not use after the expiration date supplied by the manufacturer.

**Use:** For the isolation and detection of *V. parahaemolyticus*, *V. vulnificus* and *V. cholerae*. *V. parahaemolyticus* form mauve colonies. *V. vulnificus* and *V. cholera* form green blue to turquoise blue colonies. *V. alginolyticus* form colorless colonies.

## Columbia Agar
## (Columbia Blood Agar)

**Composition** per liter:

Columbia agar base ................................................................ 950.0mL
Sheep blood ............................................................................. 50.0mL

pH 7.3 ± 0.2 at 25°C

**Columbia Agar Base:**

**Composition** per liter:

Agar ............................................................................................ 13.5g
Pancreatic digest of casein ....................................................... 12.0g
NaCl .............................................................................................. 5.0g
Peptic digest of animal tissue ..................................................... 5.0g
Beef extract ................................................................................. 3.0g
Yeast extract ............................................................................... 3.0g
Cornstarch ................................................................................... 1.0g

**Preparation of Columbia Agar Base:** Add components to distilled/deionized water and bring volume to 1.0L. Mix thoroughly. Gently heat until boiling. Autoclave for 15 min at 15 psi pressure–121°C. Cool to 45°–50°C.

**Preparation of Medium:** To 950.0mL of cooled, sterile Columbia agar base, aseptically add 50.0mL of sterile, defibrinated sheep blood. Mix thoroughly. Pour into sterile Petri dishes or distribute into sterile tubes.

**Storage/Shelf Life:** Store dehydrated media in the dark in a sealed container below 30°C. Prepared media should be stored under refrigeration (2-8°C). Media should be used within 60 days of preparation. Media should not be used if there are any signs of deterioration (shrinking, cracking, or discoloration) or contamination, or if the expiration date supplied by the manufacturer has passed.

**Use:** For the isolation and cultivation of nonfastidious and fastidious microorganisms from a variety of clinical speciments and nonclinical specimens of public health importance.

## Columbia Blood Agar

**Composition** per liter:

Columbia blood agar base ....................................................... 950.0mL
Sheep blood ............................................................................. 50.0mL

pH 7.3 ± 0.2 at 25°C

**Columbia Blood Agar Base:**

**Composition** per liter:

Agar ............................................................................................ 15.0g
Pantone ...................................................................................... 10.0g
Bitone ........................................................................................ 10.0g
NaCl .............................................................................................. 5.0g
Tryptic digest of beef heart ....................................................... 3.0g
Cornstarch ................................................................................... 1.0g

**Source:** Columbia blood agar base is available as a premixed powder from BD Diagnostic Systems.

**Preparation of Columbia Blood Agar Base:** Add components to distilled/deionized water and bring volume to 1.0L. Mix thoroughly.

Gently heat until boiling. Autoclave for 15 min at 15 psi pressure–121°C. Cool to 45°–50°C.

**Preparation of Medium:** To 950.0mL of cooled, sterile Columbia blood agar base, aseptically add 50.0mL of sterile, defibrinated sheep blood. Mix thoroughly. Pour into sterile Petri dishes or distribute into sterile tubes.

**Storage/Shelf Life:** Store dehydrated media in the dark in a sealed container below 30°C. Prepared media should be stored under refrigeration (2-8°C). Media should be used within 60 days of preparation. Media should not be used if there are any signs of deterioration (shrinking, cracking, or discoloration) or contamination, or if the expiration date supplied by the manufacturer has passed.

**Use:** With the addition of blood or other enrichments, used for the isolation and cultivation of fastidious microorganisms.

## Columbia Blood Agar Base with 1% Agar,
## HiVeg with Blood

**Composition** per liter:

Plant special peptone ................................................................ 23.3g
Agar ............................................................................................ 10.0g
NaCl .............................................................................................. 5.0g
Corn starch .................................................................................. 1.0g
Sheep blood, defibrinated ....................................................... 50.0mL

pH 7.3 ± 0.2 at 25°C

**Source:** This medium is available as a premixed powder from Hi-Media.

**Preparation of Medium:** Add components to distilled/deionized water and bring volume to 1.0L. Mix thoroughly. Gently heat until boiling. Autoclave for 15 min at 15 psi pressure–121°C. Cool to 45°–50°C. For Columbia Blood Agar: Add 5% sterile defibrinated sheep blood to sterile cool base. For Chocolate Agar: Add 10% sterile defibrinated sheep blood to sterile cool base. Heat to 80°C for 10 min with constant agitation. For Selective Medium: Add desired quantity of antimicrobial agent to sterile base. Mix thoroughly. Pour into sterile Petri dishes or distribute into sterile tubes.

**Storage/Shelf Life:** Store dehydrated media in the dark in a sealed container below 30°C. Prepared media should be stored under refrigeration (2-8°C). Media should be used within 60 days of preparation. Media should not be used if there are any signs of deterioration (shrinking, cracking, or discoloration) or contamination, or if the expiration date supplied by the manufacturer has passed.

**Use:** For the isolation and cultivation of fastidious bacteria from a variety of clinical speciments and nonclinical specimens of public health importance.

## Columbia Blood Agar Base, HiVeg with Blood

**Composition** per liter:

Plant special peptone ................................................................ 23.0g
Agar ............................................................................................ 15.0g
NaCl .............................................................................................. 5.0g
Corn starch .................................................................................. 1.0g
Sheep blood, defibrinated ....................................................... 50.0mL

pH 7.3 ± 0.2 at 25°C

**Source:** This medium is available as a premixed powder from Hi-Media.

**Preparation of Medium:** Add components to distilled/deionized water and bring volume to 950.0mL. Mix thoroughly. Gently heat until boiling. Autoclave for 15 min at 15 psi pressure–121°C. Cool to 45°–

50°C. For Columbia Blood Agar: Add 5% sterile defibrinated sheep blood to sterile cool base. For Chocolate Agar: Add 10% sterile defibrinated sheep blood to sterile cool base. Heat to 80°C for 10 min with constant agitation. For Selective Medium: Add desired quantity of antimicrobial agent to sterile base. Mix thoroughly. Pour into sterile Petri dishes or distribute into sterile tubes.

**Storage/Shelf Life:** Store dehydrated media in the dark in a sealed container below 30°C. Prepared media should be stored under refrigeration (2-8°C). Media should be used within 60 days of preparation. Media should not be used if there are any signs of deterioration (shrinking, cracking, or discoloration) or contamination, or if the expiration date supplied by the manufacturer has passed.

**Use:** For the isolation and cultivation of fastidious microorganisms from a variety of clinical speciments and nonclinical specimens of public health importance.

### Columbia Broth

**Composition** per liter:
| | |
|---|---|
| Bitone | 10.0g |
| Pancreatic digest of casein | 5.0g |
| Peptic digest of animal tissue | 5.0g |
| NaCl | 5.0g |
| Tryptic digest of beef heart | 3.0g |
| Tris(hydroxymethyl)aminomethane·HCl | 2.86g |
| Glucose | 2.5g |
| Tris(hydroxymethyl)aminomethane | 0.83g |
| $Na_2CO_3$ | 0.6g |
| L-Cysteine·HCl | 0.1g |
| $MgSO_4$, anhydrous | 0.1g |
| $FeSO_4$ | 0.02g |

pH 7.5 ± 0.2 at 25°C

**Source:** This medium is available as a premixed powder from BD Diagnostic Systems.

**Preparation of Medium:** Add components to distilled/deionized water and bring volume to 1.0L. Mix thoroughly. Gently heat until boiling. Distribute into tubes or flasks. Autoclave for 15 min at 15 psi pressure–121°C.

**Storage/Shelf Life:** Store dehydrated media in the dark in a sealed container below 30°C. Prepared media should be stored under refrigeration (2-8°C). Media should be used within 60 days of preparation. Media should not be used if there are any signs of deterioration (discoloration) or contamination, or if the expiration date supplied by the manufacturer has passed.

**Use:** For the cultivation and isolation of fastidious bacteria from clinical specimens or as a general purpose broth.

### Columbia Broth

**Composition** per liter:
| | |
|---|---|
| Pancreatic digest of casein | 10.0g |
| Peptic digest of animal tissue | 8.0g |
| NaCl | 5.0g |
| Yeast extract | 5.0g |
| Tris(hydroxymethyl) aminomethane·HCl buffer | 2.86g |
| Glucose | 2.5g |
| Tris(hydroxymethyl) aminomethane buffer | 0.83g |
| L-Cysteine·HCl·$H_2O$ | 0.1g |
| $MgSO_4$·$7H_2O$ | 0.05g |
| $FeSO_4$ | 0.012g |

pH 7.4 ± 0.2 at 25°C

**Source:** This medium is available as a premixed powder from BD Diagnostic Systems.

**Preparation of Medium:** Add components to distilled/deionized water and bring volume to 1.0L. Mix thoroughly. Gently heat until boiling. Distribute into tubes or flasks. Autoclave for 15 min at 15 psi pressure–121°C.

**Storage/Shelf Life:** Store dehydrated media in the dark in a sealed container below 30°C. Prepared media should be stored under refrigeration (2-8°C). Media should be used within 60 days of preparation. Media should not be used if there are any signs of deterioration (discoloration) or contamination, or if the expiration date supplied by the manufacturer has passed.

**Use:** For the cultivation of a wide variety of microorganisms. Used as a general purpose medium.

### Columbia Broth Base, HiVeg with Blood

**Composition** per liter:
| | |
|---|---|
| Plant peptone No. 5 | 10.0g |
| Plant special peptone | 10.0g |
| NaCl | 5.0g |
| Plant infusion | 3.0g |
| Tris(hydroxymethyl)aminomethane | 2.86 |
| Glucose | 2.5g |
| $Na_2CO_3$ | 0.6g |
| L-Cystine hydrochloride | 0.1g |
| $MgSO_4$ | 0.1g |
| $FeSO_4$ | 0.02g |
| Sheep blood, defibrinated | 50.0mL |

pH 7.5 ± 0.2 at 25°C

**Source:** This medium, without blood, is available as a premixed powder from HiMedia.

**Preparation of Medium:** Add components to distilled/deionized water and bring volume to 1.0L. Mix thoroughly. Gently heat until boiling. Distribute into tubes or flasks. Autoclave for 15 min at 15 psi pressure–121°C. Cool to 45°–50°C. Add blood and/or selective antimicrobics. Mix well.

**Storage/Shelf Life:** Store dehydrated media in the dark in a sealed container below 30°C. Prepared media should be stored under refrigeration (2-8°C). Media should be used within 60 days of preparation. Media should not be used if there are any signs of deterioration (discoloration) or contamination, or if the expiration date supplied by the manufacturer has passed.

**Use:** For the cultivation and isolation of fastidious bacteria from clinical specimens.

### Columbia Broth Base, HiVeg with SPS

**Composition** per liter:
| | |
|---|---|
| Plant peptone No. 5 | 10.0g |
| Plant special peptone | 10.0g |
| NaCl | 5.0g |
| Plant infusion | 3.0g |
| Tris (hydroxymethyl) aminomethane | 2.86 |
| Glucose | 2.5g |
| $Na_2CO_3$ | 0.6g |
| L-Cystine hydrochloride | 0.1g |
| $MgSO_4$ | 0.1g |

FeSO₄ ..................................................................0.02g

$FeSO_4$ ..................................................................0.02g

SPS (sodium polystyrene sulfonate) .........................0.1mL

pH 7.5 ± 0.2 at 25°C

**Source:** This medium, without SPS, is available as a premixed powder from HiMedia.

**Preparation of Medium:** Add components to distilled/deionized water and bring volume to 1.0L. Mix thoroughly. Gently heat until boiling. Distribute into tubes or flasks. Autoclave for 15 min at 15 psi pressure–121°C. Cool to 45°–50°C. Mix thoroughly.

**Storage/Shelf Life:** Store dehydrated media in the dark in a sealed container below 30°C. Prepared media should be stored under refrigeration (2-8°C). Media should be used within 60 days of preparation. Media should not be used if there are any signs of deterioration (discoloration) or contamination, or if the expiration date supplied by the manufacturer has passed.

**Use:** For the cultivation and isolation of fastidious bacteria from clinical specimens. For blood cultures, the SPS inhibits lysozyme activity and interferes with phagocytosis and destroys the aminoglycosides.

## Columbia CNA Agar
## (Columbia Colistin Nalidixic Acid Agar)

**Composition** per liter:

Columbia blood agar base ...........................................950.0L

Sheep blood ...............................................................50.0mL

pH 7.3 ± 0.2 at 25°C

**Source:** This medium is available as a premixed powder from BD Diagnostic Systems.

**Columbia Blood Agar Base:**

**Composition** per liter:

Agar ........................................................................13.5g

Pancreatic digest of casein ......................................12.0g

NaCl .........................................................................5.0g

Peptic digest of animal tissue ...................................5.0g

Beef extract ..............................................................3.0g

Yeast extract .............................................................3.0g

Cornstarch ...............................................................1.0g

Nalidixic acid .........................................................15.0mg

Colistin ...................................................................10.0mg

**Preparation of Columbia Blood Agar Base:** Add components to distilled/deionized water and bring volume to 1.0L. Mix thoroughly. Gently heat until boiling. Autoclave for 15 min at 15 psi pressure–121°C. Cool to 45°–50°C.

**Preparation of Medium:** To 950.0mL of cooled, sterile Columbia blood agar base, aseptically add 50.0mL of sterile, defibrinated sheep blood. Mix thoroughly. Pour into sterile Petri dishes or distribute into sterile tubes.

**Storage/Shelf Life:** Store dehydrated media in the dark in a sealed container below 30°C. Prepared media should be stored under refrigeration (2-8°C). Media should be used within 60 days of preparation. Media should not be used if there are any signs of deterioration (shrinking, cracking, or discoloration) or contamination, or if the expiration date supplied by the manufacturer has passed.

**Use:** For the selective isolation, cultivation, and differentiation of Gram-positive cocci from clinical speciments and nonclinical specimens of public health importance.

## Columbia C.N.A. Agar Base with Blood

**Composition** per liter:

Peptone, special ......................................................23.0g

Agar ........................................................................15.0g

NaCl .........................................................................5.0g

Corn starch ..............................................................1.0g

Sheep blood, defibrinated ......................................50.0mL

pH 7.3 ± 0.2 at 25°C

**Source:** This medium, without blood, is available as a premixed powder from HiMedia.

**Preparation of Medium:** Add components to distilled/deionized water and bring volume to 1.0L. Mix thoroughly. Gently heat until boiling. Autoclave for 15 min at 15 psi pressure–121°C. Cool to 45°–50°C. Aseptically add 50.0mL of sterile, defibrinated sheep blood to 950.0mL of cooled, sterile agar base. Mix thoroughly. Pour into sterile Petri dishes or distribute into sterile tubes.

**Storage/Shelf Life:** Store dehydrated media in the dark in a sealed container below 30°C. Prepared media should be stored under refrigeration (2-8°C). Media should be used within 60 days of preparation. Media should not be used if there are any signs of deterioration (shrinking, cracking, or discoloration) or contamination, or if the expiration date supplied by the manufacturer has passed.

**Use:** For the selective isolation, cultivation, and differentiation of Gram-positive cocci from clinical speciments and nonclinical specimens of public health importance.

## Columbia C.N.A. Agar Base with Blood

**Composition** per liter:

Biopeptone ...............................................................20.0g

Agar ........................................................................15.0g

NaCl .........................................................................5.0g

Tryptic digest of beef heart ......................................3.0g

Cornstarch ...............................................................1.0g

Nalidixic acid .........................................................0.015g

Colistin sulfate .......................................................0.01g

Sheep blood, defibrinated ......................................50.0mL

pH 7.3 ± 0.2 at 25°C

**Source:** This medium, without blood, is available as a premixed powder from HiMedia.

**Preparation of Medium:** Add components to distilled/deionized water and bring volume to 1.0L. Mix thoroughly. Gently heat until boiling. Autoclave for 15 min at 15 psi pressure–121°C. Cool to 45°–50°C. Aseptically add 50.0mL of sterile, defibrinated sheep blood to 950.0mL of cooled, sterile agar base. Mix thoroughly. Pour into sterile Petri dishes or distribute into sterile tubes.

**Storage/Shelf Life:** Store dehydrated media in the dark in a sealed container below 30°C. Prepared media should be stored under refrigeration (2-8°C). Media should be used within 60 days of preparation. Media should not be used if there are any signs of deterioration (shrinking, cracking, or discoloration) or contamination, or if the expiration date supplied by the manufacturer has passed.

**Use:** For the selective isolation, cultivation, and differentiation of Gram-positive cocci from clinical speciments and nonclinical specimens of public health importance.

## Columbia C.N.A. Agar Base with 1% Agar and Blood

**Composition** per liter:

Biopeptone ...............................................................20.0g

Agar ........................................................................10.0g

NaCl.................................................................................5.0g
Tryptic digest of beef heart...........................................3.0g
Corn starch......................................................................1.0g
Nalidixic acid...............................................................0.015g
Colistin sulfate.............................................................0.01g
Sheep blood, defibrinated ........................................50.0mL

pH 7.3 ± 0.2 at 25°C

**Source:** This medium, without blood, is available as a premixed powder from HiMedia.

**Preparation of Medium:** Add components to distilled/deionized water and bring volume to 1.0L. Mix thoroughly. Gently heat until boiling. Autoclave for 15 min at 15 psi pressure–121°C. Cool to 45°–50°C. Aseptically add 50.0mL of sterile, defibrinated sheep blood to 950.0mL of cooled, sterile agar base. Mix thoroughly. Pour into sterile Petri dishes or distribute into sterile tubes.

**Storage/Shelf Life:** Store dehydrated media in the dark in a sealed container below 30°C. Prepared media should be stored under refrigeration (2-8°C). Media should be used within 60 days of preparation. Media should not be used if there are any signs of deterioration (shrinking, cracking, or discoloration) or contamination, or if the expiration date supplied by the manufacturer has passed.

**Use:** For the selective isolation, cultivation, and differentiation of Gram-positive cocci from clinical speciments and nonclinical specimens of public health importance.

### Columbia CNA Agar, Modified with Sheep Blood
**Composition** per liter:
Columbia blood agar base.......................................950.0mL
Sheep blood, defibrinated ........................................50.0mL

pH 7.3 ± 0.2 at 25°C

**Source:** This medium is available as a premixed powder from BD Diagnostic Systems.

**Columbia Blood Agar Base:**
**Composition** per liter:
Agar .................................................................................13.5g
Pancreatic digest of casein............................................12.0g
NaCl..................................................................................5.0g
Peptic digest of animal tissue.........................................5.0g
Beef extract......................................................................3.0g
Yeast extract....................................................................3.0g
Cornstarch........................................................................1.0g
Nalidixic acid.................................................................5.0mg
Colistin...........................................................................10.0mg

**Preparation of Columbia Blood Agar Base:** Add components to distilled/deionized water and bring volume to 1.0L. Mix thoroughly. Gently heat until boiling. Autoclave for 15 min at 15 psi pressure–121°C. Cool to 45°–50°C.

**Preparation of Medium:** To 950.0L of cooled, sterile Columbia blood agar base, aseptically add 50.0mL of sterile, defibrinated sheep blood. Mix thoroughly. Pour into sterile Petri dishes or distribute into sterile tubes.

**Storage/Shelf Life:** Store dehydrated media in the dark in a sealed container below 30°C. Prepared media should be stored under refrigeration (2-8°C). Media should be used within 60 days of preparation. Media should not be used if there are any signs of deterioration (shrinking, cracking, or discoloration) or contamination, or if the expiration date supplied by the manufacturer has passed.

**Use:** For the selective isolation, cultivation, and differentiation of Gram-positive cocci from clinical and nonclinical materials.

### Columbia C.N.A. HiVeg Agar Base with 1% Agar
**Composition** per liter:
Plant peptone No. 5........................................................20.0g
Agar .................................................................................10.0g
NaCl..................................................................................5.0g
Plant infusion..................................................................3.0g
Corn starch......................................................................1.0g
Nalidixic acid...............................................................0.015g
Colistin sulfate.............................................................0.01g
Sheep blood, defibrinated ........................................50.0mL

pH 7.3 ± 0.2 at 25°C

**Source:** This medium, without blood, is available as a premixed powder from HiMedia.

**Preparation of Medium:** Add components to distilled/deionized water and bring volume to 1.0L. Mix thoroughly. Gently heat until boiling. Autoclave for 15 min at 15 psi pressure–121°C. Cool to 45°–50°C. Aseptically add 50.0mL of sterile, defibrinated sheep blood to 950.0mL of cooled, sterile agar base. Mix thoroughly. Pour into sterile Petri dishes or distribute into sterile tubes.

**Storage/Shelf Life:** Store dehydrated media in the dark in a sealed container below 30°C. Prepared media should be stored under refrigeration (2-8°C). Media should be used within 60 days of preparation. Media should not be used if there are any signs of deterioration (shrinking, cracking, or discoloration) or contamination, or if the expiration date supplied by the manufacturer has passed.

**Use:** For the selective isolation, cultivation, and differentiation of Gram-positive cocci from clinical speciments and nonclinical specimens of public health importance.

### Columbia C.N.A. HiVeg Agar Base with Blood
**Composition** per liter:
Plant peptone No. 5........................................................20.0g
Agar .................................................................................15.0g
NaCl..................................................................................5.0g
Plant infusion..................................................................3.0g
Cornstarch........................................................................1.0g
Nalidixic acid...............................................................0.015g
Colistin sulfate.............................................................0.01g
Sheep blood, defibrinated ........................................50.0mL

pH 7.3 ± 0.2 at 25°C

**Source:** This medium, without blood, is available as a premixed powder from HiMedia.

**Preparation of Medium:** Add components to distilled/deionized water and bring volume to 1.0L. Mix thoroughly. Gently heat until boiling. Autoclave for 15 min at 15 psi pressure–121°C. Cool to 45°–50°C. Aseptically add 50.0mL of sterile, defibrinated sheep blood to 950.0mL of cooled, sterile agar base. Mix thoroughly. Pour into sterile Petri dishes or distribute into sterile tubes.

**Storage/Shelf Life:** Store dehydrated media in the dark in a sealed container below 30°C. Prepared media should be stored under refrigeration (2-8°C). Media should be used within 60 days of preparation. Media should not be used if there are any signs of deterioration (shrinking, cracking, or discoloration) or contamination, or if the expiration date supplied by the manufacturer has passed.

**Use:** For the selective isolation, cultivation, and differentiation of Gram-positive cocci from clinical speciments and nonclinical specimens of public health importance.

## Congo Red Agar
## (CR Agar)

**Composition** per liter:

| | |
|---|---|
| GC agar base | 890.0mL |
| Hemoglobin solution | 100.0mL |
| Supplement solution | 10.0mL |
| Congo Red (0.01% solution) | 0.1mL |

pH 7.2 ± 0.2 at 25°C

### GC Agar Base:
**Composition** per 890.0mL:

| | |
|---|---|
| Agar | 10.0g |
| Pancreatic digest of casein | 7.5g |
| Peptic digest of animal tissue | 7.5g |
| NaCl | 5.0g |
| $K_2HPO_4$ | 4.0g |
| Cornstarch | 1.0g |
| $KH_2PO_4$ | 1.0g |

**Preparation of GC Agar Base:** Add components to distilled/deionized water and bring volume to 890.0mL. Mix thoroughly. Gently heat until boiling. Autoclave for 15 min at 15 psi pressure–121°C. Cool to 45°–50°C.

### Hemoglobin Solution:
**Composition** per 100.0mL:

| | |
|---|---|
| Hemoglobin | 2.0g |

**Preparation of Hemoglobin Solution:** Add hemoglobin to distilled/deionized water and bring volume to 100.0mL. Mix thoroughly. Autoclave for 15 min at 15 psi pressure–121°C. Cool to 50°C.

### Congo Red Solution:
**Composition** per 100.0mL:

| | |
|---|---|
| Congo Red | 0.01g |

**Preparation of Congo Red Solution:** Add Congo Red to 100.0mL of distilled/deionized water. Mix thoroughly. Autoclave for 15 min at 15 psi pressure–121°C.

### Supplement Solution:
**Composition** per liter:

| | |
|---|---|
| Glucose | 100.0g |
| L-Cysteine·HCl | 25.9g |
| L-Glutamine | 10.0g |
| L-Cystine | 1.1g |
| Adenine | 1.0g |
| Nicotinamide adenine dinucleotide | 0.25g |
| Vitamin $B_{12}$ | 0.1g |
| Thiamine pyrophosphate | 0.1g |
| Guanine·HCl | 0.03g |
| $Fe(NO_3)_3·6H_2O$ | 0.02g |
| *p*-Aminobenzoic acid | 0.013g |
| Thiamine·HCl | 3.0mg |

**Source:** The supplement solution IsoVitaleX® enrichment is available from BD Diagnostic Systems. This enrichment may be replaced by supplement VX from BD Diagnostic Systems.

**Preparation of Supplement Solution:** Add components to distilled/deionized water and bring volume to 1.0L. Mix thoroughly. Filter sterilize.

**Preparation of Medium:** To 890.0mL of sterile, cooled GC agar base aseptically add 100.0mL of sterile, cooled hemoglobin solution, 10.0mL of sterile supplement solution, and 0.1mL of sterile Congo Red solution. Mix thoroughly. Pour into sterile Petri dishes.

**Storage/Shelf Life:** Store dehydrated media in the dark in a sealed container below 30°C. Prepared media should be stored under refrigeration (2-8°C). Media should be used within 60 days of preparation. Media should not be used if there are any signs of deterioration (shrinking, cracking, or discoloration) or contamination, or if the expiration date supplied by the manufacturer has passed.

**Use:** For the isolation and differentiation of virulent and avirulent strains of *Shigella*, *Vibrio cholerae*, *Escherichia coli*, and *Neisseria meningitidis*. Used for the detection and differentiation of "iron-responsive" avirulent mutants. Used in the preparation of live vaccines. Used for the differentiation of sensitive *Neisseria gonorrhoeae* (no growth) from other *Neisseria* species (growth) that are resistant to Congo Red.

## Congo Red Agar
## (CR Agar)

**Composition** per liter:

| | |
|---|---|
| Soybean-casein digest agar | 890.0mL |
| Hemoglobin solution | 100.0mL |
| Supplement solution | 10.0mL |
| Congo Red (0.01% solution) | 0.1mL |

pH 7.3 ± 0.2 at 25°C

### Soybean-Casein Digest Agar:
**Composition** per 890.0mL:

| | |
|---|---|
| Pancreatic digest of casein | 17.0g |
| Agar | 15.0g |
| NaCl | 5.0g |
| Papaic digest of soybean meal | 3.0g |
| Glucose | 2.5g |
| $K_2HPO_4$ | 2.5g |

**Preparation of Soybean-Casein Digest Agar:** Add components to distilled/deionized water and bring volume to 890.0mL. Mix thoroughly. Gently heat until boiling. Autoclave for 15 min at 15 psi pressure–121°C. Cool to 45°–50°C.

### Hemoglobin Solution:
**Composition** per 100.0mL:

| | |
|---|---|
| Hemoglobin | 2.0g |

**Preparation of Hemoglobin Solution:** Add hemoglobin to distilled/deionized water and bring volume to 100.0mL. Mix thoroughly. Autoclave for 15 min at 15 psi pressure–121°C. Cool to 50°C.

### Congo Red Solution:
**Composition** per 100.0mL:

| | |
|---|---|
| Congo Red | 0.01g |

**Preparation of Congo Red Solution:** Add Congo Red to 100.0mL of distilled/deionized water. Mix thoroughly. Autoclave for 15 min at 15 psi pressure–121°C.

### Supplement Solution:
**Composition** per liter:

| | |
|---|---|
| Glucose | 100.0g |
| L-Cysteine·HCl | 25.9g |
| L-Glutamine | 10.0g |
| L-Cystine | 1.1g |
| Adenine | 1.0g |
| Nicotinamide adenine dinucleotide | 0.25g |
| Vitamin $B_{12}$ | 0.1g |
| Thiamine pyrophosphate | 0.1g |
| Guanine·HCl | 0.03g |
| $Fe(NO_3)_3·6H_2O$ | 0.02g |

p-Aminobenzoic acid........................................................0.013g
Thiamine·HCl ..................................................................3.0mg

**Preparation of Supplement Solution:** Add components to distilled/deionized water and bring volume to 1.0L. Mix thoroughly. Filter sterilize.

**Source:** The supplement solution IsoVitaleX® enrichment is available from BD Diagnostic Systems. This enrichment may be replaced by supplement VX from BD Diagnostic Systems.

**Preparation of Medium:** To 890.0mL of sterile, cooled soybean-casein digest agar, aseptically add 100.0mL of sterile, cooled hemoglobin solution, 10.0mL of sterile supplement solution, and 0.1mL of sterile Congo Red solution. Mix thoroughly. Pour into sterile Petri dishes.

**Storage/Shelf Life:** Store dehydrated media in the dark in a sealed container below 30°C. Prepared media should be stored under refrigeration (2-8°C). Media should be used within 60 days of preparation. Media should not be used if there are any signs of deterioration (shrinking, cracking, or discoloration) or contamination, or if the expiration date supplied by the manufacturer has passed.

**Use:** For the isolation and differentiation of virulent and avirulent strains of *Shigella, Vibrio cholerae, Escherichia coli,* and *Neisseria meningitidis.* Used for the detection and differentiation of "iron-responsive" avirulent mutants. Used in the preparation of live vaccines. Used for the differentiation of sensitive *Neisseria gonorrhoeae* (no growth) from other *Neisseria* species (growth) that are resistant to Congo Red.

### Congo Red BHI Agarose Medium

**Composition** per liter:
Agarose ..........................................................................15.0g
Pancreatic digest of gelatin ............................................14.5g
Brain heart, solids from infusion ......................................6.0g
Peptic digest of animal tissue...........................................6.0g
NaCl.................................................................................5.0g
Glucose ...........................................................................3.0g
Na$_2$HPO$_4$........................................................................2.5g
Congo Red .....................................................................0.075g
pH 7.4 ± 0.2 at 25°C

**Preparation of Medium:** Add components to distilled/deionized water and bring volume to 1.0L. Mix thoroughly. Gently heat and bring to boiling. Distribute into tubes or flasks. Autoclave for 15 min at 15 psi pressure–121°C. Pour into sterile Petri dishes in 20.0mL volumes.

**Storage/Shelf Life:** Store dehydrated media in the dark in a sealed container below 30°C. Prepared media should be stored under refrigeration (2-8°C). Media should be used within 60 days of preparation. Media should not be used if there are any signs of deterioration (shrinking, cracking, or discoloration) or contamination, or if the expiration date supplied by the manufacturer has passed.

**Use:** For the isolation, cultivation, and detection of virulent strains of *Yersinia enterocolitica.*

### Congo Red BHI Agarose Medium
### (CRBHO Medium)
### (BAM M41)

**Composition** per liter:
Pancreatic digest of gelatin ............................................14.5g
Agarose ..........................................................................12.0g
Brain heart, solids from infusion ......................................6.0g
Peptic digest of animal tissue...........................................6.0g
NaCl.................................................................................5.0g

Glucose ...........................................................................3.0g
Na$_2$HPO$_4$........................................................................2.5g
MgCl$_2$............................................................................1.0g
Congo Red solution ......................................................20.0mL
pH 7.4 ± 0.2 at 25°C

**Preparation of Medium:** Add components to distilled/deionized water and bring volume to 1.0L. Mix thoroughly. Gently heat and bring to boiling. Distribute into tubes or flasks. Autoclave for 15 min at 15 psi pressure–121°C. Pour into sterile Petri dishes in 20.0mL volumes.

**Congo Red Solution:**
**Composition** per 100.0mL:
Congo Red ....................................................................375.0mg

**Preparation of Congo Red Solution:** Add Congo Red to 100.0mL of distilled/deionized water. Mix thoroughly. Autoclave for 15 min at 15 psi pressure–121°C. Cool to 25°C.

**Storage/Shelf Life:** Store dehydrated media in the dark in a sealed container below 30°C. Prepared media should be stored under refrigeration (2-8°C). Media should be used within 60 days of preparation. Media should not be used if there are any signs of deterioration (shrinking, cracking, or discoloration) or contamination, or if the expiration date supplied by the manufacturer has passed.

**Use:** For the isolation, cultivation, and detection of virulent strains of *Yersinia enterocolitica.*

### Congo Red Magnesium Oxalate Agar
### (CRMOX Agar)

**Composition** per liter:
Solution 1....................................................................825.0mL
Solution 2......................................................................80.0mL
Solution 3......................................................................80.0mL
Solution 4......................................................................10.0mL
Solution 5........................................................................5.0mL
pH 7.3 ± 0.2 at 25°C

**Solution 1:**
**Composition** per 825.0mL:
Pancreatic digest of casein .............................................15.0g
Agar ...............................................................................15.0g
Papaic digest of soybean meal..........................................5.0g
NaCl.................................................................................5.0g
pH 7.3 ± 0.2 at 25°C

**Preparation of Solution 1:** Add components to distilled/deionized water and bring volume to 825.0mL. Mix thoroughly. Gently heat and bring to boiling. Autoclave for 15 min at 15 psi pressure–121°C. Do not overheat.

**Solution 2:**
**Composition** per liter:
MgCl$_2$·6H$_2$O ..............................................................50.8g

**Preparation of Solution 2:** Add MgCl$_2$·6H$_2$O to distilled/deionized water and bring volume to 1.0L. Mix thoroughly. Autoclave for 15 min at 15 psi pressure–121°C.

**Solution 3:**
**Composition** per liter:
Sodium oxalate ................................................................33.2g

**Preparation of Solution 3:** Add sodium oxalate to distilled/deionized water and bring volume to 1.0L. Mix thoroughly. Autoclave for 15 min at 15 psi pressure–121°C.

**Solution 4:**
**Composition** per 100.0mL:
D-Galactose .................................................................20.0g

**Preparation of Solution 4:** Add D-galactose to distilled/deionized water and bring volume to 100.0mL. Mix thoroughly. Filter sterilize.

**Solution 5:**
**Composition** per 10.0mL:
Congo Red ....................................................................0.1g

**Preparation of Solution 5:** Add Congo Red to distilled/deionized water and bring volume to 10.0mL. Mix thoroughly. Autoclave for 15 min at 15 psi pressure–121°C.

**Preparation of Medium:** Aseptically combine 80.0mL of sterile solution 2, 80.0mL of sterile solution 3, 10.0mL of sterile solution 4, and 5.0mL of sterile solution 5. Mix thoroughly. Warm to 50°C. Add this mixture to 825.0mL of cooled, sterile solution 1. Mix thoroughly. Pour into sterile Petri dishes.

**Storage/Shelf Life:** Store dehydrated media in the dark in a sealed container below 30°C. Prepared media should be stored under refrigeration (2-8°C). Media should be used within 60 days of preparation. Media should not be used if there are any signs of deterioration (shrinking, cracking, or discoloration) or contamination, or if the expiration date supplied by the manufacturer has passed.

**Use:** For the cultivation and identification of pathogenic serotypes of *Yersinia enterocolitica*. For the determination of whether *Yersinia* strains contain the *Yersinia* virulence plasmid.

### *Corynebacterium* Liquid Enrichment Medium
**Composition** per 2000.0mL:
Fosfomycin ...................................................................0.15g
Glucose 6-phosphate.....................................................0.03g
Solution A ................................................................985.0mL
Bovine serum ............................................................100.0mL
Nystatin solution..........................................................1.15mL
L-Cystine (1% solution) ................................................1.0mL
Egg yolk emulsion ......................................................10 eggs
pH 7.4 ± 0.2 at 25°C

**Solution A:**
**Composition** per liter:
Meat extract ..................................................................9.0g
Proteose peptone No. 3 ..................................................9.0g
NaCl..............................................................................2.7g
Glucose .........................................................................1.8g
$Na_2HPO_4 \cdot 12H_2O$.............................................1.8g
$K_2TeO_3$ (2% solution)..............................................75.0mL
L-Cystine (1% solution) ..............................................10.0mL

**Caution:** Potassium tellurite is toxic.

**Preparation of Solution A:** Add components to distilled/deionized water and bring volume to 985.0mL. Mix thoroughly. Filter sterilize.

**Egg Yolk Emulsion:**
**Composition**:
Chicken egg yolks...............................................................9
Whole chicken egg..............................................................1

**Preparation of Egg Yolk Emulsion:** Soak eggs with 1:100 dilution of saturated mercuric chloride solution for 1 min. Crack eggs and separate yolks from whites. Mix egg yolks with 1 chicken egg. Filter sterilize.

**Nystatin Solution:**
**Composition** per 10.0mL:
Nystatin.................................................................10,000U

**Preparation of Nystatin Solution:** Add nystatin to distilled/deionized water and bring volume to 10.0mL. Mix thoroughly. Filter sterilize.

**L-Cystine Solution:**
**Composition** per 10.0mL:
L-Cystine .....................................................................0.1g

**Preparation of L-Cystine Solution:** Add L-cystine to distilled/deionized water and bring volume to 10.0mL. Mix thoroughly. Filter sterilize.

**Preparation of Medium:** To 985.0mL of sterile solution A, aseptically add the remaining components. Mix thoroughly. Aseptically distribute into sterile tubes in 2.0–3.0mL volumes.

**Storage/Shelf Life:** Store dehydrated media in the dark in a sealed container below 30°C. Prepared media should be stored under refrigeration (2-8°C). Media should be used within 60 days of preparation. Media should not be used if there are any signs of deterioration (discoloration) or contamination, or if the expiration date supplied by the manufacturer has passed.

**Use:** For the isolation and cultivation of *Corynebacterium diphtheriae*.

### CPC Agar Base with Cellobiose, Colistin, and Polymyxin B
**Composition** per liter:
NaCl.............................................................................20.0g
Agar .............................................................................15.0g
Cellobiose .....................................................................15.0g
Peptic digest of animal tissue .........................................10.0g
Beef extract ....................................................................5.0g
Bromthymol Blue ..........................................................0.04g
Cresol Red.....................................................................0.04g
Cellobiose colistin polymyxin B solution............................100.0mL
pH 7.6 ± 0.2 at 25°C

**Source:** This medium, without cellobiose colistin polymyxin B solution, is available as a premixed powder from HiMedia.

**Cellobiose Colistin Polymyxin B Solution:**
**Composition** per 100.0mL:
Cellobiose .....................................................................15.0g
Colistin....................................................................1,360,000U
Polymyxin B ............................................................100,000U

**Preparation of Cellobiose Colistin Polymyxin B Solution:** Add components to distilled/deionized water and bring volume to 100.0mL. Mix thoroughly. Filter sterilize.

**Preparation of Medium:** Add components, except cellobiose colistin polymyxin B solution, to tap water and bring volume to 1.0L. Mix thoroughly. Distribute into tubes or flasks. Autoclave for 15 min at 15 psi pressure–121°C. Aseptically add 100.0mL of sterile cellobiose colistin polymyxin B solution to 900.0 mL of the cooled agar base. Mix thoroughly. Pour into sterile Petri dishes. Use within 7 days.

**Storage/Shelf Life:** Store dehydrated media in the dark in a sealed container below 30°C. Prepared media should be stored under refrigeration (2-8°C). Media should be used within 60 days of preparation. Media should not be used if there are any signs of deterioration (shrink-

ing, cracking, or discoloration) or contamination, or if the expiration date supplied by the manufacturer has passed.

**Use:** For the cultivation and identification of *Vibrio* species from foods.

## CPC HiVeg Agar Base
## with Cellobiose, Colistin, and Polymyxin B

**Composition** per liter:

| | |
|---|---|
| NaCl | 20.0g |
| Agar | 15.0g |
| Cellobiose | 15.0g |
| Plant peptone | 10.0g |
| Plant extract | 5.0g |
| Bromthymol Blue | 0.04g |
| Cresol Red | 0.04g |
| Cellobiose colistin polymyxin B solution | 100.0mL |

pH 7.6 ± 0.2 at 25°C

**Source:** This medium, without cellobiose colistin polymyxin B solution, is available as a premixed powder from HiMedia.

### Cellobiose Colistin Polymyxin B Solution:

**Composition** per 100.0mL:

| | |
|---|---|
| Cellobiose | 15.0g |
| Colistin | 1,360,000U |
| Polymyxin B | 100,000U |

**Preparation of Cellobiose Colistin Polymyxin B Solution:** Add components to distilled/deionized water and bring volume to 100.0mL. Mix thoroughly. Filter sterilize.

**Preparation of Medium:** Add components, except cellobiose colistin polymyxin B solution, to tap water and bring volume to 1.0L. Mix thoroughly. Distribute into tubes or flasks. Autoclave for 15 min at 15 psi pressure–121°C. Aseptically add 100.0mL of sterile cellobiose colistin polymyxin B solution to 900.0 mL of the cooled agar base. Mix thoroughly. Pour into sterile Petri dishes. Use within 7 days.

**Storage/Shelf Life:** Store dehydrated media in the dark in a sealed container below 30°C. Prepared media should be stored under refrigeration (2-8°C). Media should be used within 60 days of preparation. Media should not be used if there are any signs of deterioration (shrinking, cracking, or discoloration) or contamination, or if the expiration date supplied by the manufacturer has passed.

**Use:** For the cultivation and identification of *Vibrio* species from foods.

## Crossley Milk Medium

**Composition** per liter:

| | |
|---|---|
| Skim milk powder | 100.0g |
| Peptone | 10.0g |
| Bromcresol Purple | 0.1g |

pH 5.8 ± 0.2 at 25°C

**Source:** This medium is available as a premixed powder from Thermo Scientific.

**Preparation of Medium:** Add components to a very small volume of distilled/deionized water and mix to a paste. Gradually add more distilled/deionized water and bring volume to 1.0L. Distribute in 10.0mL volumes into tubes. Autoclave for 5 min at 15 psi pressure–121°C.

**Storage/Shelf Life:** Store dehydrated media in the dark in a sealed container below 30°C. Prepared media should be stored under refrigeration (2-8°C). Media should be used within 60 days of preparation.

Media should not be used if there are any signs of deterioration (discoloration) or contamination, or if the expiration date supplied by the manufacturer has passed.

**Use:** For the routine examination of canned food samples for anaerobic bacteria.

## Crystal Violet Agar

**Composition** per liter:

| | |
|---|---|
| Agar | 15.0g |
| Lactose | 10.0g |
| Proteose peptone | 5.0g |
| Beef extract | 3.0g |
| Crystal Violet | 3.3mg |

pH 6.8 ± 0.1 at 25°C

**Preparation of Medium:** Add components to distilled/deionized water and bring volume to 1.0L. Mix thoroughly. Gently heat until boiling. Distribute into tubes or flasks. Autoclave for 15 min at 15 psi pressure–121°C. Pour into sterile Petri dishes or leave in tubes.

**Storage/Shelf Life:** Store dehydrated media in the dark in a sealed container below 30°C. Prepared media should be stored under refrigeration (2-8°C). Media should be used within 60 days of preparation. Media should not be used if there are any signs of deterioration (shrinking, cracking, or discoloration) or contamination, or if the expiration date supplied by the manufacturer has passed.

**Use:** For the differentiation of pathogenic staphylococci from nonpathogenic staphylococci. Hemolytic and coagulating strains of *Staphylococcus aureus* appear as purple or yellow colonies. Nonhemolytic and noncoagulating strains of *Staphylococcus* species appear as white colonies.

## Crystal Violet Lactose Agar

**Composition** per liter:

| | |
|---|---|
| Agar | 15.0g |
| Lactose | 10.0g |
| Proteose peptone | 5.0g |
| Beef extract | 3.0g |
| Crystal Violet | 3.3mg |

pH 6.8 ± 0.2 at 25°C

**Source:** This medium is available as a premixed powder from Hi-Media.

**Preparation of Medium:** Add components to distilled/deionized water and bring volume to 1.0L. Mix thoroughly. Gently heat and bring to boiling. Distribute into tubes or flasks. Autoclave for 15 min at 15 psi pressure–121°C. Pour into sterile Petri dishes or leave in tubes.

**Storage/Shelf Life:** Store dehydrated media in the dark in a sealed container below 30°C. Prepared media should be stored under refrigeration (2-8°C). Media should be used within 60 days of preparation. Media should not be used if there are any signs of deterioration (shrinking, cracking, or discoloration) or contamination, or if the expiration date supplied by the manufacturer has passed.

**Use:** For the differentiation of pure cultures of pathogenic staphylococci.

## Crystal Violet Lactose Broth

**Composition** per liter:

| | |
|---|---|
| Lactose | 5.0g |
| Peptic digest of animal tissue | 5.0g |
| $K_2HPO_4$ | 5.0g |

KH$_2$PO$_4$............................................................................1.0g
Crystal Violet..................................................................1.43mg

pH 7.4 ± 0.2 at 25°C

**Source:** This medium is available as a premixed powder from Hi-Media.

**Preparation of Medium:** Add components to distilled/deionized water and bring volume to 1.0L. Mix thoroughly. Gently heat and bring to boiling. Distribute into tubes or flasks. Autoclave for 15 min at 15 psi pressure–121°C. Pour into sterile Petri dishes or leave in tubes.

**Storage/Shelf Life:** Store dehydrated media in the dark in a sealed container below 30°C. Prepared media should be stored under refrigeration (2-8°C). Media should be used within 60 days of preparation. Media should not be used if there are any signs of deterioration (discoloration) or contamination, or if the expiration date supplied by the manufacturer has passed.

**Use:** For the differentiation of pure cultures of pathogenic staphylococci.

### Crystal Violet Lactose HiVeg Agar

**Composition** per liter:
Agar ......................................................................................15.0g
Lactose...............................................................................10.0g
Plant peptone No. 3.............................................................5.0g
Plant extract .........................................................................3.0g
Crystal Violet ....................................................................3.3mg

pH 6.8 ± 0.2 at 25°C

**Source:** This medium is available as a premixed powder from Hi-Media.

**Preparation of Medium:** Add components to distilled/deionized water and bring volume to 1.0L. Mix thoroughly. Gently heat and bring to boiling. Distribute into tubes or flasks. Autoclave for 15 min at 15 psi pressure–121°C. Pour into sterile Petri dishes or leave in tubes.

**Storage/Shelf Life:** Store dehydrated media in the dark in a sealed container below 30°C. Prepared media should be stored under refrigeration (2-8°C). Media should be used within 60 days of preparation. Media should not be used if there are any signs of deterioration (shrinking, cracking, or discoloration) or contamination, or if the expiration date supplied by the manufacturer has passed.

**Use:** For the differentiation of pure cultures of pathogenic staphylococci.

### CTA Agar
### (Cystine Trypticase™ Agar)

**Composition** per liter:
Pancreatic digest of casein...........................................20.0g
Agar .....................................................................................14.0g
NaCl.......................................................................................5.0g
L-Cystine...............................................................................0.5g
Na$_2$SO$_3$................................................................................0.5g
Phenol Red........................................................................0.017g

pH 7.3 ± 0.2 at 25°C

**Preparation of Medium:** Add components to distilled/deionized water and bring volume to 1.0L. Mix thoroughly. Gently heat until boiling. Distribute into tubes or flasks. Autoclave for 15 min at 15 psi pressure–118°C. Pour into sterile Petri dishes or leave in tubes. Two drops of sterile rabbit serum added per tube prior to solidification enhances the recovery of *Corynebacterium diphtheriae*.

**Storage/Shelf Life:** Store dehydrated media in the dark in a sealed container below 30°C. Prepared media should be stored under refrigeration (2-8°C). Media should be used within 60 days of preparation. Media should not be used if there are any signs of deterioration (shrinking, cracking, or discoloration) or contamination, or if the expiration date supplied by the manufacturer has passed.

**Use:** For the cultivation and maintenance of a variety of fastidious microorganisms, including *Corynebacterium diphtheriae*. For carbohydrate fermentation tests in the differentiation of *Neisseria* species.

### CTA Medium
### (Cystine Trypticase™ Agar Medium)
### (Cystine Tryptic Agar)

**Composition** per liter:
Pancreatic digest of casein...........................................20.0g
NaCl.......................................................................................5.0g
Carbohydrate........................................................................5.0g
Agar .......................................................................................2.5g
L-Cystine...............................................................................0.5g
Na$_2$SO$_3$................................................................................0.5g
Phenol Red........................................................................0.017g

pH 7.3 ± 0.2 at 25°C

**Source:** The medium is available as a premixed powder from BD Diagnostic Systems.

**Preparation of Medium:** Add components to distilled/deionized water and bring volume to 1.0L. Mix thoroughly. Adjust pH to 7.3. Gently heat until boiling. Distribute into tubes or flasks. Autoclave for 15 min at 15 psi pressure–118°C. Cool tubes in an upright position. Store at room temperature.

**Storage/Shelf Life:** Store dehydrated media in the dark in a sealed container below 30°C. Prepared media should be stored under refrigeration (2-8°C). Media should be used within 60 days of preparation. Media should not be used if there are any signs of deterioration (discoloration) or contamination, or if the expiration date supplied by the manufacturer has passed.

**Use:** For the cultivation and maintenance of a variety of fastidious microorganisms. For the detection of bacterial motility. Used, with added specific carbohydrate, for fermentation reactions of fastidious microorganisms, especially *Neisseria* species, pneumococci, streptococci, and nonspore-forming anaerobes.

### CTA Medium with Yeast Extract and Rabbit Serum
### (Cystine Trypticase™ Agar Medium with Yeast Extract and Rabbit Serum)

**Composition** per liter:
Yeast extract......................................................................50.0g
Pancreatic digest of casein...........................................20.0g
NaCl.......................................................................................5.0g
Carbohydrate........................................................................5.0g
Agar .......................................................................................2.5g
L-Cystine...............................................................................0.5g
Na$_2$SO$_3$................................................................................0.5g
Phenol Red........................................................................0.017g
Rabbit serum ................................................................. 250.0mL

pH 7.3 ± 0.2 at 25°C

**Preparation of Medium:** Add components, except rabbit serum, to distilled/deionized water and bring volume to 750.0mL. Mix thoroughly. Adjust pH to 7.3. Gently heat until boiling. Autoclave for 15 min at

15 psi pressure–118°C. Cool to 50°C. Aseptically add sterile rabbit serum. Mix thoroughly. Distribute into sterile tubes. Store at room temperature. Do not refrigerate.

**Storage/Shelf Life:** Store dehydrated media in the dark in a sealed container below 30°C. Prepared media should be stored under refrigeration (2-8°C). Media should be used within 60 days of preparation. Media should not be used if there are any signs of deterioration (discoloration) or contamination, or if the expiration date supplied by the manufacturer has passed.

**Use:** For the cultivation and maintenance of fastidious microorganisms, especially mycoplasmas and related microorganisms.

## Cultivation Medium for Chlamydiae
**Composition** per 101.0mL:

IM medium ....................................................................... 90.0mL
Fetal bovine serum............................................................ 10.0mL
Amino acids, 100x ............................................................... 1.0mL

pH 7.4 ± 0.2 at 25°C

### IM medium:
**Composition** per 100.0mL:

Pancreatic digest of gelatin ................................................0.05g
Bile salts No. 3 ...................................................................0.05g
Brain heart, solids from infusion .......................................0.02g
Peptic digest of animal tissue.............................................0.02g
NaCl .................................................................................0.017g
Glucose ...............................................................................0.01g
Na$_2$HPO$_4$ .............................................................................8.0mg
Earle's balanced salts solution .......................................... 80.0mL
Fetal bovine serum, heat inactivated (2 h at 55°C).................. 20.0mL

pH 7.4 ± 0.2 at 25°C

### Earle's Balanced Salts Solution:
**Composition** per liter:

NaCl .....................................................................................6.8g
NaHCO$_3$ .................................................................................2.2g
Glucose ..................................................................................1.0g
KCl ........................................................................................0.4g
CaCl$_2$·2H$_2$O.........................................................................0.265g
MgSO$_4$·7H$_2$O .......................................................................0.2g
NaH$_2$PO$_4$·H$_2$O ...................................................................0.14g

**Preparation of Earle's Balanced Salts Solution:** Add components to distilled/deionized water and bring volume to 1.0L. Mix thoroughly. Filter sterilize.

**Preparation of IM:** Combine components. Mix thoroughly. Filter sterilize. Store at 4°–10°C.

**Preparation of Medium:** Combine components. Mix thoroughly. Filter sterilize. Store for no longer than 4 weeks at room temperature to facilitate detection of contamination. Prepare a 25 cm$^2$ flask and seed with either cells L929 (ACC 2) or HeLa (ACC 57) cells. Incubate at 37°C plus 5% CO$_2$. When a confluent layer has formed, infection can be carried out. Exchange medium with 6.0mL of IM with the addition of 0.001mg/mL cycloheximide (final concentration) and add 0.5–1.0mL of EB stock solution (thawed quickly to 37°C). Centrifuge for 1 h onto the cell layer at 1600 rpm at 20°C. Incubate at 37°C + 5% CO$_2$. Control cells daily and look for inclusions. Not all chlamydiae form well-visible inclusions; ultimately, immunofluorescence or *in situ* hybridization techniques are necessary to visualize inclusions.

**Storage/Shelf Life:** Store dehydrated media in the dark in a sealed container below 30°C. Prepared media should be stored under refrigeration (2-8°C). Media should be used within 60 days of preparation.

Media should not be used if there are any signs of deterioration (discoloration) or contamination, or if the expiration date supplied by the manufacturer has passed.

**Use:** For the screening for *Chlamydia* using cell line cultures to test for infectivity.

## CVA Medium
## (Cefoperazone Vancomycin Amphotericin Medium)
**Composition** per liter:

Agar ....................................................................................15.0g
Casein peptone....................................................................10.0g
Meat peptone ......................................................................10.0g
NaCl......................................................................................5.0g
Yeast autolysate ....................................................................2.0g
Glucose .................................................................................1.0g
NaHSO$_3$ ................................................................................0.1g
Sheep blood, defibrinated .................................................. 50.0mL
CVA antibiotic solution ..................................................... 10.0mL

pH 7.0 ± 0.2 at 25°C

### CVA Antibiotic Solution:
**Composition** per 10.0mL:

Cefoperazone ....................................................................20.0mg
Vancomycin .......................................................................10.0mg
Amphotericin B ...................................................................2.0mg

**Preparation of CVA Antibiotic Solution:** Add components to distilled/deionized water and bring volume to 10.0mL. Mix thoroughly. Filter sterilize.

**Preparation of Medium:** Add components, except CVA antibiotic solution and sheep blood, to distilled/deionized water and bring volume to 940.0mL. Mix thoroughly. Gently heat until boiling. Autoclave for 15 min at 15 psi pressure–121°C. Cool to 45°–50°C. Aseptically add sterile CVA antibiotic solution and sterile, defibrinated sheep blood. Mix thoroughly. Pour into sterile Petri dishes.

**Storage/Shelf Life:** Store dehydrated media in the dark in a sealed container below 30°C. Prepared media should be stored under refrigeration (2-8°C). Media should be used within 60 days of preparation. Media should not be used if there are any signs of deterioration (shrinking, cracking, or discoloration) or contamination, or if the expiration date supplied by the manufacturer has passed.

**Use:** For the isolation and cultivation of *Campylobacter* species from clinical specimens.

## Cycloserine Cefoxitin EggYolk Fructose Agar
**Composition** per liter:

Proteose peptone No. 2 .......................................................40.0g
Agar ....................................................................................25.0g
Fructose.................................................................................6.0g
Na$_2$HPO$_4$ ..............................................................................5.0g
NaCl.......................................................................................2.0g
KH$_2$PO$_4$ ...............................................................................1.0g
MgSO$_4$·7H$_2$O .......................................................................0.1g
Egg yolk emulsion ............................................................ 100.0mL
Antibiotic solution .............................................................. 10.0mL
Neutral Red solution ............................................................ 3.0mL
Hemin solution...................................................................... 1.0mL

### Egg Yolk Emulsion:
**Composition**:

Chicken egg yolks.................................................................... 11
Whole chicken egg .................................................................... 1

**Preparation of Egg Yolk Emulsion:** Soak eggs with 1:100 dilution of saturated mercuric chloride solution for 1 min. Crack eggs. Separate yolks from whites for 11 eggs. Mix egg yolks with 1 chicken egg.

**Antibiotic Solution:**
**Composition** per 10.0mL:

Cycloserine .................................................................0.5g
Cefoxitin ..................................................................0.016g

**Preparation of Antibiotic Solution:** Add components to distilled/deionized water and bring volume to 10.0mL. Mix thoroughly. Filter sterilize.

**Neutral Red Solution:**
**Composition** per 10.0mL:

Neutral Red ................................................................0.1g
Ethanol ..................................................................10.0mL

**Preparation of Neutral Red Solution:** Add Neutral Red to 10.0mL of ethanol. Mix thoroughly.

**Hemin Solution:**
**Composition** per 100.0mL:

Hemin ......................................................................0.5g
NaOH ($1N$ solution) ....................................................10.0mL

**Preparation of Hemin Solution:** Add hemin to 10.0mL of $1N$ NaOH solution. Mix thoroughly. Bring volume to 100.0mL with distilled/deionized water.

**Preparation of Medium:** Add components, except egg yolk emulsion and antibiotic solution, to distilled/deionized water and bring volume to 890.0mL. Mix thoroughly. Gently heat and bring to boiling. Autoclave for 15 min at 15 psi pressure–121°C. Cool to 45°–50°C. Aseptically add sterile egg yolk emulsion and antibiotic solution. Mix thoroughly. Pour into sterile Petri dishes.

**Storage/Shelf Life:** Store dehydrated media in the dark in a sealed container below 30°C. Prepared media should be stored under refrigeration (2-8°C). Media should be used within 60 days of preparation. Media should not be used if there are any signs of deterioration (shrinking, cracking, or discoloration) or contamination, or if the expiration date supplied by the manufacturer has passed.

**Use:** For the selective isolation and cultivation of *Clostridium difficile* from feces.

### Cystine Heart Agar with Rabbit Blood
**Composition** per liter:

Beef heart, solids from infusion...................................500.0g
Agar .......................................................................15.0g
Glucose ...................................................................10.0g
Proteose peptone .......................................................10.0g
NaCl .........................................................................5.0g
L-Cystine ...................................................................1.0g
Rabbit blood, defibrinated ..........................................50.0mL

pH 6.8 ± 0.2 at 25°C

**Source:** This medium is available as a premixed powder from BD Diagnostic Systems.

**Preparation of Medium:** Add components, except rabbit blood, to distilled/deionized water and bring volume to 950.0mL. Mix thoroughly. Gently heat until boiling. Autoclave for 15 min at 15 psi pressure–121°C. Cool to 50°–60°C. Aseptically add 50.0mL of sterile, defibrinated rabbit blood. Mix thoroughly. Pour into sterile Petri dishes or distribute into sterile tubes.

**Storage/Shelf Life:** Store dehydrated media in the dark in a sealed container below 30°C. Prepared media should be stored under refrigeration (2-8°C). Media should be used within 60 days of preparation. Media should not be used if there are any signs of deterioration (shrinking, cracking, or discoloration) or contamination, or if the expiration date supplied by the manufacturer has passed.

**Use:** For the cultivation and maintenance of *Francisella tularensis* and *Francisella philomiragia*. Without the hemoglobin enrichment, it supports excellent growth of Gram-negative cocci and other pathogenic microorganisms.

### Cystine HiVeg Agar Base with Hemoglobin
**Composition** per liter:

Agar .......................................................................15.0g
Plant infusion ...........................................................10.0g
Plant peptone No. 3.....................................................10.0g
Glucose ...................................................................10.0g
NaCl .........................................................................5.0g
L-Cystine ...................................................................1.0g
Hemoglobin solution ..............................................100.0mL

pH 6.8 ± 0.2 at 25°C

**Source:** This medium, without hemoglobin, is available as a premixed powder from HiMedia.

**Hemoglobin Solution:**
**Composition** per 100.0mL:

Bovine hemoglobin........................................................2.0g

**Preparation of Hemoglobin Solution:** Add bovine hemoglobin to distilled/deionized water and bring volume to 100.0mL. Mix thoroughly. Autoclave for 15 min at 15 psi pressure–121°C. Cool to 45°–50°C.

**Preparation of Medium:** Add components, except hemoglobin solution, to distilled/deionized water and bring volume to 900.0mL. Mix thoroughly. Gently heat until boiling. Autoclave for 15 min at 15 psi pressure–121°C. Cool to 45°–50°C. Add 100.0mL sterile hemoglobin solution. Mix thoroughly. Pour into sterile Petri dishes or distribute into sterile tubes.

**Storage/Shelf Life:** Store dehydrated media in the dark in a sealed container below 30°C. Prepared media should be stored under refrigeration (2-8°C). Media should be used within 60 days of preparation. Media should not be used if there are any signs of deterioration (shrinking, cracking, or discoloration) or contamination, or if the expiration date supplied by the manufacturer has passed.

**Use:** For the cultivation of Gram-negative cocci and other fastidious pathogens. For the cultivation of *Francicella tularensis*.

### Cystine Tellurite Blood Agar
**Composition** per liter:

Heart infusion agar .................................................900.0mL
$K_2TeO_3$ solution.....................................................75.0mL
Rabbit blood.............................................................25.0mL
L-Cystine ................................................................22.0mg

pH 7.4 ± 0.2 at 25°C

**Heart Infusion Agar:**
**Composition** per 900.0mL:

Beef heart, solids from infusion...................................500.0g
Agar .......................................................................20.0g
Tryptose ..................................................................10.0g
Yeast extract ...............................................................5.0g
NaCl .........................................................................5.0g

**Preparation of Heart Infusion Agar:** Add components to distilled/deionized water and bring volume to 900.0mL. Mix thoroughly. Autoclave for 15 min at 15 psi pressure–121°C. Cool to 45°–50°C.

**K₂TeO₃ Solution:**

**Composition** per 100.0mL:

$K_2TeO_3$ ....................................................................................0.3g

**Preparation of K₂TeO₃ Solution:** Add $K_2TeO_3$ to distilled/deionized water and bring volume to 100.0mL. Mix thoroughly. Autoclave for 15 min at 15 psi pressure–121°C.

**Caution:** Potassium tellurite is toxic.

**Preparation of Medium:** Add sterile $K_2TeO_3$ solution, sterile rabbit blood, and sterile, solid L-cystine to sterile, cooled heart infusion agar. Mix thoroughly. Pour into sterile Petri dishes or distribute into sterile tubes.

**Storage/Shelf Life:** Store dehydrated media in the dark in a sealed container below 30°C. Prepared media should be stored under refrigeration (2-8°C). Media should be used within 60 days of preparation. Media should not be used if there are any signs of deterioration (shrinking, cracking, or discoloration) or contamination, or if the expiration date supplied by the manufacturer has passed.

**Use:** For the isolation, differentiation, and cultivation of *Corynebacterium diphtheriae*. *Corynebacterium diphtheriae* appears as dark gray to black colonies.

## Cystine Tellurite Blood Agar

**Composition** per 120.0mL:

Heart infusion agar.................................................... 100.0mL
K₂TeO₃ solution ........................................................ 15.0mL
Sheep blood.............................................................. 5.0mL
L-Cystine ................................................................5.0mg

pH 7.4 ± 0.2 at 25°C

**Heart Infusion Agar:**

**Composition** per liter:

Beef heart, infusion from ............................................500.0g
Agar ......................................................................20.0g
Tryptose ..................................................................10.0g
Yeast extract............................................................5.0g
NaCl ......................................................................5.0g

**Preparation of Heart Infusion Agar:** Add components to distilled/deionized water and bring volume to 1.0L. Mix thoroughly. Autoclave for 15 min at 15 psi pressure–121°C. Cool to 45°–50°C.

**K₂TeO₃ Solution:**

**Composition** per 100.0mL:

$K_2TeO_3$ ....................................................................................0.3g

**Preparation of K₂TeO₃ Solution:** Add $K_2TeO_3$ to distilled/deionized water and bring volume to 100.0mL. Mix thoroughly. Autoclave for 15 min at 15 psi pressure–121°C.

**Caution:** Potassium tellurite is toxic.

**Preparation of Medium:** Add sterile $K_2TeO_3$ solution, sterile, defibrinated sheep blood, and sterile, solid L-cystine to sterile, cooled heart infusion agar. Mix thoroughly. Pour into sterile Petri dishes or distribute into sterile tubes.

**Storage/Shelf Life:** Store dehydrated media in the dark in a sealed container below 30°C. Prepared media should be stored under refrigeration (2-8°C). Media should be used within 60 days of preparation. Media should not be used if there are any signs of deterioration (shrink-

ing, cracking, or discoloration) or contamination, or if the expiration date supplied by the manufacturer has passed.

**Use:** For the isolation, differentiation, and cultivation of *Corynebacterium diphtheriae*. *Corynebacterium diphtheriae* appears as dark gray to black colonies.

## DCLS Agar
### (Deoxycholate Citrate Lactose Sucrose Agar)

**Composition** per liter:

Agar .......................................................................12.0g
Sodium citrate·3H₂O...................................................10.5g
Lactose....................................................................5.0g
Na₂S₂O₃...................................................................5.0g
Sucrose....................................................................5.0g
Pancreatic digest of casein...........................................3.5g
Peptic digest of animal tissue ........................................3.5g
Beef extract..............................................................3.0g
Sodium deoxycholate...................................................2.5g
Neutral Red...............................................................0.03g

pH 7.2 ± 0.1 at 25°C

**Source:** This medium is available as a premixed powder from BD Diagnostic Systems and Thermo Scientific.

**Preparation of Medium:** Add components to distilled/deionized water and bring volume to 1.0L. Mix thoroughly. Gently heat while stirring and bring to boiling. Do not overheat. Do not autoclave. Pour into sterile Petri dishes in 20.0mL volumes.

**Storage/Shelf Life:** Store dehydrated media in the dark in a sealed container below 30°C. Prepared media should be stored under refrigeration (2-8°C). Media should be used within 60 days of preparation. Media should not be used if there are any signs of deterioration (shrinking, cracking, or discoloration) or contamination, or if the expiration date supplied by the manufacturer has passed.

**Use:** For the selective isolation of *Salmonella* species, *Shigella* species, and *Vibrio* species from fecal specimens.

## DCLS Agar

**Composition** per liter:

Agar .......................................................................12.0g
Sodium citrate...........................................................10.0g
Proteose peptone........................................................7.0g
Lactose....................................................................5.0g
Na₂S₂O₃...................................................................5.0g
Sucrose....................................................................5.0g
Beef extract..............................................................3.0g
Sodium deoxycholate...................................................2.5g
Neutral Red...............................................................0.03

pH 7.2 ± 0.1 at 25°C

**Source:** This medium is available as a premixed powder from Hi-Media.

**Preparation of Medium:** Add components to distilled/deionized water and bring volume to 1.0L. Mix thoroughly. Gently heat while stirring and bring to boiling. Do not overheat. Do not autoclave. Pour into sterile Petri dishes or distribute into sterile tubes.

**Storage/Shelf Life:** Store dehydrated media in the dark in a sealed container below 30°C. Prepared media should be stored under refrigeration (2-8°C). Media should be used within 60 days of preparation. Media should not be used if there are any signs of deterioration (shrink-

ing, cracking, or discoloration) or contamination, or if the expiration date supplied by the manufacturer has passed.

**Use:** For the selective isolation of *Salmonella* species, *Shigella* species, and *Vibrio* species from fecal specimens.

## DCLS Agar, Hajna

**Composition** per liter:

| | |
|---|---|
| Agar | 20.0g |
| Sodium citrate | 10.0g |
| Lactose | 7.5g |
| Sucrose | 7.5g |
| Peptic digest of animal tissue | 5.0g |
| Casein enzymatic hydrolysate | 5.0g |
| NaCl | 5.0g |
| $Na_2S_2O_3$ | 5.0g |
| Plant extract | 3.0g |
| Beef extract | 3.0g |
| Sodium deoxycholate | 2.5g |
| Bromcresol Purple | 0.02g |

pH 7.2 ± 0.1 at 25°C

**Source:** This medium is available as a premixed powder from Hi-Media.

**Preparation of Medium:** Add components to distilled/deionized water and bring volume to 1.0L. Mix thoroughly. Gently heat while stirring and bring to boiling. Do not overheat. Do not autoclave. Pour into sterile Petri dishes or distribute into sterile tubes.

**Storage/Shelf Life:** Store dehydrated media in the dark in a sealed container below 30°C. Prepared media should be stored under refrigeration (2-8°C). Media should be used within 60 days of preparation. Media should not be used if there are any signs of deterioration (shrinking, cracking, or discoloration) or contamination, or if the expiration date supplied by the manufacturer has passed.

**Use:** For the selective isolation of *Salmonella* species, *Shigella* species, and *Vibrio* species from fecal specimens.

## DCLS HiVeg Agar

**Composition** per liter:

| | |
|---|---|
| Agar | 12.0g |
| Sodium citrate | 10.0g |
| Plant peptone No. 3 | 8.0g |
| Lactose | 5.0g |
| $Na_2S_2O_3$ | 5.0g |
| Sucrose | 5.0g |
| Plant extract | 3.0g |
| Synthetic detergent No. III | 1.5g |
| Neutral Red | 0.03 |

pH 7.2 ± 0.1 at 25°C

**Source:** This medium is available as a premixed powder from Hi-Media.

**Preparation of Medium:** Add components to distilled/deionized water and bring volume to 1.0L. Mix thoroughly. Gently heat while stirring and bring to boiling. Do not overheat. Do not autoclave. Pour into sterile Petri dishes or distribute into sterile tubes.

**Storage/Shelf Life:** Store dehydrated media in the dark in a sealed container below 30°C. Prepared media should be stored under refrigeration (2-8°C). Media should be used within 60 days of preparation. Media should not be used if there are any signs of deterioration (shrink-

ing, cracking, or discoloration) or contamination, or if the expiration date supplied by the manufacturer has passed.

**Use:** For the selective isolation of *Salmonella* species, *Shigella* species, and *Vibrio* species from fecal specimens.

## DCLS HiVeg Agar, Hajna

**Composition** per liter:

| | |
|---|---|
| Agar | 20.0g |
| Sodium citrate | 10.0g |
| Lactose | 7.5g |
| Sucrose | 7.5g |
| Plant peptone | 6.0g |
| Plant hydrolysate | 5.0g |
| NaCl | 5.0g |
| $Na_2S_2O_3$ | 5.0g |
| Plant extract | 3.0g |
| Yeast extract | 3.0g |
| Synthetic detergent No. III | 1.5g |
| Bromresol Purple | 0.02g |

pH 7.2 ± 0.1 at 25°C

**Source:** This medium is available as a premixed powder from Hi-Media.

**Preparation of Medium:** Add components to distilled/deionized water and bring volume to 1.0L. Mix thoroughly. Gently heat while stirring and bring to boiling. Do not overheat. Do not autoclave. Pour into sterile Petri dishes or distribute into sterile tubes.

**Storage/Shelf Life:** Store dehydrated media in the dark in a sealed container below 30°C. Prepared media should be stored under refrigeration (2-8°C). Media should be used within 60 days of preparation. Media should not be used if there are any signs of deterioration (shrinking, cracking, or discoloration) or contamination, or if the expiration date supplied by the manufacturer has passed.

**Use:** For the selective isolation of *Salmonella* species, *Shigella* species, and *Vibrio* species from fecal specimens.

## Decarboxylase Basal Medium (BAM M44)

**Composition** per liter:

| | |
|---|---|
| Peptone or gelysate | 5.0g |
| Yeast extract | 3.0g |
| Glucose | 1.0g |
| Bromcresol Purple | 0.02g |

pH 6.5 ± 0.2 at 25°C

**Preparation of Medium:** Add components to distilled/deionized water and bring volume to 1.0L. Mix thoroughly. Adjust pH so that it will be 6.5 ± 0.2 after sterilization. Distribute into 16 × 150mm screw-capped tubes in 5.0mL volumes. Autoclave medium with loosely capped tubes for 10 min at 15 psi pressure–121°C. Screw the caps on tightly for storage and after inoculation.

**Storage/Shelf Life:** Store dehydrated media in the dark in a sealed container below 30°C. Prepared media should be stored under refrigeration (2-8°C). Media should be used within 60 days of preparation. Media should not be used if there are any signs of deterioration (discoloration) or contamination, or if the expiration date supplied by the manufacturer has passed.

**Use:** For the differentiation of bacteria based on their ability to decarboxylate the amino acid. As the basal medium for arginine broth, lysine broth, and ornithine broth. Bacteria that decarboxylate arginine, lysine, or ornithine turn the medium turbid purple. The unsupplemented decarboylase basal medium is used as a control.

## Decarboxylase Basal Medium with Sodium Chloride (BAM M44)

**Composition** per liter:
| | |
|---|---|
| Peptone or gelysate | 5.0g |
| Yeast extract | 3.0g |
| Glucose | 1.0g |
| Bromcresol Purple | 0.02g |

pH 6.5 ± 0.2 at 25°C

**Preparation of Medium:** Add components to distilled/deionized water and bring volume to 1.0L. Mix thoroughly. Adjust pH so that it will be 6.5 ± 0.2 after sterilization. Distribute into 16 × 150mm screw-capped tubes in 5.0mL volumes. Autoclave medium with loosely capped tubes for 10 min at 15 psi pressure–121°C. Screw the caps on tightly for storage and after inoculation.

**Storage/Shelf Life:** Store dehydrated media in the dark in a sealed container below 30°C. Prepared media should be stored under refrigeration (2-8°C). Media should be used within 60 days of preparation. Media should not be used if there are any signs of deterioration (discoloration) or contamination, or if the expiration date supplied by the manufacturer has passed.

**Use:** For the differentiation of *Vibrio* spp. based on their ability to decarboxylate the amino acid. Used as the basal medium for arginine broth, lysine broth, and ornithine broth. Bacteria that decarboxylate arginine, lysine, or ornithine turn the medium turbid purple. The unsupplemented decarboylase basal medium is used as a control.

## Decarboxylase Base, Møller

**Composition** per liter:
| | |
|---|---|
| Amino acid | 10.0g |
| Beef extract | 5.0g |
| Peptone | 5.0g |
| Glucose | 0.5g |
| Bromcresol Purple | 0.01g |
| Cresol Red | 5.0mg |
| Pyridoxal | 5.0mg |
| Mineral oil | 200.0mL |

pH 6.0 ± 0.2 at 25°C

**Source:** This medium is available as a premixed powder from BD Diagnostic Systems.

**Preparation of Medium:** Add components, except mineral oil, to distilled/deionized water and bring volume to 1.0L. For amino acid, use L-arginine, L-lysine, or L-ornithine. Mix thoroughly. Distribute into screw-capped tubes in 5.0mL volumes. Autoclave medium and mineral oil separately for 15 min at 15 psi pressure–121°C. After inoculation, overlay medium with 1.0mL of sterile mineral oil per tube.

**Storage/Shelf Life:** Store dehydrated media in the dark in a sealed container below 30°C. Prepared media should be stored under refrigeration (2-8°C). Media should be used within 60 days of preparation. Media should not be used if there are any signs of deterioration (discoloration) or contamination, or if the expiration date supplied by the manufacturer has passed.

**Use:** For the differentiation of bacteria based on their ability to decarboxylate the amino acid. Bacteria that decarboxylate arginine, lysine, or ornithine turn the medium turbid purple.

## Decarboxylase HiVeg Agar Base

**Composition** per liter:
| | |
|---|---|
| Agar | 15.0g |
| Plant peptone | 5.0g |
| Yeast extract | 3.0g |
| Glucose | 1.0g |
| Bromcresol Purple | 0.02 |
| Amino acid solution | 100.0mL |

pH 6.5 ± 0.2 at 25°C

**Source:** This medium without amino acid is available as a premixed powder from HiMedia.

**Amino Acid Solution:**
**Composition** per 100.0mL:
| | |
|---|---|
| L-arginine, L-lysine, or L-ornithine | 10.0g |

**Preparation of Amino Acid Solution:** Add amino acid to distilled/deionized water and bring volume to 100.0mL. Mix thoroughly.

**Preparation of Medium:** Add components and bring volume to 900.0L. For amino acid, use L-arginine, L-lysine, or L-ornithine and add 100.0ml of a 10% solution. Mix thoroughly. Distribute into screw-capped tubes in 5.0mL volumes. Autoclave for 15 min at 15 psi pressure–121°C.

**Storage/Shelf Life:** Store dehydrated media in the dark in a sealed container below 30°C. Prepared media should be stored under refrigeration (2-8°C). Media should be used within 60 days of preparation. Media should not be used if there are any signs of deterioration (shrinking, cracking, or discoloration) or contamination, or if the expiration date supplied by the manufacturer has passed.

**Use:** As a basal medium for the cultivation and differentiation of bacteria based on their ability to decarboxylate amino acids. The medium is supplemented with specific L-amino acids for testing decarboxylase activity on that amino acid. Amino acids are added to a final concentration of 0.5 percent.

## Decarboxylase HiVeg Broth Base, Moeller

**Composition** per liter:
| | |
|---|---|
| Plant extract | 5.0g |
| Plant peptone | 5.0g |
| Glucose | 0.5g |
| Bromcresol Purple | 0.01g |
| Cresol Red | 5.0mg |
| Pyridoxal | 5.0mg |
| Amino acid solution | 100.0mL |

pH 6.0 ± 0.2 at 25°C

**Source:** This medium without amino acid is available as a premixed powder from HiMedia.

**Amino Acid Solution:**
**Composition** per 100.0mL:
| | |
|---|---|
| L-arginine, L-lysine, or L-ornithine | 10.0g |

**Preparation of Amino Acid Solution:** Add amino acid to distilled/deionized water and bring volume to 100.0mL. Mix thoroughly.

**Preparation of Medium:** Add components and bring volume to 900.0L. For amino acid, use L-arginine, L-lysine, or L-ornithine and add 100.0ml of a 10% solution. Mix thoroughly. Distribute into screw-capped tubes in 5.0mL volumes. Autoclave for 15 min at 15 psi pressure–121°C.

**Storage/Shelf Life:** Store dehydrated media in the dark in a sealed container below 30°C. Prepared media should be stored under refriger-

ation (2-8°C). Media should be used within 60 days of preparation. Media should not be used if there are any signs of deterioration (discoloration) or contamination, or if the expiration date supplied by the manufacturer has passed.

**Use:** As a basal medium for the differentiation of bacteria based on their ability to decarboxylate amino acids. The medium is supplemented with specific L-amino acids for testing decarboxylase activity on that amino acid. Bacteria that decarboxylate arginine, lysine, or ornithine turn the medium turbid purple.

## Decarboxylase Medium Base, Falkow

**Composition** per liter:
| | |
|---|---|
| Amino acid (arginine, lysine, or ornithine) | 5.0g |
| Peptone | 5.0g |
| Yeast extract | 3.0g |
| Glucose | 1.0g |
| Bromcresol Purple | 0.02g |
| Mineral oil | 200.0mL |

pH 6.8 ± 0.2 at 25°C

**Source:** This medium is available as a premixed powder from BD Diagnostic Systems.

**Preparation of Medium:** Add components, except mineral oil, to distilled/deionized water and bring volume to 1.0L. For amino acid, use L-arginine, L-lysine, or L-ornithine. Mix thoroughly. Distribute into screw-capped tubes in 5.0mL volumes. Autoclave medium and mineral oil separately for 15 min at 15 psi pressure–121°C. After inoculation, overlay medium with 1.0mL of sterile mineral oil per tube.

**Storage/Shelf Life:** Store dehydrated media in the dark in a sealed container below 30°C. Prepared media should be stored under refrigeration (2-8°C). Media should be used within 60 days of preparation. Media should not be used if there are any signs of deterioration (discoloration) or contamination, or if the expiration date supplied by the manufacturer has passed.

**Use:** For the differentiation of bacteria based on their ability to decarboxylate a specific amino acid. Bacteria that decarboxylate arginine, lysine, or ornithine turn the medium turbid purple.

## Decarboxylase Medium, Ornithine Modified

**Composition** per liter:
| | |
|---|---|
| L-Ornithine | 10.0g |
| Meat peptone | 5.0g |
| Yeast extract | 3.0g |
| Bromcresol Purple solution | 5.0mL |

pH 5.5 ± 0.2 at 25°C

**Bromcresol Purple Solution:**
**Composition** per 100.0mL:
| | |
|---|---|
| Bromcresol Purple | 0.2g |
| Ethanol | 50.0mL |

**Preparation of Bromcresol Purple Solution:** Add Bromcresol Purple to ethanol. Mix thoroughly. Bring volume to 100.0mL with distilled/deionized water. Mix thoroughly. Filter sterilize.

**Preparation of Medium:** Add components to distilled/deionized water and bring volume to 1.0L. Mix thoroughly. Gently heat until dissolved. Adjust pH to 5.5 with HCl or NaOH. Distribute into screw-capped tubes. Autoclave for 15 min at 15 psi pressure–121°C.

**Storage/Shelf Life:** Store dehydrated media in the dark in a sealed container below 30°C. Prepared media should be stored under refriger-

ation (2-8°C). Media should be used within 60 days of preparation. Media should not be used if there are any signs of deterioration (discoloration) or contamination, or if the expiration date supplied by the manufacturer has passed.

**Use:** For the differentiation of bacteria based on their ability to decarboxylate ornithine. Bacteria that decarboxylate ornithine turn the medium turbid purple.

## Decarboxylase Test HiVeg Medium Base (Falkow)

**Composition** per liter:
| | |
|---|---|
| Plant peptone | 5.0g |
| Yeast extract | 3.0g |
| Glucose | 1.0g |
| Bromcresol Purple | 0.02 |
| Amino acid solution | 100.0mL |

pH 6.8 ± 0.2 at 25°C

**Source:** This medium without amino acid is available as a premixed powder from HiMedia.

**Amino Acid Solution:**
**Composition** per 100.0mL:
| | |
|---|---|
| L-arginine, L-lysine, or L-ornithine | 10.0g |

**Preparation of Amino Acid Solution:** Add amino acid to distilled/deionized water and bring volume to 100.0mL. Mix thoroughly.

**Preparation of Medium:** Add components and bring volume to 900.0L. For amino acid, use L-arginine, L-lysine, or L-ornithine and add 100.0ml of a 10% solution. Mix thoroughly. Distribute into screw-capped tubes in 5.0mL volumes. Autoclave for 15 min at 15 psi pressure–121°C.

**Storage/Shelf Life:** Store dehydrated media in the dark in a sealed container below 30°C. Prepared media should be stored under refrigeration (2-8°C). Media should be used within 60 days of preparation. Media should not be used if there are any signs of deterioration (discoloration) or contamination, or if the expiration date supplied by the manufacturer has passed.

**Use:** As a basal medium for the differentiation of bacteria based on their ability to decarboxylate amino acids. The medium is supplemented with specific L-amino acids for testing decarboxylase activity on that amino acid. Amino acids are added to a final concentration of 0.5 percent. Bacteria that decarboxylate arginine, lysine, or ornithine turn the medium turbid purple.

## Demi-Fraser Broth

**Composition** per liter:
| | |
|---|---|
| NaCl | 20.0g |
| Tryptose | 10.0g |
| $Na_2HPO_4$ | 9.6g |
| Beef extract | 5.0g |
| Yeast extract | 5.0g |
| LiCl | 3.0g |
| $KH_2PO_4$ | 1.35g |
| Esculin | 1.0g |
| Acriflavin·HCl | 12.5mg |
| Nalidixic acid | 10.0mg |
| Ferric ammonium citrate supplement | 10.0mL |

pH 7.2 ± 0.2 at 25°C

**Source:** This medium is available as a premixed powder and supplement from BD Diagnostic Systems.

**Caution:** Lithium chloride is harmful. Avoid bodily contact and inhalation of vapors. On contact with skin wash with plenty of water immediately.

**Ferric Ammonium Citrate Supplement:**
**Composition** per 10.0mL:

Ferric ammonium citrate.................................................0.5g

**Preparation of Ferric Ammonium Citrate Supplement:** Add ferric ammonium citrate to distilled/deionized water and bring volume to 10.0mL. Mix thoroughly. Filter sterilize.

**Preparation of Medium:** Add components, except ferric ammonium citrate supplement, to distilled/deionized water and bring volume to 990.0mL. Mix thoroughly. Autoclave for 15 min at 15 psi pressure–121°C. Aseptically add 10.0mL of sterile ferric ammonium citrate supplement. Mix thoroughly. Aseptically distribute into sterile tubes or flasks.

**Storage/Shelf Life:** Store dehydrated media in the dark in a sealed container below 30°C. Prepared media should be stored under refrigeration (2-8°C). Media should be used within 60 days of preparation. Media should not be used if there are any signs of deterioration (discoloration) or contamination, or if the expiration date supplied by the manufacturer has passed.

**Use:** For the cultivation of *Listeria* species from food and environmental samples.

## Deoxycholate Agar

**Composition** per liter:

Agar .............................................................................16.0g
Lactose .........................................................................10.0g
NaCl ...............................................................................5.0g
Pancreatic digest of casein............................................5.0g
Peptic digest of animal tissue........................................5.0g
K$_2$HPO$_4$...........................................................................2.0g
Ferric citrate..................................................................1.0g
Sodium citrate................................................................1.0g
Sodium deoxycholate.....................................................1.0g
Neutral Red.................................................................0.033g

pH 7.3 ± 0.2 at 25°C

**Source:** This medium is available as a premixed powder from BD Diagnostic Systems.

**Preparation of Medium:** Add components to distilled/deionized water and bring volume to 1.0L. Mix thoroughly. Gently heat and bring to boiling. Do not autoclave. Cool to 45°–50°C. Pour into sterile Petri dishes.

**Storage/Shelf Life:** Store dehydrated media in the dark in a sealed container below 30°C. Prepared media should be stored under refrigeration (2-8°C). Media should be used within 60 days of preparation. Media should not be used if there are any signs of deterioration (shrinking, cracking, or discoloration) or contamination, or if the expiration date supplied by the manufacturer has passed.

**Use:** For the selective isolation, cultivation, enumeration, and differentiation of Gram-negative enteric microorganisms from a variety of clinical speciments and nonclinical specimens of public health importance. *Escherichia coli* appears as large, flat, rose-red colonies. *Enterobacter* and *Klebsiella* species appear as large, mucoid, pale colonies with a pink center. *Proteus* and *Salmonella* species appear as large, colorless to tan colonies. *Shigella* species appear as colorless to pink colonies. *Pseudomonas* species appear as irregular colorless to brown colonies.

## Deoxycholate Agar
### (Desoxycholate Agar)

**Composition** per liter:

Agar .............................................................................15.0g
Lactose .........................................................................10.0g
Peptone .........................................................................10.0g
NaCl ...............................................................................5.0g
K$_2$HPO$_4$...........................................................................2.0g
Ferric citrate..................................................................1.0g
Sodium citrate................................................................1.0g
Sodium deoxycholate.....................................................1.0g
Neutral Red.................................................................0.03g

pH 7.3 ± 0.2 at 25°C

**Source:** This medium is available as a premixed powder from Thermo Scientific and BD Diagnostic Systems.

**Preparation of Medium:** Add components to distilled/deionized water and bring volume to 1.0L. Mix thoroughly. Gently heat and bring to boiling. Do not autoclave. Cool to 50°C. Pour into sterile Petri dishes.

**Storage/Shelf Life:** Store dehydrated media in the dark in a sealed container below 30°C. Prepared media should be stored under refrigeration (2-8°C). Media should be used within 60 days of preparation. Media should not be used if there are any signs of deterioration (shrinking, cracking, or discoloration) or contamination, or if the expiration date supplied by the manufacturer has passed.

**Use:** For the selective isolation, cultivation, enumeration, and differentiation of Gram-negative enteric microorganisms from a variety of clinical speciments and nonclinical specimens of public health importance. *Escherichia coli* appears as large, flat, rose-red colonies. *Enterobacter* and *Klebsiella* species appear as large, mucoid, pale colonies with a pink center. *Proteus* and *Salmonella* species appear as large, colorless to tan colonies. *Shigella* species appear as colorless to pink colonies. *Pseudomonas* species appear as irregular colorless to brown colonies.

## Deoxycholate Agar

**Composition** per liter:

Agar .............................................................................15.0g
Peptic digest of animal tissue .......................................10.0g
Lactose .........................................................................10.0g
NaCl ...............................................................................5.0g
K$_2$HPO$_4$...........................................................................2.0g
Ferric citrate..................................................................1.0g
Sodium citrate................................................................1.0g
Sodium deoxycholate.....................................................1.0g
Neutral Red.................................................................0.03g

pH 7.5 ± 0.2 at 25°C

**Source:** This medium is available as a premixed powder from Hi-Media.

**Preparation of Medium:** Add components to distilled/deionized water and bring volume to 1.0L. Mix thoroughly. Gently heat and bring to boiling. Do not autoclave. Cool to 45°–50°C. Pour into sterile Petri dishes.

**Storage/Shelf Life:** Store dehydrated media in the dark in a sealed container below 30°C. Prepared media should be stored under refrigeration (2-8°C). Media should be used within 60 days of preparation. Media should not be used if there are any signs of deterioration (shrinking, cracking, or discoloration) or contamination, or if the expiration date supplied by the manufacturer has passed.

**Use:** For the selective isolation, cultivation, enumeration, and differentiation of Gram-negative enteric microorganisms from a variety of clinical speciments and nonclinical specimens of public health importance. *Escherichia coli* appears as large, flat, rose-red colonies. *Enterobacter* and *Klebsiella* species appear as large, mucoid, pale colonies with a pink center. *Proteus* and *Salmonella* species appear as large, colorless to tan colonies. *Shigella* species appear as colorless to pink colonies. *Pseudomonas* species appear as irregular colorless to brown colonies.

## Deoxycholate Agar, HiVeg

**Composition** per liter:

| | |
|---|---|
| Agar | 15.0g |
| Plant peptone | 10.0g |
| Lactose | 10.0g |
| NaCl | 5.0g |
| $K_2HPO_4$ | 2.0g |
| Ferric citrate | 1.0g |
| Sodium citrate | 1.0g |
| Synthetic detergent No. III | 1.0g |
| Neutral Red | 0.03g |

pH 7.5 ± 0.2 at 25°C

**Source:** This medium is available as a premixed powder from Hi-Media.

**Preparation of Medium:** Add components to distilled/deionized water and bring volume to 1.0L. Mix thoroughly. Gently heat and bring to boiling. Do not autoclave. Cool to 45°–50°C. Pour into sterile Petri dishes.

**Storage/Shelf Life:** Store dehydrated media in the dark in a sealed container below 30°C. Prepared media should be stored under refrigeration (2-8°C). Media should be used within 60 days of preparation. Media should not be used if there are any signs of deterioration (shrinking, cracking, or discoloration) or contamination, or if the expiration date supplied by the manufacturer has passed.

**Use:** For the selective isolation, cultivation, enumeration, and differentiation of Gram-negative enteric microorganisms from a variety of clinical speciments and nonclinical specimens of public health importance. *Escherichia coli* appears as large, flat, rose-red colonies. *Enterobacter* and *Klebsiella* species appear as large, mucoid, pale colonies with a pink center. *Proteus* and *Salmonella* species appear as large, colorless to tan colonies. *Shigella* species appear as colorless to pink colonies. *Pseudomonas* species appear as irregular colorless to brown colonies.

## Deoxycholate Citrate Agar

**Composition** per liter:

| | |
|---|---|
| Sodium citrate | 50.0g |
| Agar | 15.0g |
| Lactose | 10.0g |
| Beef extract | 5.0g |
| Peptone | 5.0g |
| $Na_2S_2O_3 \cdot 5H_2O$ | 5.0g |
| Sodium deoxycholate | 2.5g |
| Ferric citrate | 1.0g |
| Neutral Red | 0.025g |

pH 7.3 ± 0.2 at 25°C

**Source:** This medium is available as a premixed powder from Thermo Scientific.

**Preparation of Medium:** Add components to distilled/deionized water and bring volume to 1.0L. Mix thoroughly. Gently heat and bring

to boiling. Do not autoclave. Cool to 45°–50°C. Pour into sterile Petri dishes. Dry the agar surface before use.

**Storage/Shelf Life:** Store dehydrated media in the dark in a sealed container below 30°C. Prepared media should be stored under refrigeration (2-8°C). Media should be used within 60 days of preparation. Media should not be used if there are any signs of deterioration (shrinking, cracking, or discoloration) or contamination, or if the expiration date supplied by the manufacturer has passed.

**Use:** For the selective isolation and cultivation of enteric pathogens, especially *Salmonella* and *Shigella* species.

## Deoxycholate Citrate Agar

**Composition** per liter:

| | |
|---|---|
| Sodium citrate | 20.0g |
| Agar | 17.0g |
| Lactose | 10.0g |
| Meat, solids from infusion | 10.0g |
| Peptic digest of animal tissue | 10.0g |
| Sodium deoxycholate | 5.0g |
| Ferric citrate | 1.0g |
| Neutral Red | 0.02g |

pH 7.3 ± 0.2 at 25°C

**Source:** This medium is available as a premixed powder from BD Diagnostic Systems.

**Preparation of Medium:** Add components to distilled/deionized water and bring volume to 1.0L. Mix thoroughly. Gently heat and bring to boiling. Do not autoclave. Cool to 45°–50°C. Pour into sterile Petri dishes. Dry the agar surface before use.

**Storage/Shelf Life:** Store dehydrated media in the dark in a sealed container below 30°C. Prepared media should be stored under refrigeration (2-8°C). Media should be used within 60 days of preparation. Media should not be used if there are any signs of deterioration (shrinking, cracking, or discoloration) or contamination, or if the expiration date supplied by the manufacturer has passed.

**Use:** For the selective isolation and cultivation of enteric pathogens, especially *Salmonella* and *Shigella* species.

## Deoxycholate Citrate Agar
### (Desoxycholate Citrate Agar)

**Composition** per liter:

| | |
|---|---|
| Pork infusion | 330.0g |
| Sodium citrate | 20.0g |
| Agar | 13.5g |
| Lactose | 10.0g |
| Proteose peptone No. 3 | 10.0g |
| Sodium deoxycholate | 5.0g |
| Ferric ammonium citrate | 2.0g |
| Neutral Red | 0.02g |

pH 7.5 ± 0.2 at 25°C

**Source:** This medium is available as a premixed powder from BD Diagnostic Systems.

**Preparation of Medium:** Add components to distilled/deionized water and bring volume to 1.0L. Mix thoroughly. Gently heat and bring to boiling. Do not autoclave. Cool to 45°–50°C. Pour into sterile Petri dishes. Dry the agar surface before use.

**Storage/Shelf Life:** Store dehydrated media in the dark in a sealed container below 30°C. Prepared media should be stored under refrigeration (2-8°C). Media should be used within 60 days of preparation.

Media should not be used if there are any signs of deterioration (shrinking, cracking, or discoloration) or contamination, or if the expiration date supplied by the manufacturer has passed.

**Use:** For the selective isolation and cultivation of enteric pathogens, especially *Salmonella* and *Shigella* species.

## Deoxycholate Citrate Agar

**Composition** per liter:

| | |
|---|---|
| Sodium citrate | 20.0g |
| Agar | 13.0g |
| Proteose peptone | 10.0g |
| Heart infusion solids | 10.0g |
| Lactose | 10.0g |
| Sodium deoxycholate | 5.0g |
| Ferric ammonium citrate | 2.0g |
| Neutral Red | 0.02g |

pH 7.5 ± 0.2 at 25°C

**Source:** This medium is available as a premixed powder from Hi-Media.

**Preparation of Medium:** Add components to distilled/deionized water and bring volume to 1.0L. Mix thoroughly. Gently heat and bring to boiling. Do not autoclave. Cool to 45°–50°C. Pour into sterile Petri dishes. Dry the agar surface before use. Avoid excessive heating as it is detrimental to the medium.

**Storage/Shelf Life:** Store dehydrated media in the dark in a sealed container below 30°C. Prepared media should be stored under refrigeration (2-8°C). Media should be used within 60 days of preparation. Media should not be used if there are any signs of deterioration (shrinking, cracking, or discoloration) or contamination, or if the expiration date supplied by the manufacturer has passed.

**Use:** For the selective isolation and cultivation of enteric pathogens, especially *Salmonella* and *Shigella* species.

## Deoxycholate Citrate Agar, HiVeg

**Composition** per liter:

| | |
|---|---|
| Sodium citrate | 20.0g |
| Agar | 13.5g |
| Plant peptone No. 3 | 13.0g |
| Plant infusion | 10.0g |
| Lactose | 10.0g |
| Synthetic detergent No. III | 2.0g |
| Ferric ammonium citrate | 2.0g |
| Neutral Red | 0.02g |

pH 7.5 ± 0.2 at 25°C

**Source:** This medium is available as a premixed powder from Hi-Media.

**Preparation of Medium:** Add components to distilled/deionized water and bring volume to 1.0L. Mix thoroughly. Gently heat and bring to boiling. Do not autoclave. Cool to 45°–50°C. Pour into sterile Petri dishes. Dry the agar surface before use. Avoid excessive heating as it is detrimental to the medium.

**Storage/Shelf Life:** Store dehydrated media in the dark in a sealed container below 30°C. Prepared media should be stored under refrigeration (2-8°C). Media should be used within 60 days of preparation. Media should not be used if there are any signs of deterioration (shrinking, cracking, or discoloration) or contamination, or if the expiration date supplied by the manufacturer has passed.

**Use:** For the selective isolation and cultivation of enteric pathogens, especially *Salmonella* and *Shigella* species.

## Deoxycholate Citrate Agar, Hynes

**Composition** per liter:

| | |
|---|---|
| Agar | 12.0g |
| Lactose | 10.0g |
| Sodium citrate | 8.5g |
| $Na_2S_2O_3·5H_2O$ | 5.4g |
| Beef extract powder | 5.0g |
| Peptone | 5.0g |
| Sodium deoxycholate | 5.0g |
| Ferric citrate | 1.0g |
| Neutral Red | 0.02g |

pH 7.3 ± 0.2 at 25°C

**Source:** This medium is available as a premixed powder from Thermo Scientific.

**Preparation of Medium:** Add components to distilled/deionized water and bring volume to 1.0L. Mix thoroughly. Gently heat and bring to boiling. Do not autoclave. Cool to 45°–50°C. Pour into sterile Petri dishes. Dry the agar surface before use.

**Storage/Shelf Life:** Store dehydrated media in the dark in a sealed container below 30°C. Prepared media should be stored under refrigeration (2-8°C). Media should be used within 60 days of preparation. Media should not be used if there are any signs of deterioration (shrinking, cracking, or discoloration) or contamination, or if the expiration date supplied by the manufacturer has passed.

**Use:** For the selective isolation, cultivation, and differentiation of enteric pathogens, especially *Salmonella* and *Shigella* species. Lactose-fermenting bacteria appear as pink colonies that may or may not be surrounded by a zone of precipitated deoxycholate. Nonlactose-fermenting bacteria appear as colorless colonies that are surrounded by a clear orange-yellow zone.

## Deoxycholate Lactose Agar

**Composition** per liter:

| | |
|---|---|
| Agar | 15.0g |
| Lactose | 10.0g |
| NaCl | 5.0g |
| Pancreatic digest of casein | 5.0g |
| Peptic digest of animal tissue | 5.0g |
| Sodium citrate | 2.0g |
| Sodium deoxycholate | 0.5g |
| Neutral Red | 0.033g |

pH 7.1 ± 0.2 at 25°C

**Source:** This medium is available as a premixed powder from BD Diagnostic Systems.

**Preparation of Medium:** Add components to distilled/deionized water and bring volume to 1.0L. Mix thoroughly. Gently heat and bring to boiling. Do not autoclave. Cool to 45°–50°C. Pour into sterile Petri dishes. Dry the agar surface before use.

**Storage/Shelf Life:** Store dehydrated media in the dark in a sealed container below 30°C. Prepared media should be stored under refrigeration (2-8°C). Media should be used within 60 days of preparation. Media should not be used if there are any signs of deterioration (shrinking, cracking, or discoloration) or contamination, or if the expiration date supplied by the manufacturer has passed.

**Use:** For the selective isolation, cultivation, and differentiation of enteric pathogens, especially *Salmonella* and *Shigella* species. Lactose-fermenting bacteria appear as pink colonies that may or may not be surrounded by a zone of precipitated deoxycholate. Nonlactose-fermenting bacteria appear as colorless colonies that are surrounded by a clear orange-yellow zone. Also used for the enumeration of coliform bacteria from water, milk, and dairy products.

## Deoxycholate Lactose HiVeg Agar

**Composition** per liter:

| | |
|---|---|
| Agar | 15.0g |
| Plant special peptone | 10.0g |
| Lactose | 10.0g |
| NaCl | 5.0g |
| Sodium citrate | 2.0g |
| Synthetic detergent No. III | 0.5g |
| Neutral Red | 0.03g |

pH 7.1 ± 0.2 at 25°C

**Source:** This medium is available as a premixed powder from Hi-Media.

**Preparation of Medium:** Add components to distilled/deionized water and bring volume to 1.0L. Mix thoroughly. Gently heat and bring to boiling. Do not autoclave. Cool to 45°–50°C. Pour into sterile Petri dishes. Dry the agar surface before use.

**Storage/Shelf Life:** Store dehydrated media in the dark in a sealed container below 30°C. Prepared media should be stored under refrigeration (2-8°C). Media should be used within 60 days of preparation. Media should not be used if there are any signs of deterioration (shrinking, cracking, or discoloration) or contamination, or if the expiration date supplied by the manufacturer has passed.

**Use:** For the selective isolation, cultivation, and differentiation of enteric pathogens, especially *Salmonella* and *Shigella* species. Lactose-fermenting bacteria appear as pink colonies that may or may not be surrounded by a zone of precipitated deoxycholate. Nonlactose-fermenting bacteria appear as colorless colonies that are surrounded by a clear orange-yellow zone. Also used for the enumeration of coliform bacteria from water, milk, and dairy products.

## DEV Lactose Peptone MUG Broth

**Composition** per liter:

| | |
|---|---|
| Lactose | 10.0g |
| Meat peptone | 10.0g |
| NaCl | 5.0g |
| Tryptophan | 1.0g |
| 4-Methylumbelliferyl-β-D-glucuronide | 0.1g |
| Bromocresol Purple | 0.01g |

pH 7.2 ± 0.2 at 37°C

**Source:** This medium is available from Fluka, Sigma-Aldrich.

**Preparation of Medium:** Add components to distilled/deionized water and bring volume to 1.0L. Mix thoroughly. Distribute into test tubes that contain an inverted Durham tube in 10.0mL volumes. Autoclave for 15 min at 15 psi pressure–121°C.

**Storage/Shelf Life:** Store dehydrated media in the dark in a sealed conainer below 30°C. Prepared plates should be stored under refrigeration (2-8°C). This medium has a limited shelf life. Media should not be used if there are any signs of deterioration (shrinking, cracking, or discoloration) or contamination, or if the expiration date supplied by the manufacturer has passed.

**Use:** For the enrichment and titer determination of coliform bacteria in connection with the bacteriological examination of water. The presence of *E. coli* can be demonstrated by fluorescence in the UV and a positive indole test.

## Dextrose Agar

**Composition** per liter:

| | |
|---|---|
| Agar | 15.0g |
| Glucose | 10.0g |
| NaCl | 5.0g |
| Pancreatic digest of casein | 5.0g |
| Peptic digest of animal tissue | 5.0g |
| Beef extract | 3.0g |

pH 6.9 ± 0.2 at 25°C

**Source:** This medium is available as a premixed powder from BD Diagnostic Systems.

**Preparation of Medium:** Add components to distilled/deionized water and bring volume to 1.0L. Mix thoroughly. Gently heat and bring to boiling. Distribute into tubes or flasks. Autoclave for 15 min at 15 psi pressure–121°C. Pour into sterile Petri dishes or leave in tubes.

**Use:** For the cultivation and enumeration of microorganisms from foods. For use as a base for the preparation of blood agar.

## Dextrose Agar

**Composition** per liter:

| | |
|---|---|
| Agar | 15.0g |
| Glucose | 10.0g |
| Tryptose | 10.0g |
| NaCl | 5.0g |
| Beef extract | 3.0g |

pH 7.3 ± 0.2 at 25°C

**Source:** This medium is available as a premixed powder from BD Diagnostic Systems.

**Preparation of Medium:** Add components to distilled/deionized water and bring volume to 1.0L. Mix thoroughly. Gently heat and bring to boiling. Distribute into tubes or flasks. Autoclave for 15 min at 15 psi pressure–121°C. Pour into sterile Petri dishes or leave in tubes.

**Storage/Shelf Life:** Store dehydrated media in the dark in a sealed container below 30°C. Prepared media should be stored under refrigeration (2-8°C). Media should be used within 60 days of preparation. Media should not be used if there are any signs of deterioration (shrinking, cracking, or discoloration) or contamination, or if the expiration date supplied by the manufacturer has passed.

**Use:** For the cultivation of a wide variety of microorganisms. For use as a base for the preparation of blood agar and for general laboratory procedures.

## Dextrose Agar

**Composition** per liter:

| | |
|---|---|
| Agar | 15.0g |
| Glucose | 10.0g |
| Tryptose | 10.0g |
| NaCl | 5.0g |
| Beef extract | 3.0g |

pH 7.3 ± 0.2 at 25°C

**Source:** This medium is available as a premixed powder from Hi-Media.

**Preparation of Medium:** Add components to distilled/deionized water and bring volume to 1.0L. Mix thoroughly. Gently heat and bring to boiling. Distribute into tubes or flasks. Autoclave for 20 min at 15 psi pressure–121°C. Pour into sterile Petri dishes or leave in tubes.

**Storage/Shelf Life:** Store dehydrated media in the dark in a sealed container below 30°C. Prepared media should be stored under refrigeration (2-8°C). Media should be used within 60 days of preparation. Media should not be used if there are any signs of deterioration (shrinking, cracking, or discoloration) or contamination, or if the expiration date supplied by the manufacturer has passed.

**Use:** For the cultivation and maintenance of a wide variety of microorganisms.

### Dextrose Ascitic Fluid Semisolid Agar

**Composition** per liter:
| | |
|---|---|
| Pancreatic digest of casein | 2.66g |
| NaCl | 1.33g |
| Agar | 0.5g |
| Phenol Red | 4.8mg |
| Ascitic fluid | 50.0mL |
| Glucose solution | 15.0mL |

pH 7.4 ± 0.2 at 25°C

**Glucose Solution:**

**Composition** per 15.0mL:
| | |
|---|---|
| Glucose | 3.0g |

**Preparation of Glucose Solution:** Add glucose to distilled/deionized water and bring volume to 15.0mL. Mix thoroughly. Filter sterilize.

**Preparation of Medium:** Add components, except ascitic fluid and glucose solution, to distilled/deionized water and bring volume to 935.0mL. Mix thoroughly. Gently heat and bring to boiling. Autoclave for 15 min at 15 psi pressure–121°C. Cool to 45°–50°C. Aseptically add sterile ascitic fluid and glucose solution. Mix thoroughly. Aseptically distribute into sterile tubes.

**Storage/Shelf Life:** Store dehydrated media in the dark in a sealed container below 30°C. Prepared media should be stored under refrigeration (2-8°C). Media should be used within 60 days of preparation. Media should not be used if there are any signs of deterioration (shrinking, cracking, or discoloration) or contamination, or if the expiration date supplied by the manufacturer has passed.

**Use:** For the isolation and cultivation of microorganisms from spinal fluid.

### Dextrose Broth

**Composition** per liter:
| | |
|---|---|
| Tryptose | 10.0g |
| Glucose | 5.0g |
| NaCl | 5.0g |
| Beef extract | 3.0g |

pH 7.2 ± 0.2 at 25°C

**Source:** This medium is available as a premixed powder from BD Diagnostic Systems and Thermo Scientific.

**Preparation of Medium:** Add components to distilled/deionized water and bring volume to 1.0L. Mix thoroughly. Distribute into tubes or flasks. Autoclave for 15 min at 15 psi pressure–121°C.

**Storage/Shelf Life:** Store dehydrated media in the dark in a sealed container below 30°C. Prepared media should be stored under refriger-

ation (2-8°C). Media should be used within 60 days of preparation. Media should not be used if there are any signs of deterioration (discoloration) or contamination, or if the expiration date supplied by the manufacturer has passed.

**Use:** For the isolation and enrichment of fastidious or damaged microorganisms.

### Dextrose Broth

**Composition** per liter:
| | |
|---|---|
| Pancreatic digest of casein | 10.0g |
| Glucose | 5.0g |
| NaCl | 5.0g |

pH 7.3 ± 0.2 at 25°C

**Source:** This medium is available as a premixed powder from BD Diagnostic Systems.

**Preparation of Medium:** Add components to distilled/deionized water and bring volume to 1.0L. Mix thoroughly. Distribute into tubes or flasks. Autoclave for 15 min at 15 psi pressure–121°C.

**Storage/Shelf Life:** Store dehydrated media in the dark in a sealed container below 30°C. Prepared media should be stored under refrigeration (2-8°C). Media should be used within 60 days of preparation. Media should not be used if there are any signs of deterioration (discoloration) or contamination, or if the expiration date supplied by the manufacturer has passed.

**Use:** For the cultivation and differentiation of microorganisms based on their ability to ferment glucose. If desired, a Durham tube may be added to the test tubes to determine gas production.

### Dextrose HiVeg Agar

**Composition** per liter:
| | |
|---|---|
| Agar | 15.0g |
| Glucose | 10.0g |
| Plant hydrolysate No. 1 | 10.0g |
| NaCl | 5.0g |
| Plant extract | 3.0g |

pH 7.3 ± 0.2 at 25°C

**Source:** This medium is available as a premixed powder from HiMedia.

**Preparation of Medium:** Add components to distilled/deionized water and bring volume to 1.0L. Mix thoroughly. Gently heat and bring to boiling. Distribute into tubes or flasks. Autoclave for 20 min at 15 psi pressure–121°C. Pour into sterile Petri dishes or leave in tubes.

**Storage/Shelf Life:** Store dehydrated media in the dark in a sealed container below 30°C. Prepared media should be stored under refrigeration (2-8°C). Media should be used within 60 days of preparation. Media should not be used if there are any signs of deterioration (shrinking, cracking, or discoloration) or contamination, or if the expiration date supplied by the manufacturer has passed.

**Use:** For the cultivation and maintenance of a wide variety of microorganisms.

### Dextrose HiVeg Agar with Blood

**Composition** per liter:
| | |
|---|---|
| Agar | 15.0g |
| Glucose | 10.0g |
| Plant hydrolysate No. 1 | 10.0g |
| NaCl | 5.0g |

Plant extract ................................................................3.0g
Sheep blood, defibrinated ......................................... 50.0mL

pH 7.3 ± 0.2 at 25°C

**Source:** This medium wtihout sheep blood is available as a premixed powder from HiMedia.

**Preparation of Medium:** Add components, except sheep blood, to distilled/deionized water and bring volume to 950.0mL. Mix thoroughly. Gently heat and bring to boiling. Autoclave for 15 min at 15 psi pressure–121°C. Cool to 50°C. Aseptically add 50.0mL sterile blood. Pour into sterile Petri dishes or leave in tubes.

**Storage/Shelf Life:** Store dehydrated media in the dark in a sealed container below 30°C. Prepared media should be stored under refrigeration (2-8°C). Media should be used within 60 days of preparation. Media should not be used if there are any signs of deterioration (shrinking, cracking, or discoloration) or contamination, or if the expiration date supplied by the manufacturer has passed.

**Use:** For the cultivation and maintenance of a wide variety of microorganisms.

## Dextrose Proteose Peptone HiVeg Agar Base with Tellurite and Blood

**Composition** per liter:
Plant peptone No. 3........................................................20.0g
Agar .............................................................................15.0g
NaCl...............................................................................5.0g
Glucose ..........................................................................2.0g
Sheep blood, defibrinated ......................................... 50.0mL
Tellurite solution ...................................................... 2.0mL

pH 7.4 ± 0.2 at 25°C

**Source:** This medium, without tellurite or blood, is available as a premixed powder from HiMedia.

**Tellurite Solution:**

**Composition** per 10.0mL:
$K_2TeO_3$.........................................................................0.1g

**Preparation of Tellurite Solution:** Add $K_2TeO_3$ to distilled/deionized water and bring volume to 100.0mL. Mix thoroughly. Filter sterilize.

**Caution:** Potassium tellurite is toxic.

**Preparation of Medium:** Add components, except tellurite solution and blood, to distilled/deionized water and bring volume to 950.0mL. Mix thoroughly. Gently heat and bring to boiling. Autoclave for 15 min at 15 psi pressure–121°C. Cool to 50°C. Aseptically add 2.0mL of sterile tellurite soltuion and 50.0mL of sterile defibrinated blood. Pour into sterile Petri dishes or leave in tubes.

**Storage/Shelf Life:** Store dehydrated media in the dark in a sealed container below 30°C. Prepared media should be stored under refrigeration (2-8°C). Media should be used within 60 days of preparation. Media should not be used if there are any signs of deterioration (shrinking, cracking, or discoloration) or contamination, or if the expiration date supplied by the manufacturer has passed.

**Use:** For the cultivation and maintenance of a wide variety of microorganisms. For use as a base for the preparation of blood agar and for general laboratory procedures. For the isolation of *Corynebacterium diphtheriae*.

## Dextrose Proteose Peptone Agar Base with Tellurite and Blood

**Composition** per liter:
Proteose peptone...........................................................20.0g
Agar .............................................................................15.0g
NaCl...............................................................................5.0g
Glucose ..........................................................................2.0g
Sheep blood, defibrinated ......................................... 50.0mL
Tellurite solution ...................................................... 2.0mL

pH 7.4 ± 0.2 at 25°C

**Source:** This medium, without tellurite or blood, is available as a premixed powder from HiMedia.

**Tellurite Solution:**

**Composition** per 10.0mL:
$K_2TeO_3$.........................................................................0.1g

**Preparation of Tellurite Solution:** Add $K_2TeO_3$ to distilled/deionized water and bring volume to 100.0mL. Mix thoroughly. Filter sterilize.

**Caution:** Potassium tellurite is toxic.

**Preparation of Medium:** Add components, except tellurite solution and blood, to distilled/deionized water and bring volume to 950.0mL. Mix thoroughly. Gently heat and bring to boiling. Autoclave for 15 min at 15 psi pressure–121°C. Cool to 50°C. Aseptically add 2.0mL of sterile tellurite soltuion and 50.0mL of sterile defibrinated blood. Pour into sterile Petri dishes or leave in tubes.

**Storage/Shelf Life:** Store dehydrated media in the dark in a sealed container below 30°C. Prepared media should be stored under refrigeration (2-8°C). Media should be used within 60 days of preparation. Media should not be used if there are any signs of deterioration (shrinking, cracking, or discoloration) or contamination, or if the expiration date supplied by the manufacturer has passed.

**Use:** For the cultivation and maintenance of a wide variety of microorganisms. For use as a base for the preparation of blood agar and for general laboratory procedures. For the isolation of *Corynebacterium diphtheriae*.

## Diagnostic Sensitivity Test Agar (DST Agar)

**Composition** per liter:
Agar .............................................................................12.0g
Proteose peptone...........................................................10.0g
Veal infusion solids........................................................10.0g
NaCl...............................................................................3.0g
$Na_2HPO_4$........................................................................2.0g
Glucose ..........................................................................2.0g
Sodium acetate ...............................................................1.0g
Adenine sulfate ............................................................0.01g
Guanine·HCl ................................................................0.01g
Uracil ..........................................................................0.01g
Xanthine........................................................................0.01g
Thiamine .....................................................................0.02mg
Horse blood, defibrinated ......................................... 70.0mL

pH 7.4 ± 0.2 at 25°C

**Source:** This medium is available as a premixed powder from Thermo Scientific.

**Preparation of Medium:** Add components, except horse blood, to distilled/deionized water and bring volume to 930.0mL. Mix throughly. Gently heat and bring to boiling. Autoclave for 15 min at 15 psi pressure–121°C. Cool to 45°–50°C. Aseptically add sterile horse blood. Mix thoroughly. Pour into sterile Petri dishes or distribute into sterile tubes.

**Storage/Shelf Life:** Store dehydrated media in the dark in a sealed container below 30°C. Prepared media should be stored under refrigeration (2-8°C). Media should be used within 60 days of preparation. Media should not be used if there are any signs of deterioration (shrinking, cracking, or discoloration) or contamination, or if the expiration date supplied by the manufacturer has passed.

**Use:** For antimicrobial testing of various pathogenic microorganisms. DSTA is primarily used for susceptibility tests rather than the primary isolation of organisms from clinical samples. An essential requirement for satisfactory antimicrobial susceptibility media is that the reactive levels of thymidine and thymine must be sufficiently reduced to avoid antagonism of trimethoprim and sulphonamides. DSTA meets this requirement and in the presence of lysed horse blood (or defibrinated horse blood if the plates are stored long enough to allow some lysis of the erythrocytes) the level of thymidine will be further reduced. This is caused by the action of the enzyme thymidine phosphorylase, which is released from lysed horse erythrocytes. Thymidine is an essential growth factor for thymidine-dependent organisms and they will not grow in its absence or they will grow poorly in media containing reduced levels.

## Differential Agar for Group D Streptococci

**Composition** per liter:

| | |
|---|---|
| NaCl | 65.0g |
| Agar | 13.5g |
| Casein enzymic hydrolysate | 16.0g |
| Glucose | 10.0g |
| Brain heart infusion | 8.0g |
| Peptic digest of animal tissue | 5.0g |
| Na$_2$HPO$_4$ | 2.5g |
| Bromcresol Purple | 0.02g |

pH 7.4 ± 0.2 at 25°C

**Source:** This medium is available from HiMedia.

**Preparation of Medium:** Add components to distilled/deionized water and bring volume to 1.0L. Mix thoroughly. Gently heat and bring to boiling. Distribute into tubes or flasks. Autoclave for 15 min at 15 psi pressure–121°C. Pour into sterile Petri dishes or leave in tubes. For tubes, allow to solidify in slanted position.

**Storage/Shelf Life:** Store dehydrated media in the dark in a sealed container below 30°C. Prepared media should be stored under refrigeration (2-8°C). Media should be used within 60 days of preparation. Media should not be used if there are any signs of deterioration (shrinking, cracking, or discoloration) or contamination, or if the expiration date supplied by the manufacturer has passed.

**Use:** For the differentiation and identification of Group D streptococci.

## Differential Buffered Charcoal Yeast Extract Agar Base with Selective Supplement

**Composition** per liter:

| | |
|---|---|
| Agar | 15.0g |
| ACES buffer | 10.0g |
| Yeast extract | 10.0g |
| Charcoal, activated | 1.5g |
| L-Cysteine·HCl | 0.4g |

| | |
|---|---|
| Ferric pyrophosphate, soluble | 0.25g |
| α-Ketoglutarate | 0.2g |
| Bromcresol Purple | 0.01g |
| Bromthymol Blue | 0.01g |
| Selective supplement | 10.0mL |

pH 6.9 ± 0.2 at 25°C

**Source:** This medium, wthout selective supplement, is available as a premixed powder from HiMedia.

**Selective Supplement:**
**Composition** per 10.0mL:

| | |
|---|---|
| Vancomycin | 0.1g |
| Polymyxin B | 50,000 units |

**Preparation of Selective Supplement:** Add components to distilled/deionized water and bring volume to 10.0mL. Mix thoroughly. Filter sterilize.

**Preparation of Medium:** Add components, except selective supplement, to distilled/deionized water and bring volume to 1.0L. Mix thoroughly. Gently heat and bring to boiling. Autoclave for 15 min at 15 psi pressure–121°C. Cool to 50°C. Aseptically add 10.0mL sterile selective supplement. Mix thoroughly. Pour into sterile Petri dishes or distribute into sterile tubes.

**Storage/Shelf Life:** Store dehydrated media in the dark in a sealed container below 30°C. Prepared media should be stored under refrigeration (2-8°C). Media should be used within 60 days of preparation. Media should not be used if there are any signs of deterioration (shrinking, cracking, or discoloration) or contamination, or if the expiration date supplied by the manufacturer has passed.

**Use:** For the selective isolation and differentiation of *Legionella* species.

## Diphtheria Virulence HiVeg Agar Base with Tellurite and Diphtheria Virulence Supplement

**Composition** per liter:

| | |
|---|---|
| Plant peptone No. 3 | 20.0g |
| Agar | 15.0g |
| NaCl | 2.5g |
| Diptheria virulence supplement | 200.0mL |
| Tellurite solution | 50.0mL |

pH 7.8± 0.2 at 25°C

**Source:** This medium, without tellurite or diphtheria virulence supplement, is available as a premixed powder from HiMedia.

**Tellurite Solution:**
**Composition** per 100.0mL:

| | |
|---|---|
| K$_2$TeO$_3$ | 1.0g |

**Preparation of Tellurite Solution:** Add K$_2$TeO$_3$ to distilled/deionized water and bring volume to 100.0mL. Mix thoroughly. Filter sterilize.

**Caution:** Potassium tellurite is toxic.

**Diphtheria Virulence Supplement:**
**Composition** per 260.0mL:

| | |
|---|---|
| Horse serum | 200.0mL |
| Potassium tellurite solution | 60.0mL |

**Preparation of Diphtheria Virulence Supplement:** Aseptically combine sterile horse serum and sterile tellurite solution. Mix thoroughly.

**Preparation of Medium:** Add components, except tellurite solution and diphtheria virulence supplement, to distilled/deionized water and bring volume to 1.0L. Mix thoroughly. Gently heat and bring to boiling. Distribute into tubes or flasks. Autoclave for 15 min at 15 psi pressure–121°C. Cool to 55–60°C. Aseptically add 2.0mL of sterile diphtheria virulence supplement and 0.5mL sterile tellurite solution to each Petri dish. Quickly add 10.0mL sterile Diphtheria Virulence HiVeg Base Agar to each Petri dish. Before the medium solidifies, place a filter paper strip saturated with potent diphtheria antitoxin across the diameter of the plate. Allow the strip to sink to the bottom of the Petri plate. Inoculate the plate with a heavy inoculum across the strip.

**Storage/Shelf Life:** Store dehydrated media in the dark in a sealed container below 30°C. Prepared media should be stored under refrigeration (2-8°C). Media should be used within 60 days of preparation. Media should not be used if there are any signs of deterioration (shrinking, cracking, or discoloration) or contamination, or if the expiration date supplied by the manufacturer has passed.

**Use:** For the detection of diphtheria toxin producing strains of *Corynebacterium diphtheriae*. For testing the toxigenicity of *Corynebacterium diphtheriae*. The reaction of antitoxin forms the actual basis for the detection of the diphtheria toxin.

## Disinfectant Test Broth
### (*Staphylococcus aureus* Enrichment Broth)
**Composition** per liter:
Peptic digest of animal tissue........................................10.0g
Beef infusion....................................................................5.0g
NaCl.................................................................................5.0g
pH 6.8 ± 0.2 at 25°C

**Source:** This medium is available as a premixed powder from Hi-Media.

**Preparation of Medium:** Add components to distilled/deionized water and bring volume to 1.0L. Mix thoroughly. Distribute into tubes or flasks. Autoclave for 15 min at 15 psi pressure–121°C.

**Storage/Shelf Life:** Store dehydrated media in the dark in a sealed container below 30°C. Prepared media should be stored under refrigeration (2-8°C). Media should be used within 60 days of preparation. Media should not be used if there are any signs of deterioration (discoloration) or contamination, or if the expiration date supplied by the manufacturer has passed.

**Use:** For the determination of phenol coefficients of disinfectants.

## Disinfectant Test Broth AOAC
**Composition** per liter:
Peptic digest of animal tissue........................................10.0g
Beef extract.....................................................................5.0g
NaCl.................................................................................5.0g
pH 6.8 ± 0.2 at 25°C

**Source:** This medium is available as a premixed powder from BD Diagnostic Systems.

**Preparation of Medium:** Add components to distilled/deionized water and bring volume to 1.0L. Mix thoroughly. Distribute into tubes or flasks. Autoclave for 15 min at 13 psi pressure–118°C.

**Storage/Shelf Life:** Store dehydrated media in the dark in a sealed container below 30°C. Prepared media should be stored under refrigeration (2-8°C). Media should be used within 60 days of preparation. Media should not be used if there are any signs of deterioration (discoloration) or contamination, or if the expiration date supplied by the manufacturer has passed.

**Use:** For the determination of phenol coefficients of disinfectants.

## Disinfectant Test HiVeg Broth
**Composition** per liter:
Plant peptone ................................................................10.0g
Plant extract ...................................................................5.0g
NaCl.................................................................................5.0g
pH 6.8 ± 0.2 at 25°C

**Source:** This medium is available as a premixed powder from Hi-Media.

**Preparation of Medium:** Add components to distilled/deionized water and bring volume to 1.0L. Mix thoroughly. Distribute into tubes or flasks. Autoclave for 15 min at 15 psi pressure–121°C.

**Storage/Shelf Life:** Store dehydrated media in the dark in a sealed container below 30°C. Prepared media should be stored under refrigeration (2-8°C). Media should be used within 60 days of preparation. Media should not be used if there are any signs of deterioration (discoloration) or contamination, or if the expiration date supplied by the manufacturer has passed.

**Use:** For the determination of phenol coefficients of disinfectants.

## Disinfectant Test Medium
**Composition** per liter:
Peptic digest of animal tissue ........................................5.0g
Proteose peptone.............................................................5.0g
NaCl.................................................................................5.0g
Beef extract.....................................................................5.0g
Yeast extract....................................................................5.0g
pH 6.8 ± 0.2 at 25°C

**Source:** This medium is available as a premixed powder from Hi-Media.

**Preparation of Medium:** Add components to distilled/deionized water and bring volume to 1.0L. Mix thoroughly. Distribute into tubes or flasks. Autoclave for 15 min at 15 psi pressure–121°C.

**Storage/Shelf Life:** Store dehydrated media in the dark in a sealed container below 30°C. Prepared media should be stored under refrigeration (2-8°C). Media should be used within 60 days of preparation. Media should not be used if there are any signs of deterioration (discoloration) or contamination, or if the expiration date supplied by the manufacturer has passed.

**Use:** For the determination of phenol coefficients of disinfectants.

## DNase Agar
**Composition** per liter:
Tryptose .........................................................................20.0g
Agar ...............................................................................12.0g
NaCl.................................................................................5.0g
Deoxyribonucleic acid.....................................................2.0g
pH 7.3 ± 0.2 at 25°C

**Source:** This medium is available as a premixed powder from Thermo Scientific.

**Preparation of Medium:** Add components to distilled/deionized water and bring volume to 1.0L. Mix thoroughly. Gently heat and bring to boiling. Distribute into tubes or flasks. Autoclave for 15 min at 15 psi pressure–121°C. Pour into sterile Petri dishes or leave in tubes.

**Storage/Shelf Life:** Store dehydrated media in the dark in a sealed container below 30°C. Prepared media should be stored under refriger-

ation (2-8°C). Media should be used within 60 days of preparation. Media should not be used if there are any signs of deterioration (shrinking, cracking, or discoloration) or contamination, or if the expiration date supplied by the manufacturer has passed.

**Use:** For the differentiation of microorganisms, especially *Staphylococcus* species and *Serratia marcescens*, based on their production of deoxyribo-nuclease.

## DNase Test Agar

**Composition** per liter:

Agar ...................................................................................15.0g
Pancreatic digest of casein.................................................15.0g
NaCl......................................................................................5.0g
Papaic digest of soybean meal .............................................5.0g
Deoxyribonucleic acid .........................................................2.0g

pH 7.3 ± 0.2 at 25°C

**Source:** This medium is available as a premixed powder from BD Diagnostic Systems.

**Preparation of Medium:** Add components to distilled/deionized water and bring volume to 1.0L. Mix thoroughly. Gently heat while stirring and bring to boiling. Distribute into tubes or flasks. Autoclave for 15 min at 13 psi pressure–118°C. Pour into sterile Petri dishes or leave in tubes.

**Storage/Shelf Life:** Store dehydrated media in the dark in a sealed container below 30°C. Prepared media should be stored under refrigeration (2-8°C). Media should be used within 60 days of preparation. Media should not be used if there are any signs of deterioration (shrinking, cracking, or discoloration) or contamination, or if the expiration date supplied by the manufacturer has passed.

**Use:** For the differentiation of microorganisms, especially *Staphylococcus* species and *Serratia marcescens*, based on their production of deoxyribonuclease.

## DNase Test Agar with Methyl Green

**Composition** per liter:

Agar ...................................................................................15.0g
Pancreatic digest of casein.................................................10.0g
Peptic digest of animal tissue............................................10.0g
NaCl......................................................................................5.0g
Deoxyribonucleic acid .........................................................2.0g
Methyl Green .....................................................................0.05g

pH 7.3 ± 0.2 at 25°C

**Source:** This medium is available as a premixed powder from BD Diagnostic Systems.

**Preparation of Medium:** Add components to distilled/deionized water and bring volume to 1.0L. Mix thoroughly. Gently heat while stirring and bring to boiling. Distribute into tubes or flasks. Autoclave for 15 min at 13 psi pressure–118°C. Pour into sterile Petri dishes or leave in tubes.

**Storage/Shelf Life:** Store dehydrated media in the dark in a sealed container below 30°C. Prepared media should be stored under refrigeration (2-8°C). Media should be used within 60 days of preparation. Media should not be used if there are any signs of deterioration (shrinking, cracking, or discoloration) or contamination, or if the expiration date supplied by the manufacturer has passed.

**Use:** For the differentiation of microorganisms, especially *Staphylococcus* species and *Serratia marcescens*, based on their production of deoxyribonuclease.

## DNase Test Agar with Toluidine Blue

**Composition** per liter:

Agar ...................................................................................15.0g
Pancreatic digest of casein.................................................10.0g
Peptic digest of animal tissue ............................................10.0g
NaCl......................................................................................5.0g
Deoxyribonucleic acid .........................................................2.0g
Toluidine Blue ....................................................................0.1g

pH 7.3 ± 0.2 at 25°C

**Preparation of Medium:** Add components to distilled/deionized water and bring volume to 1.0L. Mix thoroughly. Gently heat while stirring and bring to boiling. Distribute into tubes or flasks. Autoclave for 15 min at 13 psi pressure–118°C. Pour into sterile Petri dishes or leave in tubes.

**Storage/Shelf Life:** Store dehydrated media in the dark in a sealed container below 30°C. Prepared media should be stored under refrigeration (2-8°C). Media should be used within 60 days of preparation. Media should not be used if there are any signs of deterioration (shrinking, cracking, or discoloration) or contamination, or if the expiration date supplied by the manufacturer has passed.

**Use:** For the differentiation of microorganisms, especially *Staphylococcus* species and *Serratia marcescens*, based on their production of deoxyribonuclease.

## DNase Test HiVeg Agar Base

**Composition** per liter:

Agar ...................................................................................15.0g
Plant hydrolysate ...............................................................15.0g
Papaic digest of soybean meal.............................................5.0g
NaCl......................................................................................5.0g
Deoxyribonucleic acid (DNA)..............................................2.0g

pH 7.3 ± 0.2 at 25°C

**Source:** This medium is available as a premixed powder from Hi-Media.

**Preparation of Medium:** Add components to distilled/deionized water and bring volume to 1.0L. Mix thoroughly. Gently heat while stirring and bring to boiling. Distribute into tubes or flasks. Autoclave for 15 min at 13 psi pressure–118°C. Pour into sterile Petri dishes or leave in tubes.

**Storage/Shelf Life:** Store dehydrated media in the dark in a sealed container below 30°C. Prepared media should be stored under refrigeration (2-8°C). Media should be used within 60 days of preparation. Media should not be used if there are any signs of deterioration (shrinking, cracking, or discoloration) or contamination, or if the expiration date supplied by the manufacturer has passed.

**Use:** For the differentiation of microorganisms, especially *Staphylococcus* species and *Serratia marcescens*, based on their production of deoxyribonuclease.

## DNase Test HiVeg Agar Base without DNA

**Composition** per liter:

Agar ...................................................................................15.0g
Plant hydrolysate ...............................................................15.0g
Papaic digest of soybean meal.............................................5.0g
NaCl......................................................................................5.0g

pH 6.8 ± 0.2 at 25°C

**Source:** This medium is available as a premixed powder from Hi-Media.

**Preparation of Medium:** Add components to distilled/deionized water and bring volume to 1.0L. Mix thoroughly. Distribute into tubes or flasks. Autoclave for 15 min at 15 psi pressure–121°C. Pour into sterile Petri dishes or leave in tubes

**Storage/Shelf Life:** Store dehydrated media in the dark in a sealed container below 30°C. Prepared media should be stored under refrigeration (2-8°C). Media should be used within 60 days of preparation. Media should not be used if there are any signs of deterioration (shrinking, cracking, or discoloration) or contamination, or if the expiration date supplied by the manufacturer has passed.

**Use:** As a base medium for the differentiation of microorganisms, especially *Staphylococcus* species and *Serratia marcescens*, based on their production of deoxyribonuclease.

### DNase Test HiVeg Agar with Toluidine Blue
**Composition** per liter:
| | |
|---|---|
| Plant hydrolysate No. 1 | 20.0g |
| Agar | 15.0g |
| NaCl | 5.0g |
| Deoxyribonucleic acid (DNA) | 2.0g |
| Toluidine Blue | 0.1g |

pH 7.3 ± 0.2 at 25°C

**Source:** This medium is available as a premixed powder from Hi-Media.

**Preparation of Medium:** Add components to distilled/deionized water and bring volume to 1.0L. Mix thoroughly. Gently heat while stirring and bring to boiling. Distribute into tubes or flasks. Autoclave for 15 min at 13 psi pressure–118°C. Pour into sterile Petri dishes or leave in tubes.

**Storage/Shelf Life:** Store dehydrated media in the dark in a sealed container below 30°C. Prepared media should be stored under refrigeration (2-8°C). Media should be used within 60 days of preparation. Media should not be used if there are any signs of deterioration (shrinking, cracking, or discoloration) or contamination, or if the expiration date supplied by the manufacturer has passed.

**Use:** For the differentiation of microorganisms, especially *Staphylococcus* species and *Serratia marcescens*, based on their production of deoxyribonuclease.

### Doyle and Roman Enrichment Medium
**Composition** per liter:
| | |
|---|---|
| Pancreatic digest of casein | 10.0g |
| Peptic digest of animal tissue | 10.0g |
| NaCl | 5.0g |
| Sodium succinate | 3.0g |
| Yeast extract | 2.0g |
| Glucose | 1.0g |
| NaHSO$_3$ | 0.1g |
| L-Cysteine·HCl·H$_2$O | 0.1g |
| Horse blood, lysed | 70.0mL |
| Antibiotic solution | 10.0mL |

pH 7.0 ± 0.2 at 25°C

**Antibiotic Solution:**
**Composition** per 10.0mL:
| | |
|---|---|
| Cycloheximide | 0.05g |
| Vancomycin | 0.015g |
| Trimethoprim lactate | 5.0mg |
| Polymyxin B | 200,000U |

**Preparation of Antibiotic Solution:** Add components to distilled/deionized water and bring volume to 10.0mL. Mix thoroughly. Filter sterilize.

**Caution:** Cycloheximide is toxic. Avoid skin contact or aerosol formation and inhalation.

**Preparation of Medium:** Add components, except antibiotic solution and horse blood, to distilled/deionized water and bring volume to 920.0mL. Mix thoroughly. Gently heat and bring to boiling. Autoclave for 15 min at 15 psi pressure–121°C. Cool to 45°–50°C. Aseptically add sterile antibiotic solution and horse blood. Mix thoroughly. Aseptically distribute into sterile flasks in 90.0–100.0mL volumes.

**Storage/Shelf Life:** Store dehydrated media in the dark in a sealed container below 30°C. Prepared media should be stored under refrigeration (2-8°C). Media should be used within 60 days of preparation. Media should not be used if there are any signs of deterioration (discoloration) or contamination, or if the expiration date supplied by the manufacturer has passed.

**Use:** For the cultivation and enrichment of *Campylobacter* species from foods.

### Doyle's Enrichment Broth Base with Antibiotic Solution
**Composition** per liter:
| | |
|---|---|
| Casein enzymatic hydrolysate | 10.0g |
| Peptic digest of animal tissue | 10.0g |
| NaCl | 5.0g |
| Sodium succinate | 3.0g |
| Yeast extract | 2.0g |
| Glucose | 1.0g |
| L-Cysteine·HCl | 0.1g |
| NaHSO$_3$ | 0.1g |
| Antibiotic solution | 10.0mL |

pH 7.0 ± 0.2 at 25°C

**Source:** This medium, without antibiotic solution, is available as a premixed powder from HiMedia.

**Antibiotic Solution:**
**Composition** per 10.0mL:
| | |
|---|---|
| Cycloheximide | 0.05g |
| Vancomycin | 0.015g |
| Trimethoprim lactate | 5.0mg |
| Polymyxin B | 200,000U |

**Preparation of Antibiotic Solution:** Add components to distilled/deionized water and bring volume to 10.0mL. Mix thoroughly. Filter sterilize.

**Caution:** Cycloheximide is toxic. Avoid skin contact or aerosol formation and inhalation.

**Preparation of Medium:** Add components, except antibiotic solution, to distilled/deionized water and bring volume to 990.0mL. Mix thoroughly. Gently heat and bring to boiling. Autoclave for 15 min at 15 psi pressure–121°C. Cool to 45°–50°C. Aseptically add a selective sterile antibiotic solution. Generally 50.0mL of defibrinated horse blood is also added as an enrichment. Mix thoroughly.

**Storage/Shelf Life:** Store dehydrated media in the dark in a sealed container below 30°C. Prepared media should be stored under refrigeration (2-8°C). Media should be used within 60 days of preparation. Media should not be used if there are any signs of deterioration (discoloration) or contamination, or if the expiration date supplied by the manufacturer has passed.

**Use:** For the cultivation and enrichment of *Campylobacter* species from foods.

## Doyle's Enrichment HiVeg Broth Base with Antibiotic Solution

**Composition** per liter:

| | |
|---|---|
| Plant hydrolysate | 10.0g |
| Plant peptone | 10.0g |
| NaCl | 5.0g |
| Sodium succinate | 3.0g |
| Yeast extract | 2.0g |
| Glucose | 1.0g |
| L-Cysteine·HCl | 0.1g |
| NaHSO$_3$ | 0.1g |
| Antibiotic solution | 10.0mL |

pH 7.0 ± 0.2 at 25°C

**Source:** This medium, without antibiotic solution, is available as a premixed powder from HiMedia.

**Antibiotic Solution:**
**Composition** per 10.0mL:

| | |
|---|---|
| Cycloheximide | 0.05g |
| Vancomycin | 0.015g |
| Trimethoprim lactate | 5.0mg |
| Polymyxin B | 200,000U |

**Preparation of Antibiotic Solution:** Add components to distilled/deionized water and bring volume to 10.0mL. Mix thoroughly. Filter sterilize.

**Caution:** Cycloheximide is toxic. Avoid skin contact or aerosol formation and inhalation.

**Preparation of Medium:** Add components, except antibiotic solution, to distilled/deionized water and bring volume to 990.0mL. Mix thoroughly. Gently heat and bring to boiling. Autoclave for 15 min at 15 psi pressure–121°C. Cool to 45°–50°C. Aseptically add a selective sterile antibiotic solution. Generally 50.0mL of defibrinated horse blood is also added as an enrichment. Mix thoroughly.

**Use:** For the cultivation and enrichment of *Campylobacter* species from foods.

## Dubos Broth

**Composition** per liter:

| | |
|---|---|
| Na$_2$HPO$_4$ | 2.5g |
| L-Asparagine | 2.0g |
| KH$_2$PO$_4$ | 1.0g |
| Pancreatic digest of casein | 0.5g |
| Tween™ 80 | 0.2g |
| CaCl$_2$·2H$_2$O | 0.5mg |
| CuSO$_4$ | 0.1mg |
| ZnSO$_4$·7H$_2$O | 0.1mg |
| Ferric ammonium citrate | 0.05g |
| MgSO$_4$·7H$_2$O | 0.01g |
| Bovine serum albumin or bovine serum | 20.0mL |

pH 6.5 ± 0.2 at 25°C

**Source:** This medium is available as a premixed powder from BD Diagnostic Systems.

**Preparation of Medium:** Add components, except bovine serum or bovine serum albumin, to distilled/deionized water and bring volume to 980.0mL. Mix thoroughly. Gently heat and bring to boiling. Autoclave for 15 min at 15 psi pressure–121°C. Cool to 45°–50°C. Asepti-

cally add sterile bovine serum or bovine serum albumin. Mix thoroughly. Aseptically distribute into sterile tubes.

**Storage/Shelf Life:** Store dehydrated media in the dark in a sealed container below 30°C. Prepared media should be stored under refrigeration (2-8°C). Media should be used within 60 days of preparation. Media should not be used if there are any signs of deterioration (discoloration) or contamination, or if the expiration date supplied by the manufacturer has passed.

**Use:** For the cultivation of *Mycobacterium tuberculosis* and other *Mycobacterium* species.

## Dubos Broth Base with Serum and Glycerol

**Composition** per liter:

| | |
|---|---|
| Na$_2$HPO$_4$ | 2.5g |
| L-Asparagine | 2.0g |
| KH$_2$PO$_4$ | 1.0g |
| Casein enzymatic hydrolysate | 0.5g |
| Polysorbate 80 | 0.2g |
| Ferric ammonium citrate | 0.05g |
| MgSO$_4$ | 0.01g |
| CaCl$_2$ | 0.5mg |
| CuSO$_4$ | 0.1mg |
| ZnSO$_4$ | 0.1mg |
| Glycerol | 50.0mL |
| Bovine serum or bovine albumin V | 20.0mL |

pH 6.5 ± 0.2 at 25°C

**Source:** This medium, without bovine serum or glycerol, is available as a premixed powder from HiMedia.

**Preparation of Medium:** Add components, except bovine serum, to distilled/deionized water and bring volume to 980.0mL. Mix thoroughly. Gently heat and bring to boiling. Autoclave for 15 min at 15 psi pressure–121°C. Cool to 45°–50°C. Aseptically add 20.0mL sterile bovine serum or bovine serum albumin. Mix thoroughly. Aseptically distribute into sterile tubes.

**Storage/Shelf Life:** Store dehydrated media in the dark in a sealed container below 30°C. Prepared media should be stored under refrigeration (2-8°C). Media should be used within 60 days of preparation. Media should not be used if there are any signs of deterioration (discoloration) or contamination, or if the expiration date supplied by the manufacturer has passed.

**Use:** For the cultivation of *Mycobacterium tuberculosis* and other *Mycobacterium* species.

## Dubos Oleic Agar

**Composition** per liter:

| | |
|---|---|
| Agar | 15.0g |
| Na$_2$HPO$_4$ | 2.5g |
| KH$_2$PO$_4$ | 1.0g |
| L-Asparagine | 1.0g |
| Pancreatic digest of casein | 0.5g |
| Ferric ammonium citrate | 0.05g |
| MgSO$_4$·7H$_2$O | 0.01g |
| CaCl$_2$·2H$_2$O | 0.5mg |
| CuSO$_4$ | 0.1mg |
| ZnSO$_4$·7H$_2$O | 0.1mg |
| Dubos oleic albumin complex | 20.0mL |
| Penicillin solution | 10.0mL |

pH 6.6 ± 0.2 at 25°C

**Source:** This medium is available as a premixed powder from BD Diagnostic Systems.

**Dubos Oleic Albumin Complex:**
**Composition** per 100.0mL:
Bovine serum albumin, fraction V .......................................5.0g
Oleic acid, sodium salt .......................................................0.05g
NaCl (0.85% solution) ................................................... 100.0mL

**Preparation of Dubos Oleic Albumin Complex:** Add bovine serum albumin and oleic acid to 100.0mL of NaCl solution. Mix thoroughly. Filter sterilize.

**Penicillin Solution:**
**Composition** per 10.0mL:
Penicillin .....................................................................10,000U

**Preparation of Penicillin Solution:** Add penicillin to distilled/deionized water and bring volume to 10.0mL. Mix thoroughly. Filter sterilize.

**Preparation of Medium:** Add components, except Dubos oleic albumin complex and penicillin solution, to distilled/deionized water and bring volume to 970.0mL. Mix thoroughly. Gently heat and bring to boiling. Autoclave for 15 min at 15 psi pressure–121°C. Cool to 45°–50°C. Aseptically add sterile Dubos oleic albumin complex and penicillin solution. Mix thoroughly. Pour into sterile Petri dishes or distribute into sterile tubes. Allow tubes to cool in a slanted position.

**Storage/Shelf Life:** Store dehydrated media in the dark in a sealed container below 30°C. Prepared media should be stored under refrigeration (2-8°C). Media should be used within 60 days of preparation. Media should not be used if there are any signs of deterioration (shrinking, cracking, or discoloration) or contamination, or if the expiration date supplied by the manufacturer has passed.

**Use:** For the isolation of *Mycobacterium tuberculosis* and determining its sensitivity to chemotherapeutic agents.

## Dulcitol Selenite Broth
## (Selenite-F Broth with Dulcitol)
**Composition** per liter:
NaH$_2$PO$_4$ ......................................................................10.0g
Peptic digest of animal tissue .............................................5.0g
Dulcitol ...............................................................................4.0g
HNaO$_3$Se...........................................................................4.0g
pH 7.0 ± 0.2 at 25°C

**Source:** This medium is available from HiMedia.

**Caution:** Sodium hydrogen selenite is a very toxic, corrosive agent and causes teratogenicity. Upon contact with skin, wash immediately with a lot of water.

**Preparation of Medium:** Add sodium hydrogen selenite to distilled/deionized water and bring volume to 1.0L. Mix thoroughly. Add remaining components. Mix thoroughly. Gently heat if needed to get all compoents to dissolve. Distribute into tubes or flasks. Sterilize in a boiling water bath or free flowing steam for 10 min. Excessive heating is detrimental. Do not autoclave.

**Storage/Shelf Life:** Store dehydrated media in the dark in a sealed container below 30°C. Prepared media should be stored under refrigeration (2-8°C). Media should be used within 60 days of preparation. Media should not be used if there are any signs of deterioration (discoloration) or contamination, or if the expiration date supplied by the manufacturer has passed.

**Use:** For the selective enrichment of *Salmonella* species.

## Dunkelberg Carbohydrate Medium, Modified
**Composition** per 100.0mL:
Proteose peptone No. 3 ........................................................1.5g
Carbohydrate.......................................................................1.0g
Na$_2$HPO$_4$·2H$_2$O...............................................................0.207g
Phenol Red........................................................................0.055g
NaH$_2$PO$_4$·H$_2$O...............................................................0.038g
Horse serum ........................................................................5.0mL
pH 7.4 ± 0.2 at 25°C

**Preparation of Medium:** Add components, except horse serum, to distilled/deionized water and bring volume to 95.0mL. For carbohydrate, use glucose, maltose, or starch. Mix thoroughly. Filter sterilize. Aseptically add sterile horse serum. Mix thoroughly. Aseptically distribute into sterile tubes or flasks.

**Storage/Shelf Life:** Store dehydrated media in the dark in a sealed container below 30°C. Prepared media should be stored under refrigeration (2-8°C). Media should be used within 60 days of preparation. Media should not be used if there are any signs of deterioration (discoloration) or contamination, or if the expiration date supplied by the manufacturer has passed.

**Use:** For the cultivation and differentiation of *Gardnerella vaginalis* based on its ability to ferment glucose, maltose, or starch.

## Dunkelberg Semisolid Carbohydrate
## Fermentation Medium
**Composition** per liter:
Proteose peptone No. 3 ......................................................20.0g
Carbohydrate.....................................................................10.0g
Agar ....................................................................................5.0g
Bromcresol Purple solution ................................................ 1.0mL
pH 7.4 ± 0.2 at 25°C

**Bromcresol Purple Solution:**
**Composition** per 10.0mL:
Bromcresol Purple ............................................................0.16g
Ethanol (95% solution).................................................... 10.0mL

**Preparation of Bromcresol Purple Solution:** Add Bromcresol Purple to 10.0mL of ethanol. Mix thoroughly. Filter sterilize.

**Preparation of Medium:** Add components to distilled/deionized water and bring volume to 1.0L. For carbohydrate, use glucose, maltose, or starch. Mix thoroughly. Gently heat and bring to boiling. Filter sterilize. Aseptically distribute into sterile tubes or flasks.

**Storage/Shelf Life:** Store dehydrated media in the dark in a sealed container below 30°C. Prepared media should be stored under refrigeration (2-8°C). Media should be used within 60 days of preparation. Media should not be used if there are any signs of deterioration (discoloration) or contamination, or if the expiration date supplied by the manufacturer has passed.

**Use:** For the cultivation and differentiation of *Gardnerella vaginalis* based on its ability to ferment glucose, maltose, or starch.

## *E. coli*-Coliforms Chromogenic Medium
**Composition** per liter:
Agar ..................................................................................10.0g
NaCl....................................................................................5.0g
Phosphate buffer .................................................................4.9g
Peptone ...............................................................................3.0g
Tryptophan..........................................................................1.0g

| | |
|---|---|
| Sodium pyruvate | 1.0g |
| Tergitol-7 | 0.1g |

pH 6.8 ± 0.2 at 25°C

**Source:** This medium is available from CONDA, Barcelona, Spain.

**Preparation of Medium:** Add components to distilled/deionized water and bring volume to 1.0L. Mix thoroughly. Gently heat and bring to boiling. Do not autoclave. Cool to 50°C. Pour into sterile Petri dishes.

**Storage/Shelf Life:** Store dehydrated media in the dark in a sealed container below 30°C. Prepared plates should be stored in the dark under refrigeration (2-8°C). Chromogenic media are especially light and temperature sensitive; protect from light, excessive heat, moisture, and freezing. Media should not be used if there are any signs of deterioration (shrinking, cracking, or discoloration) or contamination, or if the expiration date supplied by the manufacturer has passed.

**Use:** For the simultaneous detection of *E. coli* and other coliforms in foods and water samples.

## *E. coli* O157:H7 MUG Agar

**Composition** per liter:

| | |
|---|---|
| Casein peptone | 20.0g |
| Agar | 13.0g |
| Sorbitol | 10.0g |
| NaCl | 5.0g |
| Meat extract | 2.0g |
| $Na_2S_2O_3$ | 2.0g |
| Na-deoxycholate | 1.12g |
| Yeast extract | 1.0g |
| Ammonium ferric citrate | 0.5g |
| 4-Methylumbelliferyl-β-D-glucuronide | 0.1g |
| Bromthymol Blue | 0.025g |

pH 7.4 ± 0.2 at 37°C

**Source:** This medium is available from Fluka, Sigma-Aldrich.

**Preparation of Medium:** Add components to distilled/deionized water and bring volume to 1.0L. Mix thoroughly. Gently heat while stirring and bring to boiling. Autoclave for 15 min at 15 psi pressure–121°C. Cool to 50°C. Pour into sterile Petri dishes.

**Storage/Shelf Life:** Store dehydrated media in the dark in a sealed conainer below 30°C. Prepared plates should be stored under refrigeration (2-8°C). This medium has a limited shelf life. Media should not be used if there are any signs of deterioration (shrinking, cracking, or discoloration) or contamination, or if the expiration date supplied by the manufacturer has passed.

**Use:** For the isolation and differentiation of enterohemorrhagic (EHEC) *E. coli* O157:H7 strains from food and clinical specimens.

## *E. sakazakii* Agar

**Composition** per liter:

| | |
|---|---|
| Agar | 15.0g |
| Casein peptone | 7.0g |
| NaCl | 5.0g |
| Yeast extract | 3.0g |
| Sodium deoxycholate | 0.6g |
| X-α Glucoside | 0.15g |
| Crystal Violet | 2.0mg |

pH 7.2± 0.2 at 25°C

**Source:** This medium is available from CHROMagar, Paris, France.

**Preparation of Medium:** Add components to distilled/deionized water and bring volume to 1.0L. Mix thoroughly. Distribute into tubes

or flasks. Autoclave for 15 min at 15 psi pressure–121°C. Pour into sterile Petri dishes.

**Storage/Shelf Life:** Store dehydrated media in the dark in a sealed container below 30°C. Prepared plates should be stored in the dark under refrigeration (2-8°C). Chromogenic media are especially light and temperature sensitive; protect from light, excessive heat, moisture, and freezing. Media should not be used if there are any signs of deterioration (shrinking, cracking, or discoloration) or contamination, or if the expiration date supplied by the manufacturer has passed.

**Use:** For the detection of *Enterobacter sakazakii* (*Cronobacter* spp.) from infant formula and environmental samples of public health significance according to ISO/TS 22964 standard. *E. sakazakii* produce green to intense blue colonies.

## EC Broth
## (*Escherichia coli* Broth)
## (EC Medium)
## (BAM M49)

**Composition** per liter:

| | |
|---|---|
| Pancreatic digest of casein | 20.0g |
| Lactose | 5.0g |
| NaCl | 5.0g |
| $K_2HPO_4$ | 4.0g |
| $KH_2PO_4$ | 1.5g |
| Bile salts No.3 | 1.12g |
| Novobiocin solution | 10.0mL |

pH 6.9 ± 0.2 at 25°C

**Source:** This medium is available as a premixed powder from Thermo Scientific.

**Novobiocin Solution:**
**Composition** per 50.0mL:

| | |
|---|---|
| Novobiocin | 0.1g |

**Preparation of Novobiocin Solution:** Add novobiocin to distilled/deionized water and bring volume to 50.0mL. Mix thoroughly. Filter sterilize.

**Preparation of Medium:** Add components, except novobiocin solution, to distilled/deionized water and bring volume to 990.0mL. Mix thoroughly. Autoclave for 15 min at 15 psi pressure–121°C. Cool to 50°C. Aseptically add 10.0mL novobiocin solution. Mix thoroughly. Aseptically distribute to sterile tubes or flasks.

**Storage/Shelf Life:** Store dehydrated media in the dark in a sealed container below 30°C. Prepared media should be stored under refrigeration (2-8°C). Media should be used within 60 days of preparation. Media should not be used if there are any signs of deterioration (discoloration) or contamination, or if the expiration date supplied by the manufacturer has passed.

**Use:** A selective enrichment broth for the growth of *E. coli* O157 from food and environmental samples.

## EC Broth with MUG

**Composition** per liter:

| | |
|---|---|
| Pancreatic digest of casein | 20.0g |
| Lactose | 5.0g |
| NaCl | 5.0g |
| $K_2HPO_4$ | 4.0g |
| Bile salts mixture | 1.5g |

KH$_2$PO$_4$..................................................................1.5g
4-Methylumbelliferyl-β-D-glucuronide (MUG)............................0.05g
pH 6.9 ± 0.2 at 25°C

**Source:** This medium is available as a premixed powder from BD Diagnostic Systems.

**Preparation of Medium:** Add components to distilled/deionized water and bring volume to 1.0L. Mix thoroughly. Distribute into test tubes that contain an inverted Durham tube in 10.0mL volumes. Autoclave for 15 min at 15 psi pressure–121°C.

**Storage/Shelf Life:** Store dehydrated media in the dark in a sealed container below 30°C. Prepared media should be stored under refrigeration (2-8°C). Media should be used within 60 days of preparation. Media should not be used if there are any signs of deterioration (discoloration) or contamination, or if the expiration date supplied by the manufacturer has passed.

**Use:** For the detection of *Escherichia coli* in water and food samples by a fluorogenic procedure.

### EC HiVeg Broth
**Composition** per liter:
Plant hydrolysate no. 1.................................................20.0g
Lactose...........................................................................5.0g
NaCl.................................................................................5.0g
K$_2$HPO$_4$..........................................................................4.0g
KH$_2$PO$_4$..........................................................................1.5g
Synthetic detergent.........................................................1.5g
pH 6.9 ± 0.2 at 25°C

**Source:** This medium is available as a premixed powder from Hi-Media.

**Preparation of Medium:** Add components to distilled/deionized water and bring volume to 1.0L. Mix thoroughly. Distribute into test tubes that contain an inverted Durham tube. Autoclave for 15 min at 15 psi pressure–121°C. Cool broth as quickly as possible.

**Storage/Shelf Life:** Store dehydrated media in the dark in a sealed container below 30°C. Prepared media should be stored under refrigeration (2-8°C). Media should be used within 60 days of preparation. Media should not be used if there are any signs of deterioration (discoloration) or contamination, or if the expiration date supplied by the manufacturer has passed.

**Use:** For the cultivation and differentiation of coliform bacteria at 37°C and of *Escherichia coli* at 45.5°C. Recommended for selective enumeration of presumptive *Escherichia coli* by MPN technique. For the selective enumeration of faecal and nonfaecal coliforms in water, wastewater, and shellfish.

### EC Medium, Modified with Novobiocin
**Composition** per liter:
Tryptone.........................................................................20.0g
NaCl.................................................................................5.0g
Lactose...........................................................................5.0g
K$_2$HPO$_4$..........................................................................4.0g
KH$_2$PO$_4$..........................................................................1.5g
Bile salts........................................................................1.12g
Novobiocin supplement ...............................................10.0mL
pH 6.9 ± 0.2 at 25°C

**Source:** This medium is available as a premixed powder and supplement from BD Diagnostic Systems.

**Novobiocin Supplement:**
**Composition** per 10.0mL:
Sodium novobiocin..........................................................20.0mg

**Preparation of Novobiocin Supplement:** Add sodium novobiocin to distilled/deionized water and bring volume to 10.0mL. Mix thoroughly. Filter sterilize.

**Preparation of Medium:** Add components, except novobiocin supplement, to distilled/deionized water and bring volume to 990.0mL. Mix thoroughly. Autoclave for 15 min at 15 psi pressure–121°C. Aseptically add 10.0mL of sterile novobiocin supplement. Mix thoroughly. Aseptically distribute into sterile tubes or flasks.

**Storage/Shelf Life:** Store dehydrated media in the dark in a sealed container below 30°C. Prepared media should be stored under refrigeration (2-8°C). Media should be used within 60 days of preparation. Media should not be used if there are any signs of deterioration (discoloration) or contamination, or if the expiration date supplied by the manufacturer has passed.

**Use:** For the cultivation of *Escherichia coli* O157:H7.

### ECD Agar
**Composition** per liter:
Casein enzymic hydrolysate ..........................................20.0g
Agar ................................................................................15.0g
Yeast extract...................................................................5.0g
NaCl.................................................................................5.0g
Na$_2$HPO$_4$..........................................................................5.0g
KH$_2$PO$_4$..........................................................................1.5g
Bile salts........................................................................1.5g
pH 7.2 ± 0.2 at 25°C

**Preparation of Medium:** Add components to distilled/deionized water and bring volume to 1.0L. Mix thoroughly. Gently heat and bring to boiling. Distribute into tubes or flasks. Autoclave for 15 min at 15 psi pressure–121°C. Pour into sterile Petri dishes or leave in tubes.

**Storage/Shelf Life:** Store dehydrated media in the dark in a sealed container below 30°C. Prepared media should be stored under refrigeration (2-8°C). Media should be used within 60 days of preparation. Media should not be used if there are any signs of deterioration (shrinking, cracking, or discoloration) or contamination, or if the expiration date supplied by the manufacturer has passed.

**Use:** For the selective detection of coliforms, especially *Escherichia coli* in water and food.

### ECD Agar, Fluorocult®
### (Fluorocult *ECD Agar*)
**Composition** per liter:
Peptone from casein ......................................................20.0g
Agar ................................................................................15.0g
NaCl.................................................................................5.0g
Lactose ...........................................................................5.0g
K$_2$HPO$_4$..........................................................................4.0g
Bile salt mixture...............................................................1.5g
KH$_2$PO$_4$..........................................................................1.5g
Tryptophan.......................................................................1.0g
4-Methylumbelliferyl-ß-D-glucuronide.............................0.07g
pH 7.0 ± 0.2 at 25°C

**Source:** This medium is available from Merck.

**Preparation of Medium:** Add components to distilled/deionized water and bring volume to 1.0L. Mix thoroughly. Autoclave for 15 min at 15 psi pressure–121°C. Cool to 45°–50°C. Pour into sterile Petri dishes. The prepared medium is clear and yellowish brown.

**Storage/Shelf Life:** Store dehydrated media in the dark in a sealed container below 30°C. Prepared plates should be stored in the dark under refrigeration (2-8°C). Chromogenic media are especially light and temperature sensitive; protect from light, excessive heat, moisture, and freezing. Media should not be used if there are any signs of deterioration (shrinking, cracking, or discoloration) or contamination, or if the expiration date supplied by the manufacturer has passed.

**Use:** For the detection of *E. coli* in meats. The medium complies with the German-DIN-Norm 10110 for the examination of meat, with the regulations according to § 35 LMBG (06.00/36) for the examination of food, and with ISO Standard 6391 (1996) for the enumeration of *E. coli* in meat and meat products. The bile salt mixture of this medium largely inhibits the accompanying flora not usually found in the intestines. Using fluorescence under UV light and a positive indole reaction, *E. coli* colonies can be identified among the grown colonies.

## ECD MUG Agar

**Composition** per liter:

| | |
|---|---|
| Casein peptone | 20.0g |
| Agar | 15.0g |
| NaCl | 5.0g |
| Lactose | 5.0g |
| $K_2HPO_4$ | 4.0g |
| Bile salt mixture | 1.5g |
| $KH_2PO_4$ | 1.5g |
| Tryptophan | 1.0g |
| 4-Methylumbelliferyl-β-D-glucuronide | 0.07g |

pH 7.0 ± 0.2 at 37°C

**Source:** This medium is available from Fluka, Sigma-Aldrich.

**Preparation of Medium:** Add components to distilled/deionized water and bring volume to 1.0L. Mix thoroughly. Gently heat while stirring and bring to boiling. Autoclave for 15 min at 15 psi pressure–121°C. Cool to 50°C. Pour into sterile Petri dishes.

**Storage/Shelf Life:** Store dehydrated media in the dark in a sealed conainer below 30°C. Prepared plates should be stored under refrigeration (2-8°C). This medium has a limited shelf life. Media should not be used if there are any signs of deterioration (shrinking, cracking, or discoloration) or contamination, or if the expiration date supplied by the manufacturer has passed.

**Use:** For detection of *Escherichia coli* in a variety of specimens. The bile-salt mixture in this *E. coli* Direct Agar extensively inhibits the non-obligatory intestinal accompanying flora. Fluorescence in the UV and a positive indole test demonstrate the presence of *E. coli* in the colonies.

## Egg Yolk Agar

**Composition** per liter:

| | |
|---|---|
| Proteose peptone No. 2 | 40.0g |
| Agar | 25.0g |
| $Na_2HPO_4$ | 5.0g |
| Glucose | 2.0g |
| NaCl | 2.0g |
| $KH_2PO_4$ | 1.0g |
| $MgSO_4 \cdot 7H_2O$ | 0.1g |

| | |
|---|---|
| Egg yolk emulsion | 100.0mL |
| Hemin solution | 1.0mL |

pH 7.6 ± 0.2 at 25°C

**Hemin Solution:**
**Composition** per 100.0mL:

| | |
|---|---|
| Hemin | 0.5g |
| NaOH (1$N$ solution) | 20.0mL |

**Preparation of Hemin Solution:** Add hemin to 20.0mL of 1$N$ NaOH solution. Mix thoroughly. Bring volume to 100.0mL with distilled/deionized water.

**Egg Yolk Emulsion:**
**Composition:**

| | |
|---|---|
| Chicken egg yolks | 11 |
| Whole chicken egg | 1 |

**Preparation of Egg Yolk Emulsion:** Soak eggs with 1:100 dilution of saturated mercuric chloride solution for 1 min. Crack eggs and separate yolks from whites. Mix egg yolks with 1 chicken egg.

**Preparation of Medium:** Add components, except egg yolk emulsion, to distilled/deionized water and bring volume to 900.0mL. Mix thoroughly. Gently heat and bring to boiling. Autoclave for 15 min at 15 psi pressure–121°C. Cool to 45°–50°C. Aseptically add sterile egg yolk emulsion. Mix thoroughly. Pour into sterile Petri dishes.

**Storage/Shelf Life:** Store dehydrated media in the dark in a sealed container below 30°C. Prepared media should be stored under refrigeration (2-8°C). Media should be used within 60 days of preparation. Media should not be used if there are any signs of deterioration (shrinking, cracking, or discoloration) or contamination, or if the expiration date supplied by the manufacturer has passed.

**Use:** For the isolation, cultivation, and differentiation of *Clostridium* species and some other anaerobic bacteria.

## Egg Yolk Agar Base, HiVeg with Egg Yolk Emulsion

**Composition** per liter:

| | |
|---|---|
| Plant peptone No. 3 | 40.0g |
| Agar | 25.0g |
| $Na_2HPO_4$ | 5.0g |
| Glucose | 2.0g |
| NaCl | 2.0g |
| $KH_2PO_4$ | 1.0g |
| $MgSO_4$ | 0.1g |
| $Fe_4(P_2O_7)_3 \cdot H_2O$ | 5.0mg |
| Egg yolk emulsion | 100.0mL |

pH 7.6 ± 0.2 at 25°C

**Source:** This medium, without egg yolk emulsion, is available as a premixed powder from HiMedia.

**Egg Yolk Emulsion:**
**Composition** per liter:

| | |
|---|---|
| Egg yolks | 30.0mL |
| NaCl, 0.9% solution | 70.0mL |

**Preparation of Egg Yolk Emulsion:** Soak eggs with 1:100 dilution of saturated mercuric chloride solution for 1 min. Crack 11 eggs and separate yolks from whites. Mix egg yolks. Measure 30.0mL of egg yolk emulsion and add to 70.0mL of 0.9% sterile NaCl solution. Mix thoroughly. Warm to 45°–50°C.

**Preparation of Medium:** Add components, except egg yolk emulsion, to distilled/deionized water and bring volume to 900.0mL. Mix thoroughly. Gently heat and bring to boiling. Autoclave for 15 min at

15 psi pressure–121°C. Cool to 45°–50°C. Aseptically add sterile egg yolk emulsion. Mix thoroughly. Pour into sterile Petri dishes.

**Storage/Shelf Life:** Store dehydrated media in the dark in a sealed container below 30°C. Prepared media should be stored under refrigeration (2-8°C). Media should be used within 60 days of preparation. Media should not be used if there are any signs of deterioration (shrinking, cracking, or discoloration) or contamination, or if the expiration date supplied by the manufacturer has passed.

**Use:** For the isolation, cultivation, and differentiation of *Clostridium* species and some other anaerobic bacteria.

## Egg Yolk Agar, Modified

**Composition** per liter:

| | |
|---|---|
| Agar | 20.0g |
| Pancreatic digest of casein | 15.0g |
| Vitamin $K_1$ | 10.0g |
| NaCl | 5.0g |
| Papaic digest of soybean meal | 5.0g |
| Yeast extract | 5.0g |
| L-Cystine | 0.4g |
| Hemin | 5.0mg |
| Egg yolk emulsion | 100.0mL |

**Source:** This medium is available as a prepared medium from BD Diagnostic Systems.

**Egg Yolk Emulsion:**
**Composition**:

| | |
|---|---|
| Chicken egg yolks | 11 |
| Whole chicken egg | 1 |

**Preparation of Egg Yolk Emulsion:** Soak eggs with 1:100 dilution of saturated mercuric chloride solution for 1 min. Crack eggs and separate yolks from whites. Mix egg yolks with 1 chicken egg.

**Preparation of Medium:** Add components, except egg yolk emulsion, to distilled/deionized water and bring volume to 900.0mL. Mix thoroughly. Gently heat and bring to boiling. Autoclave for 15 min at 15 psi pressure–121°C. Cool to 45°–50°C. Aseptically add sterile egg yolk emulsion. Mix thoroughly. Pour into sterile Petri dishes.

**Storage/Shelf Life:** Store dehydrated media in the dark in a sealed container below 30°C. Prepared media should be stored under refrigeration (2-8°C). Media should be used within 60 days of preparation. Media should not be used if there are any signs of deterioration (shrinking, cracking, or discoloration) or contamination, or if the expiration date supplied by the manufacturer has passed.

**Use:** For the isolation, cultivation, and differentiation of *Clostridium* species and some other anaerobic bacteria.

## Egg Yolk Emulsion

**Composition** per 100.0mL:

| | |
|---|---|
| Sterile saline | 70.0mL |
| Egg yolk | 30.0mL |

**Source:** Sterile egg yolk emulsion is available from Fluka, Sigma-Aldrich.

**Preparation of Medium:** Use fresh eggs, less than 1 week old. Scrub the shells with soap. Let stand in a soap solution for 30 min. Rinse in running water. Soak eggs in 70% ethanol for 15 min. or soak eggs with 1:100 dilution of saturated mercuric chloride solution for 1 min. Crack eggs and separate yolks from whites, placing egg yolks into a sterile container. Use enough eggs to produce at least 30.0mL egg yolk. Homogenize by shaking. Add 0.9g NaCl to distilled/deionized water and bring volume to 100.0mL. Sterilze the saline solution by filtration or by autoclaving for 15 min at 15 psi pressure–121°C. If autoclaving is used, cool to 25°C. Aseptically add 30.0mL homogenized egg yolks to 70.0mL of sterile saline solution. Mix thoroughly.

**Storage/Shelf Life:** Store dehydrated media in the dark in a sealed container below 30°C. Prepared media should be stored under refrigeration (2-8°C). Media should be used within 60 days of preparation. Media should not be used if there are any signs of deterioration (shrinking, cracking, or discoloration) or contamination, or if the expiration date supplied by the manufacturer has passed.

**Use:** Sterile stabilized emulsion of egg yolk is recommended for use in various culture media.

## Egg Yolk Emulsion

**Composition** per 100.0mL:

| | |
|---|---|
| NaCl | 0.45g |
| Egg yolk | 50.0mL |

**Preparation of Medium:** Use fresh eggs, less than 1 week old. Scrub the shells with soap. Let stand in a soap solution for 30 min. Rinse in running water. Soak eggs in 70% ethanol for 15 min. or soak eggs with 1:100 dilution of saturated mercuric chloride solution for 1 min. Crack eggs and separate yolks from whites, placing egg yolks into a sterile container. Use enough eggs to produce at least 50.0mL egg yolk. Homogenize by shaking. Add 0.45g NaCl to distilled/deionized water and bring volume to 50.0mL. Sterilize the saline solution by filtration or by autoclaving for 15 min at 15 psi pressure–121°C. If autoclaving is used, cool to 25°C. Aseptically add 50.0mL homogenized egg yolks to 50.0mL of the sterile NaCl solution. Mix thoroughly.

**Storage/Shelf Life:** Prepare fresh. Use right away.

**Use:** Sterile stabilized emulsion of egg yolk is recommended for use in various culture media.

## Egg Yolk Emulsion, 50% (BAM M51)

**Composition** per 100.0mL:

| | |
|---|---|
| Chicken egg yolks | variable |
| NaCl (0.85% solution) | 40.0mL |

**Preparation of Egg Yolk Emulsion:** Wash fresh eggs with a stiff brush and drain. Soak eggs in 70% ethanol for 1 h. Crack eggs aseptically and separate yolks from whites. Remove egg yolks with a sterile syringe or a wide-mouth pipet. Place 50.0mL of egg yolks into a sterile container. Add 50.0mL sterile 0.85% saline.

**Storage/Shelf Life:** Prepare fresh. Use right away.

**Use:** For use in media requiring egg yolk emulsion.

## Egg Yolk Tellurite Emulsion 20%

**Composition** per 100.0mL:

| | |
|---|---|
| NaCl | 0.425g |
| $K_2TeO_3$ | 0.21g |
| Egg yolk | 20.0mL |

**Preparation of Medium:** Use fresh eggs, less than 1 week old. Scrub the shells with soap. Let stand in a soap solution for 30 min. Rinse in running water. Soak eggs in 70% ethanol for 15 min or soak eggs with 1:100 dilution of saturated mercuric chloride solution for 1 min. Crack eggs and separate yolks from whites, placing egg yolks into a sterile container. Use enough eggs to produce at least 20.0mL egg yolk. Homogenize by shaking. Add 0.45g NaCl and 0.21g $K_2TeO_3$ to distilled/deionized water and bring volume to 80.0mL. Sterilize the saline-tellurite solution by filtration or by autoclaving for 15 min at 15 psi pressure–121°C. If autoclaving is used, cool to 25°C. Aseptically add 20.0mL homogenized egg yolks to 80.0mL of the sterile saline-tellurite solution. Mix thoroughly.

**Storage/Shelf Life:** Prepare fresh. Use right away.

**Use:** For use in various culture media. It may be added directly to nutrient media for the identification of *Clostridium, Bacillus,* and *Staphylococcus* species by their lipase activity.

## Eijkman Lactose HiVeg Broth

**Composition** per liter:
Plant hydrolysate No. 1...................................................15.0g
NaCl...............................................................................5.0g
$K_2HPO_4$.......................................................................4.0g
Lactose...........................................................................3.0g
$KH_2PO_4$.......................................................................1.5g

pH 6.8 ± 0.1 at 25°C

**Source:** This medium is available as a premixed powder from Hi-Media.

**Preparation of Medium:** Add components to distilled/deionized water and bring volume to 1.0L. Mix thoroughly. Distribute into test tubes that contain an inverted Durham tube. Autoclave for 15 min at 15 psi pressure–121°C.

**Storage/Shelf Life:** Store dehydrated media in the dark in a sealed container below 30°C. Prepared media should be stored under refrigeration (2-8°C). Media should be used within 60 days of preparation. Media should not be used if there are any signs of deterioration (discoloration) or contamination, or if the expiration date supplied by the manufacturer has passed.

**Use:** For the cultivation and differentiation of *Escherichia coli* from other coliform organisms based on their ability to ferment lactose and produce gas.

## Eijkman Lactose Medium

**Composition** per liter:
Pancreatic digest of casein.............................................15.0g
$K_2HPO_4$.....................................................................10.0g
$KH_2PO_4$.......................................................................4.0g
Lactose...........................................................................3.0g
NaCl...............................................................................2.5g

pH 6.8 ± 0.1 at 25°C

**Preparation of Medium:** Add components to distilled/deionized water and bring volume to 1.0L. Mix thoroughly. Distribute into test tubes that contain an inverted Durham tube. Autoclave for 15 min at 15 psi pressure–121°C.

**Storage/Shelf Life:** Store dehydrated media in the dark in a sealed container below 30°C. Prepared media should be stored under refrigeration (2-8°C). Media should be used within 60 days of preparation. Media should not be used if there are any signs of deterioration (discoloration) or contamination, or if the expiration date supplied by the manufacturer has passed.

**Use:** For the cultivation and differentiation of *Escherichia coli* from other coliform organisms based on their ability to ferment lactose and produce gas.

## EMB Agar
## (Eosin Methylene Blue Agar)

**Composition** per liter:
Agar.............................................................................13.5g
Pancreatic digest of casein.............................................10.0g
Lactose...........................................................................5.0g
Sucrose...........................................................................5.0g
$K_2HPO_4$.......................................................................2.0g
Eosin Y...........................................................................0.4g
Methylene Blue...........................................................0.065g

pH 7.2 ± 0.2 at 25°C

**Source:** This medium is available as a premixed powder from BD Diagnostic Systems.

**Preparation of Medium:** Add components to distilled/deionized water and bring volume to 1.0L. Mix thoroughly. Gently heat and bring to boiling. Distribute into tubes or flasks. Autoclave for 15 min at 15 psi pressure–121°C. Pour into sterile Petri dishes.

**Storage/Shelf Life:** Store dehydrated media in the dark in a sealed container below 30°C. Prepared media should be stored under refrigeration (2-8°C). Media should be used within 60 days of preparation. Media should not be used if there are any signs of deterioration (shrinking, cracking, or discoloration) or contamination, or if the expiration date supplied by the manufacturer has passed.

**Use:** For the isolation, cultivation, and differentiation of Gram-negative enteric bacteria based on lactose fermentation. Bacteria that ferment lactose, especially the coliform bacterium *Escherichia coli*, appear as colonies with a green metallic sheen or blue-black to brown color. Bacteria that do not ferment lactose appear as colorless or transparent, light purple colonies.

## EMB Agar Base

**Composition** per liter:
Agar.............................................................................15.0g
Peptone........................................................................10.0g
$K_2HPO_4$.......................................................................2.0g
Eosin Y...........................................................................0.4g
Methylene Blue...........................................................0.065g

pH 7.3 ± 0.2 at 25°C

**Preparation of Medium:** Add components to distilled/deionized water and bring volume to 1.0L. Mix thoroughly. Gently heat and bring to boiling. Distribute into tubes or flasks. Autoclave for 15 min at 15 psi pressure–121°C. Pour into sterile Petri dishes.

**Storage/Shelf Life:** Store dehydrated media in the dark in a sealed container below 30°C. Prepared media should be stored under refrigeration (2-8°C). Media should be used within 60 days of preparation. Media should not be used if there are any signs of deterioration (shrinking, cracking, or discoloration) or contamination, or if the expiration date supplied by the manufacturer has passed.

**Use:** For the isolation, cultivation, and differentiation of Gram-negative enteric bacteria based on lactose fermentation. Bacteria that ferment lactose, especially the coliform bacterium *Escherichia coli*, appear as colonies with a green metallic sheen or blue-black to brown color. Bacteria that do not ferment lactose appear as colorless or transparent, light purple colonies.

## EMB Agar, Modified
## (Eosin Methylene Blue Agar, Modified)

**Composition** per liter:

| | |
|---|---|
| Agar | 15.0g |
| Lactose | 10.0g |
| Pancreatic digest of gelatin | 10.0g |
| $K_2HPO_4$ | 2.0g |
| Eosin Y | 0.4g |
| Methylene Blue | 0.065g |

pH 6.8 ± 0.2 at 25°C

**Source:** This medium is available as a premixed powder from Thermo Scientific.

**Preparation of Medium:** Add components to distilled/deionized water and bring volume to 1.0L. Mix thoroughly. Gently heat and bring to boiling. Distribute into tubes or flasks. Autoclave for 15 min at 15 psi pressure–121°C. Cool to 60°C. Shake medium to oxidize methylene blue. Pour into sterile Petri dishes. Swirl flask while pouring plates to distribute precipitate.

**Storage/Shelf Life:** Store dehydrated media in the dark in a sealed container below 30°C. Prepared media should be stored under refrigeration (2-8°C). Media should be used within 60 days of preparation. Media should not be used if there are any signs of deterioration (shrinking, cracking, or discoloration) or contamination, or if the expiration date supplied by the manufacturer has passed.

**Use:** For the isolation, cultivation, and differentiation of Gram-negative enteric bacteria based on lactose fermentation. Bacteria that ferment lactose, especially the coliform bacterium *Escherichia coli*, appear as colonies with a green metallic sheen or blue-black to brown color. Bacteria that do not ferment lactose appear as colorless or transparent, light purple colonies.

## EMB HiVeg Agar

**Composition** per liter:

| | |
|---|---|
| Agar | 13.5g |
| Plant peptone | 10.0g |
| Lactose | 5.0g |
| Sucrose | 5.0g |
| $K_2HPO_4$ | 2.0g |
| Eosin Y | 0.4g |
| Methylene Blue | 0.065g |

pH 7.2 ± 0.2 at 25°C

**Source:** This medium is available as a premixed powder from Hi-Media.

**Preparation of Medium:** Add components to distilled/deionized water and bring volume to 1.0L. Mix thoroughly. Gently heat and bring to boiling. Distribute into tubes or flasks. Autoclave for 15 min at 15 psi pressure–121°C. Pour into sterile Petri dishes.

**Storage/Shelf Life:** Store dehydrated media in the dark in a sealed container below 30°C. Prepared media should be stored under refrigeration (2-8°C). Media should be used within 60 days of preparation. Media should not be used if there are any signs of deterioration (shrinking, cracking, or discoloration) or contamination, or if the expiration date supplied by the manufacturer has passed.

**Use:** For the differential isolation of Gram-negative enteric bacilli from clinical speciments and nonclinical specimens of public health importance. For the isolation, cultivation, and differentiation of Gram-negative enteric bacteria based on lactose fermentation. Bacteria that ferment lactose, especially the coliform bacterium *Escherichia coli*,

appear as colonies with a green metallic sheen or blue-black to brown color. Bacteria that do not ferment lactose appear as colorless or transparent, light purple colonies.

## EMB HiVeg Agar, Levine

**Composition** per liter:

| | |
|---|---|
| Agar | 15.0g |
| Plant peptone | 10.0g |
| Lactose | 10.0g |
| $K_2HPO_4$ | 2.0g |
| Eosin Y | 0.4g |
| Methylene Blue | 0.065g |

pH 7.2 ± 0.2 at 25°C

**Source:** This medium is available as a premixed powder from Hi-Media.

**Preparation of Medium:** Add components to distilled/deionized water and bring volume to 1.0L. Mix thoroughly. Gently heat and bring to boiling. Distribute into tubes or flasks. Autoclave for 15 min at 15 psi pressure–121°C. Pour into sterile Petri dishes.

**Storage/Shelf Life:** Store dehydrated media in the dark in a sealed container below 30°C. Prepared media should be stored under refrigeration (2-8°C). Media should be used within 60 days of preparation. Media should not be used if there are any signs of deterioration (shrinking, cracking, or discoloration) or contamination, or if the expiration date supplied by the manufacturer has passed.

**Use:** For the isolation, enumeration, and differentiation of members of Enterobacteriaceae. For the isolation, cultivation, and differentiation of Gram-negative enteric bacteria based on lactose fermentation. Bacteria that ferment lactose, especially the coliform bacterium *Escherichia coli*, appear as colonies with a green metallic sheen or blue-black to brown color. Bacteria that do not ferment lactose appear as colorless or transparent, light purple colonies.

## EMB HiVeg Broth

**Composition** per liter:

| | |
|---|---|
| Plant peptone | 10.0g |
| Lactose | 5.0g |
| Sucrose | 5.0g |
| $K_2HPO_4$ | 2.0g |
| Eosin Y | 0.4g |
| Methylene Blue | 0.065g |

pH 7.2 ± 0.2 at 25°C

**Source:** This medium is available as a premixed powder from Hi-Media.

**Preparation of Medium:** Add components to distilled/deionized water and bring volume to 1.0L. Mix thoroughly. Distribute into test tubes that contain an inverted Durham tube. Autoclave for 15 min at 15 psi pressure–121°C.

**Storage/Shelf Life:** Store dehydrated media in the dark in a sealed container below 30°C. Prepared media should be stored under refrigeration (2-8°C). Media should be used within 60 days of preparation. Media should not be used if there are any signs of deterioration (discoloration) or contamination, or if the expiration date supplied by the manufacturer has passed.

**Use:** For the differential cultivation of Gram-negative enteric bacilli from clinical speciments and nonclinical specimens of public health importance.

## Endo Agar

**Composition** per liter:

Agar ....................................................................................15.0g
Lactose ...............................................................................10.0g
Peptic digest of animal tissue......................................10.0g
$K_2HPO_4$.............................................................................3.5g
$Na_2SO_3$.............................................................................2.5g
Basic Fuchsin.....................................................................0.5g

pH 7.4 ± 0.2 at 25°C

**Source:** This medium is available as a premixed powder from BD Diagnostic Systems.

**Caution:** Basic Fuchsin is a potential carcinogen and care must be taken to avoid inhalation of the powdered dye and contact with the skin.

**Preparation of Medium:** Add components to distilled/deionized water and bring volume to 1.0L. Mix thoroughly. Gently heat and bring to boiling. Autoclave for 15 min at 15 psi pressure–121°C. Cool to 45°–50°C. Pour into sterile Petri dishes. Swirl flask while pouring plates to keep precipitate in suspension. Protect from the light.

**Storage/Shelf Life:** Store dehydrated media in the dark in a sealed container below 30°C. Prepared media should be stored under refrigeration (2-8°C). Media should be used within 60 days of preparation. Media should not be used if there are any signs of deterioration (shrinking, cracking, or discoloration) or contamination, or if the expiration date supplied by the manufacturer has passed.

**Use:** For the selective isolation, cultivation, and differentiation of coliform and other enteric microorganisms based on their ability to ferment lactose. Lactose-fermenting bacteria appear as dark red colonies with a gold metallic sheen. Lactose-nonfermenting bacteria appear as colorless or translucent colonies.

## Endo Agar

**Composition** per liter:

Agar ....................................................................................10.0g
Lactose ...............................................................................10.0g
Peptic digest of animal tissue......................................10.0g
$K_2HPO_4$.............................................................................3.5g
$Na_2SO_3$.............................................................................2.5g
Basic Fuchsin solution ................................................. 4.0mL

pH 7.5 ± 0.2 at 25°C

**Source:** This medium is available as a premixed powder from Thermo Scientific.

**Basic Fuchsin Solution:**

**Composition** per 10.0mL:

Basic Fuchsin.....................................................................1.0g
Ethanol (95% solution) ........................................... 10.0mL

**Preparation of Basic Fuchsin Solution:** Add Basic Fuchsin to 10.0mL of ethanol. Mix thoroughly.

**Caution:** Basic Fuchsin is a potential carcinogen and care must be taken to avoid inhalation of the powdered dye and contact with the skin.

**Preparation of Medium:** Add components to distilled/deionized water and bring volume to 1.0L. Mix thoroughly. Gently heat and bring to boiling. Autoclave for 15 min at 15 psi pressure–121°C. Cool to 45°–50°C. Pour into sterile Petri dishes. Swirl flask while pouring plates to keep precipitate in suspension. Protect from the light.

**Storage/Shelf Life:** Store dehydrated media in the dark in a sealed container below 30°C. Prepared media should be stored under refrigeration (2-8°C). Media should be used within 60 days of preparation. Media should not be used if there are any signs of deterioration (shrinking, cracking, or discoloration) or contamination, or if the expiration date supplied by the manufacturer has passed.

**Use:** For the selective isolation, cultivation, and differentiation of coliform and other enteric microorganisms based on their ability to ferment lactose. Lactose-fermenting bacteria appear as dark red colonies with a gold metallic sheen. Lactose-nonfermenting bacteria appear as colorless or translucent colonies.

## Endo Agar

**Composition** per liter:

Agar ....................................................................................15.0g
Peptic digest of animal tissue .....................................10.0g
Lactose ...............................................................................10.0g
$K_2HPO_4$.............................................................................3.5g
$Na_2SO_3$ ............................................................................2.5g
Basic Fuchsin.....................................................................0.5g

pH 7.5 ± 0.2 at 25°C

**Source:** This medium is available as a premixed powder from HiMedia.

**Caution:** Basic Fuchsin is a potential carcinogen and care must be taken to avoid inhalation of the powdered dye and contact with the skin.

**Preparation of Medium:** Add components to distilled/deionized water and bring volume to 1.0L. Mix thoroughly. Gently heat and bring to boiling. Autoclave for 15 min at 15 psi pressure–121°C. Cool to 45°–50°C. Pour into sterile Petri dishes. Swirl flask while pouring plates to keep precipitate in suspension. Protect from the light.

**Storage/Shelf Life:** Store dehydrated media in the dark in a sealed container below 30°C. Prepared media should be stored under refrigeration (2-8°C). Media should be used within 60 days of preparation. Media should not be used if there are any signs of deterioration (shrinking, cracking, or discoloration) or contamination, or if the expiration date supplied by the manufacturer has passed.

**Use:** For the selective isolation, cultivation, and differentiation of coliform and other enteric microorganisms based on their ability to ferment lactose. Lactose-fermenting bacteria appear as dark red colonies with a gold metallic sheen. Lactose-nonfermenting bacteria appear as colorless or translucent colonies.

## Endo Agar Base

**Composition** per liter:

Agar ....................................................................................12.0g
Peptic digest of animal tissue .....................................10.0g
Lactose ...............................................................................10.0g
$K_2HPO_4$.............................................................................3.5g
$Na_2SO_3$ ............................................................................2.5g
Basic Fuchsin solution ................................................. 4.0mL

pH 7.5 ± 0.2 at 25°C

**Source:** This medium, without Basic Fuchsin solution, is available as a premixed powder from HiMedia.

**Basic Fuchsin Solution:**

**Composition** per 10.0mL:

Basic Fuchsin...................................................................1.0g
Ethanol (95% solution) ........................................... 10.0mL

**Preparation of Basic Fuchsin Solution:** Add Basic Fuchsin to 10.0mL of ethanol. Mix thoroughly.

**Caution:** Basic Fuchsin is a potential carcinogen and care must be taken to avoid inhalation of the powdered dye and contact with the skin.

**Preparation of Medium:** Add components to distilled/deionized water and bring volume to 1.0L. Mix thoroughly. Gently heat and bring to boiling. Autoclave for 15 min at 15 psi pressure–121°C. Cool to 45°–50°C. Pour into sterile Petri dishes. Swirl flask while pouring plates to keep precipitate in suspension. Protect from the light.

**Storage/Shelf Life:** Store dehydrated media in the dark in a sealed container below 30°C. Prepared media should be stored under refrigeration (2-8°C). Media should be used within 60 days of preparation. Media should not be used if there are any signs of deterioration (shrinking, cracking, or discoloration) or contamination, or if the expiration date supplied by the manufacturer has passed.

**Use:** For the selective isolation, cultivation, and differentiation of coliform and other enteric microorganisms based on their ability to ferment lactose. Lactose-fermenting bacteria appear as dark red colonies with a gold metallic sheen. Lactose-nonfermenting bacteria appear as colorless or translucent colonies.

## Endo Agar, Modified

**Composition** per liter:

| | |
|---|---|
| Agar | 12.5g |
| Peptic digest of animal tissue | 10.0g |
| Lactose | 10.0g |
| Na$_2$SO$_3$ | 3.3g |
| K$_2$HPO$_4$ | 2.5g |
| Basic Fuchsin | 0.3g |

pH 7.4 ± 0.2 at 25°C

**Source:** This medium is available as a premixed powder from Hi-Media.

**Caution:** Basic Fuchsin is a potential carcinogen and care must be taken to avoid inhalation of the powdered dye and contact with the skin.

**Preparation of Medium:** Add components to distilled/deionized water and bring volume to 1.0L. Mix thoroughly. Gently heat and bring to boiling. Autoclave for 15 min at 15 psi pressure–121°C. Cool to 45°–50°C. Pour into sterile Petri dishes. Swirl flask while pouring plates to keep precipitate in suspension. Protect from the light.

**Storage/Shelf Life:** Store dehydrated media in the dark in a sealed container below 30°C. Prepared media should be stored under refrigeration (2-8°C). Media should be used within 60 days of preparation. Media should not be used if there are any signs of deterioration (shrinking, cracking, or discoloration) or contamination, or if the expiration date supplied by the manufacturer has passed.

**Use:** For the selective isolation, cultivation, and differentiation of coliform and other enteric microorganisms based on their ability to ferment lactose. Lactose-fermenting bacteria appear as dark red colonies with a gold metallic sheen. Lactose-nonfermenting bacteria appear as colorless or translucent colonies.

## Endo Agar, LES
## (m-Endo Agar, LES)
## (m-LES, Endo Agar)

**Composition** per liter:

| | |
|---|---|
| Agar | 14.0g |
| Lactose | 9.4g |
| Peptones (pancreatic digest of casein 65% and yeast extract 35%) | 7.5g |
| NaCl | 3.7g |
| Pancreatic digest of casein | 3.7g |
| Peptic digest of animal tissue | 3.7g |
| K$_2$HPO$_4$ | 3.3g |
| Na$_2$SO$_3$ | 1.6g |
| Yeast extract | 1.2g |
| KH$_2$PO$_4$ | 1.0g |
| Basic Fuchsin | 0.8g |
| Sodium lauryl sulfate | 0.05g |
| Ethanol | 20.0mL |

pH 7.2 ± 0.2 at 25°C

**Source:** This medium is available as a premixed powder from BD Diagnostic Systems.

**Caution:** Basic Fuchsin is a potential carcinogen and care must be taken to avoid inhalation of the powdered dye and contact with the skin.

**Preparation of Medium:** Add ethanol to approximately 900.0mL of distilled/deionized water. Add remaining components. Bring volume to 1.0L with distilled/deionized water. Mix thoroughly. Gently heat and bring to boiling. Autoclave for 15 min at 15 psi pressure–121°C. Pour into sterile 60mm Petri dishes in 4.0mL volumes. Protect from the light.

**Storage/Shelf Life:** Store dehydrated media in the dark in a sealed container below 30°C. Prepared media should be stored under refrigeration (2-8°C). Media should be used within 60 days of preparation. Media should not be used if there are any signs of deterioration (shrinking, cracking, or discoloration) or contamination, or if the expiration date supplied by the manufacturer has passed.

**Use:** For the cultivation and enumeration of coliform bacteria by the membrane filter method.

## Endo Agar, LES
## (m-Endo Agar, LES)

**Composition** per liter:

| | |
|---|---|
| Agar | 10.0g |
| Lactose | 9.4g |
| Tryptose | 7.5g |
| NaCl | 3.7g |
| Peptone | 3.7g |
| Pancreatic digest of casein | 3.7g |
| K$_2$HPO$_4$ | 3.3g |
| Na$_2$SO$_3$ | 1.6g |
| Yeast extract | 1.2g |
| KH$_2$PO$_4$ | 1.0g |
| Sodium deoxycholate | 0.1g |
| Sodium lauryl sulfate | 0.05g |
| Basic Fuchsin solution | 8.0mL |

pH 7.2 ± 0.2 at 25°C

**Basic Fuchsin Solution:**
**Composition** per 10.0mL:

Basic Fuchsin ................................................................1.0g
Ethanol (95% solution) ...............................................10.0mL

**Caution:** Basic Fuchsin is a potential carcinogen and care must be taken to avoid inhalation of the powdered dye and contact with the skin.

**Preparation of Basic Fuchsin Solution:** Add Basic Fuchsin to 10.0mL of ethanol. Mix thoroughly.

**Preparation of Medium:** Add components to distilled/deionized water and bring volume to 1.0L. Mix thoroughly. Gently heat and bring to boiling. Autoclave for 15 min at 15 psi pressure–121°C. Cool to 45°–50°C. Pour into sterile Petri dishes. Swirl flask while pouring plates to keep precipitate in suspension. Protect from the light.

**Storage/Shelf Life:** Store dehydrated media in the dark in a sealed container below 30°C. Prepared media should be stored under refrigeration (2-8°C). Media should be used within 60 days of preparation. Media should not be used if there are any signs of deterioration (shrinking, cracking, or discoloration) or contamination, or if the expiration date supplied by the manufacturer has passed.

**Use:** For the cultivation and enumeration of coliform bacteria from water by the membrane filter method.

## Endo Agar with Sodium Chloride

**Composition** per liter:

Agar ...............................................................................12.0g
Lactose .........................................................................10.0g
Peptone, special .............................................................8.0g
NaCl ................................................................................3.0g
$Na_2SO_3$ ...........................................................................2.5g
$K_2HPO_4$ ...........................................................................2.0g
Basic Fuchsin ................................................................0.2g

pH 7.4 ± 0.2 at 25°C

**Source:** This medium is available as a premixed powder from Hi-Media.

**Caution:** Basic Fuchsin is a potential carcinogen and care must be taken to avoid inhalation of the powdered dye and contact with the skin.

**Preparation of Medium:** Add components to distilled/deionized water and bring volume to 1.0L. Mix thoroughly. Gently heat and bring to boiling. Autoclave for 15 min at 15 psi pressure–121°C. Cool to 45°–50°C. Pour into sterile Petri dishes. Swirl flask while pouring plates to keep precipitate in suspension. Protect from the light.

**Storage/Shelf Life:** Store dehydrated media in the dark in a sealed container below 30°C. Prepared media should be stored under refrigeration (2-8°C). Media should be used within 60 days of preparation. Media should not be used if there are any signs of deterioration (shrinking, cracking, or discoloration) or contamination, or if the expiration date supplied by the manufacturer has passed.

**Use:** For the selective isolation, cultivation, and differentiation of coliform and other enteric microorganisms based on their ability to ferment lactose.

## Endo Broth
## (m-Endo Broth)

**Composition** per liter:

Lactose .........................................................................12.5g
Peptone..........................................................................10.0g

NaCl ................................................................................5.0g
Pancreatic digest of casein............................................5.0g
Peptic digest of animal tissue .......................................5.0g
$K_2HPO_4$ ........................................................................4.375g
$Na_2SO_3$ ...........................................................................2.1g
Yeast extract .................................................................1.5g
$KH_2PO_4$ .......................................................................1.375g
Basic Fuchsin ..............................................................1.05g
Sodium deoxycholate.....................................................0.1g
Ethanol (95% solution) ...............................................20.0mL

pH 7.2 ± 0.1 at 25°C

**Source:** This medium is available as a premixed powder from BD Diagnostic Systems.

**Caution:** Basic Fuchsin is a potential carcinogen and care must be taken to avoid inhalation of the powdered dye and contact with the skin.

**Preparation of Medium:** Add ethanol to approximately 900.0mL of distilled/deionized water. Add remaining components. Bring volume to 1.0L with distilled/deionized water. Mix thoroughly. Gently heat and bring to boiling. Rapidly cool broth below 45°C. Do not autoclave. Use 1.8–2.0mL for each filter pad. Protect from the light. Prepare broth freshly.

**Storage/Shelf Life:** Store dehydrated media in the dark in a sealed container below 30°C. Prepared media should be stored under refrigeration (2-8°C). Media should be used within 60 days of preparation. Media should not be used if there are any signs of deterioration (discoloration) or contamination, or if the expiration date supplied by the manufacturer has passed.

**Use:** For the cultivation and enumeration of coliform bacteria from water by the membrane filter method.

## Endo HiVeg Agar

**Composition** per liter:

Agar ...............................................................................15.0g
Plant peptone ...............................................................10.0g
Lactose .........................................................................10.0g
$K_2HPO_4$ ...........................................................................3.5g
$Na_2SO_3$ ...........................................................................2.5g
Basic Fuchsin ................................................................0.5g

pH 7.5 ± 0.2 at 25°C

**Source:** This medium is available as a premixed powder from Hi-Media.

**Caution:** Basic Fuchsin is a potential carcinogen and care must be taken to avoid inhalation of the powdered dye and contact with the skin.

**Preparation of Medium:** Add components to distilled/deionized water and bring volume to 1.0L. Mix thoroughly. Gently heat and bring to boiling. Autoclave for 15 min at 15 psi pressure–121°C. Cool to 45°–50°C. Pour into sterile Petri dishes. Swirl flask while pouring plates to keep precipitate in suspension. Protect from the light.

**Storage/Shelf Life:** Store dehydrated media in the dark in a sealed container below 30°C. Prepared media should be stored under refrigeration (2-8°C). Media should be used within 60 days of preparation. Media should not be used if there are any signs of deterioration (shrinking, cracking, or discoloration) or contamination, or if the expiration date supplied by the manufacturer has passed.

**Use:** For the selective isolation, cultivation, and differentiation of coliform and other enteric microorganisms based on their ability to ferment lactose. Lactose-fermenting bacteria appear as dark red colonies with a gold metallic sheen. Lactose-nonfermenting bacteria appear as colorless or translucent colonies.

## Endo HiVeg Agar Base

**Composition** per liter:

| | |
|---|---|
| Agar | 12.0g |
| Plant peptone | 10.0g |
| Lactose | 10.0g |
| $K_2HPO_4$ | 3.5g |
| $Na_2SO_3$ | 2.5g |
| Basic Fuchsin solution | 4.0mL |

pH 7.5 ± 0.2 at 25°C

**Source:** This medium, without Basic Fuchsin solution, is available as a premixed powder from HiMedia.

**Basic Fuchsin Solution:**
**Composition** per 10.0mL:

| | |
|---|---|
| Basic Fuchsin | 1.0g |
| Ethanol (95% solution) | 10.0mL |

**Preparation of Basic Fuchsin Solution:** Add Basic Fuchsin to 10.0mL of ethanol. Mix thoroughly.

**Caution:** Basic Fuchsin is a potential carcinogen and care must be taken to avoid inhalation of the powdered dye and contact with the skin.

**Preparation of Medium:** Add components to distilled/deionized water and bring volume to 1.0L. Mix thoroughly. Gently heat and bring to boiling. Autoclave for 15 min at 15 psi pressure–121°C. Cool to 45°–50°C. Pour into sterile Petri dishes. Swirl flask while pouring plates to keep precipitate in suspension. Protect from the light.

**Storage/Shelf Life:** Store dehydrated media in the dark in a sealed container below 30°C. Prepared media should be stored under refrigeration (2-8°C). Media should be used within 60 days of preparation. Media should not be used if there are any signs of deterioration (shrinking, cracking, or discoloration) or contamination, or if the expiration date supplied by the manufacturer has passed.

**Use:** For the selective isolation, cultivation, and differentiation of coliform and other enteric microorganisms based on their ability to ferment lactose. Lactose-fermenting bacteria appear as dark red colonies with a gold metallic sheen. Lactose-nonfermenting bacteria appear as colorless or translucent colonies.

## Endo HiVeg Agar, Modified

**Composition** per liter:

| | |
|---|---|
| Agar | 12.5g |
| Plant peptone | 10.0g |
| Lactose | 10.0g |
| $Na_2SO_3$ | 3.3g |
| $K_2HPO_4$ | 2.5g |
| Basic Fuchsin | 0.3g |

pH 7.4 ± 0.2 at 25°C

**Source:** This medium is available as a premixed powder from Hi-Media.

**Caution:** Basic Fuchsin is a potential carcinogen and care must be taken to avoid inhalation of the powdered dye and contact with the skin.

**Preparation of Medium:** Add components to distilled/deionized water and bring volume to 1.0L. Mix thoroughly. Gently heat and bring to boiling. Autoclave for 15 min at 15 psi pressure–121°C. Cool to 45°–

50°C. Pour into sterile Petri dishes. Swirl flask while pouring plates to keep precipitate in suspension. Protect from the light.

**Storage/Shelf Life:** Store dehydrated media in the dark in a sealed container below 30°C. Prepared media should be stored under refrigeration (2-8°C). Media should be used within 60 days of preparation. Media should not be used if there are any signs of deterioration (shrinking, cracking, or discoloration) or contamination, or if the expiration date supplied by the manufacturer has passed.

**Use:** For the selective isolation, cultivation, and differentiation of coliform and other enteric microorganisms based on their ability to ferment lactose. Lactose-fermenting bacteria appear as dark red colonies with a gold metallic sheen. Lactose-nonfermenting bacteria appear as colorless or translucent colonies.

## Endo HiVeg Agar with NaCl

**Composition** per liter:

| | |
|---|---|
| Agar | 12.0g |
| Lactose | 10.0g |
| Plant special peptone | 8.0g |
| NaCl | 3.0g |
| $Na_2SO_3$ | 2.5g |
| $K_2HPO_4$ | 2.0g |
| Basic Fuchsin | 0.2g |

pH 7.4 ± 0.2 at 25°C

**Source:** This medium is available as a premixed powder from Hi-Media.

**Caution:** Basic Fuchsin is a potential carcinogen and care must be taken to avoid inhalation of the powdered dye and contact with the skin.

**Preparation of Medium:** Add components to distilled/deionized water and bring volume to 1.0L. Mix thoroughly. Gently heat and bring to boiling. Autoclave for 15 min at 15 psi pressure–121°C. Cool to 45°–50°C. Pour into sterile Petri dishes. Swirl flask while pouring plates to keep precipitate in suspension. Protect from the light.

**Storage/Shelf Life:** Store dehydrated media in the dark in a sealed container below 30°C. Prepared media should be stored under refrigeration (2-8°C). Media should be used within 60 days of preparation. Media should not be used if there are any signs of deterioration (shrinking, cracking, or discoloration) or contamination, or if the expiration date supplied by the manufacturer has passed.

**Use:** For the selective isolation, cultivation, and differentiation of coliform and other enteric microorganisms based on their ability to ferment lactose.

## Enrichment Broth for *Aeromonas hydrophila*

**Composition** per liter:

| | |
|---|---|
| NaCl | 5.0g |
| Maltose | 3.5g |
| Yeast extract | 3.0g |
| Bile salts No. 3 | 1.0g |
| L-Cysteine·HCl·$H_2O$ | 0.3g |
| Bromthymol Blue | 0.03g |
| Novobiocin | 5.0mg |

pH 7.0 ± 0.2 at 25°C

**Preparation of Medium:** Add components to distilled/deionized water and bring volume to 1.0L. Mix thoroughly. Distribute into tubes or flasks. Autoclave for 15 min at 15 psi pressure–121°C.

**Storage/Shelf Life:** Store dehydrated media in the dark in a sealed container below 30°C. Prepared media should be stored under refrigeration (2-8°C). Media should be used within 60 days of preparation. Media should not be used if there are any signs of deterioration (discoloration) or contamination, or if the expiration date supplied by the manufacturer has passed.

**Use:** For the cultivation and enrichment of *Aeromonas hydrophila*.

## Enrichment Broth, pH 7.3

**Composition** per liter:
| | |
|---|---|
| Pancreatic digest of casein | 17.0g |
| Yeast extract | 6.0g |
| NaCl | 5.0g |
| Papaic digest of soybean meal | 3.0g |
| Glucose | 2.5g |
| $K_2HPO_4$ | 2.5g |
| Nalidixic acid solution | 8.0mL |
| Cycloheximide solution | 5.1mL |
| Acriflavin·HCl solution | 3.0mL |

pH 7.3 ± 0.2 at 25°C

**Cycloheximide Solution:**
**Composition** per 10.0mL:
| | |
|---|---|
| Cycloheximide | 0.1g |
| Ethanol, absolute | 4.0mL |

**Preparation of Cycloheximide Solution:** Add components to distilled/deionized water and bring volume to 10.0mL. Mix thoroughly. Filter sterilize.

**Caution:** Cycloheximide is toxic. Avoid skin contact or aerosol formation and inhalation.

**Nalidixic Acid Solution:**
**Composition** per 10.0mL:
| | |
|---|---|
| Nalidixic acid | 0.05g |

**Preparation of Nalidixic Acid Solution:** Add nalidixic acid to distilled/deionized water and bring volume to 10.0mL. Mix thoroughly. Filter sterilize.

**Acriflavin·HCl Solution:**
**Composition** per 10.0mL:
| | |
|---|---|
| Acriflavin·HCl | 0.05g |

**Preparation of Acriflavin·HCl Solution:** Add acriflavin·HCl to distilled/deionized water and bring volume to 10.0mL. Mix thoroughly. Filter sterilize.

**Preparation of Medium:** Add components—except cycloheximide solution, nalidixic acid solution, and acriflavin·HCl solution—to distilled/deionized water and bring volume to 983.9mL. Mix thoroughly. Gently heat and bring to boiling. Autoclave for 15 min at 15 psi pressure–121°C. Cool to 45°–50°C. Aseptically add 5.1mL of sterile cycloheximide solution, 8.0mL of sterile nalidixic acid solution, and 3.0mL of sterile acriflavin·HCl solution. Mix thoroughly. Aseptically distribute into sterile tubes.

**Storage/Shelf Life:** Store dehydrated media in the dark in a sealed container below 30°C. Prepared media should be stored under refrigeration (2-8°C). Media should be used within 60 days of preparation. Media should not be used if there are any signs of deterioration (discoloration) or contamination, or if the expiration date supplied by the manufacturer has passed.

**Use:** For the isolation, cultivation, and enrichment of a variety of microorganisms from nondairy foods.

## Enrichment Broth, pH 7.3

**Composition** per liter:
| | |
|---|---|
| Pancreatic digest of casein | 17.0g |
| Yeast extract | 6.0g |
| NaCl | 5.0g |
| Papaic digest of soybean meal | 3.0g |
| Glucose | 2.5g |
| $K_2HPO_4$ | 2.5g |
| Nalidixic acid solution | 8.0mL |
| Cycloheximide solution | 5.1mL |
| Acriflavin·HCl solution | 2.0mL |

pH 7.3 ± 0.2 at 25°C

**Cycloheximide Solution:**
**Composition** per 10.0mL:
| | |
|---|---|
| Cycloheximide | 0.1g |
| Ethanol, absolute | 4.0mL |

**Preparation of Cycloheximide Solution:** Add components to distilled/deionized water and bring volume to 10.0mL. Mix thoroughly. Filter sterilize.

**Caution:** Cycloheximide is toxic. Avoid skin contact or aerosol formation and inhalation.

**Nalidixic Acid Solution:**
**Composition** per 10.0mL:
| | |
|---|---|
| Nalidixic acid | 0.05g |

**Preparation of Nalidixic Acid Solution:** Add nalidixic acid to distilled/deionized water and bring volume to 10.0mL. Mix thoroughly. Filter sterilize.

**Acriflavin·HCl Solution:**
**Composition** per 10.0mL:
| | |
|---|---|
| Acriflavin·HCl | 0.05g |

**Preparation of Acriflavin·HCl Solution:** Add acriflavin·HCl to distilled/deionized water and bring volume to 10.0mL. Mix thoroughly. Filter sterilize.

**Preparation of Medium:** Add components—except cycloheximide solution, nalidixic acid solution, and acriflavin·HCl solution—to distilled/deionized water and bring volume to 984.9mL. Mix thoroughly. Gently heat and bring to boiling. Autoclave for 15 min at 15 psi pressure–121°C. Cool to 45°–50°C. Aseptically add 5.1mL of cycloheximide solution, 8.0mL of nalidixic acid solution, and 2.0mL of acriflavin·HCl solution. Mix thoroughly. Aseptically distribute into sterile tubes.

**Storage/Shelf Life:** Store dehydrated media in the dark in a sealed container below 30°C. Prepared media should be stored under refrigeration (2-8°C). Media should be used within 60 days of preparation. Media should not be used if there are any signs of deterioration (discoloration) or contamination, or if the expiration date supplied by the manufacturer has passed.

**Use:** For the isolation, cultivation, and enrichment of a variety of microorganisms from milk and dairy products.

## Enrichment Broth, pH 7.3 with Pyruvate

**Composition** per liter:
| | |
|---|---|
| Pancreatic digest of casein | 17.0g |
| Yeast extract | 6.0g |
| NaCl | 5.0g |
| Papaic digest of soybean meal | 3.0g |
| Glucose | 2.5g |

K₂HPO₄.................................................................2.5g
Pyruvate solution ........................................... 11.1mL
Nalidixic acid solution ..................................... 8.0mL
Cycloheximide solution ..................................... 5.1mL
Acriflavin·HCl solution ..................................... 3.0mL

pH 7.3 ± 0.2 at 25°C

**Pyruvate Solution:**
**Composition** per 20.0mL:
Sodium pyruvate ..........................................2.0g

**Preparation of Pyruvate Solution:** Add sodium pyruvate to distilled/deionized water and bring volume to 20.0mL. Mix thoroughly. Filter sterilize.

**Cycloheximide Solution:**
**Composition** per 10.0mL:
Cycloheximide ...............................................0.1g
Ethanol, absolute.......................................... 4.0mL

**Preparation of Cycloheximide Solution:** Add components to distilled/deionized water and bring volume to 10.0mL. Mix thoroughly. Filter sterilize.

**Caution:** Cycloheximide is toxic. Avoid skin contact or aerosol formation and inhalation.

**Nalidixic Acid Solution:**
**Composition** per 10.0mL:
Nalidixic acid.................................................0.05g

**Preparation of Nalidixic Acid Solution:** Add nalidixic acid to distilled/deionized water and bring volume to 10.0mL. Mix thoroughly. Filter sterilize.

**Acriflavin·HCl Solution:**
**Composition** per 10.0mL:
Acriflavin·HCl ...............................................0.05g

**Preparation of Acriflavin·HCl Solution:** Add acriflavin·HCl to distilled/deionized water and bring volume to 10.0mL. Mix thoroughly. Filter sterilize.

**Preparation of Medium:** Add components—except pyruvate solution, cycloheximide solution, nalidixic acid solution, and acriflavin·HCl solution—to distilled/deionized water and bring volume to 972.8mL. Mix thoroughly. Gently heat and bring to boiling. Autoclave for 15 min at 15 psi pressure–121°C. Cool to 45°–50°C. Aseptically add 11.1mL of sterile pyruvate solution. Mix thoroughly. Inoculate medium and incubate at 30°C for 6 h. Aseptically add 5.1mL of sterile cycloheximide solution, 8.0mL of sterile nalidixic acid solution, and 3.0mL of sterile acriflavin·HCl solution. Mix thoroughly.

**Storage/Shelf Life:** Store dehydrated media in the dark in a sealed container below 30°C. Prepared media should be stored under refrigeration (2-8°C). Media should be used within 60 days of preparation. Media should not be used if there are any signs of deterioration (discoloration) or contamination, or if the expiration date supplied by the manufacturer has passed.

**Use:** For the isolation, cultivation, and enrichment of *Listeria* species from nondairy foods.

## Enrichment Broth, pH 7.3 with Pyruvate
**Composition** per liter:
Pancreatic digest of casein ...........................17.0g
NaCl......................................................................5.0g
Papaic digest of soybean meal ........................3.0g
Glucose ..............................................................2.5g

K₂HPO₄.................................................................2.5g
Yeast extract.....................................................6.0g
Pyruvate solution ........................................... 11.1mL
Nalidixic acid solution..................................... 8.0mL
Cycloheximide solution ..................................... 5.1mL
Acriflavin·HCl solution ..................................... 2.0mL

pH 7.3 ± 0.2 at 25°C

**Pyruvate Solution:**
**Composition** per 20.0mL:
Sodium pyruvate..........................................2.0g

**Preparation of Pyruvate Solution:** Add sodium pyruvate to distilled/deionized water and bring volume to 20.0mL. Mix thoroughly. Filter sterilize.

**Cycloheximide Solution:**
**Composition** per 10.0mL:
Cycloheximide...............................................0.1g
Ethanol, absolute..........................................4.0mL

**Preparation of Cycloheximide Solution:** Add components to distilled/deionized water and bring volume to 10.0mL. Mix thoroughly. Filter sterilize.

**Caution:** Cycloheximide is toxic. Avoid skin contact or aerosol formation and inhalation.

**Nalidixic Acid Solution:**
**Composition** per 10.0mL:
Nalidixic acid.................................................0.05g

**Preparation of Nalidixic Acid Solution:** Add nalidixic acid to distilled/deionized water and bring volume to 10.0mL. Mix thoroughly. Filter sterilize.

**Acriflavin·HCl Solution:**
**Composition** per 10.0mL:
Acriflavin·HCl ...............................................0.05g

**Preparation of Acriflavin·HCl Solution:** Add acriflavin·HCl to distilled/deionized water and bring volume to 10.0mL. Mix thoroughly. Filter sterilize.

**Preparation of Medium:** Add components—except pyruvate solution, cycloheximide solution, nalidixic acid solution, and acriflavin·HCl solution—to distilled/deionized water and bring volume to 973.8mL. Mix thoroughly. Gently heat and bring to boiling. Autoclave for 15 min at 15 psi pressure–121°C. Cool to 45°–50°C. Aseptically add 11.1mL of sterile pyruvate solution. Mix thoroughly. Inoculate medium and incubate at 30°C for 6 h. Aseptically add 5.1mL of sterile cycloheximide solution, 8.0mL of sterile nalidixic acid solution, and 2.0mL of sterile acriflavin·HCl solution. Mix thoroughly.

**Storage/Shelf Life:** Store dehydrated media in the dark in a sealed container below 30°C. Prepared media should be stored under refrigeration (2-8°C). Media should be used within 60 days of preparation. Media should not be used if there are any signs of deterioration (discoloration) or contamination, or if the expiration date supplied by the manufacturer has passed.

**Use:** For the isolation, cultivation, and enrichment of *Listeria* species from milk and dairy products.

## *Enterobacter sakazakii* Isolation Chromogenic Agar
**Composition** per liter:
Agar .................................................................15.0g
Pancreatic digest of casein...............................7.0g

NaCl ............................................................................5.0g
Yeast extract ...............................................................3.0g
Sodium desoxycholate ................................................0.6g
α-X-Glucoside ...........................................................0.15g
Crystal Violet ...........................................................2.0mg
pH 7.0 ± 0.2 at 25°C

**Source:** This medium is available from CONDA, Barcelona, Spain.

**Preparation of Medium:** Add components to distilled/deionized water and bring volume to 1.0L. Mix thoroughly. Distribute into tubes or flasks. Autoclave for 15 min at 15 psi pressure–121°C. Pour into sterile Petri dishes.

**Storage/Shelf Life:** Store dehydrated media in the dark in a sealed container below 30°C. Prepared plates should be stored in the dark under refrigeration (2-8°C). Chromogenic media are especially light and temperature sensitive; protect from light, excessive heat, moisture, and freezing. Media should not be used if there are any signs of deterioration (shrinking, cracking, or discoloration) or contamination, or if the expiration date supplied by the manufacturer has passed.

**Use:** For the isolation and presumptive identification of *Enterobacter sakazakii* from food.

## Enterococci Broth, Chromocult (Chromocult Enterococci Broth)

**Composition** per liter:
Peptone .......................................................................8.6g
NaCl ............................................................................6.4g
Tween 80 .....................................................................2.2g
NaN₃ ............................................................................0.6g
5-Bromo-4-chloro-3-indolyl-β-D-glucopyranoside (X-GLU) .......0.04
pH 7.5 ± 0.2 at 25°C

**Source:** This medium is available from Merck.

**Preparation of Medium:** Add components to distilled/deionized water and bring volume to 1.0L. Mix well. Distribute into tubes. Autoclave for 15 min at 15 psi pressure–121°C. The prepared broth is clear and yellowish.

**Caution:** Sodium azide is toxic. Azides also react with metals and disposal must be highly diluted.

**Storage/Shelf Life:** Store dehydrated media in the dark in a sealed container below 30°C. Prepared plates should be stored in the dark under refrigeration (2-8°C). Chromogenic media are especially light and temperature sensitive; protect from light, excessive heat, moisture, and freezing. Media should not be used if there are any signs of deterioration (shrinking, cracking, or discoloration) or contamination, or if the expiration date supplied by the manufacturer has passed.

**Use:** For the detection of enterococci in food and water. The sodium azide present in this medium largely inhibits the growth of the accompanying, and especially the Gram-negative microbial flora while sparing the enterococci. The substrate X-GLU (5-bromo-4-chloro-3-indolyl-β-D-glucopyranoside) is cleaved, stimulated by selected peptones, by the enzyme β-D-glucosidase which is characteristic for enterococci. This results in an intensive blue-green color of the broth. Azide, at the same time, prevents a false positive result by most other β-D-glucosidase positive bacteria. Therefore, the color change of the broth largely confirms the presence of enterococci and group D streptococci.

## Enterococci Confirmatory Agar

**Composition** per liter:
Agar ..........................................................................15.0g
Glucose .......................................................................5.0g

Pancreatic digest of casein .........................................5.0g
Yeast extract ...............................................................5.0g
NaN₃ ............................................................................0.4g
Methylene Blue .......................................................10.0mg
Enterococci confirmatory broth ..............................variable
pH 8.0 ± 0.2 at 25°C

**Source:** This medium is available as a premixed powder from BD Diagnostic Systems.

**Caution:** Sodium azide is toxic. Azides also react with metals and disposal must be highly diluted.

**Preparation of Medium:** Add components to distilled/deionized water and bring volume to 1.0L. Mix thoroughly. Gently heat and bring to boiling. Distribute into tubes. Autoclave for 15 min at 15 psi pressure–121°C. Allow tubes to cool in a slanted position. Add sufficient amount of Enterococci confirmatory broth to cover half the slant.

**Storage/Shelf Life:** Store dehydrated media in the dark in a sealed container below 30°C. Prepared media should be stored under refrigeration (2-8°C). Media should be used within 60 days of preparation. Media should not be used if there are any signs of deterioration (shrinking, cracking, or discoloration) or contamination, or if the expiration date supplied by the manufacturer has passed.

**Use:** For the identification of enterococci from water by the confirmatory test.

## Enterococci Confirmatory Broth

**Composition** per liter:
NaCl ..........................................................................65.0g
Glucose .......................................................................5.0g
Pancreatic digest of casein .........................................5.0g
Yeast extract ...............................................................5.0g
NaN₃ ............................................................................0.4g
Methylene Blue .......................................................10.0mg
Penicillin ....................................................................650U
pH 8.0 ± 0.2 at 25°C

**Source:** This medium is available as a premixed powder from BD Diagnostic Systems.

**Caution:** Sodium azide is toxic. Azides also react with metals and disposal must be highly diluted.

**Preparation of Medium:** Add components, except penicillin, to distilled/deionized water and bring volume to 1.0L. Mix thoroughly. Gently heat and bring to boiling. Autoclave for 15 min at 15 psi pressure–121°C. Cool to 25°C. Aseptically add penicillin. Mix thoroughly.

**Storage/Shelf Life:** Store dehydrated media in the dark in a sealed container below 30°C. Prepared media should be stored under refrigeration (2-8°C). Media should be used within 60 days of preparation. Media should not be used if there are any signs of deterioration (discoloration) or contamination, or if the expiration date supplied by the manufacturer has passed.

**Use:** For the identification of enterococci from water by the confirmatory test.

## Enterococci Presumptive Broth

**Composition** per liter:
Glucose .......................................................................5.0g
Pancreatic digest of casein .........................................5.0g
Yeast extract ...............................................................5.0g

NaN$_3$.................................................................0.4g
Bromthymol Blue ...........................................32.0mg

pH 8.4 ± 0.2 at 25°C

**Source:** This medium is available as a premixed powder from BD Diagnostic Systems.

**Caution:** Sodium azide is toxic. Azides also react with metals and disposal must be highly diluted.

**Preparation of Medium:** Add components to distilled/deionized water and bring volume to 1.0L. Mix thoroughly. Distribute into tubes or flasks. Autoclave for 15 min at 15 psi pressure–121°C.

**Storage/Shelf Life:** Store dehydrated media in the dark in a sealed container below 30°C. Prepared media should be stored under refrigeration (2-8°C). Media should be used within 60 days of preparation. Media should not be used if there are any signs of deterioration (discoloration) or contamination, or if the expiration date supplied by the manufacturer has passed.

**Use:** For the isolation and identification of enterococci from water by the presumptive test. Bacteria that produce acid and turn the medium yellow and turbid after incubation at 45°C are presumptive enterococci.

## Enterococcosel™ Agar

**Composition** per liter:
Pancreatic digest of casein............................................17.0g
Agar ....................................................................13.5g
Oxgall.................................................................10.0g
NaCl......................................................................5.0g
Yeast extract..........................................................5.0g
Peptic digest of animal tissue........................................3.0g
Esculin ..................................................................1.0g
Sodium citrate ........................................................1.0g
Ferric ammonium citrate.............................................0.5g
NaN$_3$................................................................0.25g

pH 7.1 ± 0.2 at 25°C

**Source:** This medium is available as a premixed powder from BD Diagnostic Systems.

**Caution:** Sodium azide is toxic. Azides also react with metals and disposal must be highly diluted.

**Preparation of Medium:** Add components to distilled/deionized water and bring volume to 1.0L. Mix thoroughly. Gently heat while stirring and bring to boiling. Distribute into tubes or flasks. Autoclave for 15 min at 15 psi pressure–121°C. Pour into sterile Petri dishes or leave in tubes.

**Storage/Shelf Life:** Store dehydrated media in the dark in a sealed container below 30°C. Prepared media should be stored under refrigeration (2-8°C). Media should be used within 60 days of preparation. Media should not be used if there are any signs of deterioration (shrinking, cracking, or discoloration) or contamination, or if the expiration date supplied by the manufacturer has passed.

**Use:** For the rapid, selective isolation, cultivation, and enumeration of fecal group D streptococci (enterococci). For the cultivation of staphylococci and *Listeria monocytogenes.*

## Enterococcosel™ Agar with Vancomycin

**Composition** per liter:
Pancreatic digest of casein............................................17.0g
Agar ....................................................................13.5g
Oxgall.................................................................10.0g

NaCl......................................................................5.0g
Yeast extract..........................................................5.0g
Peptic digest of animal tissue........................................3.0g
Esculin ..................................................................1.0g
Sodium citrate ........................................................1.0g
Ferric ammonium citrate.............................................0.5g
NaN$_3$................................................................0.25g
Vancomycin solution ..............................................10.0mL

pH 7.1 ± 0.2 at 25°C

**Source:** This medium is available as a premixed powder from BD Diagnostic Systems.

**Caution:** Sodium azide is toxic. Azides also react with metals and disposal must be highly diluted.

**Vancomycin Solution:**
**Composition** per 10.0mL:
Vancomycin ...........................................................8.0mg

**Preparation of Vancomycin Solution:** Add vancomycin to distilled/deionized water and bring volume to 10.0mL. Mix thoroughly. Filter sterilize.

**Preparation of Medium:** Add components—except vancomycin solution—to distilled/deionized water and bring volume to 990.0mL. Mix thoroughly. Gently heat and bring to boiling. Autoclave for 15 min at 15 psi pressure–121°C. Cool to 45°–50°C. Aseptically add 10.0mL of freshly prepared sterile vancomycin solution. Mix thoroughly. Pour into sterile Petri dishes or distribute into sterile tubes.

**Storage/Shelf Life:** Store dehydrated media in the dark in a sealed container below 30°C. Prepared media should be stored under refrigeration (2-8°C). Media should be used within 60 days of preparation. Media should not be used if there are any signs of deterioration (shrinking, cracking, or discoloration) or contamination, or if the expiration date supplied by the manufacturer has passed.

**Use:** For the detection of vanocmycin resistant enterococci (VRE), particularly for primary screening of asymptomatic gastrointestinal carriage of VRE.

## Enterococcosel™ Broth

**Composition** per liter:
Pancreatic digest of casein............................................17.0g
Oxgall.................................................................10.0g
NaCl......................................................................5.0g
Yeast extract..........................................................5.0g
Peptic digest of animal tissue........................................3.0g
Esculin ..................................................................1.0g
Sodium citrate ........................................................1.0g
Ferric ammonium citrate.............................................0.5g
NaN$_3$................................................................0.25g

pH 7.1 ± 0.2 at 25°C

**Source:** This medium is available as a premixed powder from BD Diagnostic Systems.

**Caution:** Sodium azide is toxic. Azides also react with metals and disposal must be highly diluted.

**Preparation of Medium:** Add components to distilled/deionized water and bring volume to 1.0L. Mix thoroughly. Gently heat while stirring until dissolved. Distribute into tubes or flasks. Autoclave for 15 min at 15 psi pressure–121°C.

**Storage/Shelf Life:** Store dehydrated media in the dark in a sealed container below 30°C. Prepared media should be stored under refrigeration (2-8°C). Media should be used within 60 days of preparation.

Media should not be used if there are any signs of deterioration (discoloration) or contamination, or if the expiration date supplied by the manufacturer has passed.

**Use:** For the cultivation and differentiation of group D streptococci (enterococci).

## *Enterococcus* Agar
## (m-*Enterococcus* Agar)
## (Azide Agar)

**Composition** per liter:

| | |
|---|---|
| Pancreatic digest of casein | 15.0g |
| Agar | 10.0g |
| Papaic digest of soybean meal | 5.0g |
| Yeast extract | 5.0g |
| $KH_2PO_4$ | 4.0g |
| Glucose | 2.0g |
| $NaN_3$ | 0.4g |
| Triphenyltetrazolium chloride | 0.1g |

pH 7.2 ± 0.2 at 25°C

**Source:** This medium is available as a premixed powder from BD Diagnostic Systems.

**Caution:** Sodium azide is toxic. Azides also react with metals and disposal must be highly diluted.

**Preparation of Medium:** Add components to distilled/deionized water and bring volume to 1.0L. Mix thoroughly. Gently heat and bring to boiling. Cool to 45°–50°C. Do not autoclave. Pour into sterile Petri dishes.

**Storage/Shelf Life:** Store dehydrated media in the dark in a sealed container below 30°C. Prepared media should be stored under refrigeration (2-8°C). Media should be used within 60 days of preparation. Media should not be used if there are any signs of deterioration (shrinking, cracking, or discoloration) or contamination, or if the expiration date supplied by the manufacturer has passed.

**Use:** For the isolation, cultivation, and enumeration of entercocci in water, sewage, and feces by the membrane filter method. For the direct plating of specimens for the detection and enumeration of fecal streptococci.

## *Enterococcus* Confirmatory HiVeg Agar
## with Penicillin

**Composition** per liter:

| | |
|---|---|
| Agar | 15.0g |
| Glucose | 5.0g |
| Plant hydrolysate | 5.0g |
| Yeast extract | 5.0g |
| $NaN_3$ | 0.4g |
| Methylene Blue | 0.01g |
| Penicillin solution | 10.0mL |

pH 8.0± 0.2 at 25°C

**Source:** This medium, without penicillin, is available as a premixed powder from HiMedia.

**Caution:** Sodium azide is toxic. Azides also react with metals and disposal must be highly diluted.

**Penicillin Solution:**

**Composition** per 10.0mL:

| | |
|---|---|
| Penicillin | 650U |

**Preparation of Penicillin Solution:** Add penicillin to distilled/deionized water and bring volume to 10.0mL. Mix thoroughly. Filter sterilize.

**Preparation of Medium:** Add components, except penicillin solution, to distilled/deionized water and bring volume to 1.0L. Mix thoroughly. Gently heat and bring to boiling. Autoclave for 15 min at 15 psi pressure–121°C. Cool to 45°–50°C. Aseptically add 10.0mL sterile penicillin solution. Mix thoroughly. Pour into sterile Petri dishes or dispense into sterile tubes.

**Storage/Shelf Life:** Store dehydrated media in the dark in a sealed container below 30°C. Prepared media should be stored under refrigeration (2-8°C). Media should be used within 60 days of preparation. Media should not be used if there are any signs of deterioration (shrinking, cracking, or discoloration) or contamination, or if the expiration date supplied by the manufacturer has passed.

**Use:** For the isolation, cultivation, and enumeration of entercocci in water, sewage, and feces by the membrane filter method. For the direct plating of specimens for the detection and enumeration of fecal streptococci.

## *Enterococcus* Confirmatory HiVeg Broth
## with Penicillin

**Composition** per liter:

| | |
|---|---|
| NaCl | 65.0g |
| Glucose | 5.0g |
| Plant hydrolysate | 5.0g |
| Yeast extract | 5.0g |
| $NaN_3$ | 0.4g |
| Methylene Blue | 0.01g |
| Penicillin solution | 10.0mL |

pH 8.0 ± 0.2 at 25°C

**Source:** This medium, without penicillin, is available as a premixed powder from HiMedia.

**Caution:** Sodium azide is toxic. Azides also react with metals and disposal must be highly diluted.

**Penicillin Solution:**

**Composition** per 10.0mL:

| | |
|---|---|
| Penicillin | 650U |

**Preparation of Penicillin Solution:** Add penicillin to distilled/deionized water and bring volume to 10.0mL. Mix thoroughly. Filter sterilize.

**Preparation of Medium:** Add components, except penicillin solution, to distilled/deionized water and bring volume to 1.0L. Mix thoroughly. Gently heat and bring to boiling. Autoclave for 15 min at 15 psi pressure–121°C. Cool to 45°–50°C. Aseptically add 10.0mL sterile penicillin solution. Mix thoroughly.

**Storage/Shelf Life:** Store dehydrated media in the dark in a sealed container below 30°C. Prepared media should be stored under refrigeration (2-8°C). Media should be used within 60 days of preparation. Media should not be used if there are any signs of deterioration (discoloration) or contamination, or if the expiration date supplied by the manufacturer has passed.

**Use:** For the cultivation of entercocci in water, sewage, and feces.

## Eosin Methylene Blue Agar, Modified, Holt-Harris and Teague (Eosin Methylene Blue Agar) (EMB Agar)

**Composition** per liter:

| | |
|---|---|
| Agar | 13.5g |
| Pancreatic digest of gelatin | 10.0g |
| Lactose | 5.0g |
| Sucrose | 5.0g |
| Dipotassium phosphate | 2.0g |
| Eosin Y | 0.4g |
| Methylene Blue | 0.065g |

pH 7.2 ± 0.2 at 25°C

**Source:** This medium is available as a premixed powder from BD Diagnostic Systems.

**Preparation of Medium:** Add components to distilled/deionized water and bring volume to 1.0L. Mix thoroughly. Gently heat and bring to boiling. Autoclave for 15 min at 15 psi pressure–121°C. Pour into sterile Petri dishes.

**Storage/Shelf Life:** Store dehydrated media in the dark in a sealed container below 30°C. Prepared media should be stored under refrigeration (2-8°C). Media should be used within 60 days of preparation. Media should not be used if there are any signs of deterioration (shrinking, cracking, or discoloration) or contamination, or if the expiration date supplied by the manufacturer has passed.

**Use:** For the isolation, cultivation, and differentiation of Gram-negative enteric bacteria based on lactose fermentation. Bacteria that ferment lactose, especially the coliform bacterium *Escherichia coli*, appear as colonies with a green metallic sheen or blue-black to brown color. Bacteria that do not ferment lactose appear as colorless or transparent, light purple colonies.

## Esculin Agar

**Composition** per liter:

| | |
|---|---|
| Agar | 15.0g |
| Pancreatic digest of casein | 13.0g |
| NaCl | 5.0g |
| Yeast extract | 5.0g |
| Heart muscle, solids from infusion | 2.0g |
| Esculin | 1.0g |
| Ferric citrate | 0.5g |

pH 7.3 ± 0.2 at 25°C

**Preparation of Medium:** Add components to distilled/deionized water and bring volume to 1.0L. Mix thoroughly. Gently heat and bring to boiling. Distribute into screw-capped tubes in 3.0mL volumes. Autoclave for 15 min at 15 psi pressure–121°C. Allow tubes to cool in a slanted position.

**Storage/Shelf Life:** Store dehydrated media in the dark in a sealed container below 30°C. Prepared media should be stored under refrigeration (2-8°C). Media should be used within 60 days of preparation. Media should not be used if there are any signs of deterioration (shrinking, cracking, or discoloration) or contamination, or if the expiration date supplied by the manufacturer has passed.

**Use:** For the cultivation and differentiation of bacteria based on their ability to hydrolyze esculin and produce $H_2S$. Bacteria that hydrolyze esculin appear as colonies surrounded by a reddish-brown to dark brown zone. Bacteria that produce $H_2S$ appear as black colonies.

## Esculin Agar, Modified CDC (BAM M53)

**Composition** per liter:

| | |
|---|---|
| Heart muscle, infusion from | 375.0g |
| Agar | 15.0g |
| Thiotone | 10.0g |
| NaCl | 5.0g |
| Esculin | 1.0g |
| Ferric citrate | 0.5g |

pH 7.0 ± 0.2 at 25°C

**Preparation of Medium:** Add components to distilled/deionized water and bring volume to 1.0L. Mix thoroughly. Gently heat and bring to boiling. Cool to 55°C. Adjust pH to 7.0. Distribute into tubes or leave in flask. Autoclave for 20 min at 15 psi pressure–121°C. Pour into sterile Petri dishes or leave in tubes. Allow tubes to cool in inclined position to produce slants.

**Storage/Shelf Life:** Store dehydrated media in the dark in a sealed container below 30°C. Prepared media should be stored under refrigeration (2-8°C). Media should be used within 60 days of preparation. Media should not be used if there are any signs of deterioration (shrinking, cracking, or discoloration) or contamination, or if the expiration date supplied by the manufacturer has passed.

**Use:** For the differentiation of *Enterobacter* spp.

## Esculin Azide Broth

**Composition** per liter:

| | |
|---|---|
| Peptic digest of animal tissue | 20.0g |
| Bile salts | 10.0g |
| Yeast extract | 5.0g |
| Esculin | 1.0g |
| Sodium citrate | 1.0g |
| Ferric ammonium citrate | 0.5g |
| $NaN_3$ | 0.25g |

pH 7.2 ± 0.2 at 25°C

**Source:** This medium is available as a premixed powder from Hi-Media.

**Caution:** Sodium azide is toxic. Azides also react with metals and disposal must be highly diluted.

**Preparation of Medium:** Add components to distilled/deionized water and bring volume to 1.0L. Mix thoroughly. Gently heat and bring to boiling. Cool to 45°–50°C. Do not autoclave.

**Storage/Shelf Life:** Store dehydrated media in the dark in a sealed container below 30°C. Prepared media should be stored under refrigeration (2-8°C). Media should be used within 60 days of preparation. Media should not be used if there are any signs of deterioration (discoloration) or contamination, or if the expiration date supplied by the manufacturer has passed.

**Use:** For the cultivation of entercocci in water, sewage, and feces.

## Esculin Azide HiVeg Broth

**Composition** per liter:

| | |
|---|---|
| Plant peptone | 25.0g |
| Synthetic detergent | 5.0g |
| Yeast extract | 5.0g |
| Esculin | 1.0g |
| Sodium citrate | 1.0g |

Ferric ammonium citrate.................................................0.5g

NaN$_3$ ...........................................................................0.25g

pH 7.2 ± 0.2 at 25°C

**Source:** This medium is available as a premixed powder from Hi-Media.

**Caution:** Sodium azide is toxic. Azides also react with metals and disposal must be highly diluted.

**Preparation of Medium:** Add components to distilled/deionized water and bring volume to 1.0L. Mix thoroughly. Gently heat and bring to boiling. Cool to 45°–50°C. Do not autoclave.

**Storage/Shelf Life:** Store dehydrated media in the dark in a sealed container below 30°C. Prepared media should be stored under refrigeration (2-8°C). Media should be used within 60 days of preparation. Media should not be used if there are any signs of deterioration (discoloration) or contamination, or if the expiration date supplied by the manufacturer has passed.

**Use:** For the cultivation of entercocci in water, sewage, and feces.

### Esculin Broth

**Composition** per liter:

Beef heart, solids from infusion.....................................500.0g

Tryptose .......................................................................10.0g

NaCl...............................................................................5.0g

Agar ...............................................................................1.0g

Esculin ...........................................................................1.0g

pH 7.0 ± 0.2 at 25°C

**Preparation of Medium:** Add components to distilled/deionized water and bring volume to 1.0L. Mix thoroughly. Gently heat and bring to boiling. Distribute into screw-capped tubes in 7.0mL volumes. Autoclave for 15 min at 15 psi pressure–121°C.

**Storage/Shelf Life:** Store dehydrated media in the dark in a sealed container below 30°C. Prepared media should be stored under refrigeration (2-8°C). Media should be used within 60 days of preparation. Media should not be used if there are any signs of deterioration (discoloration) or contamination, or if the expiration date supplied by the manufacturer has passed.

**Use:** For the cultivation and differentiation of bacteria based on their ability to hydrolyze esculin. Bacteria that hydrolyze esculin turn the medium brown-black to black.

### Esculin Iron Agar

**Composition** per liter:

Agar ...........................................................................15.0g

Esculin ..........................................................................1.0g

Ferric ammonium citrate.................................................0.5g

pH 7.1 ± 0.2 at 25°C

**Source:** This medium is available as a premixed powder from BD Diagnostic Systems.

**Preparation of Medium:** Add components to distilled/deionized water and bring volume to 1.0L. Mix thoroughly. Gently heat and bring to boiling. Distribute into tubes or flasks. Autoclave for 15 min at 15 psi pressure–121°C. Pour into sterile Petri dishes.

**Storage/Shelf Life:** Store dehydrated media in the dark in a sealed container below 30°C. Prepared media should be stored under refrigeration (2-8°C). Media should be used within 60 days of preparation. Media should not be used if there are any signs of deterioration (shrink-

ing, cracking, or discoloration) or contamination, or if the expiration date supplied by the manufacturer has passed.

**Use:** For the cultivation and identification of enterococci based on their ability to hydrolyze esculin. Used in conjunction with E agar and the membrane filter method.

### Esculin Mannitol Agar

**Composition** per liter:

Agar .............................................................................13.5g

Polypeptone™ ..............................................................10.0g

D-Mannitol ...................................................................10.0g

Pancreatic digest of casein .............................................5.0g

Yeast extract...................................................................5.0g

NaCl...............................................................................5.0g

Heart peptone.................................................................3.0g

Cornstarch......................................................................1.0g

Esculin ...........................................................................1.0g

Ferric ammonium citrate.................................................0.5g

Phenol Red..................................................................0.025g

Nalidixic acid solution ............................................... 10.0mL

Colistin solution......................................................... 10.0mL

pH 7.3 ± 0.2 at 25°C

**Nalidixic Acid Solution:**

**Composition** per 10.0mL:

Nalidixic acid.............................................................0.015g

**Preparation of Nalidixic Acid Solution:** Add nalidixic acid to distilled/deionized water and bring volume to 10.0mL. Mix thoroughly. Filter sterilize.

**Colistin Solution:**

**Composition** per 10.0mL:

Colistin...........................................................................0.01g

**Preparation of Colistin Solution:** Add colistin to distilled/deionized water and bring volume to 10.0mL. Mix thoroughly. Filter sterilize.

**Preparation of Medium:** Add components, except nalidixic acid solution and colistin solution, to distilled/deionized water and bring volume to 980.0mL. Mix thoroughly. Gently heat and bring to boiling. Autoclave for 15 min at 15 psi pressure–121°C. Cool to 45°–50°C. Aseptically add sterile nalidixic acid solution and colistin solution. Mix thoroughly. Pour into sterile Petri dishes or distribute into sterile tubes.

**Storage/Shelf Life:** Store dehydrated media in the dark in a sealed container below 30°C. Prepared media should be stored under refrigeration (2-8°C). Media should be used within 60 days of preparation. Media should not be used if there are any signs of deterioration (shrinking, cracking, or discoloration) or contamination, or if the expiration date supplied by the manufacturer has passed.

**Use:** For the selective isolation, cultivation, and differentiation of *Staphylococcus aureus* and group D streptococci based on mannitol fermentation and hydrolysis of esculin. Bacteria that ferment mannitol appear as yellow colonies surrounded by a yellow zone. Bacteria that hydrolyze esculin appear as dark brown to black colonies surrounded by a dark brown to black zone.

### ETGPA
### (Egg Tellurite Glycine Pyruvate Agar)

**Composition** per liter:

Agar .............................................................................17.0g

Glycine..........................................................................12.0g

Sodium pyruvate .................................................. 10.0g
Pancreatic digest of casein ................................. 10.0g
Beef extract .......................................................... 5.0g
LiCl ...................................................................... 5.0g
Yeast extract ........................................................ 1.0g
Egg yolk emulsion ......................................... 50.0mL
$K_2TeO_3$ solution ............................................ 10.0mL

pH 7.0 ± 0.2 at 25°C

**Source:** This medium is available as a premixed powder from BD Diagnostic Systems.

**Caution:** Lithium chloride is harmful. Avoid bodily contact and inhalation of vapors. On contact with skin wash with plenty of water immediately.

**Egg Yolk Emulsion:**
**Composition**:
Chicken egg yolks ................................................. 11
Whole chicken egg ................................................. 1

**Preparation of Egg Yolk Emulsion:** Soak egg with 1:100 dilution of saturated mercuric chloride solution for 1 min. Crack eggs and separate yolks from whites. Mix egg yolks with 1 chicken egg.

**$K_2TeO_3$ Solution:**
**Composition** per 100.0mL:
$K_2TeO_3$ ............................................................. 1.0g

**Preparation of $K_2TeO_3$ Solution:** Add $K_2TeO_3$ to distilled/deionized water and bring volume to 100.0mL. Mix thoroughly. Filter sterilize.

**Caution:** Potassium tellurite is toxic.

**Preparation of Medium:** Add components to distilled/deionized water and bring volume to 940.0mL. Mix thoroughly. Gently heat and bring to boiling. Autoclave for 15 min at 15 psi pressure–121°C. Cool to 45°–50°C. Add 10.0mL of sterile 1% tellurite solution and 50.0mL of sterile egg yolk emulsion. If desired, add sulfamethazine to a final concentration of 50.0mg/mL. Mix thoroughly but gently and pour into sterile Petri dishes.

**Storage/Shelf Life:** Store dehydrated media in the dark in a sealed container below 30°C. Prepared media should be stored under refrigeration (2-8°C). Media should be used within 60 days of preparation. Media should not be used if there are any signs of deterioration (shrinking, cracking, or discoloration) or contamination, or if the expiration date supplied by the manufacturer has passed.

**Use:** For the selective isolation and enumeration of coagulase-positive staphylococci from food, skin, soil, air, and other materials. For the differentiation and identification of staphylococci on the basis of their ability to clear egg yolk. Addition of sulfamethazine inhibits the growth of *Proteus*. Gray-black colonies surrounded by a clear zone are diagnostic for *Staphylococcus aureus*.

## Ethyl Violet Azide Broth
### (EVA Broth)

**Composition** per liter:
Pancreatic digest of casein .............................. 13.5g
Yeast extract ........................................................ 6.5g
Glucose ................................................................. 5.0g
NaCl ...................................................................... 5.0g
$K_2HPO_4$ .............................................................. 2.7g
$KH_2PO_4$ .............................................................. 2.7g

$NaN_3$ ................................................................... 0.4g
Ethyl Violet ..................................................... 0.83mg

pH 7.0 ± 0.2 at 25°C

**Source:** This medium is available as a premixed powder from BD Diagnostic Systems.

**Caution:** Sodium azide is toxic. Azides also react with metals and disposal must be highly diluted.

**Preparation of Medium:** Add components to distilled/deionized water and bring volume to 1.0L. Mix thoroughly. Gently heat and bring to boiling. Distribute into tubes in 10.0mL volumes. Autoclave for 15 min at 15 psi pressure–121°C.

**Storage/Shelf Life:** Store dehydrated media in the dark in a sealed container below 30°C. Prepared media should be stored under refrigeration (2-8°C). Media should be used within 60 days of preparation. Media should not be used if there are any signs of deterioration (discoloration) or contamination, or if the expiration date supplied by the manufacturer has passed.

**Use:** For the isolation, cultivation, and enumeration of enterococci from water and other specimens. Fecal enterococci turn the medium turbid with a purple sediment on the bottom of the tube.

## Ethyl Violet Azide Broth
### (EVA Broth)

**Composition** per liter:
Tryptose ............................................................. 20.0g
Glucose ................................................................. 5.0g
NaCl ...................................................................... 5.0g
$K_2HPO_4$ .............................................................. 2.7g
$KH_2PO_4$ .............................................................. 2.7g
$NaN_3$ ................................................................... 0.4g
Ethyl Violet ..................................................... 0.83mg

pH 7.0 ± 0.2 at 25°C

**Source:** This medium is available as a premixed powder from BD Diagnostic Systems.

**Caution:** Sodium azide is toxic. Azides also react with metals and disposal must be highly diluted.

**Preparation of Medium:** Add components to distilled/deionized water and bring volume to 1.0L. Mix thoroughly. Gently heat and bring to boiling. Distribute into tubes in 10.0mL volumes. Autoclave for 15 min at 15 psi pressure–121°C.

**Storage/Shelf Life:** Store dehydrated media in the dark in a sealed container below 30°C. Prepared media should be stored under refrigeration (2-8°C). Media should be used within 60 days of preparation. Media should not be used if there are any signs of deterioration (discoloration) or contamination, or if the expiration date supplied by the manufacturer has passed.

**Use:** For the isolation, cultivation, and enumeration of enterococci from water and other specimens. Fecal enterococci turn the medium turbid with a purple sediment on the bottom of the tube.

## Ethyl Violet Azide Broth
### (EVA Broth)

**Composition** per liter:
Tryptose ............................................................. 20.0g
Glucose ................................................................. 5.0g
NaCl ...................................................................... 5.0g
$K_2HPO_4$ .............................................................. 2.7g

KH$_2$PO$_4$..........................................................................................2.7g
NaN$_3$ ..............................................................................................0.3g
Ethyl Violet....................................................................................0.5mg

pH 6.8 ± 0.2 at 25°C

**Source:** This medium is available as a premixed powder from Thermo Scientific.

**Preparation of Medium:** Add components to distilled/deionized water and bring volume to 1.0L. Mix thoroughly. Gently heat and bring to boiling. Distribute into tubes in 10.0mL volumes. Autoclave for 15 min at 15 psi pressure–121°C.

**Caution:** Sodium azide is toxic. Azides also react with metals and disposal must be highly diluted.

**Storage/Shelf Life:** Store dehydrated media in the dark in a sealed container below 30°C. Prepared media should be stored under refrigeration (2-8°C). Media should be used within 60 days of preparation. Media should not be used if there are any signs of deterioration (discoloration) or contamination, or if the expiration date supplied by the manufacturer has passed.

**Use:** For the isolation, cultivation, and enumeration of enterococci from water and other specimens. Fecal enterococci turn the medium turbid with a purple sediment on the bottom of the tube.

## Ethyl Violet Azide HiVeg Broth
### (E.V.A. HiVeg Broth)

**Composition** per liter:
Plant hydrolysate............................................................20.0g
Glucose ..............................................................................5.0g
NaCl.....................................................................................5.0g
K$_2$HPO$_4$.........................................................................2.7g
KH$_2$PO$_4$.........................................................................2.7g
NaN$_3$ ..............................................................................0.4g
Ethyl Violet......................................................................8.3mg

pH 7.0 ± 0.2 at 25°C

**Source:** This medium is available as a premixed powder from Hi-Media.

**Caution:** Sodium azide is toxic. Azides also react with metals and disposal must be highly diluted.

**Preparation of Medium:** Add components to distilled/deionized water and bring volume to 1.0L. Mix thoroughly. Gently heat and bring to boiling. Distribute into tubes in 10.0mL volumes. Autoclave for 15 min at 15 psi pressure–121°C.

**Storage/Shelf Life:** Store dehydrated media in the dark in a sealed container below 30°C. Prepared media should be stored under refrigeration (2-8°C). Media should be used within 60 days of preparation. Media should not be used if there are any signs of deterioration (discoloration) or contamination, or if the expiration date supplied by the manufacturer has passed.

**Use:** For the isolation, cultivation, and enumeration of enterococci from water and other specimens. Fecal enterococci turn the medium turbid with a purple sediment on the bottom of the tube.

## Fastidious Anaerobe Agar
### (FAA)

**Composition** per liter:
Peptone................................................................................23.0g
Agar .....................................................................................12.0g
NaCl.......................................................................................5.0g

Glucose ................................................................................1.0g
L-Arginine ...........................................................................1.0g
Sodium pyruvate ................................................................1.0g
Soluble starch......................................................................1.0g
L-Cysteine·HCl·H$_2$O .......................................................0.5g
Sodium succinate ...............................................................0.5g
NaHCO$_3$..............................................................................0.4g
Na$_4$P$_2$O$_7$·10H$_2$O .......................................................0.25g
Sheep blood, defibrinated .............................................50.0mL
Hemin solution............................................................... 1.0mL
Vitamin K$_1$ solution...................................................... 0.1mL

pH 7.2 ± 0.2 at 25°C

**Vitamin K$_1$ Solution:**
**Composition** per 100.0mL:
Vitamin K$_1$ ........................................................................1.0g
Ethanol...............................................................................99.0mL

**Preparation of Vitamin K$_1$ Solution:** Add vitamin K$_1$ to 99.0mL of absolute ethanol. Mix thoroughly.

**Hemin Solution:**
**Composition** per 100.0mL:
Hemin ..................................................................................1.0g
NaOH (1$N$ solution)........................................................ 20.0mL

**Preparation of Hemin Solution:** Add hemin to 20.0mL of 1$N$ NaOH solution. Mix thoroughly. Bring volume to 100.0mL with distilled/deionized water.

**Preparation of Medium:** Add components, except defibrinated sheep blood, to distilled/deionized water and bring volume to 950.0mL. Mix thoroughly. Gently heat and bring to boiling. Autoclave for 15 min at 15 psi pressure–121°C. Cool to 45°–50°C. Aseptically add 50.0mL of sterile defibrinated sheep blood. Mix thoroughly. Pour into sterile Petri dishes or distribute into sterile tubes.

**Storage/Shelf Life:** Store dehydrated media in the dark in a sealed container below 30°C. Prepared media should be stored under refrigeration (2-8°C). Media should be used within 60 days of preparation. Media should not be used if there are any signs of deterioration (shrinking, cracking, or discoloration) or contamination, or if the expiration date supplied by the manufacturer has passed.

**Use:** For the cultivation of a variety of fastidious anaerobes from clinical speciments and nonclinical specimens of public health importance.

## Fastidious Anaerobe Agar, Alternative Selective
### (FAA Alternative Selective)

**Composition** per liter:
Peptone ................................................................................23.0g
Agar .....................................................................................12.0g
NaCl.......................................................................................5.0g
Glucose ................................................................................1.0g
L-Arginine ...........................................................................1.0g
Sodium pyruvate ................................................................1.0g
Soluble starch......................................................................1.0g
L-Cysteine·HCl·H$_2$O .......................................................0.5g
Sodium succinate ...............................................................0.5g
NaHCO$_3$..............................................................................0.4g
Na$_4$P$_2$O$_7$·10H$_2$O .......................................................0.25g
Sheep blood, defibrinated .............................................50.0mL
Hemin solution............................................................... 1.0mL
Vitamin K$_1$ solution...................................................... 0.1mL

pH 7.2 ± 0.2 at 25°C

**Vitamin K₁ Solution:**
**Composition** per 100.0mL:
Vitamin K₁ .................................................................1.0g
Ethanol .................................................................99.0mL

**Preparation of Vitamin K₁ Solution:** Add vitamin K₁ to 99.0mL of absolute ethanol. Mix thoroughly.

**Hemin Solution:**
**Composition** per 100.0mL:
Hemin.........................................................................1.0g
NaOH (1*N* solution)..............................................20.0mL

**Preparation of Hemin Solution:** Add hemin to 20.0mL of 1*N* NaOH solution. Mix thoroughly. Bring volume to 100.0mL with distilled/deionized water.

**Preparation of Medium:** Add components, except defibrinated sheep blood, to distilled/deionized water and bring volume to 950.0mL. Mix thoroughly. Gently heat and bring to boiling. Autoclave for 15 min at 15 psi pressure–121°C. Cool to 45°–50°C. Aseptically add 50.0mL of sterile defibrinated sheep blood. Mix thoroughly. Pour into sterile Petri dishes or distribute into sterile tubes.

**Storage/Shelf Life:** Store dehydrated media in the dark in a sealed container below 30°C. Prepared media should be stored under refrigeration (2-8°C). Media should be used within 60 days of preparation. Media should not be used if there are any signs of deterioration (shrinking, cracking, or discoloration) or contamination, or if the expiration date supplied by the manufacturer has passed.

**Use:** For the cultivation of a variety of fastidious anaerobes from clinical speciments and nonclinical specimens of public health importance.

## Fastidious Anaerobe Agar, Alternative Selective with Neomycin, Vancomycin, and Josamycin (FAA Alternative Selective Medium with Neomycin, Vancomycin, and Josamycin)
**Composition** per liter:
Peptone...................................................................23.0g
Agar ........................................................................12.0g
NaCl...........................................................................5.0g
Glucose......................................................................1.0g
L-Arginine.................................................................1.0g
Sodium pyruvate.......................................................1.0g
Soluble starch...........................................................1.0g
L-Cysteine·HCl·H₂O .................................................0.5g
Sodium succinate .....................................................0.5g
NaHCO₃......................................................................0.4g
Na₄P₂O₇·10H₂O........................................................0.25g
Neomycin...................................................................0.1g
Sheep blood, defibrinated .....................................50.0mL
Vancomycin solution..............................................10.0mL
Josamycin solution.................................................10.0mL
Hemin solution.........................................................1.0mL
Vitamin K₁ solution.................................................0.1mL
pH 7.2 ± 0.2 at 25°C

**Vitamin K₁ Solution:**
**Composition** per 100.0mL:
Vitamin K₁ .................................................................1.0g
Ethanol .................................................................99.0mL

**Preparation of Vitamin K₁ Solution:** Add vitamin K₁ to 99.0mL of absolute ethanol. Mix thoroughly.

**Hemin Solution:**
**Composition** per 100.0mL:
Hemin .........................................................................1.0g
NaOH (1*N* solution)..............................................20.0mL

**Preparation of Hemin Solution:** Add hemin to 20.0mL of 1*N* NaOH solution. Mix thoroughly. Bring volume to 100.0mL with distilled/deionized water.

**Vancomycin Solution:**
**Composition** per 10.0mL:
Vancomycin .............................................................5.0mg

**Preparation of Vancomycin Solution:** Add vancomycin to distilled/deionized water and bring volume to 10.0mL. Mix thoroughly. Filter sterilize.

**Josamycin Solution:**
**Composition** per 10.0mL:
Josamycin .................................................................3.0mg

**Preparation of Josamycin Solution:** Add josamycin to distilled/deionized water and bring volume to 10.0mL. Mix thoroughly. Filter sterilize.

**Preparation of Medium:** Add components, except defibrinated sheep blood, vancomycin solution, and josamycin solution, to distilled/deionized water and bring volume to 930.0mL. Mix thoroughly. Gently heat and bring to boiling. Autoclave for 15 min at 15 psi pressure–121°C. Cool to 45°–50°C. Aseptically add 50.0mL of sterile defibrinated sheep blood, 10.0mL vancomycin solution, and 10.0mL of josamycin solution. Mix thoroughly. Pour into sterile Petri dishes or distribute into sterile tubes.

**Storage/Shelf Life:** Store dehydrated media in the dark in a sealed container below 30°C. Prepared media should be stored under refrigeration (2-8°C). Media should be used within 60 days of preparation. Media should not be used if there are any signs of deterioration (shrinking, cracking, or discoloration) or contamination, or if the expiration date supplied by the manufacturer has passed.

**Use:** For the selective cultivation of *Fusobacterium* species from clinical speciments and nonclinical specimens of public health importance.

## Fastidious Anaerobe Agar, Selective (FAA Selective)
**Composition** per liter:
Peptone ...................................................................23.0g
Agar ........................................................................12.0g
NaCl...........................................................................5.0g
Glucose......................................................................1.0g
L-Arginine.................................................................1.0g
Sodium pyruvate.......................................................1.0g
Soluble starch...........................................................1.0g
L-Cysteine·HCl·H₂O .................................................0.5g
Sodium succinate .....................................................0.5g
NaHCO₃......................................................................0.4g
Na₄P₂O₇·10H₂O........................................................0.25g
Sheep blood, defibrinated .....................................50.0mL
Hemin solution.........................................................1.0mL
Vitamin K₁ solution.................................................0.1mL
pH 7.2 ± 0.2 at 25°C

**Vitamin K₁ Solution:**
**Composition** per 100.0mL:
Vitamin K₁ .................................................................1.0g
Ethanol.................................................................99.0mL

**Preparation of Vitamin K$_1$ Solution:** Add vitamin K$_1$ to 99.0mL of absolute ethanol. Mix thoroughly.

**Hemin Solution:**
**Composition** per 100.0mL:
Hemin......................................................................1.0g
NaOH (1*N* solution)................................................ 20.0mL

**Preparation of Hemin Solution:** Add hemin to 20.0mL of 1*N* NaOH solution. Mix thoroughly. Bring volume to 100.0mL with distilled/deionized water.

**Preparation of Medium:** Add components, except defibrinated sheep blood, to distilled/deionized water and bring volume to 950.0mL. Mix thoroughly. Gently heat and bring to boiling. Autoclave for 15 min at 15 psi pressure–121°C. Cool to 45°–50°C. Aseptically add 50.0mL of sterile defibrinated sheep blood. Mix thoroughly. Pour into sterile Petri dishes or distribute into sterile tubes.

**Storage/Shelf Life:** Store dehydrated media in the dark in a sealed container below 30°C. Prepared media should be stored under refrigeration (2-8°C). Media should be used within 60 days of preparation. Media should not be used if there are any signs of deterioration (shrinking, cracking, or discoloration) or contamination, or if the expiration date supplied by the manufacturer has passed.

**Use:** For the cultivation of a variety of fastidious anaerobes from clinical speciments and nonclinical specimens of public health importance.

### Fastidious Anaerobe Agar, Selective with Neomycin and Vancomycin (FAA Selective with Neomycin and Vancomycin)

**Composition** per liter:
Peptone.......................................................................23.0g
Agar ..........................................................................12.0g
NaCl............................................................................5.0g
Glucose........................................................................1.0g
L-Arginine....................................................................1.0g
Sodium pyruvate...........................................................1.0g
Soluble starch...............................................................1.0g
L-Cysteine·HCl·H$_2$O ...................................................0.5g
Sodium succinate..........................................................0.5g
NaHCO$_3$......................................................................0.4g
Na$_4$P$_2$O$_7$·10H$_2$O ........................................................0.25g
Neomycin.....................................................................0.1g
Sheep blood, defibrinated ........................................... 50.0mL
Vancomycin solution.................................................. 10.0mL
Hemin solution............................................................ 1.0mL
Vitamin K$_1$ solution..................................................... 0.1mL
pH 7.2 ± 0.2 at 25°C

**Vitamin K$_1$ Solution:**
**Composition** per 100.0mL:
Vitamin K$_1$ ..................................................................1.0g
Ethanol................................................................... 99.0mL

**Preparation of Vitamin K$_1$ Solution:** Add vitamin K$_1$ to 99.0mL of absolute ethanol. Mix thoroughly.

**Hemin Solution:**
**Composition** per 100.0mL:
Hemin......................................................................1.0g
NaOH (1*N* solution)................................................ 20.0mL

**Preparation of Hemin Solution:** Add hemin to 20.0mL of 1*N* NaOH solution. Mix thoroughly. Bring volume to 100.0mL with distilled/deionized water.

**Vancomycin Solution:**
**Composition** per 10.0mL:
Vancomycin..................................................................7.5mg

**Preparation of Vancomycin Solution:** Add vancomycin to distilled/deionized water and bring volume to 10.0mL. Mix thoroughly. Filter sterilize.

**Preparation of Medium:** Add components, except defibrinated sheep blood and vancomycin solution, to distilled/deionized water and bring volume to 940.0mL. Mix thoroughly. Gently heat and bring to boiling. Autoclave for 15 min at 15 psi pressure–121°C. Cool to 45°–50°C. Aseptically add 50.0mL of sterile defibrinated sheep blood and 10.0mL of vancomycin solution. Mix thoroughly. Pour into sterile Petri dishes or distribute into sterile tubes.

**Storage/Shelf Life:** Store dehydrated media in the dark in a sealed container below 30°C. Prepared media should be stored under refrigeration (2-8°C). Media should be used within 60 days of preparation. Media should not be used if there are any signs of deterioration (shrinking, cracking, or discoloration) or contamination, or if the expiration date supplied by the manufacturer has passed.

**Use:** For the selective cultivation of *Fusobacterium* species from clinical speciments and nonclinical specimens of public health importance.

### FC Agar (Fecal Coliform Agar) (m-FC Agar) (m-Fecal Coliform Agar)

**Composition** per liter:
Agar .........................................................................15.0g
Lactose......................................................................12.5g
NaCl............................................................................5.0g
Proteose peptone No. 3 ..................................................5.0g
Yeast extract.................................................................3.0g
Bile salts.....................................................................1.5g
Aniline Blue.................................................................0.1g
Rosolic acid solution................................................ 10.0mL
pH 7.4 ± 0.2 at 25°C

**Source:** This medium is available as a premixed powder from BD Diagnostic Systems.

**Rosolic Acid Solution:**
**Composition** per 100.0mL:
Rosolic acid ...............................................................1.0g

**Preparation of Rosolic Acid Solution:** Add rosolic acid to 0.2*N* NaOH and bring volume to 100.0L. Mix thoroughly.

**Preparation of Medium:** Add 10.0mL rosolic acid solution to 950.0mL distilled/deionized water. Mix thoroughly. Add other components and bring volume to 1.0L with distilled/deionized water. Mix thoroughly. Gently heat and bring to boiling with frequent mixing. Do not autoclave. Pour into sterile Petri dishes or leave in tubes.

**Storage/Shelf Life:** Store dehydrated media in the dark in a sealed container below 30°C. Prepared media should be stored under refrigeration (2-8°C). Media should be used within 60 days of preparation. Media should not be used if there are any signs of deterioration (shrinking, cracking, or discoloration) or contamination, or if the expiration date supplied by the manufacturer has passed.

**Use:** For the cultivation of fecal coliform bacteria from waters and the enumeration of coliform bacteria using the membrane filtration method.

## FC Agar
## (Fecal Coliform Agar)
## (m-FC Agar)
## (m-Fecal Coliform Agar)

**Composition** per liter:

Agar ............................................................................15.0g
Lactose ......................................................................12.5g
Tryptose ....................................................................10.0g
NaCl.............................................................................5.0g
Proteose peptone No. 3 ...............................................5.0g
Yeast extract...............................................................3.0g
Bile salts.....................................................................1.5g
Aniline Blue................................................................0.1g
Rosolic acid solution............................................. 10.0mL

pH 7.4 ± 0.2 at 25°C

**Rosolic Acid Solution:**
**Composition** per 100.0mL:
Rosolic acid....................................................................1.0g

**Preparation of Rosolic Acid Solution:** Add rosolic acid to 0.2*N* NaOH and bring volume to 100.0L. Mix thoroughly.

**Preparation of Medium:** Add 10.0mL rosolic acid solution to 950.0mL of distilled/deionized water. Mix thoroughly. Add other components and bring volume to 1.0L with distilled/deionized water. Mix thoroughly. Gently heat and bring to boiling with frequent mixing. Do not autoclave. Pour into sterile Petri dishes or leave in tubes.

**Storage/Shelf Life:** Store dehydrated media in the dark in a sealed container below 30°C. Prepared media should be stored under refrigeration (2-8°C). Media should be used within 60 days of preparation. Media should not be used if there are any signs of deterioration (shrinking, cracking, or discoloration) or contamination, or if the expiration date supplied by the manufacturer has passed.

**Use:** For the cultivation of fecal coliform bacteria from waters and the enumeration of coliform bacteria using the membrane filtration method.

## FC Broth
## (Fecal Coliform Broth)
## (m-FC Broth)
## (m-Fecal Coliform Broth)

**Composition** per liter:

Lactose ......................................................................12.5g
Tryptose ....................................................................10.0g
NaCl.............................................................................5.0g
Proteose peptone No. 3 ...............................................5.0g
Yeast extract...............................................................3.0g
Bile salts.....................................................................1.5g
Aniline Blue................................................................0.1g
Rosolic acid solution............................................. 10.0mL

pH 7.4 ± 0.2 at 25°C

**Rosolic Acid Solution:**
**Composition** per 100.0mL:
Rosolic acid....................................................................1.0g

**Preparation of Rosolic Acid Solution:** Add rosolic acid to 0.2*N* NaOH and bring volume to 100.0L. Mix thoroughly.

**Preparation of Medium:** Add 10.0mL of rosolic acid solution to 950.0mL of distilled/deionized water. Mix thoroughly. Add other components and bring volume to 1.0L with distilled/deionized water. Mix

thoroughly. Gently heat and bring to boiling with frequent mixing. Do not autoclave. Pour into sterile Petri dishes or leave in tubes.

**Storage/Shelf Life:** Store dehydrated media in the dark in a sealed container below 30°C. Prepared media should be stored under refrigeration (2-8°C). Media should be used within 60 days of preparation. Media should not be used if there are any signs of deterioration (discoloration) or contamination, or if the expiration date supplied by the manufacturer has passed.

**Use:** For the cultivation of fecal coliform bacteria from waters and the enumeration of coliform bacteria using the membrane filtration method.

## FC Broth
## (Fecal Coliform Broth)
## (m-FC Broth)
## (m-Fecal Coliform Broth)

**Composition** per liter:

Lactose........................................................................12.5g
NaCl.............................................................................5.0g
Proteose peptone No. 3 ...............................................5.0g
Yeast extract...............................................................3.0g
Bile salts.....................................................................1.5g
Aniline Blue................................................................0.1g
Rosolic acid solution............................................. 10.0mL

pH 7.4 ± 0.2 at 25°C

**Source:** This medium is available as a premixed powder from BD Diagnostic Systems.

**Rosolic Acid Solution:**
**Composition** per 100.0mL:
Rosolic acid ...................................................................1.0g

**Preparation of Rosolic Acid Solution:** Add rosolic acid to 0.2*N* NaOH and bring volume to 100.0L. Mix thoroughly.

**Preparation of Medium:** Add 10.0mL of rosolic acid solution to 950.0mL of distilled/deionized water. Mix thoroughly. Add other components and bring volume to 1.0L with distilled/deionized water. Mix thoroughly. Gently heat and bring to boiling with frequent mixing. Do not autoclave. Pour into sterile Petri dishes or leave in tubes.

**Storage/Shelf Life:** Store dehydrated media in the dark in a sealed container below 30°C. Prepared media should be stored under refrigeration (2-8°C). Media should be used within 60 days of preparation. Media should not be used if there are any signs of deterioration (discoloration) or contamination, or if the expiration date supplied by the manufacturer has passed.

**Use:** For the cultivation of fecal coliform bacteria from waters and the enumeration of coliform bacteria using the membrane filtration method.

## Fecal Coliform Agar, Modified
## (m-Fecal Coliform Agar, Modified)
## (FCIC)

**Composition** per liter:

Agar ............................................................................15.0g
Inositol ......................................................................10.0g
Tryptose ....................................................................10.0g
Proteose peptone No. 3 ...............................................5.0g
NaCl.............................................................................5.0g
Yeast extract...............................................................3.0g

Bile salts No. 3.................................................................1.5g
Aniline Blue.....................................................................0.1g
pH 7.4 ± 0.2 at 25°C

**Preparation of Medium:** Add components to distilled/deionized water and bring volume to 1.0L. Mix thoroughly. Gently heat and bring to boiling. Do not autoclave. Cool to 50°C. Adjust pH to 7.4. Pour into sterile Petri dishes in 20.0mL volumes. Allow surface of plates to dry before using.

**Use:** For the isolation, cultivation, and enumeration of *Klebsiella* species using the membrane filter method.

## Fecal Coliform Agar, Modified
**Composition** per liter:

Agar ...............................................................................15.0g
Lactose..........................................................................12.5g
Tryptose .......................................................................10.0g
Proteose peptone No. 3 ..................................................5.0g
NaCl................................................................................5.0g
Yeast extract..................................................................3.0g
Bile salts No. 3...............................................................1.5g
Aniline Blue....................................................................0.1g
pH 7.4 ± 0.2 at 25°C

**Preparation of Medium:** Add components and bring volume to 1.0L. Mix thoroughly. Gently heat and bring to boiling. Do not autoclave. Cool to 50°C. Adjust pH to 7.4. Pour into sterile Petri dishes in 20.0mL volumes. Allow surface of plates to dry before using.

**Storage/Shelf Life:** Store dehydrated media in the dark in a sealed container below 30°C. Prepared media should be stored under refrigeration (2-8°C). Media should be used within 60 days of preparation. Media should not be used if there are any signs of deterioration (shrinking, cracking, or discoloration) or contamination, or if the expiration date supplied by the manufacturer has passed.

**Use:** For the isolation, cultivation, and identification of stressed fecal coliform microorganisms based on their ability to ferment lactose. Lactose-fermenting bacteria turn the medium blue.

## Feeley Gorman HiVeg Agar
## (F-G HiVeg Agar)
**Composition** per liter:

Plant acid hydrolysate...................................................17.5g
Agar ...............................................................................17.0g
Plant extract ...................................................................3.0g
Starch .............................................................................1.5g
L-Cysteine·HCl...............................................................0.4g
Fe$_4$(P$_2$O$_7$)$_3$·H$_2$O, soluble ............................................0.25g
pH 6.9 ± 0.05 at 25°C

**Source:** This medium is available as a premixed powder from Hi-Media.

**Preparation of Medium:** Add components to distilled/deionized water and bring volume to 1.0L. Mix thoroughly. Gently heat and bring to boiling. Autoclave for 15 min at 15 psi pressure–121°C. Cool to 45°–50°C. Mix thoroughly. Adjust pH to 6.9. Pour into sterile Petri dishes or distribute into sterile tubes.

**Storage/Shelf Life:** Store dehydrated media in the dark in a sealed container below 30°C. Prepared media should be stored under refrigeration (2-8°C). Media should be used within 60 days of preparation. Media should not be used if there are any signs of deterioration (shrink-

ing, cracking, or discoloration) or contamination, or if the expiration date supplied by the manufacturer has passed.

**Use:** For the isolation and cultivation of *Legionella pneumophila*.

## F-G Agar
## (Feeley-Gorman Agar)
**Composition** per liter:

Casein, acid hydrolyzed.................................................17.5g
Agar ...............................................................................17.0g
Beef extract.....................................................................3.0g
Starch .............................................................................1.5g
L-Cysteine solution .....................................................10.0mL
Fe$_4$(P$_2$O$_7$)$_3$ solution .............................................10.0mL
pH 6.9 ± 0.05 at 25°C

**L-Cysteine Solution:**
**Composition** per 10.0mL:

L-Cysteine·HCl·H$_2$O.......................................................0.4g

**Preparation of L-Cysteine Solution:** Add L-cysteine·HCl·H$_2$O to distilled/deionized water and bring volume to 10.0mL. Mix thoroughly. Filter sterilize.

**Fe$_4$(P$_2$O$_7$)$_3$ Solution:**
**Composition** per 10.0mL:

Fe$_4$(P$_2$O$_7$)$_3$.........................................................................0.25g

**Preparation of Fe$_4$(P$_2$O$_7$)$_3$ Solution:** Add Fe$_4$(P$_2$O$_7$)$_3$ to distilled/ deionized water and bring volume to 10.0mL. Mix thoroughly. Filter sterilize.

**Preparation of Medium:** Add components, except L-cysteine solution and Fe$_4$(P$_2$O$_7$)$_3$ solution, to distilled/deionized water and bring volume to 980.0mL. Mix thoroughly. Gently heat and bring to boiling. Autoclave for 15 min at 15 psi pressure–121°C. Cool to 45°–50°C. Aseptically add 10.0mL of L-cysteine solution. Mix thoroughly. Aseptically add 10.0mL of Fe$_4$(P$_2$O$_7$)$_3$ solution. Mix thoroughly. Adjust pH to 6.9. Pour into sterile Petri dishes or distribute into sterile tubes.

**Storage/Shelf Life:** Store dehydrated media in the dark in a sealed container below 30°C. Prepared media should be stored under refrigeration (2-8°C). Media should be used within 60 days of preparation. Media should not be used if there are any signs of deterioration (shrinking, cracking, or discoloration) or contamination, or if the expiration date supplied by the manufacturer has passed.

**Use:** For the isolation and cultivation of *Legionella pneumophila*.

## F-G Agar with Selenium
## (Feeley-Gorman Agar with Selenium)
**Composition** per liter:

Casein, acid hydrolyzed.................................................17.5g
Agar ...............................................................................17.0g
Beef extract.....................................................................3.0g
Starch .............................................................................1.5g
L-Cysteine solution .....................................................10.0mL
Fe$_4$(P$_2$O$_7$)$_3$ solution .............................................10.0mL
Na$_2$SeO$_3$·5H$_2$O solution...........................................10.0mL
pH 6.9 ± 0.05 at 25°C

**L-Cysteine Solution:**
**Composition** per 10.0mL:

L-Cysteine·HCl·H$_2$O .....................................................0.4g

**Preparation of L-Cysteine Solution:** Add L-cysteine·HCl·H$_2$O to distilled/deionized water and bring volume to 10.0mL. Mix thoroughly. Filter sterilize.

**Fe$_4$(P$_2$O$_7$)$_3$ Solution:**
**Composition** per 10.0mL:
Fe$_4$(P$_2$O$_7$)$_3$ ......................................................................0.25g

**Preparation of Fe$_4$(P$_2$O$_7$)$_3$ Solution:** Add Fe$_4$(P$_2$O$_7$)$_3$ to distilled/deionized water and bring volume to 10.0mL. Mix thoroughly. Filter sterilize.

**Na$_2$SeO$_3$·5H$_2$O Solution:**
**Composition** per 10.0mL:
Na$_2$SeO$_3$·5H$_2$O ..................................................................0.01g

**Preparation of Na$_2$SeO$_3$·5H$_2$O Solution:** Add Na$_2$SeO$_3$·5H$_2$O to distilled/deionized water and bring volume to 10.0mL. Mix thoroughly. Filter sterilize.

**Preparation of Medium:** Add components—except L-cysteine solution, Fe$_4$(P$_2$O$_7$)$_3$ solution, and Na$_2$SeO$_3$·5H$_2$O solution—to distilled/deionized water and bring volume to 970.0mL. Mix thoroughly. Gently heat and bring to boiling. Autoclave for 15 min at 15 psi pressure–121°C. Cool to 45°–50°C. Aseptically add 10.0mL of sterile L-cysteine solution. Mix thoroughly. Aseptically add 10.0mL of sterile Fe$_4$(P$_2$O$_7$)$_3$ solution and 10.0mL of sterile Na$_2$SeO$_3$·5H$_2$O solution. Mix thoroughly. Adjust pH to 6.9. Pour into sterile Petri dishes or distribute into sterile tubes.

**Storage/Shelf Life:** Store dehydrated media in the dark in a sealed container below 30°C. Prepared media should be stored under refrigeration (2-8°C). Media should be used within 60 days of preparation. Media should not be used if there are any signs of deterioration (shrinking, cracking, or discoloration) or contamination, or if the expiration date supplied by the manufacturer has passed.

**Use:** For the isolation and cultivation of *Legionella pneumophila*.

## F-G Broth
## (Feeley-Gorman Broth)
**Composition** per liter:
Casein, acid hydrolyzed ................................................17.5g
Beef extract ......................................................................3.0g
Starch ................................................................................1.5g
L-Cysteine solution ................................................... 10.0mL
Fe$_4$(P$_2$O$_7$)$_3$ solution ............................................... 10.0mL
pH 6.9 ± 0.05 at 25°C

**L-Cysteine Solution:**
**Composition** per 10.0mL:
L-Cysteine·HCl·H$_2$O..................................................0.4g

**Preparation of L-Cysteine Solution:** Add L-cysteine·HCl·H$_2$O to distilled/deionized water and bring volume to 10.0mL. Mix thoroughly. Filter sterilize.

**Fe$_4$(P$_2$O$_7$)$_3$ Solution:**
**Composition** per 10.0mL:
Fe$_4$(P$_2$O$_7$)$_3$......................................................................0.25g

**Preparation of Fe$_4$(P$_2$O$_7$)$_3$ Solution:** Add Fe$_4$(P$_2$O$_7$)$_3$ to distilled/deionized water and bring volume to 10.0mL. Mix thoroughly. Filter sterilize.

**Preparation of Medium:** Add components, except L-cysteine solution and Fe$_4$(P$_2$O$_7$)$_3$ solution, to distilled/deionized water and bring vol-

ume to 980.0mL. Mix thoroughly. Gently heat and bring to boiling. Autoclave for 15 min at 15 psi pressure–121°C. Cool to 45°–50°C. Aseptically add 10.0mL of L-cysteine solution. Mix thoroughly. Aseptically add 10.0mL of Fe$_4$(P$_2$O$_7$)$_3$ solution. Mix thoroughly. Adjust pH to 6.9. Aseptically distribute into sterile tubes or flasks.

**Storage/Shelf Life:** Store dehydrated media in the dark in a sealed container below 30°C. Prepared media should be stored under refrigeration (2-8°C). Media should be used within 60 days of preparation. Media should not be used if there are any signs of deterioration (discoloration) or contamination, or if the expiration date supplied by the manufacturer has passed.

**Use:** For the cultivation of *Legionella pneumophila*.

## Fermentation Broth
## (CHO Medium)
**Composition** per liter:
Pancreatic digest of casein............................................15.0g
Yeast extract......................................................................7.0g
NaCl...................................................................................2.5g
Agar.................................................................................0.75g
Sodium thioglycolate.........................................................0.5g
L-Cystine.........................................................................0.25g
Ascorbic acid.....................................................................0.1g
Bromthymol Blue.............................................................0.01g
Carbohydrate or starch solution........................... 100.0mL
pH 7.0 ± 0.1 at 25°C

**Source:** This medium is available as a premixed powder from BD Diagnostic Systems.

**Carbohydrate Solution:**
**Composition** per 100.0mL:
Carbohydrate....................................................................6.0g

**Preparation of Carbohydrate Solution:** Add carbohydrate to distilled/deionized water and bring volume to 10.0mL. Mix thoroughly. Filter sterilize.

**Starch Solution:**
**Composition** per 100.0mL:
Starch ................................................................................2.5g

**Preparation of Starch Solution:** Add starch to distilled/deionized water and bring volume to 100.0mL. Mix thoroughly. Filter sterilize.

**Preparation of Medium:** Add components, except carbohydrate solution, to distilled/deionized water and bring volume to 900.0mL. Mix thoroughly. Distribute into tubes or flasks. Autoclave for 15 min at 15 psi pressure–121°C. Cool to 45°–50°C. Aseptically add 100.0mL of sterile carbohydrate solution. Mix thoroughly. Aseptically distribute into sterile tubes or flasks. Loosen caps on tubes. Place in an anaerobic chamber under an atmosphere of 85% N$_2$, 10% H$_2$, and 5% CO$_2$. Fasten the caps securely or maintain in an anaerobic chamber.

**Storage/Shelf Life:** Store dehydrated media in the dark in a sealed container below 30°C. Prepared media should be stored under refrigeration (2-8°C). Media should be used within 60 days of preparation. Media should not be used if there are any signs of deterioration (discoloration) or contamination, or if the expiration date supplied by the manufacturer has passed.

**Use:** For the differentiation of anaerobic bacteria based upon carbohydrate fermentation. Bacteria athat ferment the specific carbohydrates added to the medium turn the medium yellow.

## Fildes Enrichment Agar

**Composition** per liter:

Agar ..................................................................................15.0g
Peptone................................................................................5.0g
Beef extract ........................................................................3.0g
Fildes enrichment solution .............................................. 50.0mL

### Fildes Enrichment Solution:
**Composition** per 206.0mL:

Pepsin..................................................................................1.0g
NaCl (0.85% solution) ..................................................... 150.0mL
Sheep blood, defibrinated ................................................. 50.0mL
HCl....................................................................................... 6.0mL

pH 7.0–7.2 at 25°C

**Source:** Fildes enrichment solution is available from BD Diagnostic Systems.

**Preparation of Fildes Enrichment Solution:** Combine components. Mix thoroughly. Incubate at 56°C for 4 h. Bring pH to 7.0 with 20% NaOH. Adjust pH to 7.2 with HCl. Do not autoclave. Add 0.25 mL of chloroform and store at 4°C. Before use, heat to 56°C to remove chloroform.

**Preparation of Medium:** Add components, except Fildes enrichment solution, to distilled/deionized water and bring volume to 950.0mL. Mix thoroughly. Gently heat and bring to boiling. Autoclave for 15 min at 15 psi pressure–121°C. Cool to 56°C. Aseptically add 50.0mL of sterile Fildes enrichment solution. Mix thoroughly. Pour into sterile Petri dishes or distribute into sterile tubes.

**Storage/Shelf Life:** Store dehydrated media in the dark in a sealed container below 30°C. Prepared media should be stored under refrigeration (2-8°C). Media should be used within 60 days of preparation. Media should not be used if there are any signs of deterioration (shrinking, cracking, or discoloration) or contamination, or if the expiration date supplied by the manufacturer has passed.

**Use:** For the isolation and cultivation of *Haemophilus influenzae*.

## Fletcher *Leptospira* HiVeg Medium Base
## (*Leptospira* HiVeg Medium Base, Fletcher)

**Composition** per liter:

Agar ..................................................................................1.5g
NaCl....................................................................................0.5g
Plant peptone.....................................................................0.3g
Plant extract ......................................................................0.2g
Rabbit serum ................................................................. 50.0mL

pH 7.9 ± 0.1 at 25°C

**Source:** This medium, without rabbit serum, is available as a premixed powder from BD Diagnostic Systems.

**Preparation of Medium:** Add components, except rabbit serum, to distilled/deionized water and bring volume to 950.0mL. Mix thoroughly. Gently heat and bring to boiling. Autoclave for 15 min at 15 psi pressure–121°C. Cool to 50°–55°C. Aseptically add 50.0mL of sterile rabbit serum. Mix thoroughly. Aseptically distribute into sterile tubes or flasks.

**Storage/Shelf Life:** Store dehydrated media in the dark in a sealed container below 30°C. Prepared media should be stored under refrigeration (2-8°C). Media should be used within 60 days of preparation. Media should not be used if there are any signs of deterioration (shrinking, cracking, or discoloration) or contamination, or if the expiration date supplied by the manufacturer has passed.

**Use:** For the isolation, cultivation, and maintenance of cultures of *Leptospira* species.

## Fletcher Medium

**Composition** per liter:

Agar ..................................................................................1.5g
NaCl....................................................................................0.5g
Peptone ..............................................................................0.3g
Beef extract ........................................................................0.2g
Rabbit serum ................................................................. 50.0mL

pH 7.9 ± 0.1 at 25°C

**Source:** This medium is available as a premixed powder from BD Diagnostic Systems.

**Preparation of Medium:** Add components, except rabbit serum, to distilled/deionized water and bring volume to 950.0mL. Mix thoroughly. Gently heat and bring to boiling. Autoclave for 15 min at 15 psi pressure–121°C. Cool to 50°–55°C. Aseptically add 50.0mL of sterile rabbit serum. Mix thoroughly. Aseptically distribute into sterile tubes or flasks.

**Storage/Shelf Life:** Store dehydrated media in the dark in a sealed container below 30°C. Prepared media should be stored under refrigeration (2-8°C). Media should be used within 60 days of preparation. Media should not be used if there are any signs of deterioration (shrinking, cracking, or discoloration) or contamination, or if the expiration date supplied by the manufacturer has passed.

**Use:** For the isolation, cultivation, and maintenance of cultures of *Leptospira* species.

## Fletcher Medium with Fluorouracil
## (Fluorouracil *Leptospira* Medium)

**Composition** per liter:

Agar ..................................................................................1.5g
NaCl....................................................................................0.5g
Peptone ..............................................................................0.3g
Beef extract ........................................................................0.2g
Rabbit serum ................................................................. 50.0mL
Fluorouracil solution......................................................... 20.0mL

pH 7.9 ± 0.1 at 25°C

### Fluorouracil Solution:
**Composition** per 100.0mL:

Fluorouracil.......................................................................... 10.0g

**Preparation of Fluorouracil Solution:** Add fluorouracil to 50.0mL of distilled/deionized water. Add 1.0mL of 2$N$ NaOH and bring volume to 100.0mL. Gently heat to 56°C for 2 h. Adjust pH to 7.4–7.6 with NaOH. Mix thoroughly. Filter sterilize.

**Preparation of Medium:** Add components, except rabbit serum and fluorouracil solution, to distilled/deionized water and bring volume to 930.0mL. Mix thoroughly. Gently heat and bring to boiling. Autoclave for 15 min at 15 psi pressure–121°C. Cool to 50°–55°C. Aseptically add 80.0mL of sterile rabbit serum. Mix thoroughly. Aseptically distribute into sterile tubes or flasks. Immediately prior to use, add 0.1mL of fluorouracil solution per 5.0mL of medium.

**Storage/Shelf Life:** Store dehydrated media in the dark in a sealed container below 30°C. Prepared plates should be stored in the dark under refrigeration (2-8°C). Chromogenic media are especially light and temperature sensitive; protect from light, excessive heat, moisture, and freezing. Media should not be used if there are any signs of deterioration (shrinking, cracking, or discoloration) or contamination, or if the expiration date supplied by the manufacturer has passed.

**Use:** For the isolation, cultivation, and maintenance of cultures of *Leptospira* species.

## Fluid Casein Digest Soya Lecithin HiVeg Medium
**Composition** per liter:

| | |
|---|---|
| Plant hydrolysate | 20.0g |
| Soya lecithin | 5.0g |
| Polysorbate 20 | 40.0 ml |

pH 7.3 ± 0.2 at 25°C

**Source:** This medium, without polysorbate 20, is available as a pre-mixed powder from HiMedia.

**Preparation of Medium:** Add plant hydrolysate and soya lecithin to distilled/deionized water and bring volume to 960.0mL. Mix thoroughly. Heat as necessary until components are completely dissolved. Add 40.0mL polysorbate 20. Mix thoroughly. Distribute into tubes or flasks. Autoclave for 15 min at 15 psi pressure–121°C.

**Storage/Shelf Life:** Store dehydrated media in the dark in a sealed container below 30°C. Prepared media should be stored under refrigeration (2-8°C). Media should be used within 60 days of preparation. Media should not be used if there are any signs of deterioration (discoloration) or contamination, or if the expiration date supplied by the manufacturer has passed.

**Use:** This is a highly nutritional medium containing neutralizing agents for neutralizing quaternary ammonium compounds used for the sanitary examination of surfaces.

## Fluid Casein Digest Soya Lecithin Medium
**Composition** per liter:

| | |
|---|---|
| Casein enzymatic hydrolysate | 20.0g |
| Soya lecithin | 5.0g |
| Polysorbate 20 | 40.0 ml |

pH 7.3 ± 0.2 at 25°C

**Source:** This medium, without polysorbate 20, is available as a pre-mixed powder from HiMedia.

**Preparation of Medium:** Add plant hydrolysate and soya lecithin to distilled/deionized water and bring volume to 960.0mL. Mix thoroughly. Heat is necessary until components are completely dissolved. Add 40.0mL polysorbate 20. Mix thoroughly. Distribute into tubes or flasks. Autoclave for 15 min at 15 psi pressure–121°C.

**Storage/Shelf Life:** Store dehydrated media in the dark in a sealed container below 30°C. Prepared media should be stored under refrigeration (2-8°C). Media should be used within 60 days of preparation. Media should not be used if there are any signs of deterioration (discoloration) or contamination, or if the expiration date supplied by the manufacturer has passed.

**Use:** This is a highly nutritional medium containing neutralizing agents for neutralizing quaternary ammonium compounds used for the sanitary examination of surfaces.

## Fluid Lactose HiVeg Medium
**Composition** per liter:

| | |
|---|---|
| Plant peptone No. 2 | 5.0g |
| Lactose | 5.0g |
| Plant extract | 3.0g |

pH 6.9 ± 0.2 at 25°C

**Source:** This medium is available as a premixed powder from Hi-Media.

**Preparation of Medium:** Add components to distilled/deionized water and bring volume to 1.0L. Mix thoroughly. Distribute into tubes or flasks. Autoclave for 15 min at 15 psi pressure–121°C.

**Storage/Shelf Life:** Store dehydrated media in the dark in a sealed container below 30°C. Prepared media should be stored under refrigeration (2-8°C). Media should be used within 60 days of preparation. Media should not be used if there are any signs of deterioration (discoloration) or contamination, or if the expiration date supplied by the manufacturer has passed.

**Use:** As a pre-enrichment medium for the detection of coliform bacteria in water, dairy products, and foods.

## Fluid Lactose HiVeg Medium with Soya Lecithin and Polysorbate 20
**Composition** per liter:

| | |
|---|---|
| Plant peptone No. 2 | 5.0g |
| Lactose | 5.0g |
| Soya lecithin | 5.0g |
| Plant extract | 3.0g |
| Polysorbate 20 | 40 mL |

pH 6.9 ± 0.2 at 25°C

**Source:** This medium is available as a premixed powder from Hi-Media.

**Preparation of Medium:** Add components, except polysorbate 20, to distilled/deionized water and bring volume to 960.0mL. Mix thoroughly. Heat as necessary until components are completely dissolved. Add 40.0mL polysorbate 20. Mix thoroughly. Distribute into tubes or flasks. Autoclave for 15 min at 15 psi pressure–121°C.

**Storage/Shelf Life:** Store dehydrated media in the dark in a sealed container below 30°C. Prepared media should be stored under refrigeration (2-8°C). Media should be used within 60 days of preparation. Media should not be used if there are any signs of deterioration (discoloration) or contamination, or if the expiration date supplied by the manufacturer has passed.

**Use:** For the cultivation of coliform bacteria in water, dairy products, and foods.

## Fluid Selenite Cystine HiVeg Medium (Selenite Cystine HiVeg Broth)
**Composition** per liter:

| | |
|---|---|
| $Na_2HPO_4$ | 10.0g |
| Plant hydrolysate | 5.0g |
| Lactose | 4.0g |
| L-Cystine | 0.01g |
| $HNaO_3Se$ | 4.0g |

pH 7.0 ± 0.2 at 25°C

**Source:** This medium is available as a premixed powder from Hi-Media.

**Caution:** Sodium selenite is toxic and a potential teratogen and care must be taken to avoid inhalation of the powdered dye, contact with the skin, or ingestion, especially in pregnant laboratory workers.

**Preparation of Medium:** Add sodium hydrogen selenite to distilled/deionized water and bring volume to 1.0L. Add remaining components. Mix thoroughly. Gently heat. Do not autoclave. Distribute into sterile tubes in 10.0mL volumes. Sterilize for 15 min at 0 psi pressure–100°C. Do not autoclave.

**Storage/Shelf Life:** Store dehydrated media in the dark in a sealed container below 30°C. Prepared media should be stored under refrigeration (2-8°C). Media should be used within 60 days of preparation. Media should not be used if there are any signs of deterioration (discoloration) or contamination, or if the expiration date supplied by the manufacturer has passed.

**Use:** For the isolation and cultivation of *Salmonella* species from feces, dairy products, and other specimens.

## Fluid Selenite Cystine Medium
### (Selenite Cystine Broth)

**Composition** per liter:

| | |
|---|---|
| Na$_2$HPO$_4$ | 10.0g |
| Casein enzymatic hydrolysate | 5.0g |
| Lactose | 4.0g |
| L-Cystine | 0.01g |
| HNaO$_3$Se | 4.0g |

pH 7.0 ± 0.2 at 25°C

**Source:** This medium is available as a premixed powder from Hi-Media.

**Caution:** Sodium selenite is toxic and a potential teratogen and care must be taken to avoid inhalation of the powdered dye, contact with the skin, or ingestion, especially in pregnant laboratory workers.

**Preparation of Medium:** Add sodium hydrogen selenite to distilled/deionized water and bring volume to 1.0L. Add remaining components. Mix thoroughly. Gently heat. Do not autoclave. Distribute into sterile tubes in 10.0mL volumes. Sterilize for 15 min at 0 psi pressure–100°C. Do not autoclave.

**Storage/Shelf Life:** Store dehydrated media in the dark in a sealed container below 30°C. Prepared media should be stored under refrigeration (2-8°C). Media should be used within 60 days of preparation. Media should not be used if there are any signs of deterioration (discoloration) or contamination, or if the expiration date supplied by the manufacturer has passed.

**Use:** For the isolation and cultivation of *Salmonella* species from feces, dairy products, and other specimens.

## Fluid Tetrathionate HiVeg Medium without Iodine and BG
### (Tetrathionate HiVeg Broth Base without Iodine & BG)

**Composition** per liter:

| | |
|---|---|
| Na$_2$S$_2$O$_3$ | 30.0g |
| CaCO$_3$ | 10.0g |
| Plant hydrolysate | 2.5g |
| Plant peptone | 2.5g |
| Synthetic detergent | 1.0g |
| Iodine-potassium iodide solution | 20.0mL |
| Brilliant Green solution | 10.0mL |

pH 8.4 ± 0.2 at 25°C

**Source:** This medium, without iodine-potassium iodine solution and Brilliant Green solution, is available as a premixed powder from Hi-Media.

**Preparation of Tetrathionate Broth Base:** Add components to distilled/deionized water and bring volume to 1.0L. Mix thoroughly.

Gently heat and bring to boiling. A slight precipitate will remain. Do not autoclave. Cool to 25°C. Store at 4°C.

**Iodine-Potassium Iodide Solution:**
**Composition** per 20.0mL:

| | |
|---|---|
| Iodine, resublimed | 6.0g |
| KI | 5.0g |

**Preparation of Iodine-Potassium Iodide Solution:** Add KI to 5.0mL of sterile distilled/deionized water. Mix thoroughly. Add iodine. Mix thoroughly. Bring volume to 20.0mL with sterile distilled/deionized water.

**Brilliant Green Solution:**
**Composition** per 100.0mL:

| | |
|---|---|
| Brilliant Green | 0.1g |

**Preparation of Brilliant Green Solution:** Add Brilliant Green to sterile distilled/deionized water. Mix thoroughly.

**Preparation of Medium:** Add components, except iodine-potassium iodine soulution and Brilliant Green solution to distilled/deionized water and bring volume to 1.0L. Mix thoroughly. Gently heat and bring to boiling. A slight precipitate will remain. Do not autoclave. Cool to 25°C. Add 20.0mL of iodine-potassium iodide solution, and 10.0mL of Brilliant Green solution. Mix thoroughly. Aseptically distribute into tubes in 10.0mL volumes. Do not heat medium after it has been mixed.

**Storage/Shelf Life:** Store dehydrated media in the dark in a sealed container below 30°C. Prepared media should be stored under refrigeration (2-8°C). Media should be used within 60 days of preparation. Media should not be used if there are any signs of deterioration (discoloration) or contamination, or if the expiration date supplied by the manufacturer has passed.

**Use:** For the selective isolation and cultivation of *Salmonella* species from foods.

## Fluid Tetrathionate Medium without Iodine and BG
### (Tetrathionate Broth Base without Iodine & BG)

**Composition** per liter:

| | |
|---|---|
| Na$_2$S$_2$O$_3$ | 30.0g |
| CaCO$_3$ | 10.0g |
| Casein enzymatic hydrolysate | 2.5g |
| Peptic digest of animal tissue | 2.5g |
| Bile salts | 1.0g |
| Iodine-potassium iodide solution | 20.0mL |
| Brilliant Green solution | 10.0mL |

pH 8.4 ± 0.2 at 25°C

**Source:** This medium, without iodine-potassium iodine solution and Brilliant Green solution, is available as a premixed powder from Hi-Media.

**Preparation of Tetrathionate Broth Base:** Add components to distilled/deionized water and bring volume to 1.0L. Mix thoroughly. Gently heat and bring to boiling. A slight precipitate will remain. Do not autoclave. Cool to 25°C. Store at 4°C.

**Iodine-Potassium Iodide Solution:**
**Composition** per 20.0mL:

| | |
|---|---|
| Iodine, resublimed | 6.0g |
| KI | 5.0g |

**Preparation of Iodine-Potassium Iodide Solution:** Add KI to 5.0mL of sterile distilled/deionized water. Mix thoroughly. Add iodine. Mix thoroughly. Bring volume to 20.0mL with sterile distilled/deionized water.

**Brilliant Green Solution:**
**Composition** per 100.0mL:
Brilliant Green ................................................................0.1g

**Preparation of Brilliant Green Solution:** Add Brilliant Green to sterile distilled/deionized water. Mix thoroughly.

**Preparation of Medium:** Add components, except iodine-potassium iodine soulution and Brilliant Green solution, to distilled/deionized water and bring volume to 1.0L. Mix thoroughly. Gently heat and bring to boiling. A slight precipitate will remain. Do not autoclave. Cool to 25°C. Add 20.0mL of iodine-potassium iodide solution, and 10.0mL of Brilliant Green solution. Mix thoroughly. Aseptically distribute into tubes in 10.0mL volumes. Do not heat medium after it has been mixed.

**Storage/Shelf Life:** Store dehydrated media in the dark in a sealed container below 30°C. Prepared media should be stored under refrigeration (2-8°C). Media should be used within 60 days of preparation. Media should not be used if there are any signs of deterioration (discoloration) or contamination, or if the expiration date supplied by the manufacturer has passed.

**Use:** For the selective isolation and cultivation of *Salmonella* species from foods.

## Fluorocult® *E. coli* O157:H7 Agar
### (*E. coli* O157:H7 Agar, Fluorocult)
**Composition** per liter:
Peptone from casein................................................20.0g
Agar................................................................................13.0g
Sorbitol........................................................................10.0g
NaCl................................................................................5.0g
Meat extract..................................................................2.0g
$Na_2S_2O_3$......................................................................2.0g
Sodium deoxycholate..................................................1.12g
Yeast extract..................................................................1.0g
Ammonium ferric citrate ............................................0.5g
4-Methylumbelliferyl-β-D-glucuronide ......................0.1g
Bromthymol Blue......................................................0.025g
pH 7.4 ± 0.2 at 25°C

**Source:** This medium is available from Merck.

**Preparation of Medium:** Add components to distilled/deionized water and bring volume to 1.0L. Mix thoroughly. Autoclave for 15 min at 15 psi pressure–121°C. Cool to 45°–50°C. Pour into sterile Petri dishes.

**Storage/Shelf Life:** Store dehydrated media in the dark in a sealed container below 30°C. Prepared plates should be stored in the dark under refrigeration (2-8°C). Chromogenic media are especially light and temperature sensitive; protect from light, excessive heat, moisture, and freezing. Media should not be used if there are any signs of deterioration (shrinking, cracking, or discoloration) or contamination, or if the expiration date supplied by the manufacturer has passed.

**Use:** For the isolation and differentiation of enterohemorrhagic (EHEC) *Escherichia coli* O157:H7 strains from foods. In contrast to most other *E. coli* strains, *E. coli* O157:H7 shows the following characteristics: no sorbitol-cleavage capacity within 48 h and no formation of glucuronidase (MUG-negative/no fluorescence). Sodium deoxycholate inhibits the growth of the Gram-positive accompanying flora for the greater part. Sorbitol serves, together with the pH indicator Bromthymol Blue, to determine the degradation of sorbitol which, in the case of sorbitol-positive microorganisms, results in the colonies turning yellow in color. Sorbitol-negative strains, on the other hand, do not lead to any change in the color of the culture medium and thus proliferate as greenish colonies. Sodium thiosulfate and ammonium iron(III) citrate result in black-brown discoloration of the agar for colonies, in the presence of hydrogen-sulfide-forming pathogens, precipitating iron sulfide. *Proteus mirabilis* in particular, which displays biochemical properties similar to those of *E. coli* O157:H7, can thus be very easily differentiated from *E. coli* O157:H7 on account of the brownish discoloration. 4-Methylumbelliferyl-β-D-glucuronide (MUG) is converted into 4-methylumbelliferone by β-D-glucuronidase-forming pathogens; 4-methylumbelliferone fluoresces under UV light. The activity of β-D-glucuronidase is a highly specific characteristic of *E. coli*. In contrast to most *E. coli* strains, *E. coli* O157:H7 is not capable of forming β-D-glucoronidase. When irradiated with long-wave UV light, no fluorescence is formed.

## Forget Fredette Agar
**Composition** per liter:
Casein enzymatic hydrolysate ....................................17.0g
Agar................................................................................10.0g
NaCl................................................................................5.0g
Papaic digest of soybean meal....................................3.0g
Glucose..........................................................................2.5g
$K_2HPO_4$......................................................................2.5g
$NaN_3$..........................................................................0.5g
pH 7.4 ± 0.2 at 25°C

**Source:** This medium is available from HiMedia.

**Caution:** Sodium azide is toxic. Azides also react with metals and disposal must be highly diluted.

**Preparation of Medium:** Add components to distilled/deionized water and bring volume to 1.0L. Mix thoroughly. Gently heat and bring to boiling. Distribute into tubes or flasks. Autoclave for 15 min at 15 psi pressure–121°C. Pour into sterile Petri dishes or leave in tubes.

**Storage/Shelf Life:** Store dehydrated media in the dark in a sealed container below 30°C. Prepared media should be stored under refrigeration (2-8°C). Media should be used within 60 days of preparation. Media should not be used if there are any signs of deterioration (shrinking, cracking, or discoloration) or contamination, or if the expiration date supplied by the manufacturer has passed.

**Use:** For the selective isolation of anaerobic microorganisms.

## FRAG Agar
### (Fragilis Agar)
**Composition** per 1025.0mL:
L-Cysteine·HCl·$H_2O$..................................................0.5g
Basal solution........................................................995.0mL
Glucuronic acid solution........................................25.0mL
Gentamicin solution..................................................1.0mL
Hemin-vitamin $K_1$ solution..................................1.0mL
Ferric sulfate solution ..............................................1.0mL
Mineral solution........................................................1.0mL
Phenol Red (1% solution)..........................................1.0mL
Vitamin $B_{12}$ solution ........................................0.05mL
pH 7.0 ± 0.1 at 25°C

**Basal Solution:**
**Composition** per 995.0mL:
Oxgall..........................................................................20.0g
Agar..............................................................................15.4g
$K_2HPO_4$..................................................................2.26g
Yeast extract..................................................................2.0g

Pancreatic digest of casein ............................................1.4g
$(NH_4)_2HPO_4$............................................................1.0g
$K_2HPO_4$....................................................................0.9g
Papaic digest of soybean meal ....................................0.12g
NaCl............................................................................0.12g

**Preparation of Basal Solution:** Add components to distilled/deionized water and bring volume to 995.0mL. Mix thoroughly. Gently heat and bring to boiling. Autoclave for 15 min at 15 psi pressure–121°C. Cool to 45°–50°C.

**Glucuronic Acid Solution:**
**Composition** per 100.0mL:
D-Glucuronic acid........................................................40.0g

**Preparation of Glucuronic Acid Solution:** Add glucuronic acid to distilled/deionized water and bring volume to 100.0mL. Mix thoroughly. Filter sterilize.

**Gentamicin Solution:**
**Composition** per 10.0mL:
Gentamicin.................................................................0.1mg

**Preparation of Gentamicin Solution:** Add gentamicin to distilled/deionized water and bring volume to 10.0mL. Mix thoroughly. Filter sterilize.

**Vitamin $K_1$-Hemin Solution:**
**Composition** per liter:
Vitamin $K_1$ solution ...................................... 10.0mL
Hemin solution................................................ 10.0mL

**Preparation of Vitamin $K_1$-Hemin Solution:** Add components to distilled/deionized water and bring volume to 1.0L. Mix thoroughly. Filter sterilize.

**Vitamin $K_1$ Solution:**
**Composition** per 100.0mL:
Vitamin $K_1$ ...............................................................1.0g
Ethanol ................................................................ 99.0mL

**Preparation of Vitamin $K_1$ Solution:** Add vitamin $K_1$ to 99.0mL of absolute ethanol. Mix thoroughly.

**Hemin Solution:**
**Composition** per 10.0mL:
Hemin..........................................................................0.5g
NaOH...........................................................................0.4g

**Preparation of Hemin Solution:** Add hemin and NaOH to distilled/deionized water and bring volume to 10.0mL. Mix thoroughly.

**Ferric Sulfate Solution:**
**Composition** per 100.0mL:
$FeSO_4·9H_2O$...............................................................0.04g

**Preparation of Ferric Sulfate Solution:** Add $FeSO_4·9H_2O$ to distilled/deionized water and bring volume to 100.0mL. Mix thoroughly. Filter sterilize.

**Mineral Solution:**
**Composition** per 100.0mL:
NaCl............................................................................9.0g
$CaCl_2·2H_2O$...............................................................0.27g
$MgCl_2·6H_2O$..............................................................0.2g
$CoCl_2·6H_2O$...............................................................0.1g
$MnCl_2·4H_2O$..............................................................0.1g

**Preparation of Mineral Solution:** Add components to distilled/deionized water and bring volume to 100.0mL. Mix thoroughly. Filter sterilize.

**Vitamin $B_{12}$ Solution:**
**Composition** per 10.0mL:
Vitamin $B_{12}$.............................................................0.1mg

**Preparation of Vitamin $B_{12}$ Solution:** Add vitamin $B_{12}$ to distilled/deionized water and bring volume to 10.0mL. Mix thoroughly. Filter sterilize.

**Preparation of Medium:** To 995.0mL of cooled, sterile basal solution, aseptically add 0.5g of L-cysteine·HCl·H₂O, 25.0mL of sterile glucuronic acid solution, 1.0mL of sterile gentamicin solution, 1.0mL of sterile hemin-vitamin $K_1$ solution, 1.0mL of sterile ferric sulfate solution, 1.0mL of sterile mineral solution, 1.0mL of Phenol Red solution, and 0.05mL of sterile vitamin $B_{12}$ solution. Mix thoroughly. Pour into sterile Petri dishes or distribute into sterile tubes.

**Storage/Shelf Life:** Store dehydrated media in the dark in a sealed container below 30°C. Prepared media should be stored under refrigeration (2-8°C). Media should be used within 60 days of preparation. Media should not be used if there are any signs of deterioration (shrinking, cracking, or discoloration) or contamination, or if the expiration date supplied by the manufacturer has passed.

**Use:** For the isolation, cultivation, and differentiation of the *Bacteroides fragilis* group (*Bacteroides fragilis*, *Bacteroides thetaiotamicron*, *Bacteroides vulgatus*, *Bacteroides distasonis*, *Bacteriodes ovatus*, and *Bacteroides uniformis*) from clinical specimens.

## *Francisella tularensis* Isolation Medium

**Composition** per liter:
Agar ...........................................................................10.0g
Glucose .......................................................................10.0g
Pancreatic digest of casein.......................................10.0g
Peptic digest of animal tissue ..................................10.0g
L-Cysteine·HCl·H₂O ..................................................5.0g
NaCl.............................................................................5.0g
Sodium thioglycolate.................................................2.0g
Glucose .......................................................................1.0g
Thiamine·HCl ...........................................................5.0mg

pH 7.2 ± 0.2 at 25°C

**Preparation of Medium:** Add components, except agar, to distilled/deionized water and bring volume to 1.0L. Mix thoroughly. Adjust pH to 7.2. Add agar. Gently heat and bring to boiling. Autoclave for 20 min at 15 psi pressure–121°C. Cool to 45°–50°C. Pour into sterile Petri dishes.

**Storage/Shelf Life:** Store dehydrated media in the dark in a sealed container below 30°C. Prepared media should be stored under refrigeration (2-8°C). Media should be used within 60 days of preparation. Media should not be used if there are any signs of deterioration (shrinking, cracking, or discoloration) or contamination, or if the expiration date supplied by the manufacturer has passed.

**Use:** For the isolation and cultivation of *Francisella tularensis*.

## Fraser Broth

**Composition** per liter:
NaCl............................................................................20.0g
$Na_2HPO_4$...................................................................12.0g
Beef extract..................................................................5.0g
Proteose peptone..........................................................5.0g
Pancreatic digest of casein..........................................5.0g
Yeast extract.................................................................5.0g
LiCl..............................................................................3.0g

KH$_2$PO$_4$............................................................................1.35g
Esculin ..............................................................................1.0g
Fraser supplement solution .......................................... 10.0mL

pH 7.2 ± 0.2 at 25°C

**Source:** This medium is available as a premixed powder from BD Diagnostic Systems.

**Caution:** Lithium chloride is harmful. Avoid bodily contact and inhalation of vapors. On contact with skin wash with plenty of water immediately.

**Fraser Supplement Solution:**
**Composition** per 10.0mL:

Ferric ammonium citrate................................................0.5g
Acriflavine·HCl...............................................................0.25g
Nalidixic acid...................................................................0.1g
Ethanol ............................................................................ 5.0mL

**Preparation of Fraser Supplement Solution:** Add components to distilled/deionized water and bring volume to 10.0mL. Mix thoroughly. Filter sterilize.

**Preparation of Medium:** Add components, except Fraser supplement solution, to distilled/deionized water and bring volume to 990.0mL. Mix thoroughly. Gently heat and bring to boiling. Autoclave for 15 min at 15 psi pressure–121°C. Cool to 45°–50°C. Aseptically add sterile Fraser supplement solution. Mix thoroughly. Aseptically distribute into sterile tubes or flasks.

**Storage/Shelf Life:** Store dehydrated media in the dark in a sealed container below 30°C. Prepared media should be stored under refrigeration (2-8°C). Media should be used within 60 days of preparation. Media should not be used if there are any signs of deterioration (discoloration) or contamination, or if the expiration date supplied by the manufacturer has passed.

**Use:** For the isolation of *Listeria* species from food and environmental species.

## Fraser Broth

**Composition** per liter:

NaCl....................................................................................20.0g
Na$_2$HPO$_4$.........................................................................12.0g
Proteose peptone ..........................................................5.0g
Tryptone ...........................................................................5.0g
Lab Lemco powder ........................................................5.0g
LiCl......................................................................................3.0g
KH$_2$PO$_4$............................................................................1.35g
Esculin ..............................................................................1.0g
Fraser supplement solution .......................................... 10.0mL

pH 7.2 ± 0.2 at 25°C

**Source:** This medium is available as a premixed powder from Thermo Scientific.

**Caution:** Lithium chloride is harmful. Avoid bodily contact and inhalation of vapors. On contact with skin wash with plenty of water immediately.

**Fraser Supplement Solution:**
**Composition** per 10.0mL:

Ferric ammonium citrate................................................0.5g
Acriflavine·HCl...............................................................0.25g
Nalidixic acid...................................................................0.1g
Ethanol ............................................................................ 5.0mL

**Preparation of Fraser Supplement Solution:** Add components to distilled/deionized water and bring volume to 10.0mL. Mix thoroughly. Filter sterilize.

**Preparation of Medium:** Add components, except Fraser supplement solution, to distilled/deionized water and bring volume to 990.0mL. Mix thoroughly. Gently heat and bring to boiling. Autoclave for 15 min at 15 psi pressure–121°C. Cool to 45°–50°C. Aseptically add sterile Fraser supplement solution. Mix thoroughly. Aseptically distribute into sterile tubes or flasks.

**Storage/Shelf Life:** Store dehydrated media in the dark in a sealed container below 30°C. Prepared media should be stored under refrigeration (2-8°C). Media should be used within 60 days of preparation. Media should not be used if there are any signs of deterioration (discoloration) or contamination, or if the expiration date supplied by the manufacturer has passed.

**Use:** For the isolation of *Listeria* species from food and environmental species.

## Fraser HiVeg Broth Base

**Composition** per liter:

NaCl....................................................................................20.0g
Na$_2$HPO$_4$·2H$_2$O ............................................................12.0g
Plant extract No. 1 .........................................................5.0g
Plant hydrolysate ...........................................................5.0g
Plant peptone .................................................................5.0g
Yeast extract....................................................................5.0g
LiCl......................................................................................3.0g
KH$_2$PO$_4$............................................................................1.35g
Esculin ..............................................................................1.0g
Fraser supplement solution .......................................... 10.0mL

pH 7.2 ± 0.2 at 25°C

**Source:** This medium, without Fraser supplement solution, is available as a premixed powder from HiMedia.

**Caution:** Lithium chloride is harmful. Avoid bodily contact and inhalation of vapors. On contact with skin wash with plenty of water immediately.

**Fraser Supplement Solution:**
**Composition** per 10.0mL:

Ferric ammonium citrate................................................0.5g
Acriflavine·HCl ...............................................................0.25g
Nalidixic acid...................................................................0.1g
Ethanol............................................................................. 5.0mL

**Preparation of Fraser Supplement Solution:** Add components to distilled/deionized water and bring volume to 10.0mL. Mix thoroughly. Filter sterilize.

**Preparation of Medium:** Add components, except Fraser supplement solution, to distilled/deionized water and bring volume to 990.0mL. Mix thoroughly. Gently heat and bring to boiling. Autoclave for 15 min at 15 psi pressure–121°C. Cool to 45°–50°C. Aseptically add sterile Fraser supplement solution. Mix thoroughly. Aseptically distribute into sterile tubes or flasks.

**Storage/Shelf Life:** Store dehydrated media in the dark in a sealed container below 30°C. Prepared media should be stored under refrigeration (2-8°C). Media should be used within 60 days of preparation. Media should not be used if there are any signs of deterioration (discoloration) or contamination, or if the expiration date supplied by the manufacturer has passed.

**Use:** For the isolation of *Listeria* species from food and environmental species.

## Fraser Secondary Enrichment Broth

**Composition** per liter:

| | |
|---|---|
| NaCl | 20.0g |
| Na₂HPO₄ | 12.0g |
| Beef extract | 5.0g |
| Proteose peptone | 5.0g |
| Pancreatic digest of casein | 5.0g |
| Yeast extract | 5.0g |
| LiCl | 3.0g |
| KH₂PO₄ | 1.35g |
| Esculin | 1.0g |
| Acriflavin solution | 10.0mL |
| Ferric ammonium citrate solution | 10.0mL |
| Nalidixic acid solution | 1.0mL |

**Caution:** Lithium chloride is harmful. Avoid bodily contact and inhalation of vapors. On contact with skin wash with plenty of water immediately.

**Ferric Ammonium Citrate Solution:**

**Composition** per 10.0mL:

| | |
|---|---|
| Ferric ammonium citrate | 0.5g |

**Preparation of Ferric Ammonium Citrate Solution:** Add ferric ammonium citrate to distilled/deionized water and bring volume to 10.0mL. Mix thoroughly. Filter sterilize.

**Acriflavin Solution:**

**Composition** per 10.0mL:

| | |
|---|---|
| Acriflavin | 0.025g |

**Preparation of Acriflavin Solution:** Add acriflavin to distilled/deionized water and bring volume to 10.0mL. Mix thoroughly. Filter sterilize.

**Nalidixic Acid Solution:**

**Composition** per 10.0mL:

| | |
|---|---|
| Nalidixic acid | 0.04g |
| NaOH (0.1$N$ solution) | 10.0mL |

**Preparation of Nalidixic Acid Solution:** Add nalidixic acid to 10.0mL of NaOH solution. Mix thoroughly. Filter sterilize.

**Preparation of Medium:** Add components, except acriflavin solution and ferric ammonium citrate solution, to distilled/deionized water and bring volume to 980.0mL. Mix thoroughly. Gently heat and bring to boiling. Distribute into tubes in 10.0mL volumes. Autoclave for 12 min at 15 psi pressure–121°C. Cool rapidly to 25°C. Immediately prior to inoculation, aseptically add 0.1mL of sterile acriflavin solution and 0.1mL of ferric ammonium citrate solution to each tube. Mix thoroughly.

**Storage/Shelf Life:** Store dehydrated media in the dark in a sealed container below 30°C. Prepared media should be stored under refrigeration (2-8°C). Media should be used within 60 days of preparation. Media should not be used if there are any signs of deterioration (discoloration) or contamination, or if the expiration date supplied by the manufacturer has passed.

**Use:** For the isolation, cultivation, and enrichment of *Listeria monocytogenes* from foods and environmental specimens based on esculin hydrolysis. Bacteria that hydrolyze esculin appear as black colonies.

## Fraser Secondary Enrichment HiVeg Broth Base

**Composition** per liter:

| | |
|---|---|
| NaCl | 20.0g |
| Na₂HPO₄ | 12.0g |
| Plant extract | 5.0g |
| Plant hydrolysate | 5.0g |
| Plant peptone No. 3 | 5.0g |
| Yeast extract | 5.0g |
| LiCl | 3.0g |
| KH₂PO₄ | 1.35g |
| Esculin | 1.0g |
| Ferric ammonium citrate | 0.5g |
| Acriflavin solution | 10.0mL |
| Ferric ammonium citrate solution | 10.0mL |
| Nalidixic acid solution | 1.0mL |

pH 7.2 ± 0.2 at 25°C

**Source:** This medium, without acriflavin, ferric ammonim citrate, and nalidixic acid solutions, is available as a premixed powder from Hi-Media.

**Caution:** Lithium chloride is harmful. Avoid bodily contact and inhalation of vapors. On contact with skin wash with plenty of water immediately.

**Ferric Ammonium Citrate Solution:**

**Composition** per 10.0mL:

| | |
|---|---|
| Ferric ammonium citrate | 0.5g |

**Preparation of Ferric Ammonium Citrate Solution:** Add ferric ammonium citrate to distilled/deionized water and bring volume to 10.0mL. Mix thoroughly. Filter sterilize.

**Acriflavin Solution:**

**Composition** per 10.0mL:

| | |
|---|---|
| Acriflavin | 0.025g |

**Preparation of Acriflavin Solution:** Add acriflavin to distilled/deionized water and bring volume to 10.0mL. Mix thoroughly. Filter sterilize.

**Nalidixic Acid Solution:**

**Composition** per 10.0mL:

| | |
|---|---|
| Nalidixic acid | 0.04g |
| NaOH (0.1$N$ solution) | 10.0mL |

**Preparation of Nalidixic Acid Solution:** Add nalidixic acid to 10.0mL of NaOH solution. Mix thoroughly. Filter sterilize.

**Preparation of Medium:** Add components, except acriflavin solution and ferric ammonium citrate solution, to distilled/deionized water and bring volume to 980.0mL. Mix thoroughly. Gently heat and bring to boiling. Distribute into tubes in 10.0mL volumes. Autoclave for 12 min at 15 psi pressure–121°C. Cool rapidly to 25°C. Immediately prior to inoculation, aseptically add 0.1mL of sterile acriflavin solution and 0.1mL of ferric ammonium citrate solution to each tube. Mix thoroughly.

**Storage/Shelf Life:** Store dehydrated media in the dark in a sealed container below 30°C. Prepared media should be stored under refrigeration (2-8°C). Media should be used within 60 days of preparation. Media should not be used if there are any signs of deterioration (discoloration) or contamination, or if the expiration date supplied by the manufacturer has passed.

**Use:** For the isolation, cultivation, and enrichment of *Listeria monocytogenes* from foods and environmental specimens based on esculin hydrolysis. Bacteria that hydrolyze esculin appear as black colonies.

## Furunculosis Agar

Agar ................................................................15.0g
Tryptone ..........................................................10.0g
Yeast extract......................................................5.0g
NaCl ................................................................2.5g
Tyrosine............................................................1.0g

pH 7.3 ± 0.2 at 25°C

**Preparation of Medium:** Add components to distilled/deionized water and bring volume to 1.0L. Mix thoroughly. Gently heat and bring to boiling. Mix thoroughly. Distribute into tubes or flasks. Autoclave for 15 min at 15 psi pressure–121°C. Pour into sterile Petri dishes.

**Storage/Shelf Life:** Store dehydrated media in the dark in a sealed container below 30°C. Prepared media should be stored under refrigeration (2-8°C). Media should be used within 60 days of preparation. Media should not be used if there are any signs of deterioration (shrinking, cracking, or discoloration) or contamination, or if the expiration date supplied by the manufacturer has passed.

**Use:** For the cultivation amd identification based upon pigment production of *Aeromonas salmonicida*.

## Furunculosis Agar

**Composition** per liter:

Agar ................................................................15.0g
Casein enzymatic hydrolysate ...................................10.0g
Yeast extract......................................................5.0g
NaCl ................................................................2.5g
Tyrosine............................................................1.0g

pH 7.3 ± 0.2 at 25°C

**Source:** This medium is available as a premixed powder from Hi-Media.

**Preparation of Medium:** Add components to distilled/deionized water and bring volume to 1.0L. Mix thoroughly. Gently heat and bring to boiling. Mix thoroughly. Distribute into tubes or flasks. Autoclave for 15 min at 15 psi pressure–121°C. Pour into sterile Petri dishes or leave in tubes. Allow tubes to cool in a slanted position.

**Storage/Shelf Life:** Store dehydrated media in the dark in a sealed container below 30°C. Prepared media should be stored under refrigeration (2-8°C). Media should be used within 60 days of preparation. Media should not be used if there are any signs of deterioration (shrinking, cracking, or discoloration) or contamination, or if the expiration date supplied by the manufacturer has passed.

**Use:** For the cultivation amd identification based upon pigment production of *Aeromonas salmonicida*.

## Furunculosis HiVeg Agar

**Composition** per liter:

Agar ................................................................15.0g
Plant hydrolysate................................................10.0g
Yeast extract......................................................5.0g
NaCl ................................................................2.5g
Tyrosine............................................................1.0g

pH 7.3 ± 0.2 at 25°C

**Source:** This medium is available as a premixed powder from Hi-Media.

**Preparation of Medium:** Add components to distilled/deionized water and bring volume to 1.0L. Mix thoroughly. Gently heat and bring

to boiling. Mix thoroughly. Distribute into tubes or flasks. Autoclave for 15 min at 15 psi pressure–121°C. Pour into sterile Petri dishes or leave in tubes. Allow tubes to cool in a slanted position.

**Storage/Shelf Life:** Store dehydrated media in the dark in a sealed container below 30°C. Prepared media should be stored under refrigeration (2-8°C). Media should be used within 60 days of preparation. Media should not be used if there are any signs of deterioration (shrinking, cracking, or discoloration) or contamination, or if the expiration date supplied by the manufacturer has passed.

**Use:** For the cultivation amd identification based upon pigment production of *Aeromonas salmonicida*.

## *Gardnerella vaginalis* Selective Medium

**Composition** per liter:

Columbia blood agar base ....................................... 940.0mL
Rabbit or horse serum............................................. 50.0mL
Antibiotic inhibitor solution .................................... 10.0mL

pH 7.2 ± 0.2 at 25°C

**Source:** This medium is available as a premixed powder from Thermo Scientific.

**Columbia Blood Agar Base:**

**Composition** per liter:

Special peptone...................................................23.0g
Agar ................................................................10.0g
NaCl ................................................................5.0g
Starch ..............................................................1.0g

**Source:** Columbia blood agar base is available as a premixed powder from Thermo Scientific.

**Preparation of Columbia Blood Agar Base:** Add components to distilled/deionized water and bring volume to 1.0L. Mix thoroughly. Gently heat until boiling. Autoclave for 15 min at 15 psi pressure–121°C. Cool to 45°–50°C.

**Antibiotic Inhibitor Solution:**

**Composition** per 10.0mL:

Nalidixic acid.....................................................0.035g
Gentamicin sulfate ................................................4.0mg
Amphotericin B ...................................................2.0mg
Ethanol ............................................................4.0mL

**Preparation of Antibiotic Inhibitor Solution:** Add components to distilled/deionized water and bring volume to 10.0mL. Mix thoroughly. Filter sterilize.

**Preparation of Medium:** To 940.0mL of cooled, sterile Columbia blood agar base, aseptically add 50.0mL of rabbit or horse blood serum and 10.0mL of sterile antibiotic inhibitor solution. Pour into sterile Petri dishes or distribute into sterile tubes.

**Storage/Shelf Life:** Store dehydrated media in the dark in a sealed container below 30°C. Prepared media should be stored under refrigeration (2-8°C). Media should be used within 60 days of preparation. Media should not be used if there are any signs of deterioration (shrinking, cracking, or discoloration) or contamination, or if the expiration date supplied by the manufacturer has passed.

**Use:** For the selective isolation, cultivation, and differentiation of *Gardnerella vaginalis* from clinical specimens, such as the vaginal discharge of patients with vaginitis. *Gardnerella vaginalis* exhibits β-hemolysis on this medium.

## Gassner Agar
### (Water-Blue Metachrome-Yellow Lactose Agar)

**Composition** per liter:

| | |
|---|---|
| Lactose | 43.0g |
| Peptone | 14.0g |
| Agar | 13.0g |
| NaCl | 5.0g |
| Metachrome Yellow | 1.25g |
| Water Blue | 0.62g |

pH 7.2 ± 0.2 at 25°C

**Preparation of Medium:** Add components to distilled/deionized water and bring volume to 1.0L. Mix thoroughly. Gently heat and bring to boiling. Mix thoroughly. Distribute into tubes or flasks. Autoclave for 15 min at 15 psi pressure–121°C. Pour into sterile Petri dishes.

**Storage/Shelf Life:** Store dehydrated media in the dark in a sealed container below 30°C. Prepared media should be stored under refrigeration (2-8°C). Media should be used within 60 days of preparation. Media should not be used if there are any signs of deterioration (shrinking, cracking, or discoloration) or contamination, or if the expiration date supplied by the manufacturer has passed.

**Use:** For the detection and isolation of pathogenic Enterobacteriaceae. For use in the execution of the German Meat Inspection Law (Deutsches Fleischbeschaugesetz). This culture medium contains Metachrome Yellow, which primarily inhibits the accompanying Gram-positive microbial flora. It also contains lactose, which, when degraded to acid, is shown by the indicator Water Blue, which is deep blue in the acidic range and colorless in the alkaline range. The prepared culture medium is green; in the acidic pH range it becomes blue-green to blue. At alkaline pHs, however, the yellow color of the Metachrome Yellow becomes increasingly apparent.

## Gassner Lactose Agar

**Composition** per liter:

| | |
|---|---|
| Lactose | 50.0g |
| Agar | 13.0g |
| Meat peptone | 7.0g |
| NaCl | 5.0g |
| Metachrome Yellow | 1.25g |
| Water Blue | 0.625g |

pH 7.2 ± 0.2 at 25°C

**Source:** This medium is available as a premixed powder from Hi-Media.

**Preparation of Medium:** Add components to distilled/deionized water and bring volume to 1.0L. Mix thoroughly. Gently heat and bring to boiling. Mix thoroughly. Distribute into tubes or flasks. Autoclave for 15 min at 15 psi pressure–121°C. Pour into sterile Petri dishes.

**Storage/Shelf Life:** Store dehydrated media in the dark in a sealed container below 30°C. Prepared media should be stored under refrigeration (2-8°C). Media should be used within 60 days of preparation. Media should not be used if there are any signs of deterioration (shrinking, cracking, or discoloration) or contamination, or if the expiration date supplied by the manufacturer has passed.

**Use:** For the detection and isolation of pathogenic Enterobacteriaceae.

## Gassner Lactose HiVeg Agar

**Composition** per liter:

| | |
|---|---|
| Lactose | 50.0g |
| Agar | 13.0g |
| Plant peptone No. 1 | 7.0g |
| NaCl | 5.0g |
| Metachrome Yellow | 1.25g |
| Water Blue | 0.625g |

pH 7.2 ± 0.2 at 25°C

**Source:** This medium is available as a premixed powder from Hi-Media.

**Preparation of Medium:** Add components to distilled/deionized water and bring volume to 1.0L. Mix thoroughly. Gently heat and bring to boiling. Mix thoroughly. Distribute into tubes or flasks. Autoclave for 15 min at 15 psi pressure–121°C. Pour into sterile Petri dishes.

**Storage/Shelf Life:** Store dehydrated media in the dark in a sealed container below 30°C. Prepared media should be stored under refrigeration (2-8°C). Media should be used within 60 days of preparation. Media should not be used if there are any signs of deterioration (shrinking, cracking, or discoloration) or contamination, or if the expiration date supplied by the manufacturer has passed.

**Use:** For the detection and isolation of pathogenic Enterobacteriaceae.

## GBNA Medium
### (Gum Base Nalidixic Acid Medium)

**Composition** per liter:

| | |
|---|---|
| Gellan gum | 8.0g |
| Pancreatic digest of casein | 5.7g |
| NaCl | 1.7g |
| Papaic digest of soybean meal | 1.0g |
| Glucose | 0.83g |
| $K_2HPO_4$ | 0.83g |
| $MgCl_2 \cdot 6H_2O$ | 0.33g |
| Nalidixic acid | 0.05g |

pH 7.2 ± 0.2 at 25°C

**Source:** This medium is available as a premixed powder from BD Diagnostic Systems.

**Preparation of Medium:** Add components to tap water and bring volume to 1.0L. Mix thoroughly. Gently heat and bring to boiling. Distribute into tubes or flasks. Autoclave for 15 min at 15 psi pressure–121°C. Pour into sterile Petri dishes or leave in tubes.

**Storage/Shelf Life:** Store dehydrated media in the dark in a sealed container below 30°C. Prepared media should be stored under refrigeration (2-8°C). Media should be used within 60 days of preparation. Media should not be used if there are any signs of deterioration (shrinking, cracking, or discoloration) or contamination, or if the expiration date supplied by the manufacturer has passed.

**Use:** For the isolation and cultivation of *Listeria monocytogenes* from clinical speciments and nonclinical specimens of public health importance.

## GBS Agar Base, Islam
### (Group B Streptococci Agar)
### (Islam GBS Agar)

**Composition** per liter:

| | |
|---|---|
| Proteose peptone | 23.0g |
| Agar | 10.0g |
| $Na_2HPO_4$ | 5.75g |
| Soluble starch | 5.0g |
| $NaH_2PO_4$ | 1.5g |
| Horse serum, heat inactivated | 50.0mL |

pH 7.5 ± 0.1 at 25°C

**Source:** This medium is available as a premixed powder from Thermo Scientific.

**Preparation of Medium:** Add components, except horse serum, to distilled/deionized water and bring volume to 950.0mL. Mix thoroughly. Gently heat and bring to boiling. Autoclave for 15 min at 15 psi pressure–121°C. Cool to 45°–50°C. Aseptically add 50.0mL of sterile inactivated horse serum. Mix thoroughly. Pour into sterile Petri dishes or distribute into sterile tubes.

**Storage/Shelf Life:** Store dehydrated media in the dark in a sealed container below 30°C. Prepared media should be stored under refrigeration (2-8°C). Media should be used within 60 days of preparation. Media should not be used if there are any signs of deterioration (shrinking, cracking, or discoloration) or contamination, or if the expiration date supplied by the manufacturer has passed.

**Use:** For the isolation and detection of group B streptococci (GBS) in clinical specimens. The medium is designed to exploit the ability of most group B streptococci (GBS) to produce orange/red pigmented colonies when incubated under anaerobic conditions. There is a pigment-enhancing effect around a sulphonamide disc, which does not grow; the enhanced pigment effect can be seen over a radius of 10–20mm. Non-group B organisms able to grow on this medium do not produce the orange/red pigment.

## GBS Medium, Rapid
## (Group B Streptococci Medium)
## (GBS Medium Base)

| | |
|---|---|
| Starch | 80.0g |
| Proteose peptone | 23.0g |
| $Na_2HPO_4$ | 5.75g |
| $NaH_2PO_4$ | 1.5g |
| Horse serum, inactivated | 50.0mL |
| Antibiotic inhibitor solution | 10.0 mL |

pH 7.5 ± 0.2 at 25°C

**Source:** This medium is available as a premixed powder from Thermo Scientific and HiMedia.

**Antibiotic Inhibitor Solution:**
**Composition** per 10.0mL:

| | |
|---|---|
| Metronidazole | 10.0mg |
| Gentamicin | 2.0mg |

**Preparation of Antibiotic Inhibitor Solution:** Add components to distilled/deionized water and bring volume to 10.0mL. Mix thoroughly. Filter sterilize.

**Preparation of Medium:** Add components, except horse serum and antibiotic inhibitor solution, to distilled/deionized water and bring volume to 940.0mL. Mix thoroughly. Gently heat and bring to boiling. Autoclave for 15 min at 15 psi pressure–121°C. Cool to 45°–50°C. Aseptically add 50.0mL of sterile heat-inactivated horse serum and 10.0mL of sterile antibiotic inhibitor solution. Mix thoroughly. Aseptically distribute into sterile tubes or flasks. Cool to 5°C and hold at that temperature for 12 h prior to use.

**Storage/Shelf Life:** Store dehydrated media in the dark in a sealed container below 30°C. Prepared media should be stored under refrigeration (2-8°C). Media should be used within 60 days of preparation. Media should not be used if there are any signs of deterioration (discoloration) or contamination, or if the expiration date supplied by the manufacturer has passed.

**Use:** For the rapid isolation and cultivation of group B streptococci from clinical specimens.

## GC Agar Base with Blood
**Composition** per liter:

| | |
|---|---|
| Peptone, special | 15.0g |
| Agar | 10.0g |
| NaCl | 5.0g |
| $K_2HPO_4$ | 4.0g |
| $KH_2PO_4$ | 1.0g |
| Cornstarch | 1.0g |
| Blood, defibrinated | 50.0mL |

pH 7.2 ± 0.2 at 25°C

**Source:** This medium, without blood, is available as a premixed powder from HiMedia.

**Preparation of Medium:** Add components, except blood, to distilled/deionized water and bring volume to 950.0L. Mix thoroughly. Gently heat until boiling. Autoclave for 15 min at 15 psi pressure–121°C. Cool to 75°–80°C. Add 50.0mL sterile defibrinated blood with thorough mixing and maintain at 75°–80°C for 15–20 min until the medium is chocolatized. Pour into sterile Petri dishes or distribute into sterile tubes.

**Storage/Shelf Life:** Store dehydrated media in the dark in a sealed container below 30°C. Prepared media should be stored under refrigeration (2-8°C). Media should be used within 60 days of preparation. Media should not be used if there are any signs of deterioration (shrinking, cracking, or discoloration) or contamination, or if the expiration date supplied by the manufacturer has passed.

**Use:** For the isolation and cultivation of fastidious bacteria, especially *Neisseria* and *Haemophilus* species.

## GC HiVeg Agar Base with Blood
**Composition** per liter:

| | |
|---|---|
| Plant special peptone | 15.0g |
| Agar | 10.0g |
| NaCl | 5.0g |
| $K_2HPO_4$ | 4.0g |
| $KH_2PO_4$ | 1.0g |
| Cornstarch | 1.0g |
| Blood, defibrinated | 50.0mL |

pH 7.2 ± 0.2 at 25°C

**Source:** This medium, without blood, is available as a premixed powder from HiMedia.

**Preparation of Medium:** Add components, except blood, to distilled/deionized water and bring volume to 950.0L. Mix thoroughly. Gently heat until boiling. Autoclave for 15 min at 15 psi pressure–121°C. Cool to 75°–80°C. Add 50.0mL sterile defibrinated blood with thorough mixing and maintain at 75°–80°C for 15–20 min until the medium is chocolatized. Pour into sterile Petri dishes or distribute into sterile tubes.

**Storage/Shelf Life:** Store dehydrated media in the dark in a sealed container below 30°C. Prepared media should be stored under refrigeration (2-8°C). Media should be used within 60 days of preparation. Media should not be used if there are any signs of deterioration (shrinking, cracking, or discoloration) or contamination, or if the expiration date supplied by the manufacturer has passed.

**Use:** For the isolation and cultivation of fastidious bacteria, especially *Neisseria* and *Haemophilus* species.

## GC HiVeg Agar Base with Hemoglobin

**Composition** per liter:

| | |
|---|---|
| Plant special peptone | 15.0g |
| Agar | 10.0g |
| NaCl | 5.0g |
| K$_2$HPO$_4$ | 4.0g |
| KH$_2$PO$_4$ | 1.0g |
| Cornstarch | 1.0g |
| Hemoglobin solution | 100.0mL |

pH 7.2 ± 0.2 at 25°C

**Source:** This medium, without hemoglobin solution, is available as a premixed powder from HiMedia.

**Hemoglobin Solution:**

**Composition** 100.0mL:

| | |
|---|---|
| Bovine hemoglobin | 2.0g |

**Preparation of Hemoglobin Solution:** Add bovine hemoglobin to distilled/deionized water and bring volume to 100.0mL. Mix thoroughly. Autoclave for 15 min at 15 psi pressure–121°C. Cool to 45°–50°C.

**Preparation of Medium:** Add components, except hemoglobin, to distilled/deionized water and bring volume to 500.0L. Mix thoroughly. Gently heat until boiling. Autoclave for 15 min at 15 psi pressure–121°C. Cool to 45°–50°C. Add 100.0mL sterile hemoglobin solution. Mix thoroughly. Note: Antibiotics may be added to increase selectivity. Pour into sterile Petri dishes or distribute into sterile tubes.

**Storage/Shelf Life:** Store dehydrated media in the dark in a sealed container below 30°C. Prepared media should be stored under refrigeration (2-8°C). Media should be used within 60 days of preparation. Media should not be used if there are any signs of deterioration (shrinking, cracking, or discoloration) or contamination, or if the expiration date supplied by the manufacturer has passed.

**Use:** For the isolation and cultivation of fastidious bacteria, especially *Neisseria* and *Haemophilus* species.

## GC Medium, New York City Formulation

**Composition** per liter:

| | |
|---|---|
| GC agar base | 850.0mL |
| Horse blood, lysed | 100.0mL |
| Yeast autolysate supplement | 30.0mL |
| LCAT antibiotic solution | 20.0mL |

pH 7.3 ± 0.2 at 25°C

**GC Agar Base:**

**Composition** per 850.0mL:

| | |
|---|---|
| Special peptone | 15.0g |
| Agar | 10.0g |
| NaCl | 5.0g |
| K$_2$HPO$_4$ | 4.0g |
| Cornstarch | 1.0g |
| KH$_2$PO$_4$ | 1.0g |

pH 7.2 ± 0.2 at 25°C

**Preparation of GC Agar Base:** Add components of GC medium base and the hemoglobin to distilled/deionized water and bring volume to 850.0mL. Mix thoroughly. Gently heat until boiling. Autoclave for 15 min at 15 psi pressure–121°C. Cool to 45°–50°C.

**Horse Blood, Lysed:**

**Composition** per 100.0mL:

| | |
|---|---|
| Saponin | 0.5g |
| Horse blood, defibrinated | 100.0mL |

**Preparation of Horse Blood, Lysed:** Add saponin to defibrinated horse blood. Mix thoroughly. Allow blood to lyse.

**Yeast Autolysate Supplement:**

**Composition** per 30.0mL:

| | |
|---|---|
| Yeast autolysate | 10.0g |
| Glucose | 1.0g |
| NaHCO$_3$ | 0.15g |

**Preparation of Yeast Autolysate Supplement:** Add components to distilled/deionized water and bring volume to 30.0mL. Mix thoroughly. Filter sterilize.

**LCAT Antibiotic Solution:**

**Composition** per 20.0mL:

| | |
|---|---|
| Colistin | 6.0mg |
| Trimethoprim lactate | 5.0mg |
| Lincomycin | 1.0mg |
| Amphotericin B | 1.0mg |

**Preparation of LCAT Antibiotic Solution:** Add components to distilled/deionized water and bring volume to 20.0mL. Mix thoroughly. Filter sterilize.

**Preparation of Medium:** To 850.0mL of cooled sterile GC agar base, aseptically add 100.0mL of sterile lysed horse blood, 30.0mL of sterile yeast autolysate supplement, and 20.0mL of LCAT antibiotic solution. Mix thoroughly. Pour into sterile Petri dishes or distribute into sterile tubes.

**Storage/Shelf Life:** Store dehydrated media in the dark in a sealed container below 30°C. Prepared media should be stored under refrigeration (2-8°C). Media should be used within 60 days of preparation. Media should not be used if there are any signs of deterioration (shrinking, cracking, or discoloration) or contamination, or if the expiration date supplied by the manufacturer has passed.

**Use:** For the selective isolation and cultivation of fastidious microorganisms, especially *Neisseria* species.

## GCII Agar

**Composition** per liter:

| | |
|---|---|
| GCII agar base, 2X | 490.0mL |
| Hemoglobin solution | 490.0mL |
| Supplement solution | 10.0mL |

pH 7.2 ± 0.2 at 25°C

**GCII Agar Base, 2X:**

**Composition** per liter:

| | |
|---|---|
| Agar | 10.0g |
| Pancreatic digest of casein | 7.5g |
| Selected meat peptone | 7.5g |
| NaCl | 5.0g |
| K$_2$HPO$_4$ | 4.0g |
| Cornstarch | 1.0g |
| KH$_2$PO$_4$ | 1.0g |

**Source:** GCII agar base is available as a premixed powder from BD Diagnostic Systems.

**Preparation of GCII Agar Base, 2X:** Add components to distilled/deionized water and bring volume to 500.0mL. Mix thoroughly. Gently heat until boiling. Autoclave for 15 min at 15 psi pressure–121°C. Cool to 45°–50°C.

**Hemoglobin Solution:**

**Composition** per 500.0mL:

| | |
|---|---|
| Hemoglobin | 10.0g |

**Preparation of Hemoglobin Solution:** Add hemoglobin to distilled/deionized water and bring volume to 500.0mL. Mix thoroughly. Autoclave for 15 min at 15 psi pressure–121°C. Cool to 45°–50°C.

**Supplement Solution:**
**Composition** per liter:

| | |
|---|---:|
| Glucose | 100.0g |
| L-Cysteine·HCl | 25.9g |
| L-Glutamine | 10.0g |
| L-Cystine | 1.1g |
| Adenine | 1.0g |
| Nicotinamide adenine dinucleotide | 0.25g |
| Vitamin B$_{12}$ | 0.1g |
| Thiamine pyrophosphate | 0.1g |
| Guanine·HCl | 0.03g |
| Fe(NO$_3$)$_3$·6H$_2$O | 0.02g |
| *p*-Aminobenzoic acid | 0.013g |
| Thiamine·HCl | 3.0mg |

**Preparation of Supplement Solution:** Add components to distilled/deionized water and bring volume to 1.0L. Mix thoroughly. Filter sterilize.

**Preparation of Medium:** To 490.0mL of sterile GCII agar base, aseptically add 490.0mL of sterile hemoglobin solution at 45°–50°C. Mix thoroughly. Aseptically add 10.0mL of sterile supplement solution. Mix thoroughly. Pour into sterile Petri dishes or distribute into sterile tubes.

**Storage/Shelf Life:** Store dehydrated media in the dark in a sealed container below 30°C. Prepared media should be stored under refrigeration (2-8°C). Media should be used within 60 days of preparation. Media should not be used if there are any signs of deterioration (shrinking, cracking, or discoloration) or contamination, or if the expiration date supplied by the manufacturer has passed.

**Use:** For the isolation and cultivation of fastidious microorganisms, especially *Neisseria* and *Haemophilus* species, from clinical specimens.

## GCII Agar

**Composition** per liter:

| | |
|---|---:|
| GCII agar base | 950.0mL |
| Blood, defibrinated | 50.0mL |

pH 7.2 ± 0.2 at 25°C

**GCII Agar Base with Extra Agar:**
**Composition** per liter:

| | |
|---|---:|
| Agar | 12.0g |
| Pancreatic digest of casein | 7.5g |
| Selected meat peptone | 7.5g |
| NaCl | 5.0g |
| K$_2$HPO$_4$ | 4.0g |
| Cornstarch | 1.0g |
| KH$_2$PO$_4$ | 1.0g |

**Source:** GCII agar base is available as a premixed powder from BD Diagnostic Systems.

**Preparation of GCII Agar Base with Extra Agar:** Add components to distilled/deionized water and bring volume to 1.0L. Mix thoroughly. Gently heat until boiling. Autoclave for 15 min at 15 psi pressure–121°C. Cool to 45°–50°C.

**Preparation of Medium:** To 950.0mL of sterile GCII agar base, aseptically add 50.0mL of sterile defibrinated blood with thorough

mixing and maintain at 75°–80°C for 15–20 min until the medium is chocolatized. Pour into sterile Petri dishes or distribute into sterile tubes.

**Storage/Shelf Life:** Store dehydrated media in the dark in a sealed container below 30°C. Prepared media should be stored under refrigeration (2-8°C). Media should be used within 60 days of preparation. Media should not be used if there are any signs of deterioration (shrinking, cracking, or discoloration) or contamination, or if the expiration date supplied by the manufacturer has passed.

**Use:** For the isolation and cultivation of fastidious microorganisms, especially *Neisseria* and *Haemophilus* species, from clinical specimens.

## GC-Lect™ Agar

**Composition** per liter:

| | |
|---|---:|
| GCII agar base, 2X | 500.0mL |
| Hemoglobin solution | 500.0mL |
| Supplement solution | 10.0mL |
| Selective agent solution | 10.0mL |

pH 7.2 ± 0.2 at 25°C

**Source:** This medium is available as a prepared medium from BD Diagnostic Systems.

**GCII Agar Base, 2X with Extra Agar:**
**Composition** per liter:

| | |
|---|---:|
| Agar | 12.0g |
| Pancreatic digest of casein | 7.5g |
| Selected meat peptone | 7.5g |
| NaCl | 5.0g |
| K$_2$HPO$_4$ | 4.0g |
| Cornstarch | 1.0g |
| KH$_2$PO$_4$ | 1.0g |

**Source:** GCII agar base is available as a premixed powder from BD Diagnostic Systems.

**Preparation of GCII Agar Base, 2X with Extra Agar:** Add components to distilled/deionized water and bring volume to 500.0mL. Mix thoroughly. Gently heat until boiling. Autoclave for 15 min at 15 psi pressure–121°C. Cool to 45°–50°C.

**Hemoglobin Solution:**
**Composition** per 500.0mL:

| | |
|---|---:|
| Hemoglobin | 10.0g |

**Preparation of Hemoglobin Solution:** Add hemoglobin to distilled/deionized water and bring volume to 500.0mL. Mix thoroughly. Autoclave for 15 min at 15 psi pressure–121°C. Cool to 45°–50°C.

**Supplement Solution:**
**Composition** per liter:

| | |
|---|---:|
| Glucose | 100.0g |
| L-Cysteine·HCl | 25.9g |
| L-Glutamine | 10.0g |
| L-Cystine | 1.1g |
| Adenine | 1.0g |
| Nicotinamide adenine dinucleotide | 0.25g |
| Vitamin B$_{12}$ | 0.1g |
| Thiamine pyrophosphate | 0.1g |
| Guanine·HCl | 0.03g |
| Fe(NO$_3$)$_3$·6H$_2$O | 0.02g |
| *p*-Aminobenzoic acid | 0.013g |
| Thiamine·HCl | 3.0mg |

**Source:** The supplement solution (IsoVitaleX® enrichment) is available from BD Diagnostic Systems. This enrichment may be replaced by supplement VX from BD Diagnostic Systems.

**Preparation of Supplement Solution:** Add components to distilled/deionized water and bring volume to 1.0L. Mix thoroughly. Filter sterilize.

**Selective Agent Solution:**
**Composition** per 10.0mL:
Selective agents.................................................................0.017g

**Preparation of Selective Agent Solution:** Add selective agents to distilled/deionized water and bring volume to 10.0mL. Mix thoroughly. Filter sterilize.

**Preparation of Medium:** To 500.0mL of sterile GCII agar base, aseptically add 500.0mL of sterile hemoglobin solution at 45°–50°C. Mix thoroughly. Aseptically add 10.0mL of sterile supplement solution and 10.0mL of selective agents solution. Mix thoroughly. Pour into sterile Petri dishes or distribute into sterile tubes.

**Storage/Shelf Life:** Store dehydrated media in the dark in a sealed container below 30°C. Prepared media should be stored under refrigeration (2-8°C). Media should be used within 60 days of preparation. Media should not be used if there are any signs of deterioration (shrinking, cracking, or discoloration) or contamination, or if the expiration date supplied by the manufacturer has passed.

**Use:** For the isolation and cultivation of *Neisseria gonorrhoeae* from clinical specimens.

### GCA Agar with Thiamine
**Composition** per liter:
Glucose ..............................................................................25.0g
Agar ...................................................................................14.0g
Papaic digest of soybean meal ..........................................10.0g
NaCl .....................................................................................5.0g
Pancreatic digest of heart muscle........................................3.0g
Cysteine·HCl·H$_2$O................................................................1.0g
Thiamine ............................................................................0.05mg
Rabbit blood, defibrinated ................................................50.0mL
pH 6.8 ± 0.2 at 25°C

**Preparation of Medium:** Add components, except rabbit blood, to distilled/deionized water and bring volume to 950.0mL. Mix thoroughly. Gently heat and bring to boiling. Autoclave for 15 min at 15 psi pressure–121°C. Cool to 45°–50°C. Aseptically add sterile rabbit blood. Mix thoroughly. Pour into sterile Petri dishes or distribute into sterile tubes.

**Storage/Shelf Life:** Store dehydrated media in the dark in a sealed container below 30°C. Prepared media should be stored under refrigeration (2-8°C). Media should be used within 60 days of preparation. Media should not be used if there are any signs of deterioration (shrinking, cracking, or discoloration) or contamination, or if the expiration date supplied by the manufacturer has passed.

**Use:** For the isolation and cultivation of *Francisella tularensis*.

### Gelatin Agar
**Composition** per liter:
Gelatin...............................................................................30.0g
Agar ...................................................................................15.0g
Pancreatic digest of casein................................................10.0g
NaCl...................................................................................10.0g
pH 7.2 ± 0.2 at 25°C

**Preparation of Medium:** Add components to distilled/deionized water and bring volume to 1.0L. Mix thoroughly. Gently heat and bring to boiling. Distribute into tubes or flasks. Autoclave for 15 min at 15 psi pressure–121°C.

**Storage/Shelf Life:** Store dehydrated media in the dark in a sealed container below 30°C. Prepared media should be stored under refrigeration (2-8°C). Media should be used within 60 days of preparation. Media should not be used if there are any signs of deterioration (shrinking, cracking, or discoloration) or contamination, or if the expiration date supplied by the manufacturer has passed.

**Use:** For the cultivation of bacteria isolated from foods and their differentiation based on proteolytic activity.

### Gelatin Agar
### (GA Medium)
**Composition** per liter:
Solution 1.......................................................................950.0mL
Solution 2.........................................................................50.0mL
pH 7.2 ± 0.2 at 25°C

**Solution 1:**
**Composition** per 950.0mL:
Gelatin...............................................................................30.0g
Agar ...................................................................................15.0g
Pancreatic digest of casein................................................10.0g
NaCl.....................................................................................2.0g
D-Mannitol...........................................................................1.0g
Glucose ................................................................................1.0g
KNO$_3$ ..................................................................................1.0g
Sodium acetate ....................................................................1.0g
Sodium formate ...................................................................1.0g
Sodium succinate .................................................................1.0g
Yeast extract........................................................................1.0g
Sodium lactate (60% solution)........................................... 5.0mL

**Preparation of Solution 1:** Add components to distilled/deionized water and bring volume to 950.0mL. Mix thoroughly. Gently heat and bring to boiling. Autoclave for 15 min at 15 psi pressure–121°C.

**Solution 2:**
**Composition** per 50.0mL:
Na$_2$HPO$_4$.............................................................................1.0g
L-Cysteine·HCl·H$_2$O ...........................................................0.5g
Na$_2$CO$_3$·H$_2$O .....................................................................0.5g
Sucrose.................................................................................0.5g
Dithiothreitol.......................................................................0.1g
Menadione solution ......................................................... 2.0mL

**Preparation of Solution 2:** Add components to distilled/deionized water and bring volume to 50.0mL. Mix thoroughly. Filter sterilize.

**Menadione Solution:**
**Composition** per 100.0mL:
Menadione (vitamin K$_3$) .....................................................0.05g
Ethanol............................................................................. 99.0mL

**Preparation of Menadione Solution:** Add menadione to 99.0mL of absolute ethanol. Mix thoroughly.

**Preparation of Medium:** Aseptically combine sterile solution 1 with sterile solution 2. Mix thoroughly. Pour into sterile Petri dishes.

**Storage/Shelf Life:** Store dehydrated media in the dark in a sealed container below 30°C. Prepared media should be stored under refrigeration (2-8°C). Media should be used within 60 days of preparation.

Media should not be used if there are any signs of deterioration (shrinking, cracking, or discoloration) or contamination, or if the expiration date supplied by the manufacturer has passed.

**Use:** For the cultivation and differentiation of microorganisms from dental plaque based on their ability to produce gelatinase. For the differentiation of aerobic, anaerobic, and facultative microorganisms of clinical significance.

### Gelatin Medium

**Composition** per liter:

Gelatin.................................................................................4.0g

pH 7.0 ± 0.2 at 25°C

**Preparation of Medium:** Add gelatin to distilled/deionized water and bring volume to 1.0L. Mix thoroughly. Gently heat and bring to boiling. Distribute into tubes. Autoclave for 15 min at 15 psi pressure–121°C.

**Storage/Shelf Life:** Store dehydrated media in the dark in a sealed container below 30°C. Prepared media should be stored under refrigeration (2-8°C). Media should be used within 60 days of preparation. Media should not be used if there are any signs of deterioration (shrinking, cracking, or discoloration) or contamination, or if the expiration date supplied by the manufacturer has passed.

**Use:** For the cultivation and differentiation of *Nocardia* and *Streptomyces* species based on utilization of gelatin. *Nocardia asteroides* usually exhibits no growth. *Nocardia brasiliensis* shows good growth and round, compact colonies. *Streptomyces* species show varying degrees of growth.

### Gelatin Metronidazole Cadmium Medium
### (GMC Medium)

**Composition** per liter:

Solution 1 ..........................................................950.0mL
Solution 2 ............................................................50.0mL

pH 7.2 ± 0.2 at 25°C

**Solution 1:**

**Composition** per 950.0mL:

Gelatin................................................................30.0g
Agar ...................................................................15.0g
Pancreatic digest of casein................................10.0g
NaCl......................................................................2.0g
D-Mannitol...........................................................1.0g
Glucose ...............................................................1.0g
$KNO_3$ ....................................................................1.0g
Sodium acetate....................................................1.0g
Sodium formate....................................................1.0g
Sodium succinate ................................................1.0g
Yeast extract .......................................................1.0g
$CdSO_4 \cdot 8H_2O$ ......................................................0.02g
Metronidazole ...................................................0.01g
Sodium lactate (60% solution).........................5.0mL

**Preparation of Solution 1:** Add components to distilled/deionized water and bring volume to 950.0mL. Mix thoroughly. Gently heat and bring to boiling. Autoclave for 15 min at 15 psi pressure–121°C.

**Solution 2:**

**Composition** per 50.0mL:

$Na_2HPO_4$................................................................1.0g
L-Cysteine·HCl·$H_2O$............................................0.5g
$Na_2CO_3 \cdot H_2O$ .....................................................0.5g

Sucrose.................................................................0.5g
Dithiothreitol.......................................................0.1g
Menadione solution ..........................................2.0mL

**Preparation of Solution 2:** Add components to distilled/deionized water and bring volume to 50.0mL. Mix thoroughly. Filter sterilize.

**Menadione Solution:**

**Composition** per 100.0mL:

Menadione (vitamin $K_3$) .................................0.05g
Ethanol ............................................................99.0mL

**Preparation of Menadione Solution:** Add menadione to 99.0mL of absolute ethanol. Mix thoroughly.

**Preparation of Medium:** Aseptically combine sterile solution 1 with sterile solution 2. Mix thoroughly. Pour into sterile Petri dishes.

**Storage/Shelf Life:** Store dehydrated media in the dark in a sealed container below 30°C. Prepared media should be stored under refrigeration (2-8°C). Media should be used within 60 days of preparation. Media should not be used if there are any signs of deterioration (shrinking, cracking, or discoloration) or contamination, or if the expiration date supplied by the manufacturer has passed.

**Use:** For the cultivation and differentiation of microorganisms from dental plaque based on their ability to produce gelatinase. For the differentiation of aerobic, anaerobic, and facultative microorganisms of clinical significance.

### Gelatin Phosphate Salt Agar
### (GPS Agar)

**Composition** per liter:

Agar ..................................................................15.0g
Gelatin................................................................10.0g
NaCl...................................................................10.0g
$K_2HPO_4$..............................................................5.0g

pH 7.2 ± 0.2 at 25°C

**Preparation of Medium:** Add components to distilled/deionized water and bring volume to 1.0L. Mix thoroughly. Gently heat and bring to boiling. Distribute into tubes or flasks. Autoclave for 15 min at 15 psi pressure–121°C. Pour into sterile Petri dishes.

**Storage/Shelf Life:** Store dehydrated media in the dark in a sealed container below 30°C. Prepared media should be stored under refrigeration (2-8°C). Media should be used within 60 days of preparation. Media should not be used if there are any signs of deterioration (shrinking, cracking, or discoloration) or contamination, or if the expiration date supplied by the manufacturer has passed.

**Use:** For the cultivation and differentiation of *Vibrio* species from foods.

### Gelatin Phosphate Salt Broth
### (GPS Broth)

**Composition** per liter:

Gelatin................................................................10.0g
NaCl...................................................................10.0g
$K_2HPO_4$..............................................................5.0g

pH 7.2 ± 0.2 at 25°C

**Preparation of Medium:** Add components to distilled/deionized water and bring volume to 1.0L. Mix thoroughly. Gently heat and bring to boiling. Distribute into tubes or flasks. Autoclave for 15 min at 15 psi pressure–121°C.

**Use:** For the cultivation of *Vibrio* species from foods.

## Gelatin Salt Agar

**Composition** per liter:

NaCl.................................................................30.0g
Agar .................................................................15.0g
Gelatin..............................................................15.0g
Peptone..............................................................4.0g
Yeast extract........................................................1.0g

pH 7.2 ± 0.2 at 25°C

**Preparation of Medium:** Add components to distilled/deionized water and bring volume to 1.0L. Mix thoroughly. Gently heat and bring to boiling. Distribute into tubes or flasks. Autoclave for 15 min at 15 psi pressure–121°C. Pour into sterile Petri dishes or leave in tubes.

**Storage/Shelf Life:** Store dehydrated media in the dark in a sealed container below 30°C. Prepared media should be stored under refrigeration (2-8°C). Media should be used within 60 days of preparation. Media should not be used if there are any signs of deterioration (discoloration) or contamination, or if the expiration date supplied by the manufacturer has passed.

**Use:** For the cultivation and differentiation of *Vibrio* species from foods.

## Gelatin Salt Agar
## (GS Agar)
## (BAM M55)

**Composition** per liter:

NaCl.................................................................30.0g
Agar .................................................................25.0g
Gelatin..............................................................15.0g
Peptone..............................................................4.0g
Yeast extract........................................................1.0g

pH 7.2 ± 0.2 at 25°C

**Preparation of Medium:** Add components to distilled/deionized water and bring volume to 1.0L. Mix thoroughly. Gently heat and bring to boiling. Distribute into tubes or flasks. Autoclave for 15 min at 15 psi pressure–121°C. Pour into sterile Petri dishes or leave in tubes.

**Storage/Shelf Life:** Store dehydrated media in the dark in a sealed container below 30°C. Prepared media should be stored under refrigeration (2-8°C). Media should be used within 60 days of preparation. Media should not be used if there are any signs of deterioration (shrinking, cracking, or discoloration) or contamination, or if the expiration date supplied by the manufacturer has passed.

**Use:** For the cultivation and differentiation of *Vibrio* species from foods. The high concentration of agar inhibits the spreading of *V. alginolyticus* and some other *Vibrio* spp.

## Glucose Azide Broth

**Composition** per liter:

Peptic digest of animal tissue.......................................10.0g
Glucose ..............................................................5.0g
K$_2$HPO$_4$..........................................................5.0g
NaCl.................................................................5.0g
Yeast extract........................................................3.0g
KH$_2$PO$_4$..........................................................2.0g
NaN$_3$ ..............................................................0.25g
Bromcresol Purple ...................................................0.03g

pH 6.7 ± 0.2 at 25°C

**Source:** This medium is available as a premixed powder from Hi-Media.

**Caution:** Sodium azide is toxic. Azides also react with metals and disposal must be highly diluted.

**Preparation of Medium:** Add components to distilled/deionized water and bring volume to 1.0L. Mix thoroughly. Gently heat and bring to boiling. Mix thoroughly. Distribute into tubes or flasks. Autoclave for 15 min at 15 psi pressure–121°C.

**Storage/Shelf Life:** Store dehydrated media in the dark in a sealed container below 30°C. Prepared media should be stored under refrigeration (2-8°C). Media should be used within 60 days of preparation. Media should not be used if there are any signs of deterioration (discoloration) or contamination, or if the expiration date supplied by the manufacturer has passed.

**Use:** For the enumeration of fecal streptococci by the MPN technique from water and sewage.

## Glucose Azide HiVeg Broth

**Composition** per liter:

Plant peptone .......................................................10.0g
Glucose ..............................................................5.0g
K$_2$HPO$_4$..........................................................5.0g
NaCl.................................................................5.0g
Yeast extract........................................................3.0g
KH$_2$PO$_4$..........................................................2.0g
NaN$_3$ ..............................................................0.25g
Bromcresol Purple ...................................................0.03g

pH 6.7 ± 0.2 at 25°C

**Source:** This medium is available as a premixed powder from Hi-Media.

**Caution:** Sodium azide is toxic. Azides also react with metals and disposal must be highly diluted.

**Preparation of Medium:** Add components to distilled/deionized water and bring volume to 1.0L. Mix thoroughly. Gently heat and bring to boiling. Mix thoroughly. Distribute into tubes or flasks. Autoclave for 15 min at 15 psi pressure–121°C.

**Storage/Shelf Life:** Store dehydrated media in the dark in a sealed container below 30°C. Prepared media should be stored under refrigeration (2-8°C). Media should be used within 60 days of preparation. Media should not be used if there are any signs of deterioration (discoloration) or contamination, or if the expiration date supplied by the manufacturer has passed.

**Use:** For the enumeration of fecal streptococci by the MPN technique from water and sewage.

## Glucose Salt Teepol Broth
## (GSTB)

**Composition** per liter:

NaCl.................................................................30.0g
Peptone .............................................................10.0g
Glucose ..............................................................5.0g
Beef extract.........................................................3.0g
Methyl Violet........................................................2.0mg
Sodium lauryl sulfate (Teepol—0.1% solution) .......................4.0mL

pH 8.8 ± 0.2 at 25°C

**Preparation of Medium:** Add components to distilled/deionized water and bring volume to 1.0L. Mix thoroughly. Adjust pH to 8.8. Distribute into tubes or flasks. Autoclave for 15 min at 15 psi pressure–121°C.

**Storage/Shelf Life:** Store dehydrated media in the dark in a sealed container below 30°C. Prepared media should be stored under refrigeration (2-8°C). Media should be used within 60 days of preparation. Media should not be used if there are any signs of deterioration (discoloration) or contamination, or if the expiration date supplied by the manufacturer has passed.

**Use:** For the cultivation of *Vibrio* species from foods.

## Glucose Salt Teepol HiVeg Broth

**Composition** per liter:
| | |
|---|---|
| NaCl | 30.0g |
| Plant peptone | 10.0g |
| Glucose | 5.0g |
| Teepol | 4.0g |
| Plant extract | 3.0g |
| Methyl Violet | 2.0mg |

pH 8.8 ± 0.2 at 25°C

**Source:** This medium is available as a premixed powder from Hi-Media.

**Preparation of Medium:** Add components to distilled/deionized water and bring volume to 1.0L. Mix thoroughly. Adjust pH to 8.8. Distribute into tubes or flasks. Autoclave for 15 min at 15 psi pressure–121°C.

**Storage/Shelf Life:** Store dehydrated media in the dark in a sealed container below 30°C. Prepared media should be stored under refrigeration (2-8°C). Media should be used within 60 days of preparation. Media should not be used if there are any signs of deterioration (discoloration) or contamination, or if the expiration date supplied by the manufacturer has passed.

**Use:** For the cultivation of *Vibrio* species from foods.

## GN Broth, Hajna

**Composition** per liter:
| | |
|---|---|
| Pancreatic digest of casein | 10.0g |
| Peptic digest of animal tissue | 10.0g |
| NaCl | 5.0g |
| Sodium citrate | 5.0g |
| K$_2$HPO$_4$ | 4.0g |
| D-Mannitol | 2.0g |
| KH$_2$PO$_4$ | 1.5g |
| Glucose | 1.0g |
| Sodium deoxycholate | 0.5g |

pH 7.0 ± 0.2 at 25°C

**Source:** This medium is available as a premixed powder from BD Diagnostic Systems.

**Preparation of Medium:** Add components to distilled/deionized water and bring volume to 1.0L. Mix thoroughly. Gently heat and bring to boiling. Distribute into tubes or flasks. Autoclave for 15 min at 13 psi pressure–118°C.

**Storage/Shelf Life:** Store dehydrated media in the dark in a sealed container below 30°C. Prepared media should be stored under refrigeration (2-8°C). Media should be used within 60 days of preparation. Media should not be used if there are any signs of deterioration (discoloration) or contamination, or if the expiration date supplied by the manufacturer has passed.

**Use:** For the selective cultivation of *Salmonella* and *Shigella* species.

## GN Broth, Hajna

**Composition** per liter:
| | |
|---|---|
| Tryptose | 20.0g |
| NaCl | 5.0g |
| Sodium citrate | 5.0g |
| K$_2$HPO$_4$ | 4.0g |
| Mannitol | 2.0g |
| KH$_2$PO$_4$ | 1.5g |
| Glucose | 1.0g |
| Sodium deoxycholate | 0.5g |

pH 7.0 ± 0.2 at 25°C

**Source:** This medium is available as a premixed powder from Hi-Media.

**Preparation of Medium:** Add components to distilled/deionized water and bring volume to 1.0L. Mix thoroughly. Gently heat and bring to boiling. Distribute into tubes or flasks. Autoclave for 15 min at 10 psi pressure–115°C.

**Storage/Shelf Life:** Store dehydrated media in the dark in a sealed container below 30°C. Prepared media should be stored under refrigeration (2-8°C). Media should be used within 60 days of preparation. Media should not be used if there are any signs of deterioration (discoloration) or contamination, or if the expiration date supplied by the manufacturer has passed.

**Use:** For the selective cultivation of *Salmonella* and *Shigella* species.

## GN HiVeg Broth

**Composition** per liter:
| | |
|---|---|
| Plant hydrolysate No. 1 | 20.0g |
| NaCl | 5.0g |
| Sodium citrate | 5.0g |
| K$_2$HPO$_4$ | 4.0g |
| Mannitol | 2.0g |
| KH$_2$PO$_4$ | 1.5g |
| Glucose | 1.0g |
| Synthetic detergent No. III | 0.5g |

pH 7.0 ± 0.2 at 25°C

**Source:** This medium is available as a premixed powder from Hi-Media.

**Preparation of Medium:** Add components to distilled/deionized water and bring volume to 1.0L. Mix thoroughly. Gently heat and bring to boiling. Distribute into tubes or flasks. Autoclave for 15 min at 10 psi pressure–115°C.

**Storage/Shelf Life:** Store dehydrated media in the dark in a sealed container below 30°C. Prepared media should be stored under refrigeration (2-8°C). Media should be used within 60 days of preparation. Media should not be used if there are any signs of deterioration (discoloration) or contamination, or if the expiration date supplied by the manufacturer has passed.

**Use:** For the selective cultivation of *Salmonella* and *Shigella* species.

## GPVA Medium

**Composition** per liter:
| | |
|---|---|
| Agar | 15.0g |
| Yeast extract | 10.0g |
| ACES buffer (2-[(2-amino-2-oxoethyl)-amino]-ethane sulfonic acid) | 10.0g |
| Glycine | 3.0g |
| Charcoal, activated | 2.0g |

α-Ketoglutarate ............................................................1.0g
Fe$_4$(P$_2$O$_7$)$_3$·9H$_2$O ..............................................0.25g
Antibiotic inhibitor solution................................ 10.0mL
pH 6.9 ± 0.2 at 25°C

### Antibiotic Inhibitor Solution:
**Composition** per 10.0mL:
Anisomycin.................................................................0.08g
Vancomycin ...............................................................5.0mg
Polymyxin B ......................................................100,000U

**Preparation of Antibiotic Inhibitor Solution:** Add components to distilled/deionized water and bring volume to 10.0mL. Mix thoroughly. Filter sterilize.

**Preparation of Medium:** Add components, except antibiotic inhibitor solution, to distilled/deionized water and bring volume to 990.0mL. Mix thoroughly. Gently heat and bring to boiling. Autoclave for 15 min at 15 psi pressure–121°C. Cool to 45°–50°C. Adjust pH to 6.9. Aseptically add 10.0mL of sterile antibiotic inhibitor solution. Mix thoroughly. Pour into sterile Petri dishes or distribute into sterile tubes.

**Storage/Shelf Life:** Store dehydrated media in the dark in a sealed container below 30°C. Prepared media should be stored under refrigeration (2-8°C). Media should be used within 60 days of preparation. Media should not be used if there are any signs of deterioration (shrinking, cracking, or discoloration) or contamination, or if the expiration date supplied by the manufacturer has passed.

**Use:** For the isolation and cultivation of *Legionella* species from environmental waters.

## Granada Medium
**Composition** per liter:
Starch, soluble..........................................................150.0g
Proteose peptone No. 3 .............................................38.0g
NaCl............................................................................3.0g
Trimethoprim lactate................................................0.015g
Sodium phosphate buffer (0.06*M*, pH 7.4) ............ 900.0mL
Horse serum, coagulated ........................................ 100.0mL
pH 7.4 ± 0.2 at 25°C

**Preparation of Medium:** Add proteose peptone No. 3 and NaCl to 200.0mL of sodium phosphate buffer and bring to boiling. Add 400.0mL of cold sodium phosphate buffer, starch, and trimethoprim lactate. Mix thoroughly. Bring volume to 900.0mL with sodium phosphate buffer. Gently heat while stirring in a boiling water bath for exactly 20 min. Do not autoclave. Cool to 90°–95°C. Add horse serum. Mix thoroughly. Cool to 60°–65°C while stirring. Pour into sterile Petri dishes. Medium will solidify in 2–3 h.

**Storage/Shelf Life:** Store dehydrated media in the dark in a sealed container below 30°C. Prepared media should be stored under refrigeration (2-8°C). Media should be used within 60 days of preparation. Media should not be used if there are any signs of deterioration (discoloration) or contamination, or if the expiration date supplied by the manufacturer has passed.

**Use:** For the early selective isolation and cultivation of Group B streptococci from clinical specimens.

## Group A Selective Strep Agar with Sheep Blood
**Composition** per liter:
Pancreatic digest of casein.........................................14.5g
Agar ...........................................................................14.0g
NaCl.............................................................................5.0g

Papaic digest of soybean meal.....................................5.0g
Sheep blood ............................................................ 50.0mL
Growth factor solution............................................ 10.0mL
Selective agents solution......................................... 10.0mL
pH 7.4 ± 0.2 at 25°C

### Growth Factor Solution:
**Composition** per 10.0mL:
Growth factors, BBL ..................................................1.5g

**Preparation of Growth Factor Solution:** Add growth factors to distilled/deionized water and bring volume to 10.0mL. Mix thoroughly. Filter sterilize.

### Selective Agents Solution:
**Composition** per 10.0mL:
Selective agents .....................................................0.042g

**Preparation of Selective Agents Solution:** Add selective agents to distilled/deionized water and bring volume to 10.0mL. Mix thoroughly. Filter sterilize.

**Preparation of Medium:** Add components, except sheep blood, growth factor solution, and selective agents solution, to distilled/deionized water and bring volume to 930.0mL. Mix thoroughly. Gently heat and bring to boiling. Autoclave for 15 min at 15 psi pressure–121°C. Cool to 45°–50°C. Aseptically add 50.0mL of sheep blood, 10.0mL of sterile growth factor solution, and 10.0mL of sterile selective agents solution. Mix thoroughly. Pour into sterile Petri dishes or distribute into sterile tubes.

**Storage/Shelf Life:** Store dehydrated media in the dark in a sealed container below 30°C. Prepared media should be stored under refrigeration (2-8°C). Media should be used within 60 days of preparation. Media should not be used if there are any signs of deterioration (shrinking, cracking, or discoloration) or contamination, or if the expiration date supplied by the manufacturer has passed.

**Use:** For the selective cultivation and primary isolation of group A streptococci, especially *Streptococcus pyogenes,* from clinical specimens.

## *Haemophilus* Test Medium (HTM)
**Composition** per liter:
Beef infusion...........................................................300.0g
Acid hydrolysate of casein.........................................17.5g
Agar ...........................................................................17.0g
Yeast extract................................................................5.0g
Starch ..........................................................................1.5g
HTM supplement ..................................................... 10.0mL
pH 7.4 ± 0.2 at 25°C

**Source:** This medium is available as a premixed powder from Thermo Scientific.

### HTM Supplement:
**Composition** per 10.0mL:
Nicotinamide adenine dinucleotide ..............................0.03g
Hematin........................................................................0.03g

**Preparation of HTM Supplement:** Add components to distilled/deionized water and bring volume to 10.0mL. Mix thoroughly. Filter sterilize.

**Preparation of Medium:** Add components, except HTM supplement, to distilled/deionized water and bring volume to 990.0mL. Mix thoroughly. Gently heat and bring to boiling. Autoclave for 15 min at 15 psi pressure–

121°C. Cool to 45°–50°C. Aseptically add 10.0mL of sterile HTM supplement. Mix thoroughly. Pour into sterile Petri dishes or distribute into sterile tubes.

**Storage/Shelf Life:** Store dehydrated media in the dark in a sealed container below 30°C. Prepared media should be stored under refrigeration (2-8°C). Media should be used within 60 days of preparation. Media should not be used if there are any signs of deterioration (shrinking, cracking, or discoloration) or contamination, or if the expiration date supplied by the manufacturer has passed.

**Use:** For the susceptibility testing of *Haemophilus influenzae*. The medium forms part of the recommended methods of the United States National Committee for Clinical Laboratory Standards (NCCLS). *Haemophilus influenzae* require complex media for growth. These complex media have aggravated the routine susceptibility testing of *Haemophilus influenzae* because of antagonism between some essential nutrients and certain antimicrobial agents. This medium overcomes those limitations. The transparency of the medium allows zones of inhibition to be read easily through the bottom of the Petri dish. HTM contains low levels of antimicrobial antagonists, which allows testing of trimethoprim/sulphamethoxazole to be carried out.

### Half Fraser Broth

**Composition** per liter:
| | |
|---|---|
| NaCl | 20.0g |
| $Na_2HPO_4$ | 12.0g |
| Proteose peptone | 5.0g |
| Tryptone | 5.0g |
| Lab Lemco powder | 5.0g |
| LiCl | 3.0g |
| $KH_2PO_4$ | 1.35g |
| Esculin | 1.0g |
| Half Fraser supplement solution | 10.0mL |

pH 7.2 ± 0.2 at 25°C

**Source:** This medium is available as a premixed powder from Thermo Scientific.

**Caution:** Lithium chloride is harmful. Avoid bodily contact and inhalation of vapors. On contact with skin wash with plenty of water immediately.

**Half Fraser Supplement Solution:**
**Composition** per 10.0mL:
| | |
|---|---|
| Ferric ammonium citrate | 0.5g |
| Acriflavine·HCl | 0.125g |
| Nalidixic acid | 0.05g |
| Ethanol | 5.0mL |

**Preparation of Half Fraser Supplement Solution:** Add components to distilled/deionized water and bring volume to 10.0mL. Mix thoroughly. Filter sterilize.

**Preparation of Medium:** Add components, except half Fraser supplement solution, to distilled/deionized water and bring volume to 990.0mL. Mix thoroughly. Gently heat and bring to boiling. Autoclave for 15 min at 15 psi pressure–121°C. Cool to 45°–50°C. Aseptically add sterile half Fraser supplement solution. Mix thoroughly. Aseptically distribute into sterile tubes or flasks.

**Storage/Shelf Life:** Store dehydrated media in the dark in a sealed container below 30°C. Prepared media should be stored under refrigeration (2-8°C). Media should be used within 60 days of preparation. Media should not be used if there are any signs of deterioration (discol-

oration) or contamination, or if the expiration date supplied by the manufacturer has passed.

**Use:** For the isolation of *Listeria* species from food and environmental species. A primary selective enrichment broth for *Listeria* spp.

### Half Fraser Broth without Ferric Ammonium Citrate
**Composition** per liter:
| | |
|---|---|
| NaCl | 20.0g |
| $Na_2HPO_4$ | 12.0g |
| Proteose peptone | 5.0g |
| Tryptone | 5.0g |
| Lab Lemco powder | 5.0g |
| LiCl | 3.0g |
| $KH_2PO_4$ | 1.35g |
| Esculin | 1.0g |
| Half Fraser supplement solution without ferric ammonium citrate | 10.0mL |

pH 7.2 ± 0.2 at 25°C

**Source:** This medium is available as a premixed powder from Thermo Scientific.

**Caution:** Lithium chloride is harmful. Avoid bodily contact and inhalation of vapors. On contact with skin wash with plenty of water immediately.

**Half Fraser Supplement Solution without Ferric Ammonium Citrate:**
**Composition** per 10.0mL:
| | |
|---|---|
| Acriflavine·HCl | 0.125g |
| Nalidixic acid | 0.05g |
| Ethanol | 5.0mL |

**Preparation of Half Fraser Supplement Solution without Ferric Ammonium Citrate:** Add components to distilled/deionized water and bring volume to 10.0mL. Mix thoroughly. Filter sterilize.

**Preparation of Medium:** Add components, except half Fraser supplement solution without ferric ammonium citrate, to distilled/deionized water and bring volume to 990.0mL. Mix thoroughly. Gently heat and bring to boiling. Autoclave for 15 min at 15 psi pressure–121°C. Cool to 45°–50°C. Aseptically add sterile Half Fraser supplement solution without ferric ammonium citrate. Mix thoroughly. Aseptically distribute into sterile tubes or flasks.

**Storage/Shelf Life:** Store dehydrated media in the dark in a sealed container below 30°C. Prepared media should be stored under refrigeration (2-8°C). Media should be used within 60 days of preparation. Media should not be used if there are any signs of deterioration (discoloration) or contamination, or if the expiration date supplied by the manufacturer has passed.

**Use:** For the isolation of *Listeria* species from food and environmental specimens. A primary selective enrichment broth for *Listeria* spp. A pre-supplemented primary selective enrichment broth for *Listeria* spp.

### Ham's F-10 Medium
**Composition** per liter:
| | |
|---|---|
| NaCl | 7.4g |
| $NaHCO_3$ | 1.2g |
| Glucose | 1.1g |
| $NaH_2PO_4 \cdot H_2O$ | 0.29g |
| KCl | 0.28g |
| L-Arginine·HCl | 0.21g |

| | |
|---|---|
| L-Glutamine | 0.15g |
| MgSO$_4$·7H$_2$O | 0.15g |
| Sodium pyruvate | 0.11g |
| KH$_2$PO$_4$ | 0.08g |
| CaCl$_2$·2H$_2$O | 0.04g |
| L-Cystine·2HCl | 0.04g |
| L-Histidine·HCl·H$_2$O | 0.02g |
| L-Lysine·HCl | 0.02g |
| L-Asparagine-H$_2$O | 0.01g |
| L-Aspartic acid | 0.01g |
| L-Glutamic acid | 0.01g |
| L-Leucine | 0.01g |
| L-Proline | 0.01g |
| L-Serine | 0.01g |
| L-Alanine | 8.9mg |
| Glycine | 7.5mg |
| D-Phenylalanine | 5.0mg |
| L-Methionine | 4.5mg |
| Hypoxanthine | 4.1mg |
| L-Threonine | 3.6mg |
| L-Valine | 3.5mg |
| L-Isoleucine | 2.6mg |
| L-Tyrosine | 1.8mg |
| Vitamin B$_{12}$ | 1.4mg |
| Folic acid | 1.3mg |
| Phenol Red | 1.2mg |
| Thiamine·HCl | 1.0mg |
| FeSO$_4$·7H$_2$O | 0.8mg |
| Choline chloride | 0.7mg |
| D-Calcium pantothenate | 0.7mg |
| Thymidine | 0.7mg |
| Niacinamide | 0.6mg |
| L-Tryptophan | 0.6mg |
| Isoinositol | 0.5mg |
| Riboflavin | 0.4mg |
| Lipoic acid | 0.2mg |
| Pyridoxine·HCl | 0.2mg |
| ZnSO$_4$·7H$_2$O | 0.03mg |
| Biotin | 0.02mg |
| CuSO$_4$·5H$_2$O | 3.0µg |

pH 7.0 ± 0.2 at 25°C

**Preparation of Medium:** Add components to distilled/deionized water and bring volume to 1.0L. Mix thoroughly. Filter sterilize.

**Storage/Shelf Life:** Store dehydrated media in the dark in a sealed container below 30°C. Prepared media should be stored under refrigeration (2-8°C). Media should be used within 60 days of preparation. Media should not be used if there are any signs of deterioration (discoloration) or contamination, or if the expiration date supplied by the manufacturer has passed.

**Use:** For the growth of Y-1 cell cultures used in the mouse adrenal assay for heat-labile toxin of enterotoxigenic *Escherichia coli* and *Vibrio* species.

## HardyCHROM™ Blu*Ecoli*™

**Composition** per liter:
Proprietary

**Source:** This medium is available from Hardy Diagnostics.

**Preparation of Medium:** Available as prepared plates.

**Storage/Shelf Life:** Store in the dark under refrigeration (2-8°C). Chromogenic agars are especially light and temperature sensitive; protect from light, excessive heat, moisture, and freezing. Do not use after the expiration date supplied by the manufacturer.

**Use:** For screening urine specimens for *E. coli*. Colonies that are blue is confirmatory for *E. coli*.

## HardyCHROM™ Carbapenemase Agar

**Composition** per liter:
Proprietary

pH 7.5 ± 0.2 at 25°C

**Source:** This medium is available from Hardy Diagnostics.

**Preparation of Medium:** Available as prepared plates.

**Storage/Shelf Life:** Store in the dark under refrigeration (2-8°C). Chromogenic agars are especially light and temperature sensitive; protect from light, excessive heat, moisture, and freezing. Do not use after the expiration date supplied by the manufacturer.

**Use:** For the selection and differentiation of carbapenemase producing Gram-negative bacteria. Not intended for use in the identification of colonization with carbapenem-resistant bacteria in the prevention and control of such bacteria in a healthcare setting. Not intended to diagnose infections by carbapenem-resistant bacteria. Minimize exposure to light during storage and incubation.

## HardyCHROM™ ECC

**Composition** per liter:
Proprietary

**Source:** This medium is available from Hardy Diagnostics.

**Preparation of Medium:** Available as prepared plates.

**Storage/Shelf Life:** Store in the dark under refrigeration (2-8°C). Chromogenic agars are especially light and temperature sensitive; protect from light, excessive heat, moisture, and freezing. Do not use after the expiration date supplied by the manufacturer.

**Use:** For the rapid and reliable detection and differentiation of *E. coli* and other coliforms. *E. coli* can be identified as pink to violet colored colonies on the plate, while other coliform bacteria will appear as turquoise colonies.

## HardyCHROM™ ESBL

**Composition** per liter:
Proprietary

**Source:** This medium is available from Hardy Diagnostics.

**Preparation of Medium:** Available as prepared plates.

**Storage/Shelf Life:** Store in the dark under refrigeration (2-8°C). Chromogenic agars are especially light and temperature sensitive; protect from light, excessive heat, moisture, and freezing. Do not use after the expiration date supplied by the manufacturer.

**Use:** For the screening and differentiation of Extended-Spectrum Beta-Lactamase (ESBL) in Enterobacteriaceae. *Escherichia coli* produces colonies that are rose to magenta in color, with darker pink centers. *Klebsiella* and *Enterobacter* spp. produce large, dark blue colonies. *Citrobacter* spp. produce dark blue colonies often with a rose halo in the surrounding media. *Proteus* and *Morganella* spp. produce clear to light yellow colonies with golden-orange halo diffused through surrounding media. Additionally, approximately 50% of *Proteus vulgaris* isolates will produce blue-green or green colonies with a golden-orange halo.

## HardyCHROM™ HUrBi™ Urine Biplate (HardyCHROM™ HUrBi™)

**Composition** per liter:
Proprietary

**Source:** This medium is available from Hardy Diagnostics.

**Preparation of Medium:** Available as prepared plates.

**Storage/Shelf Life:** Store in the dark under refrigeration (2-8°C). Chromogenic agars are especially light and temperature sensitive; protect from light, excessive heat, moisture, and freezing. Do not use after the expiration date supplied by the manufacturer.

**Use:** For the isolation and differentiation of urinary tract pathogens. formulated to assist in characterizing Gram-positive organisms grow on on one side of the biplate and gram-negative organisms on the other side of the biplate.

## HardyCHROM™ *Listeria*

**Composition** per liter:
Proprietary

**Source:** This medium is available from Hardy Diagnostics.

**Preparation of Medium:** Available as prepared plates.

**Storage/Shelf Life:** Store in the dark under refrigeration (2-8°C). Chromogenic agars are especially light and temperature sensitive; protect from light, excessive heat, moisture, and freezing. Do not use after the expiration date supplied by the manufacturer.

**Use:** For the selective isolation of *Listeria monocytogenes* from food and environmental samples. *Listeria* species produce turquoise colored colonies and L. *monocytogenes* colonies are surrounded by a white halo.

## HardyCHROM™ MRSA

**Composition** per liter:
Proprietary

**Source:** This medium is available from Hardy Diagnostics.

**Preparation of Medium:** Available as prepared plates.

**Storage/Shelf Life:** Store in the dark under refrigeration (2-8°C). Chromogenic agars are especially light and temperature sensitive; protect from light, excessive heat, moisture, and freezing. Do not use after the expiration date supplied by the manufacturer.

**Use:** For the isolation and identification of methicillin-resistant *Staphylococcus aureus* (MRSA). MRSA strains produce deep pink to magenta colonies within 24 hours.

## HardyCHROM™ O157

**Composition** per liter:
Proprietary

**Source:** This medium is available from Hardy Diagnostics.

**Preparation of Medium:** Available as prepared plates.

**Storage/Shelf Life:** Store in the dark under refrigeration (2-8°C). Chromogenic agars are especially light and temperature sensitive; protect from light, excessive heat, moisture, and freezing. Do not use after the expiration date supplied by the manufacturer.

**Use:** For differentiating *E. coli* O157 from non-*E. coli* O157. *E. coli* O157 produces purple-pink colored colonies on the plate. Organisms

## HardyCHROM™ *sakazakii*

**Composition** per liter:
Proprietary

**Source:** This medium is available from Hardy Diagnostics.

**Preparation of Medium:** Available as prepared plates.

**Storage/Shelf Life:** Store in the dark under refrigeration (2-8°C). Chromogenic agars are especially light and temperature sensitive; protect from light, excessive heat, moisture, and freezing. Do not use after the expiration date supplied by the manufacturer.

**Use:** for For the isolation and chromogenic detection o *Cronobacter sakazakii*. *C. sakazakii* produces smooth, turquoise colonies. Other members of the Enterobacteriaceae family will produce white or colorless colonies with or without black centers.

All gram-positive bacteria and yeast will be inhibited on this medium.

## HardyCHROM™ *Salmonella*

**Composition** per liter:
Proprietary

**Source:** This medium is available from Hardy Diagnostics.

**Preparation of Medium:** Available as prepared plates.

**Storage/Shelf Life:** Store in the dark under refrigeration (2-8°C). Chromogenic agars are especially light and temperature sensitive; protect from light, excessive heat, moisture, and freezing. Do not use after the expiration date supplied by the manufacturer.

**Use:** For the isolation and differentiation of *Salmonella spp.* from other members of the family *Enterobacteriaceae*. *Salmonella* species use only one of the chromogens and will produce deep pink to magenta colored colonies. Bacteria other than *Salmonella* spp. may utilize the other chromogenic substrates and produce blue colonies. If none of the substrates are utilized, natural or white colored colonies will be present. *E. coli* will be partially inhibited.

other than *E. coli* O157 will be inhibited or appear as blue colonies.

## HardyCHROM™ SS

**Composition** per liter:
Proprietary

**Source:** This medium is available from Hardy Diagnostics.

**Preparation of Medium:** Available as prepared plates.

**Storage/Shelf Life:** Store in the dark under refrigeration (2-8°C). Chromogenic agars are especially light and temperature sensitive; protect from light, excessive heat, moisture, and freezing. Do not use after the expiration date supplied by the manufacturer.

**Use:** For the screening of stools for the isolation and differentiation of *Salmonella* and *Shigella*.

## HardyCHROM™ *Staphylococcus aureus*

**Composition** per liter:
Proprietary

**Source:** This medium is available from Hardy Diagnostics.

**Preparation of Medium:** Available as prepared plates.

**Storage/Shelf Life:** Store in the dark under refrigeration (2-8°C). Chromogenic agars are especially light and temperature sensitive; protect from light, excessive heat, moisture, and freezing. Do not use after the expiration date supplied by the manufacturer.

**Use:** For the rapid and reliable detection of *S. aureus* from both clinical and food specimens within 24 hours. *Staphylococcus aureus* can be identified as smooth, pink colored colonies on the plate. Other organisms may appear as colorless, blue or cream colonies, or will be inhibited. *Staphylococcus saprophyticus* will appear as turquoise colored colonies. *Staphylococcus epidermidis* will be inhibited.

## HardyCHROM™ UTI

**Composition** per liter:
Proprietary

**Source:** This medium is available from Hardy Diagnostics.

**Preparation of Medium:** Available as prepared plates.

**Storage/Shelf Life:** Store in the dark under refrigeration (2-8°C). Chromogenic agars are especially light and temperature sensitive; protect from light, excessive heat, moisture, and freezing. Do not use after the expiration date supplied by the manufacturer.

**Use:** For the isolation and differentiation of urinary tract pathogens, including gram-negative and gram-positive bacteria. The development of various colors, due to chromogenic substances in the medium, allows for the differentiation of microorganisms from the primary set-up of a urine specimen.

## HardyCHROM™ VRE

**Composition** per liter:
Proprietary

**Source:** This medium is available from Hardy Diagnostics.

**Preparation of Medium:** Available as prepared plates.

**Storage/Shelf Life:** Store in the dark under refrigeration (2-8°C). Chromogenic agars are especially light and temperature sensitive; protect from light, excessive heat, moisture, and freezing. Do not use after the expiration date supplied by the manufacturer.

**Use:** For the detection and differentiation of vancomycin-resistant enterococci (VRE). *Enterococcus faecalis* form red colonies with a metallic sheen and *Enterococcus faecium* produce bluecolonies.

## Heart Infusion Agar

**Composition** per liter:
Beef heart, infusion from .............................................................500.0g
Agar ...............................................................................................15.0g
Tryptose .........................................................................................10.0g
NaCl ................................................................................................5.0g
pH 7.4 ± 0.2 at 25°C

**Source:** This medium is available as a premixed powder from BD Diagnostic Systems.

**Preparation of Medium:** Add components to distilled/deionized water and bring volume to 1.0L. Mix thoroughly. Gently heat and bring to boiling. Distribute into tubes or flasks. Autoclave for 15 min at 15 psi pressure–121°C. Pour into sterile Petri dishes or leave in tubes.

**Storage/Shelf Life:** Store dehydrated media in the dark in a sealed container below 30°C. Prepared media should be stored under refrigeration (2-8°C). Media should be used within 60 days of preparation.

Media should not be used if there are any signs of deterioration (shrinking, cracking, or discoloration) or contamination, or if the expiration date supplied by the manufacturer has passed.

**Use:** For the isolation and cultivation of a wide variety of fastidious microorganisms. It can also be used as a base for the preparation of blood agar in determining hemolytic reactions. When using for blood agar, reduce volume to 950.0mL to allow for addition of 50.0mL of defibrinated horse blood.

## Heart Infusion Agar, HiVeg

**Composition** per liter:
Agar ...............................................................................................15.0g
Plant hydrolysate No. 1 ..................................................................10.0g
Plant infusion ................................................................................10.0g
NaCl ................................................................................................5.0g
pH 7.4 ± 0.2 at 25°C

**Source:** This medium is available as a premixed powder from HiMedia.

**Preparation of Medium:** Add components to distilled/deionized water and bring volume to 1.0L. Mix thoroughly. Gently heat and bring to boiling. Distribute into tubes or flasks. Autoclave for 15 min at 15 psi pressure–121°C. Pour into sterile Petri dishes or leave in tubes.

**Storage/Shelf Life:** Store dehydrated media in the dark in a sealed container below 30°C. Prepared media should be stored under refrigeration (2-8°C). Media should be used within 60 days of preparation. Media should not be used if there are any signs of deterioration (shrinking, cracking, or discoloration) or contamination, or if the expiration date supplied by the manufacturer has passed.

**Use:** For the isolation and cultivation of a wide variety of fastidious microorganisms.

## Heart Infusion Agar, HiVeg with Blood

**Composition** per liter:
Agar ...............................................................................................15.0g
Plant hydrolysate No. 1 ..................................................................10.0g
Plant infusion ................................................................................10.0g
NaCl ................................................................................................5.0g
Horse blood, defibrinated ..........................................................50.0mL
pH 7.4 ± 0.2 at 25°C

**Source:** This medium, without blood, is available as a premixed powder from HiMedia.

**Preparation of Medium:** Add components, except blood, to distilled/deionized water and bring volume to 950.0mL. Mix thoroughly. Gently heat and bring to boiling. Distribute into tubes or flasks. Autoclave for 15 min at 15 psi pressure–121°C. Cool to 50°C. Aseptically add 50.0mL defibrinated blood. Mix thoroughly. Pour into sterile Petri dishes or leave in tubes.

**Storage/Shelf Life:** Store dehydrated media in the dark in a sealed container below 30°C. Prepared media should be stored under refrigeration (2-8°C). Media should be used within 60 days of preparation. Media should not be used if there are any signs of deterioration (shrinking, cracking, or discoloration) or contamination, or if the expiration date supplied by the manufacturer has passed.

**Use:** For the isolation and cultivation of a wide variety of fastidious microorganisms. For determining hemolytic reactions.

## Heart Infusion Broth
## (HI)
## (BAM M60)

**Composition** per liter:

Agar ................................................................................15.0g
Tryptose .........................................................................10.0g
NaCl..................................................................................5.0g
Beef heart, infusion from 500.0g ................................... 1.0L

pH 7.4 ± 0.2 at 25°C

**Source:** This medium without added NaCl is available as a premixed powder from BD Diagnostic Systems.

**Preparation of Medium:** Add components to distilled/deionized water and bring volume to 1.0L. Mix thoroughly. Distribute into tubes or flasks. Autoclave for 15 min at 15 psi pressure–121°C.

**Storage/Shelf Life:** Store dehydrated media in the dark in a sealed container below 30°C. Prepared media should be stored under refrigeration (2-8°C). Media should be used within 60 days of preparation. Media should not be used if there are any signs of deterioration (discoloration) or contamination, or if the expiration date supplied by the manufacturer has passed.

**Use:** For the isolation and cultivation of a wide variety of fastidious microorganisms. For the cultivation of *Bacillus cereus, Staphylococcus aureus, Vibrio vulnificus,* and *Vibrio cholerae.* It can also be used as a base for the preparation of blood agar in determining hemolytic reactions. When using for blood broth, reduce volume to 950.0mL to allow for addition of 50.0mL of defibrinated horse blood. Blood is added aseptically after autoclaving.

## Hektoen Enteric Agar

**Composition** per liter:

Agar .................................................................................13.5g
Lactose ............................................................................12.0g
Peptic digest of animal tissue........................................12.0g
Sucrose ............................................................................12.0g
Bile salts...........................................................................9.0g
NaCl..................................................................................5.0g
$Na_2S_2O_3$ .............................................................................5.0g
Yeast extract.....................................................................3.0g
Salicin ...............................................................................2.0g
Ferric ammonium citrate..................................................1.5g
Acid Fuchsin .....................................................................0.1g
Bromthymol Blue ...........................................................0.064g

pH 7.6 ± 0.2 at 25°C

**Source:** This medium is available as a premixed powder from BD Diagnostic Systems and Thermo Scientific.

**Caution:** Acid Fuchsin is a potential carcinogen and care must be taken to avoid inhalation of the powdered dye and contact with the skin.

**Preparation of Medium:** Add components to distilled/deionized water and bring volume to 1.0L. Mix thoroughly. Gently heat while stirring until components are dissolved. Do not autoclave. Pour into sterile Petri dishes. Allow agar to solidify with the Petri dish covers partially off.

**Storage/Shelf Life:** Store dehydrated media in the dark in a sealed container below 30°C. Prepared media should be stored under refrigeration (2-8°C). Media should be used within 60 days of preparation. Media should not be used if there are any signs of deterioration (shrinking, cracking, or discoloration) or contamination, or if the expiration date supplied by the manufacturer has passed.

**Use:** For the isolation and cultivation of Gram-negative enteric microorganisms from a variety of clinical speciments and nonclinical specimens of public health importance based on lactose or sucrose fermentation and $H_2S$ production. For the isolation and differentiation of *Salmonella* and *Shigella.* Bacteria that ferment lactose or sucrose appear as yellow to orange colonies. Bacteria that produce $H_2S$ appear as colonies with black centers.

## Hektoen Enteric Agar

**Composition** per liter:

Agar ..................................................................................15.0g
Proteose peptone.............................................................12.0g
Lactose .............................................................................12.0g
Sucrose ............................................................................12.0g
Bile salts...........................................................................9.0g
NaCl..................................................................................5.0g
$Na_2S_2O_3$ .............................................................................5.0g
Yeast extract.....................................................................3.0g
Salicin ...............................................................................2.0g
Ferric ammonium citrate..................................................1.5g
Acid Fuchsin .....................................................................0.1g
Bromthymol Blue .........................................................0.065g

pH 7.5 ± 0.2 at 25°C

**Source:** This medium is available as a premixed powder from HiMedia.

**Caution:** Acid Fuchsin is a potential carcinogen and care must be taken to avoid inhalation of the powdered dye and contact with the skin.

**Preparation of Medium:** Add components to distilled/deionized water and bring volume to 1.0L. Mix thoroughly. Gently heat while stirring until components are dissolved. Do not autoclave. Pour into sterile Petri dishes. Allow agar to solidify with the Petri dish covers partially off.

**Storage/Shelf Life:** Store dehydrated media in the dark in a sealed container below 30°C. Prepared media should be stored under refrigeration (2-8°C). Media should be used within 60 days of preparation. Media should not be used if there are any signs of deterioration (shrinking, cracking, or discoloration) or contamination, or if the expiration date supplied by the manufacturer has passed.

**Use:** For the isolation and cultivation of Gram-negative enteric microorganisms from a variety of clinical speciments and nonclinical specimens of public health importance based on lactose or sucrose fermentation and $H_2S$ production. For the isolation and differentiation of *Salmonella* and *Shigella.* Bacteria that ferment lactose or sucrose appear as yellow to orange colonies. Bacteria that produce $H_2S$ appear as colonies with black centers.

## Hektoen Enteric Agar, HiVeg

**Composition** per liter:

Plant peptone No. 3..........................................................19.0g
Agar ..................................................................................15.0g
Lactose .............................................................................12.0g
Sucrose ............................................................................12.0g
NaCl..................................................................................5.0g
$Na_2S_2O_3$ .............................................................................5.0g
Yeast extract.....................................................................3.0g
Synthetic detergent No. I ..................................................2.0g
Salicin ...............................................................................2.0g
Ferric ammonium citrate..................................................1.5g

Acid Fuchsin ..................................................................0.1g
Bromthymol Blue ........................................................0.065g
pH 7.5 ± 0.2 at 25°C

**Source:** This medium is available as a premixed powder from Hi-Media.

**Caution:** Acid Fuchsin is a potential carcinogen and care must be taken to avoid inhalation of the powdered dye and contact with the skin.

**Preparation of Medium:** Add components to distilled/deionized water and bring volume to 1.0L. Mix thoroughly. Gently heat while stirring until components are dissolved. Do not autoclave. Pour into sterile Petri dishes. Allow agar to solidify with the Petri dish covers partially off.

**Storage/Shelf Life:** Store dehydrated media in the dark in a sealed container below 30°C. Prepared media should be stored under refrigeration (2-8°C). Media should be used within 60 days of preparation. Media should not be used if there are any signs of deterioration (shrinking, cracking, or discoloration) or contamination, or if the expiration date supplied by the manufacturer has passed.

**Use:** For the isolation and cultivation of Gram-negative enteric microorganisms from a variety of clinical speciments and nonclinical specimens of public health importance based on lactose or sucrose fermentation and $H_2S$ production. For the isolation and differentiation of *Salmonella* and *Shigella*. Bacteria that ferment lactose or sucrose appear as yellow to orange colonies. Bacteria that produce $H_2S$ appear as colonies with black centers.

## *Helicobacter pylori* Isolation Agar

**Composition** per liter:
Agar ..............................................................................15.0g
Bitone ...........................................................................10.0g
Pancreatic digest of casein ............................................5.0g
NaCl ...............................................................................5.0g
Peptic digest of animal tissue ........................................5.0g
Tryptic digest of beef heart ...........................................3.0g
Cornstarch .....................................................................1.0g
Horse blood, laked .....................................................35.0mL
Antibiotic inhibitor solution......................................10.0mL
pH 7.3 ± 0.2 at 25°C

**Antibiotic Inhibitor Solution:**
**Composition** per 10.0mL:
Vancomycin ..................................................................0.01g
Amphotericin B.............................................................5.0mg
Cefsulodin.....................................................................5.0mg
Trimethoprim lactate.....................................................5.0mg

**Preparation of Antibiotic Inhibitor Solution:** Add components to distilled/deionized water and bring volume to 10.0mL. Mix thoroughly. Filter sterilize.

**Preparation of Medium:** Add components, except horse blood and antibiotic inhibitor solution, to distilled/deionized water and bring volume to 955.0mL. Mix thoroughly. Gently heat and bring to boiling. Autoclave for 15 min at 15 psi pressure–121°C. Cool to 45°–50°C. Aseptically add sterile horse blood and sterile antibiotic inhibitor solution. Mix thoroughly. Pour into sterile Petri dishes or distribute into sterile tubes.

**Storage/Shelf Life:** Store dehydrated media in the dark in a sealed container below 30°C. Prepared media should be stored under refrigeration (2-8°C). Media should be used within 60 days of preparation.

Media should not be used if there are any signs of deterioration (shrinking, cracking, or discoloration) or contamination, or if the expiration date supplied by the manufacturer has passed.

**Use:** For the isolation and cultivation of *Helicobacter pylori* from clinical specimens.

## *Helicobacter pylori* Selective Medium

**Composition** per 1080.0mL:
Special peptone.............................................................23.0g
Agar ..............................................................................10.0g
NaCl ...............................................................................5.0g
Starch .............................................................................1.0g
Horse blood, laked .....................................................70.0mL
Selective supplement solution ...................................10.0mL
pH 7.3 ± 0.2 at 25°C

**Source:** This medium is available as a premixed powder from Thermo Scientific.

**Horse Blood, Laked:**
**Composition** per 100.0mL:
Horse blood, fresh.....................................................100.0mL

**Preparation of Horse Blood, Laked:** Add blood to a sterile polypropylene bottle. Freeze overnight at –20°C. Thaw at 8°C. Refreeze at –20°C. Thaw again at 8°C.

**Selective Supplement Solution:**
**Composition** per 10.0mL:
Vancomycin ................................................................10.0mg
Trimethoprim ...............................................................5.0mg
Cefsulodin.....................................................................5.0mg
Amphotericin B.............................................................5.0mg

**Preparation of Selective Supplement Solution:** Add components to distilled/deionized water and bring volume to 10.0mL. Mix thoroughly. Filter sterilize.

**Preparation of Medium:** Add components, except selective supplement solution and laked horse blood, to distilled/deionized water and bring volume to 1.0L. Mix thoroughly. Gently heat while stirring and bring to boiling. Autoclave for 15 min at 15 psi pressure–121°C. Cool to 50°C. Aseptically add 10.0mL selective supplement solution and 70.0mL sterile laked horse blood. Mix thoroughly. Pour into sterile Petri dishes.

**Storage/Shelf Life:** Store dehydrated media in the dark in a sealed container below 30°C. Prepared media should be stored under refrigeration (2-8°C). Media should be used within 60 days of preparation. Media should not be used if there are any signs of deterioration (shrinking, cracking, or discoloration) or contamination, or if the expiration date supplied by the manufacturer has passed.

**Use:** For the isolation of *Helicobacter pylori* from clinical specimens. *H. pylori* forms discrete, translucent, and non-coalescent colonies.

## Hemo ID Quad Plate with Growth Factors (*Haemophilus* Identification Quadrant Plate with Growth Factors)

**Composition** per plate:
Quadrant I .....................................................................5.0mL
Quadrant II....................................................................5.0mL
Quadrant III ..................................................................5.0mL
Quadrant IV ..................................................................5.0mL

**Quadrant I:**
**Composition** per 5.0mL:
Hemin.................................................................0.1mg
Brain heart infusion agar...................................5.0mL

**Quadrant II:**
**Composition** per 5.0mL:
Brain heart infusion agar...................................5.0mL
Supplement solution .......................................0.05mL

**Quadrant III:**
**Composition** per 5.0mL:
Hemin.................................................................0.1mg
Brain heart infusion agar...................................5.0mL
Supplement solution .......................................0.05mL

**Quadrant IV:**
**Composition** per 5.0mL:
Hemin.................................................................0.1mg
Brain heart infusion agar...................................5.0mL
Horse blood....................................................0.25mL
Supplement solution .......................................0.05mL

**Source:** The supplement solution (IsoVitaleX® enrichment) is available from BD Diagnostic Systems. This enrichment may be replaced by supplement VX from BD Diagnostic Systems.

**Preparation of Quadrant Media:** Sterilize Brain Heart Infusion Agar by autoclaving for 15 min at 15 psi pressure–121°C. Cool to 45°–50°C. Add additional components as filter sterilized solutions. Mix and distribute as 5.0mL aliquots into quadrants.

**Storage/Shelf Life:** Store dehydrated media in the dark in a sealed container below 30°C. Prepared media should be stored under refrigeration (2-8°C). Media should be used within 60 days of preparation. Media should not be used if there are any signs of deterioration (shrinking, cracking, or discoloration) or contamination, or if the expiration date supplied by the manufacturer has passed.

**Use:** For the differentiation and presumptive identification of *Haemophilus* species. The Hemo ID Quad Plate is a four-sectored plate, each with a different medium.

### Hemorrhagic Coli Agar
### (HC Agar)

**Composition** per liter:
Sorbitol.............................................................20.0g
Pancreatic digest of casein................................20.0g
Agar .................................................................15.0g
NaCl...................................................................5.0g
Bile salts No. 3..................................................1.12g
Bromcresol Purple ..........................................0.015g
pH 7.2 ± 0.2 at 25°C

**Preparation of Medium:** Add components to distilled/deionized water and bring volume to 1.0L. Mix thoroughly. Gently heat and bring to boiling. Distribute into tubes or flasks. Autoclave for 15 min at 15 psi pressure–121°C. Pour into sterile Petri dishes.

**Storage/Shelf Life:** Store dehydrated media in the dark in a sealed container below 30°C. Prepared media should be stored under refrigeration (2-8°C). Media should be used within 60 days of preparation. Media should not be used if there are any signs of deterioration (shrinking, cracking, or discoloration) or contamination, or if the expiration date supplied by the manufacturer has passed.

**Use:** For the isolation and cultivation of enterohemorrhaghic *Escherichia coli* from food.

### Hemorrhagic Coli Agar with MUG
### (HC Agar with MUG)
### (BAM M62)

**Composition** per liter:
Sorbitol.............................................................20.0g
Pancreatic digest of casein................................20.0g
Agar .................................................................15.0g
NaCl...................................................................5.0g
Bile salts No. 3..................................................1.12g
Bromcresol Purple ..........................................0.015g
MUG reagent ......................................................0.1g
pH 7.2 ± 0.2 at 25°C

**Source:** MUG reagent is available from Hach Company, Loveland, Colorado.

**Preparation of Medium:** Add components to distilled/deionized water and bring volume to 1.0L. Mix thoroughly. Gently heat and bring to boiling. Distribute into tubes or flasks. Autoclave for 15 min at 15 psi pressure–121°C. Pour into sterile Petri dishes.

**Storage/Shelf Life:** Store dehydrated media in the dark in a sealed conainer below 30°C. Prepared plates should be stored under refrigeration (2-8°C). This medium has a limited shelf life. Media should not be used if there are any signs of deterioration (shrinking, cracking, or discoloration) or contamination, or if the expiration date supplied by the manufacturer has passed.

**Use:** For the isolation and cultivation of enterohemorraghic *Escherichia coli* from food.

### *Herellea* Agar

**Composition** per liter:
Agar .................................................................16.0g
Pancreatic digest of casein................................15.0g
Lactose.............................................................10.0g
Maltose ............................................................10.0g
Enzymatic digest of soybean meal .......................5.0g
NaCl...................................................................5.0g
Bile salts...........................................................1.25g
Bromcresol Purple ............................................0.02g
pH 6.8 ± 0.2 at 25°C

**Source:** This medium is available as a premixed powder from BD Diagnostic Systems.

**Preparation of Medium:** Add components to distilled/deionized water and bring volume to 1.0L. Mix thoroughly. Gently heat and bring to boiling. Distribute into tubes or flasks. Autoclave for 15 min at 15 psi pressure–121°C. Pour into sterile Petri dishes or leave in tubes.

**Storage/Shelf Life:** Store dehydrated media in the dark in a sealed container below 30°C. Prepared media should be stored under refrigeration (2-8°C). Media should be used within 60 days of preparation. Media should not be used if there are any signs of deterioration (shrinking, cracking, or discoloration) or contamination, or if the expiration date supplied by the manufacturer has passed.

**Use:** For the isolation, cultivation, and differentiation of Gram-negative nonfermentative and fermentative bacteria. It is especially recommended for the differentiation of *Acinetobacter (Herellea)* species from *Neisseria gonorrhoeae* in urethral or vaginal specimens. Fermentative bacteria appear as yellow colonies surrounded by yellow zones. Nonfermentative bacteria, such as *Acinetobacter* species, appear as pale lavender colonies.

## HHD Medium

**Composition** per liter:

| | |
|---|---|
| Agar | 20.0g |
| Pancreatic digest of casein | 10.0g |
| Casamino acids | 3.0g |
| Fructose | 2.5g |
| KH$_2$PO$_4$ | 2.5g |
| Papaic digest of soybean meal | 1.5g |
| Tween™ 80 | 1.0g |
| Yeast extract | 1.0g |
| Bromcresol Green solution | 20.0mL |

pH 7.0 ± 0.2 at 25°C

**Bromcresol Green Solution:**

**Composition** per 30.0mL:

| | |
|---|---|
| Bromcresol Green | 0.1g |
| NaOH (0.01*N* solution) | 30.0mL |

**Preparation of Bromcresol Green Solution:** Add Bromcresol Green to 30.0mL of NaOH solution. Mix thoroughly. Filter sterilize.

**Preparation of Medium:** Add components to distilled/deionized water and bring volume to 1.0L. Mix thoroughly. Gently heat and bring to boiling. Adjust pH to 7.0. Distribute into tubes or flasks. Autoclave for 15 min at 15 psi pressure–121°C. Pour into sterile Petri dishes or leave in tubes.

**Storage/Shelf Life:** Store dehydrated media in the dark in a sealed container below 30°C. Prepared media should be stored under refrigeration (2-8°C). Media should be used within 60 days of preparation. Media should not be used if there are any signs of deterioration (shrinking, cracking, or discoloration) or contamination, or if the expiration date supplied by the manufacturer has passed.

**Use:** For the cultivation of *Salmonella* species from foods.

## HiCrome™ Aureus Agar Base with Egg Tellurite
## (Aureus Agar Base, HiCrome™)
## (HiCrome™ *Staphylococcus aureus* Agar)
## (*Staphylococcus aureus* Agar, HiCrome)

**Composition** per liter:

| | |
|---|---|
| Agar | 20.0g |
| Casein enzymic hydrolysate | 12.0g |
| Sodium pyruvate | 10.0g |
| Beef extract | 6.0g |
| LiCl | 5.0g |
| Yeast extract | 5.0g |
| Pancreatic digest of gelatin | 3.0g |
| Chromogenic mixture | 2.1g |
| Egg tellurite emulsion | 50.0mL |

pH 7.4 ± 0.2 at 25°C

**Source:** This medium is available as a premixed powder from HiMedia.

**Caution:** Lithium chloride is harmful. Avoid bodily contact and inhalation of vapors. On contact with skin, wash with plenty of water immediately.

**Egg Yolk Tellurite Emulsion:**

**Composition** per 100.0mL:

| | |
|---|---|
| Sterile saline | 64.0mL |
| Egg yolk | 30.0mL |
| Sterile potassium tellurite solution, 3.5% | 6.0mL |

**Source:** This medium is available from Fluka, Sigma-Aldrich, and HiMedia.

**Preparation of Medium:** Add components, except egg yolk tellurite emulsion, to distilled/deionized water and bring volume to 950.0mL. Mix thoroughly. Gently heat and bring to boiling. Autoclave for 15 min at 15 psi pressure–121°C. Cool to 50°C. Aseptically add 50.0mL sterile egg yolk tellurite emulsion. Mix thoroughly. Pour into sterile Petri dishes or distribute into sterile tubes.

**Storage/Shelf Life:** Store dehydrated media in the dark in a sealed container below 30°C. Prepared plates should be stored in the dark under refrigeration (2-8°C). Chromogenic media are especially light and temperature sensitive; protect from light, excessive heat, moisture, and freezing. Media should not be used if there are any signs of deterioration (shrinking, cracking, or discoloration) or contamination, or if the expiration date supplied by the manufacturer has passed.

**Use:** For the isolation and identification of staphylococci from environmental samples. For the isolation and enumeration of coagulase positive *Staphylococcus aureus*. Coagulase positive *S. aureus* gives brown-black colonies whereas *S. epidermidis* gives yellow, slightly brownish, colonies.

## HiCrome™ *Bacillus* Agar wtih Polymyxin B
## (*Bacillus cereus* HiCrome™ Agar)

**Composition** per liter:

| | |
|---|---|
| Agar | 15.0g |
| Peptic digest of animal tissue | 10.0g |
| D-Mannitol | 10.0g |
| NaCl | 10.0g |
| Chromogenic mixture | 3.2g |
| Meat extract | 1.0g |
| Phenol Red | 0.025g |
| Polymyxin B solution | 1.0mL |

pH 7.1 ± 0.2 at 25°C

**Source:** This medium, without Polymyxin B solution, is available as a premixed powder from HiMedia.

**Polymyxin B Solution:**

**Composition** per 1.0mL:

| | |
|---|---|
| Polymyxin B | 1.0mg |

**Preparation of Polymyxin B Solution:** Add Polymyxin B to distilled/deionized water and bring volume to 1.0mL. Mix thoroughly. Filter sterilize.

**Preparation of Medium:** Add components, except Polymyxin B solution, to distilled/deionized water and bring volume to 1.0L. Mix thoroughly and heat with frequent agitation until boiling. Boil until components are fully dissolved. Do not autoclave. Cool to 45°–50°C. Mix thoroughly. Aseptically add 1.0mL of Polymyxin B solution. Mix thoroughly. Pour into sterile Petri dishes.

**Storage/Shelf Life:** Store dehydrated media in the dark in a sealed container below 30°C. Prepared plates should be stored in the dark under refrigeration (2-8°C). Chromogenic media are especially light and temperature sensitive; protect from light, excessive heat, moisture, and freezing. Media should not be used if there are any signs of deterioration (shrinking, cracking, or discoloration) or contamination, or if the expiration date supplied by the manufacturer has passed.

**Use:** For the rapid identification of *Bacillus* species from a mixed culture by chromogenic method. For the enumeration of *Bacillus cereus* and *Bacillus thuringiensis* when present in large numbers in certain foodstuffs.

## HiCrome™ Coliform Agar
### (Coliform Agar, HiCrome™)

**Composition** per liter:

Agar ....................................................................................12.0g
Sodium chloride.................................................................5.0g
Peptone, special ...............................................................3.0g
$K_2HPO_4$.............................................................................3.0g
$KH_2PO_4$............................................................................1.7g
Sodium pyruvate...............................................................1.0g
Tryptophan.........................................................................1.0g
Chromogenic mixture .......................................................0.2g
Sodium lauryl sulphate .....................................................0.1g

pH 6.8 ± 0.2 at 25°C

**Source:** This medium is available as a premixed powder from Hi-Media.

**Preparation of Medium:** Add components to distilled/deionized water and bring volume to 1.0L. Mix thoroughly. Gently heat while stirring and bring to boiling. Autoclave for 15 min at 15 psi pressure–121°C. Pour into sterile Petri dishes.

**Storage/Shelf Life:** Store dehydrated media in the dark in a sealed container below 30°C. Prepared plates should be stored in the dark under refrigeration (2-8°C). Chromogenic media are especially light and temperature sensitive; protect from light, excessive heat, moisture, and freezing. Media should not be used if there are any signs of deterioration (shrinking, cracking, or discoloration) or contamination, or if the expiration date supplied by the manufacturer has passed.

**Use:** For the simultaneous detection of *Escherichia coli* and total coliforms in water and food samples. Sodium lauryl sulphate inhibits Gram-positive organisms. The chromogenic mixture contains two chromogenic substrates as Salmon-GAL and X-glucuronide. The enzyme β-D-galactosidase produced by coliforms cleaves Salmon-GAL, resulting in the salmon to red coloration of coliform colonies. The enzyme β-D-glucuronidase produced by *E. coli* cleaves X-glucuronide. *Escherichia coli* forms dark blue to violet colored colonies due to cleavage of both Salmon-GAL and X-glucuronide. The addition of tryptophan improves the indole reaction, thereby increasing detection reliability in combination with the two chromogens.

## HiCrome™ Coliform Agar with Novobiocin
### (Coliform Agar with Novobiocin, HiCrome™)

**Composition** per liter:

Agar ....................................................................................12.0g
Sodium chloride.................................................................5.0g
Peptone, special ...............................................................3.0g
$K_2HPO_4$.............................................................................3.0g
$KH_2PO_4$............................................................................1.7g
Sodium pyruvate...............................................................1.0g
Tryptophan.........................................................................1.0g
Chromogenic mixture .......................................................0.2g
Sodium lauryl sulphate .....................................................0.1g
Novobiocin.......................................................................5.0mg

pH 6.8 ± 0.2 at 25°C

**Source:** This medium is available as a premixed powder from Hi-Media.

**Preparation of Medium:** Add components to distilled/deionized water and bring volume to 1.0L. Mix thoroughly. Gently heat while stirring and bring to boiling. Autoclave for 15 min at 15 psi pressure–121°C. Pour into sterile Petri dishes.

**Storage/Shelf Life:** Store dehydrated media in the dark in a sealed container below 30°C. Prepared plates should be stored in the dark under refrigeration (2-8°C). Chromogenic media are especially light and temperature sensitive; protect from light, excessive heat, moisture, and freezing. Media should not be used if there are any signs of deterioration (shrinking, cracking, or discoloration) or contamination, or if the expiration date supplied by the manufacturer has passed.

**Use:** For the simultaneous detection of *Escherichia coli* and total coliforms in water and food samples when high numbers of Gram-positive bacteria may be present. Novobiocin and sodium lauryl sulphate inhibit Gram-positive organisms. The chromogenic mixture contains two chromogenic substrates as Salmon-GAL and X-glucuronide. The enzyme β-D-galactosidase produced by coliforms cleaves Salmon-GAL, resulting in the salmon to red coloration of coliform colonies. The enzyme β-D-glucuronidase produced by *E. coli* cleaves X-glucuronide. *Escherichia coli* forms dark blue to violet colored colonies due to cleavage of both Salmon-GAL and X-glucuronide. The addition of tryptophan improves the indole reaction, thereby increasing detection reliability in combination with the two chromogens.

## HiCrome™ Coliform Agar with SLS

**Composition** per liter:

Agar ....................................................................................12.0g
NaCl....................................................................................5.0g
$K_2HPO_4$.............................................................................3.0g
Peptone, special ...............................................................3.0g
$KH_2PO_4$............................................................................1.7g
Sodium pyruvate...............................................................1.0g
Tryptophan.........................................................................1.0g
Chromogenic mixture .......................................................0.2g
Sodium lauryl sulfate.........................................................0.1g

pH 7.1± 0.2 at 25°C

**Source:** This medium is available as a premixed powder from Hi-Media.

**Preparation of Medium:** Add components to distilled/deionized water and bring volume to 1.0L. Mix thoroughly. Gently heat and bring to boiling. Distribute into tubes or flasks. Autoclave for 15 min at 15 psi pressure–121°C. Pour into sterile Petri dishes.

**Storage/Shelf Life:** Store dehydrated media in the dark in a sealed container below 30°C. Prepared plates should be stored in the dark under refrigeration (2-8°C). Chromogenic media are especially light and temperature sensitive; protect from light, excessive heat, moisture, and freezing. Media should not be used if there are any signs of deterioration (shrinking, cracking, or discoloration) or contamination, or if the expiration date supplied by the manufacturer has passed.

**Use:** For the simultaneous detection of *Escherichia coli* and total coliforms in water and food samples.

## HiCrome™ Coliform Agar with SLS and Novobiocin

**Composition** per liter:

Agar ....................................................................................12.0g
NaCl....................................................................................5.0g
$K_2HPO_4$.............................................................................3.0g
Peptone, special ...............................................................3.0g
$KH_2PO_4$............................................................................1.7g
Sodium pyruvate...............................................................1.0g
Tryptophan.........................................................................1.0g
Chromogenic mixture .......................................................0.2g

Sodium lauryl sulfate ........................................................0.1g
Novobiocin........................................................5.0mg

pH 7.1 ± 0.2 at 25°C

**Source:** This medium, without novobiocin, is available as a premixed powder from HiMedia.

**Preparation of Medium:** Add components to distilled/deionized water and bring volume to 1.0L. Mix thoroughly. Gently heat and bring to boiling. Distribute into tubes or flasks. Autoclave for 15 min at 15 psi pressure–121°C. Pour into sterile Petri dishes.

**Storage/Shelf Life:** Store dehydrated media in the dark in a sealed container below 30°C. Prepared plates should be stored in the dark under refrigeration (2-8°C). Chromogenic media are especially light and temperature sensitive; protect from light, excessive heat, moisture, and freezing. Media should not be used if there are any signs of deterioration (shrinking, cracking, or discoloration) or contamination, or if the expiration date supplied by the manufacturer has passed.

**Use:** For the simultaneous detection of *Escherichia coli* and total coliforms in water and food samples when a high number of Gram-positive accompanying bacteria are expected.

## HiCrome™ Coliform HiVeg Agarwtih SLS

**Composition** per liter:

Agar ........................................................12.0g
NaCl ........................................................5.0g
K₂HPO₄ ........................................................3.0g
Plant special peptone ........................................................3.0g
KH₂PO₄ ........................................................1.7g
Sodium pyruvate ........................................................1.0g
Tryptophan ........................................................1.0g
Chromogenic mixture ........................................................0.2g
Sodium lauryl sulfate ........................................................0.1g

pH 7.1 ± 0.2 at 25°C

**Source:** This medium is available as a premixed powder from HiMedia.

**Preparation of Medium:** Add components to distilled/deionized water and bring volume to 1.0L. Mix thoroughly. Gently heat and bring to boiling. Distribute into tubes or flasks. Autoclave for 15 min at 15 psi pressure–121°C. Pour into sterile Petri dishes.

**Storage/Shelf Life:** Store dehydrated media in the dark in a sealed container below 30°C. Prepared plates should be stored in the dark under refrigeration (2-8°C). Chromogenic media are especially light and temperature sensitive; protect from light, excessive heat, moisture, and freezing. Media should not be used if there are any signs of deterioration (shrinking, cracking, or discoloration) or contamination, or if the expiration date supplied by the manufacturer has passed.

**Use:** For the simultaneous detection of *Escherichia coli* and total coliforms in water and food samples.

## HiCrome™ *E. coli* Agar

**Composition** per liter:

Casein enzymic hydrolysate ........................................................14.0g
Agar ........................................................12.0g
Peptone, special ........................................................5.0g
NaCl ........................................................2.4g
Bile salts mixture ........................................................1.5g
Na₂HPO₄ ........................................................1.0g

NaH₂PO₄ ........................................................0.6g
X-Glucuronide ........................................................0.075g

pH 7.1 ± 0.2 at 25°C

**Source:** This medium is available as a premixed powder from HiMedia.

**Preparation of Medium:** Add components to distilled/deionized water and bring volume to 1.0L. Mix thoroughly. Gently heat and bring to boiling. Distribute into tubes or flasks. Autoclave for 15 min at 15 psi pressure–121°C. Pour into sterile Petri dishes.

**Storage/Shelf Life:** Store dehydrated media in the dark in a sealed container below 30°C. Prepared plates should be stored in the dark under refrigeration (2-8°C). Chromogenic media are especially light and temperature sensitive; protect from light, excessive heat, moisture, and freezing. Media should not be used if there are any signs of deterioration (shrinking, cracking, or discoloration) or contamination, or if the expiration date supplied by the manufacturer has passed.

**Use:** For the simultaneous detection of *Escherichia coli* in foods without further confirmation on membrane filter or by indole reagent.

## HiCrome™ *E. coli* Agar

**Composition** per liter:

Casein enzymic hydrolysate ........................................................20.0g
Agar ........................................................15.0g
Bile salts mixture ........................................................1.50g
X-Glucuronide ........................................................7.5mg

pH 7.1 ± 0.2 at 25°C

**Source:** This medium is available as a premixed powder from HiMedia.

**Preparation of Medium:** Add components to distilled/deionized water and bring volume to 1.0L. Mix thoroughly. Gently heat and bring to boiling. Distribute into tubes or flasks. Autoclave for 15 min at 15 psi pressure–121°C. Pour into sterile Petri dishes.

**Storage/Shelf Life:** Store dehydrated media in the dark in a sealed container below 30°C. Prepared plates should be stored in the dark under refrigeration (2-8°C). Chromogenic media are especially light and temperature sensitive; protect from light, excessive heat, moisture, and freezing. Media should not be used if there are any signs of deterioration (shrinking, cracking, or discoloration) or contamination, or if the expiration date supplied by the manufacturer has passed.

**Use:** For the simultaneous detection of *Escherichia coli* in foods without further confirmation on membrane filter or by indole reagent.

## HiCrome™ *E. coli* Agar A
## (*E. coli* Agar A, HiCrome™)

**Composition** per liter:

Casein enzymic hydrolysate ........................................................14.0g
Agar ........................................................12.0g
Peptone, special ........................................................5.0g
NaCl ........................................................2.4g
Bile salts mixture ........................................................1.5g
Na₂HPO₄ ........................................................1.0g
NaH₂PO₄ ........................................................0.6g
X-Glucuronide ........................................................0.075g

pH 7.2 ± 0.2 at 25°C

**Source:** This medium is available as a premixed powder from HiMedia.

**Preparation of Medium:** Add components to distilled/deionized water and bring volume to 1.0L. Mix thoroughly. Gently heat while stirring and bring to boiling. Autoclave for 15 min at 15 psi pressure–121°C. Cool to 50°C. Pour into sterile Petri dishes.

**Storage/Shelf Life:** Store dehydrated media in the dark in a sealed container below 30°C. Prepared plates should be stored in the dark under refrigeration (2-8°C). Chromogenic media are especially light and temperature sensitive; protect from light, excessive heat, moisture, and freezing. Media should not be used if there are any signs of deterioration (shrinking, cracking, or discoloration) or contamination, or if the expiration date supplied by the manufacturer has passed.

**Use:** For the detection and enumeration of *Escherichia coli* in foods without further confirmation on membrane filter or by indole reagent. The chromogenic agent X-glucuronide used in this medium helps to detect glucuronidase activity. *E. coli* cells absorb X-glucuronide and the intracellular glucuronidase splits the bond between the chromophore and the glucuronide. The released chromophore gives coloration to the colonies. Bile salts mixture inhibits Gram-positive organisms.

## HiCrome™ *E. coli* Agar B
## (*E. coli* Agar B, HiCrome™)

**Composition** per liter:

Casein enzymic hydrolysate ...............................20.0g
Agar ...............................................................15.0g
Bile salts mixture ..............................................1.5g
X-Glucuronide ................................................0.075g

pH 7.2 ± 0.2 at 25°C

**Source:** This medium is available as a premixed powder from Hi-Media.

**Preparation of Medium:** Add components to distilled/deionized water and bring volume to 1.0L. Mix thoroughly. Gently heat while stirring and bring to boiling. Autoclave for 15 min at 15 psi pressure–121°C. Pour into sterile Petri dishes.

**Storage/Shelf Life:** Store dehydrated media in the dark in a sealed container below 30°C. Prepared plates should be stored in the dark under refrigeration (2-8°C). Chromogenic media are especially light and temperature sensitive; protect from light, excessive heat, moisture, and freezing. Media should not be used if there are any signs of deterioration (shrinking, cracking, or discoloration) or contamination, or if the expiration date supplied by the manufacturer has passed.

**Use:** For the detection and enumeration of *Escherichia coli* in foods without further confirmation on membrane filter or by indole reagent. The chromogenic agent X-glucuronide used in this medium helps to detect glucuronidase activity. *E. coli* cells absorb X-glucuronide and the intracellular glucuronidase splits the bond between the chromophore and the glucuronide. The released chromophore gives coloration to the colonies. Bile salts mixture inhibits Gram-positive organisms.

## HiCrome™ *E. coli* Agar, HiVeg

**Composition** per liter:

Plant hydrolysate....................................................14.0g
Agar ...................................................................12.0g
Plant special peptone .............................................5.0g
NaCl...................................................................2.4g
Synthetic detergent.................................................1.5g
Na$_2$HPO$_4$......................................................1.0g
NaH$_2$PO$_4$......................................................0.6g
X-Glucuronide .....................................................0.075g

pH 7.1 ± 0.2 at 25°C

**Source:** This medium is available as a premixed powder from Hi-Media.

**Preparation of Medium:** Add components to distilled/deionized water and bring volume to 1.0L. Mix thoroughly. Gently heat and bring to boiling. Distribute into tubes or flasks. Autoclave for 15 min at 15 psi pressure–121°C. Pour into sterile Petri dishes.

**Storage/Shelf Life:** Store dehydrated media in the dark in a sealed container below 30°C. Prepared plates should be stored in the dark under refrigeration (2-8°C). Chromogenic media are especially light and temperature sensitive; protect from light, excessive heat, moisture, and freezing. Media should not be used if there are any signs of deterioration (shrinking, cracking, or discoloration) or contamination, or if the expiration date supplied by the manufacturer has passed.

**Use:** For the simultaneous detection of *Escherichia coli* in foods without further confirmation on membrane filter or by indole reagent.

## HiCrome™ EC O157:H7 Agar

**Composition** per liter:

Agar ...............................................................15.0g
Casein enzymic hydrolysate .....................................8.0g
Sorbitol ...............................................................7.0g
Bile salts mixture ..................................................1.5g
Chromogenic mixture..............................................0.25g
Sodium lauryl sulfate..............................................0.1g

pH 7.1 ± 0.2 at 25°C

**Source:** This medium is available as a premixed powder from Hi-Media.

**Preparation of Medium:** Add components to distilled/deionized water and bring volume to 1.0L. Mix thoroughly. Gently heat and bring to boiling. Distribute into tubes or flasks. Autoclave for 15 min at 15 psi pressure–121°C. Pour into sterile Petri dishes.

**Storage/Shelf Life:** Store dehydrated media in the dark in a sealed container below 30°C. Prepared plates should be stored in the dark under refrigeration (2-8°C). Chromogenic media are especially light and temperature sensitive; protect from light, excessive heat, moisture, and freezing. Media should not be used if there are any signs of deterioration (shrinking, cracking, or discoloration) or contamination, or if the expiration date supplied by the manufacturer has passed.

**Use:** For the isolation and differentiation of *Escherichia coli* O157:H7 from food and recommended for selective isolation and easy detection of *Escherichia coli* O157:H7.

## HiCrome™ ECC Agar
## (ECC Agar, HiCrome™)

**Composition** per liter:

Chromogenic mixture ...............................................20.3g
Agar ...................................................................15.0g
Peptone, special ....................................................5.0g
NaCl...................................................................5.0g
Na$_2$HPO$_4$......................................................3.5g
Yeast extract.........................................................3.0g
Lactose...............................................................2.5g
KH$_2$PO$_4$.......................................................1.5g
Neutral Red..........................................................0.03g

pH 6.8 ± 0.2 at 25°C

**Source:** This medium is available as a premixed powder from Hi-Media.

**Preparation of Medium:** Add components to distilled/deionized water and bring volume to 1.0L. Mix thoroughly. Gently heat while stirring and bring to boiling. Autoclave for 15 min at 15 psi pressure–121°C. Pour into sterile Petri dishes.

**Storage/Shelf Life:** Store dehydrated media in the dark in a sealed container below 30°C. Prepared plates should be stored in the dark under refrigeration (2-8°C). Chromogenic media are especially light and temperature sensitive; protect from light, excessive heat, moisture, and freezing. Media should not be used if there are any signs of deterioration (shrinking, cracking, or discoloration) or contamination, or if the expiration date supplied by the manufacturer has passed.

**Use:** For the presumptive identification of *Escherichia coli* and other coliforms in food and environmental samples. A differential medium for presumptive identification of *E. coli* and other coliforms in food samples. The chromogenic mixture contains two chromogens as X-glucuronide and Salmon-GAL. X-glucuronide is cleaved by the enzyme β-glucuronidase produced by *E. coli*. Salmon-GAL is cleaved by the enzyme galactosidase produced by the majority of coliforms, including *E. coli*.

## HiCrome™ ECC HiVeg Agar

**Composition** per liter:

| | |
|---|---|
| Chromogenic mixture | 20.3g |
| Agar | 15.0g |
| Plant special peptone | 5.0g |
| NaCl | 5.0g |
| $Na_2HPO_4$ | 3.5g |
| Yeast extract | 3.0g |
| Lactose | 2.5g |
| $KH_2PO_4$ | 1.5g |
| Neutral Red | 0.03g |

pH 7.2 ± 0.2 at 25°C

**Source:** This medium is available as a premixed powder from Hi-Media.

**Preparation of Medium:** Add components to distilled/deionized water and bring volume to 1.0L. Mix thoroughly. Gently heat and bring to boiling. Distribute into tubes or flasks. Autoclave for 15 min at 15 psi pressure–121°C. Pour into sterile Petri dishes.

**Storage/Shelf Life:** Store dehydrated media in the dark in a sealed container below 30°C. Prepared plates should be stored in the dark under refrigeration (2-8°C). Chromogenic media are especially light and temperature sensitive; protect from light, excessive heat, moisture, and freezing. Media should not be used if there are any signs of deterioration (shrinking, cracking, or discoloration) or contamination, or if the expiration date supplied by the manufacturer has passed.

**Use:** For the presumptive identification of *Escherichia coli* and other coliforms in food and environmental samples.

## HiCrome™ ECC Selective Agar
## (ECC Selective Agar, HiCrome™)

**Composition** per liter:

| | |
|---|---|
| Agar | 10.0g |
| Peptone, special | 6.0g |
| Casein enzymic hydrolysate | 3.3g |
| NaCl | 2.0g |
| Sodium pyruvate | 1.0g |
| Tryptophan | 1.0g |
| Sorbitol | 1.0g |
| $Na_2HPO_4$ | 1.0g |

| | |
|---|---|
| $NaH_2PO_4$ | 0.6g |
| Tergitol 7® | 0.15g |
| Chromogenic mixture | 0.43g |

pH 6.8 ± 0.2 at 25°C

**Source:** This medium is available as a premixed powder from Hi-Media.

**Preparation of Medium:** Add components to distilled/deionized water and bring volume to 1.0L. Mix thoroughly and heat with frequent agitation until components are completely dissolved (approximately 35 min). Do not autoclave. Cool to 45°–50°C. Pour into sterile Petri dishes. Medium may show haziness.

**Storage/Shelf Life:** Store dehydrated media in the dark in a sealed container below 30°C. Prepared plates should be stored in the dark under refrigeration (2-8°C). Chromogenic media are especially light and temperature sensitive; protect from light, excessive heat, moisture, and freezing. Media should not be used if there are any signs of deterioration (shrinking, cracking, or discoloration) or contamination, or if the expiration date supplied by the manufacturer has passed.

**Use:** For detection of *Escherichia coli* and coliforms in water and food samples. Tergitol inhibits Gram-positive as well as some Gram-negative bacteria other than coliforms. The chromogenic mixture contains two chromogenic substrates as Salmon-GAL and X-glucuronide. The enzyme b-D-galactosidase produced by coliforms cleaves Salmon-GAL, resulting in the salmon to red coloration of coliform colonies. The enzyme β-D-glucuronidase produced by *E. coli* cleaves X-glucuronide. *E. coli* forms dark blue to violet colored colonies due to cleavage of both Salmon-GAL and X-glucuronide. The addition of tryptophan improves the indole reaction.

## HiCrome™ ECC Selective Agar
## (ECC Selective Agar, HiCrome™)

**Composition** per liter:

| | |
|---|---|
| Agar | 10.0g |
| Peptone, special | 6.0g |
| Casein enzymic hydrolysate | 3.3g |
| NaCl | 2.0g |
| Sodium pyruvate | 1.0g |
| Tryptophan | 1.0g |
| Sorbitol | 1.0g |
| $Na_2HPO_4$ | 1.0g |
| $NaH_2PO_4$ | 0.6g |
| Tergitol 7® | 0.15g |
| Chromogenic mixture | 0.43g |
| Cefsulodin | 5.0mg |

pH 6.8 ± 0.2 at 25°C

**Source:** This medium is available as a premixed powder from Hi-Media.

**Preparation of Medium:** Add components to distilled/deionized water and bring volume to 1.0L. Mix thoroughly and heat with frequent agitation until components are completely dissolved (approximately 35 min.). Do not autoclave. Cool to 45°–50°C. Pour into sterile Petri dishes. Medium may show haziness.

**Storage/Shelf Life:** Store dehydrated media in the dark in a sealed container below 30°C. Prepared plates should be stored in the dark under refrigeration (2-8°C). Chromogenic media are especially light and temperature sensitive; protect from light, excessive heat, moisture, and freezing. Media should not be used if there are any signs of deterioration (shrinking, cracking, or discoloration) or contamination, or if the expiration date supplied by the manufacturer has passed.

**Use:** For detection of *Escherichia coli* and coliforms in water and food samples using pour plate or streak plate methods. Tergitol inhibits Gram-positive as well as some Gram-negative bacteria other than coliforms. The cefsulodin inhibits *Pseudomonas* and *Aeromonas* species. The chromogenic mixture contains two chromogenic substrates as Salmon-GAL and X-glucuronide. The enzyme β-D-galactosidase produced by coliforms cleaves Salmon-GAL, resulting in the salmon to red coloration of coliform colonies. The enzyme β-D-glucuronidase produced by *E. coli* cleaves X-glucuronide. *E. coli* forms dark blue to violet colored colonies due to cleavage of both Salmon-GAL and X-glucuronide. The addition of tryptophan improves the indole reaction.

## HiCrome™ ECC Selective Agar
### (ECC Selective Agar, HiCrome™)
**Composition** per liter:

| | |
|---|---|
| Agar | 10.0g |
| Peptone, special | 6.0g |
| Casein enzymic hydrolysate | 3.3g |
| NaCl | 2.0g |
| Sodium pyruvate | 1.0g |
| Tryptophan | 1.0g |
| Sorbitol | 1.0g |
| Na$_2$HPO$_4$ | 1.0g |
| NaH$_2$PO$_4$ | 0.6g |
| Tergitol 7® | 0.15g |
| Chromogenic mixture | 0.43g |
| Cefsulodin | 10.0mg |

pH 6.8 ± 0.2 at 25°C

**Source:** This medium is available as a premixed powder from Hi-Media.

**Preparation of Medium:** Add components to distilled/deionized water and bring volume to 1.0L. Mix thoroughly and heat with frequent agitation until components are completely dissolved (approximately 35 min). Do not autoclave. Cool to 45°–50°C. Pour into sterile Petri dishes. Medium may show haziness.

**Storage/Shelf Life:** Store dehydrated media in the dark in a sealed container below 30°C. Prepared plates should be stored in the dark under refrigeration (2-8°C). Chromogenic media are especially light and temperature sensitive; protect from light, excessive heat, moisture, and freezing. Media should not be used if there are any signs of deterioration (shrinking, cracking, or discoloration) or contamination, or if the expiration date supplied by the manufacturer has passed.

**Use:** For detection of *Escherichia coli* and coliforms in water and food samples using the membrane filter technique. Tergitol inhibits Gram-positive as well as some Gram-negative bacteria other than coliforms. The cefsulodin inhibits *Pseudomonas* and *Aeromonas* species. The chromogenic mixture contains two chromogenic substrates as Salmon-GAL and X-glucuronide. The enzyme β-D-galactosidase produced by coliforms cleaves Salmon-GAL, resulting in the salmon to red coloration of coliform colonies. The enzyme β-D-glucuronidase produced by *E. coli* cleaves X-glucuronide. *E. coli* forms dark blue to violet colored colonies due to cleavage of both Salmon-GAL and X-glucuronide. The addition of tryptophan improves the indole reaction.

## HiCrome™ ECC Selective Agar Base
**Composition** per liter:

| | |
|---|---|
| Agar | 10.0g |
| Peptone, special | 6.0g |
| Casein enzymic hydrolysate | 3.3g |
| NaCl | 2.0g |

| | |
|---|---|
| Na$_2$HPO$_4$ | 1.0g |
| Sodium pyruvate | 1.0g |
| Sorbitol | 1.0g |
| Tryptophan | 1.0g |
| Sodium dihydrogen phosphate | 0.6g |
| Chromogenic mixture | 0.43g |
| Tergitol 7 | 0.15g |

pH 7.0± 0.2 at 25°C

**Source:** This medium is available as a premixed powder from Hi-Media.

**Preparation of Medium:** Add components to distilled/deionized water and bring volume to 1.0L. Mix thoroughly and heat with frequent agitation in boiling water bath or flowing steam. Boil until components are fully dissolved (approximately 35 min). Do not autoclave. Cool to 45°–50°C. Pour into sterile Petri dishes. Medium may show haziness. To inhibit *Pseudomonas* and *Aeromonas* species 5 mg cefsulodin may be added for surface and pour plate methods or 10 mg cefsulodin may be aseptically added to the medium for the membrane filter technique.

**Storage/Shelf Life:** Store dehydrated media in the dark in a sealed container below 30°C. Prepared plates should be stored in the dark under refrigeration (2-8°C). Chromogenic media are especially light and temperature sensitive; protect from light, excessive heat, moisture, and freezing. Media should not be used if there are any signs of deterioration (shrinking, cracking, or discoloration) or contamination, or if the expiration date supplied by the manufacturer has passed.

**Use:** For the detection of *Escherichia coli* and coliforms in water and food samples.

## HiCrome™ ECC Selective Agar Base, HiVeg
**Composition** per liter:

| | |
|---|---|
| Agar | 10.0g |
| Plant special peptone | 6.0g |
| Plant hydrolysate | 3.3g |
| NaCl | 2.0g |
| Sodium dihydrogen phosphate | 0.6g |
| Na$_2$HPO$_4$ | 1.0g |
| Sodium pyruvate | 1.0g |
| Sorbitol | 1.0g |
| Tryptophan | 1.0g |
| Chromogenic mixture | 0.43g |
| Tergitol 7 | 0.15g |

pH 7.0± 0.2 at 25°C

**Source:** This medium is available as a premixed powder from Hi-Media.

**Preparation of Medium:** Add components to distilled/deionized water and bring volume to 1.0L. Mix thoroughly and heat with frequent agitation in boiling water bath or flowing steam. Boil until components are fully dissolved (approximately 35 min). Do not autoclave. Cool to 45°–50°C. Pour into sterile Petri dishes. Medium may show haziness. To inhibit *Pseudomonas* and *Aeromonas* species 5 mg cefsulodin may be added for surface and pour plate methods or 10 mg cefsulodin may be aseptically added to the medium for membrane filter technique.

**Storage/Shelf Life:** Store dehydrated media in the dark in a sealed container below 30°C. Prepared plates should be stored in the dark under refrigeration (2-8°C). Chromogenic media are especially light and temperature sensitive; protect from light, excessive heat, moisture, and freezing. Media should not be used if there are any signs of deterioration (shrinking, cracking, or discoloration) or contamination, or if the expiration date supplied by the manufacturer has passed.

**Use:** For the detection of *Escherichia coli* and coliforms in water and food samples.

## HiCrome™ ECD Agar with MUG
**Composition** per liter:

Casein enzymic hydrolysate .......................................................20.0g
Agar ..............................................................................................15.0g
Lactose ..........................................................................................5.0g
NaCl ................................................................................................5.0g
K$_2$HPO$_4$ ...........................................................................................4.0g
Bile salts mixture ..........................................................................1.5g
KH$_2$PO$_4$ ..........................................................................................1.5g
L-Tryptophan.................................................................................1.0g
Chromogenic substrate ................................................................0.1g
Fluorogenic substrate....................................................................0.07g

pH 7.0 ± 0.2 at 37°C

**Source:** This medium is available as a premixed powder from Hi-Media.

**Preparation of Medium:** Add components to distilled/deionized water and bring volume to 1.0L. Mix thoroughly. Gently heat while stirring and bring to boiling. Autoclave for 15 min at 15 psi pressure–121°C. Cool to 50°C. Pour into sterile Petri dishes.

**Storage/Shelf Life:** Store dehydrated media in the dark in a sealed container below 30°C. Prepared plates should be stored in the dark under refrigeration (2-8°C). Chromogenic media are especially light and temperature sensitive; protect from light, excessive heat, moisture, and freezing. Media should not be used if there are any signs of deterioration (shrinking, cracking, or discoloration) or contamination, or if the expiration date supplied by the manufacturer has passed.

**Use:** For detection of *Escherichia coli* in a variety of specimens. Fluorescence in the UV and a positive indole test demonstrate the presence of *E. coli* in the colonies.

## HiCrome™ ECD HiVeg Agar with MUG
**Composition** per liter:

Plant hydrolysate.......................................................................20.0g
Agar ..............................................................................................15.0g
NaCl ................................................................................................5.0g
Lactose ..........................................................................................5.0g
K$_2$HPO$_4$ ...........................................................................................4.0g
KH$_2$PO$_4$ ..........................................................................................1.5g
Synthetic detergent......................................................................1.5g
Tryptophan ...................................................................................1.0g
Chromogenic substrate ................................................................0.1g
Fluorogenic substrate....................................................................0.07g

pH 7.0 ± 0.2 at 37°C

**Source:** This medium is available as a premixed powder from Hi-Media.

**Preparation of Medium:** Add components to distilled/deionized water and bring volume to 1.0L. Mix thoroughly. Gently heat while stirring and bring to boiling. Autoclave for 15 min at 15 psi pressure–121°C. Cool to 50°C. Pour into sterile Petri dishes.

**Storage/Shelf Life:** Store dehydrated media in the dark in a sealed container below 30°C. Prepared plates should be stored in the dark under refrigeration (2-8°C). Chromogenic media are especially light and temperature sensitive; protect from light, excessive heat, moisture, and freezing. Media should not be used if there are any signs of deteriora-

tion (shrinking, cracking, or discoloration) or contamination, or if the expiration date supplied by the manufacturer has passed.

**Use:** For detection of *Escherichia coli* in a variety of specimens, including foods and water. Fluorescence in the UV and a positive indole test demonstrate the presence of *E. coli* in the colonies.

## HiCrome™ Enrichment Broth Base for EC O157:H7
**Composition** per liter:

Casein enzymic hydrolysate .......................................................10.0g
Sorbitol ..........................................................................................10.0g
Bile salts mixture ..........................................................................1.5g
Chromogenic mixture ..................................................................1.3g

pH 7.1 ± 0.2 at 25°C

**Source:** This medium is available as a premixed powder from Hi-Media.

**Preparation of Medium:** Add components to distilled/deionized water and bring volume to 1.0L. Mix thoroughly and heat with frequent agitation until boiling. Boil until components are fully dissolved. Do not autoclave. Cool to 45°–50°C.

**Use:** For the enrichment culture of *Escherichia coli* O157:H7 in foods and environmental samples.

## HiCrome™ *Enterobacter sakazakii* Agar
**Composition** per liter:

Casein enzymic hydrolysate .......................................................15.0g
Agar ..............................................................................................15.0g
Chromogenic mixture ................................................................10.17g
Papaic digest of soybean meal....................................................5.0g
NaCl ................................................................................................5.0g
Na$_2$S$_2$O$_3$ ...........................................................................................1.0g
Sodium deoxycholate....................................................................0.5g

pH 7.2 ± 0.2 at 25°C

**Source:** This medium is available as a premixed powder from Hi-Media.

**Preparation of Medium:** Add components to distilled/deionized water and bring volume to 1.0L. Mix thoroughly. Gently heat and bring to boiling. Distribute into tubes or flasks. Autoclave for 15 min at 15 psi pressure–121°C. Pour into sterile Petri dishes.

**Storage/Shelf Life:** Store dehydrated media in the dark in a sealed container below 30°C. Prepared plates should be stored in the dark under refrigeration (2-8°C). Chromogenic media are especially light and temperature sensitive; protect from light, excessive heat, moisture, and freezing. Media should not be used if there are any signs of deterioration (shrinking, cracking, or discoloration) or contamination, or if the expiration date supplied by the manufacturer has passed.

**Use:** For the isolation and identification of *Enterobacter sakazakii* from food and environmental samples.

## HiCrome™ *Enterobacter sakazakii* Agar, Modified
**Composition** per liter:

Agar ..............................................................................................15.0g
Casein enzymatic hydrolysate .....................................................7.0g
NaCl ................................................................................................5.0g
Yeast extract..................................................................................3.0g
Sodium deoxycholate....................................................................0.6g
Chromogenic substrate ................................................................0.15g
Crystal Violet................................................................................0.02g

pH 7.2 ± 0.2 at 25°C

**Source:** This medium is available as a premixed powder from Hi-Media.

**Preparation of Medium:** Add components to distilled/deionized water and bring volume to 1.0L. Mix thoroughly. Gently heat and bring to boiling. Distribute into tubes or flasks. Autoclave for 15 min at 15 psi pressure–121°C. Pour into sterile Petri dishes.

**Storage/Shelf Life:** Store dehydrated media in the dark in a sealed container below 30°C. Prepared plates should be stored in the dark under refrigeration (2-8°C). Chromogenic media are especially light and temperature sensitive; protect from light, excessive heat, moisture, and freezing. Media should not be used if there are any signs of deterioration (shrinking, cracking, or discoloration) or contamination, or if the expiration date supplied by the manufacturer has passed.

**Use:** For the isolation and identification of *Enterobacter sakazakii* from food and environmental samples. Meets the formulation recommended by ISO Committee for the isolation and identification of *Enterobacter sakazakii* from milk and milk products.

## HiCrome™ Enterococci Broth
## (Enterococci HiCrome™ Broth)
**Composition** per liter:

Peptone, special ................................................10.0g
NaCl ...................................................................5.0g
Polysorbate 80 .................................................2.0g
NaH$_2$PO$_4$ ........................................................1.25g
NaN$_3$ ................................................................0.3g
Chromogenic mixture ......................................0.04g

pH 7.5 ± 0.2 at 25°C

**Source:** This medium is available as a premixed powder from Hi-Media.

**Caution:** Sodium azide is toxic. It also has a tendency to form explosive metal azides with plumbing materials. It is advisable to use enough water to flush off the disposables.

**Preparation of Medium:** Add components to distilled/deionized water and bring volume to 1.0L. Mix thoroughly. Distribute into tubes or flasks. Autoclave for 15 min at 15 psi pressure–121°C.

**Storage/Shelf Life:** Store dehydrated media in the dark in a sealed container below 30°C. Prepared plates should be stored in the dark under refrigeration (2-8°C). Chromogenic media are especially light and temperature sensitive; protect from light, excessive heat, moisture, and freezing. Media should not be used if there are any signs of deterioration (discoloration) or contamination, or if the expiration date supplied by the manufacturer has passed.

**Use:** For the rapid and easy identification and differentiation of enterococci. It contains a chromogenic substrate that aids in the detection of enterococci, especially in water samples.

## HiCrome™ Enterococci HiVeg Broth
**Composition** per liter:

Plant special peptone ......................................10.0g
NaCl ...................................................................5.0g
Polysorbate 80 .................................................2.0g
Na$_2$HPO$_4$ ........................................................1.25g
NaN$_3$ ................................................................0.3g
Chromogenic mixture ......................................0.04g

pH 7.2 ± 0.2 at 25°C

**Source:** This medium is available as a premixed powder from Hi-Media.

**Caution:** Sodium azide is toxic. It also has a tendency to form explosive metal azides with plumbing materials. It is advisable to use enough water to flush off the disposables.

**Preparation of Medium:** Add components to distilled/deionized water and bring volume to 1.0L. Mix thoroughly. Gently heat and bring to boiling. Distribute into tubes or flasks. Autoclave for 15 min at 15 psi pressure–121°C. Pour into sterile Petri dishes.

**Storage/Shelf Life:** Store dehydrated media in the dark in a sealed container below 30°C. Prepared plates should be stored in the dark under refrigeration (2-8°C). Chromogenic media are especially light and temperature sensitive; protect from light, excessive heat, moisture, and freezing. Media should not be used if there are any signs of deterioration (discoloration) or contamination, or if the expiration date supplied by the manufacturer has passed.

**Use:** For the identification and differentiation of Enterococci from water samples.

## HiCrome™ *Enterococcus faecium* Agar Base
**Composition** per liter:

Peptone, special ................................................23.0g
Agar ..................................................................15.0g
Arabinose ..........................................................10.0g
NaCl ...................................................................5.0g
Cornstarch ..........................................................1.0g
Chromogenic substrate ......................................0.1g
Phenol Red ..........................................................0.1g

pH 7.2 ± 0.2 at 25°C

**Source:** This medium is available from HiMedia.

**Preparation of Medium:** Add components to distilled/deionized water and bring volume to 1.0L. Mix thoroughly. Gently heat and bring to boiling. Distribute into tubes or flasks. Autoclave for 15 min at 15 psi pressure–121°C. Pour into sterile Petri dishes.

**Storage/Shelf Life:** Store dehydrated media in the dark in a sealed container below 30°C. Prepared plates should be stored in the dark under refrigeration (2-8°C). Chromogenic media are especially light and temperature sensitive; protect from light, excessive heat, moisture, and freezing. Media should not be used if there are any signs of deterioration (shrinking, cracking, or discoloration) or contamination, or if the expiration date supplied by the manufacturer has passed.

**Use:** For the chromogenic identification of *Enterococcus faecium* from water and sewage samples.

## HiCrome™ Improved *Salmonella* Agar
**Composition** per liter:

Agar ...................................................................12.0g
Peptone, special ................................................8.0g
Chromogenic mixture ......................................3.25g
Yeast extract ......................................................2.0g
Sodium deoxycholate .........................................1.0g

pH 7.1 ± 0.2 at 25°C

**Source:** This medium is available as a premixed powder from Hi-Media.

**Preparation of Medium:** Add components to distilled/deionized water and bring volume to 1.0L. Mix thoroughly and heat with frequent agitation until boiling. Boil until components are fully dissolved. Do not autoclave. Cool to 45°–50°C. Mix well. Pour into sterile Petri dishes.

**Storage/Shelf Life:** Store dehydrated media in the dark in a sealed container below 30°C. Prepared plates should be stored in the dark under refrigeration (2-8°C). Chromogenic media are especially light and temperature sensitive; protect from light, excessive heat, moisture, and freezing. Media should not be used if there are any signs of deterioration (shrinking, cracking, or discoloration) or contamination, or if the expiration date supplied by the manufacturer has passed.

**Use:** For the improved selective and differential medium for *Salmonella* species.

## HiCrome™ *Klebsiella* Selective Agar Base with Carbenicillin

**Composition** per liter:
| | |
|---|---|
| Agar | 15.0g |
| Peptone, special | 12.0g |
| Yeast extract | 7.0g |
| NaCl | 5.0g |
| Bile salts mixture | 1.5g |
| Chromogenic mixture | 0.2g |
| Sodium lauryl sulfate | 0.1g |
| *Klebsiella* selective supplement | 2.0mL |

pH 7.1 ± 0.2 at 25°C

**Source:** This medium, without *Klebsiella* selective supplement, is available as a premixed powder from HiMedia.

### *Klebsiella* Selective Supplement:
**Composition** per 2.0mL:
| | |
|---|---|
| Carbenicillin | 50.0mg |

**Preparation of *Klebsiella* Selective Supplement:** Add components to distilled/deionized water and bring volume to 2.0mL. Mix thoroughly. Filter sterilize.

**Preparation of Medium:** Add components to distilled/deionized water and bring volume to 1.0L. Mix thoroughly and heat with frequent agitation until boiling. Boil until components are fully dissolved. Do not autoclave. Cool to 45°–50°C. Aseptically add *Klebsiella* selective supplement. Mix well. Pour into sterile Petri dishes.

**Storage/Shelf Life:** Store dehydrated media in the dark in a sealed container below 30°C. Prepared plates should be stored in the dark under refrigeration (2-8°C). Chromogenic media are especially light and temperature sensitive; protect from light, excessive heat, moisture, and freezing. Media should not be used if there are any signs of deterioration (shrinking, cracking, or discoloration) or contamination, or if the expiration date supplied by the manufacturer has passed.

**Use:** For the selective isolation and easy detection of *Klebsiella* species from water and sewage.

## HiCrome™ *Listeria* Agar Base, Modified, with Moxalactam
### (*Listeria* HiCrome Agar Base, Modified)

**Composition** per liter:
| | |
|---|---|
| Peptone, special | 23.0g |
| Agar | 13.0g |
| Rhamnose | 10.0g |
| Chromogenic mixture | 5.13g |
| LiCl | 5.0g |
| Meat extract | 5.0g |
| NaCl | 5.0g |
| Yeast extract | 1.0g |
| Phenol Red | 0.12g |
| Moxolactam solution | 10.0mL |

pH 7.3 ± 0.2 at 25°C

**Source:** This medium is available as a premixed powder from Hi-Media.

**Caution:** Lithium chloride is harmful. Avoid bodily contact and inhalation of vapors. On contact with skin, wash with plenty of water immediately.

**Moxalactam Solution (*Listeria* Selective Supplement):**
**Composition** per 10.0mL:
| | |
|---|---|
| Moxalactam | 0.2g |

**Preparation of Moxalactam Solution (*Listeria* Selective Supplement):** Add moxalactam to distilled/deionized water and bring volume to 10.0mL. Mix thoroughly. Filter sterilize.

**Preparation of Medium:** Add components, except moxalactam solution, to distilled/deionized water and bring volume to 990.0mL. Mix thoroughly. Gently heat and bring to boiling. Distribute into tubes or flasks. Autoclave for 15 min at 15 psi pressure–121°C. Cool to 45°–50°C. Add 10.0mL moxalactam solution (*Listeria* selective supplement). Mix well. Pour into sterile Petri dishes.

**Storage/Shelf Life:** Store dehydrated media in the dark in a sealed container below 30°C. Prepared plates should be stored in the dark under refrigeration (2-8°C). Chromogenic media are especially light and temperature sensitive; protect from light, excessive heat, moisture, and freezing. Media should not be used if there are any signs of deterioration (shrinking, cracking, or discoloration) or contamination, or if the expiration date supplied by the manufacturer has passed.

**Use:** For the rapid and direct identification of *Listeria* species, specifically *Listeria monocytogenes*. A selective and differential agar medium recommended for rapid and direct identification of *Listeria* species, specifically *Listeria monocytogenes*.

## HiCrome™ MacConkey-Sorbitol Agar (MacConkey-Sorbitol Agar, HiCrome)

**Composition** per liter:
| | |
|---|---|
| Casein enzymic hydrolysate | 17.0g |
| Agar | 13.5g |
| Sorbitol | 10.0g |
| NaCl | 5.0g |
| Proteose peptone | 3.0g |
| Bile salts mixture | 1.5g |
| 5-Bromo-4-chloro-3-indolyl-β-D-glucuronidesodium salt | 0.1g |
| Neutral Red | 0.03g |
| Crystal Violet | 0.001g |

pH 7.1 ± 0.3 at 25°C

**Source:** This medium is available as a premixed powder from Hi-Media.

**Preparation of Medium:** Add components to distilled/deionized water and bring volume to 1.0L. Mix thoroughly. Gently heat while stirring and bring to boiling. Mix to completely dissolve components. Do not autoclave. Cool to 50°C. Pour into sterile Petri dishes.

**Storage/Shelf Life:** Store dehydrated media in the dark in a sealed container below 30°C. Prepared plates should be stored in the dark under refrigeration (2-8°C). Chromogenic media are especially light and temperature sensitive; protect from light, excessive heat, moisture, and freezing. Media should not be used if there are any signs of deterioration (shrinking, cracking, or discoloration) or contamination, or if the expiration date supplied by the manufacturer has passed.

**Use:** For the direct isolation and differentiation of *E. coli O157:H7* strains from foodstuffs. Recommended for selective isolation of *Escherichia coli* O157:H7 from food and animal feeds. The medium contains sorbitol instead of lactose. Enteropathogenic strains of *Escherichia coli* O157:H7 ferment lactose but do not ferment sorbitol and hence produce colorless colonies. Sorbitol fermenting strains of *Escherichia coli* produce pink-red colonies. The red color is due to production of acid from sorbitol, absorption of Neutral Red and a subsequent color change of the dye when the pH of the medium falls below 6.8. The chromogenic indicator is added to detect the presence of an enzyme β-D-glucuronidase. Strains of *Escherichia coli* possessing β-D-glucuronidase appear as blue colored colonies on the medium. Enteropathogenic strains of *Escherichia coli* O157 do not possess β-D-glucuronidase activity and thus produce colorless colonies. *Escherichia coli* fermenting sorbitol and possessing β-D-glucuronidase activity produce purple colored colonies. Most of the Gram-positive organisms are inhibited by Crystal Violet and bile salts.

## HiCrome™ MacConkey-Sorbitol Agar with Tellurite-Cefixime Supplement (MacConkey-Sorbitol Agar with Tellurite-Cefixime Supplement)

**Composition** per 1004.0mL:

| | |
|---|---|
| Casein enzymic hydrolysate | 17.0g |
| Agar | 13.5g |
| Sorbitol | 10.0g |
| NaCl | 5.0g |
| Proteose peptone | 3.0g |
| Bile salts mixture | 1.5g |
| 5-Bromo-4-chloro-3-indolyl-β-D-glucuronidesodium salt | 0.1g |
| Neutral Red | 0.03g |
| Crystal Violet | 0.001g |
| Tellurite-cefixime supplement | 4.0mL |

pH 7.1 ± 0.3 at 25°C

**Source:** This medium, without tellurite-cefixime supplement, is available as a premixed powder from HiMedia.

**Tellurite-Cefixime Supplement:**

**Composition** per 4.0mL:

| | |
|---|---|
| $K_2TeO_3$ | 5.0mg |
| Cefixime | 0.1mg |

**Preparation of Tellurite-Cefixime Supplement:** Add components to 4.0mL of distilled/deionized water. Mix thoroughly. Filter sterilize.

**Caution:** Potassium tellurite is toxic.

**Source:** This medium is available from Fluka, Sigma-Aldrich. Tellurite-cefixime supplement is available from Thermo Scientific.

**Preparation of Medium:** Add components, except tellurite-cefixime supplement, to distilled/deionized water and bring volume to 1.0L. Mix thoroughly. Gently heat while stirring and bring to boiling. Mix to completely dissolve components. Do not autoclave. Cool to 50°C. Add 4.0mL sterile tellurite-cefixime supplement. Mix thoroughly. Pour into sterile Petri dishes.

**Storage/Shelf Life:** Store dehydrated media in the dark in a sealed container below 30°C. Prepared plates should be stored in the dark under refrigeration (2-8°C). Chromogenic media are especially light and temperature sensitive; protect from light, excessive heat, moisture, and freezing. Media should not be used if there are any signs of deterioration (shrinking, cracking, or discoloration) or contamination, or if the expiration date supplied by the manufacturer has passed.

**Use:** For the direct isolation and differentiation of *E. coli* O157:H7-strains from foodstuffs. Recommended for selective isolation of *Escherichia coli* O157:H7 from food and animal feeding stuffs. The medium contains sorbitol instead of lactose. Enteropathogenic strains of *Escherichia coli* O157:H7 ferment lactose but do not ferment sorbitol and hence produce colorless colonies. Sorbitol fermenting strains of *Escherichia coli* produce pink-red colonies. The red color is due to production of acid from sorbitol, absorption of Neutral Red and a subsequent color change of the dye when the pH of the medium falls below 6.8. The chromogenic indicator is added to detect the presence of an enzyme β-D-glucuronidase. Strains of *Escherichia coli* possessing β-D-glucuronidase appear as blue colored colonies on the medium. Enteropathogenic strains of *Escherichia coli* O157 do not possess β-D-glucuronidase activity and thus produce colorless colonies. *Escherichia coli* fermenting sorbitol and possessing β-D-glucuronidase activity produce purple colored colonies. Most of the Gram-positive organisms are inhibited by Crystal Violet and bile salts. Addition of tellurite-cefixime supplement makes the medium selective. Potassium tellurite selects the serogroups O157 from other *E. coli* serogroups and inhibits *Aeromonas* species and *Providencia* species. Cefixime inhibits *Proteus* species.

## HiCrome™ M-CP Agar Base (M-CP HiCrome™ Agar Base) (Membrane *Clostridium perfringens* HiCrome™ Agar Base)

**Composition** per liter:

| | |
|---|---|
| Tryptose | 30.0g |
| Yeast extract | 20.0g |
| Agar | 15.0g |
| Sucrose | 5.0g |
| L-Cysteine·HCl·H$_2$O | 1.0g |
| MgSO$_4$·7H$_2$O | 0.1g |
| FeCl$_3$·6H$_2$O | 0.09g |
| Indoxyl-β-D-glucoside | 0.06g |
| Bromcresol Purple | 0.04g |

pH 7.6 ± 0.2 at 25°C

**Source:** This medium is available as a premixed powder from HiMedia.

**Preparation of Medium:** Add components to distilled/deionized water and bring volume to 1.0L. Mix thoroughly. Gently heat while stirring and bring to boiling. Autoclave for 15 min at 15 psi pressure–121°C. Pour into sterile Petri dishes.

**Storage/Shelf Life:** Store dehydrated media in the dark in a sealed container below 30°C. Prepared plates should be stored in the dark under refrigeration (2-8°C). Chromogenic media are especially light and temperature sensitive; protect from light, excessive heat, moisture, and freezing. Media should not be used if there are any signs of deterioration (shrinking, cracking, or discoloration) or contamination, or if the expiration date supplied by the manufacturer has passed.

**Use:** For the detection of *Clostridium perfringens*. Recommended by the Directive of the Council of the European Union 98/83/EC for isolation and enumeration of *C. perfringens* from water samples using the membrane filtration technique.

## HiCrome™ MeReSa Agar with Methicillin

**Composition** per liter:

| | |
|---|---|
| NaCl | 40.0g |
| Agar | 15.0g |

Casein enzymic hydrolysate ........................................................13.0 g
Chromogenic mixture ....................................................................5.3 g
Sodium pyruvate ...........................................................................5.0 g
Beef extract...................................................................................2.5 g
Yeast extract.................................................................................2.5 g
MRSA selective supplement....................................................... 10.0mL

pH 7.0 ± 0.2 at 25°C

**Source:** This medium, without MRSA supplement, is available as a premixed powder from HiMedia.

**MRSA Selective Supplement:**
**Composition** per 10.0mL:
Methicillin.......................................................................................4.0mg

**Preparation of MRSA Selective Supplement:** Add methicillin to distilled/deionized water and bring volume to 10.0mL. Mix thoroughly. Filter sterilize.

**Preparation of Medium:** Add components, except MRSA supplement, to distilled/deionized water and bring volume to 990.0mL. Mix thoroughly. Gently heat and bring to boiling. Distribute into tubes or flasks. Autoclave for 15 min at 15 psi pressure–121°C. Cool to 50°C. Aseptically add 10.0mL MRSA supplement. Pour into sterile Petri dishes or leave in tubes.

**Storage/Shelf Life:** Store dehydrated media in the dark in a sealed container below 30°C. Prepared plates should be stored in the dark under refrigeration (2-8°C). Chromogenic media are especially light and temperature sensitive; protect from light, excessive heat, moisture, and freezing. Media should not be used if there are any signs of deterioration (shrinking, cracking, or discoloration) or contamination, or if the expiration date supplied by the manufacturer has passed.

**Use:** For the isolation and cultivation of methicillin-resistant *Staphylococcus aureus* (MRSA).

### HiCrome™ MeReSa Agar with Oxacillin

**Composition** per liter:
NaCl.............................................................................................. 40.0g
Agar ..............................................................................................15.0g
Casein enzymic hydrolysate ........................................................13.0 g
Chromogenic mixture ....................................................................5.3 g
Sodium pyruvate ...........................................................................5.0 g
Beef extract...................................................................................2.5 g
Yeast extract.................................................................................2.5 g
MRSA selective supplement....................................................... 10.0mL

pH 7.0 ± 0.2 at 25°C

**Source:** This medium, without MRSA supplement, is available as a premixed powder from HiMedia.

**MRSA Selective Supplement:**
**Composition** per 10.0mL:
Oxacillin..........................................................................................2.0mg

**Preparation of MRSA Selective Supplement:** Add oxacillin to distilled/deionized water and bring volume to 10.0mL. Mix thoroughly. Filter sterilize.

**Preparation of Medium:** Add components, except MRSA supplement, to distilled/deionized water and bring volume to 990.0mL. Mix thoroughly. Gently heat and bring to boiling. Distribute into tubes or flasks. Autoclave for 15 min at 15 psi pressure–121°C. Cool to 50°C. Aseptically add 10.0mL MRSA supplement. Pour into sterile Petri dishes or leave in tubes.

**Storage/Shelf Life:** Store dehydrated media in the dark in a sealed container below 30°C. Prepared plates should be stored in the dark under refrigeration (2-8°C). Chromogenic media are especially light and temperature sensitive; protect from light, excessive heat, moisture, and freezing. Media should not be used if there are any signs of deterioration (shrinking, cracking, or discoloration) or contamination, or if the expiration date supplied by the manufacturer has passed.

**Use:** For the isolation and cultivation of methicillin-resistant *Staphylococcus aureus* (MRSA).

### HiCrome™ M-Lauryl Sulfate Agar

**Composition** per liter:
Peptic digest of animal tissue ....................................................40.0g
Lactose..........................................................................................30.0g
Agar ..............................................................................................10.0g
Yeast extract.................................................................................6.0g
Sodium lauryl sulfate ....................................................................1.0g
Sodium pyruvate ...........................................................................0.5g
Chromogen ....................................................................................0.2g
Phenol Red.....................................................................................0.2g

pH 7.2 ± 0.2 at 25°C

**Source:** This medium is available as a premixed powder from HiMedia.

**Preparation of Medium:** Add components to distilled/deionized water and bring volume to 1.0L. Mix thoroughly. Gently heat and bring to boiling. Distribute into tubes or flasks. Autoclave for 15 min at 15 psi pressure–121°C. Pour into sterile Petri dishes.

**Storage/Shelf Life:** Store dehydrated media in the dark in a sealed container below 30°C. Prepared plates should be stored in the dark under refrigeration (2-8°C). Chromogenic media are especially light and temperature sensitive; protect from light, excessive heat, moisture, and freezing. Media should not be used if there are any signs of deterioration (shrinking, cracking, or discoloration) or contamination, or if the expiration date supplied by the manufacturer has passed.

**Use:** For the differentiation and enumeration of *Escherichia coli* and other coliforms by the membrane filtration method.

### HiCrome™ MM Agar (HiCrome™ Miller and Mallinson Agar) (MM Agar HiCrome™)

**Composition** per liter:
Agar ..............................................................................................15.0g
Lactose..........................................................................................10.0g
Peptic digest of animal tissue ....................................................10.0g
Chromogenic mixture ....................................................................6.6g
D-Cellobiose...................................................................................3.0g
Beef extract...................................................................................2.0g
D-Trehalose ...................................................................................1.33g
D-Mannitol .....................................................................................1.2g

pH 7.6 ± 0.2 at 25°C

**Source:** This medium is available as a premixed powder from HiMedia.

**Preparation of Medium:** Add components to distilled/deionized water and bring volume to 1.0L. Mix thoroughly. Gently heat and bring to boiling. Distribute into tubes or flasks. Do not autoclave. Cool to 45°–50°C. Mix well. Pour into sterile Petri dishes.

**Storage/Shelf Life:** Store dehydrated media in the dark in a sealed container below 30°C. Prepared plates should be stored in the dark under refrigeration (2-8°C). Chromogenic media are especially light and temperature sensitive; protect from light, excessive heat, moisture, and freezing. Media should not be used if there are any signs of deterioration (shrinking, cracking, or discoloration) or contamination, or if the expiration date supplied by the manufacturer has passed.

**Use:** For the identification and differentiation of *Salmonella* and non-*Salmonella* like *Citrobacter* from water samples.

## HiCrome™ MM HiVeg Agar

**Composition** per liter:

| | |
|---|---|
| Agar | 15.0g |
| Plant peptone | 10.0g |
| Lactose | 10.0g |
| Chromogenic mixture | 6.6g |
| D-Cellobiose | 3.0g |
| Plant extract | 2.0g |
| D-Trehalose | 1.33g |
| D-Mannitol | 1.2g |

pH 7.6 ± 0.2 at 25°C

**Source:** This medium is available as a premixed powder from Hi-Media.

**Preparation of Medium:** Add components to distilled/deionized water and bring volume to 1.0L. Mix thoroughly. Gently heat and bring to boiling. Distribute into tubes or flasks. Do not autoclave. Cool to 45°–50°C. Mix well. Pour into sterile Petri dishes.

**Storage/Shelf Life:** Store dehydrated media in the dark in a sealed container below 30°C. Prepared plates should be stored in the dark under refrigeration (2-8°C). Chromogenic media are especially light and temperature sensitive; protect from light, excessive heat, moisture, and freezing. Media should not be used if there are any signs of deterioration (shrinking, cracking, or discoloration) or contamination, or if the expiration date supplied by the manufacturer has passed.

**Use:** For the identification and differentiation of *Salmonella* and non-*Salmonella* like *Citrobacter* from water samples.

## HiCrome™ MS.O157 Agar
### (MS.O157 Agar HiCrome™)

**Composition** per liter:

| | |
|---|---|
| Agar | 12.0g |
| Peptone, special | 10.0g |
| Sorbitol | 4.0g |
| Bile salt mixture | 1.0g |
| Chromogenic mixture | 0.731g |

pH 6.8 ± 0.2 at 25°C

**Source:** This medium is available from Fluka, Sigma-Aldrich.

**Preparation of Medium:** Add components to distilled/deionized water and bring volume to 1.0L. Mix thoroughly. Gently heat while stirring and bring to boiling. Do not autoclave. Cool to 50°C. Pour into sterile Petri dishes.

**Storage/Shelf Life:** Store dehydrated media in the dark in a sealed container below 30°C. Prepared plates should be stored in the dark under refrigeration (2-8°C). Chromogenic media are especially light and temperature sensitive; protect from light, excessive heat, moisture, and freezing. Media should not be used if there are any signs of deterioration (shrinking, cracking, or discoloration) or contamination, or if the expiration date supplied by the manufacturer has passed.

**Use:** For the simultaneous detection of *Escherichia coli*, *Escherichia coli O157:H7*, and coliforms in water and food samples. *Escherichia*

*coli O157:H7* gives colorless colonies because of non-fermentation of sorbitol and absence of β-glucuronidase activity, whereas other strains of *Escherichia coli* having β-glucuronidase activity and fermenting sorbitol appear as steel blue colored colonies. Some non *Escherichia coli* O157:H7 may have some colony color.

## HiCrome™ MS.O157 Agar with Tellurite
### (MS.O157 Agar with Tellurite, HiCrome™)

**Composition** 1000.25mL:

| | |
|---|---|
| Agar | 12.0g |
| Peptone, special | 10.0g |
| Sorbitol | 4.0g |
| Bile salt mixture | 1.0g |
| Chromogenic mixture | 0.731g |
| Tellurite solution | 0.25mL |

pH 6.8 ± 0.2 at 25°C

**Source:** This medium is available as a premixed powder from Hi-Media.

**Tellurite Solution:**
**Composition** per 10.0mL:

| | |
|---|---|
| $K_2TeO_3$ | 0.1g |

**Preparation of Tellurite Solution:** Add $K_2TeO_3$ to 10.0mL of distilled/deionized water. Mix thoroughly. Filter sterilize.

**Caution:** Potassium tellurite is toxic.

**Preparation of Medium:** Add components, exept tellurite solution, to distilled/deionized water and bring volume to 1.0L. Mix thoroughly. Gently heat while stirring and bring to boiling. Do not autoclave. Cool to 45°C. Aseptically add 0.25mL sterile tellurite solution. Mix thoroughly. Pour into sterile Petri dishes.

**Storage/Shelf Life:** Store dehydrated media in the dark in a sealed container below 30°C. Prepared plates should be stored in the dark under refrigeration (2-8°C). Chromogenic media are especially light and temperature sensitive; protect from light, excessive heat, moisture, and freezing. Media should not be used if there are any signs of deterioration (shrinking, cracking, or discoloration) or contamination, or if the expiration date supplied by the manufacturer has passed.

**Use:** For the simultaneous detection of *Escherichia coli*, *Escherichia coli O157:H7,* and coliforms in water and food samples. *Escherichia coli O157:H7* gives colorless colonies because of non-fermentation of sorbitol and absence of β-glucuronidase activity, whereas other strains of *Escherichia coli* having β-glucuronidase activity and fermenting sorbitol appear as steel blue colored colonies. Addition of tellurite makes the medium much more specific and selective.

## HiCrome™ RajHans Medium
### (*Salmonella* Agar)

**Composition** per liter:

| | |
|---|---|
| Agar | 13.5g |
| Casein enzymic hydrolysate | 8.0g |
| Chromogenic mixture | 7.3g |
| NaCl | 5.0g |
| Yeast extract | 5.0g |
| Peptone | 4.0g |
| Lactose | 3.0g |
| Sodium deoxycholate | 1.0g |
| Neutral Red | 0.02g |

pH 7.3 ± 0.2 at 25°C

**Preparation of Medium:** Add components to distilled/deionized water and bring volume to 1.0L. Mix thoroughly. Gently heat and bring to boiling. Distribute into tubes or flasks. Do not autoclave. Do not overheat. Cool to 50°C. Mix thoroughly. Pour into sterile Petri dishes or leave in tubes.

**Storage/Shelf Life:** Store dehydrated media in the dark in a sealed container below 30°C. Prepared plates should be stored in the dark under refrigeration (2-8°C). Chromogenic media are especially light and temperature sensitive; protect from light, excessive heat, moisture, and freezing. Media should not be used if there are any signs of deterioration (shrinking, cracking, or discoloration) or contamination, or if the expiration date supplied by the manufacturer has passed.

**Use:** For the identification and differentiation of *Salmonella* species from the members of Enterobacteriaceae, especially *Proteus* species.

## HiCrome™ RajHans Medium, Modified
### (*Salmonella* Agar, Modified)

**Composition** per liter:

| | |
|---|---|
| Agar | 12.0g |
| Casein enzymic hydrolysate | 8.0g |
| NaCl | 5.0g |
| Yeast extract | 5.0g |
| Chromogenic mixture | 4.32g |
| Peptic digest of animal tissue | 4.0g |
| Lactose | 3.0g |
| Sodium deoxycholate | 1.0g |
| Neutral Red | 0.02g |

pH 7.3 ± 0.2 at 25°C

**Preparation of Medium:** Add components to distilled/deionized water and bring volume to 1.0L. Mix thoroughly. Gently heat and bring to boiling. Distribute into tubes or flasks. Do not autoclave. Do not overheat. Cool to 50°C. Mix thoroughly. Pour into sterile Petri dishes or leave in tubes.

**Storage/Shelf Life:** Store dehydrated media in the dark in a sealed container below 30°C. Prepared plates should be stored in the dark under refrigeration (2-8°C). Chromogenic media are especially light and temperature sensitive; protect from light, excessive heat, moisture, and freezing. Media should not be used if there are any signs of deterioration (shrinking, cracking, or discoloration) or contamination, or if the expiration date supplied by the manufacturer has passed.

**Use:** For the identification and differentiation of *Salmonella* species from the members of Enterobacteriaceae, especially *Proteus* species.

## HiCrome™ Rapid Coliform Broth
### (Coliform Rapid HiCrome™ Broth)
### (Rapid Coliform HiCrome™ Broth)

**Composition** per liter:

| | |
|---|---|
| Peptone, special | 5.0g |
| NaCl | 5.0g |
| Na$_2$HPO$_4$ | 2.7g |
| KH$_2$PO$_4$ | 2.0g |
| Sorbitol | 1.0g |
| Sodium lauryl sulfate | 0.1g |
| IPTG | 0.1g |
| Chromogenic substrate | 0.08g |
| Fluorogenic substrate | 0.05g |

pH 6.8 ± 0.3 at 25°C

**Source:** This medium is available as a premixed powder from Hi-Media.

**Preparation of Medium:** Add components to distilled/deionized water and bring volume to 1.0L. Mix thoroughly. Gently heat while stirring and bring to boiling. Autoclave for 15 min at 15 psi pressure–121°C. Pour into sterile Petri dishes.

**Storage/Shelf Life:** Store dehydrated media in the dark in a sealed container below 30°C. Prepared plates should be stored in the dark under refrigeration (2-8°C). Chromogenic media are especially light and temperature sensitive; protect from light, excessive heat, moisture, and freezing. Media should not be used if there are any signs of deterioration (shrinking, cracking, or discoloration) or contamination, or if the expiration date supplied by the manufacturer has passed.

**Use:** For the detection and confirmation of *Escherichia coli* and coliforms on the basis of enzyme substrate reaction from water samples, using a combination of chromogenic and fluorogenic substrate.

## HiCrome™ Rapid Enterococci Agar
### (Enterococci Rapid HiCrome™ Agar)
### (Rapid Enterococci HiCrome™ Agar)

**Composition** per liter:

| | |
|---|---|
| Agar | 15.0g |
| Peptone special | 10.0g |
| NaCl | 5.0g |
| Polysorbate 80 | 2.0g |
| Na$_2$HPO$_4$ | 1.25g |
| NaN$_3$ | 0.3g |
| Chromogenic mixture | 0.06g |

pH 7.5 ± 0.2 at 25°C

**Source:** This medium is available as a premixed powder from Hi-Media.

**Caution:** Sodium azide is toxic. It also has a tendency to form explosive metal azides with plumbing materials. It is advisable to use enough water to flush off the disposables.

**Preparation of Medium:** Add components to distilled/deionized water and bring volume to 1.0L. Mix thoroughly. Gently heat while stirring and bring to boiling. Autoclave for 15 min at 15 psi pressure–121°C. Pour into sterile Petri dishes.

**Storage/Shelf Life:** Store dehydrated media in the dark in a sealed container below 30°C. Prepared plates should be stored in the dark under refrigeration (2-8°C). Chromogenic media are especially light and temperature sensitive; protect from light, excessive heat, moisture, and freezing. Media should not be used if there are any signs of deterioration (shrinking, cracking, or discoloration) or contamination, or if the expiration date supplied by the manufacturer has passed.

**Use:** For the rapid and easy identification and differentiation of enterococci. It contains a chromogenic substrate, which aids in the detection of enterococci, especially from water samples.

## HiCrome™ *Salmonella* Agar

**Composition** per liter:

| | |
|---|---|
| Agar | 13.0g |
| Peptic digest of animal tissue | 6.0g |
| Chromogenic mixture | 5.4g |
| Yeast extract | 2.5g |
| Bile salts mixture | 1.0g |

pH 7.3 ± 0.2 at 25°C

**Source:** This medium is available as a premixed powder from Hi-Media.

**Preparation of Medium:** Add components to distilled/deionized water and bring volume to 1.0L. Mix thoroughly. Gently heat and bring to boiling. Distribute into tubes or flasks. Do not autoclave. Cool to 45°–50°C. Mix well. Pour into sterile Petri dishes.

**Storage/Shelf Life:** Store dehydrated media in the dark in a sealed container below 30°C. Prepared plates should be stored in the dark under refrigeration (2-8°C). Chromogenic media are especially light and temperature sensitive; protect from light, excessive heat, moisture, and freezing. Media should not be used if there are any signs of deterioration (shrinking, cracking, or discoloration) or contamination, or if the expiration date supplied by the manufacturer has passed.

**Use:** For the simultaneous detection of *Escherichia coli* and *Salmonella* from food and water. For the identification, differentiation, and confirmation of enteric bacteria from specimens such as urine, water, or food that may contain large numbers of *Proteus* species as well as potentially pathogenic Gram-positive organisms.

## HiCrome™ *Salmonella* Agar
### (*Salmonella* Agar, HiCrome™)

**Composition** per liter:

| | |
|---|---|
| Agar | 13.0g |
| Peptic digest of animal tissue | 6.0g |
| Chromogenic mixture | 5.4g |
| Yeast extract | 2.5g |
| Chromogenic mix | 1.5g |
| Bile salt mixture | 1.0g |

pH 7.7 ± 0.2 at 25°C

**Source:** This medium is available as a premixed powder from Hi-Media.

**Preparation of Medium:** Add components to distilled/deionized water and bring volume to 1.0L. Mix thoroughly. Gently heat while stirring and bring to boiling. Mix to completely dissolve components. Do not autoclave. Cool to 50°C. Pour into sterile Petri dishes.

**Storage/Shelf Life:** Store dehydrated media in the dark in a sealed container below 30°C. Prepared plates should be stored in the dark under refrigeration (2-8°C). Chromogenic media are especially light and temperature sensitive; protect from light, excessive heat, moisture, and freezing. Media should not be used if there are any signs of deterioration (shrinking, cracking, or discoloration) or contamination, or if the expiration date supplied by the manufacturer has passed.

**Use:** A selective chromogenic medium used for the isolation and differentiation of *Salmonella* species from coliforms in food and water. *E. coli* and *Salmonella* are easily distinguishable due to the colony characteristics. *Salmonella* give light purple colonies with a halo. *E. coli* have a characteristic blue color. Other organisms give colorless colonies. The characteristic light purple and blue color is due to the chromogenic mixture.

## HiCrome™ *Salmonella* Chromogen Agar
### (*Salmonella* Chromogen Agar, HiCrome™)
### (Rambach Equivalent Agar)

**Composition** per liter:

| | |
|---|---|
| Agar | 15.0g |
| Peptic digest of animal tissue | 5.0g |
| NaCl | 5.0g |

| | |
|---|---|
| Yeast extract | 2.0g |
| Meat extract | 1.0g |
| Na-deoxycholate | 1.0g |
| Chromogenic mixture | 1 vial |

pH 7.3 ± 0.2 at 25°C

**Source:** This medium is available as a premixed powder from Hi-Media.

**Preparation of Medium:** Add components to distilled/deionized water and bring volume to 1.0L. Mix thoroughly. Gently heat while stirring and bring to boiling. Mix to completely dissolve components. Do not autoclave. Mix and boil in 5 min sequences for 35–40 min. Cool to 50°C. Shake gently for 30–35 min. Pour into sterile Petri dishes.

**Storage/Shelf Life:** Store dehydrated media in the dark in a sealed container below 30°C. Prepared plates should be stored in the dark under refrigeration (2-8°C). Chromogenic media are especially light and temperature sensitive; protect from light, excessive heat, moisture, and freezing. Media should not be used if there are any signs of deterioration (shrinking, cracking, or discoloration) or contamination, or if the expiration date supplied by the manufacturer has passed.

**Use:** A selective medium used for the detection of *Salmonella* species. This medium exploits a novel phenotypic characteristic of *Salmonella* spp.: the formation of acid from propylene glycol. This characteristic may be used in combination with a chromogenic indicator of β-galactosidase to differentiate *Salmonella* spp. from *Proteus* spp. and the other members of the Enterobacteriaceae. Deoxycholate is included in the plate medium as an inhibitor of Gram-positive organisms. Non-typhi *Salmonella* spp. yield distinct, bright red colonies on this medium, allowing facilitated identification and unambiguous differentiation from *Proteus* spp. Coliforms produce blue-green to blue-violet colonies. Other Enterobacteriaceae and Gram-negative bacteria such as *Proteus, Shigella, Pseudomonas, Salmonella typhi*, and *S. paratyphi* A form colorless or yellow colonies.

## HiCrome™ UTI Agar, HiVeg

**Composition** per liter:

| | |
|---|---|
| Agar | 15.0g |
| Plant peptone | 15.0g |
| Chromogenic mixture | 2.45g |

pH 7.3 ± 0.2 at 25°C

**Source:** This medium is available as a premixed powder from Hi-Media.

**Preparation of Medium:** Add components to distilled/deionized water and bring volume to 1.0L. Mix thoroughly. Gently heat and bring to boiling. Distribute into tubes or flasks. Autoclave for 15 min at 15 psi pressure–121°C. Pour into sterile Petri dishes.

**Storage/Shelf Life:** Store dehydrated media in the dark in a sealed container below 30°C. Prepared plates should be stored in the dark under refrigeration (2-8°C). Chromogenic media are especially light and temperature sensitive; protect from light, excessive heat, moisture, and freezing. Media should not be used if there are any signs of deterioration (shrinking, cracking, or discoloration) or contamination, or if the expiration date supplied by the manufacturer has passed.

**Use:** For the differentiation and enumeration of thermotolerant *E. coli* from water by the membrane filtration method. For the identification, differentiation, and confirmation of enteric bacteria from specimens such as urine, water, or food that may contain large numbers of *Proteus* species as well as potentially pathogenic Gram-positive organisms.

## HiCrome™ UTI Agar, Modified

**Composition** per liter:
Peptic digest of animal tissue......................................18.0g
Agar ..............................................................................15.0g
Chromogenic mixture ..............................................12.44g
Beef extract..................................................................6.0g
Casein enzymic hydrolysate ........................................4.0g
pH 7.2 ± 0.2 at 25°C

**Source:** This medium is available as a premixed powder from Hi-Media.

**Preparation of Medium:** Add components to distilled/deionized water and bring volume to 1.0L. Mix thoroughly. Gently heat and bring to boiling. Distribute into tubes or flasks. Autoclave for 15 min at 15 psi pressure–121°C. Pour into sterile Petri dishes.

**Storage/Shelf Life:** Store dehydrated media in the dark in a sealed container below 30°C. Prepared plates should be stored in the dark under refrigeration (2-8°C). Chromogenic media are especially light and temperature sensitive; protect from light, excessive heat, moisture, and freezing. Media should not be used if there are any signs of deterioration (shrinking, cracking, or discoloration) or contamination, or if the expiration date supplied by the manufacturer has passed.

**Use:** For the differentiation and enumeration of thermotolerant *E. coli* from water by the membrane filtration method. For the identification, differentiation, and confirmation of enteric bacteria from specimens such as urine, water, or food which may contain large numbers of *Proteus* species as well as potentially pathogenic Gram-positive organisms.

## HiCrome™ UTI Agar, Modified
## (UTI Agar, Modified HiCrome™)

**Composition** per liter:
Peptic digest of animal tissue......................................18.0g
Agar ..............................................................................15.0g
Chromogenic mixture ..............................................12.44g
Beef extract..................................................................4.0g
Casein enzymatic hydrolysate ......................................4.0g
pH 7.2 ± 0.2 at 25°C

**Source:** This medium is available as a premixed powder from Hi-Media.

**Preparation of Medium:** Add components to distilled/deionized water and bring volume to 1.0L. Mix thoroughly. Gently heat while stirring and bring to boiling. Mix to completely dissolve components. Do not autoclave. Cool to 50°C. Pour into sterile Petri dishes.

**Storage/Shelf Life:** Store dehydrated media in the dark in a sealed container below 30°C. Prepared plates should be stored in the dark under refrigeration (2-8°C). Chromogenic media are especially light and temperature sensitive; protect from light, excessive heat, moisture, and freezing. Media should not be used if there are any signs of deterioration (shrinking, cracking, or discoloration) or contamination, or if the expiration date supplied by the manufacturer has passed.

**Use:** A chromogenic medium used for detecting and identifying Enterobacteria, *Proteus* species, and other bacteria involved in urinary tract infections.

## HiFluoro™ *Pseudomonas* Agar Base

**Composition** per liter:
Pancreatic digest of gelatin ........................................18.0g
Agar ..............................................................................15.0g

$K_2SO_4$ ..............................................................................10.0g
Fluorogenic mixture...................................................2.05g
$MnCl_2$............................................................................1.4g
Cetrimide .....................................................................0.3g
Glycerol ......................................................................10.ml
pH 7.2 ± 0.2 at 25°C

**Source:** This medium, without glycerol, is available as a premixed powder from HiMedia.

**Preparation of Medium:** Add components to distilled/deionized water and bring volume to 1.0L. Mix thoroughly. Gently heat and bring to boiling. Distribute into tubes or flasks. Autoclave for 15 min at 15 psi pressure–121°C. Pour into sterile Petri dishes.

**Storage/Shelf Life:** Store dehydrated media in the dark in a sealed container below 30°C. Prepared plates should be stored in the dark under refrigeration (2-8°C). Chromogenic media are especially light and temperature sensitive; protect from light, excessive heat, moisture, and freezing. Media should not be used if there are any signs of deterioration (shrinking, cracking, or discoloration) or contamination, or if the expiration date supplied by the manufacturer has passed.

**Use:** For the selective isolation of *Pseudomonas aeruginosa* from clinical speciments and nonclinical specimens of public health importance by the fluorogenic method.

## Hi-Sensitivity Test Agar

**Composition** per liter:
Casein enzymic hydrolysate ......................................11.0g
Agar ..............................................................................8.0g
NaCl...............................................................................3.0g
Peptic digest of animal tissue ......................................3.0g
Glucose ........................................................................2.0g
$Na_2HPO_4$......................................................................2.0g
Sodium acetate..............................................................1.0g
Starch, soluble...............................................................1.0g
Magnesium glycerophosphate ......................................0.2g
Calcium gluconate.........................................................0.1g
L-Cystine hydrochloride ...........................................0.02g
L-Tryptophan ..............................................................0.02g
Adenine........................................................................0.01g
Guanine........................................................................0.01g
Uracil ...........................................................................0.01g
Xanthine.......................................................................0.01g
Calcium pantothenate ................................................3.0mg
Biotin ..........................................................................3.0mg
Nicotinamide...............................................................3.0mg
Pyridoxine hydrochloride ..........................................3.0mg
Manganese chloride ...................................................2.0mg
$ZnSO_4$ .........................................................................1.0mg
$CoSO_4$ ........................................................................1.0mg
$CuSO_4$........................................................................1.0mg
Cyanocobalamin .........................................................1.0mg
$FeSO_4$.........................................................................1.0mg
Menadione ..................................................................1.0mg
Thiamine hydrochloride.............................................0.04mg
pH 7.2 ± 0.2 at 25°C

**Source:** This medium is available as a premixed powder from Hi-Media.

**Preparation of Medium:** Add components to distilled/deionized water and bring volume to 1.0L. Mix thoroughly. Gently heat and bring

to boiling. Distribute into tubes or flasks. Autoclave for 15 min at 15 psi pressure–121°C. Pour into sterile Petri dishes.

**Storage/Shelf Life:** Store dehydrated media in the dark in a sealed container below 30°C. Prepared media should be stored under refrigeration (2-8°C). Media should be used within 60 days of preparation. Media should not be used if there are any signs of deterioration (shrinking, cracking, or discoloration) or contamination, or if the expiration date supplied by the manufacturer has passed.

**Use:** For antimicrobial susceptibility tests.

## Hi-Sensitivity Test Broth
**Composition** per liter:

| | |
|---|---|
| Casein enzymic hydrolysate | 11.0g |
| NaCl | 3.0g |
| Peptic digest of animal tissue | 3.0g |
| Glucose | 2.0g |
| $Na_2HPO_4$ | 2.0g |
| Sodium acetate | 1.0g |
| Starch, soluble | 1.0g |
| Magnesium glycerophosphate | 0.2g |
| Calcium gluconate | 0.1g |
| L-Cystine hydrochloride | 0.02g |
| L-Tryptophan | 0.02g |
| Adenine | 0.01g |
| Guanine | 0.01g |
| Uracil | 0.01g |
| Xanthine | 0.01g |
| Calcium pantothenate | 3.0mg |
| Biotin | 3.0mg |
| Nicotinamide | 3.0mg |
| Pyridoxine hydrochloride | 3.0mg |
| Manganese chloride | 2.0mg |
| $ZnSO_4$ | 1.0mg |
| $CoSO_4$ | 1.0mg |
| $CuSO_4$ | 1.0mg |
| Cyanocobalamin | 1.0mg |
| $FeSO_4$ | 1.0mg |
| Menadione | 1.0mg |
| Thiamine hydrochloride | 0.04mg |

pH 7.2 ± 0.2 at 25°C

**Source:** This medium is available as a premixed powder from Hi-Media.

**Preparation of Medium:** Add components to distilled/deionized water and bring volume to 1.0L. Mix thoroughly. Gently heat and bring to boiling. Distribute into tubes or flasks. Autoclave for 15 min at 15 psi pressure–121°C.

**Storage/Shelf Life:** Store dehydrated media in the dark in a sealed container below 30°C. Prepared media should be stored under refrigeration (2-8°C). Media should be used within 60 days of preparation. Media should not be used if there are any signs of deterioration (discoloration) or contamination, or if the expiration date supplied by the manufacturer has passed.

**Use:** For antimicrobial susceptibility testing.

## Hi-Sensitivity Test HiVeg Agar
**Composition** per liter:

| | |
|---|---|
| Plant hydrolysate | 11.0g |
| Agar | 8.0g |
| NaCl | 3.0g |

| | |
|---|---|
| Plant peptone | 3.0g |
| Glucose | 2.0g |
| $Na_2HPO_4$ | 2.0g |
| Sodium acetate | 1.0g |
| Starch, soluble | 1.0g |
| Magnesium glycerophosphate | 0.2g |
| Calcium gluconate | 0.1g |
| L-Cystine hydrochloride | 0.02g |
| L-Tryptophan | 0.02g |
| Adenine | 0.01g |
| Guanine | 0.01g |
| Uracil | 0.01g |
| Xanthine | 0.01g |
| Calcium pantothenate | 3.0mg |
| Biotin | 3.0mg |
| Nicotinamide | 3.0mg |
| Pyridoxine hydrochloride | 3.0mg |
| Manganese chloride | 2.0mg |
| $ZnSO_4$ | 1.0mg |
| $CoSO_4$ | 1.0mg |
| $CuSO_4$ | 1.0mg |
| Cyanocobalamin | 1.0mg |
| $FeSO_4$ | 1.0mg |
| Menadione | 1.0mg |
| Thiamine hydrochloride | 0.04mg |

pH 7.4 ± 0.2 at 25°C

**Source:** This medium is available as a premixed powder from Hi-Media.

**Preparation of Medium:** Add components to distilled/deionized water and bring volume to 1.0L. Mix thoroughly. Gently heat and bring to boiling. Distribute into tubes or flasks. Autoclave for 15 min at 15 psi pressure–121°C. Pour into sterile Petri dishes.

**Storage/Shelf Life:** Store dehydrated media in the dark in a sealed container below 30°C. Prepared media should be stored under refrigeration (2-8°C). Media should be used within 60 days of preparation. Media should not be used if there are any signs of deterioration (shrinking, cracking, or discoloration) or contamination, or if the expiration date supplied by the manufacturer has passed.

**Use:** For antimicrobial susceptibility tests.

## Hi-Sensitivity Test HiVeg Broth
**Composition** per liter:

| | |
|---|---|
| Plant hydrolysate | 11.0g |
| NaCl | 3.0g |
| Plant peptone | 3.0g |
| Glucose | 2.0g |
| $Na_2HPO_4$ | 2.0g |
| Sodium acetate | 1.0g |
| Starch, soluble | 1.0g |
| Magnesium glycerophosphate | 0.2g |
| Calcium gluconate | 0.1g |
| L-Cystine hydrochloride | 0.02g |
| L-Tryptophan | 0.02g |
| Adenine | 0.01g |
| Guanine | 0.01g |
| Uracil | 0.01g |
| Xanthine | 0.01g |
| Calcium pantothenate | 3.0mg |
| Biotin | 3.0mg |
| Nicotinamide | 3.0mg |
| Pyridoxine hydrochloride | 3.0mg |

| | |
|---|---|
| Manganese chloride | 2.0mg |
| ZnSO$_4$ | 1.0mg |
| CoSO$_4$ | 1.0mg |
| CuSO$_4$ | 1.0mg |
| Cyanocobalamin | 1.0mg |
| FeSO$_4$ | 1.0mg |
| Menadione | 1.0mg |
| Thiamine hydrochloride | 0.04mg |

pH 7.4 ± 0.2 at 25°C

**Source:** This medium is available as a premixed powder from Hi-Media.

**Preparation of Medium:** Add components to distilled/deionized water and bring volume to 1.0L. Mix thoroughly. Gently heat and bring to boiling. Distribute into tubes or flasks. Autoclave for 15 min at 15 psi pressure–121°C.

**Storage/Shelf Life:** Store dehydrated media in the dark in a sealed container below 30°C. Prepared media should be stored under refrigeration (2-8°C). Media should be used within 60 days of preparation. Media should not be used if there are any signs of deterioration (discoloration) or contamination, or if the expiration date supplied by the manufacturer has passed.

**Use:** For antimicrobial susceptibility testing.

## Hisitest Agar

**Composition** per liter:

| | |
|---|---|
| Casein enzymic hydrolysate | 11.0g |
| Agar | 8.0g |
| Buffer salt | 3.3g |
| Peptic digest of animal tissue | 3.0g |
| NaCl | 3.0g |
| Glucose | 2.0g |
| Starch | 1.0g |
| Nucleoside basis | 0.02g |
| Thiamine | 0.02mg |

pH 7.2 ± 0.2 at 25°C

**Source:** This medium is available as a premixed powder from Hi-Media.

**Preparation of Medium:** Add components to distilled/deionized water and bring volume to 1.0L. Mix thoroughly. Gently heat and bring to boiling. Distribute into tubes or flasks. Autoclave for 15 min at 15 psi pressure–121°C.

**Storage/Shelf Life:** Store dehydrated media in the dark in a sealed container below 30°C. Prepared media should be stored under refrigeration (2-8°C). Media should be used within 60 days of preparation. Media should not be used if there are any signs of deterioration (shrinking, cracking, or discoloration) or contamination, or if the expiration date supplied by the manufacturer has passed.

**Use:** For determination of antibiotic susceptibility of fastidious microorganisms.

## Horie Arabinose Ethyl Violet Broth (HAEB)

**Composition** per liter:

| | |
|---|---|
| NaCl | 30.0g |
| Peptone | 5.0g |
| Beef extract | 3.0g |
| Bromthymol Blue | 0.03g |
| Ethyl Violet | 1.0mg |
| Arabinose solution | 100.0mL |

pH 9.0 ± 0.2 at 25°C

**Arabinose Solution:**
**Composition** per 100.0mL:

| | |
|---|---|
| Arabinose | 5.0g |

**Preparation of Arabinose Solution:** Add arabinose to distilled/deionized water and bring volume to 100.0mL. Mix thoroughly. Filter sterilize.

**Preparation of Medium:** Add components, except arabinose solution, to distilled/deionized water and bring volume to 900.0mL. Mix thoroughly. Gently heat and bring to boiling. Adjust pH to 9.0. Autoclave for 15 min at 15 psi pressure–121°C. Cool to 45°–50°C. Aseptically add sterile arabinose solution. Mix thoroughly. Aseptically distribute into sterile tubes or flasks.

**Storage/Shelf Life:** Store dehydrated media in the dark in a sealed container below 30°C. Prepared media should be stored under refrigeration (2-8°C). Media should be used within 60 days of preparation. Media should not be used if there are any signs of deterioration (shrinking, cracking, or discoloration) or contamination, or if the expiration date supplied by the manufacturer has passed.

**Use:** For the cultivation of *Vibrio* species from foods.

## Hoyle Medium

**Composition** per 1060.0mL:

| | |
|---|---|
| Agar | 15.0g |
| Lab Lemco powder | 10.0g |
| Peptone | 10.0g |
| NaCl | 5.0g |
| Horse blood, laked | 50.0mL |
| Tellurite solution | 10.0mL |

pH 7.8 ± 0.2 at 25°C

**Source:** This medium is available as a premixed powder from Thermo Scientific.

**Horse Blood, Laked:**
**Composition** per 50.0mL:

| | |
|---|---|
| Horse blood, fresh | 50.0mL |

**Preparation of Horse Blood, Laked:** Add blood to a sterile polypropylene bottle. Freeze overnight at –20°C. Thaw at 8°C. Refreeze at –20°C. Thaw again at 8°C.

**Tellurite Solution:**
**Composition** per 100.0mL:

| | |
|---|---|
| K$_2$TeO$_3$ | 3.5g |

**Preparation of Tellurite Solution:** Add K$_2$TeO$_3$ to distilled/deionized water and bring volume to 100.0mL. Mix thoroughly. Filter sterilize.

**Caution:** Potassium tellurite is toxic.

**Preparation of Medium:** Add components, except laked horse blood and tellurite solution, to distilled/deionized water and bring volume to 1.0L. Mix thoroughly. Gently heat and bring to boiling. Autoclave for 15 min at 15 psi pressure–121°C. Cool to 50°C. Aseptically add 50.0mL sterile laked horse blood and 10.0mL sterile tellurite solution. Mix thoroughly. Pour into sterile Petri dishes or distribute into sterile tubes.

**Storage/Shelf Life:** Store dehydrated media in the dark in a sealed container below 30°C. Prepared media should be stored under refriger-

ation (2-8°C). Media should be used within 60 days of preparation. Media should not be used if there are any signs of deterioration (shrinking, cracking, or discoloration) or contamination, or if the expiration date supplied by the manufacturer has passed.

**Use:** For the isolation and differentiation of *Corynebacterium diphtheriae* strains. This medium permits very rapid growth of all types of *Corynebacterium diphtheriae*, so that diagnosis is possible after 18 hours' incubation.

### Hoyle HiVeg Medium Base

**Composition** per liter:
Agar .................................................................................15.0g
Plant extract ....................................................................10.0g
Plant peptone...................................................................10.0g
NaCl ...................................................................................5.0g
Horse blood, laked ........................................................ 50.0mL
Tellurite solution ........................................................... 10.0mL

pH 7.8 ± 0.2 at 25°C

**Source:** This medium, without tellurite solution and laked blood, is available as a premixed powder from HiMedia.

**Horse Blood, Laked:**
**Composition** per 50.0mL:
Horse blood, fresh...................................................... 50.0mL

**Preparation of Horse Blood, Laked:** Add blood to a sterile polypropylene bottle. Freeze overnight at −20°C. Thaw at 8°C. Refreeze at −20°C. Thaw again at 8°C.

**Tellurite Solution:**
**Composition** per 100.0mL:
K₂TeO₃.................................................................................3.5g

**Preparation of Tellurite Solution:** Add K₂TeO₃ to distilled/deionized water and bring volume to 100.0mL. Mix thoroughly. Filter sterilize.

**Caution:** Potassium tellurite is toxic.

**Preparation of Medium:** Add components, except laked horse blood and tellurite solution, to distilled/deionized water and bring volume to 1.0L. Mix thoroughly. Gently heat and bring to boiling. Autoclave for 15 min at 15 psi pressure–121°C. Cool to 50°C. Aseptically add 50.0mL sterile laked horse blood and 10.0mL sterile tellurite solution. Mix thoroughly. Pour into sterile Petri dishes or distribute into sterile tubes.

**Storage/Shelf Life:** Store dehydrated media in the dark in a sealed container below 30°C. Prepared media should be stored under refrigeration (2-8°C). Media should be used within 60 days of preparation. Media should not be used if there are any signs of deterioration (shrinking, cracking, or discoloration) or contamination, or if the expiration date supplied by the manufacturer has passed.

**Use:** For the isolation and differentiation of *Corynebacterium diphtheriae* strains. This medium permits very rapid growth of all types of *Corynebacterium diphtheriae*, so that diagnosis is possible after 18 hours' incubation.

### Hoyle Medium Base

**Composition** per liter:
Agar ...................................................................................15.0g
Beef extract......................................................................10.0g
Peptone.............................................................................10.0g

NaCl ....................................................................................5.0g
Blood, laked................................................................... 50.0mL
Tellurite solution ........................................................... 10.0mL

pH 7.8 ± 0.2 at 25°C

**Source:** This medium is available as a premixed powder from Thermo Scientific.

**Tellurite Solution:**
**Composition** per 100.0mL:
K₂TeO₃.................................................................................3.5g

**Caution:** Potassium tellurite is toxic.

**Preparation of Tellurite Solution:** Add K₂TeO₃ to distilled/deionized water and bring volume to 100.0mL. Mix thoroughly. Filter sterilize.

**Horse Blood, Laked:**
**Composition** per 50.0mL:
Horse blood, fresh........................................................ 50.0mL

**Preparation of Horse Blood, Laked:** Add blood to a sterile polypropylene bottle. Freeze overnight at −20°C. Thaw at 8°C. Refreeze at −20°C. Thaw again at 8°C.

**Preparation of Medium:** Add components, except laked blood, to distilled/deionized water and bring volume to 940.0mL. Mix thoroughly. Gently heat and bring to boiling. Autoclave for 15 min at 15 psi pressure–121°C. Cool to 45°–50°C. Aseptically add sterile laked blood. Mix thoroughly. Pour into sterile Petri dishes or distribute into sterile tubes.

**Storage/Shelf Life:** Store dehydrated media in the dark in a sealed container below 30°C. Prepared media should be stored under refrigeration (2-8°C). Media should be used within 60 days of preparation. Media should not be used if there are any signs of deterioration (shrinking, cracking, or discoloration) or contamination, or if the expiration date supplied by the manufacturer has passed.

**Use:** For the isolation and differentiation of *Corynebacterium diphtheriae*.

### HPC Agar
### (Heterotrophic Plate Count Agar)
### (m-HPC Agar)

**Composition** per liter:
Gelatin...............................................................................25.0g
Pancreatic digest of gelatin.............................................20.0g
Agar ...................................................................................15.0g
Glycerol ......................................................................... 10.0mL

pH 7.1 ± 0.2 at 25°C

**Source:** This medium is available from BD Diagnostic Systems.

**Preparation of Medium:** Add components, except glycerol, to distilled/deionized water and bring volume to 990.0mL. Mix thoroughly. Gently heat and bring to boiling. Add glycerol. Mix thoroughly. Autoclave for 15 min at 15 psi pressure–121°C. Cool to 45°–50°C. Pour into sterile Petri dishes.

**Storage/Shelf Life:** Store dehydrated media in the dark in a sealed container below 30°C. Prepared media should be stored under refrigeration (2-8°C). Media should be used within 60 days of preparation. Media should not be used if there are any signs of deterioration (shrinking, cracking, or discoloration) or contamination, or if the expiration date supplied by the manufacturer has passed.

**Use:** For the the cultivation and enumeration of microorganisms from potable water sources, swimming pools, and other water specimens by the membrane filter method and heterotrophic plate count technique.

## Hugh Leifson Glucose Broth

**Composition** per liter:

| | |
|---|---|
| NaCl | 30.0g |
| Glucose | 10.0g |
| Agar | 3.0g |
| Peptone | 2.0g |
| Yeast extract | 0.5g |
| Bromcresol Purple | 0.015g |

pH 7.4 ± 0.2 at 25°C

**Preparation of Medium:** Add components to distilled/deionized water and bring volume to 1.0L. Mix thoroughly. Gently heat while stirring and bring to boiling. Adjust pH to 7.4. Distibute into tubes or flasks. Autoclave for 15 min at 15 psi pressure–121°C.

**Storage/Shelf Life:** Store dehydrated media in the dark in a sealed container below 30°C. Prepared media should be stored under refrigeration (2-8°C). Media should be used within 60 days of preparation. Media should not be used if there are any signs of deterioration (discoloration) or contamination, or if the expiration date supplied by the manufacturer has passed.

**Use:** For the cultivation and differentiation of bacteria based on their ability to ferment glucose. Bacteria that ferment glucose turn the medium yellow.

## Hugh Leifson Glucose HiVeg Medium

**Composition** per liter:

| | |
|---|---|
| NaCl | 30.0g |
| Glucose | 10.0g |
| Agar | 3.0g |
| Plant peptone | 2.0g |
| Yeast extract | 0.5g |
| Bromcresol Purple | 0.015g |

pH 7.4 ± 0.2 at 25°C

**Source:** This medium is available as a premixed powder from Hi-Media.

**Preparation of Medium:** Add components to distilled/deionized water and bring volume to 1.0L. Mix thoroughly. Gently heat while stirring and bring to boiling. Adjust pH to 7.4. Distribute into tubes or flasks. Autoclave for 15 min at 15 psi pressure–121°C.

**Storage/Shelf Life:** Store dehydrated media in the dark in a sealed container below 30°C. Prepared media should be stored under refrigeration (2-8°C). Media should be used within 60 days of preparation. Media should not be used if there are any signs of deterioration (discoloration) or contamination, or if the expiration date supplied by the manufacturer has passed.

**Use:** For the cultivation and differentiation of bacteria based on their ability to ferment glucose. Bacteria that ferment glucose turn the medium yellow.

## Hugh Leifson HiVeg Medium

**Composition** per liter:

| | |
|---|---|
| Glucose | 10.0g |
| NaCl | 5.0g |

| | |
|---|---|
| Agar | 2.0g |
| Plant peptone | 2.0g |
| $K_2HPO_4$ | 0.3g |
| Bromthymol Blue | 0.05g |

pH 6.8 ± 0.2 at 25°C

**Source:** This medium is available as a premixed powder from Hi-Media.

**Preparation of Medium:** Add components to distilled/deionized water and bring volume to 1.0L. Mix thoroughly. Gently heat while stirring and bring to boiling. Adjust pH to 7.4. Distribute into tubes or flasks. Autoclave for 15 min at 15 psi pressure–121°C.

**Storage/Shelf Life:** Store dehydrated media in the dark in a sealed container below 30°C. Prepared media should be stored under refrigeration (2-8°C). Media should be used within 60 days of preparation. Media should not be used if there are any signs of deterioration (discoloration) or contamination, or if the expiration date supplied by the manufacturer has passed.

**Use:** For the cultivation and differentiation of bacteria based on their ability to ferment glucose. Bacteria that ferment glucose turn the medium yellow.

## Indole Medium

**Composition** per 200.0mL:

| | |
|---|---|
| $K_2HPO_4$ | 3.13g |
| L-Tryptophan | 1.0g |
| NaCl | 1.0g |
| $KH_2PO_4$ | 0.27g |

pH 7.2 ± 0.2 at 25°C

**Preparation of Medium:** Add components to distilled/deionized water and bring volume to 200.0mL. Mix thoroughly. Distribute into tubes or flasks. Autoclave for 15 min at 15 psi pressure–121°C.

**Storage/Shelf Life:** Store dehydrated media in the dark in a sealed container below 30°C. Prepared media should be stored under refrigeration (2-8°C). Media should be used within 60 days of preparation. Media should not be used if there are any signs of deterioration (discoloration) or contamination, or if the expiration date supplied by the manufacturer has passed.

**Use:** For the differentiation of microorganisms by means of indole production from the tryptophan test.

## Indole Medium

**Composition** per liter:

| | |
|---|---|
| Pancreatic digest of casein | 20.0g |

pH 7.3 ± 0.2 at 25°C

**Preparation of Medium:** Add pancreatic digest of casein to distilled/deionized water and bring volume to 1.0L. Mix thoroughly. Distribute into tubes or flasks. Autoclave for 15 min at 15 psi pressure–121°C.

**Storage/Shelf Life:** Store dehydrated media in the dark in a sealed container below 30°C. Prepared media should be stored under refrigeration (2-8°C). Media should be used within 60 days of preparation. Media should not be used if there are any signs of deterioration (discoloration) or contamination, or if the expiration date supplied by the manufacturer has passed.

**Use:** For the differentiation of microorganisms by means of the indole test.

## Indole Medium, CDC
## (BAM M65)

**Composition** per liter:

Pancreatic digest of casein.............................................20.0g

pH 7.3 ± 0.2 at 25°C

**Preparation of Medium:** Add pancreatic digest of casein to distilled/deionized water and bring volume to 1.0L. Mix thoroughly. Distribute into tubes or flasks. Autoclave for 15 min at 15 psi pressure–121°C.

**Storage/Shelf Life:** Store dehydrated media in the dark in a sealed container below 30°C. Prepared media should be stored under refrigeration (2-8°C). Media should be used within 60 days of preparation. Media should not be used if there are any signs of deterioration (discoloration) or contamination, or if the expiration date supplied by the manufacturer has passed.

**Use:** For the differentiation of microorganisms by means of the indole test.

## Indole Nitrate HiVeg Medium
## (Tryptone Nitrate HiVeg Medium)

**Composition** per liter:

Plant hydrolysate...........................................................20.0g
Na$_2$HPO$_4$.....................................................................2.0g
Agar .............................................................................1.0g
Glucose .........................................................................1.0g
Potassium nitrate............................................................1.0g

pH 7.2 ± 0.2 at 25°C

**Source:** This medium is available as a premixed powder from Hi-Media.

**Preparation of Medium:** Add components to distilled/deionized water and bring volume to 1.0L. Mix thoroughly. Gently heat and bring to boiling with frequent agitation. Distribute into tubes or flasks. Autoclave for 15 min at 15 psi pressure–121°C.

**Storage/Shelf Life:** Store dehydrated media in the dark in a sealed container below 30°C. Prepared media should be stored under refrigeration (2-8°C). Media should be used within 60 days of preparation. Media should not be used if there are any signs of deterioration (discoloration) or contamination, or if the expiration date supplied by the manufacturer has passed.

**Use:** For the identification of microorganisms by means of the nitrate reduction and indole tests.

## Indole Nitrate Medium
## (Trypticase™ Nitrate Broth)

**Composition** per liter:

Pancreatic digest of casein...........................................20.0g
Na$_2$HPO$_4$.....................................................................2.0g
Agar .............................................................................1.0g
Glucose .........................................................................1.0g
KNO$_3$ ...........................................................................1.0g

pH 7.2 ± 0.2 at 25°C

**Source:** This medium is available as a premixed powder from BD Diagnostic Systems.

**Preparation of Medium:** Add components to distilled/deionized water and bring volume to 1.0L. Mix thoroughly. Gently heat and bring to boiling with frequent agitation. Distribute into tubes or flasks. Autoclave for 15 min at 15 psi pressure–121°C.

**Storage/Shelf Life:** Store dehydrated media in the dark in a sealed container below 30°C. Prepared media should be stored under refrigeration (2-8°C). Media should be used within 60 days of preparation. Media should not be used if there are any signs of deterioration (discoloration) or contamination, or if the expiration date supplied by the manufacturer has passed.

**Use:** For the identification of microorganisms by means of the nitrate reduction and indole tests.

## Infection Medium
## (IM)

**Composition** per 100.0mL:

Pancreatic digest of gelatin...........................................0.05g
Bile salts No. 3.............................................................0.05g
Brain heart, solids from infusion ...................................0.02g
Peptic digest of animal tissue .......................................0.02g
NaCl............................................................................0.017g
Glucose .......................................................................0.01g
Na$_2$HPO$_4$....................................................................8.0mg
Earle's balanced salts solution .....................................80.0mL
Fetal bovine serum, heat inactivated (2 h at 55°C)...................20.0mL

pH 7.4 ± 0.2 at 25°C

**Earle's Balanced Salts Solution:**

**Composition** per liter:

NaCl.............................................................................6.8g
NaHCO$_3$.......................................................................2.2g
Glucose .........................................................................1.0g
KCl...............................................................................0.4g
CaCl$_2$·2H$_2$O ...............................................................0.265g
MgSO$_4$·7H$_2$O ...............................................................0.2g
NaH$_2$PO$_4$·H$_2$O...........................................................0.14g

**Preparation of Earle's Balanced Salts Solution:** Add components to distilled/deionized water and bring volume to 1.0L. Mix thoroughly. Filter sterilize.

**Preparation of Medium:** Combine components. Mix thoroughly. Filter sterilize. Store at 4°–10°C.

**Storage/Shelf Life:** Store dehydrated media in the dark in a sealed container below 30°C. Prepared media should be stored under refrigeration (2-8°C). Media should be used within 60 days of preparation. Media should not be used if there are any signs of deterioration (discoloration) or contamination, or if the expiration date supplied by the manufacturer has passed.

**Use:** For the screening of *Escherichia coli* for pathogenicity using the HeLa cell test for invasiveness.

## Infusion Cystine Agar Base, HiVeg

**Composition** per liter:

Agar ...........................................................................15.0g
Glucose .......................................................................10.0g
Plant infusion ..............................................................10.0g
Plant peptone No. 3.......................................................10.0g
NaCl..............................................................................5.0g
L-Cystine......................................................................1.0g

pH 7.2 ± 0.2 at 25°C

**Source:** This medium is available as a premixed powder from Hi-Media.

**Preparation of Medium:** Add components to distilled/deionized water and bring volume to 1.0L. Mix thoroughly. Gently heat and bring

to boiling with frequent agitation. Distribute into tubes or flasks. Autoclave for 15 min at 15 psi pressure–121°C. Pour into sterile Petri dishes.

**Storage/Shelf Life:** Store dehydrated media in the dark in a sealed container below 30°C. Prepared media should be stored under refrigeration (2-8°C). Media should be used within 60 days of preparation. Media should not be used if there are any signs of deterioration (shrinking, cracking, or discoloration) or contamination, or if the expiration date supplied by the manufacturer has passed.

**Use:** For the cultivation of Gram-negative cocci and other pathogenic organisms.

## Infusion Cystine Agar Base, HiVeg with Hemoglobin
**Composition** per liter:
| | |
|---|---|
| Agar | 15.0g |
| Glucose | 10.0g |
| Plant infusion | 10.0g |
| Plant peptone No. 3 | 10.0g |
| NaCl | 5.0g |
| L-Cystine | 1.0g |
| Hemoglobin solution, 2% | 100.0mL |

pH 7.2 ± 0.2 at 25°C

**Source:** This medium, without hemoglobin solution, is available as a premixed powder from HiMedia.

**Preparation of Medium:** Add components, except hemoglobin solution, to distilled/deionized water and bring volume to 900.0mL. Mix thoroughly. Gently heat and bring to boiling with frequent agitation. Distribute into tubes or flasks. Autoclave for 15 min at 15 psi pressure–121°C. Cool to 50°C. Aseptically add 100.0mL of sterile hemoglobin solution. Mix thoroughly. Pour into sterile Petri dishes.

**Storage/Shelf Life:** Store dehydrated media in the dark in a sealed container below 30°C. Prepared media should be stored under refrigeration (2-8°C). Media should be used within 60 days of preparation. Media should not be used if there are any signs of deterioration (shrinking, cracking, or discoloration) or contamination, or if the expiration date supplied by the manufacturer has passed.

**Use:** For the cultivation of Gram-negative cocci and other pathogenic organisms. With added hemoglobin it is used for cultivation of *Francisella tularensis*.

## Intracellular Growth Phase Medium (IGP Medium)
**Composition** per 100.0mL:
| | |
|---|---|
| Gentamicin sulfate | 500.0mg |
| Lysozyme | 30.0mg |
| Eagle MEM | 72.0mL |
| Dulbecco's phosphate-buffered saline | 20.0mL |
| Fetal bovine serum | 8.0mL |

pH 7.2–7.4 at 25°C

**Eagle MEM:**
**Composition** per liter:
| | |
|---|---|
| NaCl | 8.0g |
| Glucose | 1.0g |
| KCl | 0.4g |
| $CaCl_2 \cdot 2H_2O$ | 0.14g |
| $MgSO_4 \cdot 7H_2O$ | 0.1g |
| $KH_2PO_4$ | 0.06g |
| $Na_2HPO_4$ | 0.05g |
| L-Isoleucine | 0.026g |
| L-Leucine | 0.026g |
| L-Lysine | 0.026g |
| L-Threonine | 0.024g |
| L-Valine | 0.0235g |
| L-Tyrosine | 0.018g |
| L-Arginine | 0.0174g |
| L-Phenylalanine | 0.0165g |
| L-Cystine | 0.012g |
| L-Histidine | 8.0mg |
| L-Methionine | 7.5mg |
| Phenol Red | 5.0mg |
| L-Tryptophan | 4.0mg |
| Inositol | 1.8mg |
| Biotin | 1.0mg |
| Folic acid | 1.0mg |
| Calcium pantothenate | 1.0mg |
| Choline chloride | 1.0mg |
| Nicotinamide | 1.0mg |
| Pyridoxal·HCl | 1.0mg |
| Thiamine·HCl | 1.0mg |
| Riboflavin | 0.1mg |

**Preparation of Eagle MEM:** Add components to distilled/deionized water and bring volume to 1.0L. Mix thoroughly.

**Dulbecco's Phosphate-Buffered Saline:**
**Composition** per liter:
| | |
|---|---|
| NaCl | 8.0g |
| $Na_2HPO_4 \cdot 7H_2O$ | 2.16g |
| KCl | 0.2g |
| $KH_2PO_4$ | 0.2g |
| $CaCl_2$ | 0.1g |
| $MnCl_2 \cdot 6H_2O$ | 0.1g |

**Preparation of Dulbecco's Phosphate-Buffered Saline:** Add components to distilled/deionized water and bring volume to 1.0L. Mix thoroughly.

**Preparation of Medium:** Combine components. Mix thoroughly. Filter sterilize. Aseptically distribute into sterile tubes or flasks.

**Storage/Shelf Life:** Store dehydrated media in the dark in a sealed container below 30°C. Prepared media should be stored under refrigeration (2-8°C). Media should be used within 60 days of preparation. Media should not be used if there are any signs of deterioration (discoloration) or contamination, or if the expiration date supplied by the manufacturer has passed.

**Use:** For the screening of *Escherichia coli* for pathogenicity using the HeLa cell test for invasiveness.

## Ion Agar for *Ureaplasma*
**Composition** per 101.45mL:
| | |
|---|---|
| HEPES (*N*-[2-hydroxyethyl]piperazine-*N*′-[2-ethanesulfonic acid]) buffer | 1.19g |
| Ionagar No. 2 | 0.75g |
| Pancreatic digest of casein | 0.7g |
| NaCl | 0.5g |
| Beef extract | 0.3g |
| Yeast extract | 0.3g |
| Beef heart, solids from infusion | 0.2g |
| Yeast extract | 0.1g |
| Horse serum, normal sterile | 10.0mL |
| Ampicillin solution | 1.0mL |
| Urea solution | 0.25mL |

Nystatin solution ........................................................................ 0.1mL
Tripeptide solution ...................................................................... 0.1mL

pH 7.2 ± 0.2 at 25°C

**Ampicillin Solution:**
**Composition** per 10.0mL:
Ampicillin ....................................................................................1.0g

**Preparation of Ampicillin Solution:** Add ampicillin to distilled/deionized water and bring volume to 10.0mL. Mix thoroughly. Filter sterilize.

**Urea Solution:**
**Composition** per 100.0mL:
Urea ............................................................................................10.0g

**Preparation of Urea Solution:** Add urea to distilled/deionized water and bring volume to 100.0mL. Filter sterilize. Store at –20°C.

**Nystatin Solution:**
**Composition** per 1.0mL:
Nystatin ..................................................................................50,000U

**Preparation of Nystatin Solution:** Add nystatin to distilled/deionized water and bring volume to 1.0mL. Filter sterilize.

**Tripeptide Solution:**
**Composition** per 10.0mL:
Glycyl-L-histidyl-L-lysine acetate .................................................0.2mg

**Preparation of Tripeptide Solution:** Add glycyl-L-histidyl-L-lysine acetate to distilled/deionized water and bring volume to 10.0mL. Mix thoroughly. Filter sterilize. Store at –20°C.

**Preparation of Medium:** Add components—except horse serum, ampicillin solution, urea solution, nystatin solution, and tripeptide solution—to distilled/deionized water and bring volume to 90.0mL. Mix thoroughly. Gently heat and bring to boiling. Autoclave for 15 min at 15 psi pressure–121°C. Cool to 45°–50°C. Aseptically add 10.0mL of sterile horse serum, 1.0mL of sterile ampicillin solution, 0.25mL of sterile urea solution, 0.1mL of sterile nystatin solution, and 0.1mL of sterile tripeptide solution. Mix thoroughly. Pour into sterile Petri dishes.

**Storage/Shelf Life:** Store dehydrated media in the dark in a sealed container below 30°C. Prepared media should be stored under refrigeration (2-8°C). Media should be used within 60 days of preparation. Media should not be used if there are any signs of deterioration (shrinking, cracking, or discoloration) or contamination, or if the expiration date supplied by the manufacturer has passed.

**Use:** For the cultivation of *Ureaplasma* species from clinical specimens.

## Irgasan® Ticarcillin Chlorate Broth (ITC Broth)

**Composition** per liter:
MgCl$_2$·6H$_2$0 ............................................................................60.0g
Pancreatic digest of casein............................................................10.0g
NaCl............................................................................................5.0g
KClO$_4$ ........................................................................................1.0g
Yeast extract................................................................................1.0g
Malachite Green (0.2% solution)....................................................5.0mL
Irgasan solution ...........................................................................1.0mL
Ticarcillin solution ........................................................................1.0mL

pH 7.6 ± 0.2 at 25°C

**Irgasan Solution:**
**Composition** per 10.0mL:
Irgasan (triclosan) .........................................................................1.0mg

**Preparation of Irgasan Solution:** Add Irgasan to distilled/deionized water and bring volume to 10.0mL. Mix thoroughly. Filter sterilize.

**Ticarcillin Solution:**
**Composition** per 10.0mL:
Ticarcillin.....................................................................................1.0 mg

**Preparation of Ticarcillin Solution:** Add ticarcillin to distilled/deionized water and bring volume to 10.0mL. Mix thoroughly. Filter sterilize.

**Preparation of Medium:** Add components, except Irgasan solution and ticarcillin solution, to distilled/deionized water and bring volume to 998.0mL. Mix thoroughly. Autoclave for 15 min at 15 psi pressure–121°C. Cool to 45°–50°C. Adjust to pH 7.6. Aseptically add 1.0mL of Irgasan solution and 1.0mL of ticarcillin solution. Mix thoroughly. Aseptically distribute into sterile tubes or flasks.

**Storage/Shelf Life:** Store dehydrated media in the dark in a sealed container below 30°C. Prepared media should be stored under refrigeration (2-8°C). Media should be used within 60 days of preparation. Media should not be used if there are any signs of deterioration (discoloration) or contamination, or if the expiration date supplied by the manufacturer has passed.

**Use:** For the selective isolation and cultivation of *Yersinia* species.

## Iso-Sensitest Agar

**Composition** per liter:
Casein, hydrolyzed ........................................................................11.0g
Agar ...........................................................................................8.0g
Peptones......................................................................................3.0g
NaCl............................................................................................3.0g
Na$_2$HPO$_4$....................................................................................2.0g
Glucose........................................................................................2.0g
Sodium acetate..............................................................................1.0g
Soluble starch................................................................................1.0g
Magnesium glycerophosphate .........................................................0.2g
Calcium gluconate .........................................................................0.1g
L-Cysteine·HCl..............................................................................0.02g
L-Tryptophan................................................................................0.02g
Adenine.......................................................................................0.01g
Guanine.......................................................................................0.01g
Xanthine.......................................................................................0.01g
Uracil..........................................................................................0.01g
Nicotinamide.................................................................................3.0mg
Pantothenate.................................................................................3.0mg
Pyridoxine....................................................................................3.0mg
MnCl$_2$·4H$_2$O................................................................................2.0mg
CoSO$_4$ .......................................................................................1.0mg
CuSO$_4$·5H$_2$O...............................................................................1.0mg
FeSO$_4$·7H$_2$O...............................................................................1.0mg
Menadione...................................................................................1.0mg
Cyanocobalamin............................................................................1.0mg
ZnSO$_4$·7H$_2$O...............................................................................1.0mg
Biotin .........................................................................................0.3mg
Thiamine......................................................................................0.04mg

pH 7.4 ± 0.2 at 25°C

**Preparation of Medium:** Add components to distilled/deionized water and bring volume to 1.0L. Mix thoroughly. Gently heat and bring

to boiling. Distribute into tubes or flasks. Autoclave for 15 min at 15 psi pressure–121°C. Pour into sterile Petri dishes or leave in tubes.

**Storage/Shelf Life:** Store dehydrated media in the dark in a sealed container below 30°C. Prepared media should be stored under refrigeration (2-8°C). Media should be used within 60 days of preparation. Media should not be used if there are any signs of deterioration (shrinking, cracking, or discoloration) or contamination, or if the expiration date supplied by the manufacturer has passed.

**Use:** For antimicrobial susceptibility testing.

## Iso Sensitest Broth

**Composition** per liter:

| | |
|---|---|
| Casein, hydrolyzed | 11.0g |
| Peptones | 3.0g |
| NaCl | 3.0g |
| Glucose | 2.0g |
| $Na_2HPO_4$ | 2.0g |
| Sodium acetate | 1.0g |
| Soluble starch | 1.0g |
| Magnesium glycerophosphate | 0.2g |
| Calcium gluconate | 0.1g |
| L-cysteine·HCl | 0.02g |
| L-Tryptophan | 0.02g |
| Adenine | 0.01g |
| Guanine | 0.01g |
| Xanthine | 0.01g |
| Uracil | 0.01g |
| Nicotinamide | 3.0mg |
| Pantothenate | 3.0mg |
| Pyridoxine | 3.0mg |
| $MnCl_2 \cdot 4H_2O$ | 2.0mg |
| $CoSO_4$ | 1.0mg |
| $CuSO_4 \cdot 5H_2O$ | 1.0mg |
| $FeSO_4 \cdot 7H_2O$ | 1.0mg |
| Menadione | 1.0mg |
| Cyanocobalamin | 1.0mg |
| $ZnSO_4 \cdot 7H_2O$ | 1.0mg |
| Biotin | 0.3mg |
| Thiamine | 0.04mg |

pH 7.4 ± 0.2 at 25°C

**Preparation of Medium:** Add components to distilled/deionized water and bring volume to 1.0L. Mix thoroughly. Gently heat and bring to boiling. Distribute into tubes or flasks. Autoclave for 15 min at 15 psi pressure–121°C. Pour into sterile Petri dishes or leave in tubes.

**Storage/Shelf Life:** Store dehydrated media in the dark in a sealed container below 30°C. Prepared media should be stored under refrigeration (2-8°C). Media should be used within 60 days of preparation. Media should not be used if there are any signs of deterioration (discoloration) or contamination, or if the expiration date supplied by the manufacturer has passed.

**Use:** For antimicrobial susceptibility testing.

## Jones–Kendrick Pertussis Transport Medium

**Composition** per liter:

| | |
|---|---|
| Beef heart, solids from infusion | 500.0g |
| Agar | 20.0g |
| Soluble starch | 10.0g |
| Tryptose | 10.0g |
| NaCl | 5.0g |

| | |
|---|---|
| Charcoal powder, activated | 4.0g |
| Yeast extract | 3.5g |
| Penicillin solution | 10.0mL |

pH 7.4 ± 0.2 at 25°C

**Penicillin Solution:**
**Composition** per 10.0mL:

| | |
|---|---|
| Penicillin | 300U |

**Preparation of Penicillin Solution:** Add penicillin to distilled/deionized water and bring volume to 10.0mL. Mix thoroughly. Filter sterilize.

**Preparation of Medium:** Add components, except penicillin solution, starch, yeast extract, heart infusion, and agar, to water. Boil to dissolve. Add charcoal, mix well, and autoclave. Cool to 50°C, add penicillin, and dispense into small bottles as slants. Cool and seal tightly. Store at 5°C. Stable for 2 to 3 months.

**Storage/Shelf Life:** Store dehydrated media in the dark in a sealed container below 30°C. Prepared media should be stored under refrigeration (2-8°C). Media should be used within 60 days of preparation. Media should not be used if there are any signs of deterioration (shrinking, cracking, or discoloration) or contamination, or if the expiration date supplied by the manufacturer has passed.

**Use:** For the cultivation and transport of *Bordetella pertussis* between clinical isolation and laboratory cultivation.

## Jordan's Tartrate Agar

**Composition** per liter:

| | |
|---|---|
| Agar | 15.0g |
| Pancreatic digest of casein | 10.0g |
| Sodium potassium tartrate | 10.0g |
| NaCl | 5.0g |
| Phenol Red | 0.024g |

pH 7.7 ± 0.3 at 25°C

**Source:** This medium is available as a prepared medium in tubes from BD Diagnostic Systems.

**Preparation of Medium:** Add components to distilled/deionized water and bring volume to 1.0L. Mix thoroughly. Gently heat and bring to boiling. Adjust pH to 7.7. Distribute into tubes. Autoclave for 15 min at 15 psi pressure–121°C.

**Storage/Shelf Life:** Store dehydrated media in the dark in a sealed container below 30°C. Prepared media should be stored under refrigeration (2-8°C). Media should be used within 60 days of preparation. Media should not be used if there are any signs of deterioration (shrinking, cracking, or discoloration) or contamination, or if the expiration date supplied by the manufacturer has passed.

**Use:** For the differentiation and identification of members of the Enterobacteriaceae, especially *Salmonella* species, based upon the ability to utilize tartrate. Utilization of tartrate turns the medium yellow. *Salmonella enteritidis* utilizes tartrate. *Salmonella paratyphi* A does not utilize tartrate.

## Kanamycin Esculin Azide Agar

**Composition** per liter:

| | |
|---|---|
| Pancreatic digest of casein | 20.0g |
| Agar | 10.0g |
| NaCl | 5.0g |
| Yeast extract | 5.0g |
| Esculin | 1.0g |
| Sodium citrate | 1.0g |

Ferric ammonium citrate................................................0.5g
NaN₃..........................................................................0.15g
Kanamycin sulfate solution ....................................... 10.0mL
pH 7.0 ± 0.2 at 25°C

**Caution:** Sodium azide is toxic. Azides also react with metals and disposal must be highly diluted.

**Source:** This medium is available as a premixed powder from Thermo Scientific.

**Kanamycin Sulfate Solution:**
**Composition** per 10.0mL:
Kanamycin sulfate ....................................................20.0mg

**Preparation of Kanamycin Sulfate Solution:** Add kanamycin sulfate to distilled/deionized water and bring volume to 10.0mL. Mix thoroughly.

**Preparation of Medium:** Add components to distilled/deionized water and bring volume to 1.0L. Mix thoroughly. Gently heat and bring to boiling. Distribute into tubes or flasks. Autoclave for 15 min at 15 psi pressure–121°C. Pour into sterile Petri dishes or leave in tubes.

**Storage/Shelf Life:** Store dehydrated media in the dark in a sealed container below 30°C. Prepared media should be stored under refrigeration (2-8°C). Media should be used within 60 days of preparation. Media should not be used if there are any signs of deterioration (shrinking, cracking, or discoloration) or contamination, or if the expiration date supplied by the manufacturer has passed.

**Use:** For the isolation of enterococci from foods.

## Kanamycin Esculin Azide Broth

**Composition** per liter:
Pancreatic digest of casein............................................20.0g
NaCl..........................................................................5.0g
Yeast extract................................................................5.0g
Esculin .....................................................................1.0g
Sodium citrate..............................................................1.0g
Ferric ammonium citrate................................................0.5g
NaN₃..........................................................................0.15g
Kanamycin sulfate solution ....................................... 10.0mL
pH 7.0 ± 0.2 at 25°C

**Caution:** Sodium azide is toxic. Azides also react with metals and disposal must be highly diluted.

**Source:** This medium is available as a premixed powder from Thermo Scientific.

**Kanamycin Sulfate Solution:**
**Composition** per 10.0mL:
Kanamycin sulfate ....................................................0.02g

**Preparation of Kanamycin Sulfate Solution:** Add kanamycin sulfate to distilled/deionized water and bring volume to 10.0mL. Mix thoroughly.

**Preparation of Medium:** Add components to distilled/deionized water and bring volume to 1.0L. Mix thoroughly. Distribute into tubes or flasks. Autoclave for 15 min at 15 psi pressure–121°C.

**Storage/Shelf Life:** Store dehydrated media in the dark in a sealed container below 30°C. Prepared media should be stored under refrigeration (2-8°C). Media should be used within 60 days of preparation. Media should not be used if there are any signs of deterioration (discoloration) or contamination, or if the expiration date supplied by the manufacturer has passed.

**Use:** For the isolation of enterococci from foods.

## Kanamycin Esculin Azide HiVeg Agar

**Composition** per liter:
Plant hydrolysate .........................................................20.0g
Agar ..........................................................................12.0g
NaCl..........................................................................5.0g
Yeast extract................................................................5.0g
Esculin .....................................................................1.0g
Sodium citrate..............................................................1.0g
Ferric ammonium citrate................................................0.5g
NaN₃ .........................................................................0.15g
Kanamycin sulfate .......................................................0.02g
pH 7.0 ± 0.2 at 25°C

**Caution:** Sodium azide is toxic. Azides also react with metals and disposal must be highly diluted.

**Source:** This medium is available as a premixed powder from Hi-Media.

**Preparation of Medium:** Add components to distilled/deionized water and bring volume to 1.0L. Mix thoroughly. Distribute into tubes or flasks. Autoclave for 15 min at 15 psi pressure–121°C.

**Storage/Shelf Life:** Store dehydrated media in the dark in a sealed container below 30°C. Prepared media should be stored under refrigeration (2-8°C). Media should be used within 60 days of preparation. Media should not be used if there are any signs of deterioration (shrinking, cracking, or discoloration) or contamination, or if the expiration date supplied by the manufacturer has passed.

**Use:** For the isolation of enterococci from foods.

## Kanamycin Esculin Azide HiVeg Agar Base
## with Kanamycin

**Composition** per liter:
Plant hydrolysate .........................................................20.0g
Agar ..........................................................................10.0g
Yeast extract................................................................5.0g
NaCl..........................................................................5.0g
Esculin .....................................................................1.0g
Sodium citrate..............................................................1.0g
Ferric ammonium citrate................................................0.5g
NaN₃ .........................................................................0.15g
Kanamycin sulfate solution ....................................... 10.0mL
pH 7.0 ± 0.2 at 25°C

**Source:** This medium, without kanamycin sulfate solution, is available as a premixed powder from HiMedia.

**Caution:** Sodium azide is toxic. Azides also react with metals and disposal must be highly diluted.

**Kanamycin Sulfate Solution:**
**Composition** per 10.0mL:
Kanamycin sulfate ....................................................0.02g

**Preparation of Kanamycin Sulfate Solution:** Add kanamycin sulfate to distilled/deionized water and bring volume to 10.0mL. Mix thoroughly.

**Preparation of Medium:** Add components to distilled/deionized water and bring volume to 1.0L. Mix thoroughly. Distribute into tubes or flasks. Autoclave for 15 min at 15 psi pressure–121°C. Pour into sterile Petri dishes or leave in tubes.

**Use:** For the isolation of enterococci from foods.

## Kanamycin Esculin Azide HiVeg Broth

**Composition** per liter:

| | |
|---|---|
| Plant hydrolysate | 20.0g |
| NaCl | 5.0g |
| Yeast extract | 5.0g |
| Esculin | 1.0g |
| Sodium citrate | 1.0g |
| Ferric ammonium citrate | 0.5g |
| NaN$_3$ | 0.15g |
| Kanamycin sulfate | 0.02g |

pH 7.0 ± 0.2 at 25°C

**Caution:** Sodium azide is toxic. Azides also react with metals and disposal must be highly diluted.

**Source:** This medium is available as a premixed powder from Hi-Media.

**Preparation of Medium:** Add components to distilled/deionized water and bring volume to 1.0L. Mix thoroughly. Distribute into tubes or flasks. Autoclave for 15 min at 15 psi pressure–121°C.

**Storage/Shelf Life:** Store dehydrated media in the dark in a sealed container below 30°C. Prepared media should be stored under refrigeration (2-8°C). Media should be used within 60 days of preparation. Media should not be used if there are any signs of deterioration (shrinking, cracking, or discoloration) or contamination, or if the expiration date supplied by the manufacturer has passed.

**Use:** For the cultivation of enterococci from foods.

## Kanamycin Esculin Azide HiVeg Broth Base with Kanamycin

**Composition** per liter:

| | |
|---|---|
| Plant hydrolysate | 20.0g |
| NaCl | 5.0g |
| Yeast extract | 5.0g |
| Esculin | 1.0g |
| Sodium citrate | 1.0g |
| Ferric ammonium citrate | 0.5g |
| NaN$_3$ | 0.15g |
| Kanamycin sulfate solution | 10.0mL |

pH 7.0 ± 0.2 at 25°C

**Caution:** Sodium azide is toxic. Azides also react with metals and disposal must be highly diluted.

**Source:** This medium, without kanamycin sulfate solution, is available as a premixed powder from HiMedia.

**Kanamycin Sulfate Solution:**
**Composition** per 10.0mL:

| | |
|---|---|
| Kanamycin sulfate | 0.02g |

**Preparation of Kanamycin Sulfate Solution:** Add kanamycin sulfate to distilled/deionized water and bring volume to 10.0mL. Mix thoroughly.

**Preparation of Medium:** Add components to distilled/deionized water and bring volume to 1.0L. Mix thoroughly. Distribute into tubes or flasks. Autoclave for 15 min at 15 psi pressure–121°C.

**Storage/Shelf Life:** Store dehydrated media in the dark in a sealed container below 30°C. Prepared media should be stored under refrigeration (2-8°C). Media should be used within 60 days of preparation. Media should not be used if there are any signs of deterioration (discoloration) or contamination,.

**Use:** For the cultivation of enterococci from foods.

## Kanamycin Vancomycin Blood Agar (KVBA)

**Composition** per liter:

| | |
|---|---|
| Agar | 17.5g |
| Pancreatic digest of casein | 15.0g |
| Papaic digest of soybean meal | 5.0g |
| NaCl | 5.0g |
| Kanamycin | 0.1g |
| Sheep blood, defibrinated | 50.0mL |
| Vancomycin solution | 10.0mL |
| Vitamin K$_1$ solution | 1.0mL |

**Vancomycin Solution:**
**Composition** per 10.0mL:

| | |
|---|---|
| Vancomycin | 7.5mg |

**Preparation of Vancomycin Solution:** Add vancomycin to distilled/deionized water and bring volume to 10.0mL. Mix thoroughly. Filter sterilize.

**Vitamin K$_1$ Solution:**
**Composition** per 100.0mL:

| | |
|---|---|
| Vitamin K$_1$ | 1.0g |

**Preparation of Vitamin K$_1$ Solution:** Add vitamin K$_1$ to 99.0mL of absolute ethanol. Mix thoroughly. Filter sterilize.

**Preparation of Medium:** Add components, except sheep blood, vancomycin solution, and vitamin K$_1$ solution, to distilled/deionized water and bring volume to 939.0mL. Mix thoroughly. Gently heat and bring to boiling. Autoclave for 15 min at 15 psi pressure–121°C. Cool to 45°–50°C. Aseptically add sheep blood, vancomycin solution, and 1.0mL vitamin K$_1$ solution. Mix thoroughly. Pour into sterile Petri dishes or distribute into sterile tubes.

**Storage/Shelf Life:** Store dehydrated media in the dark in a sealed container below 30°C. Prepared media should be stored under refrigeration (2-8°C). Media should be used within 60 days of preparation. Media should not be used if there are any signs of deterioration (shrinking, cracking, or discoloration) or contamination, or if the expiration date supplied by the manufacturer has passed.

**Use:** For the selective isolation of anaerobes, particularly *Bacteroides*, from clinical specimens.

## Kanamycin Vancomycin Laked Blood Agar

**Composition** per liter:

| | |
|---|---|
| Agar | 17.5g |
| Pancreatic digest of casein | 15.0g |
| Papaic digest of soybean meal | 5.0g |
| NaCl | 5.0g |
| Kanamycin | 0.075g |
| Sheep blood, laked | 50.0mL |
| Vancomycin solution | 10.0mL |
| Vitamin K$_1$ solution | 1.0mL |

**Vancomycin Solution:**
**Composition** per 10.0mL:

| | |
|---|---|
| Vancomycin | 7.5mg |

**Preparation of Vancomycin Solution:** Add vancomycin to distilled/deionized water and bring volume to 10.0mL. Mix thoroughly. Filter sterilize.

**Vitamin K$_1$ Solution:**
**Composition** per 100.0mL:

| | |
|---|---|
| Vitamin K$_1$ | 1.0g |

**Preparation of Vitamin K₁ Solution:** Add vitamin K₁ to 99.0mL of absolute ethanol. Mix thoroughly. Filter sterilize.

**Preparation of Medium:** The blood is laked (hemolyzed) by freezing whole blood overnight and then thawing. Add components, except sheep blood, vancomycin solution, and vitamin K₁ solution, to distilled/deionized water and bring volume to 939.0mL. Mix thoroughly. Gently heat and bring to boiling. Autoclave for 15 min at 15 psi pressure–121°C. Cool to 45°–50°C. Aseptically add sheep blood, vancomycin solution, and 1.0mL of vitamin K₁ solution. Mix thoroughly. Pour into sterile Petri dishes or distribute into sterile tubes.

**Storage/Shelf Life:** Store dehydrated media in the dark in a sealed container below 30°C. Prepared media should be stored under refrigeration (2-8°C). Media should be used within 60 days of preparation. Media should not be used if there are any signs of deterioration (shrinking, cracking, or discoloration) or contamination, or if the expiration date supplied by the manufacturer has passed.

**Use:** For isolation of the *Bacteroides melaninogenicus* group.

## Kasai Medium

**Composition** per liter:

| | |
|---|---|
| Pancreatic digest of casein | 20.0g |
| Soluble starch | 20.0g |
| L-Cysteine·HCl·H₂O | 5.0g |
| K₂HPO₄ | 5.0g |
| NaCl | 5.0g |
| Yeast extract | 2.0g |

**Preparation of Medium:** Add components to distilled/deionized water and bring volume to 1.0L. Mix thoroughly. Gently heat and bring to boiling. Distribute into tubes or flasks. Autoclave for 15 min at 15 psi pressure–121°C.

**Storage/Shelf Life:** Store dehydrated media in the dark in a sealed container below 30°C. Prepared media should be stored under refrigeration (2-8°C). Media should be used within 60 days of preparation. Media should not be used if there are any signs of deterioration (discoloration) or contamination, or if the expiration date supplied by the manufacturer has passed.

**Use:** For the isolation and cultivation of *Leptotrichia buccalis* from saliva and plaque.

## Kelly Medium, Nonselective Modified

**Composition** per 1430.0mL:

| | |
|---|---|
| HEPES buffer (*N*-2-hydroxyethylpiperazine-*N*-2-ethanesulfonic acid) | 6.0g |
| Proteose peptone No. 2 | 5.0g |
| D-Glucose | 3.0g |
| NaHCO₃ | 2.2g |
| Pancreatic digest of casein | 1.0g |
| Yeast, autolyzed | 1.0g |
| Sodium pyruvate | 0.8g |
| Sodium citrate | 0.7g |
| *N*-Acetylglucosamine | 0.4g |
| MgCl₂·6H₂O | 0.3g |
| Gelatin solution | 200.0mL |
| Bovine serum albumin solution | 143.0mL |
| CMRL-1066 medium with glutamine, 10X | 100.0mL |
| Rabbit serum, heat inactivated | 86.0mL |
| Hemin solution | 1.0mL |

pH 7.2 ± 0.2 at 25°C

**CMRL-1066 Medium with Glutamine, 10X:**
**Composition** per liter:

| | |
|---|---|
| NaCl | 6.8g |
| NaHCO₃ | 2.2g |
| D-Glucose | 1.0g |
| KCl | 0.4g |
| L-Cysteine·HCl·H₂O | 0.26g |
| CaCl₂, anhydrous | 0.2g |
| MgSO₄·7H₂O | 0.2g |
| NaH₂PO₄·H₂O | 0.14g |
| L-Glutamine | 0.1g |
| Sodium acetate·3H₂O | 0.083g |
| L-Glutamic acid | 0.075g |
| L-Arginine·HCl | 0.07g |
| L-Lysine·HCl | 0.07g |
| L-Leucine | 0.06g |
| Glycine | 0.05g |
| Ascorbic acid | 0.05g |
| L-Proline | 0.04g |
| L-Tyrosine | 0.04g |
| L-Aspartic acid | 0.03g |
| L-Threonine | 0.03g |
| L-Alanine | 0.025g |
| L-Phenylalanine | 0.025g |
| L-Serine | 0.025g |
| L-Valine | 0.025g |
| L-Cystine | 0.02g |
| L-Histidine·HCl·H₂O | 0.02g |
| L-Isoleucine | 0.02g |
| Phenol Red | 0.02g |
| L-Methionine | 0.015g |
| Deoxyadenosine | 0.01g |
| Deoxycytidine | 0.01g |
| Deoxyguanosine | 0.01g |
| Glutathione, reduced | 0.01g |
| Thymidine | 0.01g |
| Hydroxy-L-proline | 0.01g |
| L-Tryptophan | 0.01g |
| Nicotinamide adenine dinucleotide | 7.0mg |
| Tween™ 80 | 5.0mg |
| Sodium glucuronate·H₂O | 4.2mg |
| Coenzyme A | 2.5mg |
| Cocarboxylase | 1.0mg |
| Flavin adenine dinucleotide | 1.0mg |
| Nicotinamide adenine dinucleotide phosphate | 1.0mg |
| Uridine triphosphate | 1.0mg |
| Choline chloride | 0.5mg |
| Cholesterol | 0.2mg |
| 5-Methyldeoxycytidine | 0.1mg |
| Inositol | 0.05mg |
| *p*-Aminobenzoic acid | 0.05mg |
| Niacin | 0.025mg |
| Niacinamide | 0.025mg |
| Pyridoxine | 0.025mg |
| Pyridoxal·HCl | 0.025mg |
| Biotin | 0.01mg |
| D-Calcium pantothenate | 0.01mg |
| Folic acid | 0.01mg |
| Riboflavin | 0.01mg |
| Thiamine·HCl | 0.01mg |

pH 7.2 ± 0.2 at 25°C

**Source:** This solution is available as a premixed powder from BD Diagnostics.

**Preparation of CMRL-1066 Medium with Glutamine, 10X:** Add components to distilled/deionized water and bring volume to 1.0L. Mix thoroughly. Adjust pH to 7.2. Filter sterilize.

**Gelatin Solution:**
**Composition** per 200.0mL:
Gelatin.............................................................................14.0g

**Preparation of Gelatin Solution:** Add gelatin to distilled/deionized water and bring volume to 200.0mL. Mix thoroughly. Gently heat and bring to boiling. Autoclave for 15 min at 15 psi pressure–121°C. Cool to 50°C.

**Hemin Solution:**
**Composition** per 100.0mL:
Hemin...............................................................................1.0g
NaOH (1*N* solution)..................................................20.0mL

**Preparation of Hemin Solution:** Add hemin to 20.0mL of 1*N* NaOH solution. Mix thoroughly. Bring volume to 100.0mL with distilled/deionized water.

**Bovine Serum Albumin Solution:**
**Composition** per 200.0mL:
Bovine serum albumin...............................................70.0g

**Preparation of Bovine Serum Albumin Solution:** Add bovine serum albumin to distilled/deionized water and bring volume to 200.0mL. Filter sterilize.

**Preparation of Medium:** Add components, except gelatin solution, bovine serum albumin solution, and rabbit serum, to distilled/deionized water and bring volume to 1001.0mL. Mix thoroughly. Bring pH to 7.6 with 5*N* NaOH. Filter sterilize. Aseptically add 200.0mL of sterile gelatin solution, 143.0mL of sterile bovine serum albumin, and 86.0mL of sterile heat-inactivated rabbit serum. Mix thoroughly. Aseptically dispense into sterile tubes or flasks.

**Storage/Shelf Life:** Store dehydrated media in the dark in a sealed container below 30°C. Prepared media should be stored under refrigeration (2-8°C). Media should be used within 60 days of preparation. Media should not be used if there are any signs of deterioration (discoloration) or contamination, or if the expiration date supplied by the manufacturer has passed.

**Use:** For the isolation of *Borrelia burgdorferi* and other spirochetes.

## Kelly Medium, Selective Modified
**Composition** per 1270.0mL:
Bovine serum albumin fraction V.............................50.0g
HEPES buffer (*N*-2-hydroxyethylpiperazine-
   *N*-2-ethanesulfonic acid) ........................................6.0g
Glucose .........................................................................5.0g
Neopeptone ..................................................................5.0g
NaHCO$_3$ ......................................................................2.2g
Sodium pyruvate ..........................................................0.8g
Sodium citrate ..............................................................0.7g
*N*-Acetylglucosamine...................................................0.4g
Kanamycin ................................................................8.0mg
5-Fluorouracil............................................................2.3mg
Gelatin solution.......................................................200.0mL
CMRL-1066 medium with glutamine, 10X...........100.0mL
Rabbit serum, partially hemolyzed ..........................70.0mL
pH 7.7 ± 0.2 at 25°C

**Gelatin Solution:**
**Composition** per 200.0mL:
Gelatin.............................................................................14.0g

**Preparation of Gelatin Solution:** Add gelatin to distilled/deionized water and bring volume to 1.0L. Mix thoroughly. Gently heat and bring to boiling. Autoclave for 15 min at 15 psi pressure–121°C. Cool to 50°C.

**CMRL-1066 Medium with Glutamine, 10X:**
**Composition** per liter:
NaCl..............................................................................6.8g
NaHCO$_3$........................................................................2.2g
D-Glucose......................................................................1.0g
KCl................................................................................0.4g
L-Cysteine·HCl·H$_2$O ..................................................0.26g
CaCl$_2$, anhydrous.........................................................0.2g
MgSO$_4$·7H$_2$O................................................................0.2g
NaH$_2$PO$_4$·H$_2$O.............................................................0.14g
L-Glutamine ................................................................0.1g
Sodium acetate·3H$_2$O...............................................0.083g
L-Glutamic acid.........................................................0.075g
L-Arginine·HCl............................................................0.07g
L-Lysine·HCl...............................................................0.07g
L-Leucine.....................................................................0.06g
Glycine.........................................................................0.05g
Ascorbic acid...............................................................0.05g
L-Proline......................................................................0.04g
L-Tyrosine....................................................................0.04g
L-Aspartic acid............................................................0.03g
L-Threonine.................................................................0.03g
L-Alanine...................................................................0.025g
L-Phenylalanine.........................................................0.025g
L-Serine.....................................................................0.025g
L-Valine.....................................................................0.025g
L-Cystine......................................................................0.02g
L-Histidine·HCl·H$_2$O..................................................0.02g
L-Isoleucine.................................................................0.02g
Phenol Red...................................................................0.02g
L-Methionine.............................................................0.015g
Deoxyadenosine...........................................................0.01g
Deoxycytidine..............................................................0.01g
Deoxyguanosine...........................................................0.01g
Glutathione, reduced....................................................0.01g
Thymidine....................................................................0.01g
Hydroxy-L-proline........................................................0.01g
L-Tryptophan...............................................................0.01g
Nicotinamide adenine dinucleotide...........................7.0mg
Tween™ 80.................................................................5.0mg
Sodium glucuronate·H$_2$O..........................................4.2mg
Coenzyme A...............................................................2.5mg
Cocarboxylase............................................................1.0mg
Flavin adenine dinucleotide.......................................1.0mg
Nicotinamide adenine dinucleotide phosphate ..........1.0mg
Uridine triphosphate ..................................................1.0mg
Choline chloride.........................................................0.5mg
Cholesterol.................................................................0.2mg
5-Methyldeoxycytidine...............................................0.1mg
Inositol.....................................................................0.05mg
*p*-Aminobenzoic acid..............................................0.05mg
Niacin.....................................................................0.025mg
Niacinamide...........................................................0.025mg
Pyridoxine.............................................................0.025mg

Pyridoxal·HCl ...............................................................0.025mg
Biotin ...........................................................................0.01mg
D-Calcium pantothenate ...............................................0.01mg
Folic acid.....................................................................0.01mg
Riboflavin ...................................................................0.01mg
Thiamine·HCl ..............................................................0.01mg
pH 7.2 ± 0.2 at 25°C

**Source:** This solution is available as a premixed powder from BD Diagnostics.

**Preparation of CMRL-1066 Medium with Glutamine, 10X:** Add components to distilled/deionized water and bring volume to 1.0L. Mix thoroughly. Adjust pH to 7.2. Filter sterilize.

**Preparation of Medium:** Add components, except gelatin solution, partially hemolyzed rabbit serum, kanamycin, and 5-fluorouracil, to distilled/deionized water and bring volume to 1.0L. Mix thoroughly. Bring pH to 7.6 with 5*N* NaOH. Filter sterilize. Aseptically add 200.0mL of sterile gelatin solution, 70.0mL of partially hemolyzed rabbit serum, 8.0mg of kanamycin, and 230.0mg of 5-fluorouracil. Mix thoroughly. Aseptically distribute into sterile tubes or flasks.

**Storage/Shelf Life:** Store dehydrated media in the dark in a sealed container below 30°C. Prepared media should be stored under refrigeration (2-8°C). Media should be used within 60 days of preparation. Media should not be used if there are any signs of deterioration (discoloration) or contamination, or if the expiration date supplied by the manufacturer has passed.

**Use:** For the isolation of *Borrelia burgdorferi*.

## KF Streptococcal Agar Base with Triphyenyltetrazolium Chloride
**Composition** per liter:
Agar .............................................................................15.0g
Maltose.........................................................................20.0g
Plant special peptone ...................................................10.0g
Sodium glycerophosphate.............................................10.0g
Yeast extract................................................................10.0g
NaCl...............................................................................5.0g
Lactose...........................................................................1.0g
NaN$_3$ ............................................................................0.4g
Bromcresol Purple ......................................................0.018g
2,3,5-Triphenyltetrazolium chloride solution ........................ 10.0mL
pH 7.2 ± 0.2 at 25°C

**Caution:** Sodium azide is toxic. Azides also react with metals and disposal must be highly diluted.

**2,3,5-Triphenyltetrazolium Chloride Solution:**
**Composition** per 10.0mL:
2,3,5-Triphenyltetrazolium chloride ...............................................0.1g

**Preparation of 2,3,5-Triphenyltetrazolium Chloride Solution:** Add 2,3,5-triphenyltetrazolium chloride to distilled/deionized water and bring volume to 10.0mL. Mix thoroughly. Filter sterilize.

**Preparation of Medium:** Add components, except 2,3,5-triphenyltetrazolium chloride solution, to distilled/deionized water and bring volume to 990.0mL. Mix thoroughly. Gently heat and bring to boiling. Autoclave for 10 min at 15 psi pressure–121°C. Cool to 45°–50°C. Aseptically add 2,3,5-triphenyltetrazolium chloride solution. Mix thoroughly. Pour into sterile Petri dishes or distribute into sterile tubes.

**Storage/Shelf Life:** Store dehydrated media in the dark in a sealed container below 30°C. Prepared media should be stored under refrigeration (2-8°C). Media should be used within 60 days of preparation.

Media should not be used if there are any signs of deterioration (shrinking, cracking, or discoloration) or contamination, or if the expiration date supplied by the manufacturer has passed.

**Use:** For the detection and enumeration of fecal streptococci in waters and examination of feces and other materials.

## KF Streptococcal HiVeg Agar Base
**Composition** per liter:
Agar ...........................................................................20.0g
Maltose.......................................................................20.0g
Plant special peptone .................................................10.0g
Sodium glycerophosphate...........................................10.0g
Yeast extract..............................................................10.0g
NaCl.............................................................................5.0g
Lactose.........................................................................1.0g
NaN$_3$ ...........................................................................0.4g
2,3,5-Triphenyltetrazolium chloride solution ........................ 10.0mL
pH 7.2 ± 0.2 at 25°C

**Source:** This medium, without 2,3,5-triphenyltetrazolium chloride solution, is available as a premixed powder from HiMedia.

**Caution:** Sodium azide is toxic. Azides also react with metals and disposal must be highly diluted.

**2,3,5-Triphenyltetrazolium Chloride Solution:**
**Composition** per 10.0mL:
2,3,5-Triphenyltetrazolium chloride ...............................................0.1g

**Preparation of 2,3,5-Triphenyltetrazolium Chloride Solution:** Add 2,3,5-triphenyltetrazolium chloride to distilled/deionized water and bring volume to 10.0mL. Mix thoroughly. Filter sterilize.

**Preparation of Medium:** Add components, except 2,3,5-triphenyltetrazolium chloride solution, to distilled/deionized water and bring volume to 990.0mL. Mix thoroughly. Gently heat and bring to boiling. Autoclave for 10 min at 15 psi pressure–121°C. Cool to 45°–50°C. Aseptically add 2,3,5-triphenyltetrazolium chloride solution. Mix thoroughly. Pour into sterile Petri dishes or distribute into sterile tubes.

**Storage/Shelf Life:** Store dehydrated media in the dark in a sealed container below 30°C. Prepared media should be stored under refrigeration (2-8°C). Media should be used within 60 days of preparation. Media should not be used if there are any signs of deterioration (shrinking, cracking, or discoloration) or contamination, or if the expiration date supplied by the manufacturer has passed.

**Use:** For the detection and enumeration of fecal streptococci in waters and examination of feces and other materials.

## KF Streptococcal HiVeg Broth Base with BCP and Triphyenyltetrazolium Chloride
**Composition** per liter:
Maltose .......................................................................20.0g
Proteose peptone.........................................................10.0g
Sodium glycerophosphate............................................10.0g
Yeast extract...............................................................10.0g
NaCl.............................................................................5.0g
Lactose.........................................................................1.0g
Na$_2$CO$_3$ .....................................................................0.636g
NaN$_3$ ...........................................................................0.4g
Bromcresol Purple ....................................................0.018g
2,3,5-Triphenyltetrazolium chloride solution ........................ 10.0mL
pH 7.2 ± 0.2 at 25°C

**Source:** This medium, without 2,3,5-triphenyltetrazolium chloride solution, is available as a premixed powder from HiMedia.

**Caution:** Sodium azide is toxic. Azides also react with metals and disposal must be highly diluted.

**2,3,5-Triphenyltetrazolium Chloride Solution:**
**Composition** per 10.0mL:
2,3,5-Triphenyltetrazolium chloride .................................................0.1g

**Preparation of 2,3,5-Triphenyltetrazolium Chloride Solution:** Add 2,3,5-triphenyltetrazolium chloride to distilled/deionized water and bring volume to 10.0mL. Mix thoroughly. Filter sterilize.

**Preparation of Medium:** Add components, except 2,3,5-triphenyltetrazolium chloride solution, to distilled/deionized water and bring volume to 990.0mL. Mix thoroughly. Gently heat and bring to boiling. Autoclave for 10 min at 15 psi pressure–121°C. Cool to 45°–50°C. Aseptically add 2,3,5-triphenyltetrazolium chloride solution. Mix thoroughly. Aseptically distribute into sterile tubes or flasks.

**Storage/Shelf Life:** Store dehydrated media in the dark in a sealed container below 30°C. Prepared media should be stored under refrigeration (2-8°C). Media should be used within 60 days of preparation. Media should not be used if there are any signs of deterioration (discoloration) or contamination, or if the expiration date supplied by the manufacturer has passed.

**Use:** For the selective cultivation of fecal streptococci.

## KF Streptococcal HiVeg Broth Base with Triphyenyltetrazolium Chloride

**Composition** per liter:
Maltose......................................................................................20.0g
Plant special peptone ...............................................................10.0g
Sodium glycerophosphate.........................................................10.0g
Yeast extract.............................................................................10.0g
NaCl............................................................................................5.0g
Lactose .......................................................................................1.0g
Na$_2$CO$_3$ ......................................................................................0.636g
NaN$_3$ ..........................................................................................0.4g
Phenol Red...............................................................................0.018g
2,3,5-Triphenyltetrazolium chloride solution ......................... 10.0mL
pH 7.2 ± 0.2 at 25°C

**Source:** This medium, without 2,3,5-Triphenyltetrazolium chloride solution, is available as a premixed powder from HiMedia.

**Caution:** Sodium azide is toxic. Azides also react with metals and disposal must be highly diluted.

**2,3,5-Triphenyltetrazolium Chloride Solution:**
**Composition** per 10.0mL:
2,3,5-Triphenyltetrazolium chloride .................................................0.1g

**Preparation of 2,3,5-Triphenyltetrazolium Chloride Solution:** Add 2,3,5-triphenyltetrazolium chloride to distilled/deionized water and bring volume to 10.0mL. Mix thoroughly. Filter sterilize.

**Preparation of Medium:** Add components, except 2,3,5-triphenyltetrazolium chloride solution, to distilled/deionized water and bring volume to 990.0mL. Mix thoroughly. Gently heat and bring to boiling. Autoclave for 10 min at 15 psi pressure–121°C. Cool to 45°–50°C. Aseptically add 2,3,5-triphenyltetrazolium chloride solution. Mix thoroughly. Aseptically distribute into sterile tubes or flasks.

**Storage/Shelf Life:** Store dehydrated media in the dark in a sealed container below 30°C. Prepared media should be stored under refrigeration (2-8°C). Media should be used within 60 days of preparation.

Media should not be used if there are any signs of deterioration (discoloration) or contamination, or if the expiration date supplied by the manufacturer has passed.

**Use:** For the selective cultivation of fecal streptococci.

## KF *Streptococcus* Agar

**Composition** per liter:
Agar ..........................................................................................20.0g
Maltose......................................................................................20.0g
Proteose peptone ......................................................................10.0g
Sodium glycerophosphate.........................................................10.0g
Yeast extract.............................................................................10.0g
NaCl............................................................................................5.5g
Lactose .......................................................................................1.0g
NaN$_3$ ..........................................................................................0.4g
Bromcresol Purple ..................................................................0.015g
2,3,5-Triphenyltetrazolium chloride solution ......................... 10.0mL
pH 7.2 ± 0.2 at 25°C

**Caution:** Sodium azide is toxic. Azides also react with metals and disposal must be highly diluted.

**Source:** This medium is available as a premixed powder from BD Diagnostic Systems and Thermo Scientific.

**2,3,5-Triphenyltetrazolium Chloride Solution:**
**Composition** per 10.0mL:
2,3,5-Triphenyltetrazolium chloride .................................................0.1g

**Preparation of 2,3,5-Triphenyltetrazolium Chloride Solution:** Add 2,3,5-triphenyltetrazolium chloride to distilled/deionized water and bring volume to 10.0mL. Mix thoroughly. Filter sterilize.

**Preparation of Medium:** Add components, except 2,3,5-triphenyltetrazolium chloride solution, to distilled/deionized water and bring volume to 990.0mL. Mix thoroughly. Gently heat and bring to boiling. Autoclave for 15 min at 15 psi pressure–121°C. Cool to 45°–50°C. Aseptically add 2,3,5-triphenyltetrazolium chloride solution. Mix thoroughly. Pour into sterile Petri dishes or distribute into sterile tubes.

**Storage/Shelf Life:** Store dehydrated media in the dark in a sealed container below 30°C. Prepared media should be stored under refrigeration (2-8°C). Media should be used within 60 days of preparation. Media should not be used if there are any signs of deterioration (shrinking, cracking, or discoloration) or contamination, or if the expiration date supplied by the manufacturer has passed.

**Use:** For the isolation and enumeration of enterococci.

## KF *Streptococcus* Agar

**Composition** per liter:
Agar ..........................................................................................20.0g
Maltose......................................................................................20.0g
Sodium glycerophosphate.........................................................10.0g
Yeast extract.............................................................................10.0g
NaCl............................................................................................5.0g
Pancreatic digest of casein........................................................5.0g
Peptic digest of animal tissue ....................................................5.0g
Lactose .......................................................................................1.0g
NaN$_3$ ..........................................................................................0.4g
2,3,5-Triphenyltetrazolium chloride solution ......................... 10.0mL
pH 7.2 ± 0.2 at 25°C

**Caution:** Sodium azide is toxic. Azides also react with metals and disposal must be highly diluted.

**Source:** This medium is available as a premixed powder from BD Diagnostic Systems.

## 2,3,5-Triphenyltetrazolium Chloride Solution:

**Composition** per 10.0mL:

2,3,5-Triphenyltetrazolium chloride ...............................................0.1g

**Preparation of 2,3,5-Triphenyltetrazolium Chloride Solution:** Add 2,3,5-triphenyltetrazolium chloride to distilled/deionized water and bring volume to 10.0mL. Mix thoroughly. Filter sterilize.

**Preparation of Medium:** Add components, except 2,3,5-triphenyltetrazolium chloride solution, to distilled/deionized water and bring volume to 990.0mL. Mix thoroughly. Gently heat and bring to boiling. Autoclave for 15 min at 15 psi pressure–121°C. Cool to 45°–50°C. Aseptically add 2,3,5-triphenyltetrazolium chloride solution. Mix thoroughly. Pour into sterile Petri dishes or distribute into sterile tubes.

**Storage/Shelf Life:** Store dehydrated media in the dark in a sealed container below 30°C. Prepared media should be stored under refrigeration (2-8°C). Media should be used within 60 days of preparation. Media should not be used if there are any signs of deterioration (shrinking, cracking, or discoloration) or contamination, or if the expiration date supplied by the manufacturer has passed.

**Use:** For the selective cultivation and enumeration of fecal streptococci.

## KF *Streptococcus* Broth

**Composition** per liter:

Maltose..........................................................................20.0g
Sodium glycerophosphate.............................................10.0g
Yeast extract................................................................10.0g
NaCl..............................................................................5.0g
Pancreatic digest of casein............................................5.0g
Peptic digest of animal tissue........................................5.0g
Lactose.........................................................................1.0g
$Na_2CO_3$ ..................................................................0.636g
$NaN_3$.........................................................................0.4g
Phenol Red...................................................................0.018g
2,3,5-Triphenyltetrazolium chloride solution ............. 10.0mL
pH 7.2 ± 0.2 at 25°C

**Caution:** Sodium azide is toxic. Azides also react with metals and disposal must be highly diluted.

**Source:** This medium is available as a premixed powder from BD Diagnostic Systems.

**Preparation of Medium:** Add components, except 2,3,5-triphenyltetrazolium chloride solution, to distilled/deionized water and bring volume to 990.0mL. Mix thoroughly. Gently heat and bring to boiling. Autoclave for 15 min at 15 psi pressure–121°C. Cool to 45°–50°C. Aseptically add 2,3,5-triphenyltetrazolium chloride solution. Mix thoroughly. Aseptically distribute into sterile tubes or flasks.

**Storage/Shelf Life:** Store dehydrated media in the dark in a sealed container below 30°C. Prepared media should be stored under refrigeration (2-8°C). Media should be used within 60 days of preparation. Media should not be used if there are any signs of deterioration (discoloration) or contamination, or if the expiration date supplied by the manufacturer has passed.

**Use:** For the selective cultivation of fecal streptococci.

## K-L Virulence Agar
## (Klebs-Loeffler Virulence Agar)
## (Elek Agar)
## (*Corynebacterium diphtheriae* Virulence Test Medium)

**Composition** per 1300.0mL:

K-L agar base........................................................ 1.0L
Rabbit serum ....................................................... 200.0mL
$K_2TeO_3$ solution.................................................. 100.0mL
K-L filter strips .................................................... 100
pH 7.8 ± 0.2 at 25°C

**Source:** This medium is available as a premixed powder from BD Diagnostic Systems.

### K-L Agar Base:

**Composition** per liter:

Meat peptone ........................................................20.0g
Agar .....................................................................15.0g
NaCl......................................................................2.5g

**Preparation of K-L Agar Base:** Add components to distilled/deionized water and bring volume to 1.0L. Mix thoroughly. Gently heat and bring to boiling. Autoclave for 15 min at 15 psi pressure–121°C. Cool to 50°C.

### $K_2TeO_3$ Solution:

**Composition** per 100.0mL:

$K_2TeO_3$..............................................................0.3g

**Preparation of $K_2TeO_3$ Solution:** Add $K_2TeO_3$ to distilled/deionized water and bring volume to 100.0mL. Mix thoroughly. Filter sterilize.

**Caution:** Potassium tellurite is toxic.

### K-L Filter Strips:

**Composition**:

Whatman No. 3 filter paper ................................. as needed
Diphtheria toxin solution ................................... 10.0mL

**Preparation of K-L Strips:** Cut Whatman No. 3 filter paper into 1.5cm × 7cm strips. Autoclave for 15 min at 15 psi pressure–121°C. Aseptically dip each strip into a sterile solution containing 1000U of purified diphtheria toxin/mL. Drain off excess liquid.

**Preparation of Medium:** Filter sterilize rabbit serum. To 1.0L of cooled, sterile K-L agar base, aseptically add sterile rabbit serum and sterile $K_2TeO_3$ solution. Mix thoroughly. Pour into sterile Petri dishes in 13.0mL volumes. Before the agar solidifies, aseptically add one K-L filter strip across the diameter of the plate. Allow the filter strip to sink to the bottom of the plate or press it down with sterile forceps. Allow the agar to solidify. Dry the surface of the plates by incubating at 35°C with lid of plate ajar for 2 h.

**Storage/Shelf Life:** Store dehydrated media in the dark in a sealed container below 30°C. Prepared media should be stored under refrigeration (2-8°C). Media should be used within 60 days of preparation. Media should not be used if there are any signs of deterioration (shrinking, cracking, or discoloration) or contamination, or if the expiration date supplied by the manufacturer has passed.

**Use:** For *in vitro* toxigenicity testing of *Corynebacterium diphtheriae* by the agar diffusion technique. *Corynebacterium diphtheriae* that produce toxin form white precipitin lines at approximately 45° angles from the culture streak line.

## K-L Virulence Agar
### (Klebs-Loeffler Virulence Agar)

**Composition** per 1250.0mL:

| | |
|---|---|
| K-L agar base | 1.0L |
| K-L enrichment | 200.0mL |
| $K_2TeO_3$ solution | 50.0mL |
| K-L filter strips | 100 |

pH $7.8 \pm 0.2$ at 25°C

**Source:** This medium is available as a premixed powder from BD Diagnostic Systems.

### K-L Agar Base:

**Composition** per liter:

| | |
|---|---|
| Meat peptone | 20.0g |
| Agar | 15.0g |
| NaCl | 2.5g |

**Preparation of K-L Agar Base:** Add components to distilled/deionized water and bring volume to 1.0L. Mix thoroughly. Gently heat and bring to boiling. Autoclave for 15 min at 15 psi pressure–121°C. Cool to 50°C.

### K-L Enrichment:

**Composition** per 200.0mL:

| | |
|---|---|
| Casamino acids | 4.0g |
| Glycerol | 100.0mL |
| Tween™ 80 | 100.0mL |

**Preparation of K-L Enrichment:** Combine components. Mix thoroughly. Filter sterilize.

### $K_2TeO_3$ Solution:

**Composition** per 100.0mL:

| | |
|---|---|
| $K_2TeO_3$ | 1.0g |

**Preparation of $K_2TeO_3$ Solution:** Add $K_2TeO_3$ to distilled/deionized water and bring volume to 100.0mL. Mix thoroughly. Filter sterilize.

**Caution:** Potassium tellurite is toxic.

### K-L Filter Strips:

**Composition**:

| | |
|---|---|
| Diphtheria toxin solution | 10.0mL |
| Whatman No. 3 filter paper | as needed |

**Preparation of K-L Strips:** Cut Whatman #3 filter paper into 1.5cm × 7cm strips. Autoclave for 15 min at 15 psi pressure–121°C. Aseptically dip each strip into a sterile solution containing 1000U of purified diphtheria toxin/mL. Drain off excess liquid.

**Preparation of Medium:** To 1.0L of cooled, sterile K-L agar base, aseptically add sterile K-L enrichment and sterile $K_2TeO_3$ solution. Mix thoroughly. Pour into sterile Petri dishes in 13.0mL volumes. Before the agar solidifies, aseptically add one K-L filter strip across the diameter of the plate. Allow the filter strip to sink to the bottom of the plate or press it down with sterile forceps. Allow the agar to solidify. Dry the surface of the plates by incubating at 35°C with lid of plate ajar for 2 h.

**Storage/Shelf Life:** Store dehydrated media in the dark in a sealed container below 30°C. Prepared media should be stored under refrigeration (2-8°C). Media should be used within 60 days of preparation. Media should not be used if there are any signs of deterioration (shrinking, cracking, or discoloration) or contamination, or if the expiration date supplied by the manufacturer has passed.

**Use:** For *in vitro* toxigenicity testing of *Corynebacterium diphtheriae* by the agar diffusion technique. *Corynebacterium diphtheriae* that produce toxin form white precipitin lines at approximately 45° angles from the culture streak line.

## Kleb Agar
### (m-Kleb Agar)

**Composition** per liter:

| | |
|---|---|
| Agar | 15.0g |
| Proteose peptone No. 3 | 10.0g |
| NaCl | 5.0g |
| Adonitol | 5.0g |
| Beef extract | 1.0g |
| Aniline Blue | 0.1g |
| Sodium lauryl sulfate | 0.1g |
| Phenol Red | 0.025g |
| Ethanol (95% solution) | 20.0mL |
| Carbenicillin solution | 10.0mL |

pH $7.4 \pm 0.2$ at 25°C

### Carbenicillin Solution:

**Composition** per 10.0mL:

| | |
|---|---|
| Carbenicillin | 0.05g |

**Preparation of Carbenicillin Solution:** Add carbenicillin to distilled/deionized water and bring volume to 10.0mL. Mix thoroughly. Filter sterilize.

**Preparation of Medium:** Add components, except ethanol and carbenicillin solution, to distilled/deionized water and bring volume to 970.0mL. Mix thoroughly. Gently heat and bring to boiling. Autoclave for 15 min at 15 psi pressure–121°C. Cool to 45°–50°C. Aseptically add 20.0mL of ethanol and 10.0mL of carbenicillin solution. Mix thoroughly. Pour into sterile Petri dishes or distribute into sterile tubes.

**Storage/Shelf Life:** Store dehydrated media in the dark in a sealed container below 30°C. Prepared media should be stored under refrigeration (2-8°C). Media should be used within 60 days of preparation. Media should not be used if there are any signs of deterioration (shrinking, cracking, or discoloration) or contamination, or if the expiration date supplied by the manufacturer has passed.

**Use:** For the enumeration of bacteria from waters.

## *Klebsiella* Selective Agar

**Composition** per liter:

| | |
|---|---|
| Agar | 26.0g |
| DL–Phenylalanine | 10.0g |
| L-Ornithine·HCl | 10.0g |
| Raffinose | 7.0g |
| Pancreatic digest of casein | 2.5g |
| Yeast extract | 2.5g |
| $K_2HPO_4$ | 2.0g |
| Phenol Red solution | 10.0mL |
| Carbenicillin solution | 10.0mL |

pH $5.6 \pm 0.2$ at 25°C

### Phenol Red Solution:

**Composition** per 10.0mL:

| | |
|---|---|
| Phenol Red | 0.5g |

**Preparation of Phenol Red Solution:** Add Phenol Red to 50% ethanol and bring volume to 10.0mL. Mix thoroughly.

**Carbenicillin Solution:**
**Composition** per 1.0mL:
Carbenicillin.................................................................5.0mg

**Preparation of Carbenicillin Solution:** Add carbenicillin to distilled/deionized water and bring volume to 1.0mL. Mix thoroughly. Filter sterilize.

**Preparation of Medium:** Add components, except carbenicillin solution, to distilled/deionized water and bring volume to 990.0mL. Mix thoroughly. Gently heat and bring to boiling. Autoclave for 15 min at 15 psi pressure–121°C. Cool to 45°–50°C. Aseptically add 10.0mL carbenicillin solution. Mix thoroughly. Adjust pH to 5.6–5.7 with sterile $1N$ HCl. Pour into sterile Petri dishes or distribute into sterile tubes.

**Storage/Shelf Life:** Store dehydrated media in the dark in a sealed container below 30°C. Prepared media should be stored under refrigeration (2-8°C). Media should be used within 60 days of preparation. Media should not be used if there are any signs of deterioration (shrinking, cracking, or discoloration) or contamination, or if the expiration date supplied by the manufacturer has passed.

**Use:** For the isolation and identification of *Klebsiella pneumoniae* from clinical specimens.

## Kligler Iron Agar

**Composition** per liter:
Peptone.........................................................................20.0g
Agar ..............................................................................12.0g
Lactose .........................................................................10.0g
NaCl................................................................................5.0g
Beef extract ....................................................................3.0g
Yeast extract ...................................................................3.0g
Glucose ...........................................................................1.0g
Ferric citrate ...................................................................0.3g
Na₂S₂O₃ ..........................................................................0.3g
Phenol Red....................................................................0.05g

pH 7.4 ± 0.2 at 25°C

**Source:** This medium is available as a premixed powder from BD Diagnostic Systems and Thermo Scientific.

**Preparation of Medium:** Add components to distilled/deionized water and bring volume to 1.0L. Mix thoroughly. Gently heat and bring to boiling. Distribute into tubes. Autoclave for 15 min at 15 psi pressure–121°C. Pour into sterile Petri dishes or leave in tubes.

**Storage/Shelf Life:** Store dehydrated media in the dark in a sealed container below 30°C. Prepared media should be stored under refrigeration (2-8°C). Media should be used within 60 days of preparation. Media should not be used if there are any signs of deterioration (shrinking, cracking, or discoloration) or contamination, or if the expiration date supplied by the manufacturer has passed.

**Use:** For the differentiation and identification of Enterobacteriaceae based upon sugar fermentation and hydrogen sulfide production. Sugar fermentation is indicated by the medium turning yellow. H₂S production results in the medium turning black.

## Kligler Iron Agar

**Composition** per liter:
Agar ..............................................................................15.0g
Lactose .........................................................................10.0g
Pancreatic digest of casein .........................................10.0g
Peptic digest of animal tissue......................................10.0g
NaCl................................................................................5.0g
Glucose ...........................................................................1.0g

Ferric ammonium citrate................................................0.5g
Na₂S₂O₃ ..........................................................................0.5g
Phenol Red..................................................................0.025g

pH 7.4 ± 0.2 at 25°C

**Source:** This medium is available as a premixed powder from BD Diagnostic Systems.

**Preparation of Medium:** Add components to distilled/deionized water and bring volume to 1.0L. Mix thoroughly. Gently heat and bring to boiling. Distribute into tubes or flasks. Autoclave for 15 min at 15 psi pressure–121°C. Pour into sterile Petri dishes or leave in tubes.

**Storage/Shelf Life:** Store dehydrated media in the dark in a sealed container below 30°C. Prepared media should be stored under refrigeration (2-8°C). Media should be used within 60 days of preparation. Media should not be used if there are any signs of deterioration (shrinking, cracking, or discoloration) or contamination, or if the expiration date supplied by the manufacturer has passed.

**Use:** For the differentiation and identification of Enterobacteriaceae based upon sugar fermentation and hydrogen sulfide production. Sugar fermentation is indicated by the medium turning yellow. H₂S production results in the medium turning black.

## Kligler Iron Agar
## (FDA M71)

**Composition** per liter:
Lactose .........................................................................20.0g
Agar ..............................................................................15.0g
Pancreatic digest of casein .........................................10.0g
Peptic digest of animal tissue......................................10.0g
NaCl................................................................................5.0g
Glucose ...........................................................................1.0g
Ferric ammonium citrate................................................0.5g
Na₂S₂O₃ ..........................................................................0.5g
Phenol Red..................................................................0.025g

pH 7.4 ± 0.2 at 25°C

**Preparation of Medium:** Add components to distilled/deionized water and bring volume to 1.0L. Mix thoroughly. Gently heat and bring to boiling. Distribute into tubes or flasks. Autoclave for 15 min at 15 psi pressure–121°C. Pour into sterile Petri dishes or leave in tubes.

**Storage/Shelf Life:** Store dehydrated media in the dark in a sealed container below 30°C. Prepared media should be stored under refrigeration (2-8°C). Media should be used within 60 days of preparation. Media should not be used if there are any signs of deterioration (shrinking, cracking, or discoloration) or contamination, or if the expiration date supplied by the manufacturer has passed.

**Use:** For the differentiation and identification of Enterobacteriaceae based upon sugar fermentation and hydrogen sulfide production. Sugar fermentation is indicated by the medium turning yellow. H₂S production results in the medium turning black.

## Kligler Iron Agar
## (BAM M71)

**Composition** per liter:
Lactose .........................................................................20.0g
Polypeptone ..................................................................20.0g
Agar ..............................................................................15.0g
Peptic digest of animal tissue......................................10.0g
NaCl................................................................................5.0g
Glucose ...........................................................................1.0g

Ferric ammonium citrate................................................0.5g
Na$_2$S$_2$O$_3$................................................0.5g
Phenol Red................................................0.025g

pH 7.4 ± 0.2 at 25°C

**Preparation of Medium:** Add components to distilled/deionized water and bring volume to 1.0L. Mix thoroughly. Gently heat and bring to boiling. Distribute into tubes or flasks. Autoclave for 15 min at 15 psi pressure–121°C. Pour into sterile Petri dishes or leave in tubes.

**Storage/Shelf Life:** Store dehydrated media in the dark in a sealed container below 30°C. Prepared media should be stored under refrigeration (2-8°C). Media should be used within 60 days of preparation. Media should not be used if there are any signs of deterioration (shrinking, cracking, or discoloration) or contamination, or if the expiration date supplied by the manufacturer has passed.

**Use:** For the differentiation and identification of Enterobacteriaceae based upon sugar fermentation and hydrogen sulfide production. Sugar fermentation is indicated by the medium turning yellow. H$_2$S production results in the medium turning black.

### Kligler Iron Agar with Sodium Chloride (BAM M71)

**Composition** per liter:
NaCl................................................25.0g
Lactose................................................20.0g
Polypeptone................................................20.0g
Agar................................................15.0g
Peptic digest of animal tissue................................................10.0g
NaCl................................................5.0g
Glucose................................................1.0g
Ferric ammonium citrate................................................0.5g
Na$_2$S$_2$O$_3$................................................0.5g
Phenol Red................................................0.025g

pH 7.4 ± 0.2 at 25°C

**Preparation of Medium:** Add components to distilled/deionized water and bring volume to 1.0L. Mix thoroughly. Gently heat and bring to boiling. Distribute into tubes or flasks. Autoclave for 15 min at 15 psi pressure–121°C. Pour into sterile Petri dishes or leave in tubes.

**Storage/Shelf Life:** Store dehydrated media in the dark in a sealed container below 30°C. Prepared media should be stored under refrigeration (2-8°C). Media should be used within 60 days of preparation. Media should not be used if there are any signs of deterioration (shrinking, cracking, or discoloration) or contamination, or if the expiration date supplied by the manufacturer has passed.

**Use:** For the differentiation and identification of *Vibrio* spp. based upon sugar fermentation and hydrogen sulfide production. Sugar fermentation is indicated by the medium turning yellow. H$_2$S production results in the medium turning black.

### Kligler Iron HiVeg Agar

**Composition** per liter:
Plant special peptone................................................15.0g
Lactose................................................10.0g
Agar................................................15.0g
Plant peptone No. 3................................................5.0g
NaCl................................................5.0g
Plant extract................................................3.0g
Yeast extract................................................3.0g
Glucose................................................1.0g
Na$_2$S$_2$O$_3$................................................0.3g

FeSO$_4$................................................0.2g
Phenol Red................................................0.024g

pH 7.4 ± 0.2 at 25°C

**Source:** This medium is available as a premixed powder from Hi-Media.

**Preparation of Medium:** Add components to distilled/deionized water and bring volume to 1.0L. Mix thoroughly. Gently heat and bring to boiling. Distribute into tubes. Autoclave for 15 min at 15 psi pressure–121°C. Pour into sterile Petri dishes or leave in tubes.

**Storage/Shelf Life:** Store dehydrated media in the dark in a sealed container below 30°C. Prepared media should be stored under refrigeration (2-8°C). Media should be used within 60 days of preparation. Media should not be used if there are any signs of deterioration (shrinking, cracking, or discoloration) or contamination, or if the expiration date supplied by the manufacturer has passed.

**Use:** For the differentiation and identification of Enterobacteriaceae based upon sugar fermentation and hydrogen sulfide production. Sugar fermentation is indicated by the medium turning yellow. H$_2$S production results in the medium turning black.

### Kligler Iron HiVeg Agar, Modified

**Composition** per liter:
Plant hydrolysate................................................20.0g
Agar................................................15.0g
Lactose................................................10.0g
NaCl................................................5.0g
Plant extract................................................3.0g
Yeast extract................................................3.0g
Glucose, anhydrous................................................1.0g
Na$_2$S$_2$O$_3$·5H$_2$O................................................0.3g
FeSO$_4$................................................0.2g
Phenol Red................................................0.025g

pH 7.4 ± 0.2 at 25°C

**Source:** This medium is available as a premixed powder from Hi-Media.

**Preparation of Medium:** Add components to distilled/deionized water and bring volume to 1.0L. Mix thoroughly. Gently heat and bring to boiling. Distribute into tubes. Autoclave for 15 min at 15 psi pressure–121°C. Pour into sterile Petri dishes or leave in tubes.

**Storage/Shelf Life:** Store dehydrated media in the dark in a sealed container below 30°C. Prepared media should be stored under refrigeration (2-8°C). Media should be used within 60 days of preparation. Media should not be used if there are any signs of deterioration (shrinking, cracking, or discoloration) or contamination, or if the expiration date supplied by the manufacturer has passed.

**Use:** For the differentiation and identification of Enterobacteriaceae based upon sugar fermentation and hydrogen sulfide production. Recommended for identification of *Yersinia enterocolitica*.

### Koser Citrate Medium

**Composition** per liter:
Sodium citrate................................................3.0g
NaNH$_4$HPO$_4$·4H$_2$O................................................1.5g
KH$_2$PO$_4$................................................1.0g
MgSO$_4$·7H$_2$O................................................0.2g

pH 6.7 ± 0.2 at 25°C

**Source:** This medium is available as a premixed powder from BD Diagnostic Systems.

**Preparation of Medium:** Add components to distilled/deionized water and bring volume to 1.0L. Mix thoroughly. Gently heat and bring to boiling. Distribute into tubes or flasks. Autoclave for 15 min at 15 psi pressure–121°C. Pour into sterile Petri dishes or leave in tubes.

**Storage/Shelf Life:** Store dehydrated media in the dark in a sealed container below 30°C. Prepared media should be stored under refrigeration (2-8°C). Media should be used within 60 days of preparation. Media should not be used if there are any signs of deterioration (discoloration) or contamination, or if the expiration date supplied by the manufacturer has passed.

**Use:** For the differentiation of *Escherichia coli* and *Enterobacter aerogenes* based on citrate utilization.

## Kracke Blood Culture HiVeg Medium
**Composition** per liter:
| | |
|---|---|
| NaCl | 49.0g |
| Glucose | 10.0g |
| Plant peptone No. 3 | 10.0g |
| Na$_2$HPO$_4$ | 2.0g |
| Plant infusion | 2.0g |
| Plant special infusion | 1.0g |
| Sodium citrate | 1.0g |

pH 7.4 ± 0.2 at 25°C

**Source:** This medium is available as a premixed powder from Hi-Media.

**Preparation of Medium:** Add components to distilled/deionized water and bring volume to 1.0L. Mix thoroughly. Gently heat and bring to boiling. Distribute into tubes or flasks. Autoclave for 15 min at 15 psi pressure–121°C.

**Storage/Shelf Life:** Store dehydrated media in the dark in a sealed container below 30°C. Prepared media should be stored under refrigeration (2-8°C). Media should be used within 60 days of preparation. Media should not be used if there are any signs of deterioration (discoloration) or contamination, or if the expiration date supplied by the manufacturer has passed.

**Use:** For the cultivation of pathogens in cases of bacteremia. The medium is inoculated with blood from a patient. Approximately 10-15mL of blood normally is inoculated into 50mL of medium.

## Krainsky's Asparagine Agar
**Composition** per liter:
| | |
|---|---|
| Agar | 15.0g |
| Glucose | 10.0g |
| K$_2$HPO$_4$ | 0.5g |
| L-Asparagine | 0.5g |

pH 7.0 ± 0.2 at 25°C

**Preparation of Medium:** Add components to distilled/deionized water and bring volume to 1.0L. Mix thoroughly. Gently heat and bring to boiling. Distribute into tubes or flasks. Autoclave for 15 min at 15 psi pressure–121°C. Pour into sterile Petri dishes or leave in tubes.

**Storage/Shelf Life:** Store dehydrated media in the dark in a sealed container below 30°C. Prepared media should be stored under refrigeration (2-8°C). Media should not be used within 60 days of preparation. Media should not be used if there are any signs of deterioration (shrinking, cracking, or discoloration) or contamination, or if the expiration date supplied by the manufacturer has passed.

**Use:** For the cultivation and maintenance of *Streptomyces fragmentans*.

## KRANEP Agar Base
**Composition** per liter:
| | |
|---|---|
| KSCN | 25.5g |
| Agar | 18.3g |
| Sodium pyruvate | 8.2g |
| Pancreatic digest of gelatin | 6.1g |
| LiCl | 5.1g |
| Mannitol | 5.1g |
| Beef extract | 3.7g |
| NaN$_3$ | 0.05g |
| Cycloheximide | 0.041g |
| Egg yolk emulsion | 100.0mL |

pH 6.8 ± 0.2 at 25°C

**Caution:** Sodium azide is toxic. Azides also react with metals and disposal must be highly diluted.

**Caution:** Cycloheximide is toxic. Avoid skin contact or aerosol formation and inhalation.

**Caution:** Lithium chloride is harmful. Avoid bodily contact and inhalation of vapors. On contact with skin wash with plenty of water immediately.

**Source:** This medium is available as a premixed powder from Thermo Scientific.

**Egg Yolk Emulsion:**
**Composition:**
| | |
|---|---|
| Chicken egg yolks | 11 |
| Whole chicken egg | 1 |

**Preparation of Egg Yolk Emulsion:** Soak eggs with 1:100 dilution of saturated mercuric chloride solution for 1 min. Crack eggs and separate yolks from whites. Mix egg yolks with 1 chicken egg.

**Preparation of Medium:** Add components, except egg yolk emulsion, to distilled/deionized water and bring volume to 900.0mL. Mix thoroughly. Gently heat and bring to boiling. Autoclave for 15 min at 15 psi pressure–121°C. Cool to 45°–50°C. Aseptically add 100.0mL of egg yolk emulsion. Mix thoroughly. Pour into sterile Petri dishes or distribute into sterile tubes.

**Storage/Shelf Life:** Store dehydrated media in the dark in a sealed container below 30°C. Prepared media should be stored under refrigeration (2-8°C). Media should be used within 60 days of preparation. Media should not be used if there are any signs of deterioration (shrinking, cracking, or discoloration) or contamination, or if the expiration date supplied by the manufacturer has passed.

**Use:** For the isolation and enumeration of staphylococci from foods.

## KRANEP HiVeg Agar Base with Egg Yolk Emulsion
**Composition** per liter:
| | |
|---|---|
| Potassium thiocyanate | 25.5g |
| Agar | 15.0g |
| Sodium pyruvate | 8.2g |
| LiCl | 5.1g |
| Mannitol | 5.1g |
| Plant peptone | 5.0g |
| NaCl | 5.0g |
| Plant extract | 1.5g |
| Yeast extract | 1.5g |
| NaN$_3$ | 0.05g |
| Cycloheximide | 0.041g |
| Egg yolk emulsion | 100.0mL |

pH 6.8 ± 0.2 at 25°C

**Source:** This medium, without egg yolk emulsion, is available as a premixed powder from HiMedia.

**Caution:** Sodium azide is toxic. Azides also react with metals and disposal must be highly diluted.

**Caution:** Lithium Chloride is harmful. Avoid bodily contact and inhalation of vapors. On contact with skin wash with plenty of water immediately.

**Caution:** Cycloheximide is toxic. Avoid skin contact or aerosol formation and inhalation.

**Egg Yolk Emulsion:**
**Composition** per liter:
Egg yolks ............................................. 30.0mL
NaCl, 0.9% solution.............................. 70.0mL

**Preparation of Egg Yolk Emulsion:** Soak eggs with 1:100 dilution of saturated mercuric chloride solution for 1 min. Crack 11 eggs and separate yolks from whites. Mix egg yolks. Measure 30.0mL of egg yolk emulsion and add to 70.0mL of 0.9% sterile NaCl solution. Mix thoroughly. Warm to 45°–50°C.

**Preparation of Medium:** Add components, except egg yolk emulsion, to distilled/deionized water and bring volume to 900.0mL. Mix thoroughly. Gently heat and bring to boiling. Autoclave for 15 min at 15 psi pressure–121°C. Cool to 45°–50°C. Aseptically add 100.0mL of egg yolk emulsion. Mix thoroughly. Pour into sterile Petri dishes or distribute into sterile tubes.

**Storage/Shelf Life:** Store dehydrated media in the dark in a sealed container below 30°C. Prepared media should be stored under refrigeration (2-8°C). Media should be used within 60 days of preparation. Media should not be used if there are any signs of deterioration (shrinking, cracking, or discoloration) or contamination, or if the expiration date supplied by the manufacturer has passed.

**Use:** For the isolation and enumeration of staphylococci from foods.

## Kundrant Agar

**Composition** per liter:
Agar .................................................................10.0g
Meat peptone.....................................................7.8g
Casein peptone ..................................................7.8g
Starch .................................................................4.0g
Gelatin...............................................................4.0g
NaCl ...................................................................3.0g
Yeast extract......................................................2.8g
Glucose ..............................................................1.0g
Bromcresol Purple ..........................................1.6mg
pH 6.8 ± 0.2 at 25°C

**Source:** This medium is available from HiMedia.

**Preparation of Medium:** Add components to distilled/deionized water and bring volume to 1.0L. Mix thoroughly. Gently heat and bring to boiling. Distribute into tubes or flasks. Autoclave for 15 min at 15 psi pressure–121°C.

**Storage/Shelf Life:** Store dehydrated media in the dark in a sealed container below 30°C. Prepared media should be stored under refrigeration (2-8°C). Media should be used within 60 days of preparation. Media should not be used if there are any signs of deterioration (shrinking, cracking, or discoloration) or contamination, or if the expiration date supplied by the manufacturer has passed.

**Use:** For the qualitative detection of residues from sulfonamides and other antimicrobics in animal-derived foods.

## *L. mono* Confirmatory Agar Base
### (*Listeria monocytogenes* Confirmatory Agar, Base)
**Composition** per liter:
Special peptone....................................................30.0g
Agar ......................................................................12.0g
LiCl .......................................................................10.0g
B.C. indicator........................................................ 8.6g
Yeast extract...........................................................6.0g
NaCl........................................................................5.0g
α-Methyl-D-mannoside .........................................3.0g
Na$_2$HPO$_4$, anhydrous...........................................2.5g
*Listeria mono* enrichment supplement II................10.0mL
*Listeria mono* selective supplement I .....................10.0mL
*Listeria mono* selective supplement II....................10.0mL
pH 7.2 ± 0.2 at 25°C

**Source:** This medium, without *Listeria mono* enrichment supplement II, *Listeria mono* selective supplement I, and *Listeria mono* selective supplement II, is available as a premixed powder from BioChemika.

**Caution:** LiCl is harmful. Avoid bodily contact and inhalation of vapors. On contact with skin wash with plenty of water immediately.

**Listeria mono Enrichment Supplement II Solution:**
**Composition** per 10.0mL:
L-phosphatidylinositol ....................................................1.0g

**Preparation of *Listeria mono* Enrichment Supplement II Solution:** Add L-phosphatidylinositol to distilled/deionized water and bring volume to 10.0mL. Mix thoroughly. Filter sterilize.

**Listeria mono Selective Supplement I Solution:**
**Composition** per 10.0mL:
Polymyxin B sulfate .................................................76,700U

**Preparation of *Listeria mono* Selective Supplement I Solution:** Add polymyxin B sulfate to distilled/deionized water and bring volume to 10.0mL. Mix thoroughly. Filter sterilize.

**Listeria mono Selective Supplement II Solution:**
**Composition** per 10.0mL:
Ceftazidime .............................................................20.0mg
Nalidixic acid, sodium salt .....................................20.0mg
Amphotericin B .......................................................10.0mg

**Preparation of *Listeria mono* Selective Supplement II Solution:** Add components to distilled/deionized water and bring volume to 10.0mL. Mix thoroughly. Filter sterilize.

**Preparation of Medium:** Add components, except *Listeria mono* enrichment supplement II, *Listeria mono* selective supplement I, and *Listeria mono* selective supplement II, to distilled/deionized water and bring volume to 970.0mL. Mix thoroughly. Gently heat and bring to boiling. Autoclave for 15 min at 15 psi pressure–121°C. Cool to 45°–50°C. Aseptically add 10.0mL *Listeria mono* enrichment supplement II, 10.0mL *Listeria mono* selective supplement I, and 10.0mL *Listeria mono* selective supplement II. Mix thoroughly. Pour into sterile Petri dishes or distribute into sterile tubes.

**Storage/Shelf Life:** Store dehydrated media in the dark in a sealed container below 30°C. Prepared media should be stored under refrigeration (2-8°C). Media should be used within 60 days of preparation. Media should not be used if there are any signs of deterioration (shrinking, cracking, or discoloration) or contamination, or if the expiration date supplied by the manufacturer has passed.

**Use:** For the selective and differential isolation of *Listeria monocytogenes* from clinical and food specimens.

## *L. mono* Confirmatory HiVeg Agar Base

**Composition** per liter:

Plant special peptone ...............................................30.0g
Agar .........................................................................12.0g
LiCl ..........................................................................10.0g
B.C. indicator ............................................................8.6g
Yeast extract ..............................................................6.0g
NaCl ..........................................................................5.0g
α-Methyl-D-mannoside ..............................................3.0g
Na$_2$HPO$_4$, anhydrous ..............................................2.5g
*Listeria mono* enrichment supplement II ...............10.0mL
*Listeria mono* selective supplement I ....................10.0mL
*Listeria mono* selective supplement II ...................10.0mL
pH 7.2 ± 0.2 at 25°C

**Source:** This medium, without *Listeria mono* enrichment supplement II, *Listeria mono* selective supplement I, and *Listeria mono* selective supplement II, is available as a premixed powder from HiMedia.

**Caution:** LiCl is harmful. Avoid bodily contact and inhalation of vapors. On contact with skin wash with plenty of water immediately.

*Listeria mono* **Enrichment Supplement II Solution:**
**Composition** per 10.0mL:

L-phosphatidylinositol ...............................................1.0g

**Preparation of *Listeria mono* Enrichment Supplement II Solution:** Add L-phosphatidylinositol to distilled/deionized water and bring volume to 10.0mL. Mix thoroughly. Filter sterilize.

*Listeria mono* **Selective Supplement I Solution:**
**Composition** per 10.0mL:

Polymyxin B sulfate.............................................76,700U

**Preparation of *Listeria mono* Selective Supplement I Solution:** Add polymyxin B sulfate to distilled/deionized water and bring volume to 10.0mL. Mix thoroughly. Filter sterilize.

*Listeria mono* **Selective Supplement II Solution:**
**Composition** per 10.0mL:

Ceftazidime ............................................................20.0mg
Nalidixic acid, sodium salt .....................................20.0mg
Amphotericin B ......................................................10.0mg

**Preparation of *Listeria mono* Selective Supplement II Solution:** Add components to distilled/deionized water and bring volume to 10.0mL. Mix thoroughly. Filter sterilize.

**Preparation of Medium:** Add components, except *Listeria mono* enrichment supplement II, *Listeria mono* selective supplement I, and *Listeria mono* selective supplement II, to distilled/deionized water and bring volume to 970.0mL. Mix thoroughly. Gently heat and bring to boiling. Autoclave for 15 min at 15 psi pressure–121°C. Cool to 45°–50°C. Aseptically add 10.0mL *Listeria mono* enrichment supplement II, 10.0mL *Listeria mono* selective supplement I, and 10.0mL *Listeria mono* selective supplement II. Mix thoroughly. Pour into sterile Petri dishes or distribute into sterile tubes.

**Storage/Shelf Life:** Store dehydrated media in the dark in a sealed container below 30°C. Prepared media should be stored under refrigeration (2-8°C). Media should be used within 60 days of preparation. Media should not be used if there are any signs of deterioration (shrinking, cracking, or discoloration) or contamination, or if the expiration date supplied by the manufacturer has passed.

**Use:** For the selective and differential isolation of *Listeria monocytogenes* from clinical and food specimens.

## *L. mono* Differential HiVeg Agar Base

**Composition** per liter:

Plant peptone No. 1..................................................18.0g
Agar .........................................................................15.0g
LiCl ..........................................................................10.0g
Yeast extract ............................................................10.0g
Plant hydrolysate .......................................................6.0g
NaCl ..........................................................................5.0g
Na$_2$HPO$_4$, anhydrous ..............................................2.5g
Glucose ......................................................................2.0g
Sodium pyruvate ........................................................2.0g
Magnesium glycerophosphate ...................................1.0g
MgSO$_4$ ......................................................................0.5g
Chromogenic substrate ............................................0.05g
*Listeria mono* enrichment supplement II ...............10.0mL
*Listeria mono* selective supplement I ....................10.0mL
*Listeria mono* selective supplement II ...................10.0mL
pH 7.2 ± 0.2 at 25°C

**Source:** This medium, without *Listeria mono* enrichment supplement II, *Listeria mono* selective supplement I, and *Listeria mono* selective supplement II, is available as a premixed powder from HiMedia.

**Caution:** LiCl is harmful. Avoid bodily contact and inhalation of vapors. On contact with skin wash with plenty of water immediately.

*Listeria mono* **Enrichment Supplement II Solution:**
**Composition** per 10.0mL:

L-phosphatidylinositol ...............................................1.0g

**Preparation of *Listeria mono* Enrichment Supplement II Solution:** Add L-phosphatidylinositol to distilled/deionized water and bring volume to 10.0mL. Mix thoroughly. Filter sterilize.

*Listeria mono* **Selective Supplement I Solution:**
**Composition** per 10.0mL:

Polymyxin B sulfate .............................................76,700U

**Preparation of *Listeria mono* Selective Supplement I Solution:** Add polymyxin B sulfate to distilled/deionized water and bring volume to 10.0mL. Mix thoroughly. Filter sterilize.

*Listeria mono* **Selective Supplement II Solution:**
**Composition** per 10.0mL:

Ceftazidime ............................................................20.0mg
Nalidixic acid, sodium salt .....................................20.0mg
Amphotericin B ......................................................10.0mg

**Preparation of *Listeria mono* Selective Supplement II Solution:** Add components to distilled/deionized water and bring volume to 10.0mL. Mix thoroughly. Filter sterilize.

**Preparation of Medium:** Add components, except *Listeria mono* enrichment supplement II, *Listeria mono* selective supplement I, and *Listeria mono* selective supplement II, to distilled/deionized water and bring volume to 970.0mL. Mix thoroughly. Gently heat and bring to boiling. Autoclave for 15 min at 15 psi pressure–121°C. Cool to 45°–50°C. Aseptically add 10.0mL *Listeria mono* enrichment supplement II, 10.0mL *Listeria mono* selective supplement I, and 10.0mL *Listeria mono* selective supplement II. Mix thoroughly. Pour into sterile Petri dishes or distribute into sterile tubes.

**Storage/Shelf Life:** Store dehydrated media in the dark in a sealed container below 30°C. Prepared media should be stored under refrigeration (2-8°C). Media should be used within 60 days of preparation. Media should not be used if there are any signs of deterioration (shrinking, cracking, or discoloration) or contamination, or if the expiration date supplied by the manufacturer has passed.

**Use:** For the selective and differential isolation of *Listeria monocytogenes* from clinical and food specimens.

## Lachica's Medium Base

**Composition** per liter:
Beef heart, infusion from ...........................................................500.0g
Agar ...............................................................................................15.0g
Tryptose .........................................................................................10.0g
NaCl..................................................................................................5.0g
Amylose Azure ................................................................................3.0g
Selective supplement solution .................................... 10.0mL
pH 7.4 ± 0.2 at 25°C

**Source:** This medium is available from HiMedia.

**Selective Supplement Solution:**
**Composition** per 10.0mL:
Ampicillin .....................................................................................10.0mg

**Preparation of Selective Supplement Solution:** Add ampicillin to distilled/deionized water and bring volume to 10.0mL. Mix thoroughly. Filter sterilize.

**Preparation of Medium:** Add components, except selective supplement solution, to distilled/deionized water and bring volume to 990.0mL. Mix thoroughly. Autoclave for 15 min at 15 psi pressure–121°C. Cool to 50°C. Aseptically add selective supplement solution. Mix thoroughly. Pour into Petri dishes or aseptically distribute into sterile tubes.

**Storage/Shelf Life:** Store dehydrated media in the dark in a sealed container below 30°C. Prepared media should be stored under refrigeration (2-8°C). Media should be used within 60 days of preparation. Media should not be used if there are any signs of deterioration (shrinking, cracking, or discoloration) or contamination, or if the expiration date supplied by the manufacturer has passed.

**Use:** For the isolation and cultivation of *Aeromonas hydrophila* from foods stored under different temperature conditions.

## Lactose Blue Agar
## (B.T.B. Lactose Agar, Modified)

**Composition** per liter:
Lactose ...........................................................................................15.5g
Agar .................................................................................................13.0g
NaCl..................................................................................................5.0g
Peptic digest of animal tissue............................................ 3.5g
Casein enzymic hydrolysate .......................................................3.5g
Bromthymol Blue ........................................................................0.04g
pH 7.0 ± 0.2 at 25°C

**Source:** This medium is available as a premixed powder from HiMedia.

**Preparation of Medium:** Add components to distilled/deionized water and bring volume to 1.0mL. Mix thoroughly. Gently heat and bring to boiling. Autoclave for 15 min at 15 psi pressure–121°C. Pour into sterile Petri dishes.

**Storage/Shelf Life:** Store dehydrated media in the dark in a sealed container below 30°C. Prepared media should be stored under refrigeration (2-8°C). Media should be used within 60 days of preparation. Media should not be used if there are any signs of deterioration (shrinking, cracking, or discoloration) or contamination, or if the expiration date supplied by the manufacturer has passed.

**Use:** For the differentiation of lactose-fermenting and non-fermenting bacteria belonging to Enterobacteriaceae.

## Lactose Blue HiVeg Agar
## (B.T.B. Lactose HiVeg Agar, Modified)

**Composition** per liter:
Lactose ...........................................................................................15.5g
Agar .................................................................................................13.0g
NaCl..................................................................................................5.0g
Plant extract .................................................................................. 3.5g
Plant hydrolysate ..........................................................................3.5g
Bromthymol Blue .......................................................................0.04g
pH 7.0 ± 0.2 at 25°C

**Source:** This medium is available as a premixed powder from HiMedia.

**Preparation of Medium:** Add components to distilled/deionized water and bring volume to 1.0mL. Mix thoroughly. Gently heat and bring to boiling. Autoclave for 15 min at 15 psi pressure–121°C. Pour into sterile Petri dishes.

**Storage/Shelf Life:** Store dehydrated media in the dark in a sealed container below 30°C. Prepared media should be stored under refrigeration (2-8°C). Media should be used within 60 days of preparation. Media should not be used if there are any signs of deterioration (shrinking, cracking, or discoloration) or contamination, or if the expiration date supplied by the manufacturer has passed.

**Use:** For the differentiation of lactose-fermenting and non-fermenting bacteria belonging to Enterobacteriaceae.

## Lactose HiVeg Broth

**Composition** per liter:
Plant peptone ................................................................................5.0g
Lactose............................................................................................5.0g
Plant extract.................................................................................3.0g
pH 7.0 ± 0.2 at 25°C

**Source:** This medium is available as a premixed powder from HiMedia.

**Preparation of Medium:** Add components to distilled/deionized water and bring volume to 1.0mL. Mix thoroughly. Gently heat and bring to boiling. Autoclave for 15 min at 15 psi pressure–121°C. Pour into sterile Petri dishes.

**Storage/Shelf Life:** Store dehydrated media in the dark in a sealed container below 30°C. Prepared media should be stored under refrigeration (2-8°C). Media should be used within 60 days of preparation. Media should not be used if there are any signs of deterioration (discoloration) or contamination, or if the expiration date supplied by the manufacturer has passed.

**Use:** For the detection of coliform bacteria in water, foods, and dairy products as per standard methods.

## Lactose Lecithin Agar

**Composition** per liter:
Agar .................................................................................................15.0g
Casein enzymic hydrolysate .....................................................12.65g
Lactose............................................................................................10.0g
Peptic digest of animal tissue .................................................5.5g
NaCl..................................................................................................5.5g
Yeast extract..................................................................................3.85g

| | |
|---|---|
| Pancreatic digest of heart muscle | 3.3g |
| Corn starch | 1.1g |
| Egg lecithin | 0.66g |
| L-Cysteine·HCl·H$_2$O | 0.5g |
| NaN$_3$ | 0.2g |
| Neomycin sulfate | 0.15g |
| CaCl$_2$·2H$_2$O | 0.05g |
| Bromcresol Purple | 0.025g |

pH 6.8 ± 0.2 at 25°C

**Source:** This medium is available from HiMedia.

**Caution:** Sodium azide has a tendency to form explosive metal azides with plumbing materials. It is advisable to use enough water to flush off the disposables.

**Preparation of Medium:** Add components to distilled/deionized water and bring volume to 1.0L. Mix thoroughly. Gently heat and bring to boiling. Distribute into tubes or flasks. Autoclave for 15 min at 15 psi pressure–121°C. Pour into sterile Petri dishes or leave in tubes.

**Storage/Shelf Life:** Store dehydrated media in the dark in a sealed container below 30°C. Prepared media should be stored under refrigeration (2-8°C). Media should be used within 60 days of preparation. Media should not be used if there are any signs of deterioration (shrinking, cracking, or discoloration) or contamination, or if the expiration date supplied by the manufacturer has passed.

**Use:** For the isolation and differentiation of histotoxic clostridia from clinical specimens.

## Lactose Ricinoleate Broth

**Composition** per liter:

| | |
|---|---|
| Lactose | 10.0g |
| Peptone | 5.0g |
| Sodium ricinoleate | 1.0g |

pH 7.6 ± 0.2 at 25°C

**Preparation of Medium:** Add components to distilled/deionized water and bring volume to 1.0L. Mix thoroughly. Distribute into tubes or flasks. Autoclave for 15 min at 15 psi pressure–121°C.

**Storage/Shelf Life:** Store dehydrated media in the dark in a sealed container below 30°C. Prepared media should be stored under refrigeration (2-8°C). Media should be used within 60 days of preparation. Media should not be used if there are any signs of deterioration (discoloration) or contamination.

**Use:** For the selective cultivation of members of the Enterobacteriaceae.

## Lactose Sulfite Broth Base

**Composition** per liter:

| | |
|---|---|
| Lactose | 10.0g |
| Casein enzymic hydrolysate | 5.0g |
| Yeast extract | 2.5g |
| NaCl | 2.5g |
| L-Cysteine·HCl·H$_2$O | 0.3g |
| Sodium metabisulfite solution | 50.0mL |
| Ferric ammonium citrate | 50.0mL |

pH 7.1 ± 0.2 at 25°C

**Source:** This medium is available from HiMedia.

**Selective Sodium Metabisulfite Solution:**
**Composition** per 100.0mL:

| | |
|---|---|
| Sodium metabisulfite | 1.2g |

**Preparation of Sodium Metabisulfite Solution:** Add components to distilled/deionized water and bring volume to 100.0mL. Mix thoroughly. Filter sterilize.

**Selective Ferric Ammonium Citrate Solution:**
**Composition** per 100.0mL:

| | |
|---|---|
| Ferric ammonium citrate | 1.0g |

**Preparation of Ferric Ammonium Citrate Solution:** Add components to distilled/deionized water and bring volume to 100.0mL. Mix thoroughly. Filter sterilize.

**Preparation of Medium:** Add components, except sodium metabisulfite and ferric ammonium citrate solutions, to distilled/deionized water and bring volume to 999.0mL. Mix thoroughly. Autoclave for 15 min at 15 psi pressure–121°C. Cool to 50°C. Aseptically add sodium metabisulfite and ferric ammonium citrate solutions. Mix thoroughly. Pour into Petri dishes or aseptically distribute into sterile tubes.

**Storage/Shelf Life:** Store dehydrated media in the dark in a sealed container below 30°C. Prepared media should be stored under refrigeration (2-8°C). Media should be used within 60 days of preparation. Media should not be used if there are any signs of deterioration (discoloration) or contamination, or if the expiration date supplied by the manufacturer has passed.

**Use:** For the detection and enumeration of *Clostridium perfringens* in pharmaceutical products.

## Lash Serum Medium

**Composition** per liter:

| | |
|---|---|
| Casamino acids | 14.0g |
| NaCl | 6.0g |
| Glucose | 2.0g |
| Maltose | 1.5g |
| Sodium lactate (60% solution) | 0.5g |
| KCl | 0.1g |
| CaCl$_2$·2H$_2$O | 0.1g |
| Serum solution | 500.0mL |

pH 5.8 ± 0.2 at 25°C

**Serum Solution:**
**Composition** per 500.0mL:

| | |
|---|---|
| NaHCO$_3$ | 0.1g |
| Bovine serum | 200.0mL |

**Preparation of Serum Solution:** Add components to distilled/deionized water and bring volume to 500.0mL. Mix thoroughly. Filter sterilize.

**Preparation of Medium:** Add components, except serum solution, to distilled/deionized water and bring volume to 500.0mL. Mix thoroughly. Distribute into tubes in 5.0mL volumes. Autoclave for 15 min at 15 psi pressure–121°C. Cool to 25°C. Aseptically add 5.0mL of sterile serum solution to each tube. Mix thoroughly.

**Storage/Shelf Life:** Store dehydrated media in the dark in a sealed container below 30°C. Prepared media should be stored under refrigeration (2-8°C). Media should be used within 60 days of preparation. Media should not be used if there are any signs of deterioration (discoloration) or contamination, or if the expiration date supplied by the manufacturer has passed.

**Use:** For the cultivation of *Trichomonas vaginalis* from clinical specimens.

## Lauryl Sulfate Broth
## (m-Lauryl Sulfate Broth)

**Composition** per liter:

Peptone.............................................................................39.0g
Lactose.............................................................................30.0g
Yeast extract......................................................................6.0g
Sodium lauryl sulfate.........................................................1.0g
Phenol Red........................................................................0.2g

pH 7.4 ± 0.2 at 25°C

**Source:** This medium is available as a premixed powder from Oxoid Unipath.

**Preparation of Medium:** Add components to distilled/deionized water and bring volume to 1.0L. Mix thoroughly. Distribute into bottles or flasks. Autoclave for 15 min at 15 psi pressure–121°C.

**Storage/Shelf Life:** Store dehydrated media in the dark in a sealed container below 30°C. Prepared media should be stored under refrigeration (2-8°C). Media should be used within 60 days of preparation. Media should not be used if there are any signs of deterioration (discoloration) or contamination, or if the expiration date supplied by the manufacturer has passed.

**Use:** For the cultivation and enumeration of coliform bacteria, especially *Escherichia coli*, in water by the membrane filter method.

## Lauryl Sulfate Broth
## (Lauryl Tryptose Broth)

**Composition** per liter:

Pancreatic digest of casein..............................................20.0g
Lactose...............................................................................5.0g
NaCl...................................................................................5.0g
$K_2HPO_4$..........................................................................2.75g
$KH_2PO_4$..........................................................................2.75g
Sodium lauryl sulfate.........................................................0.1g

pH 6.8 ± 0.2 at 25°C

**Source:** This medium is available as a premixed powder from BD Diagnostic Systems.

**Preparation of Medium:** Add components to distilled/deionized water and bring volume to 1.0L. Mix thoroughly. Distribute into tubes containing an inverted Durham tube in 10.0mL volumes. Autoclave for 12 min at 15 psi pressure–121°C. Cool broth quickly to 25°C. For testing water samples with 10.0mL volumes, prepare medium double strength.

**Storage/Shelf Life:** Store dehydrated media in the dark in a sealed container below 30°C. Prepared media should be stored under refrigeration (2-8°C). Media should be used within 60 days of preparation. Media should not be used if there are any signs of deterioration (discoloration) or contamination, or if the expiration date supplied by the manufacturer has passed.

**Use:** For the detection of coliform bacteria in a variety of specimens. Also, for the enumeration of coliform bacteria by the multiple-tube fermentation technique.

## Lauryl Sulfate Broth, Fluorocult
## (Fluorocult Lauryl Sulfate Broth)

**Composition** per liter:

Tryptose............................................................................20.0g
Lactose...............................................................................5.0g
NaCl...................................................................................5.0g
$K_2HPO_4$..........................................................................2.75g

$KH_2PO_4$..........................................................................2.75g
L-tryptophan........................................................................1.g
Sodium lauryl sulfate.........................................................0.1g
4-Methylumbelliferyl-ß-D-glucuronide.............................0.1g

pH 6.8 ± 0.2 at 25°C

**Source:** This medium is available from Merck.

**Preparation of Medium:** Add components to distilled/deionized water and bring volume to 1.0L. Mix thoroughly. Gently heat and bring to boiling. Cool. Distribute into test tubes containing inverted Durham tubes. Autoclave for 15 min at 15 psi pressure–121°C. The prepared broth is clear and yellowish-brown.

**Storage/Shelf Life:** Store in the dark under refrigeration (2-8°C). Chromogenic media are especially light and temperature sensitive; protect from light, excessive heat, moisture, and freezing. Do not use after the expiration date supplied by the manufacturer.

**Use:** For the detection of *E. coli* in milk. The medium complies with the German-DIN-Norm 10183 for the examination of milk and food. This medium meets the requirements of ISO/DIS 11886-2.2 (1994) for milk and milk products. The lauryl sulfate largely inhibits the growth of undesirable microbial flora. The presence of *E. coli* is indicated by fluorescence under a long wavelength UV lamp. A positive indole reaction and gas formation due to fermentation of lactose confirm the results.

## Lauryl Sulfate Broth with MUG

**Composition** per liter:

Pancreatic digest of casein..............................................20.0g
Lactose...............................................................................5.0g
NaCl...................................................................................5.0g
$K_2HPO_4$..........................................................................2.75g
$KH_2PO_4$..........................................................................2.75g
Sodium lauryl sulfate.........................................................0.1g
4-Methylumbelliferyl-β-D-glucuronide (MUG)................0.05g

pH 6.8 ± 0.2 at 25°C

**Source:** This medium is available as a premixed powder from BD Diagnostic Systems.

**Preparation of Medium:** Add components to distilled/deionized water and bring volume to 1.0L. Mix thoroughly. Distribute into tubes containing an inverted Durham tube in 10.0mL volumes. Autoclave for 12 min at 15 psi pressure–121°C. Cool broth quickly to 25°C. For testing water samples with 10.0mL volumes, prepare medium double strength.

**Storage/Shelf Life:** Store dehydrated media in the dark in a sealed conainer below 30°C. Prepared media should be stored under refrigeration (2-8°C). This medium has a limited shelf life. Media should not be used if there are any signs of deterioration (discoloration) or contamination, or if the expiration date supplied by the manufacturer has passed.

**Use:** For the detection of *Escherichia coli* in water and food samples by a fluorogenic procedure.

## Lauryl Sulfate HiVeg Broth
## (Lauryl Tryptose HiVeg Broth)

**Composition** per liter:

Plant hydrolysate No. 1....................................................20.0g
NaCl...................................................................................5.0g
Lactose...............................................................................5.0g
$K_2HPO_4$..........................................................................2.75g

KH$_2$PO$_4$.............................................................2.75g
Sodium lauryl sulfate .......................................0.1g
pH 6.8 ± 0.2 at 25°C

**Source:** This medium is available as a premixed powder from HiMedia.

**Preparation of Medium:** Add components to distilled/deionized water and bring volume to 1.0L. Mix thoroughly. Distribute into bottles or flasks. Autoclave for 15 min at 15 psi pressure–121°C.

**Storage/Shelf Life:** Store dehydrated media in the dark in a sealed container below 30°C. Prepared media should be stored under refrigeration (2-8°C). Media should be used within 60 days of preparation. Media should not be used if there are any signs of deterioration (discoloration) or contamination, or if the expiration date supplied by the manufacturer has passed.

**Use:** For the cultivation and enumeration of coliform bacteria in water, wastewater, dairy products, and other foods.

## Lauryl Tryptose Broth with MUG
## (Lauryl Sulfate Broth with MUG)
## (LST-MUG)
## (BAM M77)

**Composition** per liter:
Pancreatic digest of casein ..........................20.0g
Lactose ................................................................5.0g
NaCl ...................................................................5.0g
K$_2$HPO$_4$.............................................................2.75g
KH$_2$PO$_4$.............................................................2.75g
Sodium lauryl sulfate .......................................0.1g
4-Methylumbelliferyl-β-D-glucuronide (MUG)...........0.05g
pH 6.8 ± 0.2 at 25°C

**Source:** This medium is available as a premixed powder from BD Diagnostic Systems.

**Preparation of Medium:** Add components to distilled/deionized water and bring volume to 1.0L. Mix thoroughly. Distribute into tubes containing an inverted Durham tube in 10.0mL volumes. Autoclave for 12 min at 15 psi pressure–121°C. Cool broth quickly to 25°C. For testing water samples with 10.0mL volumes, prepare medium double strength.

**Storage/Shelf Life:** Store dehydrated media in the dark in a sealed conainer below 30°C. Prepared media should be stored under refrigeration (2-8°C). This medium has a limited shelf life. Media should not be used if there are any signs of deterioration (discoloration) or contamination, or if the expiration date supplied by the manufacturer has passed.

**Use:** For the detection of *Escherichia coli* in water and food samples by a fluorogenic procedure.

## Lauryl Tryptose Mannitol Broth
## with Tryptophan

**Composition** per liter:
Pancreatic digest of casein ..........................20.0g
Lactose ................................................................5.0g
NaCl ...................................................................5.0g
K$_2$HPO$_4$.............................................................2.75g

KH$_2$PO$_4$.............................................................2.75g
Sodium lauryl sulfate .......................................0.1g
L-Tryptophan.....................................................0.2g
pH 6.8 ± 0.2 at 25°C

**Source:** This medium is available as a premixed powder from Oxoid Unipath.

**Preparation of Medium:** Add components to distilled/deionized water and bring volume to 1.0L. Mix thoroughly. Distribute into tubes containing an inverted Durham tube in 10.0mL volumes. Autoclave for 10 min at 10 psi pressure–115°C. Cool broth quickly to 25°C.

**Storage/Shelf Life:** Store dehydrated media in the dark in a sealed container below 30°C. Prepared media should be stored under refrigeration (2-8°C). Media should be used within 60 days of preparation. Media should not be used if there are any signs of deterioration (discoloration) or contamination, or if the expiration date supplied by the manufacturer has passed.

**Use:** For the detection of *Escherichia coli* in water samples.

## LD Esculin HiVeg Agar
## (Lombard-Dowell Esculin Agar, HiVeg)

**Composition** per liter:
Agar ..................................................................20.0g
Plant hydrolysate No. 1....................................5.0g
Yeast extract ......................................................5.0g
NaCl ...................................................................2.5g
Esculin ...............................................................1.0g
Ferric citrate.....................................................0.5g
L-Cystine ...........................................................0.4g
L-Tryptophan ....................................................0.2g
Fe$_4$(P$_2$O$_7$)$_3$·H$_2$O ..............................................0.01g
Vitamin K$_1$ .......................................................0.01g
pH 7.5 ± 0.2 at 25°C

**Source:** This medium is available as a premixed powder from HiMedia.

**Preparation of Medium:** Add components to distilled/deionized water and bring volume to 1.0L. Mix thoroughly. Distribute into tubes or flasks. Autoclave for 15 min at 15 psi pressure–121°C. Pour into sterile Petri plates.

**Storage/Shelf Life:** Store dehydrated media in the dark in a sealed container below 30°C. Prepared media should be stored under refrigeration (2-8°C). Media should be used within 60 days of preparation. Media should not be used if there are any signs of deterioration (shrinking, cracking, or discoloration) or contamination, or if the expiration date supplied by the manufacturer has passed.

**Use:** For the cultivation of a wide variety of anaerobic bacteria. For the differentiation of anaerobic bacteria based on esculin hydrolysis, H$_2$S production, and catalase production. Bacteria that hydrolyze esculin appear as colonies surrounded by a red-brown to dark brown zone. Bacteria that produce H$_2$S appear as black colonies.

## LD HiVeg Agar
## (Lombard-Dowell Agar, HiVeg)

**Composition** per liter:
Agar ..................................................................20.0g
Plant hydrolysate ..............................................5.0g
Yeast extract ......................................................5.0g
NaCl ...................................................................2.5g
L-Cystine ...........................................................0.4g

L-Tryptophan.................................................................0.2g
$Na_2SO_3$ ....................................................................0.1g
Vitamin $K_1$ ...............................................................0.01g
$Fe_4(P_2O_7)_3 \cdot H_2O$..................................................0.01g

pH 7.5 ± 0.2 at 25°C

**Source:** This medium is available as a premixed powder from HiMedia.

**Preparation of Medium:** Add components to distilled/deionized water and bring volume to 1.0L. Mix thoroughly. Distribute into tubes or flasks. Autoclave for 15 min at 15 psi pressure–121°C. Pour into sterile Petri plates.

**Storage/Shelf Life:** Store dehydrated media in the dark in a sealed container below 30°C. Prepared media should be stored under refrigeration (2-8°C). Media should be used within 60 days of preparation. Media should not be used if there are any signs of deterioration (shrinking, cracking, or discoloration) or contamination, or if the expiration date supplied by the manufacturer has passed.

**Use:** For the cultivation and identification of a variety of obligate anaerobic bacteria. For the cultivation of *Bacteroides* species, *Fusobacterium* species, *Clostridium* species, and nonspore-forming Gram-positive anaerobes.

## Lead Acetate Agar

**Composition** per liter:

Agar ...........................................................................15.0g
Peptone......................................................................15.0g
Proteose peptone .......................................................5.0g
Glucose ......................................................................1.0g
Lead acetate ...............................................................0.2g
$Na_2S_2O_3$ ................................................................0.08g

pH 6.6 ± 0.2 at 25°C

**Preparation of Medium:** Add components to distilled/deionized water and bring volume to 1.0L. Mix thoroughly. Gently heat and bring to boiling. Distribute into tubes or flasks. Autoclave for 15 min at 15 psi pressure–121°C. Pour into sterile Petri dishes or leave in tubes. Allow tubes to cool in a slanted position.

**Storage/Shelf Life:** Store dehydrated media in the dark in a sealed container below 30°C. Prepared media should be stored under refrigeration (2-8°C). Media should be used within 60 days of preparation. Media should not be used if there are any signs of deterioration (shrinking, cracking, or discoloration) or contamination, or if the expiration date supplied by the manufacturer has passed.

**Use:** For the cultivation and differentiation of Gram-negative coliform bacteria based on $H_2S$ production. Bacteria that produce $H_2S$ turn the medium brown.

## Lecithin Lipase Anaerobic Agar

**Composition** per liter:

Pancreatic digest of casein....................................40.0g
Agar .........................................................................25.0g
Yeast extract ..............................................................5.0g
$Na_2HPO_4 \cdot 12H_2O$................................................5.0g
Glucose ......................................................................2.0g
NaCl............................................................................2.0g
$KH_2PO_4$....................................................................1.0g
$MgSO_4 \cdot 7H_2O$ .....................................................0.1g
Egg yolk emulsion ...............................................100.0mL

pH 7.6 ± 0.2 at 25°C

**Egg Yolk Emulsion:**
**Composition**:

Chicken egg yolks..........................................................11
Whole chicken egg ..........................................................1

**Preparation of Egg Yolk Emulsion:** Soak eggs with 1:100 dilution of saturated mercuric chloride solution for 1 min. Crack eggs and separate yolks from whites. Mix egg yolks with 1 chicken egg. Filter sterilize.

**Preparation of Medium:** Add components, except egg yolk emulsion, to distilled/deionized water and bring volume to 900.0mL. Mix thoroughly. Gently heat and bring to boiling. Autoclave for 15 min at 15 psi pressure–121°C. Cool to 45°–50°C. Aseptically add sterile egg yolk emulsion. Mix thoroughly. Pour into sterile Petri dishes or distribute into sterile tubes.

**Storage/Shelf Life:** Store dehydrated media in the dark in a sealed container below 30°C. Prepared media should be stored under refrigeration (2-8°C). Media should be used within 60 days of preparation. Media should not be used if there are any signs of deterioration (shrinking, cracking, or discoloration) or contamination, or if the expiration date supplied by the manufacturer has passed.

**Use:** For the isolation, cultivation, and differentiation of *Clostridium* species based on lecithinase production and lipase production. Bacteria that produce lecithinase appear as colonies surrounded by a zone of insoluble precipitate. Bacteria that produce lipase appear as colonies with a pearly iridescent sheen.

## Lecithin Tween™ Medium
## (LT Medium)

**Composition** per liter:

Tween™ 80.................................................................30.0g
Agar .........................................................................15.0g
Pancreatic digest of casein.....................................10.0g
Peptic digest of animal tissue ................................10.0g
NaCl............................................................................5.0g
Lecithin......................................................................5.0g
$Na_2S_2O_3 \cdot 5H_2O$ .................................................5.0g
Glycerol .....................................................................3.0g
Histidine, free base ...................................................1.0g
Glucose ......................................................................1.0g

pH 7.5 ± 0.2 at 25°C

**Antibiotic Solution:**
**Composition** per 10.0mL:

5–Fluorocytosine .......................................................0.2g
Fosfomicin.................................................................0.1g
Ticarcillin..................................................................0.1g

**Preparation of Antibiotic Solution:** Add components to distilled/deionized water and bring volume to 10.0mL. Mix thoroughly. Filter sterilize.

**Preparation of Medium:** Add components, except antibiotic solution, to distilled/deionized water and bring volume to 990.0mL. Mix thoroughly. Gently heat and bring to boiling. Autoclave for 15 min at 15 psi pressure–121°C. Cool to 45°–50°C. Aseptically add sterile antibiotic solution. Mix thoroughly. Pour into sterile Petri dishes in 20.0mL volumes.

**Storage/Shelf Life:** Store dehydrated media in the dark in a sealed container below 30°C. Prepared media should be stored under refrigeration (2-8°C). Media should be used within 60 days of preparation. Media should not be used if there are any signs of deterioration (shrink-

ing, cracking, or discoloration) or contamination, or if the expiration date supplied by the manufacturer has passed.

**Use:** For the isolation and cultivation of multiresistant lipophilic *Corynebacterium* species, especially *Corynebacterium* group JK found primarily in infections in immunocompromised hosts and patients with prosthetic valve endocarditis.

## *Legionella* Agar Base
## (*Legionella* Medium)
## (BCYEα Agar, Modified)

**Composition** per liter:

| | |
|---|---|
| Agar | 17.0g |
| Yeast extract | 10.0g |
| ACES buffer (*N*-2-acetamido-2-aminoethane sulfonic acid) | 6.0g |
| Charcoal, activated | 1.5g |
| KOH | 1.5g |
| α-Ketoglutarate | 1.0g |
| *Legionella* agar enrichment | 10.0mL |

pH 6.85–7.0 at 25°C

**Source:** This medium is available as a prepared medium from BD Diagnostic Systems.

### *Legionella* Agar Enrichment:
**Composition** per 10.0mL:

| | |
|---|---|
| L-Cysteine·HCl·H$_2$O | 0.4g |
| Fe$_4$(P$_2$O$_7$)$_3$ | 0.25g |

**Preparation of *Legionella* Agar Enrichment:** Add components to distilled/deionized water and bring volume to 10.0mL. Mix thoroughly. Filter sterilize.

**Preparation of Medium:** Add components, except *Legionella* agar enrichment, to distilled/deionized water and bring volume to 990.0mL. Mix thoroughly. Gently heat to boiling. Autoclave for 15 min at 15 psi pressure–121°C. Cool to 50° C. Add 10.0mL of sterile *Legionella* agar enrichment. Adjust pH to 6.9 at 50°C by adding 4.0–4.5mL of 1.0*N* KOH. This is a critical step. Mix thoroughly. Pour into sterile Petri dishes in 20.0mL volumes. Swirl medium while pouring to keep charcoal in suspension.

**Storage/Shelf Life:** Store dehydrated media in the dark in a sealed container below 30°C. Prepared media should be stored under refrigeration (2-8°C). Media should be used within 60 days of preparation. Media should not be used if there are any signs of deterioration (shrinking, cracking, or discoloration) or contamination, or if the expiration date supplied by the manufacturer has passed.

**Use:** For the preparation of *Legionella* agars. For the isolation and cultivation of *Legionella* species from clinical and nonclinical materials.

## *Legionella pneumophila* Medium
## (Charcoal Yeast Extract Diphasic
## Blood Culture Medium)
## (Diphasic Blood Culture Buffered
## Charcoal Yeast Extract Medium)
## (CYE-DBCM)

**Composition** per liter:

| | |
|---|---|
| Agar phase | 500.0mL |
| Broth phase | 500.0mL |

pH 6.9–7.0 at 25°C

### Agar Phase:
**Composition** per 500.0mL:

| | |
|---|---|
| Agar | 17.0g |
| Charcoal, activated | 2.0g |

**Preparation of Agar Phase:** Add components to distilled/deionized water and bring volume to 500.0mL. Mix thoroughly. Gently heat and bring to boiling. Distribute in 20.0mL volumes into 125.0mL serum bottles with aluminum crimp seals and rubber stoppers. Autoclave for 20 min at 15 psi pressure–121°C. Cool to 50°C. Swirl medium to put charcoal in suspension. Allow agar to solidify so that a slant with a 6.0cm height is formed.

### Broth Phase:
**Composition** per 500.0mL:

| | |
|---|---|
| Yeast extract | 20.0g |
| L-Cysteine·HCl·H$_2$O solution | 0.4g |
| Fe(NO$_3$)$_3$·9H$_2$O solution | 0.1g |

**Preparation of Broth Phase:** Add yeast extract to distilled/deionized water and bring volume to 480.0mL. Mix thoroughly. Autoclave for 15 min at 15 psi pressure–121°C. Cool to 25°C. Aseptically add sterile L-cysteine·HCl·H$_2$O solution and Fe(NO$_3$)$_3$·9H$_2$O solution. Mix thoroughly. Adjust pH to 6.9 with 6.0mL of sterile 1*N* KOH.

### L-Cysteine·HCl·H$_2$O Solution:
**Composition** per 10.0mL:

| | |
|---|---|
| L-Cysteine·HCl·H$_2$O | 0.04g |

**Preparation of L-Cysteine·HCl·H$_2$O Solution:** Add L-cysteine·HCl·H$_2$O to distilled/deionized water and bring volume to 10.0mL. Mix thoroughly. Filter sterilize.

### Fe(NO$_3$)$_3$·9H$_2$O Solution:
**Composition** per 10.0mL:

| | |
|---|---|
| Fe(NO$_3$)$_3$·9H$_2$O | 0.04g |

**Preparation of Fe(NO$_3$)$_3$·9H$_2$O Solution:** Add Fe(NO$_3$)$_3$·9H$_2$O to distilled/deionized water and bring volume to 10.0mL. Mix thoroughly. Filter sterilize.

**Preparation of Medium:** Add 20.0mL of sterile broth phase to 125.0mL serum bottles containing 20.0mL of solidified agar phase. Seal bottles by crimping metal caps over rubber stoppers.

**Storage/Shelf Life:** Store dehydrated media in the dark in a sealed container below 30°C. Prepared media should be stored under refrigeration (2-8°C). Media should be used within 60 days of preparation. Media should not be used if there are any signs of deterioration (shrinking, cracking, or discoloration) or contamination, or if the expiration date supplied by the manufacturer has passed.

**Use:** For the isolation and cultivation of *Legionella pneumophila* from blood cultures.

## *Legionella* Selective Agar

**Composition** per liter:

| | |
|---|---|
| Agar | 15.0g |
| ACES (2-[(2-amino-2-oxoethyl)-amino]ethane sulfonic acid) buffer | 10.0g |
| Yeast extract | 10.0g |
| Charcoal, activated | 2.0g |
| α-Ketoglutarate | 1.0g |
| L-Cysteine·HCl·H$_2$O solution | 10.0mL |
| Fe$_4$(P$_2$O$_7$)$_3$ solution | 10.0mL |
| Antibiotic solution | 10.0mL |

pH 6.85–7.0 at 25°C

**Source:** This medium is available as a prepared medium from BD Diagnostic Systems.

### L-Cysteine·HCl·H₂O Solution:
**Composition** per 10.0mL:

L-Cysteine·HCl·H₂O ................................................................0.4g

**Preparation of L-Cysteine·HCl·H₂O Solution:** Add L-cysteine·HCl·H₂O to distilled/deionized water and bring volume to 10.0mL. Mix thoroughly. Filter sterilize.

### Fe₄(P₂O₇)₃ Solution:
**Composition** per 10.0mL:

Fe₄(P₂O₇)₃ ............................................................................0.25g

**Preparation of Fe₄(P₂O₇)₃ Solution:** Add Fe₄(P₂O₇)₃ to distilled/deionized water and bring volume to 10.0mL. Mix thoroughly. Filter sterilize.

### Antibiotic Solution:
**Composition** per 10.0mL:

Anisomycin .......................................................................10.0mg

Colistin ............................................................................3.75mg

Vancomycin .......................................................................2.0mg

**Preparation of Antibiotic Solution:** Add components to distilled/deionized water and bring volume to 10.0mL. Mix thoroughly. Filter sterilize.

**Preparation of Medium:** Add components—except L-cysteine·HCl·H₂O, Fe₄(P₂O₇)₃, and antibiotic solutions—to distilled/deionized water and bring volume to 970.0mL. Mix thoroughly. Gently heat and bring to boiling. Autoclave for 15 min at 15 psi pressure–121°C. Cool to 45°–50°C. Aseptically add sterile L-cysteine·HCl·H₂O, Fe₄(P₂O₇)₃, and antibiotic solutions. Mix thoroughly. Pour into sterile Petri dishes. Swirl medium while pouring to keep charcoal in suspension.

**Storage/Shelf Life:** Store dehydrated media in the dark in a sealed container below 30°C. Prepared media should be stored under refrigeration (2–8°C). Media should be used within 60 days of preparation. Media should not be used if there are any signs of deterioration (shrinking, cracking, or discoloration) or contamination, or if the expiration date supplied by the manufacturer has passed.

**Use:** *Legionella* selective agar is used in qualitative procedures for the isolation of *Legionella* species from clinical and nonclinical specimens.

### Leifson's Deoxycholate HiVeg Agar, Modified
### (Hugh Leifson Deoxycholate HiVeg Agar, Modified)
**Composition** per liter:

Agar .................................................................................15.0g

Lactose ............................................................................10.0g

Plant extract .......................................................................5.0g

Plant peptone......................................................................5.0g

Sodium citrate ....................................................................5.0g

Na₂S₂O₃ .............................................................................5.0g

Synthetic detergent No. III...................................................2.5g

Ferric citrate ......................................................................1.0g

Neutral Red .....................................................................0.025g

pH 7.0 ± 0.2 at 25°C

**Source:** This medium is available as a premixed powder from HiMedia.

**Preparation of Medium:** Add components to distilled/deionized water and bring volume to 1.0L. Mix thoroughly. Distribute into tubes or flasks. Gently bring to boiling. Do not autoclave. Pour into sterile Petri plates.

**Storage/Shelf Life:** Store dehydrated media in the dark in a sealed container below 30°C. Prepared media should be stored under refrigeration (2–8°C). Media should be used within 60 days of preparation. Media should not be used if there are any signs of deterioration (shrinking, cracking, or discoloration) or contamination, or if the expiration date supplied by the manufacturer has passed.

**Use:** For the selective isolation and differentiation of *Salmonella* and *Shigella* species.

## Letheen Agar
**Composition** per liter:

Agar .................................................................................15.0g

Tween™ 80 ..........................................................................7.0g

Pancreatic digest of casein..................................................5.0g

Beef extract........................................................................3.0g

Glucose ..............................................................................1.0g

Lecithin..............................................................................1.0g

pH 7.0 ± 0.2 at 25°C

**Source:** This medium is available as a premixed powder from BD Diagnostic Systems.

**Preparation of Medium:** Add components to distilled/deionized water and bring volume to 1.0L. Mix thoroughly. Gently heat and bring to boiling. Distribute into tubes or flasks. Autoclave for 15 min at 15 psi pressure–121°C. Pour into sterile Petri dishes or leave in tubes.

**Storage/Shelf Life:** Store dehydrated media in the dark in a sealed container below 30°C. Prepared media should be stored under refrigeration (2–8°C). Media should be used within 60 days of preparation. Media should not be used if there are any signs of deterioration (shrinking, cracking, or discoloration) or contamination, or if the expiration date supplied by the manufacturer has passed.

**Use:** For determination of the antimicrobial activity of quaternary ammonium compounds.

## Letheen Agar, Modified
**Composition** per liter:

Agar .................................................................................20.0g

Thiotone............................................................................10.0g

Pancreatic digest of casein.................................................10.0g

Tween™ 80 ..........................................................................7.0g

NaCl ..................................................................................5.0g

Beef extract........................................................................3.0g

Yeast extract.......................................................................2.0g

Glucose ..............................................................................1.0g

Lecithin..............................................................................1.0g

NaHSO₃ ..............................................................................0.1g

pH 7.2 ± 0.2 at 25°C

**Preparation of Medium:** Add components to distilled/deionized water and bring volume to 1.0L. Mix thoroughly. Gently heat and bring to boiling. Distribute into tubes or flasks. Autoclave for 15 min at 15 psi pressure–121°C. Pour into sterile Petri dishes.

**Storage/Shelf Life:** Store dehydrated media in the dark in a sealed container below 30°C. Prepared media should be stored under refrigeration (2–8°C). Media should be used within 60 days of preparation. Media should not be used if there are any signs of deterioration (shrinking, cracking, or discoloration) or contamination, or if the expiration date supplied by the manufacturer has passed.

**Use:** For determination of the antimicrobial activity of quaternary ammonium compounds.

## Letheen HiVeg Agar

**Composition** per liter:

| | |
|---|---|
| Agar | 15.0g |
| Polysorbate 80 | 7.0g |
| Plant hydrolysate | 5.0g |
| Plant extract | 3.0g |
| Glucose | 1.0g |
| Lecithin | 1.0g |

pH 7.0 ± 0.2 at 25°C

**Source:** This medium is available as a premixed powder from HiMedia.

**Preparation of Medium:** Add components to distilled/deionized water and bring volume to 1.0L. Mix thoroughly. Gently heat and bring to boiling. Distribute into tubes or flasks. Autoclave for 15 min at 15 psi pressure–121°C. Pour into sterile Petri dishes or leave in tubes.

**Storage/Shelf Life:** Store dehydrated media in the dark in a sealed container below 30°C. Prepared media should be stored under refrigeration (2-8°C). Media should be used within 60 days of preparation. Media should not be used if there are any signs of deterioration (shrinking, cracking, or discoloration) or contamination, or if the expiration date supplied by the manufacturer has passed.

**Use:** For the determination of the antimicrobial activity of quaternary ammonium compounds.

## Letheen HiVeg Agar, Modified

**Composition** per liter:

| | |
|---|---|
| Agar | 20.0g |
| Plant peptone | 20.0g |
| Plant extract | 5.0g |
| Plant hydrolysate | 5.0g |
| NaCl | 5.0g |
| Polysorbate 80 | 5.0g |
| Yeast extract | 2.0g |
| Lecithin | 0.7g |
| NaHSO$_3$ | 0.1g |

pH 7.0 ± 0.2 at 25°C

**Source:** This medium is available as a premixed powder from HiMedia.

**Preparation of Medium:** Add components to distilled/deionized water and bring volume to 1.0L. Mix thoroughly. Gently heat and bring to boiling. Distribute into tubes or flasks. Autoclave for 15 min at 15 psi pressure–121°C. Pour into sterile Petri dishes.

**Storage/Shelf Life:** Store dehydrated media in the dark in a sealed container below 30°C. Prepared media should be stored under refrigeration (2-8°C). Media should be used within 60 days of preparation. Media should not be used if there are any signs of deterioration (shrinking, cracking, or discoloration) or contamination, or if the expiration date supplied by the manufacturer has passed.

**Use:** For the screening of cosmetic products for microbial contamination.

## Letheen HiVeg Broth, AOAC

**Composition** per liter:

| | |
|---|---|
| Plant peptone | 10.0g |
| Plant extract | 5.0g |
| Polysorbate 80 | 5.0g |
| NaCl | 5.0g |
| Lecithin | 0.7g |

pH 7.0 ± 0.2 at 25°C

**Source:** This medium is available as a premixed powder from HiMedia.

**Preparation of Medium:** Add components to distilled/deionized water and bring volume to 1.0L. Mix thoroughly. Gently heat and bring to boiling. Distribute into tubes or flasks. Autoclave for 15 min at 15 psi pressure–121°C.

**Storage/Shelf Life:** Store dehydrated media in the dark in a sealed container below 30°C. Prepared media should be stored under refrigeration (2-8°C). Media should be used within 60 days of preparation. Media should not be used if there are any signs of deterioration (discoloration) or contamination, or if the expiration date supplied by the manufacturer has passed.

**Use:** For the determination of the antimicrobial activity of quaternary ammonium compounds.

## Letheen HiVeg Broth, Modified

**Composition** per liter:

| | |
|---|---|
| Plant peptone | 20.0g |
| Plant extract | 5.0g |
| Plant hydrolysate | 5.0g |
| NaCl | 5.0g |
| Polysorbate 80 | 5.0g |
| Yeast extract | 2.0g |
| Lecithin | 0.7g |
| NaHSO$_3$ | 0.1g |

pH 7.0 ± 0.2 at 25°C

**Source:** This medium is available as a premixed powder from HiMedia.

**Preparation of Medium:** Add components to distilled/deionized water and bring volume to 1.0L. Mix thoroughly. Gently heat and bring to boiling. Distribute into tubes or flasks. Autoclave for 15 min at 15 psi pressure–121°C.

**Storage/Shelf Life:** Store dehydrated media in the dark in a sealed container below 30°C. Prepared media should be stored under refrigeration (2-8°C). Media should be used within 60 days of preparation. Media should not be used if there are any signs of deterioration (discoloration) or contamination, or if the expiration date supplied by the manufacturer has passed.

**Use:** For the screening of cosmetic products for microbial contamination.

## Levine EMB Agar
## (Levine Eosin Methylene Blue Agar)
## (Eosin Methylene Blue Agar, Levine)
## (LEMB Agar)

**Composition** per liter:

| | |
|---|---|
| Agar | 15.0g |
| Lactose | 10.0g |
| Peptone | 10.0g |
| K$_2$HPO$_4$ | 2.0g |
| Eosin Y | 0.4g |
| Methylene Blue | 0.065mg |

pH 7.1 ± 0.2 at 25°C

**Source:** This medium is available as a premixed powder from BD Diagnostic Systems.

**Preparation of Medium:** Add components to distilled/deionized water and bring volume to 1.0L. Mix thoroughly. Gently heat and bring to boiling. Distribute into tubes or flasks. Autoclave for 15 min at 15 psi pressure–121°C. Pour into sterile Petri dishes or leave in tubes.

**Storage/Shelf Life:** Store dehydrated media in the dark in a sealed container below 30°C. Prepared media should be stored under refrigeration (2-8°C). Media should be used within 60 days of preparation. Media should not be used if there are any signs of deterioration (shrinking, cracking, or discoloration) or contamination, or if the expiration date supplied by the manufacturer has passed.

**Use:** For the isolation, cultivation, and differentiation of Gram-negative enteric bacteria based on lactose fermentation. Bacteria that ferment lactose, especially the coliform bacterium *Escherichia coli*, appear as colonies with a green metallic sheen or blue-black to brown color. Bacteria that do not ferment lactose appear as colorless or transparent light purple colonies.

## LICNR Broth
### (Lysine Iron Cystine Neutral Red Broth)
**Composition** per 500.0mL:

| | |
|---|---|
| L-Lysine·HCl | 10.0g |
| Mannitol | 5.0g |
| Pancreatic digest of casein | 5.0g |
| Yeast extract | 3.0g |
| Glucose | 1.0g |
| Salicin | 1.0g |
| Ferric ammonium citrate | 0.5g |
| L-Cystine | 0.1g |
| Na$_2$S$_2$O$_3$ | 0.1g |
| Neutral Red | 0.025g |
| Novobiocin solution | 10.0mL |

pH 6.2 ± 0.2 at 25°C

**Novobiocin Solution:**
**Composition** per 10.0mL:

| | |
|---|---|
| Novobiocin | 0.015g |

**Preparation of Novobiocin Solution:** Add novobiocin to distilled/deionized water and bring volume to 10.0mL. Mix thoroughly. Filter sterilize.

**Preparation of Medium:** Add components, except novobiocin solution, to distilled/deionized water and bring volume to 990.0mL. Mix thoroughly. Gently heat and bring to boiling. Continue boiling for 2–3 min. Distribute into tubes in 10.0mL volumes. Autoclave for 15 min at 15 psi pressure–121°C. Cool to 45°–50°C. Aseptically add 0.1mL of sterile novobiocin solution to each tube. Mix thoroughly.

**Storage/Shelf Life:** Store dehydrated media in the dark in a sealed container below 30°C. Prepared media should be stored under refrigeration (2-8°C). Media should be used within 60 days of preparation. Media should not be used if there are any signs of deterioration (discoloration) or contamination, or if the expiration date supplied by the manufacturer has passed.

**Use:** For the rapid, presumptive detection of *Salmonella* in foods, food ingredients, and feed materials.

## Lim Broth
**Composition** per liter:

| | |
|---|---|
| Pancreatic digest of casein | 20.0g |
| Yeast extract | 10.0g |
| Infusion from heart, solids | 3.1g |
| Glucose | 2.0g |
| NaHCO$_3$ | 2.5g |
| NaCl | 2.0g |
| Na$_2$HPO$_4$ | 0.4g |
| Antibiotic solution | 10.0mL |

pH 7.8 ± 0.2 at 25°C

**Source:** This medium is available as a premixed powder from BD Diagnostic Systems.

**Antibiotic Solution:**
**Composition** per 10.0mL:

| | |
|---|---|
| Colistin | 0.01 g |
| Nalidixic acid | 0.015 g |

**Preparation of Antibiotic Solution:** Add components to distilled/deionized water and bring volume to 10.0mL. Mix thoroughly. Filter sterilize.

**Preparation of Medium:** Add components—except antibiotic solution—to distilled/deionized water and bring volume to 990.0mL. Mix thoroughly. Gently heat and bring to boiling. Autoclave for 15 min at 15 psi pressure–121°C. Cool to 45°–50°C. Aseptically add 10.0mL of sterile antibiotic solution. Mix thoroughly. Aseptically distribute into sterile tubes.

**Storage/Shelf Life:** Store dehydrated media in the dark in a sealed container below 30°C. Prepared media should be stored under refrigeration (2-8°C). Media should be used within 60 days of preparation. Media should not be used if there are any signs of deterioration (discoloration) or contamination, or if the expiration date supplied by the manufacturer has passed.

**Use:** For the enhanced growth of Group B streptococci and the specific isolation and cultivation of *Streptococcus agalactiae*.

## Lipovitellin Salt Mannitol Agar
**Composition** per liter:

| | |
|---|---|
| NaCl | 75.0g |
| Egg yolk | 20.0g |
| Agar | 15.0g |
| D-Mannitol | 10.0g |
| Polypeptone™ | 10.0g |
| Beef extract | 1.0g |
| Phenol Red | 0.025g |

**Preparation of Medium:** Add components to distilled/deionized water and bring volume to 1.0L. Mix thoroughly. Gently heat and bring to boiling. Distribute into tubes or flasks. Autoclave for 15 min at 15 psi pressure–121°C. Pour into sterile Petri dishes or leave in tubes.

**Storage/Shelf Life:** Store dehydrated media in the dark in a sealed container below 30°C. Prepared media should be stored under refrigeration (2-8°C). Media should be used within 60 days of preparation. Media should not be used if there are any signs of deterioration (shrinking, cracking, or discoloration) or contamination, or if the expiration date supplied by the manufacturer has passed.

**Use:** For the detection of *Staphylococcus aureus* in swimming pool water based on lipovitellin-lipase activity and mannitol fermentation. *Staphylococcus aureus* and other bacteria with lipovitellin-lipase activity attack the egg yolk and appear as colonies surrounded by an opaque zone. Bacteria that ferment mannitol appear as colonies surrounded by a yellow zone.

## Liquoid Broth

**Composition** per liter:

| | |
|---|---|
| Beef heart, infusion from | 250.0g |
| Calf brain, infusion from | 200.0g |
| Proteose peptone | 10.0g |
| NaCl | 5.0g |
| Na$_2$HPO$_4$ | 2.5g |
| Glucose | 2.0g |
| Sodium polyanethol sulfonate | 0.5g |

pH 7.4 ± 0.2 at 25°C

**Source:** This medium is available from HiMedia.

**Preparation of Medium:** Add components to distilled/deionized water and bring volume to 1.0L. Mix thoroughly. Gently heat and bring to boiling. Distribute into tubes or flasks. Autoclave for 15 min at 15 psi pressure–121°C.

**Storage/Shelf Life:** Store dehydrated media in the dark in a sealed container below 30°C. Prepared media should be stored under refrigeration (2-8°C). Media should be used within 60 days of preparation. Media should not be used if there are any signs of deterioration (discoloration) or contamination.

**Use:** For the screening of blood specimens from suspected cases of bacteremia.

## *Listeria* Chromogenic Agar

**Composition** per liter:

| | |
|---|---|
| Peptone | 18.0g |
| Agar | 13.5g |
| Yeast extract | 10.0g |
| LiCl | 10.0g |
| Tryptone | 6.0g |
| NaCl | 5.0g |
| Glucose | 2.0g |
| Na$_2$HPO$_4$ | 2.5g |
| Sodium pyruvate | 2.0g |
| Magnesium glycerophosphate X-glucoside | 1.0g |
| MgSO$_4$ | 0.5g |
| *Listeria* lipase C supplement | 10.0ml |
| *Listeria* chromogenic selective supplement | 10.0ml |

pH 7.2 ± 0.2 at 25°C

**Source:** This medium is available from CONDA, Barcelona, Spain.

### *Listeria* lipase C Supplement:
**Composition** per 10.0mL:

| | |
|---|---|
| Lipase C substrate | 10.0mL |

### *Listeria* Chromogenic Selective Supplement:
**Composition** per 10.0mL:

| | |
|---|---|
| Cyclohexamide | 50.0mg |
| Ceftazidime | 20.0mg |
| Nalidixic acid | 20.0mg |
| Polymyxin B | 76,700U |

**Preparation of *Listeria* Chromogenic Selective Supplement:** Add components to distilled/deionized water and bring volume to 10.0mL. Mix thoroughly. Filter sterilize.

**Preparation of Medium:** Add components, except *Listeria* lipase C supplement and *Listeria* chromogenic selective supplement, to distilled/deionized water and bring volume to 980.0L. Mix thoroughly. Gently heat and bring to boiling. Autoclave for 15 min at 15 psi pressure–121°C. Cool to 50°C. Aseptically add 10.0mL of *Listeria* lipase C

supplement and 10.0mL of *Listeria* chromogenic selective supplement. Mix thoroughly. Pour into sterile Petri dishes.

**Storage/Shelf Life:** Store dehydrated media in the dark in a sealed container below 30°C. Prepared media should be stored in the dark under refrigeration (2-8°C). Chromogenic media are especially light and temperature sensitive; protect from light, excessive heat, moisture, and freezing. Media should not be used if there are any signs of deterioration (shrinking, cracking, or discoloration) or contamination, or if the expiration date supplied by the manufacturer has passed.

**Use:** For the selective detection of *Listeria monocytogenes*. *L. monocytogenes* colonies have a white halo around the blue colonies.

## *Listeria* Enrichment Broth

**Composition** per liter:

| | |
|---|---|
| Pancreatic digest of casein | 17.0g |
| Yeast extract | 6.0g |
| NaCl | 5.0g |
| Papaic digest of soybean meal | 3.0g |
| Glucose | 2.5g |
| K$_2$HPO$_4$ | 2.5g |
| Cycloheximide | 0.05g |
| Nalidixic acid | 0.04g |
| Acriflavine·HCl | 0.015g |

pH 7.3 ± 0.2 at 25°C

**Caution:** Cycloheximide is toxic. Avoid skin contact or aerosol formation and inhalation.

**Preparation of Medium:** Add components to distilled/deionized water and bring volume to 1.0L. Mix thoroughly. Gently heat and bring to boiling. Distribute into tubes or flasks. Autoclave for 15 min at 15 psi pressure–121°C.

**Preparation of Medium:** Add components—except antibiotic solution—to distilled/deionized water and bring volume to 990.0mL. Mix thoroughly. Gently heat and bring to boiling. Autoclave for 15 min at 15 psi pressure–121°C. Cool to 45°–50°C. Aseptically add 10.0mL of sterile antibiotic solution. Mix thoroughly. Aseptically distribute into sterile tubes.

**Storage/Shelf Life:** Store dehydrated media in the dark in a sealed container below 30°C. Prepared media should be stored under refrigeration (2-8°C). Media should be used within 60 days of preparation. Media should not be used if there are any signs of deterioration (discoloration) or contamination, or if the expiration date supplied by the manufacturer has passed.

**Use:** For the isolation and cultivation of *Listeria monocytogenes* according to the FDA formula.

## *Listeria* Enrichment Broth, FDA
## (LEB, FDA)

**Composition** per liter:

| | |
|---|---|
| Soybean casein digest broth yeast extract | 1.0L |
| Nalidixic acid solution | 8.0mL |
| Cycloheximide solution | 5.1mL |
| Acriflavin·HCl solution | 3.0mL |

pH 7.3 ± 0.2 at 25°C

**Caution:** Cycloheximide is toxic. Avoid skin contact or aerosol formation and inhalation.

### Soybean Casein Digest Broth Yeast Extract:
**Composition** per liter:

| | |
|---|---|
| Pancreatic digest of casein | 17.0g |
| Yeast extract | 6.0g |

NaCl.................................................................5.0g
Papaic digest of soybean meal.................................3.0g
K$_2$HPO$_4$......................................................2.5g
Glucose.............................................................2.5g

**Source**: This medium is available as a premixed powder from BD Diagnostic Systems.

**Preparation of Soybean Casein Digest Broth Yeast Extract:** Add components to distilled/deionized water and bring volume to 1.0L. Mix thoroughly. Autoclave for 15 min at 15 psi pressure–121°C.

**Nalidixic Acid Solution:**
**Composition** per 100.0mL:
Nalidixic acid....................................................0.5g

**Preparation of Nalidixic Acid Solution:** Add nalidixic acid to distilled/deionized water and bring volume to 100.0mL. Mix thoroughly. Filter sterilize.

**Cycloheximide Solution:**
**Composition** per 100.0mL:
Cycloheximide...................................................1.5g
Ethanol..........................................................40.0mL

**Preparation of Cycloheximide Solution:** Add cycloheximide to 40.0mL of ethanol. Mix thoroughly. Bring volume to 100.0mL with distilled/deionized water. Filter sterilize.

**Caution:** Cycloheximide is toxic. Avoid skin contact or aerosol formation and inhalation.

**Acriflavin·HCl Solution:**
**Composition** per 100.0mL:
Acriflavin·HCl solution .......................................0.5g

**Preparation of Acriflavin·HCl Solution:** Add acriflavin·HCl solution to distilled/deionized water and bring volume to 100.0mL. Mix thoroughly. Filter sterilize.

**Preparation of Medium:** Add components—except nalidixic acid solution, acriflavin solution, and cycloheximide solution—to distilled/deionized water and bring volume to 990.0mL. Mix thoroughly. Gently heat and bring to boiling. Autoclave for 15 min at 15 psi pressure–121°C. Cool to 45°–50°C. Aseptically add 8.0mL of sterile nalidixic acid solution, 5.1mL of sterile cycloheximide solution, and 3.0mL of sterile acriflavin solution. Mix thoroughly. Pour into sterile Petri dishes or distribute into sterile tubes.

**Storage/Shelf Life:** Store dehydrated media in the dark in a sealed container below 30°C. Prepared media should be stored under refrigeration (2-8°C). Media should be used within 60 days of preparation. Media should not be used if there are any signs of deterioration (discoloration) or contamination, or if the expiration date supplied by the manufacturer has passed.

**Use:** For the isolation and enrichment of *Listeria monocytogenes* from foods.

## *Listeria* Enrichment Broth I, USDA FSIS
### (*Listeria* Primary Selective Enrichment Broth, UVM I)
### (University of Vermont I *Listeria* Primary Selective Enrichment Broth)
**Composition** per liter:
NaCl.................................................................20.0g
Na$_2$HPO$_4$.....................................................12.0g

Beef extract......................................................5.0g
Proteose peptone...............................................5.0g
Pancreatic digest of casein...................................5.0g
Yeast extract.....................................................5.0g
KH$_2$PO$_4$......................................................1.35g
Esculin............................................................1.0g
Nalidixic acid solution........................................1.0mL
Acriflavine solution............................................1.0mL

pH 7.4 ± 0.2 at 25°C

**Source:** This medium is available as a premixed powder from Oxoid Unipath.

**Nalidixic Acid Solution:**
**Composition** per 10.0mL:
Nalidixic acid....................................................0.2g
NaOH (0.1*M* solution)........................................10.0mL

**Preparation of Nalidixic Acid Solution:** Add nalidixic acid to 10.0mL of NaOH solution. Mix thoroughly. Filter sterilize.

**Acriflavine Solution:**
**Composition** per 10.0mL:
Acriflavine.......................................................0.12g

**Preparation of Acriflavine Solution:** Add acriflavine to distilled/deionized water and bring volume to 10.0mL. Mix thoroughly. Filter sterilize. Use freshly prepared solution.

**Preparation of Medium:** Add components, except nalidixic acid solution and acriflavine solution, to distilled/deionized water and bring volume to 998.0mL. Mix thoroughly. Gently heat and bring to boiling. Autoclave for 15 min at 15 psi pressure–121°C. Cool to 45°–50°C. Aseptically add 1.0mL of sterile nalidixic acid solution. Mix thoroughly. Store at 4°C. Immediately prior to use, aseptically add 1.0mL of sterile acriflavine solution. Mix thoroughly. Aseptically distribute into sterile tubes or flasks.

**Storage/Shelf Life:** Store dehydrated media in the dark in a sealed container below 30°C. Prepared media should be stored under refrigeration (2-8°C). Media should be used within 60 days of preparation. Media should not be used if there are any signs of deterioration (discoloration) or contamination, or if the expiration date supplied by the manufacturer has passed.

**Use:** For the selective isolation, cultivation, and enrichment of *Listeria monocytogenes* from food, milk, and dairy products.

## *Listeria* Enrichment Broth II, USDA FSIS
### (*Listeria* Primary Selective Enrichment Broth, UVM II)
### (University of Vermont II *Listeria* Primary Selective Enrichment Broth
**Composition** per liter:
NaCl.................................................................20.0g
Na$_2$HPO$_4$.....................................................12.0g
Beef extract......................................................5.0g
Proteose peptone...............................................5.0g
Pancreatic digest of casein...................................5.0g
Yeast extract.....................................................5.0g
KH$_2$PO$_4$......................................................1.35g
Esculin............................................................1.0g

Nalidixic acid solution ................................................. 1.0mL
Acriflavine solution ..................................................... 1.0mL

pH 7.4 ± 0.2 at 25°C

**Source:** This medium is available as a premixed powder from Oxoid Unipath.

**Nalidixic Acid Solution:**
**Composition** per 10.0mL:

Nalidixic acid .............................................................0.2g
NaOH (0.1*M* solution) ............................................ 10.0mL

**Preparation of Nalidixic Acid Solution:** Add nalidixic acid to 10.0mL of NaOH solution. Mix thoroughly. Filter sterilize.

**Acriflavine Solution:**
**Composition** per 10.0mL:

Acriflavine ...............................................................0.25g

**Preparation of Acriflavine Solution:** Add acriflavine to distilled/deionized water and bring volume to 10.0mL. Mix thoroughly. Filter sterilize. Use freshly prepared solution.

**Preparation of Medium:** Add components, except nalidixic acid solution and acriflavine solution, to distilled/deionized water and bring volume to 998.0mL. Mix thoroughly. Gently heat and bring to boiling. Autoclave for 15 min at 15 psi pressure–121°C. Cool to 45°–50°C. Aseptically add 1.0mL of sterile nalidixic acid solution. Mix thoroughly. Store at 4°C. Immediately prior to use, aseptically add 1.0mL of sterile acriflavine solution. Mix thoroughly. Aseptically distribute into sterile tubes or flasks.

**Storage/Shelf Life:** Store dehydrated media in the dark in a sealed container below 30°C. Prepared media should be stored under refrigeration (2-8°C). Media should be used within 60 days of preparation. Media should not be used if there are any signs of deterioration (discoloration) or contamination, or if the expiration date supplied by the manufacturer has passed.

**Use:** For the selective isolation, cultivation, and enrichment of *Listeria monocytogenes* from food, milk, and dairy products.

## *Listeria* Enrichment HiVeg Agar

**Composition** per liter:

Potassium thiocyanate ...............................................37.5g
Agar ........................................................................13.0g
Plant hydrolysate........................................................10.0g
Plant peptone.............................................................10.0g
NaCl..........................................................................5.0g
Glucose ......................................................................1.0g
Acriflavin hydrochloride (Trypaflavin) ........................0.01g
Thiaminium dichloride ..............................................5.0mg

pH 7.4 ± 0.2 at 25°C

**Source:** This medium is available as a premixed powder from HiMedia.

**Preparation of Medium:** Add components to distilled/deionized water and bring volume to 1.0L. Mix thoroughly. Gently heat and bring to boiling. Distribute into tubes or flasks. Autoclave for 15 min at 15 psi pressure–121°C. Pour into sterile Petri dishes.

**Storage/Shelf Life:** Store dehydrated media in the dark in a sealed container below 30°C. Prepared media should be stored under refrigeration (2-8°C). Media should be used within 60 days of preparation. Media should not be used if there are any signs of deterioration (shrink-

ing, cracking, or discoloration) or contamination, or if the expiration date supplied by the manufacturer has passed.

**Use:** For the selective isolation of *Listeria monocytogenes* from clinical specimens.

## *Listeria* Enrichment HiVeg Broth

**Composition** per liter:

Potassium thiocyanate ................................................37.5g
Plant hydrolysate ......................................................10.0g
Plant peptone ...........................................................10.0g
NaCl..........................................................................5.0g
Glucose ......................................................................1.0g
Acriflavin hydrochloride (Trypaflavin) ........................0.01g
Thiaminium dichloride ..............................................5.0mg

pH 7.4 ± 0.2 at 25°C

**Source:** This medium is available as a premixed powder from HiMedia.

**Preparation of Medium:** Add components to distilled/deionized water and bring volume to 1.0L. Mix thoroughly. Gently heat and bring to boiling. Distribute into tubes or flasks. Autoclave for 15 min at 15 psi pressure–121°C.

**Storage/Shelf Life:** Store dehydrated media in the dark in a sealed container below 30°C. Prepared media should be stored under refrigeration (2-8°C). Media should be used within 60 days of preparation. Media should not be used if there are any signs of deterioration (discoloration) or contamination, or if the expiration date supplied by the manufacturer has passed.

**Use:** For the selective enrichment of *Listeria monocytogenes* from clinical specimens.

## *Listeria* Enrichment HiVeg Broth, Modified

**Composition** per liter:

NaCl..........................................................................20.0g
Plant hydrolysate No. 1...............................................10.0g
Na$_2$HPO$_4$.................................................................9.6g
Yeast extract...............................................................5.0g
Plant extract ...............................................................5.0g
KH$_2$PO$_4$..................................................................1.35g
Esculin ......................................................................1.0g
Nalidixic acid.............................................................0.02g
Acriflavin hydrochloride (Trypaflavin) ......................0.012g

pH 7.4 ± 0.2 at 25°C

**Source:** This medium is available as a premixed powder from HiMedia.

**Preparation of Medium:** Add components to distilled/deionized water and bring volume to 1.0L. Mix thoroughly. Gently heat and bring to boiling. Distribute into tubes or flasks. Autoclave for 15 min at 15 psi pressure–121°C.

**Storage/Shelf Life:** Store dehydrated media in the dark in a sealed container below 30°C. Prepared media should be stored under refrigeration (2-8°C). Media should be used within 60 days of preparation. Media should not be used if there are any signs of deterioration (discoloration) or contamination, or if the expiration date supplied by the manufacturer has passed.

**Use:** For the selective enrichment of *Listeria* species.

## *Listeria* Enrichment HiVeg Medium Base with Acriflavine and Nalidixic Acid (UVM)

**Composition** per liter:
| | |
|---|---|
| NaCl | 20.0g |
| Na$_2$HPO$_4$ | 12.0g |
| Plant extract | 5.0g |
| Plant hydrolysate | 5.0g |
| Plant peptone No. 3 | 5.0g |
| Yeast extract | 5.0g |
| KH$_2$PO$_4$ | 1.35g |
| Esculin | 1.0g |
| Nalidixic acid solution | 1.0mL |
| Acriflavine solution | 1.0mL |

pH 7.4 ± 0.2 at 25°C

**Source:** This medium, without nalidixic acid or acriflavine solutions, is available as a premixed powder from HiMedia.

**Nalidixic Acid Solution:**
**Composition** per 10.0mL:
| | |
|---|---|
| Nalidixic acid | 0.2g |
| NaOH (0.1*M* solution) | 10.0mL |

**Preparation of Nalidixic Acid Solution:** Add nalidixic acid to 10.0mL of NaOH solution. Mix thoroughly. Filter sterilize.

**Acriflavine Solution:**
**Composition** per 10.0mL:
| | |
|---|---|
| Acriflavine | 0.12g |

**Preparation of Acriflavine Solution:** Add acriflavine to distilled/deionized water and bring volume to 10.0mL. Mix thoroughly. Filter sterilize. Use freshly prepared solution.

**Preparation of Medium:** Add components, except nalidixic acid solution and acriflavine solution, to distilled/deionized water and bring volume to 998.0mL. Mix thoroughly. Gently heat and bring to boiling. Autoclave for 15 min at 15 psi pressure–121°C. Cool to 45°–50°C. Aseptically add 1.0mL of sterile nalidixic acid solution. Mix thoroughly. Store at 4°C. Immediately prior to use, aseptically add 1.0mL of sterile acriflavine solution. Mix thoroughly. Aseptically distribute into sterile tubes or flasks.

**Storage/Shelf Life:** Store dehydrated media in the dark in a sealed container below 30°C. Prepared media should be stored under refrigeration (2-8°C). Media should be used within 60 days of preparation. Media should not be used if there are any signs of deterioration (discoloration) or contamination, or if the expiration date supplied by the manufacturer has passed.

**Use:** For the selective isolation, cultivation, and enrichment of *Listeria monocytogenes* from food, milk, and dairy products. For the selective isolation and cultivation of *Listeria monocytogenes* from clinical specimens.

## *Listeria* Identification HiVeg Agar Base with PALCAM Selective Supplement (PALCAM)

**Composition** per liter:
| | |
|---|---|
| Plant peptone | 23.0g |
| LiCl | 15.0g |
| Agar | 13.0g |
| Mannitol | 10.0g |
| NaCl | 5.0g |
| Starch | 1.0g |
| Esculin | 0.8g |
| Ammonium ferric citrate | 0.5g |
| Glucose | 0.5g |
| Phenol Red | 0.08g |
| PALCAM selective supplement | 10.0mL |

pH 7.0 ± 0.2 at 25°C

**Source:** This medium, without PALCAM selective supplement, is available as a premixed powder from HiMedia.

**Caution:** LiCl is harmful. Avoid bodily contact and inhalation of vapors. On contact with skin wash with plenty of water immediately.

**PALCAM Selective Supplement:**
**Composition** per 10.0mL:
| | |
|---|---|
| Ceftazidime | 20.0mg |
| Polymyxin B | 10.0mg |
| Acriflavine·HCl | 5.0mg |

**Preparation of PALCAM Selective Supplement:** Add components to distilled/deionized water and bring volume to 10.0mL. Mix thoroughly. Filter sterilize.

**Preparation of Medium:** Add components, except PALCAM selective supplement, to distilled/deionized water and bring volume to 990.0mL. Mix thoroughly. Gently heat and bring to boiling. Autoclave for 15 min at 15 psi pressure–121°C. Cool to 45°–50°C. Aseptically add sterile PALCAM selective supplement. Mix thoroughly. Pour into sterile Petri dishes.

**Storage/Shelf Life:** Store dehydrated media in the dark in a sealed container below 30°C. Prepared media should be stored under refrigeration (2-8°C). Media should be used within 60 days of preparation. Media should not be used if there are any signs of deterioration (shrinking, cracking, or discoloration) or contamination, or if the expiration date supplied by the manufacturer has passed.

**Use:** For the selective isolation and identification of *Listeria monocytogenes* and other *Listeria* species from foods.

## *Listeria* Identification HiVeg Broth Base with PALCAM Selective Supplement (PALCAM)

**Composition** per liter:
| | |
|---|---|
| Plant peptone | 23.0g |
| LiCl | 10.0g |
| D-Mannitol | 5.0g |
| Yeast extract | 5.0g |
| Polysorbate 80 | 2.0g |
| Soy lecithin | 1.0g |
| Esculin | 0.8g |
| Ammonium ferric citrate | 0.5g |
| Phenol Red | 0.08g |
| PALCAM selective supplement | 10.0mL |

pH 7.2 ± 0.2 at 25°C

**Source:** This medium, without PALCAM selective supplement, is available as a premixed powder from HiMedia.

**Caution:** LiCl is harmful. Avoid bodily contact and inhalation of vapors. On contact with skin wash with plenty of water immediately.

**PALCAM Selective Supplement:**
**Composition** per 10.0mL:

Ceftazidime ...................................................................20.0mg
Polymyxin B ..................................................................10.0mg
Acriflavine·HCl ..............................................................5.0mg

**Preparation of PALCAM Selective Supplement:** Add components to distilled/deionized water and bring volume to 10.0mL. Mix thoroughly. Filter sterilize.

**Preparation of Medium:** Add components, except PALCAM selective supplement, to distilled/deionized water and bring volume to 990.0mL. Mix thoroughly. Gently heat and bring to boiling. Autoclave for 15 min at 15 psi pressure–121°C. Cool to 45°–50°C. Aseptically add sterile PALCAM selective supplement. Mix thoroughly.

**Storage/Shelf Life:** Store dehydrated media in the dark in a sealed container below 30°C. Prepared media should be stored under refrigeration (2-8°C). Media should be used within 60 days of preparation. Media should not be used if there are any signs of deterioration (discoloration) or contamination, or if the expiration date supplied by the manufacturer has passed.

**Use:** For the selective cultivation of *Listeria monocytogenes* and other *Listeria* species.

## *Listeria* Oxford HiVeg Medium Base with Antibiotic Inhibitor

**Composition** per liter:

Plant special peptone ......................................................23.0g
LiCl .................................................................................15.0g
Agar ................................................................................10.0g
NaCl ................................................................................5.0g
Cornstarch ......................................................................1.0g
Esculin ............................................................................1.0g
Ammonium ferric citrate ................................................0.5g
Antibiotic inhibitor ........................................................ 10.0mL
pH 7.0 ± 0.2 at 25°C

**Source:** This medium, without antibiotic inhibitor, is available as a premixed powder from HiMedia.

**Caution:** LiCl is harmful. Avoid bodily contact and inhalation of vapors. On contact with skin wash with plenty of water immediately.

**Antibiotic Inhibitor:**
**Composition** per 10.0mL:

Cycloheximide ................................................................0.4g
Colistin sulfate ...............................................................0.02g
Fosfomycin .....................................................................0.01g
Acriflavine .....................................................................5.0mg
Cefotetan ........................................................................2.0mg
Ethanol (50% solution) .................................................. 10.0mL

**Preparation of Antibiotic Inhibitor:** Add antibiotics to 10.0mL of ethanol. Mix thoroughly. Filter sterilize.

**Caution:** Cycloheximide is toxic. Avoid skin contact or aerosol formation and inhalation.

**Preparation of Medium:** Add components, except antibiotic inhibitor, to distilled/deionized water and bring volume to 990.0mL. Mix thoroughly. Gently heat and bring to boiling. Autoclave for 15 min at 15 psi pressure–121°C. Cool to 45°–50°C. Aseptically add 10.0mL of sterile antibiotic inhibitor. Mix thoroughly. Pour into sterile Petri dishes or distribute into sterile tubes.

**Storage/Shelf Life:** Store dehydrated media in the dark in a sealed container below 30°C. Prepared media should be stored under refrigeration (2-8°C). Media should be used within 60 days of preparation. Media should not be used if there are any signs of deterioration (shrinking, cracking, or discoloration) or contamination, or if the expiration date supplied by the manufacturer has passed.

**Use:** For the isolation and cultivation of *Listeria monocytogenes* from specimens containing a mixed bacterial flora. For the isolation of *Listeria* species from pathological specimens.

## *Listeria* Oxford Medium Base with Antibiotic Inhibitor

**Composition** per liter:

Peptone, special ............................................................23.0g
LiCl .................................................................................15.0g
Agar ................................................................................10.0g
NaCl ................................................................................5.0g
Cornstarch ......................................................................1.0g
Esculin ............................................................................1.0g
Ammonium ferric citrate ................................................0.5g
Antibiotic inhibitor ........................................................ 10.0mL
pH 7.0 ± 0.2 at 25°C

**Source:** This medium, without antibiotic inhibitor, is available as a premixed powder from HiMedia.

**Caution:** LiCl is harmful. Avoid bodily contact and inhalation of vapors. On contact with skin wash with plenty of water immediately.

**Antibiotic Inhibitor:**
**Composition** per 10.0mL:

Cycloheximide ................................................................0.4g
Colistin sulfate ...............................................................0.02g
Fosfomycin .....................................................................0.01g
Acriflavine .....................................................................5.0mg
Cefotetan ........................................................................2.0mg
Ethanol (50% solution) .................................................. 10.0mL

**Preparation of Antibiotic Inhibitor:** Add antibiotics to 10.0mL of ethanol. Mix thoroughly. Filter sterilize.

**Caution:** Cycloheximide is toxic. Avoid skin contact or aerosol formation and inhalation.

**Preparation of Medium:** Add components, except antibiotic inhibitor, to distilled/deionized water and bring volume to 990.0mL. Mix thoroughly. Gently heat and bring to boiling. Autoclave for 15 min at 15 psi pressure–121°C. Cool to 45°–50°C. Aseptically add 10.0mL of sterile antibiotic inhibitor. Mix thoroughly. Pour into sterile Petri dishes or distribute into sterile tubes.

**Storage/Shelf Life:** Store dehydrated media in the dark in a sealed container below 30°C. Prepared media should be stored under refrigeration (2-8°C). Media should be used within 60 days of preparation. Media should not be used if there are any signs of deterioration (shrinking, cracking, or discoloration) or contamination, or if the expiration date supplied by the manufacturer has passed.

**Use:** For the isolation and cultivation of *Listeria monocytogenes* from specimens containing a mixed bacterial flora. For the isolation of *Listeria* species from pathological specimens.

## *Listeria* Selective HiVeg Agar

**Composition** per liter:

Potassium thiocyanate ...................................................37.5g
Agar ................................................................................13.0g

Plant hydrolysate.................................................................10.0g
Plant peptone....................................................................10.0g
NaCl...................................................................................5.0g
Glucose .............................................................................1.0g
Nalidixic acid..................................................................0.04g
Acriflavin hydrochloride (Trypaflavin) .........................0.01g
Thiaminium dichloride ...................................................5.0mg
pH 7.4 ± 0.2 at 25°C

**Source:** This medium is available as a premixed powder from HiMedia.

**Preparation of Medium:** Add components to distilled/deionized water and bring volume to 1.0L. Mix thoroughly. Gently heat and bring to boiling. Autoclave for 15 min at 15 psi pressure–121°C. Cool to 45°–50°C. Pour into sterile Petri dishes or distribute into sterile tubes.

**Storage/Shelf Life:** Store dehydrated media in the dark in a sealed container below 30°C. Prepared media should be stored under refrigeration (2-8°C). Media should be used within 60 days of preparation. Media should not be used if there are any signs of deterioration (shrinking, cracking, or discoloration) or contamination, or if the expiration date supplied by the manufacturer has passed.

**Use:** For the selective isolation and cultivation of *Listeria* species from clinical specimens.

## *Listeria* Selective HiVeg Broth Base with Antibiotic Inhibitor
**Composition** per liter:
Plant hydrolysate...........................................................17.0g
Yeast extract.....................................................................6.0g
NaCl...................................................................................5.0g
Papaic digest of soybean meal .........................................3.0g
Glucose .............................................................................2.5g
K₂HPO₄..............................................................................2.5g
Antibiotic inhibitor ....................................................... 10.0mL
pH 7.3 ± 0.2 at 25°C

**Source:** This medium, without antibiotic inhibitor, is available as a premixed powder from HiMedia.

**Antibiotic Inhibitor:**
**Composition** per 10.0mL:
Cycloheximide ..................................................................0.4g
Colistin sulfate ...............................................................0.02g
Fosfomycin ....................................................................0.01g
Acriflavine .....................................................................5.0mg
Cefotetan.........................................................................2.0mg
Ethanol (50% solution) ................................................. 10.0mL

**Preparation of Antibiotic Inhibitor:** Add antibiotics to 10.0mL of ethanol. Mix thoroughly. Filter sterilize.

**Caution:** Cycloheximide is toxic. Avoid skin contact or aerosol formation and inhalation.

**Preparation of Medium:** Add components, except antibiotic inhibitor, to distilled/deionized water and bring volume to 990.0mL. Mix thoroughly. Gently heat and bring to boiling. Autoclave for 15 min at 15 psi pressure–121°C. Cool to 45°–50°C. Aseptically add 10.0mL of sterile antibiotic inhibitor. Mix thoroughly. Distribute into sterile tubes.

**Storage/Shelf Life:** Store dehydrated media in the dark in a sealed container below 30°C. Prepared media should be stored under refrigeration (2-8°C). Media should be used within 60 days of preparation. Media should not be used if there are any signs of deterioration (discol-

oration) or contamination, or if the expiration date supplied by the manufacturer has passed.

**Use:** For the cultivation of *Listeria monocytogenes* from specimens containing a mixed bacterial flora. For the cultivation of *Listeria* species from pathological specimens.

## *Listeria* Transport Enrichment Medium
**Composition** per liter:
Sodium glycerophosphate.................................................10.0g
Agar ..................................................................................2.0g
Sodium thioglycolate ........................................................1.0g
CaCl₂..................................................................................0.1g
Nalidixic acid..................................................................0.04g
Acridine solution ......................................................... 2.0mL
pH 7.4 ± 0.2 at 25°C

**Acridine Solution:**
**Composition** per 10.0mL:
Acridine ..........................................................................0.04g

**Preparation of Acridine Solution:** Add acridine to distilled/deionized water and bring volume to 10.0mL. Mix thoroughly. Autoclave for 15 min at 15 psi pressure–121°C. Cool to 45°–50°C.

**Preparation of Medium:** Add components, except acridine solution, to distilled/deionized water and bring volume to 998.0mL. Mix thoroughly. Gently heat and bring to boiling. Autoclave for 15 min at 15 psi pressure–121°C. Cool to 45°–50°C. Aseptically add 2.0mL acridine solution. Mix thoroughly. Aseptically distribute into sterile tubes in 10.0mL volumes or fill bottles 4/5 full.

**Storage/Shelf Life:** Store dehydrated media in the dark in a sealed container below 30°C. Prepared media should be stored under refrigeration (2-8°C). Media should be used within 60 days of preparation. Media should not be used if there are any signs of deterioration (discoloration) or contamination, or if the expiration date supplied by the manufacturer has passed.

**Use:** For the maintenance—as a transport medium—t of *Listeria* species.

## LMX Broth Modified, Fluorocult (Fluorocult LMX Broth, Modified)
**Composition** per liter:
Tryptose ..............................................................................5.0
NaCl....................................................................................5.0
Sorbitol ..............................................................................1.0
Tryptophan..........................................................................1.0
K₂HPO₄...............................................................................2.7g
KH₂PO₄...............................................................................2.0g
Lauryl sulfate sodium salt................................................0.1g
1-Isopropyl-β-D-1-thio-galactopyranoside (IPTG) ........0.1g
5-Bromo-4-chloro-3-indolyl-β-D-galactopyranoside (X-GAL) ...0.08g
4-Methylumbelliferyl-β-D-glucuronide ........................0.05g
pH: 6.8 ± 0.2 at 25°C

**Source:** This medium is available from Merck.

**Preparation of Medium:** Add components to distilled/deionized water and bring volume to 1.0L. Mix thoroughly. Distribute into test tubes. Autoclave for 15 min at 15 psi pressure–121°C. The broth is clear and yellowish brown.

**Storage/Shelf Life:** Store dehydrated media in the dark in a sealed container below 30°C. Prepared media should be stored under refriger-

ation (2-8°C). Media should be used within 60 days of preparation. Media should not be used if there are any signs of deterioration (discoloration) or contamination, or if the expiration date supplied by the manufacturer has passed.

**Use:** For the simultaneous detection of total coliforms and *E. coli* in water, food, and diary products by the fluorogenic procedure. A color change of the broth from yellow to blue-green indicates the presence of coliforms. A blue fluorescence under long-wave UV light permits the rapid detection of *E.coli*. As tryptophan is added to the broth, the indole reaction is easily done by adding KOVACS reagent. The formation of a red ring additionally confirms the presence of *E. coli*.

### Loeffler Blood Serum Medium

**Composition** per liter:

Beef blood serum ..................................................................... 750.0mL
Dextrose broth............................................................................ 250.0mL

pH 7.1 ± 0.2 at 25°C

**Source:** This medium is available as a premixed powder from BD Diagnostic Systems.

**Dextrose Broth:**
**Composition** per liter:

Tryptose ....................................................................................10.0g
Glucose ........................................................................................5.0g
Sodium chloride............................................................................5.0g
Beef extract ..................................................................................3.0g

**Preparation of Dextrose Broth:** Add components to distilled/deionized water and bring volume to 1.0L. Mix thoroughly.

**Preparation of Medium:** Combine 750.0mL of beef blood serum with 250.0mL of dextrose broth. Mix thoroughly. Distribute into screw-capped tubes. Slant tubes in the autoclave. Close the autoclave door loosely. Autoclave for 10 min at 0 psi pressure–100°C. Close the autoclave door tightly. Autoclave for 15 min at 15 psi pressure–121°C.

**Storage/Shelf Life:** Store dehydrated media in the dark in a sealed container below 30°C. Prepared media should be stored under refrigeration (2-8°C). Media should be used within 60 days of preparation. Media should not be used if there are any signs of deterioration (discoloration) or contamination, or if the expiration date supplied by the manufacturer has passed.

**Use:** For the cultivation of *Corynebacterium diphtheriae*. For demonstration of pigment production and proteolysis by *Corynebacterium diphtheriae*.

### Loeffler Blood Serum Medium

**Composition** per liter:

Beef blood serum ..................................................................... 750.0mL
Dextrose broth............................................................................ 250.0mL

pH 7.1 ± 0.2 at 25°C

**Dextrose Broth:**
**Composition** per liter:

Enzymatic digest of protein .............................................................2.5g
Glucose ......................................................................................1.25g
NaCl.............................................................................................1.25g
Beef extract ..................................................................................0.75g

**Preparation of Dextrose Broth:** Add components to distilled/deionized water and bring volume to 1.0L. Mix thoroughly.

**Preparation of Medium:** Combine 750.0mL of beef blood serum with 250.0mL of dextrose broth. Mix thoroughly. Distribute into

screw-capped tubes. Slant tubes in the autoclave. Close the autoclave door loosely. Autoclave for 10 min at 0 psi pressure–100°C. Close the autoclave door tightly. Autoclave for 15 min at 15 psi pressure–121°C.

**Storage/Shelf Life:** Store dehydrated media in the dark in a sealed container below 30°C. Prepared media should be stored under refrigeration (2-8°C). Media should be used within 60 days of preparation. Media should not be used if there are any signs of deterioration (discoloration) or contamination, or if the expiration date supplied by the manufacturer has passed.

**Use:** For the cultivation of *Corynebacterium diphtheriae*. For demonstration of pigment production and proteolysis by *Corynebacterium diphtheriae*.

### Loeffler Medium

**Composition** per liter:

Beef serum ....................................................................................70.0g
Egg, dried......................................................................................7.5g
Heart muscle, solids from infusion...............................................0.72g
Glucose .........................................................................................0.71g
Peptic digest of animal tissue ......................................................0.71g
NaCl..............................................................................................0.36g

pH 7.6 ± 0.2 at 25°C

**Source:** This medium is available as a premixed powder from BD Diagnostic Systems.

**Preparation of Medium:** Add components to distilled/deionized water and bring volume to 1.0L. Mix thoroughly. Distribute into screw-capped tubes. Slant tubes in the autoclave. Close the autoclave door loosely. Autoclave for 10 min at 0 psi pressure–100°C. Close the autoclave door tightly. Autoclave for 15 min at 15 psi pressure–121°C.

**Storage/Shelf Life:** Store dehydrated media in the dark in a sealed container below 30°C. Prepared media should be stored under refrigeration (2-8°C). Media should be used within 60 days of preparation. Media should not be used if there are any signs of deterioration (discoloration) or contamination, or if the expiration date supplied by the manufacturer has passed.

**Use:** For the cultivation of *Corynebacterium diphtheriae*. For demonstration of pigment production and proteolysis by *Corynebacterium diphtheriae*.

### Loeffler Slant

**Composition** per liter:

Tryptose ........................................................................................5.0g
Glucose .........................................................................................1.0g
Beef serum ............................................................................... 750.0mL

**Preparation of Medium:** Add components to distilled/deionized water and bring volume to 1.0L. Mix thoroughly. Distribute into screw-capped tubes. Slant tubes in the autoclave. Close the autoclave door loosely. Autoclave for 10 min at 0 psi pressure–100°C. Close the autoclave door tightly. Autoclave for 15 min at 15 psi pressure–121°C.

**Storage/Shelf Life:** Store dehydrated media in the dark in a sealed container below 30°C. Prepared media should be stored under refrigeration (2-8°C). Media should be used within 60 days of preparation. Media should not be used if there are any signs of deterioration (discoloration) or contamination, or if the expiration date supplied by the manufacturer has passed.

**Use:** For the cultivation of *Corynebacterium diphtheriae*. For demonstration of pigment production and proteolysis by *Corynebacterium diphtheriae*.

## Loeffler Slant, Modified

**Composition** per liter:

Glucose ......................................................................1.0g
Peptone.....................................................................0.5g
Beef serum ........................................................ 300.0mL

pH 7.6 ± 0.2 at 25°C

**Preparation of Medium:** Add peptone and glucose to distilled/deionized water and bring volume to 100.0mL. Mix thoroughly. Add beef serum. Mix thoroughly. Adjust pH to 7.6. Distribute into screw-capped tubes in 3.0mL volumes. Slant tubes in the autoclave. Autoclave for 30 min at 0 psi pressure–100°C.

**Storage/Shelf Life:** Store dehydrated media in the dark in a sealed container below 30°C. Prepared media should be stored under refrigeration (2-8°C). Media should be used within 60 days of preparation. Media should not be used if there are any signs of deterioration (discoloration) or contamination, or if the expiration date supplied by the manufacturer has passed.

**Use:** For the cultivation of *Corynebacterium diphtheriae*. For demonstration of pigment production and proteolysis by *Corynebacterium diphtheriae*.

## Lombard-Dowell Agar
## (LD Agar)

**Composition** per liter:

Agar ........................................................................20.0g
Pancreatic digest of casein.......................................5.0g
Yeast extract.............................................................5.0g
Gelatin......................................................................4.0g
NaCl..........................................................................2.5g
Glucose .....................................................................1.0g
L-Cystine ..................................................................0.4g
L-Tryptophan ............................................................0.2g
Na$_2$SO$_3$ ....................................................................0.1g
Hemin.................................................................10.0mg
Vitamin K$_1$ solution............................................... 1.0mL

pH 7.5 ± 0.2 at 25°C

**Vitamin K$_1$ Solution:**
**Composition** per 100.0mL:

Vitamin K$_1$ .................................................................1.0g
Ethanol................................................................ 99.0mL

**Preparation of Vitamin K$_1$ Solution:** Add vitamin K$_1$ to 99.0mL of absolute ethanol. Mix thoroughly.

**Preparation of Medium:** Add hemin and L-cystine to 5.0mL of NaOH. Mix thoroughly. Add remaining components, except agar and gelatin. Bring volume to 750.0mL with distilled/deionized water. Mix thoroughly. Gently heat and bring to boiling. In a separate flask, add gelatin to 100.0mL of cold distilled/deionized water. Gently heat and bring to 70°C. Add gelatin solution to the 750.0mL of basal medium. Mix thoroughly. Add agar. Bring volume to 1.0L with distilled/deionized water. Autoclave for 15 min at 15 psi pressure–121°C. Pour into sterile Petri dishes.

**Storage/Shelf Life:** Store dehydrated media in the dark in a sealed container below 30°C. Prepared media should be stored under refrigeration (2-8°C). Media should be used within 60 days of preparation. Media should not be used if there are any signs of deterioration (shrinking, cracking, or discoloration) or contamination, or if the expiration date supplied by the manufacturer has passed.

**Use:** For the cultivation of a wide variety of anaerobic bacteria. For the differentiation of anaerobic bacteria, including *Bacteroides species, Fusobacterium species, Clostridium* species, and nonspore-forming Gram-positive anaerobes, based on gelatinase production. After incubation of plates, gelatinase activity is determined by the addition of Frazier's reagent. Bacteria that hydrolyze gelatin appear as colonies surrounded by a clear zone.

## Lombard-Dowell Bile Agar
## (LD Bile Agar)

**Composition** per liter:

Agar ........................................................................20.0g
Oxgall ....................................................................20.0g
Pancreatic digest of casein.......................................5.0g
Yeast extract.............................................................5.0g
NaCl..........................................................................2.5g
D-Glucose .................................................................1.0g
L-Cystine ..................................................................0.4g
L-Tryptophan ............................................................0.2g
Na$_2$SO$_3$ ....................................................................0.1g
Hemin.................................................................10.0mg
Bile................................................................... 200.0mL
NaOH (1$N$) ............................................................ 5.0mL
Vitamin K$_1$ solution ............................................... 1.0mL

pH 7.5 ± 0.2 at 25°C

**Vitamin K$_1$ Solution:**
**Composition** per 100.0mL:

Vitamin K$_1$ .................................................................1.0g
Ethanol................................................................ 99.0mL

**Preparation of Vitamin K$_1$ Solution:** Add vitamin K$_1$ to 99.0mL of absolute ethanol. Mix thoroughly.

**Preparation of Medium:** Add hemin and L-cystine to 5.0mL of NaOH. Mix thoroughly. Add remaining components. Bring volume to 1.0L with distilled/deionized water. Mix thoroughly. Gently heat and bring to boiling. Distribute into tubes or flasks. Pour into sterile Petri dishes.

**Storage/Shelf Life:** Store dehydrated media in the dark in a sealed container below 30°C. Prepared media should be stored under refrigeration (2-8°C). Media should be used within 60 days of preparation. Media should not be used if there are any signs of deterioration (shrinking, cracking, or discoloration) or contamination, or if the expiration date supplied by the manufacturer has passed.

**Use:** For the cultivation and identification of a variety of obligate anaerobic bacteria in the presence of 20% bile.

## Lombard-Dowell Egg Yolk Agar
## (LD Egg Yolk Agar)
## (Egg Yolk Agar, Lombard-Dowell)

**Composition** per 9100.0mL:

Na$_2$HPO$_4$·12H$_2$O....................................................5.0g
Glucose .....................................................................2.0g
LD Agar ............................................................ 9000.0mL
Egg yolk emulsion ............................................ 100.0mL
MgSO$_4$·7H$_2$O (5% solution) ................................. 0.2mL

pH 7.5 ± 0.2 at 25°C

**LD Agar:**
**Composition** per liter:

Agar ........................................................................20.0g
Pancreatic digest of casein.......................................5.0g

| | |
|---|---|
| Yeast extract | 5.0g |
| NaCl | 2.5g |
| L-Cystine | 0.4g |
| L-Tryptophan | 0.2g |
| $Na_2SO_3$ | 0.1g |
| Hemin | 10.0mg |
| NaOH (1$N$ NaOH) | 5.0mL |
| Vitamin $K_1$ solution | 1.0mL |

**Preparation of LD Agar:** Add hemin and L-cystine to 5.0mL of NaOH. Mix thoroughly. Add remaining components. Mix thoroughly. Gently heat and bring to boiling.

**Vitamin $K_1$ Solution:**
**Composition** per 100.0mL:

| | |
|---|---|
| Vitamin $K_1$ | 1.0g |
| Ethanol | 99.0mL |

**Preparation of Vitamin $K_1$ Solution:** Add vitamin $K_1$ to 99.0mL of absolute ethanol. Mix thoroughly.

**Egg Yolk Emulsion:**
**Composition**:

| | |
|---|---|
| Chicken egg yolks | 11 |
| Whole chicken egg | 1 |

**Preparation of Egg Yolk Emulsion:** Soak eggs with 1:100 dilution of saturated mercuric chloride solution for 1 min. Crack eggs and separate yolks from whites. Mix egg yolks with 1 chicken egg.

**Preparation of Medium:** Combine components, except egg yolk emulsion. Mix thoroughly. Autoclave for 15 min at 15 psi pressure–121°C. Cool to 45°–50°C. Aseptically add 100.0mL of egg yolk emulsion. Mix thoroughly. Pour into sterile Petri dishes.

**Storage/Shelf Life:** Store dehydrated media in the dark in a sealed container below 30°C. Prepared media should be stored under refrigeration (2-8°C). Media should be used within 60 days of preparation. Media should not be used if there are any signs of deterioration (shrinking, cracking, or discoloration) or contamination, or if the expiration date supplied by the manufacturer has passed.

**Use:** For the cultivation of a wide variety of anaerobic bacteria. For the differentiation of anaerobic bacteria based on lecithinase production, lipase production, and proteolytic ability. Bacteria that produce lecithinase appear as colonies surrounded by a zone of insoluble precipitate. Bacteria that produce lipase appear as colonies with a pearly iridescent sheen. Bacteria that produce proteolytic activity appear as colonies surrounded by a clear zone.

## Lombard-Dowell Esculin Agar
## (LD Esculin Agar)
## (Esculin Agar, Lombard-Dowell)
**Composition** per liter:

| | |
|---|---|
| Agar | 20.0g |
| Pancreatic digest of casein | 5.0g |
| Yeast extract | 5.0g |
| NaCl | 2.5g |
| Esculin | 1.0g |
| Ferric citrate | 0.5g |
| L-Cystine | 0.4g |
| L-Tryptophan | 0.2g |
| Hemin | 10.0mg |
| NaOH (1$N$) | 5.0mL |
| Vitamin $K_1$ solution | 1.0mL |

pH 7.5 ± 0.2 at 25°C

**Vitamin $K_1$ Solution:**
**Composition** per 100.0mL:

| | |
|---|---|
| Vitamin $K_1$ | 1.0g |
| Ethanol | 99.0mL |

**Preparation of Vitamin $K_1$ Solution:** Add vitamin $K_1$ to 99.0mL of absolute ethanol. Mix thoroughly.

**Preparation of Medium:** Add hemin and L-cystine to 5.0mL of NaOH. Mix thoroughly. Add remaining components. Bring volume to 1.0L with distilled/deionized water. Mix thoroughly. Gently heat and bring to boiling. Distribute into tubes or flasks. Autoclave for 15 min at 15 psi pressure–121°C. Pour into sterile Petri dishes.

**Storage/Shelf Life:** Store dehydrated media in the dark in a sealed container below 30°C. Prepared media should be stored under refrigeration (2-8°C). Media should be used within 60 days of preparation. Media should not be used if there are any signs of deterioration (shrinking, cracking, or discoloration) or contamination, or if the expiration date supplied by the manufacturer has passed.

**Use:** For the cultivation of a wide variety of anaerobic bacteria. For the differentiation of anaerobic bacteria based on esculin hydrolysis, $H_2S$ production, and catalase production. Bacteria that hydrolyze esculin appear as colonies surrounded by a red-brown to dark brown zone. Bacteria that produce $H_2S$ appear as black colonies.

## Lombard-Dowell Gelatin Agar
## (LD Gelatin Agar)
**Composition** per liter:

| | |
|---|---|
| Agar | 20.0g |
| Pancreatic digest of casein | 5.0g |
| Yeast extract | 5.0g |
| Gelatin | 4.0g |
| NaCl | 2.5g |
| Glucose | 1.0g |
| L-Cystine | 0.4g |
| L-Tryptophan | 0.2g |
| $Na_2SO_3$ | 0.1g |
| Hemin | 10.0mg |
| NaOH (1$N$) | 5.0mL |
| Vitamin $K_1$ solution | 1.0mL |

pH 7.5 ± 0.2 at 25°C

**Vitamin $K_1$ Solution:**
**Composition** per 100.0mL:

| | |
|---|---|
| Vitamin $K_1$ | 1.0g |
| Ethanol | 99.0mL |

**Preparation of Vitamin $K_1$ Solution:** Add vitamin $K_1$ to 99.0mL of absolute ethanol. Mix thoroughly.

**Preparation of Medium:** Add hemin and L-cystine to 5.0mL of NaOH. Mix thoroughly. Add remaining components, except agar and gelatin. Bring volume to 750.0mL with distilled/deionized water. Mix thoroughly. Gently heat and bring to boiling. In a separate flask, add gelatin to 100.0mL of cold distilled/deionized water. Gently heat and bring to 70°C. Add gelatin solution to the 750.0mL of basal medium. Mix thoroughly. Add agar. Bring volume to 1.0L with distilled/deionized water. Autoclave for 15 min at 15 psi pressure–121°C. Pour into sterile Petri dishes.

**Storage/Shelf Life:** Store dehydrated media in the dark in a sealed container below 30°C. Prepared media should be stored under refrigeration (2-8°C). Media should be used within 60 days of preparation. Media should not be used if there are any signs of deterioration (shrink-

ing, cracking, or discoloration) or contamination, or if the expiration date supplied by the manufacturer has passed.

**Use:** For the cultivation of a wide variety of anaerobic bacteria. For the differentiation of anaerobic bacteria based on gelatinase production. After incubation of plates, gelatinase activity is determined by the addition of Frazier's reagent. Bacteria that hydrolyze gelatin appear as colonies surrounded by a clear zone.

## Lombard-Dowell Neomycin Agar
### (Egg Yolk Agar with Neomycin)

**Composition** per 9100.0mL:

| | |
|---|---|
| $Na_2HPO_4 \cdot 12H_2O$ | 5.0g |
| Glucose | 2.0g |
| Neomycin sulfate | 0.1g |
| LD Agar | 9000.0mL |
| Egg yolk emulsion | 100.0mL |
| $MgSO_4 \cdot 7H_2O$ (5% solution) | 0.2mL |

pH 7.5 ± 0.2 at 25°C

### LD Agar:
**Composition** per liter:

| | |
|---|---|
| Agar | 20.0g |
| Pancreatic digest of casein | 5.0g |
| Yeast extract | 5.0g |
| NaCl | 2.5g |
| L-Cystine | 0.4g |
| L-Tryptophan | 0.2g |
| $Na_2SO_3$ | 0.1g |
| Hemin | 10.0mg |
| NaOH ($1N$ NaOH) | 5.0mL |
| Vitamin $K_1$ solution | 1.0mL |

**Preparation of LD Agar:** Add hemin and L-cystine to 5.0mL of NaOH. Mix thoroughly. Add remaining components. Mix thoroughly. Gently heat and bring to boiling.

### Vitamin $K_1$ Solution:
**Composition** per 100.0mL:

| | |
|---|---|
| Vitamin $K_1$ | 1.0g |
| Ethanol | 99.0mL |

**Preparation of Vitamin $K_1$ Solution:** Add vitamin $K_1$ to 99.0mL of absolute ethanol. Mix thoroughly.

### Egg Yolk Emulsion:
**Composition:**

| | |
|---|---|
| Chicken egg yolks | 11 |
| Whole chicken egg | 1 |

**Preparation of Egg Yolk Emulsion:** Soak eggs with 1:100 dilution of saturated mercuric chloride solution for 1 min. Crack eggs and separate yolks from whites. Mix egg yolks with 1 chicken egg.

**Preparation of Medium:** Combine components, except egg yolk emulsion and neomycin sulfate. Mix thoroughly. Autoclave for 15 min at 15 psi pressure–121°C. Cool to 45°–50°C. Aseptically add 100.0mL of egg yolk emulsion and neomycin sulfate. Mix thoroughly. Pour into sterile Petri dishes.

**Storage/Shelf Life:** Store dehydrated media in the dark in a sealed container below 30°C. Prepared media should be stored under refrigeration (2-8°C). Media should be used within 60 days of preparation. Media should not be used if there are any signs of deterioration (shrinking, cracking, or discoloration) or contamination, or if the expiration date supplied by the manufacturer has passed.

**Use:** For the selective cultivation of a wide variety of anaerobic bacteria. For the differentiation of anaerobic bacteria based on lecithinase production, lipase production, and proteolytic ability. Bacteria that produce lecithinase appear as colonies surrounded by a zone of insoluble precipitate. Bacteria that produce lipase appear as colonies with a pearly iridescent sheen. Bacteria that produce proteolytic activity appear as colonies surrounded by a clear zone.

## Lowenstein-Gruft Medium

**Composition** per 1600.0mL:

| | |
|---|---|
| Potato starch | 30.0g |
| Asparagine | 3.6g |
| $KH_2PO_4$ | 2.4g |
| Magnesium citrate | 0.6g |
| Malachite Green | 0.4g |
| $MgSO_4 \cdot 7H_2O$ | 0.24g |
| Nalidixic acid | 0.056g |
| Ribonucleic acid | 0.08mg |
| Homogenized whole egg | 1.0L |
| Glycerol | 12.0mL |
| Penicillin | 80,000U |

### Homogenized Whole Egg:
**Composition** per liter:

| | |
|---|---|
| Whole eggs | 18–24 |

**Preparation of Homogenized Whole Egg:** Use fresh eggs, less than 1 week old. Scrub the shells with soap. Let stand in a soap solution for 30 min. Rinse in running water. Soak eggs in 70% ethanol for 15 min. Break the eggs into a sterile container. Homogenize by shaking. Filter through four layers of sterile cheesecloth into a sterile graduated cylinder. Measure out 1.0L.

**Preparation of Medium:** Add glycerol to 600.0mL of distilled/deionized water. Mix thoroughly. Add remaining components, except fresh egg mixture. Mix thoroughly. Gently heat while stirring and bring to boiling. Autoclave for 15 min at 15 psi pressure–121°C. Cool to 50°C. Aseptically add 1.0L of homogenized whole egg. Mix thoroughly. Distribute into sterile screw-capped tubes. Place tubes in a slanted position. Inspissate at 85°C (moist heat) for 45 min.

**Storage/Shelf Life:** Store dehydrated media in the dark in a sealed container below 30°C. Prepared media should be stored under refrigeration (2-8°C). Media should be used within 60 days of preparation. Media should not be used if there are any signs of deterioration (discoloration) or contamination, or if the expiration date supplied by the manufacturer has passed.

**Use:** For the cultivation and differentiation of *Mycobacterium* species. *Mycobacterium tuberculosis* appears as granular, rough, dry colonies. *Mycobacterium kansasii* appears as smooth to rough photochromogenic colonies. *Mycobacterium gordonae* appears as smooth yellow-orange colonies. *Mycobacterium avium* appears as smooth, colorless colonies. *Mycobacterium smegmatis* appears as wrinkled, creamy white colonies.

## Lowenstein-Jensen Medium

**Composition** per 1600.0mL:

| | |
|---|---|
| Potato starch | 30.0g |
| Asparagine | 3.6g |
| $KH_2PO_4$ | 2.4g |
| Magnesium citrate | 0.6g |
| Malachite Green | 0.4g |
| $MgSO_4 \cdot 7H_2O$ | 0.24g |

Homogenized whole egg ................................................. 1.0L
Glycerol ........................................................ 12.0mL

**Source:** This medium is available as a prepared medium from BD Diagnostic Systems and Oxoid Unipath.

## Homogenized Whole Egg:
**Composition** per liter:
Whole eggs ..........................................................18–24

**Preparation of Homogenized Whole Egg:** Use fresh eggs, less than 1 week old. Scrub the shells with soap. Let stand in a soap solution for 30 min. Rinse in running water. Soak eggs in 70% ethanol for 15 min. Break the eggs into a sterile container. Homogenize by shaking. Filter through four layers of sterile cheesecloth into a sterile graduated cylinder. Measure out 1.0L.

**Preparation of Medium:** Add glycerol to 600.0mL of distilled/deionized water. Mix thoroughly. Add remaining components, except fresh egg mixture. Mix thoroughly. Gently heat while stirring and bring to boiling. Autoclave for 15 min at 15 psi pressure–121°C. Cool to 50°C. Aseptically add 1.0L of homogenized whole egg. Mix thoroughly. Distribute into sterile screw-capped tubes. Place tubes in a slanted position. Inspissate at 85°C (moist heat) for 45 min.

**Storage/Shelf Life:** Store dehydrated media in the dark in a sealed container below 30°C. Prepared media should be stored under refrigeration (2-8°C). Media should be used within 60 days of preparation. Media should not be used if there are any signs of deterioration (shrinking, cracking, or discoloration) or contamination, or if the expiration date supplied by the manufacturer has passed.

**Use:** For the cultivation and differentiation of *Mycobacterium* species. *Mycobacterium tuberculosis* appears as granular, rough, dry colonies. *Mycobacterium kansasii* appears as smooth to rough photochromogenic colonies. *Mycobacterium gordonae* appears as smooth yellow-orange colonies. *Mycobacterium avium* appears as smooth, colorless colonies. *Mycobacterium smegmatis* appears as wrinkled, creamy white colonies.

## LPM Agar
## (Lithium Chloride Phenylethanol Moxalactam Plating Agar)
**Composition** per liter:
Agar ........................................................15.0g
Glycine anhydride......................................10.0g
LiCl$_2$........................................................5.0g
NaCl........................................................5.0g
Pancreatic digest of casein.........................5.0g
Peptic digest of animal tissue......................5.0g
Beef extract ..............................................3.0g
Phenylethyl alcohol...................................2.5g
Moxalactam solution..............................2.0mL
pH 7.3 ± 0.2 at 25°C

**Source:** This medium is available as a premixed powder from BD Diagnostic Systems.

## Moxalactam Solution:
**Composition** per 10.0mL:
Moxalactam ..............................................0.1g

**Preparation of Moxalactam Solution:** Add moxalactam to distilled/deionized water and bring volume to 10.0mL. Mix thoroughly. Filter sterilize.

**Preparation of Medium:** Add components, except moxalactam solution, to distilled/deionized water and bring volume to 998.0mL. Mix thoroughly. Gently heat while stirring and bring to boiling. Autoclave

for 12 min at 15 psi pressure–121°C. Cool to 45°–50°C. Aseptically add 2.0mL of sterile moxalactam solution. Mix thoroughly. Pour into sterile Petri dishes or distribute into sterile tubes.

**Storage/Shelf Life:** Store dehydrated media in the dark in a sealed container below 30°C. Prepared media should be stored under refrigeration (2-8°C). Media should be used within 60 days of preparation. Media should not be used if there are any signs of deterioration (shrinking, cracking, or discoloration) or contamination, or if the expiration date supplied by the manufacturer has passed.

**Use:** For the isolation and cultivation of *Listeria monocytogenes*.

## LPM Agar
## (Lithium Phenylethanol Moxalactam Agar)
**Composition** per liter:
Agar ........................................................15.0g
Glycine anhydride......................................10.0g
Casein enzymic hydrolysate .........................5.0g
Peptic digest of animal tissue ......................5.0g
Beef extract ..............................................3.0g
LiCl........................................................5.0g
NaCl........................................................5.0g
Phenylethyl alcohol...................................2.5g
Selective supplement solution ..................10.0mL
pH 7.3 ± 0.2 at 25°C

**Source:** This medium is available from HiMedia.

## Selective Supplement Solution:
**Composition** per 10.0mL:
Moxalactam ..........................................20.0mg

**Preparation of Selective Supplement Solution:** Add moxalactam to distilled/deionized water and bring volume to 10.0mL. Mix thoroughly. Filter sterilize.

**Caution:** Lithium chloride is harmful. Avoid bodily contact and inhalation of vapors. On contact with skin, wash with plenty of water immediately.

**Preparation of Medium:** Add components, except selective supplement solution, to distilled/deionized water and bring volume to 990.0mL. Mix thoroughly. Autoclave for 15 min at 15 psi pressure–121°C. Cool to 50°C. Aseptically add selective supplement solution. Mix thoroughly. Pour into Petri dishes or aseptically distribute into sterile tubes.

**Storage/Shelf Life:** Store dehydrated media in the dark in a sealed container below 30°C. Prepared media should be stored under refrigeration (2-8°C). Media should be used within 60 days of preparation. Media should not be used if there are any signs of deterioration (shrinking, cracking, or discoloration) or contamination, or if the expiration date supplied by the manufacturer has passed.

**Use:** For the isolation and cultivation of *Listeria monocytogenes* from food and dairy products.

## LPM Agar with Esculin and Ferric Iron
**Composition** per liter:
Agar ........................................................15.0g
Glycine anhydride......................................10.0g
LiCl........................................................5.0g
NaCl........................................................5.0g
Pancreatic digest of casein.........................5.0g
Peptic digest of animal tissue ......................5.0g
Beef extract..............................................3.0g

Phenylethyl alcohol.................................................................2.5g
Esculin ......................................................................................1.0g
Ferric ammonium citrate........................................................0.5g
Moxalactam solution...........................................................2.0mL

pH 7.3 ± 0.2 at 25°C

**Moxalactam Solution:**
**Composition** per 10.0mL:
Moxalactam .............................................................................0.1g

**Preparation of Moxalactam Solution:** Add moxalactam to distilled/deionized water and bring volume to 10.0mL. Mix thoroughly. Filter sterilize.

**Preparation of Medium:** Add components, except moxalactam solution, to distilled/deionized water and bring volume to 998.0mL. Mix thoroughly. Gently heat while stirring and bring to boiling. Autoclave for 12 min at 15 psi pressure–121°C. Cool to 45°–50°C. Aseptically add 2.0mL of sterile moxalactam solution. Mix thoroughly. Pour into sterile Petri dishes or distribute into sterile tubes.

**Storage/Shelf Life:** Store dehydrated media in the dark in a sealed container below 30°C. Prepared media should be stored under refrigeration (2-8°C). Media should be used within 60 days of preparation. Media should not be used if there are any signs of deterioration (shrinking, cracking, or discoloration) or contamination, or if the expiration date supplied by the manufacturer has passed.

**Use:** For the isolation and cultivation of *Listeria monocytogenes*.

## LPM HiVeg Agar Base with Moxalactam

**Composition** per liter:
Agar .......................................................................................15.0g
Glycine anhydride.................................................................10.0g
Plant hydrolysate....................................................................5.0g
Plant peptone..........................................................................5.0g
LiC ..........................................................................................5.0g
NaCl.........................................................................................5.0g
Plant extract ...........................................................................3.0g
Phenylethyl alcohol.................................................................2.5g
Moxalactam solution...........................................................2.0mL

pH 7.3 ± 0.2 at 25°C

**Source:** This medium, without moxalactam, is available as a premixed powder from HiMedia.

**Moxalactam Solution:**
**Composition** per 10.0mL:
Moxalactam .............................................................................0.1g

**Preparation of Moxalactam Solution:** Add moxalactam to distilled/deionized water and bring volume to 10.0mL. Mix thoroughly. Filter sterilize.

**Preparation of Medium:** Add components, except moxalactam solution, to distilled/deionized water and bring volume to 998.0mL. Mix thoroughly. Gently heat while stirring and bring to boiling. Autoclave for 12 min at 15 psi pressure–121°C. Cool to 45°–50°C. Aseptically add 2.0mL of sterile moxalactam solution. Mix thoroughly. Pour into sterile Petri dishes or distribute into sterile tubes.

**Storage/Shelf Life:** Store dehydrated media in the dark in a sealed container below 30°C. Prepared media should be stored under refrigeration (2-8°C). Media should be used within 60 days of preparation. Media should not be used if there are any signs of deterioration (shrinking, cracking, or discoloration) or contamination, or if the expiration date supplied by the manufacturer has passed.

**Use:** For the isolation and cultivation of *Listeria monocytogenes*.

## LST-MUG Broth

**Composition** per liter:
Tryptose .................................................................................20.0g
Lactose....................................................................................5.0g
$K_2HPO_4$ ............................................................................2.75g
$KH_2PO_4$ ...........................................................................2.75g
NaCl.........................................................................................5.0g
L-Tryptophan ..........................................................................1.0g
Sodium lauryl sulfate..............................................................0.1g
4-Methylumbelliferyl-β-D-glucuronide ..................................0.1g

pH 6.8 ± 0.2 at 37°C

**Source:** This medium is available from Fluka, Sigma-Aldrich.

**Preparation of Medium:** Add components to distilled/deionized water and bring volume to 1.0L. Mix thoroughly. Distribute into test tubes that contain an inverted Durham tube in 10.0mL volumes. Autoclave for 15 min at 15 psi pressure–121°C.

**Storage/Shelf Life:** Store dehydrated media in the dark in a sealed conainer below 30°C. Prepared plates should be stored under refrigeration (2-8°C). This medium has a limited shelf life. Media should not be used if there are any signs of deterioration (discoloration) or contamination, or if the expiration date supplied by the manufacturer has passed.

**Use:** For the detection of *E. coli* by the fluorogenic method. The presence of *E. coli* results in fluorescence in the UV. A positive indole test provides confirmation. β-D-glucoronidase, which is produced by *E. coli*, cleaves 4-methylumbelliferyl-β-D-glucuronide to 4-methylumbelliferone and glucuronide. The fluorogen 4-methylumbelliferone can be detected under a long wavelength UV lamp.

## Lysine Agar, Selective

**Composition** per liter:
Agar .......................................................................................15.0g
L-Lysine.................................................................................10.0g
Peptone ...................................................................................5.0g
Glucose ...................................................................................3.5g
Yeast extract............................................................................3.0g
Bile salts mixture ...................................................................1.5g
Sulfapyridine...........................................................................0.3g
Bromcresol Purple ................................................................0.03g
Crystal Violet........................................................................0.001g

pH 6.8 ± 0.1 at 25°C

**Source:** This medium is available as a premixed powder from BD Diagnostic Systems.

**Preparation of Medium:** Add components to distilled/deionized water and bring volume to 1.0L. Mix thoroughly. Gently heat and bring to boiling. Distribute into tubes or flasks. Autoclave for 15 min at 15 psi pressure–121°C. Pour into sterile Petri dishes or leave in tubes.

**Storage/Shelf Life:** Store dehydrated media in the dark in a sealed container below 30°C. Prepared media should be stored under refrigeration (2-8°C). Media should be used within 60 days of preparation. Media should not be used if there are any signs of deterioration (shrinking, cracking, or discoloration) or contamination, or if the expiration date supplied by the manufacturer has passed.

**Use:** For the selective isolation and cultivation of *Salmonella* species from food by the hydrophobic grid membrane filter method.

## Lysine Arginine Iron Agar

**Composition** per liter:

| | |
|---|---|
| Agar | 15.0g |
| L-Arginine | 10.0g |
| L-Lysine | 10.0g |
| Peptone | 5.0g |
| Yeast extract | 3.0g |
| Glucose | 1.0g |
| Ferric ammonium citrate | 0.5g |
| Sodium thiosulfate | 0.04g |
| Bromcresol Purple | 0.02g |

pH 6.8 ± 0.2 at 25°C

**Preparation of Medium:** Add components to distilled/deionized water and bring volume to 1.0L. Mix thoroughly. Gently heat and bring to boiling. Adjust pH to 6.8. Distribute into screw-capped tubes in 5.0mL volumes. Autoclave for 12 min at 15 psi pressure–121°C. Allow tubes to cool in a slanted position.

**Storage/Shelf Life:** Store dehydrated media in the dark in a sealed container below 30°C. Prepared media should be stored under refrigeration (2-8°C). Media should be used within 60 days of preparation. Media should not be used if there are any signs of deterioration (shrinking, cracking, or discoloration) or contamination, or if the expiration date supplied by the manufacturer has passed.

**Use:** For the cultivation and differentiation of bacteria based on their ability to decarboxylate lysine, decarboxylate arginine, and produce $H_2S$. Bacteria that decarboxylate lysine or arginine turn the medium purple. Bacteria that produce $H_2S$ appear as black colonies.

## Lysine Arginine Iron HiVeg Agar

**Composition** per liter:

| | |
|---|---|
| Agar | 15.0g |
| L-Arginine | 10.0g |
| L-Lysine | 10.0g |
| Plant peptone | 5.0g |
| Yeast extract | 3.0g |
| Glucose | 1.0g |
| Ferric ammonium citrate | 0.5g |
| $Na_2S_2O_3$ | 0.04g |
| Bromcresol Purple | 0.02g |

pH 6.8 ± 0.2 at 25°C

**Source:** This medium is available as a premixed powder from HiMedia.

**Preparation of Medium:** Add components to distilled/deionized water and bring volume to 1.0L. Mix thoroughly. Gently heat and bring to boiling. Adjust pH to 6.8. Distribute into screw-capped tubes in 5.0mL volumes. Autoclave for 12 min at 15 psi pressure–121°C. Allow tubes to cool in a slanted position.

**Storage/Shelf Life:** Store dehydrated media in the dark in a sealed container below 30°C. Prepared media should be stored under refrigeration (2-8°C). Media should be used within 60 days of preparation. Media should not be used if there are any signs of deterioration (shrinking, cracking, or discoloration) or contamination, or if the expiration date supplied by the manufacturer has passed.

**Use:** For the cultivation and differentiation of bacteria based on their ability to decarboxylate lysine, decarboxylate arginine, and produce $H_2S$. Bacteria that decarboxylate lysine or arginine turn the medium purple.

## Lysine Broth Falkow with Sodium Chloride (BAM M44)

**Composition** per liter:

| | |
|---|---|
| L-Lysine | 5.0g |
| Peptone or gelysate | 5.0g |
| Yeast extract | 3.0g |
| Glucose | 1.0g |
| Bromcresol Purple | 0.02g |

pH 6.5 ± 0.2 at 25°C

**Preparation of Medium:** Add components to distilled/deionized water and bring volume to 1.0L. Mix thoroughly. Adjust pH so that it will be 6.5 ± 0.2 after sterilization. Distribute into 16 × 150mm screw-capped tubes in 5.0mL volumes. Autoclave medium with loosely capped tubes for 10 min at 15 psi pressure–121°C. Screw the caps on tightly for storage and after inoculation.

**Storage/Shelf Life:** Store dehydrated media in the dark in a sealed container below 30°C. Prepared media should be stored under refrigeration (2-8°C). Media should be used within 60 days of preparation. Media should not be used if there are any signs of deterioration (discoloration) or contamination, or if the expiration date supplied by the manufacturer has passed.

**Use:** For the cultivation and differentiation of *Vibrio* spp. based on their ability to decarboxylate the amino acid lysine. Bacteria that decarboxylate lysine turn the medium turbid purple.

## Lysine Iron Agar

**Composition** per liter:

| | |
|---|---|
| Agar | 13.5g |
| L-Lysine | 10.0g |
| Pancreatic digest of gelatin | 5.0g |
| Yeast extract | 3.0g |
| Glucose | 1.0g |
| Ferric ammonium citrate | 0.5g |
| $Na_2S_2O_3 \cdot 5H_2O$ | 0.04g |
| Bromcresol Purple | 0.02g |

pH 6.7 ± 0.2 at 25°C

**Source:** This medium is available as a premixed powder from BD Diagnostic Systems and Oxoid Unipath.

**Preparation of Medium:** Add components to distilled/deionized water and bring volume to 1.0L. Mix thoroughly. Gently heat while stirring and bring to boiling. Distribute into tubes in 10.0mL volumes. Autoclave for 12 min at 15 psi pressure–121°C. Allow tubes to cool in a slanted position.

**Storage/Shelf Life:** Store dehydrated media in the dark in a sealed container below 30°C. Prepared media should be stored under refrigeration (2-8°C). Media should be used within 60 days of preparation. Media should not be used if there are any signs of deterioration (shrinking, cracking, or discoloration) or contamination, or if the expiration date supplied by the manufacturer has passed.

**Use:** For the cultivation and differentiation of members of the Enterobacteriaceae based on their ability to decarboxylate lysine and to form $H_2S$. Bacteria that decarboxylate lysine turn the medium purple. Bacteria that produce $H_2S$ appear as black colonies.

## Lysine Iron Cystine HiVeg Broth Base with Novobiocin
**Composition** per liter:

| | |
|---|---|
| L-Lysine hydrochloride | 10.0g |
| Plant hydrolysate | 5.0g |

Mannitol................................................................5.0g
Yeast extract.........................................................3.0g
Glucose .................................................................1.0g
Salicin ...................................................................1.0g
Ferric ammonium citrate.......................................0.5g
Na$_2$S$_2$O$_3$ ...............................................................0.1g
L-Cystine ..............................................................0.1g
Neutral Red .......................................................0.025g
Novobiocin solution...................................... 10.0mL

pH 6.2 ± 0.2 at 25°C

**Source:** This medium, without novobiocin solution, is available as a premixed powder from HiMedia.

**Novobiocin Solution:**
**Composition** per 10.0mL:
Novobiocin...........................................................0.015g

**Preparation of Novobiocin Solution:** Add novobiocin to distilled/deionized water and bring volume to 10.0mL. Mix thoroughly. Filter sterilize.

**Preparation of Medium:** Add components, except novobiocin solution, to distilled/deionized water and bring volume to 990.0mL. Mix thoroughly. Gently heat and bring to boiling. Continue boiling for 2–3 min. Distribute into tubes in 10.0mL volumes. Autoclave for 15 min at 15 psi pressure–121°C. Cool to 45°–50°C. Aseptically add 0.1mL of sterile novobiocin solution to each tube. Mix thoroughly.

**Storage/Shelf Life:** Store dehydrated media in the dark in a sealed container below 30°C. Prepared media should be stored under refrigeration (2-8°C). Media should be used within 60 days of preparation. Media should not be used if there are any signs of deterioration (discoloration) or contamination, or if the expiration date supplied by the manufacturer has passed.

**Use:** For the rapid, presumptive detection of *Salmonella* in foods, food ingredients, and feed materials.

**Lysine Iron HiVeg Agar**

**Composition** per liter:
Agar .....................................................................15.0g
L-Lysine................................................................10.0g
Plant peptone.........................................................5.0g
Yeast extract.........................................................3.0g
Glucose .................................................................1.0g
Ferric ammonium citrate.......................................0.5g
Na$_2$S$_2$O$_3$ .............................................................0.04g
Bromcresol Purple ..............................................0.02g

pH 6.7 ± 0.2 at 25°C

**Source:** This medium is available as a premixed powder from HiMedia.

**Preparation of Medium:** Add components to distilled/deionized water and bring volume to 1.0L. Mix thoroughly. Gently heat while stirring and bring to boiling. Distribute into tubes in 10.0mL volumes. Autoclave for 12 min at 15 psi pressure–121°C. Allow tubes to cool in a slanted position.

**Storage/Shelf Life:** Store dehydrated media in the dark in a sealed container below 30°C. Prepared media should be stored under refrigeration (2-8°C). Media should be used within 60 days of preparation. Media should not be used if there are any signs of deterioration (shrinking, cracking, or discoloration) or contamination, or if the expiration date supplied by the manufacturer has passed.

**Use:** For the cultivation and differentiation of members of the Enterobacteriaceae based on their ability to decarboxylate lysine and to form H$_2$S. Bacteria that decarboxylate lysine turn the medium purple. Bacteria that produce H$_2$S appear as black colonies. For the differentiation of enteric organisms, especially *Salmonella* serotype Arizona.

**Lysine Lactose HiVeg Broth**

**Composition** per liter:
Lactose.................................................................10.0g
Plant peptone No. 2...............................................5.0g
L-Lysine .................................................................5.0g
Yeast extract.........................................................3.0g
Glucose .................................................................1.0g
Bromcresol Purple ..............................................0.02g

pH 6.8 ± 0.2 at 25°C

**Source:** This medium is available as a premixed powder from HiMedia.

**Preparation of Medium:** Add components to distilled/deionized water and bring volume to 1.0L. Mix thoroughly. Gently heat and bring to boiling. Adjust pH to 6.8. Autoclave for 15 min at 15 psi pressure–121°C. Aseptically distribute into sterile tubes in 1.0mL volumes.

**Use:** For the determination of lysine decarboxylase activity of lactose nonfermenting members of Enterobacteriaceae, especially *Salmonella* species. Lysozyme Broth
**Composition** per 1010.0mL:
Nutrient broth...................................................... 1.0L
Lysozyme solution ...................................... 10.0mL

pH 6.9 ± 0.2 at 25°C

**Nutrient Broth:**
**Composition** per liter:
Pancreatic digest of gelatin..................................5.0g
Beef extract...........................................................3.0g

**Source:** Nutrient broth is available as a premixed powder from BD Diagnostic Systems.

**Preparation of Nutrient Broth:** Add components to distilled/deionized water and bring volume to 1.0L. Mix thoroughly. Distribute into bottles in 99.0mL volumes. Autoclave for 15 min at 15 psi pressure–121°C. Cool to 25°C.

**Lysozyme Solution:**
**Composition** per 100.0mL:
Lysozyme................................................................0.1g

**Preparation of Lysozyme Solution:** Add lysozyme to distilled/deionized water and bring volume to 100.0mL. Mix thoroughly. Filter sterilize.

**Preparation of Medium:** Add 1.0mL of sterile lysozyme solution to 99.0mL of cooled, sterile nutrient broth. Mix thoroughly. Aseptically distribute into sterile tubes in 2.5mL volumes.

**Storage/Shelf Life:** Store dehydrated media in the dark in a sealed container below 30°C. Prepared media should be stored under refrigeration (2-8°C). Media should be used within 60 days of preparation. Media should not be used if there are any signs of deterioration (discoloration) or contamination, or if the expiration date supplied by the manufacturer has passed.

**Use:** For the cultivation and differentiation of *Bacillus cereus* in foods. *Bacillus cereus* is resistant to lysozyme and will grow in this medium.

## M Broth

**Composition** per liter:

| | |
|---|---|
| Pancreatic digest of casein | 12.5g |
| $K_2HPO_4$ | 5.0g |
| NaCl | 5.0g |
| Sodium citrate | 5.0g |
| Yeast extract | 5.0g |
| Mannose | 2.0g |
| $MgSO_4 \cdot 7H_2O$ | 0.8g |
| Polysorbate 80 | 0.75g |
| $FeSO_4$ | 0.04g |

pH $7.0 \pm 0.22$ at 25°C

**Source:** Available as a premixed powder from BD Diagnostic Systems.

**Preparation of Medium:** Add components to distilled/deionized water and bring volume to 1.0L. Mix thoroughly. Distribute into tubes or flasks. Autoclave for 15 min at 15 psi pressure–121°C.

**Storage/Shelf Life:** Store dehydrated media in the dark in a sealed container below 30°C. Prepared media should be stored under refrigeration (2-8°C). Media should be used within 60 days of preparation. Media should not be used if there are any signs of deterioration (discoloration) or contamination, or if the expiration date supplied by the manufacturer has passed.

**Use:** For the detection of *Salmonella* in dried foods and feeds.

## MacConkey Agar

**Composition** per liter:

| | |
|---|---|
| Pancreatic digest of gelatin | 17.0g |
| Agar | 13.5g |
| Lactose | 10.0g |
| NaCl | 5.0g |
| Bile salts | 1.5g |
| Pancreatic digest of casein | 1.5g |
| Peptic digest of animal tissue | 1.5g |
| Neutral Red | 0.03g |
| Crystal Violet | 1.0mg |

pH $7.1 \pm 0.2$ at 25°C

**Source:** This medium is available as a premixed powder from BD Diagnostic Systems.

**Preparation of Medium:** Add components to distilled/deionized water and bring volume to 1.0L. Mix thoroughly. Gently heat while stirring until boiling. Autoclave for 15 min at 15 psi pressure–121°C. Pour into sterile Petri dishes or distribute into sterile tubes.

**Storage/Shelf Life:** Store dehydrated media in the dark in a sealed container below 30°C. Prepared media should be stored under refrigeration (2-8°C). Media should be used within 60 days of preparation. Media should not be used if there are any signs of deterioration (shrinking, cracking, or discoloration) or contamination, or if the expiration date supplied by the manufacturer has passed.

**Use:** For the selective isolation, cultivation, and differentiation of coliforms and enteric pathogens based on the ability to ferment lactose. Lactose-fermenting organisms appear as red to pink colonies. Lactose-nonfermenting organisms appear as colorless or transparent colonies.

## MacConkey Agar

**Composition** per liter:

| | |
|---|---|
| Peptone | 20.0g |
| Agar | 12.0g |

---

| | |
|---|---|
| Lactose | 10.0g |
| Bile salts | 5.0g |
| NaCl | 5.0g |
| Neutral Red | 0.075g |

pH $7.4 \pm 0.2$ at 25°C

**Source:** This medium is available as a premixed powder from Oxoid Unipath.

**Preparation of Medium:** Add components to distilled/deionized water and bring volume to 1.0L. Mix thoroughly. Gently heat while stirring until boiling. Autoclave for 15 min at 15 psi pressure–121°C. Pour into sterile Petri dishes or distribute into sterile tubes.

**Storage/Shelf Life:** Store dehydrated media in the dark in a sealed container below 30°C. Prepared media should be stored under refrigeration (2-8°C). Media should be used within 60 days of preparation. Media should not be used if there are any signs of deterioration (shrinking, cracking, or discoloration) or contamination, or if the expiration date supplied by the manufacturer has passed.

**Use:** For the selective isolation, cultivation, and differentiation of coliforms and enteric pathogens based on the ability to ferment lactose. Lactose-fermenting organisms appear as red to pink colonies. Lactose-nonfermenting organisms appear as colorless or transparent colonies.

## MacConkey Agar Base, HiVeg

**Composition** per liter:

| | |
|---|---|
| Plant peptone | 17.0g |
| Agar | 13.5g |
| NaCl | 5.0g |
| Plant peptone No. 3 | 3.0g |
| Synthetic detergent | 1.5g |
| Neutral Red | 0.03g |
| Crystal Violet | 1.0mg |

pH $7.1 \pm 0.2$ at 25°C

**Source:** This medium is available as a premixed powder from Hi-Media.

**Preparation of Medium:** Add components to distilled/deionized water and bring volume to 1.0L. Mix thoroughly. Gently heat while stirring until boiling. Autoclave for 15 min at 15 psi pressure–121°C. Pour into sterile Petri dishes or distribute into sterile tubes.

**Storage/Shelf Life:** Store dehydrated media in the dark in a sealed container below 30°C. Prepared media should be stored under refrigeration (2-8°C). Media should be used within 60 days of preparation. Media should not be used if there are any signs of deterioration (shrinking, cracking, or discoloration) or contamination, or if the expiration date supplied by the manufacturer has passed.

**Use:** For the cultivation and differentiation of lactose-fermenting and nonfermenting Gram-negative bacteria. Lactose-fermenting organisms appear as red to pink colonies. Lactose-nonfermenting organisms appear as colorless or transparent colonies.

## MacConkey Agar with 0.15% Bile Salts, Crystal Violet, and Sodium Chloride, HiVeg

**Composition** per liter:

| | |
|---|---|
| Plant peptone No. 2 | 17.0g |
| Agar | 15.0g |
| Lactose | 10.0g |
| NaCl | 5.0g |
| Plant hydrolysate | 1.5g |
| Plant peptone | 1.5g |

| | |
|---|---|
| Synthetic detergent | 1.5g |
| Neutral Red | 0.03g |
| Crystal Violet | 1.0mg |

pH 7.2 ± 0.2 at 25°C

**Source:** This medium is available as a premixed powder from Hi-Media.

**Preparation of Medium:** Add components to distilled/deionized water and bring volume to 1.0L. Mix thoroughly. Gently heat while stirring until boiling. Autoclave for 15 min at 15 psi pressure–121°C. Pour into sterile Petri dishes or distribute into sterile tubes.

**Storage/Shelf Life:** Store dehydrated media in the dark in a sealed container below 30°C. Prepared media should be stored under refrigeration (2-8°C). Media should be used within 60 days of preparation. Media should not be used if there are any signs of deterioration (shrinking, cracking, or discoloration) or contamination, or if the expiration date supplied by the manufacturer has passed.

**Use:** For the selective isolation, cultivation, and differentiation of enteric pathogens, especially enterococci, in clinical specimens and in materials of sanitary importance.

## MacConkey Agar, CS

**Composition** per liter:

| | |
|---|---|
| Peptone | 17.0g |
| Agar | 13.5g |
| Lactose | 10.0g |
| NaCl | 5.0g |
| Proteose peptone | 3.0g |
| Bile salts | 1.5g |
| Neutral Red | 0.03g |
| Crystal Violet | 1.0mg |

pH 7.1 ± 0.2 at 25°C

**Source:** This medium is available as a prepared medium from BD Diagnostic Systems.

**Preparation of Medium:** Add components to distilled/deionized water and bring volume to 1.0L. Mix thoroughly. Gently heat while stirring until boiling. Autoclave for 15 min at 15 psi pressure–121°C. Pour into sterile Petri dishes or distribute into sterile tubes.

**Storage/Shelf Life:** Store dehydrated media in the dark in a sealed container below 30°C. Prepared media should be stored under refrigeration (2-8°C). Media should be used within 60 days of preparation. Media should not be used if there are any signs of deterioration (shrinking, cracking, or discoloration) or contamination, or if the expiration date supplied by the manufacturer has passed.

**Use:** For the cultivation and differentiation of lactose-fermenting and lactose-nonfermenting Gram-negative bacteria while also controlling the swarming of *Proteus* species, if present. Lactose-fermenting organisms appear as red to pink colonies. Lactose-nonfermenting organisms appear as colorless or transparent colonies.

## MacConkey Agar, Fluorocult
## (Fluorocult MacConkey Agar)

**Composition** per liter:

| | |
|---|---|
| Peptone from casein | 17.0g |
| Agar | 13.5g |
| Lactose | 10.0g |
| NaCl | 5.0g |
| Peptone from meat | 3.0g |

| | |
|---|---|
| Bile salt mixture | 1.5g |
| 4-Methylumbelliferyl-β-D-glucuronide | 0.1g |
| Neutral Red | 0.03g |
| Crystal Violet | 0.001g |

pH 7.1 ± 0.2 at 25°C

**Source:** This medium is available from Merck.

**Preparation of Medium:** Add components to distilled/deionized water and bring volume to 1.0L. Mix thoroughly. Autoclave for 15 min at 15 psi pressure–121°C. Cool to 45°–50°C. Pour into sterile Petri dishes. The plates are clear and red to red-brown.

**Storage/Shelf Life:** Store dehydrated media in the dark in a sealed container below 30°C. Prepared media should be stored in the dark under refrigeration (2-8°C). Chromogenic media are especially light and temperature sensitive; protect from light, excessive heat, moisture, and freezing. Media should not be used if there are any signs of deterioration (shrinking, cracking, or discoloration) or contamination, or if the expiration date supplied by the manufacturer has passed.

**Use:** For the isolation of *Salmonella*, *Shigella*, and coliform bacteria, in particular *E. coli*, from various materials. The bile salts and Crystal Violet largely inhibit the growth of Gram-positive microbial flora. Lactose together with the pH indicator Neutral Red are used to detect lactose-positive colonies and *E. coli* can be seen among these because of fluorescence under UV light.

## MacConkey Agar without Crystal Violet

**Composition** per liter:

| | |
|---|---|
| Agar | 12.0g |
| Lactose | 10.0g |
| Pancreatic digest of casein | 10.0g |
| Peptic digest of animal tissue | 10.0g |
| Bile salts | 5.0g |
| NaCl | 5.0g |
| Neutral Red | 0.05g |

pH 7.4 ± 0.2 at 25°C

**Source:** This medium is available as a premixed powder from BD Diagnostic Systems.

**Preparation of Medium:** Add components to distilled/deionized water and bring volume to 1.0L. Mix thoroughly. Gently heat while stirring until boiling. Autoclave for 15 min at 15 psi pressure–121°C. Pour into sterile Petri dishes or distribute into sterile tubes.

**Storage/Shelf Life:** Store dehydrated media in the dark in a sealed container below 30°C. Prepared media should be stored under refrigeration (2-8°C). Media should be used within 60 days of preparation. Media should not be used if there are any signs of deterioration (shrinking, cracking, or discoloration) or contamination, or if the expiration date supplied by the manufacturer has passed.

**Use:** For the detection of members of the Enterobacteriaceae and enterococci as well as some staphylococci. For the isolation and detection of coliforms and enteric pathogens from water and wastewater.

## MacConkey Agar without Crystal Violet with Sodium Chloride and 0.5% Sodium Taurocholate, HiVeg

**Composition** per liter:

| | |
|---|---|
| Agar | 20.0g |
| Plant peptone | 20.0g |
| Lactose | 10.0g |
| Synthetic detergent No. V | 5.0g |

NaCl .................................................................................5.0g
Neutral Red .......................................................................0.04g

pH 7.2 ± 0.2 at 25°C

**Source:** This medium is available as a premixed powder from Hi-Media.

**Preparation of Medium:** Add components to distilled/deionized water and bring volume to 1.0L. Mix thoroughly. Gently heat while stirring until boiling. Autoclave for 15 min at 15 psi pressure–121°C. Pour into sterile Petri dishes or distribute into sterile tubes.

**Storage/Shelf Life:** Store dehydrated media in the dark in a sealed container below 30°C. Prepared media should be stored under refrigeration (2-8°C). Media should be used within 60 days of preparation. Media should not be used if there are any signs of deterioration (shrinking, cracking, or discoloration) or contamination, or if the expiration date supplied by the manufacturer has passed.

**Use:** For the selective isolation, cultivation, and differentiation of *Vibrio* spp. in clinical specimens and in materials of sanitary importance.

## MacConkey Agar without Crystal Violet and Sodium Chloride with 0.5% Sodium Taurocholate, HiVeg

**Composition** per liter:

Agar ..................................................................................20.0g
Plant peptone....................................................................20.0g
Lactose .............................................................................10.0g
Synthetic detergent No. V ..................................................5.0g
Neutral Red .......................................................................0.04g

pH 7.2 ± 0.2 at 25°C

**Source:** This medium, without NaCl, is available as a premixed powder from HiMedia.

**Preparation of Medium:** Add components to distilled/deionized water and bring volume to 1.0L. Mix thoroughly. Gently heat while stirring until boiling. Autoclave for 15 min at 15 psi pressure–121°C. Pour into sterile Petri dishes or distribute into sterile tubes.

**Storage/Shelf Life:** Store dehydrated media in the dark in a sealed container below 30°C. Prepared media should be stored under refrigeration (2-8°C). Media should be used within 60 days of preparation. Media should not be used if there are any signs of deterioration (shrinking, cracking, or discoloration) or contamination, or if the expiration date supplied by the manufacturer has passed.

**Use:** For the selective isolation, cultivation, and differentiation of enteric pathogens, especially enterococci, in clinical specimens and in materials of sanitary importance.

## MacConkey Agar without Salt

**Composition** per liter:

Peptone.............................................................................20.0g
Agar ..................................................................................12.0g
Lactose .............................................................................10.0g
Bile salts............................................................................5.0g
Neutral Red .......................................................................0.075g

pH 7.4 ± 0.2 at 25°C

**Source:** This medium is available as a premixed powder from BD Diagnostic Systems and Oxoid Unipath.

**Preparation of Medium:** Add components to distilled/deionized water and bring volume to 1.0L. Mix thoroughly. Gently heat while stirring until boiling. Autoclave for 15 min at 15 psi pressure–121°C.

Pour into sterile Petri dishes or distribute into sterile tubes. Dry the surface of plates before inoculation.

**Storage/Shelf Life:** Store dehydrated media in the dark in a sealed container below 30°C. Prepared media should be stored under refrigeration (2-8°C). Media should be used within 60 days of preparation. Media should not be used if there are any signs of deterioration (shrinking, cracking, or discoloration) or contamination, or if the expiration date supplied by the manufacturer has passed.

**Use:** For the isolation and detection of coliforms and enteric pathogens from urine. Provides a low electrolyte medium on which most *Proteus* species will not swarm and therefore avoids overgrowth of the plate.

## MacConkey Agar No. 2 (MacConkey II Agar)

**Composition** per liter:

Peptone .............................................................................20.0g
Agar ..................................................................................15.0g
Lactose .............................................................................10.0g
NaCl..................................................................................5.0g
Bile salts No. 2 ..................................................................1.5g
Neutral Red .......................................................................0.05g
Crystal Violet....................................................................1.0mg

pH 7.2 ± 0.2 at 25°C

**Source:** This medium is available as a premixed powder from BD Diagnostic Systems and Oxoid Unipath.

**Preparation of Medium:** Add components to distilled/deionized water and bring volume to 1.0L. Mix thoroughly. Gently heat while stirring until boiling. Autoclave for 15 min at 15 psi pressure–121°C. Pour into sterile Petri dishes or distribute into sterile tubes.

**Storage/Shelf Life:** Store dehydrated media in the dark in a sealed container below 30°C. Prepared media should be stored under refrigeration (2-8°C). Media should be used within 60 days of preparation. Media should not be used if there are any signs of deterioration (shrinking, cracking, or discoloration) or contamination, or if the expiration date supplied by the manufacturer has passed.

**Use:** For the selective isolation, cultivation, and differentiation of enteric pathogens, especially enterococci, in clinical specimens and in materials of sanitary importance.

## MacConkey Agar No. 3

**Composition** per liter:

Peptone .............................................................................20.0g
Agar ..................................................................................15.0g
Lactose .............................................................................10.0g
NaCl..................................................................................5.0g
Bile salts No. 3 ..................................................................1.5g
Neutral Red .......................................................................0.03g
Crystal Violet....................................................................0.001g

pH 7.1 ± 0.2 at 25°C

**Source:** This medium is available as a premixed powder from Oxoid Unipath.

**Preparation of Medium:** Add components to distilled/deionized water and bring volume to 1.0L. Mix thoroughly. Gently heat while stirring until boiling. Autoclave for 15 min at 15 psi pressure–121°C. Pour into sterile Petri dishes or distribute into sterile tubes.

**Storage/Shelf Life:** Store dehydrated media in the dark in a sealed container below 30°C. Prepared media should be stored under refrigeration (2-8°C). Media should be used within 60 days of preparation.

Media should not be used if there are any signs of deterioration (shrinking, cracking, or discoloration) or contamination, or if the expiration date supplied by the manufacturer has passed.

**Use:** For the selective isolation, cultivation, and differentiation of enteric pathogens, especially *Salmonella* and *Shigella*, in clinical specimens and in foods.

## MacConkey Broth

**Composition** per liter:

Pancreatic digest of gelatin..........................................20.0g
Lactose..............................................................10.0g
Oxgall.................................................................5.0g
Bromcresol Purple.....................................................0.02g

pH 7.3 ± 0.2 at 25°C

**Source:** This medium is available as a premixed powder from BD Diagnostic Systems.

**Preparation of Medium:** Add components to distilled/deionized water and bring volume to 1.0L. If testing 10.0mL samples, prepare medium double strength. Mix thoroughly. Gently heat while stirring until boiling. Distribute into test tubes containing inverted Durham tubes. Autoclave for 15 min at 15 psi pressure–121°C.

**Storage/Shelf Life:** Store dehydrated media in the dark in a sealed container below 30°C. Prepared media should be stored under refrigeration (2-8°C). Media should be used within 60 days of preparation. Media should not be used if there are any signs of deterioration (discoloration) or contamination, or if the expiration date supplied by the manufacturer has passed.

**Use:** For the selective isolation and cultivation of coliforms in milk and water.

## MacConkey Broth

**Composition** per liter:

Peptone...............................................................20.0g
Lactose..............................................................10.0g
Bile salts.............................................................5.0g
NaCl...................................................................5.0g
Neutral Red..........................................................0.075g

pH 7.4 ± 0.2 at 25°C

**Source:** This medium is available as a premixed powder from Oxoid Unipath.

**Preparation of Medium:** Add components to distilled/deionized water and bring volume to 1.0L. If testing 10.0mL samples, prepare medium double strength. Mix thoroughly. Gently heat while stirring until boiling. Distribute into test tubes containing inverted Durham tubes. Autoclave for 15 min at 15 psi pressure–121°C.

**Storage/Shelf Life:** Store dehydrated media in the dark in a sealed container below 30°C. Prepared media should be stored under refrigeration (2-8°C). Media should be used within 60 days of preparation. Media should not be used if there are any signs of deterioration (discoloration) or contamination, or if the expiration date supplied by the manufacturer has passed.

**Use:** For the selective isolation and cultivation of coliforms in milk and water.

## MacConkey Broth, Purple

**Composition** per liter:

Peptone...............................................................20.0g
Lactose..............................................................10.0g

Bile salts.............................................................5.0g
NaCl...................................................................5.0g
Bromcresol Purple.....................................................0.01g

pH 7.4 ± 0.2 at 25°C

**Source:** This medium is available as a premixed powder or tablets from Oxoid Unipath.

**Preparation of Medium:** Add components to distilled/deionized water and bring volume to 1.0L. If testing 10.0mL samples, prepare medium double strength. Mix thoroughly. Gently heat while stirring until boiling. Distribute into test tubes containing inverted Durham tubes. Autoclave for 15 min at 15 psi pressure–121°C.

**Storage/Shelf Life:** Store dehydrated media in the dark in a sealed container below 30°C. Prepared media should be stored under refrigeration (2-8°C). Media should be used within 60 days of preparation. Media should not be used if there are any signs of deterioration (discoloration) or contamination, or if the expiration date supplied by the manufacturer has passed.

**Use:** For the selective isolation and cultivation of coliforms in milk and water.

## MacConkey Broth, Purple, with Bromcresol Purple, HiVeg

**Composition** per liter:

Plant special peptone.................................................23.0g
Lactose..............................................................10.0g
NaCl...................................................................5.0g
Synthetic detergent No. V..............................................2.0g
Bromcresol Purple.....................................................0.01g

pH 7.2 ± 0.2 at 25°C

**Source:** This medium is available as a premixed powder from Hi-Media.

**Preparation of Medium:** Add components to distilled/deionized water and bring volume to 1.0L. Mix thoroughly. Gently heat while stirring until boiling. Autoclave for 15 min at 15 psi pressure–121°C.

**Storage/Shelf Life:** Store dehydrated media in the dark in a sealed container below 30°C. Prepared media should be stored under refrigeration (2-8°C). Media should be used within 60 days of preparation. Media should not be used if there are any signs of deterioration (discoloration) or contamination, or if the expiration date supplied by the manufacturer has passed.

**Use:** For the selective isolation, cultivation, and differentiation of enteric bacteria, especially coliforms.

## MacConkey HiVeg Agar with Bromthymol Blue

**Composition** per liter:

Plant peptone........................................................17.0g
Agar.................................................................15.0g
Lactose..............................................................10.0g
NaCl...................................................................5.0g
Plant peptone No. 3....................................................3.0g
Synthetic detergent....................................................1.5g
Bromthymol Blue.......................................................0.03g

pH 7.2 ± 0.2 at 25°C

**Source:** This medium is available as a premixed powder from Hi-Media.

**Preparation of Medium:** Add components to distilled/deionized water and bring volume to 1.0L. Mix thoroughly. Gently heat while stirring until boiling. Autoclave for 15 min at 15 psi pressure–121°C. Pour into sterile Petri dishes or distribute into sterile tubes.

**Storage/Shelf Life:** Store dehydrated media in the dark in a sealed container below 30°C. Prepared media should be stored under refrigeration (2-8°C). Media should be used within 60 days of preparation. Media should not be used if there are any signs of deterioration (shrinking, cracking, or discoloration) or contamination, or if the expiration date supplied by the manufacturer has passed.

**Use:** For the selective isolation, cultivation, and differentiation of enteric bacteria.

## MacConkey HiVeg Agar with Crystal Violet and Sodium Chloride
**Composition** per liter:

Plant peptone.....................20.0g
Agar .............................15.0g
Lactose ..........................10.0g
NaCl ..............................5.0g
Synthetic detergent................1.5g
Neutral Red ......................0.05g
Crystal Violet ...................1.0mg

pH 7.2 ± 0.2 at 25°C

**Source:** This medium is available as a premixed powder from Hi-Media.

**Preparation of Medium:** Add components to distilled/deionized water and bring volume to 1.0L. Mix thoroughly. Gently heat while stirring until boiling. Autoclave for 15 min at 15 psi pressure–121°C. Pour into sterile Petri dishes or distribute into sterile tubes.

**Storage/Shelf Life:** Store dehydrated media in the dark in a sealed container below 30°C. Prepared media should be stored under refrigeration (2-8°C). Media should be used within 60 days of preparation. Media should not be used if there are any signs of deterioration (shrinking, cracking, or discoloration) or contamination, or if the expiration date supplied by the manufacturer has passed.

**Use:** For the selective isolation, cultivation, and differentiation of enteric bacteria. For the identification of Enterobacteriaceae in the presence of coliforms and lactose nonfermenters from water and sewage.

## MacConkey HiVeg Agar with 1.35% Agar, Crystal Violet, and Sodium Chloride
**Composition** per liter:

Plant peptone No. 2................17.0g
Agar .............................13.5g
Lactose ..........................10.0g
NaCl ..............................5.0g
Plant hydrolysate..................1.5g
Plant peptone......................1.5g
Sodium acetate (anhydrous)..........1.5g
Neutral Red ......................0.03g
Crystal Violet ...................1.0mg

pH 7.2 ± 0.2 at 25°C

**Source:** This medium is available as a premixed powder from Hi-Media.

**Preparation of Medium:** Add components to distilled/deionized water and bring volume to 1.0L. Mix thoroughly. Gently heat while stirring until boiling. Autoclave for 15 min at 15 psi pressure–121°C. Pour into sterile Petri dishes or distribute into sterile tubes.

**Storage/Shelf Life:** Store dehydrated media in the dark in a sealed container below 30°C. Prepared media should be stored under refrigeration (2-8°C). Media should be used within 60 days of preparation. Media should not be used if there are any signs of deterioration (shrinking, cracking, or discoloration) or contamination, or if the expiration date supplied by the manufacturer has passed.

**Use:** For the selective isolation and differentiation of lactose-fermenting and lactose-nonfermenting enteric bacteria.

## MacConkey HiVeg Agar without Crystal Violet and Sodium Chloride
**Composition** per liter:

Plant peptone .....................23.0g
Agar .............................12.0g
Lactose ..........................10.0g
Synthetic detergent ...............2.0g
Neutral Red ......................0.075g

pH 7.2 ± 0.2 at 25°C

**Source:** This medium is available as a premixed powder from Hi-Media.

**Preparation of Medium:** Add components to distilled/deionized water and bring volume to 1.0L. Mix thoroughly. Gently heat while stirring until boiling. Autoclave for 15 min at 15 psi pressure–121°C. Pour into sterile Petri dishes or distribute into sterile tubes.

**Storage/Shelf Life:** Store dehydrated media in the dark in a sealed container below 30°C. Prepared media should be stored under refrigeration (2-8°C). Media should be used within 60 days of preparation. Media should not be used if there are any signs of deterioration (shrinking, cracking, or discoloration) or contamination, or if the expiration date supplied by the manufacturer has passed.

**Use:** For the cultivation and differentiation of enteric bacteria, restricting swarming of *Proteus* species.

## MacConkey HiVeg Agar, Modified
**Composition** per liter:

Plant peptone .....................17.0g
Agar .............................13.5g
Inositol ..........................10.0g
NaCl ..............................5.0g
Plant peptone No. 3................3.0g
Synthetic detergent ...............1.5g
Neutral Red ......................0.03g
Crystal Violet ...................1.0mg

pH 7.2 ± 0.2 at 25°C

**Source:** This medium is available as a premixed powder from Hi-Media.

**Preparation of Medium:** Add components to distilled/deionized water and bring volume to 1.0L. Mix thoroughly. Gently heat while stirring until boiling. Autoclave for 15 min at 15 psi pressure–121°C. Pour into sterile Petri dishes or distribute into sterile tubes.

**Storage/Shelf Life:** Store dehydrated media in the dark in a sealed container below 30°C. Prepared media should be stored under refrigeration (2-8°C). Media should be used within 60 days of preparation. Media should not be used if there are any signs of deterioration (shrinking, cracking, or discoloration) or contamination, or if the expiration date supplied by the manufacturer has passed.

**Use:** For the isolation of *Klebsiella* species from water samples.

## MacConkey HiVeg Broth (Double Strength) with Neutral Red

**Composition** per liter:

| | |
|---|---|
| Plant peptone | 46.0g |
| Lactose | 20.0g |
| NaCl | 10.0g |
| Synthetic detergent | 4.0g |
| Neutral Red | 0.15g |

pH 7.2 ± 0.2 at 25°C

**Source:** This medium is available as a premixed powder from Hi-Media.

**Preparation of Medium:** Add components to distilled/deionized water and bring volume to 1.0L. Mix thoroughly. Distribute into tubes or leave in flasks. Gently heat while stirring until boiling. Autoclave for 15 min at 15 psi pressure–121°C.

**Storage/Shelf Life:** Store dehydrated media in the dark in a sealed container below 30°C. Prepared media should be stored under refrigeration (2-8°C). Media should be used within 60 days of preparation. Media should not be used if there are any signs of deterioration (discoloration) or contamination, or if the expiration date supplied by the manufacturer has passed.

**Use:** For the primary isolation of coliforms from large samples such as water or wastewater.

## MacConkey HiVeg Broth Purple with Bromo Cresol Purple

**Composition** per liter:

| | |
|---|---|
| Plant special peptone | 23.0g |
| Lactose | 10.0g |
| NaCl | 5.0g |
| Synthetic detergent No. V | 2.0g |
| Bromcresol Purple | 0.01g |

pH 7.2 ± 0.2 at 25°C

**Source:** This medium is available as a premixed powder from Hi-Media.

**Preparation of Medium:** Add components to distilled/deionized water and bring volume to 1.0L. Mix thoroughly. Gently heat while stirring until boiling. Autoclave for 15 min at 15 psi pressure–121°C. Pour into sterile Petri dishes or distribute into sterile tubes.

**Storage/Shelf Life:** Store dehydrated media in the dark in a sealed container below 30°C. Prepared media should be stored under refrigeration (2-8°C). Media should be used within 60 days of preparation. Media should not be used if there are any signs of deterioration (discoloration) or contamination, or if the expiration date supplied by the manufacturer has passed.

**Use:** For the presumptive identification of coliforms from water.

## MacConkey Sorbitol HiVeg Agar (Sorbitol HiVeg Agar)

**Composition** per liter:

| | |
|---|---|
| Plant peptone | 17.0g |
| Agar | 13.5g |
| D-Sorbitol | 10.0g |
| NaCl | 5.0g |
| Plant peptone No. 5 | 3.0g |

| | |
|---|---|
| Synthetic detergent No. I | 1.5g |
| Neutral Red | 0.03g |
| Crystal Violet | 1.0mg |

pH 7.2 ± 0.2 at 25°C

**Source:** This medium is available as a premixed powder from Hi-Media.

**Preparation of Medium:** Add components to distilled/deionized water and bring volume to 1.0L. Mix thoroughly. Gently heat while stirring until boiling. Autoclave for 15 min at 15 psi pressure–121°C. Pour into sterile Petri dishes or distribute into sterile tubes.

**Storage/Shelf Life:** Store dehydrated media in the dark in a sealed container below 30°C. Prepared media should be stored under refrigeration (2-8°C). Media should be used within 60 days of preparation. Media should not be used if there are any signs of deterioration (shrinking, cracking, or discoloration) or contamination, or if the expiration date supplied by the manufacturer has passed.

**Use:** For the isolation and cultivation of pathogenic *Escherichia coli*.

## M-*Aeromonas* Selective Agar Base, Havelaar

**Composition** per liter:

| | |
|---|---|
| Agar | 13.0g |
| Dextrin | 11.4g |
| Tryptose | 5.0g |
| NaCl | 3.0g |
| KCl | 2.0g |
| Yeast extract | 2.0g |
| $MgSO_4 \cdot 7H_2O$ | 0.1g |
| Sodium deoxycholate | 0.1g |
| Bromthymol Blue | 0.08g |
| $FeCl_3 \cdot 6H_2O$ | 0.06g |

pH 7.5 ± 0.2 at 25°C

**Source:** This medium is available from HiMedia.

**Preparation of Medium:** Add components to distilled/deionized water and bring volume to 1.0L. Mix thoroughly. Gently heat and bring to boiling. Distribute into tubes or flasks. Autoclave for 15 min at 15 psi pressure–121°C. Pour into sterile Petri dishes or leave in tubes.

**Storage/Shelf Life:** Store dehydrated media in the dark in a sealed container below 30°C. Prepared media should be stored under refrigeration (2-8°C). Media should be used within 60 days of preparation. Media should not be used if there are any signs of deterioration (shrinking, cracking, or discoloration) or contamination, or if the expiration date supplied by the manufacturer has passed.

**Use:** For the detection of *Aeromonas* species in water and other liquid samples by the membrane filter technique.

## m-EI Chromogenic Agar

**Composition** per liter:

| | |
|---|---|
| Yeast extract | 30.0g |
| NaCl | 15.0g |
| Agar | 15.0g |
| Peptone | 10.0g |
| Esculine | 1.0g |
| X-Glucoside | 0.75g |
| $NaN_3$ | 0.1g |
| Cycloheximide | 0.05g |

pH 7.1 ± 0.2 at 25°C

**Source:** This medium is available from CONDA, Barcelona, Spain.

**Preparation of Medium:** Add components to distilled/deionized water and bring volume to 1.0L. Mix thoroughly. Distribute into tubes or flasks. Autoclave for 15 min at 15 psi pressure–121°C. Pour into sterile Petri dishes.

**Storage/Shelf Life:** Store dehydrated media in the dark in a sealed container below 30°C. Prepared media should be stored in the dark under refrigeration (2-8°C). Chromogenic media are especially light and temperature sensitive; protect from light, excessive heat, moisture, and freezing. Media should not be used if there are any signs of deterioration (shrinking, cracking, or discoloration) or contamination, or if the expiration date supplied by the manufacturer has passed.

**Use:** For the detection and enumeration of enterococci in water using membrane filtration.

## m-EI Chromogenic Agar
## with Nalidix Acid

**Composition** per liter:

| | |
|---|---|
| Yeast extract | 30.0g |
| NaCl | 15.0g |
| Agar | 15.0g |
| Peptone | 10.0g |
| Esculine | 1.0g |
| X-Glucoside | 0.75g |
| NaN$_3$ | 0.1g |
| Cycloheximide | 0.05g |
| Nalidixic acid solution | 5.0mL |

pH 7.1 ± 0.2 at 25°C

**Source:** This medium is available from CONDA, Barcelona, Spain.

**Nalidixic Acid Solution:**

**Composition** per 5.0mL:

| | |
|---|---|
| Nalidixic acid | 0.24g |

**Preparation of Nalidixic Acid Solutiont:** Add nalidixic acid with a few drops of 0).1*N* NaOH to distilled/deionized water and bring volume to 5.0mL. Mix thoroughly. Filter sterilize.

**Preparation of Medium:** Add components, except nalidixic acid solution, to distilled/deionized water and bring volume to 995.0mL. Mix thoroughly. Gently heat to boiling. Autoclave for 15 min at 15 psi pressure–121°C. Cool to 45°–50°C. Aseptically add 5.0mL of sterile nalidixic acid solution. Mix thoroughly. Pour into sterile Petri dishes or distribute into sterile tubes.

**Storage/Shelf Life:** Store dehydrated media in the dark in a sealed container below 30°C. Prepared media should be stored in the dark under refrigeration (2-8°C). Chromogenic media are especially light and temperature sensitive; protect from light, excessive heat, moisture, and freezing. Media should not be used if there are any signs of deterioration (shrinking, cracking, or discoloration) or contamination, or if the expiration date supplied by the manufacturer has passed.

**Use:** For the selective detection and enumeration of enterococci in water using membrane filtration.

## Magnesium Oxalate Agar
## (MOX Agar)

**Composition** per liter:

| | |
|---|---|
| Pancreatic digest of casein | 15.0g |
| Agar | 15.0g |
| Papaic digest of soybean meal | 5.0g |
| NaCl | 5.0g |

| | |
|---|---|
| MgCl$_2$·6H$_2$O | 4.1g |
| Sodium oxalate | 2.68g |

pH 7.4–7.6 at 25°C

**Preparation of Medium:** Add components to distilled/deionized water and bring volume to 1.0L. Mix thoroughly. Gently heat and bring to boiling. Distribute into tubes or flasks. Autoclave for 15 min at 15 psi pressure–121°C. Pour into sterile Petri dishes or leave in tubes.

**Storage/Shelf Life:** Store dehydrated media in the dark in a sealed container below 30°C. Prepared media should be stored under refrigeration (2-8°C). Media should be used within 60 days of preparation. Media should not be used if there are any signs of deterioration (shrinking, cracking, or discoloration) or contamination, or if the expiration date supplied by the manufacturer has passed.

**Use:** For the cultivation of *Yersinia enterocolitica* from foods.

## Mannitol Egg Yolk Polymyxin Agar

**Composition** per liter:

| | |
|---|---|
| Agar | 15.0g |
| D-Mannitol | 10.0g |
| Peptone | 10.0g |
| NaCl | 10.0g |
| Beef extract | 1.0g |
| Phenol Red | 0.025g |
| Egg yolk emulsion, 50% | 50.0mL |
| Polymyxin B solution | 10.0mL |

pH 7.1 ± 0.1 at 25°C

**Source:** Available as a prepared medium from BD Diagnostic Systems.

**Egg Yolk Emulsion, 50%:**

**Composition** per 100.0mL:

| | |
|---|---|
| Chicken egg yolks | 11 |
| Whole chicken egg | 1 |
| NaCl (0.9% solution) | 80.0mL |

**Preparation of Egg Yolk Emulsion, 50%:** Soak eggs with 1:100 dilution of saturated mercuric chloride solution for 1 min. Crack eggs and separate yolks from whites. Mix egg yolks with 1 chicken egg. Beat to form emulsion. Measure 50.0mL of egg yolk emulsion and add to 50.0mL of 0.9% NaCl solution. Mix thoroughly. Filter sterilize. Warm to 45°–50°C.

**Polymyxin B Solution:**

**Composition** per 10.0mL:

| | |
|---|---|
| Polymyxin B | 100,000U |

**Preparation of Polymyxin B Solution:** Add polymyxin B to distilled/deionized water and bring volume to 10.0mL. Mix thoroughly. Filter sterilize.

**Preparation of Medium:** Add components—except egg yolk emulsion, 50%, and polymyxin B solution—to distilled/deionized water and bring volume to 940.0mL. Mix thoroughly. Gently heat and bring to boiling. Autoclave for 15 min at 15 psi pressure–121°C. Cool to 45°–50°C. Aseptically add 50.0mL of sterile egg yolk emulsion, 50%, and 10.0mL of sterile polymyxin B solution. Mix thoroughly. Pour into sterile Petri dishes.

**Storage/Shelf Life:** Store dehydrated media in the dark in a sealed container below 30°C. Prepared media should be stored under refrigeration (2-8°C). Media should be used within 60 days of preparation. Media should not be used if there are any signs of deterioration (shrinking, cracking, or discoloration) or contamination, or if the expiration date supplied by the manufacturer has passed.

**Use:** For the cultivation and enumeration of *Bacillus cereus* from foods.

## Mannitol Maltose Agar

**Composition** per liter:

| | |
|---|---|
| NaCl | 20.0g |
| Agar | 13.0g |
| D-Mannitol | 10.0g |
| Maltose | 10.0g |
| Beef extract | 5.0g |
| Papaic digest of soybean meal | 5.0g |
| Polypeptone™ | 5.0g |
| Dye stock solution, 1000X | 1.0mL |

pH 7.8 ± 0.2 at 25°C

**Dye Stock Solution, 1000X:**
**Composition** per 100.0mL:

| | |
|---|---|
| Bromthymol Blue | 4.0g |
| Cresol Red | 4.0g |
| Ethanol, 95% | 100.0mL |

**Preparation of Dye Stock Solution, 1000X:** Add Bromthymol Blue and Cresol Red to 100.0mL of ethanol. Mix thoroughly.

**Preparation of Medium:** Add components to distilled/deionized water and bring volume to 1.0L. Mix thoroughly. Adjust to pH 7.8. Gently heat and bring to boiling. Distribute into tubes or flasks. Autoclave for 15 min at 15 psi pressure–121°C. Pour into sterile Petri dishes or leave in tubes.

**Storage/Shelf Life:** Store dehydrated media in the dark in a sealed container below 30°C. Prepared media should be stored under refrigeration (2-8°C). Media should be used within 60 days of preparation. Media should not be used if there are any signs of deterioration (shrinking, cracking, or discoloration) or contamination, or if the expiration date supplied by the manufacturer has passed.

**Use:** For the cultivation of *Vibrio* species from foods.

## Mannitol Motility Test HiVeg Medium

**Composition** per liter:

| | |
|---|---|
| Plant peptone | 20.0g |
| Agar | 3.0g |
| Mannitol | 2.0g |
| Potassium nitrate | 1.0g |
| Phenol red | 0.04g |

pH 7.6 ± 0.2 at 25°C

**Source:** This medium is available as a premixed powder from HiMedia.

**Preparation of Medium:** Add components to distilled/deionized water and bring volume to 1.0L. Mix thoroughly. Gently heat while stirring and bring to boiling. Distribute into tubes or flasks. Autoclave for 15 min at 15 psi pressure–121°C. Pour into sterile Petri dishes or leave in tubes.

**Storage/Shelf Life:** Store dehydrated media in the dark in a sealed container below 30°C. Prepared media should be stored under refrigeration (2-8°C). Media should be used within 60 days of preparation. Media should not be used if there are any signs of deterioration (discoloration) or contamination, or if the expiration date supplied by the manufacturer has passed.

**Use:** For the selective isolation, cultivation, and enumeration of staphylococci from clinical and nonclinical specimens. Mannitol-utilizing organisms turn the medium yellow.

## Mannitol Salt Agar

**Composition** per liter:

| | |
|---|---|
| NaCl | 75.0g |
| Agar | 15.0g |
| D-Mannitol | 10.0g |
| Pancreatic digest of casein | 5.0g |
| Peptic digest of animal tissue | 5.0g |
| Beef extract | 1.0g |
| Phenol Red | 0.025g |

pH 7.4 ± 0.2 at 25°C

**Source:** This medium is available as a premixed powder from BD Diagnostic Systems and Oxoid Unipath.

**Preparation of Medium:** Add components to distilled/deionized water and bring volume to 1.0L. Mix thoroughly. Gently heat while stirring and bring to boiling. Distribute into tubes or flasks. Autoclave for 15 min at 15 psi pressure–121°C. Pour into sterile Petri dishes or leave in tubes.

**Storage/Shelf Life:** Store dehydrated media in the dark in a sealed container below 30°C. Prepared media should be stored under refrigeration (2-8°C). Media should be used within 60 days of preparation. Media should not be used if there are any signs of deterioration (shrinking, cracking, or discoloration) or contamination, or if the expiration date supplied by the manufacturer has passed.

**Use:** For the selective isolation, cultivation, and enumeration of staphylococci from clinical and nonclinical specimens. Mannitol-utilizing organisms turn the medium yellow.

## Mannitol Salt Agar
## (BAM M97)

**Composition** per liter:

| | |
|---|---|
| NaCl | 75.0g |
| Agar | 15.0g |
| D-Mannitol | 10.0g |
| Polypeptone | 10.0g |
| Beef extract | 1.0g |
| Phenol Red | 0.025g |

pH 7.4 ± 0.2 at 25°C

**Preparation of Medium:** Add components to distilled/deionized water and bring volume to 1.0L. Mix thoroughly. Gently heat while stirring and bring to boiling. Distribute into tubes or flasks. Autoclave for 15 min at 15 psi pressure–121°C. Pour into sterile Petri dishes or leave in tubes.

**Storage/Shelf Life:** Store dehydrated media in the dark in a sealed container below 30°C. Prepared media should be stored under refrigeration (2-8°C). Media should be used within 60 days of preparation. Media should not be used if there are any signs of deterioration (shrinking, cracking, or discoloration) or contamination, or if the expiration date supplied by the manufacturer has passed.

**Use:** For the selective isolation, cultivation, and enumeration of staphylococci. Mannitol-utilizing organisms turn the medium yellow.

## Mannitol Salt Agar
## with Egg Yolk Emulsion

**Composition** per liter:

| | |
|---|---|
| NaCl | 75.0g |
| Agar | 15.0g |
| D-Mannitol | 10.0g |
| Proteose peptone | 10.0g |

Beef extract ..................................................................1.0g
Phenol Red ..................................................................0.025g
Egg yolk emulsion ................................................. 100.0mL

pH 7.4 ± 0.2 at 25°C

**Source:** This medium is available from HiMedia.

### Egg Yolk Emulsion
**Composition** per 100.0mL:
Chicken egg yolks.............................................................. 11
Whole chicken egg.............................................................. 1
NaCl (0.9% solution) .............................................. 50.0mL

**Preparation of Egg Yolk Emulsion:** Soak eggs with 1:100 dilution of saturated mercuric chloride solution for 1 min. Crack eggs and separate yolks from whites. Mix egg yolks with 1 chicken egg. Beat to form emulsion. Measure 50.0mL of egg yolk emulsion and add to 50.0mL of 0.9% NaCl solution. Mix thoroughly. Filter sterilize. Warm to 45°–50°C.

**Preparation of Medium:** Add components, except egg yolk emulsion, to distilled/deionized water and bring volume to 900.0mL. Mix thoroughly. Autoclave for 15 min at 15 psi pressure–121°C. Cool to 50°C. Aseptically add egg yolk emulsion. Mix thoroughly. Pour into Petri dishes or aseptically distribute into sterile tubes or flasks.

**Storage/Shelf Life:** Store dehydrated media in the dark in a sealed container below 30°C. Prepared media should be stored under refrigeration (2-8°C). Media should be used within 60 days of preparation. Media should not be used if there are any signs of deterioration (shrinking, cracking, or discoloration) or contamination, or if the expiration date supplied by the manufacturer has passed.

**Use:** For the selective isolation of pathogenic staphylococci.

### Mannitol Salt Broth
**Composition** per liter:
NaCl ..................................................................75.0g
Proteose peptone ..................................................10.0g
D-Mannitol ..........................................................10.0g
Beef extract ..........................................................1.0g
Phenol Red ..........................................................0.025g

pH 7.4 ± 0.2 at 25°C

**Source:** This medium is available from HiMedia.

**Preparation of Medium:** Add components to distilled/deionized water and bring volume to 1.0L. Mix thoroughly. Gently heat and bring to boiling. Distribute into tubes or flasks. Autoclave for 15 min at 15 psi pressure–121°C. Pour into sterile Petri dishes or leave in tubes.

**Storage/Shelf Life:** Store dehydrated media in the dark in a sealed container below 30°C. Prepared media should be stored under refrigeration (2-8°C). Media should be used within 60 days of preparation. Media should not be used if there are any signs of deterioration (discoloration) or contamination, or if the expiration date supplied by the manufacturer has passed.

**Use:** For the selective isolation of presumptive pathogenic staphylococci.

### Mannitol Salt Broth
**Composition** per liter:
NaCl ..................................................................100.0g
Pancreatic digest of casein ..................................17.0g
Papaic digest of soybean meal ..........................3.0g
K₂HPO₄ ..................................................................2.5g

D-Mannitol ..................................................................2.5g
Phenol Red ..................................................................0.025g

pH 7.4 ± 0.2 at 25°C

**Source:** This medium is available as a premixed powder from BD Diagnostic Systems.

**Preparation of Medium:** Add components to distilled/deionized water and bring volume to 1.0L. Mix thoroughly. Distribute into tubes or flasks. Autoclave for 15 min at 15 psi pressure–121°C.

**Storage/Shelf Life:** Store dehydrated media in the dark in a sealed container below 30°C. Prepared media should be stored under refrigeration (2-8°C). Media should be used within 60 days of preparation. Media should not be used if there are any signs of deterioration (discoloration) or contamination, or if the expiration date supplied by the manufacturer has passed.

**Use:** For the selective isolation and cultivation of staphylococci from foods and nonclinical specimens. Mannitol-utilizing organisms turn the medium yellow.

### Mannitol Salt HiVeg Agar Base
**Composition** per liter:
NaCl ..................................................................75.0g
Agar ..................................................................15.0g
D-Mannitol ..........................................................10.0g
Plant peptone No. 3..........................................10.0g
Plant extract ..........................................................1.0g
Phenol Red ..........................................................0.025g

pH 7.4 ± 0.2 at 25°C

**Source:** This medium is available as a premixed powder from HiMedia.

**Preparation of Medium:** Add components to distilled/deionized water and bring volume to 1.0L. Mix thoroughly. Gently heat while stirring and bring to boiling. Distribute into tubes or flasks. Autoclave for 15 min at 15 psi pressure–121°C. Pour into sterile Petri dishes or leave in tubes.

**Storage/Shelf Life:** Store dehydrated media in the dark in a sealed container below 30°C. Prepared media should be stored under refrigeration (2-8°C). Media should be used within 60 days of preparation. Media should not be used if there are any signs of deterioration (shrinking, cracking, or discoloration) or contamination, or if the expiration date supplied by the manufacturer has passed.

**Use:** For the selective isolation of pathogenic staphylococci.

### Mannitol Salt HiVeg Broth
**Composition** per liter:
NaCl ..................................................................75.0g
D-Mannitol ..........................................................10.0g
Plant peptone No. 3..........................................10.0g
Plant extract ..........................................................1.0g
Phenol Red ..........................................................0.025g

pH 7.4 ± 0.2 at 25°C

**Source:** This medium is available as a premixed powder from HiMedia.

**Preparation of Medium:** Add components to distilled/deionized water and bring volume to 1.0L. Mix thoroughly. Gently heat while stirring and bring to boiling. Distribute into tubes or flasks. Autoclave for 15 min at 15 psi pressure–121°C.

**Storage/Shelf Life:** Store dehydrated media in the dark in a sealed container below 30°C. Prepared media should be stored under refriger-

ation (2-8°C). Media should be used within 60 days of preparation. Media should not be used if there are any signs of deterioration (discoloration) or contamination, or if the expiration date supplied by the manufacturer has passed.

**Use:** For the selective isolation of presumptive pathogenic staphylococci.

## Mannitol Selenite Broth
## (Selenite Mannitol Broth)

**Composition** per liter:

| | |
|---|---|
| NaH$_2$PO$_4$ | 10.0g |
| Peptic digest of animal tissue | 5.0g |
| Mannitol | 4.0g |
| NaHSeO$_3$ | 4.0g |

pH 7.1 ± 0.2 at 25°C

**Source:** This medium is available from HiMedia.

**Caution:** Sodium hydrogen selenite (sodium biselenite) is a very toxic, corrosive agent and causes teratogenicity; it should be handled with great care. If there is contact, wash immediately with lots of water.

**Preparation of Medium:** Add NaHSeO$_3$ to distilled/deionized water and bring volume to 1.0L. Mix thoroughly. Add remaining components. Mix thoroughly. Distribute into tubes or flasks. Gently heat and bring to boiling. Sterilize in a boiling water bath or free flowing steam for 10 min. Do not autoclave. Discard the prepared medium if a large amount of selenite is reduced (indicated by red precipitate at the bottom of the tube).

**Storage/Shelf Life:** Store dehydrated media in the dark in a sealed container below 30°C. Prepared media should be stored under refrigeration (2-8°C). Media should be used within 60 days of preparation. Media should not be used if there are any signs of deterioration (discoloration) or contamination, or if the expiration date supplied by the manufacturer has passed.

**Use:** For the selective enrichment of *Salmonella* spp. from clinical materials.

## Mannitol Selenite Broth with Brilliant Green

**Composition** per liter:

| | |
|---|---|
| Meat peptone | 5.0g |
| Yeast extract | 5.0g |
| Mannitol | 5.0g |
| K$_2$HPO$_4$ | 4.35g |
| KH$_2$PO$_4$ | 3.4g |
| Sodium taurocholate | 1.0g |
| Brilliant Green | 0.005g |
| Na$_2$SeO$_3$·5H$_2$O | 1.0g |

pH 7.0 ± 0.2 at 25°C

**Source:** This medium is available from HiMedia.

**Caution:** Sodium hydrogen selenite (sodium biselenite) is a very toxic, corrosive agent and causes teratogenicity; it should be handled with great care. If there is contact, wash immediately with lots of water.

**Preparation of Medium:** Add NaHSeO$_3$ to distilled/deionized water and bring volume to 1.0L. Mix thoroughly. Add remaining components. Mix thoroughly. Distribute into tubes or flasks. Gently heat and bring to boiling. Sterilize in a boiling water bath or free flowing steam for 10 min. Do not autoclave. Discard the prepared medium if a large amount of selenite is reduced (indicated by red precipitate at the bottom of the tube).

**Storage/Shelf Life:** Store dehydrated media in the dark in a sealed container below 30°C. Prepared media should be stored under refrigeration (2-8°C). Media should be used within 60 days of preparation. Media should not be used if there are any signs of deterioration (discoloration) or contamination, or if the expiration date supplied by the manufacturer has passed.

**Use:** For the enrichment of *Salmonella* spp. from feces, foodstuffs, and other materials.

## Martin-Lewis Agar

**Composition** per liter:

| | |
|---|---|
| Agar | 12.0g |
| Hemoglobin | 10.0g |
| Pancreatic digest of casein | 7.5g |
| Selected meat peptone | 7.5g |
| NaCl | 5.0g |
| K$_2$HPO$_4$ | 4.0g |
| Cornstarch | 1.0g |
| KH$_2$PO$_4$ | 1.0g |
| Supplement solution | 10.0mL |
| VCAT inhibitor | 10.0mL |

pH 7.2 ± 0.22 at 25°C

**Source:** Martin-Lewis agar is available as a prepared medium from BD Diagnostic Systems.

**Supplement Solution:**
**Composition** per liter:

| | |
|---|---|
| Glucose | 100.0g |
| L-Cysteine·HCl | 25.9g |
| L-Glutamine | 10.0g |
| L-Cystine | 1.1g |
| Adenine | 1.0g |
| Nicotinamide adenine dinucleotide | 0.25g |
| Vitamin B$_{12}$ | 0.1g |
| Thiamine pyrophosphate | 0.1g |
| Guanine·HCl | 0.03g |
| Fe(NO$_3$)$_3$·6H$_2$O | 0.02g |
| *p*-Aminobenzoic acid | 0.013g |
| Thiamine·HCl | 3.0mg |

**Source:** The supplement solution IsoVitaleX® enrichment is available from BD Diagnostic Systems. This enrichment may be replaced by supplement VX from BD Diagnostic Systems.

**Preparation of Supplement Solution:** Add components to distilled/deionized water and bring volume to 1.0L. Mix thoroughly. Filter sterilize.

**VCAT Inhibitor:**
**Composition** per 10.0mL:

| | |
|---|---|
| Colistin | 7.5mg |
| Trimethoprim lactate | 5.0mg |
| Vancomycin | 4.0mg |
| Anisomycin | 0.02g |

**Preparation of VCAT Inhibitor:** Add components to distilled/deionized water and bring volume to 10.0mL. Mix thoroughly. Filter sterilize.

**Preparation of Medium:** Add components, except supplement solution and VCAT inhibitor, to distilled/deionized water and bring volume to 980.0mL. Gently heat while stirring and bring to boiling. Autoclave for 15 min at 15 psi pressure–121°C. Cool to 45°–50°C.

Aseptically add sterile supplement solution and sterile VCAT inhibitor. Mix thoroughly. Pour into sterile Petri dishes.

**Storage/Shelf Life:** Store dehydrated media in the dark in a sealed container below 30°C. Prepared media should be stored under refrigeration (2-8°C). Media should be used within 60 days of preparation. Media should not be used if there are any signs of deterioration (shrinking, cracking, or discoloration) or contamination, or if the expiration date supplied by the manufacturer has passed.

**Use:** For the isolation and cultivation of pathogenic *Neisseria* from specimens containing mixed flora of bacteria and fungi.

## Martin-Lewis Agar, Enriched

**Composition** per liter:

| | |
|---|---|
| Agar | 12.0g |
| Pancreatic digest of casein | 7.5g |
| Selected meat peptone | 7.5g |
| NaCl | 5.0g |
| K$_2$HPO$_4$ | 4.0g |
| Cornstarch | 1.0g |
| KH$_2$PO$_4$ | 1.0g |
| *Sarcina lutea* suspension | 20.0mL |
| Horse serum, inactivated | 20.0mL |
| Supplement solution | 10.0mL |
| PCAT inhibitor | 10.0mL |

pH 7.2 ± 0.22 at 25°C

**Source:** The supplement solution (IsoVitaleX® enrichment) is available from BD Diagnostic Systems. This enrichment may be replaced by supplement VX from BD Diagnostic Systems.

### *Sarcina lutea* Suspension:

**Composition** per 20.0mL:

| | |
|---|---|
| *Sarcina lutea* FDA 1001 | 10$^6$–10$^7$ cells |

**Preparation of *Sarcina lutea* Suspension:** Aseptically wash the growth of 24-hr cultures of *Sarcina lutea* FDA 1001 cells from Thayer-Martin plates with sterile soybean casein digest broth. Standardize the suspension by adding additional sterile tryptic soy broth to yield 40% light transmission at 530nm wavelength.

### Soybean Casein Digest Broth:

**Composition** per liter:

| | |
|---|---|
| Pancreatic digest of casein | 17.0g |
| NaCl | 5.0g |
| Papaic digest of soybean meal | 3.0g |
| K$_2$HPO$_4$ | 2.5g |
| Glucose | 2.5g |

pH 7.3 ± 0.2 at 25°C

**Preparation of Soybean Casein Digest Broth:** Add components to distilled/deionized water and bring volume to 1.0L. Mix thoroughly. Distribute into tubes or flasks. Autoclave for 15 min at 15 psi pressure–121°C.

### Supplement Solution:

**Composition** per liter:

| | |
|---|---|
| Glucose | 100.0g |
| L-Cysteine·HCl | 25.9g |
| L-Glutamine | 10.0g |
| L-Cystine | 1.1g |
| Adenine | 1.0g |
| Nicotinamide adenine dinucleotide | 0.25g |
| Vitamin B$_{12}$ | 0.1g |
| Thiamine pyrophosphate | 0.1g |
| Guanine·HCl | 0.03g |
| Fe(NO$_3$)$_3$·6H$_2$O | 0.02g |
| *p*-Aminobenzoic acid | 0.013g |
| Thiamine·HCl | 3.0mg |

**Preparation of Supplement Solution:** Add components to distilled/deionized water and bring volume to 1.0L. Mix thoroughly. Filter sterilize.

### PCAT Inhibitor:

**Composition** per 10.0mL:

| | |
|---|---|
| Anisomycin | 0.02g |
| Colistin | 7.5mg |
| Trimethoprim lactate | 5.0mg |
| Penicillin G | 25,000U |

**Preparation of PCAT Inhibitor:** Add components to distilled/deionized water and bring volume to 10.0mL. Mix thoroughly. Filter sterilize.

**Preparation of Medium:** Add components—except *Sarcina lutea* suspension, horse serum, supplement solution, and PCAT inhibitor—to distilled/deionized water and bring volume to 940.0mL. Gently heat while stirring and bring to boiling. Autoclave for 15 min at 15 psi pressure–121°C. Cool to 45°–50°C. Aseptically add 20.0mL of sterile *Sarcina lutea* suspension, 20.0mL of sterile horse serum, 10.0mL of supplement solution, and 10.0mL of sterile PCAT inhibitor. Mix thoroughly. Pour into sterile Petri dishes.

**Storage/Shelf Life:** Store dehydrated media in the dark in a sealed container below 30°C. Prepared media should be stored under refrigeration (2-8°C). Media should be used within 60 days of preparation. Media should not be used if there are any signs of deterioration (shrinking, cracking, or discoloration) or contamination, or if the expiration date supplied by the manufacturer has passed.

**Use:** For the isolation and cultivation of pathogenic *Neisseria*, especially penicillinase-producing strains, from specimens containing mixed flora of bacteria and fungi.

## M-Azide HiVeg Broth Base with Triphenyltetrazolium Chloride

**Composition** per liter:

| | |
|---|---|
| Saccharose | 100.0g |
| Plant hydrolysate No. 1 | 40.0g |
| Yeast extract | 10.0g |
| K$_2$HPO$_4$ | 4.0g |
| Glucose | 2.0g |
| NaN$_3$ | 0.4g |
| Triphenyltetrazolium chloride solution | 5.0mL |

pH 7.2 ± 0.2 at 25°C

**Source:** This medium is available as a premixed powder from Hi-Media.

**Caution:** Sodium azide is toxic. Azides also react with metals and disposal must be highly diluted.

### Triphenyltetrazolium Choride Solution:

**Composition** per 5.0mL:

| | |
|---|---|
| Triphenyltetrazolium chloride | 0.1g |

**Preparation of Triphenyltetrazolium Choride Solution:** Add triphenyltetrazolium chloride to distilled/deionized water and bring volume to 5.0mL. Mix thoroughly. Filter sterilize.

**Preparation of Medium:** Add components, except triphenyltetrazolium chloride solution, to distilled/deionized water and bring volume to

1.0L.Mix thoroughly. Gently heat and bring to boiling. Distribute into tubes or flasks. Autoclave for 15 min at 15 psi pressure–121°C. Cool to 45°–50°C. Aseptically add 5.0mL triphenyltetrazolium chloride solution. Mix thoroughly. Aseptically distribute into tubes.

**Storage/Shelf Life:** Store dehydrated media in the dark in a sealed container below 30°C. Prepared media should be stored under refrigeration (2-8°C). Media should be used within 60 days of preparation. Media should not be used if there are any signs of deterioration (discoloration) or contamination, or if the expiration date supplied by the manufacturer has passed.

**Use:** For the detection and enrichment of fecal streptococci in water and sewage by the membrane filtration method.

## M-Bismuth Sulfite Broth

**Composition** per liter:

| | |
|---|---|
| Peptic digest of animal tissue | 20.0g |
| Bismuth sulfite indicator | 16.0g |
| Glucose | 10.0g |
| Plant extract | 10.0g |
| Na$_2$HPO$_4$ | 8.0g |
| FeSO$_4$ | 0.6g |
| Brilliant Green | 0.05g |

pH 7.7 ± 0.2 at 25°C

**Source:** This medium is available as a premixed powder from Hi-Media.

**Preparation of Medium:** Add components to distilled/deionized water and bring volume to 1.0L. Mix thoroughly and heat with frequent agitation until boiling. Boil for 1 min. Do not autoclave. Cool to 45°–50°C. Mix to disperse the precipitate and aseptically distribute into sterile tubes or flasks. Use 2.0–2.2mL of medium for each membrane filter.

**Storage/Shelf Life:** Store dehydrated media in the dark in a sealed container below 30°C. Prepared media should be stored under refrigeration (2-8°C). Media should be used within 60 days of preparation. Media should not be used if there are any signs of deterioration (discoloration) or contamination, or if the expiration date supplied by the manufacturer has passed.

**Use:** For the selective isolation of *Salmonella typhi* and other enteric bacilli and for the detection of *Salmonella* by the membrane filter method.

## M-Bismuth Sulfite HiVeg Broth

**Composition** per liter:

| | |
|---|---|
| Plant peptone | 20.0g |
| Bismuth sulfite indicator | 16.0g |
| Glucose | 10.0g |
| Plant extract | 10.0g |
| Na$_2$HPO$_4$ | 8.0g |
| FeSO$_4$ | 0.6g |
| Brilliant Green | 0.05g |

pH 7.7 ± 0.2 at 25°C

**Source:** This medium is available as a premixed powder from Hi-Media.

**Preparation of Medium:** Add components to distilled/deionized water and bring volume to 1.0L. Mix thoroughly and heat with frequent agitation until boiling. Boil for 1 min. Do not autoclave. Cool to 45°–50°C. Mix to disperse the precipitate and aseptically distribute into

sterile tubes or flasks. Use 2.0–2.2mL of medium for each membrane filter.

**Storage/Shelf Life:** Store dehydrated media in the dark in a sealed container below 30°C. Prepared media should be stored under refrigeration (2-8°C). Media should be used within 60 days of preparation. Media should not be used if there are any signs of deterioration (discoloration) or contamination, or if the expiration date supplied by the manufacturer has passed.

**Use:** For the selective isolation of *Salmonella typhi* and other enteric bacilli and for the detection of *Salmonella* by the membrane filter method.

## M-Brilliant Green Broth

**Composition** per liter:

| | |
|---|---|
| Proteose peptone | 20.0g |
| Lactose | 20.0g |
| Saccharose | 20.0g |
| NaCl | 10.0g |
| Yeast extract | 6.0g |
| Phenol Red | 0.16g |
| Brilliant Green | 0.025g |

pH 6.9 ± 0.2 at 25°C

**Source:** This medium is available as a premixed powder from Hi-Media.

**Preparation of Medium:** Add components to distilled/deionized water and bring volume to1.0L. Mix thoroughly. Distribute into tubes containing inverted Durham tubes, in 10.0mL amounts for testing 1.0mL or less of sample. Gently heat and bring to boiling. Do not autoclave. Cool the broth rapidly. Medium is sensitive to light.

**Storage/Shelf Life:** Store dehydrated media in the dark in a sealed container below 30°C. Prepared media should be stored under refrigeration (2-8°C). Media should be used within 60 days of preparation. Media should not be used if there are any signs of deterioration (discoloration) or contamination, or if the expiration date supplied by the manufacturer has passed.

**Use:** For the detection of coliform microorganisms in foods, dairy products, water, and wastewater, as well as in other materials of sanitary importance.

## M-Brilliant Green HiVeg Broth

**Composition** per liter:

| | |
|---|---|
| Plant peptone No. 3 | 20.0g |
| Lactose | 20.0g |
| Saccharose | 20.0g |
| NaCl | 10.0g |
| Yeast extract | 6.0g |
| Phenol Red | 0.16g |
| Brilliant Green | 0.025g |

pH 6.9 ± 0.2 at 25°C

**Source:** This medium is available as a premixed powder from Hi-Media.

**Preparation of Medium:** Add components to distilled/deionized water and bring volume to1.0L. Mix thoroughly. Distribute into tubes containing inverted Durham tubes, in 10.0mL amounts for testing 1.0mL or less of sample. Gently heat and bring to boiling. Do not autoclave. Cool the broth rapidly. Medium is sensitive to light.

**Storage/Shelf Life:** Store dehydrated media in the dark in a sealed container below 30°C. Prepared media should be stored under refriger-

ation (2-8°C). Media should be used within 60 days of preparation. Media should not be used if there are any signs of deterioration (discoloration) or contamination, or if the expiration date supplied by the manufacturer has passed.

**Use:** For the detection of coliform microorganisms in foods, dairy products, water, and wastewater, as well as in other materials of sanitary importance.

## M-Broth, HiVeg

**Composition** per liter:
| | |
|---|---|
| Plant hydrolysate | 12.5g |
| K₂HPO₄ | 5.0g |
| NaCl | 5.0g |
| Sodium citrate | 5.0g |
| Yeast extract | 5.0g |
| D-Mannose | 2.0g |
| MgSO₄ | 0.8g |
| MnCl₂ | 0.14g |
| FeSO₄ | 0.04g |
| Polysorbate 80 | 0.75mL |

pH $7.0 \pm 0.22$ at 25°C

**Source:** This medium, without polysorbate 80, is available as a premixed powder from HiMedia.

**Preparation of Medium:** Add components to distilled/deionized water and bring volume to 1.0L. Mix thoroughly. Distribute into tubes or flasks. Autoclave for 15 min at 15 psi pressure–121°C.

**Storage/Shelf Life:** Store dehydrated media in the dark in a sealed container below 30°C. Prepared media should be stored under refrigeration (2-8°C). Media should be used within 60 days of preparation. Media should not be used if there are any signs of deterioration (discoloration) or contamination, or if the expiration date supplied by the manufacturer has passed.

**Use:** For the detection of *Salmonella* in dried foods and feeds.

## McBride Agar, Modified

**Composition** per liter:
| | |
|---|---|
| Agar | 15.0g |
| Glycine anhydride | 10.0g |
| Tryptose | 10.0g |
| NaCl | 5.0g |
| Beef extract | 3.0g |
| Phenylethanol | 2.5g |
| LiCl | 0.5g |
| Cycloheximide solution | 10.0mL |

pH $7.3 \pm 0.2$ at 25°C

**Cycloheximide Solution:**
**Composition** per 10.0mL:
| | |
|---|---|
| Cycloheximide | 0.2g |

**Preparation of Cycloheximide Solution:** Add cycloheximide to distilled/deionized water and bring volume to 10.0mL. Mix thoroughly. Filter sterilize.

**Caution:** Cycloheximide is toxic. Avoid skin contact or aerosol formation and inhalation.

**Preparation of Medium:** Add components, except cycloheximide solution, to distilled/deionized water and bring volume to 990.0mL. Mix thoroughly. Gently heat and bring to boiling. Autoclave for 15 min at 15 psi pressure–121°C. Cool to 45°–50°C. Aseptically add sterile

cycloheximide solution. Mix thoroughly. Pour into sterile Petri dishes or distribute into sterile tubes.

**Storage/Shelf Life:** Store dehydrated media in the dark in a sealed container below 30°C. Prepared media should be stored under refrigeration (2-8°C). Media should be used within 60 days of preparation. Media should not be used if there are any signs of deterioration (shrinking, cracking, or discoloration) or contamination, or if the expiration date supplied by the manufacturer has passed.

**Use:** For the isolation of *Listeria monocytogenes* from dairy products.

## McBride *Listeria* Agar

**Composition** per liter:
| | |
|---|---|
| Agar | 15.0g |
| Glycine | 10.0g |
| Pancreatic digest of casein | 5.0g |
| Peptic digest of animal tissue | 5.0g |
| NaCl | 5.0g |
| Beef extract | 3.0g |
| Phenylethyl alcohol | 2.5g |
| LiCl | 0.5g |

pH $7.3 \pm 0.2$ at 25°C

**Source:** This medium is available as a premixed powder from BD Diagnostic Systems.

**Preparation of Medium:** Add components to distilled/deionized water and bring volume to 1.0L. Mix thoroughly. Gently heat while stirring and bring to boiling. Distribute into tubes or flasks. Autoclave for 15 min at 15 psi pressure–121°C. Pour into sterile Petri dishes or leave in tubes.

**Storage/Shelf Life:** Store dehydrated media in the dark in a sealed container below 30°C. Prepared media should be stored under refrigeration (2-8°C). Media should be used within 60 days of preparation. Media should not be used if there are any signs of deterioration (shrinking, cracking, or discoloration) or contamination, or if the expiration date supplied by the manufacturer has passed.

**Use:** For the selective isolation of *Listeria monocytogenes* from clinical and nonclinical specimens containing mixed flora.

## McBride *Listeria* HiVeg Agar Base with Blood and Selective Supplement

**Composition** per liter:
| | |
|---|---|
| Agar | 15.0g |
| Glycine anhydride | 10.0g |
| Plant hydrolysate No. 1 | 10.0g |
| NaCl | 5.0g |
| Plant extract | 3.0g |
| Phenyl ethanol | 2.5g |
| LiCl | 0.5g |
| Sheep blood, defibrinated | 50.0mL |
| Selective supplement solution | 10.0mL |

pH $7.3 \pm 0.2$ at 25°C

**Source:** This medium, without blood or selective supplement solution, is available as a premixed powder from HiMedia.

**Caution:** LiCl is harmful. Avoid bodily contact and inhalation of vapors. On contact with skin wash with plenty of water immediately.

**Selective Supplement Solution:**
**Composition** per 10.0mL:
| | |
|---|---|
| Cycloheximide | 0.2g |

**Preparation of Selective Supplement Solution:** Add cycloheximide to distilled/deionized water and bring volume to 10.0mL. Mix thoroughly. Filter sterilize.

**Caution:** Cycloheximide is toxic. Avoid skin contact or aerosol formation and inhalation.

**Preparation of Medium:** Add components, except blood and selective supplement solution, to distilled/deionized water and bring volume to 940.0mL. Mix thoroughly. Gently heat and bring to boiling. Distribute into tubes or flasks. Autoclave for 15 min at 15 psi pressure–121°C. Cool to 50°C. Aseptically add 50.0mL defibrinated blood and 10.0mL selective supplement solution. Mix thoroughly. Pour into sterile Petri dishes or leave in tubes.

**Storage/Shelf Life:** Store dehydrated media in the dark in a sealed container below 30°C. Prepared media should be stored under refrigeration (2-8°C). Media should be used within 60 days of preparation. Media should not be used if there are any signs of deterioration (shrinking, cracking, or discoloration) or contamination, or if the expiration date supplied by the manufacturer has passed.

**Use:** For the selective isolation of *Listeria monocytogenes* from clinical and nonclinical specimens containing mixed flora.

## McBride *Listeria* HiVeg Agar Base, Modified, with Blood and Selective Supplement
### (Modified McBride *Listeria* HiVeg Agar Base)
**Composition** per liter:
| | |
|---|---|
| Agar | 15.0g |
| Glycine anhydride | 10.0g |
| NaCl | 5.0g |
| Plant extract | 3.0g |
| Phenyl ethanol | 2.5g |
| LiCl | 0.5g |
| Sheep blood, defibrinated | 50.0mL |
| Selective supplement solution | 10.0mL |

pH 7.3 ± 0.2 at 25°C

**Source:** This medium, without blood or selective supplement solution, is available as a premixed powder from HiMedia.

**Caution:** LiCl is harmful. Avoid bodily contact and inhalation of vapors. On contact with skin wash with plenty of water immediately.

**Selective Supplement Solution:**
**Composition** per 10.0mL:
| | |
|---|---|
| Cycloheximide | 0.2g |

**Preparation of Selective Supplement Solution:** Add cycloheximide to distilled/deionized water and bring volume to 10.0mL. Mix thoroughly. Filter sterilize.

**Caution:** Cycloheximide is toxic. Avoid skin contact or aerosol formation and inhalation.

**Preparation of Medium:** Add components, except blood and selective supplement solution, to distilled/deionized water and bring volume to 940.0mL. Mix thoroughly. Gently heat and bring to boiling. Distribute into tubes or flasks. Autoclave for 15 min at 15 psi pressure–121°C. Cool to 50°C. Aseptically add 50.0mL defibrinated blood and 10.0mL selective supplement solution. Mix thoroughly. Pour into sterile Petri dishes or leave in tubes.

**Storage/Shelf Life:** Store dehydrated media in the dark in a sealed container below 30°C. Prepared media should be stored under refrigeration (2-8°C). Media should be used within 60 days of preparation. Media should not be used if there are any signs of deterioration (shrink

ing, cracking, or discoloration) or contamination, or if the expiration date supplied by the manufacturer has passed.

**Use:** For the selective isolation of *Listeria monocytogenes* from clinical and nonclinical specimens containing mixed flora.

## McCarthy Agar
**Composition** per liter:
| | |
|---|---|
| Cornstarch | 10.0g |
| Naladixic acid | 0.015g |
| Colistin | 0.01g |
| GC agar base | 1.0L |

pH 7.2 ± 0.2 at 25°C

**GC Agar Base:**
**Composition** per liter:
| | |
|---|---|
| Agar | 10.0g |
| Pancreatic digest of casein | 7.5g |
| Peptic digest of animal tissue | 7.5g |
| NaCl | 5.0g |
| K$_2$HPO$_4$ | 4.0g |
| Cornstarch | 1.0g |
| KH$_2$PO$_4$ | 1.0g |

**Preparation of GC Agar Base:** Add components to distilled/deionized water and bring volume to 1.0L. Mix thoroughly.

**Preparation of Medium:** To 1.0L of GC agar base, add the cornstarch. Gently heat while stirring to dissolve. Add the naladixic acid and colistin. Mix thoroughly. Distribute into tubes or flasks. Autoclave for 15 min at 15 psi pressure–121°C. Pour into sterile Petri dishes or leave in tubes.

**Storage/Shelf Life:** Store dehydrated media in the dark in a sealed container below 30°C. Prepared media should be stored under refrigeration (2-8°C). Media should be used within 60 days of preparation. Media should not be used if there are any signs of deterioration (shrinking, cracking, or discoloration) or contamination, or if the expiration date supplied by the manufacturer has passed.

**Use:** For the isolation and differentiation of *Gardnerella vaginalis* (*Haemophilus vaginalis, Corynebacterium vaginale*) from genitourinary specimens. Bacteria that can utilize starch appear as colonies surrounded by a clear zone.

## McClung-Toabe Agar
**Composition** per liter:
| | |
|---|---|
| Proteose peptone | 40.0g |
| Agar | 25.0g |
| Na$_2$HPO$_4$ | 5.0g |
| Glucose | 2.0g |
| NaCl | 2.0g |
| KH$_2$PO$_4$ | 1.0g |
| MgSO$_4$·7H$_2$O | 0.1g |
| Egg yolk emulsion, 50% | 100.0mL |

pH 7.3 ± 0.2 at 25°C

**Source:** This medium is available as a premixed powder from BD Diagnostic Systems.

**Egg Yolk Emulsion, 50%:**
**Composition** per 100.0mL:
| | |
|---|---|
| Chicken egg yolks | 11 |
| Whole chicken egg | 1 |
| NaCl (0.9% solution) | 50.0mL |

**Preparation of Egg Yolk Emulsion, 50%:** Soak eggs with 1:100 dilution of saturated mercuric chloride solution for 1 min. Crack eggs and separate yolks from whites. Mix egg yolks with 1 chicken egg. Beat to form emulsion. Measure 50.0mL of egg yolk emulsion and add to 50.0mL of 0.9% NaCl solution. Mix thoroughly. Filter sterilize. Warm to 45°–50°C.

**Preparation of Medium:** Add components, except egg yolk emulsion, 50%, to distilled/deionized water and bring volume to 900.0mL. Mix thoroughly. Gently heat while stirring and bring to boiling. Autoclave for 15 min at 15 psi pressure–121°C. Cool to 50°–55°C. Aseptically add 100.0mL of sterile egg yolk emulsion, 50%. Mix thoroughly. Pour into sterile Petri dishes in 15.0mL volumes.

**Storage/Shelf Life:** Store dehydrated media in the dark in a sealed container below 30°C. Prepared media should be stored under refrigeration (2-8°C). Media should be used within 60 days of preparation. Media should not be used if there are any signs of deterioration (shrinking, cracking, or discoloration) or contamination, or if the expiration date supplied by the manufacturer has passed.

**Use:** For the isolation and cultivation of *Clostridium perfringens* in foods.

## McClung-Toabe Agar, Modified

**Composition** per liter:

| | |
|---|---|
| Proteose peptone No. 2 | 40.0g |
| Agar | 20.0g |
| Na$_2$HPO$_4$ | 5.0g |
| Glucose | 2.0g |
| NaCl | 2.0g |
| KH$_2$PO$_4$ | 1.0g |
| MgSO$_4$·7H$_2$O | 0.1g |
| Egg yolk emulsion, 50% | 100.0mL |
| Hemin solution | 1.0mL |

pH 7.6 ± 0.2 at 25°C

**Egg Yolk Emulsion, 50%:**
**Composition** per 100.0mL:

| | |
|---|---|
| Chicken egg yolks | 11 |
| Whole chicken egg | 1 |
| NaCl (0.9% solution) | 50.0mL |

**Preparation of Egg Yolk Emulsion, 50%:** Soak eggs with 1:100 dilution of saturated mercuric chloride solution for 1 min. Crack eggs and separate yolks from whites. Mix egg yolks with 1 chicken egg. Beat to form emulsion. Measure 50.0mL of egg yolk emulsion and add to 50.0mL of 0.9% NaCl solution. Mix thoroughly. Filter sterilize. Warm to 45°–50°C.

**Hemin Solution:**
**Composition** per 100.0mL:

| | |
|---|---|
| Hemin | 0.5g |
| NaOH (1*N* solution) | 20.0mL |

**Preparation of Hemin Solution:** Add hemin to 20.0mL of 1*N* NaOH solution. Mix thoroughly. Bring volume to 100.0mL with distilled/deionized water.

**Preparation of Medium:** Add components, except egg yolk emulsion, 50%, to distilled/deionized water and bring volume to 900.0mL. Mix thoroughly. Gently heat while stirring and bring to boiling. Autoclave for 15 min at 15 psi pressure–121°C. Cool to 50°–55°C. Aseptically add 100.0mL of sterile egg yolk emulsion, 50%. Mix thoroughly. Pour into sterile Petri dishes in 20.0mL volumes.

**Storage/Shelf Life:** Store dehydrated media in the dark in a sealed container below 30°C. Prepared media should be stored under refrigeration (2-8°C). Media should be used within 60 days of preparation. Media should not be used if there are any signs of deterioration (shrinking, cracking, or discoloration) or contamination, or if the expiration date supplied by the manufacturer has passed.

**Use:** For the cultivation of a wide variety of anaerobic bacteria. For the differentiation of anaerobic bacteria based on lecithinase production and lipase production. Bacteria that produce lecithinase appear as colonies surrounded by a zone of insoluble precipitate. Bacteria that produce lipase appear as colonies with a pearly iridescent sheen.

## McClung-Toabe Agar, Modified

**Composition** per liter:

| | |
|---|---|
| Proteose peptone No. 2 | 40.0g |
| Agar | 20.0g |
| Na$_2$HPO$_4$ | 5.0g |
| Glucose | 2.0g |
| NaCl | 2.0g |
| KH$_2$PO$_4$ | 1.0g |
| MgSO$_4$·7H$_2$O | 0.1g |
| Neomycin | 0.1g |
| Egg yolk emulsion, 50% | 100.0mL |
| Hemin solution | 1.0mL |

pH 7.6 ± 0.2 at 25°C

**Egg Yolk Emulsion, 50%:**
**Composition** per 100.0mL:

| | |
|---|---|
| Chicken egg yolks | 11 |
| Whole chicken egg | 1 |
| NaCl (0.9% solution) | 50.0mL |

**Preparation of Egg Yolk Emulsion, 50%:** Soak eggs with 1:100 dilution of saturated mercuric chloride solution for 1 min. Crack eggs and separate yolks from whites. Mix egg yolks with 1 chicken egg. Beat to form emulsion. Measure 50.0mL of egg yolk emulsion and add to 50.0mL of 0.9% NaCl solution. Mix thoroughly. Filter sterilize. Warm to 45°–50°C.

**Hemin Solution:**
**Composition** per 100.0mL:

| | |
|---|---|
| Hemin | 0.5g |
| NaOH (1*N* solution) | 20.0mL |

**Preparation of Hemin Solution:** Add hemin to 20.0mL of 1*N* NaOH solution. Mix thoroughly. Bring volume to 100.0mL with distilled/deionized water.

**Preparation of Medium:** Add components, except egg yolk emulsion, 50%, to distilled/deionized water and bring volume to 900.0mL. Mix thoroughly. Gently heat while stirring and bring to boiling. Autoclave for 15 min at 15 psi pressure–121°C. Cool to 50°–55°C. Aseptically add 100.0mL of sterile egg yolk emulsion, 50%. Mix thoroughly. Pour into sterile Petri dishes in 20.0mL volumes.

**Storage/Shelf Life:** Store dehydrated media in the dark in a sealed container below 30°C. Prepared media should be stored under refrigeration (2-8°C). Media should be used within 60 days of preparation. Media should not be used if there are any signs of deterioration (shrinking, cracking, or discoloration) or contamination, or if the expiration date supplied by the manufacturer has passed.

**Use:** For the cultivation of *Clostridium* species. For the differentiation of *Clostridium* species based on lecithinase production and lipase production. Bacteria that produce lecithinase appear as colonies surrounded by a zone of insoluble precipitate. Bacteria that produce lipase appear as colonies with a pearly iridescent sheen.

## McClung-Toabe Agar, Modified

**Composition** per liter:

| | |
|---|---|
| Proteose peptone No. 2 | 20.0g |
| Agar | 20.0g |
| Yeast extract | 5.0g |
| Pancreatic digest of casein | 5.0g |
| NaCl | 5.0g |
| Sodium thioglycolate | 1.0g |
| Egg yolk emulsion, 50% | 80.0mL |

pH 7.6 ± 0.2 at 25°C

**Egg Yolk Emulsion, 50%:**

**Composition** per 100.0mL:

| | |
|---|---|
| Chicken egg yolks | 11 |
| Whole chicken egg | 1 |
| NaCl (0.9% solution) | 50.0mL |

**Preparation of Egg Yolk Emulsion, 50%:** Soak eggs with 1:100 dilution of saturated mercuric chloride solution for 1 min. Crack eggs and separate yolks from whites. Mix egg yolks with 1 chicken egg. Beat to form emulsion. Measure 50.0mL of egg yolk emulsion and add to 50.0mL of 0.9% NaCl solution. Mix thoroughly. Filter sterilize. Warm to 45°–50°C.

**Preparation of Medium:** Add components, except egg yolk emulsion, 50%, to distilled/deionized water and bring volume to 920.0mL. Mix thoroughly. Gently heat while stirring and bring to boiling. Autoclave for 15 min at 15 psi pressure–121°C. Cool to 50°–55°C. Aseptically add 80.0mL of sterile egg yolk emulsion, 50%. Mix thoroughly. Pour into sterile Petri dishes in 20.0mL volumes.

**Storage/Shelf Life:** Store dehydrated media in the dark in a sealed container below 30°C. Prepared media should be stored under refrigeration (2-8°C). Media should be used within 60 days of preparation. Media should not be used if there are any signs of deterioration (shrinking, cracking, or discoloration) or contamination, or if the expiration date supplied by the manufacturer has passed.

**Use:** For the cultivation of *Clostridium botulinum*.

## McClung-Toabe Egg Yolk Agar, CDC Modified
### (CDC Modified McClung-Toabe Egg Yolk Agar)

**Composition** per liter:

| | |
|---|---|
| Pancreatic digest of casein | 40.0g |
| Agar | 25.0g |
| NaHPO$_4$ | 5.0g |
| Yeast extract | 5.0g |
| D-Glucose | 2.0g |
| NaCl | 2.0g |
| Egg yolk emulsion | 100.0mL |
| MgSO$_4$ (5% solution) | 0.2mL |

pH 7.4 ± 0.2 at 25°C

**Egg Yolk Emulsion:**

**Composition**:

| | |
|---|---|
| Chicken egg yolks | 11 |
| Whole chicken egg | 1 |

**Preparation of Egg Yolk Emulsion:** Soak eggs with 1:100 dilution of saturated mercuric chloride solution for 1 min. Crack eggs. Separate yolks from whites for 11 eggs. Mix egg yolks with 1 chicken egg.

**Preparation of Medium:** Add components, except egg yolk emulsion, to distilled/deionized water and bring volume to 900.0mL. Mix thoroughly. Gently heat while stirring and bring to boiling. Autoclave for 15 min at 15 psi pressure–121°C. Cool to 60°C. Aseptically add 100.0mL of sterile egg yolk emulsion. Mix thoroughly. Pour into sterile Petri dishes in 20.0mL volumes.

**Storage/Shelf Life:** Store dehydrated media in the dark in a sealed container below 30°C. Prepared media should be stored under refrigeration (2-8°C). Media should be used within 60 days of preparation. Media should not be used if there are any signs of deterioration (shrinking, cracking, or discoloration) or contamination, or if the expiration date supplied by the manufacturer has passed.

**Use:** For the isolation, cultivation, and differentiation of anaerobic bacteria from foods. Bacteria that produce lecithinase appear as colonies surrounded by an insoluble opaque precipitate. Bacteria that produce lipase activity appear as colonies with a sheen or "pearly" surface. Bacteria that possess proteolytic activity appear as colonies surrounded by a clear zone.

## McClung-Toabe Egg Yolk Agar, CDC Modified
### (CDC Modified McClung-Toabe Egg Yolk Agar)

**Composition** per liter:

| | |
|---|---|
| Pancreatic digest of casein | 40.0g |
| Agar | 25.0g |
| Na$_2$HPO$_4$ | 5.0g |
| Yeast extract | 5.0g |
| Glucose | 2.0g |
| NaCl | 2.0g |
| KH$_2$PO$_4$ | 1.0g |
| Egg yolk emulsion, 50% | 100.0mL |
| MgSO$_4$·7H$_2$O (5% solution) | 0.2mL |

pH 7.3 ± 0.2 at 25°C

**Egg Yolk Emulsion, 50%:**

**Composition** per 100.0mL:

| | |
|---|---|
| Chicken egg yolks | 11 |
| Whole chicken egg | 1 |
| NaCl (0.9% solution) | 50.0mL |

**Preparation of Egg Yolk Emulsion, 50%:** Soak eggs with 1:100 dilution of saturated mercuric chloride solution for 1 min. Crack eggs and separate yolks from whites. Mix egg yolks with 1 chicken egg. Beat to form emulsion. Measure 50.0mL of egg yolk emulsion and add to 50.0mL of 0.9% NaCl solution. Mix thoroughly. Filter sterilize. Warm to 45°–50°C.

**Preparation of Medium:** Add components—except egg yolk emulsion, 50%—to distilled/deionized water and bring volume to 900.0mL. Mix thoroughly. Gently heat while stirring and bring to boiling. Autoclave for 15 min at 15 psi pressure–121°C. Cool to 50°–55°C. Aseptically add 100.0mL of sterile egg yolk emulsion, 50%. Mix thoroughly. Pour into sterile Petri dishes in 15.0mL volumes.

**Storage/Shelf Life:** Store dehydrated media in the dark in a sealed container below 30°C. Prepared media should be stored under refrigeration (2-8°C). Media should be used within 60 days of preparation. Media should not be used if there are any signs of deterioration (shrinking, cracking, or discoloration) or contamination, or if the expiration date supplied by the manufacturer has passed.

**Use:** For the cultivation of a wide variety of anaerobic bacteria. For the differentiation of anaerobic bacteria based on lecithinase production, lipase production, and proteolytic ability. Bacteria that produce lecithi-nase appear as colonies surrounded by a zone of insoluble precipitate. Bacteria that produce lipase appear as colonies with a pearly iridescent sheen. Bacteria that produce proteolytic activity appear as colonies sur-rounded by a clear zone.

## McClung Toabe HiVeg Agar Base wtih Egg Yolk

**Composition** per liter:
| | |
|---|---|
| Plant peptone No. 3 | 40.0g |
| Agar | 25.0g |
| Na$_2$HPO$_4$ | 5.0g |
| Glucose | 2.0g |
| NaCl | 2.0g |
| KH$_2$PO$_4$ | 1.0g |
| MgSO$_4$ | 0.1g |
| Egg yolk emulsion, 50% | 100.0mL |

pH 7.3 ± 0.2 at 25°C

**Source:** This medium, without egg yolk emulsion, is available as a premixed powder from HiMedia.

**Egg Yolk Emulsion, 50%:**

**Composition** per 100.0mL:
| | |
|---|---|
| Chicken egg yolks | 11 |
| Whole chicken egg | 1 |
| NaCl (0.9% solution) | 50.0mL |

**Preparation of Egg Yolk Emulsion, 50%:** Soak eggs with 1:100 dilution of saturated mercuric chloride solution for 1 min. Crack eggs and separate yolks from whites. Mix egg yolks with 1 chicken egg. Beat to form emulsion. Measure 50.0mL of egg yolk emulsion and add to 50.0mL of 0.9% NaCl solution. Mix thoroughly. Filter sterilize. Warm to 45°–50°C.

**Preparation of Medium:** Add components, except egg yolk emul-sion, 50%, to distilled/deionized water and bring volume to 900.0mL. Mix thoroughly. Gently heat while stirring and bring to boiling. Auto-clave for 20 min at 15 psi pressure–121°C. Cool to 50°–55°C. Asepti-cally add 100.0mL of sterile egg yolk emulsion, 50%. Mix thoroughly. Pour into sterile Petri dishes in 15.0mL volumes.

**Storage/Shelf Life:** Store dehydrated media in the dark in a sealed container below 30°C. Prepared media should be stored under refriger-ation (2-8°C). Media should be used within 60 days of preparation. Media should not be used if there are any signs of deterioration (shrink-ing, cracking, or discoloration) or contamination, or if the expiration date supplied by the manufacturer has passed.

**Use:** For the isolation and cultivation of *Clostridium perfringens* in foods.

## M-CP Agar Base

**Composition** per liter:
| | |
|---|---|
| Tryptose | 30.0g |
| Yeast extract | 20.0g |
| Agar | 15.0g |
| Sucrose | 5.0g |
| L-Cysteine·HCl·H$_2$O | 1.0g |
| MgSO$_4$·7H$_2$O | 0.1g |
| FeCl$_3$·6H$_2$O | 0.09g |
| Indoxyl β-D-glucoside | 0.06g |
| Bromcresol Purple | 0.04g |
| Selective supplement solution B | 20.0mL |
| Selective supplement solution A | 10.0mL |

pH 7.6 ± 0.2 at 25°C

**Source:** This medium is available from HiMedia.

**Selective Supplement Solution A:**

**Composition** per 10.0mL:
| | |
|---|---|
| D-Cycloserine | 400.0mg |
| Polymyxin B sulfate | 25.0mg |

**Preparation of Selective Supplement Solution A:** Add compo-nents to distilled/deionized water and bring volume to 10.0mL. Mix thoroughly. Filter sterilize.

**Selective Supplement Solution B:**

**Composition** per 20.0mL:
| | |
|---|---|
| Phenolphthalein diphosphate | 0.1g |

**Preparation of Selective Supplement Solution B:** Add phenol-phthalein diphosphate to distilled/deionized water and bring volume to 20.0mL. Mix thoroughly. Filter sterilize.

**Preparation of Medium:** Add components, except selective sup-plement solutions A and B, to distilled/deionized water and bring vol-ume to 970.0mL. Mix thoroughly. Autoclave for 15 min at 15 psi pressure–121°C. Cool to 50°C. Aseptically add selective supplement solutions A and B. Mix thoroughly. Pour into Petri dishes or aseptically distribute into sterile tubes.

**Storage/Shelf Life:** Store dehydrated media in the dark in a sealed container below 30°C. Prepared media should be stored under refriger-ation (2-8°C). Media should be used within 60 days of preparation. Media should not be used if there are any signs of deterioration (shrink-ing, cracking, or discoloration) or contamination, or if the expiration date supplied by the manufacturer has passed.

**Use:** Recommended by the Directive of the Council of the European Union 98/83/EC for the isolation and enumeration of *Clostridium per-fringens* from water samples using the membrane filtration technique.

## MD Medium

**Composition** per liter:
| | |
|---|---|
| Agar | 20.0g |
| L-Malic acid | 20.0g |
| Pancreatic digest of casein | 10.0g |
| D-Glucose | 5.0g |
| Casamino acids | 3.0g |
| Pancreatic digest of soybean meal | 1.5g |
| Tween™ 80 | 1.0g |
| Yeast extract | 1.0g |
| Bromcresol Green solution | 20.0mL |

pH 7.0 ± 0.2 at 25°C

**Bromcresol Green Solution:**

**Composition** per 30.0mL:
| | |
|---|---|
| Bromcresol Green | 0.1g |
| NaOH (0.01$N$ solution) | 30.0mL |

**Preparation of Bromcresol Green Solution:** Add Bromcresol Green to 30.0mL of NaOH solution. Mix thoroughly. Filter sterilize.

**Preparation of Medium:** Add components to distilled/deionized water and bring volume to 1.0L. Mix thoroughly. Gently heat and bring to boiling. Distribute into tubes or flasks. Adjust pH to 7.0 with 10$N$ KOH. Autoclave for 15 min at 15 psi pressure–121°C. Pour into sterile Petri dishes or leave in tubes.

**Storage/Shelf Life:** Store dehydrated media in the dark in a sealed container below 30°C. Prepared media should be stored under refriger-ation (2-8°C). Media should be used within 60 days of preparation. Media should not be used if there are any signs of deterioration (shrink-

ing, cracking, or discoloration) or contamination, or if the expiration date supplied by the manufacturer has passed.

**Use:** For the isolation and cultivation of *Salmonella* species from foods.

## M-EC Test Agar

**Composition** per liter:

| | |
|---|---|
| Agar | 15.0g |
| Lactose | 10.0g |
| NaCl | 7.5g |
| Proteose peptone | 5.0g |
| Dipotassium phosphate | 3.3g |
| Yeast extract | 3.0g |
| $KH_2PO_4$ | 1.0g |
| Sodium lauryl sulphate | 0.2g |
| Sodium deoxycholate | 0.1g |
| Bromcresol Purple | 0.08g |
| Bromphenol Red | 0.08g |

pH 7.3 ± 0.2 at 25°C

**Source:** This medium is available from HiMedia.

**Preparation of Medium:** Add components to distilled/deionized water and bring volume to 1.0L. Mix thoroughly. Gently heat and bring to boiling. Distribute into tubes or flasks. Do not autoclave. Pour into sterile Petri dishes or leave in tubes.

**Storage/Shelf Life:** Store dehydrated media in the dark in a sealed container below 30°C. Prepared media should be stored under refrigeration (2-8°C). Media should be used within 60 days of preparation. Media should not be used if there are any signs of deterioration (shrinking, cracking, or discoloration) or contamination, or if the expiration date supplied by the manufacturer has passed.

**Use:** For the detection of *Escherichia coli* in water samples using the membrane filter technique.

## Membrane *Clostridium perfringens* Medium (m-CP Medium)

**Composition** per 1040.0mL:

| | |
|---|---|
| Tryptose | 30.0g |
| Yeast extract | 20.0g |
| Agar | 15.0g |
| Sucrose | 5.0g |
| L-Cysteine·HCl·H2O | 1.0g |
| $MgSO_4$·7H2O | 0.1g |
| Bromcresol Purple | 0.04g |
| Phenolphthalein solution | 20.0mL |
| Indoxyl-β-D-glucoside solution | 8.0mL |
| Selective supplement solution | 8.0mL |
| Ferric chloride solution | 4.0mL |

pH 7.6 ± 0.2 at 25°C

**Source:** This medium is available as a premixed powder from Oxoid Unipath.

**Selective Supplement Solution:**

**Composition** per 8.0mL:

| | |
|---|---|
| D-Cycloserine | 0.8g |
| Polymyxin B sulfate | 50.0mg |

**Preparation of Selective Supplement Solution:** Add components to distilled/deionized water and bring volume to 8.0mL. Mix thoroughly. Filter sterilize.

**Indoxyl-β-D-glucoside Solution:**

**Composition** per 10.0mL:

| | |
|---|---|
| Indoxyl-β-D-glucoside | 75.0mg |

**Preparation of Indoxyl-β-D-glucoside Solution:** Add indoxyl-β-D-glucoside to distilled/deionized water and bring volume to 10.0mL. Mix thoroughly. Filter sterilize.

**Ferric Chloride Solution:**

**Composition** per 10.0mL:

| | |
|---|---|
| $FeCl_3$·6H2O | 0.45g |

**Preparation of Ferric Chloride Solution:** Add ferric chloride to distilled/deionized water and bring volume to 10.0mL. Mix thoroughly. Filter sterilize.

**Phenolphthalein Solution:**

**Composition** per 20.0mL:

| | |
|---|---|
| Phenolphthalein biphosphate tetrasodium salt | 0.15g |

**Preparation of Phenolphthalein Solution:** Add phenolphthalein biphosphate tetrasodium salt to distilled/deionized water and bring volume to 20.0mL. Mix thoroughly. Filter sterilize.

**Preparation of Medium:** Add components, except selective supplement solution, phenolphthalein solution, ferric chloride solution, and indoxyl-β-D-glucoside solution to distilled/deionized water and bring volume to 1.0L. Mix thoroughly. Distribute into tubes or flasks. Autoclave for 15 min at 15 psi pressure–121°C. Cool to 50°C. Aseptically add 8.0mL selective supplement solution, 20.0mL phenolphthalein solution, 4.0mL ferric chloride solution, and 8.0mL indoxyl-β-D-glucoside solution. Mix thoroughly. Pour into sterile Petri dishes or aseptically distribute into sterile tubes.

**Storage/Shelf Life:** Store dehydrated media in the dark in a sealed container below 30°C. Prepared media should be stored under refrigeration (2-8°C). Media should be used within 60 days of preparation. Media should not be used if there are any signs of deterioration (discoloration) or contamination, or if the expiration date supplied by the manufacturer has passed.

**Use:** For the rapid isolation and presumptive identification of *Clostridium perfringens* from food and water samples. A selective and chromogenic medium for the presumptive identification of *Clostridium perfringens,* especially from water samples. Recommended in European Council Directive 98/83/EC for testing the quality of water intended for human consumption. *C. perfringens* colonies have a characteristic opaque yellow appearance. Most other *Clostridium* spp. will appear as either purple colonies, due to the lack of sucrose fermentation, or blue/green colonies where the organism is still cleaving indoxyl-β-D-glucoside and also fermenting sucrose.

## Membrane Lactose Glucuronide Agar (MLGA)

**Composition** per liter:

| | |
|---|---|
| Peptone | 40.0g |
| Lactose | 30.0g |
| Agar | 10.0g |
| Yeast extract | 6.0g |
| Sodium lauryl sulfate | 1.0g |
| Sodium pyruvate | 0.5g |
| 5-Bromo-4-chloro-3-indoxyl-β-D-glucuronic acid | 0.2g |
| Phenol Red | 0.2g |

pH 7.4 ± 0.2 at 25°C

**Source:** This medium is available from Oxoid Unipath.

**Preparation of Medium:** Add components to distilled/deionized water and bring volume to 1.0L. Mix thoroughly. Autoclave for 15 min at 15 psi pressure–121°C. Cool to 45°–50°C. Pour into sterile Petri dishes.

**Storage/Shelf Life:** Store dehydrated media in the dark in a sealed container below 30°C. Prepared media should be stored in the dark under refrigeration (2-8°C). Chromogenic media are especially light and temperature sensitive; protect from light, excessive heat, moisture, and freezing. Media should not be used if there are any signs of deterioration (discoloration) or contamination, or if the expiration date supplied by the manufacturer has passed.

**Use:** For the direct enumeration of *E. coli* and coliforms in foods by the membrane filtration method. The chromogenic substrate 5-bromo-4-chloro-3-indoxyl-β-D-glucuronic acid (BCIG) is cleaved by the enzyme β-glucuronidase and produces a blue chromophore that builds up within the bacterial cells. In addition, the incorporation of Phenol Red detects lactose fermentation and results in yellow colonies when acid is produced. Since coliform colonies are lactose positive, they will appear yellow on this medium and as *E. coli* colonies are both lactose and β-glucuronidase positive, they will appear green.

## Membrane Lauryl Sulfate Broth

**Composition** per liter:

| | |
|---|---|
| Peptone | 39.0g |
| Lactose | 30.0g |
| Yeast extract | 6.0g |
| Sodium lauryl sulfate | 1.0g |
| Phenol Red | 0.2g |

pH 7.4 ± 0.2 at 25°C

**Preparation of Medium:** Add components to distilled/deionized water and bring volume to 1.0L. Mix thoroughly. Distribute into tubes or flasks. Autoclave for 15 min at 15 psi pressure–121°C.

**Storage/Shelf Life:** Store dehydrated media in the dark in a sealed container below 30°C. Prepared media should be stored under refrigeration (2-8°C). Media should be used within 60 days of preparation. Media should not be used if there are any signs of deterioration (discoloration) or contamination, or if the expiration date supplied by the manufacturer has passed.

**Use:** For the enumeration of coliform organisms and *Escherichia coli* in water.

## M-Endo HiVeg Agar LES

**Composition** per liter:

| | |
|---|---|
| Agar | 15.0g |
| Lactose | 9.4g |
| Plant hydrolysate No. 1 | 7.5g |
| NaCl | 3.7g |
| Plant hydrolysate | 3.7g |
| Plant peptone | 3.7g |
| K$_2$HPO$_4$ | 3.3g |
| Na$_2$SO$_3$ | 1.6g |
| Yeast extract | 1.2g |
| KH$_2$PO$_4$ | 1.0g |
| Basic Fuchsin | 0.8g |
| Synthetic detergent No. III | 0.1g |
| Sodium lauryl sulfate | 0.05g |

pH 7.2 ± 0.2 at 25°C

**Source:** This medium is available as a premixed powder from Hi-Media.

**Caution:** Basic Fuchsin is a potential carcinogen and care must be taken to avoid inhalation of the powdered dye and contact with the skin.

**Preparation of Medium:** Add ethanol to approximately 900.0mL of distilled/deionized water. Add remaining components. Bring volume to 1.0L with distilled/deionized water. Mix thoroughly. Gently heat and bring to boiling. Autoclave for 15 min at 15 psi pressure–121°C. Pour into sterile 60mm Petri dishes in 4.0mL volumes. Protect from the light.

**Storage/Shelf Life:** Store dehydrated media in the dark in a sealed container below 30°C. Prepared media should be stored under refrigeration (2-8°C). Media should be used within 60 days of preparation. Media should not be used if there are any signs of deterioration (shrinking, cracking, or discoloration) or contamination, or if the expiration date supplied by the manufacturer has passed.

**Use:** For the cultivation and enumeration of coliform bacteria by the membrane filter method.

## M-Endo HiVeg Broth

**Composition** per liter:

| | |
|---|---|
| Lactose | 25.0g |
| Plant peptone | 20.0g |
| K$_2$HPO$_4$ | 7.0g |
| Yeast extract | 6.0g |
| Na$_2$SO$_3$ | 2.5g |
| Basic Fuchsin | 1.0g |

pH 7.2 ± 0.1 at 25°C

**Source:** This medium is available as a premixed powder from Hi-Media.

**Caution:** Basic Fuchsin is a potential carcinogen and care must be taken to avoid inhalation of the powdered dye and contact with the skin.

**Preparation of Medium:** Add ethanol to approximately 900.0mL of distilled/deionized water. Add remaining components. Bring volume to 1.0L with distilled/deionized water. Mix thoroughly. Gently heat and bring to boiling. Rapidly cool broth below 45°C. Do not autoclave. Use 1.8–2.0mL for each filter pad. Protect from the light. Prepare broth freshly.

**Storage/Shelf Life:** Store dehydrated media in the dark in a sealed container below 30°C. Prepared media should be stored under refrigeration (2-8°C). Media should be used within 60 days of preparation. Media should not be used if there are any signs of deterioration (discoloration) or contamination, or if the expiration date supplied by the manufacturer has passed.

**Use:** For the cultivation and enumeration of coliform bacteria from water by the membrane filter method.

## M-Endo HiVeg Broth MF
## (MF Endo HiVeg Medium)
## (M-Coliform HiVeg Broth)

**Composition** per liter:

| | |
|---|---|
| Lactose | 12.5g |
| Plant hydrolysate No. 1 | 10.0g |
| Plant hydrolysate | 5.0g |
| Plant special peptone | 5.0g |
| NaCl | 5.0g |
| K$_2$HPO$_4$ | 4.375g |
| Na$_2$SO$_3$ | 2.1g |

Yeast extract .................................................................1.5g
KH$_2$PO$_4$ .................................................................1.375g
Basic Fuchsin ..............................................................1.05g
Synthetic detergent No. III ........................................0.1g
Sodium lauryl sulfate ................................................0.05g

pH 7.2 ± 0.1 at 25°C

**Source:** This medium is available as a premixed powder from Hi-Media.

**Caution:** Basic Fuchsin is a potential carcinogen and care must be taken to avoid inhalation of the powdered dye and contact with the skin.

**Preparation of Medium:** Add ethanol to approximately 900.0mL of distilled/deionized water. Add remaining components. Bring volume to 1.0L with distilled/deionized water. Mix thoroughly. Gently heat and bring to boiling. Rapidly cool broth below 45°C. Do not autoclave. Use 1.8–2.0mL for each filter pad. Protect from the light. Prepare broth freshly.

**Storage/Shelf Life:** Store dehydrated media in the dark in a sealed container below 30°C. Prepared media should be stored under refrigeration (2-8°C). Media should be used within 60 days of preparation. Media should not be used if there are any signs of deterioration (discoloration) or contamination, or if the expiration date supplied by the manufacturer has passed.

**Use:** For the cultivation and enumeration of coliform bacteria from water by the membrane filter method.

## M-*Enterococcus* Agar Base with Polysorbate 80 and Sodium Carbonate

**Composition** per liter:

Agar ............................................................................10.0g
Casein enzymic hydrolysate .......................................15.0g
Papaic digest of soybean meal ....................................5.0g
Yeast extract ................................................................5.0g
K$_2$HPO$_4$ ....................................................................4.0g
Glucose .......................................................................2.0g
NaN$_3$ ..........................................................................0.4g
Triphenyl tetrazolium chloride ...................................0.1g
Sodium carbonate solution ..........................................2.0mL
Polysorbate 80 ............................................................0.5mL

pH 7.2 ± 0.2 at 25°C

**Source:** This medium is available from HiMedia.

**Caution:** Sodium azide has a tendency to form explosive metal azides with plumbing materials. It is advisable to use enough water to flush off the disposables.

**Sodium Carbonate Solution:**
**Composition** per 10.0mL:
Na$_2$CO$_3$ .....................................................................1.0g

**Preparation of Sodium Carbonate Solution:** Add Na$_2$CO$_3$ to distilled/deionized water and bring volume to 10.0mL. Mix thoroughly. Filter sterilize.

**Preparation of Medium:** Add components, except polysorbate 80 and sodium carbonate solution, to distilled/deionized water and bring volume to 997.5mL. Mix thoroughly. Gently heat to dissolve components. Do not autoclave. Cool to 50°C. Add polysorbate 80 and sodium carbonate solution. Mix thoroughly. Pour into Petri dishes or aseptically distribute into sterile tubes.

**Storage/Shelf Life:** Store dehydrated media in the dark in a sealed container below 30°C. Prepared media should be stored under refrigeration (2-8°C). Media should be used within 60 days of preparation. Media should not be used if there are any signs of deterioration (shrinking, cracking, or discoloration) or contamination, or if the expiration date supplied by the manufacturer has passed.

**Use:** For the selective isolation and enumeration of enterococci from water, sewage, food, or other materials.

## M-*Enterococcus* Agar Base, Modified

**Composition** per liter:

Yeast extract ...............................................................30.0g
Pancreatic digest of gelatin ........................................10.0g
Agar ............................................................................15.0g
NaCl ............................................................................15.0g
Esculin ........................................................................1.0g
Nalidixic acid ..............................................................0.25g
NaN$_3$ ..........................................................................0.15g
Cycloheximide ............................................................0.05g
Selective supplement solution ....................................15.0mL

pH 7.1 ± 0.2 at 25°C

**Source:** This medium is available from HiMedia.

**Caution:** Sodium azide has a tendency to form explosive metal azides with plumbing materials. It is advisable to use enough water to flush off the disposables.

**Caution:** Cycloheximide is toxic. Avoid skin contact or aerosol formation and inhalation.

**Selective Supplement Solution:**
**Composition** per 20.0mL:
2,3,5-Triphenyl tetrazolium chloride ..........................0.2g

**Preparation of Selective Supplement Solution:** Add 2,3,5-triphenyl tetrazolium chloride to distilled/deionized water and bring volume to 20.0mL. Mix thoroughly. Filter sterilize.

**Preparation of Medium:** Add components, except selective supplement solution, to distilled/deionized water and bring volume to 985.0mL. Mix thoroughly. Autoclave for 15 min at 15 psi pressure–121°C. Cool to 50°C. Aseptically add selective supplement solution. Mix thoroughly. Pour into Petri dishes or aseptically distribute into sterile tubes.

**Storage/Shelf Life:** Store dehydrated media in the dark in a sealed container below 30°C. Prepared media should be stored under refrigeration (2-8°C). Media should be used within 60 days of preparation. Media should not be used if there are any signs of deterioration (shrinking, cracking, or discoloration) or contamination, or if the expiration date supplied by the manufacturer has passed.

**Use:** For the recovery of enterococci in water samples using the membrane filter technique.

## M-*Enterococcus* HiVeg Agar Base

**Composition** per liter:

Plant hydrolysate ........................................................15.0g
Agar ............................................................................10.0g
Papaic digest of soybean meal ....................................5.0g
Yeast extract ................................................................5.0g
KH$_2$PO$_4$ ....................................................................4.0g
Glucose .......................................................................2.0g

NaN$_3$ ..................................................................................0.4g
Triphenyl tetrazolium chloride.......................................0.1g
pH 7.2 ± 0.2 at 25°C

**Source:** This medium is available as a premixed powder from Hi-Media.

**Caution:** Sodium azide is toxic. Azides also react with metals and disposal must be highly diluted.

**Preparation of Medium:** Add components to distilled/deionized water and bring volume to 1.0L. Mix thoroughly. Gently heat and bring to boiling. Cool to 45°–50°C. Do not autoclave. Pour into sterile Petri dishes.

**Storage/Shelf Life:** Store dehydrated media in the dark in a sealed container below 30°C. Prepared media should be stored under refrigeration (2-8°C). Media should be used within 60 days of preparation. Media should not be used if there are any signs of deterioration (shrinking, cracking, or discoloration) or contamination, or if the expiration date supplied by the manufacturer has passed.

**Use:** For the isolation, cultivation, and enumeration of entercocci in water, sewage, and feces by the membrane filter method. For the direct plating of specimens for the detection and enumeration of fecal streptococci.

## MeReSa Agar Base with Methicillin
## (Methicillin-Resistant
### *Staphylococcus aureus* Agar)

**Composition** per liter:
Agar ...............................................................................20.0g
Casein enzymic hydrolysate .........................................10.0g
Glycine...........................................................................10.0g
Mannitol.........................................................................10.0g
NaCl ...............................................................................10.0g
Sodium pyruvate ...........................................................10.0g
LiCl...................................................................................5.0g
Beef extract......................................................................5.0g
Indicator mix..................................................................0.13g
MRSA selective supplement....................................... 10.0mL
pH 7.0 ± 0.2 at 25°C

**Source:** This medium, without MRSA selective supplement, is available as a premixed powder from HiMedia.

**Caution:** Lithium chloride is harmful. Avoid bodily contact and inhalation of vapors. On contact with skin wash with plenty of water immediately.

**MRSA Selective Supplement:**
**Composition** per 10.0mL:
Methicillin....................................................................4.0mg

**Preparation of MRSA Selective Supplement:** Add methicillin to distilled/deionized water and bring volume to 10.0mL. Mix thoroughly. Filter sterilize.

**Preparation of Medium:** Add components, except MRSA selective supplement, to distilled/deionized water and bring volume to 990.0mL. Mix thoroughly. Gently heat and bring to boiling. Distribute into tubes or flasks. Autoclave for 15 min at 15 psi pressure–121°C. Cool to 50°C. Aseptically add 10.0mL MRSA selective supplement. Pour into sterile Petri dishes or leave in tubes.

**Storage/Shelf Life:** Store dehydrated media in the dark in a sealed container below 30°C. Prepared media should be stored under refrigeration (2-8°C). Media should be used within 60 days of preparation.

Media should not be used if there are any signs of deterioration (shrinking, cracking, or discoloration) or contamination, or if the expiration date supplied by the manufacturer has passed.

**Use:** For the isolation and cultivation of methicillin-resistant *Staphylococcus aureus* (MRSA).

## MeReSa Agar Base with Oxacillin
## (Methicillin-Resistant
### *Staphylococcus aureus* Agar)

**Composition** per liter:
Agar ...............................................................................20.0g
Casein enzymic hydrolysate .........................................10.0g
Glycine...........................................................................10.0g
Mannitol.........................................................................10.0g
NaCl ...............................................................................10.0g
Sodium pyruvate ...........................................................10.0g
LiCl...................................................................................5.0g
Beef extract......................................................................5.0g
Indicator mix..................................................................0.13g
MRSA selective supplement....................................... 10.0mL
pH 7.0 ± 0.2 at 25°C

**Source:** This medium, without MRSA supplement solution, is available as a premixed powder from HiMedia.

**Caution:** Lithium chloride is harmful. Avoid bodily contact and inhalation of vapors. On contact with skin wash with plenty of water immediately.

**MRSA Selective Supplement:**
**Composition** per 10.0mL:
Oxacillin .......................................................................2.0mg

**Preparation of MRSA Selective Supplement:** Add oxacillin to distilled/deionized water and bring volume to 10.0mL. Mix thoroughly. Filter sterilize.

**Preparation of Medium:** Add components, except MRSA selective supplement, to distilled/deionized water and bring volume to 990.0mL. Mix thoroughly. Gently heat and bring to boiling. Distribute into tubes or flasks. Autoclave for 15 min at 15 psi pressure–121°C. Cool to 50°C. Aseptically add 10.0mL MRSA selective supplement. Pour into sterile Petri dishes or leave in tubes.

**Storage/Shelf Life:** Store dehydrated media in the dark in a sealed container below 30°C. Prepared media should be stored under refrigeration (2-8°C). Media should be used within 60 days of preparation. Media should not be used if there are any signs of deterioration (shrinking, cracking, or discoloration) or contamination, or if the expiration date supplied by the manufacturer has passed.

**Use:** For the isolation and cultivation of methicillin-resistant *Staphylococcus aureus* (MRSA).

## Methylene Blue Milk Medium
## (MBM Medium)

**Composition** per liter:
Skim milk, dehydrated...............................................100.0g
Methylene Blue.............................................................10.0g
pH 6.4 ± 0.2 at 25°C

**Preparation of Medium:** Add components to distilled/deionized water and bring volume to 1.0L. Mix thoroughly. Distribute into tubes or flasks. Autoclave for 20 min at 10 psi pressure–115°C.

Storage/Shelf Life: Store dehydrated media in the dark in a sealed container below 30°C. Prepared media should be stored under refrigeration (2-8°C). Media should be used within 60 days of preparation. Media should not be used if there are any signs of deterioration (discoloration) or contamination, or if the expiration date supplied by the manufacturer has passed.

Use: For the cultivation and differentiation of group D streptococci (enterococci) from other *Streptococcus* species.

## M-FC HiVeg Agar Base
## with Rosalic Acid

Composition per liter:
| | |
|---|---|
| Agar | 15.0g |
| Lactose | 12.5g |
| Plant hydrolysate No. 1 | 10.0g |
| Plant peptone No. 3 | 5.0g |
| NaCl | 5.0g |
| Yeast extract | 3.0g |
| Synthetic detergent No. I | 1.5g |
| Aniline Blue | 0.1g |
| Rosolic acid solution | 10.0mL |

pH 7.4 ± 0.2 at 25°C

Source: This medium, without rosolic acid, is available as a premixed powder from HiMedia.

Rosolic Acid Solution:
Composition per 100.0mL:
Rosolic acid .............. 1.0g

Preparation of Rosolic Acid Solution: Add rosolic acid to 0.2*N* NaOH and bring volume to 100.0L. Mix thoroughly.

Preparation of Medium: Add 10.0mL rosolic acid solution to 950.0mL of distilled/deionized water. Mix thoroughly. Add other components and bring volume to 1.0L with distilled/deionized water. Mix thoroughly. Gently heat and bring to boiling with frequent mixing. Do not autoclave. Pour into sterile Petri dishes or leave in tubes.

Storage/Shelf Life: Store dehydrated media in the dark in a sealed container below 30°C. Prepared media should be stored under refrigeration (2-8°C). Media should be used within 60 days of preparation. Media should not be used if there are any signs of deterioration (shrinking, cracking, or discoloration) or contamination, or if the expiration date supplied by the manufacturer has passed.

Use: For the detection and enumeration of fecal coliforms using the membrane filter technique.

## M-FC HiVeg Agar Base, Modified
## with Rosalic Acid

Composition per liter:
| | |
|---|---|
| Agar | 15.0g |
| Plant hydrolysate No. 1 | 10.0g |
| Inositol | 10.0g |
| Plant peptone No. 3 | 5.0g |
| NaCl | 5.0g |
| Yeast extract | 3.0g |
| Synthetic detergent No. I | 1.5g |
| Aniline blue | 0.1g |
| Rosolic acid solution | 10.0mL |

pH 7.4 ± 0.2 at 25°C

Source: This medium, without rosolic acid, is available as a premixed powder from HiMedia.

Rosolic Acid Solution:
Composition per 100.0mL:
Rosolic acid .............. 1.0g

Preparation of Rosolic Acid Solution: Add rosolic acid to 0.2*N* NaOH and bring volume to 100.0L. Mix thoroughly.

Preparation of Medium: Add 10.0mL rosolic acid solution to 950.0mL of distilled/deionized water. Mix thoroughly. Add other components and bring volume to 1.0L with distilled/deionized water. Mix thoroughly. Gently heat and bring to boiling with frequent mixing. Do not autoclave. Pour into sterile Petri dishes or leave in tubes.

Storage/Shelf Life: Store dehydrated media in the dark in a sealed container below 30°C. Prepared media should be stored under refrigeration (2-8°C). Media should be used within 60 days of preparation. Media should not be used if there are any signs of deterioration (shrinking, cracking, or discoloration) or contamination, or if the expiration date supplied by the manufacturer has passed.

Use: For the detection and enumeration of fecal coliforms using membrane filter technique.

## M-FC HiVeg Broth Base
## with Rosalic Acid

Composition per liter:
| | |
|---|---|
| Lactose | 12.5g |
| Plant hydrolysate No. 1 | 10.0g |
| Plant peptone No. 3 | 5.0g |
| NaCl | 5.0g |
| Yeast extract | 3.0g |
| Synthetic detergent No. I | 1.5g |
| Aniline blue | 0.1g |
| Rosolic acid solution | 10.0mL |

pH 7.4 ± 0.2 at 25°C

Source: This medium, without rosolic acid, is available as a premixed powder from HiMedia.

Rosolic Acid Solution:
Composition per 100.0mL:
Rosolic acid .............. 1.0g

Preparation of Rosolic Acid Solution: Add rosolic acid to 0.2*N* NaOH and bring volume to 100.0L. Mix thoroughly.

Preparation of Medium: Add 10.0mL of rosolic acid solution to 950.0mL of distilled/deionized water. Mix thoroughly. Add other components and bring volume to 1.0L with distilled/deionized water. Mix thoroughly. Gently heat and bring to boiling with frequent mixing. Do not autoclave. Pour into sterile Petri dishes or leave in tubes.

Storage/Shelf Life: Store dehydrated media in the dark in a sealed container below 30°C. Prepared media should be stored under refrigeration (2-8°C). Media should be used within 60 days of preparation. Media should not be used if there are any signs of deterioration (discoloration) or contamination, or if the expiration date supplied by the manufacturer has passed.

Use: For the cultivation of fecal coliform bacteria from waters and the enumeration of coliform bacteria using the membrane filtration method.

## M-FC Agar

Composition per liter:
| | |
|---|---|
| Agar | 15.0g |
| Tryptose | 10.0g |

Inositol ..................................................................................10.0g
Proteose peptone .....................................................................5.0g
NaCl .........................................................................................5.0g
Yeast extract ...........................................................................3.0g
Bile salts mixture ....................................................................1.5g
Aniline Blue .............................................................................0.1g
Selective supplement solution ...............................................10.0mL
Rosolic acid solution ..............................................................10.0mL
pH 7.4 ± 0.2 at 25°C

**Source:** This medium is available from HiMedia.

## Rosolic Acid Solution:
**Composition** per 10.0mL:
Rosolic acid ............................................................................0.1g

**Preparation of Rosolic Acid Solution:** Add rosoloic acid to distilled/deionized water and bring volume to 10.0mL. Mix thoroughly. Filter sterilize.

## Selective Supplement Solution:
**Composition** per 10.0mL:
Carbenicillin ...........................................................................0.05g

**Preparation of Selective Supplement Solution:** Add carbenicillin to distilled/deionized water and bring volume to 10.0mL. Mix thoroughly. Filter sterilize.

**Preparation of Medium:** Add components, except selective supplement solution, to distilled/deionized water and bring volume to 990.0mL. Mix thoroughly. Autoclave for 15 min at 15 psi pressure–121°C. Cool to 50°C. Aseptically add selective supplement solution. Mix thoroughly. Pour into Petri dishes or aseptically distribute into sterile tubes.

**Storage/Shelf Life:** Store dehydrated media in the dark in a sealed container below 30°C. Prepared media should be stored under refrigeration (2-8°C). Media should be used within 60 days of preparation. Media should not be used if there are any signs of deterioration (shrinking, cracking, or discoloration) or contamination, or if the expiration date supplied by the manufacturer has passed.

**Use:** For the rapid enumeration of *Klebsiella* using the membrane filter technique.

## M-FC Agar Base
**Composition** per liter:
Agar .......................................................................................15.0g
Lactose ...................................................................................12.5g
Tryptose ..................................................................................10.0g
Proteose peptone .....................................................................5.0g
NaCl .........................................................................................5.0g
Yeast extract ...........................................................................3.0g
Bile salts mixture ....................................................................1.5g
Aniline Blue .............................................................................0.1g
Rosolic acid solution ..............................................................10.0mL
pH 7.4 ± 0.2 at 25°C

**Source:** This medium is available from HiMedia.

## Rosolic Acid Solution:
**Composition** per 10.0mL:
Rosolic acid ............................................................................0.1g

**Preparation of Rosolic Acid Solution:** Add rosolic acid to distilled/deionized water and bring volume to 10.0mL. Mix thoroughly. Filter sterilize.

**Preparation of Medium:** Add components to distilled/deionized water and bring volume to 1.0L. Mix thoroughly. Gently heat to dissolve components. Do not autoclave. Cool to 50°C. Pour into Petri dishes or aseptically distribute into sterile tubes.

**Storage/Shelf Life:** Store dehydrated media in the dark in a sealed container below 30°C. Prepared media should be stored under refrigeration (2-8°C). Media should be used within 60 days of preparation. Media should not be used if there are any signs of deterioration (shrinking, cracking, or discoloration) or contamination, or if the expiration date supplied by the manufacturer has passed.

**Use:** For the detection and enumeration of fecal coliforms using the membrane filter technique at 44.5°C.

## Microbial Content Test Agar
**Composition** per liter:
Agar .......................................................................................15.0g
Pancreatic digest of casein .....................................................15.0g
NaCl .........................................................................................5.0g
Tween™ 80 ..............................................................................5.0g
Enzymatic hydrolysate of soybean meal .................................5.0g
Lecithin ...................................................................................0.7g
pH 7.3 ± 0.2 at 25°C

**Source:** This medium is available as a premixed powder from BD Diagnostic Systems.

**Preparation of Medium:** Add components to distilled/deionized water and bring volume to 1.0L. Mix thoroughly. Gently heat and bring to boiling. Boil for 1–2 min. Distribute into tubes or flasks. Autoclave for 15 min at 15 psi pressure–121°C. Pour into sterile Petri dishes or leave in tubes.

**Storage/Shelf Life:** Store dehydrated media in the dark in a sealed container below 30°C. Prepared media should be stored under refrigeration (2-8°C). Media should be used within 60 days of preparation. Media should not be used if there are any signs of deterioration (shrinking, cracking, or discoloration) or contamination, or if the expiration date supplied by the manufacturer has passed.

**Use:** For use in the microbial content test of water-soluble cosmetic products. Also used for determining the efficiency of sanitization of containers, equipment, and environmental surfaces.

## Microbial Content Test HiVeg Agar (Tryptone Soy HiVeg Agar with Lecithin and Tween 80)
**Composition** per liter:
Agar .......................................................................................15.0g
Plant hydrolysate ....................................................................15.0g
Papaic digest of soybean meal .................................................5.0g
NaCl .........................................................................................5.0g
Lecithin ...................................................................................0.7g
Polysorbate 80 ........................................................................5.0mL
pH 7.3 ± 0.2 at 25°C

**Source:** This medium, without polysorbate 80, is available as a premixed powder from HiMedia.

**Preparation of Medium:** Add components to distilled/deionized water and bring volume to 1.0L. Mix thoroughly. Gently heat and bring to boiling. Boil for 1–2 min. Distribute into tubes or flasks. Autoclave for 15 min at 15 psi pressure–121°C. Pour into sterile Petri dishes or leave in tubes.

**Storage/Shelf Life:** Store dehydrated media in the dark in a sealed container below 30°C. Prepared media should be stored under refrigeration (2-8°C). Media should be used within 60 days of preparation. Media should not be used if there are any signs of deterioration (shrinking, cracking, or discoloration) or contamination, or if the expiration date supplied by the manufacturer has passed.

**Use:** For use in the microbial content test of water-soluble cosmetic products. Also used for determining the efficiency of sanitization of containers, equipment, and environmental surfaces.

## Middlebrook 13A Medium

**Composition** per 112.5mL:

| | |
|---|---|
| Casein hydrolysate | 0.1g |
| Tween™ 80 | 0.02g |
| Sodium polyanetholesulfonate | 0.025g |
| Middlebrook 7H9 broth | 100.0mL |
| Middlebrook 13A enrichment | 12.5mL |
| Catalase | 36,000U |
| $^{14}$C-substrate | 125µCi (185kBq) |

pH 6.6 ± 0.2 at 25°C

**Middlebrook 7H9 Broth:**

**Composition** per liter:

| | |
|---|---|
| $Na_2HPO_4$ | 2.5g |
| $KH_2PO_4$ | 1.0g |
| Monosodium glutamate | 0.5g |
| $(NH_4)_2SO_4$ | 0.5g |
| Sodium citrate | 0.1g |
| $MgSO_4 \cdot 7H_2O$ | 0.05g |
| Ferric ammonium citrate | 0.04g |
| $CuSO_4 \cdot 5H_2O$ | 1.0mg |
| Pyridoxine | 1.0mg |
| $ZnSO_4 \cdot 7H_2O$ | 1.0mg |
| Biotin | 0.5mg |
| $CaCl_2 \cdot 2H_2O$ | 0.5mg |
| Glycerol | 2.0mL |

**Preparation of Middlebrook 7H9 Broth:** Add components to distilled/deionized water and bring volume to 1.0L. Mix thoroughly.

**Middlebrook 13A Enrichment:**

**Composition** per 20.0mL:

| | |
|---|---|
| Bovine serum albumin | 3.0g |

**Preparation of Middlebrook 13A Enrichment:** Add bovine serum albumin to distilled/deionized water and bring volume to 20.0mL. Mix thoroughly. Filter sterilize.

**Preparation of Medium:** To 100.0mL of Middlebrook 7H9 broth, add remaining components, except Middlebrook 13A enrichment. Mix thoroughly. Filter sterilize. Aseptically distribute into bottles in 4.0mL volumes. Prior to inoculation, aseptically add 0.5mL of Middlebrook 13A enrichment to each bottle. Mix thoroughly.

**Storage/Shelf Life:** Store dehydrated media in the dark in a sealed container below 30°C. Prepared media should be stored under refrigeration (2-8°C). Media should be used within 60 days of preparation. Media should not be used if there are any signs of deterioration (discoloration) or contamination, or if the expiration date supplied by the manufacturer has passed.

**Use:** For the cultivation of *Mycobacterium* species from the blood of patients suspected of having mycobacteremia.

## Middlebrook ADC Enrichment
## (Middlebrook Albumin Dextrose Catalase Enrichment)

**Composition** per 100.0mL:

| | |
|---|---|
| Bovine albumin fraction V | 5.0g |
| Glucose | 2.0g |
| Catalase | 0.003g |

**Source:** This medium is available as a prepared enrichment from BD Diagnostic Systems.

**Preparation of Enrichment:** Add components to distilled/deionized water and bring volume to 100.0mL. Mix thoroughly. Filter sterilize.

**Storage/Shelf Life:** Store dehydrated media in the dark in a sealed container below 30°C. Prepared media should be stored under refrigeration (2-8°C). Media should be used within 60 days of preparation. Media should not be used if there are any signs of deterioration (discoloration) or contamination, or if the expiration date supplied by the manufacturer has passed.

**Use:** For use as a supplement to other Middlebrook media for the isolation, cultivation, and maintenance of *Mycobacterium* species. Also used as a supplement to other Middlebrook media for determining the antimicrobial susceptibility of mycobacteria.

## Middlebrook 7H9 Broth with
## Middlebrook ADC Enrichment

**Composition** per liter:

| | |
|---|---|
| $Na_2HPO_4$ | 2.5g |
| $KH_2PO_4$ | 1.0g |
| Monosodium glutamate | 0.5g |
| $(NH_4)_2SO_4$ | 0.5g |
| Sodium citrate | 0.1g |
| $MgSO_4 \cdot 7H_2O$ | 0.05g |
| Ferric ammonium citrate | 0.04g |
| $CuSO_4 \cdot 5H_2O$ | 1.0mg |
| Pyridoxine | 1.0mg |
| $ZnSO_4 \cdot 7H_2O$ | 1.0mg |
| Biotin | 0.5mg |
| $CaCl_2 \cdot 2H_2O$ | 0.5mg |
| Middlebrook ADC enrichment | 100.0mL |
| Glycerol | 2.0mL |

pH 6.6 ± 0.2 at 25°C

**Source:** This medium is available as a premixed powder from BD Diagnostic Systems.

**Middlebrook ADC Enrichment:**

**Composition** per 100.0mL:

| | |
|---|---|
| Bovine albumin fraction V | 5.0g |
| Glucose | 2.0g |
| Catalase | 3.0mg |

**Source:** This enrichment is available as a prepared enrichment from BD Diagnostic Systems.

**Preparation of Middlebrook ADC Enrichment:** Add components to distilled/deionized water and bring volume to 100.0mL. Mix thoroughly. Filter sterilize.

**Preparation of Medium:** Add glycerol to 900.0mL of distilled/deionized water and add remaining components, except Middlebrook ADC enrichment. Mix thoroughly. Gently heat and bring to boiling. Autoclave for 15 min at 15 psi pressure–121°C. Cool to 50°–55°C. Aseptically add 100.0mL of sterile Middlebrook ADC enrichment. Mix thoroughly. Distribute into sterile tubes or flasks.

**Storage/Shelf Life:** Store dehydrated media in the dark in a sealed container below 30°C. Prepared media should be stored under refrigeration (2-8°C). Media should be used within 60 days of preparation. Media should not be used if there are any signs of deterioration (discoloration) or contamination, or if the expiration date supplied by the manufacturer has passed.

**Use:** For the isolation, cultivation, and maintenance of *Mycobacterium* species, including *Mycobacterium tuberculosis*. Also used for determining the antimicrobial susceptibility of mycobacteria.

## Middlebrook 7H9 Broth with Middlebrook OADC Enrichment

**Composition** per liter:

| | |
|---|---|
| Na$_2$HPO$_4$ | 2.5g |
| KH$_2$PO$_4$ | 1.0g |
| Monosodium glutamate | 0.5g |
| (NH$_4$)$_2$SO$_4$ | 0.5g |
| Sodium citrate | 0.1g |
| MgSO$_4$·7H$_2$O | 0.05g |
| Ferric ammonium citrate | 0.04g |
| CuSO$_4$·5H$_2$O | 1.0mg |
| Pyridoxine | 1.0mg |
| ZnSO$_4$·7H$_2$O | 1.0mg |
| Biotin | 0.5mg |
| CaCl$_2$·2H$_2$O | 0.5mg |
| Middlebrook OADC enrichment | 100.0mL |
| Glycerol | 2.0mL |

pH 6.6 ± 0.2 at 25°C

**Source:** This medium is available as a premixed powder from BD Diagnostic Systems.

**Middlebrook OADC Enrichment:**

**Composition** per 100.0mL:

| | |
|---|---|
| Bovine albumin fraction V | 5.0g |
| Glucose | 2.0g |
| NaCl | 0.85g |
| Oleic acid | 0.05g |
| Catalase | 4.0mg |

**Source:** This enrichment is available as a prepared enrichment from BD Diagnostic Systems.

**Preparation of Middlebrook OADC Enrichment:** Add components to distilled/deionized water and bring volume to 100.0mL. Mix thoroughly. Filter sterilize.

**Preparation of Medium:** Add glycerol to 900.0mL of distilled/deionized water and add remaining components, except Middlebrook OADC enrichment. Mix thoroughly. Gently heat and bring to boiling. Autoclave for 15 min at 15 psi pressure–121°C. Cool to 50°–55°C. Aseptically add 100.0mL of sterile Middlebrook OADC enrichment. Mix thoroughly. Distribute into sterile tubes or flasks.

**Storage/Shelf Life:** Store dehydrated media in the dark in a sealed container below 30°C. Prepared media should be stored under refrigeration (2-8°C). Media should be used within 60 days of preparation. Media should not be used if there are any signs of deterioration (discoloration) or contamination, or if the expiration date supplied by the manufacturer has passed.

**Use:** For the isolation, cultivation, and maintenance of *Mycobacterium* species, including *Mycobacterium tuberculosis*. Also used for determining the antimicrobial susceptibility of mycobacteria.

## Middlebrook 7H9 Broth with Middlebrook OADC Enrichment and Triton™ WR 1339

**Composition** per liter:

| | |
|---|---|
| Na$_2$HPO$_4$ | 2.5g |
| KH$_2$PO$_4$ | 1.0g |
| Monosodium glutamate | 0.5g |
| (NH$_4$)$_2$SO$_4$ | 0.5g |
| Sodium citrate | 0.1g |
| MgSO$_4$·7H$_2$O | 0.05g |
| Ferric ammonium citrate | 0.04g |
| CuSO$_4$·5H$_2$O | 1.0mg |
| Pyridoxine | 1.0mg |
| ZnSO$_4$·7H$_2$O | 1.0mg |
| Biotin | 0.5mg |
| CaCl$_2$·2H$_2$O | 0.5mg |
| Middlebrook OADC enrichment with Triton™ WR 1339 | 100.0mL |
| Glycerol | 2.0mL |

pH 6.6 ± 0.2 at 25°C

**Source:** This medium is available as a premixed powder from BD Diagnostic Systems.

**Middlebrook OADC Enrichment with Triton™ WR 1339:**

**Composition** per 100.0mL:

| | |
|---|---|
| Bovine albumin fraction V | 5.0g |
| Glucose | 2.0g |
| NaCl | 0.85g |
| Triton™ WR 1339 | 0.25g |
| Oleic acid | 0.05g |
| Catalase | 4.0mg |

**Source:** This enrichment is available as a prepared enrichment from BD Diagnostic Systems.

**Preparation of Middlebrook OADC Enrichment with Triton™ WR 1339:** Add components to distilled/deionized water and bring volume to 100.0mL. Mix thoroughly. Filter sterilize.

**Preparation of Medium:** Add glycerol to 900.0mL of distilled/deionized water and add remaining components, except Middlebrook OADC enrichment with Triton™ WR 1339. Mix thoroughly. Gently heat and bring to boiling. Autoclave for 15 min at 15 psi pressure–121°C. Cool to 50°–55°C. Aseptically add 100.0mL of sterile Middlebrook OADC enrichment with Triton™ WR 1339. Mix thoroughly. Distribute into sterile tubes or flasks.

**Storage/Shelf Life:** Store dehydrated media in the dark in a sealed container below 30°C. Prepared media should be stored under refrigeration (2-8°C). Media should be used within 60 days of preparation. Media should not be used if there are any signs of deterioration (discoloration) or contamination, or if the expiration date supplied by the manufacturer has passed.

**Use:** For the isolation, cultivation, and maintenance of *Mycobacterium* species, including *Mycobacterium tuberculosis*. Also used for determining the antimicrobial susceptibility of mycobacteria.

## Middlebrook 7H9 Broth, Supplemented

**Composition** per liter:

| | |
|---|---|
| Na$_2$HPO$_4$ | 2.5g |
| KH$_2$PO$_4$ | 1.0g |
| Monosodium glutamate | 0.5g |

(NH$_4$)$_2$SO$_4$................................................................0.5g
Tween™ 80....................................................................0.5g
Sodium citrate............................................................0.1g
MgSO$_4$·7H$_2$O.............................................................0.05g
Ferric ammonium citrate.........................................0.04g
Mycobactin J...............................................................2.0mg
CuSO$_4$·5H$_2$O..............................................................1.0mg
Pyridoxine...................................................................1.0mg
ZnSO$_4$·7H$_2$O..............................................................1.0mg
Biotin...........................................................................0.5mg
CaCl$_2$·2H$_2$O...............................................................0.5mg
Dubos oleic albumin complex ............................ 100.0mL
Glycerol ..................................................................... 2.0mL

pH 6.6 ± 0.2 at 25°C

**Source:** Mycobactin J is available from Allied Laboratories, Inc.

### Dubos Oleic Albumin Complex:
**Composition** per 100.0mL:
Bovine serum albumin, fraction V.................................5.0g
Oleic acid, sodium salt................................................0.05g
NaCl (0.85% solution) ............................................ 100.0mL

**Preparation of Dubos Oleic Albumin Complex:** Add bovine serum albumin and oleic acid to 100.0mL of NaCl solution. Mix thoroughly. Filter sterilize.

**Preparation of Medium:** Add components, except Dubos oleic albumin complex, to distilled/deionized water and bring volume to 900.0mL. Mix thoroughly. Gently heat and bring to boiling. Autoclave for 15 min at 15 psi pressure–121°C. Cool to 45°–50°C. Aseptically add sterile Dubos oleic albumin complex. Mix thoroughly. Pour into sterile Petri dishes or distribute into sterile tubes.

**Storage/Shelf Life:** Store dehydrated media in the dark in a sealed container below 30°C. Prepared media should be stored under refrigeration (2-8°C). Media should be used within 60 days of preparation. Media should not be used if there are any signs of deterioration (discoloration) or contamination, or if the expiration date supplied by the manufacturer has passed.

**Use:** For the cultivation and maintenance of *Mycobacterium avium*.

### Middlebrook 7H10 Agar with Middlebrook ADC Enrichment

**Composition** per liter:
Agar ...........................................................................15.0g
Na$_2$HPO$_4$....................................................................1.5g
KH$_2$PO$_4$.......................................................................1.5g
(NH$_4$)$_2$SO$_4$................................................................0.5g
L-Glutamic acid...........................................................0.5g
Sodium citrate............................................................0.4g
Ferric ammonium citrate.........................................0.04g
MgSO$_4$·7H$_2$O.............................................................0.025g
ZnSO$_4$·7H$_2$O..............................................................1.0mg
CuSO$_4$·5H$_2$O..............................................................1.0mg
Pyridoxine...................................................................1.0mg
Biotin...........................................................................0.5mg
CaCl$_2$·2H$_2$O...............................................................0.5mg
Malachite Green..........................................................0.25mg
Middlebrook ADC enrichment ............................ 100.0mL
Glycerol ..................................................................... 5.0mL

pH 6.6 ± 0.2 at 25°C

### Middlebrook ADC Enrichment:
**Composition** per 100.0mL:
Bovine albumin fraction V ..........................................5.0g
Glucose .......................................................................2.0g
Catalase.......................................................................0.003g

**Source:** The medium and enrichment are available as a prepared enrichment from BD Diagnostic Systems.

**Preparation of Middlebrook ADC Enrichment:** Add components to distilled/deionized water and bring volume to 100.0mL. Mix thoroughly. Filter sterilize.

**Preparation of Medium:** Add glycerol to 900.0mL of distilled/deionized water and add remaining components, except Middlebrook ADC enrichment. Mix thoroughly. Gently heat and bring to boiling. Autoclave for 15 min at 15 psi pressure–121°C. Cool to 50°–55°C. Aseptically add 100.0mL of sterile Middlebrook ADC enrichment. Mix thoroughly. Pour into sterile Petri dishes or distribute into sterile tubes.

**Storage/Shelf Life:** Store dehydrated media in the dark in a sealed container below 30°C. Prepared media should be stored under refrigeration (2-8°C). Media should be used within 60 days of preparation. Media should not be used if there are any signs of deterioration (shrinking, cracking, or discoloration) or contamination, or if the expiration date supplied by the manufacturer has passed.

**Use:** For the isolation, cultivation, and maintenance of *Mycobacterium* species, including *Mycobacterium tuberculosis*. Also used for determining the antimicrobial susceptibility of mycobacteria.

### Middlebrook 7H10 Agar with Middlebrook OADC Enrichment (Middlebrook and Cohn 7H10 Agar)

**Composition** per liter:
Agar ...........................................................................15.0g
Na$_2$HPO$_4$....................................................................1.5g
KH$_2$PO$_4$.......................................................................1.5g
(NH$_4$)$_2$SO$_4$ ...............................................................0.5g
L-Glutamic acid...........................................................0.5g
Sodium citrate............................................................0.4g
Ferric ammonium citrate.........................................0.04g
MgSO$_4$·7H$_2$O.............................................................0.025g
ZnSO$_4$·7H$_2$O..............................................................1.0mg
CuSO$_4$·5H$_2$O..............................................................1.0mg
Pyridoxine...................................................................1.0mg
Biotin...........................................................................0.5mg
CaCl$_2$·2H$_2$O...............................................................0.5mg
Malachite Green..........................................................0.25mg
Middlebrook OADC enrichment .......................... 100.0mL
Glycerol ..................................................................... 5.0mL

pH 6.6 ± 0.2 at 25°C

**Source:** This medium is available as a premixed powder from BD Diagnostic Systems.

### Middlebrook OADC Enrichment:
**Composition** per 100.0mL:
Bovine albumin fraction V ..........................................5.0g
Glucose .......................................................................2.0g
NaCl.............................................................................0.85g
Oleic acid.....................................................................0.05g
Catalase.......................................................................4.0mg

**Source:** This enrichment is available as a prepared enrichment from BD Diagnostic Systems.

**Preparation of Middlebrook OADC Enrichment:** Add components to distilled/deionized water and bring volume to 100.0mL. Mix thoroughly. Filter sterilize.

**Preparation of Medium:** Add glycerol to 900.0mL of distilled/deionized water and add remaining components, except Middlebrook OADC enrichment. Mix thoroughly. Gently heat and bring to boiling. Autoclave for 15 min at 15 psi pressure–121°C. Cool to 50°–55°C. Aseptically add 100.0mL of sterile Middlebrook OADC enrichment. Mix thoroughly. Pour into sterile Petri dishes or distribute into sterile tubes.

**Storage/Shelf Life:** Store dehydrated media in the dark in a sealed container below 30°C. Prepared media should be stored under refrigeration (2-8°C). Media should be used within 60 days of preparation. Media should not be used if there are any signs of deterioration (shrinking, cracking, or discoloration) or contamination, or if the expiration date supplied by the manufacturer has passed.

**Use:** For the isolation, cultivation, and maintenance of *Mycobacterium* species, including *Mycobacterium tuberculosis*. Also used for determining the antimicrobial susceptibility of mycobacteria.

## Middlebrook 7H10 Agar
## with Middlebrook OADC Enrichment and Hemin
### (Hemin Medium for *Mycobacterium*)

**Composition** per liter:

| | |
|---|---|
| Agar | 15.0g |
| $Na_2HPO_4$ | 1.5g |
| $KH_2PO_4$ | 1.5g |
| $(NH_4)_2SO_4$ | 0.5g |
| L-Glutamic acid | 0.5g |
| Sodium citrate | 0.4g |
| Ferric ammonium citrate | 0.04g |
| $MgSO_4 \cdot 7H_2O$ | 0.025g |
| $ZnSO_4 \cdot 7H_2O$ | 1.0mg |
| $CuSO_4 \cdot 5H_2O$ | 1.0mg |
| Pyridoxine | 1.0mg |
| Biotin | 0.5mg |
| $CaCl_2 \cdot 2H_2O$ | 0.5mg |
| Malachite Green | 0.25mg |
| Middlebrook OADC enrichment | 100.0mL |
| Glycerol | 5.0mL |
| Hemin solution | 3.9mL |

pH 6.6 ± 0.2 at 25°C

**Source:** This medium is available as a premixed powder from BD Diagnostic Systems.

**Middlebrook OADC Enrichment:**

**Composition** per 100.0mL:

| | |
|---|---|
| Bovine albumin fraction V | 5.0g |
| Glucose | 2.0g |
| NaCl | 0.85g |
| Oleic acid | 0.05g |
| Catalase | 4.0mg |

**Preparation of Middlebrook OADC Enrichment:** Add components to distilled/deionized water and bring volume to 100.0mL. Mix thoroughly. Filter sterilize.

**Hemin Solution:**

**Composition** per 100.0mL:

| | |
|---|---|
| Hemin | 1.0g |
| NaOH (1*N* solution) | 20.0mL |

**Preparation of Hemin Solution:** Add hemin to 20.0mL of 1*N* NaOH solution. Mix thoroughly. Bring volume to 100.0mL with distilled/deionized water.

**Preparation of Medium:** Add glycerol to 891.1mL of distilled/deionized water and add remaining components, except Middlebrook OADC enrichment. Mix thoroughly. Gently heat and bring to boiling. Autoclave for 15 min at 15 psi pressure–121°C. Cool to 50°–55°C. Aseptically add 100.0mL of sterile Middlebrook OADC enrichment. Mix thoroughly. Pour into sterile Petri dishes or distribute into sterile tubes.

**Storage/Shelf Life:** Store dehydrated media in the dark in a sealed container below 30°C. Prepared media should be stored under refrigeration (2-8°C). Media should be used within 60 days of preparation. Media should not be used if there are any signs of deterioration (shrinking, cracking, or discoloration) or contamination, or if the expiration date supplied by the manufacturer has passed.

**Use:** For the isolation, cultivation, and maintenance of *Mycobacterium* species, including *Mycobacterium tuberculosis*. For the cultivation and maintenance of *Mycobacterium haemophilum*. Also used for determining the antimicrobial susceptibility of mycobacteria.

## Middlebrook 7H10 Agar
## with Middlebrook OADC
### Enrichment and Triton™ WR 1339

**Composition** per liter:

| | |
|---|---|
| Agar | 15.0g |
| $Na_2HPO_4$ | 1.5g |
| $KH_2PO_4$ | 1.5g |
| $(NH_4)_2SO_4$ | 0.5g |
| L-Glutamic acid | 0.5g |
| Sodium citrate | 0.4g |
| Ferric ammonium citrate | 0.04g |
| $MgSO_4 \cdot 7H_2O$ | 0.025g |
| $ZnSO_4 \cdot 7H_2O$ | 1.0mg |
| $CuSO_4 \cdot 5H_2O$ | 1.0mg |
| Pyridoxine | 1.0mg |
| Biotin | 0.5mg |
| $CaCl_2 \cdot 2H_2O$ | 0.5mg |
| Malachite Green | 0.25mg |
| Middlebrook OADC enrichment with Triton™ WR 1339 | 100.0mL |
| Glycerol | 5.0mL |

pH 6.6 ± 0.2 at 25°C

**Source:** This medium is available as a premixed powder from BD Diagnostic Systems.

**Middlebrook OADC Enrichment with Triton™ WR 1339:**

**Composition** per 100.0mL:

| | |
|---|---|
| Bovine albumin fraction V | 5.0g |
| Glucose | 2.0g |
| NaCl | 0.85g |
| Triton™ WR 1339 | 0.25g |
| Oleic acid | 0.05g |
| Catalase | 4.0mg |

**Source:** This enrichment is available as a prepared enrichment from BD Diagnostic Systems.

**Preparation of Middlebrook OADC Enrichment with Triton™ WR 1339:** Add components to distilled/deionized water and bring volume to 100.0mL. Mix thoroughly. Filter sterilize.

**Preparation of Medium:** Add glycerol to 900.0mL of distilled/deionized water and add remaining components, except Middlebrook OADC enrichment with Triton™ WR 1339. Mix thoroughly. Gently heat and bring to boiling. Autoclave for 15 min at 15 psi pressure–121°C. Cool to 50°–55°C. Aseptically add 100.0mL of sterile Middlebrook OADC enrichment with Triton™ WR 1339. Mix thoroughly. Pour into sterile Petri dishes or distribute into sterile tubes.

**Storage/Shelf Life:** Store dehydrated media in the dark in a sealed container below 30°C. Prepared media should be stored under refrigeration (2-8°C). Media should be used within 60 days of preparation. Media should not be used if there are any signs of deterioration (shrinking, cracking, or discoloration) or contamination, or if the expiration date supplied by the manufacturer has passed.

**Use:** For the isolation, cultivation, and maintenance of *Mycobacterium* species, including *Mycobacterium tuberculosis*. Also used for determining the antimicrobial susceptibility of mycobacteria.

## Middlebrook 7H10 Agar with Streptomycin

**Composition** per liter:

| | |
|---|---|
| Agar | 15.0g |
| $Na_2HPO_4$ | 1.5g |
| $KH_2PO_4$ | 1.5g |
| $(NH_4)_2SO_4$ | 0.5g |
| L-Glutamic acid | 0.5g |
| Sodium citrate | 0.4g |
| Ferric ammonium citrate | 0.04g |
| $MgSO_4 \cdot 7H_2O$ | 0.025g |
| $ZnSO_4 \cdot 7H_2O$ | 1.0mg |
| $CuSO_4 \cdot 5H_2O$ | 1.0mg |
| Pyridoxine | 1.0mg |
| Biotin | 0.5mg |
| $CaCl_2 \cdot 2H_2O$ | 0.5mg |
| Malachite Green | 0.25mg |
| Glycerol | 5.0mL |
| Streptomycin | 100.0mg |

pH $6.6 \pm 0.2$ at 25°C

**Source:** This medium is available as a premixed powder from BD Diagnostic Systems.

**Preparation of Medium:** Add glycerol to 1.0L of distilled/deionized water and add remaining components. Mix thoroughly. Gently heat and bring to boiling. Autoclave for 15 min at 15 psi pressure–121°C. Cool to 50°–55°C. Aseptically add streptomycin. Mix thoroughly. Pour into sterile Petri dishes or distribute into sterile tubes.

**Storage/Shelf Life:** Store dehydrated media in the dark in a sealed container below 30°C. Prepared media should be stored under refrigeration (2-8°C). Media should be used within 60 days of preparation. Media should not be used if there are any signs of deterioration (shrinking, cracking, or discoloration) or contamination, or if the expiration date supplied by the manufacturer has passed.

**Use:** For the isolation, cultivation, and maintenance of *Mycobacterium kansasii*.

## Middlebrook 7H11 Agar, Selective

**Composition** per liter:

| | |
|---|---|
| Agar | 15.0g |
| $Na_2HPO_4$ | 1.5g |
| $KH_2PO_4$ | 1.5g |
| Pancreatic digest of casein | 1.0g |
| $(NH_4)_2SO_4$ | 0.5g |
| L-Glutamic acid | 0.5g |
| Sodium citrate | 0.4g |
| $MgSO_4 \cdot 7H_2O$ | 0.05g |
| Ferric ammonium citrate | 0.04g |
| Pyridoxine | 1.0mg |
| $ZnSO_4 \cdot 7H_2O$ | 1.0mg |
| $CuSO_4 \cdot 5H_2O$ | 1.0mg |
| $CaCl_2 \cdot 2H_2O$ | 0.5mg |
| Malachite Green | 0.25mg |
| D–Biotin | 0.5μg |
| Middlebrook OADC enrichment | 100.0mL |
| Antibiotic solution | 10.0mL |
| Glycerol | 5.0mL |

pH $6.6 \pm 0.2$ at 25°C

**Middlebrook OADC Enrichment:**

**Composition** per 100.0mL:

| | |
|---|---|
| Bovine albumin fraction V | 5.0g |
| Glucose | 2.0g |
| NaCl | 0.85g |
| Oleic acid | 0.05g |
| Catalase | 4.0mg |

**Source:** This enrichment is available as a prepared enrichment from BD Diagnostic Systems.

**Preparation of Middlebrook OADC Enrichment:** Add components to distilled/deionized water and bring volume to 100.0mL. Mix thoroughly. Filter sterilize.

**Antibiotic Solution:**

**Composition** per 10.0mL:

| | |
|---|---|
| Carbenicillin | 0.05mg |
| Trimethoprim lactate | 0.02mg |
| Amphotericin B | 0.01mg |
| Polymyxin B | 200,000U |

**Preparation of Antibiotic Solution:** Add components to distilled/deionized water and bring volume to 10.0mL. Mix thoroughly. Filter sterilize.

**Preparation of Medium:** Add glycerol to 890.0mL of distilled/deionized water and add remaining components, except Middlebrook OADC enrichment and antibiotic solution. Mix thoroughly. Gently heat and bring to boiling. Autoclave for 15 min at 15 psi pressure–121°C. Cool to 50°–55°C. Aseptically add 100.0mL of sterile Middlebrook OADC enrichment and 10.0mL of sterile antibiotic solution. Mix thoroughly. Pour into sterile Petri dishes or distribute into sterile tubes.

**Storage/Shelf Life:** Store dehydrated media in the dark in a sealed container below 30°C. Prepared media should be stored under refrigeration (2-8°C). Media should be used within 60 days of preparation. Media should not be used if there are any signs of deterioration (shrinking, cracking, or discoloration) or contamination, or if the expiration date supplied by the manufacturer has passed.

**Use:** For the selective isolation and cultivation of pathogenic mycobacteria from specimens potentially contaminated with bacteria and fungi.

## Middlebrook 7H11 Agar with Middlebrook ADC Enrichment (Mycobacteria 7H11 Agar with Middlebrook ADC Enrichment)

**Composition** per liter:

| | |
|---|---|
| Agar | 15.0g |
| $Na_2HPO_4$ | 1.5g |
| $KH_2PO_4$ | 1.5g |
| Pancreatic digest of casein | 1.0g |
| $(NH_4)_2SO_4$ | 0.5g |
| L-Glutamic acid | 0.5g |
| Sodium citrate | 0.4g |
| $MgSO_4 \cdot 7H_2O$ | 0.05g |
| Ferric ammonium citrate | 0.04g |
| Pyridoxine | 1.0mg |
| Malachite Green | 0.25mg |
| D-Biotin | 0.5μg |
| Middlebrook ADC enrichment | 100.0mL |
| Glycerol | 5.0mL |

pH 6.6 ± 0.2 at 25°C

**Source:** This medium is available as a premixed powder from BD Diagnostic Systems.

**Middlebrook ADC Enrichment:**
**Composition** per 100.0mL:

| | |
|---|---|
| Bovine albumin fraction V | 5.0g |
| Glucose | 2.0g |
| Catalase | 0.003g |

**Source:** This enrichment is available as a prepared enrichment from BD Diagnostic Systems.

**Preparation of Middlebrook ADC Enrichment:** Add components to distilled/deionized water and bring volume to 100.0mL. Mix thoroughly. Filter sterilize.

**Preparation of Medium:** Add glycerol to 900.0mL of distilled/deionized water and add remaining components, except Middlebrook ADC enrichment. Mix thoroughly. Gently heat and bring to boiling. Autoclave for 15 min at 15 psi pressure–121°C. Cool to 50°–55°C. Aseptically add 100.0mL of sterile Middlebrook ADC enrichment. Mix thoroughly. Pour into sterile Petri dishes or distribute into sterile tubes.

**Storage/Shelf Life:** Store dehydrated media in the dark in a sealed container below 30°C. Prepared media should be stored under refrigeration (2-8°C). Media should be used within 60 days of preparation. Media should not be used if there are any signs of deterioration (shrinking, cracking, or discoloration) or contamination, or if the expiration date supplied by the manufacturer has passed.

**Use:** For the cultivation of drug-resistant (isoniazid [INH]) strains of *Mycobacterium tuberculosis*. For the cultivation of particularly fastidious strains of tubercle bacilli that occur following treatment of tuberculosis patients with secondary antitubercular drugs. Generally, these strains fail to grow on 7H10 medium.

## Middlebrook 7H11 Agar with Middlebrook OADC Enrichment (Mycobacteria 7H11 Agar with Middlebrook OADC Enrichment)

**Composition** per liter:

| | |
|---|---|
| Agar | 15.0g |
| $Na_2HPO_4$ | 1.5g |
| $KH_2PO_4$ | 1.5g |
| Pancreatic digest of casein | 1.0g |
| $(NH_4)_2SO_4$ | 0.5g |
| L-Glutamic acid | 0.5g |
| Sodium citrate | 0.4g |
| $MgSO_4 \cdot 7H_2O$ | 0.05g |
| Ferric ammonium citrate | 0.04g |
| Pyridoxine | 1.0mg |
| Malachite Green | 0.25mg |
| D-Biotin | 0.5μg |
| Middlebrook OADC enrichment | 100.0mL |
| Glycerol | 5.0mL |

pH 6.6 ± 0.2 at 25°C

**Source:** This medium is available as a premixed powder from BD Diagnostic Systems.

**Middlebrook OADC Enrichment:**
**Composition** per 100.0mL:

| | |
|---|---|
| Bovine albumin fraction V | 5.0g |
| Glucose | 2.0g |
| NaCl | 0.85g |
| Oleic acid | 0.05g |
| Catalase | 4.0mg |

**Source:** This enrichment is available as a prepared enrichment from BD Diagnostic Systems.

**Preparation of Middlebrook OADC Enrichment:** Add components to distilled/deionized water and bring volume to 100.0mL. Mix thoroughly. Filter sterilize.

**Preparation of Medium:** Add glycerol to 900.0mL of distilled/deionized water and add remaining components, except Middlebrook OADC enrichment. Mix thoroughly. Gently heat and bring to boiling. Autoclave for 15 min at 15 psi pressure–121°C. Cool to 50°–55°C. Aseptically add 100.0mL of sterile Middlebrook OADC enrichment. Mix thoroughly. Pour into sterile Petri dishes or distribute into sterile tubes.

**Storage/Shelf Life:** Store dehydrated media in the dark in a sealed container below 30°C. Prepared media should be stored under refrigeration (2-8°C). Media should be used within 60 days of preparation. Media should not be used if there are any signs of deterioration (shrinking, cracking, or discoloration) or contamination, or if the expiration date supplied by the manufacturer has passed.

**Use:** For the cultivation of drug-resistant (isoniazid [INH]) strains of *Mycobacterium tuberculosis*. For the cultivation of particularly fastidious strains of tubercle bacilli that occur following treatment of tuberculosis patients with secondary antitubercular drugs. Generally, these strains fail to grow on 7H10 medium.

## Middlebrook 7H11 Agar with Middlebrook OADC Enrichment and Triton™ WR 1339 (Mycobacteria 7H11 Agar with Middlebrook OADC Enrichment and Triton™ WR 1339)

**Composition** per liter:

| | |
|---|---|
| Agar | 15.0g |
| $Na_2HPO_4$ | 1.5g |
| $KH_2PO_4$ | 1.5g |
| Pancreatic digest of casein | 1.0g |
| $(NH_4)_2SO_4$ | 0.5g |
| L-Glutamic acid | 0.5g |

Sodium citrate ................................................................0.4g
MgSO$_4$·7H$_2$O ..........................................................0.05g
Ferric ammonium citrate............................................0.04g
Pyridoxine...................................................................1.0mg
Malachite Green.........................................................0.25mg
D-Biotin .......................................................................0.5μg
Middlebrook OADC enrichment
   with Triton™ WR 1339......................................100.0mL
Glycerol .......................................................................5.0mL

pH 6.6 ± 0.2 at 25°C

**Source:** This medium is available as a premixed powder from BD Diagnostic Systems.

### Middlebrook OADC Enrichment with Triton™ WR 1339:

**Composition** per 100.0mL:
Bovine albumin fraction V ...........................................5.0g
Glucose .........................................................................2.0g
NaCl...............................................................................0.85g
Triton™ WR 1339 .........................................................0.25g
Oleic acid .....................................................................0.05g
Catalase.........................................................................4.0mg

**Source:** This enrichment is available as a prepared enrichment from BD Diagnostic Systems.

**Preparation of Middlebrook OADC Enrichment with Triton™ WR 1339:** Add components to distilled/deionized water and bring volume to 100.0mL. Mix thoroughly. Filter sterilize.

**Preparation of Medium:** Add glycerol to 900.0mL of distilled/deionized water and add remaining components, except Middlebrook OADC enrichment with Triton™ WR-1339. Mix thoroughly. Gently heat and bring to boiling. Autoclave for 15 min at 15 psi pressure–121°C. Cool to 50°–55°C. Aseptically add 100.0mL of sterile Middlebrook OADC enrichment with Triton™ WR-1339. Mix thoroughly. Pour into sterile Petri dishes or distribute into sterile tubes.

**Storage/Shelf Life:** Store dehydrated media in the dark in a sealed container below 30°C. Prepared media should be stored under refrigeration (2-8°C). Media should be used within 60 days of preparation. Media should not be used if there are any signs of deterioration (shrinking, cracking, or discoloration) or contamination, or if the expiration date supplied by the manufacturer has passed.

**Use:** For the cultivation of drug-resistant (isoniazid [INH]) strains of *Mycobacterium tuberculosis*. For the cultivation of particularly fastidious strains of tubercle bacilli that occur following treatment of tuberculosis patients with secondary antitubercular drugs. Generally, these strains fail to grow on 7H10 medium.

### Middlebrook 7H11 HiVeg Agar Base with Middlebrook ADC Enrichment

**Composition** per liter:
Agar ..............................................................................15.0g
Na$_2$HPO$_4$..................................................................1.5g
KH$_2$PO$_4$....................................................................1.5g
Plant hydrolysate..........................................................1.0g
L-Glutamic acid............................................................0.5g
(NH$_4$)$_2$SO$_4$.............................................................0.5g
Sodium citrate ..............................................................0.4g
MgSO$_4$........................................................................0.05g
Ferric ammonium citrate..............................................0.04g
Pyridoxine.....................................................................1.0mg

Malachite green .............................................................1.0mg
Biotin .............................................................................0.5mg
Middlebrook ADC enrichment...................................100.0mL
Glycerol .........................................................................5.0mL

pH 6.6 ± 0.2 at 25°C

**Source:** This medium, without glycerol and Middlebrook ADC enrichment, is available as a premixed powder from HiMedia.

### Middlebrook ADC Enrichment:

**Composition** per 100.0mL:
Bovine albumin fraction V ...........................................5.0g
Glucose .........................................................................2.0g
Catalase.........................................................................0.003g

**Preparation of Middlebrook ADC Enrichment:** Add components to distilled/deionized water and bring volume to 100.0mL. Mix thoroughly. Filter sterilize.

**Preparation of Medium:** Add glycerol to 900.0mL of distilled/deionized water and add remaining components, except Middlebrook ADC enrichment. Mix thoroughly. Gently heat and bring to boiling. Autoclave for 15 min at 15 psi pressure–121°C. Cool to 50°–55°C. Aseptically add 100.0mL of sterile Middlebrook ADC enrichment. Mix thoroughly. Pour into sterile Petri dishes or distribute into sterile tubes.

**Storage/Shelf Life:** Store dehydrated media in the dark in a sealed container below 30°C. Prepared media should be stored under refrigeration (2-8°C). Media should be used within 60 days of preparation. Media should not be used if there are any signs of deterioration (shrinking, cracking, or discoloration) or contamination, or if the expiration date supplied by the manufacturer has passed.

**Use:** For the cultivation of drug-resistant (isoniazid [INH]) strains of *Mycobacterium tuberculosis*. For the cultivation of particularly fastidious strains of tubercle bacilli that occur following treatment of tuberculosis patients with secondary antitubercular drugs.

### Middlebrook 7H11 HiVeg Agar Base with Middlebrook OADC Enrichment

**Composition** per liter:
Agar ..............................................................................15.0g
Na$_2$HPO$_4$..................................................................1.5g
KH$_2$PO$_4$....................................................................1.5g
Plant hydrolysate ..........................................................1.0g
L-Glutamic acid ............................................................0.5g
(NH$_4$)$_2$SO$_4$.............................................................0.5g
Sodium citrate...............................................................0.4g
MgSO$_4$........................................................................0.05g
Ferric ammonium citrate...............................................0.04g
Pyridoxine.....................................................................1.0mg
Malachite green .............................................................1.0mg
Biotin .............................................................................0.5mg
Middlebrook OADC enrichment ................................100.0mL
Glycerol .........................................................................5.0mL

pH 6.6 ± 0.2 at 25°C

**Source:** This medium, without glycerol and Middlebrook OADC enrichment, is available as a premixed powder from HiMedia.

### Middlebrook OADC Enrichment:

**Composition** per 100.0mL:
Bovine albumin fraction V ...........................................5.0g
Glucose .........................................................................2.0g
NaCl...............................................................................0.85g

Oleic acid .................................................................................0.05g
Catalase ....................................................................................4.0mg

**Preparation of Middlebrook OADC Enrichment:** Add components to distilled/deionized water and bring volume to 100.0mL. Mix thoroughly. Filter sterilize.

**Preparation of Medium:** Add glycerol to 900.0mL of distilled/deionized water and add remaining components, except Middlebrook OADC enrichment. Mix thoroughly. Gently heat and bring to boiling. Autoclave for 15 min at 15 psi pressure–121°C. Cool to 50°–55°C. Aseptically add 100.0mL of sterile Middlebrook OADC enrichment. Mix thoroughly. Pour into sterile Petri dishes or distribute into sterile tubes.

**Storage/Shelf Life:** Store dehydrated media in the dark in a sealed container below 30°C. Prepared media should be stored under refrigeration (2-8°C). Media should be used within 60 days of preparation. Media should not be used if there are any signs of deterioration (shrinking, cracking, or discoloration) or contamination, or if the expiration date supplied by the manufacturer has passed.

**Use:** For the cultivation of drug-resistant (isoniazid [INH]) strains of *Mycobacterium tuberculosis*. For the cultivation of particularly fastidious strains of tubercle bacilli that occur following treatment of tuberculosis patients with secondary antitubercular drugs.

## Middlebrook 7H12 Medium

**Composition** per 102.5mL:
Bovine serum albumin .............................................................0.5g
Casein hydrolyslate ..................................................................0.1g
Catalase .................................................................................4800U
$^{14}$C-Palmitic acid..................................................................100µCi
Middlebrook 7H9 broth ......................................................100.0mL
Antibiotic solution ..................................................................2.5mL

pH 6.8 ± 0.1 at 25°C

**Middlebrook 7H9 Broth:**

**Composition** per liter:
Na$_2$HPO$_4$ .................................................................................2.5g
KH$_2$PO$_4$ ....................................................................................1.0g
Monosodium glutamate ...........................................................0.5g
(NH$_4$)$_2$SO$_4$ ...............................................................................0.5g
Sodium citrate .........................................................................0.1g
MgSO$_4$·7H$_2$O ........................................................................0.05g
Ferric ammonium citrate........................................................0.04g
CuSO$_4$·5H$_2$O .........................................................................1.0mg
Pyridoxine ..............................................................................1.0mg
ZnSO$_4$·7H$_2$O ........................................................................1.0mg
Biotin .....................................................................................0.5mg
CaCl$_2$·2H$_2$O...........................................................................0.5mg
Glycerol ..................................................................................2.0mL

**Preparation of Middlebrook 7H9 Broth:** Add components to distilled/deionized water and bring volume to 1.0L. Mix thoroughly.

**Antibiotic Solution:**

**Composition** per 5.0mL:
Nalidixic acid ..........................................................................0.2g
Azlocillin ................................................................................0.1g
Amphotericin B........................................................................0.05g
Trimethoprim ...........................................................................0.05g
Polymyxin B .....................................................................500,000U

**Preparation of Antibiotic Solution:** Add components to distilled/deionized water and bring volume to 5.0mL. Mix thoroughly. Filter sterilize.

**Preparation of Medium:** To 100.0mL of Middlebrook 7H9 broth, add remaining components, except antibiotic solution. Mix thoroughly. Filter sterilize. Aseptically distribute into bottles in 4.0mL volumes. Prior to inoculation, aseptically add 0.1mL of antibiotic solution to each bottle. Mix thoroughly.

**Storage/Shelf Life:** Store dehydrated media in the dark in a sealed container below 30°C. Prepared media should be stored under refrigeration (2-8°C). Media should be used within 60 days of preparation. Media should not be used if there are any signs of deterioration (shrinking, cracking, or discoloration) or contamination, or if the expiration date supplied by the manufacturer has passed.

**Use:** For the cultivation of *Mycobacterium* species from the blood of patients suspected of having mycobacteremia.

## Middlebrook Medium

**Composition** per liter:
Agar .......................................................................................15.0g
Na$_2$HPO$_4$ .................................................................................1.5g
KH$_2$PO$_4$ ....................................................................................1.5g
(NH$_4$)$_2$SO$_4$ ...............................................................................0.5g
L-Glutamic acid........................................................................0.5g
Sodium citrate .........................................................................0.4g
Ferric ammonium citrate.........................................................0.04g
MgSO$_4$·7H$_2$O ......................................................................0.025g
ZnSO$_4$·7H$_2$O ........................................................................1.0mg
CuSO$_4$·5H$_2$O .........................................................................1.0mg
Pyridoxine ..............................................................................1.0mg
Biotin .....................................................................................0.5mg
CaCl$_2$·2H$_2$O...........................................................................0.5mg
Malachite Green.....................................................................0.25mg
Middlebrook OADC enrichment .........................................100.0mL
Glycerol ..................................................................................5.0mL

pH 6.6 ± 0.2 at 25°C

**Source:** This medium is available as a premixed powder from BD Diagnostic Systems.

**Middlebrook OADC Enrichment:**

**Composition** per 100.0mL:
Bovine albumin fraction V .......................................................5.0g
Glucose ....................................................................................2.0g
Catalase ................................................................................0.003g
Distilled water ....................................................................100.0mL

**Source:** This enrichment is available as a prepared enrichment from BD Diagnostic Systems.

**Preparation of Middlebrook OADC Enrichment:** Add components to distilled/deionized water and bring volume to 100.0mL. Mix thoroughly. Filter sterilize.

**Preparation of Medium:** Add glycerol to 900.0mL of distilled/deionized water and add remaining components, except Middlebrook OADC enrichment. Mix thoroughly. Gently heat and bring to boiling. Autoclave for 15 min at 15 psi pressure–121°C. Cool to 50°–55°C. Aseptically add 100.0mL of sterile Middlebrook OADC enrichment. Mix thoroughly. Pour into sterile Petri dishes or distribute into sterile tubes.

**Storage/Shelf Life:** Store dehydrated media in the dark in a sealed container below 30°C. Prepared media should be stored under refriger-

ation (2-8°C). Media should be used within 60 days of preparation. Media should not be used if there are any signs of deterioration (shrinking, cracking, or discoloration) or contamination, or if the expiration date supplied by the manufacturer has passed.

**Use:** For the isolation, cultivation, and maintenance of *Mycobacterium* species.

## Middlebrook Medium with Mycobactin

**Composition** per liter:

| | |
|---|---|
| Agar | 15.0g |
| Na$_2$HPO$_4$ | 1.5g |
| KH$_2$PO$_4$ | 1.5g |
| (NH$_4$)$_2$SO$_4$ | 0.5g |
| L-Glutamic acid | 0.5g |
| Sodium citrate | 0.4g |
| Ferric ammonium citrate | 0.04g |
| MgSO$_4$·7H$_2$O | 0.025g |
| Mycobactin J | 2.0mg |
| ZnSO$_4$·7H$_2$O | 1.0mg |
| CuSO$_4$·5H$_2$O | 1.0mg |
| Pyridoxine | 1.0mg |
| Biotin | 0.5mg |
| CaCl$_2$·2H$_2$O | 0.5mg |
| Malachite Green | 0.25mg |
| Middlebrook ADC enrichment | 100.0mL |
| Glycerol | 5.0mL |

pH 6.6 ± 0.2 at 25°C

**Source:** Mycobactin J is available from Allied Laboratories, Inc.

**Middlebrook ADC Enrichment:**

**Composition** per 100.0mL:

| | |
|---|---|
| Bovine albumin fraction V | 5.0g |
| Glucose | 2.0g |
| Catalase | 0.003g |
| Distilled water | 100.0mL |

**Source:** This medium and enrichment is available from BD Diagnostic Systems.

**Preparation of Middlebrook ADC Enrichment:** Add components to distilled/deionized water and bring volume to 100.0mL. Mix thoroughly. Filter sterilize.

**Preparation of Medium:** Add glycerol to 900.0mL of distilled/deionized water and add remaining components, except Middlebrook ADC enrichment and mycobactin. Mix thoroughly. Gently heat and bring to boiling. Autoclave for 15 min at 15 psi pressure–121°C. Cool to 50°–55°C. Aseptically add 100.0mL of sterile Middlebrook ADC enrichment and mycobactin. The mycobactin is dissolved in 2.0mL ethanol. Be sure to add all of the mycobactin; wash with additional 2.0mL ethanol if needed. Mix thoroughly. Pour into sterile Petri dishes or distribute into sterile tubes.

**Storage/Shelf Life:** Store dehydrated media in the dark in a sealed container below 30°C. Prepared media should be stored under refrigeration (2-8°C). Media should be used within 60 days of preparation. Media should not be used if there are any signs of deterioration (shrinking, cracking, or discoloration) or contamination, or if the expiration date supplied by the manufacturer has passed.

**Use:** For the cultivation of *Mycobacterium avium* subsp. *paratuberculosis.*

## Middlebrook OADC Enrichment (Middlebrook Oleic Albumin Dextrose Catalase Enrichment)

**Composition** per 100.0mL:

| | |
|---|---|
| Bovine albumin fraction V | 5.0g |
| Glucose | 2.0g |
| NaCl | 0.85g |
| Oleic acid | 0.05g |
| Catalase | 4.0mg |

**Source:** This enrichment is available as a prepared enrichment from BD Diagnostic Systems.

**Preparation of Enrichment:** Add components to distilled/deionized water and bring volume to 100.0mL. Mix thoroughly. Filter sterilize.

**Storage/Shelf Life:** Store dehydrated media in the dark in a sealed container below 30°C. Prepared media should be stored under refrigeration (2-8°C). Media should be used within 60 days of preparation. Media should not be used if there are any signs of deterioration (discoloration) or contamination, or if the expiration date supplied by the manufacturer has passed.

**Use:** For use as a supplement to other Middlebrook media for the isolation, cultivation, and maintenance of *Mycobacterium* species. Also used as a supplement to other Middlebrook media for determining the antimicrobial susceptibility of mycobacteria.

## Middlebrook OADC Enrichment with Triton™ WR 1339 (Middlebrook Oleic Albumin Dextrose Catalase Enrichment with Triton™ WR 1339)

**Composition** per 100.0mL:

| | |
|---|---|
| Bovine albumin fraction V | 5.0g |
| Glucose | 2.0g |
| NaCl | 0.85g |
| Triton™ WR 1339 | 0.25g |
| Oleic acid | 0.05g |
| Catalase | 4.0mg |

**Source:** This enrichment is available as a prepared enrichment from BD Diagnostic Systems.

**Preparation of Enrichment:** Add components to distilled/deionized water and bring volume to 100.0mL. Mix thoroughly. Filter sterilize.

**Storage/Shelf Life:** Store dehydrated media in the dark in a sealed container below 30°C. Prepared media should be stored under refrigeration (2-8°C). Media should be used within 60 days of preparation. Media should not be used if there are any signs of deterioration (discoloration) or contamination, or if the expiration date supplied by the manufacturer has passed.

**Use:** For use as a supplement to other Middlebrook media for the isolation, cultivation, and maintenance of *Mycobacterium* species. Also used as a supplement to other Middlebrook media for determining the antimicrobial susceptibility of mycobacteria.

## MIL Medium (Motility Indole Lysine Medium)

**Composition** per liter:

| | |
|---|---|
| Peptone | 10.0g |
| Pancreatic digest of casein | 10.0g |

L-Lysine·HCl ...............................................................10.0g
Yeast extract................................................................3.0g
Agar ............................................................................2.0g
Dextrose .....................................................................1.0g
Ferric ammonium citrate.............................................0.5g
Bromcresol Purple .....................................................0.02g

pH 6.6 ± 0.2 at 25°C

**Source:** This medium is available as a premixed powder and prepared medium from BD Diagnostic Systems.

**Preparation of Medium:** Add components to distilled/deionized water and bring volume to 1.0L. Mix thoroughly. Gently heat and bring to boiling. Distribute into tubes in 5.0mL volumes. Autoclave for 15 min at 15 psi pressure–121°C.

**Storage/Shelf Life:** Store dehydrated media in the dark in a sealed container below 30°C. Prepared media should be stored under refrigeration (2-8°C). Media should be used within 60 days of preparation. Media should not be used if there are any signs of deterioration (discoloration) or contamination, or if the expiration date supplied by the manufacturer has passed.

**Use:** For the cultivation and differentiation of members of the Enterobacteriaceae on the basis of motility, lysine decarboxylase activity, lysine deaminase activity, and indole production.

## Mitis Salivarius Agar

**Composition** per liter:

Sucrose.......................................................................50.0g
Agar ...........................................................................15.0g
Enzymatic digest of protein .......................................10.0g
Proteose peptone ........................................................10.0g
K₂HPO₄.......................................................................4.0g
Dextrose .....................................................................1.0g
Trypan Blue................................................................0.08g
Crystal Violet .............................................................0.8mg
Na₂TeO₃ solution ........................................................1.0mL

pH 7.0 ± 0.2 at 25°C

**Source:** This medium is available as a premixed powder from BD Diagnostic Systems.

**Na₂TeO₃ Solution:**
**Composition** per 10.0mL:
Na₂TeO₃ ......................................................................0.1g

**Preparation of Na₂TeO₃ Solution:** Add Na₂TeO₃ to 10.0mL of distilled/deionized water. Mix thoroughly. Filter sterilize.

**Caution:** Potassium tellurite is toxic.

**Preparation of Medium:** Add components to distilled/deionized water and bring volume to 999.0mL. Mix thoroughly. Gently heat and bring to boiling. Autoclave for 15 min at 15 psi pressure–121°C. Cool medium to 50°–55°C. Aseptically add 1.0mL of the sterile Na₂TeO₃ solution to the cooled basal medium. Mix thoroughly. Pour into sterile Petri dishes or distribute into sterile tubes.

**Storage/Shelf Life:** Store dehydrated media in the dark in a sealed container below 30°C. Prepared media should be stored under refrigeration (2-8°C). Media should be used within 60 days of preparation. Media should not be used if there are any signs of deterioration (shrinking, cracking, or discoloration) or contamination, or if the expiration date supplied by the manufacturer has passed.

**Use:** For the selective isolation of *Streptococcus mitis, Streptococcus salivarius,* and other viridans streptococci and enterococci.

## Mitis Salivarius HiVeg Agar Base with Tellurite

**Composition** per liter:

Sucrose.......................................................................50.0g
Agar ...........................................................................15.0g
Plant hydrolysate ........................................................15.0g
Plant peptone ..............................................................5.0g
K₂HPO₄.......................................................................4.0g
Glucose .......................................................................1.0g
Trypan Blue ................................................................0.075g
Crystal Violet ..............................................................0.8mg
Na₂TeO₃ solution .........................................................1.0mL

pH 7.0 ± 0.2 at 25°C

**Source:** This medium, without tellurite, is available as a premixed powder from HiMedia.

**Na₂TeO₃ Solution:**
**Composition** per 10.0mL:
Na₂TeO₃ ......................................................................0.1g

**Preparation of Na₂TeO₃ Solution:** Add Na₂TeO₃ to 10.0mL of distilled/deionized water. Mix thoroughly. Filter sterilize.

**Caution:** Potassium tellurite is toxic.

**Preparation of Medium:** Add components to distilled/deionized water and bring volume to 999.0mL. Mix thoroughly. Gently heat and bring to boiling. Autoclave for 15 min at 15 psi pressure–121°C. Cool medium to 50°–55°C. Aseptically add 1.0mL of the sterile Na₂TeO₃ solution to the cooled basal medium. Mix thoroughly. Pour into sterile Petri dishes or distribute into sterile tubes.

**Storage/Shelf Life:** Store dehydrated media in the dark in a sealed container below 30°C. Prepared media should be stored under refrigeration (2-8°C). Media should be used within 60 days of preparation. Media should not be used if there are any signs of deterioration (shrinking, cracking, or discoloration) or contamination, or if the expiration date supplied by the manufacturer has passed.

**Use:** For the selective isolation of *Streptococcus mitis, Streptococcus salivarius,* and other viridans streptococci and enterococci.

## MLCB Agar
## (Mannitol Lysine Crystal Violet-Brilliant Green Agar)

**Composition** per liter:

Agar ...........................................................................15.0g
Peptone ......................................................................10.0g
Yeast extract...............................................................5.0g
L-Lysine·HCl...............................................................5.0g
NaCl ...........................................................................4.0g
Na₂S₂O₃ .......................................................................4.0g
Mannitol......................................................................3.0g
Beef extract ................................................................2.0g
Ferric ammonium citrate .............................................1.0g
Crystal Violet ..............................................................0.01g
Brilliant Green ............................................................12.5mg

pH 6.8 ± 0.1 at 25°C

**Source:** This medium is available as a premixed powder from Oxoid Unipath.

**Preparation of Medium:** Add components to distilled/deionized water and bring volume to 1.0L. Mix thoroughly. Gently heat while stirring and bring to boiling. Do not autoclave. Cool to 50°C. Pour into sterile Petri dishes in 20.0mL volumes.

**Storage/Shelf Life:** Store dehydrated media in the dark in a sealed container below 30°C. Prepared media should be stored under refrigeration (2-8°C). Media should be used within 60 days of preparation. Media should not be used if there are any signs of deterioration (shrinking, cracking, or discoloration) or contamination, or if the expiration date supplied by the manufacturer has passed.

**Use:** For the selective isolation and cultivation of *Salmonella* species from fecal material and foods.

## MM10 Agar
## (Modified Medium 10 Agar)
**Composition** per liter:

| | |
|---|---|
| Base | 954.0mL |
| Dithiothreitol solution | 20.0mL |
| Sheep blood | 20.0mL |
| $Na_2CO_3$ solution | 5.0mL |
| Menadione solution | 1.0mL |

pH 7.2 ± 0.2 at 25°C

**Base**:
**Composition** per 954.0mL:

| | |
|---|---|
| Agar | 15.0g |
| Casein peptone | 2.0g |
| Glucose | 1.0g |
| Sodium formate | 1.0g |
| $KNO_3$ | 0.5g |
| Yeast extract | 0.5g |
| Hemin | 0.01g |
| Mineral salt solution 1 | 38.0mL |
| Mineral salt solution 2 | 38.0mL |
| Sodium lactate solution | 4.0mL |

**Preparation of Base:** Add components to distilled/deionized water and bring volume to 954.0mL. Mix thoroughly. Gently heat and bring to boiling. Autoclave for 15 min at 15 psi pressure–121°C. Cool to 45°–50°C.

**Mineral Salt Solution 1:**
**Composition** per 100.0mL:

| | |
|---|---|
| $K_2HPO_4$ | 0.6g |

**Preparation of Mineral Salt Solution 1:** Add $K_2HPO_4$ to distilled/deionized water and bring volume to 100.0mL. Mix thoroughly.

**Mineral Salt Solution 2:**
**Composition** per 100.0mL:

| | |
|---|---|
| NaCl | 1.2g |
| $(NH_4)_2SO_4$ | 1.2g |
| $KH_2PO_4$ | 0.6g |
| $CaCl_2$ | 0.12g |

**Preparation of Mineral Salt Solution 2:** Add components to distilled/deionized water and bring volume to 100.0mL. Mix thoroughly.

**Sodium Lactate Solution:**
**Composition** per 100.0mL:

| | |
|---|---|
| Sodium lactate | 60.0g |

**Preparation of Sodium Lactate Solution:** Add sodium lactate to distilled/deionized water and bring volume to 100.0mL. Mix thoroughly.

**Dithiothreitol Solution:**
**Composition** per 100.0mL:

| | |
|---|---|
| Dithiothreitol | 1.0g |

**Preparation of Dithiothreitol Solution:** Add dithiothreitol to distilled/deionized water and bring volume to 100.0mL. Mix thoroughly. Filter sterilize.

**Menadione Solution:**
**Composition** per 100.0mL:

| | |
|---|---|
| Vitamin $K_1$ (phytomenadione) | 0.05g |
| Ethanol (95% solution) | 100.0mL |

**Preparation of Menadione Solution:** Add vitamin $K_1$ to 100.0mL of ethanol. Mix thoroughly. Filter sterilize.

**$Na_2CO_3$ Solution:**
**Composition** per 100.0mL:

| | |
|---|---|
| $Na_2CO_3$ | 8.0g |

**Preparation of $Na_2CO_3$ Solution:** Add $Na_2CO_3$ to distilled/deionized water and bring volume to 100.0mL. Mix thoroughly. Filter sterilize.

**Preparation of Medium:** To 954.0mL of sterile cooled base, aseptically add 20.0mL of sterile dithiothreitol solution, 20.0mL of sterile, defibrinated sheep blood, 5.0mL of sterile $Na_2CO_3$ solution, and 1.0mL of sterile menadione solution. Mix thoroughly. Pour into sterile Petri dishes or distribute into sterile screw-capped tubes.

**Storage/Shelf Life:** Store dehydrated media in the dark in a sealed container below 30°C. Prepared media should be stored under refrigeration (2-8°C). Media should be used within 60 days of preparation. Media should not be used if there are any signs of deterioration (shrinking, cracking, or discoloration) or contamination, or if the expiration date supplied by the manufacturer has passed.

**Use:** For the isolation and quantitation of plaque bacteria, especially *Streptococcus mutans, Streptococcus sanguis,* and *Streptococcus salivarius.*

## Modified Bile Esculin Azide Agar
**Composition** per liter:

| | |
|---|---|
| Casein enzymic hydrolysate | 17.0g |
| Agar | 13.5g |
| Oxgall | 10.0g |
| Yeast extract | 5.0g |
| NaCl | 5.0g |
| Peptic digest of animal tissue | 3.0g |
| Sodium citrate | 1.0g |
| Esculin | 1.0g |
| Ferric ammonium citrate | 0.5g |
| $NaN_3$ | 0.25g |

pH 7.1 ± 0.2 at 25°C

**Source:** This medium is available from HiMedia.

**Caution:** Sodium azide has a tendency to form explosive metal azides with plumbing materials. It is advisable to use enough water to flush off the disposables.

**Preparation of Medium:** Add components to distilled/deionized water and bring volume to 1.0L. Mix thoroughly. Gently heat and bring to boiling. Distribute into tubes or flasks. Autoclave for 15 min at 15 psi pressure–121°C. Pour into sterile Petri dishes or leave in tubes.

**Storage/Shelf Life:** Store dehydrated media in the dark in a sealed container below 30°C. Prepared media should be stored under refrigeration (2-8°C). Media should be used within 60 days of preparation. Media should not be used if there are any signs of deterioration (shrinking, cracking, or discoloration) or contamination, or if the expiration date supplied by the manufacturer has passed.

**Use:** For the selective isolation and enumeration of group D streptococci.

## Modified Buffered Charcoal HiVeg Agar Base with Cysteine

**Composition** per liter:

Agar ..................................................................................17.0g
ACES buffer.......................................................................10.0g
Plant peptone No. 3.............................................................10.0g
Charcoal, activated...............................................................2.0g
α-Ketoglutarate monopotassium salt............................................1.0g
L-Cysteine solution ............................................................4.0mL

pH 6.9 ± 0.2 at 25°C

**Source:** This medium, without L-cysteine solution, is available as a premixed powder from HiMedia.

**L-Cysteine Solution:**

**Composition** per 10.0mL:

L-cysteine·HCl·H$_2$O............................................................0.4g

**L-Cysteine Solution:** Add L-cysteine·HCl·H$_2$O to distilled/deionized water and bring volume to 10.0mL. Mix thoroughly. Filter sterilize.

**Preparation of Medium:** Add components, except L-cysteine solution, to distilled/deionized water and bring volume to 1.0L. Mix thoroughly. Adjust medium to pH 6.9 with 1N KOH. Heat gently and bring to boiling for 1 min. Autoclave for 15 min at 15 psi pressure–121°C. Cool to 50°–55°C. Aseptically add 4.0mL of L-cysteine solution. Mix thoroughly. Pour into sterile Petri dishes with constant agitation to keep charcoal in suspension.

**Storage/Shelf Life:** Store dehydrated media in the dark in a sealed container below 30°C. Prepared media should be stored under refrigeration (2-8°C). Media should be used within 60 days of preparation. Media should not be used if there are any signs of deterioration (shrinking, cracking, or discoloration) or contamination, or if the expiration date supplied by the manufacturer has passed.

**Use:** For the isolation, cultivation, and maintenance of *Legionella pneumophila* and other *Legionella* species from environmental and clinical specimens.

## Modified *Campylobacter* Blood-Free Selective Agar Base (Modified *Campylobacter* Charcoal Differential Agar) (Modified CCDA) (BAM M30a)

**Composition** per 1012.0mL:

Agar ..................................................................................12.0g
Beef extract.......................................................................10.0g
Peptone..............................................................................10.0g
NaCl....................................................................................5.0g
Charcoal..............................................................................4.0g
Casein hydrolysate...............................................................3.0g
Yeast extract........................................................................2.0g
Sodium deoxycholate.............................................................1.0g
FeSO$_4$..............................................................................0.25g
Sodium pyruvate..................................................................0.25g
Cefoperazone solution ........................................................4.0mL

Amphotericin B solution.......................................................4.0mL
Rifampicin solution ............................................................4.0mL

pH 7.4 ± 0.2 at 25°C

**Cefoperazone Solution:**

**Composition** per 10.0mL:

Cefoperazone...................................................................0.037g

**Preparation of Cefoperazone Solution:** Add cefoperazone to distilled/deionized water and bring volume to 10.0mL. Mix thoroughly. Filter sterilize.

**Rifampicin Solution:**

**Composition** per 100.0mL:

Rifampicin .......................................................................0.25g
Ethanol, absolute................................................................50.0mL

**Preparation of Rifampicin Solution:** Add rifampicin to 50.0mL of ethanol. Mix thoroughly. Bring volume to 100.0mL with distilled/deionized water. Filter sterilize.

**Amphotericin B Solution:**

**Composition** per 10.0mL:

Amphotericin B.................................................................0.005g

**Preparation of Amphotericin B Solution:** Add amphotericin B to distilled/deionized water and bring volume to 10.0mL. Mix thoroughly. Filter sterilize. Can be stored for 1 year at –20°C.

**Preparation of Medium:** Add components, except cefoperazone solution, amphotericin B solution, and rifampicin solution, to distilled/deionized water and bring volume to 1.0L. Mix thoroughly. Gently heat and bring to boiling. Autoclave for 15 min at 15 psi pressure–121°C. Cool to 45°–50°C. Aseptically add 10.0mL of sterile cefoperazone solution, 10.0mL of sterile amphotericin B solution, and 10.0mL of sterile rifampicin solution. Mix thoroughly. Pour into sterile Petri dishes or distribute into sterile tubes.

**Storage/Shelf Life:** Store dehydrated media in the dark in a sealed container below 30°C. Prepared media should be stored under refrigeration (2-8°C). Media should be used within 60 days of preparation. Media should not be used if there are any signs of deterioration (shrinking, cracking, or discoloration) or contamination, or if the expiration date supplied by the manufacturer has passed.

**Use:** For the cultivation of *Campylobacter* species. For the recovery of injured *Campylobacter* spp. from foods.

## Modified CPLM HiVeg Medium Base with Horse Serum, Penicillin, Streptomycin, and Nystatin (*Trichomonas* Modified CPLM HiVeg Medium Base)

**Composition** per liter:

Plant peptone ....................................................................32.0g
Liver digest.......................................................................20.0g
L-Cysteine·HCl.....................................................................2.4g
Maltose................................................................................1.6g
Ringer's salt solution, 1/4X ...................................................1.0L
Horse serum.....................................................................100.0mL
Penicillin-streptomycin solution ..........................................10.0mL
Nystatin solution.................................................................10.0mL

pH 6.0 ± 0.2 at 25°C

**Ringer's Salt Solution, 1/4X:**

**Composition** per 400.0mL:

NaCl....................................................................................9.0g
KCl....................................................................................0.042g
CaCl$_2$..............................................................................0.024g

**Preparation of Ringer's Salt Solution, 1/4X:** Add components to distilled/deionized water and bring volume to 400.0mL. Mix thoroughly.

**Penicllin-Streptomycin Solution:**
**Composition** per 10.0mL:
Streptomycin ................................................................0.1g
Penicillin ...........................................................1,000,000U

**Preparation of Penicllin-Streptomycin Solution:** Add components to distilled/deionized water and bring volume to 10.0mL. Mix thoroughly. Filter sterilize

**Nystatin Solution:**
**Composition** per 10.0mL:
Nystatin ...............................................................50,000U

**Preparation of Nystatin Solution:** Add nystatin to distilled/deionized water and bring volume to 10.0mL. Mix thoroughly. Filter sterilize

**Preparation of Medium:** Add components, except horse serum, penicillin-streptomycin solution, and nystatin solution, to 1.0L of Ringer's salt solution, 1/4X. Mix thoroughly. Adjust pH to 6.0. Gently heat and bring to boiling. Autoclave for 10 min at 15 psi pressure–121°C. Cool to 25°C. Aseptically add 100.0mL of sterile, heat-inactivated horse serum, 10.0mL sterile penicllin-streptomycin solution, and 10.0mL sterile nystatin solution. Mix thoroughly. Aseptically distribute into sterile, screw-capped tubes or flasks.

**Storage/Shelf Life:** Store dehydrated media in the dark in a sealed container below 30°C. Prepared media should be stored under refrigeration (2-8°C). Media should be used within 60 days of preparation. Media should not be used if there are any signs of deterioration (shrinking, cracking, or discoloration) or contamination, or if the expiration date supplied by the manufacturer has passed.

**Use:** For the selective cultivation of *Trichomonas vaginalis*.

## Modified Differential Clostridial Broth

**Composition** per liter:
Meat extract .................................................................8.0g
Casein enzymic hydrolysate ...........................................5.0g
Meat peptone................................................................5.0g
Sodium acetate .............................................................5.0g
Yeast extract ................................................................1.0g
Starch .........................................................................1.0g
Glucose .......................................................................1.0g
L-Cysteine hydrochloride...............................................0.5g
NaHSO₃ ......................................................................0.5g
Ammonium ferric citrate ................................................0.5g
Resazurin ..................................................................0.002g
pH 7.2 ± 0.2 at 25°C

**Source:** This medium is available from HiMedia.

**Preparation of Medium:** Add components to distilled/deionized water and bring volume to 1.0L. Mix thoroughly. Gently heat and bring to boiling. Distribute into tubes or flasks. Autoclave for 15 min at 15 psi pressure–121°C.

**Storage/Shelf Life:** Store dehydrated media in the dark in a sealed container below 30°C. Prepared media should be stored under refrigeration (2-8°C). Media should be used within 60 days of preparation. Media should not be used if there are any signs of deterioration (discoloration) or contamination, or if the expiration date supplied by the manufacturer has passed.

**Use:** For the detection of *Clostridium* spp. from foods by the MPN technique.

## Modified Letheen HiVeg Agar
### (Letheen HiVeg Agar, Modified)

**Composition** per liter:
Agar ..........................................................................20.0g
Plant peptone ..............................................................20.0g
Plant extract .................................................................5.0g
Plant hydrolysate ..........................................................5.0g
NaCl............................................................................5.0g
Polysorbate 80 ..............................................................5.0g
Yeast extract ................................................................2.0g
Lecithin .......................................................................0.7g
NaHSO₃ .......................................................................0.1g
pH 7.2 ± 0.2 at 25°C

**Source:** This medium is available as a premixed powder from Hi-Media.

**Preparation of Medium:** Add components to distilled/deionized water and bring volume to 1.0L. Mix thoroughly. Gently heat and bring to boiling. Distribute into tubes or flasks. Autoclave for 15 min at 15 psi pressure–121°C. Pour into sterile Petri dishes.

**Storage/Shelf Life:** Store dehydrated media in the dark in a sealed container below 30°C. Prepared media should be stored under refrigeration (2-8°C). Media should be used within 60 days of preparation. Media should not be used if there are any signs of deterioration (shrinking, cracking, or discoloration) or contamination, or if the expiration date supplied by the manufacturer has passed.

**Use:** For screening cosmetic products for microbial contamination.

## Modified Letheen HiVeg Broth
### (Letheen HiVeg Broth, Modified)

**Composition** per liter:
Plant peptone ..............................................................20.0g
Plant extract .................................................................5.0g
Plant hydrolysate ..........................................................5.0g
NaCl............................................................................5.0g
Polysorbate 80 ..............................................................5.0g
Yeast extract ................................................................2.0g
Lecithin .......................................................................0.7g
NaHSO₃ .......................................................................0.1g
pH 7.0 ± 0.2 at 25°C

**Source:** This medium is available as a premixed powder from Hi-Media.

**Preparation of Medium:** Add components to distilled/deionized water and bring volume to 1.0L. Mix thoroughly. Gently heat and bring to boiling. Distribute into tubes or flasks. Autoclave for 15 min at 15 psi pressure–121°C.

**Storage/Shelf Life:** Store dehydrated media in the dark in a sealed container below 30°C. Prepared media should be stored under refrigeration (2-8°C). Media should be used within 60 days of preparation. Media should not be used if there are any signs of deterioration (discoloration) or contamination, or if the expiration date supplied by the manufacturer has passed.

**Use:** For the screening of cosmetic products for microbial contamination.

## Modified McBride *Listeria* HiVeg Agar Base with Blood and Cycloheximide

**Composition** per liter:

| | |
|---|---|
| Agar | 15.0g |
| Glycine anhydride | 10.0g |
| Plant hydrolysate | 5.0g |
| Plant peptone | 5.0g |
| NaCl | 5.0g |
| Plant extract | 3.0g |
| Phenyl ethanol | 2.5g |
| LiCl | 0.5g |
| Sheep blood, defibrinated | 50.0mL |
| Selective supplement solution | 10.0mL |

pH 7.3 ± 0.2 at 25°C

**Source:** This medium, without blood or selective supplement solution, is available as a premixed powder from HiMedia.

**Caution:** LiCl is harmful. Avoid bodily contact and inhalation of vapors. On contact with skin wash with plenty of water immediately.

**Caution:** Cycloheximide is toxic. Avoid skin contact or aerosol formation and inhalation.

**Selective Supplement Solution:**

**Composition** per 10.0mL:

| | |
|---|---|
| Cycloheximide | 0.2g |

**Preparation of Selective Supplement Solution:** Add cycloheximide to distilled/deionized water and bring volume to 10.0mL. Mix thoroughly. Filter sterilize.

**Preparation of Medium:** Add components, except blood and selective supplement solution, to distilled/deionized water and bring volume to 940.0mL. Mix thoroughly. Gently heat and bring to boiling. Distribute into tubes or flasks. Autoclave for 15 min at 15 psi pressure–121°C. Cool to 50°C. Aseptically add 50.0mL defibrinated blood and 10.0mL selective supplement solution. Mix thoroughly. Pour into sterile Petri dishes or leave in tubes.

**Storage/Shelf Life:** Store dehydrated media in the dark in a sealed container below 30°C. Prepared media should be stored under refrigeration (2-8°C). Media should be used within 60 days of preparation. Media should not be used if there are any signs of deterioration (shrinking, cracking, or discoloration) or contamination, or if the expiration date supplied by the manufacturer has passed.

**Use:** For the selective isolation of *Listeria monocytogenes* from clinical and nonclinical specimens containing mixed flora.

## Modified Oxford *Listeria* Selective Agar (Oxford Agar, Modified) (*Listeria* Selective Agar, Modified Oxford) (MOX Agar) (BAM M103a)

**Composition** per 1002.0mL:

| | |
|---|---|
| Special peptone | 23.0g |
| LiCl | 15.0g |
| Agar | 12.0g |
| NaCl | 5.0g |
| Cornstarch | 1.0g |
| Esculin | 1.0g |
| Ferric ammonium citrate | 0.5g |

| | |
|---|---|
| Buffered colistin methane sulfonate solution | 1.0mL |
| Buffered moxalactam solution | 2.0mL |

pH 7.2 ± 0.2 at 25°C

**Buffered Colistin Methane Sulfonate Solution:**

**Composition** per 100.0mL:

| | |
|---|---|
| Colistin methane sulfonate | 1.0g |
| Potassium phosphate buffer, 0.1*M*. pH 6.0 | 100.0mL |

**Preparation of Buffered Colistin Methane Sulfonate Solution:** Add colistin methane sulfonate to 100.0mL 0.1*M* potassium phosphate buffer. Mix thoroughly. Adjust pH to 6.0. Filter sterilize. Store at –20°C.

**Buffered Moxalactam Solution:**

**Composition** per 100.0mL:

| | |
|---|---|
| Sodium or ammonium moxalactam | 1.0g |

**Preparation of Buffered Moxalactam Solution:** Add sodium or ammonium moxalactam to distilled/deionized water and bring volume to 100.0mL. Mix thoroughly. Filter sterilize. Store at –20°C.

**Preparation of Medium:** Gradually add components, except buffered moxalactam solution, to distilled/deionized water and bring volume to 1.0L. Mix thoroughly. Adjust pH to 7.2. Gently heat and bring to boiling. Autoclave for 10 min at 15 psi pressure–121°C. Cool quickly to 46°C. Aseptically add 2.0mL of sterile moxalactam solution. Mix thoroughly. Pour into sterile Petri dishes or distribute into sterile tubes.

**Storage/Shelf Life:** Store dehydrated media in the dark in a sealed container below 30°C. Prepared media should be stored under refrigeration (2-8°C). Media should be used within 60 days of preparation. Media should not be used if there are any signs of deterioration (shrinking, cracking, or discoloration) or contamination, or if the expiration date supplied by the manufacturer has passed.

**Use:** For the isolation and cultivation of *Listeria monocytogenes* from specimens containing a mixed bacterial flora.

## Modified Rappaport Vassiliadis HiVeg Medium

**Composition** per liter:

| | |
|---|---|
| MgCl₂·6H₂O | 40.0g |
| NaCl | 8.0g |
| Papaic digest of soybean meal | 5.0g |
| KH₂PO₄ | 1.6 |
| Malachite Green | 0.04g |

pH 5.2 ± 0.2 at 25°C

**Source:** This medium is available as a premixed powder from HiMedia.

**Preparation of Medium:** Add components to distilled/deionized water and bring volume to 1.0L. Mix thoroughly. Gently heat and bring to boiling. Distribute into tubes or flasks. Autoclave for 15 min at 10 psi pressure–115°C.

**Storage/Shelf Life:** Store dehydrated media in the dark in a sealed container below 30°C. Prepared media should be stored under refrigeration (2-8°C). Media should be used within 60 days of preparation. Media should not be used if there are any signs of deterioration (discoloration) or contamination, or if the expiration date supplied by the manufacturer has passed.

**Use:** For the selective enrichment of *Salmonella* species from food and environmental specimens.

## Modified Semisolid Rappaport Vassiliadis Medium (MSRV Medium)

**Composition** per liter:

| | |
|---|---|
| MgCl₂, anhydrous | 10.93g |

MgCl$_2$, anhydrous ..................................................................10.93g
NaCl .............................................................................................7.34g
Casein hydrolysate .....................................................................4.59g
Tryptose ......................................................................................4.59g
Agar ..............................................................................................2.7g
KH$_2$PO$_4$ .....................................................................................1.47g
Malachite Green oxalate ...........................................................0.037g
Novobiocin solution ................................................................10.0mL

pH 5.2 ± 0.2 at 25°C

**Source:** This medium is available as a premixed powder from Oxoid Unipath.

**Novobiocin Solution:**
**Composition** per 10.0mL:
Novobiocin ....................................................................................0.02g

**Preparation of Novobiocin Solution:** Add novobiocin to 10.0mL of distilled/deionized water. Mix thoroughly. Filter sterilize.

**Preparation of Medium:** Add components, except novobiocin solution, to distilled/deionized water and bring volume to 990.0mL. Mix thoroughly. Gently heat to boiling. Do not autoclave. Cool to 45°–50°C. Aseptically add 10.0mL of sterile novobiocin solution. Mix thoroughly. Pour into sterile Petri dishes. Air-dry plates for at least 1 h.

**Storage/Shelf Life:** Store dehydrated media in the dark in a sealed container below 30°C. Prepared media should be stored under refrigeration (2-8°C). Media should be used within 60 days of preparation. Media should not be used if there are any signs of deterioration (shrinking, cracking, or discoloration) or contamination, or if the expiration date supplied by the manufacturer has passed.

**Use:** For the isolation and cultivation of motile *Salmonella* species from food and environmental samples.

## Modified Soyabean Bile Broth Base

**Composition** per liter:
Casein enzymic hydrolysate .........................................................17.0g
NaCl ..............................................................................................5.0g
K$_2$HPO$_4$ .......................................................................................4.0g
Papaic digest of soybean meal .....................................................3.0g
D-Glucose......................................................................................2.5g
Bile salts mixture .........................................................................1.5g
Selective supplement solution ................................................10.0mL

pH 7.3 ± 0.2 at 25°C

**Source:** This medium is available from HiMedia.

**Selective Supplement Solution:**
**Composition** per 10.0mL:
Novobiocin....................................................................................10.0mg

**Preparation of Selective Supplement Solution:** Add novobiocin to distilled/deionized water and bring volume to 10.0mL. Mix thoroughly. Filter sterilize.

**Preparation of Medium:** Add components, except selective supplement solution, to distilled/deionized water and bring volume to 990.0mL. Mix thoroughly. Autoclave for 15 min at 15 psi pressure–121°C. Cool to 50°C. Aseptically add selective supplement solution. Mix thoroughly. Pour into Petri dishes or aseptically distribute into sterile tubes.

**Storage/Shelf Life:** Store dehydrated media in the dark in a sealed container below 30°C. Prepared media should be stored under refrigeration (2-8°C). Media should be used within 60 days of preparation. Media should not be used if there are any signs of deterioration (discoloration) or contamination, or if the expiration date supplied by the manufacturer has passed.

**Use:** For the detection of *Escherichia coli* O157:H7 from food.

## Moeller Decarboxylase Broth

**Composition** per liter:
Amino acid....................................................................................10.0g
Peptic digest of animal tissue .......................................................5.0g
Beef extract...................................................................................5.0g
Glucose..........................................................................................0.5g
Bromcresol Purple .......................................................................0.01g
Cresol Red ..................................................................................5.0mg
Pyridoxal....................................................................................5.0mg

pH 6.0 ± 0.2 at 25°C

**Source:** This medium is available as a premixed powder from BD Diagnostic Systems.

**Preparation of Medium:** Add components to distilled/deionized water and bring volume to 1.0L. Use L-lysine, L-arginine, or L-ornithine. Mix thoroughly. Gently heat until dissolved. Distribute into screw-capped tubes in 5.0mL volumes. Autoclave for 15 min at 15 psi pressure–121°C. A slight precipitate may form in the ornithine broth.

**Storage/Shelf Life:** Store dehydrated media in the dark in a sealed container below 30°C. Prepared media should be stored under refrigeration (2-8°C). Media should be used within 60 days of preparation. Media should not be used if there are any signs of deterioration (discoloration) or contamination, or if the expiration date supplied by the manufacturer has passed.

**Use:** For the differentiation of Gram-negative enteric bacteria based on the production of arginine dihydrolase, lysine decarboxylase, or ornithine decarboxylase.

## Moeller Decarboxylase HiVeg Broth Base (Decarboxylase Broth Base, Moeller)

**Composition** per liter:
Plant extract .................................................................................5.0g
Plant peptone ................................................................................5.0g
Glucose..........................................................................................0.5g
Bromcresol Purple .......................................................................0.01g
Pyridoxal....................................................................................5.0mg
Cresol Red ..................................................................................5.0mg

pH 6.0 ± 0.2 at 25°C

**Source:** This medium is available as a premixed powder from HiMedia.

**Preparation of Medium:** Add components to distilled/deionized water and bring volume to 1.0L. Mix thoroughly. Gently heat until dissolved. Distribute into screw-capped tubes in 5.0mL volumes. Autoclave for 15 min at 15 psi pressure–121°C.

**Storage/Shelf Life:** Store dehydrated media in the dark in a sealed container below 30°C. Prepared media should be stored under refrigeration (2-8°C). Media should be used within 60 days of preparation. Media should not be used if there are any signs of deterioration (discoloration) or contamination, or if the expiration date supplied by the manufacturer has passed.

**Use:** With the addition of amino acid solutions, this medium is used for the differentiation of Gram-negative enteric bacteria based on the production of amino acid decarboxylation reactions.

## Moeller Decarboxylase HiVeg Broth Arginine HCl
### (Decarboxylase Broth Base, Moeller with Arginine)

**Composition** per liter:

L-Arginine hydrochloride.............................................10.0g
Plant extract ..............................................................5.0g
Plant peptone.............................................................5.0g
Glucose ......................................................................0.5g
Bromcresol Purple .....................................................0.01g
Pyridoxal...................................................................5.0mg
Cresol Red.................................................................5.0mg

pH 6.0 ± 0.2 at 25°C

**Source:** This medium is available as a premixed powder from Hi-Media.

**Preparation of Medium:** Add components to distilled/deionized water and bring volume to 1.0L. Mix thoroughly. Gently heat until dissolved. Distribute into screw-capped tubes in 5.0mL volumes. Autoclave for 15 min at 15 psi pressure–121°C.

**Storage/Shelf Life:** Store dehydrated media in the dark in a sealed container below 30°C. Prepared media should be stored under refrigeration (2-8°C). Media should be used within 60 days of preparation. Media should not be used if there are any signs of deterioration (discoloration) or contamination, or if the expiration date supplied by the manufacturer has passed.

**Use:** For the differentiation of Gram-negative enteric bacteria based on the production of arginine dihydrolase.

## Moeller Decarboxylase HiVeg Broth with Lysine HCl
### (Decarboxylase Broth Base, Moeller with Lysine)

**Composition** per liter:

L-Lysine hydrochloride ...............................................10.0g
Plant extract ..............................................................5.0g
Plant peptone.............................................................5.0g
Glucose ......................................................................0.5g
Bromcresol Purple .....................................................0.01g
Pyridoxal...................................................................5.0mg
Cresol Red.................................................................5.0mg

pH 6.0 ± 0.2 at 25°C

**Source:** This medium is available as a premixed powder from Hi-Media.

**Preparation of Medium:** Add components to distilled/deionized water and bring volume to 1.0L. Mix thoroughly. Gently heat until dissolved. Distribute into screw-capped tubes in 5.0mL volumes. Autoclave for 15 min at 15 psi pressure–121°C.

**Storage/Shelf Life:** Store dehydrated media in the dark in a sealed container below 30°C. Prepared media should be stored under refrigeration (2-8°C). Media should be used within 60 days of preparation. Media should not be used if there are any signs of deterioration (discoloration) or contamination, or if the expiration date supplied by the manufacturer has passed.

**Use:** For the differentiation of Gram-negative enteric bacteria based on the production of lysine decarboxylase.

## Moeller Decarboxylase HiVeg Broth with Ornithine HCl

**Composition** per liter:

L-Ornithine hydrochloride ..........................................10.0g
Plant extract ..............................................................5.0g
Plant peptone.............................................................5.0g
Glucose ......................................................................0.5g
Bromcresol Purple .....................................................0.01g
Pyridoxal...................................................................5.0mg
Cresol Red.................................................................5.0mg

pH 6.0 ± 0.2 at 25°C

**Source:** This medium is available as a premixed powder from Hi-Media.

**Preparation of Medium:** Add components to distilled/deionized water and bring volume to 1.0L. Mix thoroughly. Gently heat until dissolved. Distribute into screw-capped tubes in 5.0mL volumes. Autoclave for 15 min at 15 psi pressure–121°C.

**Storage/Shelf Life:** Store dehydrated media in the dark in a sealed container below 30°C. Prepared media should be stored under refrigeration (2-8°C). Media should be used within 60 days of preparation. Media should not be used if there are any signs of deterioration (discoloration) or contamination, or if the expiration date supplied by the manufacturer has passed.

**Use:** For the differentiation of Gram-negative enteric bacteria based on the production of ornithine decarboxylase.

## Moeller KCN Broth Base

**Composition** per liter:

$Na_2HPO_4$.................................................................5.64g
NaCl...........................................................................5.0g
Pancreatic digest of casein...........................................1.5g
Peptic digest of animal tissue .......................................1.5g
$KH_2PO_4$..................................................................0.225g
KCN solution ............................................................. 0.15mL

pH 7.6 ± 0.2 at 25°C

**Source:** This medium is available as a premixed powder from BD Diagnostic Systems.

**KCN Solution:**
**Composition** per 100.0mL:

KCN..........................................................................0.5g

**Preparation of KCN Solution:** Add KCN to 100.0mL of cold distilled/deionized water. Mix thoroughly and cap. Do not mouth pipette.

**Caution:** Cyanide is toxic.

**Preparation of Medium:** Add components, except KCN solution, to distilled/deionized water and bring volume to 1.0L. Mix thoroughly. Autoclave for 15 min at 15 psi pressure–121°C. Cool to room temperature. Prior to use, add 0.15mL of KCN solution. Mix thoroughly. Aseptically distribute into sterile tubes.

**Storage/Shelf Life:** Store dehydrated media in the dark in a sealed container below 30°C. Prepared media should be stored under refrigeration (2-8°C). Media should be used within 60 days of preparation. Media should not be used if there are any signs of deterioration (discoloration) or contamination, or if the expiration date supplied by the manufacturer has passed.

**Use:** For the differentiation of Gram-negative enteric bacteria on the basis of their ability to grow in the presence of cyanide.

## Monsur Agar
### (Taurocholate Tellurite Gelatin Agar)

**Composition** per liter:

| | |
|---|---|
| Gelatin | 30.0g |
| Agar | 15.0g |
| Casein peptone | 10.0g |
| NaCl | 10.0g |
| Sodium taurocholate | 5.0g |
| $Na_2CO_3 \cdot H_2O$ | 1.0g |
| $K_2TeO_3$ solution | 10.0mL |

pH 8.5 ± 0.2 at 25°C

**$K_2TeO_3$ Solution:**

**Composition** per 10.0mL:

| | |
|---|---|
| $K_2TeO_3$ | 0.02g |

**Preparation of $K_2TeO_3$ Solution:** Add $K_2TeO_3$ to 10.0mL of distilled/deionized water. Mix thoroughly. Filter sterilize.

**Caution:** Potassium tellurite is toxic.

**Preparation of Medium:** Add components, except $K_2TeO_3$ solution, to distilled/deionized water and bring volume to 990.0mL. Mix thoroughly. Gently heat and bring to boiling. Autoclave for 15 min at 15 psi pressure–121°C. Cool to 45°–50°C. Add 10.0mL of sterile $K_2TeO_3$ solution. Mix thoroughly. Pour into sterile Petri dishes or distribute into sterile tubes.

**Storage/Shelf Life:** Store dehydrated media in the dark in a sealed container below 30°C. Prepared media should be stored under refrigeration (2-8°C). Media should be used within 60 days of preparation. Media should not be used if there are any signs of deterioration (shrinking, cracking, or discoloration) or contamination, or if the expiration date supplied by the manufacturer has passed.

**Use:** For the isolation of *Vibrio cholerae* from fecal specimens.

## MOX HiVeg Agar

**Composition** per liter:

| | |
|---|---|
| Agar | 15.0g |
| Plant hydrolysate | 15.0g |
| Papaic digest of soybean meal | 5.0g |
| NaCl | 5.0g |
| $MnCl_2$ | 4.067g |
| Sodium oxalate | 2.68g |

pH 7.5 ± 0.2 at 25°C

**Source:** This medium is available as a premixed powder from Hi-Media.

**Preparation of Medium:** Add components to distilled/deionized water and bring volume to 1.0L. Mix thoroughly. Gently heat and bring to boiling. Distribute into tubes or flasks. Autoclave for 15 min at 15 psi pressure–121°C. Pour into sterile Petri dishes or leave in tubes.

**Storage/Shelf Life:** Store dehydrated media in the dark in a sealed container below 30°C. Prepared media should be stored under refrigeration (2-8°C). Media should be used within 60 days of preparation. Media should not be used if there are any signs of deterioration (shrinking, cracking, or discoloration) or contamination, or if the expiration date supplied by the manufacturer has passed.

**Use:** For the cultivation of *Yersinia enterocolitica* from foods.

## MRS Agar
### (DeMan, Rogosa, Sharpe Agar)

**Composition** per liter:

| | |
|---|---|
| Glucose | 18.5g |
| Agar | 13.5g |
| Pancreatic digest of gelatin | 10.0g |
| Beef extract | 8.0g |
| Yeast extract | 4.0g |
| Sodium acetate | 3.0g |
| $K_2HPO_4$ | 2.0g |
| Ammonium citrate | 2.0g |
| Polysorbate 80 | 1.0g |
| $MgSO_4 \cdot 7H_2O$ | 0.2g |
| $MnSO_4 \cdot 4H_2O$ | 0.05g |

pH 6.2 ± 0.2 at 25°C

**Source:** This medium is available as a premixed powder from BD Diagnostic Systems.

**Preparation of Medium:** Add components to distilled/deionized water and bring volume to 1.0L. Mix thoroughly. Gently heat while stirring and bring to boiling. Distribute into tubes or flasks. Autoclave for 15 min at 15 psi pressure–121°C. Pour into sterile Petri dishes or leave in tubes.

**Storage/Shelf Life:** Store dehydrated media in the dark in a sealed container below 30°C. Prepared media should be stored under refrigeration (2-8°C). Media should be used within 60 days of preparation. Media should not be used if there are any signs of deterioration (shrinking, cracking, or discoloration) or contamination, or if the expiration date supplied by the manufacturer has passed.

**Use:** For the isolation and cultivation of *Lactobacillus* species from clinical specimens, foods, and dairy products.

## MRSA Agar

**Composition** per liter:

| | |
|---|---|
| Growth Factors | 78.0g |
| Agar | 12.5g |
| Peptones | 11.0g |
| Chromogenic substrate | 1.9g |
| Cefoxitin supplement | 10.0mL |

pH 7.2 ± 0.2 at 25°C

**Source:** This medium is available from CONDA, Barcelona, Spain.

**Cefoxitin Supplement:**
**Composition** per 10.0mL:

| | |
|---|---|
| Cefoxitin | 4.0mg |

**Preparation of Cefoxitin Supplement:** Add cefoxitin to distilled/deionized water and bring volume to 10.0mL. Mix thoroughly. Filter sterilize.

**Preparation of Medium:** Add components, except cefoxitin supplement, to distilled/deionized water and bring volume to 990.0mL. Mix thoroughly. Gently heat to boiling. Autoclave for 15 min at 15 psi pressure–121°C. Cool to 45°–50°C. Aseptically add 10.0mL of sterile cefoxitin supplement. Mix thoroughly. Pour into sterile Petri dishes or distribute into sterile tubes.

**Storage/Shelf Life:** Store dehydrated media in the dark in a sealed container below 30°C. Prepared media should be stored under refrigeration (2-8°C). Media should be used within 60 days of preparation. Media should not be used if there are any signs of deterioration (shrinking, cracking, or discoloration) or contamination, or if the expiration date supplied by the manufacturer has passed.

**Use:** For the detection of methicillin resistant *Staphylococcus aureus* (MRSA) in clinical specimens.

## MRSASelect™

**Composition** per liter:

Proprietary

**Source:** Available from BioRad

**Preparation of Medium:** Prepared plates.

**Storage/Shelf Life:** Prepared plates should be stored under refrigeration (2-8°C). Media should not be used if there are any signs of deterioration (shrinking, cracking, or discoloration) or contamination, or if the expiration date supplied by the manufacturer has passed.

**Use:** For the rapid screening of nasal specimens for MRSA (methicillin-resistant *Staphylococcus aureus*). MRSA strains form pink colonies.

## MRVP Broth
## (Methyl Red- Voges-Proskauer Broth)

**Composition** per liter:

Glucose ................................................................5.0g
KH$_2$PO$_4$.................................................................5.0g
Pancreatic digest of casein...............................3.5g
Peptic digest of animal tissue.............................3.5g
pH 6.9 ± 0.2 at 25°C

**Source:** Available as a premixed powder from BD Diagnostic Systems and as a prepared medium from BD Diagnostic Systems.

**Preparation of Medium:** Add components to distilled/deionized water and bring volume to 1.0L. Mix thoroughly. Distribute into tubes or flasks. Autoclave for 15 min at 15 psi pressure–121°C.

**Storage/Shelf Life:** Store dehydrated media in the dark in a sealed container below 30°C. Prepared media should be stored under refrigeration (2-8°C). Media should be used within 60 days of preparation. Media should not be used if there are any signs of deterioration (discoloration) or contamination, or if the expiration date supplied by the manufacturer has passed.

**Use:** For the differentiation of bacteria based on acid production (Methyl Red test) and acetoin production (Voges-Proskauer reaction).

## MRVP Broth
## (Methyl Red-Voges-Proskauer Broth)
## (BAM M104 Medium 1)

**Composition** per liter:

Buffered peptone-water powder .......................................7.0g
Glucose ..................................................................5.0g
KH$_2$PO$_4$.................................................................5.0g
pH 7.0 ± 0.2 at 25°C

**Source:** Buffered peptone-water powder is available from BD Diagnostic Systems.

**Preparation of Medium:** Add components to distilled/deionized water and bring volume to 1.0L. Mix thoroughly. Distribute into tubes or flasks. Autoclave for 15 min at 15 psi pressure–121°C.

**Storage/Shelf Life:** Store dehydrated media in the dark in a sealed container below 30°C. Prepared media should be stored under refrigeration (2-8°C). Media should be used within 60 days of preparation. Media should not be used if there are any signs of deterioration (discoloration) or contamination, or if the expiration date supplied by the manufacturer has passed.

**Use:** For the differentiation of bacteria based on acid production (Methyl Red test) and acetoin production (Voges-Proskauer reaction).

## M-Slanetz *Enterococcus* Broth Base
## with Triphenyltetrazolium Chloride

**Composition** per liter:

Sucrose...........................................................100.0g
Casein enzymic hydrolysate .............................25.0g
Peptic digest of animal tissue ...........................15.0g
Yeast extract.....................................................10.0g
K$_2$HPO$_4$.............................................................4.0g
Glucose .............................................................2.0g
NaN$_3$ ..................................................................0.4g
Triphenyltetrazolium chloride solution ....................5.0mL
pH 7.2 ± 0.2 at 25°C

**Source:** This medium, without triphenyltetrazolium chloride solution, is available as a premixed powder from HiMedia.

**Caution:** Sodium azide is toxic. Azides also react with metals and disposal must be highly diluted.

**Triphenyltetrazolium Choride Solution:**

**Composition** per 5.0mL:

Triphenyltetrazolium chloride ...........................0.1g

**Preparation of Triphenyltetrazolium Choride Solution:** Add triphenyltetrazolium chloride to distilled/deionized water and bring volume to 5.0mL. Mix thoroughly. Filter sterilize.

**Preparation of Medium:** Add components, except triphenyltetrazolium chloride solution, to distilled/deionized water and bring volume to 1.0L.Mix thoroughly. Gently heat and bring to boiling. Distribute into tubes or flasks. Autoclave for 15 min at 15 psi pressure–121°C. Cool to 45°–50°C. Aseptically add 5.0mL triphenyltetrazolium chloride solution. Mix thoroughly.

**Storage/Shelf Life:** Store dehydrated media in the dark in a sealed container below 30°C. Prepared media should be stored under refrigeration (2-8°C). Media should be used within 60 days of preparation. Media should not be used if there are any signs of deterioration (discoloration) or contamination, or if the expiration date supplied by the manufacturer has passed.

**Use:** For the isolation, cultivation, and enumeration of entercocci in water, sewage, and feces by the membrane filter method. For the direct plating of specimens for the detection and enumeration of fecal streptococci. For the isolation and detection of enterococci using the membrane filter technique.

## M-Slanetz *Enterococcus* HiVeg Broth Base with Triphenyltetrazolium Chloride

**Composition** per liter:

Sucrose............................................................100.0g
Plant hydrolysate .............................................25.0g
Plant peptone ...................................................15.0g
Yeast extract.....................................................10.0g
K$_2$HPO$_4$.............................................................4.0g
Glucose .............................................................2.0g
NaN$_3$ ..................................................................0.4g
Triphenyltetrazolium chloride solution ....................5.0mL
pH 7.2 ± 0.2 at 25°C

**Source:** This medium, without triphenyltetrazolium chloride solution, is available as a premixed powder from HiMedia.

**Caution:** Sodium azide is toxic. Azides also react with metals and disposal must be highly diluted.

**Triphenyltetrazolium Choride Solution:**
**Composition** per 5.0mL:
Triphenyltetrazolium chloride.........................................................0.1g

**Preparation of Triphenyltetrazolium Choride Solution:** Add triphenyltetrazolium chloride to distilled/deionized water and bring volume to 5.0mL. Mix thoroughly. Filter sterilize.

**Preparation of Medium:** Add components, except triphenyltetrazolium chloride solution, to distilled/deionized water and bring volume to 1.0L. Mix thoroughly. Gently heat and bring to boiling. Distribute into tubes or flasks. Autoclave for 15 min at 15 psi pressure–121°C. Cool to 45°–50°C. Aseptically add 5.0mL triphenyltetrazolium chloride solution. Mix thoroughly.

**Storage/Shelf Life:** Store dehydrated media in the dark in a sealed container below 30°C. Prepared media should be stored under refrigeration (2-8°C). Media should be used within 60 days of preparation. Media should not be used if there are any signs of deterioration (discoloration) or contamination, or if the expiration date supplied by the manufacturer has passed.

**Use:** For the cultivation and enumeration of entercocci in water, sewage, and feces by the membrane filter method. For the isolation and detection of enterococci using the membrane filter technique.

### M-*Staphylococcus* HiVeg Broth

**Composition** per liter:
NaCl.........................................................................................75.0g
Plant hydrolysate......................................................................10.0g
Mannitol...................................................................................10.0g
$K_2HPO_4$...................................................................................5.0g
Yeast extract..............................................................................2.5g
Lactose.......................................................................................2.0g
$NaN_3$ .....................................................................................0.049g
pH 7.0 ± 0.2 at 25°C

**Source:** This medium is available as a premixed powder from Hi-Media.

**Caution:** Sodium azide is toxic. Azides also react with metals and disposal must be highly diluted.

**Preparation of Medium:** Add components to distilled/deionized water and bring volume to 1.0L. Mix thoroughly. Distribute into tubes or flasks. Autoclave for 15 min at 15 psi pressure–121°C.

**Storage/Shelf Life:** Store dehydrated media in the dark in a sealed container below 30°C. Prepared media should be stored under refrigeration (2-8°C). Media should be used within 60 days of preparation. Media should not be used if there are any signs of deterioration (discoloration) or contamination, or if the expiration date supplied by the manufacturer has passed.

**Use:** For the cultivation and enumeration of pathogenic and enterotoxigenic staphylococci by the membrane filter method.

### M-Tetrathionate HiVeg Broth Base with Iodine

**Composition** per liter:
$Na_2S_2O_3$ ...............................................................................30.0g
Plant peptone No. 3....................................................................5.0g
Synthetic detergent.....................................................................1.0g
Iodine solution .......................................................................20.0mL

**Source:** This medium, without iodine solution, is available as a premixed powder from HiMedia.

**Iodine Solution:**
**Composition** per 20.0mL:
Iodine ........................................................................................6.0g
KI ...............................................................................................5.0g

**Preparation of Iodine Solution:** Add iodine and KI to distilled/deionized water and bring volume to 20.0mL. Mix thoroughly.

**Preparation of Medium:** Add components, except iodine solution, to distilled/deionized water and bring volume to 980.0mL. Mix thoroughly. Gently heat and bring to boiling. Do not autoclave. Cool to 40°C. Add 20.0mL of iodine solution. Mix thoroughly. Distribute into tubes in 10.0mL volumes. Use medium the same day it is prepared.

**Storage/Shelf Life:** Store dehydrated media in the dark in a sealed container below 30°C. Prepared media should be stored under refrigeration (2-8°C). Media should be used within 60 days of preparation. Media should not be used if there are any signs of deterioration (discoloration) or contamination, or if the expiration date supplied by the manufacturer has passed.

**Use:** For the selective isolation and enrichment of *Salmonella typhi* and other salmonellae from fecal specimens, sewage, and other specimens.

### Mucate Broth

**Composition** per liter:
Mucic acid .................................................................................10.0g
Peptone .....................................................................................10.0g
Bromthymol Blue .....................................................................0.024g
pH 7.4 ± 0.1 at 25°C.

**Preparation of Medium:** Add components to distilled/deionized water and bring volume to 1.0L. Mix thoroughly. Add 5N NaOH while stirring until mucic acid dissolves. Distribute into screw-capped tubes in 5.0mL volumes. Autoclave for 10 min at 15 psi pressure–121°C.

**Storage/Shelf Life:** Store dehydrated media in the dark in a sealed container below 30°C. Prepared media should be stored under refrigeration (2-8°C). Media should be used within 60 days of preparation. Media should not be used if there are any signs of deterioration (discoloration) or contamination, or if the expiration date supplied by the manufacturer has passed.

**Use:** For the isolation and cultivation of enterovirulent *Escherichia coli* and *Shigella* species.

### Mucate Control Broth

**Composition** per liter:
Peptone ......................................................................................10.0g
Bromthymol Blue .....................................................................0.024g
pH 7.4 ± 0.1 at 25°C

**Preparation of Medium:** Add components to distilled/deionized water and bring volume to 1.0L. Mix thoroughly. Distribute into screw-capped tubes in 5.0mL volumes. Autoclave for 10 min at 15 psi pressure–121°C.

**Storage/Shelf Life:** Store dehydrated media in the dark in a sealed container below 30°C. Prepared media should be stored under refrigeration (2-8°C). Media should be used within 60 days of preparation. Media should not be used if there are any signs of deterioration (discoloration) or contamination, or if the expiration date supplied by the manufacturer has passed.

**Use:** For the isolation and cultivation of enterovirulent *Escherichia coli* and *Shigella* species.

## Mucate Control HiVeg Broth

**Composition** per liter:
Plant peptone................................................................10.0g
Bromthymol Blue ...........................................................0.024g

pH 7.4 ± 0.2 at 25°C

**Source:** This medium is available as a premixed powder from Hi-Media.

**Preparation of Medium:** Add components to distilled/deionized water and bring volume to 1.0L. Mix thoroughly. Distribute into screw-capped tubes in 5.0mL volumes. Autoclave for 10 min at 15 psi pressure–121°C.

**Storage/Shelf Life:** Store dehydrated media in the dark in a sealed container below 30°C. Prepared media should be stored under refrigeration (2-8°C). Media should be used within 60 days of preparation. Media should not be used if there are any signs of deterioration (discoloration) or contamination, or if the expiration date supplied by the manufacturer has passed.

**Use:** For the isolation and cultivation of enterovirulent *Escherichia coli* and *Shigella* species.

## Mucate HiVeg Broth

**Composition** per liter:
Plant peptone................................................................10.0g
Mucic acid.....................................................................10.0g
Bromthymol Blue ...........................................................0.024g

pH 7.4 ± 0.2 at 25°C.

**Source:** This medium is available as a premixed powder from Hi-Media.

**Preparation of Medium:** Add components to distilled/deionized water and bring volume to 1.0L. Mix thoroughly. Add 5*N* NaOH while stirring until mucic acid dissolves. Distribute into screw-capped tubes in 5.0mL volumes. Autoclave for 10 min at 15 psi pressure–121°C.

**Storage/Shelf Life:** Store dehydrated media in the dark in a sealed container below 30°C. Prepared media should be stored under refrigeration (2-8°C). Media should be used within 60 days of preparation. Media should not be used if there are any signs of deterioration (discoloration) or contamination, or if the expiration date supplied by the manufacturer has passed.

**Use:** For the isolation and cultivation of enterovirulent *Escherichia coli* and *Shigella* species.

## MUD SF Broth Base

**Composition** per liter:
Tryptose .......................................................................40.0g
KH$_2$PO$_4$....................................................................10.0g
D-Galactose...................................................................2.0g
Tween 80 (polysorbate 80)..............................................1.5g
4-Methylumbelliferyl-β-D-glucoside (MUD) ...............0.15g
Selective supplement solution A.....................................10.0mL
Selective supplement solution B......................................1.0mL

pH 7.5 ± 0.2 at 25°C

**Source:** This medium is available from HiMedia.

**Selective Supplement Solution A:**
**Composition** per 10.0mL:
Thallium acetate.............................................................2.0g
Nalidixic acid................................................................0.25g

**Preparation of Selective Supplement Solution A:** Add components to distilled/deionized water and bring volume to 10.0mL. Mix thoroughly. Filter sterilize.

**Selective Supplement Solution B:**
**Composition** per 20.0mL:
2,3,5,Triphenyltetrazolium chloride ...............................0.2g

**Preparation of Selective Supplement Solution B:** Add 0.2g of 2,3,5,triphenyltetrazolium chloride to distilled/deionized water and bring volume to 20.0mL. Mix thoroughly. Filter sterilize.

**Preparation of Medium:** Add components, except selective supplement solutions A and B, to distilled/deionized water and bring volume to 989.0mL. Mix thoroughly. Autoclave for 15 min at 15 psi pressure–121°C. Cool to 50°C. Aseptically add selective supplement solutions A and B. Mix thoroughly. Aseptically distribute into sterile tubes or flasks.

**Storage/Shelf Life:** Store dehydrated media in the dark in a sealed container below 30°C. Prepared media should be stored in the dark under refrigeration (2-8°C). Chromogenic media are especially light and temperature sensitive; protect from light, excessive heat, moisture, and freezing. Media should not be used if there are any signs of deterioration (discoloration) or contamination, or if the expiration date supplied by the manufacturer has passed.

**Use:** For the detection and enumeration of intestinal enterococci in surface and wastewater in accordance with ISO committee under ISO 7899-1:1998.

## Mueller-Hinton Agar

**Composition** per liter:
Beef infusion.................................................................300.0g
Acid hydrolysate of casein.............................................17.5g
Agar .............................................................................17.0g
Starch ...........................................................................1.5g

pH 7.4 ± 0.2 at 25°C

**Source:** This medium is available as a premixed powder from BD Diagnostic Systems and Oxoid Unipath.

**Preparation of Medium:** Add components to distilled/deionized water and bring to 1.0L. Mix thoroughly. Gently heat and bring to boiling. Distribute into tubes or flasks. Autoclave for 15 min at 15 psi pressure–121°C. Pour into sterile Petri dishes or leave in tubes.

**Use:** For the isolation of pathogenic *Neisseria* species. For antimicrobial susceptibility testing of a variety of bacterial species.

## Mueller-Hinton Agar with Sodium Chloride (BAM M107)

**Composition** per liter:
Beef infusion from..........................................................300.0g
NaCl..............................................................................30.0g
Acid hydrolysate of casein.............................................17.5g
Agar .............................................................................17.0g
Starch ...........................................................................1.5g

pH 7.3 ± 0.2 at 25°C

**Source:** This medium without NaCl is available as a premixed powder from BD Diagnostics and Oxoid Unipath.

**Preparation of Medium:** Add components to distilled/deionized water and bring to 1.0L. Mix thoroughly. Gently heat and bring to boiling. Distribute into tubes or flasks. Autoclave for 15 min at 11 psi pressure–116°C. Pour into sterile Petri dishes or leave in tubes.

**Storage/Shelf Life:** Store dehydrated media in the dark in a sealed container below 30°C. Prepared media should be stored under refrigeration (2-8°C). Media should be used within 60 days of preparation. Media should not be used if there are any signs of deterioration (shrinking, cracking, or discoloration) or contamination, or if the expiration date supplied by the manufacturer has passed.

**Use**: For antimicrobial susceptibility testing of a variety of halophilic *Vibrio* spp.

## Mueller-Hinton Broth

**Composition** per liter:
Acid hydrolysate of casein...............................................17.5g
Beef extract ...................................................................3.0g
Starch ...........................................................................1.5g

pH 7.3 ± 0.1 at 25°C

**Source:** This medium is available as a premixed powder from BD Diagnostic Systems and Oxoid Unipath.

**Preparation of Medium:** Add components to distilled/deionized water and bring to 1.0L. Mix thoroughly. Gently heat and bring to boiling. Distribute into tubes or flasks. Autoclave for 10 min at 10 psi pressure–115°C. Do not overheat.

**Storage/Shelf Life:** Store dehydrated media in the dark in a sealed container below 30°C. Prepared media should be stored under refrigeration (2-8°C). Media should be used within 60 days of preparation. Media should not be used if there are any signs of deterioration (discoloration) or contamination, or if the expiration date supplied by the manufacturer has passed.

**Use:** For the cultivation of a wide variety of microorganisms. For antimicrobial susceptibility testing.

## Mueller-Hinton Chocolate Agar

**Composition** per liter:
Beef infusion...............................................................300.0g
Acid hydrolysate of casein...........................................17.5g
Agar ............................................................................17.0g
Starch ...........................................................................1.5g
Sheep blood................................................................50.0mL

pH 7.4 ± 0.2 at 25°C

**Preparation of Medium:** Add components, except sheep blood, to distilled/deionized water and bring volume to 950.0mL. Mix thoroughly. Gently heat and bring to boiling. Autoclave for 15 min at 15 psi pressure–121°C. Cool to 45°–50°C. Aseptically add sterile sheep blood. Mix thoroughly. Gently heat to 70°C for 10 min. Pour into sterile Petri dishes or distribute into sterile tubes.

**Storage/Shelf Life:** Store dehydrated media in the dark in a sealed container below 30°C. Prepared media should be stored under refrigeration (2-8°C). Media should be used within 60 days of preparation. Media should not be used if there are any signs of deterioration (shrinking, cracking, or discoloration) or contamination, or if the expiration date supplied by the manufacturer has passed.

**Use:** For the cultivation and maintenance of *Neisseria gonorrhoeae* and *Neisseria meningitidis*. For antimicrobial susceptibility testing of fastidious microorganisms.

## Mueller-Hinton II Agar

**Composition** per liter:
Acid hydrolysate of casein...........................................17.5g
Agar ............................................................................17.0g
Beef extract ..................................................................2.0g
Starch ...........................................................................1.5g

pH 7.3 ± 0.1 at 25°C

**Source:** This medium is available as a premixed powder from BD Diagnostic Systems.

**Preparation of Medium:** Add components to distilled/deionized water and bring to 1.0L. Mix thoroughly. Gently heat and bring to boiling. Distribute into tubes or flasks. Autoclave for 15 min at 15 psi pressure–121°C. Pour into sterile Petri dishes or leave in tubes.

**Storage/Shelf Life:** Store dehydrated media in the dark in a sealed container below 30°C. Prepared media should be stored under refrigeration (2-8°C). Media should be used within 60 days of preparation. Media should not be used if there are any signs of deterioration (shrinking, cracking, or discoloration) or contamination, or if the expiration date supplied by the manufacturer has passed.

**Use:** For antimicrobial disc diffusion susceptibility testing by the Bauer-Kirby method of a variety of bacteria. This medium supplemented with 5% sheep blood is recommended for use in antimicrobial susceptibility testing of *Streptococcus pneumoniae* and *Haemophilus influenzae*.

## Mueller-Hinton HiVeg Agar

**Composition** per liter:
Plant infusion .............................................................300.0g
Plant acid hydrolysate.................................................17.5g
Agar ............................................................................17.0g
Starch ...........................................................................1.5g

pH 7.3 ± 0.2 at 25°C

**Source:** This medium is available as a premixed powder from Hi-Media.

**Preparation of Medium:** Add components to distilled/deionized water and bring to 1.0L. Mix thoroughly. Gently heat and bring to boiling. Distribute into tubes or flasks. Autoclave for 15 min at 15 psi pressure–121°C. Pour into sterile Petri dishes or leave in tubes.

**Storage/Shelf Life:** Store dehydrated media in the dark in a sealed container below 30°C. Prepared media should be stored under refrigeration (2-8°C). Media should be used within 60 days of preparation. Media should not be used if there are any signs of deterioration (shrinking, cracking, or discoloration) or contamination, or if the expiration date supplied by the manufacturer has passed.

**Use:** For the cultivation of *Neisseria* species and for determination of susceptibility of microorganisms to antimicrobial agents.

## Mueller-Hinton HiVeg Agar No. 2

**Composition** per liter:
Plant hydrolysate .........................................................17.5g
Agar ............................................................................17.0g
Plant infusion................................................................2.0g
Starch, soluble..............................................................1.5g

pH 7.3 ± 0.2 at 25°C

**Source:** This medium is available as a premixed powder from Hi-Media.

**Preparation of Medium:** Add components to distilled/deionized water and bring to 1.0L. Mix thoroughly. Gently heat and bring to boiling. Distribute into tubes or flasks. Autoclave for 15 min at 15 psi pressure–121°C. Pour into sterile Petri dishes or leave in tubes.

**Storage/Shelf Life:** Store dehydrated media in the dark in a sealed container below 30°C. Prepared media should be stored under refrigeration (2-8°C). Media should be used within 60 days of preparation. Media should not be used if there are any signs of deterioration (shrinking, cracking, or discoloration) or contamination, or if the expiration date supplied by the manufacturer has passed.

**Use:** For the isolation of pathogenic *Neisseria* species. For testing susceptibility of common and rapidly growing bacteria using antimicrobial discs by the Bauer-Kirby method.

## Mueller-Hinton HiVeg Broth

**Composition** per liter:

Plant acid hydrolysate ....................................................17.5g
Plant infusion .................................................................2.0g
Starch ...........................................................................1.5g

pH 7.3 ± 0.1 at 25°C

**Source:** This medium is available as a premixed powder from Hi-Media.

**Preparation of Medium:** Add components to distilled/deionized water and bring to 1.0L. Mix thoroughly. Gently heat and bring to boiling. Distribute into tubes or flasks. Autoclave for 10 min at 10 psi pressure–115°C. Do not overheat.

**Storage/Shelf Life:** Store dehydrated media in the dark in a sealed container below 30°C. Prepared media should be stored under refrigeration (2-8°C). Media should be used within 60 days of preparation. Media should not be used if there are any signs of deterioration (discoloration) or contamination, or if the expiration date supplied by the manufacturer has passed.

**Use:** For the cultivation of a wide variety of microorganisms. For antimicrobial susceptibility testing.

## Mueller-Tellurite Medium

**Composition** per liter:

Casamino acids .............................................................20.0g
Agar ...............................................................................20.0g
Casein.............................................................................5.0g
$KH_2PO_4$.............................................................................0.3g
$MgSO_4 \cdot 7H_2O$ ............................................................0.1g
L-Tryptophan .................................................................0.05g
Mueller-tellurite serum ............................................25.0mL

pH 7.4 ± 0.1 at 25°C

**Mueller-Tellurite Serum:**

**Composition** per 100.0mL:

$K_2TeO_3$ solution ...........................................................0.4g
Calcium pantothenate .................................................0.2mg
Horse or beef serum, sterile ....................................50.0mL
Sodium lactate solution..............................................40.0mL
Ethyl alcohol ..............................................................10.0mL

**Preparation of Mueller-Tellurite Serum:** Add calcium pantothenate to 1.0mL of distilled/deionized water. Autoclave for 15 min at 15 psi pressure–121°C. Add $K_2TeO_3$ to 1.0mL of sterile distilled/deionized water. To 40.0mL of cooled, sterile sodium lactate solution, add filter-sterilized ethanol, sterile calcium pantothenate solution, sterile serum, and $K_2TeO_3$ solution. Mix thoroughly.

**Sodium Lactate Solution:**

**Composition** per 100.0mL:

Lactic acid (85% solution)........................................50.0mL
Phenol Red solution (0.2g in 50% ethanol) ...............0.1mL

**Preparation of Sodium Lactate Solution:** Add lactic acid to distilled/deionized water and bring volume to 100.0mL. Add 0.1mL of Phenol Red solution. Add enough 40% NaOH solution to adjust pH to 7.0. Gently heat and bring to boiling for 5 min. Add more NaOH solution to retain red color, if necessary. Autoclave for 15 min at 15 psi pressure–121°C. Cool to 50°C.

**Caution:** Potassium tellurite is toxic.

**Preparation of Medium:** Add components to distilled/deionized water and bring volume to 975.0mL. Gently heat and bring to boiling. Autoclave for 15 min at 15 psi pressure–121°C. Cool quickly to 50°C. Aseptically add 25.0mL of Mueller-Tellurite serum. Mix thoroughly. Distribute into sterile Petri dishes. Allow the surface of the plates to dry by partially removing the covers during solidification.

**Storage/Shelf Life:** Store dehydrated media in the dark in a sealed container below 30°C. Prepared media should be stored under refrigeration (2-8°C). Media should be used within 60 days of preparation. Media should not be used if there are any signs of deterioration (discoloration) or contamination, or if the expiration date supplied by the manufacturer has passed.

**Use:** For the isolation, cultivation, and differentiation of *Corynebacterium diphtheriae*.

## MUG Bromcresol Purple Broth with Lactose

**Composition** per liter:

Casein enzymic hydrolysate .......................................17.0g
Lactose..........................................................................10.0g
NaCl................................................................................5.0g
Papaic digest of soybean meal ......................................3.0g
Tryptophan.....................................................................1.0g
Bromcresol Purple .........................................................0.02g
4-Methylumbelliferyl-β-D-glucuronide
(MUG)......................................................................0.01g

pH 7.0 ± 0.2 at 25°C

**Source:** This medium is available from HiMedia.

**Preparation of Medium:** Add components to distilled/deionized water and bring volume to 1.0L. Mix thoroughly. Gently heat and bring to boiling. Distribute into tubes with inverted Durham tubes. Autoclave for 20 min at 10 psi pressure–115°C.

**Storage/Shelf Life:** Store dehydrated media in the dark in a sealed container below 30°C. Prepared media should be stored in the dark under refrigeration (2-8°C). Chromogenic media are especially light and temperature sensitive; protect from light, excessive heat, moisture, and freezing. Media should not be used if there are any signs of deterioration (discoloration) or contamination, or if the expiration date supplied by the manufacturer has passed.

**Use:** For the identification of *Escherichia coli* and coliform bacteria from water samples by a fluorogenic assay method.

## MUG EC O157 Agar

**Composition** per liter:

Casein peptone.............................................................20.0g
Agar ..............................................................................13.0g

Sorbitol..................................................................10.0g
NaCl.........................................................................5.0g
Meat extract ..........................................................2.0g
Na$_2$S$_2$O$_3$ .................................................................2.0g
Sodium deoxycholate...........................................1.12g
Yeast extract...........................................................1.0g
Ferric ammonium citrate.......................................0.5g
4-Methylumbellifery-lβ-D-glucuronide (MUG) ...........0.1g
Bromthymol Blue ..............................................0.025g

pH 7.4 ± 0.2 at 25°C

**Source:** This medium is available from HiMedia.

**Preparation of Medium:** Add components to distilled/deionized water and bring volume to 1.0L. Mix thoroughly. Gently heat and bring to boiling. Distribute into tubes or flasks. Autoclave for 15 min at 15 psi pressure–121°C. Pour into sterile Petri dishes or leave in tubes.

**Storage/Shelf Life:** Store dehydrated media in the dark in a sealed container below 30°C. Prepared media should be stored in the dark under refrigeration (2-8°C). Chromogenic media are especially light and temperature sensitive; protect from light, excessive heat, moisture, and freezing. Media should not be used if there are any signs of deterioration (discoloration) or contamination, or if the expiration date supplied by the manufacturer has passed.

**Use:** For the isolation and differentiation of enterohemorrhagic *Escherichia coli* O157:H7 from foodstuffs, water, and clinical samples by a fluorogenic method.

## MUG EC O157 Agar, Modified

**Composition** per liter:
Peptic digest of animal tissue.......................................20.0g
Sorbitol.......................................................................20.0g
Agar ...........................................................................12.0g
NaCl.............................................................................5.0g
Bile salts.....................................................................1.12g
4-Methylumbelliferyl-β-D-glucuronide (MUG) ...........0.05g
Bromcresol Purple .....................................................0.01g

pH 7.2 ± 0.2 at 25°C

**Source:** This medium is available from HiMedia.

**Preparation of Medium:** Add components to distilled/deionized water and bring volume to 1.0L. Mix thoroughly. Gently heat and bring to boiling. Distribute into tubes or flasks. Autoclave for 15 min at 15 psi pressure–121°C. Pour into sterile Petri dishes or leave in tubes.

**Storage/Shelf Life:** Store dehydrated media in the dark in a sealed container below 30°C. Prepared media should be stored in the dark under refrigeration (2-8°C). Chromogenic media are especially light and temperature sensitive; protect from light, excessive heat, moisture, and freezing. Media should not be used if there are any signs of deterioration (discoloration) or contamination, or if the expiration date supplied by the manufacturer has passed.

**Use:** For the isolation and differentiation of enterohemorrhagic *Escherichia coli* O157:H7 from foodstuffs, water, and clinical samples by a fluorogenic method.

## MUG EC Broth

**Composition** per liter:
Casein enzymatic hydrolysate ..........................20.0g
Lactose ........................................................................5.0g
NaCl.............................................................................5.0g
K$_2$HPO$_4$......................................................................4.0g

KH$_2$PO$_4$ ........................................................................1.5g
Bile salts mixture ........................................................1.5g
4-Methylumbelliferyl-β-D-glucuronide (MUG) ...........0.05g

pH 6.9 ± 0.2 at 25°C

**Source:** This medium is available as a premixed powder from HiMedia.

**Preparation of Medium:** Add components to distilled/deionized water and bring volume to 1.0L. Mix thoroughly. Distribute into test tubes that contain an inverted Durham tube in 10.0mL volumes. Autoclave for 15 min at 15 psi pressure–121°C.

**Storage/Shelf Life:** Store dehydrated media in the dark in a sealed container below 30°C. Prepared media should be stored in the dark under refrigeration (2-8°C). Chromogenic media are especially light and temperature sensitive; protect from light, excessive heat, moisture, and freezing. Media should not be used if there are any signs of deterioration (discoloration) or contamination, or if the expiration date supplied by the manufacturer has passed.

**Use:** For the detection of *Escherichia coli* in water and food samples by a fluorogenic procedure.

## MUG EC Broth, Modified

**Composition** per liter:
Casein enzymic hydrolysate .......................................40.0g
Salicin .........................................................................1.0g
Triton X-100 .................................................................1.0g
4-Methylumbelliferyl-β-D-glucuronide (MUG)............0.1g

pH 6.9 ± 0.2 at 25°C

**Source:** This medium is available from HiMedia.

**Preparation of Medium:** Add components to distilled/deionized water and bring volume to 1.0L. Mix thoroughly. Gently heat and bring to boiling. Distribute into tubes or flasks. Autoclave for 15 min at 15 psi pressure–121°C.

**Storage/Shelf Life:** Store dehydrated media in the dark in a sealed container below 30°C. Prepared media should be stored in the dark under refrigeration (2-8°C). Chromogenic media are especially light and temperature sensitive; protect from light, excessive heat, moisture, and freezing. Media should not be used if there are any signs of deterioration (discoloration) or contamination, or if the expiration date supplied by the manufacturer has passed.

**Use:** For the detection and enumeration of *Escherichia coli* in surface and waste water by the miniaturized method (MPN) in accordance with the ISO committee under ISO 9308-3:1998.

## MUG EC HiVeg Broth

**Composition** per liter:
Plant hydrolysate .....................................................20.0g
Lactose.........................................................................5.0g
NaCl.............................................................................5.0g
K$_2$HPO$_4$......................................................................4.0g
KH$_2$PO$_4$ ........................................................................1.5g
Synthetic detergent No. I ..........................................1.5g
4-Methylumbelliferyl-β-D-glucuronide (MUG) ...........0.05g

pH 6.9 ± 0.2 at 25°C

**Source:** This medium is available as a premixed powder from HiMedia.

**Preparation of Medium:** Add components to distilled/deionized water and bring volume to 1.0L. Mix thoroughly. Distribute into test tubes that contain an inverted Durham tube in 10.0mL volumes. Autoclave for 15 min at 15 psi pressure–121°C.

**Storage/Shelf Life:** Store dehydrated media in the dark in a sealed container below 30°C. Prepared media should be stored in the dark under refrigeration (2-8°C). Chromogenic media are especially light and temperature sensitive; protect from light, excessive heat, moisture, and freezing. Media should not be used if there are any signs of deterioration (discoloration) or contamination, or if the expiration date supplied by the manufacturer has passed.

**Use:** For the detection of *Escherichia coli* in water and food samples by a fluorogenic procedure.

## MUG Lauryl Sulfate Broth

**Composition** per liter:
Casein enzymic hydrolysate .......................................................20.0g
Lactose .....................................................................................5.0g
NaCl .........................................................................................5.0g
K$_2$HPO$_4$ ...................................................................................2.75g
KH$_2$PO$_4$ ...................................................................................2.75g
Sodium lauryl sulfate .................................................................0.1g
4-Methylumbelliferyl-β-D-glucuronide (MUG) ...........................0.05g
pH 6.8 ± 0.2 at 25°C

**Source:** This medium is available from HiMedia.

**Preparation of Medium:** Add components to distilled/deionized water and bring volume to 1.0L. Mix thoroughly. Gently heat and bring to boiling. Distribute into tubes or flasks. Autoclave for 15 min at 15 psi pressure–121°C.

**Storage/Shelf Life:** Store dehydrated media in the dark in a sealed container below 30°C. Prepared media should be stored in the dark under refrigeration (2-8°C). Chromogenic media are especially light and temperature sensitive; protect from light, excessive heat, moisture, and freezing. Media should not be used if there are any signs of deterioration (discoloration) or contamination, or if the expiration date supplied by the manufacturer has passed.

**Use:** For the detection of coliform bacteria in water and food specimens by a fluorogenic procedure.

## MUG Lauryl Sulfate Broth, Modified

**Composition** per liter:
Casein enzymatic hydrolysate ...................................................20.0g
Lactose .....................................................................................5.0g
NaCl .........................................................................................5.0g
K$_2$HPO$_4$ ...................................................................................2.75g
KH$_2$PO$_4$ ...................................................................................2.75g
Sodium lauryl sulfate .................................................................0.1g
4-Methylumbelliferyl-β-D-glucuronide (MUG) ...........................0.05g
pH 6.8 ± 0.2 at 25°C

**Source:** This medium is available as a premixed powder from HiMedia.

**Preparation of Medium:** Add components to distilled/deionized water and bring volume to 1.0L. Mix thoroughly. Distribute into tubes containing an inverted Durham tube in 10.0mL volumes. Autoclave for 12 min at 15 psi pressure–121°C. Cool broth quickly to 25°C. For testing water samples with 10.0mL volumes, prepare medium double strength.

**Storage/Shelf Life:** Store dehydrated media in the dark in a sealed container below 30°C. Prepared media should be stored in the dark under refrigeration (2-8°C). Chromogenic media are especially light and

temperature sensitive; protect from light, excessive heat, moisture, and freezing. Media should not be used if there are any signs of deterioration (discoloration) or contamination, or if the expiration date supplied by the manufacturer has passed.

**Use:** For the detection of *Escherichia coli* in water and food samples by a fluorogenic procedure.

## MUG Lauryl Sulfate HiVeg Broth, Modified

**Composition** per liter:
Plant hydrolysate .....................................................................20.0g
Lactose .....................................................................................5.0g
NaCl .........................................................................................5.0g
K$_2$HPO$_4$ ...................................................................................2.75g
KH$_2$PO$_4$ ...................................................................................2.75g
Sodium lauryl sulfate .................................................................0.1g
4-Methylumbelliferyl-β-D-glucuronide (MUG) ...........................0.05g
pH 6.8 ± 0.2 at 25°C

**Source:** This medium is available as a premixed powder from HiMedia.

**Preparation of Medium:** Add components to distilled/deionized water and bring volume to 1.0L. Mix thoroughly. Distribute into tubes containing an inverted Durham tube in 10.0mL volumes. Autoclave for 12 min at 15 psi pressure–121°C. Cool broth quickly to 25°C. For testing water samples with 10.0mL volumes, prepare medium double strength.

**Storage/Shelf Life:** Store dehydrated media in the dark in a sealed container below 30°C. Prepared media should be stored in the dark under refrigeration (2-8°C). Chromogenic media are especially light and temperature sensitive; protect from light, excessive heat, moisture, and freezing. Media should not be used if there are any signs of deterioration (discoloration) or contamination, or if the expiration date supplied by the manufacturer has passed.

**Use:** For the detection of *Escherichia coli* in water and food samples by a fluorogenic procedure.

## MUG MacConkey HiVeg Agar

**Composition** per liter:
Plant peptone ...........................................................................20.0g
Agar .........................................................................................15.0g
Lactose .....................................................................................10.0g
NaCl .........................................................................................5.0g
Synthetic detergent No. I ..........................................................1.5g
4-Methylumbelliferyl-β-D-glucuronide (MUG) ...........................0.1g
Neutral Red ..............................................................................0.03g
Crystal Violet ...........................................................................1.0mg
pH 7.4 ± 0.2 at 25°C

**Source:** This medium is available as a premixed powder from HiMedia.

**Preparation of Medium:** Add components to distilled/deionized water and bring volume to 1.0L. Mix thoroughly. Gently heat while stirring until boiling. Autoclave for 15 min at 15 psi pressure–121°C. Pour into sterile Petri dishes or distribute into sterile tubes.

**Storage/Shelf Life:** Store dehydrated media in the dark in a sealed container below 30°C. Prepared media should be stored in the dark under refrigeration (2-8°C). Chromogenic media are especially light and temperature sensitive; protect from light, excessive heat, moisture, and freezing. Media should not be used if there are any signs of deterioration (discoloration) or contamination, or if the expiration date supplied by the manufacturer has passed.

**Use:** For the selective isolation, cultivation, and differentiation of coliforms and enteric pathogens based on the ability to ferment lactose by a fluorogenic procedure.

## MUG Nutrient Agar

**Composition** per liter:

| | |
|---|---|
| Agar | 15.0g |
| Peptic digest of animal tissue | 5.0g |
| NaCl | 5.0g |
| Beef extract | 1.5g |
| Yeast extract | 1.5g |
| 4-Methylumbelliferyl-β-D-glucuronide (MUG) | 0.1g |

pH 7.4 ± 0.2 at 25°C

**Source:** This medium is available from HiMedia.

**Preparation of Medium:** Add components to distilled/deionized water and bring volume to 1.0L. Mix thoroughly. Gently heat and bring to boiling. Distribute into tubes or flasks. Autoclave for 15 min at 15 psi pressure–121°C. Pour into sterile Petri dishes or leave in tubes.

**Storage/Shelf Life:** Store dehydrated media in the dark in a sealed container below 30°C. Prepared media should be stored in the dark under refrigeration (2-8°C). Chromogenic media are especially light and temperature sensitive; protect from light, excessive heat, moisture, and freezing. Media should not be used if there are any signs of deterioration (discoloration) or contamination, or if the expiration date supplied by the manufacturer has passed.

**Use:** For the detection of *Escherichia coli* in water and food samples by a fluorogenic procedure.

## MUG Sorbitol Agar

**Composition** per liter:

| | |
|---|---|
| Peptic digest of animal tissue | 17.0g |
| Agar | 13.5g |
| D-Sorbitol | 10.0g |
| NaCl | 5.0g |
| Proteose peptone | 3.0g |
| Bile salts mixture | 1.5g |
| 4-Methylumbelliferyl-β-D-glucuronide (MUG) | 0.1g |
| Neutral Red | 0.03g |
| Crystal Violet | 0.001g |

pH 7.1 ± 0.2 at 25°C

**Source:** This medium is available from HiMedia.

**Preparation of Medium:** Add components to distilled/deionized water and bring volume to 1.0L. Mix thoroughly. Gently heat and bring to boiling. Distribute into tubes or flasks. Autoclave for 15 min at 15 psi pressure–121°C. Pour into sterile Petri dishes or leave in tubes.

**Storage/Shelf Life:** Store dehydrated media in the dark in a sealed container below 30°C. Prepared media should be stored in the dark under refrigeration (2-8°C). Chromogenic media are especially light and temperature sensitive; protect from light, excessive heat, moisture, and freezing. Media should not be used if there are any signs of deterioration (shrinking, cracking, or discoloration) or contamination, or if the expiration date supplied by the manufacturer has passed.

**Use:** For the isolation and identification of enteropathogenic *Escherichia coli* associated with infant diarrhea by fluorogenic method.

## MUG Violet Red Agar

**Composition** per liter:

| | |
|---|---|
| Agar | 15.0g |
| Lactose | 10.0g |
| Peptic digest of animal tissue | 7.0g |
| NaCl | 5.0g |
| Yeast extract | 3.0g |
| Synthetic detergent No. I | 1.5g |
| 4-Methylumbelliferyl-β-D-glucuronide (MUG) | 0.1g |
| Neutral Red | 0.03g |
| Crystal Violet | 2.0mg |

pH 7.4 ± 0.2 at 25°C

**Source:** This medium is available as a premixed powder from Hi-Media.

**Preparation of Medium:** Add components to distilled/deionized water and bring volume to 1.0L. Mix thoroughly. Gently heat while stirring and bring to boiling. Distribute into tubes or flasks. Do not autoclave. Pour immediately into sterile Petri dishes or leave in tubes.

**Storage/Shelf Life:** Store dehydrated media in the dark in a sealed container below 30°C. Prepared media should be stored in the dark under refrigeration (2-8°C). Chromogenic media are especially light and temperature sensitive; protect from light, excessive heat, moisture, and freezing. Media should not be used if there are any signs of deterioration (shrinking, cracking, or discoloration) or contamination, or if the expiration date supplied by the manufacturer has passed.

**Use:** For the differentiation of *Escherichia coli* from dairy products and other foods by a fluorogenic procedure based on their ability to produce β-glucuronidase.

## MUG Violet Red HiVeg Agar

**Composition** per liter:

| | |
|---|---|
| Agar | 15.0g |
| Lactose | 10.0g |
| Plant peptone | 7.0g |
| NaCl | 5.0g |
| Yeast extract | 3.0g |
| Synthetic detergent No. I | 1.5g |
| 4-Methylumbelliferyl-β-D-glucuronide (MUG) | 0.1g |
| Neutral Red | 0.03g |
| Crystal Violet | 2.0mg |

pH 7.4 ± 0.2 at 25°C

**Source:** This medium is available as a premixed powder from Hi-Media.

**Preparation of Medium:** Add components to distilled/deionized water and bring volume to 1.0L. Mix thoroughly. Gently heat while stirring and bring to boiling. Distribute into tubes or flasks. Do not autoclave. Pour immediately into sterile Petri dishes or leave in tubes.

**Storage/Shelf Life:** Store dehydrated media in the dark in a sealed container below 30°C. Prepared media should be stored in the dark under refrigeration (2-8°C). Chromogenic media are especially light and temperature sensitive; protect from light, excessive heat, moisture, and freezing. Media should not be used if there are any signs of deterioration (shrinking, cracking, or discoloration) or contamination, or if the expiration date supplied by the manufacturer has passed.

**Use:** For the differentiation of *Escherichia coli* from dairy products and other foods by a fluorogenic procedure based on their ability to produce β-glucuronidase.

## MWY Medium
### (Wadowsky and Yee Medium, Modified)

**Composition** per liter:

| | |
|---|---|
| Agar | 13.0g |
| Yeast extract | 10.0g |
| Glycine | 3.0g |
| ACES buffer (2-[(2-amino-2-oxoethyl)-amino]-ethane sulfonic acid) | 2.0g |
| Charcoal, activated | 2.0g |
| α-Ketoglutarate | 0.2g |
| $Fe_4(P_2O_7)_3 \cdot 9H_2O$ | 0.05g |
| Bromcresol Purple | 0.01g |
| Bromcresol Blue | 0.01g |
| Antibiotic inhibitor | 10.0mL |
| L-Cysteine·HCl·H$_2$O solution | 10.0mL |

pH 6.9 ± 0.2 at 25°C

**Antibiotic Inhibitor:**
**Composition** per 10.0mL:

| | |
|---|---|
| Anisomycin | 0.16g |
| Cefamandole | 4.0mg |
| Vancomycin | 1.0mg |
| Polymyxin B | 130,000U |

**Preparation of Antibiotic Inhibitor:** Add components to distilled/deionized water and bring volume to 10.0mL. Mix thoroughly. Filter sterilize.

**L-Cysteine·HCl·H$_2$O Solution:**
**Composition** per 10.0mL:

| | |
|---|---|
| L-Cysteine·HCl·H$_2$O | 0.08g |

**Preparation of L-Cysteine·HCl·H$_2$O Solution:** Add L-cysteine·HCl·H$_2$O to distilled/deionized water and bring volume to 10.0mL. Mix thoroughly. Filter sterilize.

**Preparation of Medium:** Add components, except L-cysteine solution and antibiotic inhibitor, to distilled/deionized water and bring volume to 980.0mL. Mix thoroughly. Adjust medium to pH 6.9 with 1*N* KOH. Heat gently and bring to boiling for 1 min. Autoclave for 15 min at 15 psi pressure–121°C. Cool to 50°–55°C. Add 10.0mL of the sterile L-cysteine·HCl·H$_2$O solution and 10.0mL of the sterile antibiotic solution. Mix thoroughly. Pour into sterile Petri dishes with constant agitation to keep charcoal in suspension.

**Storage/Shelf Life:** Store dehydrated media in the dark in a sealed container below 30°C. Prepared media should be stored under refrigeration (2-8°C). Media should be used within 60 days of preparation. Media should not be used if there are any signs of deterioration (discoloration) or contamination, or if the expiration date supplied by the manufacturer has passed.

**Use:** For the selective isolation and cultivation of *Legionella pneumophila* and other *Legionella* species.

## Mycobactosel™ L-J Medium

**Composition** per liter:

| | |
|---|---|
| Potato flour | 30.0g |
| L-Asparagine | 3.6g |
| KH$_2$PO$_4$, anhydrous | 2.5g |
| Sodium citrate | 0.6g |
| MgSO$_4$·7H$_2$O | 0.24g |
| Homogenized whole egg | 1.0L |
| Malachite Green solution | 20.0mL |
| Glycerol | 12.0mL |
| Antibiotic solution | 10.0mL |

pH 7.0 ± 0.2 at 25°C

**Source:** This medium is available as a prepared medium from BD Diagnostic Systems.

**Homogenized Whole Egg:**
**Composition** per liter:

| | |
|---|---|
| Whole eggs | 18–24 |

**Preparation of Whole Egg:** Use fresh eggs, less than 1 week old. Scrub the shells with soap. Let stand in a soap solution for 30 min. Rinse in running water. Soak eggs in 70% ethanol for 15 min. Break the eggs into a sterile container. Homogenize by shaking. Filter through four layers of sterile cheesecloth into a sterile graduated cylinder. Measure out 1.0L.

**Malachite Green Solution:**
**Composition** per 20.0mL:

| | |
|---|---|
| Malachite Green | 0.4g |

**Preparation of Malachite Green Solution:** Add Malachite Green to sterile distilled/deionized water and bring volume to 20.0mL in a sterile container. Mix thoroughly.

**Antibiotic Solution:**
**Composition** per 10.0mL:

| | |
|---|---|
| Cycloheximide | 0.64g |
| Nalidixic acid | 0.056g |
| Lincomycin | 3.2mg |

**Preparation of Antibiotic Solution:** Add components to distilled/deionized water and bring volume to 10.0mL. Mix thoroughly. Filter sterilize.

**Caution:** Cycloheximide is toxic. Avoid skin contact or aerosol formation and inhalation.

**Preparation of Medium:** Add components—except whole egg, Malachite Green solution, and antibiotic solution—to distilled/deionized water and bring volume to 600.0mL. Mix thoroughly. Autoclave for 30 min at 15 psi pressure–121°C. Cool to room temperature. Add the homogenized whole egg, Malachite Green solution, and antibiotic solution. Distribute into sterile tubes in 8.0mL volumes. Coagulate medium in a slanted position at 85°C (moist heat) for 50 min.

**Storage/Shelf Life:** Store dehydrated media in the dark in a sealed container below 30°C. Prepared media should be stored under refrigeration (2-8°C). Media should be used within 60 days of preparation. Media should not be used if there are any signs of deterioration (discoloration) or contamination, or if the expiration date supplied by the manufacturer has passed.

**Use:** For the isolation and cultivation of *Mycobacterium* species from clinical specimens.

## *Mycoplasma* HiVeg Broth Base
### with Crystal Violet and Tellurite
### (PPLO HiVeg Broth Base with CV)

**Composition** per liter:

| | |
|---|---|
| Plant peptone | 10.0g |
| Plant infusion | 6.0g |
| NaCl | 5.0g |
| Crystal Violet | 0.01g |
| Chapman tellurite solution | 2.85mL |
| Ascitic fluid | 250.0mL |

pH 7.8 ± 0.2 at 25°C

**Source:** This medium, without tellurite, is available as a premixed powder from HiMedia.

**Chapman Tellurite Solution:**
**Composition** per 100.0mL:
K₂TeO₃.....................................................................................1.0g

$K_2TeO_3$......................................................................................1.0g

**Preparation of Chapman Tellurite Solution:** Add $K_2TeO_3$ to distilled/deionized water and bring volume to 100.0mL. Mix thoroughly. Filter sterilize.

**Caution:** Potassium tellurite is toxic.

**Preparation of Medium:** Add components, except ascitic fluid and Chapman tellurite solution, to distilled/deionized water and bring volume to 747.15mL. Mix thoroughly. Autoclave for 15 min at 15 psi pressure–121°C. Cool to less than 37°C. Aseptically add sterile ascitic fluid and 2.85mL of Chapman tellurite solution. Mix thoroughly. Aseptically distribute into sterile tubes or flasks.

**Storage/Shelf Life:** Store dehydrated media in the dark in a sealed container below 30°C. Prepared media should be stored under refrigeration (2-8°C). Media should be used within 60 days of preparation. Media should not be used if there are any signs of deterioration (discoloration) or contamination, or if the expiration date supplied by the manufacturer has passed.

**Use:** For the isolation of *Mycoplasma* species from clinical specimens.

## Mycoplasma HiVeg Broth Base without Crystal Violet and with Ascitic Fluid (PPLO HiVeg Broth Base without CV)

**Composition** per liter:
Plant peptone...........................................................................10.0g
Plant infusion ...........................................................................6.0g
NaCl..........................................................................................5.0g
Ascitic fluid........................................................................ 250.0mL
pH 7.8 ± 0.2 at 25°C

**Source:** This medium, without ascitic fluid, is available as a premixed powder from HiMedia.

**Preparation of Medium:** Add components, except ascitic fluid, to distilled/deionized water and bring volume to 750.0mL. Mix thoroughly. Autoclave for 15 min at 15 psi pressure–121°C. Cool to less than 37°C. Aseptically add sterile ascitic fluid. If desired, 0.5g of thallium acetate or 100,000U of penicillin may be added for a more selective medium. Mix thoroughly. Aseptically distribute into sterile tubes or flasks.

**Storage/Shelf Life:** Store dehydrated media in the dark in a sealed container below 30°C. Prepared media should be stored under refrigeration (2-8°C). Media should be used within 60 days of preparation. Media should not be used if there are any signs of deterioration (discoloration) or contamination, or if the expiration date supplied by the manufacturer has passed.

**Use:** For the enrichment of pleuro-pneumonia-like organisms (PPLOs) and *Mycoplasma* species from clinical specimens.

## *Mycoplasma pneumoniae* Isolation Medium

**Composition** per 1200.0mL:
Beef heart for infusion ...........................................................50.0g
Peptone....................................................................................10.0g
NaCl...........................................................................................5.0g
Water ................................................................................. 900.0mL

Yeast extract solution........................................................ 100.0mL
α-Gamma horse serum, unheated........................................ 200.0mL
pH 7.6–7.8 at 25°C

**Yeast Extract Solution:**
**Composition** per 10.0mL:
Yeast, active, dry, Baker's....................................................250.0g

**Preparation of Yeast Extract Solution:** Add yeast to 1.0L of distilled/deionized water. Mix thoroughly. Gently heat and bring to boiling. Filter through Whatman #2 filter paper. Adjust the pH of the filtrate to 8.0 with NaOH. Distribute into tubes in 10.0mL volumes. Autoclave for 15 min at 15 psi pressure–121°C. Store at –20°C.

**Preparation of Medium:** Add components, except yeast extract solution and α-gamma horse serum, to distilled/deionized water and bring volume to 990.0mL. Mix thoroughly. Gently heat and bring to boiling. Autoclave for 15 min at 15 psi pressure–121°C. Cool to 45°–50°C. Aseptically add sterile yeast extract solution and α-gamma horse serum. Mix thoroughly. Aseptically distribute into sterile tubes.

**Storage/Shelf Life:** Store dehydrated media in the dark in a sealed container below 30°C. Prepared media should be stored under refrigeration (2-8°C). Media should be used within 60 days of preparation. Media should not be used if there are any signs of deterioration (discoloration) or contamination, or if the expiration date supplied by the manufacturer has passed.

**Use:** For the isolation and cultivation of *Mycoplasma pneumoniae*.

## Mycoplasmal Agar

**Composition** per liter:
Papaic digest of soy meal .......................................................20.0g
Agarose....................................................................................10.0g
NaCl...........................................................................................5.0g
Phenol Red (2% solution).................................................... 1.0mL
pH 7.3 ± 0.2 at 25°C

**Preparation of Medium:** Add components, except agarose, to distilled/deionized water and bring volume to 1.0L. Mix thoroughly. Adjust pH to 7.3 with 1*N* NaOH. Add agarose. Mix thoroughly. Gently heat and bring to boiling. Distribute into tubes or flasks. Autoclave for 15 min at 15 psi pressure–121°C. Pour into sterile Petri dishes or leave in tubes.

**Storage/Shelf Life:** Store dehydrated media in the dark in a sealed container below 30°C. Prepared media should be stored under refrigeration (2-8°C). Media should be used within 60 days of preparation. Media should not be used if there are any signs of deterioration (shrinking, cracking, or discoloration) or contamination, or if the expiration date supplied by the manufacturer has passed.

**Use:** For the isolation and cultivation of human mycoplasmas and ureaplasmas.

## Mycosel™ Agar (Cycloheximide Chloramphenicol Agar)

**Composition** per liter:
Agar .......................................................................................15.5g
Papaic digest of soybean meal................................................10.0g
Glucose ...................................................................................10.0g
Cycloheximide...........................................................................0.4g
Chloramphenicol......................................................................0.05g
pH 6.9 ± 0.2 at 25°C

**Caution:** Cycloheximide is toxic. Avoid skin contact or aerosol formation and inhalation.

**Preparation of Medium:** Add components to distilled/deionized water and bring volume to 1.0L. Mix thoroughly. Gently heat while stirring and bring to boiling. Autoclave for 15 min at 14 psi pressure–118°C. Avoid overheating. Pour into sterile Petri dishes or distribute into sterile tubes.

**Storage/Shelf Life:** Store dehydrated media in the dark in a sealed container below 30°C. Prepared media should be stored under refrigeration (2-8°C). Media should be used within 60 days of preparation. Media should not be used if there are any signs of deterioration (shrinking, cracking, or discoloration) or contamination, or if the expiration date supplied by the manufacturer has passed.

**Use:** For the selective isolation of pathogenic fungi from specimens with other fungi and bacteria.

## *Neisseria meningitidis* Medium

**Composition** per liter:

Beef infusion ...................................................................300.0g
Acid hydrolysate of casein................................................17.5g
Agar ....................................................................................17.0g
Starch ...................................................................................1.5g
Antibiotic solution ........................................................ 10.0mL

pH 7.4 ± 0.2 at 25°C

**Antibiotic Solution:**

**Composition** per 10.0mL:

Vancomycin .......................................................................3.0mg
Colistin...............................................................................7.5mg
Nystatin.........................................................................12,500U

**Preparation of Antibiotic Solution:** Add components to distilled/deionized water and bring volume to 10.0mL. Mix thoroughly. Filter sterilize.

**Preparation of Medium:** Add components, except antibiotic solution, to distilled/deionized water and bring volume to 990.0mL. Mix thoroughly. Gently heat and bring to boiling. Autoclave for 15 min at 15 psi pressure–121°C. Cool to 45°–50°C. Aseptically add sterile antibiotic solution. Mix thoroughly. Pour into sterile Petri dishes or distribute into sterile tubes.

**Storage/Shelf Life:** Store dehydrated media in the dark in a sealed container below 30°C. Prepared media should be stored under refrigeration (2-8°C). Media should be used within 60 days of preparation. Media should not be used if there are any signs of deterioration (discoloration) or contamination, or if the expiration date supplied by the manufacturer has passed.

**Use:** For the selective isolation and cultivation of *Neisseria meningitidis.*

## Neomycin Blood Agar

**Composition** per liter:

Pancreatic digest of casein...............................................14.5g
Agar ...................................................................................14.0g
Papaic digest of soybean meal ...........................................5.0g
NaCl......................................................................................5.0g
Growth factors ....................................................................1.5g
Sheep blood, defibrinated ...............................................50.0mL
Neomycin solution..........................................................10.0mL

pH 7.3 ± 0.2 at 25°C

**Source:** This medium is available as a premixed powder from BD Diagnostic Systems.

**Neomycin Solution:**

**Composition** per 10.0mL:

Neomycin sulfate .............................................................0.03g

**Preparation of Neomycin Solution:** Add neomycin sulfate to distilled/deionized water and bring volume to 10.0mL. Mix thoroughly. Filter sterilize.

**Preparation of Medium:** Add components, except sheep blood and neomycin solution, to distilled/deionized water and bring volume to 940.0mL. Mix thoroughly. Gently heat and bring to boiling. Autoclave for 15 min at 15 psi pressure–121°C. Cool to 45°–50°C. Aseptically add sterile sheep blood and sterile neomycin solution. Mix thoroughly. Pour into sterile Petri dishes or distribute into sterile tubes.

**Storage/Shelf Life:** Store dehydrated media in the dark in a sealed container below 30°C. Prepared media should be stored under refrigeration (2-8°C). Media should be used within 60 days of preparation. Media should not be used if there are any signs of deterioration (shrinking, cracking, or discoloration) or contamination, or if the expiration date supplied by the manufacturer has passed.

**Use:** For the isolation and cultivation of group A streptococci (*Streptococcus pyogenes)* and group B streptococci (*Streptococcus agalactiae*) from throat cultures and other clinical specimens.

## New York City Medium

**Composition** per liter:

NYC basal medium....................................................... 640.0mL
Horse blood cells .......................................................... 200.0mL
Horse plasma, citrated ................................................. 120.0mL
Yeast dialysate ............................................................... 25.0mL
Glucose solution ............................................................ 10.0mL
Antibiotic VCNT solution .............................................. 5.0mL

pH 7.4 ± 0.2 at 25°C

**NYC Basal Medium:**

**Composition** per 640.0mL:

Solution 1..................................................................... 400.0mL
Solution 3..................................................................... 200.0mL
Solution 2....................................................................... 40.0mL

**Preparation of NYC Basal Medium:** Combine solution 1, solution 2, and solution 3. Mix thoroughly. Autoclave for 15 min at 15 psi pressure–121°C. Cool to 45°–50°C.

**Solution 1:**

**Composition** per 400.0mL:

Agar ..................................................................................20.0g

**Preparation of Solution 1:** Add agar to distilled/deionized water and bring volume to 400.0mL. Mix thoroughly. Melt agar in autoclave for 10 min at 0 psi pressure–100°C. Cool to 45°–50°C.

**Solution 2:**

**Composition** per 40.0mL:

Cornstarch...........................................................................1.0g

**Preparation of Solution 2:** Add cornstarch to distilled/deionized water and bring volume to 40.0mL. Mix thoroughly. Warm to 45°–50°C.

**Solution 3:**

**Composition** per 200.0mL:

Proteose peptone No. 3 .....................................................15.0g
NaCl......................................................................................5.0g
$K_2HPO_4$............................................................................4.0g
$KH_2PO_4$............................................................................1.0g

**Preparation of Solution 3:** Add components to distilled/deionized water and bring volume to 1.0L. Mix thoroughly. Gently heat and bring to boiling. Cool to 45°–50°C.

**Horse Blood Cells:**
**Composition** per 200.0mL:
Horse blood cells, sedimented ....................................................... 6.0mL

**Preparation of Horse Blood Cells:** Cow blood may be used instead of horse blood but do not use sheep blood. Use cells freshly packed by sedimentation. Do not pack by centrifugation. Aseptically add 6.0mL of sedimented blood cells to 200.0mL of sterile distilled/deionized water. Mix thoroughly.

**Horse Plasma, Citrated:**
**Composition** per 6.0L:
Horse blood ........................................................................... 5400.0mL
Citrate solution ....................................................................... 600.0mL

**Preparation of Horse Plasma, Citrated:** Place 600.0mL of sterile citrate solution into a receiving bottle. Draw horse blood to the 6.0L mark. Allow cells to sediment out. Aseptically remove plasma.

**Citrate Solution:**
**Composition** per liter:
Sodium citrate ............................................................................150.0g
NaCl ..........................................................................................81.13g

**Preparation of Citrate Solution:** Add components to distilled/deionized water and bring volume to 1.0L. Mix thoroughly. Filter sterilize.

**Glucose Solution:**
**Composition** per 10.0mL:
D-Glucose ....................................................................................5.0g

**Preparation of Glucose Solution:** Add D-glucose to distilled/deionized water and bring volume to 10.0mL. Mix thoroughly. Autoclave for 10 min at 10 psi pressure–115°C. Cool to 45°–50°C.

**Yeast Dialysate:**
**Composition** per 2500.0mL:
Baker's yeast, fresh ..................................................................908.0g

**Preparation of Yeast Dialysate:** Add fresh baker's yeast to 2500.0mL of distilled/deionized water. Mix thoroughly. Autoclave for 10 min at 15 psi pressure–121°C. Cool to 25°C. Put into dialysis tubing. Dialyze against 2.0L of distilled/deionized water for 48 hr at 4°C.

**Antibiotic VCNT Solution:**
**Composition** per 5.0mL:
Colistin ......................................................................................7.5mg
Trimethorprim lactate ................................................................3.0mg
Vancomycin·HCl.........................................................................2.0mg
Nystatin .....................................................................................12.5U

**Preparation of Antibiotic Solution:** Add components to distilled/deionized water and bring volume to 5.0mL. Mix thoroughly. Filter sterilize.

**Preparation of Medium:** Have all solutions prepared and at 45°–50°C. Aseptically combine components. Mix thoroughly. Pour into sterile Petri dishes.

**Storage/Shelf Life:** Store dehydrated media in the dark in a sealed container below 30°C. Prepared media should be stored under refrigeration (2-8°C). Media should be used within 60 days of preparation. Media should not be used if there are any signs of deterioration (shrinking, cracking, or discoloration) or contamination, or if the expiration date supplied by the manufacturer has passed.

**Use:** For the isolation and cultivation of pathogenic *Neisseria* species. Used as a transport medium for urogenital and other clinical specimens. For the isolation and presumptive identification of Mycoplasmatales, including large-colony species (*Mycoplasma pneumoniae*) and T–mycoplasmas from urogenital specimens.

## New York City Medium, Modified

**Composition** per liter:
NYC basal medium.................................................................. 840.0mL
α-Gamma horse serum (Flow Labs)...................................... 120.0mL
Yeast dialysate ........................................................................ 25.0mL
Glucose solution ...................................................................... 10.0mL
Antibiotic LCNT solution......................................................... 5.0mL
pH 7.4 ± 0.2 at 25°C

**NYC Basal Medium:**
**Composition** per 840.0mL:
Horse blood ............................................................................ 5400.0mL
Solution 1................................................................................. 600.0mL
Solution 3................................................................................. 200.0mL
Solution 2................................................................................. 40.0mL

**Preparation of NYC Basal Medium:** Combine solution 1, solution 2, and solution 3. Mix thoroughly. Autoclave for 15 min at 15 psi pressure–121°C. Cool to 45°–50°C.

**Solution 1:**
**Composition** per 600.0mL:
Agar ...........................................................................................20.0g

**Preparation of Solution 1:** Add agar to distilled/deionized water and bring volume to 600.0mL. Mix thoroughly. Melt agar in autoclave for 10 min at 0 psi pressure–100°C. Cool to 45°–50°C.

**Solution 2:**
**Composition** per 40.0mL:
Cornstarch....................................................................................1.0g

**Preparation of Solution 2:** Add cornstarch to distilled/deionized water and bring volume to 40.0mL. Mix thoroughly. Warm to 45°–50°C.

**Solution 3:**
**Composition** per 200.0mL:
Proteose peptone No. 3 ...............................................................15.0g
NaCl............................................................................................5.0g
$K_2HPO_4$......................................................................................4.0g
$KH_2PO_4$......................................................................................1.0g

**Preparation of Solution 3:** Add components to distilled/deionized water and bring volume to 200.0mL. Mix thoroughly. Gently heat and bring to boiling. Cool to 45°–50°C.

**Glucose Solution:**
**Composition** per 10.0mL:
D-Glucose....................................................................................5.0g

**Preparation of Glucose Solution:** Add glucose to distilled/deionized water and bring volume to 10.0mL. Mix thoroughly. Autoclave for 10 min at 10 psi pressure–115°C. Cool to 45°–50°C.

**Yeast Dialysate:**
**Composition** per 2500.0mL:
Baker's yeast, fresh....................................................................908.0g

**Preparation of Yeast Dialysate:** Add fresh Baker's yeast to 2500.0mL of distilled/deionized water. Mix thoroughly. Autoclave for 10 min at 15 psi pressure–121°C. Cool to 25°C. Put into dialysis tubing. Dialyze against 2.0L of distilled/deionized water for 48 hr at 4°C.

**Antibiotic LCNT Solution:**
**Composition** per 5.0mL:

Colistin................................................................7.5mg
Lincomycin·HCl ...................................................4.0mg
Trimethorprim lactate ...........................................3.0mg
Nystatin............................................................12.5U

**Preparation of Antibiotic LCNT Solution:** Add the components to distilled/deionized water and bring volume to 5.0mL. Mix thoroughly. Filter sterilize the solution.

**Preparation of Medium:** Have all solutions prepared and at 45°–50°C. Aseptically combine components. Mix thoroughly. Pour into sterile Petri dishes.

**Storage/Shelf Life:** Store dehydrated media in the dark in a sealed container below 30°C. Prepared media should be stored under refrigeration (2-8°C). Media should be used within 60 days of preparation. Media should not be used if there are any signs of deterioration (shrinking, cracking, or discoloration) or contamination, or if the expiration date supplied by the manufacturer has passed.

**Use:** For the isolation and cultivation of pathogenic *Neisseria* species. Used as a transport medium for urogenital and other clinical specimens. For the isolation and presumptive identification of Mycoplasmatales, including large-colony species (*Mycoplasma pneumoniae*) and T–mycoplasmas from urogenital specimens.

## Nitrate Agar

**Composition** per liter:

Agar ...................................................................12.0g
Peptone................................................................5.0g
Beef extract..........................................................3.0g
KNO$_3$ ................................................................1.0g

pH 6.8 ± 0.2 at 25°C

**Preparation of Medium:** Add components to distilled/deionized water and bring volume to 1.0L. Mix thoroughly. Gently heat and bring to boiling. Distribute into tubes. Autoclave for 15 min at 15 psi pressure–121°C. Allow tubes to cool in a slanted position.

**Storage/Shelf Life:** Store dehydrated media in the dark in a sealed container below 30°C. Prepared media should be stored under refrigeration (2-8°C). Media should be used within 60 days of preparation. Media should not be used if there are any signs of deterioration (shrinking, cracking, or discoloration) or contamination, or if the expiration date supplied by the manufacturer has passed.

**Use:** For the differentiation of aerobic and facultative Gram-negative microorganisms based on their ability to reduce nitrate. Test for nitrates with sulfanilic acid and α-naphthylamine reagents. Bacteria that reduce nitrate to nitrite turn the reagents red or pink.

## Nitrate Broth

**Composition** per liter:

Peptone................................................................5.0g
Beef extract..........................................................3.0g
KNO$_3$ ................................................................1.0g

pH 7.0 ± 0.2 at 25°C

**Source:** This medium is available as a premixed powder from BD Diagnostic Systems.

**Preparation of Medium:** Add components to distilled/deionized water and bring volume to 1.0L. Mix thoroughly. Distribute into tubes or flasks. Autoclave for 15 min at 15 psi pressure–121°C.

**Storage/Shelf Life:** Store dehydrated media in the dark in a sealed container below 30°C. Prepared media should be stored under refrigeration (2-8°C). Media should be used within 60 days of preparation. Media should not be used if there are any signs of deterioration (discoloration) or contamination, or if the expiration date supplied by the manufacturer has passed.

**Use:** For the differentiation of aerobic and facultative Gram-negative microorganisms based on their ability to reduce nitrate. Test for nitrates with sulfanilic acid and α-naphthylamine reagents. Bacteria that reduce nitrate to nitrite turn the reagents red or pink.

## Nitrate Broth

**Composition** per liter:

Pancreatic digest of gelatin.......................................20.0g
KNO$_3$ ................................................................2.0g

pH 7.2 ± 0.2 at 25°C

**Source:** This medium is available as a premixed powder from BD Diagnostic Systems.

**Preparation of Medium:** Add components to distilled/deionized water and bring volume to 1.0L. Mix thoroughly. Distribute into tubes or flasks. Autoclave for 15 min at 15 psi pressure–121°C.

**Storage/Shelf Life:** Store dehydrated media in the dark in a sealed container below 30°C. Prepared media should be stored under refrigeration (2-8°C). Media should be used within 60 days of preparation. Media should not be used if there are any signs of deterioration (discoloration) or contamination, or if the expiration date supplied by the manufacturer has passed.

**Use:** For the differentiation of aerobic and facultative Gram-negative microorganisms based on their ability to reduce nitrate. Test for nitrates with sulfanilic acid and α-naphthylamine reagents. Bacteria that reduce nitrate to nitrite turn the reagents red or pink.

## Nutrient Agar
## (BAM M113)

**Composition** per liter:

Agar ...................................................................15.0g
Pancreatic digest of gelatin........................................5.0g
Beef extract..........................................................3.0g

pH 6.8 ± 0.2 at 25°C

**Source:** This medium is available as a premixed powder from BD Diagnostic Systems.

**Preparation of Medium:** Add components to distilled/deionized water and bring volume to 1.0L. Mix thoroughly. Gently heat while stirring and bring to boiling. Distribute into tubes or flasks. Autoclave for 15 min at 15 psi pressure–121°C. Pour into sterile Petri dishes or leave in tubes.

**Storage/Shelf Life:** Store dehydrated media in the dark in a sealed container below 30°C. Prepared media should be stored under refrigeration (2-8°C). Media should be used within 60 days of preparation. Media should not be used if there are any signs of deterioration (shrinking, cracking, or discoloration) or contamination, or if the expiration date supplied by the manufacturer has passed.

**Use:** For the cultivation of a wide variety of bacteria and for the enumeration of organisms in water, sewage, feces, and other materials. For the cultivation of *Bacillus cereus*.

## Nutrient Agar 1.5%, HiVeg with Ascitic Fluid

**Composition** per liter:

| | |
|---|---|
| Agar | 15.0g |
| NaCl | 8.0g |
| Plant peptone | 5.0g |
| Plant extract | 3.0g |
| Ascitic fluid | 250.0mL |

pH 7.4 ± 0.2 at 25°C

**Source:** This medium, without ascitic fluid, is available as a premixed powder from HiMedia.

**Preparation of Medium:** Add components, except ascitic fluid, to distilled/deionized water and bring volume to 750.0mL. Mix thoroughly. Autoclave for 15 min at 15 psi pressure–121°C. Cool to less than 37°C. Aseptically add sterile ascitic fluid. If desired, 0.5g of thallium acetate or 100,000U of penicillin may be added for a more selective medium. Mix thoroughly. Aseptically distribute into sterile tubes or flasks.

**Storage/Shelf Life:** Store dehydrated media in the dark in a sealed container below 30°C. Prepared media should be stored under refrigeration (2-8°C). Media should be used within 60 days of preparation. Media should not be used if there are any signs of deterioration (shrinking, cracking, or discoloration) or contamination, or if the expiration date supplied by the manufacturer has passed.

**Use:** For the enrichment of pleuro-pneumonia-like organisms (PPLOs) and *Mycoplasma* species from clinical specimens.

## N-Z Amine A™ Glycerol Agar

**Composition** per liter:

| | |
|---|---|
| Agar | 15.0g |
| N-Z Amine A™ | 5.0g |
| Beef extract | 1.0g |
| Glycerol | 70.0mL |

pH 6.5–7.0 at 25°C

**Preparation of Medium:** Add components to distilled/deionized water and bring volume to 1.0L. Mix thoroughly. Gently heat and bring to boiling. Distribute into tubes or flasks. Autoclave for 15 min at 15 psi pressure–121°C. Pour into sterile Petri dishes or leave in tubes.

**Storage/Shelf Life:** Store dehydrated media in the dark in a sealed container below 30°C. Prepared media should be stored under refrigeration (2-8°C). Media should be used within 60 days of preparation. Media should not be used if there are any signs of deterioration (shrinking, cracking, or discoloration) or contamination, or if the expiration date supplied by the manufacturer has passed.

**Use:** For the isolation and cultivation of *Actinomadura* species.

## O157:H7(+) Plating Medium

**Composition** per liter:

| | |
|---|---|
| Agar | 15.0g |
| Sorbitol | 12.0g |
| Salicin | 10.0g |
| Inositol | 10.0g |
| Peptone | 10.0g |
| Adonitol | 8.0g |
| NaCl | 5.0g |
| Tryptone | 5.0g |
| Proteose peptone | 3.0g |
| Bile salts No. 3 | 1.25g |
| Indoxyl-β-D-galactopyranoside | 0.12g |
| 5-Bromo-4-chloro-3-indoxyl-β-D-galactopyranoside | 0.12g |
| Phenol Red | 0.1g |
| Isopropyl-β-D-thiogalactopyranoside | 0.1g |
| Novobiocin solution | 1.0mL |
| Tellurite solution | 1.0mL |

pH 6.8 ± 0.2 at 25°C

**Source:** This medium is available as a premixed powder from BIO-SYNTH International, Inc.

**Novobiocin Solution:**

**Composition** per 10.0mL:

| | |
|---|---|
| Novobiocin | 0.1g |

**Preparation of Novobiocin Solution:** Add novobiocin to distilled/deionized water and bring volume to 10.0mL. Mix thoroughly. Filter sterilize.

**Tellurite Solution:**

**Composition** per 10.0mL:

| | |
|---|---|
| $K_2TeO_3$ | 0.01g |

**Preparation of Tellurite Solution:** Add $K_2TeO_3$ to distilled/deionized water and bring volume to 10.0mL. Mix thoroughly. Filter sterilize.

**Caution:** Potassium tellurite is toxic.

**Preparation of Medium:** Add components, except novobiocin solution and tellurite solution, to distilled/deionized water and bring volume to 998.0mL. Mix thoroughly. Gently heat while stirring and bring to boiling. Autoclave for 15 min at 15 psi pressure–121°C. Cool to 50°C. Aseptially add 1.0mL sterile novobiocin solution and 1.0 sterile tellurite solution. Mix thouroughly. Pour into sterile Petri dishes.

**Storage/Shelf Life:** Store dehydrated media in the dark in a sealed container below 30°C. Prepared media should be stored under refrigeration (2-8°C). Media should be used within 60 days of preparation. Media should not be used if there are any signs of deterioration (shrinking, cracking, or discoloration) or contamination, or if the expiration date supplied by the manufacturer has passed.

**Use:** For the detection of *Escherichia coli* O157:H7. *E.coli* O157:H7 grow with blue-black colonies and *E.coli* non-O157 with green-yellow colonies.

## O157:H7 ID Agar

**Composition** per liter:
Proprietary

**Source:** This medium is available from bioMérieux.

**Storage/Shelf Life:** Store in the dark under refrigeration (2-8°C). Chromogenic media are especially light and temperature sensitive; protect from light, excessive heat, moisture, and freezing. Do not use after the expiration date supplied by the manufacturer.

**Use:** A new chromogenic medium for the detection of *Escherichia coli* O157:H7. In accordance with USDA, BAM, and ISO 16654 standards, O157:H7 ID Agar allows the detection and presumptive identification of *E. coli* O157:H7.

## Ogawa TB Medium

**Composition** per 300.0mL:

| | |
|---|---|
| $KH_2PO_4$ | 1.0g |
| Homogenized whole egg | 200.0mL |
| Glycerol | 6.0mL |
| Malachite Green (2% solution) | 6.0mL |

pH 6.5 ± 0.2 at 25°C

**Homogenized Whole Egg:**
**Composition** per liter:
Whole eggs ...................................................................................18–24

**Preparation of Homogenized Whole Egg:** Use fresh eggs, less than 1 week old. Scrub the shells with soap. Let stand in a soap solution for 30 min. Rinse in running water. Soak eggs in 70% ethanol for 15 min. Break the eggs into a sterile container. Homogenize by shaking. Filter through four layers of sterile cheesecloth into a sterile graduated cylinder. Measure out 1.0L.

**Preparation of Medium:** Add components, except homogenized whole egg, to distilled/deionized water and bring volume to 100.0mL. Mix thoroughly. Autoclave for 15 min at 15 psi pressure–121°C. Cool to 45°–50°C. Aseptically add 200.0mL of sterile homogenized whole egg. Mix thoroughly. Aseptically distribute into sterile screw-capped tubes in 7.0mL volumes. Inspissate at 85°–90°C (moist heat) for 60 min.

**Storage/Shelf Life:** Store dehydrated media in the dark in a sealed container below 30°C. Prepared media should be stored under refrigeration (2-8°C). Media should be used within 60 days of preparation. Media should not be used if there are any signs of deterioration (discoloration) or contamination, or if the expiration date supplied by the manufacturer has passed.

**Use:** For the isolation and cultivation of *Mycobacterium* species, except for *Mycobacterium leprae*.

## Oleic Albumin Complex

**Composition** per liter:
Bovine albumin fraction V ...........................................................50.0g
NaCl.................................................................................................8.5g
Oleic acid .................................................................................... 0.6mL

**Preparation of Medium:** Add components to distilled/deionized water and bring volume to 1.0L. Mix thoroughly. Filter sterilize.

**Storage/Shelf Life:** Store dehydrated media in the dark in a sealed container below 30°C. Prepared media should be stored under refrigeration (2-8°C). Media should be used within 60 days of preparation. Media should not be used if there are any signs of deterioration (discoloration) or contamination, or if the expiration date supplied by the manufacturer has passed.

**Use:** For use in media employed for the cultivation of mycobacteria.

## ONE Broth-*Listeria*
## (Oxoid Novel Enrichment (ONE) Broth-*Listeria)*

**Composition** per 1005.0mL:
Peptone..........................................................................................28.0g
Salt mix .........................................................................................10.0g
Carbohydrate mix ...........................................................................6.0g
ONE Broth-*Listeria* selective supplement (proprietary) ........... 5.0mL
pH 7.4 ± 0.2 at 25°C

**Source:** This medium is available as a premixed powder from Oxoid Unipath.

**Preparation of Medium:** Add components, except ONE Broth-*Listeria* selective supplement, to distilled/deionized water and bring volume to 1.0L. Mix thoroughly. Gently heat while stirring and bring to boiling. Autoclave for 15 min at 15 psi pressure–121°C. Cool to 50°C. Aseptically add 5.0mL ONE Broth-*Listeria* selective supplement. Mix thoroughly. Pour into sterile Petri dishes.

**Storage/Shelf Life:** Store dehydrated media in the dark in a sealed container below 30°C. Prepared media should be stored under refrigeration (2-8°C). Media should be used within 60 days of preparation. Media should not be used if there are any signs of deterioration (discoloration) or contamination, or if the expiration date supplied by the manufacturer has passed.

**Use:** For the selective enrichment of *Listeria* spp. from food and environmental samples. A selective enrichment broth for *Listeria* species from food samples in 24 h.

## Önöz *Salmonella* Agar

**Composition** per liter:
Agar ..............................................................................................15.0g
Sucrose...........................................................................................13.0g
Lactose.......................................................................................... 11.5g
Trisodium citrate–5,5–hydrate.........................................................9.3g
Meat peptone ..................................................................................6.8g
Beef extract.....................................................................................6.0g
L–Phenylalanine .............................................................................5.0g
$Na_2S_2O_3 \cdot 5H_2O$ ...........................................................................4.25g
Bile salt mixture...........................................................................3.825g
Yeast extract...................................................................................3.0g
$Na_2HPO_4 \cdot 2H_2O$...........................................................................1.0g
Ferric citrate...................................................................................0.5g
Metachrome Yellow.......................................................................0.47g
$MgSO_4 \cdot 7H_2O$..............................................................................0.4g
Aniline Blue..................................................................................0.25g
Neutral Red.................................................................................0.022g
Brilliant Green .........................................................................0.00166g
pH 7.1 ± 0.2 at 25°C

**Preparation of Medium:** Add components to distilled/deionized water and bring volume to 1.0L. Mix thoroughly. Gently heat and bring to boiling. Distribute into tubes or flasks. Autoclave for 15 min at 15 psi pressure–121°C. Pour into sterile Petri dishes or leave in tubes.

**Storage/Shelf Life:** Store dehydrated media in the dark in a sealed container below 30°C. Prepared media should be stored under refrigeration (2-8°C). Media should be used within 60 days of preparation. Media should not be used if there are any signs of deterioration (discoloration) or contamination, or if the expiration date supplied by the manufacturer has passed.

**Use:** For the isolation and cultivation of *Salmonella* from feces.

## ONPG Broth

**Composition** per liter:
Peptone water....................................................................... 750.0mL
ONPG solution...................................................................... 250.0mL
pH 7.2–7.4 at 25°C

**ONPG Solution:**
**Composition** per 250.0mL:
ONPG (*o*-nitrophenyl-β-D-galactopyranoside)..............................1.5g
Sodium phosphate buffer (0.01*M*, pH 7.5) ........................... 250.0mL

**Preparation of ONPG Solution:** Add ONPG to 250.0mL of sodium phosphate buffer. Mix thoroughly. Filter sterilize.

**Peptone Water:**
**Composition** per 750.0mL:
Peptone ...........................................................................................7.5g
NaCl...............................................................................................3.75g

**Preparation of Peptone Water:** Add components to distilled/deionized water and bring volume to 750.0mL. Mix thoroughly. Gently heat and bring to boiling. Adjust pH to 8.0–8.4. Continue boiling for 10 min. Filter through Whatman #1 filter paper. Readjust pH of filtrate to 7.2–7.4. Autoclave for 20 min at 10 psi pressure–115°C. Cool to 25°C.

**Preparation of Medium:** Aseptically combine the sterile ONPG solution with the cooled, sterile peptone water. Mix thoroughly. Aseptically distribute into tubes in 2.5–3.0mL volumes. Store at 4°C for up to 1 month.

**Storage/Shelf Life:** Store dehydrated media in the dark in a sealed container below 30°C. Prepared media should be stored under refrigeration (2-8°C). Media should be used within 60 days of preparation. Media should not be used if there are any signs of deterioration (discoloration) or contamination, or if the expiration date supplied by the manufacturer has passed.

**Use:** For the differentiation of a variety of Gram-negative bacteria based on production of β-galactosidase. For the differentiation of lactose-delayed bacteria from lactose-negative bacteria. For the differentiation of *Pseudomonas cepacia* (positive) and *Pseudomonas maltophila* (positive) from other *Pseudomonas* species (negative). Bacteria that produce β-galactosidase turn the medium yellow.

## OR Indicator Agar
## (Oxidation-Reduction Indicator Agar)

**Composition** per liter:

Agar ....................................................................15.0g
Sodium glycerol phosphate.................................10.0g
Sodium thioglycolate ............................................1.7g
CaCl$_2$·2H$_2$O.........................................................0.1g
Methylene Blue.....................................................6.0mg

**Preparation of Medium:** Add components to distilled/deionized water and bring volume to 1.0L. Mix thoroughly. Gently heat and bring to boiling. Distribute into tubes or flasks. Autoclave for 15 min at 15 psi pressure–121°C. Pour into sterile Petri dishes or leave in tubes.

**Storage/Shelf Life:** Store dehydrated media in the dark in a sealed container below 30°C. Prepared media should be stored under refrigeration (2-8°C). Media should be used within 60 days of preparation. Media should not be used if there are any signs of deterioration (shrinking, cracking, or discoloration) or contamination, or if the expiration date supplied by the manufacturer has passed.

**Use:** For use as an indicator of oxygen-free conditions in anaerobic culture chambers.

## Orange Serum Agar

**Composition** per liter:

Agar ....................................................................15.5g
Orange serum.......................................................10.0g
Pancreatic digest of casein.................................10.0g
Glucose .................................................................4.0g
Yeast extract.........................................................3.0g
K$_2$HPO$_4$.................................................................2.5g

pH 5.5 ± 0.2 at 25°C

**Source:** This medium is available as a premixed powder from BD Diagnostic Systems.

**Preparation of Medium:** Add components to distilled/deionized water and bring volume to 1.0L. Mix thoroughly. Gently heat and bring to boiling. Distribute into tubes or flasks. Autoclave for 10 min at 15 psi pressure–121°C. Pour into sterile Petri dishes or leave in tubes.

**Storage/Shelf Life:** Store dehydrated media in the dark in a sealed container below 30°C. Prepared media should be stored under refrigeration (2-8°C). Media should be used within 60 days of preparation. Media should not be used if there are any signs of deterioration (shrinking, cracking, or discoloration) or contamination, or if the expiration date supplied by the manufacturer has passed.

**Use:** For the enumeration and cultivation of microorganisms from citrus juice and other products. For the cultivation of lactobacilli, pathogenic fungi, and other aciduric microorganisms.

## Orange Serum Broth Concentrate 10X

**Composition** per liter:

Pancreatic digest of casein.................................10.0g
Glucose .................................................................4.0g
Yeast extract.........................................................3.0g
K$_2$HPO$_4$.................................................................2.5g
Orange serum concentrate ............................. 100.0mL

pH 5.6 ± 0.2 at 25°C

**Source:** This medium is available as a premixed solution from BD Diagnostic Systems.

**Preparation of Medium:** Add components to distilled/deionized water and bring volume to 1.0L. Mix thoroughly. Distribute into tubes or flasks. Autoclave for 15 min at 15 psi pressure–121°C.

**Storage/Shelf Life:** Store dehydrated media in the dark in a sealed container below 30°C. Prepared media should be stored under refrigeration (2-8°C). Media should be used within 60 days of preparation. Media should not be used if there are any signs of deterioration (discoloration) or contamination, or if the expiration date supplied by the manufacturer has passed.

**Use:** For the cultivation and enumeration of microorganisms associated with the spoilage of citrus products. For the cultivation of lactobacilli, other aciduric microorganisms, and pathogenic fungi.

## Orange Serum HiVeg Agar

**Composition** per liter:

Agar ....................................................................17.0g
Plant hydrolysate ...............................................10.0g
Orange serum (solids from 200mL) .....................9.0g
Glucose .................................................................4.0g
Yeast extract.........................................................3.0g
K$_2$HPO$_4$.................................................................2.5g

pH 5.5 ± 0.2 at 25°C

**Source:** This medium is available as a premixed powder from HiMedia.

**Preparation of Medium:** Add components to distilled/deionized water and bring volume to 1.0L. Mix thoroughly. Gently heat and bring to boiling. Distribute into tubes or flasks. Autoclave for 15 min at 15 psi pressure–121°C. Pour into sterile Petri dishes or leave in tubes.

**Storage/Shelf Life:** Store dehydrated media in the dark in a sealed container below 30°C. Prepared media should be stored under refrigeration (2-8°C). Media should be used within 60 days of preparation. Media should not be used if there are any signs of deterioration (shrinking, cracking, or discoloration) or contamination, or if the expiration date supplied by the manufacturer has passed.

**Use:** For the cultivation and enumeration of microorganisms associated with the spoilage of citrus products. For the cultivation of lactobacilli, other aciduric microorganisms, and pathogenic fungi.

## Orange Serum HiVeg Broth

**Composition** per liter:

| | |
|---|---|
| Plant hydrolysate | 10.0g |
| Orange serum (solids from 200mL) | 9.0g |
| Glucose | 4.0g |
| Yeast extract | 3.0g |
| K$_2$HPO$_4$ | 2.5g |

pH 5.5 ± 0.2 at 25°C

**Source:** This medium is available as a premixed powder from Hi-Media.

**Preparation of Medium:** Add components to distilled/deionized water and bring volume to 1.0L. Mix thoroughly. Gently heat and bring to boiling. Distribute into tubes or flasks. Autoclave for 15 min at 15 psi pressure–121°C.

**Storage/Shelf Life:** Store dehydrated media in the dark in a sealed container below 30°C. Prepared media should be stored under refrigeration (2-8°C). Media should be used within 60 days of preparation. Media should not be used if there are any signs of deterioration (discoloration) or contamination, or if the expiration date supplied by the manufacturer has passed.

**Use:** For the cultivation of lactobacilli, other aciduric microorganisms, and pathogenic fungi.

## Oxacillin Resistance Screening Agar Base

**Composition** per liter:

| | |
|---|---|
| NaCl | 55.0g |
| Agar | 12.5g |
| Peptic digest of animal tissue | 11.8g |
| Mannitol | 10.0g |
| Yeast extract | 9.0g |
| LiCl | 5.0g |
| Aniline Blue | 0.2g |

pH 7.2 ± 0.2 at 25°C

**Source:** This medium is available from HiMedia.

**Caution:** Lithium chloride is harmful. Avoid bodily contact and inhalation of vapors. On contact with skin, wash with plenty of water immediately.

**Selective Supplement Solution:**

**Composition** per 10.0mL:

| | |
|---|---|
| Oxacillin | 2.0mg |
| Polymyxin B | 50,000U |

**Preparation of Selective Supplement Solution:** Add components to distilled/deionized water and bring volume to 10.0mL. Mix thoroughly. Filter sterilize.

**Preparation of Medium:** Add components, except selective supplement solution, to distilled/deionized water and bring volume to 990.0mL. Mix thoroughly. Autoclave for 15 min at 15 psi pressure–121°C. Cool to 50°C. Aseptically add selective supplement solution. Mix thoroughly. Pour into Petri dishes or aseptically distribute into sterile tubes.

**Storage/Shelf Life:** Store dehydrated media in the dark in a sealed container below 30°C. Prepared media should be stored under refrigeration (2-8°C). Media should be used within 60 days of preparation. Media should not be used if there are any signs of deterioration (shrinking, cracking, or discoloration) or contamination, or if the expiration date supplied by the manufacturer has passed.

**Use:** For the screening of oxacillin-resistant microorganisms.

## Oxford Agar
### (*Listeria* Selective Agar, Oxford)

**Composition** per liter:

| | |
|---|---|
| Special peptone | 23.0g |
| LiCl | 15.0g |
| Agar | 10.0g |
| NaCl | 5.0g |
| Cornstarch | 1.0g |
| Esculin | 1.0g |
| Ferric ammonium citrate | 0.5g |
| Antibiotic inhibitor | 10.0mL |

pH 7.0 ± 0.2 at 25°C

**Source:** This medium is available as a premixed powder from Oxoid Unipath.

**Caution:** Lithium chloride is harmful. Avoid bodily contact and inhalation of vapors. On contact with skin, wash with plenty of water immediately.

**Antibiotic Inhibitor:**

**Composition** per 10.0mL:

| | |
|---|---|
| Cycloheximide | 0.4g |
| Colistin sulfate | 0.02g |
| Fosfomycin | 0.01g |
| Acriflavine | 5.0mg |
| Cefotetan | 2.0mg |
| Ethanol (50% solution) | 10.0mL |

**Preparation of Antibiotic Inhibitor:** Add antibiotics to 10.0mL of ethanol. Mix thoroughly. Filter sterilize.

**Caution:** Cycloheximide is toxic. Avoid skin contact or aerosol formation and inhalation.

**Preparation of Medium:** Add components, except antibiotic inhibitor, to distilled/deionized water and bring volume to 990.0mL. Mix thoroughly. Gently heat and bring to boiling. Autoclave for 15 min at 15 psi pressure–121°C. Cool to 45°–50°C. Aseptically add 10.0mL of sterile antibiotic inhibitor. Mix thoroughly. Pour into sterile Petri dishes or distribute into sterile tubes.

**Storage/Shelf Life:** Store dehydrated media in the dark in a sealed container below 30°C. Prepared media should be stored under refrigeration (2-8°C). Media should be used within 60 days of preparation. Media should not be used if there are any signs of deterioration (shrinking, cracking, or discoloration) or contamination, or if the expiration date supplied by the manufacturer has passed.

**Use:** For the isolation and cultivation of *Listeria monocytogenes* from specimens containing a mixed bacterial flora.

## Oxford Agar, Modified
### (*Listeria* Selective Agar, Modified Oxford)
### (MOX Agar)

**Composition** per liter:

| | |
|---|---|
| Special peptone | 23.0g |
| LiCl | 15.0g |
| Agar | 12.0g |
| NaCl | 5.0g |
| Cornstarch | 1.0g |
| Esculin | 1.0g |

Ferric ammonium citrate.................................................0.5g
Antibiotic inhibitor ................................................ 10.0mL

pH 7.0 ± 0.2 at 25°C

**Caution:** Lithium chloride is harmful. Avoid bodily contact and inhalation of vapors. On contact with skin, wash with plenty of water immediately.

**Antibiotic Inhibitor:**
**Composition** per 10.0mL:
Moxalactam ...............................................................0.015g
Colistin sulfate ............................................................0.01g

**Preparation of Antibiotic Inhibitor:** Add components to distilled/deionized water and bring volume to 10.0mL. Mix thoroughly. Filter sterilize.

**Preparation of Medium:** Add components, except antibiotic inhibitor, to distilled/deionized water and bring volume to 990.0mL. Mix thoroughly. Gently heat and bring to boiling. Autoclave for 10 min at 15 psi pressure–121°C. Cool to 45°–50°C. Aseptically add 10.0mL of sterile antibiotic inhibitor. Mix thoroughly. Pour into sterile Petri dishes or distribute into sterile tubes.

**Storage/Shelf Life:** Store dehydrated media in the dark in a sealed container below 30°C. Prepared media should be stored under refrigeration (2-8°C). Media should be used within 60 days of preparation. Media should not be used if there are any signs of deterioration (shrinking, cracking, or discoloration) or contamination, or if the expiration date supplied by the manufacturer has passed.

**Use:** For the isolation and cultivation of *Listeria monocytogenes* from specimens containing a mixed bacterial flora.

## Oxford Medium
### (BAM M118)

**Composition** per liter:
Special peptone ............................................................23.0g
LiCl .............................................................................15.0g
Agar ............................................................................10.0g
NaCl..............................................................................5.0g
Cornstarch .....................................................................1.0g
Esculin ..........................................................................1.0g
Ferric ammonium citrate................................................0.5g
Antibiotic inhibitor ................................................ 10.0mL

pH 7.0 ± 0.2 at 25°C

**Source:** This medium is available as a premixed powder from Oxoid Unipath.

**Caution:** Lithium chloride is harmful. Avoid bodily contact and inhalation of vapors. On contact with skin, wash with plenty of water immediately.

**Supplement Mix:**
**Composition** per 10.0mL:
Cycloheximide ..............................................................0.4g
Colistin sulfate ............................................................0.02g
Fosfomycin .................................................................0.01g
Acriflavine .................................................................5.0mg
Cefotetan....................................................................2.0mg
Ethanol (50% solution) ......................................... 10.0mL

**Preparation of Supplement Mix:** Add components to 10.0mL of ethanol. Mix thoroughly. Filter sterilize.

**Caution:** Cycloheximide is toxic. Avoid skin contact or aerosol formation and inhalation.

**Preparation of Medium:** Add components, except supplement mix, to distilled/deionized water and bring volume to 990.0mL. Mix thoroughly. Gently heat and bring to boiling. Autoclave for 15 min at 15 psi pressure–121°C. Cool to 50°C. Aseptically add 10.0mL of sterile supplement mix. Mix thoroughly. Pour into sterile Petri dishes or distribute into sterile tubes.

**Storage/Shelf Life:** Store dehydrated media in the dark in a sealed container below 30°C. Prepared media should be stored under refrigeration (2-8°C). Media should be used within 60 days of preparation. Media should not be used if there are any signs of deterioration (discoloration) or contamination, or if the expiration date supplied by the manufacturer has passed.

**Use**: For the isolation and cultivation of *Listeria monocytogenes* from specimens containing a mixed bacterial flora.

## Oxidation-Fermentation Medium
### (OF Medium)

**Composition** per liter:
NaCl..............................................................................5.0g
Agar ..............................................................................2.5g
Pancreatic digest of casein.............................................2.0g
K$_2$HPO$_4$..............................................................0.3g
Bromthymol Blue ........................................................0.03g
Carbohydrate solution............................................ 100.0mL

pH 6.8 ± 0.1 at 25°C

**Source:** This medium is available as a premixed powder from BD Diagnostic Systems.

**Carbohydrate Solution:**
**Composition** per 100.0mL:
Carbohydrate................................................................10.0g

**Preparation of Carbohydrate Solution:** Add carbohydrate to distilled/deionized water and bring volume to 100.0mL. Mix thoroughly. Filter sterilize.

**Preparation of Medium:** Add components, except carbohydrate solution, to distilled/deionized water and bring volume to 900.0mL. Mix thoroughly. Gently heat and bring to boiling. Autoclave for 15 min at 15 psi pressure–121°C. Cool to 45°–50°C. Aseptically add 100.0mL of sterile carbohydrate solution. Mix thoroughly. Distribute into sterile tubes.

**Storage/Shelf Life:** Store dehydrated media in the dark in a sealed container below 30°C. Prepared media should be stored under refrigeration (2-8°C). Media should be used within 60 days of preparation. Media should not be used if there are any signs of deterioration (discoloration) or contamination, or if the expiration date supplied by the manufacturer has passed.

**Use:** For differentiating Gram-negative bacteria based upon determining the oxidative and fermentative metabolism of carbohydrates.

## Oxidation-Fermentation Medium, Hugh-Leifson's
### (Hugh-Leifson's Oxidation Fermentation Medium)
**Composition** per liter:
NaCl..............................................................................5.0g
Agar ..............................................................................3.0g
Peptone .........................................................................2.0g
K$_2$HPO$_4$..............................................................0.3g

Carbohydrate solution.................................................100.0mL
Bromthymol Blue solution (0.2% )...........................15.0mL
pH 7.1 ± 0.2 at 25°C

**Carbohydrate Solution:**
**Composition** per 100.0mL:
Carbohydrate...............................................................10.0g

**Preparation of Carbohydrate Solution:** Add carbohydrate to distilled/deionized water and bring volume to 100.0mL. Mix thoroughly. Filter sterilize.

**Preparation of Medium:** Add components, except carbohydrate solution, to distilled/deionized water and bring volume to 900.0mL. Mix thoroughly. Gently heat and bring to boiling. Autoclave for 15 min at 15 psi pressure–121°C. Cool to 45°–50°C. Aseptically add 100.0mL of sterile carbohydrate solution. Mix thoroughly. Distribute into sterile tubes.

**Storage/Shelf Life:** Store dehydrated media in the dark in a sealed container below 30°C. Prepared media should be stored under refrigeration (2-8°C). Media should be used within 60 days of preparation. Media should not be used if there are any signs of deterioration (discoloration) or contamination, or if the expiration date supplied by the manufacturer has passed.

**Use:** For differentiating Gram-negative bacteria, such as *Vibrio* species, based upon determining the oxidative and fermentative metabolism of carbohydrates. Bacteria that ferment the carbohydrate turn the medium yellow.

## Oxidation-Fermentation Medium, King's
## (King's OF Medium)
**Composition** per liter:
Agar .......................................................................3.0g
Pancreatic digest of casein....................................2.0g
Carbohydrate solution...........................................100.0mL
Phenol Red (1.5% solution) ...................................2.0mL
Carbohydrate solution...........................................100.0mL
pH to 7.3 ± 0.2

**Preparation of Medium:** Add components, except carbohydrate solution, to distilled/deionized water and bring volume to 900.0mL. Mix thoroughly. Gently heat and bring to boiling. Autoclave for 15 min at 15 psi pressure–121°C. Cool to 45°–50°C. Aseptically add 100.0mL of sterile carbohydrate solution. Mix thoroughly. Pour into sterile Petri dishes or distribute into sterile tubes.

**Carbohydrate Solution:**
**Composition** per 100.0mL:
Carbohydrate...............................................................10.0g

**Preparation of Carbohydrate Solution:** Add carbohydrate to distilled/deionized water and bring volume to 100.0mL. Mix thoroughly. Filter sterilize.

**Preparation of Medium:** Add components, except carbohydrate solution, to distilled/deionized water and bring volume to 900.0mL. Mix thoroughly. Gently heat and bring to boiling. Autoclave for 15 min at 15 psi pressure–121°C. Cool to 45°–50°C. Aseptically add 100.0mL of sterile carbohydrate solution. Mix thoroughly. Distribute into sterile tubes.

**Storage/Shelf Life:** Store dehydrated media in the dark in a sealed container below 30°C. Prepared media should be stored under refrigeration (2-8°C). Media should be used within 60 days of preparation. Media should not be used if there are any signs of deterioration (discol-

oration) or contamination, or if the expiration date supplied by the manufacturer has passed.

**Use:** For differentiating bacteria based upon determining the oxidative and fermentative metabolism of carbohydrates. Bacteria that ferment the carbohydrate turn the medium yellow.

## Oxidative-Fermentative Medium
## (OF Medium)
**Composition** per liter:
NaCl.........................................................................5.0g
Agar .........................................................................2.0g
Pancreatic digest of casein.......................................2.0g
$K_2HPO_4$..................................................................0.3g
Bromthymol Blue .....................................................0.08g
Carbohydrate solution...............................................100.0mL
pH 6.8 ± 0.1 at 25°C

**Source:** This medium is available as a premixed powder from BD Diagnostic Systems and Oxoid Unipath.

**Carbohydrate Solution:**
**Composition** per 100.0mL:
Carbohydrate...............................................................10.0g

**Preparation of Carbohydrate Solution:** Add carbohydrate to distilled/deionized water and bring volume to 100.0mL. Mix thoroughly. Filter sterilize.

**Preparation of Medium:** Add components, except carbohydrate solution, to distilled/deionized water and bring volume to 900.0mL. Mix thoroughly. Gently heat and bring to boiling. Autoclave for 15 min at 15 psi pressure–121°C. Cool to 45°–50°C. Aseptically add 100.0mL of sterile carbohydrate solution. Mix thoroughly. Pour into sterile Petri dishes or distribute into sterile tubes.

**Use:** For differentiating bacteria based upon determining the oxidative and fermentative metabolism of carbohydrates. Bacteria that ferment the carbohydrate turn the medium yellow.

**Preparation of Medium:** Add components to distilled/deionized water and bring volume to 1.0L. Mix thoroughly. Adjust pH to 6.5. Autoclave for 15 min at 15 psi pressure–121°C. Cool to 55°–60°C. Readjust pH to 7.1. Mix thoroughly. Pour into 50mm × 12mm Petri dishes in 3.0mL volumes.

**Storage/Shelf Life:** Store dehydrated media in the dark in a sealed container below 30°C. Prepared media should be stored under refrigeration (2-8°C). Media should be used within 60 days of preparation. Media should not be used if there are any signs of deterioration (discoloration) or contamination, or if the expiration date supplied by the manufacturer has passed.

**Use:** For the cultivation and estimation of numbers of *Pseudomonas aeruginosa* in water by the membrane filter method. For the detection of presence and absence of coliform bacteria in water from treatment plants or distribution systems.

## PA-C Agar
## (mPA-C Agar)
**Composition** per liter:
Agar ..........................................................................12.0g
L-Lysine·HCl..............................................................5.0g
NaCl...........................................................................5.0g
$Na_2S_2O_3$ ..................................................................5.0g
Yeast extract...............................................................2.0g
$MgSO_4 \cdot 7H_2O$ .........................................................1.5g

| | |
|---|---|
| Lactose | 1.25g |
| Sucrose | 1.25g |
| Xylose | 1.25g |
| Ferric ammonium citrate | 0.8g |
| Phenol Red | 0.08g |
| Nalidixic acid | 0.037g |
| Kanamycin | 8.0mg |

pH 7.2 ± 0.1 at 25°C

**Source:** This medium is available as a premixed powder from BD Diagnostic Systems.

**Preparation of Medium:** Add components to distilled/deionized water and bring volume to 1.0L. Mix thoroughly. Gently heat and bring to boiling. Distribute into tubes or flasks. Autoclave for 15 min at 15 psi pressure–121°C. Pour into sterile Petri dishes or leave in tubes.

**Storage/Shelf Life:** Store dehydrated media in the dark in a sealed container below 30°C. Prepared media should be stored under refrigeration (2-8°C). Media should be used within 60 days of preparation. Media should not be used if there are any signs of deterioration (shrinking, cracking, or discoloration) or contamination, or if the expiration date supplied by the manufacturer has passed.

**Use:** For the selective recovery and enumeration of *Pseudomonas aeruginosa* from water samples.

## PALCAM Agar
### (Polymyxin Acriflavine Lithium Chloride Ceftazidime Esculin Mannitol Agar)

**Composition** per liter:

| | |
|---|---|
| Peptone | 23.0g |
| LiCl$_2$ | 15.0g |
| Agar | 10.0g |
| Mannitol | 10.0g |
| NaCl | 5.0g |
| Yeast extract | 3.0g |
| Starch | 1.0g |
| Esculin | 0.8g |
| Ferric ammonium citrate | 0.5g |
| Glucose | 0.5g |
| Phenol Red | 0.08g |
| PALCAM selective supplement | 10.0mL |

pH 7.2 ± 0.2 at 25°C

**Source:** This medium is available as a premixed powder from Oxoid Unipath.

### PALCAM Selective Supplement:
**Composition** per 10.0mL:

| | |
|---|---|
| Ceftazidime | 20.0mg |
| Polymyxin B | 10.0mg |
| Acriflavine·HCl | 5.0mg |

**Preparation of PALCAM Selective Supplement:** Add components to distilled/deionized water and bring volume to 10.0mL. Mix thoroughly. Filter sterilize.

**Preparation of Medium:** Add components, except PALCAM selective supplement, to distilled/deionized water and bring volume to 990.0mL. Mix thoroughly. Gently heat and bring to boiling. Autoclave for 15 min at 15 psi pressure–121°C. Cool to 45°–50°C. Aseptically add sterile PALCAM selective supplement. Mix thoroughly.

**Storage/Shelf Life:** Store dehydrated media in the dark in a sealed container below 30°C. Prepared media should be stored under refrigeration (2-8°C). Media should be used within 60 days of preparation.

Media should not be used if there are any signs of deterioration (shrinking, cracking, or discoloration) or contamination, or if the expiration date supplied by the manufacturer has passed.

**Use:** For the selective isolation, cultivation, and differentiation of *Listeria monocytogenes* and other *Listeria* species from foods.

## PALCAM Agar with Egg Yolk Emulsion
### (Polymyxin Acriflavine Lithium Chloride Ceftazidime Esculin Mannitol Agar with Egg Yolk Emulsion)

**Composition** per liter:

| | |
|---|---|
| Peptone | 23.0g |
| LiCl$_2$ | 15.0g |
| Agar | 10.0g |
| Mannitol | 10.0g |
| NaCl | 5.0g |
| Yeast extract | 3.0g |
| Starch | 1.0g |
| Esculin | 0.8g |
| Ferric ammonium citrate | 0.5g |
| Glucose | 0.5g |
| Phenol Red | 0.08g |
| Egg yolk emulsion | 25.0mL |
| PALCAM selective supplement | 10.0mL |

pH 7.2 ± 0.2 at 25°C

**Source:** This medium is available as a premixed powder from Oxoid Unipath.

### PALCAM Selective Supplement:
**Composition** per 10.0mL:

| | |
|---|---|
| Ceftazidime | 20.0mg |
| Polymyxin B | 10.0mg |
| Acriflavine·HCl | 5.0mg |

**Preparation of PALCAM Selective Supplement:** Add components to distilled/deionized water and bring volume to 10.0mL. Mix thoroughly. Filter sterilize.

### Egg Yolk Emulsion:
**Composition**:

| | |
|---|---|
| Chicken egg yolks | 11 |
| Whole chicken egg | 1 |

**Preparation of Egg Yolk Emulsion:** Soak eggs with 1:100 dilution of saturated mercuric chloride solution for 1 min. Crack eggs and separate yolks from whites. Mix egg yolks with 1 chicken egg.

**Preparation of Medium:** Add components, except PALCAM selective supplement and egg yolk emulsion, to distilled/deionized water and bring volume to 990.0mL. Mix thoroughly. Gently heat and bring to boiling. Autoclave for 15 min at 15 psi pressure–121°C. Cool to 45°–50°C. Aseptically add sterile PALCAM selective supplement and egg yolk emulsion. Mix thoroughly. Pour into sterile Petri dishes or distribute into sterile tubes.

**Storage/Shelf Life:** Store dehydrated media in the dark in a sealed container below 30°C. Prepared media should be stored under refrigeration (2-8°C). Media should be used within 60 days of preparation. Media should not be used if there are any signs of deterioration (shrinking, cracking, or discoloration) or contamination, or if the expiration date supplied by the manufacturer has passed.

**Use:** For the selective isolation, cultivation, and differentiation of *Listeria monocytogenes* and other *Listeria* species from foods The addition of egg yolk emulsion aids in the recovery of damaged *Listeria*.

## Park and Sanders Enrichment Broth

**Composition** per 1010.0mL:

Basal medium ................................................................ 1.0L
Supplement A .......................................................... 5.0mL
Supplement B .......................................................... 5.0mL

pH 7.0 ± 0.2 at 25°C

**Basal Medium:**
**Composition** per liter:

Pancreatic digest of casein ......................................... 10.0g
Peptic digest of animal tissue ..................................... 10.0g
NaCl ............................................................................. 5.0g
Yeast extract ................................................................ 2.0g
Glucose ........................................................................ 1.0g
Sodium pyruvate ....................................................... 0.25g
NaHSO$_3$ .................................................................... 0.1g
Horse blood, lysed .................................................. 50.0mL

**Preparation of Basal Medium:** Add components, except horse blood, to distilled/deionized water and bring volume to 950.0mL. Mix thoroughly. Gently heat and bring to boiling. Autoclave for 15 min at 15 psi pressure–121°C. Cool to 25°C. Aseptically add sterile horse blood. Mix thoroughly.

**Supplement A:**
**Composition** per 5.0mL:

Vancomycin ................................................................ 0.01g
Trimethoprim lactate ................................................. 0.01g

**Preparation of Supplement A:** Add components to distilled/deionized water and bring volume to 5.0mL. Mix thoroughly. Filter sterilize.

**Supplement B:**
**Composition** per 5.0mL:

Cefoperazone ........................................................... 0.032g
Cycloheximide ............................................................. 0.1g

**Preparation of Supplement B:** Add components to distilled/deionized water and bring volume to 5.0mL. Mix thoroughly. Filter sterilize.

**Caution:** Cycloheximide is toxic. Avoid skin contact or aerosol formation and inhalation.

**Preparation of Medium:** To 1.0L of cooled, sterile basal medium, aseptically add 5.0mL of sterile supplement A. Mix thoroughly. Aseptically distribute into flasks in 100.0mL volumes. Inoculate medium with food samples. Incubate at 31°–32°C for 4 h to recover and resuscitate injured cells. Aseptically add 0.5mL of supplement B to each 100.0mL of medium. Incubate cultures at 37°C for 2 h.

**Storage/Shelf Life:** Store dehydrated media in the dark in a sealed container below 30°C. Prepared media should be stored under refrigeration (2-8°C). Media should be used within 60 days of preparation. Media should not be used if there are any signs of deterioration (discoloration) or contamination, or if the expiration date supplied by the manufacturer has passed.

**Use:** For the cultivation and enrichment of *Campylobacter* species from foods.

## Park and Sanders Enrichment HiVeg Broth Base with Horse Blood and Selective Antibiotics

**Composition** per liter:

Plant hydrolysate ....................................................... 10.0g
Plant peptone ............................................................. 10.0g

NaCl ............................................................................. 5.0g
Yeast extract ................................................................ 2.0g
Glucose ........................................................................ 1.0g
Sodium pyruvate ....................................................... 0.25g
Sodium biselenite ........................................................ 0.1g
Horse blood, lysed .................................................. 50.0mL
Supplement A .......................................................... 5.0mL
Supplement B .......................................................... 5.0mL

pH 7.0 ± 0.2 at 25°C

**Source:** This medium, without horse blood and selective antibiotic supplements, is available as a premixed powder from HiMedia.

**Supplement A:**
**Composition** per 5.0mL:

Vancomycin ................................................................ 0.01g
Trimethoprim lactate ................................................. 0.01g

**Preparation of Supplement A:** Add components to distilled/deionized water and bring volume to 5.0mL. Mix thoroughly. Filter sterilize.

**Supplement B:**
**Composition** per 5.0mL:

Cefoperazone ........................................................... 0.032g
Cycloheximide ............................................................. 0.1g

**Preparation of Supplement B:** Add components to distilled/deionized water and bring volume to 5.0mL. Mix thoroughly. Filter sterilize.

**Caution:** Cycloheximide is toxic. Avoid skin contact or aerosol formation and inhalation.

**Preparation of Medium:** Add components, except horse blood, supplement A and supplement B, to distilled/deionized water and bring volume to 940.0mL. Mix thoroughly. Gently heat and bring to boiling. Autoclave for 15 min at 15 psi pressure–121°C. Cool to 25°C. Aseptically add 50.0mL sterile horse blood. Mix thoroughly. Aseptically add 5.0mL of sterile supplement A. Mix thoroughly. Aseptically distribute into flasks in 100.0mL volumes. Inoculate medium with food samples. Incubate at 31°–32°C for 4 h to recover and resuscitate injured cells. Aseptically add 0.5mL of supplement B to each 100.0mL of medium. Incubate cultures at 37°C for 2 h.

**Storage/Shelf Life:** Store dehydrated media in the dark in a sealed container below 30°C. Prepared media should be stored under refrigeration (2-8°C). Media should be used within 60 days of preparation. Media should not be used if there are any signs of deterioration (discoloration) or contamination, or if the expiration date supplied by the manufacturer has passed.

**Use:** For the cultivation and enrichment of *Campylobacter* species from foods.

## *Pasteurella haemolytica* Selective Medium

**Composition** per 1010.0mL:

Tryptose agar with peptic digest of blood ..................... 1.0L
Antibiotic solution .................................................. 10.0mL

pH 7.2 ± 0.2 at 25°C

**Tryptose Agar with Peptic Digest of Blood:**
**Composition** per liter:

Agar ........................................................................... 15.0g
Pancreatic digest of casein ......................................... 10.0g
Peptic digest of animal tissue ..................................... 10.0g

NaCl ................................................................................5.0g
Glucose ...........................................................................1.0g
Peptic digest of blood ..............................................50.0mL

**Preparation of Tryptose Agar with Peptic Digest of Blood:**
Add components to distilled/deionized water and bring volume to
950.0mL. Mix thoroughly. Gently heat and bring to boiling. Autoclave
for 15 min at 15 psi pressure–121°C. Cool to 45°–50°C. Aseptically
add peptic digest of blood. Mix thoroughly.

**Antibiotic Solution:**
**Composition** per 10.0mL:
Actidione (cycloheximide) ..............................................0.1g
Novobiocin................................................................2.0mg
Neomycin...................................................................1.5mg

**Preparation of Antibiotic Solution:** Add components to distilled/
deionized water and bring volume to 10.0mL. Mix thoroughly. Filter
sterilize.

**Caution:** Cycloheximide is toxic. Avoid skin contact or aerosol formation and inhalation.

**Preparation of Medium:** To 1.0L of cooled, sterile tryptose agar
with peptic digest of blood, aseptically add 10.0mL of sterile antibiotic
solution. Mix thoroughly. Pour into sterile Petri dishes or distribute into
sterile tubes.

**Storage/Shelf Life:** Store dehydrated media in the dark in a sealed
container below 30°C. Prepared media should be stored under refrigeration (2-8°C). Media should be used within 60 days of preparation.
Media should not be used if there are any signs of deterioration (discoloration) or contamination, or if the expiration date supplied by the
manufacturer has passed.

**Use:** For the selective cultivation of *Pasteurella haemolytica*.

### *Pasteurella multocida* Selective Medium
**Composition** per 1020.0mL:
Tryptose agar with peptic digest of blood.....................1.0L
Antibiotic solution .....................................................10.0mL
K$_2$TeO$_3$ solution .........................................................10.0mL
pH 7.2 ± 0.2 at 25°C

**Tryptose Agar with Peptic Digest of Blood:**
**Composition** per liter:
Agar ..............................................................................15.0g
Pancreatic digest of casein............................................10.0g
Peptic digest of animal tissue........................................10.0g
NaCl ................................................................................5.0g
Glucose ...........................................................................1.0g
Peptic digest of blood ..................................................50.0mL

**Preparation of Tryptose Agar with Peptic Digest of Blood:**
Add components to distilled/deionized water and bring volume to
950.0mL. Mix thoroughly. Gently heat and bring to boiling. Autoclave
for 15 min at 15 psi pressure–121°C. Cool to 45°–50°C. Aseptically
add peptic digest of blood. Mix thoroughly.

**Antibiotic Solution:**
**Composition** per 10.0mL:
Actidione (cycloheximide) ..............................................0.1g
Novobiocin...............................................................0.01g
Erythrocin ...................................................................5.0mg

**Preparation of Antibiotic Solution:** Add components to distilled/
deionized water and bring volume to 10.0mL. Mix thoroughly. Filter
sterilize.

**Caution:** Cycloheximide is toxic. Avoid skin contact or aerosol formation and inhalation.

**K$_2$TeO$_3$ Solution:**
**Composition** per 10.0mL:
K$_2$TeO$_3$...........................................................................5.0mg

**Preparation of K$_2$TeO$_3$ Solution:** Add K$_2$TeO$_3$ to distilled/deion-
ized water and bring volume to 10.0mL. Mix thoroughly. Filter sterilize.

**Caution:** Potassium tellurite is toxic.

**Preparation of Medium:** To 1.0L of cooled, sterile tryptose agar with
peptic digest of blood, aseptically add 10.0mL of sterile antibiotic solution
and 10.0mL of sterile K$_2$TeO$_3$ solution. Mix thoroughly. Pour into sterile
Petri dishes or distribute into sterile tubes.

**Storage/Shelf Life:** Store dehydrated media in the dark in a sealed
container below 30°C. Prepared media should be stored under refrigeration (2-8°C). Media should be used within 60 days of preparation.
Media should not be used if there are any signs of deterioration (discoloration) or contamination, or if the expiration date supplied by the
manufacturer has passed.

**Use:** For the selective cultivation of *Pasteurella multocida*.

### PE-2 HiVeg Medium
**Composition** per liter:
Plant peptone ...............................................................20.0g
Yeast extract....................................................................3.0g
Bromcresol Purple ........................................................0.04g
Alaska seed peas.........................................................variable
pH 6.8 ± 0.2 at 25°C

**Source:** This medium is available as a premixed powder from Hi-
Media.

**Preparation of Medium:** Add components, except alaska seed
peas, to distilled/deionized water and bring volume to 1.0L. Mix thoroughly. Gently heat and bring to boiling. Distribute into screw-cap
tubes. Add 8–10 untreated Alaska seed peas per tube and let the tubes
stand for 1 h to permit hydration. Autoclave for 15 min at 15 psi pressure–121°C.

**Storage/Shelf Life:** Store dehydrated media in the dark in a sealed
container below 30°C. Prepared media should be stored under refrigeration (2-8°C). Media should be used within 60 days of preparation.
Media should not be used if there are any signs of deterioration (discoloration) or contamination, or if the expiration date supplied by the
manufacturer has passed.

**Use:** For the cultivation of *Clostridium botulinum* from foods.

### PE2 Medium
**Composition** per liter:
Peptone ..........................................................................20.0g
Yeast extract....................................................................3.0g
Alaska seed peas........................................................416–520
Bromcresol Purple solution ........................................2.0mL

**Bromcresol Purple Solution:**
**Composition** per 100.0mL:
Bromcresol Purple ..........................................................2.0g
Ethanol.........................................................................10.0mL

**Preparation of Bromcresol Purple Solution:** Add Bromcresol
Purple to 10.0mL of ethanol. Mix thoroughly. Bring volume to
100.0mL with distilled/deionized water. Filter sterilize.

**Preparation of Medium:** Add components, except Alaska seed peas, to distilled/deionized water and bring volume to 1.0L. Mix thoroughly. Gently heat until dissolved. Add 8–10 Alaska seed peas to each of 18 × 150 mm screw-capped tubes. Distribute the broth into each tube in 19.0mL volumes. Allow tubes to stand for 1 h. Autoclave for 15 min at 15 psi pressure–121°C.

**Storage/Shelf Life:** Store dehydrated media in the dark in a sealed container below 30°C. Prepared media should be stored under refrigeration (2-8°C). Media should be used within 60 days of preparation. Media should not be used if there are any signs of deterioration (discoloration) or contamination, or if the expiration date supplied by the manufacturer has passed.

**Use:** For the cultivation of *Clostridium botulinum* from foods.

### Peptone Iron Agar
**Composition** per liter:
Agar .............................................................15.0g
Peptone..........................................................15.0g
Proteose peptone ..............................................5.0g
Sodium glycerophosphate...................................1.0g
Ferric ammonium citrate....................................0.5g
$Na_2S_2O_3$ .......................................................0.08g
pH 6.7 ± 0.2 at 25°C

**Source:** This medium is available as a premixed powder from BD Diagnostic Systems.

**Preparation of Medium:** Add components to distilled/deionized water and bring volume to 1.0L. Mix thoroughly. Gently heat and bring to boiling. Distribute into tubes. Autoclave for 15 min at 15 psi pressure–121°C. Allow tubes to cool in an upright position.

**Storage/Shelf Life:** Store dehydrated media in the dark in a sealed container below 30°C. Prepared media should be stored under refrigeration (2-8°C). Media should be used within 60 days of preparation. Media should not be used if there are any signs of deterioration (shrinking, cracking, or discoloration) or contamination, or if the expiration date supplied by the manufacturer has passed.

**Use:** For the cultivation and differentiation of microorganisms based on their ability to produce $H_2S$. Microorganisms that produce $H_2S$ turn the medium black.

### Peptone Iron HiVeg Agar
**Composition** per liter:
Agar .............................................................15.0g
Plant peptone...................................................15.0g
Plant peptone No. 3............................................5.0g
Sodium glycerophosphate...................................1.0g
Ferric ammonium citrate....................................0.5g
$Na_2S_2O_3$ .......................................................0.08g
pH 6.7 ± 0.2 at 25°C

**Source:** This medium is available as a premixed powder from HiMedia.

**Preparation of Medium:** Add components to distilled/deionized water and bring volume to 1.0L. Mix thoroughly. Gently heat and bring to boiling. Distribute into tubes. Autoclave for 15 min at 15 psi pressure–121°C. Allow tubes to cool in an upright position.

**Storage/Shelf Life:** Store dehydrated media in the dark in a sealed container below 30°C. Prepared media should be stored under refrigeration (2-8°C). Media should be used within 60 days of preparation. Media should not be used if there are any signs of deterioration (shrink-

ing, cracking, or discoloration) or contamination, or if the expiration date supplied by the manufacturer has passed.

**Use:** For the cultivation and differentiation of microorganisms based on their ability to produce $H_2S$. Microorganisms that produce $H_2S$ turn the medium black.

### Peptone Starch Dextrose Agar (PSD Agar) (Dunkelberg Agar)
**Composition** per liter:
Proteose peptone No. 3.....................................20.0g
Agar .............................................................15.0g
Soluble starch.................................................10.0g
Glucose ...........................................................2.0g
$Na_2HPO_4$ .......................................................1.0g
$NaH_2PO_4$ .......................................................1.0g
pH 6.8 ± 0.2 at 25°C

**Preparation of Medium:** Add starch to approximately 100.0mL of cold distilled/deionized water. Mix thoroughly. Add starch solution to 400.0mL of boiling distilled/deionized water. Add remaining components. Mix thoroughly. Bring volume to 1.0L with distilled/deionized water. Autoclave for 12 min at 8 psi pressure–112°C. Pour into sterile Petri dishes or distribute into screw-capped tubes.

**Storage/Shelf Life:** Store dehydrated media in the dark in a sealed container below 30°C. Prepared media should be stored under refrigeration (2-8°C). Media should be used within 60 days of preparation. Media should not be used if there are any signs of deterioration (shrinking, cracking, or discoloration) or contamination, or if the expiration date supplied by the manufacturer has passed.

**Use:** For the selective isolation and cultivation of *Gardnerella vaginalis*.

### Perfringens HiVeg Agar Base (O.P.S.P.) with Antibiotics
**Composition** per liter:
Agar .............................................................15.0g
Plant hydrolysate ............................................15.0g
Plant extract No. 2 ............................................7.0g
Papaic digest of soybean meal............................5.0g
Yeast extract....................................................5.0g
Tris buffer .......................................................1.5g
Ferric ammonium citrate....................................1.0g
$Na_2S_2O_5$ .......................................................1.0g
Antibiotic inhibitor ........................................ 10.0mL
pH 7.3 ± 0.2 at 25°C

**Source:** This medium, without antibiotic inhibitor, is available as a premixed powder from HiMedia.

**Antibiotic Inhibitor:**
**Composition** per 10.0mL:
Sodium sulfadiazine...........................................0.1g
Oleandomycin phosphate...................................0.5mg
Polymyxin B ............................................... 10,000U

**Preparation of Antibiotic Inhibitor:** Add components to distilled/deionized water and bring volume to 10.0mL. Mix thoroughly. Filter sterilize.

**Preparation of Medium:** Add components, except antibiotic inhibitor, to distilled/deionized water and bring volume to 990.0mL. Mix thoroughly. Gently heat and bring to boiling. Autoclave for 15 min at 15 psi pressure–121°C. Cool to 45°–50°C. Aseptically add sterile antibiotic inhibitor. Mix thoroughly. Pour into sterile Petri dishes or distribute into sterile tubes.

**Storage/Shelf Life:** Store dehydrated media in the dark in a sealed container below 30°C. Prepared media should be stored under refrigeration (2-8°C). Media should be used within 60 days of preparation. Media should not be used if there are any signs of deterioration (shrinking, cracking, or discoloration) or contamination, or if the expiration date supplied by the manufacturer has passed.

**Use:** For the presumptive identification and enumeration of *Clostridium perfringens* in foods.

## Perfringens HiVeg Agar Base with Egg Yolk and Antibiotics (T.S.C./S.F.P. HiVeg Agar Base)

**Composition** per liter:

| | |
|---|---|
| Agar | 15.0g |
| Plant hydrolysate No. 1 | 15.0g |
| Papaic digest of soybean meal | 5.0g |
| Plant extract | 5.0g |
| Yeast extract | 5.0g |
| $Na_2S_2O_5$ | 1.0g |
| Ferric ammonium citrate | 1.0g |
| Egg yolk emulsion | 25.0mL |
| Perfringens SFP supplement | 4.0mL |
| Perfringens TSC supplement | 4.0mL |

pH 7.6 ± 0.2 at 25°C

**Source:** This medium, without egg yolk emulsion, perfringens SFP supplement, and perfringens TSC supplement, is available as a premixed powder from HiMedia.

### Egg Yolk Emulsion:

**Composition** per 100.0mL:

| | |
|---|---|
| Chicken egg yolks | 9 |
| Whole chicken egg | 1 |
| NaCl (0.9% solution) | 25.0mL |

**Preparation of Egg Yolk Emulsion:** Soak eggs with 1:100 dilution of saturated mercuric chloride solution for 1 min. Crack eggs and separate yolks from whites. Mix egg yolks with 1 chicken egg. Beat to form emulsion. Measure 50.0mL of egg yolk emulsion and add to 50.0mL of 0.9% NaCl solution. Mix thoroughly. Filter sterilize. Warm to 45°–50°C.

### Perfringens SFP Supplement:

**Composition** per 10.0mL:

| | |
|---|---|
| Kanamycin sulfate | 30.0mg |
| Polymyxin B | 75,000U |

**Preparation of Perfringens SFP Supplement:** Add components to distilled/deionized water and bring volume to 10.0mL. Mix thoroughly. Filter sterilize.

### Perfringens TSC Supplement:

**Composition** per 10.0mL:

| | |
|---|---|
| D-Cycloserine | 1.0g |

**Preparation of Perfringens TSC Supplement:** Add D-cycloserine to distilled/deionized water and bring volume to 10.0mL. Mix thoroughly. Filter sterilize.

**Preparation of Medium:** Add components, except perfringens SFP supplement, egg yolk emulsion, and perfringens TSC supplement, to distilled/deionized water and bring volume to 975mL. Mix thoroughly. Gently heat and bring to boiling. Distribute into tubes or flasks. Auto-

clave for 15 min at 15 psi pressure–121°C. Cool to 55°C. Aseptically add 25.0mL egg yolk emulsion, 4.0mL perfringens SFP supplement, and 4.00mL perfringens TSC supplement. Mix thoroughly. Pour into sterile Petri dishes or distribute into sterile tubes.

**Storage/Shelf Life:** Store dehydrated media in the dark in a sealed container below 30°C. Prepared media should be stored under refrigeration (2-8°C). Media should be used within 60 days of preparation. Media should not be used if there are any signs of deterioration (shrinking, cracking, or discoloration) or contamination, or if the expiration date supplied by the manufacturer has passed.

**Use:** For the cultivation, enumeration, and presumptive identification of *Clostridium perfringens* from foods.

## Petragnani Medium

**Composition** per 2398.0mL:

| | |
|---|---|
| Skim milk | 100.0g |
| Potato flour | 36.4g |
| L-Asparagine | 5.1g |
| Pancreatic digest of casein | 5.1g |
| Malachite Green | 1.2g |
| Whole egg | 1277.0mL |
| Egg yolk | 121.0mL |
| Glycerol | 60.0mL |

pH 7.0 ± 0.2 at 25°C

**Source:** This medium is available as a prepared medium from BD Diagnostic Systems.

**Preparation of Medium:** Add components—except whole egg, egg yolk, and glycerol—to distilled/deionized water and bring volume to 940.0mL. Mix thoroughly. Add glycerol. Gently heat while stirring and bring to boiling. Autoclave for 15 min at 15 psi pressure–121°C. Cool to 45°–50°C. Scrub the eggshells with soap. Let stand in a soap solution for 30 min. Rinse in running water. Soak eggs in 70% ethanol for 15 min. Break the eggs into a sterile container. Homogenize by shaking. Filter through four layers of sterile cheesecloth into a sterile graduated cylinder. Measure out 1277.0mL. Add separated egg yolks to another sterile container. Measure out 121.0mL. Aseptically add homogenized whole egg and egg yolk to cooled sterile basal medium. Mix thoroughly. Aseptically distribute into sterile tubes. Inspissate at 85°–90°C (moist heat) for 45 min.

**Storage/Shelf Life:** Store dehydrated media in the dark in a sealed container below 30°C. Prepared media should be stored under refrigeration (2-8°C). Media should be used within 60 days of preparation. Media should not be used if there are any signs of deterioration (discoloration) or contamination, or if the expiration date supplied by the manufacturer has passed.

**Use:** For the isolation and cultivation of *Mycobacterium* species from clinical specimens. For the cultivation and maintenance of *Mycobacterium smegmatis*.

## Petragnani Medium

**Composition** per 2285.0mL:

| | |
|---|---|
| Potato | 500.0g |
| Potato flour | 36.0g |
| Malachite Green | 1.2g |
| Whole egg | 1200.0mL |
| Whole milk | 900.0mL |
| Egg yolk | 115.0mL |
| Glycerol | 70.0mL |

pH 7.2 ± 0.2 at 25°C

**Source:** This medium is available as a prepared medium from BD Diagnostic Systems.

**Preparation of Medium:** Peel and dice potato. Add potato to 500.0mL of distilled/deionized water. Gently heat and bring to boiling. Continue boiling for 30 min. Filter solids through two layers of cheesecloth. Combine potato solids with remaining components, except whole egg, egg yolk, and glycerol. Mix thoroughly. Add glycerol. Gently heat while stirring and bring to boiling. Autoclave for 15 min at 15 psi pressure–121°C. Cool to 45°–50°C. Scrub the eggshells with soap. Let stand in a soap solution for 30 min. Rinse in running water. Soak eggs in 70% ethanol for 15 min. Break the eggs into a sterile container. Homogenize by shaking. Filter through four layers of sterile cheesecloth into a sterile graduated cylinder. Measure out 1200.0mL. Add separated egg yolks to another sterile container. Measure out 115.0mL. Aseptically add homogenized whole egg and egg yolk to cooled sterile basal medium. Mix thoroughly. Aseptically distribute into sterile tubes. Inspissate at 85°–90°C (moist heat) for 45 min.

**Storage/Shelf Life:** Store dehydrated media in the dark in a sealed container below 30°C. Prepared media should be stored under refrigeration (2-8°C). Media should be used within 60 days of preparation. Media should not be used if there are any signs of deterioration (discoloration) or contamination, or if the expiration date supplied by the manufacturer has passed.

**Use:** For the isolation and cultivation of *Mycobacterium* species from clinical specimens. For the cultivation and maintenance of *Mycobacterium smegmatis*.

## Pfizer Selective *Enterococcus* Agar (PSE Agar)

| | |
|---|---|
| Peptone C | 17.0g |
| Agar | 15.0g |
| Bile | 10.0g |
| NaCl | 5.0g |
| Yeast extract | 5.0g |
| Peptone B | 3.0g |
| Esculin | 1.0g |
| Sodium citrate | 1.0g |
| Ferric ammonium citrate | 0.5g |
| NaN$_3$ | 0.25g |

pH 7.1 ± 0.2 at 25°C

**Caution:** Sodium azide is toxic. Azides also react with metals and disposal must be highly diluted.

**Preparation of Medium:** Add components to distilled/deionized water and bring volume to 1.0L. Mix thoroughly. Gently heat and bring to boiling. Distribute into tubes or flasks. Autoclave for 15 min at 15 psi pressure–121°C. Pour into sterile Petri dishes or leave in tubes.

**Storage/Shelf Life:** Store dehydrated media in the dark in a sealed container below 30°C. Prepared media should be stored under refrigeration (2-8°C). Media should be used within 60 days of preparation. Media should not be used if there are any signs of deterioration (shrinking, cracking, or discoloration) or contamination, or if the expiration date supplied by the manufacturer has passed.

**Use:** For the selective isolation, cultivation, and enumeration of *Enterococcus* species by the multiple tube technique.

## Pfizer Selective *Enterococcus* HiVeg Agar

**Composition** per liter:

| | |
|---|---|
| Plant hydrolysate | 21.0g |
| Agar | 15.0g |
| Plant peptone | 6.0g |
| NaCl | 5.0g |
| Yeast extract | 5.0g |
| Synthetic detergent | 3.0g |
| Esculin | 1.0g |
| Sodium citrate | 1.0g |
| Ferric ammonium citrate | 0.5g |
| NaN$_3$ | 0.25g |

pH 7.1 ± 0.2 at 25°C

**Source:** This medium is available as a premixed powder from HiMedia.

**Caution:** Sodium azide is toxic. Azides also react with metals and disposal must be highly diluted.

**Preparation of Medium:** Add components to distilled/deionized water and bring volume to 1.0L. Mix thoroughly. Gently heat and bring to boiling. Distribute into tubes or flasks. Autoclave for 15 min at 15 psi pressure–121°C. Pour into sterile Petri dishes or leave in tubes.

**Storage/Shelf Life:** Store dehydrated media in the dark in a sealed container below 30°C. Prepared media should be stored under refrigeration (2-8°C). Media should be used within 60 days of preparation. Media should not be used if there are any signs of deterioration (shrinking, cracking, or discoloration) or contamination, or if the expiration date supplied by the manufacturer has passed.

**Use:** For the selective isolation, cultivation, and enumeration of *Enterococcus* species by the multiple tube technique.

## Pfizer TB Medium Base with Glycerol, Egg Yolk, Glucose, and Malachite Green

**Composition** per liter:

| | |
|---|---|
| Agar | 15.0g |
| Potato starch | 15.0g |
| Casein acid hydrolysate | 10.0g |
| K$_2$HPO$_4$ | 3.5g |
| Beef extract | 3.0g |
| L-Asparagine | 3.0g |
| Citric acid | 0.1g |
| Ferric ammonium citrate | 0.1g |
| MgSO$_4$ | 0.015g |
| Egg yolk emulsion | 100.0mL |
| Glycerol | 40.0mL |
| Malachite Green solution | 13.0mL |
| Glucose solution | 1.0mL |

pH 7.0 ± 0.2 at 35°C

**Source:** This medium, without glycerol, glucose solution, Malachite Green solution, and egg yolk emulsion, is available as a premixed powder from HiMedia.

**Egg Yolk Emulsion:**

**Composition** per 100.0mL:

| | |
|---|---|
| Chicken egg yolks | 9 |
| Whole chicken egg | 1 |
| NaCl (0.9% solution) | 25.0mL |

**Preparation of Egg Yolk Emulsion:** Soak eggs with 1:100 dilution of saturated mercuric chloride solution for 1 min. Crack eggs and separate yolks from whites. Mix egg yolks with 1 chicken egg. Beat to form emulsion. Measure 50.0mL of egg yolk emulsion and add to 50.0mL of 0.9% NaCl solution. Mix thoroughly. Filter sterilize. Warm to 45°–50°C.

**Glucose Solution:**
**Composition** per 100.0mL:
Glucose ..............................................................................20.0g

**Preparation of Glucose Solution:** Add glucose to distilled/deionized water and bring volume to 100.0mL. Mix thoroughly. Filter sterilize.

**Malachite Green Solution:**
**Composition** per 20.0mL:
Malachite Green ................................................................0.2g

**Preparation of Malachite Green Solution:** Add Malachite Green to distilled/deionized water and bring volume to 100.0mL. Mix thoroughly. Filter sterilize.

**Preparation of Medium:** Add glycerol and the other components, except glucose solution, egg yolk emulsion, and Malachite Green solution, to distilled/deionized water and bring volume to 1.0L. Mix thoroughly. Gently heat and bring to boiling. Distribute into tubes or flasks. Autoclave for 15 min at 12 psi pressure–118°C. Cool to 55°C. Aseptically add 100.0mL egg yolk emulsion, 1.0mL glucose solution, and 13.0mL Malachite Green solution. Mix thoroughly. Aseptically dispense into sterile tubes. Allow to solidify as slants.

**Storage/Shelf Life:** Store dehydrated media in the dark in a sealed container below 30°C. Prepared media should be stored under refrigeration (2-8°C). Media should be used within 60 days of preparation. Media should not be used if there are any signs of deterioration (shrinking, cracking, or discoloration) or contamination, or if the expiration date supplied by the manufacturer has passed.

**Use:** For the cultivation of *Mycobacterium tuberculosis*.

## PGT Medium

**Composition** per liter:
Casamino acids ..............................................................30.0g
L-Glutamic acid ................................................................0.5g
MgSO$_4$·7H$_2$O ..............................................................0.45g
Maltose..............................................................................0.2g
L-Cystine ..........................................................................0.2g
DL-Tryptophan..................................................................0.1g
Solution 3 ...................................................................100.0mL
Solution 2......................................................................2.0mL
Calcium pantothenate (0.1% solution).........................0.5mL
pH 6.8 ± 0.2 at 25°C

**Solution 3:**
**Composition** per 500.0mL:
Maltose..........................................................................200.0g
CaCl$_2$................................................................................1.5g
Calcium pantothenate (0.1% solution).........................3.0mL
FeSO$_4$ (1% in 1*N* HCl) ..................................................0.2mL

**Preparation of Solution 3:** Add components to distilled/deionized water and bring volume to 500.0mL. Mix thoroughly. Autoclave for 15 min at 7 psi pressure–111°C. Cool to 45°–50°C.

**Solution 2:**
**Composition** per 100.0mL:
β-Alanine ........................................................................0.115g
Nicotinic acid ................................................................0.115g
CuSO$_4$·5H$_2$O ..................................................................0.05g
ZnSO$_4$·7H$_2$O ................................................................0.045g
MnCl$_2$·4H$_2$O..................................................................0.015g
Pimelic acid....................................................................7.5mg
HCl, concentrated ..........................................................3.0mL

**Preparation of Solution 2:** Add components to distilled/deionized water and bring volume to 100.0mL. Mix thoroughly.

**Preparation of Medium:** Add components, except solution 3, to distilled/deionized water and bring volume to 900.0mL. Mix thoroughly. Adjust pH to 6.8 with 50% KOH. Gently heat and bring to boiling. Autoclave for 15 min at 15 psi pressure–121°C. Cool to 45°–50°C. Aseptically add sterile solution 3. Mix thoroughly. Aseptically distribute into sterile tubes or flasks.

**Storage/Shelf Life:** Store dehydrated media in the dark in a sealed container below 30°C. Prepared media should be stored under refrigeration (2-8°C). Media should be used within 60 days of preparation. Media should not be used if there are any signs of deterioration (discoloration) or contamination, or if the expiration date supplied by the manufacturer has passed.

**Use:** For the cultivation of *Corynebacterium diphtheriae*.

## Phenethyl Alcohol Agar
## (Phenylethanol Agar)
## (Phenylethyl Alcohol Agar)

**Composition** per liter:
Agar ................................................................................15.0g
Pancreatic digest of casein............................................15.0g
NaCl....................................................................................5.0g
Papaic digest of soybean meal........................................5.0g
β-Phenethyl alcohol..........................................................2.5g
Blood.............................................................................. 50.0mL
pH 7.3 ± 0.2 at 25°C

**Source:** This medium is available as a premixed powder from BD Diagnostic Systems.

**Preparation of Medium:** Add components, except blood, to distilled/deionized water and bring volume to 950.0mL. Mix thoroughly. Gently heat and bring to boiling. Autoclave for 15 min at 13 psi pressure–118°C. Cool to 45°–50°C. Aseptically add sterile defibrinated blood. Mix thoroughly. Pour into sterile Petri dishes or distribute into sterile tubes.

**Storage/Shelf Life:** Store dehydrated media in the dark in a sealed container below 30°C. Prepared media should be stored under refrigeration (2-8°C). Media should be used within 60 days of preparation. Media should not be used if there are any signs of deterioration (shrinking, cracking, or discoloration) or contamination, or if the expiration date supplied by the manufacturer has passed.

**Use:** For the selective isolation of Gram-positive bacteria, particularly Gram-positive cocci, from specimens with a mixed flora. Do not use for the observation of hemolytic reactions.

## Phenol Red Agar

**Composition** per liter:
Agar ................................................................................15.0g
Pancreatic digest of casein............................................10.0g
NaCl....................................................................................5.0g
Phenol Red......................................................................0.018g
Carbohydrate solution.................................................. 20.0mL
pH 7.4 ± 0.2 at 25°C

**Source:** This medium is available as a premixed powder from BD Diagnostic Systems.

**Carbohydrate Solution:**
**Composition** per 20.0mL:
Carbohydrate..............................................................5.0–10.0g

**Preparation of Carbohydrate Solution:** Add carbohydrate to distilled/deionized water and bring volume to 20.0mL. Mix thoroughly. Filter sterilize.

**Preparation of Medium:** Add components, except carbohydrate solution, to distilled/deionized water and bring volume to 980.0mL. Mix thoroughly. Adjust pH to 7.4 if necessary. Autoclave for 15 min at 15 psi pressure–121°C. Cool to 45°–50°C. Aseptically add 20.0mL of sterile carbohydrate solution. Pour into sterile Petri dishes or distribute into sterile tubes. Allow tubes to cool in a slanted position.

**Storage/Shelf Life:** Store dehydrated media in the dark in a sealed container below 30°C. Prepared media should be stored under refrigeration (2-8°C). Media should be used within 60 days of preparation. Media should not be used if there are any signs of deterioration (shrinking, cracking, or discoloration) or contamination, or if the expiration date supplied by the manufacturer has passed.

**Use:** For the determination of fermentation reactions. Bacteria that can ferment the added carbohydrate turn the medium yellow.

## Phenol Red Agar

**Composition** per liter:

| | |
|---|---|
| Agar | 15.0g |
| Proteose peptone No. 3 | 10.0g |
| NaCl | 5.0g |
| Beef extract | 1.0g |
| Phenol Red | 0.025g |
| Carbohydrate solution | 20.0mL |

pH 7.4 ± 0.2 at 25°C

**Source:** This medium is available as a premixed powder from BD Diagnostic Systems.

**Carbohydrate Solution:**
**Composition** per 20.0mL:

| | |
|---|---|
| Carbohydrate | 5.0–10.0g |

**Preparation of Carbohydrate Solution:** Add carbohydrate to distilled/deionized water and bring volume to 20.0mL. Mix thoroughly. Filter sterilize.

**Preparation of Medium:** Add components, except carbohydrate solution, to distilled/deionized water and bring volume to 980.0mL. Mix thoroughly. Adjust pH to 7.4 if necessary. Autoclave for 15 min at 15 psi pressure–121°C. Cool to 45°–50°C. Aseptically add 20.0mL of sterile carbohydrate solution. Pour into sterile Petri dishes or distribute into sterile tubes. Allow tubes to cool in a slanted position.

**Storage/Shelf Life:** Store dehydrated media in the dark in a sealed container below 30°C. Prepared media should be stored under refrigeration (2-8°C). Media should be used within 60 days of preparation. Media should not be used if there are any signs of deterioration (shrinking, cracking, or discoloration) or contamination, or if the expiration date supplied by the manufacturer has passed.

**Use:** For the determination of fermentation reactions. Bacteria that can ferment the added carbohydrate turn the medium yellow.

## Phenol Red Broth

**Composition** per liter:

| | |
|---|---|
| Pancreatic digest of casein | 10.0g |
| NaCl | 5.0g |
| Phenol Red | 0.018g |
| Carbohydrate solution | 20.0mL |

pH 7.4 ± 0.2 at 25°C

**Source:** This medium is available as a premixed powder from BD Diagnostic Systems.

**Carbohydrate Solution:**
**Composition** per 20.0mL:

| | |
|---|---|
| Carbohydrate | 5.0–10.0g |

**Preparation of Carbohydrate Solution:** Add carbohydrate to distilled/deionized water and bring volume to 20.0mL. Mix thoroughly. Filter sterilize.

**Preparation of Medium:** Add components, except carbohydrate solution, to distilled/deionized water and bring volume to 980.0mL. Mix thoroughly. Adjust pH to 7.4 if necessary. Distribute into tubes containing an inverted Durham tube. Fill each tube with 9.8mL of medium. Autoclave for 15 min at 13 psi pressure–118°C. Cool to 45°–50°C. Aseptically add 0.2mL of sterile carbohydrate solution to each tube.

**Storage/Shelf Life:** Store dehydrated media in the dark in a sealed container below 30°C. Prepared media should be stored under refrigeration (2-8°C). Media should be used within 60 days of preparation. Media should not be used if there are any signs of deterioration (discoloration) or contamination, or if the expiration date supplied by the manufacturer has passed.

**Use:** For the determination of fermentation reactions in the differentiation of microorganisms. Fermentation is determined by the production of acid—broth turns yellow—and formation of gas—bubble trapped in Durham tube.

## Phenol Red Glucose Broth (BAM M122)

**Composition** per liter:

| | |
|---|---|
| Proteose peptone No. 3 | 10.0g |
| Glucose | 5.0g |
| NaCl | 5.0g |
| Beef extract | 1.0g |
| Phenol Red solution | 7.2mL |

pH 7.4 ± 0.2 at 25°C

**Source:** This medium is available as a premixed powder from BD Diagnostics.

**Phenol Red Solution:**
**Composition** per 10.0mL:

| | |
|---|---|
| Phenol Red | 0.025g |

**Preparation of Phenol Red Solution:** Add Phenol Red to distilled/deionized water and bring volume to 10.0mL. Mix thoroughly.

**Preparation of Medium:** Add components to distilled/deionized water and bring volume to 1.0L. Mix thoroughly. Adjust pH to 7.4 if necessary. Distribute into tubes containing an inverted Durham tube. Fill each tube with 10.0mL of medium. Autoclave for 10 min at 13 psi pressure–118°C.

**Storage/Shelf Life:** Store dehydrated media in the dark in a sealed container below 30°C. Prepared media should be stored under refrigeration (2-8°C). Media should be used within 60 days of preparation. Media should not be used if there are any signs of deterioration (discoloration) or contamination, or if the expiration date supplied by the manufacturer has passed.

**Use:** For determination of the ability of a microorganism to ferment glucose. Fermentation is determined by the production of acid—broth turns yellow—and formation of gas—bubble trapped in Durham tube.

## Phenol Red Glucose Broth, HiVeg

**Composition** per liter:
Plant peptone No. 3.................................................................10.0g
Glucose ......................................................................................5.0g
NaCl............................................................................................5.0g
Plant extract ..............................................................................1.0g
Phenol Red..............................................................................0.018g

pH 7.3 ± 0.2 at 25°C

**Source:** This medium is available as a premixed powder from Hi-Media.

**Preparation of Medium:** Add components to distilled/deionized water and bring volume to 1.0L. Mix thoroughly. Adjust pH to 7.3 if necessary. Distribute into tubes containing an inverted Durham tube. Fill each tube with 10.0mL of medium. Autoclave for 15 min at 13 psi pressure–118°C.

**Storage/Shelf Life:** Store dehydrated media in the dark in a sealed container below 30°C. Prepared media should be stored under refrigeration (2-8°C). Media should be used within 60 days of preparation. Media should not be used if there are any signs of deterioration (discoloration) or contamination, or if the expiration date supplied by the manufacturer has passed.

**Use:** For determination of the ability of a microorganism to ferment glucose. Fermentation is determined by the production of acid—broth turns yellow—and formation of gas—bubble trapped in Durham tube.

## Phenol Red Glucose HiVeg Agar

**Composition** per liter:
Agar ........................................................................................15.0g
Plant peptone No. 3.................................................................10.0g
Glucose ....................................................................................10.0g
NaCl............................................................................................5.0g
Plant extract ..............................................................................1.0g
Phenol Red..............................................................................0.025g

pH 7.4 ± 0.2 at 25°C

**Source:** This medium is available as a premixed powder from Hi-Media.

**Preparation of Medium:** Add components to distilled/deionized water and bring volume to 1.0L. Mix thoroughly. Gently heat and bring to boiling. Distribute into tubes or flasks. Autoclave for 15 min at 13 psi pressure–118°C. Pour into sterile Petri dishes or leave in tubes. Allow tubes to cool in a slanted position.

**Storage/Shelf Life:** Store dehydrated media in the dark in a sealed container below 30°C. Prepared media should be stored under refrigeration (2-8°C). Media should be used within 60 days of preparation. Media should not be used if there are any signs of deterioration (shrinking, cracking, or discoloration) or contamination, or if the expiration date supplied by the manufacturer has passed.

**Use:** For the determination of the ability of a microorganism to ferment glucose. Fermentation is determined by the production of acid—medium turns yellow.

## Phenol Red HiVeg Agar Base with Carbohydrate

**Composition** per liter:
Agar ........................................................................................15.0g
Plant peptone No. 3.................................................................10.0g
NaCl............................................................................................5.0g

Plant extract ..............................................................................1.0g
Phenol Red..............................................................................0.025g
Carbohydrate solution.......................................................... 20.0mL

pH 7.4 ± 0.2 at 25°C

**Source:** This medium, without carbohydrate, is available as a premixed powder from HiMedia.

**Carbohydrate Solution:**
**Composition** per 20.0mL:
Carbohydrate.................................................................... 5.0–10.0g

**Preparation of Carbohydrate Solution:** Add carbohydrate to distilled/deionized water and bring volume to 20.0mL. Mix thoroughly. Filter sterilize.

**Preparation of Medium:** Add components, except carbohydrate solution, to distilled/deionized water and bring volume to 980.0mL. Mix thoroughly. Adjust pH to 7.4 if necessary. Autoclave for 15 min at 15 psi pressure–121°C. Cool to 45°–50°C. Aseptically add 20.0mL of sterile carbohydrate solution. Pour into sterile Petri dishes or distribute into sterile tubes. Allow tubes to cool in a slanted position.

**Storage/Shelf Life:** Store dehydrated media in the dark in a sealed container below 30°C. Prepared media should be stored under refrigeration (2-8°C). Media should be used within 60 days of preparation. Media should not be used if there are any signs of deterioration (shrinking, cracking, or discoloration) or contamination, or if the expiration date supplied by the manufacturer has passed.

**Use:** For the determination of fermentation reactions. Bacteria that can ferment the added carbohydrate turn the medium yellow.

## Phenol Red Tartrate Agar

**Composition** per liter:
Agar ........................................................................................15.0g
Peptone ...................................................................................10.0g
Potassium tartrate..................................................................10.0g
NaCl............................................................................................5.0g
Phenol Red..............................................................................0.024g

pH 7.6 ± 0.2 at 25°C

**Source:** This medium is available as a premixed powder from BD Diagnostic Systems.

**Preparation of Medium:** Add components to cold distilled/deionized water and bring volume to 1.0L. Mix thoroughly. Gently heat and bring to boiling. Distribute into tubes or flasks. Autoclave for 15 min at 13 psi pressure–118°C. Pour into sterile Petri dishes or leave in tubes. Allow tubes to cool in an upright position.

**Storage/Shelf Life:** Store dehydrated media in the dark in a sealed container below 30°C. Prepared media should be stored under refrigeration (2-8°C). Media should be used within 60 days of preparation. Media should not be used if there are any signs of deterioration (shrinking, cracking, or discoloration) or contamination, or if the expiration date supplied by the manufacturer has passed.

**Use:** For the differentiation of Gram-negative bacteria of the intestinal groups, particularly members of the *Salmonella* (paratyphoid) group based on their ability to ferment tartrate.

## Phenol Red Tartrate Broth

**Composition** per liter:
Pancreatic digest of casein................................................10.0g
Potassium tartrate..................................................................10.0g
Agar ............................................................................................5.0g

NaCl ................................................................................5.0g
Phenol Red .....................................................................0.024g
pH 7.6 ± 0.2 at 25°C

**Preparation of Medium:** Add components to distilled/deionized water and bring volume to 1.0L. Mix thoroughly. Distribute into tubes or flasks. Autoclave for 15 min at 15 psi pressure–121°C.

**Storage/Shelf Life:** Store dehydrated media in the dark in a sealed container below 30°C. Prepared media should be stored under refrigeration (2-8°C). Media should be used within 60 days of preparation. Media should not be used if there are any signs of deterioration (discoloration) or contamination, or if the expiration date supplied by the manufacturer has passed.

**Use:** For the differentiation of Gram-negative bacteria of the intestinal groups, particularly members of the *Salmonella* (paratyphoid) group based on their ability to ferment tartrate.

## Phenolphthalein Phosphate HiVeg Agar

**Composition** per liter:
Agar ................................................................................15.0g
Plant peptone .................................................................5.0g
NaCl ................................................................................5.0g
Plant extract ...................................................................3.0g
Sodium phenolphthalein phosphate ...............................0.012g
pH 7.4 ± 0.2 at 25°C

**Source:** This medium is available as a premixed powder from Hi-Media.

**Preparation of Medium:** Add components to distilled/deionized water and bring volume to 1.0L. Mix thoroughly. Gently heat and bring to boiling. Distribute into tubes or flasks. Autoclave for 15 min at 15 psi pressure–21°C.

**Storage/Shelf Life:** Store dehydrated media in the dark in a sealed container below 30°C. Prepared media should be stored under refrigeration (2-8°C). Media should be used within 60 days of preparation. Media should not be used if there are any signs of deterioration (shrinking, cracking, or discoloration) or contamination, or if the expiration date supplied by the manufacturer has passed.

**Use:** For the identification of phosphatase-positive *Staphylococcus aureus*.

## Phenylalanine Agar
## (Phenylalanine Deaminase Medium)

**Composition** per liter:
Agar ................................................................................12.0g
NaCl ................................................................................5.0g
Yeast extract ...................................................................3.0g
DL-Phenylalanine ...........................................................2.0g
Na$_2$HPO$_4$ ................................................................1.0g
pH 7.3 ± 0.2 at 25°C

**Source:** This medium is available as a premixed powder from BD Diagnostic Systems.

**Preparation of Medium:** Add components to distilled/deionized water and bring volume to 1.0L. Mix thoroughly. Gently heat while stirring and bring to boiling. Distribute into tubes or flasks. Autoclave for 10 min at 15 psi pressure–121°C. Pour into sterile Petri dishes or leave in tubes.

**Storage/Shelf Life:** Store dehydrated media in the dark in a sealed container below 30°C. Prepared media should be stored under refriger-

ation (2-8°C). Media should be used within 60 days of preparation. Media should not be used if there are any signs of deterioration (shrinking, cracking, or discoloration) or contamination, or if the expiration date supplied by the manufacturer has passed.

**Use:** For the differentiation of enteric Gram-negative bacilli on the basis of their ability to produce phenylpyruvic acid from phenylalanine. After appropriate incubation of bacteria, ferric chloride reagent is added on the agar. Formation of a green color in 1–5 min indicates the production of phenylpyruvic acid.

## Phenylethanol Agar

**Composition** per liter:
Agar ................................................................................15.0g
Tryptose ..........................................................................10.0g
NaCl ................................................................................5.0g
Beef extract .....................................................................3.0g
Phenylethanol .................................................................2.5g
pH 7.3 ± 0.2 at 25°C

**Source:** This medium is available as a premixed powder from BD Diagnostic Systems.

**Preparation of Medium:** Add components to distilled/deionized water and bring volume to 1.0L. Mix thoroughly. Gently heat and bring to boiling. Distribute into tubes or flasks. Autoclave for 15 min at 15 psi pressure–121°C. Pour into sterile Petri dishes or leave in tubes.

**Storage/Shelf Life:** Store dehydrated media in the dark in a sealed container below 30°C. Prepared media should be stored under refrigeration (2-8°C). Media should be used within 60 days of preparation. Media should not be used if there are any signs of deterioration (shrinking, cracking, or discoloration) or contamination, or if the expiration date supplied by the manufacturer has passed.

**Use:** For the isolation of staphylococci and streptococci from specimens containing a mixed flora.

## Phenylethanol Blood Agar

**Composition** per liter:
Agar ................................................................................15.0g
Tryptose ..........................................................................10.0g
NaCl ................................................................................5.0g
Beef extract .....................................................................3.0g
Phenylethanol .................................................................2.5g
Blood, defibrinated .........................................................50.0mL
pH 7.3 ± 0.2 at 25°C

**Preparation of Medium:** Add components, except blood, to distilled/deionized water and bring volume to 950.0mL. Mix thoroughly. Gently heat and bring to boiling. Autoclave for 15 min at 13 psi pressure–118°C. Cool to 45°–50°C. Aseptically add sterile defibrinated blood. Mix thoroughly. Pour into sterile Petri dishes or distribute into sterile tubes.

**Storage/Shelf Life:** Store dehydrated media in the dark in a sealed container below 30°C. Prepared media should be stored under refrigeration (2-8°C). Media should be used within 60 days of preparation. Media should not be used if there are any signs of deterioration (shrinking, cracking, or discoloration) or contamination, or if the expiration date supplied by the manufacturer has passed.

**Use:** For the isolation of staphylococci and streptococci from specimens containing a mixed flora.

## Phenylethyl Alcohol HiVeg Agar

**Composition** per liter:

| | |
|---|---|
| Agar | 15.0g |
| Plant hydrolysate | 15.0g |
| Papaic digest of soybean meal | 5.0g |
| NaCl | 5.0g |
| Phenylethyl alcohol | 2.5g |

pH 7.3 ± 0.2 at 25°C

**Source:** This medium is available as a premixed powder from HiMedia.

**Preparation of Medium:** Add components to distilled/deionized water and bring volume to 1.0L. Mix thoroughly. Gently heat and bring to boiling. Distribute into tubes or flasks. Autoclave for 15 min at 15 psi pressure–121°C. Pour into sterile Petri dishes or leave in tubes.

**Storage/Shelf Life:** Store dehydrated media in the dark in a sealed container below 30°C. Prepared media should be stored under refrigeration (2-8°C). Media should be used within 60 days of preparation. Media should not be used if there are any signs of deterioration (shrinking, cracking, or discoloration) or contamination, or if the expiration date supplied by the manufacturer has passed.

**Use:** For the isolation of staphylococci and streptococci from specimens containing a mixed flora.

## Phytone™ Yeast Extract Agar

**Composition** per liter:

| | |
|---|---|
| Glucose | 40.0g |
| Agar | 17.0g |
| Papaic digest of soybean meal | 10.0g |
| Yeast extract | 5.0g |
| Chloramphenicol | 0.05g |
| Streptomycin | 0.03g |

pH 6.6 ± 0.2 at 25°C

**Source:** This medium is available as a premixed powder from BD Diagnostic Systems.

**Preparation of Medium:** Add components to distilled/deionized water and bring volume to 1.0L. Mix thoroughly. Gently heat and bring to boiling. Distribute into tubes or flasks. Autoclave for 15 min at 15 psi pressure–121°C. Pour into sterile Petri dishes or leave in tubes.

**Storage/Shelf Life:** Store dehydrated media in the dark in a sealed container below 30°C. Prepared media should be stored under refrigeration (2-8°C). Media should be used within 60 days of preparation. Media should not be used if there are any signs of deterioration (shrinking, cracking, or discoloration) or contamination, or if the expiration date supplied by the manufacturer has passed.

**Use:** For the selective isolation of dermatophytes, particularly *Trichophyton verrucosum,* and other pathogenic fungi, from clinical specimens.

## Pike Streptococcal Broth

**Composition** per liter:

| | |
|---|---|
| Pancreatic digest of casein | 10.0g |
| Tryptose | 10.0g |
| Yeast extract | 10.0g |
| Glucose | 0.2g |
| NaN$_3$ | 0.065g |
| Crystal Violet | 2.0mg |
| Rabbit blood, defibrinated | 50.0mL |

pH 7.4 ± 0.2 at 25°C

**Caution:** Sodium azide is toxic. Azides also react with metals and disposal must be highly diluted.

**Preparation of Medium:** Add components, except rabbit blood, to distilled/deionized water and bring volume to 950.0mL. Mix thoroughly. Gently heat and bring to boiling. Distribute into flasks in 100.0mL volumes. Autoclave for 15 min at 15 psi pressure–121°C. Cool to 45°–50°C. Aseptically add 5.0mL of sterile rabbit blood to each flask. Mix thoroughly.

**Storage/Shelf Life:** Store dehydrated media in the dark in a sealed container below 30°C. Prepared media should be stored under refrigeration (2-8°C). Media should be used within 60 days of preparation. Media should not be used if there are any signs of deterioration (discoloration) or contamination, or if the expiration date supplied by the manufacturer has passed.

**Use:** For the isolation and enrichment of hemolytic streptococci from throat swabs and other clinical specimens. After incubation of bacteria for 18–24 h in this medium, they may be isolated by streaking the culture onto blood agar plates.

## Pike Streptococcal HiVeg Broth Base with Blood

**Composition** per liter:

| | |
|---|---|
| Yeast extract | 10.0g |
| Plant hydrolysate | 10.0g |
| Plant hydrolysate No. 1 | 10.0g |
| Glucose | 0.2g |
| NaN$_3$ | 0.065g |
| Crystal Violet | 2.0mg |
| Rabbit blood, defibrinated | 50.0mL |

pH 7.4 ± 0.2 at 25°C

**Caution:** Sodium azide is toxic. Azides also react with metals and disposal must be highly diluted.

**Preparation of Medium:** Add components, except rabbit blood, to distilled/deionized water and bring volume to 950.0mL. Mix thoroughly. Gently heat and bring to boiling. Distribute into flasks in 100.0mL volumes. Autoclave for 15 min at 15 psi pressure–121°C. Cool to 45°–50°C. Aseptically add 5.0mL of sterile rabbit blood to each flask. Mix thoroughly.

**Storage/Shelf Life:** Store dehydrated media in the dark in a sealed container below 30°C. Prepared media should be stored under refrigeration (2-8°C). Media should be used within 60 days of preparation. Media should not be used if there are any signs of deterioration (discoloration) or contamination, or if the expiration date supplied by the manufacturer has passed.

**Use:** For the isolation and enrichment of hemolytic streptococci from throat swabs and other clinical specimens. After incubation of bacteria for 18–24 h in this medium, they may be isolated by streaking the culture onto blood agar plates.

## PKU Test Agar
### (Phenylketonuria Test Agar)

**Composition** per liter:

| | |
|---|---|
| Agar | 15.0g |
| K$_2$HPO$_4$ | 15.0g |
| Glucose | 10.0g |
| KH$_2$PO$_4$ | 5.0g |
| (NH$_4$)Cl | 2.5g |
| (NH$_4$)NO$_3$ | 0.5g |
| Asparagine | 0.5g |

DL-Alanine......................................................................0.5g
L-Glutamic acid.............................................................0.5g
Na$_2$SO$_4$.......................................................................0.5g
MgSO$_4$·7H$_2$O.............................................................0.05g
FeCl$_3$..........................................................................5.0mg
MnCl$_2$·4H$_2$O............................................................5.0mg
β-2-Thienylalanine.....................................................3.3mg
CaCl$_2$·2H$_2$O............................................................2.5mg
*Bacillus subtilis* spore suspension..........................10.0mL
pH 7.0 ± 0.2 at 25°C

**Source:** This medium is available as a premixed powder from BD Diagnostic Systems.

**Preparation of Medium:** Add components to distilled/deionized water and bring volume to 1.0L. Mix thoroughly. Gently heat and bring to boiling. Continue boiling for 5 min. Do not autoclave. Cool to 50°C. Add 10.0mL of a suspension of *Bacillus subtilis* ATCC 6633 spores. Mix thoroughly. Pour into sterile Petri dishes or other containers.

**Storage/Shelf Life:** Store dehydrated media in the dark in a sealed container below 30°C. Prepared media should be stored under refrigeration (2-8°C). Media should be used within 60 days of preparation. Media should not be used if there are any signs of deterioration (shrinking, cracking, or discoloration) or contamination, or if the expiration date supplied by the manufacturer has passed.

**Use:** For the determination of phenylalanine concentrations in serum or urine. Used in the Guthrie-modified bacterial-inhibition assay procedure for screening newborn infants for phenylketonuria (PKU).

## PKU Test Agar
## (Phenylketonuria Test Agar)

**Composition** per liter:
K$_2$HPO$_4$....................................................................15.0g
Agar ...........................................................................13.5g
Glucose .........................................................................5.0g
KH$_2$PO$_4$......................................................................5.0g
(NH$_4$)Cl.......................................................................2.5g
L-Asparagine ...............................................................0.5g
L-Glutamic acid.............................................................0.5g
Na$_2$SO$_4$.......................................................................0.5g
(NH$_4$)NO$_3$...................................................................0.5g
L-Alanine ....................................................................0.25g
MgSO$_4$·7H$_2$O.............................................................0.05g
FeCl$_3$ ........................................................................0.005g
MnSO$_4$......................................................................0.005g
CaCl$_2$·2H$_2$O...............................................................2.5mg
β-2-Thienylalanine solution.......................................1.0mL
pH 6.9 ± 0.2 at 25°C

**Source:** This medium is available as a premixed powder from BD Diagnostic Systems.

**β-2-Thienylalanine Solution:**
**Composition** per 100.0mL:
β-2-Thienylalanine......................................................0.33g

**Preparation of β-2-Thienylalanine Solution:** Add β-2-thienylalanine to distilled/deionized water and bring volume to 100.0mL. Mix thoroughly. Filter sterilize.

**Preparation of Medium:** Add components, except β-2-thienylalanine solution, to distilled/deionized water and bring volume to 1.0L. Mix thoroughly. Gently heat and bring to boiling. Continue boiling for 5 min. Do not autoclave. Cool to 50°C. Add 10.0mL of a suspension of

*Bacillus subtilis* ATCC 6633 spores and 1.0mL of sterile β-2-thienylalanine solution. Mix thoroughly. Pour into sterile Petri dishes or other containers.

**Storage/Shelf Life:** Store dehydrated media in the dark in a sealed container below 30°C. Prepared media should be stored under refrigeration (2-8°C). Media should be used within 60 days of preparation. Media should not be used if there are any signs of deterioration (shrinking, cracking, or discoloration) or contamination, or if the expiration date supplied by the manufacturer has passed.

**Use:** For the determination of phenylalanine concentrations in serum or urine. Used in the Guthrie-modified bacterial-inhibition assay procedure for screening newborn infants for phenylketonuria (PKU).

## PL Agar

**Composition** per liter:
Agar ...........................................................................15.0g
Mannitol........................................................................7.5g
L-Arabinose .................................................................5.0g
Peptone .........................................................................5.0g
NaCl..............................................................................5.0g
Lysine............................................................................2.0g
Yeast extract.................................................................2.0g
Bile salts No. 2..............................................................1.0g
Inositol ..........................................................................1.0g
Phenol Red....................................................................0.08g
pH 7.4 ± 0.2 at 25°C

**Preparation of Medium:** Add components to distilled/deionized water and bring volume to 1.0L. Mix thoroughly. Gently heat and bring to boiling. Adjust pH to 7.4. Distribute into tubes or flasks. Autoclave for 15 min at 10 psi pressure–115°C. Pour into sterile Petri dishes or leave in tubes.

**Storage/Shelf Life:** Store dehydrated media in the dark in a sealed container below 30°C. Prepared media should be stored under refrigeration (2-8°C). Media should be used within 60 days of preparation. Media should not be used if there are any signs of deterioration (shrinking, cracking, or discoloration) or contamination, or if the expiration date supplied by the manufacturer has passed.

**Use:** For the isolation and cultivation of *Pleisomonas shigelloides* from foods.

## PL HiVeg Agar

**Composition** per liter:
Agar ...........................................................................15.0g
Mannitol........................................................................7.5g
Plant peptone ................................................................5.0g
L-Arabinose..................................................................5.0g
NaCl..............................................................................5.0g
Yeast extract.................................................................2.0g
Lysine............................................................................2.0g
Inositol ..........................................................................1.0g
Synthetic detergent .......................................................1.0g
Phenol Red....................................................................0.08g
pH 7.4 ± 0.2 at 25°C

**Source:** This medium is available as a premixed powder from HiMedia.

**Preparation of Medium:** Add components to distilled/deionized water and bring volume to 1.0L. Mix thoroughly. Gently heat and bring to boiling. Adjust pH to 7.4. Distribute into tubes or flasks. Autoclave for 15 min at 10 psi pressure–115°C. Pour into sterile Petri dishes or leave in tubes.

**Storage/Shelf Life:** Store dehydrated media in the dark in a sealed container below 30°C. Prepared media should be stored under refrigeration (2-8°C). Media should be used within 60 days of preparation. Media should not be used if there are any signs of deterioration (shrinking, cracking, or discoloration) or contamination, or if the expiration date supplied by the manufacturer has passed.

**Use:** For the isolation and cultivation of *Pleisomonas shigelloides* from foods.

## Plate Count Agar
## (ATCC Medium 1048)

**Composition** per liter:
Agar ........................................................................15.0g
Pancreatic digest of casein .......................................5.0g
Yeast extract............................................................2.5g
Glucose ...................................................................1.0g

pH 7.0 ± 0.2 at 25°C

**Source:** This medium is available as a premixed powder from BD Diagnostic Systems and Oxoid Unipath.

**Preparation of Medium:** Add components to distilled/deionized water and bring volume to 1.0L. Mix thoroughly. Gently heat and bring to boiling. Distribute into tubes or flasks. Autoclave for 15 min at 15 psi pressure–121°C. Pour into sterile Petri dishes or leave in tubes.

**Storage/Shelf Life:** Store dehydrated media in the dark in a sealed container below 30°C. Prepared media should be stored under refrigeration (2-8°C). Media should be used within 60 days of preparation. Media should not be used if there are any signs of deterioration (shrinking, cracking, or discoloration) or contamination, or if the expiration date supplied by the manufacturer has passed.

**Use:** For the enumeration of bacteria in milk, water, food, and dairy products.

## Plate Count Agar, HiVeg
## (Standard Methods Agar, HiVeg)

**Composition** per liter:
Agar .......................................................................15.0g
Plant hydrolysate......................................................5.0g
Yeast extract............................................................2.5g
Glucose ...................................................................1.0g

pH 7.0 ± 0.2 at 25°C

**Source:** This medium is available as a premixed powder from Oxoid Unipath.

**Preparation of Medium:** Add components to distilled/deionized water and bring volume to 1.0L. Mix thoroughly. Gently heat while stirring and bring to boiling. Distribute into tubes or flasks. Autoclave for 15 min at 15 psi pressure–121°C. Pour into sterile Petri dishes or leave in tubes.

**Storage/Shelf Life:** Store dehydrated media in the dark in a sealed container below 30°C. Prepared media should be stored under refrigeration (2-8°C). Media should be used within 60 days of preparation. Media should not be used if there are any signs of deterioration (shrinking, cracking, or discoloration) or contamination, or if the expiration date supplied by the manufacturer has passed.

**Use:** For the enumeration of viable bacteria in milk and dairy products. Also used for the estimation of the number of live heterotrophic bacteria in water.

## PLET Agar

**Composition** per liter:
Beef heart, solids from infusion.................................500.0g
Agar .......................................................................15.0g
Tryptose .................................................................10.0g
NaCl.........................................................................5.0g
Ethylenediamine tetracetic acid (EDTA).....................0.3g
Thallous acetate .....................................................0.04g
Antibiotic inhibitor ............................................... 10.0mL

**Antibiotic Inhibitor:**
**Composition** per 10.0mL:
Lysozyme...........................................................300,000U
Polymyxin...........................................................30,000U

**Preparation of Antibiotic Inhibitor:** Add components to distilled/deionized water and bring volume to 10.0mL. Mix thoroughly. Filter sterilize.

**Preparation of Medium:** Add components, except antibiotic inhibitor, to distilled/deionized water and bring volume to 990.0mL. Mix thoroughly. Gently heat and bring to boiling. Autoclave for 15 min at 15 psi pressure–121°C. Cool to 50°C. Aseptically add sterile antibiotic inhibitor. Mix thoroughly. Pour into sterile Petri dishes or distribute into sterile tubes.

**Storage/Shelf Life:** Store dehydrated media in the dark in a sealed container below 30°C. Prepared media should be stored under refrigeration (2-8°C). Media should be used within 60 days of preparation. Media should not be used if there are any signs of deterioration (shrinking, cracking, or discoloration) or contamination, or if the expiration date supplied by the manufacturer has passed.

**Use:** For the selective isolation and cultivation of *Bacillus anthracis*.

## Polymyxin *Staphylococcus* Medium

**Composition** per liter:
Agar .......................................................................15.0g
Pancreatic digest of gelatin.......................................5.0g
Beef extract.............................................................3.0g
Lecithin...................................................................0.7g
Polymyxin..............................................................0.075g
Tween™ 80 ......................................................... 10.2mL

pH 6.8 ± 0.2 at 25°C

**Preparation of Medium:** Add components to distilled/deionized water and bring volume to 1.0L. Mix thoroughly. Gently heat and bring to boiling. Distribute into tubes or flasks. Autoclave for 15 min at 15 psi pressure–121°C. Pour into sterile Petri dishes.

**Storage/Shelf Life:** Store dehydrated media in the dark in a sealed container below 30°C. Prepared media should be stored under refrigeration (2-8°C). Media should be used within 60 days of preparation. Media should not be used if there are any signs of deterioration (shrinking, cracking, or discoloration) or contamination, or if the expiration date supplied by the manufacturer has passed.

**Use:** For the selective isolation and cultivation of pathogenic, coagulase-positive *Staphylococcus aureus*. *Proteus* species will grow on this medium but appear as translucent colonies.

## Porcine Heart Agar

**Composition** per liter:

Porcine heart, infusion from ......................................375.0g
Agar .............................................................................15.0g
Papaic digest of soybean meal ......................................6.5g
Glucose ..........................................................................5.0g
NaCl ...............................................................................5.0g
Proteose peptone No. 3 ..................................................5.0g
Yeast extract..................................................................3.5g
Sheep blood, defibrinated ..........................................50.0mL

pH 7.2 ± 0.2 at 25°C

**Source:** This medium is available as a premixed powder from BD Diagnostic Systems.

**Preparation of Medium:** Add components, except sheep blood, to distilled/deionized water and bring volume to 950.0mL. Mix thoroughly. Gently heat and bring to boiling. Autoclave for 15 min at 15 psi pressure–121°C. Cool to 45°–50°C. Aseptically add sterile sheep blood. Mix thoroughly. Pour into sterile Petri dishes.

**Storage/Shelf Life:** Store dehydrated media in the dark in a sealed container below 30°C. Prepared media should be stored under refrigeration (2-8°C). Media should be used within 60 days of preparation. Media should not be used if there are any signs of deterioration (shrinking, cracking, or discoloration) or contamination, or if the expiration date supplied by the manufacturer has passed.

**Use:** For determination of the sensitivity of microorganisms using the disc plate technique.

## Pork Plasma Fibrinogen Overlay Agar

**Composition** per plate:

Baird-Parker agar, modified....................................... 15.0mL
Pork plasma fibrinogen overlay agar ......................... 8.0mL

pH 7.0 ± 0.1 at 25°C

**Baird-Parker Agar, Modified:**

**Composition** per liter:

Agar ..............................................................................17.0g
Glycine...........................................................................12.0g
Sodium pyruvate ...........................................................10.0g
Pancreatic digest of casein ...........................................10.0g
Beef extract ....................................................................5.0g
LiCl .................................................................................5.0g
Yeast extract...................................................................1.0g
Chapman tellurite solution...........................................1.0mL

pH 7.0 ± 0.2 at 25°C

**Chapman Tellurite Solution:**

**Composition** per 100.0mL:

$K_2TeO_3$.............................................................................1.0g

**Preparation of Chapman Tellurite Solution:** Add $K_2TeO_3$ to distilled/deionized water and bring volume to 100.0mL. Mix thoroughly. Filter sterilize.

**Preparation of Baird-Parker Agar, Modified:** Add components, except Chapman tellurite solution, to distilled/deionized water and bring volume to 999.0mL. Mix thoroughly. Gently heat and bring to boiling. Autoclave for 15 min at 15 psi pressure–121°C. Cool to 45–50°C. Aseptically add 1.0mL of sterile Chapman tellurite solution. Mix thoroughly but gently.

## Pork Plasma Fibrinogen Overlay Agar:

**Composition** per 100.5mL:

Agar solution .............................................................. 50.0mL
Bovine fibrinogen solution .......................................... 47.5mL
Pork plasma ................................................................. 2.5mL
Trypsin inhibitor solution ........................................... 0.5mL

**Agar Solution:**

**Composition** per 50.0mL:

Agar ................................................................................0.7g

**Preparation of Agar Solution:** Add agar to distilled/deionized water and bring volume to 50.0mL. Mix thoroughly. Autoclave for 15 min at 15 psi pressure–121°C. Cool to 45°–50°C.

**Bovine Fibrinogen Solution:**

**Composition** per 50.0mL:

Bovine fibrinogen, fraction I .......................................0.4g
Sodium phosphate buffer
    (0.05*M* solution, pH 7.0) ...................................... 50.0mL

**Preparation of Bovine Fibrinogen Solution:** Grind bovine fibrinogen in a mortar to a fine powder. Add bovine fibrinogen to 50.0mL of sodium phosphate buffer. Mix thoroughly on a magnetic stirrer for 30 min. Filter through Whatman #1 filter paper. Filter sterilize.

**Pork Plasma:**

**Composition** per 10.0mL:

Pork plasma-EDTA................................................... 10.0mL

**Preparation of Pork Plasma:** Filter sterilize fresh or rehydrated commercial pork plasma-EDTA.

**Trypsin Inhibitor Solution:**

**Composition** per 5.0mL:

Trypsin inhibitor ..........................................................0.015g
Sodium phosphate buffer (0.05*M* solution, pH 7.0) .................. 5.0mL

**Preparation of Trypsin Inhibitor Solution:** Add trypsin inhibitor to 5.0mL of sodium phosphate buffer. Mix thoroughly. Filter sterilize.

**Preparation of Pork Plasma Fibrinogen Overlay Agar:** Aseptically combine 50.0mL of cooled, sterile agar solution, 47.5mL of sterile bovine fibrinogen solution, 2.5mL of sterile pork plasma, and 0.5mL of trypsin inhibitor solution. Mix thoroughly. Maintain at 45°–50°C but use within 1 h.

**Caution:** Potassium tellurite is toxic.

**Preparation of Medium:** Pour cooled, sterile Baird-Parker agar, modified, into sterile Petri dishes in 15.0mL volumes. Allow agar to solidify. Overlay each plate with 8.0mL of sterile pork plasma fibrinogen overlay agar.

**Storage/Shelf Life:** Store dehydrated media in the dark in a sealed container below 30°C. Prepared media should be stored under refrigeration (2-8°C). Media should be used within 60 days of preparation. Media should not be used if there are any signs of deterioration (shrinking, cracking, or discoloration) or contamination, or if the expiration date supplied by the manufacturer has passed.

**Use:** For the cultivation of *Staphylococcus aureus* from foods.

## Potassium Cyanide Broth

**Composition** per liter:

$Na_2HPO_4$.............................................................................5.64g
NaCl ...............................................................................5.0g

Proteose peptone No. 3 ...................................................3.0g
KH$_2$PO$_4$...........................................................................0.225g
KCN solution ............................................. 15.0mL
pH 7.6 ± 0.2 at 25°C

**KCN Solution:**
**Composition** per 100.0mL:
KCN ....................................................................................0.5g

**Preparation of KCN Solution:** Add KCN to distilled/deionized water and bring volume to 100.0mL. Mix thoroughly.

**Caution:** Cyanide is toxic.

**Preparation of Medium:** Add components, except KCN solution, to distilled/deionized water and bring volume to 985.0mL. Mix thoroughly. Gently heat and bring to boiling. Autoclave for 15 min at 15 psi pressure–121°C. Cool to 25°C. Aseptically add 15.0mL of KCN solution. Mix thoroughly. Distribute into sterile screw-capped tubes or flasks in 1.0–1.5mL volumes. Close caps tightly.

**Storage/Shelf Life:** Store dehydrated media in the dark in a sealed container below 30°C. Prepared media should be stored under refrigeration (2-8°C). Media should be used within 60 days of preparation. Media should not be used if there are any signs of deterioration (discoloration) or contamination, or if the expiration date supplied by the manufacturer has passed.

**Use:** For the cultivation and differentiation of urease-negative, Gram-negative enteric bacteria. *Salmonella* species and *Shigella* species are nonmotile in this medium. *Proteus* species are motile in this medium.

## Potassium Cyanide HiVeg Broth Base with KCN
**Composition** per liter:
Na$_2$HPO$_4$........................................................................5.64g
NaCl....................................................................................5.0g
Plant peptone No. 3............................................................3.0g
KH$_2$PO$_4$...........................................................................0.225g
KCN solution ............................................. 15.0mL
pH 7.6 ± 0.2 at 25°C

**Source:** This medium, without potassium cyanide solution, is available as a premixed powder from HiMedia.

**KCN Solution:**
**Composition** per 100.0mL:
KCN ....................................................................................0.5g

**Preparation of KCN Solution:** Add KCN to distilled/deionized water and bring volume to 100.0mL. Mix thoroughly.

**Caution:** Cyanide is toxic.

**Preparation of Medium:** Add components, except KCN solution, to distilled/deionized water and bring volume to 985.0mL. Mix thoroughly. Gently heat and bring to boiling. Autoclave for 15 min at 15 psi pressure–121°C. Cool to 25°C. Aseptically add 15.0mL of KCN solution. Mix thoroughly. Distribute into sterile screw-capped tubes or flasks in 1.0–1.5mL volumes. Close caps tightly.

**Storage/Shelf Life:** Store dehydrated media in the dark in a sealed container below 30°C. Prepared media should be stored under refrigeration (2-8°C). Media should be used within 60 days of preparation. Media should not be used if there are any signs of deterioration (discoloration) or contamination, or if the expiration date supplied by the manufacturer has passed.

**Use:** For the cultivation and differentiation of urease-negative, Gram-negative enteric bacteria. *Salmonella* species and *Shigella* species are nonmotile in this medium. *Proteus* species are motile in this medium.

## Potassium Tellurite Agar
**Composition** per liter:
Beef heart, solids from infusion................................500.0g
Agar ................................................................................15.0g
Tryptose ..........................................................................10.0g
NaCl..................................................................................5.0g
Blood, defibrinated ..................................................... 50.0mL
K$_2$TeO$_3$ solution..................................................... 20.0mL
pH 6.0 ± 0.2 at 25°C

**K$_2$TeO$_3$ Solution:**
**Composition** per 20.0mL:
K$_2$TeO$_3$.............................................................................0.5g

**Preparation of K$_2$TeO$_3$ Solution:** Add K$_2$TeO$_3$ to distilled/deionized water and bring volume to 20.0mL. Mix thoroughly. Filter sterilize.

**Caution:** Potassium tellurite is toxic.

**Preparation of Medium:** Add components, except K$_2$TeO$_3$ solution, to distilled/deionized water and bring volume to 930.0mL. Mix thoroughly. Gently heat and bring to boiling. Autoclave for 15 min at 15 psi pressure–121°C. Cool to 45°–50°C. Aseptically add sterile K$_2$TeO$_3$ solution and 50.0mL of blood. Rabbit or sheep blood may be used. Mix thoroughly. Pour into sterile Petri dishes or distribute into sterile tubes. Allow tubes to cool in a slanted position.

**Storage/Shelf Life:** Store dehydrated media in the dark in a sealed container below 30°C. Prepared media should be stored under refrigeration (2-8°C). Media should be used within 60 days of preparation. Media should not be used if there are any signs of deterioration (shrinking, cracking, or discoloration) or contamination, or if the expiration date supplied by the manufacturer has passed.

**Use:** For the cultivation and differentiation of *Enterococcus faecalis*. *Enterococcus faecalis* appears as black colonies.

## PPLO Broth with Crystal Violet
**Composition** per liter:
Beef heart, infusion from ...........................................50.0g
Peptone ............................................................................10.0g
NaCl....................................................................................5.0g
Crystal Violet ...................................................................0.01g
Ascitic fluid .............................................................. 250.0mL
Chapman tellurite solution.......................................2.85mL
pH 7.8 ± 0.2 at 25°C

**Source:** This medium is available as a premixed powder from BD Diagnostic Systems.

**Chapman Tellurite Solution:**
**Composition** per 100.0mL:
K$_2$TeO$_3$.............................................................................1.0g

**Preparation of Chapman Tellurite Solution:** Add K$_2$TeO$_3$ to distilled/deionized water and bring volume to 100.0mL. Mix thoroughly. Filter sterilize.

**Caution:** Potassium tellurite is toxic.

**Preparation of Medium:** Add components, except ascitic fluid and Chapman tellurite solution, to distilled/deionized water and bring volume to 747.15mL. Mix thoroughly. Autoclave for 15 min at 15 psi pressure–121°C. Cool to less than 37°C. Aseptically add sterile ascitic fluid and 2.85mL of Chapman tellurite solution. Mix thoroughly. Aseptically distribute into sterile tubes or flasks.

**Storage/Shelf Life:** Store dehydrated media in the dark in a sealed container below 30°C. Prepared media should be stored under refriger-

ation (2-8°C). Media should be used within 60 days of preparation. Media should not be used if there are any signs of deterioration (discoloration) or contamination, or if the expiration date supplied by the manufacturer has passed.

**Use:** For the isolation of *Mycoplasma* species from clinical specimens.

## PPLO Broth without Crystal Violet

**Composition** per liter:

Beef heart, infusion from ........................................................50.0g
Peptone.....................................................................................10.0g
NaCl............................................................................................5.0g
Thallium acetate (optional)......................................................0.5g
Penicillin (optional) ........................................................100,000U
Ascitic fluid.......................................................................250.0mL

pH 7.8 ± 0.2 at 25°C

**Source:** This medium is available as a premixed powder from BD Diagnostic Systems.

**Preparation of Medium:** Add components, except ascitic fluid, to distilled/deionized water and bring volume to 750.0mL. Mix thoroughly. Autoclave for 15 min at 15 psi pressure–121°C. Cool to less than 37°C. Aseptically add sterile ascitic fluid. If desired, 0.5g of thallium acetate or 100,000U of penicillin may be added for a more selective medium. Mix thoroughly. Aseptically distribute into sterile tubes or flasks.

**Storage/Shelf Life:** Store dehydrated media in the dark in a sealed container below 30°C. Prepared media should be stored under refrigeration (2-8°C). Media should be used within 60 days of preparation. Media should not be used if there are any signs of deterioration (discoloration) or contamination, or if the expiration date supplied by the manufacturer has passed.

**Use:** For the enrichment of pleuro-pneumonia-like organisms (PPLOs) and *Mycoplasma* species from clinical specimens.

## PPLO Broth without Crystal Violet with Calf Serum, Fresh Yeast Extract, and Sodium Acetate

**Composition** per liter:

Beef heart, infusion from ........................................................50.0g
Peptone.....................................................................................10.0g
Sodium acetate .........................................................................9.0g
NaCl............................................................................................5.0g
Yeast extract solution, fresh...............................................250.0mL
Calf serum............................................................................100.0mL

pH 7.8 ± 0.2 at 25°C

**Yeast Extract Solution:**
**Composition** per 300.0mL:

Baker's yeast, live, pressed, starch-free.................................75.0g

**Preparation of Yeast Extract Solution:** Add the live Baker's yeast to 300.0mL of distilled/deionized water. Autoclave for 90 min at 15 psi pressure–121°C. Allow to stand. Remove supernatant solution. Adjust pH to 6.6–6.8. Filter sterilize.

**Preparation of Medium:** Add components, except fresh yeast extract solution and calf serum, to distilled/deionized water and bring volume to 550.0mL. Mix thoroughly. Autoclave for 15 min at 15 psi pressure–121°C. Cool to 45°–50°C. Aseptically add sterile fresh yeast extract solution and calf serum. Mix thoroughly. Aseptically distribute into sterile tubes or flasks.

**Storage/Shelf Life:** Store dehydrated media in the dark in a sealed container below 30°C. Prepared media should be stored under refrigeration (2-8°C). Media should be used within 60 days of preparation. Media should not be used if there are any signs of deterioration (discoloration) or contamination, or if the expiration date supplied by the manufacturer has passed.

**Use:** For the cultivation and maintenance of *Mycoplasma* species.

## PPLO Broth without Crystal Violet with Horse Serum

**Composition** per liter:

Beef heart, infusion from ........................................................50.0g
Peptone ....................................................................................10.0g
NaCl............................................................................................5.0g
Horse serum, inactivated ....................................................200.0mL

pH 7.8 ± 0.2 at 25°C

**Preparation of Medium:** Add components, except horse serum, to distilled/deionized water and bring volume to 800.0mL. Mix thoroughly. Autoclave for 15 min at 15 psi pressure–121°C. Cool to 45°–50°C. Aseptically add sterile horse serum. Mix thoroughly. Aseptically distribute into sterile tubes or flasks.

**Storage/Shelf Life:** Store dehydrated media in the dark in a sealed container below 30°C. Prepared media should be stored under refrigeration (2-8°C). Media should be used within 60 days of preparation. Media should not be used if there are any signs of deterioration (discoloration) or contamination, or if the expiration date supplied by the manufacturer has passed.

**Use:** For the cultivation and maintenance of *Acholeplasma* species and *Mycoplasma* species.

## PPLO Broth without Crystal Violet with Horse Serum and Fresh Yeast Extract

**Composition** per liter:

Beef heart, solids from infusion.............................................50.0g
Peptone .....................................................................................10.0g
NaCl............................................................................................5.0g
Yeast extract solution, fresh...............................................250.0mL
Horse serum ........................................................................200.0mL

pH 7.8 ± 0.2 at 25°C

**Yeast Extract Solution:**
**Composition** per 300.0mL:

Baker's yeast, live, pressed, starch-free.................................75.0g

**Preparation of Yeast Extract Solution:** Add the live Baker's yeast to 300.0mL of distilled/deionized water. Autoclave for 90 min at 15 psi pressure–121°C. Allow to stand. Remove supernatant solution. Adjust pH to 6.6–6.8. Filter sterilize.

**Preparation of Medium:** Add components, except fresh yeast extract solution and horse serum, to distilled/deionized water and bring volume to 550.0mL. Mix thoroughly. Autoclave for 15 min at 15 psi pressure–121°C. Cool to 45°–50°C. Aseptically add sterile fresh yeast extract solution and horse serum. Mix thoroughly. Aseptically distribute into sterile tubes or flasks.

**Storage/Shelf Life:** Store dehydrated media in the dark in a sealed container below 30°C. Prepared media should be stored under refrigeration (2-8°C). Media should be used within 60 days of preparation. Media should not be used if there are any signs of deterioration (discol-

oration) or contamination, or if the expiration date supplied by the manufacturer has passed.

**Use:** For the cultivation and maintenance of *Mycoplasma putrefaciens*.

## PPLO Broth without Crystal Violet with Horse Serum, Glucose, and Fresh Yeast Extract
**Composition** per liter:

Beef heart, infusion from ...............................................50.0g
Peptone.............................................................................10.0g
Glucose ..............................................................................5.0g
NaCl....................................................................................5.0g
Yeast extract solution, fresh......................................250.0mL
Horse serum .................................................................200.0mL

pH 7.8 ± 0.2 at 25°C

**Yeast Extract Solution:**
**Composition** per 300.0mL:
Baker's yeast, live, pressed, starch-free......................75.0g

**Preparation of Yeast Extract Solution:** Add the live Baker's yeast to 300.0mL of distilled/deionized water. Autoclave for 90 min at 15 psi pressure–121°C. Allow to stand. Remove supernatant solution. Adjust pH to 6.6–6.8. Filter sterilize.

**Preparation of Medium:** Add components, except fresh yeast extract solution and horse serum, to distilled/deionized water and bring volume to 550.0mL. Mix thoroughly. Autoclave for 15 min at 15 psi pressure–121°C. Cool to 45°–50°C. Aseptically add sterile fresh yeast extract solution and horse serum. Mix thoroughly. Aseptically distribute into sterile tubes or flasks.

**Storage/Shelf Life:** Store dehydrated media in the dark in a sealed container below 30°C. Prepared media should be stored under refrigeration (2-8°C). Media should be used within 60 days of preparation. Media should not be used if there are any signs of deterioration (discoloration) or contamination, or if the expiration date supplied by the manufacturer has passed.

**Use:** For the cultivation and maintenance of *Mycoplasma putrefaciens*, *Mycoplasma collis*, and *Mycoplasma cricetuli*.

## PPLO Broth without Crystal Violet with Sodium Acetate, Fresh Yeast Extract, and Calf Serum
**Composition** per liter:

Beef heart, infusion from ...............................................50.0g
Peptone.............................................................................10.0g
NaCl....................................................................................5.0g
Sodium acetate.................................................................1.0g
Calf serum..................................................................100.0mL
Yeast extract solution, fresh........................................50.0mL

pH 7.8 ± 0.2 at 25°C

**Yeast Extract Solution:**
**Composition** per 300.0mL:
Baker's yeast, live, pressed, starch-free......................75.0g

**Preparation of Yeast Extract Solution:** Add the live Baker's yeast to 300.0mL of distilled/deionized water. Autoclave for 90 min at 15 psi pressure–121°C. Allow to stand. Remove supernatant solution. Adjust pH to 6.6–6.8. Filter sterilize.

**Preparation of Medium:** Add components, except fresh yeast extract solution and calf serum, to distilled/deionized water and bring volume to 850.0mL. Mix thoroughly. Autoclave for 15 min at 15 psi pressure–121°C. Cool to 25°C. Aseptically add sterile fresh yeast extract solution and calf serum. Mix thoroughly. Aseptically distribute into sterile tubes or flasks.

**Storage/Shelf Life:** Store dehydrated media in the dark in a sealed container below 30°C. Prepared media should be stored under refrigeration (2-8°C). Media should be used within 60 days of preparation. Media should not be used if there are any signs of deterioration (discoloration) or contamination, or if the expiration date supplied by the manufacturer has passed.

**Use:** For the cultivation and maintenance of *Acholeplasma laidlawii.*

## PPLO Broth with Penicillin
**Composition** per 1010.0mL:
Pancreatic digest of casein................................................7.0g
NaCl....................................................................................5.0g
Beef extract........................................................................3.0g
Yeast extract.......................................................................3.0g
Beef heart, solids from infusion.........................................2.0g
*Mycoploasma* supplement solution........................... 300.0mL
Penicillin solution ...................................................... 10.0mL

pH 7.8 ± 0.2 at 25°C

**Source:** This medium is available as a premixed powder from BD Diagnostic Systems.

### *Mycoplasma* Supplement Solution:
**Composition** per 300.0mL:
Horse serum ............................................................... 200.0mL
Yeast extract (fresh autolysate)................................. 100.0mL
Thallium acetate...........................................................50.0 mg

**Preparation of *Mycoplasma* Supplement Solution:** Combine components. Mix thoroughly. Filter sterilize.

### Penicllin Solution:
**Composition** per 10.0mL:
Penicillin.................................................................... 500,000U

**Preparation of Penicllin Solution:** Add penicillin to distilled/deionized water and bring volume to 10.0mL. Mix thoroughly. Filter sterilize.

**Preparation of Medium:** Add components, except *Mycoplasma* supplement solution and penicllin solution, to distilled/deionized water and bring volume to 700.0mL. Mix thoroughly. Gently heat and bring to boiling. Boil for 1 min. Autoclave for 15 min at 15 psi pressure–121°C. Cool to 45°–50°C. Aseptically add sterile *Mycoplasma* supplement solution and penicllin solution. Mix thoroughly.

**Storage/Shelf Life:** Store dehydrated media in the dark in a sealed container below 30°C. Prepared media should be stored under refrigeration (2-8°C). Media should be used within 60 days of preparation. Media should not be used if there are any signs of deterioration (discoloration) or contamination, or if the expiration date supplied by the manufacturer has passed.

**Use:** For the selective cultivation of *Mycoplasma* species.

## PPT Agar, 1*M*
**Composition** per liter:
NaCl..................................................................................58.4g
Agar .................................................................................18.0g
Proteose peptone No. 3 .................................................10.0g
Pancreatic digest of casein ...........................................10.0g

**Preparation of Medium:** Add components to distilled/deionized water and bring volume to 1.0L. Mix thoroughly. Gently heat and bring

to boiling. Distribute into tubes or flasks. Autoclave for 15 min at 15 psi pressure–121°C. Pour into sterile Petri dishes or leave in tubes.

**Use:** For the cultivation of *Pseudomonas* species.

## Pre-Enrichment HiVeg Broth Base with Magnesium Sulfate and Calcium Chloride

**Composition** per liter:

| | |
|---|---|
| Yeast extract | 20.0g |
| Plant special peptone | 10.0g |
| $Na_2HPO_4$ | 7.1g |
| KCl | 1.0g |
| NaCl | 1.0g |
| Magnesium sulfate solution | 10.0mL |
| Calcium chloride soltuion | 10.0mL |

pH 8.3 ± 0.2 at 25°C

**Source:** This medium, without magnesium sulfate and calcium chloride solutions, is available as a premixed powder from HiMedia.

### Magnesium Sulfate Solution:

**Composition** per 10.0mL:

| | |
|---|---|
| $MgSO_4$ | 0.01g |

**Preparation of Magnesium Sulfate Solution:** Add $MgSO_4$ to distilled/deionized water and bring volume to 10.0mL. Mix thoroughly. Filter sterilize.

### Calcium Chloride Solution:

**Composition** per 10.0mL:

| | |
|---|---|
| $CaCl_2$ | 0.01g |

**Preparation of Calcium Chloride Solution:** Add $CaCl_2$ to distilled/deionized water and bring volume to 10.0mL. Mix thoroughly. Filter sterilize.

**Preparation of Medium:** Add components, except magnesium chloride solution and calcium chloride solution, to distilled/deionized water and bring volume to 980.0mL. Mix thoroughly. Gently heat while stirring and bring to boiling. Distribute into tubes or flasks. Autoclave for 15 min at 15 psi pressure–121°C. Cool to 45°–50°C. Aseptically add 10.0mL of sterile magnesium sulfate solution and 10.0mL of sterile calcium chloride solution. Mix thoroughly. Aseptically distribute to sterile tubes or flasks.

**Storage/Shelf Life:** Store dehydrated media in the dark in a sealed container below 30°C. Prepared media should be stored under refrigeration (2-8°C). Media should be used within 60 days of preparation. Media should not be used if there are any signs of deterioration (shrinking, cracking, or discoloration) or contamination, or if the expiration date supplied by the manufacturer has passed.

**Use:** For the isolation, enrichment, and cultivation of *Yersinia enterocolitica* from foods.

## Preenrichment Medium
## (PEM)

**Composition** per liter:

| | |
|---|---|
| Yeast extract | 20.0g |
| Special peptone | 10.0g |
| $Na_2HPO_4$ | 7.1g |
| NaCl | 1.0g |
| KCl | 1.0g |
| $MgSO_4 \cdot 7H_2O$ solution | 10.0mL |
| $CaCl_2 \cdot 2H_2O$ solution | 10.0mL |

pH 8.3 ± 0.2 at 25°C

### $MgSO_4 \cdot 7H_2O$ Solution:

**Composition** per 10.0mL:

| | |
|---|---|
| $MgSO_4 \cdot 7H_2O$ | 0.01g |

**Preparation of $MgSO_4 \cdot 7H_2O$ Solution:** Add $MgSO_4 \cdot 7H_2O$ to distilled/deionized water and bring volume to 10.0mL. Mix thoroughly. Filter sterilize.

### $CaCl_2 \cdot 2H_2O$ Solution:

| | |
|---|---|
| $CaCl_2 \cdot 2H_2O$ | 0.01g |

**Preparation of $CaCl_2 \cdot 2H_2O$ Solution:** Add the $CaCl_2 \cdot 2H_2O$ to distilled/deionized water and bring volume to 10.0mL. Mix thoroughly. Filter sterilize.

**Preparation of Medium:** Add components, except $MgSO_4 \cdot 7H_2O$ solution and $CaCl_2 \cdot 2H_2O$ solution, to distilled/deionized water and bring volume to 980.0mL. Mix thoroughly. Adjust pH to 8.3. Gently heat and bring to boiling. Autoclave for 15 min at 15 psi pressure–121°C. Cool to 45°–50°C. Aseptically add sterile $MgSO_4 \cdot 7H_2O$ solution and $CaCl_2 \cdot 2H_2O$ solution. Mix thoroughly. Aseptically distribute into sterile tubes.

**Storage/Shelf Life:** Store dehydrated media in the dark in a sealed container below 30°C. Prepared media should be stored under refrigeration (2-8°C). Media should be used within 60 days of preparation. Media should not be used if there are any signs of deterioration (discoloration) or contamination, or if the expiration date supplied by the manufacturer has passed.

**Use:** For the isolation and enrichment of *Yersinia enterocolitica* from foods.

## Presence-Absence Broth
## (P-A Broth)

**Composition** per liter:

| | |
|---|---|
| Pancreatic digest of casein | 10.0g |
| Lactose | 7.5g |
| Pancreatic digest of gelatin | 5.0g |
| Beef extract | 3.0g |
| NaCl | 2.5g |
| $K_2HPO_4$ | 1.375g |
| $KH_2PO_4$ | 1.375g |
| Sodium lauryl sulfate | 0.05g |
| Bromcresol Purple | 8.5mg |

pH 6.8 ± 0.2 at 25°C

**Source:** This medium is available as a premixed powder from BD Diagnostic Systems.

**Preparation of Medium:** Add components to distilled/deionized water and bring volume to 333.0mL. Mix thoroughly. Distribute into screw-capped 250.0mL milk dilution bottles in 50.0mL volumes. Autoclave for 15 min at 15 psi pressure–121°C.

**Storage/Shelf Life:** Store dehydrated media in the dark in a sealed container below 30°C. Prepared media should be stored under refrigeration (2-8°C). Media should be used within 60 days of preparation. Media should not be used if there are any signs of deterioration (discoloration) or contamination, or if the expiration date supplied by the manufacturer has passed.

**Use:** For the detection of coliform bacteria in water from treatment plants or distribution systems using the presence-absence coliform test.

## Preston Enrichment Broth

**Composition** per liter:

| | |
|---|---|
| Beef extract | 10.0g |
| Peptone | 10.0g |

NaCl .................................................................................5.0g
Horse blood, lysed ........................................... 50.0mL
Antibiotic solution ........................................... 10.0mL

pH 7.5 ± 0.2 at 25°C

**Antibiotic Solution:**
**Composition** per 10.0mL:
Cycloheximide ..................................................................0.1g
Rifampicin ......................................................................0.01g
Trimethoprim lactate .....................................................0.01g
Polymyxin B ..................................................................5000U

**Preparation of Antibiotic Solution:** Add components to distilled/deionized water and bring volume to 10.0mL. Mix thoroughly. Filter sterilize.

**Caution:** Cycloheximide is toxic. Avoid skin contact or aerosol formation and inhalation.

**Preparation of Medium:** Add components, except horse blood and antibiotic solution, to distilled/deionized water and bring volume to 940.0mL. Mix thoroughly. Gently heat and bring to boiling. Autoclave for 15 min at 15 psi pressure–121°C. Cool to 45°–50°C. Aseptically add sterile horse blood and antibiotic solution. Mix thoroughly. Aseptically distribute into sterile tubes or flasks.

**Storage/Shelf Life:** Store dehydrated media in the dark in a sealed container below 30°C. Prepared media should be stored under refrigeration (2-8°C). Media should be used within 60 days of preparation. Media should not be used if there are any signs of deterioration (discoloration) or contamination, or if the expiration date supplied by the manufacturer has passed.

**Use:** For the isolation and enrichment of *Campylobacter* species from foods.

## Preston HiVeg Agar Base with Horse Blood and Antibiotics

**Composition** per liter:
Agar ...............................................................................12.0g
Plant extract ..................................................................10.0g
Plant peptone.................................................................10.0g
NaCl .................................................................................5.0g
Horse blood, lysed ........................................... 50.0mL
Antibiotic solution ........................................... 10.0mL

pH 7.5 ± 0.2 at 25°C

**Source:** This medium, without horse blood and antibiotics, is available as a premixed powder from HiMedia.

**Antibiotic Solution:**
**Composition** per 10.0mL:
Cycloheximide ..................................................................0.1g
Rifampicin ......................................................................0.01g
Trimethoprim lactate.......................................................0.01g
Polymyxin B ..................................................................5000U

**Preparation of Antibiotic Solution:** Add components to distilled/deionized water and bring volume to 10.0mL. Mix thoroughly. Filter sterilize.

**Caution:** Cycloheximide is toxic. Avoid skin contact or aerosol formation and inhalation.

**Preparation of Medium:** Add components, except horse blood and antibiotic solution, to distilled/deionized water and bring volume to 940.0mL. Mix thoroughly. Gently heat and bring to boiling. Autoclave for 15 min at 15 psi pressure–121°C. Cool to 45°–50°C. Aseptically

add sterile horse blood and antibiotic solution. Mix thoroughly. Aseptically distribute into sterile tubes or flasks.

**Storage/Shelf Life:** Store dehydrated media in the dark in a sealed container below 30°C. Prepared media should be stored under refrigeration (2-8°C). Media should be used within 60 days of preparation. Media should not be used if there are any signs of deterioration (shrinking, cracking, or discoloration) or contamination, or if the expiration date supplied by the manufacturer has passed.

**Use:** For the isolation and enrichment of *Campylobacter* species from foods.

## Presumpto Media

**Composition** per plate:
Quadrant I ......................................................... 5.0mL
Quadrant II ........................................................ 5.0mL
Quadrant III ....................................................... 5.0mL
Quadrant IV ....................................................... 5.0mL

**Quadrant I:**
**Composition** per 5.0mL:
Lombard-Dowell agar.......................................... 5.0mL

**Quadrant II:**
**Composition** per 5.0mL:
Lombard-Dowell bile agar.................................... 5.0mL

**Quadrant III:**
**Composition** per 5.0mL:
Lombard-Dowell egg yolk agar............................. 5.0mL

**Quadrant IV:**
**Composition** per 5.0mL:
Lombard-Dowell esculin agar .............................. 5.0mL

**Preparation of Quadrant Media:** Sterilize Lombard-Dowell Agar by autoclaving for 15 min at 15 psi pressure–121°C. Cool to 45°–50°C. Add additional components as filter sterilized solutions. Mix and distribute as 5.0mL aliquots into quadrants.

**Storage/Shelf Life:** Store dehydrated media in the dark in a sealed container below 30°C. Prepared media should be stored under refrigeration (2-8°C). Media should be used within 60 days of preparation. Media should not be used if there are any signs of deterioration (shrinking, cracking, or discoloration) or contamination, or if the expiration date supplied by the manufacturer has passed.

**Use:** For the differentiation and presumptive identification of anaerobic bacteria. The Presumpto media is a plate with four sectors, each containing a different medium.

## Proteose Agar

**Composition** per liter:
Agar ...............................................................................15.0g
Proteose peptone No. 3 ...................................................15.0g
Yeast extract....................................................................7.5g
Casamino acids ................................................................5.0g
K$_2$HPO$_4$....................................................................5.0g
(NH$_4$)$_2$SO$_4$ ...........................................................1.5g
Starch, soluble.................................................................1.0g

pH 9.0 ± 0.2 at 25°C

**Preparation of Medium:** Add components to distilled/deionized water and bring volume to 1.0L. Mix thoroughly. Gently heat and bring to boiling. Distribute into tubes in 10.0mL volumes. Autoclave for 15 min at 15 psi pressure–121°C. Allow tubes to cool in a slanted position.

**Storage/Shelf Life:** Store dehydrated media in the dark in a sealed container below 30°C. Prepared media should be stored under refrigeration (2-8°C). Media should be used within 60 days of preparation. Media should not be used if there are any signs of deterioration (shrinking, cracking, or discoloration) or contamination, or if the expiration date supplied by the manufacturer has passed.

**Use:** For the cultivation of *Vibrio* species from foods.

## Proteose HiVeg Agar

**Composition** per liter:

| | |
|---|---|
| Agar | 15.0g |
| Plant peptone No. 3 | 15.0g |
| K$_2$HPO$_4$ | 5.0g |
| Plant acid hydrolysate | 5.0g |
| Yeast extract | 7.5g |
| (NH$_4$)$_2$SO$_4$ | 1.5g |
| Starch, soluble | 1.0g |

pH 9.0 ± 0.2 at 25°C

**Source:** This medium is available as a premixed powder from Hi-Media.

**Preparation of Medium:** Add components to distilled/deionized water and bring volume to 1.0L. Mix thoroughly. Gently heat while stirring and bring to boiling. Distribute into tubes or flasks. Autoclave for 10 min at 15 psi pressure–121°C. Pour into sterile Petri dishes or leave in tubes.

**Storage/Shelf Life:** Store dehydrated media in the dark in a sealed container below 30°C. Prepared media should be stored under refrigeration (2-8°C). Media should be used within 60 days of preparation. Media should not be used if there are any signs of deterioration (shrinking, cracking, or discoloration) or contamination, or if the expiration date supplied by the manufacturer has passed.

**Use:** For the cultivation and maintenance of *Vibrio* species from foods.

## *Pseudomonas* Agar P

**Composition** per liter:

| | |
|---|---|
| Proteose peptone No. 3 | 20.0g |
| Agar | 15.0g |
| Glycerol | 10.0g |
| K$_2$HPO$_4$ | 10.0g |
| MgCl$_2$·6H$_2$O | 1.4g |

pH 7.0 ± 0.2 at 25°C

**Source:** This medium is available as a premixed powder from BD Diagnostic Systems.

**Preparation of Medium:** Add components to distilled/deionized water and bring volume to 1.0L. Mix thoroughly. Gently heat and bring to boiling. Distribute into tubes or flasks. Autoclave for 15 min at 15 psi pressure–121°C. Pour into sterile Petri dishes or leave in tubes.

**Storage/Shelf Life:** Store dehydrated media in the dark in a sealed container below 30°C. Prepared media should be stored under refrigeration (2-8°C). Media should be used within 60 days of preparation. Media should not be used if there are any signs of deterioration (shrinking, cracking, or discoloration) or contamination, or if the expiration date supplied by the manufacturer has passed.

**Use:** For the isolation, cultivation, and differentiation of *Pseudomonas aeruginosa* on the basis of pigment production.

## *Pseudomonas* Isolation Agar

**Composition** per liter:

| | |
|---|---|
| Peptone | 20.0g |
| Agar | 13.6g |
| K$_2$SO$_4$ | 10.0g |
| MgCl$_2$·6H$_2$O | 1.4g |
| Irgasan® (triclosan) | 0.025g |
| Glycerol | 20.0mL |

pH 7.0 ± 0.2 at 25°C

**Source:** This medium is available as a premixed powder from BD Diagnostic Systems.

**Preparation of Medium:** Add components to distilled/deionized water and bring volume to 1.0L. Mix thoroughly. Gently heat and bring to boiling. Distribute into tubes or flasks. Autoclave for 15 min at 15 psi pressure–121°C. Pour into sterile Petri dishes or leave in tubes.

**Storage/Shelf Life:** Store dehydrated media in the dark in a sealed container below 30°C. Prepared media should be stored under refrigeration (2-8°C). Media should be used within 60 days of preparation. Media should not be used if there are any signs of deterioration (shrinking, cracking, or discoloration) or contamination, or if the expiration date supplied by the manufacturer has passed.

**Use:** For the isolation and cultivation of *Pseudomonas* species.

## *Pseudomonas* Isolation HiVeg Agar Base with Glycerol

**Composition** per liter:

| | |
|---|---|
| Plant peptone | 20.0g |
| Agar | 13.6g |
| K$_2$SO$_4$ | 10.0g |
| MnCl$_2$ | 1.4g |
| Triclosan (Irgasan®) | 0.025g |
| Glycerol | 10.0mL |

pH 7.0 ± 0.2 at 25°C

**Source:** This medium, without glycerol, is available as a premixed powder from HiMedia.

**Preparation of Medium:** Add glycerol and then other components to distilled/deionized water and bring volume to 1.0L. Mix thoroughly. Gently heat and bring to boiling. Distribute into tubes or flasks. Autoclave for 15 min at 15 psi pressure–121°C. Pour into sterile Petri dishes or leave in tubes.

**Storage/Shelf Life:** Store dehydrated media in the dark in a sealed container below 30°C. Prepared media should be stored under refrigeration (2-8°C). Media should be used within 60 days of preparation. Media should not be used if there are any signs of deterioration (shrinking, cracking, or discoloration) or contamination, or if the expiration date supplied by the manufacturer has passed.

**Use:** For the selective isolation and identification of *Pseudomonas aeruginosa* from clinical and nonclinical specimens.

## PSTA Enrichment HiVeg Broth Base

**Composition** per liter:

| | |
|---|---|
| Tris hydroxymethyl aminomethane | 3.0g |
| Plant peptone | 1.0g |
| Sucrose | 1.0g |
| NaN$_3$ | 0.192g |
| Brilliant Green | 0.0125g |

pH 7.0 ± 0.2 at 25°C

**Source:** This medium is available as a premixed powder from Hi-Media.

**Preparation of Medium:** Add components to distilled/deionized water and bring volume to 1.0L. Mix thoroughly. Gently heat and bring to boiling. Distribute into tubes or flasks. Autoclave for 15 min at 15 psi pressure–121°C. Pour into sterile Petri dishes or leave in tubes.

**Storage/Shelf Life:** Store dehydrated media in the dark in a sealed container below 30°C. Prepared media should be stored under refrigeration (2-8°C). Media should be used within 60 days of preparation. Media should not be used if there are any signs of deterioration (discoloration) or contamination, or if the expiration date supplied by the manufacturer has passed.

**Use:** For the secondary enrichment of *Yersinia enterocolitica* from foods.

## Purple Agar

**Composition** per liter:

| | |
|---|---|
| Agar | 15.0g |
| Proteose peptone No. 3 | 10.0g |
| NaCl | 5.0g |
| Beef extract | 1.0g |
| Bromcresol Purple | 0.02g |
| Carbohydrate solution | 20.0mL |

pH 6.8 ± 0.2 at 25°C

**Source:** This medium is available as a premixed powder from BD Diagnostic Systems.

**Carbohydrate Solution:**
**Composition** per 20.0mL:

| | |
|---|---|
| Carbohydrate | 10.0g |

**Preparation of Carbohydrate Solution:** Add carbohydrate to distilled/deionized water and bring volume to 20.0mL. For expensive carbohydrates, 5.0g may be used instead of 10.0g. Mix thoroughly. Filter sterilize.

**Preparation of Medium:** Add components, except carbohydrate solution, to distilled/deionized water and bring volume to 980.0mL. Mix thoroughly. Gently heat and bring to boiling. Distribute into tubes in 9.8mL volumes. Autoclave for 15 min at 15 psi pressure–121°C. Cool to 45°–50°C. Aseptically add 0.2mL of sterile carbohydrate solution to each tube. Mix thoroughly. Allow tubes to cool in a slanted position.

**Storage/Shelf Life:** Store dehydrated media in the dark in a sealed container below 30°C. Prepared media should be stored under refrigeration (2-8°C). Media should be used within 60 days of preparation. Media should not be used if there are any signs of deterioration (shrinking, cracking, or discoloration) or contamination, or if the expiration date supplied by the manufacturer has passed.

**Use:** For the preparation of carbohydrate media used in fermentation studies for the identification of bacteria, especially members of the Enterobacteriaceae. Bacteria that can ferment the carbohydrate turn the medium yellow.

## Purple Broth
## (Purple Carbohydrate Broth)

**Composition** per liter:

| | |
|---|---|
| Proteose peptone No. 3 | 10.0g |
| NaCl | 5.0g |
| Beef extract | 1.0g |
| Bromcresol Purple | 0.015g |
| Carbohydrate solution | 20.0mL |

pH 6.8 ± 0.2 at 25°C

**Source:** This medium is available as a premixed powder from BD Diagnostic Systems.

**Carbohydrate Solution:**
**Composition** per 20.0mL:

| | |
|---|---|
| Carbohydrate | 10.0g |

**Preparation of Carbohydrate Solution:** Add carbohydrate to distilled/deionized water and bring volume to 20.0mL. For expensive carbohydrates, 5.0g may be used instead of 10.0g. Mix thoroughly. Filter sterilize.

**Preparation of Medium:** Add components, except carbohydrate solution, to distilled/deionized water and bring volume to 980.0mL. Mix thoroughly. Gently heat and bring to boiling. Distribute into tubes in 9.8mL volumes. Autoclave for 15 min at 15 psi pressure–121°C. Cool to 25°C. Aseptically add 0.2mL of sterile carbohydrate solution to each tube. Mix thoroughly.

**Storage/Shelf Life:** Store dehydrated media in the dark in a sealed container below 30°C. Prepared media should be stored under refrigeration (2-8°C). Media should be used within 60 days of preparation. Media should not be used if there are any signs of deterioration (discoloration) or contamination, or if the expiration date supplied by the manufacturer has passed.

**Use:** For the preparation of carbohydrate media used in fermentation studies for the identification of bacteria, especially members of the Enterobacteriaceae. Bacteria that can ferment the carbohydrate turn the medium yellow.

## Purple Carbohydrate Fermentation Broth Base
## (BAM M130a)

**Composition** per liter:

| | |
|---|---|
| Pancreatic digest of gelatin | 10.0g |
| NaCl | 5.0g |
| Bromcresol Purple | 0.02g |
| Carbohydrate solution | 100.0mL |

pH 6.8 ± 0.2 at 25°C

**Source:** This medium is available as a premixed powder from BD Diagnostics.

**Carbohydrate Solution:**
**Composition** per 100.0mL:

| | |
|---|---|
| Carbohydrate | 5.0g |

**Preparation of Carbohydrate Solution:** Add carbohydrate to distilled/deionized water and bring volume to 100.0mL. Mix thoroughly. Filter sterilize.

**Preparation of Medium:** Add components, except carbohydrate solution, to distilled/deionized water and bring volume to 900.0mL. Mix thoroughly. Adjust pH to 7.4 if necessary. Distribute into tubes containing an inverted Durham tube. Fill each tube with 9.0mL of medium. Autoclave for 15 min at 15 psi pressure–121°C. Aseptically add 1.0mL of sterile carbohydrate solution to each tube.

**Storage/Shelf Life:** Store dehydrated media in the dark in a sealed container below 30°C. Prepared media should be stored under refrigeration (2-8°C). Media should be used within 60 days of preparation. Media should not be used if there are any signs of deterioration (discoloration) or contamination, or if the expiration date supplied by the manufacturer has passed.

**Use:** For the preparation of liquid fermentation media. Bacteria that can ferment the carbohydrate turn the medium yellow.

## Purple Carbohydrate Fermentation Broth Base with Esculin
## (BAM M130a)

**Composition** per liter:

Pancreatic digest of gelatin ............................................................. 10.0g
NaCl ............................................................................................... 5.0g
Esculin ........................................................................................... 5.0g
Bromcresol Purple .........................................................................0.02g
Carbohydrate solution ................................................................ 10.0mL

pH 6.8 ± 0.2 at 25°C

**Source:** This medium, without esculin, is available as a premixed powder from BD Diagnostics.

**Carbohydrate Solution:**
**Composition** per 100.0mL:

Carbohydrate .................................................................................5.0g

**Preparation of Carbohydrate Solution:** Add carbohydrate to distilled/deionized water and bring volume to 100.0mL. Mix thoroughly. Filter sterilize.

**Preparation of Medium:** Add components, except carbohydrate solution and esculin, to distilled/deionized water and bring volume to 900.0mL. Mix thoroughly. Adjust pH to 7.4 if necessary. Distribute into tubes containing an inverted Durham tube. Fill each tube with 9.0mL of medium. Add 0.05g esculin to each tube. Autoclave for 15 min at 10 psi pressure–115°C. Aseptically add 1.0mL of sterile carbohydrate solution to each tube.

**Storage/Shelf Life:** Store dehydrated media in the dark in a sealed container below 30°C. Prepared media should be stored under refrigeration (2-8°C). Media should be used within 60 days of preparation. Media should not be used if there are any signs of deterioration (discoloration) or contamination, or if the expiration date supplied by the manufacturer has passed.

**Use:** For the preparation of liquid fermentation media, e.g., for *Listeria* spp. and *Enterococcus* spp. Bacteria that can ferment the carbohydrate turn the medium yellow.

## Purple HiVeg Agar Base with Carbohydrate

**Composition** per liter:

Agar ............................................................................................. 15.0g
Plant special peptone ................................................................... 10.0g
NaCl ............................................................................................... 5.0g
Plant extract .................................................................................. 1.0g
Bromcresol Purple .........................................................................0.02g
Carbohydrate solution ................................................................ 20.0mL

pH 6.8 ± 0.2 at 25°C

**Source:** This medium, without carbohydrate, is available as a premixed powder from HiMedia.

**Carbohydrate Solution:**
**Composition** per 20.0mL:

Carbohydrate ............................................................................... 10.0g

**Preparation of Carbohydrate Solution:** Add carbohydrate to distilled/deionized water and bring volume to 20.0mL. For expensive carbohydrates, 5.0g may be used instead of 10.0g. Mix thoroughly. Filter sterilize.

**Preparation of Medium:** Add components, except carbohydrate solution, to distilled/deionized water and bring volume to 980.0mL. Mix thoroughly. Gently heat and bring to boiling. Distribute into tubes in 9.8mL volumes. Autoclave for 15 min at 15 psi pressure–121°C. Cool to

45°–50°C. Aseptically add 0.2mL of sterile carbohydrate solution to each tube. Mix thoroughly. Allow tubes to cool in a slanted position.

**Storage/Shelf Life:** Store dehydrated media in the dark in a sealed container below 30°C. Prepared media should be stored under refrigeration (2-8°C). Media should be used within 60 days of preparation. Media should not be used if there are any signs of deterioration (shrinking, cracking, or discoloration) or contamination, or if the expiration date supplied by the manufacturer has passed.

**Use:** For the preparation of carbohydrate media used in fermentation studies for the identification of bacteria, especially members of the Enterobacteriaceae. Bacteria that can ferment the carbohydrate turn the medium yellow.

## Purple HiVeg Broth Base with Carbohydrate

**Composition** per liter:

Plant special peptone ................................................................... 10.0g
NaCl ............................................................................................... 5.0g
Bromcresol Purple .........................................................................0.02g
Carbohydrate solution ................................................................ 20.0mL

pH 6.8 ± 0.2 at 25°C

**Source:** This medium, without carbohydrate, is available as a premixed powder from HiMedia.

**Carbohydrate Solution:**
**Composition** per 20.0mL:

Carbohydrate ............................................................................... 10.0g

**Preparation of Carbohydrate Solution:** Add carbohydrate to distilled/deionized water and bring volume to 20.0mL. For expensive carbohydrates, 5.0g may be used instead of 10.0g. Mix thoroughly. Filter sterilize.

**Preparation of Medium:** Add components, except carbohydrate solution, to distilled/deionized water and bring volume to 980.0mL. Mix thoroughly. Gently heat and bring to boiling. Distribute into tubes in 9.8mL volumes. Autoclave for 15 min at 15 psi pressure–121°C. Cool to 25°C. Aseptically add 0.2mL of sterile carbohydrate solution to each tube. Mix thoroughly.

**Storage/Shelf Life:** Store dehydrated media in the dark in a sealed container below 30°C. Prepared media should be stored under refrigeration (2-8°C). Media should be used within 60 days of preparation. Media should not be used if there are any signs of deterioration (discoloration) or contamination, or if the expiration date supplied by the manufacturer has passed.

**Use:** For the preparation of carbohydrate media used in fermentation studies for the identification of bacteria, especially members of the Enterobacteriaceae. Bacteria that can ferment the carbohydrate turn the medium yellow.

## Purple Lactose Agar

**Composition** per liter:

Agar ............................................................................................. 10.0g
Lactose ......................................................................................... 10.0g
Peptone ......................................................................................... 5.0g
Beef extract ................................................................................... 3.0g
Bromcresol Purple .......................................................................0.025g

pH 6.8 ± 0.1 at 25°C

**Source:** This medium is available as a premixed powder from BD Diagnostic Systems.

**Preparation of Medium:** Add components to distilled/deionized water and bring volume to 1.0L. Mix thoroughly. Gently heat and bring to boiling. Distribute into tubes or flasks. Autoclave for 15 min at 15 psi pressure–121°C. Pour into sterile Petri dishes or leave in tubes. Allow tubes to cool in a slanted position.

**Storage/Shelf Life:** Store dehydrated media in the dark in a sealed container below 30°C. Prepared media should be stored under refrigeration (2-8°C). Media should be used within 60 days of preparation. Media should not be used if there are any signs of deterioration (shrinking, cracking, or discoloration) or contamination, or if the expiration date supplied by the manufacturer has passed.

**Use:** For the detection and differentiation of members of the Enterobacteriaceae. Bacteria that can ferment lactose turn the medium yellow.

## Purple Serum Agar Base

**Composition** per liter:
| | |
|---|---|
| Agar | 20.0g |
| Lactose | 20.0g |
| Peptone | 20.0g |
| NaCl | 5.0g |
| Bromcresol Purple | 0.03g |
| Phenol Red | 0.024g |

pH 7.6 ± 0.2 at 25°C

**Preparation of Medium:** Add components to distilled/deionized water and bring volume to 1.0L. Mix thoroughly. Gently heat and bring to boiling. Distribute into tubes or flasks. Autoclave for 15 min at 15 psi pressure–121°C. Pour into sterile Petri dishes or leave in tubes.

**Storage/Shelf Life:** Store dehydrated media in the dark in a sealed container below 30°C. Prepared media should be stored under refrigeration (2-8°C). Media should be used within 60 days of preparation. Media should not be used if there are any signs of deterioration (shrinking, cracking, or discoloration) or contamination, or if the expiration date supplied by the manufacturer has passed.

**Use:** For the cultivation and differentiation of Gram-negative bacteria isolated from the urinary tract. Bacteria that can ferment lactose turn the medium yellow.

## PV Blood Agar
### (Paromomycin Vancomycin Blood Agar)

**Composition** per liter:
| | |
|---|---|
| Agar | 20.0g |
| Pancreatic digest of casein | 15.0g |
| NaCl | 5.0g |
| Papaic digest of soybean meal | 5.0g |
| Yeast extract | 5.0g |
| L-Cystine | 0.4g |
| Paromomycin | 0.1g |
| Vancomycin | 7.5mg |
| Hemin | 5.0mg |
| Sheep blood, defibrinated | 50.0mL |
| Vitamin K$_1$ solution | 10.0mL |

pH 7.5 ± 0.2 at 25°C

**Vitamin K$_1$ Solution:**
**Composition** per 10.0mL:
| | |
|---|---|
| Vitamin K$_1$ | 0.01g |
| Ethanol | 10.0mL |

**Preparation of Vitamin K$_1$ Solution:** Add vitamin K$_1$ to 10.0mL of absolute ethanol. Mix thoroughly. Filter sterilize.

**Preparation of Medium:** Add components—except vitamin K$_1$ solution, sheep blood, paromomycin, and vancomycin—to distilled/deionized water and bring volume to 940.0mL. Mix thoroughly. Gently heat and bring to boiling for 1 min. Autoclave for 15 min at 15 psi pressure–121°C. Cool to 50°–55°C. Aseptically add the sterile vitamin K$_1$ solution, sheep blood, vancomycin, and paromomycin. Mix thoroughly. Pour into sterile Petri dishes.

**Storage/Shelf Life:** Store dehydrated media in the dark in a sealed container below 30°C. Prepared media should be stored under refrigeration (2-8°C). Media should be used within 60 days of preparation. Media should not be used if there are any signs of deterioration (shrinking, cracking, or discoloration) or contamination, or if the expiration date supplied by the manufacturer has passed.

**Use:** For the selective cultivation of fastidious anaerobic bacteria.

## Pyrazinamidase Agar
### (Pyrazinamide Medium)

**Composition** per liter:
| | |
|---|---|
| Pancreatic digest of casein | 15.0g |
| Agar | 15.0g |
| Papaic digest of soybean meal | 5.0g |
| NaCl | 5.0g |
| Yeast extract | 3.0g |
| Pyrazinecarboxamide | 1.0g |
| Tris(hydroxymethyl)amino- methane maleate buffer (0.2$M$, pH 6.0) | 1.0L |

pH 6.0 ± 0.2 at 25°C

**Preparation of Medium:** Combine components. Mix thoroughly. Gently heat and bring to boiling. Distribute into tubes in 5.0mL volumes. Autoclave for 15 min at 15 psi pressure–121°C. Allow tubes to cool in a slanted position.

**Storage/Shelf Life:** Store dehydrated media in the dark in a sealed container below 30°C. Prepared media should be stored under refrigeration (2-8°C). Media should be used within 60 days of preparation. Media should not be used if there are any signs of deterioration (shrinking, cracking, or discoloration) or contamination, or if the expiration date supplied by the manufacturer has passed.

**Use:** For the cultivation, differentiation, and maintenance of pathogenic *Yersinia* species. Bacteria that produce pyrazinamidase turn the medium pink.

## Pyrazinamidase Agar
### (BAM M131)

**Composition** per liter:
| | |
|---|---|
| Pancreatic digest of casein | 11.25g |
| Agar | 11.25g |
| Papaic digest of soybean meal | 3.75g |
| NaCl | 3.75g |
| Yeast extract | 3.0g |
| Pyrazine-carboxamide | 1.0g |
| Tris maleate, 0.2$M$, pH6.0 | 1.0L |

pH 6.0 ± 0.2 at 25°C

**Preparation of Medium:** Add components to 0.2$M$ tris maleate and bring volume to 1.0L. Mix thoroughly. Gently heat and bring to boiling. Distribute into tubes or flasks. Autoclave for 15 min at 15 psi pressure–121°C. Pour into sterile Petri dishes or leave in tubes. For slants allow tubes to cool in an inclined position.

**Storage/Shelf Life:** Store dehydrated media in the dark in a sealed container below 30°C. Prepared media should be stored under refriger-

ation (2-8°C). Media should be used within 60 days of preparation. Media should not be used if there are any signs of deterioration (shrinking, cracking, or discoloration) or contamination, or if the expiration date supplied by the manufacturer has passed.

**Use:** For the cultivation of *Yersinia* spp.

## Pyruvic Acid Egg Medium

**Composition** per 1640.0mL:

| | |
|---|---|
| $KH_2PO_4$ | 11.4g |
| D-Glucose | 10.0g |
| $Na_2HPO_4$ | 6.0g |
| Pyruvic acid | 3.0g |
| $MgSO_4 \cdot 7H_2O$ | 0.3g |
| Malachite Green | 0.125g |
| Egg, homogenized whole | 1.0L |
| Penicillin solution | 10.0mL |

**Source:** This medium is available as a prepared medium from Oxoid Unipath.

**Penicillin Solution:**

**Composition** per 10.0mL:

| | |
|---|---|
| Penicillin G | 100,000U |

**Preparation of Penicillin Solution:** Add penicillin G to distilled/deionized water and bring volume to 10.0mL. Mix thoroughly. Filter sterilize.

**Homogenized Whole Egg:**

**Composition** per liter:

| | |
|---|---|
| Whole eggs | 18–24 |

**Preparation of Homogenized Whole Egg:** Use fresh eggs, less than 1 week old. Scrub the shells with soap. Let stand in a soap solution for 30 min. Rinse in running water. Soak eggs in 70% ethanol for 15 min. Break the eggs into a sterile container. Homogenize by shaking. Filter through four layers of sterile cheesecloth into a sterile graduated cylinder. Measure out 1.0L.

**Preparation of Medium:** Add components, except homogenized whole egg and penicillin solution, to distilled/deionized water and bring volume to 630.0mL. Mix thoroughly. Autoclave for 15 min at 15 psi pressure–121°C. Cool to 45°–50°C. Aseptically add homogenized whole egg and penicillin solution to cooled sterile basal medium. Mix thoroughly. Aseptically distribute into sterile tubes. Inspissate at 85°–90°C (moist heat) for 45 min.

**Storage/Shelf Life:** Store dehydrated media in the dark in a sealed container below 30°C. Prepared media should be stored under refrigeration (2-8°C). Media should be used within 60 days of preparation. Media should not be used if there are any signs of deterioration (discoloration) or contamination, or if the expiration date supplied by the manufacturer has passed.

**Use:** For the isolation and cultivation of *Mycobacterium* species, especially ones that are drug resistant and difficult to grow.

## PYS Agar
## (Peptone Yeast Extract Salt Agar)

**Composition** per liter:

| | |
|---|---|
| Agar | 15.0g |
| Peptone | 15.0g |
| NaCl | 5.0g |
| Yeast extract | 5.0g |

pH 7.2–7.4 at 25°C

**Preparation of Medium:** Add components to tap water and bring volume to 1.0L. Mix thoroughly. Gently heat and bring to boiling. Distribute into tubes or flasks. Autoclave for 15 min at 15 psi pressure–121°C. Pour into sterile Petri dishes or leave in tubes.

**Use:** For the cultivation and maintenance of *Actinomadura madurae.*

## R2A Agar

**Composition** per liter:

| | |
|---|---|
| Agar | 15.0g |
| Yeast extract | 0.5g |
| Acid hydrolysate of casein | 0.5g |
| Glucose | 0.5g |
| Soluble starch | 0.5g |
| $K_2HPO_4$ | 0.3g |
| Sodium pyruvate | 0.3g |
| Pancreatic digest of casein | 0.25g |
| Peptic digest of animal tissue | 0.25g |
| $MgSO_4$, anhydrous | 0.024g |

pH 7.2 ± 0.2 at 25°C

**Source:** This medium is available as a premixed powder from BD Diagnostic Systems.

**Preparation of Medium:** Add components to distilled/deionized water and bring volume to 1.0L. Mix thoroughly. Gently heat with mixing and bring to boiling. Distribute into tubes or flasks. Autoclave for 15 min at 15 psi pressure–121°C. Do not overheat. Pour into sterile Petri dishes or leave in tubes.

**Storage/Shelf Life:** Store dehydrated media in the dark in a sealed container below 30°C. Prepared media should be stored under refrigeration (2-8°C). Media should be used within 60 days of preparation. Media should not be used if there are any signs of deterioration (shrinking, cracking, or discoloration) or contamination, or if the expiration date supplied by the manufacturer has passed.

**Use:** For use in standard methods for pour plate, spread plate, and membrane filter analysis to enumerate heterotrophic bacteria from waters.

## R2A Agar

**Composition** per liter:

| | |
|---|---|
| Agar | 15.0g |
| Yeast extract | 0.5g |
| Acid hydrolysate of casein | 0.5g |
| Glucose | 0.5g |
| Soluble starch | 0.5g |
| $K_2HPO_4$ | 0.3g |
| Sodium pyruvate | 0.3g |
| Pancreatic digest of casein | 0.25g |
| Peptic digest of animal tissue | 0.25g |
| $MgSO_4$, anhydrous | 0.024g |

pH 7.2 ± 0.2 at 25°C

**Source:** This medium is available as a premixed powder from BD Diagnostic Systems.

**Preparation of Medium:** Add components to distilled/deionized water and bring volume to 1.0L. Mix thoroughly. Gently heat with mixing and bring to boiling. Distribute into tubes or flasks. Autoclave for 15 min at 15 psi pressure–121°C. Do not overheat. Pour into sterile Petri dishes or leave in tubes.

**Storage/Shelf Life:** Store dehydrated media in the dark in a sealed container below 30°C. Prepared media should be stored under refriger-

ation (2-8°C). Media should be used within 60 days of preparation. Media should not be used if there are any signs of deterioration (shrinking, cracking, or discoloration) or contamination, or if the expiration date supplied by the manufacturer has passed.

**Use:** For use in standard methods for pour plate, spread plate, and membrane filter analysis to enumerate heterotrophic bacteria from potable waters.

## Rainbow Agar O157

**Composition** per liter:
Proprietary.

**Source:** This medium is available as a premixed powder from Biolog Inc.

**Preparation:** Suspend 60.0g of the proprietary mixture in distilled/deionized water and bring volume to 1.0L. Mix thoroughly. Gently heat and bring to boiling. Autoclave for 15 min at 15 psi pressure–121°C. Cool to 45°C–50°C. Mix thoroughly. Pour into sterile Petri dishes or distribute into sterile tubes. The final medium should be clear and virtually colorless. No pH adjustment is needed. The final pH should be pH 7.9–8.3. To increase the selectivity of the medium, a sterile solution containing 0.8mg potassium tellurite and 10mg novobiocin can be added. Caution must be used because tellurite is toxic.

**Storage/Shelf Life:** Store dehydrated media in the dark in a sealed container below 30°C. Prepared media should be stored in the dark under refrigeration (2-8°C). Chromogenic media are especially light and temperature sensitive; protect from light, excessive heat, moisture, and freezing. Media should not be used if there are any signs of deterioration (shrinking, cracking, or discoloration) or contamination, or if the expiration date supplied by the manufacturer has passed.

**Use**: For the detection, isolation, and presumptive identification of verotoxin-producing strains of *Escherichia coli*, particularly serotype O157:H7. The medium contains chromogenic substrates that are specific for two *E. coli*-associated enzymes: β-galactosidase (a blue-black chromogenic substrate) and β-glucuronidase (a red chromogenic substrate). The distinctive black or gray coloration of *E. coli* O157:H7 colonies is easily viewed by laying the Petri plate against a white background. When O157 is surrounded by pink or magenta non-toxigenic colonies, it may have a bluish hue. The addition of selective agents improves performance. *E. coli* O157:H7 colony coloration will be slightly bluer with these selective agents added. Tellurite is highly selective for *E. coli* O157:H7 and can reduce background flora considerably. Novobiocin inhibits *Proteus* swarming and the growth of tellurite-reducing bacteria. Rare strains of O157:H7 are tellurite sensitive.

## Rainbow Agar *Salmonella*

**Composition** per liter:
Proprietary

pH 7.2–7.6 at 25°C

**Source:** This medium is available from Biolog.

**Preparation of Medium:** Add components to distilled/deionized water and bring volume to 990.0mL. Mix thoroughly. Add 10.0mL 35% glycerol. Stir until components are evenly dispersed. Gently heat and boil. Autoclave for 10 min at 15 psi pressure–121°C. Do not exceed 10 min. Cool agar to 45°C–50°C before pouring plates.

**Storage/Shelf Life:** Store dehydrated media in the dark in a sealed container below 30°C. Prepared media should be stored in the dark under refrigeration (2-8°C). Chromogenic media are especially light and temperature sensitive; protect from light, excessive heat, moisture, and

freezing. Media should not be used if there are any signs of deterioration (shrinking, cracking, or discoloration) or contamination, or if the expiration date supplied by the manufacturer has passed.

**Use:** As a selective, chromogenic medium to aid in the detection and isolation of H$_2$S-producing *Salmonella* species. Black colonies are formed by even weak H$_2$S-producing strains.

## RajHans Medium
## (HiCrome™ *Salmonella* Medium, Modified)

**Composition** per liter:
| | |
|---|---|
| Agar | 12.0 g |
| Casein enzymic hydrolysate | 8.0 g |
| Yeast extract | 5.0 g |
| NaCl | 5.0 g |
| Chromogenic mixture | 4.32 g |
| Peptic digest of animal tissue | 4.0 g |
| Lactose | 3.0 g |
| Sodium deoxycholate | 1.0 g |
| Neutral Red | 0.02 g |

pH 7.3 ± 0.2 at 25°C

**Source:** This medium is available as a premixed powder from Hi-Media.

**Preparation of Medium:** Add components to distilled/deionized water and bring volume to 1.0L. Mix thoroughly. Gently heat while stirring and bring to boiling. Mix to completely dissolve components. Do not autoclave. Cool to 50°C. Mix thoroughly. Pour into sterile Petri dishes.

**Storage/Shelf Life:** Store dehydrated media in the dark in a sealed container below 30°C. Prepared media should be stored in the dark under refrigeration (2-8°C). Chromogenic media are especially light and temperature sensitive; protect from light, excessive heat, moisture, and freezing. Media should not be used if there are any signs of deterioration (shrinking, cracking, or discoloration) or contamination, or if the expiration date supplied by the manufacturer has passed.

**Use:** A selective chromgenic medium used for the isolation and differentiation of *Salmonella* species from the members of Enterobacteriaceae, especially *Proteus* species.

## RajHans Medium, HiVeg
## (HiCrome™ *Salmonella* Medium, Modified)

**Composition** per liter:
| | |
|---|---|
| Agar | 12.0g |
| Propylene glycol | 10.0g |
| Plant special peptone | 8.0g |
| Yeast extract | 2.0g |
| B.C. indicator | 2.0g |
| Sodium deoxycholate | 1.0g |

pH 7.3 ± 0.2 at 25°C

**Source:** This medium is available as a premixed powder from Hi-Media.

**Preparation of Medium:** Add components to distilled/deionized water and bring volume to 1.0L. Mix thoroughly. Gently heat while stirring and bring to boiling. Mix to completely dissolve components. Do not autoclave. Cool to 50°C. Mix thoroughly. Pour into sterile Petri dishes.

**Storage/Shelf Life:** Store dehydrated media in the dark in a sealed container below 30°C. Prepared media should be stored in the dark under refrigeration (2-8°C). Chromogenic media are especially light and temperature sensitive; protect from light, excessive heat, moisture, and freezing. Media should not be used if there are any signs of deterioration (shrinking, cracking, or discoloration) or contamination, or if the expiration date supplied by the manufacturer has passed.

**Use:** A selective chromgenic medium used for the isolation and differentiation of *Salmonella* species from the members of Enterobacteriaceae, especially *Proteus* species.

## Rambach® Agar

**Composition** per liter:

| | |
|---|---|
| Agar | 15.0g |
| Polypropylene glycol | 10.5g |
| Peptone | 8.0g |
| NaCl | 5.0g |
| Chromogenic mix | 1.5g |
| Na-desoxycholate | 1.0g |

pH 7.4 ± 0.2 at 25°C

**Source:** Rambach Agar is available from CHROMagar Microbiology.

**Preparation of Medium:** Add components to distilled/deionized water and bring volume to 1.0L. Mix thoroughly. Heat in a boiling water bath or in a current of steam, while shaking from time to time. The medium is totally suspended, if no visual particles stick to the glass wall. The medium should not be heat treated further. Complete dissolution with shaking in 5-min sequences is approximately 35–40 minutes. Do not autoclave. Do not overheat. Cool as fast as possible to 45°–50°C while gently shaking from time to time. Pour into sterile Petri dishes. To prevent any precipitate or clotting of the chromogenic mix in the plates, place Petri dishes during pouring procedure on a cool (max. 25°C) surface. The plates are opaque and pink.

**Storage/Shelf Life:** Store dehydrated media in the dark in a sealed container below 30°C. Prepared media should be stored in the dark under refrigeration (2-8°C). Chromogenic media are especially light and temperature sensitive; protect from light, excessive heat, moisture, and freezing. Media should not be used if there are any signs of deterioration (shrinking, cracking, or discoloration) or contamination, or if the expiration date supplied by the manufacturer has passed.

**Use:** For the detection of enteric bacteria, including coliforms and *Salmonella* spp. Sodium desoxycholate inhibits the accompanying Gram-positive flora. This medium enables *Salmonella* spp. to be differentiated unambiguously from other bacteria. *Salmonella* spp. form a characteristic red color. In order to differentiate coliforms from Salmonellae, the medium contains a chromogene indicating the presence of β-galactosidase splitting, a characteristic of coliforms. Coliform microorganisms grow as blue-green or blue-violet colonies. Other Enterobacteriaceae and Gram-negative bacteria, such as *Proteus, Pseudomonas, Shigella, S. typhi, and S. parathyphi* A, grow as colorless-yellow colonies.

## RambaCHROM™ CTX

**Composition** per liter:

| | |
|---|---|
| Agar | 15.0g |
| Peptone and yeast extract | 8.0g |
| NaCl | 5.0g |
| Chromogenic mix | 4.8g |
| CTX supplement solution (freshly prepared) | 2.0mL |

pH 7.2± 0.2 at 25°C

**Source:** Available as prepared plates from Gibson Bioscience, Lexington Kentucky.

**CTX Supplement Solution:**
**Composition** per 10.0mL:

| | |
|---|---|
| Ethanol | 5.0mL |
| CHROMagar™ CTX supplement | 1.25g |

**Storage/Shelf Life:** Store in the dark. Chromogenic agars are especially light and temperature sensitive; protect from light, excessive heat, moisture, and freezing. Prepared media plates can be kept for one day at ambient temperature. Plates can be stored at least one week under refrigeration (2-8°C) if properly prepared and protected from light and dehydration. Do not use after the expiration date supplied by the manufacturer.

**Use:** For the detection of bacteria with ESBL CTX-M-type resistance. *E.coli* that produce CTX-M–type â-lactamases (CTX-Ms) form blue colonies. Other enterobacteriaceae CTX-M form mauve colonies.

## RambaCHROM™ *E. coli*

**Composition** per liter:

| | |
|---|---|
| Agar | 15.0g |
| Chromogenic mix | 9.0g |
| Peptone and yeast extract | 8.3g |
| Sodium chloride | 5.0g |

pH 6.0± 0.2 at 25°C

**Source:** Available as prepared plates from Gibson Bioscience, Lexington Kentucky.

**Storage/Shelf Life:** Store in the dark. Chromogenic agars are especially light and temperature sensitive; protect from light, excessive heat, moisture, and freezing. Prepared media plates can be kept for one day at ambient temperature. Plates can be stored at least one week under refrigeration (2-8°C) if properly prepared and protected from light and dehydration. Do not use after the expiration date supplied by the manufacturer.

**Use:** For the detection and enumeration of *E. coli*.

## RambaCHROM™ ECC

**Composition** per liter:

| | |
|---|---|
| Agar | 15.0g |
| Peptone and yeast extract | 8.0g |
| NaCl | 5.0g |
| Chromogenic mix | 4.8g |

pH 7.2± 0.2 at 25°C

**Source:** Available as prepared plates from Gibson Bioscience, Lexington Kentucky.

**Storage/Shelf Life:** Store in the dark. Chromogenic agars are especially light and temperature sensitive; protect from light, excessive heat, moisture, and freezing. Prepared media plates can be kept for one day at ambient temperature. Plates can be stored at least one week under refrigeration (2-8°C) if properly prepared and protected from light and dehydration. Do not use after the expiration date supplied by the manufacturer.

**Use:** For the detection and enumeration of *Escherichia coli* and coliforms. *E. coli* form blue colonies. Other fecal coliforms form mauve colonies.

## RambaCHROM™ ESBL

**Composition** per liter:

| | |
|---|---|
| Peptone and yeast extract | 17.0g |
| Agar | 15.0g |

Chromogenic mix ......................................................1.0g
EBSL selective supplement ......................................0.57g

pH 7.0± 0.2 at 25°C

**Source:** Available as prepared plates from Gibson Bioscience, Lexington Kentucky.

**Storage/Shelf Life:** Store in the dark. Chromogenic agars are especially light and temperature sensitive; protect from light, excessive heat, moisture, and freezing. Prepared media plates can be kept for one day at ambient temperature. Plates can be stored at least one week under refrigeration (2-8°C) if properly prepared and protected from light and dehydration. Do not use after the expiration date supplied by the manufacturer.

**Use:** For the detection of Gram-negative bacteria producing extended spectrum beta-lactamase. Extended spectrum β-lactam resistant strains of *E. coli* form dark pink to reddish colonies. Extended spectrum β-lactam resistant strains of *Klebsiella, Enterobacter*, and *Citrobacter* form metallic blue colonies. Extended spectrum β-lactam resistant strains of *Proteus* colonies with a brown halo.

## RambaCHROM™ KPC

**Composition** per liter:
Peptone and yeast extract ...............................................17.0g
Agar ...............................................................................15.0g
Chromogenic mix ...........................................................1.0g
KPC selective supplement ..............................................0.4g

pH 7.0± 0.2 at 25°C

**Source:** Available as prepared plates from Gibson Bioscience, Lexington Kentucky.

**Storage/Shelf Life:** Store in the dark. Chromogenic agars are especially light and temperature sensitive; protect from light, excessive heat, moisture, and freezing. Prepared media plates can be kept for one day at ambient temperature. Plates can be stored at least one week under refrigeration (2-8°C) if properly prepared and protected from light and dehydration. Do not use after the expiration date supplied by the manufacturer.

**Use:** For the detection of Gram-negative bacteria with a reduced susceptibility to most carbapenem antimicrobic agents. Carbapenem resistant strains of *E. coli* form dark pink to reddish colonies. Carbapenem resistant strains of *Klebsiella, Enterobacter*, and *Citrobacter* form blue colonies. Carbapenem resistant strains of *Pseudomonas* form cream translucent colonies.

## RambaCHROM™ *Listeria*

**Composition** per liter:
Proprietary

**Source:** Available as prepared plates from Gibson Bioscience, Lexington Kentucky.

**Storage/Shelf Life:** Store in the dark. Chromogenic agars are especially light and temperature sensitive; protect from light, excessive heat, moisture, and freezing. Prepared media plates can be kept for one day at ambient temperature. Plates can be stored at least one week under refrigeration (2-8°C) if properly prepared and protected from light and dehydration. Do not use after the expiration date supplied by the manufacturer.

**Use:** For the differentiation and presumptive identification of *Listeria monocytogenes,* which form blue colonies surrounded by white halos.

## RambaCHROM™ MRSA

**Composition** per liter:
Chrompeptone..................................................................40.0g
NaCl................................................................................25.0g
Agar ...............................................................................14.0g
Chromogenic mix ...........................................................0.5g
Inhibitory agents ...........................................................0.07g
Cefoxitin ......................................................................6.0mg

**Source:** Available as prepared plates from Gibson Bioscience, Lexington Kentucky.

**Storage/Shelf Life:** Store in the dark. Chromogenic agars are especially light and temperature sensitive; protect from light, excessive heat, moisture, and freezing. Prepared media plates can be kept for one day at ambient temperature. Plates can be stored at least one week under refrigeration (2-8°C) if properly prepared and protected from light and dehydration. Do not use after the expiration date supplied by the manufacturer.

**Use:** For the qualitative direct detection of nasal colonization by methicillin resistant *Staphylococcus aureus* (MRSA) to aid in the prevention and control of MRSA infections in healthcare settings.

## RambaCHROM™ *Salmonella*

**Composition** per liter:
Agar ...............................................................................15.0g
Chromogenic and selective mix....................................12.9g
Peptone and yeast extract................................................7.0g

pH 7.6± 0.2 at 25°C

**Source:** Available as prepared plates from Gibson Bioscience, Lexington Kentucky.

**Storage/Shelf Life:** Store in the dark. Chromogenic agars are especially light and temperature sensitive; protect from light, excessive heat, moisture, and freezing. Prepared media plates can be kept for one day at ambient temperature. Plates can be stored at least one week under refrigeration (2-8°C) if properly prepared and protected from light and dehydration. Do not use after the expiration date supplied by the manufacturer.

**Use:** For the detection and isolation of *Salmonella* spp., including *S. typhi* and *S. paratyphi* in clinical specimens. *Salmonella* including *S. typhi* form mauve colonies.

## RambaCHROM™ *Staph aureus*

**Composition** per liter:
Peptone and yeast extract................................................40.0g
Salts................................................................................25.0g
Agar ...............................................................................15.0g
Chromogenic mix ...........................................................2.5g

pH 6.9± 0.2 at 25°C

**Source:** Available as prepared plates from Gibson Bioscience, Lexington Kentucky.

**Storage/Shelf Life:** Store in the dark. Chromogenic agars are especially light and temperature sensitive; protect from light, excessive heat, moisture, and freezing. Prepared media plates can be kept for one day at ambient temperature. Plates can be stored at least one week under refrigeration (2-8°C) if properly prepared and protected from light and dehydration. Do not use after the expiration date supplied by the manufacturer.

**Use:** For the isolation and direct differentiation of *Staphylococcus aureus* in clinical specimens. S. aureua forms pink to mauve colonies.

## RambaCHROM™ *Vibrio*

**Composition** per liter:

Salts..................................................................51.4g

Agar ..................................................................15.0g

Peptone and yeast extract..............................8.0g

Chromogenic mix ...........................................0.3g

pH 9.0± 0.2 at 25°C

**Source:** Available as prepared plates from Gibson Bioscience, Lexington Kentucky.

**Storage/Shelf Life:** Store in the dark. Chromogenic agars are especially light and temperature sensitive; protect from light, excessive heat, moisture, and freezing. Prepared media plates can be kept for one day at ambient temperature. Plates can be stored at least one week under refrigeration (2-8°C) if properly prepared and protected from light and dehydration. Do not use after the expiration date supplied by the manufacturer.

**Use:** For the isolation and detection of *V. parahaemolyticus, V. vulnificus* and *V. cholerae. V. parahaemolyticus* form mauve colonies. *V. vulnificus and V. cholera* form green blue to turquoise blue colonies. *V. alginolyticus* form colorless colonies.

## RambaCHROM™ VRE

**Composition** per liter:

Proprietary

**Source:** Available as prepared plates from Gibson Bioscience, Lexington Kentucky.

**Storage/Shelf Life:** Store in the dark. Chromogenic agars are especially light and temperature sensitive; protect from light, excessive heat, moisture, and freezing. Prepared media plates can be kept for one day at ambient temperature. Plates can be stored at least one week under refrigeration (2-8°C) if properly prepared and protected from light and dehydration. Do not use after the expiration date supplied by the manufacturer.

**Use:** For the differentiation and presumptive identification of vancomycin resistant *Enterococcus* (*Enterococcus faecalis/E. facecium*). Vancomycin resistant *Enterococcus* strains form rose to mauve colonies.

## RAPID´*E. coli* 2 Agar

**Composition** per liter:

Proprietary

**Source:** This medium is available from Biorad.

**Storage/Shelf Life:** Prepared plates should be stored in the dark under refrigeration (2-8°C). Chromogenic agars are especially light and temperature sensitive; protect from light, excessive heat, moisture, and freezing. Media should not be used if there are any signs of deterioration (shrinking, cracking, or discoloration) or contamination, or if the expiration date supplied by the manufacturer has passed.

**Use:** For the direct enumeration of *E. coli* and coliforms in foods. Selectivity and electivity are based on the detection of glucuronidase and galactosidase activities. Hydrolysis of chromogenic substrate results in purple to pink *E. coli* colonies (gluc+/gal+) and blue-green coliform colonies (gluc-/gal+). RAPID´*E. coli* 2 agar is AFNOR validated according to ISO 16140 protocol to enumerate *E. coli* and coliforms on the same plate at 37°C, without any further confirmation of characteristic colonies.

## RAPID´*Enterococcus* Agar

**Composition** per liter:

Proprietary

**Source:** This medium is available from Biorad.

**Storage/Shelf Life:** Prepared plates should be stored in the dark under refrigeration (2-8°C). Chromogenic agars are especially light and temperature sensitive; protect from light, excessive heat, moisture, and freezing. Media should not be used if there are any signs of deterioration (shrinking, cracking, or discoloration) or contamination, or if the expiration date supplied by the manufacturer has passed.

**Use:** A selective chromogenic culture medium for the direct enumeration, without confirmation, of enterococci in water and in food products. The cleavage of the chromogenic substrate by glucosidase activity of Enterococci leads to specific blue colonies. RAPID´*Enterococcus* totally inhibits growth of Gram-negative flora and that of practically all Gram-positive bacteria other than Enterococci, due to the combined action of temperature and selective media.

## Rapid Fermentation Medium

**Composition** per liter:

Pancreatic digest of casein.........................20.0g

NaCl................................................................5.0g

Agar................................................................3.5g

L-Cystine........................................................0.5g

Na$_2$SO$_3$.........................................................0.5g

Phenol Red..................................................0.017g

pH 7.3 ± 0.2 at 25°C

**Source:** This medium is available as a premixed powder from BD Diagnostic Systems.

**Preparation of Medium:** Add components to distilled/deionized water and bring volume to 1.0L. Mix thoroughly. Distribute into tubes or flasks. Autoclave for 15 min at 15 psi pressure–121°C.

**Storage/Shelf Life:** Store dehydrated media in the dark in a sealed container below 30°C. Prepared media should be stored under refrigeration (2-8°C). Media should be used within 60 days of preparation. Media should not be used if there are any signs of deterioration (discoloration) or contamination, or if the expiration date supplied by the manufacturer has passed.

**Use:** For the differentiation of *Neisseria* species isolated from clinical specimens.

## Rapid HiColiform Agar

**Composition** per liter:

Agar................................................................15.0g

Peptone, special.............................................5.0g

NaCl................................................................5.0g

K$_2$HPO$_4$.........................................................2.7g

KH$_2$PO$_4$.........................................................2.0g

Sorbitol...........................................................1.0g

Sodium lauryl sulfate.....................................0.1g

1-Isopropyl- ß-D-1-thiogalactopyranoside.....0.1g

Chromogenic mixture...................................0.08g

Fluorogenic mixture......................................0.05g

pH 7.2 ± 0.2 at 25°C

**Source:** This medium is available as a premixed powder from Hi-Media.

**Preparation of Medium:** Add components to distilled/deionized water and bring volume to 1.0L. Mix thoroughly. Gently heat and bring to boiling. Distribute into tubes or flasks. Autoclave for 15 min at 15 psi pressure–121°C. Pour into sterile Petri dishes or leave in tubes.

**Storage/Shelf Life:** Store dehydrated media in the dark in a sealed container below 30°C. Prepared media should be stored in the dark under refrigeration (2-8°C). Chromogenic media are especially light and temperature sensitive; protect from light, excessive heat, moisture, and freezing. Media should not be used if there are any signs of deterioration (shrinking, cracking, or discoloration) or contamination, or if the expiration date supplied by the manufacturer has passed.

**Use:** For the detection and confirmation of *Escherichia coli* and total coliforms on the basis of enzyme substrate reaction from water samples, using a combination of chromogenic and fluorogenic substrates.

### Rapid HiColiform Broth
**Composition** per liter:

| | |
|---|---|
| Peptone, special | 5.0g |
| NaCl | 5.0g |
| K$_2$HPO$_4$ | 2.7g |
| KH$_2$PO$_4$ | 2.0g |
| Sorbitol | 1.0g |
| Sodium lauryl sulfate | 0.1g |
| IPTG | 0.1g |
| Chromogenic substrate | 0.08g |
| Fluorogenic substrate | 0.05g |

pH 6.8 ± 0.2 at 25°C

**Source:** This medium is available from HiMedia.

**Preparation of Medium:** Add components to distilled/deionized water and bring volume to 1.0L. Mix thoroughly. Gently heat and bring to boiling. Distribute into tubes or flasks. Autoclave for 15 min at 15 psi pressure–121°C.

**Storage/Shelf Life:** Store dehydrated media in the dark in a sealed container below 30°C. Prepared media should be stored in the dark under refrigeration (2-8°C). Chromogenic media are especially light and temperature sensitive; protect from light, excessive heat, moisture, and freezing. Media should not be used if there are any signs of deterioration (shrinking, cracking, or discoloration) or contamination, or if the expiration date supplied by the manufacturer has passed.

**Use:** For the detection and confirmation of *Escherichia coli* and total coliforms from water samples, using a combination of chromogenic and fluorogenic substrates.

### Rapid HiColiform HiVeg Agar
**Composition** per liter:

| | |
|---|---|
| Agar | 15.0g |
| Plant special peptone | 5.0g |
| NaCl | 5.0g |
| K$_2$HPO$_4$ | 2.7g |
| KH$_2$PO$_4$ | 2.0g |
| Sorbitol | 1.0g |
| 1-Isopropyl-ß-D-1-thiogalactopyranoside | 0.1g |
| Sodium lauryl sulfate | 0.1g |
| Chromogenic mixture | 0.08g |
| Fluorogenic mixture | 0.05g |

pH 7.2 ± 0.2 at 25°C

**Source:** This medium is available as a premixed powder from HiMedia.

**Preparation of Medium:** Add components to distilled/deionized water and bring volume to 1.0L. Mix thoroughly. Gently heat and bring to boiling. Distribute into tubes or flasks. Autoclave for 15 min at 15 psi pressure–121°C. Pour into sterile Petri dishes or leave in tubes.

**Storage/Shelf Life:** Store dehydrated media in the dark in a sealed container below 30°C. Prepared media should be stored in the dark under refrigeration (2-8°C). Chromogenic media are especially light and temperature sensitive; protect from light, excessive heat, moisture, and freezing. Media should not be used if there are any signs of deterioration (shrinking, cracking, or discoloration) or contamination, or if the expiration date supplied by the manufacturer has passed.

**Use:** For the detection and confirmation of *Escherichia coli* and total coliforms on the basis of enzyme substrate reaction from water samples, using a combination of chromogenic and fluorogenic substrates.

### Rapid HiColiform HiVeg Agar
**Composition** per liter:

| | |
|---|---|
| Agar | 15.0g |
| Plant special peptone | 5.0g |
| NaCl | 5.0g |
| K$_2$HPO$_4$ | 2.7g |
| KH$_2$PO$_4$ | 2.0g |
| Sorbitol | 1.0g |
| IPTG (Isopropyl-β-D-thiogalactopyranoside) | 0.1g |
| Sodium lauryl sulfate | 0.1g |
| Chromogenic substrate | 0.08g |
| Fluorogenic substrate | 0.05g |

pH 7.2 ± 0.2 at 25°C

**Source:** This medium is available as a premixed powder from HiMedia.

**Preparation of Medium:** Add components to distilled/deionized water and bring volume to 1.0L. Mix thoroughly. Gently heat and bring to boiling. Distribute into tubes or flasks. Autoclave for 15 min at 15 psi pressure–121°C. Pour into sterile Petri dishes or leave in tubes.

**Storage/Shelf Life:** Store dehydrated media in the dark in a sealed container below 30°C. Prepared media should be stored in the dark under refrigeration (2-8°C). Chromogenic media are especially light and temperature sensitive; protect from light, excessive heat, moisture, and freezing. Media should not be used if there are any signs of deterioration (shrinking, cracking, or discoloration) or contamination, or if the expiration date supplied by the manufacturer has passed.

**Use:** For the detection and confirmation of *Escherichia coli* and total coliforms on the basis of enzyme substrate reaction from water samples, using a combination of chromogenic and fluorogenic substrates.

### Rapid HiColiform HiVeg Broth
**Composition** per liter:

| | |
|---|---|
| Plant special peptone | 5.0g |
| NaCl | 5.0g |
| K$_2$HPO$_4$ | 2.7g |
| KH$_2$PO$_4$ | 2.0g |
| Sorbitol | 1.0g |
| IPTG (Isopropyl-β-D-thiogalactopyranoside) | 0.1g |
| Sodium lauryl sulfate | 0.1g |
| Chromogenic substrate | 0.08g |
| Fluorogenic substrate | 0.05g |

pH 7.2 ± 0.2 at 25°C

**Source:** This medium is available as a premixed powder from Hi-Media.

**Preparation of Medium:** Add components to distilled/deionized water and bring volume to 1.0L. Mix thoroughly. Gently heat and bring to boiling. Distribute into tubes or flasks. Autoclave for 15 min at 15 psi pressure–121°C.

**Storage/Shelf Life:** Store dehydrated media in the dark in a sealed container below 30°C. Prepared media should be stored in the dark under refrigeration (2-8°C). Chromogenic media are especially light and temperature sensitive; protect from light, excessive heat, moisture, and freezing. Media should not be used if there are any signs of deterioration (shrinking, cracking, or discoloration) or contamination, or if the expiration date supplied by the manufacturer has passed.

**Use:** For the detection and confirmation of *Escherichia coli* and total coliforms on the basis of enzyme substrate reaction from water samples, using a combination of chromogenic and fluorogenic substrates.

## Rapid HiEnterococci Agar

**Composition** per liter:

| | |
|---|---|
| Agar | 15.0g |
| Peptone, special | 10.0g |
| NaCl | 5.0g |
| Polysorbate 80 | 2.0g |
| $NaN_3$ | 0.3g |
| $Na_2HPO_4$ | 1.25g |
| Chromogenic mixture | 0.06g |

pH 7.5 ± 0.2 at 25°C

**Source:** This medium is available from HiMedia.

**Caution:** Sodium azide has a tendency to form explosive metal azides with plumbing materials. It is advisable to use enough water to flush off the disposables.

**Storage/Shelf Life:** Store dehydrated media in the dark in a sealed container below 30°C. Prepared media should be stored in the dark under refrigeration (2-8°C). Chromogenic media are especially light and temperature sensitive; protect from light, excessive heat, moisture, and freezing. Media should not be used if there are any signs of deterioration (shrinking, cracking, or discoloration) or contamination, or if the expiration date supplied by the manufacturer has passed.

**Preparation of Medium:** Add components to distilled/deionized water and bring volume to 1.0L. Mix thoroughly. Gently heat and bring to boiling. Distribute into tubes or flasks. Autoclave for 15 min at 15 psi pressure–121°C. Pour into sterile Petri dishes or leave in tubes.

**Use:** For the rapid and easy identification and differentiation of enterococci from water samples.

## RAPID′L. mono Medium
## (BAM M131a)

**Composition** per liter:

| | |
|---|---|
| Peptones | 30.0g |
| Agar B, proprietary | 13.0g |
| D-Xylose | 10.0g |
| LiCl | 9.0g |
| Meat extract | 5.0g |
| Yeast extract | 1.0g |
| Phenol Red | 0.12g |
| Selective supplement, proprietary | 20.0g |
| Chromogenic substrate, proprietary | 1.0mL |

pH 7.3 ± 0.1 at 25°C

**Source:** This medium is available from Biorad.

**Storage/Shelf Life:** Prepared plates should be stored in the dark under refrigeration (2-8°C). Chromogenic media are especially light and temperature sensitive; protect from light, excessive heat, moisture, and freezing. Media should not be used if there are any signs of deterioration (discoloration) or contamination, or if the expiration date supplied by the manufacturer has passed.

**Use:** A selective chromogenic culture medium for the detection and differentiation of *Listeria* spp., including *L. ivanovii* and *L. monocytogenes*.

## Rappaport Broth, Modified
## (Rap Broth, Modified)

**Composition** per 250.2mL:

| | |
|---|---|
| Solution A | 155.0mL |
| Solution C | 53.0mL |
| Solution B | 40.0mL |
| Solution D | 1.6mL |
| Solution E | 0.6mL |

**Solution A:**
**Composition** per liter:

| | |
|---|---|
| Pancreatic digest of casein | 10.0g |

**Preparation of Solution A:** Add pancreatic digest of casein to distilled/deionized water and bring volume to 1.0L. Mix thoroughly.

**Solution B:**
**Composition** per liter:

| | |
|---|---|
| $Na_2HPO_4$ | 9.5g |

**Preparation of Solution B:** Add $Na_2HPO_4$ to distilled/deionized water and bring volume to 1.0L. Mix thoroughly.

**Solution C:**
**Composition** per 100.0mL:

| | |
|---|---|
| $MgCl_2 \cdot 6H_2O$ | 40.0g |

**Preparation of Solution C:** Add $MgCl_2 \cdot 6H_2O$ to distilled/deionized water and bring volume to 100.0mL. Mix thoroughly. Autoclave for 15 min at 15 psi pressure–121°C. Cool to 25°C.

**Solution D:**
**Composition** per 100.0mL:

| | |
|---|---|
| Malachite Green | 0.2g |

**Preparation of Solution D:** Add Malachite Green to sterile distilled/deionized water and bring volume to 100.0mL. Mix thoroughly. Do not sterilize.

**Solution E:**
**Composition** per 10.0mL:

| | |
|---|---|
| Carbenicillin | 0.01g |

**Preparation of Solution E:** Add carbenicillin to distilled/deionized water and bring volume to 10.0mL. Mix thoroughly. Filter sterilize.

**Preparation of Medium:** Combine 155.0mL of solution A and 40.0mL of solution B. Mix thoroughly. Autoclave for 15 min at 15 psi pressure–121°C. Cool to 45°–50°C. Aseptically add 53.0mL of sterile solution C, 1.6mL of solution D, and 0.6mL of sterile solution E. Mix thoroughly. Aseptically distribute into sterile tubes or flasks.

**Storage/Shelf Life:** Store dehydrated media in the dark in a sealed container below 30°C. Prepared media should be stored under refrigeration (2-8°C). Media should be used within 60 days of preparation. Media should not be used if there are any signs of deterioration (discol-

oration) or contamination, or if the expiration date supplied by the manufacturer has passed.

**Use:** For the isolation and cultivation of *Yersinia enterocolitica* from foods.

## Rappaport-Vassiliadis Enrichment Broth (RV Enrichment Broth)

**Composition** per liter:

| | |
|---|---|
| NaCl | 8.0g |
| Papaic digest of soybean meal | 5.0g |
| $KH_2PO_4$ | 1.6g |
| Magnesium chloride solution | 100.0mL |
| Malachite Green solution | 10.0mL |

pH 5.2 ± 0.2 at 25°C

**Source:** This medium is available as a premixed powder from Oxoid Unipath.

### Magnesium Chloride Solution:

**Composition** per 100.0mL:

| | |
|---|---|
| $MgCl_2 \cdot 6H_2O$ | 40.0g |

**Preparation of Magnesium Chloride Solution:** Add 40.0g of $MgCl_2 \cdot 6H_2O$ to distilled/deionized water and bring volume to 100.0mL. Mix thoroughly. Autoclave for 15 min at 15 psi pressure–121°C. Cool to 45°–50°C.

### Malachite Green Solution:

**Composition** per 10.0mL:

| | |
|---|---|
| Malachite Green oxalate | 0.04g |

**Preparation of Malachite Green Solution:** Add Malachite Green to distilled/deionized water and bring volume to 10.0mL. Mix thoroughly. Autoclave for 15 min at 15 psi pressure–121°C. Cool to 45°–50°C.

**Preparation of Medium:** Add components to distilled/deionized water and bring volume to 1.0L. Mix thoroughly. Distribute into tubes in 10.0mL volumes. Autoclave for 15 min at 10 psi pressure–115°C.

**Storage/Shelf Life:** Store dehydrated media in the dark in a sealed container below 30°C. Prepared media should be stored under refrigeration (2-8°C). Media should be used within 60 days of preparation. Media should not be used if there are any signs of deterioration (discoloration) or contamination, or if the expiration date supplied by the manufacturer has passed.

**Use:** For the isolation and cultivation of *Salmonella* species from food and environmental specimens.

## Rappaport-Vassiliadis Medium Semisolid, Modified with Novobiocin (MSRV)

**Composition** per liter:

| | |
|---|---|
| $MgCl_2$ | 10.93g |
| NaCl | 7.34g |
| Enzymatic digest of casein | 4.59g |
| Casein acid hydrolysate | 4.59g |
| Agar | 2.7g |
| $KH_2PO_4$ | 1.4 g |
| Malachite Green oxalate | 0.037g |
| Novobiocin solution | 1.0mL |

pH 5.6 ± 0.2 at 25°C

### Novobiocin Solution:

**Composition** per 10.0mL:

| | |
|---|---|
| Novobiocin | 0.2g |

**Preparation of Novobiocin Solution:** Add novobiocin to distilled/deionized water and bring volume to 10.0mL. Mix thoroughly. Filter sterilize.

**Preparation of Medium:** Add components, except novobiocin solutiont, to distilled/deionized water and bring volume to 990.0mL. Mix thoroughly. Gently heat and bring to boiling. Distribute into tubes or flasks. Do not autoclave. Do not overheat. Cool to 50°C. Aseptically add 1.0mL novobiocin solution. Pour into sterile Petri dishes or leave in tubes.

**Storage/Shelf Life:** Store dehydrated media in the dark in a sealed container below 30°C. Prepared media should be stored under refrigeration (2-8°C). Media should be used within 60 days of preparation. Media should not be used if there are any signs of deterioration (discoloration) or contamination, or if the expiration date supplied by the manufacturer has passed.

**Use:** For the rapid and sensitive isolation of motile *Salmonella* spp. from food products following pre-enrichment or selective enrichment. The semisolid medium allows motility to be detected as halos of growth around the original point of inoculation. Recommended by the European Chocolate Manufacturer's Association. For the isolation of *Salmonella* spp. (other than *S. typhi* and *S. partyphi* type A) from stool specimens with high sensitivity and specificity.

## Rappaport-Vassiliadis R10 Broth

**Composition** per liter:

| | |
|---|---|
| $MgCl_2$, anhydrous | 13.4g |
| NaCl | 7.2g |
| Papaic digest of soybean meal | 4.54g |
| $KH_2PO_4$ | 1.45g |
| Malachite Green oxalate | 0.036g |

pH 5.1 ± 0.2 at 25°C

**Preparation of Medium:** Add components to distilled/deionized water and bring volume to 1.0L. Mix thoroughly. Distribute into screw-capped tubes in 10.0mL volumes. Autoclave for 15 min at 10 psi pressure–116°C.

**Storage/Shelf Life:** Store dehydrated media in the dark in a sealed container below 30°C. Prepared media should be stored under refrigeration (2-8°C). Media should be used within 60 days of preparation. Media should not be used if there are any signs of deterioration (discoloration) or contamination, or if the expiration date supplied by the manufacturer has passed.

**Use:** For the isolation and cultivation of *Salmonella* species from food and environmental specimens.

## Rappaport-Vassiliadis Soy Peptone Broth (RVS Broth)

**Composition** per liter:

| | |
|---|---|
| $MgCl_2$, anhydrous | 13.58g |
| NaCl | 7.2g |
| Papaic digest of soybean meal | 4.5g |
| $KH_2PO_4$ | 1.26g |
| $K_2HPO_4$ | 0.18g |
| Malachite Green | 0.036g |

pH 5.2 ± 0.2 at 25°C

**Source:** This medium is available as a premixed powder from Oxoid Unipath.

**Preparation of Medium:** Add components to distilled/deionized water and bring volume to 1.0L. Mix thoroughly. Distribute into screw-capped tubes in 10.0mL volumes. Autoclave for 15 min at 10 psi pressure–115°C.

**Storage/Shelf Life:** Store dehydrated media in the dark in a sealed container below 30°C. Prepared media should be stored under refrigeration (2-8°C). Media should be used within 60 days of preparation. Media should not be used if there are any signs of deterioration (discoloration) or contamination, or if the expiration date supplied by the manufacturer has passed.

**Use:** For the isolation and cultivation of *Salmonella* species from food and environmental specimens.

## Reduced Salt Solution Medium
## (RSS Medium)

**Composition** per liter:
| | |
|---|---:|
| $CaCl_2 \cdot H_2O$ | 20.0g |
| $NaHCO_3$ | 10.0g |
| Dithiothreitol | 2.0g |
| $MgSO_4 \cdot 7H_2O$ | 2.0g |
| $K_2HPO_4$ | 1.0g |
| $KH_2PO_4$ | 1.0g |
| NaCl | 0.2g |

pH 9.2 ± 0.2 at 25°C

**Preparation of Medium:** Add components to distilled/deionized water and bring volume to 1.0L. Mix thoroughly. Distribute into screw-capped tubes or flasks. Autoclave for 15 min at 15 psi pressure–121°C.

**Storage/Shelf Life:** Store dehydrated media in the dark in a sealed container below 30°C. Prepared media should be stored under refrigeration (2-8°C). Media should be used within 60 days of preparation. Media should not be used if there are any signs of deterioration (discoloration) or contamination, or if the expiration date supplied by the manufacturer has passed.

**Use:** For the transport and isolation of bacteria from dental plaque, especially *Streptococcus mutans, Streptococcus sanguis*, and *Lactobacillus* species.

## Reduced Transport Fluid

**Composition** per liter:
| | |
|---|---:|
| $(NH_4)_2SO_4$ | 9.0g |
| NaCl | 9.0g |
| $K_2HPO_4$ | 4.5g |
| $KH_2PO_4$ | 4.5g |
| $Na_2CO_3$ | 4.0g |
| EDTA (ethylenediamine tetraacetic acid) | 3.8g |
| Dithiothreitol | 2.0g |
| $MgSO_4 \cdot 7H_2O$ | 1.8g |

pH 8.0 ± 0.2 at 25°C

**Preparation of Medium:** Add components to distilled/deionized water and bring volume to 1.0L. Mix thoroughly. Filter sterilize. Aseptically distribute into sterile tubes with rubber stoppers.

**Storage/Shelf Life:** Store dehydrated media in the dark in a sealed container below 30°C. Prepared media should be stored under refrigeration (2-8°C). Media should be used within 60 days of preparation. Media should not be used if there are any signs of deterioration (discol-oration) or contamination, or if the expiration date supplied by the manufacturer has passed.

**Use:** For the transport and isolation of bacteria from dental plaque, especially *Streptococcus mutans* and *Streptococcus sanguis*. Also used for the cultivation of a variety of Gram-positive bacteria from the oral cavity, especially streptococci, actinomycetes, lactobacilli, clostridia, *Bacteroides* species, *Fusobacterium* species, and *Veillonella* species.

## Reduced Transport Fluid

**Composition** per liter:
| | |
|---|---:|
| Stock mineral salt solution No. 1 | 75.0mL |
| Stock mineral salt solution No. 2 | 75.0mL |
| Dithiothreitol (1% solution) | 20.0mL |
| Ethylenediamine tetraacetic acid (1*M* solution) | 10.0mL |
| $Na_2CO_3$ (8% solution) | 5.0mL |
| Resazurin (0.1% solution) | 1.0mL |

pH 8.0 ± 0.2 at 25°C

**Stock Mineral Salt Solution No. 1:**
**Composition** per 100.0mL:
| | |
|---|---:|
| $K_2HPO_4$ | 0.6g |

**Preparation of Stock Mineral Salt Solution No. 1:** Add $K_2HPO_4$ to distilled/deionized water and bring volume to 100.0mL. Mix thoroughly.

**Stock Mineral Salt Solution No. 2:**
**Composition** per 100.0mL:
| | |
|---|---:|
| NaCl | 1.2g |
| $(NH_4)_2SO_4$ | 1.2g |
| $K_2HPO_4$ | 0.6g |
| $MgSO_4 \cdot 7H_2O$ | 0.25g |

**Preparation of Stock Mineral Salt Solution No. 2:** Add components to distilled/deionized water and bring volume to 100.0mL. Mix thoroughly.

**Preparation of Medium:** Add components to distilled/deionized water and bring volume to 1.0L. Mix thoroughly. Filter sterilize. Aseptically distribute into sterile tubes with rubber stoppers.

**Storage/Shelf Life:** Store dehydrated media in the dark in a sealed container below 30°C. Prepared media should be stored under refrigeration (2-8°C). Media should be used within 60 days of preparation. Media should not be used if there are any signs of deterioration (discoloration) or contamination, or if the expiration date supplied by the manufacturer has passed.

**Use:** For the transport and isolation of bacteria from dental plaque, especially *Streptococcus mutans* and *Streptococcus sanguis*. Also used for the cultivation of a variety of Gram-positive bacteria from the oral cavity, especially streptococci, actinomycetes, lactobacilli, clostrida, *Bacteroides*, Fusobacteria, and *Veillonela*.

## Regan-Lowe Charcoal Agar
## (Regan-Lowe Medium)

**Composition** per liter:
| | |
|---|---:|
| Agar | 12.0g |
| Beef extract | 10.0g |
| Pancreatic digest of gelatin | 10.0g |
| Soluble starch | 10.0g |
| NaCl | 5.0g |
| Charcoal | 4.0g |
| Niacin | 0.01g |

Horse blood, defibrinated ....................................................... 100.0mL
Cephalexin solution ............................................................... 10.0mL
pH 7.4 ± 0.2 at 25°C

**Source:** This medium is available as a premixed powder from BD Diagnostic Systems.

**Cephalexin Solution:**
**Composition** per 10.0mL:
Cephalexin ....................................................................................0.04g

**Preparation of Cephalexin Solution:** Add cephalexin to distilled/deionized water and bring volume to 10.0mL. Mix thoroughly. Filter sterilize.

**Preparation of Medium:** Add components, except horse blood and cephalexin solution, to distilled/deionized water and bring volume to 890.0mL. Mix thoroughly. Gently heat and bring to boiling. Autoclave for 15 min at 15 psi pressure–121°C. Cool to 45°–50°C. Aseptically add sterile horse blood and sterile cephalexin solution. Mix thoroughly. Pour into sterile Petri dishes or distribute into sterile tubes. Swirl medium while dispensing to keep charcoal in suspension.

**Storage/Shelf Life:** Store dehydrated media in the dark in a sealed container below 30°C. Prepared media should be stored under refrigeration (2-8°C). Media should be used within 60 days of preparation. Media should not be used if there are any signs of deterioration (shrinking, cracking, or discoloration) or contamination, or if the expiration date supplied by the manufacturer has passed.

**Use:** For the selective isolation and cultivation of *Bordetella pertussis* and *Bordetella parapertussis* from clinical specimens.

### Regan-Lowe Semisolid Transport Medium
**Composition** per liter:
Agar ....................................................................................6.0g
Beef extract................................................................................5.0g
Pancreatic digest of gelatin .......................................................5.0g
Soluble starch..............................................................................5.0g
NaCl.............................................................................................2.5g
Charcoal......................................................................................2.0g
Niacin ........................................................................................0.01g
Horse blood, defibrinated ....................................................... 100.0mL
Cephalexin solution ............................................................... 10.0mL
pH 7.4 ± 0.2 at 25°C

**Cephalexin Solution:**
**Composition** per 10.0mL:
Cephalexin ....................................................................................0.04g

**Preparation of Cephalexin Solution:** Add cephalexin to distilled/deionized water and bring volume to 10.0mL. Mix thoroughly. Filter sterilize.

**Preparation of Medium:** Add components, except horse blood and cephalexin solution, to distilled/deionized water and bring volume to 890.0mL. Mix thoroughly. Gently heat and bring to boiling. Autoclave for 15 min at 15 psi pressure–121°C. Cool to 45°–50°C. Aseptically add sterile horse blood and sterile cephalexin solution. Mix thoroughly. Aseptically distribute into small, sterile, screw-capped tubes. Fill tubes half-full. Swirl medium while dispensing to keep charcoal in suspension.

**Storage/Shelf Life:** Store dehydrated media in the dark in a sealed container below 30°C. Prepared media should be stored under refrigeration (2-8°C). Media should be used within 60 days of preparation. Media should not be used if there are any signs of deterioration (discoloration) or contamination, or if the expiration date supplied by the manufacturer has passed.

**Use:** For the transport of *Bordetella pertussis* and *Bordetella parapertussis* isolated from clinical specimens.

### Reinforced Clostridial Agar
**Composition** per liter:
Agar ..................................................................................13.5g
Beef extract................................................................................10.0g
Pancreatic digest of casein.......................................................10.0g
NaCl............................................................................................5.0g
Glucose .......................................................................................5.0g
Yeast extract...............................................................................3.0g
Sodium acetate...........................................................................3.0g
Soluble starch..............................................................................1.0g
L-Cysteine·HCl·H₂O ..................................................................0.5g
pH 6.8 ± 0.2 at 25°C

**Source:** This medium is available as a premixed powder from BD Diagnostic Systems and Oxoid Unipath.

**Preparation of Medium:** Add components to distilled/deionized water and bring volume to 1.0L. Mix thoroughly. Gently heat and bring to boiling. Distribute into tubes or flasks. Autoclave for 15 min at 10 psi pressure–115°C. Pour into sterile Petri dishes or leave in tubes.

**Storage/Shelf Life:** Store dehydrated media in the dark in a sealed container below 30°C. Prepared media should be stored under refrigeration (2-8°C). Media should be used within 60 days of preparation. Media should not be used if there are any signs of deterioration (shrinking, cracking, or discoloration) or contamination, or if the expiration date supplied by the manufacturer has passed.

**Use:** For the cultivation and enumeration of *Clostridium* species, *Bifidobacterium* species, other anaerobes (e.g., lactobacilli), and facultative organisms from clinical specimens and foods.

### Reinforced Clostridial Agar with Tween™
### (LMG Medium 146)
**Composition** per liter:
Agar ..................................................................................13.5g
Beef extract................................................................................10.0g
Pancreatic digest of casein.......................................................10.0g
NaCl............................................................................................5.0g
Glucose .......................................................................................5.0g
Yeast extract...............................................................................3.0g
Sodium acetate...........................................................................3.0g
Tween™ 80.................................................................................1.0g
Soluble starch..............................................................................1.0g
L-Cysteine·HCl·H₂O ..................................................................0.5g
pH 6.8 ± 0.2 at 25°C

**Preparation of Medium:** Add components to distilled/deionized water and bring volume to 1.0L. Mix thoroughly. Gently heat and bring to boiling. Distribute into tubes or flasks. Autoclave for 15 min at 10 psi pressure–115°C. Pour into sterile Petri dishes or leave in tubes.

**Storage/Shelf Life:** Store dehydrated media in the dark in a sealed container below 30°C. Prepared media should be stored under refrigeration (2-8°C). Media should be used within 60 days of preparation. Media should not be used if there are any signs of deterioration (shrinking, cracking, or discoloration) or contamination, or if the expiration date supplied by the manufacturer has passed.

**Use:** For the cultivation and maintenance of *Bifidobacterium merycicum*.

## Reinforced Clostridial HiVeg Agar

**Composition** per liter:

| | |
|---|---|
| Agar | 13.5g |
| Plant extract | 10.0g |
| Plant hydrolysate | 10.0g |
| Glucose | 5.0g |
| NaCl | 5.0g |
| Yeast extract | 3.0g |
| Sodium acetate | 3.0g |
| Starch, soluble | 1.0g |
| L-Cysteine·HCl | 0.5g |

pH 6.8 ± 0.2 at 25°C

**Source:** This medium is available as a premixed powder from Hi-Media.

**Preparation of Medium:** Add components to distilled/deionized water and bring volume to 1.0L. Mix thoroughly. Gently heat and bring to boiling. Distribute into tubes or flasks. Autoclave for 15 min at 10 psi pressure–115°C. Pour into sterile Petri dishes or leave in tubes.

**Storage/Shelf Life:** Store dehydrated media in the dark in a sealed container below 30°C. Prepared media should be stored under refrigeration (2-8°C). Media should be used within 60 days of preparation. Media should not be used if there are any signs of deterioration (shrinking, cracking, or discoloration) or contamination, or if the expiration date supplied by the manufacturer has passed.

**Use:** For the cultivation and enumeration of clostridia and other anaerobes.

## Reinforced Clostridial HiVeg Broth

**Composition** per liter:

| | |
|---|---|
| Plant extract | 10.0g |
| Plant hydrolysate | 10.0g |
| Glucose | 5.0g |
| NaCl | 5.0g |
| Sodium acetate | 3.0g |
| Yeast extract | 3.0g |
| Starch, soluble | 1.0g |
| Agar | 0.5g |
| L-Cysteine·HCl | 0.5g |

pH 6.8 ± 0.2 at 25°C

**Source:** This medium is available as a premixed powder from Hi-Media.

**Preparation of Medium:** Add components to distilled/deionized water and bring volume to 1.0L. Mix thoroughly. Gently heat and bring to boiling. Distribute into tubes or flasks. Autoclave for 15 min at 10 psi pressure–115°C.

**Storage/Shelf Life:** Store dehydrated media in the dark in a sealed container below 30°C. Prepared media should be stored under refrigeration (2-8°C). Media should be used within 60 days of preparation. Media should not be used if there are any signs of deterioration (discoloration) or contamination, or if the expiration date supplied by the manufacturer has passed.

**Use:** For the cultivation and enumeration of clostridia and other anaerobes.

## Reinforced Clostridial Medium

**Composition** per liter:

| | |
|---|---|
| Tryptose | 10.0g |
| Beef extract | 10.0g |

| | |
|---|---|
| Glucose | 5.0g |
| NaCl | 5.0g |
| Yeast extract | 3.0g |
| Sodium acetate | 3.0g |
| Soluble starch | 1.0g |
| L-Cysteine·HCl·H₂O | 0.5g |
| Agar | 0.5g |

pH 6.8 ± 0.2 at 25°C

**Source:** This medium is available as a premixed powder from BD Diagnostic Systems and Oxoid Unipath.

**Preparation of Medium:** Add components to distilled/deionized water and bring volume to 1.0L. Mix thoroughly. Gently heat and bring to boiling. Distribute into tubes or flasks. Autoclave for 15 min at 10 psi pressure–115°C. Pour into sterile Petri dishes or leave in tubes.

**Storage/Shelf Life:** Store dehydrated media in the dark in a sealed container below 30°C. Prepared media should be stored under refrigeration (2-8°C). Media should be used within 60 days of preparation. Media should not be used if there are any signs of deterioration (discoloration) or contamination, or if the expiration date supplied by the manufacturer has passed.

**Use:** For the nonselective cultivation and enumeration of *Clostridium* species, other anaerobes such as lactobacilli, and facultative organisms from clinical specimens and foods.

## Rippey-Cabelli Agar
### (RC Agar)

**Composition** per liter:

| | |
|---|---|
| Agar | 15.0g |
| Meat peptone | 5.0g |
| Trehalose | 5.0g |
| NaCl | 3.0g |
| KCl | 2.0g |
| Yeast extract | 2.0g |
| Bromthymol Blue | 0.44g |
| MgSO₄·7H₂O | 0.2g |
| FeCl₃·6H₂O | 0.1g |
| Sodium deoxycholate | 0.1g |
| Ampicillin solution | 10.0mL |
| Ethanol | 10.0mL |

pH 8.0 ± 0.2 at 25°C

**Ampicillin Solution:**
**Composition** per 10.0mL:

| | |
|---|---|
| Ampicillin | 0.02g |

**Preparation of Ampicillin Solution:** Add ampicillin to distilled/deionized water and bring volume to 10.0mL. Mix thoroughly. Filter sterilize.

**Preparation of Medium:** Add components—except sodium deoxycholate, ampicillin solution, and ethanol—to distilled/deionized water and bring volume to 980.0mL. Mix thoroughly. Gently heat and bring to boiling. Autoclave for 15 min at 15 psi pressure–121°C. Cool to 45°–50°C. Aseptically add sodium deoxycholate, 10.0mL of sterile ampicillin solution, and 10.0mL of ethanol. Mix thoroughly. Pour into sterile Petri dishes or distribute into sterile tubes.

**Storage/Shelf Life:** Store dehydrated media in the dark in a sealed container below 30°C. Prepared media should be stored under refrigeration (2-8°C). Media should be used within 60 days of preparation. Media should not be used if there are any signs of deterioration (shrink-

ing, cracking, or discoloration) or contamination, or if the expiration date supplied by the manufacturer has passed.

**Use:** For the isolation, cultivation, and differentiation of *Aeromonas* species and *Plesiomonas* species from water samples using the membrane filter method. This medium differentiates bacteria on the basis of trehalose fermentation. Bacteria that ferment trehalose turn the medium yellow.

## Rippey-Cabelli HiVeg Agar Base with Ethanol and Ampicillin

**Composition** per liter:

| | |
|---|---|
| Agar | 15.0g |
| Plant hydrolysate No. 1 | 5.0g |
| Trehalose | 5.0g |
| NaCl | 3.0g |
| KCl | 2.0g |
| Yeast extract | 2.0g |
| MgSO$_4$ | 0.2g |
| Iron (III) chloride | 0.1g |
| Bromthymol Blue | 0.04g |
| Ampicillin solution | 10.0mL |
| Ethanol | 10.0mL |

pH 8.0 ± 0.2 at 25°C

**Source:** This medium, without ampicillin solution and ethanol, is available as a premixed powder from HiMedia.

**Ampicillin Solution:**
**Composition** per 10.0mL:

| | |
|---|---|
| Ampicillin | 0.02g |

**Preparation of Ampicillin Solution:** Add ampicillin to distilled/deionized water and bring volume to 10.0mL. Mix thoroughly. Filter sterilize.

**Preparation of Medium:** Add components—except ampicillin solution and ethanol—to distilled/deionized water and bring volume to 990.0mL. Mix thoroughly. Gently heat and bring to boiling. Autoclave for 15 min at 15 psi pressure–121°C. Cool to 45°–50°C. Aseptically add 10.0mL of sterile ampicillin solution, and 10.0mL of ethanol. Mix thoroughly. Pour into sterile Petri dishes or distribute into sterile tubes.

**Storage/Shelf Life:** Store dehydrated media in the dark in a sealed container below 30°C. Prepared media should be stored under refrigeration (2-8°C). Media should be used within 60 days of preparation. Media should not be used if there are any signs of deterioration (shrinking, cracking, or discoloration) or contamination, or if the expiration date supplied by the manufacturer has passed.

**Use:** For the isolation, cultivation, and differentiation of *Aeromonas* species and *Plesiomonas* species from water samples using the membrane filter method. For the differential and selective isolation of *Aeromonas hydrophila* species from water samples

## Rogosa SL Agar
### (Rogosa Selective *Lactobacillus* Agar)

**Composition** per liter:

| | |
|---|---|
| Agar | 15.0g |
| Sodium acetate | 15.0g |
| Glucose | 10.0g |
| Pancreatic digest of casein | 10.0g |
| K$_2$HPO$_4$ | 6.0g |
| Yeast extract | 5.0g |
| Arabinose | 5.0g |

| | |
|---|---|
| Sucrose | 5.0g |
| Ammonium citrate | 2.0g |
| Sorbitan monooleate | 1.0g |
| MgSO$_4$·7H$_2$O | 0.57g |
| MnSO$_4$·7H$_2$O | 0.12g |
| FeSO$_4$·H$_2$O | 0.03g |
| Acetic acid, glacial | 1.32mL |

pH 5.4 ± 0.2 at 25°C

**Source:** This medium is available as a premixed powder from BD Diagnostic Systems.

**Preparation of Medium:** Add components, except glacial acetic acid, to distilled/deionized water and bring volume to 998.7mL. Mix thoroughly. Gently heat and bring to boiling. Add glacial acetic acid. Mix thoroughly. Gently heat while stirring and bring to 90°–100°C for 2–3 min. Do not autoclave. Pour into sterile Petri dishes or distribute into sterile tubes.

**Storage/Shelf Life:** Store dehydrated media in the dark in a sealed container below 30°C. Prepared media should be stored under refrigeration (2-8°C). Media should be used within 60 days of preparation. Media should not be used if there are any signs of deterioration (shrinking, cracking, or discoloration) or contamination, or if the expiration date supplied by the manufacturer has passed.

**Use:** For the isolation, cultivation, and enumeration of lactobacilli, especially from feces, saliva, vaginal specimens, and dairy products.

## Rogosa SL Broth
### (Rogosa Selective *Lactobacillus* Broth)

**Composition** per liter:

| | |
|---|---|
| Sodium acetate | 15.0g |
| Glucose | 10.0g |
| Pancreatic digest of casein | 10.0g |
| K$_2$HPO$_4$ | 6.0g |
| Yeast extract | 5.0g |
| Arabinose | 5.0g |
| Sucrose | 5.0g |
| Ammonium citrate | 2.0g |
| Sorbitan monooleate | 1.0g |
| MgSO$_4$·7H$_2$O | 0.57g |
| MnSO$_4$·7H$_2$O | 0.12g |
| FeSO$_4$·H$_2$O | 0.03g |
| Acetic acid, glacial | 1.32mL |

pH 5.4 ± 0.2 at 25°C

**Source:** This medium is available as a premixed powder from BD Diagnostic Systems.

**Preparation of Medium:** Add components, except glacial acetic acid, to distilled/deionized water and bring volume to 998.7mL. Mix thoroughly. Gently heat and bring to boiling. Add glacial acetic acid. Mix thoroughly. Gently heat while stirring and bring to 90°–100°C for 2–3 min. Do not autoclave. Aseptically distribute into sterile tubes.

**Storage/Shelf Life:** Store dehydrated media in the dark in a sealed container below 30°C. Prepared media should be stored under refrigeration (2-8°C). Media should be used within 60 days of preparation. Media should not be used if there are any signs of deterioration (discoloration) or contamination, or if the expiration date supplied by the manufacturer has passed.

**Use:** For the isolation, cultivation, and enumeration of lactobacilli, especially from feces, saliva, vaginal specimens, and dairy products.

## Rogosa SL HiVeg Agar

**Composition** per liter:

| | |
|---|---|
| Agar | 15.0g |
| Sodium acetate | 15.0g |
| Glucose | 10.0g |
| Plant hydrolysate No. 1 | 10.0g |
| $KH_2PO_4$ | 6.0g |
| Arabinose | 5.0g |
| Saccharose | 5.0g |
| Yeast extract | 5.0g |
| Ammonium citrate | 2.0g |
| $MgSO_4$ | 0.57g |
| $MnSO_4$ | 0.12g |
| $FeSO_4$ | 0.03g |
| Acetic acid, glacial | 1.32mL |
| Polysorbate 80 | 1.0mL |

pH 5.4 ± 0.2 at 25°C

**Source:** This medium, without acetic acid or polysorbate 80, is available as a premixed powder from HiMedia.

**Preparation of Medium:** Add components to distilled/deionized water and bring volume to 1.0L. Mix thoroughly. Gently heat and bring to boiling. Distribute into tubes or flasks. Gently heat to 90–100°C. Hold at temperature for 2–3 min. Do not autoclave. Cool to 50°C. Pour into sterile Petri dishes or leave in tubes.

**Storage/Shelf Life:** Store dehydrated media in the dark in a sealed container below 30°C. Prepared media should be stored under refrigeration (2-8°C). Media should be used within 60 days of preparation. Media should not be used if there are any signs of deterioration (shrinking, cracking, or discoloration) or contamination, or if the expiration date supplied by the manufacturer has passed.

**Use:** For the isolation, cultivation, and enumeration of lactobacilli, especially from feces, saliva, vaginal specimens, and dairy products. For the cultivation of oral and fecal lactobacilli.

## Rogosa SL HiVeg Broth

**Composition** per liter:

| | |
|---|---|
| Sodium acetate | 15.0g |
| Glucose | 10.0g |
| Plant hydrolysate No. 1 | 10.0g |
| $KH_2PO_4$ | 6.0g |
| Arabinose | 5.0g |
| Saccharose | 5.0g |
| Yeast extract | 5.0g |
| Ammonium citrate | 2.0g |
| $MgSO_4$ | 0.57g |
| $MnSO_4$ | 0.12g |
| $FeSO_4$ | 0.03g |
| Acetic acid, glacial | 1.32mL |
| Polysorbate 80 | 1.0mL |

pH 5.4 ± 0.2 at 25°C

**Source:** This medium, without acetic acid or polysorbate 80, is available as a premixed powder from HiMedia.

**Preparation of Medium:** Add components to distilled/deionized water and bring volume to 1.0L. Mix thoroughly. Gently heat and bring to boiling. Distribute into tubes or flasks. Gently heat to 90–100°C. Hold at temperature for 2–3 min. Do not autoclave.

**Storage/Shelf Life:** Store dehydrated media in the dark in a sealed container below 30°C. Prepared media should be stored under refrigeration (2-8°C). Media should be used within 60 days of preparation.

Media should not be used if there are any signs of deterioration (discoloration) or contamination, or if the expiration date supplied by the manufacturer has passed.

**Use:** For the isolation, cultivation, and enumeration of lactobacilli, especially from feces, saliva, vaginal specimens, and dairy products.

## RS HiVeg Medium Base with Novobiocin (Rimler-Shotts Medium)

**Composition** per liter:

| | |
|---|---|
| Agar | 13.5g |
| $Na_2S_2O_3$ | 6.8g |
| L-Ornithine hydrochloride | 6.5g |
| L-Lysine hydrochloride | 5.0g |
| NaCl | 5.0g |
| Maltose | 3.5g |
| Yeast extract | 3.0g |
| Synthetic detergent No. III | 1.0g |
| Ferric ammonium citrate | 0.8g |
| L-Cysteine·HCl | 0.3g |
| Bromthymol Blue | 0.03g |
| Novobiocin solution | 10.0mL |

pH 7.0 ± 0.2 at 25°C

**Source:** This medium, without novobiocin, is available as a premixed powder from HiMedia.

**Novobiocin Solution:**

**Composition** per 10.0mL:

| | |
|---|---|
| Novobiocin | 5.0mg |

**Preparation of Novobiocin Solution:** Add novobiocin to distilled/deionized water and bring volume to 10.0mL. Mix thoroughly. Filter sterilize.

**Preparation of Medium:** Add components, except novobiocin solution, to distilled/deionized water and bring volume to 990.0mL. Mix thoroughly. Gently heat and bring to boiling. Autoclave for 15 min at 15 psi pressure–121°C. Cool to 45°–50°C. Aseptically add sterile novobiocin solution. Mix thoroughly. Pour into sterile Petri dishes or distribute into sterile tubes.

**Storage/Shelf Life:** Store dehydrated media in the dark in a sealed container below 30°C. Prepared media should be stored under refrigeration (2-8°C). Media should be used within 60 days of preparation. Media should not be used if there are any signs of deterioration (shrinking, cracking, or discoloration) or contamination, or if the expiration date supplied by the manufacturer has passed.

**Use:** For the selective isolation, cultivation, and presumptive identification of *Aeromonas hydrophila* and other Gram-negative bacteria based on their ability to decarboxylate lysine and ornithine, ferment maltose, and produce $H_2S$. Maltose-fermenting bacteria appear as yellow colonies. Bacteria that produce lysine or ornithine decarboxylase turn the medium greenish-yellow to yellow. Bacteria that produce $H_2S$ appear as colonies with black centers.

## *S. aureus* ID

**Composition** per liter:
Proprietary

**Source:** This medium is available from bioMérieux.

**Storage/Shelf Life:** Prepared plates should be stored under refrigeration (2-8°C). Media should not be used if there are any signs of deterioration (shrinking, cracking, or discoloration) or contamination, or if the expiration date supplied by the manufacturer has passed.

**Use:** For the direct identification of *Staphylococcus aureus* and the selective isolation of staphylococci. Direct identification of *S. aureus* is based on the spontaneous green coloration of α-glucosidase-producing colonies.

## SA Agar

**Composition** per liter:

| | |
|---|---|
| Agar | 15.0g |
| Pancreatic digest of casein | 10.0g |
| NaCl | 5.0g |
| Starch, soluble | 1.0g |
| Ampicillin | 0.01g |
| Phenol Red | 0.018g |

pH 7.4 ± 0.2 at 25°C

**Preparation of Medium:** Add components, except ampicillin, to distilled/deionized water and bring volume to 1.0L. Mix thoroughly. Gently heat and bring to boiling. Autoclave for 15 min at 15 psi pressure–121°C. Cool to 45°–50°C. Aseptically add ampicillin. Mix thoroughly. Pour into sterile Petri dishes.

**Storage/Shelf Life:** Store dehydrated media in the dark in a sealed container below 30°C. Prepared media should be stored under refrigeration (2-8°C). Media should be used within 60 days of preparation. Media should not be used if there are any signs of deterioration (shrinking, cracking, or discoloration) or contamination, or if the expiration date supplied by the manufacturer has passed.

**Use:** For the isolation, cultivation, and differentiation, based on starch hydrolysis, of *Aeromonas hydrophila* from foods. After inoculation of plates and growth of cultures, starch hydrolysis is determined by flooding each plate with 5.0mL of Lugol's iodine solution.

## SA Agar, Modified
## (Lachica's Medium)

**Composition** per liter:

| | |
|---|---|
| Beef heart, solids from infusion | 500.0g |
| Agar | 15.0g |
| Tryptose | 10.0g |
| NaCl | 5.0g |
| Amylose Azure | 3.0g |
| Ampicillin | 0.01mg |

pH 7.4 ± 0.2 at 25°C

**Preparation of Medium:** Add components to distilled/deionized water and bring volume to 1.0L. Mix thoroughly. Gently heat and bring to boiling. Distribute into tubes or flasks. Autoclave for 15 min at 15 psi pressure–121°C. Pour into sterile Petri dishes.

**Storage/Shelf Life:** Store dehydrated media in the dark in a sealed container below 30°C. Prepared media should be stored under refrigeration (2-8°C). Media should be used within 60 days of preparation. Media should not be used if there are any signs of deterioration (shrinking, cracking, or discoloration) or contamination, or if the expiration date supplied by the manufacturer has passed.

**Use:** For the isolation and cultivation of *Aeromonas hydrophila* from foods. *Aeromonas hydrophila* appears as colonies surrounded by a light halo on a light blue background.

## SA HiVeg Agar Base with Ampicillin

**Composition** per liter:

| | |
|---|---|
| Agar | 15.0g |
| Plant hydrolysate | 10.0g |
| Starch, soluble | 10.0g |
| NaCl | 5.0g |
| Phenol Red | 0.025g |
| Ampicillin solution | 10.0mL |

pH 7.4 ± 0.2 at 25°C

**Source:** This medium, without ampicillin, is available as a premixed powder from HiMedia.

**Ampicillin Solution:**
**Composition** per 10.0mL:

| | |
|---|---|
| Ampicillin | 0.01g |

**Preparation of Ampicillin Solution:** Add ampicillin to distilled/deionized water and bring volume to 10.0mL. Mix thoroughly. Filter sterilize.

**Preparation of Medium:** Add components, except ampicillin solution, to distilled/deionized water and bring volume to 990.0mL. Mix thoroughly. Gently heat and bring to boiling. Autoclave for 15 min at 15 psi pressure–121°C. Cool to 45°C. Aseptically add sterile ampicillin solution. Mix thoroughly. Pour into sterile Petri dishes or distribute into sterile tubes.

**Storage/Shelf Life:** Store dehydrated media in the dark in a sealed container below 30°C. Prepared media should be stored under refrigeration (2-8°C). Media should be used within 60 days of preparation. Media should not be used if there are any signs of deterioration (shrinking, cracking, or discoloration) or contamination, or if the expiration date supplied by the manufacturer has passed.

**Use:** For the isolation, cultivation, and differentiation of *Aeromonas hydrophilia* from foods based on starch hydrolysis.

## Sakazakii DHL Agar

**Composition** per liter:

| | |
|---|---|
| Agar | 15.0g |
| Casein enzymic hydrolysate | 10.0g |
| Meat peptone | 10.0g |
| Lactose | 10.0g |
| Sucrose | 10.0g |
| Meat extract | 3.0g |
| Na$_2$S$_2$O$_3$ | 2.0g |
| Sodium deoxycholate | 1.5g |
| Sodium citrate | 1.0g |
| Ammonium iron (III) citrate | 1.0g |
| L-Cysteine·HCl·H$_2$O | 0.2g |
| Neutral Red | 0.03g |

pH 7.2 ± 0.2 at 25°C

**Source:** This medium is available from HiMedia.

**Preparation of Medium:** Add components to distilled/deionized water and bring volume to 1.0L. Mix thoroughly. Gently heat and bring to boiling. Distribute into tubes or flasks. Autoclave for 15 min at 15 psi pressure–121°C. Pour into sterile Petri dishes or leave in tubes.

**Storage/Shelf Life:** Store dehydrated media in the dark in a sealed container below 30°C. Prepared media should be stored under refrigeration (2-8°C). Media should be used within 60 days of preparation. Media should not be used if there are any signs of deterioration (shrinking, cracking, or discoloration) or contamination, or if the expiration date supplied by the manufacturer has passed.

**Use:** For the detection and isolation of pathogenic Enterobacteriaceae from all types of specimens.

## *Salmonella* **Chromogen Agar** (**Rambach Equivalent Agar**)

**Composition** per liter:

Agar .......................................................................15.0g
Peptone......................................................................5.0g
NaCl..........................................................................5.0g
Yeast extract.............................................................2.0g
Meat extract .............................................................1.0g
Sodium deoxycholate................................................1.0g

pH 7.3 ± 0.2 at 25°C

**Source:** This medium is available from Fluka, Sigma-Aldrich.

**Preparation of Medium:** Add components to distilled/deionized water and bring volume to 1.0L. Mix thoroughly. Gently heat while stirring and bring to boiling. Autoclave for 15 min at 15 psi pressure–121°C. Pour into sterile Petri dishes.

**Storage/Shelf Life:** Store dehydrated media in the dark in a sealed container below 30°C. Prepared media should be stored in the dark under refrigeration (2-8°C). Chromogenic media are especially light and temperature sensitive; protect from light, excessive heat, moisture, and freezing. Media should not be used if there are any signs of deterioration (shrinking, cracking, or discoloration) or contamination, or if the expiration date supplied by the manufacturer has passed.

**Use:** A differential diagnostic agar for the detection of *Salmonella* in food, including the isolation and enumeration of *Salmonella* from bivalves.

## *Salmonella* **Chromogenic Agar** (**OSCM**)

**Composition** per liter:

Chromogenic mix ....................................................28.0g
Agar .......................................................................12.0g
Special peptone .....................................................10.0g
Selective supplement solution ............................ 10.0mL

pH 7.2 ± 0.2 at 25°C

**Source:** This medium is available as a premixed powder from Oxoid Unipath.

**Selective Supplement Solution:**

**Composition** per 10.0mL:

Cefsulodin................................................................12.0mg
Novobiocin................................................................5.0mg

**Preparation of Selective Supplement Solution:** Add components to distilled/deionized water and bring volume to 10.0mL. Mix thoroughly. Filter sterilize.

**Preparation of Medium:** Add components to distilled/deionized water and bring volume to 1.0mL Mix thoroughly. Gently heat while stirring and bring to boiling. Do not autoclave. Cool quickly to 50°C. Mix thoroughly. Pour into sterile Petri dishes.

**Storage/Shelf Life:** Store dehydrated media in the dark in a sealed container below 30°C. Prepared media should be stored in the dark under refrigeration (2-8°C). Chromogenic media are especially light and temperature sensitive; protect from light, excessive heat, moisture, and freezing. Media should not be used if there are any signs of deterioration (shrinking, cracking, or discoloration) or contamination, or if the expiration date supplied by the manufacturer has passed.

**Use:** For the identification of *Salmonella* species and differentiation of *Salmonella* spp. and other organisms in the family Enterobacteriaceae. This medium combines two chromogens for the detection of *Salmonella* spp., 5-bromo-6-chloro-3-indolyl caprylate (Magenta-caprylate) and 5-bromo-4-chloro-3-indolyl-β-D galactopyranoside (X-gal). X-gal is a substrate for the enzyme β-D-galactosidase. Hydrolysis of the chromogen, Magenta-caprylate, by lactose-negative *Salmonella* species results in magenta colonies. The addition of the selective supplement solution increases the selectivity of the medium. Novobiocin inhibits *Proteus* growth and cefsulodin inhibits growth of pseudomonads.

## *Salmonella* **Chromogenic Agar**

**Composition** per liter:

SAgar .....................................................................12.8g
Sodium citrate...........................................................8.5g
Chromogenic mixture (X-gal + Magenta-caprylate)....................5.81g
Casein peptone ........................................................5.0g
Beef Extract .............................................................5.0g

pH 7.2 ± 0.2 at 25°C

**Source:** This medium is available from CONDA, Barcelona, Spain.

**Preparation of Medium:** Add components to distilled/deionized water and bring volume to 1.0L. Mix thoroughly. Gently heat and bring to boiling. Do not autoclave. Cool to 50°C. Pour into sterile Petri dishes.

**Storage/Shelf Life:** Store dehydrated media in the dark in a sealed container below 30°C. Prepared media should be stored in the dark under refrigeration (2-8°C). Chromogenic media are especially light and temperature sensitive; protect from light, excessive heat, moisture, and freezing. Media should not be used if there are any signs of deterioration (shrinking, cracking, or discoloration) or contamination, or if the expiration date supplied by the manufacturer has passed.

**Use:** For the isolation of *Salmonella* from foods, waters, and clinical samples.

## *Salmonella* **Differential Agar** (**RajHans Medium**)

**Composition** per liter:

Agar .......................................................................12.0g
Propylene glycol .....................................................10.0g
Peptone, special ........................................................8.0g
B.C. indicator............................................................2.0g
Yeast extract.............................................................2.0g
Sodium deoxycholate................................................1.0g

pH 7.3 ± 0.2 at 25°C

**Source:** This medium is available as a premixed powder from Hi-Media.

**Preparation of Medium:** Add components to distilled/deionized water and bring volume to 1.0L. Mix thoroughly. Gently heat while stirring and bring to boiling. Mix to completely dissolve components. Do not autoclave. Cool to 50°C. Mix thoroughly. Pour into sterile Petri dishes.

**Storage/Shelf Life:** Store dehydrated media in the dark in a sealed container below 30°C. Prepared media should be stored under refrigeration (2-8°C). Media should be used within 60 days of preparation. Media should not be used if there are any signs of deterioration (shrinking, cracking, or discoloration) or contamination, or if the expiration date supplied by the manufacturer has passed.

**Use:** A selective chromgenic medium used for the isolation and differentiation of *Salmonella* species from the members of Enterobacteriaceae, especially *Proteus* species.

## *Salmonella* Differential HiVeg Agar, Modified
### (*Salmonella* Differential Agar, Modified, HiVeg)
**Composition** per liter:

| | |
|---|---|
| Propylene glycol | 10.0g |
| Plant special peptone | 8.0g |
| NaCl | 5.0g |
| Yeast extract | 3.0g |
| B.C. indicator | 2.0g |
| Synthetic detergent | 1.0g |

pH 7.3 ± 0.2 at 25°C

**Source:** This medium is available as a premixed powder from Hi-Media.

**Preparation of Medium:** Add components to distilled/deionized water and bring volume to 1.0L. Mix thoroughly. Gently heat while stirring and bring to boiling. Mix to completely dissolve components. Do not autoclave. Cool to 50°C. Mix thoroughly. Pour into sterile Petri dishes.

**Storage/Shelf Life:** Store dehydrated media in the dark in a sealed container below 30°C. Prepared media should be stored under refrigeration (2-8°C). Media should be used within 60 days of preparation. Media should not be used if there are any signs of deterioration (shrinking, cracking, or discoloration) or contamination, or if the expiration date supplied by the manufacturer has passed.

**Use:** A selective chromogenic medium used for the isolation and differentiation of *Salmonella* species from the members of Enterobacteriaceae, especially *Proteus* species.

## *Salmonella* HiVeg Agar, ONOZ
**Composition** per liter:

| | |
|---|---|
| Agar | 15.0g |
| Sucrose | 13.0g |
| Lactose | 11.5g |
| Na$_3$-citrate·5H$_2$O | 9.3g |
| Plant peptone | 8.625g |
| Plant extract No. 1 | 6.0g |
| L-Phenylalanine | 5.0g |
| Na$_2$S$_2$O$_3$·5H$_2$O | 4.25g |
| Yeast extract | 3.0g |
| Synthetic detergent No. 1 | 2.0g |
| Na$_2$HPO$_4$·2H$_2$O | 1.0g |
| Ferric citrate | 0.5g |
| Metachrome Yellow | 0.47g |
| MgSO$_4$ | 0.4g |
| Aniline Blue | 0.25g |
| Neutral Red | 0.022g |
| Brilliant Green | 1.66mg |

pH 7.4 ± 0.2 at 25°C

**Preparation of Medium:** Add components to distilled/deionized water and bring volume to 1.0L. Mix thoroughly. Distribute into tubes or flasks. Gently heat and bring to boiling. Do not autoclave. Cool to 50°C. Pour into sterile Petri dishes or leave in tubes.

**Storage/Shelf Life:** Store dehydrated media in the dark in a sealed container below 30°C. Prepared media should be stored under refrigeration (2-8°C). Media should be used within 60 days of preparation. Media should not be used if there are any signs of deterioration (shrink-

ing, cracking, or discoloration) or contamination, or if the expiration date supplied by the manufacturer has passed.

**Use:** For the cultivation and maintenance *Salmonella* spp. from clinical specimens.

## *Salmonella* Rapid Test Elective Medium
**Composition** per liter:

| | |
|---|---|
| Tryptone | 10.0g |
| Na$_2$HPO$_4$ | 9.0g |
| Sodium chloride | 5.0g |
| Casein | 5.0g |
| KH$_2$PO$_4$ | 1.5g |
| Malachite Green | 0.0025g |

pH 6.5 ± 0.2 at 25°C

**Preparation of Medium:** Add components to distilled/deionized water and bring volume to 1.0L. Mix thoroughly. Autoclave for 15 min at 15 psi pressure–121°C.

**Storage/Shelf Life:** Store dehydrated media in the dark in a sealed container below 30°C. Prepared media should be stored under refrigeration (2-8°C). Media should be used within 60 days of preparation. Media should not be used if there are any signs of deterioration (discoloration) or contamination, or if the expiration date supplied by the manufacturer has passed.

**Use:** For the Oxoid *Salmonella* Rapid Test, which is for the presumptive detection of motile *Salmonella* in foods and environmental samples.

## *Salmonella* Rapid Test Elective Medium, 2X
**Composition** per liter:

| | |
|---|---|
| Tryptone | 20.0g |
| Na$_2$HPO$_4$ | 18.0g |
| Sodium chloride | 10.0g |
| KH$_2$PO$_4$ | 3.0g |
| Casein | 10.0g |
| Malachite Green | 0.005g |

pH 6.5 ± 0.2 at 25°C

**Preparation of Medium:** Add components to distilled/deionized water and bring volume to 1.0L. Mix thoroughly. Autoclave for 15 min at 15 psi pressure–121°C.

**Storage/Shelf Life:** Store dehydrated media in the dark in a sealed container below 30°C. Prepared media should be stored under refrigeration (2-8°C). Media should be used within 60 days of preparation. Media should not be used if there are any signs of deterioration (discoloration) or contamination, or if the expiration date supplied by the manufacturer has passed.

**Use:** Use as described in the Oxoid *Salmonella* Rapid Test, which is for the presumptive detection of motile *Salmonella* in foods and environmental samples.

## *Salmonella Shigella* Agar
### (SS Agar)
**Composition** per liter:

| | |
|---|---|
| Agar | 13.5g |
| Lactose | 10.0g |
| Bile salts | 8.5g |
| Na$_2$S$_2$O$_3$ | 8.5g |
| Sodium citrate | 8.5g |
| Beef extract | 5.0g |

Pancreatic digest of casein............................................2.5g
Peptic digest of animal tissue.......................................2.5g
Ferric citrate...............................................................1.0g
Neutral Red...............................................................0.025g
Brilliant Green .........................................................0.33mg

pH 7.0 ± 0.2 at 25°C

**Source:** This medium is available as a premixed powder from BD Diagnostic Systems and Oxoid Unipath.

**Preparation of Medium:** Add components to distilled/deionized water and bring volume to 1.0L. Mix thoroughly. Gently heat while stirring and bring to boiling. Do not autoclave. Cool to 45°–50°C. Pour into sterile Petri dishes in 20.0mL volumes. Allow the surface of the plates to dry before inoculation.

**Storage/Shelf Life:** Store dehydrated media in the dark in a sealed container below 30°C. Prepared media should be stored under refrigeration (2-8°C). Media should be used within 60 days of preparation. Media should not be used if there are any signs of deterioration (shrinking, cracking, or discoloration) or contamination, or if the expiration date supplied by the manufacturer has passed.

**Use:** For the selective isolation and differentiation of pathogenic enteric bacilli, especially those belonging to the genus *Salmonella*. This medium is not recommended for the primary isolation of *Shigella* species. Lactose-fermenting bacteria such as *Escherichia coli* or *Klebsiella pneumoniae* appear as small pink or red colonies. Lactose-nonfermenting bacteria—such as *Salmonella* species, *Proteus* species, and *Shigella* species—appear as colorless colonies. Production of $H_2S$ by *Salmonella* species turns the center of the colonies black.

## *Salmonella Shigella* Agar, Modified
### (SS Agar, Modified)

**Composition** per liter:
Agar ........................................................................12.0g
Lactose ....................................................................10.0g
Sodium citrate .........................................................10.0g
$Na_2S_2O_3$ .............................................................8.5g
Bile salts ...................................................................5.5g
Beef extract ...............................................................5.0g
Peptone.....................................................................5.0g
Ferric citrate ..............................................................1.0g
Neutral Red ..............................................................0.025g
Brilliant Green .......................................................0.33mg

pH 7.3 ± 0.2 at 25°C

**Source:** This medium is available as a premixed powder from Oxoid Unipath.

**Preparation of Medium:** Add components to distilled/deionized water and bring volume to 1.0L. Mix thoroughly. Gently heat while stirring and bring to boiling. Do not autoclave. Cool to 45°–50°C. Pour into sterile Petri dishes in 20.0mL volumes. Allow the surface of the plates to dry before inoculation.

**Storage/Shelf Life:** Store dehydrated media in the dark in a sealed container below 30°C. Prepared media should be stored under refrigeration (2-8°C). Media should be used within 60 days of preparation. Media should not be used if there are any signs of deterioration (shrinking, cracking, or discoloration) or contamination, or if the expiration date supplied by the manufacturer has passed.

**Use:** For the selective isolation and differentiation of pathogenic enteric bacilli, especially those belonging to the genus *Salmonella*. This medium provides better growth of *Shigella* species. Lactose-fer-

menting bacteria such as *Escherichia coli* or *Klebsiella pneumoniae* appear as small pink or red colonies. Lactose-nonfermenting bacteria—such as *Salmonella* species, *Proteus* species, and *Shigella* species—appear as colorless colonies. Production of $H_2S$ by *Salmonella* species turns the center of the colonies black.

## Salt Broth, Modified

**Composition** per liter:
NaCl..........................................................................65.0g
Enzymatic digest of animal tissue ..............................10.0g
Heart digest...............................................................10.0g
Glucose .....................................................................1.0g
Bromcresol Purple .....................................................0.016g

pH 7.2 ± 0.2 at 25°C

**Source:** Available as a prepared medium from BD Diagnostic Systems.

**Preparation of Medium:** Add components to distilled/deionized water and bring volume to 1.0L. Mix thoroughly. Distribute into tubes or flasks. Autoclave for 15 min at 15 psi pressure–121°C.

**Storage/Shelf Life:** Store dehydrated media in the dark in a sealed container below 30°C. Prepared media should be stored under refrigeration (2-8°C). Media should be used within 60 days of preparation. Media should not be used if there are any signs of deterioration (discoloration) or contamination, or if the expiration date supplied by the manufacturer has passed.

**Use:** For the differentiation of the enterococcal group D streptococci from nonenterococcal group D streptococci based on salt tolerance.

## Salt Meat Broth

**Composition** per liter:
NaCl........................................................................100.0g
Neutral ox-heart tissue ..............................................30.0g
Beef extract ..............................................................10.0g
Peptone ....................................................................10.0g

pH 7.6 ± 0.2 at 25°C

**Source:** This medium is available as tablets from Oxoid Unipath.

**Preparation of Medium:** Add components to distilled/deionized water and bring volume to 1.0L. Mix thoroughly. Distribute into tubes or flasks. Autoclave for 15 min at 15 psi pressure–121°C.

**Storage/Shelf Life:** Store dehydrated media in the dark in a sealed container below 30°C. Prepared media should be stored under refrigeration (2-8°C). Media should be used within 60 days of preparation. Media should not be used if there are any signs of deterioration (discoloration) or contamination, or if the expiration date supplied by the manufacturer has passed.

**Use:** For the isolation and cultivation of staphylococci from specimens with a mixed flora such as fecal specimens, especially during the investigation of staphylococcal food poisoning.

## Salt Polymyxin Broth
### (SPB)

**Composition** per liter:
NaCl..........................................................................20.0g
Pancreatic digest of casein.........................................10.0g
Yeast extract...............................................................3.0g
Polymyxin B ........................................................250,000U

pH 8.8 ± 0.2 at 25°C

**Preparation of Medium:** Add components to distilled/deionized water and bring volume to 1.0L. Mix thoroughly. Adjust pH to 8.8. Distribute into tubes or flasks. Autoclave for 10 min at 10 psi pressure–115°C.

**Storage/Shelf Life:** Store dehydrated media in the dark in a sealed container below 30°C. Prepared media should be stored under refrigeration (2-8°C). Media should be used within 60 days of preparation. Media should not be used if there are any signs of deterioration (discoloration) or contamination, or if the expiration date supplied by the manufacturer has passed.

**Use:** For the isolation and cultivation of *Vibrio* species from foods.

### Salt Polymyxin HiVeg Broth Base with Polymyxin
**Composition** per liter:
| | |
|---|---|
| NaCl | 20.0g |
| Plant hydrolysate | 10.0g |
| Yeast extract | 3.0g |
| Selective supplement | 10.0mL |

pH 8.8 ± 0.2 at 25°C

**Source:** This medium, without selective supplement, is available as a premixed powder from HiMedia.

**Selective Supplement:**
**Composition** per 10.0mL:
| | |
|---|---|
| Polymyxin B sulfate | 100,000U |

**Preparation of Selective Supplement:** Add polymycin B sulfate to distilled/deionized water and bring volume to 10.0mL. Mix thoroughly. Filter sterilize.

**Preparation of Medium:** Add components, except selective supplement, to distilled/deionized water and bring volume to 990.0mL. Mix thoroughly. Gently heat and bring to boiling. Distribute into tubes or flasks. Mix thoroughly. Adjust pH to 8.8. Distribute into tubes or flasks. Autoclave for 10 min at 10 psi pressure–115°C. Cool to 50°C. Aseptically add 10.0mL sterile selective supplement. Mix thoroughly. Pour into sterile Petri dishes or aseptically distribute into tubes.

**Storage/Shelf Life:** Store dehydrated media in the dark in a sealed container below 30°C. Prepared media should be stored under refrigeration (2-8°C). Media should be used within 60 days of preparation. Media should not be used if there are any signs of deterioration (discoloration) or contamination, or if the expiration date supplied by the manufacturer has passed.

**Use:** For the isolation and cultivation of *Vibrio* species from foods.

### Salt Tolerance Medium
**Composition** per liter:
| | |
|---|---|
| Beef heart, infusion from | 500.0g |
| NaCl | 65.0g |
| Tryptose | 10.0g |
| Glucose | 1.0g |
| Indicator solution | 1.0mL |

pH 7.4 ± 0.2 at 25°C

**Indicator Solution:**
**Composition** per 100.0mL:
| | |
|---|---|
| Bromcresol Purple | 1.6g |
| Ethanol (95% solution) | 100.0mL |

**Preparation of Indicator Solution:** Add Bromcresol Purple to ethanol. Mix thoroughly.

**Preparation of Medium:** Add components to distilled/deionized water and bring volume to 1.0L. Mix thoroughly. Distribute into tubes or flasks. Autoclave for 15 min at 15 psi pressure–121°C.

**Storage/Shelf Life:** Store dehydrated media in the dark in a sealed container below 30°C. Prepared media should be stored under refrigeration (2-8°C). Media should be used within 60 days of preparation. Media should not be used if there are any signs of deterioration (discoloration) or contamination, or if the expiration date supplied by the manufacturer has passed.

**Use:** For the cultivation of salt-tolerant *Streptococcus* species and other salt-tolerant Gram-positive cocci. For the differentiation of Gram-positive cocci based on salt tolerance.

### Salt Tolerance Medium
**Composition** per liter:
| | |
|---|---|
| NaCl | 60.0g |
| Peptone | 5.0g |
| Yeast extract | 2.0g |
| Beef extract | 1.0g |

pH 7.4 ± 0.2 at 25°C

**Preparation of Medium:** Add components to distilled/deionized water and bring volume to 1.0L. Mix thoroughly. Distribute into tubes or flasks. Autoclave for 15 min at 15 psi pressure–121°C.

**Storage/Shelf Life:** Store dehydrated media in the dark in a sealed container below 30°C. Prepared media should be stored under refrigeration (2-8°C). Media should be used within 60 days of preparation. Media should not be used if there are any signs of deterioration (discoloration) or contamination, or if the expiration date supplied by the manufacturer has passed.

**Use:** For the differentiation of *Aeromonas* and *Plesiomonas* species based on salt tolerance.

### SA*Select*™ Medium
**Composition** per liter:
Proprietary

**Source:** This medium is available from Bio-Rad.

**Use:** For the detection of *Staphylococcus aureus* and other *Staphylococcus* species from various clinical specimens: suppurations, wounds, respiratory secretion, nasal swab, and blood culture. *S. aureus* form pink to orange colonies. *S. epidermidis* form white to faint pink small colonies. *S. saprophyticus* form blue to turquoise colonies. *S. intermedius* form purple to grey colonies.

### SBG Enrichment Broth
### (Selenite Brilliant Green
### Enrichment Broth)
**Composition** per liter:
| | |
|---|---|
| D-Mannitol | 5.0g |
| Peptone | 5.0g |
| Yeast extract | 5.0g |
| $Na_2SeO_3 \cdot 5H_2O$ | 4.0g |
| $K_2HPO_4$ | 2.65g |
| $KH_2PO_4$ | 1.02g |
| Sodium taurocholate | 1.0g |
| Brilliant Green | 5.0mg |

pH 7.2 ± 0.2 at 25°C

**Source:** This medium is available as a premixed powder from BD Diagnostic Systems.

**Preparation of Medium:** Add components to distilled/deionized water and bring volume to 1.0L. Mix thoroughly. Gently heat and bring to boiling. Continue boiling for 5–10 min. Do not autoclave. Distribute into sterile tubes or flasks.

**Storage/Shelf Life:** Store dehydrated media in the dark in a sealed container below 30°C. Prepared media should be stored under refrigeration (2-8°C). Media should be used within 60 days of preparation. Media should not be used if there are any signs of deterioration (discoloration) or contamination, or if the expiration date supplied by the manufacturer has passed.

**Use:** For the selective isolation of *Salmonella* species, especially from eggs and egg products.

## SBG Sulfa Enrichment
**Composition** per liter:

| | |
|---|---|
| D-Mannitol | 5.0g |
| Peptone | 5.0g |
| Yeast extract | 5.0g |
| $Na_2SeO_3 \cdot 5H_2O$ | 4.0g |
| $K_2HPO_4$ | 2.65g |
| $KH_2PO_4$ | 1.02g |
| Sodium taurocholate | 1.0g |
| Sodium sulfapyridine | 0.5g |
| Brilliant Green | 5.0mg |

pH 7.2 ± 0.2 at 25°C

**Source:** This medium is available as a premixed powder from BD Diagnostic Systems.

**Preparation of Medium:** Add components to distilled/deionized water and bring volume to 1.0L. Mix thoroughly. Gently heat and bring to boiling. Continue boiling for 5–10 min. Do not autoclave. Distribute into sterile tubes or flasks.

**Storage/Shelf Life:** Store dehydrated media in the dark in a sealed container below 30°C. Prepared media should be stored under refrigeration (2-8°C). Media should be used within 60 days of preparation. Media should not be used if there are any signs of deterioration (discoloration) or contamination, or if the expiration date supplied by the manufacturer has passed.

**Use:** For the selective isolation of *Salmonella* species, especially from eggs and egg products.

## Schaedler HiVeg Agar with Blood
**Composition** per liter:

| | |
|---|---|
| Agar | 15.0g |
| Glucose | 5.83g |
| Plant hydrolysate | 5.67g |
| Plant peptone No. 3 | 5.0g |
| Yeast extract | 5.0g |
| Tris(hydroxymethyl)aminomethane buffer | 3.0g |
| NaCl | 1.67g |
| Papaic digest of soybean meal | 1.0g |
| $K_2HPO_4$ | 0.83g |
| L-Cystine | 0.4g |
| $Fe_4(P_2O_7)_3 \cdot H_2O$ | 0.01g |
| Sheep blood, defibrinated | 50.0mL |

pH 7.6 ± 0.2 at 25°C

**Source:** This medium, without blood, is available as a premixed powder from HiMedia.

**Preparation of Medium:** Add components, except sheep blood, to distilled/deionized water and bring volume to 950.0mL. Mix thoroughly. Gently heat and bring to boiling. Autoclave for 15 min at 15 psi pressure–121°C. Cool to 45°–50°C. Aseptically add 50.0mL of sterile sheep blood. Mix thoroughly.

**Storage/Shelf Life:** Store dehydrated media in the dark in a sealed container below 30°C. Prepared media should be stored under refrigeration (2-8°C). Media should be used within 60 days of preparation. Media should not be used if there are any signs of deterioration (shrinking, cracking, or discoloration) or contamination, or if the expiration date supplied by the manufacturer has passed.

**Use:** For the enumeration of various aerobic and anaerobic bacterial species present in the gastrointestinal tract.

## Schaedler Agar
## (Schaedler Anaerobic Agar)
**Composition** per liter:

| | |
|---|---|
| Agar | 13.5g |
| Glucose | 5.83g |
| Pancreatic digest of casein | 5.7g |
| Proteose peptone No. 3 | 5.0g |
| Yeast extract | 5.0g |
| Tris(hydroxymethyl)aminomethane buffer | 3.0g |
| NaCl | 1.65g |
| Papaic digest of soybean meal | 1.0g |
| $K_2HPO_4$ | 0.83g |
| L-Cystine | 0.4g |
| Hemin | 0.01g |

pH 7.6 ± 0.2 at 25°C

**Source:** This medium is available as a premixed powder from BD Diagnostic Systems and Oxoid Unipath.

**Preparation of Medium:** Add components to distilled/deionized water and bring volume to 1.0L. Mix thoroughly. Gently heat and bring to boiling. Distribute into tubes or flasks. Autoclave for 15 min at 15 psi pressure–121°C. Pour into sterile Petri dishes or leave in tubes.

**Storage/Shelf Life:** Store dehydrated media in the dark in a sealed container below 30°C. Prepared media should be stored under refrigeration (2-8°C). Media should be used within 60 days of preparation. Media should not be used if there are any signs of deterioration (shrinking, cracking, or discoloration) or contamination, or if the expiration date supplied by the manufacturer has passed.

**Use:** For the isolation, cultivation, and enumeration of anaerobic and aerobic microorganisms.

## Schaedler Agar
**Composition** per liter:

| | |
|---|---|
| Agar | 13.5g |
| Pancreatic digest of casein | 8.2g |
| Glucose | 5.8g |
| Yeast extract | 5.0g |
| Tris(hydroxymethyl)aminomethane buffer | 3.0g |
| Peptic digest of animal tissue | 2.5g |
| NaCl | 1.7g |
| Papaic digest of soybean meal | 1.0g |
| $K_2HPO_4$ | 0.8g |

L-Cystine .................................................................0.4g
Hemin..................................................................0.01g

pH 7.6 ± 0.2 at 25°C

**Source:** This medium is available as a premixed powder from BD Diagnostic Systems.

**Preparation of Medium:** Add components to distilled/deionized water and bring volume to 1.0L. Mix thoroughly. Gently heat and bring to boiling. Distribute into tubes or flasks. Autoclave for 15 min at 15 psi pressure–121°C. Pour into sterile Petri dishes or leave in tubes.

**Storage/Shelf Life:** Store dehydrated media in the dark in a sealed container below 30°C. Prepared media should be stored under refrigeration (2-8°C). Media should be used within 60 days of preparation. Media should not be used if there are any signs of deterioration (shrinking, cracking, or discoloration) or contamination, or if the expiration date supplied by the manufacturer has passed.

**Use:** For the isolation, cultivation, and enumeration of anaerobic and aerobic microorganisms.

## Schaedler Agar with Vitamin K₁ and Sheep Blood

**Composition** per liter:

Agar ...............................................................13.5g
Pancreatic digest of casein....................................8.2g
Glucose ..........................................................5.8g
Yeast extract....................................................5.0g
Tris(hydroxymethyl)aminomethane buffer....................3.0g
Peptic digest of animal tissue.................................2.5g
Papaic digest of soybean meal ...............................1.0g
NaCl..............................................................1.7g
K₂HPO₄..........................................................0.8g
L-Cystine ........................................................0.4g
Hemin...........................................................0.01g
Sheep blood, defibrinated ...................................50.0mL
Vitamin K₁ solution ...........................................1.0mL

pH 7.6 ± 0.2 at 25°C

**Vitamin K₁ Solution:**
**Composition** per 10.0mL:
Vitamin K₁ .......................................................5.0g
Ethanol, absolute............................................10.0mL

**Preparation of Vitamin K₁ Solution:** Add vitamin K₁ to ethanol. Mix thoroughly.

**Preparation of Medium:** Add components, except sheep blood, to distilled/deionized water and bring volume to 950.0mL. Mix thoroughly. Gently heat and bring to boiling. Autoclave for 15 min at 15 psi pressure–121°C. Cool to 45°–50°C. Aseptically add sterile sheep blood. Mix thoroughly. Pour into sterile Petri dishes or distribute into sterile tubes.

**Storage/Shelf Life:** Store dehydrated media in the dark in a sealed container below 30°C. Prepared media should be stored under refrigeration (2-8°C). Media should be used within 60 days of preparation. Media should not be used if there are any signs of deterioration (shrinking, cracking, or discoloration) or contamination, or if the expiration date supplied by the manufacturer has passed.

**Use:** For the recovery of fastidious anaerobic bacteria such as *Bacteroides* species.

## Schaedler CNA Agar with Vitamin K₁ and Sheep Blood

**Composition** per liter:

Agar ...............................................................13.5g
Pancreatic digest of casein....................................8.2g
Glucose ..........................................................5.8g
Yeast extract....................................................5.0g
Tris(hydroxymethyl)aminomethane buffer....................3.0g
Peptic digest of animal tissue .................................2.5g
Papaic digest of soybean meal ...............................1.0g
NaCl..............................................................1.7g
K₂HPO₄..........................................................0.8g
L-Cystine ........................................................0.4g
Hemin ...........................................................0.01g
Colistin..........................................................0.01g
Nalidixic acid...................................................0.01g
Sheep blood, defibrinated ...................................50.0mL
Vitamin K₁ solution ...........................................1.0mL

pH 7.6 ± 0.2 at 25°C

**Vitamin K₁ Solution:**
**Composition** per 10.0mL:
Vitamin K₁ .......................................................5.0g
Ethanol, absolute............................................10.0mL

**Preparation of Vitamin K₁ Solution:** Add vitamin K₁ to ethanol. Mix thoroughly.

**Preparation of Medium:** Add components, except sheep blood, to distilled/deionized water and bring volume to 950.0mL. Mix thoroughly. Gently heat and bring to boiling. Autoclave for 15 min at 15 psi pressure–121°C. Cool to 45°–50°C. Aseptically add sterile sheep blood. Mix thoroughly. Pour into sterile Petri dishes or distribute into sterile tubes.

**Storage/Shelf Life:** Store dehydrated media in the dark in a sealed container below 30°C. Prepared media should be stored under refrigeration (2-8°C). Media should be used within 60 days of preparation. Media should not be used if there are any signs of deterioration (shrinking, cracking, or discoloration) or contamination, or if the expiration date supplied by the manufacturer has passed.

**Use:** For the selective isolation of anaerobic, Gram-positive cocci, especially *Peptococcus* species and *Peptostreptococcus* species.

## Schaedler HiVeg Agar with Blood

**Composition** per liter:

Agar ...............................................................15.0g
Glucose ..........................................................5.83g
Plant hydrolysate ..............................................5.67g
Plant peptone No. 3............................................5.0g
Yeast extract....................................................5.0g
Tris(hydroxymethyl)aminomethane buffer....................3.0g
NaCl..............................................................1.67g
Papaic digest of soybean meal ...............................1.0g
K₂HPO₄..........................................................0.83g
L-Cystine ........................................................0.4g
Fe₄(P₂O₇)₃·H₂O................................................0.01g
Sheep blood, defibrinated ...................................50.0mL

pH 7.6 ± 0.2 at 25°C

**Source:** This medium, without blood, is available as a premixed powder from HiMedia.

**Preparation of Medium:** Add components, except sheep blood, to distilled/deionized water and bring volume to 950.0mL. Mix thoroughly. Gently heat and bring to boiling. Autoclave for 15 min at 15 psi pressure–121°C. Cool to 45°–50°C. Aseptically add 50.0mL of sterile sheep blood. Mix thoroughly.

**Storage/Shelf Life:** Store dehydrated media in the dark in a sealed container below 30°C. Prepared media should be stored under refrigeration (2-8°C). Media should be used within 60 days of preparation. Media should not be used if there are any signs of deterioration (shrinking, cracking, or discoloration) or contamination, or if the expiration date supplied by the manufacturer has passed.

**Use:** For the enumeration of various aerobic and anaerobic bacterial species present in the gastrointestinal tract.

## Schaedler HiVeg Broth

**Composition** per liter:

| | |
|---|---|
| Glucose | 5.83g |
| Plant hydrolysate | 5.67g |
| Plant peptone No. 3 | 5.0g |
| Tris(hydroxymethyl)aminomethane buffer | 3.0g |
| Yeast extract | 5.0g |
| NaCl | 1.67g |
| Papaic digest of soybean meal | 1.0g |
| $K_2HPO_4$ | 0.83g |
| L-Cystine | 0.4g |
| $Fe_4(P_2O_7)_3 \cdot H_2O$ | 0.01g |
| Sheep blood, defibrinated | 50.0mL |

pH 7.6 ± 0.2 at 25°C

**Source:** This medium, without blood, is available as a premixed powder from HiMedia.

**Preparation of Medium:** Add components, except sheep blood, to distilled/deionized water and bring volume to 950.0mL. Mix thoroughly. Gently heat and bring to boiling. Autoclave for 15 min at 15 psi pressure–121°C. Cool to 45°–50°C. Aseptically add 50.0mL of sterile sheep blood. Mix thoroughly. Aseptically distribute to tubes or flasks.

**Storage/Shelf Life:** Store dehydrated media in the dark in a sealed container below 30°C. Prepared media should be stored under refrigeration (2-8°C). Media should be used within 60 days of preparation. Media should not be used if there are any signs of deterioration (discoloration) or contamination, or if the expiration date supplied by the manufacturer has passed.

**Use:** For the enumeration of various aerobic and anaerobic bacterial species present in the gastrointestinal tract.

## Schaedler KV Agar
## with Vitamin $K_1$ and Sheep Blood

**Composition** per liter:

| | |
|---|---|
| Agar | 13.5g |
| Pancreatic digest of casein | 8.2g |
| Glucose | 5.8g |
| Yeast extract | 5.0g |
| Tris(hydroxymethyl)aminomethane buffer | 3.0g |
| Peptic digest of animal tissue | 2.5g |
| NaCl | 1.7g |
| Papaic digest of soybean meal | 1.0g |
| $K_2HPO_4$ | 0.8g |
| L-Cystine | 0.4g |
| Hemin | 0.01g |
| Kanamycin | 0.01g |

| | |
|---|---|
| Vancomycin | 7.5mg |
| Sheep blood, defibrinated | 50.0 mL |
| Vitamin $K_1$ solution | 1.0mL |

pH 7.6 ± 0.2 at 25°C

**Vitamin $K_1$ Solution:**
**Composition** per 10.0mL:

| | |
|---|---|
| Vitamin $K_1$ | 5.0g |
| Ethanol, absolute | 10.0mL |

**Preparation of Vitamin $K_1$ Solution:** Add vitamin $K_1$ to ethanol. Mix thoroughly.

**Preparation of Medium:** Add components, except sheep blood, to distilled/deionized water and bring volume to 950.0mL. Mix thoroughly. Gently heat and bring to boiling. Autoclave for 15 min at 15 psi pressure–121°C. Cool to 45°–50°C. Aseptically add sterile sheep blood. Mix thoroughly. Pour into sterile Petri dishes or distribute into sterile tubes.

**Storage/Shelf Life:** Store dehydrated media in the dark in a sealed container below 30°C. Prepared media should be stored under refrigeration (2-8°C). Media should be used within 60 days of preparation. Media should not be used if there are any signs of deterioration (shrinking, cracking, or discoloration) or contamination, or if the expiration date supplied by the manufacturer has passed.

**Use:** For the selective isolation of Gram-negative anaerobic bacteria.

## Schleifer-Krämer Agar
## (SK Agar)

**Composition** per liter:

| | |
|---|---|
| Agar | 13.0g |
| Glycerol | 10.0g |
| Sodium pyruvate | 10.0g |
| Pancreatic digest of casein | 10.0g |
| Beef extract | 5.0g |
| Yeast extract | 3.0g |
| Potassium isothiocyanate | 2.25g |
| LiCl | 2.0g |
| $Na_2HPO_4 \cdot 2H_2O$ | 0.9g |
| $NaH_2PO_4 \cdot H_2O$ | 0.6g |
| Glycine | 0.5g |
| $NaN_3$ solution | 10.0mL |

pH 7.2 ± 0.2 at 25°C

**$NaN_3$ Solution:**
**Composition** per 10.0mL:

| | |
|---|---|
| $NaN_3$ | 0.045g |

**Preparation of $NaN_3$ Solution:** Add $NaN_3$ to distilled/deionized water and bring volume to 10.0mL. Mix thoroughly. Filter sterilize.

**Preparation of Medium:** Add components, except $NaN_3$ solution, to distilled/deionized water and bring volume to 990.0mL. Mix thoroughly. Adjust pH to 7.2. Gently heat and bring to boiling. Autoclave for 15 min at 15 psi pressure–121°C. Cool to 45°–50°C. Aseptically add sterile $NaN_3$ solution. Mix thoroughly. Pour into sterile Petri dishes or distribute into sterile tubes.

**Storage/Shelf Life:** Store dehydrated media in the dark in a sealed container below 30°C. Prepared media should be stored under refrigeration (2-8°C). Media should be used within 60 days of preparation. Media should not be used if there are any signs of deterioration (shrinking, cracking, or discoloration) or contamination, or if the expiration date supplied by the manufacturer has passed.

**Use:** For the isolation and cultivation of *Staphylococcus* species.

## Schuberts Arginine Broth

**Composition** per liter:
Casein enzymic hydrolysate ...........................................17.0g
L-Arginine monohydrochloride ........................................10.0g
NaCl..............................................................................5.0g
Papaic digest of soabean meal .........................................3.0g
D-Glucose......................................................................0.5g
Cresol Red....................................................................0.01g
Bromothymol Blue ....................................................0.0075g
Brilliant Green .......................................................0.00038g

pH 7.0 ± 0.2 at 25°C

**Source:** This medium is available from HiMedia.

**Preparation of Medium:** Add components to distilled/deionized water and bring volume to 1.0L. Mix thoroughly. Gently heat and bring to boiling. Distribute into tubes or flasks. Autoclave for 15 min at 15 psi pressure–121°C. pH 7.2 ± 0.2 at 25°C

**Storage/Shelf Life:** Store dehydrated media in the dark in a sealed container below 30°C. Prepared media should be stored under refrigeration (2-8°C). Media should be used within 60 days of preparation. Media should not be used if there are any signs of deterioration (discoloration) or contamination, or if the expiration date supplied by the manufacturer has passed.

**Use:** For the isolation of chlorine-damaged *Pseudomonas aeruginosa* from swimming pool water.

## SDS HiVeg Agar with Polymyxin B
### (Sodium Dodecyl Sulfate Polymyxin Sucrose HiVeg Agar)

**Composition** per liter:
NaCl.............................................................................20.0g
Agar .............................................................................15.0g
Sucrose.........................................................................15.0g
Plant peptone No. 3.......................................................10.0g
Plant extract ..................................................................5.0g
Sodium dodecyl sulfate...................................................1.0g
Bromthymol Blue .........................................................0.04g
Cresol Red....................................................................0.04g
Polymyxin B solution ................................................. 1.0mL

pH 7.6 ± 0.2 at 25°C

**Source:** This medium, without polymyxin B solution, is available as a premixed powder from HiMedia.

**Polymyxin B Solution:**
**Composition** per 1.0mL:
Polymyxin B .................................................................1.0mg

**Preparation of Polymyxin B Solution:** Add polymyxin B to distilled/deionized water and bring volume to 1.0mL. Mix thoroughly. Filter sterilize.

**Preparation of Medium:** Add components, except polymyxin B solution, to distilled/deionized water and bring volume to 1.0L. Mix thoroughly. Gently heat and bring to boiling. Distribute into tubes or flasks. Autoclave for 15 min at 15 psi pressure–121°C. Cool to 45°–50°C. Aseptically add 1.0mL of polymyxin B solution. Mix thoroughly. Pour into sterile Petri dishes.

**Storage/Shelf Life:** Store dehydrated media in the dark in a sealed container below 30°C. Prepared media should be stored under refrigeration (2-8°C). Media should be used within 60 days of preparation.

Media should not be used if there are any signs of deterioration (shrinking, cracking, or discoloration) or contamination, or if the expiration date supplied by the manufacturer has passed.

**Use:** For the enrichment, isolation, and enumeration of *Vibrio vulnificus* from seafood.

## Selenite Broth
### (Selenite Broth, Lactose)
### (Selenite F Enrichment Medium)
### (Sodium Biselenite Medium)
### (Sodium Hydrogen Selenite Medium)

**Composition** per liter:
Na$_2$HPO$_4$.............................................................................10.0g
Pancreatic digest of casein..............................................5.0g
Lactose..........................................................................4.0g
NaHSeO$_3$·5H$_2$O ...........................................................4.0g

pH 7.0 ± 0.2 at 25°C

**Source:** This medium is available as a premixed powder from BD Diagnostic Systems and a prepared medium from Oxoid Unipath.

**Caution:** Sodium biselenite is toxic and a potential teratogen and care must be taken to avoid inhalation of the powdered dye, contact with the skin, or ingestion, especially in pregnant laboratory workers.

**Preparation of Medium:** Add components to distilled/deionized water and bring volume to 1.0L. Mix thoroughly. Gently heat and bring to boiling. Do not autoclave. Distribute into sterile tubes in 10.0mL volumes.

**Storage/Shelf Life:** Store dehydrated media in the dark in a sealed container below 30°C. Prepared media should be stored under refrigeration (2-8°C). Media should be used within 60 days of preparation. Media should not be used if there are any signs of deterioration (discoloration) or contamination, or if the expiration date supplied by the manufacturer has passed.

**Use:** For the isolation and enrichment of *Salmonella* species from clinical specimens and food products.

## Selenite Broth Base, Mannitol

**Composition** per liter:
Na$_2$HPO$_4$.............................................................................10.0g
Peptone .........................................................................5.0g
Mannitol........................................................................4.0g
NaHSeO$_3$·5H$_2$O ...........................................................4.0g

pH 7.1 ± 0.2 at 25°C

**Source:** This medium is available as a premixed powder from Oxoid Unipath.

**Caution:** Sodium selenite is toxic and a potential teratogen and care must be taken to avoid inhalation of the powdered dye, contact with the skin, or ingestion, especially in pregnant laboratory workers.

**Preparation of Medium:** Add components to distilled/deionized water and bring volume to 1.0L. Mix thoroughly. Gently heat. Do not autoclave. Distribute into sterile tubes in 10.0mL volumes. Sterilize for 10 min at 0 psi pressure–100°C.

**Storage/Shelf Life:** Store dehydrated media in the dark in a sealed container below 30°C. Prepared media should be stored under refrigeration (2-8°C). Media should be used within 60 days of preparation. Media should not be used if there are any signs of deterioration (discoloration) or contamination, or if the expiration date supplied by the manufacturer has passed.

**Use:** For the isolation and cultivation of *Salmonella typhi* and *Salmonella paratyphi*.

## Selenite Cystine Broth

**Composition** per liter:

| | |
|---|---|
| Na$_2$HPO$_4$ | 10.0g |
| Pancreatic digest of casein | 5.0g |
| Lactose | 4.0g |
| Na$_2$SeO$_3$·5H$_2$O | 4.0g |
| L-Cystine | 0.02g |

pH 7.0 ± 0.2 at 25°C

**Source:** This medium is available as a premixed powder from BD Diagnostic Systems and Oxoid Unipath.

**Caution:** Sodium selenite is toxic and a potential teratogen and care must be taken to avoid inhalation of the powdered dye, contact with the skin, or ingestion, especially in pregnant laboratory workers.

**Preparation of Medium:** Add components to distilled/deionized water and bring volume to 1.0L. Mix thoroughly. Gently heat. Do not autoclave. Distribute into sterile tubes in 10.0mL volumes. Sterilize for 15 min at 0 psi pressure–100°C.

**Storage/Shelf Life:** Store dehydrated media in the dark in a sealed container below 30°C. Prepared media should be stored under refrigeration (2-8°C). Media should be used within 60 days of preparation. Media should not be used if there are any signs of deterioration (discoloration) or contamination, or if the expiration date supplied by the manufacturer has passed.

**Use:** For the isolation and cultivation of *Salmonella* species from feces, dairy products, and other specimens.

## Selenite Cystine Broth
## (BAM M134)

**Composition** per liter:

| | |
|---|---|
| Na$_2$HPO$_4$ | 5.5g |
| Polypeptone | 5.0g |
| KH$_2$PO$_4$ | 4.5g |
| Lactose | 4.0g |
| Na$_2$SeO$_3$·5H$_2$O | 4.0g |
| L-Cystine | 0.01g |

pH 7.0 ± 0.2 at 25°C

**Caution:** Sodium selenite is toxic and a potential teratogen and care must be taken to avoid inhalation of the powdered dye, contact with the skin, or ingestion, especially in pregnant laboratory workers.

**Preparation of Medium:** Add components to distilled/deionized water and bring volume to 1.0L. Mix thoroughly. Gently heat. Do not autoclave. Distribute into sterile tubes in 10.0mL volumes. Sterilize for 10 min at 0 psi pressure–100°C in flowing steam.

**Storage/Shelf Life:** Store dehydrated media in the dark in a sealed container below 30°C. Prepared media should be stored under refrigeration (2-8°C). Media should be used within 60 days of preparation. Media should not be used if there are any signs of deterioration (discoloration) or contamination, or if the expiration date supplied by the manufacturer has passed.

**Use:** For the isolation and cultivation of *Salmonella* species from feces, dairy products, and other specimens.

## Selenite F Broth

**Composition** per liter:

| | |
|---|---|
| KH$_2$PO$_4$ | 7.0g |
| Pancreatic digest of casein | 5.0g |
| Lactose | 4.0g |
| Na$_2$SeO$_3$·5H$_2$O | 4.0g |
| Na$_2$HPO$_4$ | 3.0g |

pH 7.0 ± 0.2 at 25°C

**Source:** This medium is available as a premixed powder from BD Diagnostic Systems.

**Caution:** Sodium selenite is toxic and a potential teratogen and care must be taken to avoid inhalation of the powdered dye, contact with the skin, or ingestion, especially in pregnant laboratory workers.

**Preparation of Medium:** Add components to distilled/deionized water and bring volume to 1.0L. Mix thoroughly. Gently heat. Do not autoclave. Distribute into sterile tubes in 10.0mL volumes. Sterilize for 30 min at 0 psi pressure–100°C.

**Storage/Shelf Life:** Store dehydrated media in the dark in a sealed container below 30°C. Prepared media should be stored under refrigeration (2-8°C). Media should be used within 60 days of preparation. Media should not be used if there are any signs of deterioration (discoloration) or contamination, or if the expiration date supplied by the manufacturer has passed.

**Use:** For the isolation and cultivation of *Salmonella* species from feces, dairy products, and other specimens.

## Semisolid IMRV HiVeg Medium Base
## with Novobiocin

**Composition** per liter:

| | |
|---|---|
| Plant hydrolysate | 13.5g |
| Plant peptone | 13.5g |
| MnCl$_2$ | 10.91g |
| Saccharose | 7.5g |
| Agar | 2.7g |
| KH$_2$PO$_4$ | 1.47g |
| Na$_2$S$_2$O$_3$ | 0.8g |
| Lactose | 0.5g |
| FeNH$_4$(SO$_4$)$_2$·12H$_2$O | 0.2g |
| Bromcresol Purple | 0.08g |
| Malachite Green | 0.037g |
| Novobiocin solution | 1.0mL |

pH 5.6 ± 0.2 at 25°C

**Source:** This medium, without novobiocin, is available as a premixed powder from HiMedia.

**Novobiocin Solution:**
**Composition** per 10.0mL:

| | |
|---|---|
| Novobiocin | 0.1g |

**Novobiocin Solution:** Add novobiocin to distilled/deionized water and bring volume to 10.0mL. Mix thoroughly. Filter sterilize.

**Preparation of Medium:** Add components, except novobiocin solutiont, to distilled/deionized water and bring volume to 999.0mL. Mix thoroughly. Gently heat and bring to boiling. Distribute into tubes or flasks. Do not autoclave. Do not overheat. Cool to 50°C. Aseptically add 1.0mL novobiocin solution. Pour into sterile Petri dishes or leave in tubes.

**Storage/Shelf Life:** Store dehydrated media in the dark in a sealed container below 30°C. Prepared media should be stored under refriger-

ation (2-8°C). Media should be used within 60 days of preparation. Media should not be used if there are any signs of deterioration (discoloration) or contamination, or if the expiration date supplied by the manufacturer has passed.

**Use:** For the isolation of motile *Salmonella* spp. from foods

## Semisolid RV HiVeg Medium Base with Novobiocin

**Composition** per liter:

| | |
|---|---|
| MnCl₂, anhydrous | 10.93g |
| NaCl | 7.34g |
| Plant hydrolysate | 4.6g |
| Plant hydrolysate No. 1 | 4.6g |
| Agar | 2.7g |
| Malachite Green | 0.037g |
| Novobiocin solution | 1.0mL |

pH 5.4 ± 0.2 at 25°C

**Source:** This medium, without novobiocin, is available as a premixed powder from HiMedia.

**Novobiocin Solution:**

**Composition** per 10.0mL:

| | |
|---|---|
| Novobiocin | 0.1g |

**Preparation of Novobiocin Solution:** Add novobiocin to distilled/deionized water and bring volume to 10.0mL. Mix thoroughly. Filter sterilize.

**Preparation of Medium:** Add components, except novobiocin solution, to distilled/deionized water and bring volume to 999.0mL. Mix thoroughly. Gently heat and bring to boiling. Distribute into tubes or flasks. Do not autoclave. Do not overheat. Cool to 50°C. Aseptically add 1.0mL novobiocin solution. Pour into sterile Petri dishes or leave in tubes.

**Storage/Shelf Life:** Store dehydrated media in the dark in a sealed container below 30°C. Prepared media should be stored under refrigeration (2-8°C). Media should be used within 60 days of preparation. Media should not be used if there are any signs of deterioration (discoloration) or contamination, or if the expiration date supplied by the manufacturer has passed.

**Use:** For the isolation of motile *Salmonella* spp. from foods

## Sensitest Agar

**Composition** per liter:

| | |
|---|---|
| Pancreatic digest of casein | 11.0g |
| Agar | 8.0g |
| Buffer salts | 3.3g |
| Peptone | 3.0g |
| NaCl | 3.0g |
| Glucose | 2.0g |
| Starch | 1.0g |
| Nucleoside bases | 0.02g |
| Thiamine | 0.02mg |

pH 7.4 ± 0.2 at 25°C

**Source:** This medium is available as a premixed powder from Oxoid Unipath.

**Preparation of Medium:** Add components to distilled/deionized water and bring volume to 1.0L. Mix thoroughly. Gently heat and bring to boiling. Distribute into tubes or flasks. Autoclave for 15 min at 15 psi pressure–121°C. Pour into sterile Petri dishes.

**Storage/Shelf Life:** Store dehydrated media in the dark in a sealed container below 30°C. Prepared media should be stored under refrigeration (2-8°C). Media should be used within 60 days of preparation. Media should not be used if there are any signs of deterioration (shrinking, cracking, or discoloration) or contamination, or if the expiration date supplied by the manufacturer has passed.

**Use:** For the performance of antibiotic sensitivity assays.

## Sensitivity Test HiVeg Medium with Blood Serum

**Composition** per liter:

| | |
|---|---|
| Agar | 15.0g |
| Glucose | 10.0g |
| Veal, infusion from | 10.0g |
| Plant peptone No. 3 | 10.0g |
| NaCl | 3.0g |
| Na₂HPO₄ | 2.0g |
| Sodium acetate | 1.0g |
| Guanine | 0.01g |
| Uracil | 0.01g |
| Xanthine | 0.01g |
| Adenine sulfate | 0.01g |
| Sterile bovine or sheep blood | 50.0mL |

pH 7.3 ± 0.2 at 25°C

**Source:** This medium, without blood, is available as a premixed powder from HiMedia.

**Preparation of Medium:** Add components, except blood, to distilled/deionized water and bring volume to 950.0mL. Mix thoroughly. Gently heat and bring to boiling. Autoclave for 15 min at 15 psi pressure–121°C. Cool to 45°–50°C. Aseptically add 50.0mL of sterile bovine blood or sheep blood. Mix thoroughly. Pour into sterile Petri dishes.

**Storage/Shelf Life:** Store dehydrated media in the dark in a sealed container below 30°C. Prepared media should be stored under refrigeration (2-8°C). Media should be used within 60 days of preparation. Media should not be used if there are any signs of deterioration (shrinking, cracking, or discoloration) or contamination, or if the expiration date supplied by the manufacturer has passed.

**Use:** For antimicroibal sensitivity testing of sulfonamides and other antimicrobics.

## *Serratia* Differential Medium (SD Medium)

**Composition** per 102.0mL:

| | |
|---|---|
| Solution A | 92.0mL |
| Solution B | 10.0mL |

pH 6.7 ± 0.2 at 25°C

**Solution A:**

**Composition** per 92.0mL:

| | |
|---|---|
| Yeast extract | 1.0g |
| L-Ornithine | 1.0g |
| NaCl | 0.5g |
| Agar | 0.4g |
| Irgasan inhibitor | 1.0mL |
| Indicator solution | 1.0mL |

**Preparation of Solution A:** Add components to distilled/deionized water and bring volume to 92.0mL. Mix thoroughly. Adjust pH to 6.7 with 1*N* NaOH.

## Irgasan Inhibitor:

**Composition** per 100.0mL:

Irgasan-DP-300 (4,2′, 4′-trichloro-2-hydroxydiphenylether).........0.1g
NaOH (1*N* solution)................................................................. 10.0mL

**Preparation of Irgasan Inhibitor:** Add irgasan to 10.0mL of NaOH solution. Mix thoroughly. Gently heat to dissolve. Bring volume to 100.0mL with distilled/deionized water.

## Indicator Solution:

**Composition** per 100.0mL:

Bromthymol Blue ........................................................................0.2g
Phenol Red.................................................................................0.1g

**Preparation of Indicator Solution:** Add components to 50.0mL of distilled/deionized water. Mix thoroughly for 1 h. Bring volume to 100.0mL with distilled/deionized water.

## Solution B:

**Composition** per 10.0mL:

L-Arabinose ...............................................................................1.0g

**Preparation of Solution B:** Add L-arabinose to distilled/deionized water and bring volume to 10.0mL. Mix thoroughly.

**Preparation of Medium:** Combine 92.0mL of solution A with 10.0mL of solution B. Mix thoroughly. Distribute into tubes. Autoclave for 15 min at 15 psi pressure–121°C. Allow tubes to cool in an upright position.

**Storage/Shelf Life:** Store dehydrated media in the dark in a sealed container below 30°C. Prepared media should be stored under refrigeration (2-8°C). Media should be used within 60 days of preparation. Media should not be used if there are any signs of deterioration (discoloration) or contamination, or if the expiration date supplied by the manufacturer has passed.

**Use:** For the cultivation and differentiation of *Serratia* species based on the fermentation of arabinose and production of ornithine decarboxylase. *Serratia marcescens* changes the medium to purple throughout the tube. *Serratia liquefaciens* changes the medium to a band of purple at the top of the tube with a green/yellow butt. *Serratia rubidaea* changes the medium to yellow throughout the tube.

## Serum Glucose Agar, Farrell Modified

**Composition** per 1086.9mL:

Agar .........................................................................................15.0g
Peptone....................................................................................10.0g
Beef extract ..............................................................................5.0g
NaCl ..........................................................................................5.0g
Serum-glucose solution...........................................................60.0mL
Bacitracin solution ..................................................................12.5mL
Cycloheximide solution ...........................................................10.0mL
Nystatin solution .......................................................................2.0mL
Polymyxin B solution ................................................................1.0mL
Nalidixic acid solution ..............................................................1.0mL
Vancomycin solution.................................................................0.4mL

pH 7.3 ± 0.2 at 25°C

## Serum-Glucose Solution:

**Composition** per 60.0mL:

D-Glucose ................................................................................10.0g
Serum (inactivated at 56°C, 30 min) .......................................50.0mL

**Preparation of Serum-Glucose Solution:** Add glucose to 50.0mL of heat-inactivated serum. Horse serum or ox serum may be used. Mix thoroughly. Filter sterilize.

## Bacitracin Solution:

**Composition** per 12.5mL:

Bacitracin..............................................................................25,000U

**Preparation of Bacitracin Solution:** Add Bacitracin to distilled/deionized water and bring volume to 12.5mL. Mix thoroughly. Filter sterilize.

## Cycloheximide Solution:

**Composition** per 100.0mL:

Cycloheximide...........................................................................1.0g
Acetone .....................................................................................5.0mL

**Preparation of Cycloheximide Solution:** Add cycloheximide to 5.0mL of acetone. Mix thoroughly. Bring volume to 100.0mL with distilled/deionized water. Mix thoroughly. Filter sterilize.

## Nystatin Solution:

**Composition** per 5.0mL:

Nystatin.................................................................................250,000U

**Preparation of Nystatin Solution:** Add nystatin to distilled/deionized water and bring volume to 5.0mL. Mix thoroughly. Filter sterilize.

## Polymyxin B Solution:

**Composition** per 2.0mL:

Polymyxin B ..........................................................................10,000U

**Preparation of Polymyxin B Solution:** Add polymyxin B to distilled/deionized water and bring volume to 2.0mL. Mix thoroughly. Filter sterilize.

## Nalidixic Acid Solution:

**Composition** per 2.0mL:

Nalidixic acid.............................................................................0.1g
NaOH (0.5*N* solution)...............................................................2.0mL

**Preparation of Nalidixic Acid Solution:** Add nalidixic acid to 2.0mL of NaOH solution. Mix thoroughly. Immediately before use, add 1.0mL of this stock solution to 9.0mL of distilled/deionized water. Mix thoroughly. Filter sterilize.

## Vancomycin Solution:

**Composition** per 1.0mL:

Vancomycin .............................................................................0.05g

**Preparation of Vancomycin Solution:** Add vancomycin to distilled/deionized water and bring volume to 1.0mL. Mix thoroughly. Filter sterilize.

**Preparation of Medium:** Add components—except serum-glucose solution, bacitracin solution, cycloheximide solution, nystatin solution, polymyxin B solution, nalidixic acid solution, and vancomycin solution—to distilled/deionized water and bring volume to 1.0L. Mix thoroughly. Gently heat and bring to boiling. Autoclave for 15 min at 10 psi pressure–115°C. Cool to 50°C. Aseptically add 60.0mL of sterile serum-glucose solution, 12.5mL of sterile bacitracin solution, 10.0mL of sterile cycloheximide solution, 2.0mL of sterile nystatin solution, 1.0mL of sterile polymyxin B solution, 1.0mL of sterile nalidixic acid solution, and 0.4mL of sterile vancomycin solution. Mix thoroughly. Pour into sterile Petri dishes or distribute into sterile tubes. Allow tubes to cool in a slanted position.

**Storage/Shelf Life:** Store dehydrated media in the dark in a sealed container below 30°C. Prepared media should be stored under refrigeration (2-8°C). Media should be used within 60 days of preparation. Media should not be used if there are any signs of deterioration (shrinking, cracking, or discoloration) or contamination, or if the expiration date supplied by the manufacturer has passed.

**Use:** For the selective isolation and cultivation of *Brucella* species.

## Serum Tellurite Agar

**Composition** per liter:

| | |
|---|---|
| Agar | 20.0g |
| Pancreatic digest of casein | 10.0g |
| Peptic digest of animal tissue | 10.0g |
| NaCl | 5.0g |
| Glucose | 2.0g |
| Lamb serum | 50.0mL |
| Chapman tellurite solution | 10.0mL |

pH 7.5 ± 0.2 at 25°C

**Source:** This medium is available as a premixed powder from BD Diagnostic Systems.

### Chapman Tellurite Solution:

**Composition** per 100.0mL:

| | |
|---|---|
| $K_2TeO_3$ | 1.0g |

**Preparation of Chapman Tellurite Solution:** Add $K_2TeO_3$ to distilled/deionized water and bring volume to 100.0mL. Mix thoroughly. Filter sterilize.

**Preparation of Medium:** Add components, except lamb serum and Chapman tellurite solution, to distilled/deionized water and bring volume to 940.0mL. Mix thoroughly. Gently heat and bring to boiling. Autoclave for 15 min at 15 psi pressure–121°C. Cool to 45°–50°C. Aseptically add sterile lamb serum and 10.0mL of sterile Chapman tellurite solution. Mix thoroughly. Pour into sterile Petri dishes or distribute into sterile tubes.

**Storage/Shelf Life:** Store dehydrated media in the dark in a sealed container below 30°C. Prepared media should be stored under refrigeration (2-8°C). Media should be used within 60 days of preparation. Media should not be used if there are any signs of deterioration (shrinking, cracking, or discoloration) or contamination, or if the expiration date supplied by the manufacturer has passed.

**Use:** For the isolation and cultivation of *Corynebacterium* species, especially in the laboratory diagnosis of diphtheria.

## Seven H11 Agar
## (Selective 7H11 Agar)

**Composition** per 1010.0mL:

| | |
|---|---|
| Agar | 13.5g |
| $KH_2PO_4$ | 1.5g |
| $Na_2HPO_4$ | 1.5g |
| Pancreatic digest of casein | 1.0g |
| NaCl | 0.85g |
| Monosodium glutamate | 0.5g |
| $(NH_4)_2SO_4$ | 0.5g |
| Sodium citrate | 0.4g |
| $MgSO_4 \cdot 7H_2O$ | 0.05g |
| Ferric ammonium citrate | 0.04g |
| $CuSO_4 \cdot 5H_2O$ | 1.0mg |
| Pyridoxine | 1.0mg |
| $ZnSO_4 \cdot 7H_2O$ | 1.0mg |
| Biotin | 0.5mg |
| $CaCl_2 \cdot 2H_2O$ | 0.5mg |
| Malachite Green | 0.25mg |
| Middlebrook OADC enrichment | 100.0mL |

| | |
|---|---|
| Antibiotic inhibitor | 10.0mL |
| Glycerol | 5.0mL |

pH 6.6 ± 0.2 at 25°C

**Source:** This medium is available as a prepared medium from BD Diagnostic Systems.

### Middlebrook OADC Enrichment:

**Composition** per liter:

| | |
|---|---|
| Bovine albumin fraction V | 5.0g |
| Glucose | 2.0g |
| NaCl | 0.85g |
| Catalase | 3.0mg |
| Oleic acid | 0.06mL |

**Preparation of Middlebrook OADC Enrichment:** Add components to distilled/deionized water and bring volume to 100.0mL. Mix thoroughly. Filter sterilize.

### Antibiotic Inhibitor:

**Composition** per 10.0mL:

| | |
|---|---|
| Carbenicillin | 0.05g |
| Trimethoprim lactate | 0.02g |
| Amphotericin B | 0.01g |
| Polymyxin B | 200,000U |

**Preparation of Antibiotic Inhibitor:** Add components to distilled/deionized water and bring volume to 10.0mL. Mix thoroughly. Filter sterilize.

**Preparation of Medium:** Add glycerol to 900.0mL of distilled/deionized water. Mix thoroughly. Add remaining components, except Middlebrook OADC enrichment and antibiotic inhibitor. Mix thoroughly. Gently heat. Do not boil. Autoclave for 10 min at 15 psi pressure–121°C. Cool to 50°–55°C. Aseptically add 100.0mL of sterile Middlebrook OADC enrichment and 10.0mL of sterile antibiotic solution. Mix thoroughly. Pour into sterile Petri dishes or distribute into sterile tubes.

**Storage/Shelf Life:** Store dehydrated media in the dark in a sealed container below 30°C. Prepared media should be stored under refrigeration (2-8°C). Media should be used within 60 days of preparation. Media should not be used if there are any signs of deterioration (shrinking, cracking, or discoloration) or contamination, or if the expiration date supplied by the manufacturer has passed.

**Use:** For the isolation and cultivation of *Mycobacterium* species from specimens with a mixed flora.

## Seven-Hour Fecal Coliform Agar
## (Seven-Hour FC Agar)
## (m-Seven-Hour Fecal Coliform Agar)

**Composition** per liter:

| | |
|---|---|
| Agar | 15.0g |
| Lactose | 10.0g |
| NaCl | 7.5g |
| D-Mannitol | 5.0g |
| Proteose peptone No. 3 | 5.0g |
| Yeast extract | 3.0g |
| Bromcresol Purple | 0.35g |
| Phenol Red | 0.3g |
| Sodium lauryl sulfate | 0.2g |
| Sodium deoxycholate | 0.1g |

pH 7.3 ± 0.1 at 25°C

**Preparation of Medium:** Add components to distilled/deionized water and bring volume to 1.0L. Mix thoroughly. Gently heat and bring

to boiling. Continue boiling for 5 min. Cool to 55°–60°C. Adjust pH to 7.3 with 0.1$N$ NaOH. Cool to 45°–50°C. Pour into sterile Petri dishes with tight-fitting lids in 5.0mL volumes. Store at 2°–10°C.

**Storage/Shelf Life:** Store dehydrated media in the dark in a sealed container below 30°C. Prepared media should be stored under refrigeration (2-8°C). Media should be used within 60 days of preparation. Media should not be used if there are any signs of deterioration (shrinking, cracking, or discoloration) or contamination, or if the expiration date supplied by the manufacturer has passed.

**Use:** For the rapid estimation of the bacteriological quality of water using the membrane filter method.

## SF Broth
### (*Streptococcus faecalis* Broth)
**Composition** per liter:
| | |
|---|---|
| Pancreatic digest of casein | 20.0g |
| Glucose | 5.0g |
| NaCl | 5.0g |
| K$_2$HPO$_4$ | 4.0g |
| KH$_2$PO$_4$ | 1.5g |
| NaN$_3$ | 0.5g |
| Bromcresol Purple | 0.032g |

pH 6.9 ± 0.2 at 25°C

**Source:** This medium is available as a premixed powder from BD Diagnostic Systems.

**Preparation of Medium:** Add components to distilled/deionized water and bring volume to 1.0L. Mix thoroughly. Distribute into tubes or flasks. Autoclave for 15 min at 15 psi pressure–121°C.

**Storage/Shelf Life:** Store dehydrated media in the dark in a sealed container below 30°C. Prepared media should be stored under refrigeration (2-8°C). Media should be used within 60 days of preparation. Media should not be used if there are any signs of deterioration (discoloration) or contamination, or if the expiration date supplied by the manufacturer has passed.

**Use:** For the cultivation and differentiation of group D enterococci (*Streptococcus faecalis* and *Streptococcus faecium*) from group D non-enterococci and from other *Streptococcus* species. Group D enterococci turn the medium turbid and yellow-brown.

## SF HiVeg Broth
**Composition** per liter:
| | |
|---|---|
| Plant hydrolysate | 20.0g |
| Glucose | 5.0g |
| NaCl | 5.0g |
| K$_2$HPO$_4$ | 4.0g |
| KH$_2$PO$_4$ | 1.5g |
| NaN$_3$ | 0.5g |
| Bromcresol Purple | 0.032 |

pH 6.9 ± 0.2 at 25°C

**Source:** This medium is available as a premixed powder from Hi-Media.

**Preparation of Medium:** Add components to distilled/deionized water and bring volume to 1.0L. Mix thoroughly. Distribute into tubes or flasks. Autoclave for 15 min at 15 psi pressure–121°C.

**Storage/Shelf Life:** Store dehydrated media in the dark in a sealed container below 30°C. Prepared media should be stored under refrigeration (2-8°C). Media should be used within 60 days of preparation. Media should not be used if there are any signs of deterioration (discol-

oration) or contamination, or if the expiration date supplied by the manufacturer has passed.

**Use:** For the cultivation and differentiation of group D enterococci (*Streptococcus faecalis* and *Streptococcus faecium*) from group D non-enterococci and from other *Streptococcus* species. Group D enterococci turn the medium turbid and yellow-brown.

## SFP Agar
### (Shahidi-Ferguson Perfringens Agar)
**Composition** per 2020.0mL:
| | |
|---|---|
| Basal layer | 1010.0mL |
| Cover layer | 1010.0mL |

**Source:** This medium is available as a premixed powder from BD Diagnostic Systems and Oxoid Unipath.

**Basal Layer:**
**Composition** per 1010.0mL:
| | |
|---|---|
| Agar | 20.0g |
| Tryptose | 15.0g |
| Papaic digest of soybean meal | 5.0g |
| Yeast extract | 5.0g |
| Ferric ammonium citrate | 1.0g |
| NaHSO$_3$ | 1.0g |
| Egg yolk emulsion, 50% | 100.0mL |
| Antibiotic inhibitor | 10.0mL |

pH 7.6 ± 0.2 at 25°C

**Egg Yolk Emulsion, 50%:**
**Composition** per 100.0mL:
| | |
|---|---|
| Chicken egg yolks | 11 |
| Whole chicken egg | 1 |
| NaCl (0.9% solution) | 50.0mL |

**Preparation of Egg Yolk Emulsion, 50%:** Soak eggs with 1:100 dilution of saturated mercuric chloride solution for 1 min. Crack eggs and separate yolks from whites. Mix egg yolks with 1 chicken egg. Beat to form emulsion. Measure 50.0mL of egg yolk emulsion and add to 50.0mL of 0.9% NaCl solution. Mix thoroughly. Filter sterilize. Warm to 45°–50°C.

**Antibiotic Inhibitor:**
**Composition** per 10.0mL:
| | |
|---|---|
| Kanamycin | 0.012g |
| Polymyxin B sulfate | 30,000U |

**Preparation of Antibiotic Inhibitor:** Add components to distilled/deionized water and bring volume to 10.0mL. Mix thoroughly. Filter sterilize.

**Preparation of Basal Layer:** Add components—except egg yolk emulsion, 50%, and antibiotic inhibitor—to distilled/deionized water and bring volume to 990.0mL. Mix thoroughly. Gently heat and bring to boiling. Autoclave for 15 min at 15 psi pressure–121°C. Cool to 45°–50°C. Aseptically add sterile egg yolk emulsion, 50%, and antibiotic inhibitor. Mix thoroughly. Pour into sterile Petri dishes in 10.0mL volumes.

**Cover Layer:**
**Composition** per 1010.0mL:
| | |
|---|---|
| Agar | 20.0g |
| Tryptose | 15.0g |
| Papaic digest of soybean meal | 5.0g |
| Yeast extract | 5.0g |
| Ferric ammonium citrate | 1.0g |

NaHSO$_3$ ................................................................1.0g
Antibiotic inhibitor ........................................... 10.0mL

pH 7.6 ± 0.2 at 25°C

**Preparation of Cover Layer:** Add components—except antibiotic inhibitor—to distilled/deionized water and bring volume to 1.0L. Mix thoroughly. Gently heat and bring to boiling. Autoclave for 15 min at 15 psi pressure–121°C. Cool to 45°–50°C. Aseptically add sterile antibiotic inhibitor. Mix thoroughly.

**Preparation of Medium:** Prepare and dispense basal layer into sterile Petri dishes in 10.0mL volumes. Incubate overnight to dry plates and test for sterility. Inoculate plates using 0.1mL volume. Spread inoculum over suface of agar. Aseptically add 10.0mL of cover layer to each plate. Incubate at 37°C under 90% N$_2$ + 10% CO$_2$.

**Storage/Shelf Life:** Store dehydrated media in the dark in a sealed container below 30°C. Prepared media should be stored under refrigeration (2-8°C). Media should be used within 60 days of preparation. Media should not be used if there are any signs of deterioration (shrinking, cracking, or discoloration) or contamination, or if the expiration date supplied by the manufacturer has passed.

**Use:** For the isolation and enumeration of *Clostridium perfringens* from foods. *Clostridium perfringens* appears as black colonies surrounded by a precipitate.

## S.F.P. HiVeg Agar Base with Egg Yolk and Antibiotics

**Composition** per liter:
Basal layer................................................................500.0mL
Cover layer...............................................................500.0mL

**Basal Layer:**
**Composition** per liter:
Agar ......................................................................20.0g
Plant hydrolysate No. 1.................................................15.0g
Yeast extract............................................................5.0g
Papaic digest of soybean meal ..........................................5.0g
NaHSO$_3$ ...................................................................1.0g
Ferric ammonium citrate...................................................1.0g
Egg yolk emulsion, 50%....................................................100.0mL
Antibiotic inhibitor .......................................................10.0mL

pH 7.6 ± 0.2 at 25°C

**Source:** This medium, without egg yolk emulsion and antibiotic inhibitor, is available as a premixed powder from HiMedia.

**Egg Yolk Emulsion, 50%:**
**Composition** per 100.0mL:
Chicken egg yolks............................................................11
Whole chicken egg.............................................................1
NaCl (0.9% solution) .........................................................50.0mL

**Preparation of Egg Yolk Emulsion, 50%:** Soak eggs with 1:100 dilution of saturated mercuric chloride solution for 1 min. Crack eggs and separate yolks from whites. Mix egg yolks with 1 chicken egg. Beat to form emulsion. Measure 50.0mL of egg yolk emulsion and add to 50.0mL of 0.9% NaCl solution. Mix thoroughly. Filter sterilize. Warm to 45°–50°C.

**Antibiotic Inhibitor:**
**Composition** per 10.0mL:
Kanamycin ...................................................................0.012g
Polymyxin B sulfate..........................................................30,000U

**Preparation of Antibiotic Inhibitor:** Add components to distilled/deionized water and bring volume to 10.0mL. Mix thoroughly. Filter sterilize.

**Preparation of Basal Layer:** Add components—except egg yolk emulsion, 50%, and antibiotic inhibitor—to distilled/deionized water and bring volume to 990.0mL. Mix thoroughly. Gently heat and bring to boiling. Autoclave for 15 min at 15 psi pressure–121°C. Cool to 45°–50°C. Aseptically add sterile egg yolk emulsion, 50%, and antibiotic inhibitor. Mix thoroughly. Pour into sterile Petri dishes in 10.0mL volumes.

**Cover Layer:**
**Composition** per 1010.0mL:
Agar ......................................................................20.0g
Tryptose .................................................................15.0g
Papaic digest of soybean meal ..........................................5.0g
Yeast extract............................................................5.0g
Ferric ammonium citrate.................................................1.0g
NaHSO$_3$ ...................................................................1.0g
Antibiotic inhibitor .......................................................10.0mL

pH 7.6 ± 0.2 at 25°C

**Antibiotic Inhibitor:**
**Composition** per 10.0mL:
Kanamycin..................................................................0.012g
Polymyxin B sulfate .......................................................30,000U

**Preparation of Antibiotic Inhibitor:** Add components to distilled/deionized water and bring volume to 10.0mL. Mix thoroughly. Filter sterilize.

**Preparation of Cover Layer:** Add components—except antibiotic inhibitor—to distilled/deionized water and bring volume to 1.0L. Mix thoroughly. Gently heat and bring to boiling. Autoclave for 15 min at 15 psi pressure–121°C. Cool to 45°–50°C. Aseptically add sterile antibiotic inhibitor. Mix thoroughly.

**Preparation of Medium:** Prepare and dispense basal layer into sterile Petri dishes in 10.0mL volumes. Incubate overnight to dry plates and test for sterility. Inoculate plates using 0.1mL volume. Spread inoculum over suface of agar. Aseptically add 10.0mL of cover layer to each plate. Incubate at 37°C under 90% N$_2$ + 10% CO$_2$.

**Storage/Shelf Life:** Store dehydrated media in the dark in a sealed container below 30°C. Prepared media should be stored under refrigeration (2-8°C). Media should be used within 60 days of preparation. Media should not be used if there are any signs of deterioration (shrinking, cracking, or discoloration) or contamination, or if the expiration date supplied by the manufacturer has passed.

**Use:** For the isolation and enumeration of *Clostridium perfringens* from foods. *Clostridium perfringens* appears as black colonies surrounded by a precipitate.

## Sheep Blood Agar (BAM M135)

**Composition** per liter:
Proteose peptone ........................................................15.0g
Agar ......................................................................12.0g
Liver digest ..............................................................2.5g
Yeast extract............................................................5.0g
NaCl......................................................................5.0g
Sheep blood, defibrinated ...............................................50.0mL

pH 7.3 ± 0.2 at 25°C

**Source:** This medium is available as a premixed powder from Oxoid Unipath.

**Preparation of Medium:** Add components, except sheep blood, to distilled/deionized water and bring volume to 950.0mL. Mix thoroughly. Heat with frequent agitation and boil for 1 min to completely dissolve. Autoclave for 15 min at 15 psi pressure–121°C. Cool to 45°–46°C. Aseptically add 50.0mL of sterile, defibrinated sheep blood. Mix thoroughly and pour into sterile Petri dishes.

**Storage/Shelf Life:** Store dehydrated media in the dark in a sealed container below 30°C. Prepared media should be stored under refrigeration (2-8°C). Media should be used within 60 days of preparation. Media should not be used if there are any signs of deterioration (shrinking, cracking, or discoloration) or contamination, or if the expiration date supplied by the manufacturer has passed.

**Use:** For the isolation, cultivation, and detection of hemolytic activity of streptococci and other fastidious microorganisms.

### Sheep Blood Agar

**Composition** per liter:

| | |
|---|---|
| Tryptone | 14.0g |
| Agar | 12.0g |
| NaCl | 5.0g |
| Peptone, neutralized | 4.5g |
| Yeast extract | 4.5g |
| Sheep blood, defibrinated | 70.0mL |

pH 7.4 ± 0.2 at 25°C

**Source:** This medium is available as a premixed powder from Oxoid Unipath.

**Preparation of Medium:** Add components, except sheep blood, to distilled/deionized water and bring volume to 930.0mL. Mix thoroughly. Heat with frequent agitation and boil for 1 min to completely dissolve. Autoclave for 15 min at 15 psi pressure–121°C. Cool to 50°C. Aseptically add 70.0mL of sterile, defibrinated sheep blood. Mix thoroughly and pour into sterile Petri dishes.

**Storage/Shelf Life:** Store dehydrated media in the dark in a sealed container below 30°C. Prepared media should be stored under refrigeration (2-8°C). Media should be used within 60 days of preparation. Media should not be used if there are any signs of deterioration (shrinking, cracking, or discoloration) or contamination, or if the expiration date supplied by the manufacturer has passed.

**Use:** For the isolation, cultivation, and detection of hemolytic activity of streptococci and other fastidious microorganisms. Specifically formulated to give maximum recovery and improved hemolytic reactions with sheep blood.

### *Shigella* Broth

**Composition** per liter:

| | |
|---|---|
| Pancreatic digest of casein | 20.0g |
| NaCl | 5.0g |
| $K_2HPO_4$ | 2.0g |
| $KH_2PO_4$ | 2.0g |
| Glucose | 1.0g |
| Novobiocin solution | 11.1mL |
| Tween™ 80 | 1.5mL |

pH 7.0 ± 0.2 at 25°C

**Novobiocin Solution:**
**Composition** per liter:

| | |
|---|---|
| Novobiocin | 0.05g |

**Preparation of Novobiocin Solution:** Add novobiocin to distilled/deionized water and bring volume to 1.0L. Mix thoroughly. Filter sterilize.

**Preparation of Medium:** Add components, except novobiocin solution, to distilled/deionized water and bring volume to 988.9mL. Mix thoroughly. Gently heat and bring to boiling. Autoclave for 15 min at 15 psi pressure–121°C. Cool to 45°–50°C. Aseptically add sterile novobiocin solution. Mix thoroughly. Aseptically distribute into sterile tubes.

**Storage/Shelf Life:** Store dehydrated media in the dark in a sealed container below 30°C. Prepared media should be stored under refrigeration (2-8°C). Media should be used within 60 days of preparation. Media should not be used if there are any signs of deterioration (discoloration) or contamination, or if the expiration date supplied by the manufacturer has passed.

**Use:** For the isolation and cultivation of *Shigella* species from food.

### *Shigella* HiVeg Broth Base with Novobiocin

**Composition** per liter:

| | |
|---|---|
| Plant hydrolysate | 20.0g |
| NaCl | 5.0g |
| $K_2HPO_4$ | 2.0g |
| $KH_2PO_4$ | 2.0g |
| Glucose | 1.0g |
| Polysorbate 80 | 1.5 ml |
| Novobiocin solution | 11.1mL |

pH 7.0 ± 0.2 at 25°C

**Source:** This medium, without novobiocin, is available as a premixed powder from HiMedia.

**Novobiocin Solution:**
**Composition** per liter:

| | |
|---|---|
| Novobiocin | 0.05g |

**Preparation of Novobiocin Solution:** Add novobiocin to distilled/deionized water and bring volume to 1.0L. Mix thoroughly. Filter sterilize.

**Preparation of Medium:** Add components, except novobiocin solution, to distilled/deionized water and bring volume to 988.9mL. Mix thoroughly. Gently heat and bring to boiling. Autoclave for 15 min at 15 psi pressure–121°C. Cool to 45°–50°C. Aseptically add sterile novobiocin solution. Mix thoroughly. Aseptically distribute into sterile tubes.

**Storage/Shelf Life:** Store dehydrated media in the dark in a sealed container below 30°C. Prepared media should be stored under refrigeration (2-8°C). Media should be used within 60 days of preparation. Media should not be used if there are any signs of deterioration (discoloration) or contamination, or if the expiration date supplied by the manufacturer has passed.

**Use:** For the isolation and cultivation of *Shigella* species from food.

### SIM HiVeg Medium

**Composition** per liter:

| | |
|---|---|
| Plant peptone | 30.0g |
| Agar | 3.0g |
| Plant extract | 3.0g |

HiVeg peptonized iron ...................................................................0.2g

Na$_2$S$_2$O$_3$ ...................................................................................0.025g

pH 7.3 ± 0.2 at 25°C

**Source:** This medium is available as a premixed powder from Hi-Media.

**Preparation of Medium:** Add components to distilled/deionized water and bring volume to 1.0L. Mix thoroughly. Gently heat and bring to boiling. Distribute into tubes in 15.0mL volumes. Autoclave for 15 min at 15 psi pressure–121°C. Allow tubes to cool in an upright position.

**Storage/Shelf Life:** Store dehydrated media in the dark in a sealed container below 30°C. Prepared media should be stored under refrigeration (2-8°C). Media should be used within 60 days of preparation. Media should not be used if there are any signs of deterioration (discoloration) or contamination, or if the expiration date supplied by the manufacturer has passed.

**Use:** For the determination of hydrogen sulfide production, indole formation, and motility of enteric bacilli.

## SIM Medium

**Composition** per liter:

Peptone.......................................................................................30.0g

Agar ...........................................................................................3.0g

Beef extract.................................................................................3.0g

Peptonized iron ..........................................................................0.2g

Na$_2$S$_2$O$_3$·5H$_2$O .....................................................................0.025g

pH 7.3 ± 0.2 at 25°C

**Source:** This medium is available as a premixed powder from BD Diagnostic Systems.

**Preparation of Medium:** Add components to distilled/deionized water and bring volume to 1.0L. Mix thoroughly. Gently heat and bring to boiling. Distribute into tubes in 15.0mL volumes. Autoclave for 15 min at 15 psi pressure–121°C. Allow tubes to cool in an upright position.

**Storage/Shelf Life:** Store dehydrated media in the dark in a sealed container below 30°C. Prepared media should be stored under refrigeration (2-8°C). Media should be used within 60 days of preparation. Media should not be used if there are any signs of deterioration (discoloration) or contamination, or if the expiration date supplied by the manufacturer has passed.

**Use:** For the differentiation of members of the Enterobacteriaceae based on H$_2$S production, indole production, and motility.

## SIM Medium

**Composition** per liter:

Pancreatic digest of casein............................................................20.0g

Peptic digest of animal tissue.........................................................6.1g

Agar ...........................................................................................3.5g

Fe(NH$_4$)$_2$(SO$_4$)$_2$·6H$_2$O ........................................................0.2g

Na$_2$S$_2$O$_3$·5H$_2$O ........................................................................0.2g

pH 7.3 ± 0.2 at 25°C

**Source:** This medium is available as a premixed powder from BD Diagnostic Systems and Oxoid Unipath.

**Preparation of Medium:** Add components to distilled/deionized water and bring volume to 1.0L. Mix thoroughly. Gently heat and bring to boiling. Distribute into tubes in 15.0mL volumes. Autoclave for 15 min at 15 psi pressure–121°C. Allow tubes to cool in an upright position.

**Storage/Shelf Life:** Store dehydrated media in the dark in a sealed container below 30°C. Prepared media should be stored under refrigeration (2-8°C). Media should be used within 60 days of preparation. Media should not be used if there are any signs of deterioration (discoloration) or contamination, or if the expiration date supplied by the manufacturer has passed.

**Use:** For the differentiation of members of the Enterobacteriaceae based on H$_2$S production, indole production, and motility.

## SIM Motility Medium
### (BAM M137)

**Composition** per liter:

Pancreatic digest of casein............................................................20.0g

Peptic digest of animal tissue .......................................................6.1g

Agar ...........................................................................................3.5g

Fe(NH$_4$)$_2$(SO$_4$)$_2$·6H$_2$O .................................................................0.2g

Na$_2$S$_2$O$_3$·5H$_2$O ........................................................................0.2g

pH 7.3 ± 0.2 at 25°C

**Source:** This medium is available as a premixed powder from BD Diagnostic Systems and Oxoid Unipath.

**Preparation of Medium:** Add components to distilled/deionized water and bring volume to 1.0L. Mix thoroughly. Gently heat and bring to boiling. Distribute into tubes in 15.0mL volumes. Autoclave for 15 min at 15 psi pressure–121°C. Allow tubes to cool in an upright position.

**Storage/Shelf Life:** Store dehydrated media in the dark in a sealed container below 30°C. Prepared media should be stored under refrigeration (2-8°C). Media should be used within 60 days of preparation. Media should not be used if there are any signs of deterioration (discoloration) or contamination, or if the expiration date supplied by the manufacturer has passed.

**Use:** For the differentiation of members of the Enterobacteriaceae based on H$_2$S production, indole production, and motility.

## Simmons' Citrate Agar
### (Citrate Agar)

**Composition** per liter:

Agar ...........................................................................................15.0g

NaCl...........................................................................................5.0g

Sodium citrate.............................................................................2.0g

K$_2$HPO$_4$....................................................................................1.0g

(NH$_4$)H$_2$PO$_4$..............................................................................1.0g

MgSO$_4$·7H$_2$O ...........................................................................0.2g

Bromthymol Blue ........................................................................0.08g

pH 6.9 ± 0.2 at 25°C

**Source:** This medium is available as a premixed powder from BD Diagnostic Systems and Oxoid Unipath.

**Preparation of Medium:** Add components to distilled/deionized water and bring volume to 1.0L. Mix thoroughly. Gently heat while stirring and bring to boiling. Distribute into tubes or flasks. Autoclave for 15 min at 15 psi pressure–121°C. Pour into sterile Petri dishes or leave in tubes.

**Storage/Shelf Life:** Store dehydrated media in the dark in a sealed container below 30°C. Prepared media should be stored under refrigeration (2-8°C). Media should be used within 60 days of preparation.

Media should not be used if there are any signs of deterioration (shrinking, cracking, or discoloration) or contamination, or if the expiration date supplied by the manufacturer has passed.

**Use:** For the differentiation of Gram-negative bacteria on the basis of citrate utilization. Bacteria that can utilize citrate as sole carbon source turn the medium blue.

## Skirrow *Brucella* Medium
**Composition** per liter:

| | |
|---|---|
| Blood agar base No. 2 | 940.0mL |
| Horse blood, lysed defibrinated | 50.0mL |
| Antibiotic solution | 10.0mL |

pH 7.4 ± 0.2 at 25°C

### Blood Agar Base No. 2
**Composition** per 940.0mL:

| | |
|---|---|
| Proteose peptone | 15.0g |
| Agar | 12.0g |
| NaCl | 5.0g |
| Yeast extract | 5.0g |
| Liver digest | 2.5g |

pH 7.4 ± 0.2 at 25°C

**Preparation of Blood Agar Base No. 2:** Add components to distilled/deionized water and bring volume to 940.0mL. Mix thoroughly. Gently heat while stirring and bring to boiling. Autoclave for 15 min at 15 psi pressure–121°C. Cool to 45°–50°C.

### Antibiotic Solution:
**Composition** per 10.0mL:

| | |
|---|---|
| Vancomycin | 0.01g |
| Trimethoprim | 5.0mg |
| Polymyxin B | 2500U |

**Preparation of Antibiotic Solution:** Add components to distilled/deionized water and bring volume to 10.0mL. Mix thoroughly. Filter sterilize.

**Preparation of Medium:** To 940.0mL of sterile cooled blood agar base No. 2, aseptically add 50.0mL of sterile, lysed defibrinated horse blood and 10.0mL of sterile antibiotic solution. Pour into sterile Petri dishes or distribute into sterile tubes.

**Storage/Shelf Life:** Store dehydrated media in the dark in a sealed container below 30°C. Prepared media should be stored under refrigeration (2-8°C). Media should be used within 60 days of preparation. Media should not be used if there are any signs of deterioration (discoloration) or contamination, or if the expiration date supplied by the manufacturer has passed.

**Use:** For the selective isolation and cultivation of *Campylobacter* species.

## Slanetz and Bartley, HiVeg
## (Slanetz and Bartley HiVeg Medium)
**Composition** per liter:

| | |
|---|---|
| Plant hydrolysate No. 1 | 20.0g |
| Agar | 15.0g |
| Yeast extract | 5.0g |
| Na$_2$HPO$_4$ | 4.0g |
| Glucose | 2.0g |
| NaN$_3$ | 0.4g |
| 2,3,5-Triphenyltetrazolium chloride | 0.1g |

pH 7.2 ± 0.2 at 25°C

**Source:** This medium is available as a premixed powder from Hi-Media.

**Caution:** Sodium azide is toxic. Azides also react with metals and disposal must be highly diluted.

**Preparation of Medium:** Add components to distilled/deionized water and bring volume to 1.0L. Mix thoroughly. Gently heat and bring to boiling to dissolve the medium completely. Excessive heating should be avoided. Do not autoclave. Pour into sterile Petri dishes or leave in tubes.

**Storage/Shelf Life:** Store dehydrated media in the dark in a sealed container below 30°C. Prepared media should be stored under refrigeration (2-8°C). Media should be used within 60 days of preparation. Media should not be used if there are any signs of deterioration (shrinking, cracking, or discoloration) or contamination, or if the expiration date supplied by the manufacturer has passed.

**Use:** For the detection and enumeration of fecal streptococci by the membrane filter technique.

## Slanetz and Bartley Medium
**Composition** per liter:

| | |
|---|---|
| Tryptose | 20.0g |
| Agar | 10.0g |
| Yeast extract | 5.0g |
| Na$_2$HPO$_4$ 2H$_2$O | 4.0g |
| Glucose | 2.0g |
| NaN$_3$ | 0.4g |
| Tetrazolium chloride | 0.1g |

pH 7.2 ± 0.2 at 25°C

**Source:** This medium is available as a premixed powder from Oxoid Unipath.

**Preparation of Medium:** Add components to distilled/deionized water and bring volume to 1.0L. Mix thoroughly. Gently heat and bring to boiling. Distribute into tubes or flasks. Autoclave for 15 min at 15 psi pressure–121°C. Pour into sterile Petri dishes.

**Storage/Shelf Life:** Store dehydrated media in the dark in a sealed container below 30°C. Prepared media should be stored under refrigeration (2-8°C). Media should be used within 60 days of preparation. Media should not be used if there are any signs of deterioration (shrinking, cracking, or discoloration) or contamination, or if the expiration date supplied by the manufacturer has passed.

**Use:** For the detection and enumeration of enterococci by the membrane filter method.

## SM ID2 Agar
**Composition** per liter:
Proprietary

**Source:** This medium is available from bioMérieux

**Storage/Shelf Life:** Prepared plates should be stored under refrigeration (2-8°C). Chromogenic media are especially light and temperature sensitive; protect from light, excessive heat, moisture, and freezing. Media should not be used if there are any signs of deterioration (shrinking, cracking, or discoloration) or contamination, or if the expiration date supplied by the manufacturer has passed.

**Use:** A chromogenic medium for the selective isolation and detection of *Salmonella. S. typhi, S. paratyphi,* and most lactose(+) *Salmonella* present pale pink to mauve colonies. Other organisms are either inhibited, colorless, or pale blue in appearance.

## Snyder Agar

**Composition** per liter:

Glucose ............................................................20.0g
Agar ..................................................................16.0g
Pancreatic digest of casein ...............................13.5g
Yeast extract.......................................................6.5g
NaCl....................................................................5.0g
Bromcresol Green ..............................................0.02g

pH 4.8 ± 0.2 at 25°C

**Source:** This medium is available as a premixed powder from BD Diagnostic Systems.

**Preparation of Medium:** Add components to distilled/deionized water and bring volume to 1.0L. Mix thoroughly. Gently heat and bring to boiling. Distribute into tubes in 10.0mL volumes. Autoclave for 15 min at 13 psi pressure–118°C. Do not overheat. Pour into sterile Petri dishes or leave in tubes.

**Storage/Shelf Life:** Store dehydrated media in the dark in a sealed container below 30°C. Prepared media should be stored under refrigeration (2-8°C). Media should be used within 60 days of preparation. Media should not be used if there are any signs of deterioration (shrinking, cracking, or discoloration) or contamination, or if the expiration date supplied by the manufacturer has passed.

**Use:** For the cultivation and enumeration of lactobacilli in saliva and indication of dental caries activity.

## Snyder Test Agar

**Composition** per liter:

Agar ..................................................................20.0g
Glucose ............................................................20.0g
Tryptose ...........................................................20.0g
NaCl....................................................................5.0g
Bromcresol Green ..............................................0.02g

pH 4.8 ± 0.2 at 25°C

**Source:** This medium is available as a premixed powder from BD Diagnostic Systems.

**Preparation of Medium:** Add components to distilled/deionized water and bring volume to 1.0L. Mix thoroughly. Gently heat and bring to boiling. Distribute into tubes in 10.0mL volumes. Autoclave for 15 min at 13 psi pressure–118°C. Do not overheat. Pour into sterile Petri dishes or leave in tubes.

**Storage/Shelf Life:** Store dehydrated media in the dark in a sealed container below 30°C. Prepared media should be stored under refrigeration (2-8°C). Media should be used within 60 days of preparation. Media should not be used if there are any signs of deterioration (shrinking, cracking, or discoloration) or contamination, or if the expiration date supplied by the manufacturer has passed.

**Use:** For the cultivation and enumeration of lactobacilli in saliva and indication of dental caries activity.

## Snyder Test HiVeg Agar
## (BCG-Glucose Agar, HiVeg)

**Composition** per liter:

Agar ..................................................................20.0g
Glucose ............................................................20.0g
Plant peptone....................................................20.0g
NaCl....................................................................5.0g
Bromcresol Green ..............................................0.02

pH 4.8 ± 0.2 at 25°C

**Source:** This medium is available as a premixed powder from Hi-Media.

**Preparation of Medium:** Add components to distilled/deionized water and bring volume to 1.0L. Mix thoroughly. Gently heat and bring to boiling. Distribute into tubes in 10.0mL volumes. Autoclave for 15 min at 13 psi pressure–118°C. Do not overheat. Pour into sterile Petri dishes or leave in tubes.

**Storage/Shelf Life:** Store dehydrated media in the dark in a sealed container below 30°C. Prepared media should be stored under refrigeration (2-8°C). Media should be used within 60 days of preparation. Media should not be used if there are any signs of deterioration (shrinking, cracking, or discoloration) or contamination, or if the expiration date supplied by the manufacturer has passed.

**Use:** For the cultivation and enumeration of lactobacilli in saliva and indication of dental caries activity.

## Sodium Dodecyl Sulfate
## Polymyxin Sucrose Agar

**Composition** per liter:

NaCl..................................................................20.0g
Agar ..................................................................15.0g
Sucrose..............................................................15.0g
Proteose peptone ..............................................10.0g
Beef extract.........................................................5.0g
Sodium lauryl sulfate ..........................................1.0g
Bromthymol Blue ..............................................0.04g
Cresol Red ........................................................0.04g
Polymyxin B solution ...................................... 10.0mL

pH 7.6 ± 0.2 at 25°C

**Polymyxin B Solution:**
**Composition** per 10.0mL:

Polymyxin B sulfate ......................................... 100,000U

**Preparation of Polymyxin B Solution:** Add polymyxin B sulfate to distilled/deionized water and bring volume to 10.0mL. Mix thoroughly. Filter sterilize.

**Preparation of Medium:** Add components, except polymyxin B solution, to distilled/deionized water and bring volume to 990.0mL. Mix thoroughly. Gently heat and bring to boiling. Autoclave for 15 min at 15 psi pressure–121°C. Cool to 45°–50°C. Aseptically add sterile polymyxin B solution. Mix thoroughly. Pour immediately into sterile Petri dishes or distribute into sterile tubes.

**Storage/Shelf Life:** Store dehydrated media in the dark in a sealed container below 30°C. Prepared media should be stored under refrigeration (2-8°C). Media should be used within 60 days of preparation. Media should not be used if there are any signs of deterioration (shrinking, cracking, or discoloration) or contamination, or if the expiration date supplied by the manufacturer has passed.

**Use:** For the isolation and cultivation of *Vibrio* species from foods.

## Sodium Hippurate Broth
## (Hippurate Broth)

**Composition** per liter:

Beef heart, solids from infusion............................500.0g
Tryptose ...........................................................10.0g
Sodium hippurate..............................................10.0g
NaCl....................................................................5.0g

pH 7.4 ± 0.2 at 25°C

**Source:** Heart infusion broth is available as a premixed powder from BD Diagnostic Systems.

**Preparation of Medium:** Add components to distilled/deionized water and bring volume to 1.0L. Mix thoroughly. Gently heat and bring to boiling. Distribute into screw-capped tubes or flasks. Autoclave for 15 min at 15 psi pressure–121°C. Tighten caps to prevent drying.

**Storage/Shelf Life:** Store dehydrated media in the dark in a sealed container below 30°C. Prepared media should be stored under refrigeration (2-8°C). Media should be used within 60 days of preparation. Media should not be used if there are any signs of deterioration (discoloration) or contamination, or if the expiration date supplied by the manufacturer has passed.

**Use:** For the identification and differentiation of β-hemolytic streptococci based on hippurate hydrolysis. After inoculation and incubation, tubes are treated with FeCl$_3$ reagent. A heavy precipitate remaining after 10–15 min indicates that hippurate has been hydrolyzed.

### Sorbitol HiVeg Agar
### (Sorbitol MacConkey HiVeg Agar)
**Composition** per liter:

| | |
|---|---|
| Plant peptone | 17.0g |
| Agar | 13.5g |
| D-Sorbitol | 10.0g |
| Plant peptone No. 3 | 3.0g |
| Synthetic detergent | 1.5g |
| NaCl | 5.0g |
| Neutral Red | 0.03g |
| Crystal Violet | 1.0mg |

pH 7.1 ± 0.2 at 25°C

**Preparation of Medium:** Add components to distilled/deionized water and bring volume to 1.0L. Mix thoroughly. Gently heat while stirring and bring to boiling. Autoclave for 15 min at 15 psi pressure–121°C. Avoid overheating. Mix thoroughly. Pour into sterile Petri dishes.

**Storage/Shelf Life:** Store dehydrated media in the dark in a sealed container below 30°C. Prepared media should be stored under refrigeration (2-8°C). Media should be used within 60 days of preparation. Media should not be used if there are any signs of deterioration (shrinking, cracking, or discoloration) or contamination, or if the expiration date supplied by the manufacturer has passed.

**Use:** For the isolation and identification of enteropathogenic *Escherichia coli* strains associated with infant diarrhea.

### Sorbitol HiVeg Agar with Blood
### (Sorbitol MacConkey HiVeg Agar)
**Composition** per liter:

| | |
|---|---|
| Plant peptone | 17.0g |
| Agar | 13.5g |
| D-Sorbitol | 10.0g |
| Plant peptone No. 3 | 3.0g |
| Synthetic detergent | 1.5g |
| NaCl | 5.0g |
| Neutral Red | 0.03g |
| Crystal Violet | 1.0mg |
| Sterile bovine or sheep blood | 50.0mL |

pH 7.3 ± 0.2 at 25°C

**Source:** This medium, without blood, is available as a premixed powder from HiMedia.

**Preparation of Medium:** Add components, except blood, to distilled/deionized water and bring volume to 950.0mL. Mix thoroughly. Gently heat and bring to boiling. Autoclave for 15 min at 15 psi pressure–121°C. Cool to 45°–50°C. Aseptically add 50.0mL of sterile bovine blood or sheep blood. Mix thoroughly. Pour into sterile Petri dishes.

**Storage/Shelf Life:** Store dehydrated media in the dark in a sealed container below 30°C. Prepared media should be stored under refrigeration (2-8°C). Media should be used within 60 days of preparation. Media should not be used if there are any signs of deterioration (shrinking, cracking, or discoloration) or contamination, or if the expiration date supplied by the manufacturer has passed.

**Use:** For the isolation and identification of a wide variety of pathogenic bacteria.

### Sorbitol Iron HiVeg Agar
**Composition** per liter:

| | |
|---|---|
| Agar | 20.0g |
| Plant peptone No. 3 | 15.0g |
| NaCl | 5.0g |
| Plant extract | 3.0g |
| D-Sorbitol | 2.0g |
| Ferric ammonium citrate | 0.5g |
| Na$_2$S$_2$O$_3$ | 0.5g |
| Phenol Red | 0.03g |

pH 7.6 ± 0.2 at 25°C

**Source:** This medium is available as a premixed powder from HiMedia.

**Preparation of Medium:** Add components to distilled/deionized water and bring volume to 1.0L. Mix thoroughly. Gently heat and bring to boiling. Distribute into tubes or flasks. Autoclave for 15 min at 15 psi pressure–121°C. Pour into sterile Petri dishes or leave in tubes.

**Storage/Shelf Life:** Store dehydrated media in the dark in a sealed container below 30°C. Prepared media should be stored under refrigeration (2-8°C). Media should be used within 60 days of preparation. Media should not be used if there are any signs of deterioration (shrinking, cracking, or discoloration) or contamination, or if the expiration date supplied by the manufacturer has passed.

**Use:** For the isolation and cultivation of coliform bacteria. For the identification and differentiation of enteropathogenic strains of *Escherichia coli* that do not ferment sorbitol.

### Sorbitol MacConkey Agar
### (MacConkey Agar with Sorbitol)
**Composition** per liter:

| | |
|---|---|
| Peptone | 20.0g |
| Agar | 15.0g |
| Sorbitol | 10.0g |
| NaCl | 5.0g |
| Bile salts No. 3 | 1.5g |
| Neutral Red | 0.03g |
| Crystal Violet | 1.0mg |

pH 7.1 ± 0.2 at 25°C

**Source:** This medium is available as a premixed powder from BD Diagnostic Systems and Oxoid Unipath.

**Preparation of Medium:** Add components to distilled/deionized water and bring volume to 1.0L. Mix thoroughly. Gently heat and bring

to boiling. Distribute into tubes or flasks. Autoclave for 15 min at 15 psi pressure–121°C. Pour into sterile Petri dishes or leave in tubes.

**Storage/Shelf Life:** Store dehydrated media in the dark in a sealed container below 30°C. Prepared media should be stored under refrigeration (2-8°C). Media should be used within 60 days of preparation. Media should not be used if there are any signs of deterioration (shrinking, cracking, or discoloration) or contamination, or if the expiration date supplied by the manufacturer has passed.

**Use:** For the isolation and cultivation of pathogenic *Escherichia coli*.

### Sorbitol MacConkey Agar (MacConkey Agar with Sorbitol) (BAM M139)

**Composition** per liter:

| | |
|---|---|
| Peptone or gelysate | 17.0g |
| Agar | 13.5g |
| Sorbitol | 10.0g |
| NaCl | 5.0g |
| Proteose peptone No. 3 or polypeptone | 3.0g |
| Bile salts, purified | 1.5g |
| Neutral Red | 0.03g |
| Crystal Violet | 1.0mg |

pH 7.1 ± 0.2 at 25°C

**Preparation of Medium:** Add components to distilled/deionized water and bring volume to 1.0L. Mix thoroughly. Gently heat and bring to boiling. Distribute into tubes or flasks. Autoclave for 15 min at 15 psi pressure–121°C. Pour into sterile Petri dishes or leave in tubes.

**Storage/Shelf Life:** Store dehydrated media in the dark in a sealed container below 30°C. Prepared media should be stored under refrigeration (2-8°C). Media should be used within 60 days of preparation. Media should not be used if there are any signs of deterioration (shrinking, cracking, or discoloration) or contamination, or if the expiration date supplied by the manufacturer has passed.

**Use:** For the isolation and cultivation of pathogenic *Escherichia coli*.

### Sorbitol MacConkey Agar with BCIG (SMAC with BCIG)

**Composition** per liter:

| | |
|---|---|
| Peptone | 20.0g |
| Agar | 15.0g |
| Sorbitol | 10.0g |
| NaCl | 5.0g |
| Proteose peptone | 3.0g |
| Bile salts mixture | 1.5g |
| 5-Bromo-4-chloro-3-indolyl-β-D-glucuronide sodium salt | 0.1g |
| Neutral Red | 0.03g |

pH 7.1 ± 0.2 at 25°C

**Source:** This medium is available from Oxoid Unipath.

**Preparation of Medium:** Add components to distilled/deionized water and bring volume to 1.0L. Mix thoroughly. Gently heat while stirring and bring to boiling. Autoclave for 15 min at 15 psi pressure–121°C. Pour into sterile Petri dishes.

**Storage/Shelf Life:** Store dehydrated media in the dark in a sealed container below 30°C. Prepared media should be stored under refrigeration (2-8°C). Media should be used within 60 days of preparation. Media should not be used if there are any signs of deterioration (shrink-

ing, cracking, or discoloration) or contamination, or if the expiration date supplied by the manufacturer has passed.

**Use:** A selective and differential medium for the detection of *Escherichia coli* O157 incorporating the chromogen 5-bromo-4-chloro-3-indolyl-β-D-glucuronide (BCIG). The medium combines two different screening mechanisms for the detection of *E. coli* O157, the failure to ferment sorbitol and the absence of β-glucuronidase activity. The non-sorbitol-fermenting and β-glucuronidase-negative *E. coli* O157 will appear as straw-colored colonies. Organisms with β-glucuronidase activity will cleave the substrate, leading to a distinct blue-green coloration of the colonies. The intestinal tract of ruminants is the prime reservoir of *E. coli* O157 and other enterohemorrhagic *E. coli* (EHEC) strains; therefore meats derived from cattle, sheep, goat, and deer can be expected to be contaminated. Foods implicated in human illness related to *E. coli* O157 include meats, dairy products, vegetables, salads, apple juice, and water.

### Soybean HiVeg Broth Base with Novobiocin

**Composition** per liter:

| | |
|---|---|
| Plant hydrolysate | 17.0g |
| NaCl | 5.0g |
| $K_2HPO_4$ | 4.0g |
| Papaic digest of soybean meal | 3.0g |
| Glucose | 2.5g |
| Synthetic detergent No. I | 1.12g |
| Selective solution | 10.0mL |

pH 7.3 ± 0.2 at 25°C

**Source:** This medium, without selective novobiocin supplement, is available as a premixed powder from HiMedia.

**Selective Solution:**
**Composition** per 10.0mL:

| | |
|---|---|
| Novobiocin | 0.1g |

**Preparation of Selective Solution:** Add novobiocin to distilled/deionized water and bring volume to 10.0mL. Mix thoroughly. Filter sterilize.

**Preparation of Medium:** Add components, except selective solution, to distilled/deionized water and bring volume to 990.0mL. Mix thoroughly. Gently heat and bring to boiling. Mix thoroughly. Autoclave for 15 min at 15 psi pressure–121°C. Cool to 45°–50°C. Aseptically add 10.0mL selective solution. Mix thoroughly. Distribute into tubes or flasks. Autoclave for 5 min at 15 psi pressure–121°C. Pour into sterile Petri dishes.

**Storage/Shelf Life:** Store dehydrated media in the dark in a sealed container below 30°C. Prepared media should be stored under refrigeration (2-8°C). Media should be used within 60 days of preparation. Media should not be used if there are any signs of deterioration (discoloration) or contamination, or if the expiration date supplied by the manufacturer has passed.

**Use:** For the enrichment and isolation of *Escherichia coli* O157:H7 from foods.

### Soybean HiVeg Medium with 0.1% Agar (Tryptone Soy HiVeg Broth with 0.1% Agar)

**Composition** per liter:

| | |
|---|---|
| Plant hydrolysate | 17.0g |
| NaCl | 5.0g |
| Papaic digest of soybean meal | 3.0g |

K_2HPO_4 ............................................................2.5g
Agar ...................................................................1.0g

pH 7.3 ± 0.2 at 25°C

**Source:** This medium is available as a premixed powder from Hi-Media.

**Preparation of Medium:** Add components to distilled/deionized water and bring volume to 1.0L. Mix thoroughly. Gently heat and bring to boiling. Distribute into tubes or flasks. Autoclave for 15 min at 15 psi pressure–121°C. Pour into sterile Petri dishes or leave in tubes.

**Storage/Shelf Life:** Store dehydrated media in the dark in a sealed container below 30°C. Prepared media should be stored under refrigeration (2-8°C). Media should be used within 60 days of preparation. Media should not be used if there are any signs of deterioration (discoloration) or contamination, or if the expiration date supplied by the manufacturer has passed.

**Use:** For the cultivation of anaerobes from root canals, blood, and other clinical specimens.

## Soybean HiVeg Medium with 0.1% Agar with Glucose (Tryptone Soy HiVeg Broth with 0.1% Agar)

**Composition** per liter:

Plant hydrolysate ...............................................17.0g
NaCl ...................................................................5.0g
Papaic digest of soybean meal .........................3.0g
Glucose ..............................................................2.5g
K_2HPO_4 ............................................................2.5g
Agar ...................................................................1.0g

pH 7.3 ± 0.2 at 25°C

**Source:** This medium is available as a premixed powder from Hi-Media.

**Preparation of Medium:** Add components to distilled/deionized water and bring volume to 1.0L. Mix thoroughly. Gently heat and bring to boiling. Distribute into tubes or flasks. Autoclave for 15 min at 15 psi pressure–121°C.

**Storage/Shelf Life:** Store dehydrated media in the dark in a sealed container below 30°C. Prepared media should be stored under refrigeration (2-8°C). Media should be used within 60 days of preparation. Media should not be used if there are any signs of deterioration (shrinking, cracking, or discoloration) or contamination, or if the expiration date supplied by the manufacturer has passed.

**Use:** For the cultivation of anaerobes from root canals, blood, and other clinical specimens and for determining glucose fermentation.

## Special Infusion Broth, HiVeg with Blood

**Composition** per liter:

Plant infusion ....................................................10.0g
Plant peptone No. 3 ..........................................10.0g
Plant special infusion .........................................7.5g
NaCl ...................................................................5.0g
Na_2HPO_4 ..........................................................2.5g
Glucose ..............................................................2.0g
Blood, defibrinated .........................................50.0mL

pH 7.4 ± 0.2 at 25°C

**Source:** This medium is available as a premixed powder from Hi-Media.

**Preparation of Medium:** Add components, except blood, to distilled/deionized water and bring volume to 950.0L. Mix thoroughly. Gently heat until boiling. Autoclave for 15 min at 15 psi pressure–121°C. Cool to 50°C. Add 50.0mL sterile defibrinated blood. Mix thoroughly. Aseptically distribute into sterile tubes.

**Storage/Shelf Life:** Store dehydrated media in the dark in a sealed container below 30°C. Prepared media should be stored under refrigeration (2-8°C). Media should be used within 60 days of preparation. Media should not be used if there are any signs of deterioration (discoloration) or contamination, or if the expiration date supplied by the manufacturer has passed.

**Use:** For the cultivation of a variety of fastidious and nonfastidious aerobic and anaerobic microorganisms, including streptococci. For the propagation of pathogenic cocci and other fastidious organisms associated with blood culture work and allied pathological investigations.

## Specimen Preservative Medium

**Composition** per liter:

NaCl ...................................................................5.0g
Sodium citrate·2H_2O ..........................................5.0g
(NH_4)_2HPO_4 .....................................................4.0g
KH_2PO_4 .............................................................2.0g
Yeast extract .......................................................1.0g
Sodium deoxycholate ..........................................0.5g
MgSO_4·7H_2O .....................................................0.4g
Glycerol ........................................................300.0mL

pH 7.0 ± 0.2 at 25°C

**Preparation of Medium:** Add components, except glycerol, to distilled/deionized water and bring volume to 700.0mL. Mix thoroughly. Gently heat and bring to boiling. Add 300.0mL of glycerol. Mix thoroughly. Distribute into tubes or flasks. Autoclave for 10 min at 11 psi pressure–116°C.

**Storage/Shelf Life:** Store dehydrated media in the dark in a sealed container below 30°C. Prepared media should be stored under refrigeration (2-8°C). Media should be used within 60 days of preparation. Media should not be used if there are any signs of deterioration (discoloration) or contamination, or if the expiration date supplied by the manufacturer has passed.

**Use:** For the preservation of viable microorganisms in stool specimens. For the transport of fecal material.

## Spore Strip Broth

**Composition** per liter:

Spore strip broth ................................................9.0g

**Preparation of Medium:** Add 9.0g of spore strip broth powder (a mixture of glucose, buffer salts, growth factors, and Bromthymol Blue) to distilled/deionized water and bring volume to 1.0L. Mix thoroughly. Distribute into tubes or flasks. Autoclave for 15 min at 15 psi pressure–121°C.

**Storage/Shelf Life:** Store dehydrated media in the dark in a sealed container below 30°C. Prepared media should be stored under refrigeration (2-8°C). Media should be used within 60 days of preparation. Media should not be used if there are any signs of deterioration (discoloration) or contamination, or if the expiration date supplied by the manufacturer has passed.

**Use:** For the recovery of spores of *Bacillus stearothermophilus* on spore strips used to determine the sterilization efficiency of autoclaves.

## Spray's Fermentation Medium

**Composition** per 1100.0mL:

| | |
|---|---|
| Neopeptone | 10.0g |
| Pancreatic digest of casein | 10.0g |
| Agar | 2.0g |
| Sodium thioglycolate | 0.025g |
| Carbohydrate solution | 110.0mL |

pH 7.4 ± 0.1 at 25°C

**Carbohydrate Solution:**

**Composition** per 200.0mL:

| | |
|---|---|
| Carbohydrate | 20.0g |

**Preparation of Carbohydrate Solution:** Add carbohydrate to distilled/deionized water and bring volume to 200.0mL. Glucose or glycerol may be used. Mix thoroughly. Filter sterilize.

**Preparation of Medium:** Add components, except agar and carbohydrate solution, to distilled/deionized water and bring volume to 990.0mL. Mix thoroughly. Adjust pH to 7.4. Add agar. Gently heat and bring to boiling. Distribute into tubes in 9.0mL volumes. Autoclave for 15 min at 15 psi pressure–121°C. Cool to 25°C. Immediately prior to use heat tubes in a boiling water bath for 10 min. Cool to 45°C. Aseptically add 1.0mL of sterile carbohydrate solution. Mix thoroughly.

**Storage/Shelf Life:** Store dehydrated media in the dark in a sealed container below 30°C. Prepared media should be stored under refrigeration (2-8°C). Media should be used within 60 days of preparation. Media should not be used if there are any signs of deterioration (discoloration) or contamination, or if the expiration date supplied by the manufacturer has passed.

**Use:** For the differentiation of *Clostridium perfringens* based on carbohydrate fermentation patterns.

## SPS Agar

### (Sulfite Polymyxin Sulfadiazine Agar)

**Composition** per liter:

| | |
|---|---|
| Pancreatic digest of casein | 15.0g |
| Agar | 13.9g |
| Yeast extract | 10.0g |
| Ferric citrate | 0.5g |
| Na$_2$SO$_3$ | 0.5g |
| Sulfadiazine | 0.12g |
| Polymyxin sulfate | 0.01g |

pH 7.0 ± 0.2 at 25°C

**Source:** This medium is available as a premixed powder from BD Diagnostic Systems.

**Preparation of Medium:** Add components to distilled/deionized water and bring volume to 1.0L. Mix thoroughly. Gently heat while stirring and bring to boiling. Distribute into tubes or flasks. Autoclave for 15 min at 13 psi pressure–118°C. Pour into sterile Petri dishes or leave in tubes.

**Storage/Shelf Life:** Store dehydrated media in the dark in a sealed container below 30°C. Prepared media should be stored under refrigeration (2-8°C). Media should be used within 60 days of preparation. Media should not be used if there are any signs of deterioration (shrinking, cracking, or discoloration) or contamination, or if the expiration date supplied by the manufacturer has passed.

**Use:** For the isolation and detection of *Clostridium perfringens* and *Clostridium botulinum* in foods and other materials.

## SPS HiVeg Agar

**Composition** per liter:

| | |
|---|---|
| Plant hydrolysate | 15.0g |
| Agar | 13.9g |
| Yeast extract | 10.0g |
| Polysorbate 80 | 0.5g |
| Na$_2$SO$_3$ | 0.5g |
| Sulphadiazine | 0.12g |
| Polymyxin B sulfate | 0.01g |

pH 7.0 ± 0.2 at 25°C

**Source:** This medium is available as a premixed powder from Hi-Media.

**Preparation of Medium:** Add components to distilled/deionized water and bring volume to 1.0L. Mix thoroughly. Gently heat while stirring and bring to boiling. Distribute into tubes or flasks. Autoclave for 15 min at 13 psi pressure–118°C. Pour into sterile Petri dishes or leave in tubes.

**Storage/Shelf Life:** Store dehydrated media in the dark in a sealed container below 30°C. Prepared media should be stored under refrigeration (2-8°C). Media should be used within 60 days of preparation. Media should not be used if there are any signs of deterioration (shrinking, cracking, or discoloration) or contamination, or if the expiration date supplied by the manufacturer has passed.

**Use:** For the isolation and detection of *Clostridium perfringens* and *Clostridium botulinum* in foods and other materials.

## SPS HiVeg Agar, Modified

**Composition** per liter:

| | |
|---|---|
| Agar | 15.0g |
| Plant hydrolysate | 15.0g |
| Yeast extract | 10.0g |
| Na$_2$SO$_3$ | 0.5g |
| Ferric citrate | 0.5g |
| Sulphadiazine | 0.12g |
| Na-thioglycollate | 0.1g |
| Sorbitan monooleate | 0.05g |
| Polymyxin B sulfate | 0.01g |

pH 7.0 ± 0.2 at 25°C

**Source:** This medium is available as a premixed powder from Hi-Media.

**Preparation of Medium:** Add components to distilled/deionized water and bring volume to 1.0L. Mix thoroughly. Gently heat while stirring and bring to boiling. Distribute into tubes or flasks. Autoclave for 15 min at 13 psi pressure–118°C. Pour into sterile Petri dishes or leave in tubes.

**Storage/Shelf Life:** Store dehydrated media in the dark in a sealed container below 30°C. Prepared media should be stored under refrigeration (2-8°C). Media should be used within 60 days of preparation. Media should not be used if there are any signs of deterioration (shrinking, cracking, or discoloration) or contamination, or if the expiration date supplied by the manufacturer has passed.

**Use:** For the isolation and detection of *Clostridium perfringens* and *Clostridium botulinum* in foods and other materials.

## SS Agar, HiVeg

### (*Salmonella Shigella* Agar, HiVeg)

**Composition** per liter:

| | |
|---|---|
| Agar | 15.0g |
| Plant peptone | 11.5g |

Lactose ........................................................................10.0g
Sodium citrate .............................................................10.0g
Na$_2$S$_2$O$_3$ .........................................................................8.5g
Plant extract ..................................................................5.0g
Synthetic detergent No. I ..............................................2.0g
Ferric citrate .................................................................1.0g
Neutral Red .............................................................0.025g
Brilliant Green .........................................................0.33mg

pH 7.0 ± 0.2 at 25°C

**Source:** This medium is available as a premixed powder from Hi-Media.

**Preparation of Medium:** Add components to distilled/deionized water and bring volume to 1.0L. Mix thoroughly. Gently heat while stirring and bring to boiling. Do not autoclave. Cool to 45°–50°C. Pour into sterile Petri dishes in 20.0mL volumes. Allow the surface of the plates to dry before inoculation.

**Storage/Shelf Life:** Store dehydrated media in the dark in a sealed container below 30°C. Prepared media should be stored under refrigeration (2-8°C). Media should be used within 60 days of preparation. Media should not be used if there are any signs of deterioration (shrinking, cracking, or discoloration) or contamination, or if the expiration date supplied by the manufacturer has passed.

**Use:** For the selective isolation and differentiation of pathogenic enteric bacilli, especially those belonging to the genus *Salmonella*. This medium is not recommended for the primary isolation of *Shigella* species. Lactose-fermenting bacteria such as *Escherichia coli* or *Klebsiella pneumoniae* appear as small pink or red colonies. Lactose-nonfermenting bacteria—such as *Salmonella* species, *Proteus* species, and *Shigella* species—appear as colorless colonies. Production of H$_2$S by *Salmonella* species turns the center of the colonies black.

## SS Deoxycholate Agar
## (*Salmonella Shigella* Deoxycholate Agar)
## (SSDC)

**Composition** per liter:
Agar ...........................................................................13.5g
Lactose ........................................................................10.0g
Sodium deoxycholate .................................................10.0g
Bile salts.......................................................................8.5g
Na$_2$S$_2$O$_3$ .........................................................................8.5g
Sodium citrate ...............................................................8.5g
Beef extract ...................................................................5.0g
Pancreatic digest of casein............................................2.5g
Peptic digest of animal tissue........................................2.5g
CaCl$_2$·2H$_2$O.................................................................1.0g
Ferric citrate .................................................................1.0g
Neutral Red .............................................................0.025g
Brilliant Green .........................................................0.33mg

pH 7.0 ± 0.2 at 25°C

**Preparation of Medium:** Add components to distilled/deionized water and bring volume to 1.0L. Mix thoroughly. Gently heat while stirring and bring to boiling. Do not autoclave. Cool to 45°–50°C. Pour into sterile Petri dishes in 20.0mL volumes. Allow the surface of the plates to dry before inoculation.

**Storage/Shelf Life:** Store dehydrated media in the dark in a sealed container below 30°C. Prepared media should be stored under refrigeration (2-8°C). Media should not be used if there are any signs of deterioration (shrink-

ing, cracking, or discoloration) or contamination, or if the expiration date supplied by the manufacturer has passed.

**Use:** For the isolation and cultivation of *Yersinia enterocolitica* from foods.

## ST Holding Medium
## (m-ST Holding Medium)

**Composition** per liter:
KH$_2$PO$_4$......................................................................3.0g
Tris(hydroxymethyl)aminomethane buffer.....................3.0g
Sulfanilamide................................................................1.5g
NaH$_2$PO$_4$·H$_2$O............................................................0.1g
Ethanol (95% solution) ............................................ 10.0mL

pH 8.6 ± 0.2 at 25°C

**Preparation of Medium:** Dissolve the sulfanilamide in the ethanol. Add all components to distilled/deionized water and bring volume to 1.0L. Mix thoroughly. Autoclave for 15 min at 15 psi pressure–121°C. Distribute in 1.8mL volumes to sterile Petri dishes with tight-fitting lids and an absorbent filter.

**Storage/Shelf Life:** Store dehydrated media in the dark in a sealed container below 30°C. Prepared media should be stored under refrigeration (2-8°C). Media should be used within 60 days of preparation. Media should not be used if there are any signs of deterioration (discoloration) or contamination, or if the expiration date supplied by the manufacturer has passed.

**Use:** For the cultivation and enumeration of coliform bacteria by the delayed-incubation total coliform procedure. For use as a holding or transport medium to keep coliform bacteria viable between sampling and laboratory culture.

## Standard Fluid Medium 10B
## (Shepard's M10 Medium)

**Composition** per 102.5mL:
Base solution..................................................... 70.0mL
Horse serum, unheated....................................... 20.0mL
Fresh yeast extract solution ............................... 10.0mL
Penicillin solution ............................................... 1.0mL
CVA enrichment ................................................. 0.5mL
L-Cysteine·HCl·H$_2$O solution............................. 0.5mL
Urea solution...................................................... 0.4mL
Phenol Red solution............................................ 0.1mL

pH 6.0 ± 0.2 at 25°C

**Base Solution:**
**Composition** per 70.0mL:
Beef heart, solids from infusion....................................5.0g
Peptone .........................................................................1.0g
NaCl................................................................................0.5g

**Preparation of Base Solution:** Add components to distilled/deionized water and bring volume to 70.0mL. Mix thoroughly. Adjust pH to 5.5 with 2*N* HCl. Autoclave for 15 min at 15 psi pressure–121°C. Cool to 45°–50°C.

**Fresh Yeast Extract Solution:**
**Composition** per 100.0mL:
Baker's yeast, live, pressed, starch-free,......................25.0g

**Preparation of Fresh Yeast Extract Solution:** Add the live Baker's yeast to 100.0mL of distilled/deionized water. Autoclave for 90 min at 15 psi pressure–121°C. Allow to stand. Remove supernatant solution. Adjust pH to 6.6–6.8.

**Penicillin Solution:**

**Composition** per 10.0mL:

Penicillin G .........................................................1,000,000U

**Preparation of Penicillin Solution:** Add penicillin to distilled/deionized water and bring volume to 10.0mL. Mix thoroughly. Filter sterilize.

**CVA Enrichment:**

**Composition** per liter:

Glucose .................................................................100.0g
L-Cysteine·HCl·H₂O.................................................25.9g
L-Glutamine.............................................................10.0g
L-Cystine·2HCl..........................................................1.0g
Adenine.....................................................................1.0g
Nicotinamide adenine dinucleotide .........................0.25g
Cocarboxylase...........................................................0.1g
Guanine·HCl............................................................0.03g
Fe(NO₃)₃ .................................................................0.02g
*p*-Aminobenzoic acid...........................................0.013g
Vitamin B₁₂..............................................................0.01g
Thiamine·HCl .........................................................3.0mg

**Preparation of CVA Enrichment:** Add components to distilled/deionized water and bring volume to 1.0L. Mix thoroughly. Filter sterilize.

**L-Cysteine·HCl·H₂O Solution:**

**Composition** per 10.0mL:

L-Cysteine·HCl·H₂O .................................................0.2g

**Preparation of L-Cysteine·HCl·H₂O Solution:** Add L-cysteine·HCl·H₂O to distilled/deionized water and bring volume to 10.0mL. Mix thoroughly. Filter sterilize.

**Urea Solution:**

**Composition** per 10.0mL:

Urea.........................................................................1.0g

**Preparation of Urea Solution:** Add urea to distilled/deionized water and bring volume to 10.0mL. Mix thoroughly. Filter sterilize.

**Phenol Red Solution:**

**Composition** per 10.0mL:

Phenol Red................................................................0.1g

**Preparation of Phenol Red Solution:** Add Phenol Red to distilled/deionized water and bring volume to 10.0mL. Mix thoroughly. Autoclave for 15 min at 15 psi pressure–121°C.

**Preparation of Medium:** To 70.0mL of cooled, sterile base solution, aseptically add 20.0mL of sterile horse serum, 10.0mL of sterile fresh yeast extract solution, 1.0mL of sterile penicillin solution, 0.5mL of sterile CVA enrichment, 0.5mL of sterile L-cysteine·HCl·H₂O solution, 0.4mL of sterile urea solution, and 0.1mL of sterile Phenol Red solution. Mix thoroughly. Aseptically distribute into sterile tubes or flasks.

**Storage/Shelf Life:** Store dehydrated media in the dark in a sealed container below 30°C. Prepared media should be stored under refrigeration (2-8°C). Media should be used within 60 days of preparation. Media should not be used if there are any signs of deterioration (discoloration) or contamination, or if the expiration date supplied by the manufacturer has passed.

**Use:** For the isolation and cultivation of *Ureaplasma urealyticum* from clinical specimens.

## Standard Methods Agar
## (Tryptone Glucose Yeast Agar)
## (Plate Count Agar)

**Composition** per liter:

Agar .......................................................................15.0g
Pancretic digest of casein..........................................5.0g
Yeast extract.............................................................2.5g
Glucose ....................................................................1.0g

pH 7.0 ± 0.1 at 25°C

**Source:** Available as a premixed powder from BD Diagnostic Systems.

**Preparation of Medium:** Add components to distilled/deionized water and bring volume to 1.0L. Mix thoroughly. Gently heat and bring to boiling. Distribute into tubes or flasks. Autoclave for 15 min at 15 psi pressure–121°C. Pour into sterile Petri dishes or leave in tubes.

**Storage/Shelf Life:** Store dehydrated media in the dark in a sealed container below 30°C. Prepared media should be stored under refrigeration (2-8°C). Media should be used within 60 days of preparation. Media should not be used if there are any signs of deterioration (shrinking, cracking, or discoloration) or contamination, or if the expiration date supplied by the manufacturer has passed.

**Use:** For the cultivation and enumeration by microbial plate counts of microorganisms isolated from milk and dairy products, foods, water, and other specimens.

## Standard Methods Agar, HiVeg

**Composition** per liter:

Agar .......................................................................15.0g
Plant peptone ...........................................................5.0g
Yeast extract.............................................................2.5g
Glucose ....................................................................1.0g

pH 7.0 ± 0.1 at 25°C

**Source:** This medium is available as a premixed powder from HiMedia.

**Preparation of Medium:** Add components to distilled/deionized water and bring volume to 1.0L. Mix thoroughly. Gently heat and bring to boiling. Distribute into tubes or flasks. Autoclave for 15 min at 15 psi pressure–121°C. Pour into sterile Petri dishes or leave in tubes.

**Storage/Shelf Life:** Store dehydrated media in the dark in a sealed container below 30°C. Prepared media should be stored under refrigeration (2-8°C). Media should be used within 60 days of preparation. Media should not be used if there are any signs of deterioration (shrinking, cracking, or discoloration) or contamination, or if the expiration date supplied by the manufacturer has passed.

**Use:** For the cultivation and enumeration by microbial plate counts of microorganisms isolated from milk and dairy products, foods, water, and other specimens.

## Standard Methods Agar
## with Lecithin and Polysorbate 80

**Composition** per liter:

Agar .......................................................................15.0g
Pancreatic digest of casein.........................................5.0g
Polysorbate 80 ..........................................................5.0g
Yeast extract.............................................................2.5g

Glucose ........................................................................1.0g
Lecithin .......................................................................0.7g

pH 7.0 ± 0.2 at 25°C

**Source:** This medium is available as a premixed powder from BD Diagnostic Systems.

**Preparation of Medium:** Add components to distilled/deionized water and bring volume to 1.0L. Mix thoroughly. Gently heat and bring to boiling. Distribute into tubes or flasks. Autoclave for 15 min at 15 psi pressure–121°C. Pour into sterile Petri dishes or leave in tubes.

**Storage/Shelf Life:** Store dehydrated media in the dark in a sealed container below 30°C. Prepared media should be stored under refrigeration (2-8°C). Media should be used within 60 days of preparation. Media should not be used if there are any signs of deterioration (shrinking, cracking, or discoloration) or contamination, or if the expiration date supplied by the manufacturer has passed.

**Use:** For determination of the sterility of surfaces.

## *Staphylococcus* Agar No. 110

**Composition** per liter:

NaCl........................................................................75.0g
Gelatin.....................................................................30.0g
Agar .......................................................................15.0g
D-Mannitol................................................................10.0g
Pancreatic digest of casein ............................................10.0g
K$_2$HPO$_4$................................................................5.0g
Yeast extract...............................................................2.5g
Lactose .....................................................................2.0g

pH 7.0 ± 0.2 at 25°C

**Source:** This medium is available as a premixed powder from BD Diagnostic Systems and Oxoid Unipath.

**Preparation of Medium:** Add components to distilled/deionized water and bring volume to 1.0L. Mix thoroughly. Gently heat and bring to boiling. Distribute into tubes or flasks. Autoclave for 15 min at 15 psi pressure–121°C. Pour into sterile Petri dishes or leave in tubes. Swirl flask while pouring plates to disperse precipitate.

**Storage/Shelf Life:** Store dehydrated media in the dark in a sealed container below 30°C. Prepared media should be stored under refrigeration (2-8°C). Media should not be used if there are any signs of deterioration (shrinking, cracking, or discoloration) or contamination, or if the expiration date supplied by the manufacturer has passed.

**Use:** For the isolation and enumeration of staphylococci from clinical and nonclinical specimens.

## *Staphylococcus aureus* Enrichment HiVeg Broth

**Composition** per liter:

Plant peptone...........................................................10.0g
Plant infusion ...........................................................5.0g
NaCl .......................................................................5.0g

pH 6.8 ± 0.2 at 25°C

**Source:** This medium is available as a premixed powder from Hi-Media.

**Preparation of Medium:** Add components to distilled/deionized water and bring volume to 1.0L. Mix thoroughly. Gently heat and bring to boiling. Distribute into tubes or flasks. Autoclave for 15 min at 15 psi pressure–121°C.

**Storage/Shelf Life:** Store dehydrated media in the dark in a sealed container below 30°C. Prepared media should be stored under refrigeration (2-8°C). Media should be used within 60 days of preparation. Media should not be used if there are any signs of deterioration (discoloration) or contamination, or if the expiration date supplied by the manufacturer has passed.

**Use:** For the cultivation of *Staphylococcus aureus*.

## *Staphylococcus* Broth (m-*Staphylococcus* Broth)

**Composition** per liter:

NaCl........................................................................75.0g
Mannitol...................................................................10.0g
Pancreatic digest of casein.............................................10.0g
K$_2$HPO$_4$................................................................5.0g
Yeast extract...............................................................2.5g
Lactose.....................................................................2.0g

pH 7.0 ± 0.2 at 25°C

**Source:** This medium is available as a premixed powder from BD Diagnostic Systems.

**Preparation of Medium:** Add components to distilled/deionized water and bring volume to 1.0L. Mix thoroughly. Distribute into tubes or flasks. Autoclave for 15 min at 15 psi pressure–121°C.

**Storage/Shelf Life:** Store dehydrated media in the dark in a sealed container below 30°C. Prepared media should be stored under refrigeration (2-8°C). Media should be used within 60 days of preparation. Media should not be used if there are any signs of deterioration (discoloration) or contamination, or if the expiration date supplied by the manufacturer has passed.

**Use:** For the cultivation and enumeration of pathogenic and enterotoxigenic staphylococci by the membrane filter method. Also, when used in conjunction with Lipovitellin-salt-mannitol agar, for the detection of *Staphylococcus aureus* in swimming pool water.

## *Staphylococcus* Medium

**Composition** per liter:

Agar .......................................................................15.0g
Peptone ...................................................................6.0g
Pancreatic digest of casein............................................4.0g
Yeast extract...............................................................3.0g
Beef extract................................................................1.5g
Glucose ....................................................................1.0g

pH 6.6 ± 0.2 at 25°C

**Preparation of Medium:** Add components to distilled/deionized water and bring volume to 1.0L. Mix thoroughly. Gently heat and bring to boiling. Distribute into tubes or flasks. Autoclave for 15 min at 15 psi pressure–121°C. Pour into sterile Petri dishes or leave in tubes.

**Storage/Shelf Life:** Store dehydrated media in the dark in a sealed container below 30°C. Prepared media should be stored under refrigeration (2-8°C). Media should be used within 60 days of preparation. Media should not be used if there are any signs of deterioration (shrinking, cracking, or discoloration) or contamination, or if the expiration date supplied by the manufacturer has passed.

**Use:** For the cultivation and maintenance of *Staphylococcus aureus*. For the enumeration of pathogenic and enterotoxigenic staphylococci by the membrane filter method.

## *Staphylococcus/Streptococcus* Selective Medium

**Composition** per 1060.0mL:

Columbia blood agar base.................................................. 1.0L
Horse blood, defibrinated ............................................. 50.0mL
Antibiotic inhibitor ........................................................ 10.0mL

pH 7.3 ± 0.2 at 25°C

### Columbia Blood Agar Base:

**Composition** per liter:

Special peptone....................................................................23.0g
Agar ......................................................................................10.0g
NaCl........................................................................................5.0g
Starch .....................................................................................1.0g

**Source:** Columbia blood agar base is available as a premixed powder from Oxoid Unipath.

**Preparation of Columbia Blood Agar Base:** Add components to distilled/deionized water and bring volume to 1.0L. Mix thoroughly. Gently heat and bring to boiling. Autoclave for 15 min at 15 psi pressure–121°C. Cool to 45°–50°C.

### Antibiotic Inhibitor:

**Composition** per 10.0mL:

Nalidixic acid......................................................................0.015g
Colistin sulfate ....................................................................0.01g
Ethanol (95% solution) .................................................... 10.0mL

**Preparation of Antibiotic Inhibitor:** Add components to 10.0mL ethanol. Mix thoroughly. Filter sterilize.

**Preparation of Medium:** To 1.0L of cooled sterile Columbia blood agar base, aseptically add sterile horse blood and sterile antibiotic inhibitor. Mix thoroughly. Pour into sterile Petri dishes or distribute into sterile tubes.

**Storage/Shelf Life:** Store dehydrated media in the dark in a sealed container below 30°C. Prepared media should be stored under refrigeration (2-8°C). Media should be used within 60 days of preparation. Media should not be used if there are any signs of deterioration (shrinking, cracking, or discoloration) or contamination, or if the expiration date supplied by the manufacturer has passed.

**Use:** For the selective isolation of *Staphylococcus aureus* and streptococci from clinical specimens or foods.

## Starch Agar with Bromcresol Purple

**Composition** per liter:

Agar .....................................................................................15.0g
Cornstarch...........................................................................10.0g
Meat peptone.......................................................................10.0g
Bromcresol Purple solution ............................................. 1.2mL

pH 6.8 ± 0.2 at 25°C

### Bromcresol Purple Solution:

**Composition** per 10.0mL:

Bromcresol Purple ..............................................................0.16g
Ethanol (95% solution) .................................................... 10.0mL

**Preparation of Bromcresol Purple Solution:** Add Bromcresol Purple to 10.0mL of 95% ethanol. Mix thoroughly.

**Preparation of Medium:** Add components to distilled/deionized water and bring volume to 1.0L. Mix thoroughly. Gently heat and bring to boiling. Distribute into tubes or flasks. Autoclave for 15 min at 15 psi pressure–121°C. Pour into sterile Petri dishes or leave in tubes.

**Storage/Shelf Life:** Store dehydrated media in the dark in a sealed container below 30°C. Prepared media should be stored under refriger-

ation (2-8°C). Media should be used within 60 days of preparation. Media should not be used if there are any signs of deterioration (shrinking, cracking, or discoloration) or contamination, or if the expiration date supplied by the manufacturer has passed.

**Use:** For the differentiation of *Gardnerella vaginalis (Haemophilus vaginalis, Corynebacterium vaginale)* from other microorganisms found in the genitourinary tract, with the exception of some strains of *Streptococcus* and *Lactobacillus*. Differentiation is based on starch hydrolysis. Bacteria that can hydrolyze starch appear as colonies surrounded by a yellow zone.

## Starch Agar with Bromcresol Purple

**Composition** per liter:

Solution 1............................................................................ 200.0mL
Solution 2.............................................................................. 20.0mL

pH 7.8 ± 0.2 at 25°C

### Solution 1:

**Composition** per 200.0mL:

Heart infusion agar ............................................................ 5.0mL
Bromcresol Purple solution ............................................. 0.2mL

**Preparation of Solution 1:** Add components to distilled/deionized water and bring volume to 200.0mL. Mix thoroughly. Gently heat while stirring and bring to boiling.

### Heart Infusion Agar:

**Composition** per liter:

Beef heart, solids from infusion.......................................500.0g
Agar ......................................................................................15.0g
Tryptose ...............................................................................10.0g
NaCl........................................................................................5.0g

**Preparation of Heart Infusion Agar:** Add components to distilled/deionized water and bring volume to 1.0L. Mix thoroughly. Gently heat and bring to boiling.

### Bromcresol Purple Solution:

**Composition** per 10.0mL:

Bromcresol Purple ..............................................................0.16g
Ethanol (95% solution) .................................................... 10.0mL

**Preparation of Bromcresol Purple Solution:** Add Bromcresol Purple to 10.0mL of ethanol. Mix thoroughly.

### Solution 2:

**Composition** per 20.0mL:

Starch ....................................................................................0.4g

**Preparation of Solution 2:** Add starch to distilled/deionized water and bring volume to 20.0mL. Mix thoroughly. Gently heat while stirring and bring to boiling.

**Preparation of Medium:** Combine solution 1 and solution 2. Mix thoroughly. Autoclave for 15 min at 15 psi pressure–121°C. Pour into sterile Petri dishes or distribute into sterile tubes.

**Storage/Shelf Life:** Store dehydrated media in the dark in a sealed container below 30°C. Prepared media should be stored under refrigeration (2-8°C). Media should be used within 60 days of preparation. Media should not be used if there are any signs of deterioration (shrinking, cracking, or discoloration) or contamination, or if the expiration date supplied by the manufacturer has passed.

**Use:** For the differentiation of *Gardnerella vaginalis (Haemophilus vaginalis, Corynebacterium vaginale)* from other microorganisms found in the genitourinary tract, with the exception of some strains of *Streptococcus* and *Lactobacillus*. Differentiation is based on starch hydrolysis. Bacteria that can hydrolyze starch appear as colonies surrounded by a yellow zone.

## Sterility Test Broth
### (USP Alternative Thioglycolate Medium)
**Composition** per liter:

| | |
|---|---|
| Pancreatic digest of casein | 15.0g |
| Glucose | 5.0g |
| Yeast extract | 5.0g |
| NaCl | 2.5g |
| L-Cystine | 0.5g |
| Sodium thioglycolate | 0.5g |

pH 7.1 ± 0.2 at 25°C

**Source:** This medium is available as a premixed powder from BD Diagnostic Systems.

**Preparation of Medium:** Add components to distilled/deionized water and bring volume to 1.0L. Mix thoroughly. Gently heat and bring to boiling. Distribute into tubes or flasks. Autoclave for 15 min at 15 psi pressure–121°C. Cool to 25°C. If not used immediately, prior to inoculation heat tubes in a boiling water bath for 5–10 min. Cool to 25°C.

**Storage/Shelf Life:** Store dehydrated media in the dark in a sealed container below 30°C. Prepared media should be stored under refrigeration (2-8°C). Media should be used within 60 days of preparation. Media should not be used if there are any signs of deterioration (discoloration) or contamination, or if the expiration date supplied by the manufacturer has passed.

**Use:** As an alternate medium, instead of fluid thioglycolate broth, for testing the sterility of a variety of specimens.

## StrepB Carrot Broth™
**Composition** per liter:

| | |
|---|---|
| Proteose peptone No. 3 | 25.0g |
| Soluble sStarch | 20.0g |
| Selective agents | 12.2g |
| Morpholinepropanesulfonic acid (MOPS) | 11.0g |
| Na$_2$HPO$_4$ | 8.5g |
| Glucose | 2.5g |
| Sodium pyruvate | 1.0g |
| MgSO$_4$ | 20.0g |
| StrepB carrot broth tiles with growth promoting factors | variable |

pH 7.4 ± 0.1 at 25°C

**Source:** This medium is available from Hardy Diagnostics.

**Preparation:** This medium is supplied as a prepared broth in tubes. The StrepB carrot broth tile is added to a tube just prior to inoculation with a vaginal swab. The tile must remain submerged in the broth.

**Storage/Shelf Life:** Store dehydrated media in the dark in a sealed container below 30°C. Prepared media should be stored under refrigeration (2-8°C). Media should be used within 60 days of preparation. Media should not be used if there are any signs of deterioration (discoloration) or contamination, or if the expiration date supplied by the manufacturer has passed.

**Use:** For detecting the presence of Group B *Streptococcus* infections in pregnant women. This new screening test is an improvement over conventional methods, by increasing sensitivity, and decreasing turnaround time, while lowering overall cost. Tubes show an orange to red color change, typical of group B streptococci. The production of orange, red, or brick red pigment is a unique characteristic of hemolytic

Group B streptococci due to reaction with substrates such as starch, proteose peptone, serum, and folate pathway inhibitors.

## Strep ID Quad Plate
**Composition** per liter:

| | |
|---|---|
| Quadrant I | 5.0mL |
| Quadrant II | 5.0mL |
| Quadrant III | 5.0mL |
| Quadrant IV | 5.0mL |

**Source:** Available as a prepared medium from BD Diagnostic Systems.

### Quadrant I:
**Composition** per 5.0mL:

| | |
|---|---|
| Bacitracin | 0.5mg |
| TSA II agar | 5.0mL |

### Quadrant II:
**Composition** per 5.0mL:

| | |
|---|---|
| TSA II agar | 5.0mL |
| Sheep blood, defibrinated | 0.25mL |

### Quadrant III:
**Composition** per 5.0mL:

| | |
|---|---|
| Bile esculin agar | 5.0mL |

### Quadrant IV:
**Composition** per 5.0mL:

| | |
|---|---|
| Blood agar base with 6.5% NaCl | 5.0mL |

**Preparation of Quadrant Media:** Sterilize agars by autoclaving for 15 min at 15 psi pressure–121°C. Cool to 45°–50°C. Add additional components as filter sterilized solutions. Mix and distribute as 5.0mL aliquots into quadrants.

**Storage/Shelf Life:** Store dehydrated media in the dark in a sealed container below 30°C. Prepared media should be stored under refrigeration (2-8°C). Media should be used within 60 days of preparation. Media should not be used if there are any signs of deterioration (shrinking, cracking, or discoloration) or contamination, or if the expiration date supplied by the manufacturer has passed.

**Use:** For the differentiation and presumptive identification of streptococci. The Strep (*Streptococcus*) ID (Identification) Quad Plate is a four-sectored plate, each containing a different medium.

## *Streptococcal* Growth Medium
**Composition** per liter:

| | |
|---|---|
| Beef heart, solids from infusion | 500.0g |
| Tryptose | 10.0g |
| NaCl | 5.0g |
| Glucose | 1.0g |
| Bromcresol Purple solution | 1.0mL |

pH 7.4 ± 0.2 at 25°C

**Bromcresol Purple Solution:**
**Composition** per 10.0mL:

| | |
|---|---|
| Bromcresol Purple | 0.16g |
| Ethanol (95% solution) | 10.0mL |

**Preparation of Bromcresol Purple Solution:** Add Bromcresol Purple to 10.0mL of ethanol. Mix thoroughly.

**Preparation of Medium:** Add components to distilled/deionized water and bring volume to 1.0L. Mix thoroughly. Distribute into tubes in 5.0mL volumes. Autoclave for 15 min at 15 psi pressure–121°C.

**Storage/Shelf Life:** Store dehydrated media in the dark in a sealed container below 30°C. Prepared media should be stored under refrigeration (2-8°C). Media should be used within 60 days of preparation. Media should not be used if there are any signs of deterioration (discoloration) or contamination, or if the expiration date supplied by the manufacturer has passed.

**Use:** For the cultivation of *Streptococcus* species and other Gram-positive cocci. Growth in this medium turns the indicator yellow and the solution turbid.

## *Streptococcus agalactiae* Selective HiVeg Agar Base with Blood and *Staphylococcus* B toxin

**Composition** per liter:

| | |
|---|---|
| Agar | 13.0g |
| Plant peptone | 10.0g |
| NaCl | 5.0g |
| Plant extract No. 1 | 5.0g |
| Esculin | 1.0g |
| Thallous sulfate | 0.333g |
| Crystal Violet | 1.3g |
| Sheep blood, defibrinated | 50.0mL |
| *Staphylococcus* B toxin | 25.0mL |

pH 7.0 ± 0.2 at 25°C

**Source:** This medium, without blood or toxin, is available as a premixed powder from HiMedia.

**Preparation of Medium:** Add components, except sheep blood and staphylococcal toxin, to distilled/deionized water and bring volume to 925.0mL. Mix thoroughly. Gently heat and bring to boiling. Autoclave for 15 min at 15 psi pressure–121°C. Cool to 45°–50°C. Aseptically add 50.0mL of sterile sheep blood and 25.0mL of staphylococcal toxin. Mix thoroughly. Pour into sterile Petri dishes or leave in tubes.

**Storage/Shelf Life:** Store dehydrated media in the dark in a sealed container below 30°C. Prepared media should be stored under refrigeration (2-8°C). Media should be used within 60 days of preparation. Media should not be used if there are any signs of deterioration (shrinking, cracking, or discoloration) or contamination, or if the expiration date supplied by the manufacturer has passed.

**Use:** For the selective cultivation of *Streptococcus agalactiae*.

## *Streptococcus* Blood Agar, Selective

**Composition** per liter:

| | |
|---|---|
| Agar | 15.0g |
| Pancreatic digest of casein | 10.0g |
| Beef extract | 6.7g |
| Nucleic acid | 6.0g |
| NaCl | 5.0g |
| Sheep blood, defibrinated | 50.0mL |
| Maltose solution | 10.0mL |
| Antibiotic inhibitor | 10.0mL |

pH 7.3 ± 0.2 at 25°C

**Maltose Solution:**

**Composition** per 10.0mL:

| | |
|---|---|
| Maltose | 0.25–5.0g |

**Preparation of Maltose Solution:** Add maltose to distilled/deionized water and bring volume to 10.0mL. Mix thoroughly. Filter sterilize.

**Antibiotic Inhibitor:**

**Composition** per 10.0mL:

| | |
|---|---|
| Polymyxin B sulfate | 0.02g |
| Neomycin sulfate | 0.01g |

**Preparation of Antibiotic Inhibitor:** Add components to distilled/deionized water and bring volume to 10.0mL. Mix thoroughly. Filter sterilize.

**Preparation of Medium:** Add components—except sheep blood, maltose solution, and antibiotic inhibitor—to distilled/deionized water and bring volume to 930.0mL. Mix thoroughly. Gently heat and bring to boiling. Autoclave for 15 min at 15 psi pressure–121°C. Cool to 45°–50°C. Aseptically add sterile sheep blood, sterile maltose solution, and sterile antibiotic inhibitor. Mix thoroughly. Pour into sterile Petri dishes or distribute into sterile tubes.

**Storage/Shelf Life:** Store dehydrated media in the dark in a sealed container below 30°C. Prepared media should be stored under refrigeration (2-8°C). Media should be used within 60 days of preparation. Media should not be used if there are any signs of deterioration (shrinking, cracking, or discoloration) or contamination, or if the expiration date supplied by the manufacturer has passed.

**Use:** For the isolation and cultivation of group A hemolytic *Streptococcus* species from the human respiratory tract.

## *Streptococcus* Enrichment HiVeg Broth (SE HiVeg Broth)

**Composition** per liter:

| | |
|---|---|
| Plant hydrolysate | 26.0g |
| Yeast extract | 6.0g |
| NaCl | 5.0g |
| Synthetic detergent | 3.0g |
| Esculin | 1.0g |
| Sodium citrate | 1.0g |
| Ferric ammonium citrate | 0.5g |
| $NaN_3$ | 0.25g |

pH 7.0 ± 0.2 at 25°C

**Source:** This medium is available as a premixed powder from HiMedia.

**Caution:** Sodium azide is toxic. Azides also react with metals and disposal must be highly diluted.

**Preparation of Medium:** Add components to distilled/deionized water and bring volume to 1.0L. Mix thoroughly. Gently heat and bring to boiling. Distribute into tubes or flasks. Autoclave for 15 min at 15 psi pressure–121°C.

**Storage/Shelf Life:** Store dehydrated media in the dark in a sealed container below 30°C. Prepared media should be stored under refrigeration (2-8°C). Media should be used within 60 days of preparation. Media should not be used if there are any signs of deterioration (discoloration) or contamination, or if the expiration date supplied by the manufacturer has passed.

**Use:** For the selective isolation, cultivation, and enumeration of streptococci from specimens containing a mixed flora.

## *Streptococcus* Selection HiVeg Agar

**Composition** per liter:

| | |
|---|---|
| Agar | 15.0g |
| Plant hydrolysate | 15.0g |
| Glucose | 5.0g |
| Papaic digest of soybean meal | 5.0g |

NaCl ............................................................................4.0g
Sodium citrate ...........................................................1.0g
L-Cystine ..................................................................0.2g
NaN$_3$ .....................................................................0.2g
Na$_2$SO$_3$ ...............................................................0.2g
Crystal Violet ..........................................................0.2mg

pH 7.4 ± 0.2 at 25°C

**Source:** This medium is available as a premixed powder from Hi-Media.

**Preparation of Medium:** Add components to distilled/deionized water and bring volume to 1.0L. Mix thoroughly. Gently heat and bring to boiling. Distribute into tubes or flasks. Autoclave for 15 min at 15 psi pressure–121°C.

**Storage/Shelf Life:** Store dehydrated media in the dark in a sealed container below 30°C. Prepared media should be stored under refrigeration (2-8°C). Media should be used within 60 days of preparation. Media should not be used if there are any signs of deterioration (shrinking, cracking, or discoloration) or contamination, or if the expiration date supplied by the manufacturer has passed.

**Use:** For the selective isolation and enumeration of all types of streptococci including group A beta hemolytic strains.

### *Streptococcus* Selection HiVeg Agar with Cycloheximide

**Composition** per liter:
Agar ........................................................................15.0g
Plant hydrolysate......................................................15.0g
Glucose ....................................................................5.0g
Papaic digest of soybean meal .................................5.0g
NaCl.........................................................................4.0g
Sodium citrate ..........................................................1.0g
NaN$_3$ .....................................................................0.2g
Na$_2$SO$_3$ ...............................................................0.2g
L-Cystine ..................................................................0.2g
Crystal Violet ...........................................................0.2g
Cycloheximide solution ........................................ 10.0mL

pH 7.4 ± 0.2 at 25°C

**Source:** This medium is available as a premixed powder from Hi-Media.

**Cycloheximide Solution:**
**Composition** per 10.0mL:
Cycloheximide .........................................................0.01g

**Preparation of Cycloheximide Solution:** Add cycloheximide to distilled/deionized water and bring volume to 10.0mL. Mix thoroughly. Filter sterilize.

**Preparation of Medium:** Add components, except cycloheximide solution, to distilled/deionized water and bring volume to 990.0mL. Mix thoroughly. Gently heat and bring to boiling. Autoclave for 15 min at 15 psi pressure–121°C. Cool to 50°C. Aseptically add 10.0mL cycloheximide solution. Mix thoroughly. Pour into sterile Petri dishes or aseptically distribute into tubes.

**Storage/Shelf Life:** Store dehydrated media in the dark in a sealed container below 30°C. Prepared media should be stored under refrigeration (2-8°C). Media should be used within 60 days of preparation. Media should not be used if there are any signs of deterioration (shrinking, cracking, or discoloration) or contamination, or if the expiration date supplied by the manufacturer has passed.

**Use:** For the selective isolation and enumeration of all types of streptococci including group A beta hemolytic strains.

### *Streptococcus* Selection HiVeg Broth

**Composition** per liter:
Plant hydrolysate .....................................................15.0g
Glucose ....................................................................5.0g
Papaic digest of soybean meal..................................5.0g
NaCl..........................................................................4.0g
Sodium citrate ...........................................................1.0g
L-Cystine...................................................................0.2g
NaN$_3$ .....................................................................0.2g
Na$_2$SO$_3$ ...............................................................0.2g
Crystal Violet............................................................0.2mg

pH 7.4 ± 0.2 at 25°C

**Source:** This medium is available as a premixed powder from Hi-Media.

**Preparation of Medium:** Add components to distilled/deionized water and bring volume to 1.0L. Mix thoroughly. Gently heat and bring to boiling. Distribute into tubes or flasks. Autoclave for 15 min at 15 psi pressure–121°C.

**Storage/Shelf Life:** Store dehydrated media in the dark in a sealed container below 30°C. Prepared media should be stored under refrigeration (2-8°C). Media should be used within 60 days of preparation. Media should not be used if there are any signs of deterioration (discoloration) or contamination, or if the expiration date supplied by the manufacturer has passed.

**Use:** For the selective cultivation of streptococci including group A beta hemolytic strains.

### *Streptococcus* Selective Medium

**Composition** per liter:
Special peptone........................................................23.0g
Agar ......................................................................10.0g
NaCl.........................................................................5.0g
Starch .......................................................................1.0g
Horse blood, defibrinated ...................................... 50.0mL
Antibiotic inhibitor ................................................ 10.0mL

pH 7.3± 0.2 at 25°C

**Source:** This medium is available as a premixed powder from Oxoid Unipath.

**Antibiotic Inhibitor:**
**Composition** per 10.0mL:
Colistin sulfate.......................................................10.0mg
Oxolinic acid...........................................................5.0mg

**Preparation of Antibiotic Inhibitor:** Add components to distilled/deionized water and bring volume to 10.0mL. Mix thoroughly. Filter sterilize.

**Preparation of Medium:** Add components, except horse blood and antibiotic inhibitor, to distilled/deionized water and bring volume to 940.0mL. Mix thoroughly. Gently heat and bring to boiling. Autoclave for 15 min at 15 psi pressure–121°C. Cool to 45°–50°C. Aseptically add sterile horse blood and sterile antibiotic inhibitor. Mix thoroughly. Pour into sterile Petri dishes or distribute into sterile tubes.

**Storage/Shelf Life:** Store dehydrated media in the dark in a sealed container below 30°C. Prepared media should be stored under refrigeration (2-8°C). Media should be used within 60 days of preparation. Media should not be used if there are any signs of deterioration (shrink-

ing, cracking, or discoloration) or contamination, or if the expiration date supplied by the manufacturer has passed.

**Use:** For the selective isolation of streptococci from clinical specimens or foodstuffs.

## Streptosel™ Broth

**Composition** per liter:

| | |
|---|---|
| Pancreatic digest of casein | 15.0g |
| Glucose | 5.0g |
| Papaic digest of soybean meal | 5.0g |
| NaCl | 4.0g |
| Sodium citrate | 1.0g |
| L-Cystine | 0.2g |
| $Na_2SO_3$ | 0.2g |
| $NaN_3$ | 0.2g |
| Crystal Violet | 0.2mg |

pH 7.4 ± 0.2 at 25°C

**Source:** This medium is available as a premixed powder from BD Diagnostic Systems.

**Caution:** Sodium azide is toxic. Azides also react with metals and disposal must be highly diluted.

**Preparation of Medium:** Add components to distilled/deionized water and bring volume to 1.0L. Mix thoroughly. Distribute into tubes or flasks. Autoclave for 15 min at 13 psi pressure–118°C.

**Storage/Shelf Life:** Store dehydrated media in the dark in a sealed container below 30°C. Prepared media should be stored under refrigeration (2-8°C). Media should be used within 60 days of preparation. Media should not be used if there are any signs of deterioration (discoloration) or contamination, or if the expiration date supplied by the manufacturer has passed.

**Use:** For the selective isolation and cultivation of streptococci from specimens containing a mixed flora.

## Stuart Transport Medium

**Composition** per liter:

| | |
|---|---|
| Sodium glycerophosphate | 10.0g |
| Sodium thioglycolate | 1.0g |
| $CaCl_2 \cdot 2H_2O$ | 0.1g |
| Methylene Blue | 2.0mg |

pH 7.4 ± 0.2 at 25°C

**Preparation of Medium:** Add components to distilled/deionized water and bring volume to 1.0L. Mix thoroughly. Gently heat and bring to boiling. Distribute into 7.0mL screw-capped tubes. Fill tubes to capacity. Autoclave for 15 min at 15 psi pressure–121°C.

**Storage/Shelf Life:** Store dehydrated media in the dark in a sealed container below 30°C. Prepared media should be stored under refrigeration (2-8°C). Media should be used within 60 days of preparation. Media should not be used if there are any signs of deterioration (discoloration) or contamination, or if the expiration date supplied by the manufacturer has passed.

**Use:** For the preservation of *Neisseria* species and other fastidious organisms during their transport from clinic to laboratory.

## Stuart Transport Medium, Modified

**Composition** per liter:

| | |
|---|---|
| Sodium glycerophosphate | 10.0g |
| Agar | 5.0g |

| | |
|---|---|
| L-Cysteine·HCl·$H_2O$ | 0.5g |
| Sodium thioglycolate | 0.5g |
| $CaCl_2 \cdot 2H_2O$ | 0.1g |
| Methylene Blue | 1.0mg |

pH 7.4 ± 0.2 at 25°C

**Source:** This medium is available as a premixed powder from Oxoid Unipath.

**Preparation of Medium:** Add components to distilled/deionized water and bring volume to 1.0L. Mix thoroughly. Gently heat and bring to boiling. Distribute into 7.0mL screw-capped tubes. Fill tubes to capacity. Autoclave for 15 min at 15 psi pressure–121°C.

**Storage/Shelf Life:** Store dehydrated media in the dark in a sealed container below 30°C. Prepared media should be stored under refrigeration (2-8°C). Media should be used within 60 days of preparation. Media should not be used if there are any signs of deterioration (discoloration) or contamination, or if the expiration date supplied by the manufacturer has passed.

**Use:** For the preservation of *Neisseria* species and other fastidious organisms during their transport from clinic to laboratory.

## Sucrose Phosphate Glutamate Transport Medium

**Composition** per liter:

| | |
|---|---|
| Sucrose | 75.0g |
| $Na_2HPO_4$ | 1.22g |
| Glutamic acid | 0.72g |
| $K_2HPO_4$ | 0.52g |
| Bovine serum | 50.0mL |
| Antibiotic inhibitor | 10.0mL |

pH 7.4–7.6 at 25°C

**Antibiotic Inhibitor:**

**Composition** per 10.0mL:

| | |
|---|---|
| Vancomycin | 0.1g |
| Streptomycin | 0.05g |
| Nystatin | 25000U |

**Preparation of Antibiotic Inhibitor:** Add components to distilled/deionized water and bring volume to 10.0mL. Mix thoroughly. Filter sterilize.

**Preparation of Medium:** Add components, except bovine serum and antibiotic inhibitor, to distilled/deionized water and bring volume to 940.0mL. Mix thoroughly. Gently heat and bring to boiling. Adjust pH to 7.4–7.6. Autoclave for 15 min at 15 psi pressure–121°C. Cool to 45°–50°C. Aseptically add sterile bovine serum and sterile antibiotic inhibitor. Mix thoroughly. Aseptically distribute into sterile tubes or flasks.

**Storage/Shelf Life:** Store dehydrated media in the dark in a sealed container below 30°C. Prepared media should be stored under refrigeration (2-8°C). Media should be used within 60 days of preparation. Media should not be used if there are any signs of deterioration (discoloration) or contamination, or if the expiration date supplied by the manufacturer has passed.

**Use:** For the maintenance of *Chlamydia* species during transport.

## Sucrose Phosphate Transport Medium

**Composition** per liter:

| | |
|---|---|
| Sucrose | 68.5g |
| $K_2HPO_4$ | 2.1g |
| $KH_2PO_4$ | 1.1g |

Bovine serum ................................................................ 50.0mL
Antibiotic inhibitor ........................................................ 10.0mL
pH 7.0 ± 0.2 at 25°C

**Antibiotic Inhibitor:**
**Composition** per 10.0mL:
Vancomycin ......................................................................0.1g
Streptomycin ..................................................................0.05g
Nystatin ......................................................................25,000U

**Preparation of Antibiotic Inhibitor:** Add components to distilled/deionized water and bring volume to 10.0mL. Mix thoroughly. Filter sterilize.

**Preparation of Medium:** Add components, except bovine serum and antibiotic inhibitor, to distilled/deionized water and bring volume to 940.0mL. Mix thoroughly. Gently heat and bring to boiling. Adjust pH to 7.0. Autoclave for 15 min at 15 psi pressure–121°C. Cool to 45°–50°C. Aseptically add sterile bovine serum and sterile antibiotic inhibitor. Mix thoroughly. Aseptically distribute into sterile tubes or flasks.

**Storage/Shelf Life:** Store dehydrated media in the dark in a sealed container below 30°C. Prepared media should be stored under refrigeration (2-8°C). Media should be used within 60 days of preparation. Media should not be used if there are any signs of deterioration (discoloration) or contamination, or if the expiration date supplied by the manufacturer has passed.

**Use:** For the maintenance of *Chlamydia* species during transport.

### Sucrose Teepol Tellurite Agar
### (STT Agar)

**Composition** per liter:
Agar ................................................................................20.0g
Beef extract ......................................................................1.0g
Peptone............................................................................1.0g
Sucrose............................................................................1.0g
NaCl................................................................................0.5g
Bromthymol Blue (0.2% solution)................................2.5mL
Tellurite solution .........................................................2.5mL
Sodium lauryl sulfate
    (Teepol, 0.1% solution) ...........................................0.2mL
pH 8.0 ± 0.2 at 25°C

**Tellurite Solution:**
**Composition** per 100.0mL:
$K_2TeO_3$..........................................................................0.05g

**Preparation of Tellurite Solution:** Add the $K_2TeO_3$ to distilled/deionized water and bring the volume to 100.0mL. Mix thoroughly. Filter sterilize. Use freshly prepared solution.

**Caution:** Potassium tellurite is toxic.

**Preparation of Medium:** Add components to distilled/deionized water and bring volume to 1.0L. Mix thoroughly. Gently heat and bring to boiling. Do not autoclave. Pour into sterile Petri dishes.

**Storage/Shelf Life:** Store dehydrated media in the dark in a sealed container below 30°C. Prepared media should be stored under refrigeration (2-8°C). Media should be used within 60 days of preparation. Media should not be used if there are any signs of deterioration (shrinking, cracking, or discoloration) or contamination, or if the expiration date supplied by the manufacturer has passed.

**Use:** For the selective isolation, cultivation, and differentiation of *Vibrio* species based on their ability to ferment sucrose. *Vibrio chol-*

*erae* appears as flat yellow colonies. *Vibrio parahaemolyticus* appears as elevated green-yellow mucoid colonies.

### SUPERCARBA Medium

**Composition** per liter:
Peptone ...........................................................................17.0g
Lactose ............................................................................15.0g
Agar ................................................................................11.0g
Meat extract ......................................................................3.0g
Yeast extract......................................................................3.0g
Sodium deoxycholate........................................................1.0g
Sodium thiosulfate ............................................................1.0g
Bromothymol blue ..........................................................0.08g
Crystal violet...................................................................5.0mg
$ZnSO_4$ ..........................................................................70.0µg
KPC selective supplement solution ..............................10.0mL
pH 7.3± 0.2 at 25°C

**KPC Selective Supplement Solution**
**Composition** per 10.0mL:
Cloxacillin.....................................................................0.25mg
Ertapenem......................................................................0.25µg

**Preparation of KPC Selective Supplement Solution:** Add components to distilled/deionized water and bring volume to 10.0mL. Mix thoroughly to obtain a homogenous suspension with an opaque yellowish appearance. Must be freshly prepared.

**Preparation of Medium:** Add components, except KPC selective supplement solution, to distilled/deionized water and bring volume to 990mL. Mix thoroughly. Gently heat and bring to boiling. Autoclave for 15 min at 15 psi pressure–121°C. Cool to 50°C. Pour into sterile Petri dishes. Aseptically add 10.0mL of KPC selective supplement solution. Mix thoroughly. Pour into sterile Petri dishes.

**Storage/Shelf Life:** Store in the dark. Chromogenic agars are especially light and temperature sensitive; protect from light, excessive heat, moisture, and freezing. Prepared media plates can be kept for one day at ambient temperature. Plates can be stored at least one week under refrigeration (2-8°C) if properly prepared and protected from light and dehydration.

**Use:** For the detection of carbapenemase producing enterobacteria.

### SXT Blood Agar

**Composition** per liter:
Pancreatic digest of casein................................................14.5g
Agar ................................................................................14.0g
NaCl................................................................................5.0g
Papaic digest of soybean meal...........................................5.0g
Growth factor, BBL ..........................................................1.5g
Sulfamethoxazole ..........................................................0.024g
Trimethoprim ................................................................1.25mg
Sheep blood, defibrinated ............................................. 50.0mL
pH 7.3 ± 0.2 at 25°C

**Source:** This medium is available as a premixed powder from BD Diagnostic Systems.

**Preparation of Medium:** Add components, except defibrinated sheep blood, to distilled/deionized water and bring volume to 950.0mL. Mix thoroughly. Gently heat and bring to boiling. Autoclave for 15 min at 15 psi pressure–121°C. Cool to 45°–50°C. Aseptically add 50.0mL of defibrinated sheep blood. Mix thoroughly. Pour into sterile Petri dishes or distribute into sterile tubes.

**Storage/Shelf Life:** Store dehydrated media in the dark in a sealed container below 30°C. Prepared media should be stored under refrigeration (2-8°C). Media should be used within 60 days of preparation. Media should not be used if there are any signs of deterioration (shrinking, cracking, or discoloration) or contamination, or if the expiration date supplied by the manufacturer has passed.

**Use:** For the selective isolation of Lancefield group A and group B streptococci from throat cultures and other clinical specimens.

## Syncase Broth

**Composition** per liter:

| | |
|---|---|
| Casamino acids | 20.0g |
| $K_2HPO_4$ | 8.71g |
| Yeast extract | 6.0g |
| NaCl | 2.5g |

pH 8.5 ± 0.2 at 25°C

**Preparation of Medium:** Add components to distilled/deionized water and bring volume to 1.0L. Mix thoroughly. Adjust pH to 8.5. Distribute into tubes or flasks. Autoclave for 15 min at 15 psi pressure–121°C.

**Storage/Shelf Life:** Store dehydrated media in the dark in a sealed container below 30°C. Prepared media should be stored under refrigeration (2-8°C). Media should be used within 60 days of preparation. Media should not be used if there are any signs of deterioration (discoloration) or contamination, or if the expiration date supplied by the manufacturer has passed.

**Use:** For the cultivation of heat-labile, toxin-producing *Escherichia coli* from foods.

## Synthetic Broth, AOAC (Synthetic Broth, Association of Official Analytical Chemists)

**Composition** per liter:

| | |
|---|---|
| $Na_2HPO_4$ | 4.0g |
| NaCl | 3.0g |
| $K_2HPO_4$ | 1.5g |
| L-Glutamic acid | 1.3g |
| DL-Valine | 1.0g |
| L-Lysine | 0.85g |
| L-Leucine | 0.8g |
| DL-Serine | 0.61g |
| DL-Threonine | 0.5g |
| L-Aspartic acid | 0.45g |
| DL-Isoleucine | 0.44g |
| DL-Alanine | 0.43g |
| L-Arginine | 0.4g |
| DL-Methionine | 0.37g |
| DL-Histidine | 0.3g |
| DL-Phenylalanine | 0.26g |
| L-Tyrosine | 0.21g |
| KCl | 0.2g |
| Aminoacetic acid | 0.06g |
| L-Cystine | 0.05g |
| $MgSO_4$ | 0.05g |
| L-Proline | 0.05g |
| DL-Tryptophan | 0.05g |
| Nicotinamide | 0.01g |
| Thiamine·HCl | 0.01g |

pH 7.1 ± 0.1 at 25°C

**Source:** This medium is available as a premixed powder from BD Diagnostic Systems.

**Preparation of Medium:** Add components to distilled/deionized water and bring volume to 1.0L. Mix thoroughly. Gently heat and bring to boiling. Distribute into tubes or flasks. Autoclave for 20 min at 15 psi pressure–121°C.

**Storage/Shelf Life:** Store dehydrated media in the dark in a sealed container below 30°C. Prepared media should be stored under refrigeration (2-8°C). Media should be used within 60 days of preparation. Media should not be used if there are any signs of deterioration (discoloration) or contamination, or if the expiration date supplied by the manufacturer has passed.

**Use:** For the determination of phenol coefficients of disinfectants.

## T 7Agar Base (m-T7 Agar Base)

**Composition** per liter:

| | |
|---|---|
| Lactose | 20.0g |
| Agar | 15.0g |
| Polyoxyethylene ether W-1 | 5.0g |
| Yeast extract | 3.0g |
| Pancreatic digest of casein | 2.5g |
| Peptic digest of animal tissue | 2.5g |
| Sodium heptadecyl sulfate | 0.1g |
| Bromthymol Blue | 0.1g |
| Bromcresol Purple | 0.1g |

pH 7.4 ± 0.2 at 25°C

**Source:** This medium is available as a premixed powder from BD Diagnostic Systems.

**Preparation of Medium:** Add components to distilled/deionized water and bring volume to 1.0L. Mix thoroughly. Gently heat while stirring and bring to boiling. Distribute into tubes or flasks. Autoclave for 15 min at 15 psi pressure–121°C. Cool to 45°–50°C. The medium may be made more selective by adding 1.0mg of penicillin G per liter. Pour into sterile Petri dishes or leave in tubes.

**Storage/Shelf Life:** Store dehydrated media in the dark in a sealed container below 30°C. Prepared media should be stored under refrigeration (2-8°C). Media should be used within 60 days of preparation. Media should not be used if there are any signs of deterioration (shrinking, cracking, or discoloration) or contamination, or if the expiration date supplied by the manufacturer has passed.

**Use:** For the selective recovery and differential identification of injured coliform microorganisms from chlorinated water by the membrane filter method. For rapid estimation of the bacteriological quality of water using the membrane filter method.

## Tarshis Blood Agar

**Composition** per 1050.0mL:

| | |
|---|---|
| Beef heart infusion | 500.0g |
| Agar | 15.0g |
| Meat peptone | 10.0g |
| NaCl | 5.0g |
| Penicillin G, sterile | 100,000U |
| Sheep blood, sterile | 300.0mL |
| Glycerol | 10.0mL |

pH 6.6 ± 0.2 at 25°C

**Preparation of Medium:** Add components, except sheep blood and penicillin G, to distilled/deionized water and bring volume to 750.0mL.

Mix thoroughly. Gently heat and bring to boiling. Autoclave for 15 min at 15 psi pressure–121°C. Cool to 45°–50°C. Aseptically add sterile sheep blood and sterile penicillin G. Mix thoroughly. Pour into sterile Petri dishes or distribute into sterile tubes.

**Storage/Shelf Life:** Store dehydrated media in the dark in a sealed container below 30°C. Prepared media should be stored under refrigeration (2-8°C). Media should be used within 60 days of preparation. Media should not be used if there are any signs of deterioration (shrinking, cracking, or discoloration) or contamination, or if the expiration date supplied by the manufacturer has passed.

**Use:** For the isolation and cultivation of *Mycobacterium tuberculosis*.

## TAT Broth Base
## (Trypticase™ Azolectin Tween™ Broth Base)

**Composition** per liter:

| | |
|---|---|
| Pancreatic digest of casein | 20.0g |
| Lecithin | 5.0g |
| Polysorbate 20 (Tween™ 20) | 40.0mL |

pH 7.2 ± 0.2 at 25°C

**Source:** This medium is available as a premixed powder from BD Diagnostic Systems.

**Preparation of Medium:** Add pancreatic digest of casein and lecithin to distilled/deionized water and bring volume to 960.0mL. Add the Tween™ 20. Mix thoroughly. Gently heat and bring to 48°–50°C for 30 min. Distribute into tubes or flasks. Autoclave for 15 min at 15 psi pressure–121°C.

**Use:** For the isolation of Gram-negative organisms from topical drugs and cosmetics.

## TAT HiVeg Broth Base with Polysorbate

**Composition** per liter:

| | |
|---|---|
| Plant hydrolysate | 20.0g |
| Azolectin | 5.0g |
| Polysorbate 20 (Tween™ 20) | 40.0mL |

pH 7.2 ± 0.2 at 25°C

**Source:** This medium, without polysorbate, is available as a premixed powder from HiMedia.

**Preparation of Medium:** Add plant hydrolysate and azolectin to distilled/deionized water and bring volume to 960.0mL. Add 40.0mL Tween™ 20. Mix thoroughly. Gently heat and bring to 48°–50°C for 30 min. Distribute into tubes or flasks. Autoclave for 15 min at 15 psi pressure–121°C.

**Storage/Shelf Life:** Store dehydrated media in the dark in a sealed container below 30°C. Prepared media should be stored under refrigeration (2-8°C). Media should be used within 60 days of preparation. Media should not be used if there are any signs of deterioration (discoloration) or contamination, or if the expiration date supplied by the manufacturer has passed.

**Use:** For the isolation of Gram-negative organisms from topical drugs and cosmetics.

## TB Broth Base

**Composition** per liter:

| | |
|---|---|
| Proteose peptone | 4.0g |
| Na$_2$HPO$_4$ | 2.5g |
| Yeast extract | 2.0g |

| | |
|---|---|
| Sodium citrate | 1.5g |
| KH$_2$PO$_4$ | 1.0g |
| MgSO$_4$·7H$_2$O | 0.6g |
| Polysorbate 80 | 0.5g |
| Bovine albumin solution | 50.0mL |
| Glucose solution | 10.0mL |
| Glycerol | 5.0mL |

pH 7.0 ± 0.2 at 25°C

**Source:** This medium is available from HiMedia.

**Glucose Solution:**
**Composition** per 10.0mL:

| | |
|---|---|
| Glucose | 5.0g |

**Preparation of Glucose Solution:** Add glucose to distilled/deionized water and bring volume to 10.0mL. Mix thoroughly. Filter sterilize.

**Bovine Albumin Solution:**
**Composition** per 50.0mL:

| | |
|---|---|
| Bovine serum albumin | 5.0g |

**Preparation of Bovine Albumin Solution:** Add bovine serum albuin to distilled/deionized water and bring volume to 50.0mL. Mix thoroughly. Filter sterilize.

**Preparation of Medium:** Add components, except bovine albumin and glucose solution, to distilled/deionized water and bring volume to 960.0mL. Mix thoroughly. Autoclave for 15 min at 15 psi pressure–121°C. Cool to 50°C. Aseptically add glucose solution and bovine albumin solution. Mix thoroughly. Aseptically distribute into sterile tubes or flasks.

**Storage/Shelf Life:** Store dehydrated media in the dark in a sealed container below 30°C. Prepared media should be stored under refrigeration (2-8°C). Media should be used within 60 days of preparation. Media should not be used if there are any signs of deterioration (discoloration) or contamination, or if the expiration date supplied by the manufacturer has passed.

**Use:** For the cultivation of *Mycobacterium tuberculosis*.

## TBX Agar
## (Tryptone Bile X-glucuronide Agar)

**Composition** per liter:

| | |
|---|---|
| Peptone | 20.0g |
| Agar | 15.0g |
| Bile salts | 1.5g |
| X-β-D-glucuronide | 0.075g |

pH 7.2 ± 0.2 at 25°C

**Source:** This medium is available from Fluka, Sigma-Aldrich.

**Preparation of Medium:** Add components to distilled/deionized water and bring volume to 1.0L. Mix thoroughly. Gently heat while stirring and bring to boiling. Autoclave for 15 min at 15 psi pressure–121°C. Pour into sterile Petri dishes.

**Storage/Shelf Life:** Store dehydrated media in the dark in a sealed container below 30°C. Prepared media should be stored under refrigeration (2-8°C). Media should be used within 60 days of preparation. Media should not be used if there are any signs of deterioration (shrinking, cracking, or discoloration) or contamination, or if the expiration date supplied by the manufacturer has passed.

**Use:** For the detection and enumeration of *E. coli* in foodstuffs, animal food, and water without further confirmation. *E. coli* colonies are colored blue-green. The presence of the enzyme β-D-glucuronidase differentiates most *E. coli* sp. from other coliforms. *E. coli* absorbs the chromogenic substrate 5-bromo-4-chloro-3-indolyl-β-D-glucuronide. The enzyme β-glucuronidase splits the bond between the chromophore 5-bromo-4-chloro-3-indolyl and the β-D-glucuronide. Growth of accompanying Gram-positive flora is largely inhibited by the use of bile salts.

## TBX Chromogenic Agar
### (Tryptone Bile X-Glucuronide Agar)
**Composition** per liter:

| | |
|---|---|
| Casein peptone | 20.0g |
| Agar | 15.0g |
| X-β-D-Glucuronide | 0.075g |

pH 7.2 ± 0.2 at 25°C

**Source:** This medium is available from CONDA, Barcelona, Spain.

**Preparation of Medium:** Add components to distilled/deionized water and bring volume to 1.0L. Mix thoroughly. Distribute into tubes or flasks. Autoclave for 15 min at 15 psi pressure–121°C. Pour into sterile Petri dishes.

**Storage/Shelf Life:** Store dehydrated media in the dark in a sealed container below 30°C. Prepared media should be stored in the dark under refrigeration (2-8°C). Chromogenic media are especially light and temperature sensitive; protect from light, excessive heat, moisture, and freezing. Media should not be used if there are any signs of deterioration (shrinking, cracking, or discoloration) or contamination, or if the expiration date supplied by the manufacturer has passed.

**Use:** For the detection and enumeration of *E. coli* in foods. *E. coli* colonies are blue-green on this medium.

## TCBS Agar
### (Thiosulfate Citrate Bile Salt Sucrose Agar)
**Composition** per liter:

| | |
|---|---|
| Sucrose | 20.0g |
| Agar | 14.0g |
| NaCl | 10.0g |
| Sodium citrate | 10.0g |
| $Na_2S_2O_3$ | 10.0g |
| Yeast extract | 5.0g |
| Pancreatic digest of casein | 5.0g |
| Peptic digest of animal tissue | 5.0g |
| Oxgall | 5.0g |
| Sodium cholate | 3.0g |
| Ferric citrate | 1.0g |
| Thymol Blue | 0.04g |
| Bromthymol Blue | 0.04g |

pH 8.6 ± 0.2 at 25°C

**Source:** This medium is available as a premixed powder from BD Diagnostic Systems.

**Preparation of Medium:** Add components to distilled/deionized water and bring volume to 1.0L. Mix thoroughly. Gently heat while stirring and bring to boiling. Do not autoclave. Cool to 45°–50°C. Pour into sterile Petri dishes or distribute into sterile tubes.

**Storage/Shelf Life:** Store dehydrated media in the dark in a sealed container below 30°C. Prepared media should be stored under refrigeration (2-8°C). Media should be used within 60 days of preparation. Media should not be used if there are any signs of deterioration (shrinking, cracking, or discoloration) or contamination, or if the expiration date supplied by the manufacturer has passed.

**Use:** For the selective isolation of *Vibrio cholerae* and *Vibrio parahaemolyticus* from a variety of clinical and nonclinical specimens.

## TCBS HiVeg Agar
**Composition** per liter:

| | |
|---|---|
| Sucrose | 20.0g |
| Agar | 15.0g |
| Plant peptone No. 3 | 15.0g |
| NaCl | 10.0g |
| Sodium citrate | 10.0g |
| $Na_2S_2O_3$ | 10.0g |
| Yeast extract | 6.0g |
| Synthetic detergent No. II | 2.0g |
| Ferric citrate | 1.0g |
| Thymol Blue | 0.04g |
| Bromthymol Blue | 0.04g |

pH 8.6 ± 0.2 at 25°C

**Source:** This medium is available as a premixed powder from Hi-Media.

**Preparation of Medium:** Add components to distilled/deionized water and bring volume to 1.0L. Mix thoroughly. Gently heat while stirring and bring to boiling. Do not autoclave. Cool to 45°–50°C. Pour into sterile Petri dishes or distribute into sterile tubes.

**Storage/Shelf Life:** Store dehydrated media in the dark in a sealed container below 30°C. Prepared media should be stored under refrigeration (2-8°C). Media should be used within 60 days of preparation. Media should not be used if there are any signs of deterioration (shrinking, cracking, or discoloration) or contamination, or if the expiration date supplied by the manufacturer has passed.

**Use:** For the selective isolation of *Vibrio cholerae* and *Vibrio parahaemolyticus* from a variety of clinical and nonclinical specimens. For the cultivation of enteropathogenic vibrios causing food poisoning.

## TCBS HiVeg Agar (Selective)
**Composition** per liter:

| | |
|---|---|
| Sucrose | 20.0g |
| Agar | 15.0g |
| Plant special peptone | 14.5g |
| NaCl | 10.0g |
| Sodium citrate | 10.0g |
| $Na_2S_2O_3$ | 10.0g |
| Yeast extract | 5.0g |
| Synthetic detergent No. II | 2.0g |
| Synthetic detergent No. IV | 1.5g |
| Ferric citrate | 1.0g |
| BromthymolBlue | 0.04g |
| Thymol Blue | 0.04g |

pH 8.6 ± 0.2 at 25°C

**Source:** This medium is available as a premixed powder from Hi-Media.

**Preparation of Medium:** Add components to distilled/deionized water and bring volume to 1.0L. Mix thoroughly. Gently heat while stirring and bring to boiling. Do not autoclave. Cool to 45°–50°C. Pour into sterile Petri dishes or distribute into sterile tubes.

**Storage/Shelf Life:** Store dehydrated media in the dark in a sealed container below 30°C. Prepared media should be stored under refrigeration (2-8°C). Media should be used within 60 days of preparation. Media should not be used if there are any signs of deterioration (shrink-

ing, cracking, or discoloration) or contamination, or if the expiration date supplied by the manufacturer has passed.

**Use:** For the selective isolation of *Vibrio cholerae* and other enteropathogenic vibrios.

# TCH Medium
## (Thiophene 2 Carboxylic Acid Hydrazide Medium)
**Composition** per 1105.0mL:

Thiophene-2-carboxylic acid hydrazide ................................1.1mg
Middlebrook 7H10 agar base.............................................. 1.0L
OADC enrichment ......................................................... 100.0mL
Glycerol ........................................................................ 5.0mL

pH 6.6 ± 0.2 at 25°C

### Middlebrook 7H10 Agar Base:
**Composition** per liter:

Agar ..............................................................................15.0g
$Na_2HPO_4$.....................................................................1.5g
$KH_2PO_4$.......................................................................1.5g
$(NH_4)_2SO_4$.................................................................0.5g
L-Glutamic acid ............................................................0.5g
Sodium citrate ..............................................................0.4g
Ferric ammonium citrate...............................................0.04g
$MgSO_4 \cdot 7H_2O$ .......................................................0.025g
$ZnSO_4 \cdot 7H_2O$ ........................................................1.0mg
$CuSO_4 \cdot 5H_2O$ ........................................................1.0mg
Pyridoxine ....................................................................1.0mg
Biotin ...........................................................................0.5mg
$CaCl_2 \cdot 2H_2O$...........................................................0.5mg
Malachite Green............................................................0.25mg

**Preparation of Middlebrook 7H10 Agar Base:** Add glycerol to 900.0mL of distilled/deionized water and add remaining components. Mix thoroughly. Gently heat and bring to boiling.

### Middlebrook OADC Enrichment:
**Composition** per 100.0mL:

Bovine albumin fraction V ............................................5.0g
Glucose .........................................................................2.0g
NaCl..............................................................................0.85g
Oleic acid .....................................................................0.05g
Catalase ........................................................................4.0mg

**Source:** This enrichment is available as a prepared enrichment from BD Diagnostic Systems.

**Preparation of Middlebrook OADC Enrichment:** Add components to distilled/deionized water and bring volume to 100.0mL. Mix thoroughly. Filter sterilize.

**Preparation for Medium:** Combine components. Mix thoroughly. Distribute into tubes or flasks. Autoclave for 15 min at 15 psi pressure–121°C. Pour into sterile Petri dishes or leave in tubes.

**Storage/Shelf Life:** Store dehydrated media in the dark in a sealed container below 30°C. Prepared media should be stored under refrigeration (2-8°C). Media should be used within 60 days of preparation. Media should not be used if there are any signs of deterioration (shrinking, cracking, or discoloration) or contamination, or if the expiration date supplied by the manufacturer has passed.

**Use:** For the differentiation of *Mycobacterium* species based on sensitivity to TCH. *Mycobacterium bovis* is inhibited by TCH. *Mycobacterium tuberculosis* and other mycobacteria are generally resistant to low concentrations of TCH. This distinguishes *Mycobacterium bovis* from other nonchromogenic, slow-growing mycobacteria.

# Teepol Broth, Enriched
## (m-Teepol Broth, Enriched)
**Composition** per liter:

Peptone .........................................................................40.0g
Lactose..........................................................................30.0g
Yeast extract.................................................................6.0g
Phenol Red....................................................................0.2g
Sodium lauryl sulfate
  (Teepol, 0.1% solution) ..............................................4.0mL

pH 7.4 ± 0.2 at 25°C

**Preparation of Medium:** Add components to distilled/deionized water and bring volume to 1.0L. Mix thoroughly. Distribute into tubes or flasks. Autoclave for 15 min at 15 psi pressure–121°C.

**Storage/Shelf Life:** Store dehydrated media in the dark in a sealed container below 30°C. Prepared media should be stored under refrigeration (2-8°C). Media should be used within 60 days of preparation. Media should not be used if there are any signs of deterioration (discoloration) or contamination, or if the expiration date supplied by the manufacturer has passed.

**Use:** For the enumeration of coliform organisms and *Escherichia coli* in water by the membrane filter method.

# Teepol HiVeg Broth
**Composition** per liter:

Plant peptone ................................................................20.0g
Lactose..........................................................................10.0g
NaCl..............................................................................5.0g
Teepol ...........................................................................1.0g
Phenol Red....................................................................0.02g

pH 7.6 ± 0.2 at 25°C

**Source:** This medium is available as a premixed powder from Hi-Media.

**Preparation of Medium:** Add components to distilled/deionized water and bring volume to 1.0L. Mix thoroughly. Distribute into tubes or flasks. Autoclave for 15 min at 15 psi pressure–121°C.

**Storage/Shelf Life:** Store dehydrated media in the dark in a sealed container below 30°C. Prepared media should be stored under refrigeration (2-8°C). Media should be used within 60 days of preparation. Media should not be used if there are any signs of deterioration (discoloration) or contamination, or if the expiration date supplied by the manufacturer has passed.

**Use:** For the enumeration of coliform organisms and *Escherichia coli* in water by the membrane filter method. For the selective isolation and identification of enteric, lactose-fermenting bacteria.

# Tellurite Glycine Agar
**Composition** per liter:

Agar ..............................................................................17.5g
Pancreatic digest of casein............................................10.0g
Glycine..........................................................................10.0g
Yeast extract.................................................................6.5g
D-Mannitol....................................................................5.0g
$K_2HPO_4$......................................................................5.0g
LiCl................................................................................5.0g

Enzymatic hydrolysate of soybean meal ........................................3.5g
Chapman tellurite solution......................................................10.0mL
pH 7.2 ± 0.2 at 25°C

**Source:** This medium is available as a premixed powder from BD Diagnostic Systems.

**Caution:** Lithium chloride is harmful. Avoid bodily contact and inhalation of vapors. On contact with skin wash with plenty of water immediately.

**Chapman Tellurite Solution:**
**Composition** per 100.0mL:
$K_2TeO_3$...............................................................................1.0g

**Preparation of Chapman Tellurite Solution:** Add $K_2TeO_3$ to distilled/deionized water and bring volume to 100.0mL. Mix thoroughly. Filter sterilize.

**Caution:** Potassium tellurite is toxic.

**Preparation of Medium:** Add components, except Chapman tellurite solution, to distilled/deionized water and bring volume to 990.0mL. Mix thoroughly. Gently heat and bring to boiling. Autoclave for 15 min at 15 psi pressure–121°C. Cool to 50°–55°C. Aseptically add 10.0mL of sterile Chapman tellurite solution. Mix thoroughly. Pour into sterile Petri dishes or distribute into sterile tubes. Allow the surface of the plates to dry before inoculating.

**Storage/Shelf Life:** Store dehydrated media in the dark in a sealed container below 30°C. Prepared media should be stored under refrigeration (2-8°C). Media should be used within 60 days of preparation. Media should not be used if there are any signs of deterioration (shrinking, cracking, or discoloration) or contamination, or if the expiration date supplied by the manufacturer has passed.

**Use:** For the isolation and cultivation of coagulase-positive staphylococci.

## Tellurite Glycine Agar

**Composition** per liter:
Agar .......................................................................................16.0g
Pancreatic digest of casein......................................................10.0g
Glycine...................................................................................10.0g
Yeast extract.............................................................................5.0g
D-Mannitol................................................................................5.0g
$K_2HPO_4$................................................................................5.0g
LiCl...........................................................................................5.0g
Chapman tellurite solution......................................................20.0mL
pH 7.2 ± 0.2 at 25°C

**Source:** This medium is available as a premixed powder from BD Diagnostic Systems.

**Chapman Tellurite Solution:**
**Composition** per 100.0mL:
$K_2TeO_3$...............................................................................1.0g

**Preparation of Chapman Tellurite Solution:** Add $K_2TeO_3$ to distilled/deionized water and bring volume to 100.0mL. Mix thoroughly. Filter sterilize.

**Caution:** Potassium tellurite is toxic.

**Preparation of Medium:** Add components, except Chapman tellurite solution, to distilled/deionized water and bring volume to 980.0mL. Mix thoroughly. Gently heat and bring to boiling. Autoclave for 15 min at 15 psi pressure–121°C. Cool to 50°–55°C. Aseptically add 20.0mL of sterile Chapman tellurite solution. Mix thoroughly. Pour into sterile Petri

dishes or distribute into sterile tubes. Allow the surface of the plates to dry before inoculating.

**Storage/Shelf Life:** Store dehydrated media in the dark in a sealed container below 30°C. Prepared media should be stored under refrigeration (2-8°C). Media should be used within 60 days of preparation. Media should not be used if there are any signs of deterioration (shrinking, cracking, or discoloration) or contamination, or if the expiration date supplied by the manufacturer has passed.

**Use:** For the quantitative detection of coagulase-positive staphylococci from foods and other sources.

## Tergitol 7 Agar

**Composition** per liter:
Lactose...................................................................................20.0g
Agar .......................................................................................13.0g
Peptone...................................................................................10.0g
Yeast extract.............................................................................6.0g
Meat extract..............................................................................5.0g
Tergitol-7.................................................................................0.1g
Bromthymol Blue......................................................................0.05g
TTC solution..........................................................................5.0mL
pH 7.2 ± 0.2 at 25°C

**Source:** This medium is available as a premixed powder from Oxoid Unipath.

**TTC Solution:**
**Composition** per 100.0mL:
Triphenyltetrazolium chloride....................................................0.05g

**Preparation of TTC Solution:** Add triphenyltetrazolium chloride to distilled/deionized water and bring volume to 100.0mL. Mix thoroughly. Filter sterilize.

**Preparation of Medium:** Add components to distilled/deionized water and bring volume to 995.0mL. Mix thoroughly. Gently heat and bring to boiling. Autoclave for 15 min at 15 psi pressure–121°C. Cool to 50°C. Aseptically add 5.0mL of sterile TTC solution. Mix thoroughly. Pour into sterile Petri dishes or distribute into sterile tubes.

**Storage/Shelf Life:** Store dehydrated media in the dark in a sealed container below 30°C. Prepared media should be stored under refrigeration (2-8°C). Media should be used within 60 days of preparation. Media should not be used if there are any signs of deterioration (shrinking, cracking, or discoloration) or contamination, or if the expiration date supplied by the manufacturer has passed.

**Use:** For the detection and enumeration of coliforms. Lactose-fermenting bacteria appear as yellow colonies. Lactose-nonfermenting bacteria appear as blue colonies.

## Tergitol 7 Agar

**Composition** per liter:
Agar .......................................................................................15.0g
Lactose...................................................................................10.0g
Yeast extract.............................................................................3.0g
Pancreatic digest of casein........................................................2.5g
Peptic digest of animal tissue ....................................................2.5g
Tergitol 7..................................................................................0.1g
Bromthymol Blue....................................................................25.0mg
TTC solution..........................................................................3.0mL
pH 6.9 ± 0.2 at 25°C

**Source:** This medium is available as a premixed powder from BD Diagnostic Systems.

**TTC Solution:**

**Composition** per 100.0mL:

Triphenyltetrazolium chloride.........................................1.0g

**Preparation of TTC Solution:** Add triphenyltetrazolium chloride to distilled/deionized water and bring volume to 100.0mL. Mix thoroughly. Filter sterilize.

**Preparation of Medium:** Add components to distilled/deionized water and bring volume to 997.0mL. Mix thoroughly. Gently heat and bring to boiling. Autoclave for 15 min at 15 psi pressure–121°C. Cool to 50°C. Aseptically add 3.0mL of sterile TTC solution. Mix thoroughly. Pour into sterile Petri dishes or distribute into sterile tubes.

**Storage/Shelf Life:** Store dehydrated media in the dark in a sealed container below 30°C. Prepared media should be stored under refrigeration (2-8°C). Media should be used within 60 days of preparation. Media should not be used if there are any signs of deterioration (shrinking, cracking, or discoloration) or contamination, or if the expiration date supplied by the manufacturer has passed.

**Use:** For the selective isolation and differentiation of coliform bacteria based on lactose fermentation. Lactose-fermenting bacteria appear as yellow colonies. Lactose-nonfermenting bacteria appear as blue colonies.

## Tergitol 7 Agar H

**Composition** per liter:

| | |
|---|---|
| Agar | 15.0g |
| Lactose | 10.0g |
| Yeast extract | 3.0g |
| Pancreatic digest of casein | 2.5g |
| Peptic digest of animal tissue | 2.5g |
| Ferric ammonium citrate | 0.5g |
| Na$_2$S$_2$O$_3$ | 0.5g |
| Tergitol 7 | 0.1g |
| Bromthymol Blue | 0.025g |

pH 7.2 ± 0.2

**Preparation of Medium:** Add components to distilled/deionized water and bring volume to 1.0L. Mix thoroughly. Gently heat and bring to boiling. Distribute into tubes or flasks. Autoclave for 15 min at 15 psi pressure–121°C. Pour into sterile Petri dishes or leave in tubes.

**Storage/Shelf Life:** Store dehydrated media in the dark in a sealed container below 30°C. Prepared media should be stored under refrigeration (2-8°C). Media should be used within 60 days of preparation. Media should not be used if there are any signs of deterioration (shrinking, cracking, or discoloration) or contamination, or if the expiration date supplied by the manufacturer has passed.

**Use:** For the selective isolation and differentiation of enteric bacteria from urine.

## Tergitol 7 Broth

**Composition** per liter:

| | |
|---|---|
| Lactose | 10.0g |
| Yeast extract | 3.0g |
| Pancreatic digest of casein | 2.5g |
| Peptic digest of animal tissue | 2.5g |
| Tergitol 7 | 0.1g |
| Bromthymol Blue | 25.0mg |
| TTC solution | 3.0mL |

pH 6.9 ± 0.2 at 25°C

**Source:** This medium is available as a premixed powder from BD Diagnostic Systems.

**TTC Solution:**

**Composition** per 100.0mL:

Triphenyltetrazolium chloride .........................................1.0g

**Preparation of TTC Solution:** Add triphenyltetrazolium chloride to distilled/deionized water and bring volume to 100.0mL. Mix thoroughly. Filter sterilize.

**Preparation of Medium:** Add components to distilled/deionized water and bring volume to 997.0mL. Mix thoroughly. Gently heat while stirring and bring to boiling. Autoclave for 15 min at 15 psi pressure–121°C. Cool to 25°C. Aseptically add 3.0mL of sterile TTC solution. Mix thoroughly.

**Storage/Shelf Life:** Store dehydrated media in the dark in a sealed container below 30°C. Prepared media should be stored under refrigeration (2-8°C). Media should be used within 60 days of preparation. Media should not be used if there are any signs of deterioration (discoloration) or contamination, or if the expiration date supplied by the manufacturer has passed.

**Use:** For the isolation and cultivation of coliforms, *Salmonella,* and other enteric bacteria.

## Tergitol 7 HiVeg Agar Base with TTC

**Composition** per liter:

| | |
|---|---|
| Agar | 15.0g |
| Lactose | 10.0g |
| Plant peptone No. 3 | 5.0g |
| Yeast extract | 3.0g |
| Sodium heptadecyl sulfate | 0.1g |
| Bromthymol Blue | 0.025g |
| TTC solution | 5.0mL |

pH 7.2 ± 0.2 at 25°C

**Source:** This medium, without TTC solution, is available as a premixed powder from HiMedia.

**TTC Solution:**

**Composition** per 100.0mL:

Triphenyltetrazolium chloride .........................................0.05g

**Preparation of TTC Solution:** Add triphenyltetrazolium chloride to distilled/deionized water and bring volume to 100.0mL. Mix thoroughly. Filter sterilize.

**Preparation of Medium:** Add components to distilled/deionized water and bring volume to 995.0mL. Mix thoroughly. Gently heat and bring to boiling. Autoclave for 15 min at 15 psi pressure–121°C. Cool to 50°C. Aseptically add 5.0mL of sterile TTC solution. Mix thoroughly. Pour into sterile Petri dishes or distribute into sterile tubes.

**Storage/Shelf Life:** Store dehydrated media in the dark in a sealed container below 30°C. Prepared media should be stored under refrigeration (2-8°C). Media should be used within 60 days of preparation. Media should not be used if there are any signs of deterioration (shrinking, cracking, or discoloration) or contamination, or if the expiration date supplied by the manufacturer has passed.

**Use:** For the detection and enumeration of coliforms. Lactose-fermenting bacteria appear as yellow colonies. Lactose-nonfermenting bacteria appear as blue colonies. For the selective enumeration and identification of coliform organisms, as per Indian Standard published by BIS.

## Tergitol 7 HiVeg Agar H

**Composition** per liter:

| | |
|---|---|
| Agar | 15.0g |
| Lactose | 10.0g |
| Plant peptone No. 3 | 5.0g |
| Yeast extract | 3.0g |
| Ferric ammonium citrate | 0.5g |
| $Na_2S_2O_3$ | 0.5g |
| Tergitol 7 | 0.1g |
| Bromthymol Blue | 0.025g |

pH 7.2 ± 0.2

**Preparation of Medium:** Add components to distilled/deionized water and bring volume to 1.0L. Mix thoroughly. Gently heat and bring to boiling. Distribute into tubes or flasks. Autoclave for 15 min at 15 psi pressure–121°C. Pour into sterile Petri dishes or leave in tubes.

**Storage/Shelf Life:** Store dehydrated media in the dark in a sealed container below 30°C. Prepared media should be stored under refrigeration (2-8°C). Media should be used within 60 days of preparation. Media should not be used if there are any signs of deterioration (shrinking, cracking, or discoloration) or contamination, or if the expiration date supplied by the manufacturer has passed.

**Use:** For the selective isolation and differentiation of enteric bacteria from urine.

## Tergitol 7 HiVeg Broth

**Composition** per liter:

| | |
|---|---|
| Lactose | 10.0g |
| Plant peptone No. 3 | 5.0g |
| Yeast extract | 3.0g |
| Tergitol 7 | 0.1g |
| Bromthymol Blue | 0.025g |

pH 6.9 ± 0.2

**Source:** This medium is available as a premixed powder from Hi-Media.

**Preparation of Medium:** Add components to distilled/deionized water and bring volume to 1.0L. Mix thoroughly. Gently heat and bring to boiling. Distribute into tubes or flasks. Autoclave for 15 min at 15 psi pressure–121°C.

**Storage/Shelf Life:** Store dehydrated media in the dark in a sealed container below 30°C. Prepared media should be stored under refrigeration (2-8°C). Media should be used within 60 days of preparation. Media should not be used if there are any signs of deterioration (discoloration) or contamination, or if the expiration date supplied by the manufacturer has passed.

**Use:** For the selective and differential medium for detection and enumeration of coliforms.

## Tetrathionate Broth

**Composition** per liter:

| | |
|---|---|
| $Na_2S_2O_3$ | 40.7g |
| $CaCO_3$ | 25.0g |
| NaCl | 4.5g |
| Peptone | 4.5g |
| Yeast extract | 1.8g |
| Beef extract | 0.9g |
| Iodine solution | 20.0mL |

**Iodine Solution:**

**Composition** per 20.0mL:

| | |
|---|---|
| Iodine | 6.0g |
| KI | 5.0g |

**Preparation of Iodine Solution:** Add iodine and KI to distilled/deionized water and bring volume to 20.0mL. Mix thoroughly.

**Preparation of Medium:** Add components, except iodine solution, to distilled/deionized water and bring volume to 980.0mL. Mix thoroughly. Gently heat and bring to boiling. Do not autoclave. Cool to 40°C. Add 20.0mL of iodine solution. Mix thoroughly. Distribute into tubes in 10.0mL volumes. Use medium the same day it is prepared.

**Storage/Shelf Life:** Store dehydrated media in the dark in a sealed container below 30°C. Prepared media should be stored under refrigeration (2-8°C). Media should be used within 60 days of preparation. Media should not be used if there are any signs of deterioration (discoloration) or contamination, or if the expiration date supplied by the manufacturer has passed.

**Use:** For the selective isolation and enrichment of *Salmonella typhi* and other salmonellae from fecal specimens, sewage, and other specimens.

## Tetrathionate Broth
## (FDA M145)

**Composition** per 1030.0mL:

| | |
|---|---|
| Tetrathionate broth base | 1.0L |
| Iodine-potassium iodide solution | 20.0mL |
| Brilliant Green solution | 10.0mL |

pH 8.4 ± 0.2 at 25°C

**Tetrathionate Broth Base:**

**Composition** per liter:

| | |
|---|---|
| $Na_2S_2O_3 \cdot 5H_2O$ | 30.0g |
| $CaCO_3$ | 10.0g |
| Polypeptone™ | 5.0g |
| Bile salts | 1.0g |

**Preparation of Tetrathionate Broth Base:** Add components to distilled/deionized water and bring volume to 1.0L. Mix thoroughly. Gently heat and bring to boiling. A slight precipitate will remain. Do not autoclave. Cool to 25°C. Store at 4°C.

**Iodine-Potassium Iodide Solution:**

**Composition** per 20.0mL:

| | |
|---|---|
| Iodine, resublimed | 6.0g |
| KI | 5.0g |

**Preparation of Iodine-Potassium Iodide Solution:** Add KI to 5.0mL of sterile distilled/deionized water. Mix thoroughly. Add iodine. Mix thoroughly. Bring volume to 20.0mL with sterile distilled/deionized water.

**Brilliant Green Solution:**

**Composition** per 100.0mL:

| | |
|---|---|
| Brilliant Green | 0.1g |

**Preparation of Brilliant Green Solution:** Add Brilliant Green to sterile distilled/deionized water and bring volume to 100.0mL. Mix thoroughly.

**Preparation of Medium:** Combine 1.0L of tetrathionate broth base, 20.0mL of iodine-potassium iodide solution, and 10.0mL of Brilliant Green solution. Mix thoroughly. Aseptically distribute into tubes in 10.0mL volumes. Do not heat medium after it has been mixed.

**Storage/Shelf Life:** Store dehydrated media in the dark in a sealed container below 30°C. Prepared media should be stored under refriger-

ation (2-8°C). Media should be used within 60 days of preparation. Media should not be used if there are any signs of deterioration (discoloration) or contamination, or if the expiration date supplied by the manufacturer has passed.

**Use:** For the selective isolation and cultivation of *Salmonella* species from foods.

## Tetrathionate Broth
### (TT Broth)

**Composition** per liter:

| | |
|---|---|
| Na$_2$S$_2$O$_3$ | 30.0g |
| CaCO$_3$ | 10.0g |
| Proteose peptone | 5.0g |
| Bile salts | 1.0g |
| Iodine solution | 20.0mL |

pH 8.4± 0.2 at 25°C

**Source:** This medium is available as a premixed powder from BD Diagnostic Systems.

**Iodine Solution:**

**Composition** per 20.0mL:

| | |
|---|---|
| Iodine | 6.0g |
| KI | 5.0g |

**Preparation of Iodine Solution:** Add iodine and KI to distilled/deionized water and bring volume to 20.0mL. Mix thoroughly.

**Preparation of Medium:** Add components, except iodine solution, to distilled/deionized water and bring volume to 980.0mL. Mix thoroughly. Gently heat and bring to boiling. Do not autoclave. Cool to 40°C. Add 20.0mL of iodine solution. Mix thoroughly. Distribute into tubes in 10.0mL volumes. Use medium the same day it is prepared.

**Storage/Shelf Life:** Store dehydrated media in the dark in a sealed container below 30°C. Prepared media should be stored under refrigeration (2-8°C). Media should be used within 60 days of preparation. Media should not be used if there are any signs of deterioration (discoloration) or contamination, or if the expiration date supplied by the manufacturer has passed.

**Use:** For the selective isolation and enrichment of *Salmonella typhi* and other salmonellae from infectious material.

## Tetrathionate Broth
### (m-Tetrathionate Broth)
### (m-TT Broth)

**Composition** per liter:

| | |
|---|---|
| Na$_2$S$_2$O$_3$ | 30.0g |
| CaCO$_3$ | 10.0g |
| Pancreatic digest of casein | 2.5g |
| Peptic digest of animal tissue | 2.5g |
| Iodine–iodide solution | 20.0mL |

pH 8.0 ± 0.2 at 25°C

**Iodine-Iodide Solution:**

**Composition** per 20.0mL:

| | |
|---|---|
| Iodine | 6.0g |
| KI | 5.0g |

**Preparation of Iodine-Iodide Solution:** Add iodine and KI to distilled/deionized water and bring volume to 20.0mL. Mix thoroughly.

**Preparation of Medium:** Add components, except iodine-iodide solution, to distilled/deionized water and bring volume to 980.0mL. Mix thoroughly. Gently heat and bring to boiling. Do not autoclave.

Cool to 40°C. Add 20.0mL of iodine-iodide solution. Mix thoroughly. Distribute into tubes in 10.0mL volumes. Use medium the same day it is prepared.

**Storage/Shelf Life:** Store dehydrated media in the dark in a sealed container below 30°C. Prepared media should be stored under refrigeration (2-8°C). Media should be used within 60 days of preparation. Media should not be used if there are any signs of deterioration (discoloration) or contamination, or if the expiration date supplied by the manufacturer has passed.

**Use:** For the selective isolation in the membrane filter method of *Salmonella* species from feces, urine, foods, and other specimens of sanitary importance.

## Tetrathionate Broth
### (m-Tetrathionate Broth)

**Composition** per liter:

| | |
|---|---|
| Na$_2$S$_2$O$_3$ | 30.0g |
| Proteose peptone | 5.0g |
| Bile salts | 1.0g |
| Iodine solution | 20.0mL |

pH 8.0 ± 0.2 at 25°C

**Source:** This medium is available as a premixed powder from BD Diagnostic Systems.

**Iodine Solution:**

**Composition** per 20.0mL:

| | |
|---|---|
| Iodine | 6.0g |
| KI | 5.0g |

**Preparation of Iodine Solution:** Add iodine and KI to distilled/deionized water and bring volume to 20.0mL. Mix thoroughly.

**Preparation of Medium:** Add components, except iodine solution, to distilled/deionized water and bring volume to 980.0mL. Mix thoroughly. Gently heat and bring to boiling. Do not autoclave. Cool to 40°C. Add 20.0mL of iodine solution. Mix thoroughly. Use medium the same day it is prepared.

**Storage/Shelf Life:** Store dehydrated media in the dark in a sealed container below 30°C. Prepared media should be stored under refrigeration (2-8°C). Media should be used within 60 days of preparation. Media should not be used if there are any signs of deterioration (discoloration) or contamination, or if the expiration date supplied by the manufacturer has passed.

**Use:** For the enrichment of *Salmonella* species in the membrane filter method prior to placing the filter on selective media such as Brilliant Green broth.

## Tetrathionate Broth, Hajna
### (TT Broth, Hajna)

**Composition** per liter:

| | |
|---|---|
| Na$_2$S$_2$O$_3$ | 38.0g |
| CaCO$_3$ | 25.0g |
| Casein/meat peptone (50/50) | 18.0g |
| NaCl | 5.0g |
| D-Mannitol | 2.5g |
| Yeast extract | 2.0g |
| Glucose | 0.5g |
| Sodium deoxycholate | 0.5g |
| Brilliant Green | 0.01g |
| Iodine solution | 40.0mL |

pH 7.5–7.8 at 25°C

**Source:** This medium is available as a premixed powder from BD Diagnostic Systems.

**Iodine Solution:**
**Composition** per 40.0mL:

| | |
|---|---|
| KI | 8.0g |
| Iodine | 5.0g |

**Preparation of Iodine Solution:** Add iodine and KI to distilled/deionized water and bring volume to 40.0mL. Mix thoroughly.

**Preparation of Medium:** Add components, except iodine solution, to distilled/deionized water and bring volume to 960.0mL. Mix thoroughly. Gently heat and bring to boiling. Do not autoclave. Cool to 40°C. Add 40.0mL of iodine solution. Mix thoroughly. Distribute into tubes in 10.0mL volumes. Use medium the same day it is prepared.

**Storage/Shelf Life:** Store dehydrated media in the dark in a sealed container below 30°C. Prepared media should be stored under refrigeration (2-8°C). Media should be used within 60 days of preparation. Media should not be used if there are any signs of deterioration (discoloration) or contamination, or if the expiration date supplied by the manufacturer has passed.

**Use:** For the isolation of *Salmonella* species, except *Salmonella typhi*, and *Arizona* species from fecal specimens, urine, food samples, and other specimens of sanitary significance.

## Tetrathionate Broth with Novobiocin

**Composition** per liter:

| | |
|---|---|
| $Na_2S_2O_3$ | 38.0g |
| $CaCO_3$ | 25.0g |
| Casein/meat peptone (50/50) | 18.0g |
| NaCl | 5.0g |
| Yeast extract | 2.0g |
| D-Mannitol | 0.5g |
| Glucose | 0.5g |
| Sodium deoxycholate | 0.5g |
| Brilliant Green | 0.01g |
| Novobiocin | 4.0mg |
| Iodine solution | 40.0mL |

pH 7.5–7.8 at 25°C

**Iodine Solution:**
**Composition** per 40.0mL:

| | |
|---|---|
| KI | 8.0g |
| Iodine | 5.0g |

**Preparation of Iodine Solution:** Add iodine and KI to distilled/deionized water and bring volume to 40.0mL. Mix thoroughly.

**Preparation of Medium:** Add components, except iodine solution, to distilled/deionized water and bring volume to 960.0mL. Mix thoroughly. Gently heat and bring to boiling. Do not autoclave. Cool to 40°C. Add 40.0mL of iodine solution. Mix thoroughly. Distribute into tubes in 10.0mL volumes. Use medium the same day it is prepared.

**Storage/Shelf Life:** Store dehydrated media in the dark in a sealed container below 30°C. Prepared media should be stored under refrigeration (2-8°C). Media should be used within 60 days of preparation. Media should not be used if there are any signs of deterioration (discoloration) or contamination, or if the expiration date supplied by the manufacturer has passed.

**Use:** For the isolation of *Salmonella* species, except *Salmonella typhi*, and *Arizona* species from fecal specimens and other specimens of sanitary importance. Novobiocin suppresses the growth of *Proteus* species.

## Tetrathionate Broth, USA
## (TT Broth, USA)

| | |
|---|---|
| $Na_2S_2O_3$ | 30.0g |
| $CaCO_3$ | 10.0g |
| Casein peptone | 2.5g |
| Meat peptone | 2.5g |
| Bile salts | 1.0g |
| Iodine-iodide solution | 20.0mL |

**Source:** This medium is available as a premixed powder from Oxoid Unipath.

**Iodine–Iodide Solution:**
**Composition** per 20.0mL:

| | |
|---|---|
| Iodine | 6.0g |
| KI | 5.0g |

**Preparation of Iodine–Iodide Solution:** Add iodine and KI to distilled/deionized water and bring volume to 20.0mL. Mix thoroughly.

**Preparation of Medium:** Add components, except iodine solution, to distilled/deionized water and bring volume to 980.0mL. Mix thoroughly. Gently heat and bring to boiling. Do not autoclave. Cool to 40°C. Add 20.0mL of iodine solution. Mix thoroughly. Distribute into tubes in 10.0mL volumes. Use medium the same day it is prepared.

**Storage/Shelf Life:** Store dehydrated media in the dark in a sealed container below 30°C. Prepared media should be stored under refrigeration (2-8°C). Media should be used within 60 days of preparation. Media should not be used if there are any signs of deterioration (discoloration) or contamination, or if the expiration date supplied by the manufacturer has passed.

**Use:** For the selective enrichment of *Salmonella* species from feces, urine, foods, and other specimens of sanitary importance.

## Tetrathionate Crystal Violet Enhancement Broth

**Composition** per liter:

| | |
|---|---|
| Potassium tetrathionate | 20.0g |
| Casein/meat peptone (50/50) | 8.6g |
| NaCl | 6.4g |
| Crystal Violet | 0.005g |

pH 6.5 ± 0.2 at 25°C

**Preparation of Medium:** Add components to distilled/deionized water and bring volume to 1.0L. Mix thoroughly. Distribute into tubes or flasks. Autoclave for 15 min at 15 psi pressure–121°C.

**Storage/Shelf Life:** Store dehydrated media in the dark in a sealed container below 30°C. Prepared media should be stored under refrigeration (2-8°C). Media should be used within 60 days of preparation. Media should not be used if there are any signs of deterioration (discoloration) or contamination, or if the expiration date supplied by the manufacturer has passed.

**Use:** For the isolation of *Salmonella* species, except *Salmonella typhi*, and *Arizona* species from fecal specimens, urine, food samples, and other specimens of sanitary significance.

## Tetrathionate HiVeg Broth Base, Hajna with Iodine
## (TT HiVeg Broth Base)

**Composition** per liter:

| | |
|---|---|
| $Na_2S_2O_3$ | 38.0g |
| $CaCO_3$ | 25.0g |
| Plant special peptone | 18.0g |

NaCl .................................................................... 5.0g
D-Mannitol ....................................................... 2.5g
Yeast extract .................................................... 2.0g
Glucose ............................................................ 0.5g
Synthetic detergent No. III .............................. 0.5g
Brilliant Green .............................................. 0.01g
Iodine solution ........................................... 40.0mL

pH 7.6 ± 0.2 at 25°C

**Source:** This medium, without iodine, is available as a premixed powder from HiMedia.

**Iodine Solution:**

**Composition** per 40.0mL:

KI ...................................................................... 8.0g
Iodine ............................................................... 5.0g

**Preparation of Iodine Solution:** Add iodine and KI to distilled/deionized water and bring volume to 40.0mL. Mix thoroughly.

**Preparation of Medium:** Add components, except iodine solution, to distilled/deionized water and bring volume to 960.0mL. Mix thoroughly. Gently heat and bring to boiling. Do not autoclave. Cool to 40°C. Add 40.0mL of iodine solution. Mix thoroughly. Distribute into tubes in 10.0mL volumes. Use medium the same day it is prepared.

**Storage/Shelf Life:** Store dehydrated media in the dark in a sealed container below 30°C. Prepared media should be stored under refrigeration (2-8°C). Media should be used within 60 days of preparation. Media should not be used if there are any signs of deterioration (discoloration) or contamination, or if the expiration date supplied by the manufacturer has passed.

**Use:** For the isolation of *Salmonella* species, except *Salmonella typhi*, and *Arizona* species from fecal specimens, urine, food samples, and other specimens of sanitary significance.

## Tetrazolium Tolerance Agar
## (TTC Agar)

**Composition** per liter:

Pancreatic digest of casein ............................ 15.0g
Agar ................................................................ 15.0g
Triphenyltetrazolium chloride ......................... 10.0g
Papaic digest of soybean meal ......................... 5.0g
NaCl .................................................................. 5.0g

pH 7.3 ± 0.2 at 25°C

**Preparation of Medium:** Add components to distilled/deionized water and bring volume to 1.0L. Mix thoroughly. Gently heat and bring to boiling. Distribute into tubes or flasks. Autoclave for 15 min at 15 psi pressure–121°C. Do not overheat. Pour into sterile Petri dishes or leave in tubes.

**Storage/Shelf Life:** Store dehydrated media in the dark in a sealed container below 30°C. Prepared media should be stored under refrigeration (2-8°C). Media should be used within 60 days of preparation. Media should not be used if there are any signs of deterioration (shrinking, cracking, or discoloration) or contamination, or if the expiration date supplied by the manufacturer has passed.

**Use:** For the differentiation of bacteria based upon the ability to tolerate and grow in the presence of tetrazolium. *Streptococcus faecalis* (enterococci) rapidly reduces tetrazolium.

## Thayer-Martin Agar, Modified
## (MTM II)
## (Modified Thayer-Martin Agar)

**Composition** per liter:

Agar ................................................................ 12.0g
Hemoglobin ..................................................... 10.0g
Pancreatic digest of casein .............................. 7.5g
Selected meat peptone ..................................... 7.5g
NaCl .................................................................. 5.0g
K$_2$HPO$_4$ ....................................................... 4.0g
Cornstarch ........................................................ 1.0g
KH$_2$PO$_4$ ....................................................... 1.0g
CNVT inhibitor ............................................. 10.0mL
Supplement solution ..................................... 10.0mL

pH 7.2 ± 0.2 at 25°C

**CNVT Inhibitor:**

**Composition** per 10.0mL:

Colistin sulfate .............................................. 7.5mg
Trimethoprim lactate ..................................... 5.0mg
Vancomycin ................................................... 3.0mg
Nystatin ...................................................... 12,500U

**Preparation of CNVT Inhibitor:** Add components to distilled/deionized water and bring volume to 10.0mL. Mix thoroughly. Filter sterilize.

**Supplement Solution:**

**Composition** per liter:

Glucose .......................................................... 100.0g
L-Cysteine·HCl ............................................... 25.9g
L-Glutamine .................................................... 10.0g
L-Cystine .......................................................... 1.1g
Adenine ............................................................ 1.0g
Nicotinamide adenine dinucleotide ................. 0.25g
Vitamin B$_{12}$ ..................................................... 0.1g
Thiamine pyrophosphate .................................. 0.1g
Guanine·HCl .................................................... 0.03g
Fe(NO$_3$)$_3$·6H$_2$O .............................................. 0.02g
*p*-Aminobenzoic acid ................................... 0.013g
Thiamine·HCl ................................................. 3.0mg

**Source:** The supplement solution IsoVitaleX® enrichment is available from BD Diagnostic Systems. This enrichment may be replaced by supplement VX from BD Diagnostic Systems.

**Preparation of Supplement Solution:** Add components to distilled/deionized water and bring volume to 1.0L. Mix thoroughly. Filter sterilize.

**Preparation of Medium:** Add components, except CNVT inhibitor and supplement solution, to distilled/deionized water and bring volume to 990.0mL. Mix thoroughly. Gently heat and bring to boiling. Distribute into tubes or flasks. Autoclave for 15 min at 15 psi pressure–121°C. Cool to 45°–50°C. Aseptically add 10.0mL of sterile CNVT inhibitor and 10.0mL of sterile supplement solution. Mix thoroughly. Pour into sterile Petri dishes or distribute into sterile tubes.

**Storage/Shelf Life:** Store dehydrated media in the dark in a sealed container below 30°C. Prepared media should be stored under refrigeration (2-8°C). Media should be used within 60 days of preparation. Media should not be used if there are any signs of deterioration (shrinking, cracking, or discoloration) or contamination, or if the expiration date supplied by the manufacturer has passed.

**Use:** For the isolation of *Neisseria* species from specimens containing mixed flora of bacteria and fungi.

## Thayer-Martin HiVeg Medium Base with Hemoglobin and Vitox Supplement

**Composition** per liter:

| | |
|---|---|
| Plant special peptone | 23.0g |
| Agar | 13.0g |
| NaCl | 5.0g |
| Starch | 1.0g |
| Hemoglobin solution | 250.0mL |
| Vitox supplement | 10.0mL |

pH 7.0 ± 0.2 at 25°C

**Source:** This medium, without hemoglobin or Vitox supplement, is available as a premixed powder from HiMedia.

**Hemoglobin Solution:**

**Composition** per 250.0mL:

| | |
|---|---|
| Hemoglobin | 5.0g |

**Preparation of Hemoglobin Solution:** Add hemoglobin to distilled/deionized water and bring volume to 250.0mL. Mix thoroughly. Autoclave for 15 min at 15 psi pressure–121°C. Cool to 45°–50°C.

**Vitox Supplement:**

**Composition** per 10.0mL:

| | |
|---|---|
| Glucose | 2.0g |
| L-Cysteine·HCl | 0.518g |
| L-Glutamine | 0.2g |
| L-Cystine | 0.022g |
| Adenine sulfate | 0.01g |
| Nicotinamide adenine dinucleotide | 5.0mg |
| Cocarboxylase | 2.0mg |
| Guanine·HCl | 0.6mg |
| $Fe(NO_3)_3 \cdot 6H_2O$ | 0.4mg |
| *p*-Aminobenzoic acid | 0.26mg |
| Vitamin $B_{12}$ | 0.2mg |
| Thiamine·HCl | 0.06mg |

**Preparation of Vitox Supplement:** Add components to distilled/deionized water and bring volume to 10.0mL. Mix thoroughly. Filter sterilize.

**Preparation of Medium:** Add components, except hemoblobin and Vitox supplement, to distilled/deionized water and bring volume to 740.0mL. Mix thoroughly. Gently heat until boiling. Autoclave for 15 min at 15 psi pressure–121°C. Cool to 45°–50°C. Aseptically add 250.0mL of sterile hemoglobin solution and 10.0mL of sterile Vitox supplement. Mix thoroughly. Pour into sterile Petri dishes or distribute into sterile tubes.

**Storage/Shelf Life:** Store dehydrated media in the dark in a sealed container below 30°C. Prepared media should be stored under refrigeration (2-8°C). Media should be used within 60 days of preparation. Media should not be used if there are any signs of deterioration (shrinking, cracking, or discoloration) or contamination, or if the expiration date supplied by the manufacturer has passed.

**Use:** For the isolation and cultivation of fastidious microorganisms, especially *Neisseria* species. For the selective isolation of gonococci from pathological specimens.

## Thayer-Martin Medium

**Composition** per liter:

| | |
|---|---|
| GC agar base | 740.0mL |
| Hemoglobin solution | 250.0mL |
| Vitox supplement | 10.0mL |

pH 7.3 ± 0.2 at 25°C

**GC Agar Base:**

**Composition** per 740.0mL:

| | |
|---|---|
| Special peptone | 15.0g |
| Agar | 10.0g |
| NaCl | 5.0g |
| $K_2HPO_4$ | 4.0g |
| Cornstarch | 1.0g |
| $KH_2PO_4$ | 1.0g |

pH 7.2 ± 0.2 at 25°C

**Preparation of GC Agar Base:** Add components of GC medium base and the hemoglobin to distilled/deionized water and bring volume to 740.0mL. Mix thoroughly. Gently heat until boiling. Autoclave for 15 min at 15 psi pressure–121°C. Cool to 45°–50°C.

**Hemoglobin Solution:**

**Composition** per 250.0mL:

| | |
|---|---|
| Hemoglobin | 5.0g |

**Preparation of Hemoglobin Solution:** Add hemoglobin to distilled/deionized water and bring volume to 250.0mL. Mix thoroughly. Autoclave for 15 min at 15 psi pressure–121°C. Cool to 45°–50°C.

**Vitox Supplement:**

**Composition** per 10.0mL:

| | |
|---|---|
| Glucose | 2.0g |
| L-Cysteine·HCl | 0.518g |
| L-Glutamine | 0.2g |
| L-Cystine | 0.022g |
| Adenine sulfate | 0.01g |
| Nicotinamide adenine dinucleotide | 5.0mg |
| Cocarboxylase | 2.0mg |
| Guanine·HCl | 0.6mg |
| $Fe(NO_3)_3 \cdot 6H_2O$ | 0.4mg |
| *p*-Aminobenzoic acid | 0.26mg |
| Vitamin $B_{12}$ | 0.2mg |
| Thiamine·HCl | 0.06mg |

**Preparation of Vitox Supplement:** Add components to distilled/deionized water and bring volume to 10.0mL. Mix thoroughly. Filter sterilize.

**Preparation of Medium:** To 740.0mL of cooled sterile GC agar base, aseptically add 250.0mL of sterile hemoglobin solution and 10.0mL of sterile Vitox supplement. Mix thoroughly. Pour into sterile Petri dishes or distribute into sterile tubes.

**Storage/Shelf Life:** Store dehydrated media in the dark in a sealed container below 30°C. Prepared media should be stored under refrigeration (2-8°C). Media should be used within 60 days of preparation. Media should not be used if there are any signs of deterioration (shrinking, cracking, or discoloration) or contamination, or if the expiration date supplied by the manufacturer has passed.

**Use:** For the isolation and cultivation of fastidious microorganisms, especially *Neisseria* species.

## Thayer-Martin Medium

**Composition** per liter:

| | |
|---|---|
| Hemoglobin | 10.0g |
| GC medium base | 980.0mL |
| CNVT inhibitor | 10.0mL |
| Supplement B | 10.0mL |

pH 7.3 ± 0.2 at 25°C

**Source:** This medium is available as a prepared medium in tubes from BD Diagnostic Systems.

**GC Medium Base:**

**Composition** per 980.0mL:

| | |
|---|---|
| Proteose peptone No. 3 | 15.0g |
| Agar | 10.0g |
| NaCl | 5.0g |
| $K_2HPO_4$ | 4.0g |
| Cornstarch | 1.0g |
| $KH_2PO_4$ | 1.0g |

pH 7.2 ± 0.2 at 25°C

**Preparation of GC Medium Base:** Add components of GC medium base and the hemoglobin to distilled/deionized water and bring volume to 1.0L. Mix thoroughly. Gently heat until boiling. Autoclave for 15 min at 15 psi pressure–121°C. Cool to 45°–50°C.

**CNVT Inhibitor:**

**Composition** per 10.0mL:

| | |
|---|---|
| Colistin sulfate | 7.5mg |
| Trimethoprim lactate | 5.0mg |
| Vancomycin | 3.0mg |
| Nystatin | 12,500U |

**Preparation of CNVT Inhibitor:** Add components to distilled/deionized water and bring volume to 10.0mL. Mix thoroughly. Filter sterilize.

**Preparation of Medium:** To 980.0mL of cooled sterile GC medium base, aseptically add 10.0mL of sterile CNVT inhibitor and 10.0mL of sterile supplement B. Mix thoroughly. Pour into sterile Petri dishes or distribute into sterile tubes.

**Storage/Shelf Life:** Store dehydrated media in the dark in a sealed container below 30°C. Prepared media should be stored under refrigeration (2-8°C). Media should be used within 60 days of preparation. Media should not be used if there are any signs of deterioration (shrinking, cracking, or discoloration) or contamination, or if the expiration date supplied by the manufacturer has passed.

**Use:** For the isolation and cultivation of fastidious microorganisms, especially *Neisseria* species.

## Thayer-Martin Medium, Modified
## (Modified Thayer-Martin Agar)

**Composition** per liter:

| | |
|---|---|
| GC agar base | 720.0mL |
| Hemoglobin solution | 250.0mL |
| GC supplement | 30.0mL |

pH 7.3 ± 0.2 at 25°C

**GC Agar Base:**

**Composition** per 720.0mL:

| | |
|---|---|
| Special peptone | 15.0g |
| Agar | 10.0g |
| NaCl | 5.0g |
| $K_2HPO_4$ | 4.0g |
| Cornstarch | 1.0g |
| $KH_2PO_4$ | 1.0g |

pH 7.2 ± 0.2 at 25°C

**Preparation of GC Agar Base:** Add components of GC medium base to distilled/deionized water and bring volume to 720.0mL. Mix thoroughly. Gently heat until boiling. Autoclave for 15 min at 15 psi pressure–121°C. Cool to 45°–50°C.

**Hemoglobin Solution:**

**Composition** per 250.0mL:

| | |
|---|---|
| Hemoglobin | 5.0g |

**Preparation of Hemoglobin Solution:** Add hemoglobin to distilled/deionized water and bring volume to 250.0mL. Mix thoroughly. Autoclave for 15 min at 15 psi pressure–121°C. Cool to 45°–50°C.

**GC Supplement:**

**Composition** per 30.0mL:

| | |
|---|---|
| Yeast autolysate | 10.0g |
| Glucose | 1.5g |
| $NaHCO_3$ | 0.15g |
| Colistin sulfate | 7.5mg |
| Trimethoprim lactate | 5.0mg |
| Vancomycin | 3.0mg |
| Nystatin | 12,500U |

**Preparation of GC Supplement:** Add components to distilled/deionized water and bring volume to 30.0mL. Mix thoroughly. Filter sterilize.

**Preparation of Medium:** To 720.0mL of cooled sterile GC agar base, aseptically add 250.0mL of sterile hemoglobin solution and 30.0mL of sterile GC supplement. Mix thoroughly. Pour into sterile Petri dishes or distribute into sterile tubes.

**Storage/Shelf Life:** Store dehydrated media in the dark in a sealed container below 30°C. Prepared media should be stored under refrigeration (2-8°C). Media should be used within 60 days of preparation. Media should not be used if there are any signs of deterioration (shrinking, cracking, or discoloration) or contamination, or if the expiration date supplied by the manufacturer has passed.

**Use:** For the selective isolation and cultivation of fastidious microorganisms, especially *Neisseria* species.

## Thayer-Martin Medium, Modified
## (Modified Thayer-Martin Agar)

**Composition** per liter:

| | |
|---|---|
| GC agar base | 730.0mL |
| Hemoglobin solution | 250.0mL |
| Vitox supplement | 10.0mL |
| VCNT antibiotic solution | 10.0mL |

pH 7.3 ± 0.2 at 25°C

**GC Agar Base:**

**Composition** per 730.0mL:

| | |
|---|---|
| Special peptone | 15.0g |
| Agar | 10.0g |
| NaCl | 5.0g |
| $K_2HPO_4$ | 4.0g |
| Cornstarch | 1.0g |
| $KH_2PO_4$ | 1.0g |

pH 7.2 ± 0.2 at 25°C

**Preparation of GC Agar Base:** Add components of GC medium base in to distilled/deionized water and bring volume to 730.0mL. Mix thoroughly. Gently heat until boiling. Autoclave for 15 min at 15 psi pressure–121°C. Cool to 45°–50°C.

## Hemoglobin Solution:
**Composition** per 250.0mL:

Hemoglobin ............................................................5.0g

**Preparation of Hemoglobin Solution:** Add hemoglobin to distilled/deionized water and bring volume to 250.0mL. Mix thoroughly. Autoclave for 15 min at 15 psi pressure–121°C. Cool to 45°–50°C.

## Vitox Supplement:
**Composition** per 10.0mL:

Glucose .................................................................2.0g
L-Cysteine·HCl ....................................................0.518g
L-Glutamine.........................................................0.2g
L-Cystine .............................................................0.022g
Adenine sulfate ...................................................0.01g
Nicotinamide adenine dinucleotide .....................5.0mg
Cocarboxylase......................................................2.0mg
Guanine·HCl ........................................................0.6mg
$Fe(NO_3)_3$·$6H_2O$ ................................................0.4mg
*p*-Aminobenzoic acid.........................................0.26mg
Vitamin $B_{12}$ ........................................................0.2mg
Thiamine·HCl .......................................................0.06mg

**Preparation of Vitox Supplement:** Add components to distilled/deionized water and bring volume to 10.0mL. Mix thoroughly. Filter sterilize.

## VCNT Antibiotic Solution:
**Composition** per 10.0mL:

Colistin methane sulfonate......................................7.5mg
Trimethoprim lactate...............................................5.0mg
Vancomycin ............................................................3.0mg
Nystatin.................................................................12,500U

**Preparation of VCNT Antibiotic Solution:** Add components to distilled/deionized water and bring volume to 10.0mL. Mix thoroughly. Filter sterilize.

**Preparation of Medium:** To 730.0mL of cooled, sterile GC agar base, aseptically add 250.0mL of sterile hemoglobin solution, 10.0mL of sterile Vitox supplement, and 10.0mL of VCNT antibiotic solution. Mix thoroughly. Pour into sterile Petri dishes or distribute into sterile tubes.

**Storage/Shelf Life:** Store dehydrated media in the dark in a sealed container below 30°C. Prepared media should be stored under refrigeration (2-8°C). Media should be used within 60 days of preparation. Media should not be used if there are any signs of deterioration (shrinking, cracking, or discoloration) or contamination, or if the expiration date supplied by the manufacturer has passed.

**Use:** For the selective isolation and cultivation of fastidious microorganisms, especially *Neisseria* species.

## Thayer-Martin Medium, Selective
**Composition** per liter:

GC agar base...........................................................730.0mL
Hemoglobin solution................................................250.0mL
Vitox supplement ....................................................10.0mL
VCN antibiotic solution ...........................................10.0mL

pH 7.3 ± 0.2 at 25°C

## GC Agar Base:
**Composition** per 730.0mL:

Special peptone......................................................15.0g
Agar ......................................................................10.0g
NaCl.......................................................................5.0g
$K_2HPO_4$................................................................4.0g
Cornstarch..............................................................1.0g
$KH_2PO_4$................................................................1.0g

pH 7.2 ± 0.2 at 25°C

**Preparation of GC Agar Base:** Add components of GC medium base to distilled/deionized water and bring volume to 730.0mL. Mix thoroughly. Gently heat until boiling. Autoclave for 15 min at 15 psi pressure–121°C. Cool to 45°–50°C.

## Hemoglobin Solution:
**Composition** per 250.0mL:

Hemoglobin ............................................................5.0g

**Preparation of Hemoglobin Solution:** Add hemoglobin to distilled/deionized water and bring volume to 250.0mL. Mix thoroughly. Autoclave for 15 min at 15 psi pressure–121°C. Cool to 45°–50°C.

## Vitox Supplement:
**Composition** per 10.0mL:

Glucose .................................................................2.0g
L-Cysteine·HCl ....................................................0.518g
L-Glutamine.........................................................0.2g
L-Cystine .............................................................0.022g
Adenine sulfate ...................................................0.01g
Nicotinamide adenine dinucleotide .....................5.0mg
Cocarboxylase......................................................2.0mg
Guanine·HCl ........................................................0.6mg
$Fe(NO_3)_3$·$6H_2O$ ................................................0.4mg
*p*-Aminobenzoic acid.........................................0.26mg
Vitamin $B_{12}$ ........................................................0.2mg
Thiamine·HCl .......................................................0.06mg

**Preparation of Vitox Supplement:** Add components to distilled/deionized water and bring volume to 10.0mL. Mix thoroughly. Filter sterilize.

## VCN Antibiotic Solution:
**Composition** per 10.0mL:

Colistin methane sulfonate ......................................7.5mg
Vancomycin ............................................................3.0mg
Nystatin.................................................................12,500U

**Preparation of VCN Antibiotic Solution:** Add components to distilled/deionized water and bring volume to 10.0mL. Mix thoroughly. Filter sterilize.

**Preparation of Medium:** To 730.0mL of cooled, sterile GC agar base, aseptically add 250.0mL of sterile hemoglobin solution, 10.0mL of sterile Vitox supplement, and 10.0mL of VCN antibiotic solution. Mix thoroughly. Pour into sterile Petri dishes or distribute into sterile tubes.

**Storage/Shelf Life:** Store dehydrated media in the dark in a sealed container below 30°C. Prepared media should be stored under refrigeration (2-8°C). Media should be used within 60 days of preparation. Media should not be used if there are any signs of deterioration (shrinking, cracking, or discoloration) or contamination, or if the expiration date supplied by the manufacturer has passed.

**Use:** For the selective isolation and cultivation of fastidious microorganisms, especially *Neisseria* species.

## Thayer-Martin Selective Agar

**Composition** per liter:

| | |
|---|---|
| Agar | 12.0g |
| Hemoglobin | 10.0g |
| Pancreatic digest of casein | 7.5g |
| Selected meat peptone | 7.5g |
| NaCl | 5.0g |
| $K_2HPO_4$ | 4.0g |
| Cornstarch | 1.0g |
| $KH_2PO_4$ | 1.0g |
| Supplement solution | 10.0mL |
| VCN inhibitor | 10.0mL |

pH $7.2 \pm 0.2$ at 25°C

**Source:** This medium is available as a premixed powder from BD Diagnostic Systems.

### Supplement Solution:

**Composition** per liter:

| | |
|---|---|
| Glucose | 100.0g |
| L-Cysteine·HCl | 25.9g |
| L-Glutamine | 10.0g |
| L-Cystine | 1.1g |
| Adenine | 1.0g |
| Nicotinamide adenine dinucleotide | 0.25g |
| Vitamin $B_{12}$ | 0.1g |
| Thiamine pyrophosphate | 0.1g |
| Guanine·HCl | 0.03g |
| $Fe(NO_3)_3·6H_2O$ | 0.02g |
| *p*-Aminobenzoic acid | 0.013g |
| Thiamine·HCl | 3.0mg |

**Source:** The supplement solution IsoVitaleX® enrichment is available from BD Diagnostic Systems. This enrichment may be replaced by supplement VX from BD Diagnostic Systems.

**Preparation of Supplement Solution:** Add components to distilled/deionized water and bring volume to 1.0L. Mix thoroughly. Filter sterilize.

### VCN Inhibitor:

**Composition** per 10.0mL:

| | |
|---|---|
| Colistin | 7.5mg |
| Vancomycin | 3.0mg |
| Nystatin | 12,500U |

**Preparation of VCN Inhibitor:** Add components to distilled/deionized water and bring volume to 10.0mL. Mix thoroughly. Filter sterilize.

**Preparation of Medium:** Add components, except supplement solution and VCN inhibitor, to distilled/deionized water and bring volume to 980.0mL. Mix thoroughly. Gently heat and bring to boiling. Autoclave for 15 min at 15 psi pressure–121°C. Cool to 45°–50°C. Aseptically add sterile VCN inhibitor and sterile supplement solution. Mix thoroughly. Pour into sterile Petri dishes or distribute into sterile tubes.

**Storage/Shelf Life:** Store dehydrated media in the dark in a sealed container below 30°C. Prepared media should be stored under refrigeration (2-8°C). Media should be used within 60 days of preparation. Media should not be used if there are any signs of deterioration (shrinking, cracking, or discoloration) or contamination, or if the expiration date supplied by the manufacturer has passed.

**Use:** For the selective isolation of *Neisseria gonorrhoeae* and *Neisseria meningitidis* from specimens containing mixed flora of bacteria and fungi.

## Thioglycolate Bile Broth

**Composition** per 1050.0mL:

| | |
|---|---|
| Pancreatic digest of casein | 15.0g |
| Glucose | 5.5g |
| Yeast extract | 5.0g |
| NaCl | 2.5g |
| Agar | 0.75g |
| L-Cystine | 0.5g |
| Sodium thioglycolate | 0.5g |
| Bile solution | 50.0mL |

pH $7.1 \pm 0.2$ at 25°C

### Bile Solution:

**Composition** per 100.0mL:

| | |
|---|---|
| Oxgall | 40.0g |
| Sodium deoxycholate | 2.0g |

**Preparation of Bile Solution:** Add components to distilled/deionized water and bring volume to 100.0mL. Mix thoroughly. Filter sterilize.

**Preparation of Medium:** Add components, except bile solution, to distilled/deionized water and bring volume to 1.0L. Mix thoroughly. Gently heat and bring to boiling. Distribute into tubes in 10.0mL volumes. Autoclave for 15 min at 15 psi pressure–121°C. Cool to 45°–50°C. Aseptically add 0.5mL of sterile bile solution to each tube. Mix thoroughly.

**Storage/Shelf Life:** Store dehydrated media in the dark in a sealed container below 30°C. Prepared media should be stored under refrigeration (2-8°C). Media should be used within 60 days of preparation. Media should not be used if there are any signs of deterioration (discoloration) or contamination, or if the expiration date supplied by the manufacturer has passed.

**Use:** For the cultivation of *Bacteroides fragilis* and *Clostridium perfringens* from clinical specimens.

## Thioglycolate Broth USP, Alternative

**Composition** per liter:

| | |
|---|---|
| Pancreatic digest of casein | 15.0g |
| Glucose | 5.5g |
| Yeast extract | 5.0g |
| NaCl | 2.5g |
| L-Cystine | 0.5g |
| Sodium thioglycolate | 0.5g |

pH $7.1 \pm 0.2$ at 25°C

**Source:** This medium is available as a premixed powder from Oxoid Unipath.

**Preparation of Medium:** Add components to distilled/deionized water and bring volume to 1.0L. Mix thoroughly. Distribute into tubes or flasks. Autoclave for 15 min at 15 psi pressure–121°C. Prepare freshly or boil and cool the medium just before use.

**Storage/Shelf Life:** Store dehydrated media in the dark in a sealed container below 30°C. Prepared media should be stored under refrigeration (2-8°C). Media should not be used if there are any signs of deterioration (discoloration) or contamination, or if the expiration date supplied by the manufacturer has passed.

**Use:** For the cultivation of both aerobic and anaerobic organisms in the performance of sterility tests of turbid or viscous specimens.

## Thioglycolate Gelatin Medium

**Composition** per liter:

| | |
|---|---|
| Gelatin | 50.0g |
| Pancreatic digest of casein | 15.0g |
| Yeast extract | 5.0g |
| NaCl | 2.5g |
| Glucose | 2.0g |
| Agar | 0.75g |
| L-Cystine | 0.25g |
| Na$_2$SO$_3$ | 0.1g |
| Thioglycollic acid | 0.3mL |

pH 7.0 ± 0.2 at 25°C

**Source:** This medium is available as a premixed powder from BD Diagnostic Systems.

**Preparation of Medium:** Add components to distilled/deionized water and bring volume to 1.0L. Mix thoroughly. Gently heat and bring to 50°C. Let stand 5 min. Gently heat and bring to boiling. Distribute into tubes or flasks. Autoclave for 15 min at 15 psi pressure–121°C.

**Storage/Shelf Life:** Store dehydrated media in the dark in a sealed container below 30°C. Prepared media should be stored under refrigeration (2-8°C). Media should be used within 60 days of preparation. Media should not be used if there are any signs of deterioration (shrinking, cracking, or discoloration) or contamination, or if the expiration date supplied by the manufacturer has passed.

**Use:** For the determination of gelatin liquefaction by aerobes, microaerophiles, and anaerobes without special incubation.

## Thioglycolate HiVeg Agar

**Composition** per liter:

| | |
|---|---|
| Agar | 20.0g |
| Plant hydrolysate | 15.0g |
| Glucose | 5.5g |
| Yeast extract | 5.0g |
| NaCl | 2.5g |
| L-Cystine | 0.5g |
| Na-thioglycolate | 0.5g |
| Resazurin | 1.0mg |

pH 7.1 ± 0.2 at 25°C

**Source:** This medium is available as a premixed powder from Hi-Media.

**Preparation of Medium:** Add components to distilled/deionized water and bring volume to 1.0L. Mix thoroughly. Distribute into tubes or flasks. Autoclave for 15 min at 15 psi pressure–121°C. Prepare freshly or boil and cool the medium just before use. Pour into sterile Petri dishes or distribute into sterile tubes.

**Storage/Shelf Life:** Store dehydrated media in the dark in a sealed container below 30°C. Prepared media should be stored under refrigeration (2-8°C). Media should be used within 60 days of preparation. Media should not be used if there are any signs of deterioration (shrinking, cracking, or discoloration) or contamination, or if the expiration date supplied by the manufacturer has passed.

**Use:** For the cultivation of facultative and anaerobic organisms. For the performance of sterility tests of turbid or viscous specimens.

## Thioglycolate HiVeg Medium without Indicator

**Composition** per liter:

| | |
|---|---|
| Plant hydrolysate | 17.0g |
| Glucose | 6.0g |
| Papaic digest of soybean meal | 3.0g |
| NaCl | 2.5g |
| Agar | 0.7g |
| Na-thioglycolate | 0.5g |
| L-Cystine | 0.25g |
| Na$_2$SO$_3$ | 0.1g |

pH 7.1 ± 0.2 at 25°C

**Source:** This medium is available as a premixed powder from Hi-Media.

**Preparation of Medium:** Add components to distilled/deionized water and bring volume to 1.0L. Mix thoroughly. Distribute into tubes or flasks. Autoclave for 15 min at 15 psi pressure–121°C. Prepare freshly or boil and cool the medium just before use.

**Storage/Shelf Life:** Store dehydrated media in the dark in a sealed container below 30°C. Prepared media should be stored under refrigeration (2-8°C). Use freshly prepared media. Media should not be used if there are any signs of deterioration (discoloration) or contamination, or if the expiration date supplied by the manufacturer has passed.

**Use:** For the cultivation of facultative and anaerobic organisms. For the performance of sterility tests of turbid or viscous specimens.

## Thioglycolate Medium, Brewer

**Composition** per liter:

| | |
|---|---|
| Glucose | 5.0g |
| Peptone | 5.0g |
| NaCl | 5.0g |
| Yeast extract | 2.0g |
| Sodium thioglycolate | 1.1g |
| Agar | 1.0g |
| Beef extract | 1.0g |
| Methylene Blue | 2.0mg |

pH 7.2 ± 0.2 at 25°C

**Source:** This medium is available as a premixed powder from Oxoid Unipath.

**Preparation of Medium:** Add components to distilled/deionized water and bring volume to 1.0L. Mix thoroughly. Gently heat and bring to boiling. Distribute into tubes or flasks. Autoclave for 15 min at 15 psi pressure–121°C.

**Storage/Shelf Life:** Store dehydrated media in the dark in a sealed container below 30°C. Prepared media should be stored under refrigeration (2-8°C). Use freshly prepared media. Media should not be used if there are any signs of deterioration (discoloration) or contamination, or if the expiration date supplied by the manufacturer has passed.

**Use:** For determination of the sterility of solutions containing mercurial preservatives.

## Thioglycolate Medium, Fluid
## (Fluid Thioglycolate Medium)
## (FTG)
## (BAM M146)

**Composition** per liter:

| | |
|---|---|
| Pancreatic digest of casein | 15.0g |
| Glucose | 5.0g |
| Yeast extract | 5.0g |
| NaCl | 2.5g |
| Agar | 0.75g |
| L-cystine | 0.5g |

Sodium thioglycolate ...............................................................0.5g
Resazurin solution.............................................................. 1.0mL

pH 7.1 ± 0.2 at 25°C

**Resazurin Solution:**
**Composition** per 10.0mL:

Na-resazurin...........................................................................10.0mg

**Preparation of Resazurin Solution:** Add Na-resazurin to 10.0mL of distilled/deionized water. Mix thoroughly. Prepare freshly.

**Preparation of Medium:** Add components, except sodium thioglycolate and resazurin solution, to distilled/deionized water and bring volume to 1.0L. Mix thoroughly. Gently heat and bring to boiling. Add 0.5g sodium thioglycolate. Mix thoroughly. Adjust pH to 7.1. Add 1.0mL resazurin solution. Mix thoroughly. Distribute into tubes or flasks. Autoclave for 20 min at 15 psi pressure–121°C.

**Storage/Shelf Life:** Store dehydrated media in the dark in a sealed container below 30°C. Prepared media should be stored under refrigeration (2–8°C). Use freshly prepared media. Media should not be used if there are any signs of deterioration (discoloration) or contamination, or if the expiration date supplied by the manufacturer has passed.

**Use**: For the cultivation of both aerobic and anaerobic organisms in the performance of sterility tests.

## Thioglycolate Medium without Glucose
**Composition** per liter:

Pancreatic digest of casein............................................15.0g
Yeast extract.......................................................................5.0g
NaCl.....................................................................................2.5g
Agar .....................................................................................0.75g
L-Cystine .............................................................................0.25g
Methylene Blue...................................................................2.0mg
Thioglycolic acid ............................................................ 0.3mL

pH 7.2 ± 0.2 at 25°C

**Source:** This medium is available as a premixed powder from BD Diagnostic Systems.

**Preparation of Medium:** Add components to distilled/deionized water and bring volume to 1.0L. Mix thoroughly. Gently heat and bring to boiling. Distribute into tubes or flasks. Autoclave for 15 min at 15 psi pressure–121°C. If medium becomes oxidized before use (Methylene Blue turns blue), heat in a boiling water bath to expel absorbed $O_2$. Cool to 25°C.

**Storage/Shelf Life:** Store dehydrated media in the dark in a sealed container below 30°C. Prepared media should be stored under refrigeration (2–8°C). Use freshly prepared media. Media should not be used if there are any signs of deterioration (discoloration) or contamination, or if the expiration date supplied by the manufacturer has passed.

**Use:** For the cultivation of anaerobic, microaerophilic, and aerobic microorganisms. For use in sterility testing of a variety of specimens.

## Thioglycolate Medium without Glucose and Indicator
**Composition** per liter:

Pancreatic digest of casein............................................15.0g
Yeast extract.......................................................................5.0g
NaCl.....................................................................................2.5g
Agar .....................................................................................0.75g
L-Cystine .............................................................................0.25g
Thioglycolic acid ............................................................ 0.3mL

pH 7.2 ± 0.2 at 25°C

**Source:** This medium is available as a premixed powder from BD Diagnostic Systems.

**Preparation of Medium:** Add components to distilled/deionized water and bring volume to 1.0L. Mix thoroughly. Gently heat and bring to boiling. Distribute into tubes or flasks. Autoclave for 15 min at 15 psi pressure–121°C. If medium becomes oxidized before use, heat in a boiling water bath to expel absorbed $O_2$. Cool to 25°C.

**Storage/Shelf Life:** Store dehydrated media in the dark in a sealed container below 30°C. Prepared media should be stored under refrigeration (2–8°C). Use freshly prepared media. Media should not be used if there are any signs of deterioration (discoloration) or contamination, or if the expiration date supplied by the manufacturer has passed.

**Use:** For the cultivation of anaerobic, microaerophilic, and aerobic microorganisms. For use in sterility testing of a variety of specimens.

## Thioglycolate Medium without Indicator
**Composition** per liter:

Pancreatic digest of casein...........................................15.0g
Yeast extract......................................................................5.0g
Glucose ..............................................................................5.0g
NaCl....................................................................................2.5g
Agar ....................................................................................0.75g
Sodium thioglycolate.......................................................0.5g
L-Cystine ...........................................................................0.25g

pH 7.2 ± 0.2 at 25°C

**Source:** This medium is available as a premixed powder from BD Diagnostic Systems.

**Preparation of Medium:** Add components to distilled/deionized water and bring volume to 1.0L. Mix thoroughly. Gently heat and bring to boiling. Distribute into tubes or flasks. Autoclave for 15 min at 15 psi pressure–121°C. If medium becomes oxidized before use, heat in a boiling water bath to expel absorbed $O_2$. Cool to 25°C.

**Storage/Shelf Life:** Store dehydrated media in the dark in a sealed container below 30°C. Prepared media should be stored under refrigeration (2–8°C). Use freshly prepared media. Media should not be used if there are any signs of deterioration (discoloration) or contamination, or if the expiration date supplied by the manufacturer has passed.

**Use:** For the cultivation of anaerobic, microaerophilic, and aerobic microorganisms. For use in sterility testing of a variety of specimens.

## Thioglycolate Medium without Indicator
**Composition** per liter:

Pancreatic digest of casein...........................................17.0g
Glucose ..............................................................................6.0g
Papaic digest of soybean meal......................................3.0g
NaCl....................................................................................2.5g
Agar ....................................................................................0.7g
Sodium thioglycolate.......................................................0.5g
L-Cystine ...........................................................................0.25g
$Na_2SO_3$...............................................................................0.1g

pH 7.0 ± 0.2 at 25°C

**Source:** This medium is available as a premixed powder from Oxoid Unipath.

**Preparation of Medium:** Add components to distilled/deionized water and bring volume to 1.0L. Mix thoroughly. Distribute into tubes or flasks. Autoclave for 15 min at 15 psi pressure–121°C. Prepare freshly or boil and cool the medium just before use.

**Storage/Shelf Life:** Store dehydrated media in the dark in a sealed container below 30°C. Prepared media should be stored under refrigeration (2-8°C). Use freshly prepared media. Media should not be used if there are any signs of deterioration (discoloration) or contamination, or if the expiration date supplied by the manufacturer has passed.

**Use:** For the growth of aerobic and anaerobic microorganisms in diagnostic bacteriology.

## Thioglycolate Medium without Indicator-135C
**Composition** per liter:

| | |
|---|---|
| Pancreatic digest of casein | 17.0g |
| Glucose | 6.0g |
| Papaic digest of soybean meal | 3.0g |
| NaCl | 2.5g |
| Agar | 0.7g |
| Sodium thioglycolate | 0.5g |
| $Na_2SO_3$ | 0.1g |
| L-Cystine | 0.25g |

pH 7.0 ± 0.2 at 25°C

**Source:** This medium is available as a premixed powder from BD Diagnostic Systems.

**Preparation of Medium:** Add components to distilled/deionized water and bring volume to 1.0L. Mix thoroughly. Gently heat while stirring and bring to boiling. Distribute into tubes or flasks, filling them half full. For maintenance of cultures, a small quantity of $CaCO_3$ may be added to tubes before adding medium. Autoclave for 15 min at 13 psi pressure–118°C. Prepare freshly or boil and cool the medium just before use. Store prepared medium at 2°–8°C in the dark.

**Storage/Shelf Life:** Store dehydrated media in the dark in a sealed container below 30°C. Prepared media should be stored under refrigeration (2-8°C). Use freshly prepared media. Media should not be used if there are any signs of deterioration (discoloration) or contamination, or if the expiration date supplied by the manufacturer has passed.

**Use:** For the isolation and cultivation of a wide variety of microorganisms, particularly obligate anaerobes, from clinical specimens and other materials.

## Thioglycolate Medium, USP
**Composition** per liter:

| | |
|---|---|
| Pancreatic digest of casein | 15.0g |
| Glucose | 5.5g |
| Yeast extract | 5.0g |
| NaCl | 2.5g |
| Agar | 0.5g |
| L-cystine | 0.5g |
| Sodium thioglycolate | 0.5g |
| Resazurin | 1.0mg |

pH 7.1 ± 0.2 at 25°C

**Source:** This medium is available as a premixed powder from Oxoid Unipath.

**Preparation of Medium:** Add components to distilled/deionized water and bring volume to 1.0L. Mix thoroughly. Gently heat and bring to boiling. Distribute into tubes or flasks. Autoclave for 15 min at 15 psi pressure–121°C.

**Storage/Shelf Life:** Store dehydrated media in the dark in a sealed container below 30°C. Prepared media should be stored under refrigeration (2-8°C). Use freshly prepared media. Media should not be used

if there are any signs of deterioration (discoloration) or contamination, or if the expiration date supplied by the manufacturer has passed.

**Use:** For the cultivation of both aerobic and anaerobic organisms in the performance of sterility tests.

## Tinsdale Agar
**Composition** per 1100.0mL:

| | |
|---|---|
| Proteose peptone | 20.0g |
| Agar | 15.0g |
| NaCl | 5.0g |
| Yeast extract | 5.0g |
| L-Cystine | 0.24g |
| Tinsdale supplement | 150.0mL |

pH 7.4 ± 0.2 at 25°C

**Source:** This medium is available as a premixed powder from BD Diagnostic Systems and Oxoid Unipath.

**Tinsdale Supplement:**
**Composition** per 100.0mL:

| | |
|---|---|
| $Na_2S_2O_3$ | 0.43g |
| $K_2TeO_3$ | 0.35g |
| Serum | 100.0mL |

**Caution:** Potassium tellurite is toxic.

**Preparation of Tinsdale Supplement:** Add $Na_2S_2O_3$ and $K_2TeO_3$ to serum. Mix thoroughly. Filter sterilize.

**Preparation of Medium:** Add components, except Tinsdale supplement, to distilled/deionized water and bring volume to 1.0L. Mix thoroughly. Gently heat and bring to boiling. Autoclave for 15 min at 15 psi pressure–121°C. Cool to 50°–55°C. Aseptically add 100.0mL of sterile Tinsdale supplement. Mix thoroughly. Pour into sterile Petri dishes or distribute into sterile tubes.

**Storage/Shelf Life:** Store dehydrated media in the dark in a sealed container below 30°C. Prepared media should be stored under refrigeration (2-8°C). Media should be used within 60 days of preparation. Media should not be used if there are any signs of deterioration (shrinking, cracking, or discoloration) or contamination, or if the expiration date supplied by the manufacturer has passed.

**Use:** For the primary isolation and identification of *Corynebacterium diphtheriae*.

## Tinsdale HiVeg Agar Base with Tinsdale Supplement
**Composition** per liter:

| | |
|---|---|
| Plant peptone | 20.0g |
| Agar | 15.0g |
| NaCl | 5.0g |
| $Na_2S_2O_3$ | 0.43g |
| L-Cystine | 0.24g |
| Tinsdale supplement | 150.0mL |

pH 7.4 ± 0.2 at 25°C

**Source:** This medium is available as a premixed powder from Hi-Media.

**Tinsdale Supplement:**
**Composition** per 100.0mL:

| | |
|---|---|
| $Na_2S_2O_3$ | 0.43g |
| $K_2TeO_3$ | 0.35g |
| Serum | 100.0mL |

**Caution:** Potassium tellurite is toxic.

**Preparation of Tinsdale Supplement:** Add $Na_2S_2O_3$ and $K_2TeO_3$ to serum. Mix thoroughly. Filter sterilize.

**Preparation of Medium:** Add components, except Tinsdale supplement, to distilled/deionized water and bring volume to 1.0L. Mix thoroughly. Gently heat and bring to boiling. Autoclave for 15 min at 15 psi pressure–121°C. Cool to 50°–55°C. Aseptically add 100.0mL of sterile Tinsdale supplement. Mix thoroughly. Pour into sterile Petri dishes or distribute into sterile tubes.

**Storage/Shelf Life:** Store dehydrated media in the dark in a sealed container below 30°C. Prepared media should be stored under refrigeration (2-8°C). Media should be used within 60 days of preparation. Media should not be used if there are any signs of deterioration (shrinking, cracking, or discoloration) or contamination, or if the expiration date supplied by the manufacturer has passed.

**Use:** For the primary isolation and identification of *Corynebacterium diphtheriae*.

## TMAO HiVeg Medium
### (Trimethylamine-*N*-Oxide HiVeg Medium)
**Composition** per liter:

| | |
|---|---|
| Plant extract | 10.0g |
| Plant peptone | 10.0g |
| NaCl | 5.0g |
| Agar | 2.0g |
| Trimethylamine-*N*-oxide | 1.0g |
| Yeast extract | 1.0g |

pH 7.5 ± 0.2 at 25°C

**Source:** This medium is available as a premixed powder from Hi-Media.

**Preparation of Medium:** Add components to distilled/deionized water and bring volume to 1.0L. Mix thoroughly. Gently heat and bring to boiling. Distribute into screw-capped tubes in 4.0mL volumes. Autoclave for 15 min at 15 psi pressure–121°C. Allow tubes to cool in an upright position.

**Storage/Shelf Life:** Store dehydrated media in the dark in a sealed container below 30°C. Prepared media should be stored under refrigeration (2-8°C). Media should be used within 60 days of preparation. Media should not be used if there are any signs of deterioration (discoloration) or contamination, or if the expiration date supplied by the manufacturer has passed.

**Use:** For the cultivation and differentiation of *Campylobacter* species from foods. *Campylobacter jejuni* and *Campylobacter coli* will not grow.

## TN HiVeg Agar
**Composition** per liter:

| | |
|---|---|
| Agar | 15.0g |
| Plant hydrolysate | 10.0g |
| NaCl | 10.0g |

pH 7.2 ± 0.2 at 25°C

**Source:** This medium is available as a premixed powder from Hi-Media.

**Preparation of Medium:** Add components to distilled/deionized water and bring volume to 1.0L. Mix thoroughly. Gently heat and bring to boiling. Distribute into tubes or flasks. Autoclave for 15 min at 15 psi pressure–121°C. Do not overheat. Pour into sterile Petri dishes or leave in tubes.

**Storage/Shelf Life:** Store dehydrated media in the dark in a sealed container below 30°C. Prepared media should be stored under refrigeration (2-8°C). Media should be used within 60 days of preparation. Media should not be used if there are any signs of deterioration (shrinking, cracking, or discoloration) or contamination, or if the expiration date supplied by the manufacturer has passed.

**Use:** For the isolation and cultivation of vibrios from food samples.

## Todd-Hewitt Broth
**Composition** per liter:

| | |
|---|---|
| Beef heart, infusion from | 500.0g |
| Neopeptone | 20.0g |
| $Na_2CO_3$ | 2.5g |
| Glucose | 2.0g |
| NaCl | 2.0g |
| $Na_2HPO_4$ | 0.4g |

pH 7.8 ± 0.2 at 25°C

**Source:** This medium is available as a premixed powder from BD Diagnostic Systems.

**Preparation of Medium:** Add components to distilled/deionized water and bring volume to 1.0L. Mix thoroughly. Distribute into tubes or flasks. Autoclave for 15 min at 15 psi pressure–121°C.

**Storage/Shelf Life:** Store dehydrated media in the dark in a sealed container below 30°C. Prepared media should be stored under refrigeration (2-8°C). Media should be used within 60 days of preparation. Media should not be used if there are any signs of deterioration (discoloration) or contamination, or if the expiration date supplied by the manufacturer has passed.

**Use:** For the cultivation of group A streptococci used in serological typing, and for the cultivation of a variety of pathogenic microorganisms.

## Todd-Hewitt Broth
**Composition** per liter:

| | |
|---|---|
| Pancreatic digest of casein | 20.0g |
| Infusion from 450.0g fat-free minced meat | 10.0g |
| Glucose | 2.0g |
| $NaHCO_3$ | 2.0g |
| NaCl | 2.0g |
| $Na_2HPO_4$ | 0.4g |

pH 7.8 ± 0.2 at 25°C

**Source:** This medium is available as a premixed powder from Oxoid Unipath.

**Preparation of Medium:** Add components to distilled/deionized water and bring volume to 1.0L. Mix thoroughly. Distribute into tubes or flasks. Autoclave for 10 min at 10 psi pressure–115°C.

**Storage/Shelf Life:** Store dehydrated media in the dark in a sealed container below 30°C. Prepared media should be stored under refrigeration (2-8°C). Media should be used within 60 days of preparation. Media should not be used if there are any signs of deterioration (discoloration) or contamination, or if the expiration date supplied by the manufacturer has passed.

**Use:** For the cultivation of group A streptococci used in serological typing, and for the cultivation of a variety of pathogenic microorganisms.

## Todd-Hewitt Broth
## (ATCC Medium 235)

**Composition** per liter:

| | |
|---|---|
| Peptone | 20.0g |
| Beef heart, solids from infusion | 3.1g |
| $Na_2CO_3$ | 2.5g |
| Glucose | 2.0g |
| NaCl | 2.0g |
| $Na_2HPO_4$ | 0.4g |

pH 7.8 ± 0.2 at 25°C

**Source:** This medium is available as a premixed powder from BD Diagnostic Systems.

**Preparation of Medium:** Add components to distilled/deionized water and bring volume to 1.0L. Mix thoroughly. Distribute into tubes or flasks. Autoclave for 15 min at 15 psi pressure–121°C.

**Storage/Shelf Life:** Store dehydrated media in the dark in a sealed container below 30°C. Prepared media should be stored under refrigeration (2-8°C). Media should be used within 60 days of preparation. Media should not be used if there are any signs of deterioration (discoloration) or contamination, or if the expiration date supplied by the manufacturer has passed.

**Use:** For the cultivation of group A streptococci used in serological typing, and for the cultivation of a variety of pathogenic microorganisms.

## Todd-Hewitt Broth, Modified

**Composition** per liter:

| | |
|---|---|
| Neopeptone | 20.0g |
| Glucose | 2.0g |
| $NaHCO_3$ | 2.0g |
| NaCl | 2.0g |
| $Na_2HPO_4$ | 0.4g |
| Beef heart infusion | 1.0L |

pH 7.8 ± 0.2 at 25°C

**Preparation of Medium:** Add components to distilled/deionized water and bring volume to 1.0L. Mix thoroughly. Distribute into tubes or flasks. Autoclave for 10 min at 10 psi pressure–115°C.

**Storage/Shelf Life:** Store dehydrated media in the dark in a sealed container below 30°C. Prepared media should be stored under refrigeration (2-8°C). Media should be used within 60 days of preparation. Media should not be used if there are any signs of deterioration (discoloration) or contamination, or if the expiration date supplied by the manufacturer has passed.

**Use:** For the cultivation of streptococci for serological identification.

## Todd-Hewitt HiVeg Broth

**Composition** per liter:

| | |
|---|---|
| Plant peptone | 20.0g |
| Plant special infusion | 10.0g |
| $Na_2CO_3$ | 2.5g |
| NaCl | 2.0g |
| Glucose | 2.0g |
| $Na_2HPO_4$ | 0.4g |

pH 7.8 ± 0.2 at 25°C

**Source:** This medium is available as a premixed powder from Hi-Media.

**Preparation of Medium:** Add components to distilled/deionized water and bring volume to 1.0L. Mix thoroughly. Distribute into tubes or flasks. Autoclave for 15 min at 15 psi pressure–1°C.

**Storage/Shelf Life:** Store dehydrated media in the dark in a sealed container below 30°C. Prepared media should be stored under refrigeration (2-8°C). Media should be used within 60 days of preparation. Media should not be used if there are any signs of deterioration (discoloration) or contamination, or if the expiration date supplied by the manufacturer has passed.

**Use:** For the cultivation of group A streptococci used in serological typing, and for the cultivation of a variety of pathogenic microorganisms.

## Toluidine Blue DNA Agar

**Composition** per liter:

| | |
|---|---|
| Agar | 10.0g |
| NaCl | 10.0g |
| Tris(hydroxymethyl)aminomethane buffer | 6.1g |
| Deoxyribonucleic acid | 0.3g |
| Toluidine Blue O | 0.083g |
| $CaCl_2$, anhydrous | 1.1mg |

pH 9.0 ± 0.2 at 25°C

**Preparation of Medium:** Add tris(hydroxymethyl)aminomethane buffer to distilled/deionized water and bring volume to 1.0L. Mix thoroughly. Adjust pH to 9.0. Add the remaining components, except Toluidine Blue O. Mix thoroughly. Gently heat and bring to boiling. Add Toluidine Blue O. Mix thoroughly. If used the same day, sterilization is not necessary. Cool to 50°C. Pour into sterile Petri dishes or distribute into sterile tubes.

**Storage/Shelf Life:** Store dehydrated media in the dark in a sealed container below 30°C. Prepared media should be stored under refrigeration (2-8°C). Media should be used within 60 days of preparation. Media should not be used if there are any signs of deterioration (shrinking, cracking, or discoloration) or contamination, or if the expiration date supplied by the manufacturer has passed.

**Use:** For the cultivation and differentiation of *Staphylococcus aureus* from foods.

## Toluidine Blue DNA Agar

**Composition** per liter:

| | |
|---|---|
| Agar | 10.0g |
| NaCl | 10.0g |
| Tris(hydroxymethyl)aminomethane buffer | 6.1g |
| Deoxyribonucleic acid (DNA) | 0.3g |
| Toluidine Blue O | 0.083g |
| $CaCl_2$, anhydrous | 1.1mg |

pH 7.3 ± 0.2 at 25°C

**Preparation of Medium:** Add components, except Toluidine Blue O, to distilled/deionized water and bring volume to 1.0L. Mix thoroughly. Gently heat and bring to boiling. Add Toluidine Blue O. Mix thoroughly. Medium does not have to be sterilized if used immediately. Pour into sterile Petri dishes or distribute into sterile tubes. Allow tubes to cool in a slanted position.

**Storage/Shelf Life:** Store dehydrated media in the dark in a sealed container below 30°C. Prepared media should be stored under refrigeration (2-8°C). Media should be used within 60 days of preparation. Media should not be used if there are any signs of deterioration (shrink-

ing, cracking, or discoloration) or contamination, or if the expiration date supplied by the manufacturer has passed.

**Use:** For the cultivation and differentiation of bacteria based on their production of deoxyribonuclease (DNase). Bacteria that produce DNase turn the medium pink.

## TPEY Agar
### (Tellurite Polymyxin Egg Yolk Agar)

**Composition** per liter:

NaCl.................................................................20.0g
Agar .................................................................15.5g
Pancreatic digest of casein.......................................10.0g
Yeast extract.......................................................5.0g
D-Mannitol..........................................................5.0g
LiCl ................................................................2.0g
Egg yolk emulsion (30% solution) .......................... 100.0mL
Chapman tellurite solution...................................... 10.0mL
Polymyxin B solution ............................................ 0.4mL

pH 7.1 ± 0.2 at 25°C

**Source:** This medium is available as a premixed powder from BD Diagnostic Systems.

**Egg Yolk Emulsion (30% Solution):**

**Composition** per 100.0mL:

NaCl..................................................................0.6g
Egg yolk .......................................................... 30.0mL

**Preparation of Egg Yolk Emulsion (30% Solution):** Add NaCl and egg yolk to distilled/deionized water and bring volume to 100.0mL. Mix thoroughly. Filter sterilize.

**Chapman Tellurite Solution:**

**Composition** per 100.0mL:

$K_2TeO_3$.............................................................1.0g

**Preparation of Chapman Tellurite Solution:** Add $K_2TeO_3$ to distilled/deionized water and bring volume to 100.0mL. Mix thoroughly. Filter sterilize.

**Polymyxin B Solution:**

**Composition** per 100.0mL:

Polymyxin B ........................................................1.0g

**Preparation of Polymyxin B Solution:** Add polymyxin B to distilled/deionized water and bring volume to 100.0mL. Mix thoroughly. Filter sterilize.

**Caution:** Potassium tellurite is toxic.

**Preparation of Medium:** Add components—except 30% egg yolk emulsion, Chapman tellurite solution, and polymyxin B solution—to distilled/deionized water and bring volume to 890.0mL. Mix thoroughly. Gently heat and bring to boiling. Autoclave for 15 min at 15 psi pressure–121°C. Cool to 45°–50°C. Aseptically add 100.0mL of sterile 30% egg yolk emulsion, 10.0mL of sterile Chapman tellurite solution, and 0.4mL of sterile polymyxin B solution. Mix thoroughly. Pour into sterile Petri dishes or distribute into sterile tubes.

**Storage/Shelf Life:** Store dehydrated media in the dark in a sealed container below 30°C. Prepared media should be stored under refrigeration (2-8°C). Media should be used within 60 days of preparation. Media should not be used if there are any signs of deterioration (shrinking, cracking, or discoloration) or contamination, or if the expiration date supplied by the manufacturer has passed.

**Use:** For the recovery of staphylococci from foods and other materials.

## TPEY HiVeg Agar Base
### with Egg Yolk, Tellurite, and Polymyxin B

**Composition** per liter:

NaCl.................................................................20.0g
Agar .................................................................18.0g
Plant hydrolysate .................................................10.0g
D-Mannitol..........................................................5.0g
Yeast extract.......................................................5.0g
LiCl ................................................................2.0g
Egg yolk emulsion (30% solution) ...................... 100.0mL
Chapman tellurite solution...................................... 10.0mL
Polymyxin B solution ............................................ 0.4mL

pH 7.1 ± 0.2 at 25°C

**Source:** This medium, without egg yolk, tellurite, and polymyxin B, is available as a premixed powder from HiMedia.

**Egg Yolk Emulsion (30% Solution):**

**Composition** per 100.0mL:

NaCl..................................................................0.6g
Egg yolk .......................................................... 30.0mL

**Preparation of Egg Yolk Emulsion (30% Solution):** Add NaCl and egg yolk to distilled/deionized water and bring volume to 100.0mL. Mix thoroughly. Filter sterilize.

**Chapman Tellurite Solution:**

**Composition** per 100.0mL:

$K_2TeO_3$.............................................................1.0g

**Preparation of Chapman Tellurite Solution:** Add $K_2TeO_3$ to distilled/deionized water and bring volume to 100.0mL. Mix thoroughly. Filter sterilize.

**Polymyxin B Solution:**

**Composition** per 100.0mL:

Polymyxin B ........................................................1.0g

**Preparation of Polymyxin B Solution:** Add polymyxin B to distilled/deionized water and bring volume to 100.0mL. Mix thoroughly. Filter sterilize.

**Caution:** Potassium tellurite is toxic.

**Preparation of Medium:** Add components—except 30% egg yolk emulsion, Chapman tellurite solution, and polymyxin B solution—to distilled/deionized water and bring volume to 890.0mL. Mix thoroughly. Gently heat and bring to boiling. Autoclave for 15 min at 15 psi pressure–121°C. Cool to 45°–50°C. Aseptically add 100.0mL of sterile 30% egg yolk emulsion, 10.0mL of sterile Chapman tellurite solution, and 0.4mL of sterile polymyxin B solution. Mix thoroughly. Pour into sterile Petri dishes or distribute into sterile tubes.

**Storage/Shelf Life:** Store dehydrated media in the dark in a sealed container below 30°C. Prepared media should be stored under refrigeration (2-8°C). Media should be used within 60 days of preparation. Media should not be used if there are any signs of deterioration (shrinking, cracking, or discoloration) or contamination, or if the expiration date supplied by the manufacturer has passed.

**Use:** For the recovery of staphylococci from foods and other materials.

## Transgrow Medium

**Composition** per liter:

GC agar base.................................................... 730.0mL
Hemoglobin solution ........................................... 250.0mL

Vitox supplement ............................................................... 10.0mL
VCN antibiotic solution .................................................... 10.0mL
pH 7.3 ± 0.2 at 25°C

**GC Agar Base:**
**Composition** per 730.0mL:
Special peptone ......................................................................15.0g
Agar ........................................................................................20.0g
NaCl .........................................................................................5.0g
$K_2HPO_4$.................................................................................4.0g
Cornstarch ..............................................................................1.0g
$KH_2PO_4$................................................................................1.0g
pH 7.2 ± 0.2 at 25°C

**Preparation of GC Agar Base:** Add components of GC medium base and the hemoglobin to distilled/deionized water and bring volume to 730.0mL. Mix thoroughly. Gently heat until boiling. Autoclave for 15 min at 15 psi pressure–121°C. Cool to 45°–50°C.

**Hemoglobin Solution:**
**Composition** per 250.0mL:
Hemoglobin ............................................................................5.0g

**Preparation of Hemoglobin Solution:** Add hemoglobin to distilled/deionized water and bring volume to 250.0mL. Mix thoroughly. Autoclave for 15 min at 15 psi pressure–121°C. Cool to 45°–50°C.

**Vitox Supplement:**
**Composition** per 10.0mL:
Glucose ...................................................................................2.0g
L-Cysteine·HCl.....................................................................0.518g
L-Glutamine ...........................................................................0.2g
L-Cystine...............................................................................0.022g
Adenine sulfate .....................................................................0.01g
Nicotinamide adenine dinucleotide .................................5.0mg
Cocarboxylase........................................................................2.0mg
Guanine·HCl ..........................................................................0.6mg
$Fe(NO_3)_3·6H_2O$ ...............................................................0.4mg
*p*-Aminobenzoic acid.........................................................0.26mg
Vitamin $B_{12}$ .........................................................................0.2mg
Thiamine·HCl ......................................................................0.06mg

**Preparation of Vitox Supplement:** Add components to distilled/deionized water and bring volume to 10.0mL. Mix thoroughly. Filter sterilize.

**VCN Antibiotic Solution:**
**Composition** per 10.0mL:
Colistin methane sulfonate......................................................7.5mg
Vancomycin .............................................................................3.0mg
Nystatin...............................................................................12,500U

**Preparation of VCN Antibiotic Solution:** Add components to distilled/deionized water and bring volume to 10.0mL. Mix thoroughly. Filter sterilize.

**Preparation of Medium:** To 730.0mL of cooled, sterile GC agar base, aseptically add 250.0mL of sterile hemoglobin solution, 10.0mL of sterile Vitox supplement, and 10.0mL of VCN antibiotic solution. Mix thoroughly. Pour into sterile Petri dishes or distribute into sterile tubes.

**Storage/Shelf Life:** Store dehydrated media in the dark in a sealed container below 30°C. Prepared media should be stored under refrigeration (2-8°C). Media should be used within 60 days of preparation. Media should not be used if there are any signs of deterioration (discoloration) or contamination, or if the expiration date supplied by the manufacturer has passed.

**Use:** For the cultivation and transport of fastidious microorganisms, especially *Neisseria* species.

**Transgrow Medium**

**Composition** per liter:
GC medium base................................................................. 730.0mL
Hemoglobin solution ......................................................... 250.0mL
Supplement B..................................................................... 10.0mL
VCNT antibiotic solution .................................................. 10.0mL
pH 7.3 ± 0.2 at 25°C

**GC Medium Base:**
**Composition** per 730.0mL:
Proteose peptone No. 3 .........................................................15.0g
Agar ........................................................................................20.0g
NaCl .........................................................................................5.0g
$K_2HPO_4$.................................................................................4.0g
Glucose ...................................................................................1.5g
Cornstarch ..............................................................................1.0g
$KH_2PO_4$................................................................................1.0g
pH 7.2 ± 0.2 at 25°C

**Preparation of GC Medium Base:** Add components to distilled/deionized water and bring volume to 730.0mL. Mix thoroughly. Gently heat until boiling. Autoclave for 15 min at 15 psi pressure–121°C. Cool to 45°–50°C.

**Hemoglobin Solution:**
**Composition** per 250.0mL:
Hemoglobin ..........................................................................10.0g

**Preparation of Hemoglobin Solution:** Add hemoglobin to distilled/deionized water and bring volume to 250.0mL. Mix thoroughly. Autoclave for 15 min at 15 psi pressure–121°C. Cool to 45°–50°C.

**Supplement B:**
**Composition** per 10.0mL:
Supplement B contains yeast concentrate, glutamine, coenzyme, cocarboxylase, hematin, and growth factors.

**Preparation of Supplement B:** Add components to distilled/deionized water and bring volume to 10.0mL. Mix thoroughly. Filter sterilize.

**Source:** Supplement B is available as a premixed powder from BD Diagnostic Systems.

**VCNT Antibiotic Solution:**
**Composition** per 10.0mL:
Colistin methane sulfonate .....................................................7.5mg
Trimethoprim lactate................................................................5.0mg
Vancomycin ..............................................................................3.0mg
Nystatin................................................................................12,500U

**Preparation of VCNT Antibiotic Solution:** Add components to distilled/deionized water and bring volume to 10.0mL. Mix thoroughly. Filter sterilize.

**Preparation of Medium:** To 730.0mL of cooled, sterile GC medium base, aseptically add 250.0mL of sterile hemoglobin solution, 10.0mL of sterile supplement B, and 10.0mL of sterile VCNT antibiotic solution. Mix thoroughly. Pour into sterile Petri dishes or distribute into sterile tubes.

**Storage/Shelf Life:** Store dehydrated media in the dark in a sealed container below 30°C. Prepared media should be stored under refrigeration (2-8°C). Media should be used within 60 days of preparation. Media should not be used if there are any signs of deterioration (discol-

oration) or contamination, or if the expiration date supplied by the manufacturer has passed.

**Use:** For the cultivation and transport of fastidious microorganisms, especially *Neisseria* species.

## Transgrow Medium with Trimethoprim
**Composition** per liter:

| | |
|---|---|
| Agar | 20.0g |
| Hemoglobin | 10.0g |
| Pancreatic digest of casein | 7.5g |
| Selected meat peptone | 7.5g |
| NaCl | 5.0g |
| $K_2HPO_4$ | 4.0g |
| Glucose | 1.5g |
| Cornstarch | 1.0g |
| $KH_2PO_4$ | 1.0g |
| Supplement solution | 10.0mL |
| VCNT inhibitor | 10.0mL |

pH 6.7 ± 0.2 at 25°C

**Source:** This medium is available as a prepared medium from BD Diagnostic Systems.

**Supplement Solution:**
**Composition** per liter:

| | |
|---|---|
| Glucose | 100.0g |
| L-Cysteine·HCl | 25.9g |
| L-Glutamine | 10.0g |
| L-Cystine | 1.1g |
| Adenine | 1.0g |
| Nicotinamide adenine dinucleotide | 0.25g |
| Vitamin $B_{12}$ | 0.1g |
| Thiamine pyrophosphate | 0.1g |
| Guanine·HCl | 0.03g |
| $Fe(NO_3)_3·6H_2O$ | 0.02g |
| *p*-Aminobenzoic acid | 0.013g |
| Thiamine·HCl | 3.0mg |

**Source:** The supplement solution (IsoVitaleX® enrichment) is available from BD Diagnostic Systems. This enrichment may be replaced by supplement VX from BD Diagnostic Systems.

**Preparation of Supplement Solution:** Add components to distilled/deionized water and bring volume to 1.0L. Mix thoroughly. Filter sterilize.

**VCNT Inhibitor:**
**Composition** per 10.0mL:

| | |
|---|---|
| Colistin | 7.5mg |
| Trimethoprim lactate | 5.0mg |
| Vancomycin | 3.0mg |
| Nystatin | 12,500U |

**Preparation of VCNT Inhibitor:** Add components to distilled/deionized water and bring volume to 10.0mL. Mix thoroughly. Filter sterilize.

**Preparation of Medium:** Add components, except supplement solution and VCNT inhibitor, to distilled/deionized water and bring volume to 980.0mL. Mix thoroughly. Gently heat and bring to boiling. Autoclave for 15 min at 15 psi pressure–121°C. Cool to 45°–50°C under 5–30% $CO_2$. Aseptically add 10.0mL of sterile supplement solution and 10.0mL of sterile VCNT inhibitor. Mix thoroughly. Aseptically distribute under 5–30% $CO_2$ into sterile screw-capped tubes.

**Storage/Shelf Life:** Store dehydrated media in the dark in a sealed container below 30°C. Prepared media should be stored under refrigeration (2-8°C). Media should be used within 60 days of preparation. Media should not be used if there are any signs of deterioration (discoloration) or contamination, or if the expiration date supplied by the manufacturer has passed.

**Use:** For the transportation and recovery of pathogenic *Neisseria* species.

## Transgrow Medium without Trimethoprim
**Composition** per liter:

| | |
|---|---|
| Agar | 20.0g |
| Hemoglobin | 10.0g |
| Pancreatic digest of casein | 7.5g |
| Selected meat peptone | 7.5g |
| NaCl | 5.0g |
| $K_2HPO_4$ | 4.0g |
| Glucose | 1.5g |
| Cornstarch | 1.0g |
| $KH_2PO_4$ | 1.0g |
| Supplement solution | 10.0mL |
| VCN inhibitor | 10.0mL |

pH 6.7 ± 0.2 at 25°C

**Source:** This medium is available as a prepared medium from BD Diagnostic Systems.

**Supplemement Solution:**
**Composition** per liter:

| | |
|---|---|
| Glucose | 100.0g |
| L-Cysteine·HCl | 25.9g |
| L-Glutamine | 10.0g |
| L-Cystine | 1.1g |
| Adenine | 1.0g |
| Nicotinamide adenine dinucleotide | 0.25g |
| Vitamin $B_{12}$ | 0.1g |
| Thiamine pyrophosphate | 0.1g |
| Guanine·HCl | 0.03g |
| $Fe(NO_3)_3·6H_2O$ | 0.02g |
| *p*-Aminobenzoic acid | 0.013g |
| Thiamine·HCl | 3.0mg |

**Source:** The supplement solution IsoVitaleX® enrichment is available from BD Diagnostic Systems. This enrichment may be replaced by supplement VX from BD Diagnostic Systems.

**Preparation of Supplement Solution:** Add components to distilled/deionized water and bring volume to 1.0L. Mix thoroughly. Filter sterilize.

**VCN Inhibitor:**
**Composition** per 10.0mL:

| | |
|---|---|
| Colistin | 7.5mg |
| Vancomycin | 3.0mg |
| Nystatin | 12,500U |

**Preparation of VCN Inhibitor:** Add components to distilled/deionized water and bring volume to 10.0mL. Mix thoroughly. Filter sterilize.

**Preparation of Medium:** Add components, except supplement solution and VCN inhibitor, to distilled/deionized water and bring volume to 980.0mL. Mix thoroughly. Gently heat and bring to boiling. Autoclave for 15 min at 15 psi pressure–121°C. Cool to 45°–50°C under 5–30% $CO_2$. Aseptically add 10.0mL of sterile supplement solution and 10.0mL of sterile VCN inhibitor. Mix thoroughly. Aseptically distribute under 5–30% $CO_2$ into sterile screw-capped tubes.

**Storage/Shelf Life:** Store dehydrated media in the dark in a sealed container below 30°C. Prepared media should be stored under refrigeration (2-8°C). Media should be used within 60 days of preparation. Media should not be used if there are any signs of deterioration (discoloration) or contamination, or if the expiration date supplied by the manufacturer has passed.

**Use:** For the transportation and recovery of pathogenic *Neisseria* species.

## Transport Medium

**Composition** per liter:
| | |
|---|---|
| Sodium glycerophosphate | 10.0g |
| Agar | 3.0g |
| Sodium thioglycolate | 1.0g |
| $CaCl_2·2H_2O$ | 0.1g |
| Methylene Blue | 2.0mg |

pH 7.3 ± 0.2 at 25°C

**Source:** This medium is available as a premixed powder from BD Diagnostic Systems.

**Preparation of Medium:** Add components to distilled/deionized water and bring volume to 1.0L. Mix thoroughly. Gently heat while stirring and bring to boiling. Distribute into screw-capped tubes or vials. Fill tubes nearly to capacity. Leave only enough space so that when a small swab is introduced the tube does not overflow. Autoclave for 10 min at 15 psi pressure–121°C. Tighten caps on tubes.

**Storage/Shelf Life:** Store dehydrated media in the dark in a sealed container below 30°C. Prepared media should be stored under refrigeration (2-8°C). Media should be used within 60 days of preparation. Media should not be used if there are any signs of deterioration (discoloration) or contamination, or if the expiration date supplied by the manufacturer has passed.

**Use:** For the transportation of swab specimens for the recovery of a wide variety of microorganisms, including *Neisseria gonorrhoeae*.

## Transport Medium Stuart

**Composition** per liter:
| | |
|---|---|
| Sodium glycerophosphate | 10.0g |
| Agar | 3.0g |
| Sodium thioglycolate | 0.9g |
| $CaCl_2·2H_2O$ | 0.1g |
| Methylene Blue | 2.0mg |

pH 7.3 ± 0.2 at 25°C

**Source:** This medium is available as a premixed powder from BD Diagnostic Systems.

**Preparation of Medium:** Add components to distilled/deionized water and bring volume to 1.0L. Mix thoroughly. Gently heat while stirring and bring to boiling. Distribute into screw-capped tubes or vials. Fill tubes nearly to capacity. Leave only enough space so that when a small swab is introduced the tube does not overflow. Autoclave for 10 min at 15 psi pressure–121°C. Tighten caps on tubes.

**Storage/Shelf Life:** Store dehydrated media in the dark in a sealed container below 30°C. Prepared media should be stored under refrigeration (2-8°C). Media should be used within 60 days of preparation. Media should not be used if there are any signs of deterioration (discoloration) or contamination, or if the expiration date supplied by the manufacturer has passed.

**Use:** For the transportation of swab specimens for the recovery of a wide variety of microorganisms, including *Neisseria gonorrhoeae*.

## *Treponema* Isolation Medium

**Composition** per liter:
| | |
|---|---|
| Solution A | 450.0mL |
| Spirolate broth | 450.0mL |
| Rabbit serum, inactivated at 56°C for 30 min | 100.0mL |

pH 7.4 ± 0.2 at 25°C

### Solution A:
**Composition** per 450.0mL:
| | |
|---|---|
| Agar | 8.0g |
| Asparagine | 0.25g |
| Sodium thioglycolate | 0.25g |
| Pancreatic digest of casein | 0.25g |
| Brain heart infusion broth | 450.0mL |

**Preparation of Solution A:** Combine components. Mix thoroughly. Gently heat and bring to boiling. Autoclave for 15 min at 15 psi pressure–121°C. Cool to 45°–50°C.

### Brain Heart Infusion Broth:
**Composition** per liter:
| | |
|---|---|
| Pancreatic digest of gelatin | 14.5g |
| Brain heart, solids from infusion | 6.0g |
| Peptic digest of animal tissue | 6.0g |
| NaCl | 5.0g |
| Casein | 5.0g |
| Glucose | 3.0g |
| $Na_2HPO_4$ | 2.5g |

**Preparation of Brain Heart Infusion Broth:** Add components to distilled/deionized water and bring volume to 1.0L. Mix thoroughly.

### Spirolate Broth:
**Composition** per liter:
| | |
|---|---|
| Pancreatic digest of casein | 15.0g |
| Glucose | 5.0g |
| Yeast extract | 5.0g |
| NaCl | 2.5g |
| L-Cysteine·$HCl·H_2O$ | 1.0g |
| Sodium thioglycolate | 0.5g |
| Palmitic acid | 0.05g |
| Stearic acid | 0.05g |
| Oleic acid | 0.05g |
| Linoleic acid | 0.05g |

**Preparation of Spirolate Broth:** Add components to distilled/deionized water and bring volume to 1.0L. Mix thoroughly. Autoclave for 15 min at 15 psi pressure–121°C. Cool to 25°C.

**Preparation of Medium:** Combine 450.0mL of sterile solution A, 450.0mL of sterile spirolate broth, and 100.0mL of rabbit serum. Mix thoroughly. Aseptically distribute into sterile tubes or flasks.

**Storage/Shelf Life:** Store dehydrated media in the dark in a sealed container below 30°C. Prepared media should be stored under refrigeration (2-8°C). Media should be used within 60 days of preparation. Media should not be used if there are any signs of deterioration (discoloration) or contamination, or if the expiration date supplied by the manufacturer has passed.

**Use:** For the isolation and cultivation of oral, genital, and fecal treponemes.

## *Treponema* Isolation Medium

**Composition** per liter:

| | |
|---|---|
| Beef heart, solids from infusion | 20.0g |
| Ionagar No. 2 | 7.2g |
| K$_2$HPO$_4$ | 2.0g |
| Arabinose | 0.8g |
| Glucose | 0.8g |
| Maltose | 0.8g |
| Polypeptone™ | 0.8g |
| Pyruvate | 0.8g |
| Starch, soluble | 0.8g |
| Sucrose | 0.8g |
| Cysteine·HCl | 0.68g |
| (NH$_4$)$_2$SO$_4$ | 0.6g |
| Serine | 0.4g |
| Tryptose | 0.4g |
| Yeast extract | 0.4g |
| NaCl | 0.2g |
| Rumen fluid | 500.0mL |
| Rabbit serum-cocarboxylase solution | 100.0mL |

pH 7.2 ± 0.2 at 25°C

**Rabbit Serum-Cocarboxylase Solution:**

**Composition** per liter:

| | |
|---|---|
| Rabbit serum, heat inactivated | 100.0mL |
| Cocarboxylase solution | 1.0mL |

**Preparation of Rabbit Serum-Cocarboxylase Solution:** Heat rabbit serum at 56°C for 1 h. Add 1.0mL of cocarboxylase solution. Mix thoroughly.

**Cocarboxylase Solution:**

**Composition** per 1.0mL:

| | |
|---|---|
| Cocarboxylase | 0.5g |

**Preparation of Cocarboxylase Solution:** Add cocarboxylase to 1.0mL of distilled/deionized water. Mix thoroughly. Filter sterilize.

**Preparation of Medium:** Add components, except rumen fluid and rabbit serum-cocarboxylase solution, to distilled/deionized water and bring volume to 400.0mL. Mix thoroughly. Gently heat and bring to boiling. Autoclave for 15 min at 15 psi pressure–121°C. Cool to 45°–50°C. Aseptically add 500.0mL of sterile rumen fluid and 100.0mL of sterile rabbit serum-cocarboxylase solution. Mix thoroughly. Pour into sterile Petri dishes or distribute into sterile tubes.

**Storage/Shelf Life:** Store dehydrated media in the dark in a sealed container below 30°C. Prepared media should be stored under refrigeration (2-8°C). Media should be used within 60 days of preparation. Media should not be used if there are any signs of deterioration (discoloration) or contamination, or if the expiration date supplied by the manufacturer has passed.

**Use:** For the isolation of oral treponemes.

## *Treponema macrodentium* Medium

**Composition** per liter:

| | |
|---|---|
| Glucose | 1.0g |
| Nicotinamide | 0.4g |
| Spermine·4HCl | 0.15g |
| Sodium isobutyrate | 0.02g |
| Carboxylase | 5.0mg |
| PPLO agar | 900.0mL |
| Bovine serum | 100.0mL |

pH 7.0 ± 0.2 at 25°C

**PPLO Agar:**

**Composition** per 900.0mL:

| | |
|---|---|
| Beef heart, infusion from | 50.0g |
| Agar | 14.0g |
| Peptone | 10.0g |
| NaCl | 5.0g |

pH 7.8 ± 0.2 at 25°C

**Preparation of PPLO Agar:** Add components to distilled/deionized water and bring volume to 900.0mL. Mix thoroughly.

**Preparation of Medium:** Combine components, except bovine serum. Mix thoroughly. Autoclave for 15 min at 15 psi pressure–121°C. Cool to 45°–50°C. Aseptically add sterile bovine serum. Mix thoroughly. Aseptically distribute into sterile tubes or flasks.

**Storage/Shelf Life:** Store dehydrated media in the dark in a sealed container below 30°C. Prepared media should be stored under refrigeration (2-8°C). Media should be used within 60 days of preparation. Media should not be used if there are any signs of deterioration (shrinking, cracking, or discoloration) or contamination, or if the expiration date supplied by the manufacturer has passed.

**Use:** For the isolation and cultivation of *Treponema macrodentium*.

## *Treponema* Medium

**Composition** per liter:

| | |
|---|---|
| Pancreatic digest of casein | 30.0g |
| Ionagar No. 2 | 8.0g |
| Glucose | 5.0g |
| Yeast extract | 5.0g |
| NaCl | 2.5g |
| Cysteine·HCl | 0.75g |
| Horse serum, inactivated | 100.0mL |

pH 7.4 ± 0.2 at 25°C

**Preparation of Medium:** Add components, except horse serum, to distilled/deionized water and bring volume to 900.0mL. Mix thoroughly. Gently heat and bring to boiling. Autoclave for 15 min at 15 psi pressure–121°C. Cool to 45°–50°C. Aseptically add 100.0mL of sterile horse serum. Mix thoroughly. Distribute into sterile tubes.

**Use:** For the isolation and cultivation of oral treponemes.

## *Treponema* Medium

**Composition** per liter:

| | |
|---|---|
| Spirolate agar | 900.0mL |
| Rabbit serum, inactivated at 56°C for 30 min | 100.0mL |

**Spirolate Agar:**

**Composition** per liter:

| | |
|---|---|
| Pancreatic digest of casein | 15.0g |
| Agar | 14.0g |
| Glucose | 5.0g |
| Yeast extract | 5.0g |
| NaCl | 2.5g |
| L-Cysteine·HCl·H$_2$O | 1.0g |
| Sodium thioglycolate | 0.5g |
| Palmitic acid | 0.05g |
| Stearic acid | 0.05g |
| Oleic acid | 0.05g |
| Linoleic acid | 0.05g |

**Preparation of Spirolate Agar:** Add components to distilled/deionized water and bring volume to 1.0L. Mix thoroughly. Gently heat and bring to boiling. Autoclave for 15 min at 15 psi pressure–121°C. Cool to 45°–50°C.

**Preparation of Medium:** To 900.0mL of cooled, sterile spirolate agar, aseptically add 100.0mL of rabbit serum. Mix thoroughly. Aseptically distribute into sterile tubes or flasks.

**Storage/Shelf Life:** Store dehydrated media in the dark in a sealed container below 30°C. Prepared media should be stored under refrigeration (2-8°C). Media should be used within 60 days of preparation. Media should not be used if there are any signs of deterioration (shrinking, cracking, or discoloration) or contamination, or if the expiration date supplied by the manufacturer has passed.

**Use:** For the isolation of oral treponemes.

## *Treponema* Medium

**Composition** per liter:
| | |
|---|---|
| Spirolate agar | 675.0mL |
| Brain heart infusion broth | 225.0mL |
| Rabbit serum, inactivated at 56°C for 30 min | 100.0mL |

pH 7.0–7.2 ± 0.2 at 25°C

**Spirolate Agar:**
**Composition** per 675.0mL:
| | |
|---|---|
| Pancreatic digest of casein | 15.0g |
| Ionagar No. 2 | 8.0g |
| Glucose | 5.0g |
| Yeast extract | 5.0g |
| NaCl | 2.5g |
| L-Cysteine·HCl·H$_2$O | 1.0g |
| Sodium thioglycolate | 0.5g |
| Palmitic acid | 0.05g |
| Stearic acid | 0.05g |
| Oleic acid | 0.05g |
| Linoleic acid | 0.05g |

**Preparation of Spirolate Agar:** Add components to distilled/deionized water and bring volume to 675.0mL. Mix thoroughly. Autoclave for 15 min at 15 psi pressure–121°C. Cool to 45°–50°C.

**Brain Heart Infusion Broth:**
**Composition** per liter:
| | |
|---|---|
| Pancreatic digest of gelatin | 14.5g |
| Brain heart, solids from infusion | 6.0g |
| Peptic digest of animal tissue | 6.0g |
| NaCl | 5.0g |
| Casein | 5.0g |
| Glucose | 3.0g |
| Na$_2$HPO$_4$ | 2.5g |

**Preparation of Brain Heart Infusion Broth:** Add components to distilled/deionized water and bring volume to 1.0L. Mix thoroughly.

**Preparation of Medium:** Aseptically combine 675.0mL of cooled, sterile spirolate agar, 225.0mL of cooled, sterile brain heart infusion broth, and 100.0mL of rabbit serum. Mix thoroughly. Pour into sterile Petri dishes or distribute into sterile tubes.

**Storage/Shelf Life:** Store dehydrated media in the dark in a sealed container below 30°C. Prepared media should be stored under refrigeration (2-8°C). Media should be used within 60 days of preparation. Media should not be used if there are any signs of deterioration (shrinking, cracking, or discoloration) or contamination, or if the expiration date supplied by the manufacturer has passed.

**Use:** For the isolation of oral treponemes.

## *Treponema* Medium

**Composition** per liter:
| | |
|---|---|
| Solution A | 440.0mL |
| Spirolate broth | 440.0mL |
| Rabbit serum, inactivated at 56°C for 30 min | 100.0mL |
| Mucin solution | 20.0mL |

pH 7.8 ± 0.2 at 25°C

**Solution A:**
**Composition** per 440.0mL:
| | |
|---|---|
| Ionagar No. 2 | 8.0g |
| Brain heart infusion broth | 440.0mL |

**Brain Heart Infusion Broth:**
**Composition** per liter:
| | |
|---|---|
| Pancreatic digest of gelatin | 14.5g |
| Brain heart, solids from infusion | 6.0g |
| Peptic digest of animal tissue | 6.0g |
| NaCl | 5.0g |
| Casein | 5.0g |
| Glucose | 3.0g |
| Na$_2$HPO$_4$ | 2.5g |

**Preparation of Brain Heart Infusion Broth:** Add components to distilled/deionized water and bring volume to 1.0L. Mix thoroughly.

**Preparation of Solution A:** Add 8.0g of ionagar to 440.0mL of brain heart infusion broth. Mix thoroughly. Gently heat and bring to boiling. Autoclave for 15 min at 15 psi pressure–121°C. Cool to 45°–50°C.

**Spirolate Broth:**
**Composition** per liter:
| | |
|---|---|
| Pancreatic digest of casein | 15.0g |
| Glucose | 5.0g |
| Yeast extract | 5.0g |
| NaCl | 2.5g |
| L-Cysteine·HCl·H$_2$O | 1.0g |
| Sodium thioglycolate | 0.5g |
| Palmitic acid | 0.05g |
| Stearic acid | 0.05g |
| Oleic acid | 0.05g |
| Linoleic acid | 0.05g |

**Preparation of Spirolate Broth:** Add components to distilled/deionized water and bring volume to 1.0L. Mix thoroughly. Autoclave for 15 min at 15 psi pressure–121°C. Cool to 25°.

**Mucin Solution:**
**Composition** per 20.0mL:
| | |
|---|---|
| Mucin | 0.2g |

**Preparation of Mucin Solution:** Add mucin to distilled/deionized water and bring volume to 20.0mL. Mix thoroughly. Filter sterilize.

**Preparation of Medium:** Aseptically combine 440.0mL of solution A, 440.0mL of spirolate broth, 100.0mL of rabbit serum, and 20.0mL of mucin solution. Mix thoroughly. Aseptically distribute into sterile tubes or flasks.

**Storage/Shelf Life:** Store dehydrated media in the dark in a sealed container below 30°C. Prepared media should be stored under refrigeration (2-8°C). Media should be used within 60 days of preparation. Media should not be used if there are any signs of deterioration (shrinking, cracking, or discoloration) or contamination, or if the expiration date supplied by the manufacturer has passed.

**Use:** For the isolation of intestinal treponemes.

## *Treponema* Medium

**Composition** per liter:

| | |
|---|---|
| Agar | 13.0g |
| Glucose | 1.4g |
| Cysteine·HCl | 0.64g |
| $(NH_4)_2SO_4$ | 0.5g |
| Polypeptone™ | 0.5g |
| Starch, soluble | 0.5g |
| Yeast extract | 0.5g |
| Resazurin | 1.6mg |
| Salts solution | 500.0mL |
| Bovine rumen fluid | 280.0mL |

pH 7.2–7.5 at 25°C

**Salts Solution:**

**Composition** per liter:

| | |
|---|---|
| $NaHCO_3$ | 10.0g |
| NaCl | 2.0g |
| $K_2HPO_4$ | 1.0g |
| $KH_2PO_4$ | 1.0g |
| $CaCl_2$ | 0.2g |
| $MgSO_4$ | 0.2g |
| CoCl | 3.4mg |
| $MnSO_4$ | 3.4mg |
| $NaMoO_4$ | 3.4mg |

**Preparation of Salts Solution:** Add components to distilled/deionized water and bring volume to 1.0L. Mix thoroughly.

**Preparation of Medium:** Add components, except bovine rumen fluid, to distilled/deionized water and bring volume to 720.0mL. Mix thoroughly. Gently heat and bring to boiling. Autoclave for 15 min at 15 psi pressure–121°C. Cool to 45°–50°C. Aseptically add bovine rumen fluid. Mix thoroughly. Pour into sterile Petri dishes or distribute into sterile tubes.

**Storage/Shelf Life:** Store dehydrated media in the dark in a sealed container below 30°C. Prepared media should be stored under refrigeration (2-8°C). Media should be used within 60 days of preparation. Media should not be used if there are any signs of deterioration (shrinking, cracking, or discoloration) or contamination, or if the expiration date supplied by the manufacturer has passed.

**Use:** For the isolation of intestinal treponemes.

## *Treponema* Medium

**Composition** per liter:

| | |
|---|---|
| Cysteine·HCl·$H_2O$ | 1.0g |
| Glucose | 1.0g |
| Nicotinamide | 0.4g |
| Spermidine·4HCl | 0.15g |
| Sodium isobutyrate | 0.02g |
| Thiamine pyrophosphate | 5.0mg |
| PPLO broth | 900.0mL |
| Rabbit serum, inactivated | 100.0mL |

pH 7.8 ± 0.2 at 25°C

**PPLO Broth:**

**Composition** per 900.0mL:

| | |
|---|---|
| Beef heart, infusion from solids | 50.0g |
| Peptone | 10.0g |
| NaCl | 5.0g |

**Preparation of PPLO Broth:** Add components to distilled/deionized water and bring volume to 900.0mL. Mix thoroughly.

**Preparation of Medium:** Combine components, except rabbit serum. Mix thoroughly. Filter sterilize. Aseptically add sterile rabbit serum. Mix thoroughly. Aseptically distribute into sterile tubes or flasks.

**Storage/Shelf Life:** Store dehydrated media in the dark in a sealed container below 30°C. Prepared media should be stored under refrigeration (2-8°C). Media should be used within 60 days of preparation. Media should not be used if there are any signs of deterioration (discoloration) or contamination, or if the expiration date supplied by the manufacturer has passed.

**Use:** For the cultivation of oral treponemes. For the cultivation of *Treponema denticola, Treponema macrodentium,* and *Treponema oralis.*

## *Treponema* Medium, Prereduced

**Composition** per liter:

| | |
|---|---|
| Agar | 1.6g |
| Glucose | 1.4g |
| Cysteine·HCl·$H_2O$ | 0.64g |
| $(NH_4)_2SO_4$ | 0.5g |
| Polypeptone™ | 0.5g |
| Starch, soluble | 0.5g |
| Yeast extract | 0.5g |
| Resazurin | 1.6mg |
| Salts solution | 500.0mL |
| Bovine rumen fluid | 280.0mL |

pH 7.2–7.5 at 25°C

**Salts Solution:**

**Composition** per liter:

| | |
|---|---|
| $NaHCO_3$ | 10.0g |
| NaCl | 2.0g |
| $K_2HPO_4$ | 1.0g |
| $KH_2PO_4$ | 1.0g |
| $CaCl_2$ | 0.2g |
| $MgSO_4$ | 0.2g |
| CoCl | 3.4mg |
| $MnSO_4$ | 3.4mg |
| $NaMoO_4$ | 3.4mg |

**Preparation of Salts Solution:** Add components to distilled/deionized water and bring volume to 1.0L. Mix thoroughly.

**Preparation of Medium:** Add components, except bovine rumen fluid, to distilled/deionized water and bring volume to 720.0mL. Mix thoroughly. Gently heat and bring to boiling. Autoclave for 15 min at 15 psi pressure–121°C. Cool to 45°–50°C. Aseptically add 280.0mL of sterile bovine rumen fluid. Mix thoroughly. Aseptically and anaerobically distribute into sterile tubes or flasks under 100% $N_2$.

**Storage/Shelf Life:** Store dehydrated media in the dark in a sealed container below 30°C. Prepared media should be stored under refrigeration (2-8°C). Media should not be used if there are any signs of deterioration (shrinking, cracking, or discoloration) or contamination, or if the expiration date supplied by the manufacturer has passed.

**Use:** For the cultivation of fecal and intestinal treponemes.

## Trimethylamine *N*-Oxide Medium (TMAO Medium)

**Composition** per liter:

| | |
|---|---|
| Beef extract | 10.0g |
| Peptone | 10.0g |
| NaCl | 5.0g |

| | |
|---|---|
| Agar | 2.0g |
| Trimethylamine *N*-oxide | 1.0g |
| Yeast extract | 1.0g |

pH 7.5 ± 0.2 at 25°C

**Source:** This medium is available as a premixed powder from Oxoid Unipath.

**Preparation of Medium:** Add components to distilled/deionized water and bring volume to 1.0L. Mix thoroughly. Gently heat and bring to boiling. Distribute into screw-capped tubes in 4.0mL volumes. Autoclave for 15 min at 15 psi pressure–121°C. Allow tubes to cool in an upright position.

**Storage/Shelf Life:** Store dehydrated media in the dark in a sealed container below 30°C. Prepared media should be stored under refrigeration (2-8°C). Media should be used within 60 days of preparation. Media should not be used if there are any signs of deterioration (discoloration) or contamination, or if the expiration date supplied by the manufacturer has passed.

**Use:** For the cultivation and differentiation of *Campylobacter* species from foods. *Campylobacter jejuni* and *Campylobacter coli* will not grow.

## Triple Sugar Iron Agar
### (TSI Agar)

**Composition** per liter:

| | |
|---|---|
| Peptone | 20.0g |
| Agar | 12.0g |
| Lactose | 10.0g |
| Sucrose | 10.0g |
| NaCl | 5.0g |
| Beef extract | 3.0g |
| Yeast extract | 3.0g |
| Glucose | 1.0g |
| Ferric citrate | 0.3g |
| $Na_2S_2O_3$ | 0.3g |
| Phenol Red | 0.025g |

pH 7.4 ± 0.2 at 25°C

**Source:** This medium is available as a premixed powder from BD Diagnostic Systems and Oxoid Unipath.

**Preparation of Medium:** Add components to distilled/deionized water and bring volume to 1.0L. Mix thoroughly. Gently heat and bring to boiling. Distribute into tubes or flasks. Autoclave for 15 min at 15 psi pressure–121°C. Allow tubes to cool in a slanted position to form a 1.0-inch butt.

**Storage/Shelf Life:** Store dehydrated media in the dark in a sealed container below 30°C. Prepared media should be stored under refrigeration (2-8°C). Media should be used within 60 days of preparation. Media should not be used if there are any signs of deterioration (shrinking, cracking, or discoloration) or contamination, or if the expiration date supplied by the manufacturer has passed.

**Use:** For the differentiation of members of the Enterobacteriaceae based on their fermentation of lactose, sucrose, and glucose and the production of $H_2S$.

## Triple Sugar Iron Agar
### (TSI Agar)

**Composition** per liter:

| | |
|---|---|
| Agar | 13.0g |
| Pancreatic digest of casein | 10.0g |

| | |
|---|---|
| Peptic digest of animal tissue | 10.0g |
| Lactose | 10.0g |
| Sucrose | 10.0g |
| NaCl | 5.0g |
| Glucose | 1.0g |
| $Fe(NH_4)_2(SO_4)_2 \cdot 6H_2O$ | 0.2g |
| $Na_2S_2O_3$ | 0.2g |
| Phenol Red | 0.025g |

pH 7.3 ± 0.2 at 25°C

**Source:** This medium is available as a premixed powder from BD Diagnostic Systems.

**Preparation of Medium:** Add components to distilled/deionized water and bring volume to 1.0L. Mix thoroughly. Gently heat and bring to boiling. Distribute into tubes or flasks. Autoclave for 15 min at 15 psi pressure–121°C. Allow tubes to cool in a slanted position to form a 1.0-inch butt.

**Storage/Shelf Life:** Store dehydrated media in the dark in a sealed container below 30°C. Prepared media should be stored under refrigeration (2-8°C). Media should be used within 60 days of preparation. Media should not be used if there are any signs of deterioration (shrinking, cracking, or discoloration) or contamination, or if the expiration date supplied by the manufacturer has passed.

**Use:** For the differentiation of members of the Enterobacteriaceae based on their fermentation of lactose, sucrose, and glucose and the production of $H_2S$.

## Triple Sugar Iron Agar
### (TSI Agar)
### (BAM M149 Medium 2)

**Composition** per liter:

| | |
|---|---|
| Peptone | 15.0g |
| Agar | 12.0g |
| Lactose | 10.0g |
| Sucrose | 10.0g |
| Proteose peptone | 5.0g |
| NaCl | 5.0g |
| Beef extract | 3.0g |
| Yeast extract | 3.0g |
| Glucose | 1.0g |
| $Na_2S_2O_3$ | 0.3g |
| $FeSO_4$ | 0.2g |
| Phenol Red | 0.024g |

pH 7.4 ± 0.2 at 25°C

**Preparation of Medium:** Add components to distilled/deionized water and bring volume to 1.0L. Mix thoroughly. Gently heat and bring to boiling. Distribute into tubes or flasks. Autoclave for 15 min at 15 psi pressure–121°C. Allow tubes to cool in a slanted position to form a 1.0-inch butt.

**Storage/Shelf Life:** Store dehydrated media in the dark in a sealed container below 30°C. Prepared media should be stored under refrigeration (2-8°C). Media should be used within 60 days of preparation. Media should not be used if there are any signs of deterioration (shrinking, cracking, or discoloration) or contamination, or if the expiration date supplied by the manufacturer has passed.

**Use:** For the differentiation of members of the Enterobacteriaceae based on their fermentation of lactose, sucrose, and glucose and the production of $H_2S$.

## Triple Sugar Iron Agar, HiVeg

**Composition** per liter:

| | |
|---|---|
| Agar | 12.0g |
| Plant hydrolysate | 10.0g |
| Plant peptone | 10.0g |
| Lactose | 10.0g |
| Sucrose | 10.0g |
| NaCl | 5.0g |
| Plant extract | 3.0g |
| Yeast extract | 3.0g |
| Glucose | 1.0g |
| $Na_2S_2O_3$ | 0.3g |
| $FeSO_4$ | 0.2g |
| Phenol Red | 0.024g |

pH 7.4 ± 0.2 at 25°C

**Source:** This medium is available as a premixed powder from Hi-Media.

**Preparation of Medium:** Add components to distilled/deionized water and bring volume to 1.0L. Mix thoroughly. Gently heat and bring to boiling. Distribute into tubes or flasks. Autoclave for 15 min at 15 psi pressure–121°C. Allow tubes to cool in a slanted position to form a 1.0-inch butt.

**Storage/Shelf Life:** Store dehydrated media in the dark in a sealed container below 30°C. Prepared media should be stored under refrigeration (2-8°C). Media should be used within 60 days of preparation. Media should not be used if there are any signs of deterioration (shrinking, cracking, or discoloration) or contamination, or if the expiration date supplied by the manufacturer has passed.

**Use:** For the differentiation of members of the Enterobacteriaceae based on their fermentation of lactose, sucrose, and glucose and the production of $H_2S$.

## Tryptic Soy Agar with Magnesium Sulfate

**Composition** per liter:

| | |
|---|---|
| Agar | 15.0g |
| Pancreatic digest of casein | 15.0g |
| NaCl | 5.0g |
| Pancreatic digest of soybean meal | 5.0g |
| $MgSO_4 \cdot 7H_2O$ | 1.5g |

pH 7.3 ± 0.2 at 25°C

**Preparation of Medium:** Add components to distilled/deionized water and bring volume to 1.0L. Mix thoroughly. Gently heat and bring to boiling. Autoclave for 15 min at 15 psi pressure–121°C. Pour into sterile Petri dishes in 20.0mL volumes.

**Storage/Shelf Life:** Store dehydrated media in the dark in a sealed container below 30°C. Prepared media should be stored under refrigeration (2-8°C). Media should be used within 60 days of preparation. Media should not be used if there are any signs of deterioration (shrinking, cracking, or discoloration) or contamination, or if the expiration date supplied by the manufacturer has passed.

**Use:** For the cultivation of *Escherichia coli* from foods.

## Tryptic Soy Agar with Magnesium Sulfate and Sodium Chloride

**Composition** per liter:

| | |
|---|---|
| Pancreatic digest of casein | 50.0g |
| NaCl | 30.0g |
| Agar | 15.0g |

| | |
|---|---|
| Pancreatic digest of soybean meal | 5.0g |
| $MgSO_4 \cdot 7H_2O$ | 1.5g |

pH 7.3 ± 0.2 at 25°C

**Preparation of Medium:** Add components to distilled/deionized water and bring volume to 1.0L. Mix thoroughly. Gently heat and bring to boiling. Autoclave for 15 min at 15 psi pressure–121°C. Pour into sterile Petri dishes in 20.0mL volumes.

**Storage/Shelf Life:** Store dehydrated media in the dark in a sealed container below 30°C. Prepared media should be stored under refrigeration (2-8°C). Media should be used within 60 days of preparation. Media should not be used if there are any signs of deterioration (shrinking, cracking, or discoloration) or contamination, or if the expiration date supplied by the manufacturer has passed.

**Use:** For the cultivation of *Vibrio* species from foods.

## Tryptic Soy Agar with 0.6% Yeast Extract

**Composition** per liter:

| | |
|---|---|
| Agar | 15.0g |
| Pancreatic digest of casein | 15.0g |
| Yeast extract | 6.0g |
| Pancreatic digest of soybean meal | 5.0g |
| NaCl | 5.0g |

pH 7.0–7.5 at 25°C

**Source:** This medium is available as a premixed powder from BD Diagnostic Systems.

**Preparation of Medium:** Add components to distilled/deionized water and bring volume to 1.0L. Mix thoroughly. Gently heat and bring to boiling. Distribute into tubes or flasks. Autoclave for 15 min at 15 psi pressure–121°C. Pour into sterile Petri dishes or leave in tubes.

**Storage/Shelf Life:** Store dehydrated media in the dark in a sealed container below 30°C. Prepared media should be stored under refrigeration (2-8°C). Media should be used within 60 days of preparation. Media should not be used if there are any signs of deterioration (shrinking, cracking, or discoloration) or contamination, or if the expiration date supplied by the manufacturer has passed.

**Use:** For the isolation and cultivation of *Listeria monocytogenes* from foods.

## Tryptic Soy Broth with 0.001*M* Calcium Chloride
### (ATCC Medium 1380)

**Composition** per liter:

| | |
|---|---|
| Pancreatic digest of casein | 18.0g |
| Papaic digest of soybean meal | 6.0g |
| NaCl | 6.0g |
| $CaCl_2$ solution | 10.0mL |

pH 7.3 ± 0.2 at 25°C

**Source:** This medium, without $CaCl_2$, is available as a premixed powder from BD Diagnostic Systems.

**Calcium Chloride Solution:**
**Composition** per 10.0mL:

| | |
|---|---|
| $CaCl_2$ | 0.111g |

**Preparation of Calcium Chloride Solution:** Add $CaCl_2$ to distilled/deionized water and bring volume to 10.0mL. Mix thoroughly. Filter sterilize.

**Preparation of Medium:** Add components, except CaCl$_2$ solution, to distilled/deionized water and bring volume to 990mL. Mix thoroughly. Gently heat and bring to boiling. Autoclave for 15 min at 15 psi pressure–121°C. Aseptically add 10.0mL CaCl$_2$ solution. Mix thoroughly. Distribute into tubes or flasks.

**Storage/Shelf Life:** Store dehydrated media in the dark in a sealed container below 30°C. Prepared media should be stored under refrigeration (2-8°C). Media should be used within 60 days of preparation. Media should not be used if there are any signs of deterioration (discoloration) or contamination, or if the expiration date supplied by the manufacturer has passed.

**Use:** For the isolation and cultivation of *Bronchothrix thermospacta*.

## Trypticase™ Glucose Extract Agar

**Composition** per liter:
| | |
|---|---|
| Agar | 15.0g |
| Pancreatic digest of casein | 5.0g |
| Beef extract | 3.0g |
| Glucose | 1.0g |

pH 7.0 ± 0.2 at 25°C

**Source:** This medium is available as a premixed powder from BD Diagnostic Systems.

**Preparation of Medium:** Add components to distilled/deionized water and bring volume to 1.0L. Mix thoroughly. Gently heat and bring to boiling. Distribute into tubes or flasks. Autoclave for 15 min at 15 psi pressure–121°C. Pour into sterile Petri dishes or leave in tubes.

**Storage/Shelf Life:** Store dehydrated media in the dark in a sealed container below 30°C. Prepared media should be stored under refrigeration (2-8°C). Media should be used within 60 days of preparation. Media should not be used if there are any signs of deterioration (shrinking, cracking, or discoloration) or contamination, or if the expiration date supplied by the manufacturer has passed.

**Use:** For the enumeration of bacteria in water, milk, and other specimens.

## Trypticase™ Novobiocin Broth (TN Broth)

**Composition** per liter:
| | |
|---|---|
| Pancreatic digest of casein | 17.0g |
| NaCl | 5.0g |
| Papaic digest of soybean meal | 3.0g |
| K$_2$HPO$_4$ | 2.5g |
| Glucose | 2.5g |
| Bile salts No. 3 | 1.5g |
| K$_2$HPO$_4$ | 1.5g |
| Novobiocin solution | 10.0mL |

pH 7.3 ± 0.2 at 25°C

**Novobiocin Solution:**
**Composition** per 10.0mL:
| | |
|---|---|
| Novobiocin | 0.02g |

**Preparation of Novobiocin Solution:** Add novobiocin to distilled/deionized water and bring volume to 10.0mL. Mix thoroughly. Filter sterilize.

**Preparation of Medium:** Add components, except novobiocin solution, to distilled/deionized water and bring volume to 990.0mL. Mix thoroughly. Gently heat and bring to boiling. Autoclave for 15 min at 15 psi pressure–121°C. Cool to 45°–50°C. Aseptically add sterile novobiocin

solution. Mix thoroughly. Pour into sterile Petri dishes or distribute into sterile tubes.

**Storage/Shelf Life:** Store dehydrated media in the dark in a sealed container below 30°C. Prepared media should be stored under refrigeration (2-8°C). Media should be used within 60 days of preparation. Media should not be used if there are any signs of deterioration (discoloration) or contamination, or if the expiration date supplied by the manufacturer has passed.

**Use:** For the cultivation of verotoxin-producing *Escherichia coli*.

## Trypticase™ Peptone Glucose Yeast Extract Broth, Buffered

**Composition** per liter:
| | |
|---|---|
| Pancreatic digest of casein | 50.0g |
| Yeast extract | 20.0g |
| Na$_2$HPO$_4$ | 5.0g |
| Peptone | 5.0g |
| Glucose | 4.0g |
| Sodium thioglycolate | 1.0g |

pH 7.3 ± 0.2 at 25°C

**Preparation of Medium:** Add components to distilled/deionized water and bring volume to 1.0L. Mix thoroughly. Gently heat until dissolved. Adjust pH to 7.3. Distribute into tubes in 15.0mL volumes. Autoclave for 8 min at 15 psi pressure–121°C.

**Storage/Shelf Life:** Store dehydrated media in the dark in a sealed container below 30°C. Prepared media should be stored under refrigeration (2-8°C). Media should be used within 60 days of preparation. Media should not be used if there are any signs of deterioration (discoloration) or contamination, or if the expiration date supplied by the manufacturer has passed.

**Use:** For the isolation and cultivation of *Clostridium perfringens* from foods.

## Trypticase™ Soy Agar

**Composition** per liter:
| | |
|---|---|
| Pancreatic digest of casein | 17.0g |
| Agar | 15.0g |
| NaCl | 5.0g |
| Papaic digest of soybean meal | 3.0g |
| K$_2$HPO$_4$ | 2.5g |
| Glucose | 2.5g |

pH 7.3 ± 0.2 at 25°C

**Preparation of Medium:** Add components to distilled/deionized water and bring volume to 1.0L. Mix thoroughly. Gently heat and bring to boiling. Distribute into tubes or flasks. Autoclave for 15 min at 15 psi pressure–121°C. Pour into sterile Petri dishes or leave in tubes.

**Storage/Shelf Life:** Store dehydrated media in the dark in a sealed container below 30°C. Prepared media should be stored under refrigeration (2-8°C). Media should be used within 60 days of preparation. Media should not be used if there are any signs of deterioration (shrinking, cracking, or discoloration) or contamination, or if the expiration date supplied by the manufacturer has passed.

**Use:** For the cultivation and maintenance of a wide variety of heterotrophic microorganisms. For the cultivation of a wide variety of fastidious and nonfastidious microorganisms from clinical and nonclinical specimens. Also used for the rapid estimation of the bacteriological quality of water.

## Trypticase™ Soy Agar
### (Tryptic Soy Agar)
### (Soybean Casein Digest Agar)

**Composition** per liter:

Pancreatic digest of casein ...................................15.0g
Agar ..................................................................15.0g
Papaic digest of soybean meal ..............................5.0g
NaCl ...................................................................5.0g

pH 7.3 ± 0.2 at 25°C

**Source:** This medium is available as a premixed powder from BD Diagnostic Systems.

**Preparation of Medium:** Add components to distilled/deionized water and bring volume to 1.0L. Mix thoroughly. Gently heat and bring to boiling. Distribute into tubes or flasks. Autoclave for 15 min at 15 psi pressure–121°C. Do not overheat. Pour into sterile Petri dishes or leave in tubes.

**Storage/Shelf Life:** Store dehydrated media in the dark in a sealed container below 30°C. Prepared media should be stored under refrigeration (2-8°C). Media should be used within 60 days of preparation. Media should not be used if there are any signs of deterioration (shrinking, cracking, or discoloration) or contamination, or if the expiration date supplied by the manufacturer has passed.

**Use:** For the isolation and cultivation of a wide variety of fastidious as well as nonfastidious microorganisms

## Trypticase™ Soy Agar
### with Human Blood

**Composition** per liter:

Pancreatic digest of casein ...................................15.0g
Agar ..................................................................15.0g
Papaic digest of soybean meal ..............................5.0g
NaCl ...................................................................5.0g
Human blood, defibrinated ...............................50.0mL

pH 7.3 ± 0.2 at 25°C

**Preparation of Medium:** Add components, except human blood, to distilled/deionized water and bring volume to 950.0mL. Mix thoroughly. Gently heat and bring to boiling. Autoclave for 15 min at 15 psi pressure–121°C. Cool to 45°–50°C. Aseptically add sterile human blood. Mix thoroughly. Pour into sterile Petri dishes in 17.0mL volumes or distribute into sterile tubes.

**Storage/Shelf Life:** Store dehydrated media in the dark in a sealed container below 30°C. Prepared media should be stored under refrigeration (2-8°C). Media should be used within 60 days of preparation. Media should not be used if there are any signs of deterioration (shrinking, cracking, or discoloration) or contamination, or if the expiration date supplied by the manufacturer has passed.

**Use:** For the cultivation of a wide variety of fastidious microorganisms. For the observation of hemolytic reactions of a variety of bacteria. May be used to perform the CAMP test for the presumptive identification of group B streptococci (*Streptococcus agalactiae*).

## Trypticase™ Soy Agar
### with Lecithin and Polysorbate 80
### (Microbial Content Test Agar)

**Composition** per liter:

Pancreatic digest of casein ...................................15.0g
Agar ..................................................................15.0g
Papaic digest of soybean meal ..............................5.0g
NaCl ...................................................................5.0g
Polysorbate 80 (Tween™ 80) .................................5.0g
Lecithin ..............................................................0.7g

pH 7.3 ± 0.2 at 25°C

**Source:** This medium is available as a premixed powder from BD Diagnostic Systems.

**Preparation of Medium:** Add components to distilled/deionized water and bring volume to 1.0L. Mix thoroughly. Gently heat and bring to boiling. Distribute into tubes or flasks. Autoclave for 15 min at 13 psi pressure–118°C. Cool to 45°–50°C. Pour into sterile Petri dishes in 17.0mL volumes or leave in tubes.

**Storage/Shelf Life:** Store dehydrated media in the dark in a sealed container below 30°C. Prepared media should be stored under refrigeration (2-8°C). Media should be used within 60 days of preparation. Media should not be used if there are any signs of deterioration (shrinking, cracking, or discoloration) or contamination, or if the expiration date supplied by the manufacturer has passed.

**Use:** For the detection and enumeration of microorganisms in replicate plating techniques. Also used for the detection and enumeration of microorganisms present on surfaces of sanitary importance.

## Trypticase™ Soy Agar-Magnesium
### Sulfate- Sodium Chloride Agar
### (TSAMS)
### (BAM M152a)

**Composition** per liter:

NaCl .................................................................20.0g
Trypticase peptone .............................................15.0g
Agar ..................................................................15.0g
Phytone™ peptone ................................................5.0g
MgSO$_4$·7H$_2$O .......................................................1.5g

pH 7.3 ± 0.2 at 25°C

**Preparation of Medium:** Add components to distilled/deionized water and bring volume to 1.0L. Mix thoroughly. Gently heat and bring to boiling. Autoclave for 15 min at 15 psi pressure–121°C. Pour into sterile Petri dishes in 20.0mL volumes.

**Storage/Shelf Life:** Store dehydrated media in the dark in a sealed container below 30°C. Prepared media should be stored under refrigeration (2-8°C). Media should be used within 60 days of preparation. Media should not be used if there are any signs of deterioration (shrinking, cracking, or discoloration) or contamination, or if the expiration date supplied by the manufacturer has passed.

**Use**: For the cultivation of *Vibrio* species from foods.

## Trypticase™ Soy Agar, Modified
### (TSA II™)

**Composition** per liter:

Pancreatic digest of casein ...................................14.5g
Agar ..................................................................14.0g
Papaic digest of soybean meal ..............................5.0g
NaCl ...................................................................5.0g
Growth factors (BBL) ..........................................1.5g

pH 7.3 ± 0.2 at 25°C

**Source:** This medium is available as a premixed powder from BD Diagnostic Systems.

**Preparation of Medium:** Add components to distilled/deionized water and bring volume to 1.0L. Mix thoroughly. Gently heat while stirring and bring to boiling. Distribute into tubes or flasks. Autoclave for 15 min at 15 psi pressure–121°C. Do not overheat. Pour into sterile Petri dishes or leave in tubes. For blood plates, 50.0–100.0mL of sterile defibrinated sheep blood may be added to sterile medium that has been melted and cooled to 45°–50°C.

**Storage/Shelf Life:** Store dehydrated media in the dark in a sealed container below 30°C. Prepared media should be stored under refrigeration (2-8°C). Media should be used within 60 days of preparation. Media should not be used if there are any signs of deterioration (shrinking, cracking, or discoloration) or contamination, or if the expiration date supplied by the manufacturer has passed.

**Use:** Used as a base that is supplemented. For the cultivation of fastidious microorganisms. When supplemented with sheep blood, this medium is useful for the observation of hemolytic reactions of a variety of bacteria. It may be used to perform the CAMP test for the presumptive identification of group B streptococci (*Streptococcus agalactiae*).

## Trypticase™ Soy Agar, Modified with Horse Serum

**Composition** per liter:

Pancreatic digest of casein............................................17.0g
Agar ..............................................................................15.0g
NaCl................................................................................5.0g
Yeast extract...................................................................4.0g
Papaic digest of soybean meal .......................................3.0g
K$_2$HPO$_4$............................................................................2.5g
Horse serum .............................................................. 100.0mL

pH 7.3 ± 0.2 at 25°C

**Preparation of Medium:** Add components, except horse serum, to distilled/deionized water and bring volume to 900.0mL. Mix thoroughly. Gently heat and bring to boiling. Autoclave for 15 min at 15 psi pressure–121°C. Cool to 45°–50°C. Aseptically add sterile horse serum. Mix thoroughly. Pour into sterile Petri dishes or distribute into sterile tubes.

**Storage/Shelf Life:** Store dehydrated media in the dark in a sealed container below 30°C. Prepared media should be stored under refrigeration (2-8°C). Media should be used within 60 days of preparation. Media should not be used if there are any signs of deterioration (shrinking, cracking, or discoloration) or contamination, or if the expiration date supplied by the manufacturer has passed.

**Use:** For the cultivation and maintenance of *Simonsiella* species, *Alysiella* species, and *Moraxella* species.

## Trypticase™ Soy Agar with Sheep Blood (Tryptic Soy Blood Agar) (TSA Blood Agar)

**Composition** per liter:

Pancreatic digest of casein............................................15.0g
Agar ..............................................................................15.0g
Papaic digest of soybean meal .......................................5.0g
NaCl................................................................................5.0g
Sheep blood, defibrinated ........................................ 50.0mL

pH 7.3 ± 0.2 at 25°C

**Preparation of Medium:** Add components, except sheep blood, to distilled/deionized water and bring volume to 950.0mL. Mix thoroughly. Gently heat and bring to boiling. Autoclave for 15 min at 15 psi pressure–121°C. Cool to 45°–50°C. Aseptically add sterile sheep

blood. Mix thoroughly. Pour into sterile Petri dishes in 17.0mL volumes or distribute into sterile tubes.

**Storage/Shelf Life:** Store dehydrated media in the dark in a sealed container below 30°C. Prepared media should be stored under refrigeration (2-8°C). Media should be used within 60 days of preparation. Media should not be used if there are any signs of deterioration (shrinking, cracking, or discoloration) or contamination, or if the expiration date supplied by the manufacturer has passed.

**Use:** For the cultivation of a wide variety of fastidious microorganisms. For the observation of hemolytic reactions of a variety of bacteria. May be used to perform the CAMP test for the presumptive identification of group B streptococci (*Streptococcus agalactiae*).

## Trypticase™ Soy Agar with Sheep Blood, Formate, and Fumarate

**Composition** per liter:

Pancreatic digest of casein............................................14.5g
Agar ..............................................................................14.0g
Papaic digest of soybean meal .......................................5.0g
NaCl................................................................................5.0g
Sucrose...........................................................................2.0g
Growth factors................................................................1.5g
Sheep blood, defibrinated ........................................ 50.0mL
Formate-fumarate solution...................................... 13.0mL

pH 7.3 ± 0.2 at 25°C

**Source:** Growth Factors are available as a premixed powder from BD Diagnostic Systems.

**Formate-Fumarate Solution:**
**Composition** per 100.0mL:

Sodium formate ..............................................................6.0g
Fumaric acid ...................................................................6.0g

**Preparation of Formate-Fumarate Solution:** Add components to distilled/deionized water and bring volume to 100.0mL. Mix thoroughly. Adjust pH to 7.0. Filter sterilize.

**Preparation of Medium:** Add components, except sheep blood, to distilled/deionized water and bring volume to 950.0mL. Mix thoroughly. Gently heat and bring to boiling. Autoclave for 15 min at 15 psi pressure–121°C. Cool to 45°–50°C. Aseptically add sterile sheep blood. Mix thoroughly. Pour into sterile Petri dishes. Prior to inoculation, aseptically spread 0.2mL of sterile formate-fumarate solution on each plate.

**Storage/Shelf Life:** Store dehydrated media in the dark in a sealed container below 30°C. Prepared media should be stored under refrigeration (2-8°C). Media should be used within 60 days of preparation. Media should not be used if there are any signs of deterioration (shrinking, cracking, or discoloration) or contamination, or if the expiration date supplied by the manufacturer has passed.

**Use:** For the isolation of *Streptococcus pneumoniae* from a variety of clinical specimens.

## Trypticase™ Soy Agar with Sheep Blood and Gentamicin (TSA II™ with Sheep Blood and Gentamicin)

**Composition** per liter:

Pancreatic digest of casein............................................14.5g
Agar ..............................................................................14.0g
Papaic digest of soybean meal .......................................5.0g
NaCl................................................................................5.0g

Growth factors ................................................................1.5g
Sheep blood, defibrinated .......................................... 50.0mL
Gentamicin solution.................................................... 10.0mL

pH 7.3 ± 0.2 at 25°C

**Source:** This medium is available as a premixed powder from BD Diagnostic Systems.

**Gentamicin Solution:**
**Composition** per 10.0mL:
Gentamicin.................................................................2.5mg

**Preparation of Gentamicin Solution:** Add gentamicin to distilled/deionized water and bring volume to 10.0mL. Mix thoroughly. Filter sterilize.

**Preparation of Medium:** Add components, except sheep blood and gentamicin solution, to distilled/deionized water and bring volume to 940.0mL. Mix thoroughly. Gently heat and bring to boiling. Autoclave for 15 min at 15 psi pressure–121°C. Cool to 45°–50°C. Aseptically add sterile sheep blood and sterile gentamicin solution. Mix thoroughly. Pour into sterile Petri dishes or distribute into sterile tubes.

**Storage/Shelf Life:** Store dehydrated media in the dark in a sealed container below 30°C. Prepared media should be stored under refrigeration (2-8°C). Media should be used within 60 days of preparation. Media should not be used if there are any signs of deterioration (shrinking, cracking, or discoloration) or contamination, or if the expiration date supplied by the manufacturer has passed.

**Use:** For the isolation of *Streptococcus pneumoniae* from a variety of clinical specimens.

## Trypticase™ Soy Agar with Sheep Blood, Sucrose, and Tetracycline

**Composition** per liter:
Pancreatic digest of casein..........................................14.5g
Agar ..........................................................................14.0g
Papaic digest of soybean meal ......................................5.0g
NaCl...........................................................................5.0g
Sucrose.......................................................................2.0g
Growth factors ............................................................1.5g
Sheep blood, defibrinated .......................................... 50.0mL
Tetracycline solution................................................. 10.0mL

pH 7.3 ± 0.2 at 25°C

**Source:** Growth factors are available as a premixed powder from BD Diagnostic Systems.

**Tetracycline Solution:**
**Composition** per 10.0mL:
Tetracycline................................................................0.5mg

**Preparation of Tetracycline Solution:** Add tetracycline to distilled/deionized water and bring volume to 10.0mL. Mix thoroughly. Filter sterilize.

**Preparation of Medium:** Add components, except sheep blood and tetracycline solution, to distilled/deionized water and bring volume to 940.0mL. Mix thoroughly. Gently heat and bring to boiling. Autoclave for 15 min at 15 psi pressure–121°C. Cool to 45°–50°C. Aseptically add sterile sheep blood and sterile tetracycline solution. Mix thoroughly. Pour into sterile Petri dishes or distribute into sterile tubes.

**Storage/Shelf Life:** Store dehydrated media in the dark in a sealed container below 30°C. Prepared media should be stored under refrigeration (2-8°C). Media should be used within 60 days of preparation.

Media should not be used if there are any signs of deterioration (shrinking, cracking, or discoloration) or contamination, or if the expiration date supplied by the manufacturer has passed.

**Use:** For the isolation of *Streptococcus pneumoniae* from a variety of clinical specimens.

## Trypticase™ Soy Agar with Sheep Blood and Vancomycin (ATCC Medium 1976)

**Composition** per liter:
Pancreatic digest of casein.........................................14.5g
Agar ..........................................................................14.0g
Papaic digest of soybean meal......................................5.0g
NaCl...........................................................................5.0g
Growth factors ............................................................1.5g
Sheep blood, defibrinated .......................................... 50.0mL
Vancomycin solution ................................................. 10.0mL

pH 7.3 ± 0.2 at 25°C

**Source:** This medium is available as a premixed powder from BD Diagnostic Systems.

**Vancomycin Solution:**
**Composition** per 10.0mL:
Vancomycin .................................................................4.0mg

**Preparation of Vancomycin Solution:** Add vancomycin to distilled/deionized water and bring volume to 10.0mL. Mix thoroughly. Filter sterilize.

**Preparation of Medium:** Add components, except sheep blood and vancomycin solution, to distilled/deionized water and bring volume to 940.0mL. Mix thoroughly. Gently heat and bring to boiling. Autoclave for 15 min at 15 psi pressure–121°C. Cool to 50°–55°C. Aseptically add sterile sheep blood and sterile vancomycin solution. Mix thoroughly. Pour into sterile Petri dishes or distribute into sterile tubes.

**Storage/Shelf Life:** Store dehydrated media in the dark in a sealed container below 30°C. Prepared media should be stored under refrigeration (2-8°C). Media should be used within 60 days of preparation. Media should not be used if there are any signs of deterioration (shrinking, cracking, or discoloration) or contamination, or if the expiration date supplied by the manufacturer has passed.

**Use:** For the isolation of *Streptococcus and Enterococcus* spp. from a variety of clinical specimens.

## Trypticase™ Soy Agar with Yeast Extract (TSAYE)

**Composition** per liter:
Pancreatic digest of casein..........................................17.0g
Agar ..........................................................................15.0g
Yeast extract...............................................................6.0g
NaCl...........................................................................5.0g
Papaic digest of soybean meal ......................................3.0g
$K_2HPO_4$....................................................................2.5g
Glucose ......................................................................2.5g

pH 7.3 ± 0.2 at 25°C

**Preparation of Medium:** Add components to distilled/deionized water and bring volume to 1.0L. Mix thoroughly. Gently heat and bring to boiling. Distribute into tubes or flasks. Autoclave for 15 min at 15 psi pressure–121°C. Pour into sterile Petri dishes or leave in tubes.

**Storage/Shelf Life:** Store dehydrated media in the dark in a sealed container below 30°C. Prepared media should be stored under refrigeration (2-8°C). Media should be used within 60 days of preparation. Media should not be used if there are any signs of deterioration (shrinking, cracking, or discoloration) or contamination, or if the expiration date supplied by the manufacturer has passed.

**Use:** For the cultivation and maintenance of a wide variety of heterotrophic microorganisms. For the isolation and cultivation of *Listeria monocytogenes* from foods.

## Trypticase™ Soy Broth with 0.1% Agar
**Composition** per liter:

| | |
|---|---|
| Pancreatic digest of casein | 17.0g |
| NaCl | 5.0g |
| Papaic digest of soybean meal | 3.0g |
| K$_2$HPO$_4$ | 2.5g |
| Glucose | 2.5g |
| Agar | 1.0g |

pH 7.3 ± 0.2 at 25°C

**Source:** This medium is available as a premixed powder from BD Diagnostic Systems.

**Preparation of Medium:** Add components to distilled/deionized water and bring volume to 1.0L. Mix thoroughly. Distribute into tubes or flasks. Autoclave for 15 min at 15 psi pressure–121°C.

**Storage/Shelf Life:** Store dehydrated media in the dark in a sealed container below 30°C. Prepared media should be stored under refrigeration (2-8°C). Media should be used within 60 days of preparation. Media should not be used if there are any signs of deterioration (shrinking, cracking, or discoloration) or contamination, or if the expiration date supplied by the manufacturer has passed.

**Use:** For the cultivation of anaerobic microorganisms. For the cultivation of microorganisms isolated from root canals and other clinical specimens.

## Trypticase™ Soy Broth with Ferrous Sulfate
## (Tryptic Soy Broth with Ferrous Sulfate)
## (BAM M186)
**Composition** per liter:

| | |
|---|---|
| Pancreatic digest of casein | 17.0g |
| NaCl | 5.0g |
| Papaic digest of soybean meal | 3.0g |
| K$_2$HPO$_4$ | 2.5g |
| Glucose | 2.5g |
| FeSO$_4$ | 35.0mg |

pH 7.3 ± 0.2 at 25°C

**Preparation of Medium:** Add components to distilled/deionized water and bring volume to 1.0L. Mix thoroughly. Gently heat until dissolved. Adjust pH to 7.3. Distribute into tubes in 10.0mL volumes. Autoclave for 15 min at 15 psi pressure–121°C.

**Storage/Shelf Life:** Store dehydrated media in the dark in a sealed container below 30°C. Prepared media should be stored under refrigeration (2-8°C). Media should be used within 60 days of preparation. Media should not be used if there are any signs of deterioration (discoloration) or contamination, or if the expiration date supplied by the manufacturer has passed.

**Use**: For the detection of *Salmonella* spp. from foods.

## Trypticase™ Soy Broth with Glycerol
**Composition** per liter:

| | |
|---|---|
| Pancreatic digest of casein | 17.0g |
| NaCl | 15.0g |
| Papaic digest of soybean meal | 3.0g |
| K$_2$HPO$_4$ | 2.5g |
| Glycerol | 240.0mL |

pH 7.3 ± 0.2 at 25°C

**Preparation of Medium:** Add components to distilled/deionized water and bring volume to 1.0L. Mix thoroughly. Gently heat until dissolved. Adjust pH to 7.3. Distribute into tubes in 10.0mL volumes. Autoclave for 15 min at 15 psi pressure–121°C.

**Use:** For the cultivation and maintenance of a wide variety of microorgani Trypticase™ Soy Broth, Modified
**Composition** per 1000.2mL:

| | |
|---|---|
| Pancreatic digest of casein | 17.0g |
| NaCl | 15.0g |
| K$_2$HPO$_4$ | 4.0g |
| Papaic digest of soybean meal | 3.0g |
| Glucose | 2.5g |
| Bile salts No. 3 | 1.5g |
| Novobiocin solution | 0.2mL |

pH 7.3 ± 0.2 at 25°C

**Novobiocin Solution:**
**Composition** per liter:

| | |
|---|---|
| Novobiocin | 0.05g |

**Preparation of Novobiocin Solution:** Add novobiocin to distilled/deionized water and bring volume to 1.0L. Mix thoroughly. Filter sterilize.

**Preparation of Medium:** Add components, except novobiocin solution, to distilled/deionized water and bring volume to 1.0L. Mix thoroughly. Gently heat and bring to boiling. Autoclave for 15 min at 15 psi pressure–121°C. Cool to 45°–50°C. Aseptically add sterile novobiocin solution. Mix thoroughly. Aseptically distribute into sterile tubes.

**Storage/Shelf Life:** Store dehydrated media in the dark in a sealed container below 30°C. Prepared media should be stored under refrigeration (2-8°C). Media should be used within 60 days of preparation. Media should not be used if there are any signs of deterioration (discoloration) or contamination, or if the expiration date supplied by the manufacturer has passed.

**Use:** For the isolation and cultivation of *Shigella* species from food.

## Trypticase™ Soy Broth, Modified
## (mTSB)
## (BAM M156)
**Composition** per 1000.2mL:

| | |
|---|---|
| Pancreatic digest of casein | 17.0g |
| NaCl | 5.0g |
| K$_2$HPO$_4$ | 4.0g |
| Papaic digest of soybean meal | 3.0g |
| Glucose | 2.5g |
| Bile salts No. 3 | 1.5g |
| Novobiocin solution | 0.2mL |

pH 7.3 ± 0.2 at 25°C

**Novobiocin Solution:**
**Composition** per liter:

| | |
|---|---|
| Novobiocin | 0.05g |

**Preparation of Novobiocin Solution:** Add novobiocin to distilled/deionized water and bring volume to 1.0L. Mix thoroughly. Filter sterilize.

**Preparation of Medium:** Add components, except novobiocin solution, to distilled/deionized water and bring volume to 1.0L. Mix thoroughly. Gently heat and bring to boiling. Autoclave for 15 min at 15 psi pressure–121°C. Cool to 45°–50°C. Aseptically add sterile novobiocin solution. Mix thoroughly. Aseptically distribute into sterile tubes.

**Storage/Shelf Life:** Store dehydrated media in the dark in a sealed container below 30°C. Prepared media should be stored under refrigeration (2-8°C). Media should be used within 60 days of preparation. Media should not be used if there are any signs of deterioration (discoloration) or contamination, or if the expiration date supplied by the manufacturer has passed.

**Use:** For the isolation and cultivation of *Shigella* species from food.

## Trypticase™ Soy Broth with Sodium Chloride and Sodium Pyruvate

**Composition** per liter:
| | |
|---|---|
| NaCl | 100.0g |
| Pancreatic digest of casein | 17.0g |
| Sodium pyruvate | 10.0g |
| Papaic digest of soybean meal | 3.0g |
| Glucose | 2.5g |
| K₂HPO₄ | 2.5g |

$$\text{pH } 7.3 \pm 0.2 \text{ at } 25°C$$

**Preparation of Medium:** Add components to distilled/deionized water and bring volume to 1.0L. Mix thoroughly. Gently heat until dissolved. Adjust pH to 7.3. Distribute into tubes in 10.0mL volumes. Autoclave for 15 min at 15 psi pressure–121°C.

**Storage/Shelf Life:** Store dehydrated media in the dark in a sealed container below 30°C. Prepared media should be stored under refrigeration (2-8°C). Media should be used within 60 days of preparation. Media should not be used if there are any signs of deterioration (discoloration) or contamination, or if the expiration date supplied by the manufacturer has passed.

**Use:** For the isolation and cultivation of *Staphylococcus aureus* from foods.

## Trypticase™ Soy Polymyxin Broth

**Composition** per 1006.67mL:
| | |
|---|---|
| Pancreatic digest of casein | 17.0g |
| NaCl | 5.0g |
| Papaic digest of soybean meal | 3.0g |
| K₂HPO₄ | 2.5g |
| Glucose | 2.5g |
| Polymyxin B solution | 6.67mL |

$$\text{pH } 7.3 \pm 0.2 \text{ at } 25°C$$

**Polymyxin B Solution:**
**Composition** per 10.0mL:
| | |
|---|---|
| Polymyxin B | 0.015g |

**Preparation of Polymyxin B Solution:** Add polymyxin B to distilled/deionized water and bring volume to 10.0mL. Mix thoroughly. Filter sterilize.

**Preparation of Medium:** Add components, except polymyxin B solution, to distilled/deionized water and bring volume to 1.0L. Mix thoroughly. Gently heat and bring to boiling. Distribute into tubes in 15.0mL volumes. Autoclave for 15 min at 15 psi pressure–121°C. Cool

to 45°–50°C. Aseptically add 0.1mL of sterile polymyxin B solution to each tube. Mix thoroughly.

**Storage/Shelf Life:** Store dehydrated media in the dark in a sealed container below 30°C. Prepared media should be stored under refrigeration (2-8°C). Media should be used within 60 days of preparation. Media should not be used if there are any signs of deterioration (discoloration) or contamination, or if the expiration date supplied by the manufacturer has passed.

**Use:** For the isolation and cultivation of *Bacillus cereus* from foods.

## Trypticase™ Soy Sheep Blood Agar (Tryptic Soy Blood Agar) (TSA Blood Agar) (BAM M159)

**Composition** per 1050.0mL:
| | |
|---|---|
| Pancreatic digest of casein | 15.0g |
| Agar | 15.0g |
| Papaic digest of soybean meal | 5.0g |
| NaCl | 5.0g |
| Sheep blood, defibrinated | 50.0mL |

$$\text{pH } 7.3 \pm 0.2 \text{ at } 25°C$$

**Preparation of Medium:** Add components, except sheep blood, to distilled/deionized water and bring volume to 1.0L. Mix thoroughly. Gently heat and bring to boiling. Autoclave for 15 min at 15 psi pressure–121°C. Cool to 45°–50°C. Aseptically add 50.0mL sterile sheep blood. Mix thoroughly. Pour into sterile Petri dishes in 20.0mL volumes or distribute into sterile tubes.

**Storage/Shelf Life:** Store dehydrated media in the dark in a sealed container below 30°C. Prepared media should be stored under refrigeration (2-8°C). Media should be used within 60 days of preparation. Media should not be used if there are any signs of deterioration (shrinking, cracking, or discoloration) or contamination, or if the expiration date supplied by the manufacturer has passed.

**Use:** For the cultivation of a wide variety of fastidious microorganisms. For the observation of hemolytic reactions of a variety of bacteria. May be used to perform the CAMP test for the presumptive identification of group B streptococci (*Streptococcus agalactiae*).

## Trypticase™ Soy Tryptose Broth

**Composition** per liter:
| | |
|---|---|
| Pancreatic digest of casein | 13.5g |
| Peptic digest of animal tissue | 5.0g |
| NaCl | 5.0g |
| Yeast extract | 3.0g |
| Glucose | 1.75g |
| Papaic digest of soybean meal | 1.5g |
| K₂HPO₄ | 1.25g |

$$\text{pH } 7.2 \pm 0.2 \text{ at } 25°C$$

**Preparation of Medium:** Add components to distilled/deionized water and bring volume to 1.0L. Mix thoroughly. Distribute into tubes in 5.0mL volumes. Autoclave for 15 min at 15 psi pressure–121°C.

**Storage/Shelf Life:** Store dehydrated media in the dark in a sealed container below 30°C. Prepared media should be stored under refrigeration (2-8°C). Media should be used within 60 days of preparation. Media should not be used if there are any signs of deterioration (discoloration) or contamination, or if the expiration date supplied by the manufacturer has passed.

**Use:** For the enrichment of *Salmonella* species from foods.

## Trypticase™ Tellurite Agar Base

**Composition** per liter:

Agar ..................................................................20.0g
Pancreatic digest of casein.........................................10.0g
Peptic digest of animal tissue......................................10.0g
NaCl..................................................................5.0g
Glucose ..............................................................2.0g
Serum.............................................................. 50.0mL
Chapman tellurite solution..................................... 10.0mL

pH 7.5 ± 0.2 at 25°C

**Source:** This medium is available as a premixed powder from BD Diagnostic Systems.

### Chapman Tellurite Solution:

**Composition** per 100.0mL:

K₂TeO₃........................................................1.0g

$K_2TeO_3$........................................................1.0g

**Preparation of Chapman Tellurite Solution:** Add $K_2TeO_3$ to distilled/deionized water and bring volume to 100.0mL. Mix thoroughly. Filter sterilize.

**Caution:** Potassium tellurite is toxic.

**Preparation of Medium:** Add components, except serum and Chapman tellurite solution, to distilled/deionized water and bring volume to 940.0mL. Mix thoroughly. Gently heat and bring to boiling. Autoclave for 15 min at 15 psi pressure–121°C. Cool to 45°–50°C. Aseptically add sterile serum and sterile Chapman tellurite solution. Sheep serum, rabbit serum, or human serum may be used. Mix thoroughly. Pour into sterile Petri dishes.

**Storage/Shelf Life:** Store dehydrated media in the dark in a sealed container below 30°C. Prepared media should be stored under refrigeration (2-8°C). Media should be used within 60 days of preparation. Media should not be used if there are any signs of deterioration (shrinking, cracking, or discoloration) or contamination, or if the expiration date supplied by the manufacturer has passed.

**Use:** For the selective isolation of microorganisms from clinical specimens, especially from the nose, throat, and vagina.

## Tryptone Bile Agar

**Composition** per liter:

Pancreatic digest of casein.........................................20.0g
Agar ................................................................15.0g
Bile salts No. 3......................................................1.5g

pH 7.2 ± 0.2 at 25°C

**Source:** This medium is available as a premixed powder from BD Diagnostic Systems and Oxoid Unipath.

**Preparation of Medium:** Add components to distilled/deionized water and bring volume to 1.0L. Mix thoroughly. Gently heat and bring to boiling. Distribute into tubes or flasks. Autoclave for 15 min at 15 psi pressure–121°C. Pour into sterile Petri dishes.

**Storage/Shelf Life:** Store dehydrated media in the dark in a sealed container below 30°C. Prepared media should be stored under refrigeration (2-8°C). Media should be used within 60 days of preparation. Media should not be used if there are any signs of deterioration (shrinking, cracking, or discoloration) or contamination, or if the expiration date supplied by the manufacturer has passed.

**Use:** For the isolation and enumeration of *Escherichia coli* from foods.

## Tryptone Bile Glucuronide Agar, Harlequin (Harlequin TBGA)

**Composition** per liter:

Tryptone.............................................................20.0g
Agar .................................................................15.0g
Bile salts No. 3......................................................1.5g
X-glucuronide.....................................................0.075g

pH 7.2 ± 0.2 at 25°C

**Source:** This medium is available from lab m.

**Preparation of Medium:** Add components to distilled/deionized water and bring volume to 1.0L. Mix thoroughly. Allow to soak for 10 min. Autoclave for 15 min at 15 psi pressure–121°C. Cool to 45°–50°C. Pour into sterile Petri dishes.

**Storage/Shelf Life:** Store dehydrated media in the dark in a sealed container below 30°C. Prepared media should be stored in the dark under refrigeration (2-8°C). Chromogenic media are especially light and temperature sensitive; protect from light, excessive heat, moisture, and freezing. Media should not be used if there are any signs of deterioration (shrinking, cracking, or discoloration) or contamination, or if the expiration date supplied by the manufacturer has passed.

**Use:** For the simple enumeration of *E. coli* without the need for membranes or pre-incubation The medium has been modified by the addition of a chromogenic substrate to detect the β-glucuronidase, which is highly specific for *E. coli*. The advantage of the chromogenic substrate is that it requires no UV lamp to visualize the reaction, and it is concentrated within the colony, facilitating easier enumeration in the presence of other organisms, or when large numbers are present on the plate.

## Tryptone Bile X-glucuronide Agar, Chromocult (Chromocult® TBX ) (ChromocultTryptone Bile X-glucuronide Agar)

**Composition** per liter:

Peptone ..............................................................20.0g
Agar .................................................................15.0g
Bile salts No. 3......................................................1.5g
X-β-D-glucuronide .............................................0.075g

pH 7.2 ± 0.2 at 25°C

**Source:** This medium is available from Merck.

**Preparation of Medium:** Add components to distilled/deionized water and bring volume to 1.0L. Mix thoroughly. Autoclave for 15 min at 15 psi pressure–121°C. Cool to 45°–50°C. Pour into sterile Petri dishes.

**Storage/Shelf Life:** Store dehydrated media in the dark in a sealed container below 30°C. Prepared media should be stored in the dark under refrigeration (2-8°C). Chromogenic media are especially light and temperature sensitive; protect from light, excessive heat, moisture, and freezing. Media should not be used if there are any signs of deterioration (shrinking, cracking, or discoloration) or contamination, or if the expiration date supplied by the manufacturer has passed.

**Use:** For the differentiation of *E. coli* from other coliforms. The presence of the enzyme β-D-glucuronidase differentiates most *E. coli* spp. from other coliforms. *E. coli* absorbs the chromogenic substrate 5-bromo-4-chloro-3-indolyl-β-D-glucuronide (X-β-D-glucuronide). The enzyme β-glucuronidase splits the bond between the chromophore 5-bromo-4-chloro-3-indolyle and the β-D-glucuronide. *E. coli* colonies are colored blue-green. Growth of accompanying Gram-positive flora is largely inhibited by the use of bile salts. The prepared medium is clear and yellowish.

## Tryptone Broth

**Composition** per liter:
Pancreatic digest of casein ............................................................10.0g
Glucose ........................................................................................5.0g
K$_2$HPO$_4$.....................................................................................1.25g
Yeast extract.................................................................................1.0g
Bromcresol Purple solution .........................................................2.0mL

### Bromcresol Purple Solution:
**Composition** per 100.0mL:
Bromcresol Purple .......................................................................2.0g
Ethanol ....................................................................................... 10.0mL

**Preparation of Bromcresol Purple Solution:** Add Bromcresol Purple to 10.0mL of ethanol. Mix thoroughly. Bring volume to 100.0mL with distilled/deionized water.

**Preparation of Medium:** Add components to distilled/deionized water and bring volume to 1.0L. Mix thoroughly. Distribute into screw-capped tubes in 10.0mL volumes. Autoclave for 20 min at 15 psi pressure–121°C.

**Storage/Shelf Life:** Store dehydrated media in the dark in a sealed container below 30°C. Prepared media should be stored under refrigeration (2-8°C). Media should be used within 60 days of preparation. Media should not be used if there are any signs of deterioration (discoloration) or contamination, or if the expiration date supplied by the manufacturer has passed.

**Use:** For the cultivation of *Salmonella* species from foods.

## Tryptone Broth, 1%
### (ATCC Medium 274)
**Composition** per liter:
Pancreatic digest of casein ............................................................1.0g
pH 7.2 ± 0.2 at 25°C

**Preparation of Medium:** Add pancreatic digest of casein to distilled/deionized water and bring volume to 1.0L. Mix thoroughly. Distribute into tubes or flasks. Autoclave for 15 min at 15 psi pressure–121°C.

**Storage/Shelf Life:** Store dehydrated media in the dark in a sealed container below 30°C. Prepared media should be stored under refrigeration (2-8°C). Media should be used within 60 days of preparation. Media should not be used if there are any signs of deterioration (discoloration) or contamination, or if the expiration date supplied by the manufacturer has passed.

**Use:** For the cultivation and maintenance of *Escherichia coli* and *Pseudomonas* species. For the identification and confirmation of *Vibrio cholerae* in foods.

## Tryptone Glucose Beef Extract Agar
### (Tryptone Glucose Extract Agar)
**Composition** per liter:
Agar .............................................................................................15.0g
Pancreatic digest of casein ............................................................5.0g
Beef extract...................................................................................3.0g
Glucose ........................................................................................1.0g
pH 7.0 ± 0.2 at 25°C

**Source:** This medium is available as a premixed powder from BD Diagnostic Systems and Oxoid Unipath.

**Preparation of Medium:** Add components to distilled/deionized water and bring volume to 1.0L. Mix thoroughly. Gently heat and bring to boiling. Distribute into tubes or flasks. Autoclave for 15 min at 15 psi pressure–121°C. Cool to 45°–50°C. If the dilution of the specimen is greater than 1:10, add 10.0mL of sterile 10% skim milk solution. Mix thoroughly. Pour into sterile Petri dishes or leave in tubes.

**Use:** For the enumeration of bacteria by the standard plate count procedure. For the cultivation and enumeration of bacteria from milk and dairy products. For the detection of thermophilic organisms.

## Tryptone Glucose Extract HiVeg Agar
### (Tryptone Glucose Yeast Extract HiVeg Agar)
**Composition** per liter:
Agar .............................................................................................15.0g
Plant hydrolysate .........................................................................5.0g
Yeast extract.................................................................................3.0g
Glucose ........................................................................................1.0g
pH 7.0 ± 0.2 at 25°C

**Source:** This medium is available as a premixed powder from Hi-Media.

**Preparation of Medium:** Add components to distilled/deionized water and bring volume to 1.0L. Mix thoroughly. Gently heat and bring to boiling. Distribute into tubes or flasks. Autoclave for 15 min at 15 psi pressure–121°C.

**Storage/Shelf Life:** Store dehydrated media in the dark in a sealed container below 30°C. Prepared media should be stored under refrigeration (2-8°C). Media should be used within 60 days of preparation. Media should not be used if there are any signs of deterioration (shrinking, cracking, or discoloration) or contamination, or if the expiration date supplied by the manufacturer has passed.

**Use:** For the enumeration of bacteria by the standard plate count procedure. For the cultivation and enumeration of bacteria from milk and dairy products.

## Tryptone Glucose HiVeg Agar
**Composition** per liter:
Plant hydrolysate .........................................................................20.0g
Glucose ........................................................................................5.0g
Agar .............................................................................................3.5g
Bromthymol Blue .........................................................................0.01g
pH 7.0 ± 0.2 at 25°C

**Source:** This medium is available as a premixed powder from Hi-Media.

**Preparation of Medium:** Add components to distilled/deionized water and bring volume to 1.0L. Mix thoroughly. Gently heat and bring to boiling. Distribute into tubes or flasks. Autoclave for 15 min at 15 psi pressure–121°C.

**Storage/Shelf Life:** Store dehydrated media in the dark in a sealed container below 30°C. Prepared media should be stored under refrigeration (2-8°C). Media should be used within 60 days of preparation. Media should not be used if there are any signs of deterioration (shrinking, cracking, or discoloration) or contamination, or if the expiration date supplied by the manufacturer has passed.

**Use:** For the enumeration of bacteria by the standard plate count procedure. For the cultivation and enumeration of bacteria from milk and dairy products.

## Tryptone Glucose Yeast Extract HiVeg Broth

**Composition** per liter:

| | |
|---|---|
| Plant hydrolysate | 10.0g |
| Glucose | 5.0g |
| K₂HPO₄ | 1.25g |
| Yeast extract | 1.0g |

$$pH\ 6.8 \pm 0.2\ at\ 25°C$$

**Source:** This medium is available as a premixed powder from Hi-Media.

**Preparation of Medium:** Add components to distilled/deionized water and bring volume to 1.0L. Mix thoroughly. Gently heat and bring to boiling. Distribute into tubes or flasks. Autoclave for 15 min at 15 psi pressure–121°C.

**Storage/Shelf Life:** Store dehydrated media in the dark in a sealed container below 30°C. Prepared media should be stored under refrigeration (2-8°C). Media should be used within 60 days of preparation. Media should not be used if there are any signs of deterioration (discoloration) or contamination, or if the expiration date supplied by the manufacturer has passed.

**Use:** For the enumeration of bacteria by the standard plate count procedure. For the cultivation and enumeration of bacteria from milk and dairy products.

## Tryptone Peptone Glucose Yeast Extract HiVeg Broth Base without Trypsin

**Composition** per liter:

| | |
|---|---|
| Plant hydrolysate | 50.0g |
| Yeast extract | 20.0g |
| Plant peptone | 5.0g |
| Glucose | 4.0g |
| Na-thioglycolate | 1.0g |

$$pH\ 7.0 \pm 0.2\ at\ 25°C$$

**Source:** This medium is available as a premixed powder from Hi-Media.

**Preparation of Medium:** Add components to distilled/deionized water and bring volume to 1.0L. Mix thoroughly. Gently heat and bring to boiling. Distribute into tubes or flasks. Autoclave for 15 min at 15 psi pressure–121°C.

**Storage/Shelf Life:** Store dehydrated media in the dark in a sealed container below 30°C. Prepared media should be stored under refrigeration (2-8°C). Media should be used within 60 days of preparation. Media should not be used if there are any signs of deterioration (discoloration) or contamination, or if the expiration date supplied by the manufacturer has passed.

**Use:** For the enumeration of bacteria by the standard plate count procedure. For the cultivation and enumeration of bacteria from milk and dairy products.

## Tryptone Phosphate Brain Heart Infusion Yeast Extract Agar (TPBY)

**Composition** per liter:

| | |
|---|---|
| Pancreatic digest of casein | 20.0g |
| Agar | 15.0g |
| NaCl | 5.14g |

| | |
|---|---|
| K₂HPO₄ | 2.0g |
| KH₂PO₄ | 2.0g |
| Yeast extract | 1.0g |
| Oxgall | 0.5g |
| Pancreatic digest of gelatin | 0.4g |
| Brain heart, solids from infusion | 0.16g |
| Peptic digest of animal tissue | 0.16g |
| Glucose | 0.08g |
| Na₂HPO₄ | 0.06g |
| Tween™ 80 | 1.5mL |

$$pH\ 7.0 \pm 0.2\ at\ 25°C$$

**Preparation of Medium:** Add components to distilled/deionized water and bring volume to 1.0L. Mix thoroughly. Gently heat and bring to boiling. Adjust pH to 7.0. Distribute into tubes or flasks. Autoclave for 15 min at 15 psi pressure–121°C. Pour into sterile Petri dishes or leave in tubes.

**Storage/Shelf Life:** Store dehydrated media in the dark in a sealed container below 30°C. Prepared media should be stored under refrigeration (2-8°C). Media should be used within 60 days of preparation. Media should not be used if there are any signs of deterioration (shrinking, cracking, or discoloration) or contamination, or if the expiration date supplied by the manufacturer has passed.

**Use:** For the cultivation of coliform bacteria, such as *Escherichia coli*, from foods.

## Tryptone Phosphate Broth

**Composition** per liter:

| | |
|---|---|
| Pancreatic digest of casein | 20.0g |
| NaCl | 5.0g |
| K₂HPO₄ | 2.0g |
| KH₂PO₄ | 2.0g |
| Tween™ 80 | 1.5mL |

$$pH\ 7.0 \pm 0.2\ at\ 25°C$$

**Preparation of Medium:** Add components to distilled/deionized water and bring volume to 1.0L. Mix thoroughly. Distribute into tubes or flasks. Autoclave for 15 min at 15 psi pressure–121°C.

**Storage/Shelf Life:** Store dehydrated media in the dark in a sealed container below 30°C. Prepared media should be stored under refrigeration (2-8°C). Media should be used within 60 days of preparation. Media should not be used if there are any signs of deterioration (discoloration) or contamination, or if the expiration date supplied by the manufacturer has passed.

**Use:** For the cultivation of coliform bacteria, such as *Escherichia coli*, from foods.

## Tryptone Phosphate HiVeg Broth

**Composition** per liter:

| | |
|---|---|
| Plant hydrolysate | 20.0g |
| NaCl | 5.0g |
| K₂HPO₄ | 2.0g |
| KH₂PO₄ | 2.0g |
| Polysorbate 80 | 1.5g |

$$pH\ 7.0 \pm 0.2\ at\ 25°C$$

**Source:** This medium is available as a premixed powder from Hi-Media.

**Preparation of Medium:** Add components to distilled/deionized water and bring volume to 1.0L. Mix thoroughly. Distribute into tubes or flasks. Autoclave for 15 min at 15 psi pressure–121°C.

**Storage/Shelf Life:** Store dehydrated media in the dark in a sealed container below 30°C. Prepared media should be stored under refrigeration (2-8°C). Media should be used within 60 days of preparation. Media should not be used if there are any signs of deterioration (discoloration) or contamination, or if the expiration date supplied by the manufacturer has passed.

**Use:** For the cultivation of enteropathogenic *Escherichia coli*. For the enrichment and cultivation of enteropathogenic *Escherichia coli* from suspected food samples.

### Tryptone Soy HiVeg Broth with 0.1% Agar (Soybean HiVeg Medium with 0.1% Agar)

**Composition** per liter:
| | |
|---|---|
| Plant hydrolysate | 17.0g |
| NaCl | 5.0g |
| Papaic digest of soybean meal | 3.0g |
| Glucose | 2.5g |
| K$_2$HPO$_4$ | 2.5g |
| Agar | 1.0g |

pH 7.3 ± 0.2 at 25°C

**Source:** This medium is available as a premixed powder from Hi-Media.

**Preparation of Medium:** Add components to distilled/deionized water and bring volume to 1.0L. Mix thoroughly. Gently heat and bring to boiling. Distribute into tubes or flasks. Autoclave for 15 min at 15 psi pressure–121°C.

**Storage/Shelf Life:** Store dehydrated media in the dark in a sealed container below 30°C. Prepared media should be stored under refrigeration (2-8°C). Media should be used within 60 days of preparation. Media should not be used if there are any signs of deterioration (discoloration) or contamination, or if the expiration date supplied by the manufacturer has passed.

**Use:** For the cultivation of anaerobes from root canals, blood, and other clinical specimens and for determining glucose fermentation.

### Tryptone SoyHiVeg Broth without Glucose (Soybean HiVeg Medium)

**Composition** per liter:
| | |
|---|---|
| Plant hydrolysate | 17.0g |
| NaCl | 5.0g |
| Papaic digest of soybean meal | 3.0g |
| K$_2$HPO$_4$ | 2.5g |

pH 7.3 ± 0.2 at 25°C

**Source:** This medium is available as a premixed powder from Hi-Media.

**Preparation of Medium:** Add components to distilled/deionized water and bring volume to 1.0L. Mix thoroughly. Gently heat and bring to boiling. Distribute into tubes or flasks. Autoclave for 15 min at 15 psi pressure–121°C.

**Storage/Shelf Life:** Store dehydrated media in the dark in a sealed container below 30°C. Prepared media should be stored under refrigeration (2-8°C). Media should be used within 60 days of preparation. Media should not be used if there are any signs of deterioration (discol-

oration) or contamination, or if the expiration date supplied by the manufacturer has passed.

**Use:** For the cultivation of a wide variety of microorganisms. For sterility testing.

### Tryptone Soy HiVeg Broth with 10% Sodium Chloride and 1% Sodium Pyruvate

**Composition** per liter:
| | |
|---|---|
| NaCl | 105.0g |
| Plant hydrolysate | 17.0g |
| Sodium pyruvate | 10.0g |
| Papaic digest of soybean meal | 3.0g |
| Glucose | 2.5g |
| K$_2$HPO$_4$ | 2.5g |

pH 7.3 ± 0.2 at 25°C

**Source:** This medium is available as a premixed powder from Hi-Media.

**Preparation of Medium:** Add components to distilled/deionized water and bring volume to 1.0L. Mix thoroughly. Gently heat until dissolved. Adjust pH to 7.3. Distribute into tubes in 10.0mL volumes. Autoclave for 15 min at 15 psi pressure–121°C.

**Storage/Shelf Life:** Store dehydrated media in the dark in a sealed container below 30°C. Prepared media should be stored under refrigeration (2-8°C). Media should be used within 60 days of preparation. Media should not be used if there are any signs of deterioration (discoloration) or contamination, or if the expiration date supplied by the manufacturer has passed.

**Use:** For the isolation and cultivation of *Staphylococcus aureus* from foods.

### Tryptone Water Broth, HiVeg

**Composition** per liter:
| | |
|---|---|
| Plant hydrolysate | 10.0g |
| Glucose | 5.0g |
| K$_2$HPO$_4$ | 1.25g |
| Yeast extract | 1.0g |
| Bromcresol Purple | 0.04g |

pH 7.3 ± 0.2 at 25°C

**Source:** This medium is available as a premixed powder from Hi-Media.

**Preparation of Medium:** Add components to tap water and bring volume to 1.0L. Mix thoroughly. Gently heat and bring to boiling. Distribute into tubes or flasks. Autoclave for 15 min at 15 psi pressure–121°C.

**Storage/Shelf Life:** Store dehydrated media in the dark in a sealed container below 30°C. Prepared media should be stored under refrigeration (2-8°C). Media should be used within 60 days of preparation. Media should not be used if there are any signs of deterioration (discoloration) or contamination, or if the expiration date supplied by the manufacturer has passed.

**Use:** For the cultivation of *Salmonella* species from foods.

### Tryptone with Sodium Chloride Broth

**Composition** per liter:
| | |
|---|---|
| Pancreatic digest of casein | 8.0g |
| NaCl | 0.5g |

**Preparation of Medium:** Add components to distilled/deionized water and bring volume to 1.0L. Mix thoroughly. Distribute into tubes or flasks. Autoclave for 15 min at 15 psi pressure–121°C.

**Storage/Shelf Life:** Store dehydrated media in the dark in a sealed container below 30°C. Prepared media should be stored under refrigeration (2-8°C). Media should be used within 60 days of preparation. Media should not be used if there are any signs of deterioration (discoloration) or contamination, or if the expiration date supplied by the manufacturer has passed.

**Use:** For the cultivation and maintenance of fastidious aerobic and facultative microorganisms such as *Escherichia coli* and *Pseudomonas* species.

## Tryptone Yeast Extract Agar

**Composition** per liter:
Pancreatic digest of casein...........................................................10.0g
Agar ........................................................................................2.0g
Yeast extract...........................................................................1.0g
Bromcresol Purple ................................................................0.04g
Carbohydrate solution............................................... 100.0mL

pH 7.0 ± 0.2 at 25°C

**Carbohydrate Solution:**
**Composition** per 100.0mL:
Carbohydrate..........................................................................10.0g

**Preparation of Carbohydrate Solution:** Add carbohydrate to distilled/deionized water and bring volume to 100.0mL. Glucose or mannitol may be used. Mix thoroughly. Filter sterilize.

**Preparation of Medium:** Add components, except carbohydrate solution, to distilled/deionized water and bring volume to 900.0mL. Mix thoroughly. Adjust pH to 7.0. Gently heat and bring to boiling. Distribute into tubes in 13.5mL volumes. Autoclave for 20 min at 10 psi pressure–115°C. Cool to 45°–50°C. Aseptically add 1.5mL of carbohydrate solution to each tube. Mix thoroughly. Solidify agar quickly by placing tubes in ice water.

**Storage/Shelf Life:** Store dehydrated media in the dark in a sealed container below 30°C. Prepared media should be stored under refrigeration (2-8°C). Media should be used within 60 days of preparation. Media should not be used if there are any signs of deterioration (shrinking, cracking, or discoloration) or contamination, or if the expiration date supplied by the manufacturer has passed.

**Use:** For the cultivation and differentiation of *Staphylococcus aureus* based on glucose and mannitol fermentation. Bacteria that ferment the added carbohydrate turn the medium yellow.

## Tryptone Yeast Extract HiVeg Agar with Carbohydrate

**Composition** per liter:
Agar ........................................................................................12.0g
Plant hydrolysate.....................................................................6.0g
Yeast extract powder...............................................................3.0g
Carbohydrate solution............................................... 100.0mL

pH 7.0 ± 0.2 at 25°C

**Source:** This medium, without carbohydrate solution, is available as a premixed powder from HiMedia.

**Carbohydrate Solution:**
**Composition** per 100.0mL:
Carbohydrate..........................................................................10.0g

**Preparation of Carbohydrate Solution:** Add carbohydrate to distilled/deionized water and bring volume to 100.0mL. Glucose or mannitol may be used. Mix thoroughly. Filter sterilize.

**Preparation of Medium:** Add components, except carbohydrate solution, to distilled/deionized water and bring volume to 900.0mL. Mix thoroughly. Adjust pH to 7.0. Gently heat and bring to boiling. Distribute into tubes in 13.5mL volumes. Autoclave for 20 min at 10 psi pressure–115°C. Cool to 45°–50°C. Aseptically add 1.5mL of carbohydrate solution to each tube. Mix thoroughly. Solidify agar quickly by placing tubes in ice water.

**Storage/Shelf Life:** Store dehydrated media in the dark in a sealed container below 30°C. Prepared media should be stored under refrigeration (2-8°C). Media should be used within 60 days of preparation. Media should not be used if there are any signs of deterioration (shrinking, cracking, or discoloration) or contamination, or if the expiration date supplied by the manufacturer has passed.

**Use:** For the cultivation and differentiation of *Staphylococcus aureus* based on glucose and mannitol fermentation. Bacteria that ferment the added carbohydrate turn the medium yellow.

## Tryptose Agar (BAM M167)

**Composition** per liter:
Tryptose ..................................................................................20.0g
Agar ........................................................................................15.0g
NaCl.........................................................................................5.0g
Glucose ....................................................................................1.0g

pH 7.2 ± 0.2 at 25°C

**Preparation of Medium:** Add components to distilled/deionized water and bring volume to 1.0L. Mix thoroughly. Autoclave for 15 min at 15 psi pressure–121°C. Cool to 45°–50°C. Pour into sterile Petri dishes or leave in tubes. For slants allow tubes to cool in an inclined position.

**Storage/Shelf Life:** Store dehydrated media in the dark in a sealed container below 30°C. Prepared media should be stored under refrigeration (2-8°C). Media should be used within 60 days of preparation. Media should not be used if there are any signs of deterioration (shrinking, cracking, or discoloration) or contamination, or if the expiration date supplied by the manufacturer has passed.

**Use:** For the cultivation of a variety of bacteria for serology.

## Tryptose Agar with Citrate

**Composition** per liter:
Agar ........................................................................................15.0g
Pancreatic digest of casein...................................................10.0g
Peptic digest of animal tissue ..............................................10.0g
Sodium citrate.......................................................................10.0g
NaCl.........................................................................................5.0g
Glucose ....................................................................................1.0g

pH 7.2 ± 0.2 at 25°C

**Preparation of Medium:** Add components to distilled/deionized water and bring volume to 1.0L. Mix thoroughly. Gently heat and bring to boiling. Distribute into tubes or flasks. Autoclave for 15 min at 15 psi pressure–121°C. Pour into sterile Petri dishes or leave in tubes.

**Storage/Shelf Life:** Store dehydrated media in the dark in a sealed container below 30°C. Prepared media should be stored under refrigeration (2-8°C). Media should be used within 60 days of preparation. Media should not be used if there are any signs of deterioration (shrink-

ing, cracking, or discoloration) or contamination, or if the expiration date supplied by the manufacturer has passed.

**Use:** For the cultivation and maintenance of fastidious aerobic and facultative microorganisms, including *Brucella* species and streptococci.

## Tryptose Agar, HiVeg

**Composition** per liter:

| | |
|---|---|
| Plant hydrolysate No. 1 | 20.0g |
| Agar | 15.0g |
| NaCl | 5.0g |
| Glucose | 1.0g |

pH 7.2 ± 0.2 at 25°C

**Source:** This medium is available as a premixed powder from HiMedia.

**Preparation of Medium:** Add components to distilled/deionized water and bring volume to 1.0L. Mix thoroughly. Gently heat and bring to boiling. Distribute into tubes or flasks. Autoclave for 15 min at 15 psi pressure–121°C. Pour into sterile Petri dishes or leave in tubes.

**Storage/Shelf Life:** Store dehydrated media in the dark in a sealed container below 30°C. Prepared media should be stored under refrigeration (2-8°C). Media should be used within 60 days of preparation. Media should not be used if there are any signs of deterioration (shrinking, cracking, or discoloration) or contamination, or if the expiration date supplied by the manufacturer has passed.

**Use:** For the cultivation and maintenance of fastidious aerobic and facultative microorganisms. For the isolation, cultivation, and differentiation of *Brucella*, sreptococci, and pneumococci.

## Tryptose Blood Agar Base

**Composition** per liter:

| | |
|---|---|
| Agar | 15.0g |
| Tryptose | 10.0g |
| NaCl | 5.0g |
| Beef extract | 3.0g |

**Preparation of Medium:** Add components to distilled/deionized water and bring volume to 1.0L. Mix thoroughly. Gently heat and bring to boiling. Distribute into tubes. Autoclave for 15 min at 15 psi pressure–121°C. Allow tubes to cool in a slanted position to obtain a 4–5.0cm slant and a 2–3.0cm butt.

**Storage/Shelf Life:** Store dehydrated media in the dark in a sealed container below 30°C. Prepared media should be stored under refrigeration (2-8°C). Media should be used within 60 days of preparation. Media should not be used if there are any signs of deterioration (shrinking, cracking, or discoloration) or contamination, or if the expiration date supplied by the manufacturer has passed.

**Use:** For the cultivation and enumeration of *Salmonella* species from foods.

## Tryptose Blood Agar Base, HiVeg with Sheep Blood

**Composition** per liter:

| | |
|---|---|
| Agar | 15.0g |
| Plant hydrolysate No. 1 | 10.0g |
| NaCl | 5.0g |
| Plant extract | 3.0g |
| Sheep blood, defibrinated | 70.0mL |

pH 7.2 ± 0.2 at 25°C

**Source:** This medium, without blood, is available as a premixed powder from HiMedia.

**Preparation of Medium:** Add components, except sheep blood, to distilled/deionized water and bring volume to 930.0mL. Mix thoroughly. Gently heat and bring to boiling. Autoclave for 15 min at 15 psi pressure–121°C. Cool to 45°–50°C. Aseptically add sterile sheep blood. Mix thoroughly. Pour into sterile Petri dishes in 17.0mL volumes or distribute into sterile tubes.

**Storage/Shelf Life:** Store dehydrated media in the dark in a sealed container below 30°C. Prepared media should be stored under refrigeration (2-8°C). Media should be used within 60 days of preparation. Media should not be used if there are any signs of deterioration (shrinking, cracking, or discoloration) or contamination, or if the expiration date supplied by the manufacturer has passed.

**Use:** For the cultivation and maintenance of a wide variety of fastidious microorganisms. For the isolation of fastidious organisms and determining hemolytic reactions.

## Tryptose Blood Agar Base with Yeast Extract

**Composition** per liter:

| | |
|---|---|
| Agar | 15.0g |
| Tryptose | 10.0g |
| NaCl | 5.0g |
| Beef extract | 3.0g |
| Yeast extract | 1.0g |
| Sheep blood, defibrinated | 50.0mL |

pH 7.3 ± 0.2 at 25°C

**Source:** This medium is available as a premixed powder from BD Diagnostic Systems.

**Preparation of Medium:** Add components, except sheep blood, to distilled/deionized water and bring volume to 950.0mL. Mix thoroughly. Gently heat and bring to boiling. Autoclave for 15 min at 15 psi pressure–121°C. Cool to 45°–50°C. Aseptically add sterile sheep blood. Mix thoroughly. Pour into sterile Petri dishes in 17.0mL volumes or distribute into sterile tubes.

**Storage/Shelf Life:** Store dehydrated media in the dark in a sealed container below 30°C. Prepared media should be stored under refrigeration (2-8°C). Media should be used within 60 days of preparation. Media should not be used if there are any signs of deterioration (shrinking, cracking, or discoloration) or contamination, or if the expiration date supplied by the manufacturer has passed.

**Use:** For the cultivation and maintenance of a wide variety of fastidious microorganisms.

## Tryptose Blood Agar Base with Yeast Extract, HiVeg

**Composition** per liter:

| | |
|---|---|
| Agar | 15.0g |
| Plant hydrolysate No. 1 | 10.0g |
| NaCl | 5.0g |
| Plant extract | 3.0g |
| Yeast extract | 1.0g |
| Sheep blood, defibrinated | 70.0mL |

pH 7.2 ± 0.2 at 25°C

**Source:** This medium, without blood, is available as a premixed powder from HiMedia.

**Preparation of Medium:** Add components, except sheep blood, to distilled/deionized water and bring volume to 930.0mL. Mix thoroughly. Gently heat and bring to boiling. Autoclave for 15 min at 15 psi pressure–121°C. Cool to 45°–50°C. Aseptically add sterile sheep blood. Mix thoroughly. Pour into sterile Petri dishes in 17.0mL volumes or distribute into sterile tubes.

**Storage/Shelf Life:** Store dehydrated media in the dark in a sealed container below 30°C. Prepared media should be stored under refrigeration (2-8°C). Media should be used within 60 days of preparation. Media should not be used if there are any signs of deterioration (shrinking, cracking, or discoloration) or contamination, or if the expiration date supplied by the manufacturer has passed.

**Use:** For the cultivation and maintenance of a wide variety of fastidious microorganisms. For the isolation of fastidious organisms and determining hemolytic reactions.

## Tryptose Broth
## (BAM M167)

**Composition** per liter:

Tryptose ......................................................................20.0g
NaCl...............................................................................5.0g
Glucose ........................................................................1.0g

pH 7.2 ± 0.2 at 25°C

**Preparation of Medium:** Add components to distilled/deionized water and bring volume to 1.0L. Mix thoroughly. Distribute into tubes or flasks. Autoclave for 15 min at 15 psi pressure–121°C.

**Storage/Shelf Life:** Store dehydrated media in the dark in a sealed container below 30°C. Prepared media should be stored under refrigeration (2-8°C). Media should be used within 60 days of preparation. Media should not be used if there are any signs of deterioration (discoloration) or contamination, or if the expiration date supplied by the manufacturer has passed.

**Use:** For the cultivation of a variety of bacteria for serology.

## Tryptose Broth with Citrate

**Composition** per liter:

Pancreatic digest of casein...............................................10.0g
Peptic digest of animal tissue.........................................10.0g
Sodium citrate ................................................................10.0g
NaCl.................................................................................5.0g
Glucose ............................................................................1.0g
Thiamine·HCl ................................................................5.0mg

pH 7.2 ± 0.2 at 25°C

**Preparation of Medium:** Add components to distilled/deionized water and bring volume to 1.0L. Mix thoroughly. Distribute into tubes or flasks. Autoclave for 15 min at 15 psi pressure–121°C.

**Storage/Shelf Life:** Store dehydrated media in the dark in a sealed container below 30°C. Prepared media should be stored under refrigeration (2-8°C). Media should be used within 60 days of preparation. Media should not be used if there are any signs of deterioration (discoloration) or contamination, or if the expiration date supplied by the manufacturer has passed.

**Use:** For the isolation and cultivation of a variety of fastidious aerobic microorganisms, especially *Brucella* species, from clinical sources and dairy products.

## Tryptose Broth, HiVeg

**Composition** per liter:

Plant hydrolysate No. 1.................................................20.0g
NaCl.................................................................................5.0g
Glucose ...........................................................................1.0g

pH 7.2 ± 0.2 at 25°C

**Source:** This medium is available as a premixed powder from Hi-Media.

**Preparation of Medium:** Add components to distilled/deionized water and bring volume to 1.0L. Mix thoroughly. Distribute into tubes or flasks. Autoclave for 15 min at 15 psi pressure–121°C.

**Storage/Shelf Life:** Store dehydrated media in the dark in a sealed container below 30°C. Prepared media should be stored under refrigeration (2-8°C). Media should be used within 60 days of preparation. Media should not be used if there are any signs of deterioration (discoloration) or contamination, or if the expiration date supplied by the manufacturer has passed.

**Use:** For the cultivation of fastidious aerobic and facultative microorganisms, including streptococci.

## Tryptose Cycloserine Glucose HiVeg Agar Base
## with Cycloserine

**Composition** per liter:

Agar ..............................................................................20.0g
Plant hydrolysate No. 1.................................................15.0g
Papaic digest of soybean meal........................................5.0g
Yeast extract....................................................................5.0g
Ferric ammonium citrate..................................................1.0g
Cycloserine solution .................................................... 10.0mL

pH 7.6 ± 0.2 at 25°C

**Source:** This medium, without cycloserine, is available as a premixed powder from HiMedia.

**Cycloserine Solution:**
**Composition** per 10.0mL:
D-Cycloserine ..................................................................0.4g

**Preparation of Cycloserine Solution:** Add D-cycloserine to distilled/deionized water and bring volume to 10.0mL. Mix thoroughly. Filter sterilize.

**Preparation of Medium:** Add components, except cycloserine solution, to distilled/deionized water and bring volume to 990.0mL. Mix thoroughly. Gently heat and bring to boiling. Autoclave for 15 min at 15 psi pressure–121°C. Cool to 45°–50°C. Aseptically add sterile cycloserine solution. Mix thoroughly. Pour into sterile Petri dishes or distribute into sterile tubes.

**Storage/Shelf Life:** Store dehydrated media in the dark in a sealed container below 30°C. Prepared media should be stored under refrigeration (2-8°C). Media should be used within 60 days of preparation. Media should not be used if there are any signs of deterioration (shrinking, cracking, or discoloration) or contamination, or if the expiration date supplied by the manufacturer has passed.

**Use:** For the isolation and cultivation of *Clostridium* species, especially *Clostridium botulinum*, from foods.

## Tryptose Cycloserine Dextrose Agar

**Composition** per liter:

Agar ..............................................................................20.0g
Tryptose .........................................................................15.0g
Pancreatic digest of soybean meal...................................5.0g
Yeast extract....................................................................5.0g
Ferric ammonium citrate..................................................1.0g
Cycloserine solution .................................................... 10.0mL

pH 7.6 ± 0.2 at 25°C

**Cycloserine Solution:**
**Composition** per 10.0mL:
D-Cycloserine ..................................................................0.4g

**Preparation of Cycloserine Solution:** Add D-cycloserine to distilled/deionized water and bring volume to 10.0mL. Mix thoroughly. Filter sterilize.

**Preparation of Medium:** Add components, except cycloserine solution, to distilled/deionized water and bring volume to 990.0mL. Mix thoroughly. Gently heat and bring to boiling. Autoclave for 15 min at 15 psi pressure–121°C. Cool to 45°–50°C. Aseptically add sterile cycloserine solution. Mix thoroughly. Pour into sterile Petri dishes or distribute into sterile tubes.

**Storage/Shelf Life:** Store dehydrated media in the dark in a sealed container below 30°C. Prepared media should be stored under refrigeration (2-8°C). Media should be used within 60 days of preparation. Media should not be used if there are any signs of deterioration (shrinking, cracking, or discoloration) or contamination, or if the expiration date supplied by the manufacturer has passed.

**Use:** For the isolation and cultivation of *Clostridium* species, especially *Clostridium botulinum*, from foods.

## Tryptose Sulfite Cycloserine Agar
## (TSC Agar)

**Composition** per liter:
| | |
|---|---|
| Tryptose | 15.0g |
| Agar | 14.0g |
| Pancreatic digest of soybean meal | 5.0g |
| Yeast extract | 5.0g |
| Ferric ammonium citrate | 1.0g |
| $Na_2S_2O_5$ | 1.0g |
| Cycloserine solution | 10.0mL |

pH 7.6 ± 0.2 at 25°C

**Cycloserine Solution:**
**Composition** per 10.0mL:
| | |
|---|---|
| D-Cycloserine | 0.4g |

**Preparation of Cycloserine Solution:** Add cycloserine to distilled/deionized water and bring volume to 10.0mL. Mix thoroughly. Filter sterilize.

**Preparation of Medium:** Add components, except cycloserine solution, to distilled/deionized water and bring volume to 990.0mL. Mix thoroughly. Gently heat and bring to boiling. Autoclave for 15 min at 15 psi pressure–121°C. Cool to 45°–50°C. Aseptically add sterile cycloserine solution. Mix thoroughly. Pour into sterile Petri dishes.

**Storage/Shelf Life:** Store dehydrated media in the dark in a sealed container below 30°C. Prepared media should be stored under refrigeration (2-8°C). Media should be used within 60 days of preparation. Media should not be used if there are any signs of deterioration (shrinking, cracking, or discoloration) or contamination, or if the expiration date supplied by the manufacturer has passed.

**Use:** For the presumptive identification and enumeration of *Clostridium perfringens*.

## Tryptose Sulfite Cycloserine Agar
## (TSC Agar)

**Composition** per liter:
| | |
|---|---|
| Tryptose | 15.0g |
| Agar | 14.0g |
| Beef extract | 5.0g |
| Pancreatic digest of soybean meal | 5.0g |
| Yeast extract | 5.0g |
| Ferric ammonium citrate | 1.0g |

| | |
|---|---|
| $Na_2S_2O_5$ | 1.0g |
| Egg yolk emulsion | 50.0mL |
| Cycloserine solution | 10.0mL |

pH 7.6 ± 0.2 at 25°C

**Source:** This medium is available as a premixed powder from Oxoid Unipath.

**Egg Yolk Emulsion:**
**Composition**:
| | |
|---|---|
| Chicken egg yolks | 11 |
| Whole chicken egg | 1 |

**Preparation of Egg Yolk Emulsion:** Soak eggs with 1:100 dilution of saturated mercuric chloride solution for 1 min. Crack eggs and separate yolks from whites. Mix egg yolks with 1 chicken egg.

**Cycloserine Solution:**
**Composition** per 10.0mL:
| | |
|---|---|
| D-Cycloserine | 0.4g |

**Preparation of Cycloserine Solution:** Add cycloserine to distilled/deionized water and bring volume to 10.0mL. Mix thoroughly. Filter sterilize.

**Preparation of Medium:** Add components, except cycloserine solution and egg yolk emulsion, to distilled/deionized water and bring volume to 940.0mL. Mix thoroughly. Gently heat and bring to boiling. Autoclave for 15 min at 15 psi pressure–121°C. Cool to 45°–50°C. Aseptically add sterile cycloserine solution and egg yolk emulsion. Mix thoroughly. Pour into sterile Petri dishes.

**Storage/Shelf Life:** Store dehydrated media in the dark in a sealed container below 30°C. Prepared media should be stored under refrigeration (2-8°C). Media should be used within 60 days of preparation. Media should not be used if there are any signs of deterioration (shrinking, cracking, or discoloration) or contamination, or if the expiration date supplied by the manufacturer has passed.

**Use:** For the presumptive identification and enumeration of *Clostridium perfringens*.

## Tryptose Sulfite Cycloserine Agar
## with Polymyxin and Kanamycin

**Composition** per liter:
| | |
|---|---|
| Tryptose | 15.0g |
| Agar | 14.0g |
| Beef extract | 5.0g |
| Pancreatic digest of soybean meal | 5.0g |
| Yeast extract | 5.0g |
| Ferric ammonium citrate | 1.0g |
| $Na_2S_2O_5$ | 1.0g |
| Antibiotic solution | 10.0mL |

pH 7.6 ± 0.2 at 25°C

**Antibiotic Solution:**
**Composition** per 10.0mL:
| | |
|---|---|
| D-Cycloserine | 0.4g |
| Polymyxin B sulfate | 0.03g |
| Kanamycin sulfate | 0.012g |

**Preparation of Antibiotic Solution:** Add components to distilled/deionized water and bring volume to 10.0mL. Mix thoroughly. Filter sterilize.

**Preparation of Medium:** Add components, except antibiotic solution, to distilled/deionized water and bring volume to 990.0mL. Mix thoroughly. Gently heat and bring to boiling. Autoclave for 15 min at

15 psi pressure–121°C. Cool to 45°–50°C. Aseptically add sterile antibiotic solution. Mix thoroughly. Pour into sterile Petri dishes.

**Storage/Shelf Life:** Store dehydrated media in the dark in a sealed container below 30°C. Prepared media should be stored under refrigeration (2-8°C). Media should be used within 60 days of preparation. Media should not be used if there are any signs of deterioration (shrinking, cracking, or discoloration) or contamination, or if the expiration date supplied by the manufacturer has passed.

**Use:** For the isolation and enumeration of *Clostridium perfringens* from foods and clinical specimens.

## Tryptose Sulfite Cycloserine Agar without Egg Yolk (TSC Agar without Egg Yolk)

**Composition** per liter:

| | |
|---|---|
| Tryptose | 15.0g |
| Agar | 14.0g |
| Beef extract | 5.0g |
| Pancreatic digest of soybean meal | 5.0g |
| Yeast extract | 5.0g |
| Ferric ammonium citrate | 1.0g |
| $Na_2S_2O_5$ | 1.0g |
| Cycloserine solution | 10.0mL |

pH 7.6 ± 0.2 at 25°C

**Cycloserine Solution:**
**Composition** per 10.0mL:

| | |
|---|---|
| D-Cycloserine | 0.4g |

**Preparation of Cycloserine Solution:** Add cycloserine to distilled/deionized water and bring volume to 10.0mL. Mix thoroughly. Filter sterilize.

**Preparation of Medium:** Add components, except cycloserine solution, to distilled/deionized water and bring volume to 990.0mL. Mix thoroughly. Gently heat and bring to boiling. Autoclave for 15 min at 15 psi pressure–121°C. Cool to 45°–50°C. Aseptically add sterile cycloserine solution. Mix thoroughly. Pour into sterile Petri dishes.

**Storage/Shelf Life:** Store dehydrated media in the dark in a sealed container below 30°C. Prepared media should be stored under refrigeration (2-8°C). Media should be used within 60 days of preparation. Media should not be used if there are any signs of deterioration (shrinking, cracking, or discoloration) or contamination, or if the expiration date supplied by the manufacturer has passed.

**Use:** For the presumptive identification and enumeration of *Clostridium perfringens*.

## TSC Agar, Fluorocult (Fluorocult TSC Agar) (Fluorocult Tryptose Sulfite Cycloserine Agar) (Tryptose Sulfite Cycloserine Agar, Fluorocult)

**Composition** per liter:

| | |
|---|---|
| Agar | 15.0g |
| Tryptose | 15.0g |
| Peptone from soymeal | 5.0g |
| Yeast extract | 5.0g |
| $Na_2S_2O_5$ | 1.0g |
| Ammonium ferric citrate | 1.0g |
| D-Cycloserine | 0.2g |
| 4-Methylumbelliferyl-phosphate disodium salt | 50.0mg |

**Source:** This medium is available from Merck.

**Preparation of Medium:** Add components to distilled/deionized water and bring volume to 1.0L. Mix thoroughly. Autoclave for 15 min at 15 psi pressure–121°C. Cool to 45°–50°C. Pour into sterile Petri dishes.

**Storage/Shelf Life:** Store dehydrated media in the dark in a sealed container below 30°C. Prepared media should be stored in the dark under refrigeration (2-8°C). Chromogenic media are especially light and temperature sensitive; protect from light, excessive heat, moisture, and freezing. Media should not be used if there are any signs of deterioration (shrinking, cracking, or discoloration) or contamination, or if the expiration date supplied by the manufacturer has passed.

**Use:** For the isolation and enumeration of the vegetative and spore forms of *Clostridium perfringens* in foodstuffs. The culture medium complies with the recommendations of the International Organization for Standardization (ISO) (1978) and the DIN Norm 10165 for the examination of meat and meat products. It also conforms with the APHA recommendations for the examination of foods (1992). D-Cycloserine inhibits the accompanying bacterial flora and causes the colonies that develop to remain smaller. 4-Methylumbelliferyl-phosphate (MUP) is a fluorogenic substrate for the alkaline and acid phosphatase. The acid phosphatase is a highly specific indicator for *C. perfringens*. The acid phosphatase splits the fluorogenic substrate MUP forming 4-methylumbelliferone, which can be identified as fluorescence in long-wave UV light. Thus a strong suggestion for the presence of *C. perfringens* can be obtained.

## TSN Agar (Trypticase™ Sulfite Neomycin Agar)

**Composition** per liter:

| | |
|---|---|
| Pancreatic digest of casein | 15.0g |
| Agar | 13.5g |
| Yeast extract | 10.0g |
| $Na_2SO_3$ | 1.0g |
| Ferric citrate | 0.5g |
| Neomycin sulfate | 0.05g |
| Polymyxin sulfate | 0.02g |
| Buffered thioglycolate solution | 50.0mL |

pH 7.2 ± 0.2 at 25°C

**Source:** This medium is available as a premixed powder from BD Diagnostic Systems.

**Buffered Thioglycolate Solution:**
**Composition** per 50.0mL:

| | |
|---|---|
| Buffer solution | 35.0mL |
| Sodium thioglycolate solution | 15.0mL |

**Preparation of Buffered Thioglycolate Solution:** Combine components. Mix thoroughly. Autoclave for 15 min at 15 psi pressure–121°C. Cool to 45°–50°C.

**Buffer Solution:**
**Composition** per 100.0mL:

| | |
|---|---|
| $Na_2CO_3$ | 28.0g |
| $K_2HPO_4$ | 5.7g |

**Preparation of Buffer Solution:** Add components to distilled/deionized water and bring volume to 100.0mL. Mix thoroughly.

**Thioglycolate Solution:**
**Composition** per 100.0mL:

| | |
|---|---|
| Sodium thioglycolate | 13.3g |

**Preparation of Thioglycolate Solution:** Add sodium thioglycolate to distilled/deionized water and bring volume to 100.0mL. Mix thoroughly.

**Preparation of Medium:** Add components, except buffered thioglycolate solution, to distilled/deionized water and bring volume to 950.0mL. Mix thoroughly. Gently heat and bring to boiling. Autoclave for 12 min at 13 psi pressure–118°C. Do not overheat. Cool to 45°–50°C. Aseptically add buffered thioglycolate solution. Mix thoroughly. Pour into sterile Petri dishes or distribute into sterile tubes.

**Storage/Shelf Life:** Store dehydrated media in the dark in a sealed container below 30°C. Prepared media should be stored under refrigeration (2-8°C). Media should be used within 60 days of preparation. Media should not be used if there are any signs of deterioration (shrinking, cracking, or discoloration) or contamination, or if the expiration date supplied by the manufacturer has passed.

**Use:** For the selective isolation of *Clostridium perfringens*.

## U Agar Plates
## (*Ureaplasma* Agar Plates)
## (MES Agar)

**Composition** per 100.2mL:

| | |
|---|---|
| Base agar | 65.0mL |
| Horse serum | 20.0mL |
| Yeast dialysate | 10.0mL |
| MES (2-*N*-morpholinoethane sulfonic acid) buffer solution | 3.0mL |
| Penicillin solution | 2.0mL |
| Urea solution | 0.2mL |

pH 5.5 ± 0.2 at 25°C

**Base Agar:**

**Composition** per liter:

| | |
|---|---|
| Papaic digest of soybean meal | 20.0g |
| Agarose | 10.0g |
| NaCl | 5.0g |
| Phenol Red (2% solution) | 1.0mL |

**Preparation of Base Agar:** Add components to distilled/deionized water and bring volume to 1.0L. Mix thoroughly. Gently heat and bring to boiling. Adjust pH to 7.3. Autoclave for 15 min at 15 psi pressure–121°C. Cool to 45°–50°C.

**Yeast Dialysate:**

**Composition** per 10.0mL:

| | |
|---|---|
| Yeast, active dried | 450.0g |

**Preparation of Yeast Dialysate:** Add active, dried yeast to distilled/deionized water and bring volume to 1250.0mL. Gently heat and bring to 40°C. Autoclave for 15 min at 15 psi pressure–121°C. Put into dialysis tubing. Dialyze against 1.0L of distilled/deionized water for 2 days at 4°C. Discard tubing and its contents. Autoclave dialysate for 15 min at 15 psi pressure–121°C. Store at –20°C.

**MES Buffer Solution:**

**Composition** per 100.0mL:

| | |
|---|---|
| MES (2-*N*-morpholinoethane sulfonic acid) buffer | 19.52g |

**Preparation of MES Buffer Solution:** Add MES buffer to distilled/deionized water and bring volume to 100.0mL. Mix thoroughly. Adjust pH to 5.5. Filter sterilize.

**Penicillin Solution:**

**Composition** per 10.0mL:

| | |
|---|---|
| Penicillin | 100,000U |

**Preparation of Penicillin Solution:** Add penicillin to distilled/deionized water and bring volume to 10.0mL. Mix thoroughly. Filter sterilize.

**Urea Solution:**

**Composition** per 100.0mL:

| | |
|---|---|
| Urea | 6.0g |

**Preparation of Urea Solution:** Add urea to distilled/deionized water and bring volume to 100.0mL. Mix thoroughly. Filter sterilize.

**Preparation of Medium:** To 65.0mL of cooled, sterile base agar, aseptically add 10.0mL of sterile yeast dialysate, 20.0mL of horse serum, 2.0mL of sterile penicillin solution, 3.0mL of sterile MES buffer solution, and 0.2mL of sterile urea solution. Mix thoroughly. Pour into 10mm × 35mm Petri dishes in 5.0mL volumes. Allow plates to stand overnight at 25°C to remove excess surface moisture.

**Storage/Shelf Life:** Store dehydrated media in the dark in a sealed container below 30°C. Prepared media should be stored under refrigeration (2-8°C). Media should be used within 60 days of preparation. Media should not be used if there are any signs of deterioration (shrinking, cracking, or discoloration) or contamination, or if the expiration date supplied by the manufacturer has passed.

**Use:** For the isolation and cultivation of *Ureaplasma* species.

## U Broth
## (*Ureaplasma* Broth)

**Composition** per 99.5mL:

| | |
|---|---|
| Base agar | 65.0mL |
| Horse serum | 20.0mL |
| Yeast dialysate | 10.0mL |
| Penicillin solution | 2.0mL |
| MES (2-*N*-morpholinoethane sulfonic acid) buffer solution | 1.0mL |
| Na$_2$SO$_3$ solution | 1.0mL |
| Urea solution | 0.5mL |

pH 5.5 ± 0.2 at 25°C

**Base Agar:**

**Composition** per liter:

| | |
|---|---|
| Papaic digest of soybean meal | 20.0g |
| Agarose | 10.0g |
| NaCl | 5.0g |
| Phenol Red (2% solution) | 1.0mL |

**Preparation of Base Agar:** Add components to distilled/deionized water and bring volume to 1.0L. Mix thoroughly. Gently heat and bring to boiling. Adjust pH to 7.3. Autoclave for 15 min at 15 psi pressure–121°C. Cool to 45°–50°C.

**Yeast Dialysate:**

**Composition** per 10.0mL:

| | |
|---|---|
| Yeast, active dried | 450.0g |

**Preparation of Yeast Dialysate:** Add active, dried yeast to distilled/deionized water and bring volume to 1250.0mL. Gently heat and bring to 40°C. Autoclave for 15 min at 15 psi pressure–121°C. Put into dialysis tubing. Dialyze against 1.0L of distilled/deionized water for 2 days at 4°C. Discard tubing and its contents. Autoclave dialysate for 15 min at 15 psi pressure–121°C. Store at –20°C.

**Penicillin Solution:**

**Composition** per 10.0mL:

| | |
|---|---|
| Penicillin | 100,000U |

**Preparation of Penicillin Solution:** Add penicillin to distilled/deionized water and bring volume to 10.0mL. Mix thoroughly. Filter sterilize.

**MES Buffer Solution:**
**Composition** per 100.0mL:
MES (2-*N*-morpholinoethane
  sulfonic acid) buffer ..........................................19.52g

**Preparation of MES Buffer Solution:** Add MES buffer to distilled/deionized water and bring volume to 100.0mL. Mix thoroughly. Adjust pH to 5.5. Filter sterilize.

**Na₂SO₃ Solution:**
**Composition** per 10.0mL:
Na₂SO₃..............................................................0.126g

**Preparation of Na₂SO₃ Solution:** Add Na₂SO₃ to distilled/deionized water and bring volume to 10.0mL. Mix thoroughly. Filter sterilize.

**Urea Solution:**
**Composition** per 100.0mL:
Urea.....................................................................6.0g

**Preparation of Urea Solution:** Add urea to distilled/deionized water and bring volume to 100.0mL. Mix thoroughly. Filter sterilize.

**Preparation of Medium:** To 65.0mL of cooled, sterile base agar, aseptically add 10.0mL of sterile yeast dialysate, 20.0mL of horse serum, 2.0mL of sterile penicillin solution, 1.0mL of sterile MES buffer solution, 1.0mL of sterile Na₂SO₃ solution, and 0.5mL of sterile urea solution. Mix thoroughly. Pour into 10mm × 35mm Petri dishes in 5.0mL volumes. Allow plates to stand overnight at 25°C to remove excess surface moisture. Use within 48 h.

**Storage/Shelf Life:** Store dehydrated media in the dark in a sealed container below 30°C. Prepared media should be stored under refrigeration (2-8°C). Media should be used within 60 days of preparation. Media should not be used if there are any signs of deterioration (discoloration) or contamination, or if the expiration date supplied by the manufacturer has passed.

**Use:** For the isolation and cultivation of *Ureaplasma urealyticum.*

## U9 Broth
### (Urease Color Test Medium)
**Composition** per 101.6mL:
U9 base ....................................................... 95.0mL
Horse serum, unheated................................... 5.0mL
Penicillin G solution ...................................... 1.0mL
Urea solution.................................................. 0.5mL
Phenol Red solution ...................................... 0.1mL
pH 6.0 ± 0.2 at 25°C

**U9 Base:**
**Composition** per 100.0mL:
NaCl.................................................................0.63g
Pancreatic digest of casein............................0.425g
Papaic digest of soybean meal .....................0.075g
K₂HPO₄...........................................................0.063g
Glucose ...........................................................0.063g
KH₂PO₄............................................................0.02g

**Preparation of U9 Base:** Add components to distilled/deionized water and bring volume to 100.0mL. Mix thoroughly. Adjust pH to 5.5 with 1*N* HCl. Autoclave for 15 min at 15 psi pressure–121°C. Cool to 45°–50°C.

**Penicillin G Solution:**
**Composition** per 10.0mL:
Penicillin G.....................................................0.63g

**Preparation of Penicillin G Solution:** Add penicillin G to distilled/deionized water and bring volume to 10.0mL. Mix thoroughly. Filter sterilize.

**Urea Solution:**
**Composition** per 30.0mL:
Urea...................................................................3.0g

**Preparation of Urea Solution:** Add urea to distilled/deionized water and bring volume to 30.0mL. Mix thoroughly. Filter sterilize.

**Phenol Red Solution:**
**Composition** per 10.0mL:
Phenol Red........................................................0.1g

**Preparation of Phenol Red Solution:** Add Phenol Red to distilled/deionized water and bring volume to 10.0mL. Mix thoroughly. Filter sterilize.

**Preparation of Medium:** To 95.0mL of cooled, sterile U9 base, aseptically add 5.0mL of sterile horse serum, 1.0mL of sterile penicillin G solution, 0.5mL of sterile urea solution, and 0.1mL of sterile Phenol Red solution. Mix thoroughly. Aseptically distribute into sterile tubes or flasks.

**Storage/Shelf Life:** Store dehydrated media in the dark in a sealed container below 30°C. Prepared media should be stored under refrigeration (2-8°C). Media should be used within 60 days of preparation. Media should not be used if there are any signs of deterioration (discoloration) or contamination, or if the expiration date supplied by the manufacturer has passed.

**Use:** For the isolation and identification of T-strain mycoplasmas from clinical specimens, especially *Ureaplasma urealyticum.* T-mycoplasmas are the only members of the *Mycoplasma* group known to contain urease. Bacteria with urease activity turn the medium dark pink.

## U9 Broth with Amphotericin B
**Composition** per 101.6mL:
U9 base ....................................................... 95.0mL
Horse serum, unheated................................... 5.0mL
Antibiotic solution ........................................ 1.0mL
Urea solution.................................................. 0.5mL
Phenol Red solution....................................... 0.1mL
pH 6.0 ± 0.2 at 25°C

**U9 Base:**
**Composition** per 100.0mL:
NaCl.................................................................0.63g
Pancreatic digest of casein............................0.425g
Papaic digest of soybean meal.......................0.075g
K₂HPO₄...........................................................0.063g
Glucose ...........................................................0.063g
KH₂PO₄............................................................0.02g

**Preparation of U9 Base:** Add components to distilled/deionized water and bring volume to 100.0mL. Mix thoroughly. Adjust pH to 5.5 with 1*N* HCl. Autoclave for 15 min at 15 psi pressure–121°C. Cool to 45°–50°C.

**Antibiotic Solution:**
**Composition** per 10.0mL:
Penicillin G.....................................................0.63g
Amphotericin B ..............................................2.5mg

**Preparation of Antibiotic Solution:** Add penicillin G and amphotericin B to distilled/deionized water and bring volume to 10.0mL. Mix thoroughly. Filter sterilize.

**Urea Solution:**
**Composition** per 30.0mL:
Urea.........................................................................3.0g

**Preparation of Urea Solution:** Add urea to distilled/deionized water and bring volume to 30.0mL. Mix thoroughly. Filter sterilize.

**Phenol Red Solution:**
**Composition** per 10.0mL:
Phenol Red.................................................................0.1g

**Preparation of Phenol Red Solution:** Add Phenol Red to distilled/deionized water and bring volume to 10.0mL. Mix thoroughly. Filter sterilize.

**Preparation of Medium:** To 95.0mL of cooled, sterile U9 base, aseptically add 5.0mL of sterile horse serum, 1.0mL of sterile antibiotic solution, 0.5mL of sterile urea solution, and 0.1mL of sterile Phenol Red solution. Mix thoroughly. Aseptically distribute into sterile tubes or flasks.

**Storage/Shelf Life:** Store dehydrated media in the dark in a sealed container below 30°C. Prepared media should be stored under refrigeration (2-8°C). Media should be used within 60 days of preparation. Media should not be used if there are any signs of deterioration (discoloration) or contamination, or if the expiration date supplied by the manufacturer has passed.

**Use:** For the isolation and identification of T-strain mycoplasmas from clinical specimens, especially *Ureaplasma urealyticum*. T-mycoplasmas are the only members of the *Mycoplasma* group known to contain urease. Bacteria with urease activity turn the medium dark pink.

## U9B Broth

**Composition** per 102.1mL:
U9 base ...............................................................95.0mL
Horse serum, unheated......................................... 5.0mL
Penicillin G solution ............................................ 1.0mL
Urea solution.........................................................0.5mL
L-Cysteine·HCl·H$_2$O solution ............................. 0.5mL
Phenol Red solution .............................................0.1mL

pH 6.0 ± 0.2 at 25°C

**U9 Base:**
**Composition** per 100.0mL:
NaCl......................................................................0.63g
Pancreatic digest of casein..................................0.425g
Papaic digest of soybean meal............................0.075g
K$_2$HPO$_4$................................................................0.063g
Glucose ...............................................................0.063g
KH$_2$PO$_4$..................................................................0.02g

**Preparation of U9 Base:** Add components to distilled/deionized water and bring volume to 100.0mL. Mix thoroughly. Adjust pH to 5.5 with 1$N$ HCl. Autoclave for 15 min at 15 psi pressure–121°C. Cool to 45°–50°C.

**Penicillin G Solution:**
**Composition** per 10.0mL:
Penicillin G .........................................................0.63 g

**Preparation of Penicillin G Solution:** Add penicillin G to distilled/deionized water and bring volume to 10.0mL. Mix thoroughly. Filter sterilize.

**Urea Solution:**
**Composition** per 30.0mL:
Urea.........................................................................3.0g

**Preparation of Urea Solution:** Add urea to distilled/deionized water and bring volume to 30.0mL. Mix thoroughly. Filter sterilize.

**L-Cysteine·HCl·H$_2$O Solution:**
**Composition** per 50.0mL:
L-Cysteine·HCl·H$_2$O .............................................1.0g

**Preparation of L-Cysteine·HCl·H$_2$O Solution:** Add L-cysteine·HCl·H$_2$O to distilled/deionized water and bring volume to 50.0mL. Mix thoroughly. Filter sterilize.

**Phenol Red Solution:**
**Composition** per 10.0mL:
Phenol Red.................................................................0.1g

**Preparation of Phenol Red Solution:** Add Phenol Red to distilled/deionized water and bring volume to 10.0mL. Mix thoroughly. Filter sterilize.

**Preparation of Medium:** To 95.0mL of cooled, sterile U9 base, aseptically add 5.0mL of sterile horse serum, 1.0mL of sterile penicillin G solution, 0.5mL of sterile urea solution, 0.5mL of sterile L-cysteine·HCl·H$_2$O solution, and 0.1mL of sterile Phenol Red solution. Mix thoroughly. Aseptically distribute into sterile tubes or flasks.

**Storage/Shelf Life:** Store dehydrated media in the dark in a sealed container below 30°C. Prepared media should be stored under refrigeration (2-8°C). Media should be used within 60 days of preparation. Media should not be used if there are any signs of deterioration (discoloration) or contamination, or if the expiration date supplied by the manufacturer has passed.

**Use:** For the isolation and identification of T-strain mycoplasmas from clinical specimens, especially *Ureaplasma urealyticum*. T-mycoplasmas are the only members of the *Mycoplasma* group known to contain urease. Bacteria with urease activity turn the medium dark pink.

## U9C Broth

**Composition** per 102.0mL:
U9C base.............................................................. 90.0mL
Horse serum, unheated........................................ 10.0mL
Penicillin G solution ............................................ 1.0mL
Urea solution........................................................ 0.3mL
L-Cysteine·HCl·H$_2$O solution ............................. 0.5mL
GHL tripeptide solution ....................................... 0.1mL
Phenol Red solution............................................. 0.1mL

pH 6.0 ± 0.2 at 25°C

**U9C Base:**
**Composition** per 100.0mL:
NaCl......................................................................0.85g
Pancreatic digest of casein....................................0.25g
Papaic digest of soybean meal...............................0.15g
K$_2$HPO$_4$..................................................................0.12g
Glucose ...................................................................0.12g
MgCl$_2$·6H$_2$O ..........................................................0.2g
Yeast extract............................................................0.1g
KH$_2$PO$_4$..................................................................0.02g

**Preparation of U9C Base:** Add components to distilled/deionized water and bring volume to 100.0mL. Mix thoroughly. Adjust pH to 5.5 with 2$N$ HCl. Autoclave for 15 min at 15 psi pressure–121°C. Cool to 45°–50°C.

**Penicillin G Solution:**
**Composition** per 10.0mL:
Penicillin G ........................................................................0.63g

**Preparation of Penicillin G Solution:** Add penicillin G to distilled/deionized water and bring volume to 10.0mL. Mix thoroughly. Filter sterilize.

**Urea Solution:**
**Composition** per 30.0mL:
Urea.........................................................................................3.0g

**Preparation of Urea Solution:** Add urea to distilled/deionized water and bring volume to 30.0mL. Mix thoroughly. Filter sterilize.

**L-Cysteine·HCl·H$_2$O Solution:**
**Composition** per 50.0mL:
L-Cysteine·HCl·H$_2$O ........................................................1.0g

**Preparation of L-Cysteine·HCl·H$_2$O Solution:** Add L-cysteine·HCl·H$_2$O to distilled/deionized water and bring volume to 50.0mL. Mix thoroughly. Filter sterilize.

**GHL Tripeptide Solution:**
**Composition** per 10.0mL:
GHL tripeptide..................................................................0.2mg

**Preparation of GHL Tripeptide Solution:** Add GHL tripeptide (glycyl-L-histidyl-L-lysine acetate) to distilled/deionized water and bring volume to 10.0mL. Mix thoroughly. Filter sterilize.

**Phenol Red Solution:**
**Composition** per 10.0mL:
Phenol Red...........................................................................0.1g

**Preparation of Phenol Red Solution:** Add Phenol Red to distilled/deionized water and bring volume to 10.0mL. Mix thoroughly. Filter sterilize.

**Preparation of Medium:** To 90.0mL of cooled, sterile U9C base, aseptically add 10.0mL of sterile horse serum, 1.0mL of sterile penicillin G solution, 0.3mL of sterile urea solution, 0.5mL of sterile L-cysteine·HCl·H$_2$O solution, 0.1mL of sterile GHL tripeptide solution, and 0.1mL of sterile Phenol Red solution. Mix thoroughly. Aseptically distribute into sterile tubes or flasks.

**Storage/Shelf Life:** Store dehydrated media in the dark in a sealed container below 30°C. Prepared media should be stored under refrigeration (2-8°C). Media should be used within 60 days of preparation. Media should not be used if there are any signs of deterioration (discoloration) or contamination, or if the expiration date supplied by the manufacturer has passed.

**Use:** For the isolation and identification of T-strain mycoplasmas from clinical specimens, especially *Ureaplasma urealyticum*. T-mycoplasmas are the only members of the *Mycoplasma* group known to contain urease. Bacteria with urease activity turn the medium dark pink.

## Urea Agar
### (Urease Test Agar)
### (Urea Agar Base, Christensen)

**Composition** per liter:
Urea......................................................................................20.0g
Agar .....................................................................................15.0g
NaCl........................................................................................5.0g
KH$_2$PO$_4$..............................................................................2.0g
Peptone..................................................................................1.0g

Glucose ..................................................................................1.0g
Phenol Red........................................................................0.012g
pH 6.8 ± 0.2 at 25°C

**Source:** This medium is available as a premixed powder from BD Diagnostic Systems.

**Preparation of Medium:** Add components, except agar, to distilled/deionized water and bring volume to 100.0mL. Mix thoroughly. Filter sterilize. Add agar to distilled/deionized water and bring volume to 900.0mL. Mix thoroughly. Gently heat and bring to boiling. Autoclave for 15 min at 15 psi pressure–121°C. Cool to 50°C. Aseptically add the 100.0mL of sterile basal medium. Mix thoroughly. Distribute into sterile tubes. Allow tubes to solidify in a slanted position.

**Storage/Shelf Life:** Store dehydrated media in the dark in a sealed container below 30°C. Prepared media should be stored under refrigeration (2-8°C). Media should be used within 60 days of preparation. Media should not be used if there are any signs of deterioration (shrinking, cracking, or discoloration) or contamination, or if the expiration date supplied by the manufacturer has passed.

**Use:** For the differentiation of a variety of microorganisms, especially members of the Enterobacteriaceae, aerobic actinomycetes, streptococci, and nonfermenting Gram-negative bacteria, on the basis of urease production.

## Urea Agar Base

**Composition** per liter:
Agar .....................................................................................15.0g
NaCl........................................................................................5.0g
Na$_2$HPO$_4$.............................................................................1.2g
Peptone..................................................................................1.0g
Glucose ..................................................................................1.0g
KH$_2$PO$_4$..............................................................................0.8g
Phenol Red........................................................................0.012g
Urea solution.................................................................. 50.0mL
pH 6.8 ± 0.2 at 25°C

**Source:** This medium is available as a premixed powder from Oxoid Unipath.

**Urea Solution:**
**Composition** per 100.0mL:
Urea......................................................................................40.0g

**Preparation of Urea Solution:** Add urea to distilled/deionized water and bring volume to 100.0mL. Mix thoroughly. Filter sterilize.

**Preparation of Medium:** Add components, except urea solution, to distilled/deionized water and bring volume to 950.0mL. Mix thoroughly. Gently heat and bring to boiling. Autoclave for 20 min at 10 psi pressure–115°C. Cool to 50°C. Aseptically add 50.0mL of sterile urea solution. Mix thoroughly. Pour into sterile Petri dishes or distribute into sterile tubes. Allow tubes to solidify in a slanted position.

**Storage/Shelf Life:** Store dehydrated media in the dark in a sealed container below 30°C. Prepared media should be stored under refrigeration (2-8°C). Media should be used within 60 days of preparation. Media should not be used if there are any signs of deterioration (shrinking, cracking, or discoloration) or contamination, or if the expiration date supplied by the manufacturer has passed.

**Use:** For the detection of *Proteus* species based on rapid urease activity and the identification of other members of the Enterobacteriaceae based on urease activity. Urease-positive bacteria turn the medium pink.

## Urea Broth 10B for *Ureaplasma urealyticum*

**Composition** per 100.5mL:

| | |
|---|---|
| PPLO broth without Crystal Violet | 70.0mL |
| Horse serum, unheated | 20.0mL |
| Fresh yeast extract solution | 10.0mL |
| L-Cysteine·HCl·H$_2$O solution | 0.5mL |
| CVA enrichment | 0.5mL |
| Urea solution | 0.4mL |
| Phenol Red | 0.1mL |

### PPLO Broth without Crystal Violet:

**Composition** per 900.0mL:

| | |
|---|---|
| Beef heart, solids from infusion | 16.1g |
| Peptone | 3.25g |
| NaCl | 1.61g |

**Preparation of PPLO Broth without Crystal Violet:** Add components to distilled/deionized water and bring volume to 900.0mL. Adjust pH to 5.5 with 2*N* HCl. Autoclave for 15 min at 15 psi pressure–121°C. Cool to 37°C.

### Fresh Yeast Extract Solution:

**Composition** per 100.0mL:

| | |
|---|---|
| Baker's yeast live, pressed, starch-free | 25.0g |

**Preparation of Fresh Yeast Extract Solution:** Add the live Baker's yeast to 100.0mL of distilled/deionized water. Autoclave for 90 min at 15 psi pressure–121°C. Allow to stand. Remove supernatant solution. Adjust pH to 6.6–6.8.

### L-Cysteine·HCl·H$_2$O Solution:

**Composition** per 50.0mL:

| | |
|---|---|
| L-Cysteine·HCl·H$_2$O | 1.0g |

**Preparation of L-Cysteine·HCl·H$_2$O Solution:** Add L-cysteine·HCl·H$_2$O to distilled/deionized water and bring volume to 50.0mL. Mix thoroughly. Filter sterilize.

### CVA Enrichment:

**Composition** per liter:

| | |
|---|---|
| Glucose | 100.0g |
| L-Cysteine·HCl·H$_2$O | 25.9g |
| L-Glutamine | 10.0g |
| Adenine | 1.0g |
| L-Cystine·2HCl | 1.0g |
| Nicotinamide adenine dinucleotide | 0.25g |
| Cocarboxylase | 0.1g |
| Guanine·HCl | 0.03g |
| Fe(NO$_3$)$_3$ | 0.02g |
| Vitamin B$_{12}$ | 0.01g |
| *p*-Aminobenzoic acid | 0.013g |
| Thiamine·HCl | 3.0mg |

**Preparation of CVA Enrichment:** Add components to distilled/deionized water and bring volume to 1.0L. Mix thoroughly. Filter sterilize.

### Urea Solution:

**Composition** per 30.0mL:

| | |
|---|---|
| Urea | 3.0g |

**Preparation of Urea Solution:** Add urea to distilled/deionized water and bring volume to 30.0mL. Mix thoroughly. Filter sterilize.

**Preparation of Medium:** Aseptically combine the components, except the PPLO broth without Crystal Violet. Aseptically add this mixture to the cooled, sterile PPLO broth without Crystal Violet. Mix thoroughly. Aseptically distribute into sterile tubes or flasks.

**Storage/Shelf Life:** Store dehydrated media in the dark in a sealed container below 30°C. Prepared media should be stored under refrigeration (2-8°C). Media should be used within 60 days of preparation. Media should not be used if there are any signs of deterioration (discoloration) or contamination, or if the expiration date supplied by the manufacturer has passed.

**Use:** For the cultivation and maintenance of *Ureaplasma urealyticum* and other *Ureaplasma* species. Urease-positive bacteria turn the medium peach orange.

## Urea Broth Base

**Composition** per liter:

| | |
|---|---|
| NaCl | 5.0g |
| Na$_2$HPO$_4$ | 1.2g |
| Peptone | 1.0g |
| Glucose | 1.0g |
| KH$_2$PO$_4$ | 0.8g |
| Phenol Red | 0.012g |
| Urea solution | 50.0mL |

pH 6.8 ± 0.2 at 25°C

**Source:** This medium is available as a premixed powder from Oxoid Unipath.

### Urea Solution:

**Composition** per 100.0mL:

| | |
|---|---|
| Urea | 40.0g |

**Preparation of Urea Solution:** Add urea to distilled/deionized water and bring volume to 100.0mL. Mix thoroughly. Filter sterilize.

**Preparation of Medium:** Add components, except urea solution, to distilled/deionized water and bring volume to 950.0mL. Mix thoroughly. Autoclave for 20 min at 10 psi pressure–115°C. Cool to 50°C. Aseptically add 50.0mL of sterile urea solution. Mix thoroughly. Aseptically distribute into sterile tubes or flasks.

**Storage/Shelf Life:** Store dehydrated media in the dark in a sealed container below 30°C. Prepared media should be stored under refrigeration (2-8°C). Media should be used within 60 days of preparation. Media should not be used if there are any signs of deterioration (discoloration) or contamination, or if the expiration date supplied by the manufacturer has passed.

**Use:** For the differentiation of members of the Enterobacteriaceae based on urease production. Urease-positive bacteria turn the medium pink.

## Urea HiVeg Agar Base Autoclavable with Urea
## (Christensen HiVeg Agar Autoclavable)

**Composition** per liter:

| | |
|---|---|
| Agar | 15.0g |
| NaCl | 5.0g |
| Na$_2$HPO$_4$ | 1.2g |
| Glucose | 1.0g |
| Plant peptone | 1.0g |
| KH$_2$PO$_4$ | 0.8g |
| Phenol Red | 0.012g |
| Urea solution | 50.0mL |

pH 6.8 ± 0.2 at 25°C

**Source:** This medium, without urea solution, is available as a premixed powder from HiMedia.

**Urea Solution:**
**Composition** per 100.0mL:

Urea ................................................................... 40.0g

**Preparation of Urea Solution:** Add urea to distilled/deionized water and bring volume to 100.0mL. Mix thoroughly. Filter sterilize.

**Preparation of Medium:** Add components, except urea solution, to distilled/deionized water and bring volume to 950.0mL. Mix thoroughly. Gently heat and bring to boiling. Autoclave for 20 min at 10 psi pressure–115°C. Cool to 50°C. Aseptically add 50.0mL of sterile urea solution. Mix thoroughly. Pour into sterile Petri dishes or distribute into sterile tubes. Allow tubes to solidify in a slanted position.

**Storage/Shelf Life:** Store dehydrated media in the dark in a sealed container below 30°C. Prepared media should be stored under refrigeration (2-8°C). Media should be used within 60 days of preparation. Media should not be used if there are any signs of deterioration (shrinking, cracking, or discoloration) or contamination, or if the expiration date supplied by the manufacturer has passed.

**Use:** For the detection of *Proteus* species based on rapid urease activity and the identification of other members of the Enterobacteriaceae based on urease activity. Urease-positive bacteria turn the medium pink. For the detection of urease production, particularly by *Proteus vulgaris*, micrococci, and paracolon organisms.

## Urea R Broth
## (Urea Rapid Broth)

**Composition** per liter:

Urea ..................................................................... 20.0g
Yeast extract .......................................................... 0.1g
Na$_2$HPO$_4$ ....................................................... 0.095g
KH$_2$PO$_4$ ......................................................... 0.091g
Phenol Red ........................................................... 0.01g

pH 6.9 ± 0.2 at 25°C

**Source:** This medium is available as a prepared medium from BD Diagnostic Systems.

**Preparation of Medium:** Add components to distilled/deionized water and bring volume to 1.0L. Mix thoroughly. Filter sterilize. Aseptically distribute into sterile tubes or flasks.

**Storage/Shelf Life:** Store dehydrated media in the dark in a sealed container below 30°C. Prepared media should be stored under refrigeration (2-8°C). Media should be used within 60 days of preparation. Media should not be used if there are any signs of deterioration (discoloration) or contamination, or if the expiration date supplied by the manufacturer has passed.

**Use:** For the differentiation of members of the Enterobacteriaceae based on the rapid detection of urease activity. Urease-positive bacteria turn the medium cerise.

## Urea Test Broth

**Composition** per liter:

Urea ..................................................................... 20.0g
Na$_2$HPO$_4$ ......................................................... 9.5g
KH$_2$PO$_4$ ........................................................... 9.1g
Yeast extract .......................................................... 0.1g
Phenol Red ........................................................... 0.01g
Urea solution ....................................................... 100.0mL

**Urea Solution:**
**Composition** per 100.0mL:

Urea ..................................................................... 20.0g

**Preparation of Urea Solution:** Add urea to distilled/deionized water and bring volume to 100.0mL. Mix thoroughly. Filter sterilize.

**Preparation of Medium:** Add components, except urea solution, to distilled/deionized water and bring volume to 900.0mL. Mix thoroughly. Autoclave for 15 min at 15 psi pressure–121°C. Cool to 45°–50°C. Aseptically add sterile urea solution. Mix thoroughly. Aseptically distribute into sterile tubes in 3.0mL volumes.

**Storage/Shelf Life:** Store dehydrated media in the dark in a sealed container below 30°C. Prepared media should be stored under refrigeration (2-8°C). Media should be used within 60 days of preparation. Media should not be used if there are any signs of deterioration (discoloration) or contamination, or if the expiration date supplied by the manufacturer has passed.

**Use:** For the cultivation and differentiation of members of the Enterobacteriaceae and aerobic actinomycetes based on their production of urease. Bacteria that produce urease turn the medium bright red.

## Urinary Tract Infections Chromogenic Agar
## (UTIC Agar)

**Composition** per liter:

Peptones .............................................................. 16.0g
Agar ..................................................................... 16.0g
Growth factors ..................................................... 13.0g
Tryptophan ............................................................ 2.0g
Chromogenic substrate ........................................... 0.5g

pH 7.2 ± 0.2 at 25°C

**Source:** This medium is available from CONDA, Barcelona, Spain.

**Preparation of Medium:** Add components to distilled/deionized water and bring volume to 1.0L. Mix thoroughly. Distribute into tubes or flasks. Autoclave for 15 min at 15 psi pressure–121°C. Pour into sterile Petri dishes.

**Storage/Shelf Life:** Store dehydrated media in the dark in a sealed container below 30°C. Prepared media should be stored in the dark under refrigeration (2-8°C). Chromogenic media are especially light and temperature sensitive; protect from light, excessive heat, moisture, and freezing. Media should not be used if there are any signs of deterioration (shrinking, cracking, or discoloration) or contamination, or if the expiration date supplied by the manufacturer has passed.

**Use:** For the presumptive detection and differentiation of bacteria that cause urinary tract infections.

## Urogenital *Mycoplasma* Broth Base

**Composition** per liter:

Heart infusion powder ............................................. 8.0g
Casein enzymatic hydrolysate .................................. 8.0g
Yeast extract .......................................................... 4.0g
NaCl ..................................................................... 3.5g
Arginine hydrochloride ............................................ 5.0g
Cysteine hydrochloride ........................................... 0.1g
Phenol Red ........................................................... 0.05g
Horse serum ......................................................... 50.0ml
Urea solution ........................................................ 10.0mL
Vitamin solution ................................................... 10.0mL
Selective supplement solution ................................ 10.0mL

pH 6.3 ± 0.2 at 25°C

**Source:** This medium is available from HiMedia.

**Selective Supplement Solution:**
**Composition** per 10.0mL:

Penicillin .................................................................5.0mg
Amphotericin B.......................................................1.0mg
Penicillin .........................................................100,000U

**Preparation of Selective Supplement Solution:** Add components to distilled/deionized water and bring volume to 10.0mL. Mix thoroughly. Filter sterilize.

**Urea Solution:**
**Composition** per 10.0mL:

Urea.........................................................................0.5g

**Preparation of Urea Solution:** Add urea to distilled/deionized water and bring volume to 10.0mL. Mix thoroughly. Filter sterilize.

**Vitamain Solution:**
**Composition** per 10.0mL:

Glucose ....................................................................2.0g
L-Cysteine·HCl.....................................................0.518g
L-Glutamine ...........................................................0.2g
L-Cystine .............................................................0.022g
Adenine sulfate ....................................................0.02g
Nicotinamide adenine dinucleotide .....................5.0mg
Cocarboxylase......................................................2.0mg
Guanine·HCl ..........................................................0.6g
$Fe(NO_3)_3·6H_2O$........................................................0.4mg
*p*-Aminobenzoic acid.......................................0.26mg
Vitamin $B_{12}$ .........................................................0.2mg
Thiamine·HCl .....................................................0.06mg

**Preparation of Vitamin solution:** Add components to distilled/deionized water and bring volume to 10.0mL. Mix thoroughly. Filter sterilize.

**Preparation of Medium:** Add components, except vitamin solution, urea solution, horse serum, and selective supplement solution, to distilled/deionized water and bring volume to 920.0mL. Mix thoroughly. Autoclave for 15 min at 15 psi pressure–121°C. Cool to 50°C. Aseptically add vitamin solution, urea solution, horse serum, and selective supplement solution. Mix thoroughly. Aseptically distribute into sterile tubes.

**Storage/Shelf Life:** Store dehydrated media in the dark in a sealed container below 30°C. Prepared media should be stored under refrigeration (2-8°C). Media should be used within 60 days of preparation. Media should not be used if there are any signs of deterioration (discoloration) or contamination, or if the expiration date supplied by the manufacturer has passed.

**Use:** For the selective culture of *Mycoplasma hominis* and *Ureaplasma urealyticum.*

## UVM *Listeria* Enrichment Broth
### (University of Vermont *Listeria* Enrichment Broth)
**Composition** per liter:

NaCl.......................................................................20.0g
$Na_2HPO_4$ ..................................................................9.6g
Pancreatic digest of casein...................................5.0g
Peptic digest of animal tissue...............................5.0g
Beef extract ...........................................................5.0g
Yeast extract..........................................................5.0g
$KH_2PO_4$...................................................................1.35g

Esculin ...................................................................1.0g
Nalidixic acid.......................................................40.0mg
Acriflavine·HCl ...................................................12.0mg
pH 7.2 ± 0.2 at 25°C

**Preparation of Medium:** Add components to distilled/deionized water and bring volume to 1.0L. Mix thoroughly. Distribute into tubes or flasks. Autoclave for 15 min at 15 psi pressure–121°C.

**Storage/Shelf Life:** Store dehydrated media in the dark in a sealed container below 30°C. Prepared media should be stored under refrigeration (2-8°C). Media should be used within 60 days of preparation. Media should not be used if there are any signs of deterioration (discoloration) or contamination, or if the expiration date supplied by the manufacturer has passed.

**Use:** For the selective isolation of *Listeria monocytogenes.*

## UVM Modified *Listeria* Enrichment Broth
### (University of Vermont Modified *Listeria* Enrichment Broth)
**Composition** per liter:

NaCl.......................................................................20.0g
$Na_2HPO_4$ ..................................................................9.6g
Pancreatic digest of casein...................................5.0g
Peptic digest of animal tissue...............................5.0g
Beef extract ...........................................................5.0g
Yeast extract..........................................................5.0g
$KH_2PO_4$...................................................................1.35g
Esculin ...................................................................1.0g
Nalidixic acid.......................................................20.0mg
Acriflavine·HCl ...................................................12.0mg
pH 7.2 ± 0.2 at 25°C

**Source:** This medium is available as a premixed powder from BD Diagnostic Systems.

**Preparation of Medium:** Add components to distilled/deionized water and bring volume to 1.0L. Mix thoroughly. Distribute into tubes or flasks. Autoclave for 15 min at 15 psi pressure–121°C.

**Storage/Shelf Life:** Store dehydrated media in the dark in a sealed container below 30°C. Prepared media should be stored under refrigeration (2-8°C). Media should be used within 60 days of preparation. Media should not be used if there are any signs of deterioration (discoloration) or contamination, or if the expiration date supplied by the manufacturer has passed.

**Use:** For the selective isolation of *Listeria monocytogenes.*

## V Agar
**Composition** per liter:

Agar .....................................................................13.5g
Pancreatic digest of casein.................................12.0g
Peptone ................................................................10.0g
Peptic digest of animal tissue...............................5.0g
NaCl.......................................................................5.0g
Beef extract ...........................................................3.0g
Yeast extract..........................................................3.0g
Cornstarch.............................................................1.0g
Human blood, anticoagulated ...........................50.0mL
pH 7.4 ± 0.2 at 25°C

**Source:** This medium is available as a prepared medium from BD Diagnostic Systems.

**Preparation of Medium:** Add components, except human blood, to distilled/deionized water and bring volume to 950.0mL. Mix thoroughly. Gently heat and bring to boiling. Distribute into tubes or flasks. Autoclave for 15 min at 15 psi pressure–121°C. Cool to 50°C. Aseptically add 50.0mL of human blood. Mix thoroughly. Pour into sterile Petri dishes or leave in tubes.

**Storage/Shelf Life:** Store dehydrated media in the dark in a sealed container below 30°C. Prepared media should be stored under refrigeration (2-8°C). Media should be used within 60 days of preparation. Media should not be used if there are any signs of deterioration (shrinking, cracking, or discoloration) or contamination, or if the expiration date supplied by the manufacturer has passed.

**Use:** For the isolation and differentiation of *Gardnerella vaginalis* from clinical specimens. Plates are incubated under an atmosphere with 3–10% $CO_2$. *Gardnerella vaginalis* appears as small white colonies with diffuse β-hemolysis.

### V-8™ Agar

**Composition** per liter:
| | |
|---|---|
| Agar | 15.0g |
| $CaCO_3$ | 4.0g |
| V-8 canned vegetable juice | 200.0mL |

pH 7.3 ± 0.2 at 25°C

**Preparation of Medium:** Add components to distilled/deionized water and bring volume to 1.0L. Mix thoroughly. Gently heat and bring to boiling. Distribute into tubes or flasks. Autoclave for 15 min at 15 psi pressure–121°C. Pour into sterile Petri dishes or leave in tubes.

**Storage/Shelf Life:** Store dehydrated media in the dark in a sealed container below 30°C. Prepared media should be stored under refrigeration (2-8°C). Media should be used within 60 days of preparation. Media should not be used if there are any signs of deterioration (shrinking, cracking, or discoloration) or contamination, or if the expiration date supplied by the manufacturer has passed.

**Use:** For the isolation and cultivation of *Actinomadura* species.

### VACC Agar
### (Remel VACC Agar)

**Composition** per liter:
| | |
|---|---|
| Casein peptone | 15.0g |
| NaCl | 5.0g |
| Soy peptone | 5.0g |
| Vancomycin | 10.0mg |
| Amphotericin B | 2.0mg |
| Ceftazidime | 2.0mg |
| Clindamycin | 1.0mg |
| Sheep blood | 50.0mL |
| Agar | 15.0g |

pH 7.3 ± 0.2 at 25°C

**Source:** This medium is available from Remel and Oxoid.

**Preparation of Medium:** Available as prepared plates.

**Storage/Shelf Life:** Prepared plates should be stored in the dark under refrigeration (2-8°C). Chromogenic agars are especially light and temperature sensitive; protect from light, excessive heat, moisture, and freezing. Media should not be used if there are any signs of deterioration (shrinking, cracking, or discoloration) or contamination, or if the expiration date supplied by the manufacturer has passed.

**Use:** For the primary isolation, selection, and differentiation of Enterobacteriaceae that produce ESBL (extended-spectrum beta-lactamase).

### *Veillonella* Agar

**Composition** per liter:
| | |
|---|---|
| Agar | 15.0g |
| Pancreatic digest of casein | 5.0g |
| Yeast extract | 3.0g |
| Sodium thioglycolate | 0.75g |
| Vancomycin | 7.5mg |
| Basic Fuchsin | 2.0mg |
| Sodium lactate (60% solution) | 21.0mL |

pH 7.5± 0.2 at 25°C

**Source:** This medium is available as a premixed powder from BD Diagnostic Systems.

**Caution:** Basic Fuchsin is a potential carcinogen and care must be taken to avoid inhalation of the powdered dye and contact with the skin.

**Preparation of Medium:** Add components to distilled/deionized water and bring volume to 1.0L. Mix thoroughly. Gently heat and bring to boiling. Distribute into tubes or flasks. Autoclave for 15 min at 15 psi pressure–121°C. Pour into sterile Petri dishes or leave in tubes.

**Storage/Shelf Life:** Store dehydrated media in the dark in a sealed container below 30°C. Prepared media should be stored under refrigeration (2-8°C). Media should be used within 60 days of preparation. Media should not be used if there are any signs of deterioration (shrinking, cracking, or discoloration) or contamination, or if the expiration date supplied by the manufacturer has passed.

**Use:** For the isolation and cultivation of *Veillonella* species.

### *Veillonella* HiVeg Agar Base with Lactate

**Composition** per liter:
| | |
|---|---|
| Agar | 15.0g |
| Plant hydrolysate | 5.0g |
| Yeast extract | 3.0g |
| Na-thioglycolate | 0.75g |
| Basic Fuchsin | 2.0mg |
| Sodium lactate (60% solution) | 21.0mL |

pH 7.5± 0.2 at 25°C

**Source:** This medium, without lactate solution, is available as a premixed powder from HiMedia.

**Caution:** Basic Fuchsin is a potential carcinogen and care must be taken to avoid inhalation of the powdered dye and contact with the skin.

**Preparation of Medium:** Add components to distilled/deionized water and bring volume to 1.0L. Mix thoroughly. Gently heat and bring to boiling. Distribute into tubes or flasks. Autoclave for 15 min at 15 psi pressure–121°C. Mix thoroughly. Pour into sterile Petri dishes or leave in tubes.

**Storage/Shelf Life:** Store dehydrated media in the dark in a sealed container below 30°C. Prepared media should be stored under refrigeration (2-8°C). Media should be used within 60 days of preparation. Media should not be used if there are any signs of deterioration (shrinking, cracking, or discoloration) or contamination, or if the expiration date supplied by the manufacturer has passed.

**Use:** For the selective isolation and cultivation of *Veillonella* species.

## *Veillonella* HiVeg Agar Base with Lactate and Vancomycin

**Composition** per liter:

| | |
|---|---|
| Agar | 15.0g |
| Plant hydrolysate | 5.0g |
| Yeast extract | 3.0g |
| Na-thioglycolate | 0.75g |
| Basic Fuchsin | 2.0mg |
| Sodium lactate (60% solution) | 21.0mL |
| Selective supplement solution | 10.0mL |

pH 7.5± 0.2 at 25°C

**Source:** This medium, without lactate solution, is available as a pre-mixed powder from HiMedia.

**Caution:** Basic Fuchsin is a potential carcinogen and care must be taken to avoid inhalation of the powdered dye and contact with the skin.

**Selective Supplement Solution:**
**Composition** per 10.0mL:

| | |
|---|---|
| Vancomycin | 7.5mg |

**Preparation of Selective Supplement Solution:** Add vancomycin to distilled/deionized water and bring volume to 10.0mL. Mix thoroughly. Filter sterilize.

**Preparation of Medium:** Add components, except selective supplement solution, to distilled/deionized water and bring volume to 990.0mL. Mix thoroughly. Gently heat and bring to boiling. Autoclave for 15 min at 15 psi pressure–121°C. Cool to 45°–50°C. Aseptically add 10.0mL sterile selective supplement solution. Mix thoroughly. Pour into sterile Petri dishes or aseptically distribute into sterile tubes or flasks.

**Storage/Shelf Life:** Store dehydrated media in the dark in a sealed container below 30°C. Prepared media should be stored under refrigeration (2-8°C). Media should be used within 60 days of preparation. Media should not be used if there are any signs of deterioration (shrinking, cracking, or discoloration) or contamination, or if the expiration date supplied by the manufacturer has passed.

**Use:** For the selective isolation and cultivation of *Veillonella* species.

## *Vibrio* Agar

**Composition** per liter:

| | |
|---|---|
| Sucrose | 20.0g |
| Agar | 15.0g |
| NaCl | 10.0g |
| Sodium citrate·2H$_2$O | 10.0g |
| Na$_2$S$_2$O$_3$·5H$_2$O | 6.5g |
| Oxgall | 5.0g |
| Yeast extract | 5.0g |
| Pancreatic digest of casein | 4.0g |
| Proteose peptone | 3.0g |
| Sodium deoxycholate | 1.0g |
| Sodium lauryl sulfate | 0.2g |
| Water Blue | 0.2g |
| Cresol Red | 0.02g |

pH 8.5 ± 0.2 at 25°C

**Preparation of Medium:** Add components to distilled/deionized water and bring volume to 1.0L. Mix thoroughly. Adjust pH to 8.5. Gently heat and bring to boiling. Do not autoclave. Pour into sterile Petri dishes or distribute into sterile tubes.

**Storage/Shelf Life:** Store dehydrated media in the dark in a sealed container below 30°C. Prepared media should be stored under refrigeration (2-8°C). Media should be used within 60 days of preparation. Media should not be used if there are any signs of deterioration (shrinking, cracking, or discoloration) or contamination, or if the expiration date supplied by the manufacturer has passed.

**Use:** For the isolation and cultivation of the *Vibrio cholerae*.

## *Vibrio parahaemolyticus* Agar (VP Agar)

**Composition** per liter:

| | |
|---|---|
| Agar | 20.0g |
| NaCl | 20.0g |
| Sucrose | 20.0g |
| Sodium citrate | 10.0g |
| Na$_2$S$_2$O$_3$·5H$_2$O | 10.0g |
| Peptone | 10.0g |
| Sodium taurocholate | 5.0g |
| Yeast extract | 5.0g |
| Sodium lauryl sulfate | 0.2g |
| Bromthymol Blue | 0.04g |
| Thymol Blue | 0.04g |

pH 8.6 ± 0.2 at 25°C

**Preparation of Medium:** Add components to distilled/deionized water and bring volume to 1.0L. Mix thoroughly. Gently heat and bring to boiling. Do not autoclave. Pour into sterile Petri dishes.

**Storage/Shelf Life:** Store dehydrated media in the dark in a sealed container below 30°C. Prepared media should be stored under refrigeration (2-8°C). Media should be used within 60 days of preparation. Media should not be used if there are any signs of deterioration (shrinking, cracking, or discoloration) or contamination, or if the expiration date supplied by the manufacturer has passed.

**Use:** For the isolation, cultivation, enumeration, and presumptive identification of coliforms in milk, food, and other specimens of sanitary significance. For the enumeration of bacteria in cheese, especially *Pseudomonas fragi, Pseudomonas viscosa,* and *Alcaligenes metalcaligenes.* Sucrose-fermenting bacteria appear as yellow colonies with pale yellow peripheries. Sucrose-nonfermenting bacteria appear as mucoid, green colonies with a dark green center.

## *Vibrio parahaemolyticus* Sucrose Agar (VPSA)

**Composition** per liter:

| | |
|---|---|
| NaCl | 30.0g |
| Agar | 15.0g |
| Sucrose | 10.0g |
| Yeast extract | 7.0g |
| Tryptose | 5.0g |
| Pancreatic digest of casein | 5.0g |
| Bile salts No. 3 | 1.5g |
| Bromthymol Blue | 0.025g |

pH 8.6 ± 0.2 at 25°C

**Preparation of Medium:** Add components to distilled/deionized water and bring volume to 1.0L. Mix thoroughly. Gently heat and bring to boiling. Do not autoclave. Cool to 50°C. Pour into sterile Petri dishes in 20.0mL volumes. Allow plates to dry before using.

**Storage/Shelf Life:** Store dehydrated media in the dark in a sealed container below 30°C. Prepared media should be stored under refriger-

ation (2-8°C). Media should be used within 60 days of preparation. Media should not be used if there are any signs of deterioration (shrinking, cracking, or discoloration) or contamination, or if the expiration date supplied by the manufacturer has passed.

**Use:** For the isolation, cultivation, and differentiation of *Vibrio parahaemolyticus* from seafood. *Vibrio parahaemolyticus* and *Vibrio vulnificus* appear as blue to green colonies. Other *Vibrio* species appear as yellow colonies.

## *Vibrio parahaemolyticus* Sucrose HiVeg Agar
**Composition** per liter:
NaCl ........................................................................30.0g
Agar .......................................................................15.0g
Sucrose ..................................................................10.0g
Yeast extract ...........................................................7.0g
Plant hydrolysate .....................................................5.0g
Plant hydrolysate No. 1 ............................................5.0g
Synthetic detergent No. I .........................................1.5g
Bromthymol Blue ................................................0.025g

pH 8.6 ± 0.2 at 25°C

**Source:** This medium is available as a premixed powder from Hi-Media.

**Preparation of Medium:** Add components to distilled/deionized water and bring volume to 1.0L. Mix thoroughly. Gently heat and bring to boiling. Do not autoclave. Cool to 50°C. Pour into sterile Petri dishes in 20.0mL volumes. Allow plates to dry before using.

**Storage/Shelf Life:** Store dehydrated media in the dark in a sealed container below 30°C. Prepared media should be stored under refrigeration (2-8°C). Media should be used within 60 days of preparation. Media should not be used if there are any signs of deterioration (shrinking, cracking, or discoloration) or contamination, or if the expiration date supplied by the manufacturer has passed.

**Use:** For the isolation, cultivation, and differentiation of *Vibrio parahaemolyticus* from seafood. *Vibrio parahaemolyticus* and *Vibrio vulnificus* appear as blue to green colonies. Other *Vibrio* species appear as yellow colonies.

## *Vibrio vulnificus* Agar (VVA) (BAM M190)
**Composition** per liter:
NaCl ........................................................................30.0g
Agar .......................................................................25.0g
Peptone ...................................................................20.0g
Cellobiose solution .............................................100.0mL
Dye solution ........................................................10.0mL

pH 8.2 ± 0.2 at 25°C

**Dye Solution:**
**Composition** per 100.0mL:
Bromthymol Blue ......................................................0.6g
Ethanol, 70% ......................................................100.0mL

**Preparation of Dye Solution:** Add Bromthymol Blue to 100.0mL of 70% ethanol. Mix thoroughly.

**Cellobiose Solution:**
**Composition** per 100.0mL:
Cellobiose .............................................................10.0g

**Preparation of Cellobiose Solution:** Add cellobiose to distilled/deionized water and bring volume to 100.0mL. Mix thoroughly. Gently heat while mixing to dissolve the cellobiose. Cool. Filter sterilize.

**Preparation of Medium:** Add components, except cellobiose solution, to distilled/deionized water and bring volume to 900.0mL. Mix thoroughly. Gently heat until dissolved. Adjust pH to 8.2. Autoclave for 15 min at 15 psi pressure–121°C. Cool to 50°C. Aseptically add 100.0mL sterile cellobiose solution. Mix thoroughly and pour into sterile Petri dishes. Final color of medium should be light blue.

**Storage/Shelf Life:** Store dehydrated media in the dark in a sealed container below 30°C. Prepared media should be stored under refrigeration (2-8°C). Media should be used within 60 days of preparation. Media should not be used if there are any signs of deterioration (shrinking, cracking, or discoloration) or contamination, or if the expiration date supplied by the manufacturer has passed.

**Use:** For the detection of *Vibrio vulnificus* from seafoods.

## Violet Peptone Bile Lactose Broth
**Composition** per liter:
Lactose ...................................................................10.0g
Peptone ...................................................................10.0g
Bile salts ..................................................................5.0g
Gentian Violet .......................................................0.04g

pH 7.6 ± 0.2 at 25°C

**Preparation of Medium:** Add components to distilled/deionized water and bring volume to 1.0L. Mix thoroughly. Gently heat and bring to boiling. Distribute into tubes or flasks. Autoclave for 15 min at 15 psi pressure–121°C. Pour into sterile Petri dishes or leave in tubes.

**Storage/Shelf Life:** Store dehydrated media in the dark in a sealed container below 30°C. Prepared media should be stored under refrigeration (2-8°C). Media should be used within 60 days of preparation. Media should not be used if there are any signs of deterioration (discoloration) or contamination, or if the expiration date supplied by the manufacturer has passed.

**Use:** For the selective cultivation of members of the Enterobacteriaceae.

## Violet Red Bile Agar
**Composition** per liter:
Agar .......................................................................15.0g
Lactose ...................................................................10.0g
Glucose ...................................................................10.0g
Pancreatic digest of gelatin .......................................7.0g
NaCl ........................................................................5.0g
Yeast extract ...........................................................3.0g
Bile salts ..................................................................1.5g
Neutral Red ...........................................................0.03g
Crystal Violet .......................................................2.0mg

pH 7.4 ± 0.2 at 25°C

**Preparation of Medium:** Add components to distilled/deionized water and bring volume to 1.0L. Mix thoroughly. Gently heat while stirring and bring to boiling. Distribute into tubes or flasks. Autoclave for 15 min at 15 psi pressure–121°C. Pour immediately into sterile Petri dishes or leave in tubes.

**Storage/Shelf Life:** Store dehydrated media in the dark in a sealed container below 30°C. Prepared media should be stored under refrigeration (2-8°C). Media should be used within 60 days of preparation. Media should not be used if there are any signs of deterioration (shrink-

ing, cracking, or discoloration) or contamination, or if the expiration date supplied by the manufacturer has passed.

**Use:** For the isolation and cultivation of members of the Enterobacteriaceae from brined vegetables. For the enumeration of members of the Enterobacteriaceae from brined vegetables by the pour plate technique.

## Violet Red Bile Agar
### (VRB Agar)

**Composition** per liter:

| | |
|---|---|
| Agar | 15.0g |
| Lactose | 10.0g |
| Pancreatic digest of gelatin | 7.0g |
| NaCl | 5.0g |
| Yeast extract | 3.0g |
| Bile salts | 1.5g |
| Neutral Red | 0.03g |
| Crystal Violet | 2.0mg |

pH 7.4 ± 0.2 at 25°C

**Source:** This medium is available as a premixed powder from BD Diagnostic Systems and Oxoid Unipath.

**Preparation of Medium:** Add components to distilled/deionized water and bring volume to 1.0L. Mix thoroughly. Gently heat while stirring and bring to boiling. Distribute into tubes or flasks. Autoclave for 15 min at 15 psi pressure–121°C. Pour immediately into sterile Petri dishes or leave in tubes.

**Storage/Shelf Life:** Store dehydrated media in the dark in a sealed container below 30°C. Prepared media should be stored under refrigeration (2-8°C). Media should be used within 60 days of preparation. Media should not be used if there are any signs of deterioration (shrinking, cracking, or discoloration) or contamination, or if the expiration date supplied by the manufacturer has passed.

**Use:** For the detection of coliform bacteria in water and food.

## Violet Red Bile Agar, HiVeg

**Composition** per liter:

| | |
|---|---|
| Agar | 15.0g |
| Lactose | 10.0g |
| Plant peptone | 7.0g |
| NaCl | 5.0g |
| Yeast extract | 3.0g |
| Synthetic detergent No. I | 1.5g |
| Neutral Red | 0.03g |
| Crystal Violet | 2.0mg |

pH 7.4 ± 0.2 at 25°C

**Source:** This medium is available as a premixed powder from Hi-Media.

**Preparation of Medium:** Add components to distilled/deionized water and bring volume to 1.0L. Mix thoroughly. Gently heat and bring to boiling. Distribute into tubes or flasks. Boil to dissolve components completely. Do not autoclave. Cool to 45°C. Pour into sterile Petri dishes or leave in tubes.

**Storage/Shelf Life:** Store dehydrated media in the dark in a sealed container below 30°C. Prepared media should be stored under refrigeration (2-8°C). Media should be used within 60 days of preparation. Media should not be used if there are any signs of deterioration (shrinking, cracking, or discoloration) or contamination, or if the expiration date supplied by the manufacturer has passed.

**Use:** For the detection of coliform bacteria in water and food.

## Violet Red Bile Agar with MUG

**Composition** per liter:

| | |
|---|---|
| Agar | 15.0g |
| Lactose | 10.0g |
| Pancreatic digest of gelatin | 7.0g |
| NaCl | 5.0g |
| Yeast extract | 3.0g |
| Bile salts | 1.5g |
| MUG (4-methylumbelliferyl-β-D-glucuronide) | 0.1g |
| Neutral Red | 0.03g |
| Crystal Violet | 2.0mg |

pH 7.4 ± 0.2 at 25°C

**Source:** This medium is available as a premixed powder from BD Diagnostic Systems.

**Preparation of Medium:** Add components to distilled/deionized water and bring volume to 1.0L. Mix thoroughly. Gently heat while stirring and bring to boiling. Distribute into tubes or flasks. Autoclave for 15 min at 15 psi pressure–121°C. Pour immediately into sterile Petri dishes or leave in tubes.

**Storage/Shelf Life:** Store dehydrated media in the dark in a sealed container below 30°C. Prepared media should be stored in the dark under refrigeration (2-8°C). Chromogenic media are especially light and temperature sensitive; protect from light, excessive heat, moisture, and freezing. Media should not be used if there are any signs of deterioration (shrinking, cracking, or discoloration) or contamination, or if the expiration date supplied by the manufacturer has passed.

**Use:** For the differentiation of *Escherichia coli* from dairy products and other foods based on their ability to produce β-glucuronidase.

## Violet Red Bile Glucose Agar

**Composition** per liter:

| | |
|---|---|
| Agar | 12.0g |
| Glucose | 10.0g |
| Peptone | 7.0g |
| NaCl | 5.0g |
| Yeast extract | 3.0g |
| Bile salts No. 3 | 1.5g |
| Neutral Red | 0.03g |
| Crystal Violet | 2.0mg |

pH 7.4 ± 0.2 at 25°C

**Source:** This medium is available as a premixed powder from Oxoid Unipath.

**Preparation of Medium:** Add components to distilled/deionized water and bring volume to 1.0L. Mix thoroughly. Gently heat and bring to boiling. Do not autoclave. Pour into sterile Petri dishes or distribute into sterile tubes.

**Storage/Shelf Life:** Store dehydrated media in the dark in a sealed container below 30°C. Prepared media should be stored under refrigeration (2-8°C). Media should be used within 60 days of preparation. Media should not be used if there are any signs of deterioration (shrinking, cracking, or discoloration) or contamination, or if the expiration date supplied by the manufacturer has passed.

**Use:** For the detection and enumeration of Enterobacteriaceae from foods.

## Violet Red Glucose HiVeg Agar with Lactose

**Composition** per liter:

| | |
|---|---|
| Agar | 15.0g |
| Lactose monohydrate | 9.5g |
| Glucose monohydrate | 9.09g |
| Plant peptone | 7.0g |
| NaCl | 5.0g |
| Yeast extract | 3.0g |
| Synthetic detergent No. I | 1.5g |
| Neutral Red | 0.03g |
| Crystal Violet | 2.0mg |

pH 7.4 ± 0.2 at 25°C

**Source:** This medium is available as a premixed powder from Hi-Media.

**Preparation of Medium:** Add components to distilled/deionized water and bring volume to 1.0L. Mix thoroughly. Gently heat and bring to boiling. Do not autoclave. Pour into sterile Petri dishes or distribute into sterile tubes.

**Storage/Shelf Life:** Store dehydrated media in the dark in a sealed container below 30°C. Prepared media should be stored under refrigeration (2-8°C). Media should be used within 60 days of preparation. Media should not be used if there are any signs of deterioration (shrinking, cracking, or discoloration) or contamination, or if the expiration date supplied by the manufacturer has passed.

**Storage/Shelf Life:** Store dehydrated media in the dark in a sealed container below 30°C. Prepared media should be stored under refrigeration (2-8°C). Media should be used within 60 days of preparation. Media should not be used if there are any signs of deterioration (shrinking, cracking, or discoloration) or contamination, or if the expiration date supplied by the manufacturer has passed.

**Use:** For the detection and enumeration of Enterobacteriaceae from raw foods.

## Violet Red Glucose HiVeg Agar without Lactose

**Composition** per liter:

| | |
|---|---|
| Agar | 12.0g |
| Glucose | 10.0g |
| Plant peptone | 7.0g |
| NaCl | 5.0g |
| Yeast extract | 3.0g |
| Synthetic detergent No. I | 1.5g |
| Neutral Red | 0.03g |
| Crystal Violet | 2.0mg |

pH 7.4 ± 0.2 at 25°C

**Source:** This medium is available as a premixed powder from Hi-Media.

**Preparation of Medium:** Add components to distilled/deionized water and bring volume to 1.0L. Mix thoroughly. Gently heat and bring to boiling. Do not autoclave. Pour into sterile Petri dishes or distribute into sterile tubes.

**Storage/Shelf Life:** Store dehydrated media in the dark in a sealed container below 30°C. Prepared media should be stored under refrigeration (2-8°C). Media should be used within 60 days of preparation. Media should not be used if there are any signs of deterioration (shrinking, cracking, or discoloration) or contamination, or if the expiration date supplied by the manufacturer has passed.

**Use:** For the detection and enumeration of Enterobacteriaceae from raw foods.

## Violet Red HiVeg Agar

**Composition** per liter:

| | |
|---|---|
| Agar | 15.0g |
| Lactose | 10.0g |
| Plant peptone | 7.0g |
| NaCl | 5.0g |
| Yeast extract | 3.0g |
| Synthetic detergent No. I | 1.5g |
| Neutral Red | 0.03g |
| Crystal Violet | 2.0mg |

pH 7.4 ± 0.2 at 25°C

**Source:** This medium is available as a premixed powder from Hi-Media.

**Preparation of Medium:** Add components to distilled/deionized water and bring volume to 1.0L. Mix thoroughly. Gently heat and bring to boiling. Do not autoclave. Pour into sterile Petri dishes or distribute into sterile tubes.

**Storage/Shelf Life:** Store dehydrated media in the dark in a sealed container below 30°C. Prepared media should be stored under refrigeration (2-8°C). Media should be used within 60 days of preparation. Media should not be used if there are any signs of deterioration (shrinking, cracking, or discoloration) or contamination, or if the expiration date supplied by the manufacturer has passed.

**Use:** For the detection and enumeration of Enterobacteriaceae from foods. Recommended by the ISO Committee for selective isolation and enumeration of coli-aerogenes bacteria in water. For the detection and enumeration of coliforms from water and food.

## Violet Red HiVeg Agar (1.2%)

**Composition** per liter:

| | |
|---|---|
| Agar | 12.0g |
| Lactose | 10.0g |
| Plant peptone | 7.0g |
| NaCl | 5.0g |
| Yeast extract | 3.0g |
| Synthetic detergent No. I | 1.5g |
| Neutral Red | 0.03g |
| Crystal Violet | 2.0mg |

pH 7.4 ± 0.2 at 25°C

**Source:** This medium is available as a premixed powder from Hi-Media.

**Preparation of Medium:** Add components to distilled/deionized water and bring volume to 1.0L. Mix thoroughly. Gently heat and bring to boiling. Do not autoclave. Pour into sterile Petri dishes or distribute into sterile tubes.

**Storage/Shelf Life:** Store dehydrated media in the dark in a sealed container below 30°C. Prepared media should be stored under refrigeration (2-8°C). Media should be used within 60 days of preparation. Media should not be used if there are any signs of deterioration (shrinking, cracking, or discoloration) or contamination, or if the expiration date supplied by the manufacturer has passed.

**Use:** For the detection and enumeration of Enterobacteriaceae from foods. Recommended by the ISO Committee for selective isolation and enumeration of coli-aerogenes bacteria in water. For the detection and enumeration of coliforms from water and food.

## Violet Red HiVeg Broth

**Composition** per liter:

| | |
|---|---|
| Plant peptone | 7.0g |
| NaCl | 5.0g |
| Yeast extract | 3.0g |
| Lactose | 1.5g |
| Synthetic detergent No. 1 | 1.5g |
| Neutral Red | 0.03g |
| Crystal Violet | 2.0mg |

pH 7.4 ± 0.2 at 25°C

**Source:** This medium is available as a premixed powder from Hi-Media.

**Preparation of Medium:** Add components to distilled/deionized water and bring volume to 1.0L. Mix thoroughly. Gently heat and bring to boiling. Do not autoclave.

**Storage/Shelf Life:** Store dehydrated media in the dark in a sealed container below 30°C. Prepared media should be stored under refrigeration (2-8°C). Media should be used within 60 days of preparation. Media should not be used if there are any signs of deterioration (discoloration) or contamination, or if the expiration date supplied by the manufacturer has passed.

**Use:** For the isolation and detection of coliforms from water, milk, and other foods.

## VMGII Medium (Viability-Preserving Microbiostatic Medium)

**Composition** per 1100.0mL:

| | |
|---|---|
| Solution 1 | 900.0mL |
| Solution 2 | 100.0mL |
| Salt stock solution | 100.0mL |

**Solution 1:**

**Composition** per 900.0mL:

| | |
|---|---|
| Noble agar | 0.1g |

**Preparation of Solution 1:** Add agar to distilled/deionized water and bring volume to 900.0mL. Mix thoroughly. Gently heat and bring to boiling. Cool to 45°–50°C.

**Solution 2:**

**Composition** per 100.0mL:

| | |
|---|---|
| Charcoal, bacteriological | 10.0g |
| Gelatin peptone | 10.0g |
| Meat peptone | 1.0g |
| Cysteine·HCl | 0.5g |
| Thioglycolic acid | 0.5mL |

**Preparation of Solution 2:** Add components to distilled/deionized water and bring volume to 100.0mL. Mix thoroughly.

**Stock Salt Solution:**

**Composition** per liter:

| | |
|---|---|
| Sodium glycerophosphate | 100.0g |
| NaCl | 10.0g |
| KCl | 4.2g |
| CaCl$_2$·6H$_2$O | 2.4g |
| MgSO$_4$·7H$_2$O | 1.0g |
| Phenylmercuric acetate | 0.03g |

**Preparation of Stock Salt Solution:** Add phenylmercuric acetate to approximately 800.0mL of distilled/deionized water. Gently heat. Add remaining components. Bring volume to 1.0L with distilled/deionized water.

**Preparation of Medium:** To 900.0mL of cooled solution 1, add 100.0mL of solution 2 and 100.0mL of stock salt solution. Mix thoroughly. Distribute into screw-capped tubes. Autoclave for 15 min at 15 psi pressure–121°C.

**Storage/Shelf Life:** Store dehydrated media in the dark in a sealed container below 30°C. Prepared media should be stored under refrigeration (2-8°C). Media should be used within 60 days of preparation. Media should not be used if there are any signs of deterioration (discoloration) or contamination, or if the expiration date supplied by the manufacturer has passed.

**Use:** For the isolation and cultivation of oral streptococci, including *Streptococcus mutans* and *Streptococcus sanguis*, and nonspore-forming bacteria, including *Lactobacillus* species from human dental plaque.

## Vogel and Johnson Agar

**Composition** per liter:

| | |
|---|---|
| Agar | 16.0g |
| Pancreatic digest of casein | 10.0g |
| D-Mannitol | 10.0g |
| Glycine | 10.0g |
| Yeast extract | 5.0g |
| K$_2$HPO$_4$ | 5.0g |
| LiCl | 5.0g |
| Phenol Red | 0.025g |
| K$_2$TeO$_3$ solution | 20.0mL |

pH 7.2 ± 0.2 at 25°C

**Source:** This medium is available as a premixed powder from BD Diagnostic Systems and Oxoid Unipath.

**Caution:** Lithium chloride is harmful. Avoid bodily contact and inhalation of vapors. On contact with skin wash with plenty of water immediately.

**K$_2$TeO$_3$ Solution:**

**Composition** per 100.0mL:

| | |
|---|---|
| K$_2$TeO$_3$ | 1.0g |

**Preparation of K$_2$TeO$_3$ Solution:** Add K$_2$TeO$_3$ to distilled/deionized water and bring volume to 100.0mL. Mix thoroughly. Filter sterilize.

**Caution:** Potassium tellurite is toxic.

**Preparation of Medium:** Add components, except K$_2$TeO$_3$ solution, to distilled/deionized water and bring volume to 980.0mL. Mix thoroughly. Gently heat and bring to boiling. Autoclave for 15 min at 15 psi pressure–121°C. Cool to 45°–50°C. Aseptically add 20.0mL of sterile K$_2$TeO$_3$ solution. Mix thoroughly. Pour into sterile Petri dishes or distribute into sterile tubes.

**Storage/Shelf Life:** Store dehydrated media in the dark in a sealed container below 30°C. Prepared media should be stored under refrigeration (2-8°C). Media should be used within 60 days of preparation. Media should not be used if there are any signs of deterioration (shrinking, cracking, or discoloration) or contamination, or if the expiration date supplied by the manufacturer has passed.

**Use:** For the detection of coagulase-positive *Staphylococcus aureus*.

## Vogel-Johnson Agar Base, HiVeg

**Composition** per liter:

| | |
|---|---|
| Agar | 16.0g |
| K$_2$HPO$_4$ | 5.0g |

| | |
|---|---|
| Glycine | 10.0g |
| Plant hydrolysate | 10.0g |
| Mannitol | 10.0g |
| LiCl | 5.0g |
| Yeast extract | 5.0g |
| Phenol Red | 0.025g |
| $K_2TeO_3$ solution | 20.0mL |

pH 7.2 ± 0.2 at 25°C

**Source:** This medium, without tellurite, is available as a premixed powder from HiMedia.

**Caution:** Lithium chloride is harmful. Avoid bodily contact and inhalation of vapors. On contact with skin wash with plenty of water immediately.

**$K_2TeO_3$ Solution:**
**Composition** per 100.0mL:

| | |
|---|---|
| $K_2TeO_3$ | 1.0g |

**Preparation of $K_2TeO_3$ Solution:** Add $K_2TeO_3$ to distilled/deionized water and bring volume to 100.0mL. Mix thoroughly. Filter sterilize.

**Caution:** Potassium tellurite is toxic.

**Preparation of Medium:** Add components, except $K_2TeO_3$ solution, to distilled/deionized water and bring volume to 980.0mL. Mix thoroughly. Gently heat and bring to boiling. Autoclave for 15 min at 15 psi pressure–121°C. Cool to 45°–50°C. Aseptically add 20.0mL of sterile $K_2TeO_3$ solution. Mix thoroughly. Pour into sterile Petri dishes or distribute into sterile tubes.

**Storage/Shelf Life:** Store dehydrated media in the dark in a sealed container below 30°C. Prepared media should be stored under refrigeration (2-8°C). Media should be used within 60 days of preparation. Media should not be used if there are any signs of deterioration (shrinking, cracking, or discoloration) or contamination, or if the expiration date supplied by the manufacturer has passed.

**Use:** For the detection of coagulase-positive *Staphylococcus aureus*.

## VP Broth, Modified, Smith, Gordon, and Clark

**Composition** per liter:

| | |
|---|---|
| Proteose peptone | 7.0g |
| Glucose | 5.0g |
| NaCl | 5.0g |

**Preparation of Medium:** Add components to distilled/deionized water and bring volume to 1.0L. Mix thoroughly. Distribute into tubes in 5.0mL volumes. Autoclave for 15 min at 15 psi pressure–121°C.

**Storage/Shelf Life:** Store dehydrated media in the dark in a sealed container below 30°C. Prepared media should be stored under refrigeration (2-8°C). Media should be used within 60 days of preparation. Media should not be used if there are any signs of deterioration (discoloration) or contamination, or if the expiration date supplied by the manufacturer has passed.

**Use:** For the isolation and cultivation of *Bacillus cereus* from foods.

## VP HiVeg Medium

**Composition** per liter:

| | |
|---|---|
| Agar | 20.0g |
| NaCl | 20.0g |
| Sucrose | 20.0g |
| Plant peptone | 10.0g |

| | |
|---|---|
| Sodium citrate | 10.0g |
| $Na_2S_2O_3$ | 10.0g |
| Synthetic detergent No. V | 5.0g |
| Yeast extract | 5.0g |
| Sodium lauryl sulfate | 0.2g |
| Bromthymol Blue | 0.04g |
| Thymol Blue | 0.04g |

pH 6.9 ± 0.2 at 25°C

**Source:** This medium is available as a premixed powder from Hi-Media.

**Preparation of Medium:** Add components to distilled/deionized water and bring volume to 1.0L. Mix thoroughly. Adjust pH to 6.9. Distribute into tubes in 3.0mL volumes. Autoclave for 15 min at 15 psi pressure–121°C.

**Storage/Shelf Life:** Store dehydrated media in the dark in a sealed container below 30°C. Prepared media should be stored under refrigeration (2-8°C). Media should be used within 60 days of preparation. Media should not be used if there are any signs of deterioration (shrinking, cracking, or discoloration) or contamination, or if the expiration date supplied by the manufacturer has passed.

**Use:** For the cultivation and differentiation of bacteria based on their ability to produce acetoin.

## VP Medium (Voges-Proskauer Medium)

**Composition** per liter:

| | |
|---|---|
| Peptone | 7.0g |
| $K_2HPO_4$ | 5.0g |
| Glucose | 5.0g |

pH 6.9 ± 0.2 at 25°C

**Preparation of Medium:** Add components to distilled/deionized water and bring volume to 1.0L. Mix thoroughly. Adjust pH to 6.9. Distribute into tubes in 3.0mL volumes. Autoclave for 15 min at 15 psi pressure–121°C.

**Storage/Shelf Life:** Store dehydrated media in the dark in a sealed container below 30°C. Prepared media should be stored under refrigeration (2-8°C). Media should be used within 60 days of preparation. Media should not be used if there are any signs of deterioration (discoloration) or contamination, or if the expiration date supplied by the manufacturer has passed.

**Use:** For the cultivation and differentiation of bacteria based on their ability to produce acetoin.

## VRB Agar, Fluorocult (Fluorocult VRB Agar)

**Composition** per liter:

| | |
|---|---|
| Agar | 13.0g |
| Lactose | 10.0g |
| Peptone from meat | 7.0g |
| NaCl | 5.0g |
| Yeast extract | 3.0g |
| Bile salts mixture | 1.5g |
| 4-Methylumbelliferyl-β-D-glucuronide | 0.1g |
| Neutral Red | 0.03g |
| Crystal Violet | 0.002g |

pH 7.4 ± 0.2 at 25°C

**Source:** This medium is available from Merck.

**Preparation of Medium:** Add components to distilled/deionized water and bring volume to 1.0L. Mix thoroughly. Heat in a boiling water bath or in free flowing steam with frequent stirring until completely dissolved. Do not boil for more than 2 min. Do not autoclave. Do not overheat. Pour into sterile Petri dishes. The plates are clear and dark red.

**Storage/Shelf Life:** Store dehydrated media in the dark in a sealed container below 30°C. Prepared media should be stored in the dark under refrigeration (2-8°C). Chromogenic media are especially light and temperature sensitive; protect from light, excessive heat, moisture, and freezing. Media should not be used if there are any signs of deterioration (shrinking, cracking, or discoloration) or contamination, or if the expiration date supplied by the manufacturer has passed.

**Use:** For the detection and enumeration of coliform bacteria, in particular *E. coli.* Crystal Violet and bile salts largely inhibit the growth of Gram-positive accompanying bacterial flora. Lactose-positive colonies show a color change to red of the pH indicator. *E. coli* colonies show a fluorescence under UV light. Lactose-negative Enterobacteriaceae are colorless. Lactose-positive colonies are red and often surrounded by a turbid zone due to the precipitation of bile acids. Light blue fluorescing colonies denote *E. coli.*

## VRB MUG Agar
### (Violet Red Bile Lactose MUG Agar)
**Composition** per liter:

| | |
|---|---|
| Agar | 13.0g |
| Lactose | 10.0g |
| Meat peptone | 7.0g |
| NaCl | 5.0g |
| Yeast extract | 3.0g |
| Bile salts mixture | 1.5g |
| 4-Methylumbelliferyl-β-D-glucuronide | 0.1g |
| Neutral Red | 0.03g |
| Crystal Violet | 0.002g |

pH 7.4 ± 0.2 at 37°C

**Source:** This medium is available from Fluka, Sigma-Aldrich.

**Preparation of Medium:** Add components to distilled/deionized water and bring volume to 1.0L. Mix thoroughly. Gently heat while stirring and bring to boiling. Autoclave for 15 min at 15 psi pressure–121°C. Cool to 50°C. Pour into sterile Petri dishes.

**Storage/Shelf Life:** Store dehydrated media in the dark in a sealed container below 30°C. Prepared media should be stored in the dark under refrigeration (2-8°C). Chromogenic media are especially light and temperature sensitive; protect from light, excessive heat, moisture, and freezing. Media should not be used if there are any signs of deterioration (shrinking, cracking, or discoloration) or contamination, or if the expiration date supplied by the manufacturer has passed.

**Use:** For the detection and enumeration of coliform bacteria, in particular *E. coli.* Gram-positive accompanying flora are extensively inhibited by Crystal Violet and bile salts. A color change to red indicates lactose-positive colonies, within which *E. coli* can be demonstrated by fluorescence in the UV.

## VRE Agar
**Composition** per 1004.0mL:

| | |
|---|---|
| Tryptone | 20.0g |
| Agar | 10.0g |
| Yeast extract | 5.0g |
| NaCl | 5.0g |
| Sodium citrate | 1.0g |
| Esculin | 1.0g |
| Ferric ammonium citrate | 0.5g |
| NaN₃ | 0.15g |
| Selective supplement solution | 4.0mL |

pH 7.0 ± 0.2 at 25°C

**Source:** This medium is available as a premixed powder from Oxoid Unipath.

**Selective Supplement Solution:**
**Composition** per 4.0mL:

| | |
|---|---|
| Meropenum | 1.0mg |
| Vancomycin | 6.0mg |

**Preparation of Selective Supplement Solution:** Add components to distilled/deionized water and bring volume to 4.0mL. Mix thoroughly. Filter sterilize.

**Preparation of Medium:** Add components, except selective supplement solution, to distilled/deionized water and bring volume to 1.0L. Mix thoroughly. Gently heat while stirring and bring to boiling. Autoclave for 15 min at 15 psi pressure–121°C. Cool to 50°C. Aseptially add 4.0mL selective supplement solution. Mix thoroughly . Pour into sterile Petri dishes.

**Storage/Shelf Life:** Store dehydrated media in the dark in a sealed container below 30°C. Prepared media should be stored under refrigeration (2-8°C). Media should be used within 60 days of preparation. Media should not be used if there are any signs of deterioration (shrinking, cracking, or discoloration) or contamination, or if the expiration date supplied by the manufacturer has passed.

**Use:** For the isolation of vancomycin resistant enterococci (VRE) from clinical samples. Nonresistant enterococci containing the *Van C* genes will not grow on this medium. The selective supplement suppresses growth of Gram-negative bacteria and *E. gallinarum.* The medium contains an indicator system to detect the growth of esculin-hydrolyzing organisms. Enterococci produce black zones around the colonies from the formation of black iron phenolic compounds derived from esculin-hydrolyis products and ferrous iron.

## VRE Agar
**Composition** per 1004.0mL:

| | |
|---|---|
| Tryptone | 20.0g |
| Agar | 10.0g |
| Yeast extract | 5.0g |
| NaCl | 5.0g |
| Sodium citrate | 1.0g |
| Esculin | 1.0g |
| Ferric ammonium citrate | 0.5g |
| NaN₃ | 0.15g |
| Selective supplement solution | 4.0mL |

pH 7.0 ± 0.2 at 25°C

**Source:** This medium is available as a premixed powder from Oxoid Unipath.

**Selective Supplement Solution:**
**Composition** per 4.0mL:

| | |
|---|---|
| Gentamicin | 512.0mg |

**Preparation of Selective Supplement Solution:** Add gentamicin to distilled/deionized water and bring volume to 4.0mL. Mix thoroughly. Filter sterilize.

**Preparation of Medium:** Add components, except selective supplement solution, to distilled/deionized water and bring volume to

1.0L. Mix thoroughly. Gently heat while stirring and bring to boiling. Autoclave for 15 min at 15 psi pressure–121°C. Cool to 50°C. Aseptially add 4.0mL selective supplement solution. Mix thoroughly . Pour into sterile Petri dishes.

**Storage/Shelf Life:** Store dehydrated media in the dark in a sealed container below 30°C. Prepared media should be stored under refrigeration (2-8°C). Media should be used within 60 days of preparation. Media should not be used if there are any signs of deterioration (shrinking, cracking, or discoloration) or contamination, or if the expiration date supplied by the manufacturer has passed.

**Use:** For the isolation of high-level aminoglycoside-resistant enterococci (HLARE) from clinical samples. Nonresistant enterococci containing the *Van* C genes will not grow on this medium.The selective supplement suppresses growth of Gram-negative bacteria and *E. gallinarum*. The medium contains an indicator system to detect the growth of esculin-hydrolyzing organisms. Enterococci produce black zones around the colonies from the formation of black iron phenolic compounds derived from esculin hydrolyis products and ferrous iron.

### VRE Broth

**Composition** per 1004.0mL:

| | |
|---|---|
| Calf brain infusion solids | 12.5g |
| Proteose peptone | 10.0g |
| Beef heart infusion solids | 5.0g |
| NaCl | 5.0g |
| $Na_2HPO_4$ | 2.5g |
| Glucose | 2.0g |
| Selective supplement solution | 4.0mL |

pH 7.4 ± 0.2 at 25°C

**Source:** This medium is available as a premixed powder from Oxoid Unipath.

**Selective Supplement Solution:**
**Composition** per 4.0mL:

| | |
|---|---|
| Meropenum | 2.0mg |

**Preparation of Selective Supplement Solution:** Add meropenum to distilled/deionized water and bring volume to 4.0mL. Mix thoroughly. Filter sterilize.

**Preparation of Medium:** Add components, except selective supplement solution, to distilled/deionized water and bring volume to 1.0L. Mix thoroughly. Gently heat while stirring and bring to boiling. Autoclave for 15 min at 15 psi pressure–121°C. Cool to 50°C. Aseptially add 4.0mL selective supplement solution. Mix thoroughly. Aseptically distribute into sterile tubes.

**Storage/Shelf Life:** Store dehydrated media in the dark in a sealed container below 30°C. Prepared media should be stored under refrigeration (2-8°C). Media should be used within 60 days of preparation. Media should not be used if there are any signs of deterioration (discoloration) or contamination, or if the expiration date supplied by the manufacturer has passed.

**Use:** For the isolation of high-level aminoglycoside-resistant enterococci (HLARE) from clinical samples. Nonresistant enterococci will not grow on this medium. The selective supplement suppresses growth of Gram-negative bacteria and *E. gallinarum*.

### VRE*Select*™ Medium

**Composition** per liter:
Proprietary

**Source:** This medium is available from Bio-Rad.

**Use:** For the detection of vancomycin resistant enterococci from rectal swabs and fecal specimens. *Enterococcus faecium* form pink colonies. *E. faecalis* form blue colonies. Confirmation of *E. faecalis* as catalase negative and vancomycin sensitivity testing is required.

### Wagatsuma Agar

**Composition** per 1050.0mL:

| | |
|---|---|
| NaCl | 70.0g |
| Agar | 15.0g |
| Mannitol | 10.0g |
| Peptone | 10.0g |
| $K_2HPO_4$ | 5.0g |
| Yeast extract | 3.0g |
| Crystal Violet | 1.0mg |
| Red blood cells | 50.0mL |

pH 8.0 ± 0.2 at 25°C

**Red Blood Cells:**
**Composition** per 100.0mL:

| | |
|---|---|
| Blood, human or rabbit | 100.0mL |

**Preparation of Red Blood Cells:** Mix freshly drawn human or rabbit blood with anticoagulant and an equal volume of sterile 0.85% saline solution. Centrifuge cells at 4000 × g at 4°C for 15 min. Pour off saline and wash two more times with sterile saline. After last wash, pour off saline and resuspend cells to their original volume.

**Preparation of Medium:** Add components, except blood, to distilled/deionized water and bring volume to 1.0L. Mix thoroughly. Adjust pH to 8.0. Place in a steam bath for 30 min. Do not autoclave. Cool to 45°–50°C. Add 50.0mL of washed red blood cells. Mix thoroughly. Pour into sterile Petri dishes. Dry plates before using.

**Storage/Shelf Life:** Store dehydrated media in the dark in a sealed container below 30°C. Prepared media should be stored under refrigeration (2-8°C). Media should be used within 60 days of preparation. Media should not be used if there are any signs of deterioration (shrinking, cracking, or discoloration) or contamination, or if the expiration date supplied by the manufacturer has passed.

**Use:** For the cultivation and detection of thermostable hemolysin of *Vibrio parahaemolyticus* by the Kanagawa reaction.

### Wagatsuma HiVeg Agar Base with Red Blood Cells

**Composition** per 1050.0mL:

| | |
|---|---|
| NaCl | 70.0g |
| Agar | 15.0g |
| Plant peptone | 10.0g |
| Mannitol | 10.0g |
| $K_2HPO_4$ | 5.0g |
| Yeast extract | 3.0g |
| Crystal Violet | 1.0mg |
| Red blood cells | 50.0mL |

pH 8.0 ± 0.2 at 25°C

**Source:** This medium, without red blood cells, is available as a premixed powder from HiMedia.

**Red Blood Cells:**
**Composition** per 100.0mL:

| | |
|---|---|
| Blood, human or rabbit | 100.0mL |

**Preparation of Red Blood Cells:** Mix freshly drawn human or rabbit blood with anticoagulant and an equal volume of sterile 0.85% saline solution. Centrifuge cells at 4000 × g at 4°C for 15 min. Pour off saline and wash two more times with sterile saline. After last wash, pour off saline and resuspend cells to their original volume.

**Preparation of Medium:** Add components, except blood, to distilled/deionized water and bring volume to 1.0L. Mix thoroughly. Adjust pH to 8.0. Place in a steam bath for 30 min. Do not autoclave. Cool to 45°–50°C. Add 50.0mL of washed red blood cells. Mix thoroughly. Pour into sterile Petri dishes. Dry plates before using.

**Storage/Shelf Life:** Store dehydrated media in the dark in a sealed container below 30°C. Prepared media should be stored under refrigeration (2-8°C). Media should be used within 60 days of preparation. Media should not be used if there are any signs of deterioration (shrinking, cracking, or discoloration) or contamination, or if the expiration date supplied by the manufacturer has passed.

**Use:** For the cultivation and detection of thermostable hemolysin of *Vibrio parahaemolyticus* by the Kanagawa reaction.

## Wallenstein Medium
**Composition** per 4.225L:
Malachite Green.............................................................0.75g
Egg yolk emulsion ......................................................3.125L
Glycerol ....................................................................100.0mL
pH 6.75 ± 0.2 at 25°C

**Egg Yolk Emulsion:**
**Composition**:
Chicken egg yolks................................................................66
Whole chicken egg................................................................6

**Preparation of Egg Yolk Emulsion:** Soak eggs with 1:100 dilution of saturated mercuric chloride solution for 1 min. Crack eggs and separate yolks from whites. Mix egg yolks with 6 chicken eggs.

**Preparation of Medium:** Add components to distilled/deionized water and bring volume to 1.0L. Mix thoroughly. Distribute into tubes. Autoclave for 15 min at 15 psi pressure–121°C.

**Storage/Shelf Life:** Store dehydrated media in the dark in a sealed container below 30°C. Prepared media should be stored under refrigeration (2-8°C). Media should be used within 60 days of preparation. Media should not be used if there are any signs of deterioration (discoloration) or contamination, or if the expiration date supplied by the manufacturer has passed.

**Use:** For the isolation of *Mycobacterium* species other than *Mycobacterium leprae*.

## Water Agar
## (Tap Water Agar)
**Composition** per liter:
Agar ..........................................................................15.0g
Tap water......................................................................1.0L

**Preparation of Medium:** Add agar to 1.0L of tap water. Mix thoroughly. Gently heat and bring to boiling. Autoclave for 15 min at 15 psi pressure–121°C. Pour into sterile Petri dishes.

**Storage/Shelf Life:** Store dehydrated media in the dark in a sealed container below 30°C. Prepared media should be stored under refrigeration (2-8°C). Media should be used within 60 days of preparation. Media should not be used if there are any signs of deterioration (shrinking, cracking, or discoloration) or contamination, or if the expiration date supplied by the manufacturer has passed.

**Use:** For the cultivation and differentiation of fungi and aerobic actinomycetes based on filament and aerial hyphae morphology.

## Wesley Broth
**Composition** per liter:
Tryptose ......................................................................20.0g
Bicine (*N*,*N*-bis-2-[hydroxyethyl]glycine) buffer.......................10.0g
NaCl..............................................................................5.0g
Yeast extract.................................................................2.5g
Agar ..............................................................................1.0g
FeSO$_4$......................................................................0.25g
Na$_2$S$_2$O$_5$.........................................................0.25g
Sodium pyruvate.........................................................0.25g
Antibiotic solution .....................................................10.0mL
Alkaline hematin solution...........................................6.25mL

**Antibiotic Solution:**
**Composition** per 10.0mL:
Rifampin ..................................................................0.025g
Cefsulodin...............................................................6.25mg
Polymyxin B sulfate .............................................20,000U

**Preparation of Antibiotic Solution:** Add components to distilled/deionized water and bring volume to 10.0mL. Mix thoroughly. Filter sterilize.

**Alkaline Hematin Solution:**
**Composition** per 10.0mL:
Hemin ......................................................................0.032g
NaOH (0.15*N* solution)...........................................10.0mL

**Preparation of Hemin Solution:** Add hemin to 10.0mL of NaOH solution. Mix thoroughly. Autoclave for 30 min at 5 psi pressure–108°C. Cool to 25°C.

**Preparation of Medium:** Add components, except antibiotic solution and alkaline hematin solution, to distilled/deionized water and bring volume to 983.75mL. Mix thoroughly. Gently heat and bring to boiling. Autoclave for 15 min at 15 psi pressure–121°C. Cool to 45°–50°C. Aseptically add 10.0mL of sterile antibiotic solution and 6.25mL of sterile alkaline hematin solution. Mix thoroughly. Aseptically distribute into sterile tubes. Use medium immediately or store overnight at 4°C.

**Storage/Shelf Life:** Store dehydrated media in the dark in a sealed container below 30°C. Prepared media should be stored under refrigeration (2-8°C). Media should be used within 60 days of preparation. Media should not be used if there are any signs of deterioration (discoloration) or contamination, or if the expiration date supplied by the manufacturer has passed.

**Use:** For the enrichment of *Campylobacter* species from foods.

## Wesley Broth Base with Antibiotics and Hematin
**Composition** per liter:
Tryptose ......................................................................20.0g
Bicine..........................................................................10.0g
NaCl..............................................................................5.0g
Yeast extract.................................................................2.5g
Agar ..............................................................................1.0g
FeSO$_4$......................................................................0.25g
Na$_2$S$_2$O$_3$.........................................................0.25g
Sodium pyruvate.........................................................0.25g
Antibiotic solution .....................................................10.0mL
Alkaline hematin solution...........................................6.25mL
pH 8.0 ± 0.2 at 25°C

**Source:** This medium, without antibiotic and alkaline hematin solutions, is available as a premixed powder from HiMedia.

## Antibiotic Solution:
**Composition** per 10.0mL:

Rifampin ..................................................................0.025g
Cefsulodin..............................................................6.25mg
Polymyxin B sulfate..............................................20,000U

**Preparation of Antibiotic Solution:** Add components to distilled/
deionized water and bring volume to 10.0mL. Mix thoroughly. Filter
sterilize.

## Alkaline Hematin Solution:
**Composition** per 10.0mL:

Hemin......................................................................0.032g
NaOH (0.15*N* solution)........................................... 10.0mL

**Preparation of Hemin Solution:** Add hemin to 10.0mL of NaOH
solution. Mix thoroughly. Autoclave for 30 min at 5 psi pressure–108°C.
Cool to 25°C.

**Preparation of Medium:** Add components, except antibiotic solution and alkaline hematin solution, to distilled/deionized water and
bring volume to 983.75mL. Mix thoroughly. Gently heat and bring to
boiling. Autoclave for 15 min at 15 psi pressure–121°C. Cool to 45°–
50°C. Aseptically add 10.0mL of sterile antibiotic solution and 6.25mL
of sterile alkaline hematin solution. Mix thoroughly. Aseptically distribute into sterile tubes. Use medium immediately or store overnight
at 4°C.

**Storage/Shelf Life:** Store dehydrated media in the dark in a sealed
container below 30°C. Prepared media should be stored under refrigeration (2-8°C). Media should be used within 60 days of preparation.
Media should not be used if there are any signs of deterioration (discoloration) or contamination, or if the expiration date supplied by the
manufacturer has passed.

**Use:** For the enrichment of *Campylobacter* species from foods.

## Wesley HiVeg Broth Base with Antibiotics and Hematin
**Composition** per liter:

Plant hydrolysate No. 1.............................................20.0g
Bicine ......................................................................10.0g
NaCl...........................................................................5.0g
Yeast extract..............................................................2.5g
Agar ..........................................................................1.0g
FeSO₄.......................................................................0.25g
Na₂S₂O₃....................................................................0.25g
Sodium pyruvate .....................................................0.25g
Antibiotic solution ................................................. 10.0mL
Alkaline hematin solution...................................... 6.25mL
pH 8.0 ± 0.2 at 25°C

**Source:** This medium, without antibiotic and alkaline hematin solutions, is available as a premixed powder from HiMedia.

## Antibiotic Solution:
**Composition** per 10.0mL:

Rifampin ..................................................................0.025g
Cefsulodin..............................................................6.25mg
Polymyxin B sulfate..............................................20,000U

**Preparation of Antibiotic Solution:** Add components to distilled/
deionized water and bring volume to 10.0mL. Mix thoroughly. Filter
sterilize.

## Alkaline Hematin Solution:
**Composition** per 10.0mL:

Hemin......................................................................0.032g
NaOH (0.15*N* solution)........................................... 10.0mL

**Preparation of Hemin Solution:** Add hemin to 10.0mL of NaOH
solution. Mix thoroughly. Autoclave for 30 min at 5 psi pressure–108°C.
Cool to 25°C.

**Preparation of Medium:** Add components, except antibiotic solution and alkaline hematin solution, to distilled/deionized water and
bring volume to 983.75mL. Mix thoroughly. Gently heat and bring to
boiling. Autoclave for 15 min at 15 psi pressure–121°C. Cool to 45°–
50°C. Aseptically add 10.0mL of sterile antibiotic solution and 6.25mL
of sterile alkaline hematin solution. Mix thoroughly. Aseptically distribute into sterile tubes. Use medium immediately or store overnight
at 4°C.

**Storage/Shelf Life:** Store dehydrated media in the dark in a sealed
container below 30°C. Prepared media should be stored under refrigeration (2-8°C). Media should be used within 60 days of preparation.
Media should not be used if there are any signs of deterioration (discoloration) or contamination, or if the expiration date supplied by the
manufacturer has passed.

**Use:** For the enrichment of *Campylobacter* species from foods.

## Wilkins-Chalgren Agar
**Composition** per liter:

Agar ........................................................................15.0g
Gelatin peptone........................................................10.0g
Pancreatic digest of casein......................................10.0g
NaCl...........................................................................5.0g
Yeast extract..............................................................5.0g
Glucose .....................................................................1.0g
L-Arginine .................................................................1.0g
Sodium pyruvate ......................................................1.0g
Hemin ......................................................................5.0mg
Vitamin K₁ (menadione)..........................................0.5mg
pH 7.1 ± 0.2 at 25°C

**Source:** This medium is available as a premixed powder from BD Diagnostic Systems.

**Preparation of Medium:** Add components to distilled/deionized
water and bring volume to 1.0L. Mix thoroughly. Gently heat and bring
to boiling. Distribute into tubes or flasks. Autoclave for 15 min at 15
psi pressure–121°C. Cool to 50°–55°C. Add antibiotic to be assayed;
varying concentrations of antibiotics are used. Mix thoroughly. Pour
into sterile Petri dishes or leave in tubes.

**Storage/Shelf Life:** Store dehydrated media in the dark in a sealed
container below 30°C. Prepared media should be stored under refrigeration (2-8°C). Media should be used within 60 days of preparation.
Media should not be used if there are any signs of deterioration (shrinking, cracking, or discoloration) or contamination, or if the expiration
date supplied by the manufacturer has passed.

**Use:** For the cultivation and maintenance of anaerobic bacteria. For
standardized antimicrobic susceptibility testing to determine the minimum inhibitory concentrations of antimicrobics for anaerobic bacteria.

## Wilkins-Chalgren Anaerobe Agar with GN Supplement
**Composition** per liter:

Agar ........................................................................10.0g
Pancreatic digest of casein......................................10.0g
Gelatin peptone........................................................10.0g
NaCl...........................................................................5.0g
Yeast extract..............................................................5.0g

Glucose ................................................................1.0g
L-Arginine ...........................................................1.0g
Sodium pyruvate ..................................................1.0g
Hemin....................................................................5.0mg
Menadione ...........................................................0.5mg
Defibrinated blood ..............................................50.0mL
GN anaerobe selective supplement.....................20.0mL

pH 7.1 ± 0.2 at 25°C

**Source:** This medium is available as a premixed powder from Oxoid Unipath.

## GN Anaerobe Selective Supplement

**Composition** per 20.0mL:
Nalidixic acid.....................................................10.0mg
Hemin....................................................................5.0mg
Sodium succinate .................................................2.5mg
Vancomycin ..........................................................2.5mg
Menadione ............................................................0.5mg

**Preparation of GN Anaerobe Selective Supplement:** Add components to distilled/deionized water and bring volume to 20.0mL. Mix thoroughly. Filter sterilize.

**Preparation of Medium:** Add components, except defibrinated blood and GN anaerobe selective supplement, to distilled/deionized water and bring volume to 900.0mL. Mix thoroughly. Gently heat and bring to boiling. Distribute into tubes or flasks. Autoclave for 15 min at 15 psi pressure–121°C. Cool to 50°–55°C. Aseptically add 20.0mL of GN anaerobe selective supplement and 50.0mL of defibrinated blood. Bring volume to 1.0L with distilled/deionized water. Mix thoroughly. Pour into sterile Petri dishes or leave in tubes.

**Storage/Shelf Life:** Store dehydrated media in the dark in a sealed container below 30°C. Prepared media should be stored under refrigeration (2-8°C). Media should be used within 60 days of preparation. Media should not be used if there are any signs of deterioration (shrinking, cracking, or discoloration) or contamination, or if the expiration date supplied by the manufacturer has passed.

**Use:** For the selective isolation of Gram-negative anaerobes.

## Wilkins-Chalgren Anaerobe Broth (Anaerobe Broth, MIC)

**Composition** per liter:
Pancreatic digest of casein ...............................10.0g
Gelatin peptone ..................................................10.0g
NaCl.......................................................................5.0g
Yeast extract.........................................................5.0g
Glucose ..................................................................1.0g
L-Arginine .............................................................1.0g
Sodium pyruvate ...................................................1.0g
Hemin.....................................................................5.0mg
Menadione .............................................................0.5mg

pH 7.1 ± 0.2 at 25°C

**Source:** This medium is available as a premixed powder from BD Diagnostic Systems and Oxoid Unipath.

**Preparation of Medium:** Add components to distilled/deionized water and bring volume to 1.0L. Mix thoroughly. Distribute into tubes or flasks. Autoclave for 15 min at 15 psi pressure–121°C.

**Storage/Shelf Life:** Store dehydrated media in the dark in a sealed container below 30°C. Prepared media should be stored under refrigeration (2-8°C). Media should be used within 60 days of preparation. Media should not be used if there are any signs of deterioration (shrink-

ing, cracking, or discoloration) or contamination, or if the expiration date supplied by the manufacturer has passed.

**Use:** For the cultivation and antimicrobial susceptibility (MIC) testing of anaerobic bacteria.

## Wilkins-Chalgren Anaerobic HiVeg Agar Base with Blood

**Composition** per liter:
Agar ......................................................................10.0g
Plant hydrolysate ...............................................10.0g
Plant peptone .....................................................10.0g
NaCl.......................................................................5.0g
Yeast extract.........................................................5.0g
L-Arginine .............................................................1.0g
Glucose ..................................................................1.0g
Sodium pyruvate ...................................................1.0g
$Fe_4(P_2O_7)_3 \cdot H_2O$ ...............................................5.0mg
Menadione .............................................................0.5mg
Defibrinated blood .............................................. 50.0mL

pH 7.1 ± 0.2 at 25°C

**Source:** This medium, without blood, is available as a premixed powder from HiMedia.

**Preparation of Medium:** Add components, except defibrinated blood, to distilled/deionized water and bring volume to 950.0mL. Mix thoroughly. Gently heat and bring to boiling. Distribute into tubes or flasks. Autoclave for 15 min at 15 psi pressure–121°C. Cool to 50°–55°C. Aseptically add 50.0mL of defibrinated blood. Mix thoroughly. Pour into sterile Petri dishes or leave in tubes.

**Storage/Shelf Life:** Store dehydrated media in the dark in a sealed container below 30°C. Prepared media should be stored under refrigeration (2-8°C). Media should be used within 60 days of preparation. Media should not be used if there are any signs of deterioration (shrinking, cracking, or discoloration) or contamination, or if the expiration date supplied by the manufacturer has passed.

**Use:** For the cultivation of nonsporulating anaerobes. For the cultivation and maintenance of anaerobic bacteria. For standardized antimicrobic susceptibility testing to determine the minimum inhibitory concentrations of antimicrobics for anaerobic bacteria.

## Wilkins-Chalgren Anaerobic HiVeg Agar Base with Blood and Nonspore Anaerobic Supplement

**Composition** per liter:
Agar ......................................................................10.0g
Plant hydrolysate ...............................................10.0g
Plant peptone .....................................................10.0g
NaCl.......................................................................5.0g
Yeast extract.........................................................5.0g
L-Arginine .............................................................1.0g
Glucose ..................................................................1.0g
Sodium pyruvate ...................................................1.0g
$Fe_4(P_2O_7)_3 \cdot H_2O$ ...............................................5.0mg
Menadione .............................................................0.5mg
Defibrinated blood ..............................................50.0mL
Nonspore anaerobic supplement...........................

pH 7.1 ± 0.2 at 25°C

**Source:** This medium, without blood or nonspore anaerobic supplement, is available as a premixed powder from HiMedia.

## Nonspore Anaerobic Supplement:

**Composition** per 10.0mL:

| | |
|---|---|
| Sodium pyruvate | 1.0g |
| Nalidixic acid | 10.0mg |
| Ferric pyrophosphate, soluble | 5.0mg |
| Menadione | 0.5mg |

**Preparation of Nonspore Anaerobic Supplement:** Add components to distilled/deionized water and bring volume to 10.0mL. Mix thoroughly. Filter sterilize.

**Preparation of Medium:** Add components, except defibrinated blood and nonspore anaerobic supplement, to distilled/deionized water and bring volume to 950.0mL. Mix thoroughly. Gently heat and bring to boiling. Distribute into tubes or flasks. Autoclave for 15 min at 15 psi pressure–121°C. Cool to 50°–55°C. Aseptically add 50.0mL of defibrinated blood and 10.0mL nonspore anaerobic supplement. Mix thoroughly. Pour into sterile Petri dishes or leave in tubes.

**Storage/Shelf Life:** Store dehydrated media in the dark in a sealed container below 30°C. Prepared media should be stored under refrigeration (2-8°C). Media should be used within 60 days of preparation. Media should not be used if there are any signs of deterioration (shrinking, cracking, or discoloration) or contamination, or if the expiration date supplied by the manufacturer has passed.

**Use:** For the cultivation of nonsporulating anaerobes. For the cultivation and maintenance of anaerobic bacteria. For standardized antimicrobic susceptibility testing to determine the minimum inhibitory concentrations of antimicrobics for anaerobic bacteria.

## Wilkins-Chalgren Anaerobic HiVeg Broth Base with Blood

**Composition** per liter:

| | |
|---|---|
| Plant hydrolysate | 10.0g |
| Plant peptone | 10.0g |
| Yeast extract | 5.0g |
| NaCl | 5.0g |
| Sodium pyruvate | 1.0g |
| Glucose | 1.0g |
| L-Arginine | 1.0g |
| $Fe_4(P_2O_7)_3 \cdot H_2O$ | 5.0mg |
| Menadione | 0.5mg |
| Defibrinated blood | 50.0mL |

pH 7.1 ± 0.2 at 25°C

**Source:** This medium, without blood, is available as a premixed powder from HiMedia.

**Preparation of Medium:** Add components, except defibrinated blood, to distilled/deionized water and bring volume to 950.0mL. Mix thoroughly. Gently heat and bring to boiling. Distribute into tubes or flasks. Autoclave for 15 min at 15 psi pressure–121°C. Cool to 50°–55°C. Aseptically add 50.0mL of defibrinated blood. Mix thoroughly. Aseptically distribute into tubes or leave in flasks.

**Storage/Shelf Life:** Store dehydrated media in the dark in a sealed container below 30°C. Prepared media should be stored under refrigeration (2-8°C). Media should be used within 60 days of preparation. Media should not be used if there are any signs of deterioration (discoloration) or contamination, or if the expiration date supplied by the manufacturer has passed.

**Use:** For the cultivation of nonsporulating anaerobes. For the cultivation and maintenance of anaerobic bacteria. For standardized antimicrobic susceptibility testing to determine the minimum inhibitory concentrations of antimicrobics for anaerobic bacteria.

## Wilson Blair Base

**Composition** per liter:

| | |
|---|---|
| Agar | 30.0g |
| Proteose peptone No. 3 | 10.0g |
| Glucose | 10.0g |
| Beef extract | 5.0g |
| NaCl | 5.0g |
| Selective reagent | 70.0mL |
| Brilliant Green (1% solution) | 4.0mL |

pH 7.3 ± 0.2 at 25°C

### Selective Reagent:

**Composition** per 320.2mL:

| | |
|---|---|
| Solution 1 | 100.0mL |
| Solution 2 | 100.0mL |
| Solution 3 | 100.0mL |
| Solution 4 | 20.2mL |

**Preparation of Selective Reagent:** Combine 100.0mL of solution 1, 100.0mL of solution 2, 100.0mL of solution 3, and 20.2mL of solution 4. Mix thoroughly. Gently heat to boiling until a slate-grey color develops. Cool to 50°C.

### Solution 1:

**Composition** per 100.0mL:

| | |
|---|---|
| $NaHSO_3$ | 40.0g |

**Preparation of Solution 1:** Add $NaHSO_3$ to 100.0mL of distilled/deionized water. Mix thoroughly.

### Solution 2:

**Composition** per 100.0mL:

| | |
|---|---|
| $NaH_2PO_4$ | 21.0g |

**Preparation of Solution 2:** Add $NaH_2PO_4$ to 100.0mL of distilled/deionized water. Mix thoroughly.

### Solution 3:

**Composition** per 100.0mL:

| | |
|---|---|
| Bismuth ammonium citrate | 12.5g |

**Preparation of Solution 3:** Add bismuth ammonium citrate to 100.0mL of distilled/deionized water. Mix thoroughly.

### Solution 4:

**Composition** per 20.2mL:

| | |
|---|---|
| $FeSO_4$ | 0.96g |

**Preparation of Solution 4:** Add $FeSO_4$ to 20.0mL of distilled/deionized water. Add 0.2mL of concentrated HCl. Mix thoroughly.

**Preparation of Medium:** Add components, except selective reagent and Brilliant Green solution, to distilled/deionized water and bring volume to 976.0mL. Mix thoroughly. Gently heat and bring to boiling. Distribute into tubes or flasks. Autoclave for 15 min at 15 psi pressure–121°C. Cool to 50°C. Aseptically add selective reagent and Brilliant Green solution. Mix thoroughly. Pour into sterile Petri dishes or leave in tubes.

**Storage/Shelf Life:** Store dehydrated media in the dark in a sealed container below 30°C. Prepared media should be stored under refrigeration (2-8°C). Media should be used within 60 days of preparation. Media should not be used if there are any signs of deterioration (shrinking, cracking, or discoloration) or contamination, or if the expiration date supplied by the manufacturer has passed.

**Use:** For the isolation and cultivation of *Salmonella*, especially *Salmonella typhi*.

## Wilson Blair HiVeg Agar with Brilliant Green and Selective Reagent

**Composition** per liter:

Agar .................................................................20.0g
Plant peptone.....................................................10.0g
Bismuth sulfite indicator........................................8.0g
Glucose .............................................................5.0g
Plant extract ......................................................5.0g
Na$_2$HPO$_4$.........................................................4.0g
FeSO$_4$..............................................................0.3g
Brilliant Green .................................................0.025g
Selective reagent ........................................... 70.0mL

pH 7.3 ± 0.2 at 25°C

**Source:** This medium, without selective reagent, is available as a premixed powder from HiMedia.

**Selective Reagent:**

**Composition** per 320.2mL:

Solution 1.................................................... 100.0mL
Solution 2.................................................... 100.0mL
Solution 3.................................................... 100.0mL
Solution 4...................................................... 20.2mL

**Preparation of Selective Reagent:** Combine 100.0mL of solution 1, 100.0mL of solution 2, 100.0mL of solution 3, and 20.2mL of solution 4. Mix thoroughly. Gently heat to boiling until a slate-grey color develops. Cool to 50°C.

**Solution 1:**

**Composition** per 100.0mL:

NaHSO$_3$ .............................................................40.0g

**Preparation of Solution 1:** Add NaHSO$_3$ to 100.0mL of distilled/deionized water. Mix thoroughly.

**Solution 2:**

**Composition** per 100.0mL:

NaH$_2$PO$_4$.............................................................21.0g

**Preparation of Solution 2:** Add NaH$_2$PO$_4$ to 100.0mL of distilled/deionized water. Mix thoroughly.

**Solution 3:**

**Composition** per 100.0mL:

Bismuth ammonium citrate..........................................12.5g

**Preparation of Solution 3:** Add bismuth ammonium citrate to 100.0mL of distilled/deionized water. Mix thoroughly.

**Solution 4:**

**Composition** per 20.2mL:

FeSO$_4$.................................................................0.96g

**Preparation of Solution 4:** Add FeSO$_4$ to 20.0mL of distilled/deionized water. Add 0.2mL of concentrated HCl. Mix thoroughly.

**Preparation of Medium:** Add components, except selective reagent and Brilliant Green solution, to distilled/deionized water and bring volume to 976.0mL. Mix thoroughly. Gently heat and bring to boiling. Distribute into tubes or flasks. Autoclave for 15 min at 15 psi pressure–121°C. Cool to 50°C. Aseptically add selective reagent and Brilliant Green solution. Mix thoroughly. Pour into sterile Petri dishes or leave in tubes.

**Storage/Shelf Life:** Store dehydrated media in the dark in a sealed container below 30°C. Prepared media should be stored under refrigeration (2-8°C). Media should be used within 60 days of preparation. Media should not be used if there are any signs of deterioration (shrinking, cracking, or discoloration) or contamination, or if the expiration date supplied by the manufacturer has passed.

**Use:** For the isolation and cultivation of *Salmonella*, especially *Salmonella typhi*.

## Wilson Blair HiVeg Agar Base with Selective Reagent and Brilliant Green

**Composition** per liter:

Agar .................................................................30.0g
Glucose .............................................................10.0g
Plant special peptone ............................................10.0g
Plant extract ......................................................5.0g
NaCl ................................................................5.0g
Selective reagent................................................ 70.0mL
Brilliant Green (1% solution) ................................... 4.0mL

pH 7.3 ± 0.2 at 25°C

**Source:** This medium, without selective reagent or Brilliant Green, is available as a premixed powder from HiMedia.

**Selective Reagent:**

**Composition** per 100.0mL:

Solution 1.................................................... 100.0mL
Solution 2.................................................... 100.0mL
Solution 3.................................................... 100.0mL
Solution 4...................................................... 20.2mL

**Preparation of Selective Reagent:** Combine 100.0mL of solution 1, 100.0mL of solution 2, 100.0mL of solution 3, and 20.2mL of solution 4. Mix thoroughly. Gently heat to boiling until a slate-grey color develops. Cool to 50°C.

**Solution 1:**

**Composition** per 100.0mL:

NaHSO$_3$ .............................................................40.0g

**Preparation of Solution 1:** Add NaHSO$_3$ to 100.0mL of distilled/deionized water. Mix thoroughly.

**Solution 2:**

**Composition** per 100.0mL:

NaH$_2$PO$_4$.............................................................21.0g

**Preparation of Solution 2:** Add NaH$_2$PO$_4$ to 100.0mL of distilled/deionized water. Mix thoroughly.

**Solution 3:**

**Composition** per 100.0mL:

Bismuth ammonium citrate..........................................12.5g

**Preparation of Solution 3:** Add bismuth ammonium citrate to 100.0mL of distilled/deionized water. Mix thoroughly.

**Solution 4:**

**Composition** per 20.2mL:

FeSO$_4$.................................................................0.96g

**Preparation of Solution 4:** Add FeSO$_4$ to 20.0mL of distilled/deionized water. Add 0.2mL of concentrated HCl. Mix thoroughly.

**Preparation of Medium:** Add components, except selective reagent and Brilliant Green solution, to distilled/deionized water and bring volume to 976.0mL. Mix thoroughly. Gently heat and bring to boiling. Distribute into tubes or flasks. Autoclave for 15 min at 15 psi pressure–121°C. Cool to 50°C. Aseptically add selective reagent and Brilliant Green solution. Mix thoroughly. Pour into sterile Petri dishes or leave in tubes.

**Storage/Shelf Life:** Store dehydrated media in the dark in a sealed container below 30°C. Prepared media should be stored under refrigeration (2-8°C). Media should be used within 60 days of preparation. Media should not be used if there are any signs of deterioration (shrinking, cracking, or discoloration) or contamination, or if the expiration date supplied by the manufacturer has passed.

**Use:** For the isolation and cultivation of *Salmonella*, especially *Salmonella typhi*.

## XL Agar Base
### (Xylose Lysine Agar Base)

**Composition** per liter:

| | |
|---|---|
| Agar | 13.5g |
| Lactose | 7.5g |
| Sucrose | 7.5g |
| L-Lysine | 5.0g |
| NaCl | 5.0g |
| Xylose | 3.5g |
| Yeast extract | 3.0g |
| Phenol Red | 0.08g |
| Thiosulfate-citrate solution | 20.0mL |

pH 7.5 ± 0.2 at 25°C

**Source:** This medium is available as a premixed powder from BD Diagnostic Systems.

**Thiosulfate-Citrate Solution:**

**Composition** per 100.0mL:

| | |
|---|---|
| $Na_2S_2O_3$ | 34.0g |
| Ferric ammonium citrate | 4.0g |

**Preparation of Thiosulfate-Citrate Solution:** Add components to distilled/deionized water and bring volume to 100.0mL. Mix thoroughly.

**Preparation of Medium:** Add components, except thiosulfate-citrate solution, to distilled/deionized water and bring volume to 980.0mL. Mix thoroughly. Gently heat while stirring and bring to boiling. Distribute into tubes or flasks. Autoclave for 10 min at 14 psi pressure–118°C. Cool to 55°C. Aseptically add 20.0 mL of the sterile thiosulfate-citrate solution. Mix thoroughly. Pour into sterile Petri dishes or leave in tubes.

**Storage/Shelf Life:** Store dehydrated media in the dark in a sealed container below 30°C. Prepared media should be stored under refrigeration (2-8°C). Media should be used within 60 days of preparation. Media should not be used if there are any signs of deterioration (shrinking, cracking, or discoloration) or contamination, or if the expiration date supplied by the manufacturer has passed.

**Use:** For the isolation, cultivation, and differentiation of enteric pathogens. Nonfermenting xylose/lactose/sucrose bacteria appear as red colonies. Xylose-fermenting, lysine-decarboxylating bacteria appear as red colonies. Xylose-fermenting, lysine-nondecarboxylating bacteria appear as opaque yellow colonies. Lactose- or sucrose-fermenting bacteria appear as yellow colonies.

## XL Agar Base

**Composition** per liter:

| | |
|---|---|
| Agar | 15 g |
| Lactose | 7.5g |

| | |
|---|---|
| Sucrose | 7.5g |
| L-Lysine | 5.0g |
| NaCl | 5.0g |
| Xylose | 3.75g |
| Yeast extract | 3.0g |
| Phenol Red | 0.08g |
| Thiosulfate-citrate solution | 20.0mL |

pH 7.4 ± 0.2 at 25°C

**Source:** This medium is available as a premixed powder from BD Diagnostic Systems.

**Thiosulfate-Citrate Solution:**

**Composition** per 100.0mL:

| | |
|---|---|
| $Na_2S_2O_3$ | 34.0g |
| Ferric ammonium citrate | 4.0g |

**Preparation of Thiosulfate-Citrate Solution:** Add components to distilled/deionized water and bring volume to 100.0mL. Mix thoroughly.

**Preparation of Medium:** Add components, except thiosulfate-citrate solution, to distilled/deionized water and bring volume to 980.0mL. Mix thoroughly. Gently heat while stirring and bring to boiling. Distribute into tubes or flasks. Autoclave for 10 min at 14 psi pressure–118°C. Cool to 55°C. Aseptically add 20.0 mL of the sterile thiosulfate-citrate solution. Mix thoroughly. Pour into sterile Petri dishes or leave in tubes.

**Storage/Shelf Life:** Store dehydrated media in the dark in a sealed container below 30°C. Prepared media should be stored under refrigeration (2-8°C). Media should be used within 60 days of preparation. Media should not be used if there are any signs of deterioration (shrinking, cracking, or discoloration) or contamination, or if the expiration date supplied by the manufacturer has passed.

**Use:** For the isolation, cultivation, and differentiation of enteric pathogens. Nonfermenting xylose/lactose/sucrose bacteria appear as red colonies. Xylose-fermenting, lysine-decarboxylating bacteria appear as red colonies. Xylose-fermenting, lysine-nondecarboxylating bacteria appear as opaque yellow colonies. Lactose- or sucrose-fermenting bacteria appear as yellow colonies.

## XLD Agar
### (Xylose Lysine Deoxycholate Agar)

**Composition** per liter:

| | |
|---|---|
| Agar | 13.5g |
| Lactose | 7.5g |
| Sucrose | 7.5g |
| $Na_2S_2O_3$ | 6.8g |
| L-Lysine | 5.0g |
| NaCl | 5.0g |
| Xylose | 3.5g |
| Yeast extract | 3.0g |
| Sodium desoxycholate | 2.5g |
| Ferric ammonium citrate | 0.8g |
| Phenol Red | 0.08g |

pH 7.5 ± 0.2 at 25°C

**Source:** This medium is available as a premixed powder from BD Diagnostic Systems and Oxoid Unipath.

**Preparation of Medium:** Add components to distilled/deionized water and bring volume to 1.0L. Mix thoroughly. Gently heat and bring to boiling. Do not overheat. Distribute into tubes or flasks. Autoclave for 15 min at 15 psi pressure–121°C. Pour into sterile Petri dishes or leave in tubes. Plates should be poured as soon as possible to avoid precipitation.

**Storage/Shelf Life:** Store dehydrated media in the dark in a sealed container below 30°C. Prepared media should be stored under refrigeration (2-8°C). Media should be used within 60 days of preparation. Media should not be used if there are any signs of deterioration (shrinking, cracking, or discoloration) or contamination, or if the expiration date supplied by the manufacturer has passed.

**Use:** For the isolation and differentiation of enteric pathogens, especially *Shigella* and *Providencia* species. Nonfermenting xylose/lactose/sucrose bacteria appear as red colonies. Xylose-fermenting, lysine-decarboxylating bacteria appear as red colonies. Xylose-fermenting, lysine-nondecarboxylating bacteria appear as opaque yellow colonies. Lactose- or sucrose-fermenting bacteria appear as yellow colonies.

## XLD Agar
## (Xylose Lysine Deoxycholate Agar)
## (BAM M179)

**Composition** per liter:

| | |
|---|---|
| Agar | 15.0g |
| Lactose | 7.5g |
| Sucrose | 7.5g |
| $Na_2S_2O_3$ | 6.8g |
| L-Lysine | 5.0g |
| NaCl | 5.0g |
| Xylose | 3.75g |
| Yeast extract | 3.0g |
| Sodium deoxycholate | 2.5g |
| Ferric ammonium citrate | 0.8g |
| Phenol Red | 0.08g |

pH 7.5 ± 0.2 at 25°C

**Source:** This medium is available as a premixed powder from BD Diagnostic Systems and Oxoid.

**Preparation of Medium:** Add components to distilled/deionized water and bring volume to 1.0L. Mix thoroughly. Gently heat and bring to boiling. Do not overheat. Distribute into tubes or flasks. Autoclave for 15 min at 15 psi pressure–121°C. Pour into sterile Petri dishes or leave in tubes. Plates should be poured as soon as possible to avoid precipitation.

**Storage/Shelf Life:** Store dehydrated media in the dark in a sealed container below 30°C. Prepared media should be stored under refrigeration (2-8°C). Media should be used within 60 days of preparation. Media should not be used if there are any signs of deterioration (shrinking, cracking, or discoloration) or contamination, or if the expiration date supplied by the manufacturer has passed.

**Use:** For the isolation and differentiation of enteric pathogens, especially *Shigella* and *Providencia* species. Nonfermenting xylose/lactose/sucrose bacteria appear as red colonies. Xylose-fermenting, lysine-decarboxylating bacteria appear as red colonies. Xylose-fermenting, lysine-nondecarboxylating bacteria appear as opaque yellow colonies. Lactose- or sucrose-fermenting bacteria appear as yellow colonies.

## XLD Agar, HiVeg
## (Xylose Lysine Deoxycholate HiVeg Agar)

**Composition** per liter:

| | |
|---|---|
| Agar | 15.0g |
| Lactose | 7.5g |

| | |
|---|---|
| Sucrose | 7.5g |
| $Na_2S_2O_3$ | 6.8g |
| L-Lysine | 5.0g |
| NaCl | 5.0g |
| Yeast extract | 4.0g |
| Xylose | 3.5g |
| Synthetic detergent No. III | 1.5g |
| Ferric ammonium citrate | 0.8g |
| Phenol Red | 0.08g |
| Selective supplement solution | 4.6mL |

pH 7.4 ± 0.2 at 25°C

**Source:** This medium is available as a premixed powder from Hi-Media.

**Selective Supplement Solution:**
**Composition** per 100.0mL:

| | |
|---|---|
| Tergitol™ 4 | Proprietary |

**Preparation of Selective Supplement Solution:** Available as premixed solution.

**Preparation of Medium:** Add components to distilled/deionized water and bring volume to 1.0L. Mix thoroughly. Gently heat and bring to boiling. Do not overheat. Distribute into tubes or flasks. Do not autoclave. Pour into sterile Petri dishes or leave in tubes. Plates should be poured as soon as possible to avoid precipitation.

**Storage/Shelf Life:** Store dehydrated media in the dark in a sealed container below 30°C. Prepared media should be stored under refrigeration (2-8°C). Media should be used within 60 days of preparation. Media should not be used if there are any signs of deterioration (shrinking, cracking, or discoloration) or contamination, or if the expiration date supplied by the manufacturer has passed.

**Use:** For the isolation and differentiation of enteric pathogens, especially *Shigella* and *Providencia* species. Nonfermenting xylose/lactose/sucrose bacteria appear as red colonies. Xylose-fermenting, lysine-decarboxylating bacteria appear as red colonies. Xylose-fermenting, lysine-nondecarboxylating bacteria appear as opaque yellow colonies. Lactose- or sucrose-fermenting bacteria appear as yellow colonies.

## XLT4 HiVeg Agar Base

**Composition** per liter:

| | |
|---|---|
| Agar | 18.0g |
| Lactose | 7.5g |
| Saccharose | 7.5g |
| $Na_2S_2O_3$ | 6.8g |
| L-Lysine | 5.0g |
| NaCl | 5.0g |
| Xylose | 3.75g |
| Yeast extract | 3.0g |
| Plant peptone No. 3 | 1.6g |
| Ferric ammonium citrate | 0.8g |
| Phenol Red | 0.08g |
| Selective supplement solution | 4.6mL |

pH 7.4 ± 0.2 at 25°C

**Source:** This medium, without selective supplement solution, is available as a premixed powder from HiMedia.

**Selective Supplement Solution:**
**Composition** per 100.0mL:

| | |
|---|---|
| Tergitol™ 4 | Proprietary |

**Preparation of Selective Supplement Solution:** Available as premixed solution.

**Preparation of Medium:** Add components to distilled/deionized water and bring volume to 1.0L. Mix thoroughly. Gently heat and bring to boiling. Do not overheat. Distribute into tubes or flasks. Autoclave for 15 min at 15 psi pressure–121°C. Pour into sterile Petri dishes or leave in tubes. Plates should be poured as soon as possible to avoid precipitation.

**Storage/Shelf Life:** Store dehydrated media in the dark in a sealed container below 30°C. Prepared media should be stored under refrigeration (2-8°C). Media should be used within 60 days of preparation. Media should not be used if there are any signs of deterioration (shrinking, cracking, or discoloration) or contamination, or if the expiration date supplied by the manufacturer has passed.

**Use:** For the isolation and differentiation of enteric pathogens, especially *Shigella* and *Providencia* species.

## Yeast Extract HiVeg Agar

**Composition** per liter:

Agar ..................................................................15.0g
Plant peptone...................................................5.0g
Yeast extract....................................................3.0g

pH 7.2 ± 0.2 at 25°C

**Source:** This medium is available as a premixed powder from HiMedia.

**Preparation of Medium:** Add components to distilled/deionized water and bring volume to 1.0L. Mix thoroughly. Gently heat and bring to boiling. Distribute into tubes or flasks. Autoclave for 15 min at 15 psi pressure–121°C. Pour into sterile Petri dishes or leave in tubes.

**Storage/Shelf Life:** Store dehydrated media in the dark in a sealed container below 30°C. Prepared media should be stored under refrigeration (2-8°C). Media should be used within 60 days of preparation. Media should not be used if there are any signs of deterioration (shrinking, cracking, or discoloration) or contamination, or if the expiration date supplied by the manufacturer has passed.

**Use:** For the enumeration of microorganisms in potable and freshwater samples. A highly nutritive medium recommended for plate count of microorganisms in water.

## Yeast HiVeg Agar
### (Antibiotic HiVeg Assay Medium No. 4)

**Composition** per liter:

Agar ..................................................................15.0g
Plant peptone...................................................6.0g
Yeast extract....................................................3.0g
Plant extract ....................................................1.5g
Glucose ............................................................1.0g

pH 6.6 ± 0.05 at 25°C

**Source:** This medium is available from HiMedia.

**Preparation of Medium:** Add components to distilled/deionized water and bring volume to 1.0L. Mix thoroughly. Gently heat and bring to boiling. Distribute into tubes or flasks. Autoclave for 15 min at 15 psi pressure–121°C. Pour into sterile Petri dishes or leave in tubes.

**Storage/Shelf Life:** Store dehydrated media in the dark in a sealed container below 30°C. Prepared media should be stored under refrigeration (2-8°C). Media should be used within 60 days of preparation. Media should not be used if there are any signs of deterioration (shrink-

ing, cracking, or discoloration) or contamination, or if the expiration date supplied by the manufacturer has passed.

**Use:** For antibiotic assay testing.

## *Yersinia* Isolation HiVeg Agar

**Composition** per liter:

Agar ..................................................................15.0g
Plant peptone...................................................15.0g
Lactose.............................................................10.0g
Sodium citrate..................................................10.0g
Plant extract No. 1 ..........................................8.5g
$Na_2S_2O_3$...............................................................8.5g
Yeast extract....................................................5.0g
Synthetic detergent No. II................................3.0g
Synthetic detergent No. III ..............................2.0g
$CaCl_2$................................................................1.0g
Ferric citrate....................................................1.0g
Neutral Red......................................................0.025g
Brilliant Green..................................................0.3mg

pH 7.3 ± 0.2 at 25°C

**Preparation of Medium:** Add components to distilled/deionized water and bring volume to 1.0L. Mix thoroughly. Heat to boiling. Do not autoclave. Pour into sterile Petri dishes.

**Storage/Shelf Life:** Store dehydrated media in the dark in a sealed container below 30°C. Prepared media should be stored under refrigeration (2-8°C). Media should be used within 60 days of preparation. Media should not be used if there are any signs of deterioration (shrinking, cracking, or discoloration) or contamination, or if the expiration date supplied by the manufacturer has passed.

**Use:** For the isolation and characterization of *Yersinia enterocolitica* from fecal specimens and enumeration of *Yersinia enterocolitica* from water and other liquid specimens.

## *Yersinia* Selective Agar Base

**Composition** per liter:

Mannitol............................................................20.0g
Peptone ............................................................17.0g
Agar ..................................................................12.5g
Proteose peptone .............................................3.0g
Yeast extract....................................................2.0g
Sodium pyruvate..............................................2.0g
NaCl..................................................................1.0g
Sodium desoxycholate .....................................0.5g
$MgSO_4·7H_2O$....................................................0.01g
Neutral Red......................................................0.03g
Crystal Violet...................................................1.0mg
Selective supplement .......................................6.0mL

pH 7.4 ± 0.2 at 25°C

**Source:** This medium is available as a premixed powder from BD Diagnostic Systems and Oxoid Unipath.

**Selective Supplement:**
**Composition** per 6.0mL:

Cefsulodin........................................................15.0mg
Irgasan.............................................................4.0mg
Novobiocin.......................................................2.5mg
Ethanol.............................................................2.0mL

**Preparation of Selective Supplement:** Aseptically add components to 4.0mL of distilled/deionized water and 2.0mL of ethanol. Mix thoroughly.

**Preparation of Medium:** Add components to distilled/deionized water and bring volume to 1.0L. Mix thoroughly. Gently heat and bring to boiling. Distribute into tubes or flasks. Autoclave for 15 min at 15 psi pressure–121°C. Cool to 50°C. Aseptically add selective supplement. Mix thoroughly. Pour into sterile Petri dishes or leave in tubes.

**Storage/Shelf Life:** Store dehydrated media in the dark in a sealed container below 30°C. Prepared media should be stored under refrigeration (2-8°C). Media should be used within 60 days of preparation. Media should not be used if there are any signs of deterioration (shrinking, cracking, or discoloration) or contamination, or if the expiration date supplied by the manufacturer has passed.

**Use:** For the isolation and enumeration of *Yersinia enterocolitica* from food and clinical specimens.

### *Yersinia* Selective HiVeg Agar Base with Selective Supplement

**Composition** per liter:

| | |
|---|---:|
| Plant special peptone | 20.0g |
| Mannitol | 20.0g |
| Agar | 12.5g |
| Sodium pyruvate | 2.0g |
| Yeast extract | 2.0g |
| NaCl | 1.0g |
| Synthetic detergent No. III | 0.5g |
| Neutral Red | 0.03g |
| MgSO$_4$ | 0.01g |
| Crystal Violet | 1.0mg |
| Selective supplement | 6.0mL |

pH 7.4 ± 0.2 at 25°C

**Source:** This medium, without selective supplement, is available as a premixed powder from HiMedia.

**Selective Supplement:**
**Composition** per 6.0mL:

| | |
|---|---:|
| Cefsulodin | 15.0mg |
| Irgasan | 4.0mg |
| Novobiocin | 2.5mg |
| Ethanol | 2.0mL |

**Preparation of Selective Supplement:** Aseptically add components to 4.0mL of distilled/deionized water and 2.0mL of ethanol. Mix thoroughly.

**Preparation of Medium:** Add components to distilled/deionized water and bring volume to 1.0L. Mix thoroughly. Gently heat and bring to boiling. Distribute into tubes or flasks. Autoclave for 15 min at 15 psi pressure–121°C. Cool to 50°C. Aseptically add selective supplement. Mix thoroughly. Pour into sterile Petri dishes or leave in tubes.

**Storage/Shelf Life:** Store dehydrated media in the dark in a sealed container below 30°C. Prepared media should be stored under refrigeration (2-8°C). Media should be used within 60 days of preparation. Media should not be used if there are any signs of deterioration (shrinking, cracking, or discoloration) or contamination, or if the expiration date supplied by the manufacturer has passed.

**Use:** For the isolation and enumeration of *Yersinia enterocolitica* from food and clinical specimens.

## ABY Agar
## (Acid Bismuth Yeast Agar)

**Composition** per liter:

| | |
|---|---|
| Agar | 20.0g |
| Glucose | 20.0g |
| $Bi_2(SO_3)_2$ | 8.0g |
| $(NH_4)_2SO_4$ | 3.0g |
| $KH_2PO_4$ | 3.0g |
| $MgSO_4 \cdot 7H_2O$ | 0.25g |
| $CaCl_2 \cdot 2H_2O$ | 0.25g |
| Biotin | 10.0μg |

pH 7.2 ± 0.2 at 25°C

**Preparation of Medium:** Add components to distilled/deionized water and bring volume to 1.0L. Mix thoroughly. Gently heat and bring to boiling. Distribute into tubes or flasks. Autoclave for 15 min at 15 psi pressure–121°C. Cool tubes in a slanted position.

**Storage/Shelf Life:** Store dehydrated media in the dark in a sealed container below 30°C. Prepared media should be stored under refrigeration (2-8°C). Media should be used within 60 days of preparation. Media should not be used if there are any signs of deterioration (shrinking, cracking, or discoloration) or contamination, or if the expiration date supplied by the manufacturer has passed.

**Use:** For the selective isolation and differentiation of *Candida albicans* from other *Candida* species. *Candida albicans* and *Candida tropicalis* colonies appear as smooth, brownish-black round colonies. Other *Candida* species are differentially pigmented or produce diffusible pigments. Usually used in conjunction with BiGGY agar to differentiate further *Candida*; on BiGGY agar, *Candida albicans* appears as brown to black colonies with no pigment diffusion and no sheen, whereas *Candida tropicalis* appears as dark brown colonies with black centers, black pigment diffusion, and a sheen.

## Acetate Ascospore Agar
## (Ascospore Agar)

**Composition** per liter:

| | |
|---|---|
| Agar | 30.0g |
| Potassium acetate | 10.0g |
| Yeast extract | 2.5g |
| Glucose | 1.0g |

pH 6.4 ± 0.2 at 25°C

**Source:** This medium is available from HiMedia.

**Preparation of Medium:** Add components to distilled/deionized water and bring volume to 1.0L. Mix thoroughly. Gently heat and bring to boiling. Distribute into tubes or flasks. Autoclave for 15 min at 15 psi pressure–121°C. Pour into sterile Petri dishes or leave in tubes.

**Storage/Shelf Life:** Store dehydrated media in the dark in a sealed container below 30°C. Prepared media should be stored under refrigeration (2-8°C). Media should be used within 60 days of preparation. Media should not be used if there are any signs of deterioration (shrinking, cracking, or discoloration) or contamination, or if the expiration date supplied by the manufacturer has passed.

**Use:** For the enrichment of ascosporogenous yeasts and their production of ascospores.

## Antibiotic Assay Medium No. 4
## (Yeast Beef Agar)

**Composition** per liter:

| | |
|---|---|
| Agar | 15.0g |
| Peptone | 6.0g |
| Yeast extract | 3.0g |
| Beef extract | 1.5g |
| Glucose | 1.0g |

pH 6.6 ± 0.2 at 25°C

**Source:** This medium is available as a premixed powder from Hi-Media.

**Preparation of Medium:** Add components to distilled/deionized water and bring volume to 1.0L. Mix thoroughly. Gently heat and bring to boiling. Distribute into tubes or flasks. Autoclave for 15 min at 15 psi pressure–121°C. Pour into sterile Petri dishes or leave in tubes.

**Storage/Shelf Life:** Store dehydrated media in the dark in a sealed container below 30°C. Prepared media should be stored under refrigeration (2-8°C). Media should be used within 60 days of preparation. Media should not be used if there are any signs of deterioration (shrinking, cracking, or discoloration) or contamination, or if the expiration date supplied by the manufacturer has passed.

**Use:** For antibiotic assay testing.

## Antibiotic Assay Medium No. 12
## (Nystatin Assay Agar)

**Composition** per liter:

| | |
|---|---|
| Agar | 25.0g |
| Peptone | 10.0g |
| Glucose | 10.0g |
| NaCl | 10.0g |
| Yeast extract | 5.0g |
| Beef extract | 2.5g |

pH 6.0 ± 0.1 at 25°C

**Source:** This medium is available as a premixed powder from Hi-Media.

**Preparation of Medium:** Add components to distilled/deionized water and bring volume to 1.0L. Mix thoroughly. Gently heat and bring to boiling. Distribute into tubes or flasks. Autoclave for 15 min at 15 psi pressure–121°C. Pour into sterile Petri dishes.

**Storage/Shelf Life:** Store dehydrated media in the dark in a sealed container below 30°C. Prepared media should be stored under refrigeration (2-8°C). Media should be used within 60 days of preparation. Media should not be used if there are any signs of deterioration (shrinking, cracking, or discoloration) or contamination, or if the expiration date supplied by the manufacturer has passed.

**Use:** For antibiotic assay effectiveness testing. For the assay of antifungal agents such as amphotericin and nystatin.

## Antibiotic Assay Medium No. 13

**Composition** per liter:

| | |
|---|---|
| Glucose | 20.0g |
| Peptone | 10.0g |

pH 5.6 ± 0.2 at 25°C

**Source:** This medium is available as a premixed powder from Hi-Media.

**Preparation of Medium:** Add components to distilled/deionized water and bring volume to 1.0L. Mix thoroughly. Gently heat and bring to boiling. Distribute into tubes or flasks. Autoclave for 15 min at 15 psi pressure–121°C.

**Storage/Shelf Life:** Store dehydrated media in the dark in a sealed container below 30°C. Prepared media should be stored under refrigeration (2-8°C). Media should be used within 60 days of preparation.

Media should not be used if there are any signs of deterioration (discoloration) or contamination, or if the expiration date supplied by the manufacturer has passed.

**Use:** For testing the effectivness of antibiotics on yeast and molds.

### Antibiotic Assay Medium No. 19

| | |
|---|---|
| Agar | 23.5g |
| Glucose | 10.0g |
| NaCl | 10.0g |
| Peptone | 9.4g |
| Yeast extract | 4.7g |
| Beef extract | 2.4g |

pH 6.1 ± 0.2 at 25°C

**Source:** This medium is available as a premixed powder from Hi-Media.

**Preparation of Medium:** Add components to distilled/deionized water and bring volume to 1.0L. Mix thoroughly. Gently heat and bring to boiling. Distribute into tubes or flasks. Autoclave for 15 min at 15 psi pressure–121°C. Pour into sterile Petri dishes.

**Storage/Shelf Life:** Store dehydrated media in the dark in a sealed container below 30°C. Prepared media should be stored under refrigeration (2-8°C). Media should be used within 60 days of preparation. Media should not be used if there are any signs of deterioration (shrinking, cracking, or discoloration) or contamination, or if the expiration date supplied by the manufacturer has passed.

**Use:** For assaying the mycostatic activity of pharmaceutical preparations. For seed agar for the plate assay to test the effectiveness of nystatin, amphotericin B, and natamycin.

### Antibiotic Assay Medium F

**Composition** per liter:

| | |
|---|---|
| Agar | 23.5g |
| Glucose | 10.0g |
| NaCl | 10.0g |
| Peptone | 9.4g |
| Yeast extract | 4.7g |
| Beef extract | 2.4g |

pH 6.0 ± 0.2 at 25°C

**Preparation of Medium:** Add components to distilled/deionized water and bring volume to 1.0L. Mix thoroughly. Gently heat and bring to boiling. Distribute into tubes or flasks. Autoclave for 15 min at 15 psi pressure–121°C. Pour into sterile Petri dishes.

**Storage/Shelf Life:** Store dehydrated media in the dark in a sealed container below 30°C. Prepared media should be stored under refrigeration (2-8°C). Media should be used within 60 days of preparation. Media should not be used if there are any signs of deterioration (shrinking, cracking, or discoloration) or contamination, or if the expiration date supplied by the manufacturer has passed.

**Use:** For the microbiological assay of nystatin using *Saccharomyces cerevisiae* or *Candida tropicalis*.

### Antibiotic HiVeg Assay Medium No. 1
### (Antibiotic HiVeg Assay Medium - A)
### (Seed HiVeg Agar)

**Composition** per liter:

| | |
|---|---|
| Agar | 15.0g |
| Plant peptone | 6.0g |
| Plant hydrolysate | 4.0g |

| | |
|---|---|
| Yeast extract | 3.0g |
| Plant extract | 1.5g |
| Glucose | 1.0g |

pH 6.6 ± 0.2 at 25°C

**Source:** This medium is available as a premixed powder from Hi-Media.

**Preparation of Medium:** Add components to distilled/deionized water and bring volume to 1.0L. Mix thoroughly. Gently heat and bring to boiling. Distribute into tubes or flasks. Autoclave for 15 min at 15 psi pressure–121°C. Pour into sterile Petri dishes or leave in tubes.

**Storage/Shelf Life:** Store dehydrated media in the dark in a sealed container below 30°C. Prepared media should be stored under refrigeration (2-8°C). Media should be used within 60 days of preparation. Media should not be used if there are any signs of deterioration (shrinking, cracking, or discoloration) or contamination, or if the expiration date supplied by the manufacturer has passed.

**Use:** For antibiotic assay testing. Widely employed as seed agar in the preparation of plates for microbiological agar diffusion antibiotic assays.

### Antibiotic HiVeg Assay Medium No. 12
### (Nystatin HiVeg Assay Agar)

**Composition** per liter:

| | |
|---|---|
| Agar | 25.0g |
| Plant peptone | 10.0g |
| Glucose | 10.0g |
| NaCl | 10.0g |
| Yeast extract | 5.0g |
| Plant extract | 2.5g |

pH 6.0 ± 0.1 at 25°C

**Source:** This medium is available as a premixed powder from Hi-Media.

**Preparation of Medium:** Add components to distilled/deionized water and bring volume to 1.0L. Mix thoroughly. Gently heat and bring to boiling. Distribute into tubes or flasks. Autoclave for 15 min at 15 psi pressure–121°C. Pour into sterile Petri dishes.

**Storage/Shelf Life:** Store dehydrated media in the dark in a sealed container below 30°C. Prepared media should be stored under refrigeration (2-8°C). Media should be used within 60 days of preparation. Media should not be used if there are any signs of deterioration (shrinking, cracking, or discoloration) or contamination, or if the expiration date supplied by the manufacturer has passed.

**Use:** For antibiotic assay effectiveness testing. For the assay of antifungal antibiotics like amphotericin and nystatin.

### Antibiotic HiVeg Assay Medium No. 13

**Composition** per liter:

| | |
|---|---|
| Glucose | 20.0g |
| Plant peptone | 10.0g |

pH 5.6 ± 0.2 at 25°C

**Source:** This medium is available as a premixed powder from Hi-Media.

**Preparation of Medium:** Add components to distilled/deionized water and bring volume to 1.0L. Mix thoroughly. Gently heat and bring to boiling. Distribute into tubes or flasks. Autoclave for 15 min at 15 psi pressure–121°C.

**Storage/Shelf Life:** Store dehydrated media in the dark in a sealed container below 30°C. Prepared media should be stored under refrigeration (2-8°C). Media should be used within 60 days of preparation. Media should not be used if there are any signs of deterioration (discoloration) or contamination, or if the expiration date supplied by the manufacturer has passed.

**Use:** For testing the effectivness of antibiotics on yeast and molds.

## Antibiotic HiVeg Assay Medium No. 19
### (Antibiotic HiVeg Assay Medium G)

| | |
|---|---|
| Agar | 23.5g |
| Glucose | 10.0g |
| NaCl | 10.0g |
| Plant peptone | 9.4g |
| Yeast extract | 4.7g |
| Plant extract | 2.4g |

pH 6.1 ± 0.2 at 25°C

**Source:** This medium is available as a premixed powder from Hi-Media.

**Preparation of Medium:** Add components to distilled/deionized water and bring volume to 1.0L. Mix thoroughly. Gently heat and bring to boiling. Distribute into tubes or flasks. Autoclave for 15 min at 15 psi pressure–121°C. Pour into sterile Petri dishes.

**Storage/Shelf Life:** Store dehydrated media in the dark in a sealed container below 30°C. Prepared media should be stored under refrigeration (2-8°C). Media should be used within 60 days of preparation. Media should not be used if there are any signs of deterioration (shrinking, cracking, or discoloration) or contamination, or if the expiration date supplied by the manufacturer has passed.

**Use:** For assaying the mycostatic activity of pharmaceutical preparations. For seed agar for the plate assay to test the effectiveness of nystatin, amphotericin B, and natamycin.

## Antibiotic Medium 13
### (Sabouraud Liquid Broth, Modified)
### (Fluid Sabouraud Medium)

**Composition** per liter:

| | |
|---|---|
| Glucose | 20.0g |
| Pancreatic digest of casein | 5.0g |
| Peptic digest of animal tissue | 5.0g |

pH 5.7 ± 0.1 at 25°C

**Source:** This medium is available as a premixed powder from BD Diagnostic Systems.

**Preparation of Medium:** Add components to distilled/deionized water and bring volume to 1.0L. Mix thoroughly. Gently heat and bring to boiling. Distribute into tubes or flasks. Autoclave for 15 min at 15 psi pressure–121°C. Pour into sterile Petri dishes.

**Storage/Shelf Life:** Store dehydrated media in the dark in a sealed container below 30°C. Prepared media should be stored under refrigeration (2-8°C). Media should be used within 60 days of preparation. Media should not be used if there are any signs of deterioration ( discoloration) or contamination, or if the expiration date supplied by the manufacturer has passed.

**Use:** For testing the effectivness of antibiotics on yeast and molds.

## Antibiotic Medium 19
### (Nystatin Assay Agar)

**Composition** per liter:

| | |
|---|---|
| Agar | 23.5g |
| Glucose | 10.0g |
| NaCl | 10.0g |
| Pancreatic digest of gelatin | 9.4g |
| Yeast extract | 4.7g |
| Beef extract | 2.4g |

pH 6.1 ± 0.2 at 25°C

**Source:** This medium is available as a premixed powder from BD Diagnostic Systems.

**Preparation of Medium:** Add components to distilled/deionized water and bring volume to 1.0L. Mix thoroughly. Gently heat and bring to boiling. Distribute into tubes or flasks. Autoclave for 15 min at 15 psi pressure–121°C. Pour into sterile Petri dishes.

**Storage/Shelf Life:** Store dehydrated media in the dark in a sealed container below 30°C. Prepared media should be stored under refrigeration (2-8°C). Media should be used within 60 days of preparation. Media should not be used if there are any signs of deterioration (shrinking, cracking, or discoloration) or contamination, or if the expiration date supplied by the manufacturer has passed.

**Use:** For assaying the mycostatic activity of pharmaceutical preparations. For seed agar for the plate assay to test the effectiveness of nystatin, amphotericin B, and natamycin.

## Antibiotic Medium 20

**Composition** per liter:

| | |
|---|---|
| Glucose | 11.0g |
| Pancreatic digest of casein | 10.0g |
| Yeast extract | 6.5g |
| Pancreatic digest of gelatin | 5.0g |
| $K_2HPO_4$ | 3.68g |
| NaCl | 3.5g |
| Beef extract | 1.5g |
| $KH_2PO_4$ | 1.32g |

pH 6.6 ± 0.2 at 25°C

**Preparation of Medium:** Add components to distilled/deionized water and bring volume to 1.0L. Mix thoroughly. Gently heat and bring to boiling. Distribute into tubes or flasks. Autoclave for 15 min at 15 psi pressure–121°C. Pour into sterile Petri dishes.

**Storage/Shelf Life:** Store dehydrated media in the dark in a sealed container below 30°C. Prepared media should be stored under refrigeration (2-8°C). Media should be used within 60 days of preparation. Media should not be used if there are any signs of deterioration ( discoloration) or contamination, or if the expiration date supplied by the manufacturer has passed.

**Use:** For assaying the mycostatic activity of pharmaceutical preparations.

## Antifungal Assay Agar

**Composition** per liter:

| | |
|---|---|
| Glucose | 50.0g |
| Agar | 15.0g |
| Sodium citrate | 4.5g |
| Pancreatic digest of casein | 4.0g |
| Citric acid | 1.0g |
| $K_2HPO_4$ | 0.55g |

KCl.................................................................................0.425g
CaCl$_2$·2H$_2$O...................................................................0.125g
MgSO$_4$·7H$_2$O.................................................................0.125g
Inositol.........................................................................0.025g
MnSO$_4$·4H$_2$O....................................................................2.5mg
Niacin..............................................................................2.5mg
Caclium pantothenate .........................................2.5mg
FeCl$_3$..................................................................................2.5mg
Pyridoxine hydrochloride .................................0.25mg
Thiamine ....................................................................0.25mg
Biotin ......................................................................0.008mg

pH 5.5 ± 0.2 at 25°C

**Source:** This medium is available as a premixed powder from Sigma Aldrich.

**Preparation of Medium:** Add components to distilled/deionized water and bring volume to 1.0L. Mix thoroughly. Gently heat and bring to boiling. Distribute into tubes or flasks. Autoclave for 15 min at 15 psi pressure–121°C. Pour into sterile Petri dishes.

**Storage/Shelf Life:** Store dehydrated media in the dark in a sealed container below 30°C. Prepared media should be stored under refrigeration (2-8°C). Media should be used within 60 days of preparation. Media should not be used if there are any signs of deterioration (shrinking, cracking, or discoloration) or contamination, or if the expiration date supplied by the manufacturer has passed.

**Use:** For assaying antifungal activity of pharmaceutical products and other materials by cylinder plate or disc method.

## Antifungal Assay HiVeg Agar

**Composition** per liter:

Glucose ........................................................................50.0g
Agar ..............................................................................15.0g
Sodium citrate.............................................................4.5g
Plant hydrolysate.........................................................4.0g
Citric acid.....................................................................1.0g
K$_2$HPO$_4$.........................................................................0.55g
KCl................................................................................0.425g
CaCl$_2$·2H$_2$O...................................................................0.125g
MgSO$_4$·7H$_2$O.................................................................0.125g
Inositol.........................................................................0.025g
MnSO$_4$·4H$_2$O....................................................................2.5mg
Niacin...............................................................................2.5mg
Calcium pantothenate ........................................2.5mg
FeCl$_3$ ...............................................................................2.5mg
Pyridoxine hydrochloride .................................0.25mg
Thiamine ....................................................................0.25mg
Biotin ........................................................................0.008mg

pH 5.5 ± 0.2 at 25°C

**Source:** This medium is available as a premixed powder from Hi-Media.

**Preparation of Medium:** Add components to distilled/deionized water and bring volume to 1.0L. Mix thoroughly. Gently heat and bring to boiling. Distribute into tubes or flasks. Autoclave for 15 min at 15 psi pressure–121°C. Pour into sterile Petri dishes.

**Storage/Shelf Life:** Store dehydrated media in the dark in a sealed container below 30°C. Prepared media should be stored under refrigeration (2-8°C). Media should be used within 60 days of preparation. Media should not be used if there are any signs of deterioration (shrinking, cracking, or discoloration) or contamination, or if the expiration date supplied by the manufacturer has passed.

**Use:** For assaying antifungal activity of pharmaceutical products and other materials by cylinder plate or disc method.

## Antimycotic Sensitivity Test Agar

**Composition** per liter:

Agar ..............................................................................25.0g
Glucose ........................................................................20.0g
Pancreatic digest of casein.......................................19.0g
Sodium citrate.............................................................10.0g
Yeast extract...............................................................10.0g
Na$_2$HPO$_4$..........................................................................1.0g

pH 6.0 ± 0.2 at 25°C

**Source:** This medium is available as a premixed powder from Hi-Media.

**Preparation of Medium:** Add components to distilled/deionized water and bring volume to 1.0L. Mix thoroughly. Gently heat and bring to boiling. Distribute into tubes or flasks. Autoclave for 15 min at 15 psi pressure–121°C. Pour into sterile Petri dishes.

**Storage/Shelf Life:** Store dehydrated media in the dark in a sealed container below 30°C. Prepared media should be stored under refrigeration (2-8°C). Media should be used within 60 days of preparation. Media should not be used if there are any signs of deterioration (shrinking, cracking, or discoloration) or contamination, or if the expiration date supplied by the manufacturer has passed.

**Use:** For testing antimycotic sensitivity by the diffusion method.

## *Aspergillus* Differentiation Medium Base with Chloramphenicol

**Composition** per liter:

Yeast extract...............................................................20.0g
Agar ..............................................................................15.0g
Peptic digest of animal tissue .................................10.0g
Ferric ammonium citrate............................................0.5g
Chloramphenicol...........................................................0.1g
Dichloran ...................................................................2.0mg

pH 6.3 ± 0.2 at 25°C

**Source:** This medium is available from HiMedia.

**Preparation of Medium:** Add components to distilled/deionized water and bring volume to 1.0L. Mix thoroughly. Gently heat and bring to boiling. Autoclave for 15 min at 15 psi pressure–121°C. Pour into Petri dishes or leave in tubes.

**Storage/Shelf Life:** Store dehydrated media in the dark in a sealed container below 30°C. Prepared media should be stored under refrigeration (2-8°C). Media should be used within 60 days of preparation. Media should not be used if there are any signs of deterioration (shrinking, cracking, or discoloration) or contamination, or if the expiration date supplied by the manufacturer has passed.

**Use:** For the detction of aflatoxin producing *Aspergillus* spp. from foods.

## BiGGY Agar
## (Bismuth Sulfite Glucose Glycerin Yeast Extract Agar)
## (Nickerson Medium)

**Composition** per liter:

Agar ..............................................................................16.0g
Glucose ........................................................................10.0g

Glycine...................................................................................10.0g
Bismuth ammonium citrate...................................................5.0g
Na$_2$SO$_3$.................................................................................3.0g
Yeast extract..........................................................................1.0g
pH 6.8 ± 0.2 at 25°C

**Source:** This medium is available as a premixed powder from Oxoid Unipath and BD Diagnostic Systems.

**Preparation of Medium:** Add components to distilled/deionized water and bring volume to 1.0L. Mix thoroughly and heat with frequent agitation until boiling. Distribute into tubes or flasks. Do not autoclave. Cool to approximately 45°–50°C. If desired, add 2mg/L of neomycin sulfate. Swirl to disperse the insoluble material and pour into sterile Petri dishes.

**Storage/Shelf Life:** Store dehydrated media in the dark in a sealed container below 30°C. Prepared media should be stored under refrigeration (2-8°C). Media should be used within 60 days of preparation. Media should not be used if there are any signs of deterioration (shrinking, cracking, or discoloration) or contamination, or if the expiration date supplied by the manufacturer has passed.

**Use:** For the detection, isolation, and presumptive identification of *Candida* species. Addition of neomycin helps inhibit bacterial species. *Candida albicans* appears as brown to black colonies with no pigment diffusion and no sheen. *Candida tropicalis* appears as dark brown colonies with black centers, black pigment diffusion, and a sheen. *Candida krusei* appears as shiny, wrinkled, brown to black colonies with yellow pigment diffusion. *Candida pseudotropicalis* appears as flat, shiny red to brown colonies with no pigment diffusion. *Candida parakrusei* appears as flat, shiny, wrinkled, dark reddish-brown colonies with light reddish-brown peripheries and a yellow fringe. *Candida stellatoidea* appears as flat dark brown colonies with a light fringe.

## Bird Seed Agar
### (*Guizotia abyssinica* Creatinine Agar)
### (Niger Seed Agar)/(Staib Agar)
**Composition** per liter:

Agar ....................................................................................15.0g
Glucose ...............................................................................15.0g
Creatinine .............................................................................5.0g
KH$_2$PO$_4$.................................................................................3.0g
Biphenyl................................................................................1.0g
Chloramphenicol...................................................................0.5g
*Guizotia abyssinica* seed (niger seed) extract......................1000.0mL
pH 6.7 ± 0.2 at 25°C

**Preparation of Medium:** Prepare seed extract by grinding 50.0g of *Guizotia abyssinica* seed in 1.0L of distilled/deionized water. Boil for 30 min. Filter through cheesecloth and filter paper. Add remaining components to seed filtrate. Mix thoroughly and heat with frequent agitation until boiling. Distribute into flasks or tubes. Autoclave for 25 min at 15 psi pressure–110°C.

**Storage/Shelf Life:** Store dehydrated media in the dark in a sealed container below 30°C. Prepared media should be stored under refrigeration (2-8°C). Media should be used within 60 days of preparation. Media should not be used if there are any signs of deterioration (shrinking, cracking, or discoloration) or contamination, or if the expiration date supplied by the manufacturer has passed.

**Use:** For the selective isolation and differentiation of *Cryptococcus neoformans* from other *Cryptococcus* species and other yeasts.

## Blood Glucose Cystine Agar
**Composition** per 100.0mL:

Nutrient agar .....................................................................85.0mL
Glucose cystine solution ...................................................10.0mL
Human blood, fresh ............................................................5.0mL
pH 6.8 ± 0.2 at 25°C

**Nutrient Agar:**
**Composition** per liter:

Agar ....................................................................................15.0g
Pancreatic digest of gelatin..................................................5.0g
Beef extract..........................................................................3.0g

**Source:** Nutrient agar is available as a premixed powder from BD Diagnostic Systems.

**Preparation of Nutrient Agar:** Add components to distilled/deionized water and bring volume to 1.0L. Mix thoroughly. Gently heat while stirring and bring to boiling. Distribute into tubes or flasks. Autoclave for 15 min at 15 psi pressure–121°C. Cool to 45°–50°C.

**Glucose Cystine Solution:**
**Composition** per 50.0mL:

Glucose ...............................................................................12.5g
L-Cystine·HCl ......................................................................0.5g

**Preparation of Glucose Cystine Solution:** Add components to distilled/deionized water and bring volume to 50.0mL. Mix thoroughly. Filter sterilize.

**Preparation of Medium:** To 85.0mL of cooled, sterile agar solution, aseptically add 10.0mL of sterile glucose cystine solution and 5.0mL of human blood. Mix thoroughly. Pour into sterile Petri dishes or distribute into sterile tubes.

**Use:** For the cultivation of various fungi.

## Bonner-Addicott Medium
**Composition** per liter:

Agar ....................................................................................25.0g
Glucose ...............................................................................20.0g
Ca(NO$_3$)$_2$·4H$_2$O ..............................................................0.236g
KNO$_3$ ................................................................................0.081g
KCl......................................................................................0.065g
MgSO$_4$·7H$_2$O ...................................................................0.036g
KH$_2$PO$_4$.............................................................................0.012g
Ferric tartrate .....................................................................1.0mg

**Preparation of Medium:** Add components to distilled/deionized water and bring volume to 1.0L. Mix thoroughly. Gently heat and bring to boiling. Distribute into tubes or flasks. Autoclave for 15 min at 15 psi pressure–121°C. Pour into sterile Petri dishes or leave in tubes.

**Storage/Shelf Life:** Store dehydrated media in the dark in a sealed container below 30°C. Prepared media should be stored under refrigeration (2-8°C). Media should be used within 60 days of preparation. Media should not be used if there are any signs of deterioration (shrinking, cracking, or discoloration) or contamination, or if the expiration date supplied by the manufacturer has passed.

**Use:** For the cultivation of a variety of fungi.

## Brain Heart CC Agar
### (Brain Heart Cycloheximide
### Chloramphenicol Agar)
**Composition** per liter:

Pancreatic digest of casein.................................................16.0g
Agar ....................................................................................13.5g

Brain heart, solids from infusion ...................................................8.0g
Peptic digest of animal tissue.........................................5.0g
NaCl...............................................................................5.0g
Na$_2$HPO$_4$.........................................................................2.5g
Glucose ...........................................................................2.0g
Cycloheximide ...............................................................0.5g
Chloramphenicol............................................................0.05g

pH 7.4 ± 0.2 at 25°C

**Source:** This medium is available as a premixed powder from BD Diagnostic Systems.

**Caution:** Cycloheximide is toxic. Avoid skin contact or aerosol formation and inhalation.

**Preparation of Medium:** Add components to distilled/deionized water and bring volume to 1.0L. Mix thoroughly. Distribute into tubes or flasks while shaking to distribute precipitate. Autoclave for 15 min at 15 psi pressure–118°C.

**Storage/Shelf Life:** Store dehydrated media in the dark in a sealed container below 30°C. Prepared media should be stored under refrigeration (2-8°C). Media should be used within 60 days of preparation. Media should not be used if there are any signs of deterioration (shrinking, cracking, or discoloration) or contamination, or if the expiration date supplied by the manufacturer has passed.

**Use:** For the selective isolation of fastidious pathogenic fungi such as *Histoplasma capsulatum* and *Blastomyces dermatiditis* from specimens heavily contaminated with bacteria and other fungi. It may also be used as a base supplemented with sheep blood and gentamicin for enrichment and additional selectivity.

### Brain Heart CC Agar, HiVeg
### (Brain Heart Cycloheximide
### Chloramphenicol Agar, HiVeg)

**Composition** per liter:
Agar ................................................................................15.0g
Plant infusion .................................................................10.0g
Plant peptone No. 3.........................................................10.0g
Plant special infusion .......................................................7.5g
NaCl...............................................................................5.0g
Na$_2$HPO$_4$ ........................................................................2.5g
Glucose ...........................................................................2.0g
Cycloheximide ...............................................................0.5g
Chloramphenicol............................................................0.05g

pH 7.4 ± 0.2 at 25°C

**Source:** This medium is available as a premixed powder from HiMedia.

**Caution:** Cycloheximide is toxic. Avoid skin contact or aerosol formation and inhalation.

**Preparation of Medium:** Add components to distilled/deionized water and bring volume to 1.0L. Mix thoroughly. Distribute into tubes or flasks while shaking to distribute precipitate. Autoclave for 15 min at 15 psi pressure–118°C.

**Storage/Shelf Life:** Store dehydrated media in the dark in a sealed container below 30°C. Prepared media should be stored under refrigeration (2-8°C). Media should be used within 60 days of preparation. Media should not be used if there are any signs of deterioration (shrinking, cracking, or discoloration) or contamination, or if the expiration date supplied by the manufacturer has passed.

**Use:** For the selective isolation of fastidious pathogenic fungi such as *Histoplasma capsulatum* and *Blastomyces dermatiditis* from specimens heavily contaminated with bacteria and other fungi. This medium may also be used as a base supplemented with sheep blood and gentamicin for enrichment and additional selectivity.

### Brain Heart Infusion Agar

**Composition** per liter:
Pancreatic digest of casein.........................................16.0g
Agar ...............................................................................13.5g
Brain heart, solids from infusion ...................................8.0g
Peptic digest of animal tissue .......................................5.0g
NaCl...............................................................................5.0g
Glucose ...........................................................................2.0g
Na$_2$HPO$_4$.........................................................................2.5g

pH 7.4 ± 0.2 at 25°C

**Source:** This medium is available as a premixed powder from Oxoid Unipath and BD Diagnostic Systems.

**Preparation of Medium:** Add components to distilled/deionized water and bring volume to 1.0L. Mix thoroughly. Distribute into tubes or flasks while shaking to distribute precipitate. Autoclave for 15 min at 15 psi pressure–121°C.

**Storage/Shelf Life:** Store dehydrated media in the dark in a sealed container below 30°C. Prepared media should be stored under refrigeration (2-8°C). Media should be used within 60 days of preparation. Media should not be used if there are any signs of deterioration (shrinking, cracking, or discoloration) or contamination, or if the expiration date supplied by the manufacturer has passed.

**Use:** For the cultivation of a wide variety of fastidious microorganisms, including bacteria, yeasts, and molds. With the addition of 10% sheep blood, it is used for the isolation and cultivation of many fungal species, including systemic fungi, from clinical speciments and nonclinical specimens of public health importance. The addition of gentamicin and chloramphenicol with 10% sheep blood produces a selective medium used for the isolation of pathogenic fungi from specimens heavily contaminated with bacteria and saprophytic fungi. It is recommended for the isolation of *Histoplasma capsulatum* and other pathogenic fungi, including *Coccidioides immitis*.

### Brain Heart Infusion Agar
### (BAM M24 Medium 2)

**Composition** per liter:
Agar ...............................................................................15.0g
Pancreatic digest of gelatin..........................................14.5g
Brain heart, solids from infusion ...................................6.0g
Peptic digest of animal tissue .......................................6.0g
NaCl...............................................................................5.0g
Glucose ...........................................................................3.0g
Na$_2$HPO$_4$.........................................................................2.5g

pH 7.4 ± 0.2 at 25°C

**Source:** This medium is available as a premixed powder from BD Diagnostic Systems.

**Preparation of Medium:** Add components to distilled/deionized water and bring volume to 1.0L. Mix thoroughly. Distribute into tubes or flasks while shaking to distribute precipitate. Autoclave for 15 min at 15 psi pressure–121°C. Mix thoroughly. Pour into sterile Petri dishes.

**Storage/Shelf Life:** Store dehydrated media in the dark in a sealed container below 30°C. Prepared media should be stored under refriger-

ation (2-8°C). Media should be used within 60 days of preparation. Media should not be used if there are any signs of deterioration (shrinking, cracking, or discoloration) or contamination, or if the expiration date supplied by the manufacturer has passed.

**Use:** For the cultivation of a wide variety of fastidious microorganisms, including yeasts and molds.

## Brain Heart Infusion Agar with 1% Agar, HiVeg
**Composition** per liter:
| | |
|---|---|
| Agar | 15.0g |
| Plant infusion | 10.0g |
| Plant peptone No. 3 | 10.0g |
| Plant special infusion | 7.5g |
| NaCl | 5.0g |
| Na_2HPO_4 | 2.5g |
| Glucose | 2.0g |

pH 7.4 ± 0.2 at 25°C

**Source:** This medium is available as a premixed powder from Hi-Media.

**Preparation of Medium:** Add components to distilled/deionized water and bring volume to 1.0L. Mix thoroughly. Gently heat and bring to boiling. Distribute into tubes or flasks. Autoclave for 15 min at 15 psi pressure–121°C. Pour into sterile Petri dishes or leave in tubes.

**Storage/Shelf Life:** Store dehydrated media in the dark in a sealed container below 30°C. Prepared media should be stored under refrigeration (2-8°C). Media should be used within 60 days of preparation. Media should not be used if there are any signs of deterioration (shrinking, cracking, or discoloration) or contamination, or if the expiration date supplied by the manufacturer has passed.

**Use:** For the cultivation of a variety of fastidious pathogenic yeasts and molds.

## Brain Heart Infusion Agar with 1% Agar, HiVeg
**Composition** per liter:
| | |
|---|---|
| Agar | 10.0g |
| Plant infusion | 10.0g |
| Plant peptone No. 3 | 10.0g |
| Plant special infusion | 7.5g |
| NaCl | 5.0g |
| Na_2HPO_4 | 2.5g |
| Glucose | 2.0g |

pH 7.4 ± 0.2 at 25°C

**Source:** This medium is available as a premixed powder from Hi-Media.

**Preparation of Medium:** Add components to distilled/deionized water and bring volume to 1.0L. Mix thoroughly. Gently heat and bring to boiling. Distribute into tubes or flasks. Autoclave for 15 min at 15 psi pressure–121°C. Pour into sterile Petri dishes or leave in tubes.

**Storage/Shelf Life:** Store dehydrated media in the dark in a sealed container below 30°C. Prepared media should be stored under refrigeration (2-8°C). Media should be used within 60 days of preparation. Media should not be used if there are any signs of deterioration (shrinking, cracking, or discoloration) or contamination, or if the expiration date supplied by the manufacturer has passed.

**Use:** For the cultivation of a variety of fastidious pathogenic yeasts and molds.

## Brain Heart Infusion Agar with 1% Agar, HiVeg with Penicillin
**Composition** per liter:
| | |
|---|---|
| Agar | 15.0g |
| Plant infusion | 10.0g |
| Plant peptone No. 3 | 10.0g |
| Plant special infusion | 7.5g |
| NaCl | 5.0g |
| Na_2HPO_4 | 2.5g |
| Glucose | 2.0g |
| Penicillin solution | 2.0mL |

pH 7.4 ± 0.2 at 25°C

**Source:** This medium is available as a premixed powder from Hi-Media.

**Penicillin Solution:**
**Composition** per 2.0mL:
| | |
|---|---|
| Penicillin | 0.1g |

**Preparation of Penicillin Solution:** Add penicillin to ethanol and bring volume to 2.0mL. Mix thoroughly. Filter sterilize.

**Preparation of Medium:** Add components to distilled/deionized water and bring volume to 1.0L. Mix thoroughly. Gently heat and bring to boiling. Autoclave for 15 min at 15 psi pressure–121°C. Cool to 45–50°C. Aseptically add 2.0mL penicillin solution. Mix thoroughly. Pour into sterile Petri dishes or leave in tubes.

**Storage/Shelf Life:** Store dehydrated media in the dark in a sealed container below 30°C. Prepared media should be stored under refrigeration (2-8°C). Media should be used within 60 days of preparation. Media should not be used if there are any signs of deterioration (shrinking, cracking, or discoloration) or contamination, or if the expiration date supplied by the manufacturer has passed.

**Use:** For the cultivation of a variety of fastidious pathogenic yeasts and molds.

## Brain Heart Infusion Agar with Chloramphenicol
**Composition** per liter:
| | |
|---|---|
| Pancreatic digest of casein | 16.0g |
| Agar | 13.5g |
| Brain heart, solids from infusion | 8.0g |
| Peptic digest of animal tissue | 5.0g |
| NaCl | 5.0g |
| Glucose | 2.0g |
| Na_2HPO_4 | 2.5g |
| Sheep blood, defibrinated | 50.0mL |
| Chloramphenicol solution | 10.0mL |

pH 7.4 ± 0.2 at 25°C

**Chloramphenicol Solution:**
**Composition** per 10.0mL:
| | |
|---|---|
| Chloramphenicol | 0.05g |

**Preparation of Chloramphenicol Solution:** Add chloramphenicol to distilled/deionized water and bring volume to 10.0mL. Mix thoroughly. Filter sterilize.

**Preparation of Medium:** Add components, except chloramphenicol solution and sheep blood, to distilled/deionized water and bring volume to 940.0mL. Mix thoroughly. Gently heat and bring to boiling. Autoclave for 15 min at 15 psi pressure–121°C. Cool to 45°–50°C. Aseptically add sterile chloramphenicol solution and sheep blood. Mix thoroughly. Pour into sterile Petri dishes or distribute into sterile tubes.

**Storage/Shelf Life:** Store dehydrated media in the dark in a sealed container below 30°C. Prepared media should be stored under refrigeration (2-8°C). Media should be used within 60 days of preparation. Media should not be used if there are any signs of deterioration (shrinking, cracking, or discoloration) or contamination, or if the expiration date supplied by the manufacturer has passed.

**Use:** For the isolation and cultivation of a wide variety of fungal species, especially systemic fungi, from clinical speciments and nonclinical specimens of public health importance. For the selective isolation of pathogenic fungi from specimens heavily contaminated with bacteria and saprophytic fungi. For the maintenance of fungal species on slant cultures.

## Brain Heart Infusion Agar
## with Penicillin and Streptomycin

**Composition** per liter:

| | |
|---|---|
| Pancreatic digest of casein | 16.0g |
| Agar | 13.5g |
| Brain heart, solids from infusion | 8.0g |
| Peptic digest of animal tissue | 5.0g |
| NaCl | 5.0g |
| Glucose | 2.0g |
| Na$_2$HPO$_4$ | 2.5g |
| Streptomycin | 40.0mg |
| Penicillin | 20,000U |
| Sheep blood, defibrinated | 50.0mL |

pH 7.4 ± 0.2 at 25°C

**Preparation of Medium:** Add components, except sheep blood, to distilled/deionized water and bring volume to 950.0mL. Mix thoroughly and while stirring bring to a boil for 1 min to completely dissolve. Autoclave for 15 min at 15 psi pressure–121°C. Cool to 50°C. Aseptically add 50.0mL of defibrinated sheep blood. Mix thoroughly. Pour into sterile Petri dishes while agitating gently to distribute the precipitate through the medium.

**Storage/Shelf Life:** Store dehydrated media in the dark in a sealed container below 30°C. Prepared media should be stored under refrigeration (2-8°C). Media should be used within 60 days of preparation. Media should not be used if there are any signs of deterioration (shrinking, cracking, or discoloration) or contamination, or if the expiration date supplied by the manufacturer has passed.

**Use:** For the isolation and cultivation of a wide variety of fungal species, especially systemic fungi, from clinical speciments and nonclinical specimens of public health importance. For the selective isolation of pathogenic fungi from specimens heavily contaminated with bacteria and saprophytic fungi. For the maintenance of fungal species on slant cultures.

## Brain Heart Infusion Agar with 10% Sheep Blood, Gentamicin, and Chloramphenicol

**Composition** per liter:

| | |
|---|---|
| Pancreatic digest of casein | 16.0g |
| Agar | 13.5g |
| Brain heart, solids from infusion | 8.0g |
| Peptic digest of animal tissue | 5.0g |
| NaCl | 5.0g |
| Glucose | 2.0g |
| Na$_2$HPO$_4$ | 2.5g |
| Sheep blood, defibrinated | 100.0mL |
| Antibiotic solution | 10.0mL |

pH 7.4 ± 0.2 at 25°C

**Antibiotic Solution:**

**Composition** per 10.0mL:

| | |
|---|---|
| Chloramphenicol | 0.05g |
| Gentamicin | 0.05g |

**Preparation of Antibiotic Solution:** Add components to distilled/deionized water and bring volume to 10.0mL. Mix thoroughly. Filter sterilize.

**Preparation of Medium:** Add components, except antibiotic solution and sheep blood, to distilled/deionized water and bring volume to 890.0mL. Mix thoroughly. Gently heat and bring to boiling. Autoclave for 15 min at 15 psi pressure–121°C. Cool to 45°–50°C. Aseptically add sterile antibiotic solution and sheep blood. Mix thoroughly. Pour into sterile Petri dishes or distribute into sterile tubes.

**Storage/Shelf Life:** Store dehydrated media in the dark in a sealed container below 30°C. Prepared media should be stored under refrigeration (2-8°C). Media should be used within 60 days of preparation. Media should not be used if there are any signs of deterioration (shrinking, cracking, or discoloration) or contamination, or if the expiration date supplied by the manufacturer has passed.

**Use:** For the isolation and cultivation of a wide variety of fungal species, especially systemic fungi, from clinical speciments and nonclinical specimens of public health importance. For the selective isolation of pathogenic fungi from specimens heavily contaminated with bacteria and saprophytic fungi. For the maintenance of fungal species on slant cultures.

## Brilliance™ *Candida* Agar

**Composition** per liter:

| | |
|---|---|
| Chromogenic mix | 13.6g |
| Agar | 13.6g |
| Peptone | 4.0g |
| Selective supplement solution | 10.0mL |

pH 6.0 ± 0.2 at 25°C

**Source:** This medium is available as a premixed powder from Oxoid Unipath.

**Selective Supplement Solution:**

**Composition** per 10.0mL:

| | |
|---|---|
| Chloramphenicol | 0.5g |

**Preparation of Selective Supplement Solution:** Add chloramphenicol to distilled/deionized water and bring volume to 10.0mL. Mix thoroughly. Filter sterilize.

**Preparation of Medium:** Add components, except selective supplement solution, to distilled/deionized water and bring volume to 1.0L. Mix thoroughly. Add 10.0mL selective supplement solution. Gently heat while stirring and bring to boiling. Do not autoclave. Cool to 45°C. Mix thoroughly. Pour into sterile Petri dishes.

**Storage/Shelf Life:** Store in the dark under refrigeration (2-8°C). Chromogenic agars are especially light and temperature sensitive; protect from light, excessive heat, moisture, and freezing. Do not use after the expiration date supplied by the manufacturer.

**Use:** For the rapid isolation and identification of clinically important *Candida* species. The green color of *Candida albicans* and *Candida dubliniensis* is caused by the same chromogenic reaction as the dark blue color of *Candida tropicalis*. *Candida glabrata, Candida kefyr, Candida parapsilosis,* and *Candida lusitaniae* appear as a variety of beige/brown/yellow colors, due to the mixture of natural pigmentation and some alkaline phosphatase activity.

# Caffeic Acid Ferric Citrate Test Medium
## (CAFC Test Medium)
## (Caffeic Acid Agar)

**Composition** per liter:

| | |
|---|---|
| Agar | 20.0g |
| $(NH_4)_2SO_4$ | 5.0g |
| Glucose | 5.0g |
| Yeast extract | 2.0g |
| $K_2HPO_4$ | 0.8g |
| $MgSO_4 \cdot 3H_2O$ | 0.7g |
| Caffeic acid $\cdot 1/2H_2O$ | 0.18g |
| Chloramphenicol | 0.05g |
| Ferric citrate solution | 4.0mL |

pH $6.5 \pm 0.2$ at 25°C

**Ferric Citrate Solution:**

**Composition** per 20.0mL:

| | |
|---|---|
| Ferric citrate | 100.0mg |

**Preparation of Ferric Citrate Solution:** Add ferric citrate to 20.0mL of distilled/deionized water. Mix thoroughly.

**Preparation of Medium:** Add components, except chloramphenicol, to distilled/deionized water and bring volume to 1.0L. Mix thoroughly. Heat to boiling. Autoclave for 15 min at 15 psi pressure–121°C. Cool to 45°–50°C. Aseptically add 0.05g of chloramphenicol. Mix thoroughly. Pour into sterile Petri dishes.

**Storage/Shelf Life:** Store dehydrated media in the dark in a sealed container below 30°C. Prepared media should be stored under refrigeration (2-8°C). Media should be used within 60 days of preparation. Media should not be used if there are any signs of deterioration (shrinking, cracking, or discoloration) or contamination, or if the expiration date supplied by the manufacturer has passed.

**Use:** For the isolation and presumptive identification of *Cryptococcus neoformans*. *Cryptococcus neoformans* appears as dark brown colonies. All other *Cryptococcus* species appear as light brown or nonpigmented colonies.

# Canavine-Glycine Bromthymol Blue Agar

**Composition** per liter:

| | |
|---|---|
| Agar | 20.0g |
| Glycine | 10.0g |
| $K_2HPO_4$ | 1.0g |
| $MgSO_4 \cdot 7H_2O$ | 1.0g |
| Bromothymol Blue | 0.08g |
| L-Canavanine sulfate | 0.03g |
| Thiamine hydrochloride | 1.0mg |

pH $5.8 \pm 0.2$ at 25°C

**Preparation of Medium:** Add components to distilled/deionized water and bring volume to 1.0L. Mix thoroughly. Gently heat and bring to boiling. Autoclave for 15 min at 15 psi pressure–121°C. Pour into sterile Petri dishes.

**Storage/Shelf Life:** Store dehydrated media in the dark in a sealed container below 30°C. Prepared media should be stored under refrigeration (2-8°C). Media should be used within 60 days of preparation. Media should not be used if there are any signs of deterioration (shrinking, cracking, or discoloration) or contamination, or if the expiration date supplied by the manufacturer has passed.

**Use:** For the differentiation of *Cryptococcus neoformans* from *Cryptococcus gatti*. Growth and a color change to medium blue indicates *C.*

*gattii*; no growth or minimal growth that is medium yellow or green may be interpreted as *C. neoformans* var. *neoformans*.

# *Candida* Agar

**Composition** per liter:

| | |
|---|---|
| Agar | 20.0g |
| Glucose | 10.0g |
| Peptic digest of animal tissue | 5.0g |
| Yeast extract | 3.0g |
| Malt extract | 3.0g |
| Aniline Blue | 0.1g |

pH $6.2 \pm 0.2$ at 25°C

**Source:** This medium is available from HiMedia.

**Preparation of Medium:** Add components to distilled/deionized water and bring volume to 1.0L. Mix thoroughly. Distribute into tubes or flasks. Autoclave for 15 min at 15 psi pressure–121°C. Pour into Petri dishes or leave in tubes.

**Storage/Shelf Life:** Store dehydrated media in the dark in a sealed container below 30°C. Prepared media should be stored under refrigeration (2-8°C). Media should be used within 60 days of preparation. Media should not be used if there are any signs of deterioration (shrinking, cracking, or discoloration) or contamination, or if the expiration date supplied by the manufacturer has passed.

**Use:** For the isolation and differentiation of *Candida albicans*.

# *Candida* BCG Agar Base
## (*Candida* Bromcresol Green Agar Base)

**Composition** per liter:

| | |
|---|---|
| Glucose | 40.0g |
| Agar | 15.0g |
| Peptone | 10.0g |
| Yeast extract | 1.0g |
| Bromcresol Green | 0.02g |
| Neomycin solution | 10.0mL |

pH $6.1 \pm 0.1$ at 25°C

**Source:** This medium is available as a premixed powder from BD Diagnostic Systems.

**Neomycin Solution:**
**Composition** per 10.0mL:

| | |
|---|---|
| Neomycin | 0.5g |

**Preparation of Neomycin Solution:** Add neomycin to distilled/deionized water and bring volume to 10.0mL. Mix thoroughly. Filter sterilize.

**Preparation of Medium:** Add components, except neomycin solution, to distilled/deionized water and bring volume to 1.0L. Mix thoroughly and heat gently until boiling. Autoclave for 15 min at 15 psi pressure–121°C. Cool to 50°–55°C. Aseptically add 10.0mL of sterile neomycin solution. Mix thoroughly. Pour into sterile Petri dishes or leave in tubes.

**Storage/Shelf Life:** Store dehydrated media in the dark in a sealed container below 30°C. Prepared media should be stored under refrigeration (2-8°C). Media should be used within 60 days of preparation. Media should not be used if there are any signs of deterioration (shrinking, cracking, or discoloration) or contamination, or if the expiration date supplied by the manufacturer has passed.

**Use:** For the selective isolation and identification of *Candida* species. It is a highly differential medium that is used for demonstrating morphological and biochemical reactions characterizing different *Candida* species. *Candida albicans* appears as blunt conical colonies with smooth edges and yellow to blue-green color. *Candida stellatoidea* appears as convex colonies with smooth edges and yellow to green color. *Candida tropicalis* appears as convex colonies with wavy edges and yellow-green to green color with a dark blue-green base. *Candida pseudotropicalis* appears as convex, shiny colonies with smooth edges and green color with a light green edge. *Candida krusei* appears as low conical colonies with spreading edges and blue-green color. *Candida stellatoidea* appears as convex colonies with smooth edges and yellow to green color.

### *Candida* BCG HiVeg Agar Base with Neomycin

**Composition** per liter:

| | |
|---|---|
| Glucose | 40.0g |
| Agar | 15.0g |
| Plant peptone | 10.0g |
| Yeast extract | 1.0g |
| Bromcresol Green | 0.02g |
| Neomycin solution | 10.0mL |

pH 6.1 ± 0.1 at 25°C

**Source:** This medium, without neomycin solution, is available as a premixed powder from HiMedia.

**Neomycin Solution:**

**Composition** per 10.0mL:

| | |
|---|---|
| Neomycin | 0.5g |

**Preparation of Neomycin Solution:** Add neomycin to distilled/deionized water and bring volume to 10.0mL. Mix thoroughly. Filter sterilize.

**Preparation of Medium:** Add components, except neomycin solution, to distilled/deionized water and bring volume to 990.0mL. Mix thoroughly and heat gently until boiling. Autoclave for 15 min at 15 psi pressure–121°C. Cool to 50°C. Aseptically add 10.0mL of sterile neomycin solution. Mix thoroughly. Pour into sterile Petri dishes or leave in tubes.

**Storage/Shelf Life:** Store dehydrated media in the dark in a sealed container below 30°C. Prepared media should be stored under refrigeration (2-8°C). Media should be used within 60 days of preparation. Media should not be used if there are any signs of deterioration (shrinking, cracking, or discoloration) or contamination, or if the expiration date supplied by the manufacturer has passed.

**Use:** For the selective isolation and identification of *Candida* species. It is a highly differential medium that is used for demonstrating morphological and biochemical reactions characterizing different *Candida* species. *Candida albicans* appears as blunt conical colonies with smooth edges and yellow to blue-green color. *Candida stellatoidea* appears as convex colonies with smooth edges and yellow to green color. *Candida tropicalis* appears as convex colonies with wavy edges and yellow-green to green color with a dark blue-green base. *Candida pseudotropicalis* appears as convex, shiny colonies with smooth edges and green color with a light green edge. *Candida krusei* appears as low conical colonies with spreading edges and blue-green color. *Candida stellatoidea* appears as convex colonies with smooth edges and yellow to green color.

### *Candida* Chromogenic Agar

**Composition** per liter:

| | |
|---|---|
| Glucose | 20.0g |
| Peptone | 10.0g |

| | |
|---|---|
| Chloramphenicol | 0.5g |
| Chromogenic mix | 0.4g |

pH 7.0 ± 0.2 at 25°C

**Source:** This medium is available from CONDA, Barcelona, Spain.

**Preparation of Medium:** Add components to distilled/deionized water and bring volume to 1.0L. Mix thoroughly. Gently heat and bring to boiling. Do not autoclave. Cool to 50°C. Pour into sterile Petri dishes.

**Storage/Shelf Life:** Store in the dark under refrigeration (2-8°C). Chromogenic agars are especially light and temperature sensitive; protect from light, excessive heat, moisture, and freezing. Do not use after the expiration date supplied by the manufacturer.

**Use:** For the rapid isolation and identification of clinically important *Candida* species. *Candida tropicalis* colonies appear dark blue. *Candida albicans* colonies appear green. *Candida krusei* colonies are pink-purple.

### *Candida* Diagnostic Agar (CDA)

**Composition** per liter:

| | |
|---|---|
| Glucose | 40.0g |
| Agar | 15.0g |
| Peptone, mycological | 10.0g |
| Ammonium 4-{2-[4-(2-acetamido-2-deoxy-β-D-glucopyranosyloxy)-3-methoxy- phenyl]-vinyl}-1-(propan-3-yl-oate)-quinolium bromide | 0.32g |

pH 6.9 ± 0.2 at 25°C

**Source:** This medium is available from PCR Diagnostics.

**Preparation of Medium:** Add components to distilled/deionized water and bring volume to 1.0L. Mix thoroughly and heat with frequent agitation until boiling. Boil until components are fully dissolved. Do not autoclave. Cool to 45°–50°C. Pour into sterile Petri dishes.

**Storage/Shelf Life:** Store in the dark under refrigeration (2-8°C). Chromogenic agars are especially light and temperature sensitive; protect from light, excessive heat, moisture, and freezing. Do not use after the expiration date supplied by the manufacturer.

**Use:** For the rapid isolation and identification of *Candida* species. *Candida albicans* and *Candida dubliniensis* produce white colonies with deep-red spots on a yellow transparent background. Colonies of *Candida tropicalis* and *Candida kefyr* are uniformly pink, and colonies of other *Candida* spp., including *Candida glabrata* and *Candida parapsilosis*, appear white.

### *Candida* Isolation Agar

**Composition** per liter:

| | |
|---|---|
| Agar | 20.0g |
| Glucose | 10.0g |
| Peptone | 5.0g |
| Yeast extract | 3.0g |
| Malt extract | 3.0g |
| Aniline Blue | 0.1g |

pH 5.9 ± 0.5 at 25°C

**Source:** This medium is available as a premixed powder from BD Diagnostic Systems.

**Preparation of Medium:** Add components to distilled/deionized water and bring volume to 1.0L. Mix thoroughly. Gently heat and bring to boiling. Distribute into tubes or flasks. Autoclave for 15 min at 15 psi pressure–121°C. Pour into sterile Petri dishes or leave in tubes.

**Storage/Shelf Life:** Store dehydrated media in the dark in a sealed container below 30°C. Prepared media should be stored under refrigeration (2-8°C). Media should be used within 60 days of preparation. Media should not be used if there are any signs of deterioration (shrinking, cracking, or discoloration) or contamination, or if the expiration date supplied by the manufacturer has passed.

**Use:** For the isolation and differentiation of *Candida albicans. Candida albicans* turns the medium blue.

### CandiSelect 4™

**Composition** per liter:
Proprietary

**Source:** Available from BioRad.

**Preparation of Medium:** Preprepared plates.

**Storage/Shelf Life:** Prepared media should be stored under refrigeration (2-8°C). Media should not be used if there are any signs of deterioration (shrinking, cracking, or discoloration) or contamination, or if the expiration date supplied by the manufacturer has passed.

**Use:** For the direct identification of *Candida albicans* and for the presumptive identification of *Candida tropicalis, Candida glabrata*, and *Candida krusei. C. albicans* form pink to purple colones; other *Candida* species form turquoise colonies.

### Carbon Assimilation Medium

**Composition** per liter:
| | |
|---|---|
| Agar solution | 500.0mL |
| Mineral base medium | 500.0mL |

pH 6.5 ± 0.1 at 25°C

**Agar Solution:**

**Composition** per liter:
| | |
|---|---|
| Agar | 32.0g |

**Preparation of Agar Solution:** Add agar to distilled/deionized water and bring volume to 1.0L. Mix thoroughly. Gently heat and bring to boiling. Autoclave for 15 min at 15 psi pressure–121°C. Cool to 45°–50°C.

**Mineral Base Medium:**

**Composition** per 500.0mL:
| | |
|---|---|
| Carbohydrate | 10.0g |
| NaCl | 5.0g |
| $NH_4HPO_4$ | 1.0g |
| $K_2HPO_4$ | 1.0g |
| $MgSO_4 \cdot 7H_2O$, anhydrous | 0.1g |

**Preparation of Mineral Base Medium:** Add components to distilled/deionized water and bring volume to 500.0mL. Mix thoroughly. Gently heat until dissolved. Filter sterilize. Warm to 45°–50°C.

**Preparation of Medium:** Combine 500.0mL of cooled, sterile agar solution and 500.0mL of sterile mineral base medium. Mix thoroughly. Aseptically distribute into sterile tubes. Allow tubes to cool in a slanted position.

**Use:** For the cultivation and differentiation of microorganisms based on their ability to utilize a particular carbon source.

### Carbon Assimilation Medium, Auxanographic Method for Yeast Identification

**Composition** per liter:
| | |
|---|---|
| Noble agar | 20.0g |
| $(NH_4)_2SO_4$ | 0.5g |
| $KH_2PO_4$ | 0.1g |
| $MgSO_4 \cdot 7H_2O$ | 0.05g |
| NaCl | 0.01g |
| $CaCl_2 \cdot 2H_2O$ | 0.01g |
| DL-Methionine | 2.0mg |
| DL-Tryptophan | 2.0mg |
| L-Histidine·HCl | 1.0mg |
| Inositol | 0.2mg |
| KI | 0.01mg |
| $H_3BO_3$ | 0.05mg |
| $ZnSO_4 \cdot 7H_2O$ | 0.04mg |
| $MnSO_4 \cdot 4H_2O$ | 0.04mg |
| Thiamine·HCl | 0.04mg |
| Pyroxidine·HCl | 0.04mg |
| Niacin | 0.04mg |
| Calcium pantothenate | 0.04mg |
| *p*-Aminobenzoic acid | 0.02mg |
| Riboflavin | 0.02mg |
| $FeCl_3$ | 0.02mg |
| $Na_2MoO_4 \cdot 4H_2O$ | 0.02mg |
| $CuSO_4 \cdot 5H_2O$ | 4.0µg |
| Folic acid | 0.2µg |
| Biotin | 0.2µg |

pH 4.5 ± 0.2 at 25°C

**Preparation of Medium:** Add components to distilled/deionized water and bring volume to 1.0L. Mix thoroughly. Gently heat and bring to boiling. Distribute into screw-capped tubes in 20.0mL volumes. Autoclave for 15 min at 15 psi pressure–121°C.

**Storage/Shelf Life:** Store dehydrated media in the dark in a sealed container below 30°C. Prepared media should be stored under refrigeration (2-8°C). Media should be used within 60 days of preparation. Media should not be used if there are any signs of deterioration (shrinking, cracking, or discoloration) or contamination, or if the expiration date supplied by the manufacturer has passed.

**Use:** For carbohydrate assimilation tests by the auxanographic method for the identification of yeasts.

### Casein Hydrolysate Yeast Extract Salts HiVeg Broth Base with Tracer Salts (CAYES)

**Composition** per liter:
| | |
|---|---|
| Plant acid hydrolysate | 20.0g |
| $K_2HPO_4$ | 8.71g |
| Yeast extract | 6.0g |
| NaCl | 2.5g |
| Tracer salts solution | 1.0mL |

pH 7.0 ± 0.2 at 25°C

**Source:** This medium, without tracer salts solution, is available as a premixed powder from HiMedia.

**Tracer Salts Solution:**

**Composition** per 10.0mL:
| | |
|---|---|
| $MgSO_4$ | 0.5g |
| $MnCl_2$ | 0.05g |
| $FeCl_3$ | 0.05g |
| Sulfuric acid, 1*N* | 10.0mL |

**Preparation of Tracer Salts Solution:** Add components to 0.1*N* sulfuric acid and bring volume to 10.0mL. Mix thoroughly. Filter sterilize.

**Preparation of Medium:** Add components to distilled/deionized water and bring volume to 1.0L. Mix thoroughly. Gently heat and bring to boiling. Distribute into tubes or flasks. Autoclave for 15 min at 15 psi pressure–121°C.

**Storage/Shelf Life:** Store dehydrated media in the dark in a sealed container below 30°C. Prepared media should be stored under refrigeration (2-8°C). Media should be used within 60 days of preparation. Media should not be used if there are any signs of deterioration (discoloration) or contamination, or if the expiration date supplied by the manufacturer has passed.

**Use:** For agar dilution susceptibility tests with imidazole antifungal agents.

## Casein Yeast Extract Glucose Agar (CYG Agar)

**Composition** per liter:

| | |
|---|---|
| Agar | 20.0g |
| Glucose | 5.0g |
| Casein hydrolysate | 5.0g |
| Yeast extract | 5.0g |

pH 7.0 ± 0.2 at 25°C

**Preparation of Medium:** Add components to distilled/deionized water and bring volume to 1.0L. Mix thoroughly. Gently heat and bring to boiling. Distribute into tubes or flasks. Autoclave for 15 min at 15 psi pressure–121°C.

**Storage/Shelf Life:** Store dehydrated media in the dark in a sealed container below 30°C. Prepared media should be stored under refrigeration (2-8°C). Media should be used within 60 days of preparation. Media should not be used if there are any signs of deterioration (shrinking, cracking, or discoloration) or contamination, or if the expiration date supplied by the manufacturer has passed.

**Use:** For agar dilution susceptibility tests with imidazole antifungal agents.

## Casein Yeast Extract Glucose Broth (CYG Broth)

**Composition** per liter:

| | |
|---|---|
| Casein hydrolysate | 5.0g |
| Glucose | 5.0g |
| Yeast extract | 5.0g |

pH 7.0 ± 0.2 at 25°C

**Preparation of Medium:** Add components to distilled/deionized water and bring volume to 1.0L. Mix thoroughly. Gently heat and bring to boiling. Distribute into tubes or flasks. Autoclave for 15 min at 15 psi pressure–121°C.

**Storage/Shelf Life:** Store dehydrated media in the dark in a sealed container below 30°C. Prepared media should be stored under refrigeration (2-8°C). Media should be used within 60 days of preparation. Media should not be used if there are any signs of deterioration (discoloration) or contamination, or if the expiration date supplied by the manufacturer has passed.

**Use:** For agar dilution susceptibility tests with imidazole antifungal agents.

## Christensen's Urea Agar

**Composition** per liter:

| | |
|---|---|
| Agar | 15.0g |
| NaCl | 5.0g |
| $KH_2PO_4$ | 2.0g |
| Peptone | 1.0g |
| Glucose | 1.0g |
| Phenol Red | 0.012g |
| Urea solution | 100.0mL |

pH 6.8 ± 0.1 at 25°C

**Urea Solution:**
**Composition** per 100.0mL:

| | |
|---|---|
| Urea | 20.0g |

**Preparation of Urea:** Add urea to 100.0mL of distilled/deionized water. Mix thoroughly. Filter sterilize.

**Preparation of Medium:** Add components, except urea solution, to distilled/deionized water and bring volume to 900.0mL. Mix thoroughly. Gently heat and bring to boiling. Autoclave for 15 min at 15 psi pressure–121°C. Cool to 50–55°C. Aseptically add 100.0mL of sterile urea solution. Mix thoroughly. Pour into Petri dishes or distribute into sterile tubes. Allow tubes to solidify in a slanted position.

**Storage/Shelf Life:** Store dehydrated media in the dark in a sealed container below 30°C. Prepared media should be stored under refrigeration (2-8°C). Media should be used within 60 days of preparation. Media should not be used if there are any signs of deterioration (shrinking, cracking, or discoloration) or contamination, or if the expiration date supplied by the manufacturer has passed.

**Use:** For the differentiation of a variety of fungi on the basis of urease production.

## CHROMagar™ *Candida*

**Composition** per liter:

| | |
|---|---|
| Glucose | 20.0g |
| Agar | 15.0g |
| Peptone | 10.0g |
| Chromogenic mix | 2.0g |
| Chloramphenicol | 0.5g |

**Source:** CHROMagar *Candida* is available from CHROMagar Microbiology. Prepared medium is also available from BD Diagnostic Systems.

**Preparation of Medium:** Add components to distilled/deionized water and bring volume to 1.0L. Mix thoroughly. Gently heat in a boiling water bath or steam bath. Shake periodically during heating to dissolve components. Heat long enough with shaking every 5 min to ensure complete dissolution. Do not overheat. Cool to 45–50°C. Pour into sterile Petri dishes.

**Storage/Shelf Life:** Store in the dark. Chromogenic agars are especially light and temperature sensitive; protect from light, excessive heat, moisture, and freezing. Prepared media plates can be kept for one day at ambient temperature. Plates can be stored at least one week under refrigeration (2-8°C) if properly prepared and protected from light and dehydration. Do not use after the expiration date supplied by the manufacturer.

**Use:** For the differentiation of *Candida* spp. Specific *Candida* spp. give characteristic color reactions, e.g., *Candida albicans* produce distinctive green colonies and *Candida tropicalis* produce distinctive dark blue-gray colonies.

## CHROMagar™ *Malassezia* Medium

**Composition** per liter:

| | |
|---|---|
| Peptones and extracts | 38.0g |

Agar .......................................................................15.0g
Chromogenic mix .................................................2.8g
Chloramphenicol.................................................0.5g
Tween™ 40 ..................................................... 10.0mL

**Source:** This medium is available from CHROMagar, Paris, France.

**Preparation of Medium:** Add components, to distilled/deionized water and bring volume to 1.0L. Mix thoroughly. Gently heat and bring to boiling. Do not autoclave. Continue heating until complete fusion of the agar grains has taken place (large bubbles replacing foam). Cool to 50°C. Pour into sterile Petri dishes.

**Storage/Shelf Life:** Store in the dark. Chromogenic agars are especially light and temperature sensitive; protect from light, excessive heat, moisture, and freezing. Prepared media plates can be kept for one day at ambient temperature. Plates can be stored at least one week under refrigeration (2-8°C) if properly prepared and protected from light and dehydration. Do not use after the expiration date supplied by the manufacturer.

**Use:** For the cultivation of *Malassezia* species.

## CHROMagar™ *Malassezia* Medium
**Composition** per liter:

Peptones and extracts............................................38.0g
Agar .......................................................................15.0g
Chromogenic mix .................................................2.8g
Chloramphenicol.................................................0.5g
Glycerol ................................................................1.0g
Tween™ 60 ...........................................................0.5g
pH 6.1± 0.2 at 25°C

**Source:** This medium is available from CHROMagar, Paris, France.

**Preparation of Medium:** Add components, to distilled/deionized water and bring volume to 1.0L. Mix thoroughly. Gently heat and bring to boiling. Do not autoclave. Continue heating until complete fusion of the agar grains has taken place (large bubbles replacing foam). Cool to 50°C. Pour into sterile Petri dishes.

**Storage/Shelf Life:** Store in the dark. Chromogenic agars are especially light and temperature sensitive; protect from light, excessive heat, moisture, and freezing. Prepared media plates can be kept for one day at ambient temperature. Plates can be stored at least one week under refrigeration (2-8°C) if properly prepared and protected from light and dehydration. Do not use after the expiration date supplied by the manufacturer.

**Use:** For the cultivation of *Malassezia* species.

## chromID™ *Candida*
**Composition** per liter:
Proprietary.

**Source:** This medium is available from bioMérieux.

**Preparation of Medium:** Available as a prepared medium

**Storage/Shelf Life:** Prepared media should be stored in the dark under refrigeration (2-8°C). Chromogenic agars are especially light and temperature sensitive; protect from light, excessive heat, moisture, and freezing. Media should not be used if there are any signs of deterioration (shrinking, cracking, or discoloration) or contamination, or if the expiration date supplied by the manufacturer has passed.

**Use:** For the selective isolation of yeasts and the direct identification of *Candida albicans*. *C. albicans* form blue colonies. *C. tropicalis, C. kefyr,* and *C. lustrianiae form pink colonies.*

## Chromogenic *Candida* Agar
**Composition** per liter:

Chromogenic mix ...............................................13.6g
Agar .....................................................................13.6g
Peptone .................................................................4.0g
Selective supplement solution ........................... 10.0mL
pH 6.0 ± 0.2 at 25°C

**Source:** This medium is available from Oxoid Unipath.

**Selective Supplement Solution:**
**Composition** per 10.0mL:
Chloramphenicol...............................................500.0mg

**Preparation of Selective Supplement Solution:** Add chloramphenicol to distilled/deionized water and bring volume to 10.0mL. Mix thoroughly. Filter sterilize.

**Preparation of Medium:** Add components to distilled/deionized water and bring volume to 1.0mL. Mix thoroughly. Gently heat while stirring and bring to boiling. Do not autoclave. Cool to 45°C. Pour into sterile Petri dishes.

**Storage/Shelf Life:** Store dehydrated media in the dark in a sealed container below 30°C. Prepared media should be stored in the dark under refrigeration (2-8°C). Chromogenic agars are especially light and temperature sensitive; protect from light, excessive heat, moisture, and freezing. Media should not be used if there are any signs of deterioration (shrinking, cracking, or discoloration) or contamination, or if the expiration date supplied by the manufacturer has passed.

**Use:** For the rapid isolation and identification of clinically important *Candida* species. The medium incorporates two chromogens that indicate the presence of the target enzymes: X-NAG (5-bromo-4-chloro-3-indolyl N acetyl ß-D-glucosaminide) detects the activity of hexosaminidase. BCIP (5-bromo-6-chloro-3-indolyl phosphate p-toluidine salt) detects alkaline phosphatase activity. An opaque agent has been incorporated into the formulation to improve the color definition on the agar. The broad-spectrum antibacterial agent chloramphenicol is added to the agar to inhibit bacterial growth on the plates.

## CN Screen Medium
### (*Cryptococcus neoformans* Screen Medium)
**Composition** per liter:

Agar .......................................................................15.0g
$K_2HPO_4$.................................................................4.0g
$MgSO_4 \cdot 7H_2O$ ....................................................2.5g
Glucose ...............................................................1.25g
Asparagine ...........................................................1.0g
Glutamine ............................................................1.0g
Glycine.................................................................1.0g
Thiamine·HCl ......................................................1.0g
Tryptophan...........................................................1.0g
EDTA ...................................................................0.6g
Biotin .................................................................0.51g
Dihydroxyphenylalanine (Dopa) ........................0.2g
Phenol Red...........................................................0.2g
pH 5.5–5.6 ± 0.2 at 25°C

**Preparation of Medium:** Add components to distilled/deionized water and bring volume to 1.0L. Mix thoroughly. Gently heat until boiling. Distribute into tubes or flasks. Autoclave for 15 min at 15 psi pressure–121°C.

**Storage/Shelf Life:** Store dehydrated media in the dark in a sealed container below 30°C. Prepared media should be stored under refrigeration (2-8°C). Media should be used within 60 days of preparation. Media should not be used if there are any signs of deterioration (shrinking, cracking, or discoloration) or contamination, or if the expiration date supplied by the manufacturer has passed.

**Use:** For the screening of yeast isolates for the presumptive identification of *Cryptococcus neoformans*. *Cryptococcus neoformans* forms black colonies.

## Colorex™ *Candida*

**Composition** per liter:

| | |
|---|---|
| Glucose | 20.0g |
| Agar | 15.0g |
| Peptone | 10.0g |
| Chromogenic mix | 2.0g |
| Chloramphenicol | 0.5g |

**Source:** Available as prepared plates from E&O Laboratories, Bonnybridge Scotland.

**Storage/Shelf Life:** Store in the dark. Chromogenic agars are especially light and temperature sensitive; protect from light, excessive heat, moisture, and freezing. Prepared media plates can be kept for one day at ambient temperature. Plates can be stored at least one week under refrigeration (2-8°C) if properly prepared and protected from light and dehydration. Do not use after the expiration date supplied by the manufacturer.

**Use:** For the differentiation of *Candida* spp. Specific *Candida* spp. give characteristic color reactions, e.g., *Candida albicans* produce distinctive green colonies and *Candida tropicalis* produce distinctive dark blue-gray colonies.

## Cornmeal Agar

**Composition** per liter:

| | |
|---|---|
| Agar | 15.0g |
| Cornmeal, solids from infusion | 2.0g |

pH 5.6–6.0 at 25°C

**Source:** This medium is available as a premixed powder from BD Diagnostic Systems and Oxoid Unipath.

**Preparation of Medium:** Add components to distilled/deionized water and bring volume to 1.0L. Mix thoroughly. Gently heat until boiling. Distribute into tubes or flasks. Autoclave for 15 min at 15 psi pressure–121°C. Pour into sterile Petri dishes or leave in tubes.

**Storage/Shelf Life:** Store dehydrated media in the dark in a sealed container below 30°C. Prepared media should be stored under refrigeration (2-8°C). Media should be used within 60 days of preparation. Media should not be used if there are any signs of deterioration (shrinking, cracking, or discoloration) or contamination, or if the expiration date supplied by the manufacturer has passed.

**Use:** For the cultivation and maintenance of fungi.

## Cornmeal Agar

**Composition** per liter:

| | |
|---|---|
| Agar | 15.0g |
| Corn meal extract (from 50g whole maize) | 2.0g |

pH 6.0 ± 0.2 at 25°C

**Preparation of Medium:** Add components to distilled/deionized water and bring volume to 1.0L. Mix thoroughly. Gently heat and bring to boiling. Autoclave for 15 min at 15 psi pressure–121°C. Pour into sterile Petri dishes.

**Storage/Shelf Life:** Store dehydrated media in the dark in a sealed container below 30°C. Prepared media should be stored under refrigeration (2-8°C). Media should be used within 60 days of preparation. Media should not be used if there are any signs of deterioration (shrinking, cracking, or discoloration) or contamination, or if the expiration date supplied by the manufacturer has passed.

**Use:** For chlamydospore production by *Candida albicans* and for the maintenance of fungal cultures.

## Cornmeal Agar with 1% Dextrose

**Composition** per liter:

| | |
|---|---|
| Agar | 15.0g |
| Cornmeal, solids from infusion | 2.0g |
| Glucose | 1.0g |

pH 5.6–6.0 at 25°C

**Source:** This medium is available as a premixed powder from BD Diagnostic Systems and Oxoid Unipath.

**Preparation of Medium:** Add components to distilled/deionized water and bring volume to 1.0L. Mix thoroughly. Gently heat until boiling. Distribute into tubes or flasks. Autoclave for 15 min at 15 psi pressure–121°C. Pour into sterile Petri dishes or leave in tubes.

**Storage/Shelf Life:** Store dehydrated media in the dark in a sealed container below 30°C. Prepared media should be stored under refrigeration (2-8°C). Media should be used within 60 days of preparation. Media should not be used if there are any signs of deterioration (shrinking, cracking, or discoloration) or contamination, or if the expiration date supplied by the manufacturer has passed.

**Use:** For the cultivation and maintenance of fungi. For the differentiation of *Trichophyton rubrum* based upon red pigment production.

## Cornmeal Agar with Dextrose and Tween

**Composition** per liter:

| | |
|---|---|
| Agar | 15.0g |
| Cornmeal, solids from infusion | 2.0g |
| Glucose | 2.0g |
| Tween™ 80 | 1.0g |

pH 5.6–6.0 at 25°C

**Source:** This medium is available as a premixed powder from BD Diagnostic Systems.

**Preparation of Medium:** Add components to distilled/deionized water and bring volume to 1.0L. Mix thoroughly. Gently heat until boiling. Distribute into tubes or flasks. Autoclave for 15 min at 15 psi pressure–121°C. Pour into sterile Petri dishes or leave in tubes.

**Storage/Shelf Life:** Store dehydrated media in the dark in a sealed container below 30°C. Prepared media should be stored under refrigeration (2-8°C). Media should be used within 60 days of preparation. Media should not be used if there are any signs of deterioration (shrinking, cracking, or discoloration) or contamination, or if the expiration date supplied by the manufacturer has passed.

**Use:** For the cultivation of fungi. For the differentiation of *Candida* species based upon mycelia appearance.

## Cremophor EL Agar

**Composition** per liter:

Glucose ...............................................................40.0g
Agar .....................................................................15.0g
Mycological peptone............................................10.0g
Cremophor EL ................................................. 10.0mL

pH 5.8 ± 0.2 at 25°C

**Source:** This medium is available from Sigma.

**Preparation of Medium:** Add components to distilled/deionized water and bring volume to 1.0L. Mix thoroughly. Gently heat and bring to boiling. Autoclave for 15 min at 15 psi pressure–121°C. Pour into sterile Petri dishes.

**Storage/Shelf Life:** Store dehydrated media in the dark in a sealed container below 30°C. Prepared media should be stored under refrigeration (2-8°C). Media should be used within 60 days of preparation. Media should not be used if there are any signs of deterioration (shrinking, cracking, or discoloration) or contamination, or if the expiration date supplied by the manufacturer has passed.

**Use:** For the cultivation of *Malassezia* species.

## Czapek Dox Agar

**Composition** per liter:

Sucrose................................................................30.0g
Agar .....................................................................15.0g
NaNO$_3$..................................................................3.0g
K$_2$HPO$_4$................................................................1.0g
MgSO$_4$·7H$_2$O .......................................................0.5g
KCl........................................................................0.5g
FeSO$_4$·7H$_2$O.........................................................0.01g

pH 7.3 ± 0.2 at 25°C

**Preparation of Medium:** Add components to distilled/deionized water and bring volume to 1.0L. Mix thoroughly. Distribute into tubes or flasks. Autoclave for 15 min at 15 psi pressure–121°C. Pour into sterile Petri dishes or leave in tubes.

**Storage/Shelf Life:** Store dehydrated media in the dark in a sealed container below 30°C. Prepared media should be stored under refrigeration (2-8°C). Media should be used within 60 days of preparation. Media should not be used if there are any signs of deterioration (shrinking, cracking, or discoloration) or contamination, or if the expiration date supplied by the manufacturer has passed.

**Use:** For the cultivation of actinomycetes and fungi.

## Czapek Dox Agar, Modified

**Composition** per liter:

Sucrose................................................................30.0g
Agar .....................................................................12.0g
NaNO$_3$..................................................................2.0g
Magnesium glycerophosphate .............................0.5g
KCl........................................................................0.5g
K$_2$SO$_4$................................................................0.35g
FeSO$_4$ ..................................................................0.01g

pH 6.8 ± 0.2 at 25°C

**Source:** This medium is available as a premixed powder from Oxoid Unipath and HiMedia.

**Preparation of Medium:** Add components to distilled/deionized water and bring volume to 1.0L. Mix thoroughly. Distribute into tubes or flasks. Autoclave for 15 min at 15 psi pressure–121°C. Pour into sterile Petri dishes or leave in tubes.

**Storage/Shelf Life:** Store dehydrated media in the dark in a sealed container below 30°C. Prepared media should be stored under refrigeration (2-8°C). Media should be used within 60 days of preparation. Media should not be used if there are any signs of deterioration (shrinking, cracking, or discoloration) or contamination, or if the expiration date supplied by the manufacturer has passed.

**Use:** For the cultivation and maintenance of numerous fungal species. For chlamydospore production by *Candida albicans*.

## Dermasel Agar Base

**Composition** per liter:

Glucose ...............................................................20.0g
Agar .....................................................................14.5g
Papaic digest of soybean meal............................10.0g
Antibiotic inhibitor ......................................... 10.0mL

pH 6.8–7.0 at 25°C

**Source:** This medium is available as a premixed powder from Oxoid Unipath.

### Antibiotic Inhibitor:

**Composition** per 10.0mL:

Cycloheximide.......................................................0.4g
Chloramphenicol..................................................0.05g
Acetone ............................................................ 10.0mL

**Preparation of Antibiotic Inhibitor:** Add cycloheximide and chloramphenicol to 10.0mL of acetone. Mix thoroughly.

**Caution:** Cycloheximide is toxic. Avoid skin contact or aerosol formation and inhalation.

**Preparation of Medium:** Add components to distilled/deionized water and bring volume to 990.0mL. Mix thoroughly. Gently heat and bring to boiling. Do not overheat. Add antibiotic inhibitor. Mix thoroughly. Autoclave for 10 min at 15 psi pressure–121°C. Pour into sterile Petri dishes.

**Storage/Shelf Life:** Store dehydrated media in the dark in a sealed container below 30°C. Prepared media should be stored under refrigeration (2-8°C). Media should be used within 60 days of preparation. Media should not be used if there are any signs of deterioration (shrinking, cracking, or discoloration) or contamination, or if the expiration date supplied by the manufacturer has passed.

**Use:** For the isolation and cultivation of dermatophytic fungi isolated from hair, nails, or skin scrapings.

## Dermatophyte Milk Agar

**Composition** per liter:

Skim milk powder................................................40.0g
Glucose ...............................................................20.0g
Agar .....................................................................15.0g
Bromcresol Purple ..............................................16.0mb
Ethanol............................................................... 1.0mL

**Source:** This medium is available from Hardy Diagnostics.

**Preparation of Medium:** Available as prepared plates.

**Storage/Shelf Life:** Store dehydrated media in the dark in a sealed container below 30°C. Prepared media should be stored under refrigeration (2-8°C). Media should be used within 60 days of preparation. Media should not be used if there are any signs of deterioration (shrinking, cracking, or discoloration) or contamination, or if the expiration date supplied by the manufacturer has passed.

**Use:** For the identification of *Trichophyton* and other dermatophytes. Development of a violet-purple color around the fungal growth is indicative of an alkaline reaction. *Trichophyton mentagrophytes* produces diffuse growth and an alkanline reaction. *T. Rubrum* produces restricted growth and no color change in the medium—the colonies are red. *Microsporum persicolor* grows rapidly and diffusely but does not demonstrate an alkaline reaction.

## Dermatophyte Test Medium

**Composition** per liter:

Agar ............................................................................20.0g
Enzymatic digest of soybean meal..............................10.0g
Glucose ......................................................................10.0g
Cycloheximide............................................................0.5g
Phenol Red .................................................................0.2g
Selective supplement solution ................................ 10.0mL
pH 5.5 ± 0.2 at 25°C

**Source:** This medium is available from Acumedia, Neogen Corp.

**Caution:** Cycloheximide is toxic. Avoid skin contact or aerosol formation and inhalation.

**Selective Supplement Solution:**

**Composition** per 10.0mL:

Gentamicin .................................................................0.1g
Chlortetracycline .......................................................0.1g

**Preparation of Selective Supplement Solution:** Add components to distilled/deionized water and bring volume to 10.0mL. Mix thoroughly. Filter sterilize.

**Storage/Shelf Life:** Store dehydrated media in the dark in a sealed container below 30°C. Prepared media should be stored under refrigeration (2-8°C). Media should be used within 60 days of preparation. Media should not be used if there are any signs of deterioration (shrinking, cracking, or discoloration) or contamination, or if the expiration date supplied by the manufacturer has passed.

**Preparation of Medium:** Add components, except selective supplement solution, to distilled/deionized water and bring volume to 990.0mL. Mix thoroughly. Distribute into tubes or flasks. Gently heat and bring to boiling. Autoclave for 15 min at 15 psi pressure–121°C. Cool to 50°C. Aseptically add 10.0mL selective supplement solution. Mix thoroughly. Pour into sterile Petri dishes or leave in tubes.

**Use:** For the selective isolation of dermatophytic fungi.

## Dermatophyte Test Medium Base

**Composition** per liter:

Agar ............................................................................20.0g
Glucose ......................................................................10.0g
Papaic digest of soybean meal ...................................10.0g
Cycloheximide............................................................0.5g
Phenol Red .................................................................0.2g
Gentamycin sulfate .....................................................0.1g
Chlortetracycline........................................................0.1g
pH 5.5 ± 0.2 at 25°C

**Source:** This medium is available as a premixed powder from BD Diagnostic Systems.

**Caution:** Cycloheximide is toxic. Avoid skin contact or aerosol formation and inhalation.

**Preparation of Medium:** Add components, except gentamycin sulfate and chlortetracycline, to distilled/deionized water and bring vol-

ume to 1.0L. Mix thoroughly. Gently heat while stirring and bring to boiling. Autoclave for 15 min at 15 psi pressure–121°C. Cool to 45°–50°C. Aseptically add gentamycin sulfate and chlortetracycline. Mix thoroughly. Pour into sterile Petri dishes.

**Storage/Shelf Life:** Store dehydrated media in the dark in a sealed container below 30°C. Prepared media should be stored under refrigeration (2-8°C). Media should be used within 60 days of preparation. Media should not be used if there are any signs of deterioration (shrinking, cracking, or discoloration) or contamination, or if the expiration date supplied by the manufacturer has passed.

**Use:** For the selective isolation and cultivation of pathogenic fungi from cutaneous sources.

## Dextrose Agar

**Composition** per liter:

Agar ............................................................................15.0g
Glucose ......................................................................10.0g
Tryptose .....................................................................10.0g
NaCl............................................................................5.0g
Beef extract................................................................3.0g
pH 7.3 ± 0.2 at 25°C

**Source:** This medium is available as a premixed powder from Hi-Media.

**Preparation of Medium:** Add components to distilled/deionized water and bring volume to 1.0L. Mix thoroughly. Gently heat and bring to boiling. Distribute into tubes or flasks. Autoclave for 20 min at 15 psi pressure–121°C. Pour into sterile Petri dishes or leave in tubes.

**Storage/Shelf Life:** Store dehydrated media in the dark in a sealed container below 30°C. Prepared media should be stored under refrigeration (2-8°C). Media should be used within 60 days of preparation. Media should not be used if there are any signs of deterioration (shrinking, cracking, or discoloration) or contamination, or if the expiration date supplied by the manufacturer has passed.

**Use:** For the cultivation and maintenance of a wide variety of microorganisms.

## Dextrose HiVeg Agar Base, Emmons
### (Sabouraud Glucose HiVeg Agar Base, Modified)

**Composition** per liter:

Glucose ......................................................................20.0g
Agar ............................................................................17.0g
Plant special peptone .................................................10.0g
pH 6.9 ± 0.2 at 25°C

**Source:** This medium is available as a premixed powder from Hi-Media.

**Preparation of Medium:** Add components to tap water and bring volume to 1.0L. Mix thoroughly. Gently heat and bring to boiling. Distribute into tubes or flasks. Autoclave for 15 min at 15 psi pressure–121°C. Pour into sterile Petri dishes or leave in tubes.

**Storage/Shelf Life:** Store dehydrated media in the dark in a sealed container below 30°C. Prepared media should be stored under refrigeration (2-8°C). Media should be used within 60 days of preparation. Media should not be used if there are any signs of deterioration (shrinking, cracking, or discoloration) or contamination, or if the expiration date supplied by the manufacturer has passed.

**Use:** For the cultivation of dermatophytes and other pathogenic and nonpathogenic fungi from clinical speciments and nonclinical specimens of public health importance. For the cultivation of yeast and filamentous fungi.

## ESP Myco Medium

**Composition** per liter:
Proprietary.

**Source:** This medium is available from BD Diagnostic Systems and Trek Diagnositics; Cleveland, OH.

**Storage/Shelf Life:** Prepared medium should be stored under refrigeration (2-8°C). Use before the expiration date supplied by the manufacturer has passed.

**Use:** This medium is used with ESP Culture System II (TREK Diagnostic Systems) for the detection of mycobacterial growth. It is used for the detection and antibiotic susceptibility testing of *Mycobacterium tuberculosis*. The medium is composed of Middlebrook 7H9 broth enriched with glycerol, casitone, and cellulose sponge disks and OADC enrichment

## Fermentation Broth (CHO Medium)

**Composition** per liter:

| | |
|---|---|
| Pancreatic digest of casein | 15.0g |
| Yeast extract | 7.0g |
| NaCl | 2.5g |
| Agar | 0.75g |
| Sodium thioglycolate | 0.5g |
| L-Cystine | 0.25g |
| Ascorbic acid | 0.1g |
| Bromthymol Blue | 0.01g |
| Carbohydrate or starch solution | 100.0mL |

pH 7.0 ± 0.1 at 25°C

**Source:** This medium is available as a premixed powder from BD Diagnostic Systems.

**Carbohydrate Solution:**
**Composition** per 100.0mL:
Carbohydrate ............................................................................6.0g

**Preparation of Carbohydrate Solution:** Add carbohydrate to distilled/deionized water and bring volume to 10.0mL. Mix thoroughly. Filter sterilize.

**Starch Solution:**
**Composition** per 100.0mL:
Starch ........................................................................................2.5g

**Preparation of Starch Solution:** Add starch to distilled/deionized water and bring volume to 100.0mL. Mix thoroughly. Filter sterilize.

**Preparation of Medium:** Add components, except carbohydrate solution, to distilled/deionized water and bring volume to 900.0mL. Mix thoroughly. Distribute into tubes or flasks. Autoclave for 15 min at 15 psi pressure–121°C. Cool to 45°–50°C. Aseptically add 100.0mL of sterile carbohydrate solution. Mix thoroughly. Aseptically distribute into sterile tubes or flasks. Loosen caps on tubes. Place in an anaerobic chamber under an atmosphere of 85% $N_2$, 10% $H_2$, and 5% $CO_2$. Fasten the caps securely or maintain in an anaerobic chamber.

**Storage/Shelf Life:** Store dehydrated media in the dark in a sealed container below 30°C. Prepared media should be stored under refrigeration (2-8°C). Media should be used within 60 days of preparation. Media should not be used if there are any signs of deterioration (discol-

oration) or contamination, or if the expiration date supplied by the manufacturer has passed.

**Use:** For the differentiation of fungi based upon carbohydrate fermentation. Fungi that ferment the specific carbohydrates added to the medium turn the medium yellow.

## Fluconazole Testing Medium

**Composition** per liter:

| | |
|---|---|
| Glucose | 19.98g |
| $(NH_4)_2SO_4$ | 4.99g |
| $KH_2PO_4$ | 1.99g |
| $MgSO_4 \cdot 7H_2O$ anhydrous | 0.99g |
| L-Glutamine | 0.58g |
| NaCl | 0.2g |
| $CaCl_2 \cdot 2H_2O$ | 0.2g |
| L-Lysine monohydrochloride | 0.073g |
| L-Isoleucine | 0.052g |
| L-Leucine | 0.052g |
| Threonine | 0.0476g |
| Valine | 0.047g |
| L-Arginine monohydrochloride | 0.042g |
| L-Histidine | 0.023g |
| Tryptophan | 0.02g |
| DL-Methionine | 0.0189g |
| Inositol | 0.00397g |
| $ZnSO_4 \cdot 7H_2O$ | 0.0014g |
| $H_3BO_3$ | 0.00099g |
| Nicotinic acid | 0.00079g |
| Pyridoxine hydrochloride | 0.00079g |
| Calcium D-pantothenic acid | 0.00079g |
| Aneurine hydrochloride | 0.00079g |
| $MnSO_4 \cdot 2H_2O$ | 0.00079g |
| $Na_2MoO_4 \cdot 2H_2O$ | 0.00047g |
| *p*-Amino benzoic acid (PABA) | 0.000395g |
| Riboflavin | 0.000395g |
| $FeCl_3$ | 0.000395g |
| Folic acid | 0.000395g |
| KI | 0.0002g |
| $CuSO_4 \cdot 5H_2O$ | 0.00012g |
| Biotin crystalline | 0.000004g |
| Agar solution | 100.0mL |

**Source:** This medium is available from HiMedia.

**Agar Solution:**
**Composition** per 100.0mL:
Agar ........................................................................................10.0g

**Preparation of Agar Solution:** Add agar to distilled/deionized water and bring volume to 100.0mL. Mix thoroughly. Adjust pH to 7.5 with phosphate buffer. Autoclave for 10 min at 10 psi pressure–115 °C. Cool to 50°C.

**Preparation of Medium:** Add components, except agar solution, to distilled/deionized water and bring volume to 900.0mL. Mix thoroughly. Filter sterilize. Aseptically add agar solution. Mix thoroughly. Pour into Petri dishes or aseptically distribute into sterile tubes.

**Storage/Shelf Life:** Store dehydrated media in the dark in a sealed container below 30°C. Prepared media should be stored under refrigeration (2-8°C). Media should be used within 60 days of preparation. Media should not be used if there are any signs of deterioration (discoloration) or contamination, or if the expiration date supplied by the manufacturer has passed.

**Use:** For fluconazole susceptibility testing using *Candida* species.

## Fungi Kimmig HiVeg Agar Base

**Composition** per liter:

Agar .................................................................................15.0g
NaCl.................................................................................11.4g
Glucose ...........................................................................10.0g
Plant peptone.....................................................................9.3g
Plant hydrolysate................................................................4.3g
Glycerol ......................................................................... 5.0mL

pH 6.5 ± 0.2 at 25°C

**Source:** This medium without glycerol, is available as a premixed powder from HiMedia.

**Preparation of Medium:** Add components to distilled/deionized water and bring volume to 1.0L. Mix thoroughly. Gently heat and bring to boiling. Mix thoroughly. Distribute into tubes or flasks. Autoclave for 15 min at 15 psi pressure–121°C. Pour into sterile Petri dishes.

**Storage/Shelf Life:** Store dehydrated media in the dark in a sealed container below 30°C. Prepared media should be stored under refrigeration (2-8°C). Media should be used within 60 days of preparation. Media should not be used if there are any signs of deterioration (shrinking, cracking, or discoloration) or contamination, or if the expiration date supplied by the manufacturer has passed.

**Use:** For the cultivation, isolation, identification, and strain preservation of fungi.

## Fungi Kimmig Selective Agar

**Composition** per liter:

Glucose ...........................................................................19.0g
Agar .................................................................................15.0g
Peptone.............................................................................15.0g
NaCl....................................................................................1.0g
Glycerol ......................................................................... 5.0mL
Selective solution......................................................... 10.0mL

pH 6.5 ± 0.2 at 25°C

**Selective Solution:**

**Composition** per 10.0mL:

Cycloheximide .....................................................................0.4g
Streptomycin ...................................................................0.04mg
Penicillin ....................................................................40,000U

**Preparation of Selective Solution:** Add components to distilled/deionized water and bring volume to 10.0mL. Mix thoroughly. Filter sterilize.

**Caution:** Cycloheximide is toxic. Avoid skin contact or aerosol formation and inhalation.

**Preparation of Medium:** Add components, except selective solution, to distilled/deionized water and bring volume to 990.0mL. Mix thoroughly. Gently heat and bring to boiling. Mix thoroughly. Autoclave for 15 min at 15 psi pressure–121°C. Cool to 45°–50°C. Aseptically add 10.0mL selective solution. Mix thoroughly. Distribute into tubes or flasks. Autoclave for 5 min at 15 psi pressure–121°C. Pour into sterile Petri dishes.

**Storage/Shelf Life:** Store dehydrated media in the dark in a sealed container below 30°C. Prepared media should be stored under refrigeration (2-8°C). Media should be used within 60 days of preparation. Media should not be used if there are any signs of deterioration (shrinking, cracking, or discoloration) or contamination, or if the expiration date supplied by the manufacturer has passed.

**Use:** For the cultivation, isolation, identification, and strain preservation of pathogenic fungi.

## Fungi Kimmig Selective Agar

**Composition** per liter:

Glucose ...........................................................................19.0g
Agar .................................................................................15.0g
Peptone.............................................................................15.0g
NaCl....................................................................................1.0g
Glycerol ......................................................................... 5.0mL
Selective solution......................................................... 10.0mL

pH 6.5 ± 0.2 at 25°C

**Selective Solution:**

**Composition** per 10.0mL:

Novobiocin........................................................................0.1g
Colistin............................................................................0.08g

**Preparation of Selective Solution:** Add components to distilled/deionized water and bring volume to 10.0mL. Mix thoroughly. Filter sterilize.

**Preparation of Medium:** Add components, except selective solution, to distilled/deionized water and bring volume to 990.0mL. Mix thoroughly. Gently heat and bring to boiling. Mix thoroughly. Autoclave for 15 min at 15 psi pressure–121°C. Cool to 45°–50°C. Aseptically add 10.0mL selective solution. Mix thoroughly. Distribute into tubes or flasks.

**Storage/Shelf Life:** Store dehydrated media in the dark in a sealed container below 30°C. Prepared media should be stored under refrigeration (2-8°C). Media should be used within 60 days of preparation. Media should not be used if there are any signs of deterioration (shrinking, cracking, or discoloration) or contamination, or if the expiration date supplied by the manufacturer has passed.

**Use:** For the cultivation, isolation, identification, and strain preservation of pathogenic fungi.

## HardyCHROM™ *Candida*

**Composition** per liter:
Proprietary

**Source:** This medium is available from Hardy Diagnostics.

**Preparation of Medium:** Available as prepared plates.

**Storage/Shelf Life:** Store in the dark under refrigeration (2-8°C). Chromogenic agars are especially light and temperature sensitive; protect from light, excessive heat, moisture, and freezing. Do not use after the expiration date supplied by the manufacturer.

**Use:** For the isolation and differentiation of clinically important yeast species. This medium is especially useful in detecting mixed yeast infections. *Candida glabrata* produces smooth pink colonies, often with a darker mauve center. *C. tropicalis* produces smooth medium blue to dark metallic blue colonies, with a blue halo. *C. krusei* produces large rough or crenated pink to medium pink colonies. *C. albicans* produces smooth emerald green to metallic green colonies.

## HiCrome™ *Candida* Agar

**Composition** per liter:

Agar .................................................................................15.0g
Peptic digest of animal tissue ..........................................15.0g
Chromogenic mixture ....................................................11.22g

K$_2$HPO$_4$...................................................................1.0g
Chloramphenicol.......................................................0.5g

pH 6.9 ± 0.2 at 25°C

**Source:** This medium is available as a premixed powder from Hi-Media.

**Preparation of Medium:** Add components to distilled/deionized water and bring volume to 1.0L. Mix thoroughly and heat with frequent agitation until boiling. Boil until components are fully dissolved. Do not autoclave. Cool to 45°–50°C. Pour into sterile Petri dishes.

**Storage/Shelf Life:** Store dehydrated media in the dark in a sealed container below 30°C. Prepared plates should be stored in the dark under refrigeration (2-8°C). Chromogenic media are especially light and temperature sensitive; protect from light, excessive heat, moisture, and freezing. Media should not be used if there are any signs of deterioration (shrinking, cracking, or discoloration) or contamination, or if the expiration date supplied by the manufacturer has passed.

**Use:** For the rapid isolation and identification of *Candida* species from mixed cultures.

## HiCrome™ *Candida* Agar, HiVeg

**Composition** per liter:

Agar ................................................................15.0g
Plant peptone.....................................................15.0g
Chromogenic mixture ............................................11.2g
K$_2$HPO$_4$...................................................................1.0g
Chloramphenicol...................................................0.5g

pH 6.9 ± 0.2 at 25°C

**Source:** This medium is available as a premixed powder from Hi-Media.

**Preparation of Medium:** Add components to distilled/deionized water and bring volume to 1.0L. Mix thoroughly and heat with frequent agitation until boiling. Boil until components are fully dissolved. Do not autoclave. Cool to 45°–50°C. Pour into sterile Petri dishes.

**Storage/Shelf Life:** Store dehydrated media in the dark in a sealed container below 30°C. Prepared plates should be stored in the dark under refrigeration (2-8°C). Chromogenic media are especially light and temperature sensitive; protect from light, excessive heat, moisture, and freezing. Media should not be used if there are any signs of deterioration (shrinking, cracking, or discoloration) or contamination, or if the expiration date supplied by the manufacturer has passed.

**Use:** For the rapid isolation and identification of *Candida* species from mixed cultures.

## HiCrome™ *Candida* Differential Agar Base with *Candida* Selective Supplement

**Composition** per liter:

Agar ................................................................15.0g
Peptone, special ..................................................15.0g
Yeast extract........................................................4.0g
Chromogenic mixture .............................................7.220g
K$_2$HPO$_4$...................................................................1.0g
Chloramphenicol...................................................0.5g
Gentamicin solution............................................... 1.0mL

pH 7.2 ± 0.2 at 25°C

**Source:** This medium is available as a premixed powder from Hi-Media.

*Candida* **Selective Supplement:**
**Composition** per 10.0mL:
Gentamicin....................................................................0.1g

**Preparation of *Candida* Selective Supplement:** Add gentamicin to distilled/deionized water and bring volume to 10.0mL. Mix thoroughly. Filter sterilize.

**Preparation of Medium:** Add components, except *Candida* selective supplement, to distilled/deionized water and bring volume to 990.0mL. Mix thoroughly. Gently heat and bring to boiling. Distribute into tubes or flasks. Do not autoclave. Do not overheat. Cool to 50°C. Aseptically add 10.0mL *Candida* selective supplement. Mix thoroughly. Pour into sterile Petri dishes or leave in tubes.

**Storage/Shelf Life:** Store dehydrated media in the dark in a sealed container below 30°C. Prepared plates should be stored in the dark under refrigeration (2-8°C). Chromogenic media are especially light and temperature sensitive; protect from light, excessive heat, moisture, and freezing. Media should not be used if there are any signs of deterioration (shrinking, cracking, or discoloration) or contamination, or if the expiration date supplied by the manufacturer has passed.

**Use:** For the rapid isolation and identification of *Candida* species from mixed cultures.

## HiCrome™ *Candida* Differential Agar Base, Modified with *Candida* Selective Supplement

**Composition** per liter:

Agar ................................................................18.0g
Glucose ............................................................10.0g
Peptic digest of animal tissue ....................................5.0g
Malt extract........................................................3.0g
Yeast extract.......................................................3.0g
Chromogenic mixture ...............................................3.0g
Chloramphenicol...................................................0.05g
Gentamicin solution............................................... 1.0mL

pH 7.2 ± 0.2 at 25°C

**Source:** This medium is available as a premixed powder from Hi-Media.

*Candida* **Selective Supplement:**
**Composition** per 10.0mL:
Gentamicin....................................................................0.1g

**Preparation of *Candida* Selective Supplement:** Add gentamicin to distilled/deionized water and bring volume to 10.0mL. Mix thoroughly. Filter sterilize.

**Preparation of Medium:** Add components, except *Candida* selective supplement, to distilled/deionized water and bring volume to 990.0mL. Mix thoroughly. Gently heat and bring to boiling. Distribute into tubes or flasks. Do not autoclave. Do not overheat. Cool to 50°C. Aseptically add 10.0mL *Candida* selective supplement. Mix thoroughly. Pour into sterile Petri dishes or leave in tubes.

**Storage/Shelf Life:** Store dehydrated media in the dark in a sealed container below 30°C. Prepared plates should be stored in the dark under refrigeration (2-8°C). Chromogenic media are especially light and temperature sensitive; protect from light, excessive heat, moisture, and freezing. Media should not be used if there are any signs of deterioration (shrinking, cracking, or discoloration) or contamination, or if the expiration date supplied by the manufacturer has passed.

**Use:** For the rapid isolation and identification of *Candida* species from mixed cultures.

## HiCrome™ *Candida* HiVeg Agar Base, Modified wtih *Candida* Selective Supplement

**Composition** per liter:

| | |
|---|---|
| Agar | 18.0g |
| Glucose | 10.0g |
| Plant peptone | 5.0g |
| Malt extract | 3.0g |
| Yeast extract | 3.0g |
| Chromogenic mixture | 3.0g |
| Chloramphenicol | 0.05g |
| Gentamicn solution | 1.0mL |

pH 7.2 ± 0.2 at 25°C

**Source:** This medium is available as a premixed powder from Hi-Media.

### *Candida* Selective Supplement:
**Composition** per 10.0mL:

| | |
|---|---|
| Gentamicin | 0.1g |

**Preparation of *Candida* Selective Supplement:** Add gentamicin to distilled/deionized water and bring volume to 10.0mL. Mix thoroughly. Filter sterilize.

**Preparation of Medium:** Add components, except *Candida* selective supplement, to distilled/deionized water and bring volume to 990.0mL. Mix thoroughly. Gently heat and bring to boiling. Distribute into tubes or flasks. Do not autoclave. Do not overheat. Cool to 50°C. Aseptically add 10.0mL *Candida* selective supplement. Mix thoroughly. Pour into sterile Petri dishes or leave in tubes.

**Storage/Shelf Life:** Store dehydrated media in the dark in a sealed container below 30°C. Prepared plates should be stored in the dark under refrigeration (2-8°C). Chromogenic media are especially light and temperature sensitive; protect from light, excessive heat, moisture, and freezing. Media should not be used if there are any signs of deterioration (shrinking, cracking, or discoloration) or contamination, or if the expiration date supplied by the manufacturer has passed.

**Use:** For the rapid isolation and identification of *Candida* species from mixed cultures.

## Inhibitory Mold Agar

**Composition** per liter:

| | |
|---|---|
| Agar | 15.0g |
| Glucose | 5.0g |
| Yeast extract | 5.0g |
| Pancreatic digest of casein | 3.0g |
| $Na_2HPO_4$ | 2.0g |
| Peptic digest of animal tissue | 2.0g |
| Starch | 2.0g |
| Dextrin | 1.0g |
| $MgSO_4 \cdot 7H_2O$ | 0.8g |
| Chloramphenicol | 0.125g |
| $FeSO_4$ | 0.04g |
| NaCl | 0.04g |
| $MnSO_4$ | 0.16g |

pH 6.7 ± 0.2 at 25°C

**Source:** This medium is available as a premixed powder from BD Diagnostic Systems.

**Preparation of Medium:** Add components to distilled/deionized water and bring volume to 1.0L. Mix thoroughly. Gently heat and bring to boiling with frequent agitation. Distribute into tubes or flasks. Auto-

clave for 15 min at 15 psi pressure–121°C. Pour into sterile Petri dishes or leave in tubes.

**Storage/Shelf Life:** Store dehydrated media in the dark in a sealed container below 30°C. Prepared media should be stored under refrigeration (2-8°C). Media should be used within 60 days of preparation. Media should not be used if there are any signs of deterioration (shrinking, cracking, or discoloration) or contamination, or if the expiration date supplied by the manufacturer has passed.

**Use:** For the isolation of pathogenic fungi.

## Kimmig's Agar

**Composition** per liter:

| | |
|---|---|
| Agar | 15.0g |
| Glucose | 10.0g |
| Pancreatic digest of gelatin | 9.5g |
| Beef extract | 5.5g |
| NaCl | 5.0g |
| Peptone | 5.0g |
| Glycerol | 5.0mL |

pH 6.9 ± 0.2 at 35°C

**Preparation of Medium:** Add glycerol and then other components to distilled/deionized water and bring volume to 1.0L. Mix thoroughly. Gently heat and bring to boiling. Distribute into tubes or flasks. Autoclave for 15 min at 15 psi pressure–121°C. Pour into sterile Petri dishes or leave in tubes.

**Storage/Shelf Life:** Store dehydrated media in the dark in a sealed container below 30°C. Prepared media should be stored under refrigeration (2-8°C). Media should be used within 60 days of preparation. Media should not be used if there are any signs of deterioration (shrinking, cracking, or discoloration) or contamination, or if the expiration date supplied by the manufacturer has passed.

**Use:** For the assay of fungistatic agents. For agar dilution testing of antifungal agents. For the cultivation and preservation of various fungi.

## Kimmig Fungi HiVeg Agar Base with Kimmig Supplement

**Composition** per liter:

| | |
|---|---|
| Glucose | 19.0g |
| Plant peptone | 15.0g |
| Agar | 15.0g |
| NaCl | 1.0g |
| Cycloheximide | 0.4g |
| Kimmig selective supplement | 10.0mL |
| Glycerol | 5.0mL |

pH 6.9 ± 0.2 at 35°C

**Source:** This medium, without glycerol or Kimmig selective supplement, is available as a premixed powder from HiMedia.

**Caution:** Cycloheximide is toxic. Avoid skin contact or aerosol formation and inhalation.

### Kimmig Selective Supplement:
**Composition** per 10.0mL:

| | |
|---|---|
| Novobiocin | 200.0mg |
| Colistin sulfate | 80.0mg |

**Preparation of Kimmig Selective Supplement:** Add components to distilled/deionized water and bring volume to 10.0mL. Mix thoroughly. Filter sterilize.

**Preparation of Medium:** Add glycerol and then other components, except Kimmig selective supplement, to distilled/deionized water and bring volume to 990.0L. Mix thoroughly. Gently heat and bring to boiling. Autoclave for 15 min at 15 psi pressure–121°C. Cool to 50°C. Aseptically add 10.0mL sterile Kimmig selective supplement. Mix thoroughly Pour into sterile Petri dishes or leave in tubes.

**Storage/Shelf Life:** Store dehydrated media in the dark in a sealed container below 30°C. Prepared media should be stored under refrigeration (2-8°C). Media should be used within 60 days of preparation. Media should not be used if there are any signs of deterioration (shrinking, cracking, or discoloration) or contamination, or if the expiration date supplied by the manufacturer has passed.

**Use:** For the assay of fungistatic agents. For agar dilution testing of antifungal agents. For the cultivation and preservation of various fungi.

## Kimmig Fungi HiVeg Agar Base with George Kimmig Supplement

**Composition** per liter:

| | |
|---|---|
| Glucose | 19.0g |
| Plant peptone | 15.0g |
| Agar | 15.0g |
| NaCl | 1.0g |
| Cycloheximide | 0.4g |
| George Kimmig selective supplement | 10.0mL |
| Glycerol | 5.0mL |

pH 6.9 ± 0.2 at 35°C

**Source:** This medium, without glycertol or George Kimmig selective supplement, is available as a premixed powder from HiMedia.

**Caution:** Cycloheximide is toxic. Avoid skin contact or aerosol formation and inhalation.

**George Kimmig Selective Supplement:**

**Composition** per 10.0mL:

| | |
|---|---|
| Penicillin G | 40,000U |
| Streptomycin | 40,000U |

**Preparation of George Kimmig Selective Supplement:** Add components to distilled/deionized water and bring volume to 10.0mL. Mix thoroughly. Filter sterilize.

**Preparation of Medium:** Add glycerol and then other components, except George Kimmig selective supplement, to distilled/deionized water and bring volume to 990.0L. Mix thoroughly. Gently heat and bring to boiling. Autoclave for 15 min at 15 psi pressure–121°C. Cool to 50°C. Aseptically add 10.0mL sterile George Kimmig selective supplement. Mix thoroughly Pour into sterile Petri dishes or leave in tubes.

**Storage/Shelf Life:** Store dehydrated media in the dark in a sealed container below 30°C. Prepared media should be stored under refrigeration (2-8°C). Media should be used within 60 days of preparation. Media should not be used if there are any signs of deterioration (shrinking, cracking, or discoloration) or contamination, or if the expiration date supplied by the manufacturer has passed.

**Use:** For the assay of fungistatic agents. For agar dilution testing of antifungal agents. For the cultivation and preservation of various fungi.

## Lactirmel Agar
## (Borelli's Medium)

**Composition** per liter:

| | |
|---|---|
| Agar | 15.0g |
| Honey | 10.0g |

| | |
|---|---|
| Skim milk powder (Dutch Jug skimmed milk powder) | 7.0g |
| Cornmeal extract (from 50g whole maize) | 2.0g |
| Chloramphenicol | 250.0mg |

pH 6.0 ± 0.2 at 25°C

**Source:** This medium is available as a premixed powder from BD Diagnostic Systems.

**Preparation of Medium:** Add skim milk to distilled/deionized water and bring volume to approximately 150mL. Mix thoroughly until it forms a paste. Add honey and other components. Bring volume to 1.0L. Gently bring to boil and mix thoroughly. Do not adjust pH. Autoclave for 10 min at 10 psi pressure–115°C. Allow to stand for 5 min. Mix thoroughly. Pour into sterile Petri dishes or distribute into sterile tubes.

**Storage/Shelf Life:** Store dehydrated media in the dark in a sealed container below 30°C. Prepared media should be stored under refrigeration (2-8°C). Media should be used within 60 days of preparation. Media should not be used if there are any signs of deterioration (shrinking, cracking, or discoloration) or contamination, or if the expiration date supplied by the manufacturer has passed.

**Use:** For the differential identification of dermatophytes based upon the production of pigment by *Trichophyton rubrum*.

## Leeming and Notman Agar

**Composition** per liter:

| | |
|---|---|
| Agar | 15.0g |
| Peptone | 10.0g |
| Glucose | 10.0g |
| Ox bile | 8.0g |
| Yeast extract | 2.0g |
| Glycerol monostearate | 0.5g |
| Olive oil | 20.0mL |
| Glycerol | 10.0mL |
| Tween™ 60 | 5.0mL |

pH 6.0 ± 0.2 at 25°C

**Preparation of Medium:** Add components to distilled/deionized water and bring volume to 1.0L. Mix thoroughly. Gently heat and bring to boiling. Autoclave for 15 min at 15 psi pressure–121°C. Pour into sterile Petri dishes.

**Storage/Shelf Life:** Store dehydrated media in the dark in a sealed container below 30°C. Prepared media should be stored under refrigeration (2-8°C). Media should be used within 60 days of preparation. Media should not be used if there are any signs of deterioration (shrinking, cracking, or discoloration) or contamination, or if the expiration date supplied by the manufacturer has passed.

**Use:** For the cultivation of *Malassezia* species.

## Littman Oxgall Agar

**Composition** per liter:

| | |
|---|---|
| Agar | 20.0g |
| Oxgall | 15.0g |
| Glucose | 10.0g |
| Peptone | 10.0g |
| Crystal Violet | 0.01g |
| Streptomycin solution | 10.0mL |

pH 6.5 ± 0.2 at 25°C

**Source:** This medium is available as a premixed powder from BD Diagnostic Systems.

**Streptomycin Solution:**
**Composition** per 10.0mL:
Streptomycin ..................................................................0.03g

**Preparation of Streptomycin Solution:** Add streptomycin to distilled/deionized water and bring volume to 10.0mL. Mix thoroughly. Filter sterilize.

**Preparation of Medium:** Add components, except streptomycin solution, to distilled/deionized water and bring volume to 990.0mL. Mix thoroughly. Gently heat and bring to boiling. Autoclave for 15 min at 15 psi pressure–121°C. Cool to 45°–50°C. Aseptically add sterile streptomycin solution. Mix thoroughly. Pour into sterile Petri dishes or distribute into sterile tubes. Allow tubes to cool in a slanted position.

**Storage/Shelf Life:** Store dehydrated media in the dark in a sealed container below 30°C. Prepared media should be stored under refrigeration (2-8°C). Media should be used within 60 days of preparation. Media should not be used if there are any signs of deterioration (shrinking, cracking, or discoloration) or contamination, or if the expiration date supplied by the manufacturer has passed.

**Use:** For the selective isolation and cultivation of fungi, especially dermatophytes.

### Littman Oxgall HiVeg Agar Base with Streptomycin

**Composition** per liter:
Agar ..........................................................................20.0g
Plant peptone...............................................................20.0g
Glucose .......................................................................10.0g
Synthetic detergent No. II ...........................................5.0g
Crystal Violet...............................................................0.01g
Streptomycin solution ............................................10.0mL

pH 7.0 ± 0.2 at 25°C

**Source:** This medium, without streptomycin, is available as a pre-mixed powder from HiMedia.

**Streptomycin Solution:**
**Composition** per 10.0mL:
Streptomycin ..................................................................0.03g

**Preparation of Streptomycin Solution:** Add streptomycin to distilled/deionized water and bring volume to 10.0mL. Mix thoroughly. Filter sterilize.

**Preparation of Medium:** Add components, except streptomycin solution, to distilled/deionized water and bring volume to 990.0mL. Mix thoroughly. Gently heat and bring to boiling. Autoclave for 15 min at 15 psi pressure–121°C. Cool to 45°–50°C. Aseptically add sterile streptomycin solution. Mix thoroughly. Pour into sterile Petri dishes or distribute into sterile tubes. Allow tubes to cool in a slanted position.

**Storage/Shelf Life:** Store dehydrated media in the dark in a sealed container below 30°C. Prepared media should be stored under refrigeration (2-8°C). Media should be used within 60 days of preparation. Media should not be used if there are any signs of deterioration (shrinking, cracking, or discoloration) or contamination, or if the expiration date supplied by the manufacturer has passed.

**Use:** For the selective isolation and cultivation of fungi, especially dermatophytes.

### Littman Oxgall HiVeg Broth Base with Streptomycin

**Composition** per liter:
Plant peptone...............................................................20.0g
Glucose .......................................................................10.0g

Synthetic detergent No. II............................................5.0g
Crystal violet................................................................0.01g
Streptomycin solution .............................................10.0mL

pH 7.0 ± 0.2 at 25°C

**Source:** This medium, without streptomycin, is available as a pre-mixed powder from HiMedia.

**Streptomycin Solution:**
**Composition** per 10.0mL:
Streptomycin ..................................................................0.03g

**Preparation of Streptomycin Solution:** Add streptomycin to distilled/deionized water and bring volume to 10.0mL. Mix thoroughly. Filter sterilize.

**Preparation of Medium:** Add components, except streptomycin solution, to distilled/deionized water and bring volume to 990.0mL. Mix thoroughly. Gently heat and bring to boiling. Autoclave for 15 min at 15 psi pressure–121°C. Cool to 45°–50°C. Aseptically add sterile streptomycin solution. Mix thoroughly. Aseptically distribute into sterile tubes.

**Storage/Shelf Life:** Store dehydrated media in the dark in a sealed container below 30°C. Prepared media should be stored under refrigeration (2-8°C). Media should be used within 60 days of preparation. Media should not be used if there are any signs of deterioration (shrinking, cracking, or discoloration) or contamination, or if the expiration date supplied by the manufacturer has passed.

**Use:** For the selective cultivation of fungi, especially dermatophytes.

### Malt Extract Agar for Yeasts and Molds (MEAYM) (BAM M182)

**Composition** per liter:
Agar ..........................................................................20.0g
Glucose .......................................................................20.0g
Malt extract.................................................................20.0g
Peptone ........................................................................1.0g

pH 5.4 ± 0.2 at 25°C

**Preparation of Medium:** Add components to distilled/deionized water and bring volume to 1.0L. Mix thoroughly. Gently heat and bring to boiling. Distribute into tubes or flasks. Autoclave for 15 min at 15 psi pressure–121°C. Pour into sterile Petri dishes or leave in tubes.

**Storage/Shelf Life:** Store dehydrated media in the dark in a sealed container below 30°C. Prepared media should be stored under refrigeration (2-8°C). Media should be used within 60 days of preparation. Media should not be used if there are any signs of deterioration (shrinking, cracking, or discoloration) or contamination, or if the expiration date supplied by the manufacturer has passed.

**Use:** For the isolation, cultivation, and identification of heat-resistant filamentous fungi (molds) from foods. Recommended for identification of *Aspergillus* spp. and *Penicillium* spp.

### Modified Fungal Agar Base (Modified Inhibitory Mold Agar)

**Composition** per liter:
Glucose .......................................................................20.0g
Agar ..........................................................................15.0g
Casein enzymic hydrolysate ........................................2.5g
Peptic digest of animal tissue ......................................2.5g
Yeast extract................................................................5.0g
$Na_2HPO_4$..................................................................3.5g

KH$_2$PO$_4$.................................................................................3.4g

NH$_4$Cl .................................................................................1.4g

NaCO$_3$.................................................................................1.0g

Chloramphenicol................................................................0.1g

MgSO$_4$·7H$_2$O.................................................................0.06g

Polysorbate 80.......................................................... 20.0mL

pH 7.0 ± 0.2 at 25°C

**Source:** This medium is available from HiMedia.

**Preparation of Medium:** Add components, except polysorbate 80, to distilled/deionized water and bring volume to 980.0mL. Mix thoroughly. Gently heat and bring to boiling. Add polysorbate 80. Distribute into tubes or flasks. Autoclave for 15 min at 15 psi pressure–121°C. Mix thoroughly. Pour into Petri dishes or aseptically distribute into sterile tubes.

**Storage/Shelf Life:** Store dehydrated media in the dark in a sealed container below 30°C. Prepared media should be stored under refrigeration (2-8°C). Media should be used within 60 days of preparation. Media should not be used if there are any signs of deterioration (shrinking, cracking, or discoloration) or contamination, or if the expiration date supplied by the manufacturer has passed.

**Use:** For the detection and enumeration of molds in cosmetics and toiletries.

## Modified Fungal HiVeg Agar Base
## (Modified Inhibitory Mold HiVeg Agar Base)

**Composition** per liter:

Glucose .................................................................................20.0g

Agar .......................................................................................15.0g

Yeast extract........................................................................5.0g

Na$_2$HPO$_4$.............................................................................3.5g

KH$_2$PO$_4$.................................................................................3.4g

Plant hydrolysate................................................................2.5g

Plant peptone........................................................................2.5g

NH$_4$Cl .................................................................................1.4g

Na$_2$CO$_3$ .................................................................................1.0g

Chloramphenicol................................................................0.1g

MgSO$_4$ .................................................................................0.06g

Polysorbate 80.......................................................... 20.0mL

pH 7.0 ± 0.2 at 25°C

**Source:** This medium, without polysorbate 80, is available as a premixed powder from HiMedia.

**Preparation of Medium:** Add components to distilled/deionized water and bring volume to 1.0L. Mix thoroughly. Gently heat and bring to boiling with frequent agitation. Distribute into tubes or flasks. Autoclave for 15 min at 15 psi pressure–121°C. Pour into sterile Petri dishes or leave in tubes.

**Storage/Shelf Life:** Store dehydrated media in the dark in a sealed container below 30°C. Prepared media should be stored under refrigeration (2-8°C). Media should be used within 60 days of preparation. Media should not be used if there are any signs of deterioration (shrinking, cracking, or discoloration) or contamination, or if the expiration date supplied by the manufacturer has passed.

**Use:** For the isolation of pathogenic fungi.

## Mycobiotic Agar
## (Cycloheximide Chloramphenicol Agar)

**Composition** per liter:

Agar .......................................................................................15.0g

Enzymatic hydrolysate of soybean meal .....................10.0g

Glucose .................................................................................10.0g

Cycloheximide....................................................................0.5g

Chloramphenicol................................................................0.05g

pH 6.5 ± 0.2 at 25°C

**Source:** This medium is available as a premixed powder from BD Diagnostic Systems.

**Caution:** Cycloheximide is toxic. Avoid skin contact or aerosol formation and inhalation.

**Preparation of Medium:** Add components to distilled/deionized water and bring volume to 1.0L. Mix thoroughly. Gently heat and bring to boiling. Distribute into tubes or flasks. Autoclave for 15 min at 15 psi pressure–121°C. Cool tubes quickly in a slanted position.

**Storage/Shelf Life:** Store dehydrated media in the dark in a sealed container below 30°C. Prepared media should be stored under refrigeration (2-8°C). Media should be used within 60 days of preparation. Media should not be used if there are any signs of deterioration (shrinking, cracking, or discoloration) or contamination, or if the expiration date supplied by the manufacturer has passed.

**Use:** For the selective isolation and cultivation of pathogenic fungi.

## NigerSeed Agar
## (Bird Seed Agar)

**Composition** per liter:

Agar .......................................................................................15.0g

Glucose .................................................................................15.0g

Creatinine.............................................................................5.0g

KH$_2$PO$_4$.................................................................................3.0g

Biphenyl................................................................................1.0g

Chloramphenicol................................................................0.5g

*Guizotia abyssinica* seed (niger seed) extract...................... 1000.0mL

pH 6.7 ± 0.2 at 25°C

**Preparation of Medium:** Prepare seed extract by grinding 50.0g of *Guizotia abyssinica* seed in 1.0L of distilled/deionized water. Boil for 30 min. Filter through cheesecloth and filter paper. Add remaining components to seed filtrate. Mix thoroughly and heat with frequent agitation until boiling. Distribute into flasks or tubes. Autoclave for 25 min at 15 psi pressure–110°C.

**Storage/Shelf Life:** Store dehydrated media in the dark in a sealed container below 30°C. Prepared media should be stored under refrigeration (2-8°C). Media should be used within 60 days of preparation. Media should not be used if there are any signs of deterioration (shrinking, cracking, or discoloration) or contamination, or if the expiration date supplied by the manufacturer has passed.

**Use:** For the selective isolation and differentiation of *Cryptococcus neoformans* from other *Cryptococcus* species and other yeasts. *Crytococcus* species, notably *C. neoformans* and *C. gattii*, form tan to brown colonies whereas other yeasts form beige or cream colored colonies.

## Pablum Cereal Agar

**Composition** per liter:

Pablum cereal, precooked...........................................................100.0g

Agar .......................................................................................18.0g

Chloramphenicol................................................................0.05g

**Preparation of Medium:** Add components to distilled/deionized water and bring volume to 1.0L. Mix thoroughly. Gently heat and bring to boiling. Distribute into tubes or flasks. Autoclave for 15 min at 15 psi pressure–121°C. Pour into sterile Petri dishes or leave in tubes.

**Storage/Shelf Life:** Store dehydrated media in the dark in a sealed container below 30°C. Prepared media should be stored under refrigeration (2-8°C). Media should be used within 60 days of preparation. Media should not be used if there are any signs of deterioration (shrinking, cracking, or discoloration) or contamination, or if the expiration date supplied by the manufacturer has passed.

**Use:** For the cultivation of dematiaceous fungi and stimulation of spore formation.

## Pagano Levin Agar

**Composition** per liter:

Glucose ........................................................................40.0g
Agar .............................................................................15.0g
Peptone........................................................................10.0g
Yeast extract ..................................................................1.0g
Neomycin .......................................................................0.5g
2,3,5-Triphenyltetrazolium chloride .............................0.1g

pH 6.0 ± 0.1 at 25°C

**Source:** This medium is available as a premixed powder from BD Diagnostic Systems.

**Preparation of Medium:** Add components, except neomycin and 2,3,5-triphenyltetrazolium chloride, to distilled/deionized water and bring volume to 1.0L. Mix thoroughly. Gently heat and bring to boiling. Autoclave for 15 min at 15 psi pressure–121°C. Cool to 45°–50°C. Aseptically add neomycin and 2,3,5-triphenyltetrazolium chloride. Mix thoroughly. Pour into sterile Petri dishes or distribute into sterile tubes. Allow tubes to cool in a slanted position.

**Storage/Shelf Life:** Store dehydrated media in the dark in a sealed container below 30°C. Prepared media should be stored under refrigeration (2-8°C). Media should be used within 60 days of preparation. Media should not be used if there are any signs of deterioration (shrinking, cracking, or discoloration) or contamination, or if the expiration date supplied by the manufacturer has passed.

**Use:** For the isolation, cultivation, and differentiation of *Candida* species. *Candida albicans* appears as smooth, shiny, cream-light pink colonies.

## Potato Dextrose Agar
## (PDA Agar)

**Composition** per liter:

Glucose ........................................................................20.0g
Agar .............................................................................15.0g
Potato, infusion from .....................................................4.0g
Tartaric acid solution...................................................14.0mL

pH 5.6 ± 0.2 at 25°C

**Source:** This medium is available as a premixed powder from BD Diagnostic Systems and Oxoid Unipath.

**Tartaric Acid Solution:**

**Composition** per 50.0mL:

Tartaric acid ...................................................................5.0g

**Preparation of Tartaric Acid Solution:** Add tartaric acid to distilled/deionized water and bring volume to 50.0mL. Mix thoroughly. Filter sterilize.

**Preparation of Medium:** Add components to distilled/deionized water and bring volume to 986.0mL. Mix thoroughly. Gently heat and bring to boiling. Distribute into tubes or flasks. Autoclave for 15 min at 15 psi pressure–121°C. Cool to 45°–50°C. Aseptically add 14.0mL of sterile tartaric acid solution. Mix thoroughly. If medium is to be used for the enumeration of yeasts and molds in butter, adjust pH to 3.5. Pour into sterile Petri dishes or distribute into sterile tubes.

**Storage/Shelf Life:** Store dehydrated media in the dark in a sealed container below 30°C. Prepared media should be stored under refrigeration (2-8°C). Media should be used within 60 days of preparation. Media should not be used if there are any signs of deterioration (shrinking, cracking, or discoloration) or contamination, or if the expiration date supplied by the manufacturer has passed.

**Use:** For the cultivation and enumeration of yeasts and molds. For the enumeration of yeasts and molds in butter by the plate count method.

## Potato Dextrose Agar
## (PDA Agar)

**Composition** per liter:

Agar .............................................................................20.0g
Glucose ........................................................................20.0g
Potato infusion ...........................................................200.0mL

pH 5.6 ± 0.2 at 25°C

**Potato Infusion:**
**Composition** per 10.0mL:

Potatoes, unpeeled and sliced ........................200.0g

**Preparation of Potato Infusion:** Add potato slices to 1.0L of distilled/deionized water. Gently heat and bring to boiling. Continue boiling for 30 min. Filter through cheesecloth. Reserve filtrate.

**Preparation of Medium:** Add components to distilled/deionized water and bring volume to 1.0L. Mix thoroughly. Gently heat and bring to boiling. Distribute into tubes or flasks. Autoclave for 15 min at 15 psi pressure–121°C. Pour into sterile Petri dishes or leave in tubes.

**Storage/Shelf Life:** Store dehydrated media in the dark in a sealed container below 30°C. Prepared media should be stored under refrigeration (2-8°C). Media should be used within 60 days of preparation. Media should not be used if there are any signs of deterioration (shrinking, cracking, or discoloration) or contamination, or if the expiration date supplied by the manufacturer has passed.

**Use:** For the cultivation and enumeration of yeasts and filamentous fungi (molds) from foods.

## Potato Flakes Agar

**Composition** per liter:

Potato flakes................................................................20.0g
Agar .............................................................................15.0g
Glucose ........................................................................10.0g

**Preparation of Medium:** Add components to distilled/deionized water and bring volume to 1.0L. Mix thoroughly. Gently heat and bring to boiling. Distribute into tubes or flasks. Autoclave for 15 min at 15 psi pressure–121°C. Pour into sterile Petri dishes or leave in tubes.

**Storage/Shelf Life:** Store dehydrated media in the dark in a sealed container below 30°C. Prepared media should be stored under refrigeration (2-8°C). Media should be used within 60 days of preparation. Media should not be used if there are any signs of deterioration (shrinking, cracking, or discoloration) or contamination, or if the expiration date supplied by the manufacturer has passed.

**Use:** For the cultivation and induction of sporulation in all fungi.

## Rose Bengal Chloramphenicol Agar

**Composition** per liter:

Agar ....................................................................15.0g
Glucose ...............................................................10.0g
Papaic digest of soybean meal .............................5.0g
KH$_2$PO$_4$............................................................1.0g
MgSO$_4$·7H$_2$O ...................................................0.5g
Rose Bengal .........................................................0.05g
Chloramphenicol solution............................ 10.0mL

pH 7.0 ± 0.2 at 25°C

**Source:** This medium is available as a premixed powder from BD Diagnostic Systems and Oxoid Unipath.

**Chloramphenicol Solution:**
**Composition** per 10.0mL:
Chloramphenicol.....................................................0.1g

**Preparation of Chloramphenicol Solution:** Add chloramphenicol to distilled/deionized water and bring volume to 10.0mL. Mix thoroughly. Filter sterilize.

**Preparation of Medium:** Add components, except chloramphenicol solution, to distilled/deionized water and bring volume to 990.0mL. Mix thoroughly. Gently heat and bring to boiling. Autoclave for 15 min at 15 psi pressure–121°C. Cool to 45°C. Aseptically add sterile chloramphenicol solution. Mix thoroughly. Pour into sterile Petri dishes or distribute into sterile tubes.

**Storage/Shelf Life:** Store dehydrated media in the dark in a sealed container below 30°C. Prepared media should be stored under refrigeration (2-8°C). Media should be used within 60 days of preparation. Media should not be used if there are any signs of deterioration (shrinking, cracking, or discoloration) or contamination, or if the expiration date supplied by the manufacturer has passed.

**Use:** For the selective isolation, cultivation, and enumeration of yeasts and molds from environmental specimens and foods.

## Rose Bengal Chloramphenicol HiVeg Agar

**Composition** per liter:

Agar ....................................................................15.5g
Glucose ...............................................................10.0g
Plant peptone No. 4..............................................5.0g
KH$_2$PO$_4$............................................................1.0g
MgSO$_4$ ...............................................................0.5g
Rose Bengal .........................................................0.05g
Chloramphenicol solution............................ 10.0mL

pH 7.2 ± 0.2 at 25°C

**Source:** This medium, without chloramphenicol, is available as a premixed powder from HiMedia.

**Chloramphenicol Solution:**
**Composition** per 10.0mL:
Chloramphenicol.....................................................0.1g

**Preparation of Chloramphenicol Solution:** Add chloramphenicol to distilled/deionized water and bring volume to 10.0mL. Mix thoroughly. Filter sterilize.

**Preparation of Medium:** Add components, except chloramphenicol solution, to distilled/deionized water and bring volume to 990.0mL. Mix thoroughly. Gently heat and bring to boiling. Autoclave for 15 min at 15 psi pressure–121°C. Cool to 45°C. Aseptically add sterile chloramphenicol solution. Mix thoroughly. Pour into sterile Petri dishes or distribute into sterile tubes.

**Storage/Shelf Life:** Store dehydrated media in the dark in a sealed container below 30°C. Prepared media should be stored under refrigeration (2-8°C). Media should be used within 60 days of preparation. Media should not be used if there are any signs of deterioration (shrinking, cracking, or discoloration) or contamination, or if the expiration date supplied by the manufacturer has passed.

**Use:** For the selective isolation, cultivation, and enumeration of yeasts and molds from environmental specimens and foods.

## SABHI Agar
## (Sabouraud Glucose and Brain Heart Infusion Agar)

**Composition** per liter:

Glucose ...............................................................21.0g
Agar ....................................................................15.0g
Pancreatic digest of casein..................................10.5g
Peptic digest of animal tissue ..............................5.0g
Brain heart, solids from infusion ........................4.0g
NaCl.....................................................................2.5g
Na$_2$HPO$_4$............................................................1.25g

pH 6.8 ± 0.2 at 25°C

**Source:** This medium is available as a premixed powder from BD Diagnostic Systems.

**Preparation of Medium:** Add components to distilled/deionized water and bring volume to 1.0L. Mix thoroughly. Gently heat and bring to boiling. Distribute into tubes or flasks. Autoclave for 15 min at 15 psi pressure–121°C. Pour into sterile Petri dishes in 20.0mL volumes or leave in tubes.

**Storage/Shelf Life:** Store dehydrated media in the dark in a sealed container below 30°C. Prepared media should be stored under refrigeration (2-8°C). Media should be used within 60 days of preparation. Media should not be used if there are any signs of deterioration (shrinking, cracking, or discoloration) or contamination, or if the expiration date supplied by the manufacturer has passed.

**Use:** For the cultivation of dermatophytes and other pathogenic and nonpathogenic fungi from clinical and nonclinical specimens.

## SABHI Agar

**Composition** per liter:

Beef heart, infusion from....................................125.0g
Calf brains, infusion from...................................100.0g
Glucose ...............................................................21.0g
Agar ....................................................................15.0g
Neopeptone .........................................................5.0g
Proteose peptone.................................................5.0g
NaCl.....................................................................2.5g
Na$_2$HPO$_4$............................................................1.25g
Chloromycetin solution .............................. 1.0mL

pH 7.0 ± 0.2 at 25°C

**Source:** This medium is available as a premixed powder from BD Diagnostic Systems.

**Chloromycetin Solution:**
**Composition** per 10.0mL:
Chloromycetin .....................................................1.0g

**Preparation of Chloromycetin Solution:** Add chloromycetin to distilled/deionized water and bring volume to 10.0mL. Mix thoroughly. Filter sterilize.

**Preparation of Medium:** Add components, except chloromycetin solution, to distilled/deionized water and bring volume to 999.0mL. Mix thoroughly. Gently heat and bring to boiling. Autoclave for 15 min at 15 psi pressure–121°C. Cool to 45°–50°C. Aseptically add 1.0mL of sterile chloromycetin solution. Mix thoroughly. Aseptically distribute into sterile tubes in 5.0mL volumes.

**Storage/Shelf Life:** Store dehydrated media in the dark in a sealed container below 30°C. Prepared media should be stored under refrigeration (2-8°C). Media should be used within 60 days of preparation. Media should not be used if there are any signs of deterioration (shrinking, cracking, or discoloration) or contamination, or if the expiration date supplied by the manufacturer has passed.

**Use:** For the cultivation of dermatophytes and other pathogenic and nonpathogenic fungi from clinical and nonclinical specimens.

## SABHI Agar, Modified
**Composition** per liter:
| | |
|---|---|
| Beef heart, infusion from | 62.5g |
| Calf brain, infusion from | 50.0g |
| Glucose | 20.5g |
| Brain heart infusion broth | 18.6g |
| Agar | 7.5g |
| Neopeptone | 5.0g |
| Pancreatic digest of gelatin | 2.5g |
| NaCl | 1.25g |
| $Na_2HPO_4$ | 0.625g |

pH 6.8 ± 0.2 at 25°C

**Preparation of Medium:** Dissolve, then autoclave at 121°C for 15 min. Cool to 50°C and add 1.0mL of sterile chloramphenicol solution (100.0mg/mL). Mix well and dispense into sterile tubes. Slant and allow to harden. Refrigerate until needed.

**Use:** For the cultivation of dermatophytes and other pathogenic and nonpathogenic fungi from clinical and nonclinical specimens.

## SABHI Blood Agar
**Composition** per liter:
| | |
|---|---|
| Beef heart, infusion from | 125.0g |
| Calf brains, infusion from | 100.0g |
| Glucose | 21.0g |
| Agar | 15.0g |
| Neopeptone | 5.0g |
| Proteose peptone | 5.0g |
| NaCl | 2.5g |
| $Na_2HPO_4$ | 1.25g |
| Blood | 100.0mL |
| Chloromycetin solution | 1.0mL |

pH 7.0 ± 0.2 at 25°C

**Source:** This medium is available as a premixed powder from BD Diagnostic Systems.

**Chloromycetin Solution:**
**Composition** per 10.0mL:
| | |
|---|---|
| Chloromycetin | 1.0g |

**Preparation of Chloromycetin Solution:** Add chloromycetin to distilled/deionized water and bring volume to 10.0mL. Mix thoroughly. Filter sterilize.

**Preparation of Medium:** Add components, except blood and chloromycetin solution, to distilled/deionized water and bring volume to 899.0mL. Mix thoroughly. Gently heat and bring to boiling. Autoclave for 15 min at 15 psi pressure–121°C. Cool to 45°–50°C. Aseptically add 100.0mL of sterile blood and 1.0mL of sterile chloromycetin solution. Sheep blood or human blood may be used. Mix thoroughly. Aseptically distribute into sterile tubes in 5.0mL volumes.

**Use:** For the cultivation of dermatophytes and other pathogenic and nonpathogenic fungi from clinical and nonclinical specimens. Blood enhances the recovery of *Blastomyces dermatitidis* and *Histoplasma capsulatum* and their conversion to the yeast phase.

## SABHI HiVeg Agar Base with Chloramphenicol
**Composition** per liter:
| | |
|---|---|
| Glucose | 21.0g |
| Agar | 15.0g |
| Plant infusion | 5.14g |
| Plant peptone No. 3 | 5.0g |
| Plant special peptone | 5.0g |
| Plant special infusion | 4.11g |
| NaCl | 2.5g |
| $Na_2HPO_4$ | 1.25g |
| Chloramphenicol solution | 10.0mL |

pH 7.0 ± 0.2 at 25°C

**Source:** This medium, without chloramphenicol, is available as a premixed powder from HiMedia.

**Chloramphenicol Solution:**
**Composition** per 10.0mL:
| | |
|---|---|
| Chloramphenicol | 0.1g |

**Preparation of Chloramphenicol Solution:** Add chloramphenicol to distilled/deionized water and bring volume to 10.0mL. Mix thoroughly. Filter sterilize.

**Preparation of Medium:** Add components, except chloramphenicol solution, to distilled/deionized water and bring volume to 990.0mL. Mix thoroughly. Gently heat and bring to boiling. Autoclave for 15 min at 15 psi pressure–121°C. Cool to 45°–50°C. Aseptically add 10.0mL of sterile chloramphenicol solution. Mix thoroughly. Pour into Petri dishes or aseptically distribute into sterile tubes.

**Use:** For the cultivation of dermatophytes and other pathogenic and nonpathogenic fungi from clinical and nonclinical specimens.

## SABHI HiVeg Agar Base with Chloromycetin
**Composition** per liter:
| | |
|---|---|
| Glucose | 21.0g |
| Agar | 15.0g |
| Plant infusion | 5.14g |
| Plant peptone No. 3 | 5.0g |
| Plant special peptone | 5.0g |
| Plant special infusion | 4.11g |
| NaCl | 2.5g |
| $Na_2HPO_4$ | 1.25g |
| Chloromycetin solution | 1.0mL |

pH 7.0 ± 0.2 at 25°C

**Source:** This medium, without chloromycetin, is available as a premixed powder from HiMedia.

**Chloromycetin Solution:**
**Composition** per 10.0mL:
| | |
|---|---|
| Chloromycetin | 1.0g |

**Preparation of Chloromycetin Solution:** Add chloromycetin to distilled/deionized water and bring volume to 10.0mL. Mix thoroughly. Filter sterilize.

**Preparation of Medium:** Add components, except chloromycetin solution, to distilled/deionized water and bring volume to 999.0mL. Mix thoroughly. Gently heat and bring to boiling. Autoclave for 15 min at 15 psi pressure–121°C. Cool to 45°–50°C. Aseptically add 1.0mL of sterile chloromycetin solution. Mix thoroughly. Aseptically distribute into sterile tubes in 5.0mL volumes.

**Storage/Shelf Life:** Store dehydrated media in the dark in a sealed container below 30°C. Prepared media should be stored under refrigeration (2-8°C). Media should be used within 60 days of preparation. Media should not be used if there are any signs of deterioration (shrinking, cracking, or discoloration) or contamination, or if the expiration date supplied by the manufacturer has passed.

**Use:** For the cultivation of dermatophytes and other pathogenic and nonpathogenic fungi from clinical and nonclinical specimens.

## SABHI HiVeg Agar Base with Blood and Chloromycetin

**Composition** per liter:

| | |
|---|---|
| Glucose | 21.0g |
| Agar | 15.0g |
| Plant infusion | 5.14g |
| Plant peptone No. 3 | 5.0g |
| Plant special peptone | 5.0g |
| Plant special infusion | 4.11g |
| NaCl | 2.5g |
| $Na_2HPO_4$ | 1.25g |
| Blood | 100.0mL |
| Chloromycetin solution | 1.0mL |

pH 7.0 ± 0.2 at 25°C

**Source:** This medium, without blood and chloromycetin, is available as a premixed powder from HiMedia.

**Chloromycetin Solution:**

**Composition** per 10.0mL:

| | |
|---|---|
| Chloromycetin | 1.0g |

**Preparation of Chloromycetin Solution:** Add chloromycetin to distilled/deionized water and bring volume to 10.0mL. Mix thoroughly. Filter sterilize.

**Preparation of Medium:** Add components, except blood and chloromycetin solution, to distilled/deionized water and bring volume to 899.0mL. Mix thoroughly. Gently heat and bring to boiling. Autoclave for 15 min at 15 psi pressure–121°C. Cool to 45°–50°C. Aseptically add 100.0mL of sterile blood and 1.0mL of sterile chloromycetin solution. Sheep blood or human blood may be used. Mix thoroughly. Aseptically distribute into sterile tubes in 5.0mL volumes.

**Storage/Shelf Life:** Store dehydrated media in the dark in a sealed container below 30°C. Prepared media should be stored under refrigeration (2-8°C). Media should be used within 60 days of preparation. Media should not be used if there are any signs of deterioration (shrinking, cracking, or discoloration) or contamination, or if the expiration date supplied by the manufacturer has passed.

**Use:** For the cultivation of dermatophytes and other pathogenic and nonpathogenic fungi from clinical and nonclinical specimens. Blood enhances the recovery of *Blastomyces dermatitidis* and *Histoplasma capsulatum* and their conversion to the yeast phase.

## Sabouraud Agar with CCG and 3% Sodium Chloride

**Composition** per 3031.5mL:

| | |
|---|---|
| Glucose | 120.0g |
| NaCl | 90.0g |
| Agar | 45.0g |
| Peptone | 30.0g |
| Chloramphenicol solution | 15.0mL |
| Cycloheximide solution | 15.0mL |
| Gentamicin solution | 1.5mL |

**Chloramphenicol Solution:**

**Composition** per 15.0mL:

| | |
|---|---|
| Chloramphenicol | 0.15g |

**Preparation of Chloramphenicol Solution:** Add chloramphenicol to distilled/deionized water and bring volume to 15.0mL. Mix thoroughly. Filter sterilize.

**Cycloheximide Solution:**

**Composition** per 15.0mL:

| | |
|---|---|
| Cycloheximide | 0.3g |

**Preparation of Cycloheximide Solution:** Add cycloheximide to distilled/deionized water and bring volume to 15.0mL. Mix thoroughly. Filter sterilize.

**Caution:** Cycloheximide is toxic. Avoid skin contact or aerosol formation and inhalation.

**Gentamicin Solution:**

**Composition** per 10.0mL:

| | |
|---|---|
| Gentamicin | 0.4g |

**Preparation of Gentamicin Solution:** Add gentamicin to distilled/deionized water and bring volume to 10.0mL. Mix thoroughly. Filter sterilize.

**Preparation of Medium:** Add components—except chloramphenicol solution, cycloheximide solution, and gentamicin solution—to distilled/deionized water and bring volume to 3.0L. Mix thoroughly. Gently heat and bring to boiling. Autoclave for 15 min at 15 psi pressure–121°C. Cool to 45°–50°C. Aseptically add 15.0mL of sterile chloramphenicol solution, 15.0mL of sterile cycloheximide solution, and 1.5mL of sterile gentamicin solution. Mix thoroughly. Aseptically distribute into sterile tubes. Allow tubes to cool in a slanted position.

**Storage/Shelf Life:** Store dehydrated media in the dark in a sealed container below 30°C. Prepared media should be stored under refrigeration (2-8°C). Media should be used within 60 days of preparation. Media should not be used if there are any signs of deterioration (shrinking, cracking, or discoloration) or contamination, or if the expiration date supplied by the manufacturer has passed.

**Use:** For the selective isolation and cultivation of fungi from specimens with a mixed flora.

## Sabouraud Agar with CCG and 5% Sodium Chloride

**Composition** per 3031.5mL:

| | |
|---|---|
| NaCl | 150.0g |
| Glucose | 120.0g |
| Agar | 45.0g |
| Peptone | 30.0g |
| Chloramphenicol solution | 15.0mL |
| Cycloheximide solution | 15.0mL |
| Gentamicin solution | 1.5mL |

## Chloramphenicol Solution:
**Composition** per 15.0mL:
Chloramphenicol ................................................................0.15g

**Preparation of Chloramphenicol Solution:** Add chloramphenicol to distilled/deionized water and bring volume to 15.0mL. Mix thoroughly. Filter sterilize.

## Cycloheximide Solution:
**Composition** per 15.0mL:
Cycloheximide ......................................................................0.3g

**Preparation of Cycloheximide Solution:** Add cycloheximide to distilled/deionized water and bring volume to 15.0mL. Mix thoroughly. Filter sterilize.

**Caution:** Cycloheximide is toxic. Avoid skin contact or aerosol formation and inhalation.

## Gentamicin Solution:
**Composition** per 10.0mL:
Gentamicin ..........................................................................0.4g

**Preparation of Gentamicin Solution:** Add gentamicin to distilled/deionized water and bring volume to 10.0mL. Mix thoroughly. Filter sterilize.

**Preparation of Medium:** Add components—except chloramphenicol solution, cycloheximide solution, and gentamicin solution—to distilled/deionized water and bring volume to 3.0L. Mix thoroughly. Gently heat and bring to boiling. Autoclave for 15 min at 15 psi pressure–121°C. Cool to 45°–50°C. Aseptically add 15.0mL of sterile chloramphenicol solution, 15.0mL of sterile cycloheximide solution, and 1.5mL of sterile gentamicin solution. Mix thoroughly. Aseptically distribute into sterile tubes. Allow tubes to cool in a slanted position.

**Storage/Shelf Life:** Store dehydrated media in the dark in a sealed container below 30°C. Prepared media should be stored under refrigeration (2-8°C). Media should be used within 60 days of preparation. Media should not be used if there are any signs of deterioration (shrinking, cracking, or discoloration) or contamination, or if the expiration date supplied by the manufacturer has passed.

**Use:** For the selective isolation and cultivation of fungi from specimens with a mixed flora.

## Sabouraud Cycloheximide Chloramphenicol HiVeg Agar

**Composition** per liter:
Glucose ..............................................................................20.0g
Agar ..................................................................................15.0g
Plant peptone......................................................................10.0g
Cycloheximide......................................................................0.5g
Chloramphenicol...................................................................0.04g
pH 6.8 ± 0.2 at 25°C

**Source:** This medium is available as a premixed powder from Hi-Media.

**Caution:** Cycloheximide is very toxic. Avoid skin contact or aerosol formation and inhalation.

**Preparation of Medium:** Add components to distilled/deionized water and bring volume to 1.0L. Mix thoroughly. Gently heat and bring to boiling. Distribute into tubes or flasks. Autoclave for 15 min at 15 psi pressure–121°C. Pour into sterile Petri dishes or leave in tubes.

**Storage/Shelf Life:** Store dehydrated media in the dark in a sealed container below 30°C. Prepared media should be stored under refrigeration (2-8°C). Media should be used within 60 days of preparation. Media should not be used if there are any signs of deterioration (shrinking, cracking, or discoloration) or contamination, or if the expiration date supplied by the manufacturer has passed.

**Use:** For the isolation and cultivation of pathogenic fungi.

## Saboraud Dextrose Agar (SDA)

**Composition** per liter:
Glucose ..............................................................................40.0g
Agar ..................................................................................15.0g
Mycological peptone ............................................................10.0g
pH 5.6 ± 0.2 at 25°C

**Preparation of Medium:** Add components to distilled/deionized water and bring volume to 1.0L. Mix thoroughly. Gently heat and bring to boiling. Distribute into tubes or flasks. Autoclave for 15 min at 15 psi pressure–121°C. Pour into sterile Petri dishes or leave in tubes.

**Storage/Shelf Life:** Store dehydrated media in the dark in a sealed container below 30°C. Prepared media should be stored under refrigeration (2-8°C). Media should be used within 60 days of preparation. Media should not be used if there are any signs of deterioration (shrinking, cracking, or discoloration) or contamination, or if the expiration date supplied by the manufacturer has passed.

**Use:** For the cultivation of yeast and filamentous fungi. For the cultivation of pathogenic and nonpathogenic fungi, especially dermatophytes, particularly *Malassezia* species. The medium may be made more selective for fungi by the addition of chloramphenicol. Fluconozole (final concentration of 8–16mg per mL) may also be added to test for antibiotic sensitivity.

## Sabouraud Glucose Agar (Saboraud Dextrose Agar) (SabDex, 2%)

**Composition** per liter:
Glucose ..............................................................................20.0g
Agar ..................................................................................15.0g
Pancreatic digest of casein....................................................5.0g
Peptic digest of animal tissue ...............................................5.0g
pH 5.6 ± 0.2 at 25°C

**Preparation of Medium:** Add components to distilled/deionized water and bring volume to 1.0L. Mix thoroughly. Gently heat and bring to boiling. Distribute into tubes or flasks. Autoclave for 15 min at 15 psi pressure–121°C. Pour into sterile Petri dishes or leave in tubes.

**Storage/Shelf Life:** Store dehydrated media in the dark in a sealed container below 30°C. Prepared media should be stored under refrigeration (2-8°C). Media should be used within 60 days of preparation. Media should not be used if there are any signs of deterioration (shrinking, cracking, or discoloration) or contamination, or if the expiration date supplied by the manufacturer has passed.

**Use:** For the cultivation of yeast and filamentous fungi. For the cultivation of pathogenic and nonpathogenic fungi, especially dermatophytes. The medium may be made more selective for fungi by the addition of chloramphenicol. Fluconozole (final concentration of 8–16mg per mL) may also be added to test for antibiotic sensitivity.

## Sabouraud Glucose Agar
## (Saboraud Dextrose Agar)
## (SabDex, 4%)

**Composition** per liter:

Glucose ...................................................................40.0g
Agar ......................................................................15.0g
Pancreatic digest of casein....................................5.0g
Peptic digest of animal tissue................................5.0g

pH 5.6 ± 0.2 at 25°C

**Source**: This medium is available as a premixed powder from BD Diagnostic Systems and Oxoid Unipath.

**Preparation of Medium:** Add components to distilled/deionized water and bring volume to 1.0L. Mix thoroughly. Gently heat and bring to boiling. Distribute into tubes or flasks. Autoclave for 15 min at 15 psi pressure–121°C. Pour into sterile Petri dishes or leave in tubes.

**Storage/Shelf Life:** Store dehydrated media in the dark in a sealed container below 30°C. Prepared media should be stored under refrigeration (2-8°C). Media should be used within 60 days of preparation. Media should not be used if there are any signs of deterioration (shrinking, cracking, or discoloration) or contamination, or if the expiration date supplied by the manufacturer has passed.

**Use:** For the cultivation of yeast and filamentous fungi. For the cultivation of pathogenic and nonpathogenic fungi, especially dermatophytes. The medium may be made more selective for fungi by the addition of chloramphenicol. Fluconozole (final concentration of 8–16mg per mL) may also be added to test for antibiotic sensitivity.

## Sabouraud Glucose Agar, Emmons

**Composition** per liter:

Glucose ...................................................................20.0g
Agar ......................................................................17.0g
Pancreatic digest of casein....................................5.0g
Peptic digest of animal tissue................................5.0g

pH 6.9 ± 0.2 at 25°C

**Source:** This medium is available as a premixed powder from BD Diagnostic Systems.

**Preparation of Medium:** Add components to distilled/deionized water and bring volume to 1.0L. Mix thoroughly. Gently heat and bring to boiling. Distribute into tubes or flasks. Autoclave for 15 min at 13 psi pressure–118°C. Pour into sterile Petri dishes or leave in tubes.

**Storage/Shelf Life:** Store dehydrated media in the dark in a sealed container below 30°C. Prepared media should be stored under refrigeration (2-8°C). Media should be used within 60 days of preparation. Media should not be used if there are any signs of deterioration (shrinking, cracking, or discoloration) or contamination, or if the expiration date supplied by the manufacturer has passed.

**Use:** For the cultivation of dermatophytes and other pathogenic and nonpathogenic fungi from clinical and nonclinical specimens. For the cultivation of yeast and filamentous fungi.

## Sabouraud Glucose Agar, HiVeg

**Composition** per liter:

Glucose ...................................................................40.0g
Agar ......................................................................15.0g
Plant peptone No. 4..............................................10.0g
Selective supplement ......................................... 10.0mL

pH 6.9 ± 0.2 at 25°C

**Source:** This medium is available as a premixed powder from Hi-Media.

**Selective Supplement:**
**Composition** per 10.0mL:
Cycloheximide..................................................0.5g
Chloramphenicol.............................................0.04g

**Preparation of Selective Supplement:** Add components to distilled/deionized water and bring volume to 10.0mL. Mix thoroughly. Filter sterilize.

**Caution:** Cycloheximide is very toxic. Avoid skin contact or aerosol formation and inhalation.

**Preparation of Medium:** Add components, except selective supplement, to distilled/deionized water and bring volume to 990.0mL. Mix thoroughly. Gently heat and bring to boiling. Distribute into tubes or flasks. Autoclave for 15 min at 15 psi pressure–121°C. Cool to 50°C. Aseptically add 10.0mL sterile selective supplement. Mix thoroughly. Pour into sterile Petri dishes or aseptically distribute into tubes.

**Storage/Shelf Life:** Store dehydrated media in the dark in a sealed container below 30°C. Prepared media should be stored under refrigeration (2-8°C). Media should be used within 60 days of preparation. Media should not be used if there are any signs of deterioration (shrinking, cracking, or discoloration) or contamination, or if the expiration date supplied by the manufacturer has passed.

**Use:** For the cultivation of dermatophytes and other pathogenic and nonpathogenic fungi from clinical and nonclinical specimens. For the cultivation of yeast and filamentous fungi.

## Sabouraud Glucose HiVeg Broth
## (Sabouraud Liquid HiVeg Medium)

**Composition** per liter:

Glucose ...................................................................20.0g
Plant special peptone .............................................10.0g

pH 5.6 ± 0.2 at 25°C

**Source:** This medium is available as a premixed powder from Hi-Media.

**Preparation of Medium:** Add components to distilled/deionized water and bring volume to 1.0L. Mix thoroughly. Distribute into tubes or flasks. Autoclave for 15 min at 15 psi pressure–121°C.

**Storage/Shelf Life:** Store dehydrated media in the dark in a sealed container below 30°C. Prepared media should be stored under refrigeration (2-8°C). Media should be used within 60 days of preparation. Media should not be used if there are any signs of deterioration (discoloration) or contamination, or if the expiration date supplied by the manufacturer has passed.

**Use:** For the cultivation of pathogenic and nonpathogenic fungi, especially dermatophytes. The medium may be made more selective for fungi by the addition of chloramphenicol.

## Sensitest Agar

**Composition** per liter:

Pancreatic digest of casein...................................11.0g
Agar ........................................................................8.0g
Buffer salts..............................................................3.3g
Peptone ...................................................................3.0g
NaCl.........................................................................3.0g
Glucose ...................................................................2.0g
Starch ......................................................................1.0g

Nucleoside bases.................................................................0.02g
Thiamine ............................................................................0.02mg

pH 7.4 ± 0.2 at 25°C

**Source:** This medium is available as a premixed powder from Oxoid Unipath.

**Preparation of Medium:** Add components to distilled/deionized water and bring volume to 1.0L. Mix thoroughly. Gently heat and bring to boiling. Distribute into tubes or flasks. Autoclave for 15 min at 15 psi pressure–121°C. Pour into sterile Petri dishes.

**Storage/Shelf Life:** Store dehydrated media in the dark in a sealed container below 30°C. Prepared media should be stored under refrigeration (2-8°C). Media should be used within 60 days of preparation. Media should not be used if there are any signs of deterioration (shrinking, cracking, or discoloration) or contamination, or if the expiration date supplied by the manufacturer has passed.

**Use:** For the performance of antibiotic sensitivity assays.

### Soil Extract Agar

**Composition** per liter:

Soil ....................................................................................500.0g
Agar .....................................................................................15.0g
Glucose ..................................................................................2.0g
Yeast extract............................................................................1.0g
KH$_2$PO$_4$.........................................................................0.5g

**Preparation of Medium:** Add 500.0g of garden soil to 1.0L of tap water. Autoclave for 3 h at 15 psi pressure–121°C. Filter through Whatman #2 filter paper. Add remaining components to filtrate. Bring volume to 1.0L with tap water. Gently heat and bring to boiling. Distribute into tubes in 7.0mL volumes. Autoclave for 15 min at 15 psi pressure–121°C. Allow tubes to cool in a slanted position.

**Storage/Shelf Life:** Store dehydrated media in the dark in a sealed container below 30°C. Prepared media should be stored under refrigeration (2-8°C). Media should be used within 60 days of preparation. Media should not be used if there are any signs of deterioration (shrinking, cracking, or discoloration) or contamination, or if the expiration date supplied by the manufacturer has passed.

**Use:** For the cultivation and identification of *Histoplasma capsulatum*, *Blastomyces dermatitidis*, and *Bacillus* species based on the formation of typical conidia.

### Staib Agar

**Composition** per liter:

Agar .....................................................................................15.0g
Glucose .................................................................................15.0g
Creatinine...............................................................................5.0g
KH$_2$PO$_4$.........................................................................3.0g
Biphenyl..................................................................................1.0g
Chloramphenicol......................................................................0.5g
*Guizotia abyssinica* seed (niger seed) extract...................... 1000.0mL

pH 6.7 ± 0.2 at 25°C

**Preparation of Medium:** Prepare seed extract by grinding 50.0g of *Guizotia abyssinica* seed in 1.0L of distilled/deionized water. Boil for 30 min. Filter through cheesecloth and filter paper. Add remaining components to seed filtrate. Mix thoroughly and heat with frequent agitation until boiling. Distribute into flasks or tubes. Autoclave for 25 min at 15 psi pressure–110°C.

**Storage/Shelf Life:** Store dehydrated media in the dark in a sealed container below 30°C. Prepared media should be stored under refriger-

ation (2-8°C). Media should be used within 60 days of preparation. Media should not be used if there are any signs of deterioration (shrinking, cracking, or discoloration) or contamination, or if the expiration date supplied by the manufacturer has passed.

**Use:** For the selective isolation and differentiation of *Cryptococcus neoformans* from other *Cryptococcus* species and other yeasts.

### TOC Agar
### (Tween™ 80 Oxgall Caffeic Acid Agar)

**Composition** per liter:

Agar .....................................................................................20.0g
Oxgall ..................................................................................10.0g
Caffeic acid.............................................................................0.3g
Tween™ 80 ........................................................................ 10.0mL

**Source:** This medium is available as a prepared medium from BD Diagnostic Systems.

**Preparation of Medium:** Add components to distilled/deionized water and bring volume to 1.0L. Mix thoroughly. Gently heat and bring to boiling. Autoclave for 15 min at 15 psi pressure–121°C. Pour into sterile Petri dishes.

**Storage/Shelf Life:** Store dehydrated media in the dark in a sealed container below 30°C. Prepared media should be stored under refrigeration (2-8°C). Media should be used within 60 days of preparation. Media should not be used if there are any signs of deterioration (shrinking, cracking, or discoloration) or contamination, or if the expiration date supplied by the manufacturer has passed.

**Use:** For the differentiation and identification of *Candida albicans* and *Cryptococcus neoformans*. *Cryptococcus albicans* produces germ tubes and chlamydospores when grown on this medium. *Cryptococcus neoformans* appears as tan to brown colonies.

### Tomato Juice Agar
### (Tomato Juice Yeast Extract Medium)

**Composition** per liter:

Skim milk.............................................................................100.0g
Yeast extract............................................................................5.0g
Tomato juice, filtered........................................................... 100.0mL

**Preparation of Medium:** Add components to distilled/deionized water and bring volume to 1.0L. Mix thoroughly. Distribute into tubes or flasks. Autoclave for 15 min at 15 psi pressure–121°C.

**Storage/Shelf Life:** Store dehydrated media in the dark in a sealed container below 30°C. Prepared media should be stored under refrigeration (2-8°C). Media should be used within 60 days of preparation. Media should not be used if there are any signs of deterioration (shrinking, cracking, or discoloration) or contamination, or if the expiration date supplied by the manufacturer has passed.

**Use:** For the cultivation of yeasts and fungi. To promote ascospore formation for the identification of yeasts and fungi.

### *Trichophyton* Agar 1

**Composition** per liter:

Glucose .................................................................................40.0g
Agar .....................................................................................15.0g
Vitamin assay casamino acids ..................................................2.5g
KH$_2$PO$_4$.........................................................................1.8g
MgSO$_4$·7H$_2$O.................................................................0.1g

pH 6.8 ± 0.2 at 25°C

**Source:** This medium is available as a premixed powder from BD Diagnostic Systems.

**Preparation of Medium:** Add components to distilled/deionized water and bring volume to 1.0L. Mix thoroughly. Gently heat and bring to boiling. Distribute into tubes. Autoclave for 15 min at 15 psi pressure–121°C. Allow tubes to cool in a slanted position.

**Storage/Shelf Life:** Store dehydrated media in the dark in a sealed container below 30°C. Prepared media should be stored under refrigeration (2-8°C). Media should be used within 60 days of preparation. Media should not be used if there are any signs of deterioration (shrinking, cracking, or discoloration) or contamination, or if the expiration date supplied by the manufacturer has passed.

**Use:** For the differentiation of the *Trichophyton* species.

## *Trichophyton* **Agar 2**

**Composition** per liter:

| | |
|---|---|
| Glucose | 40.0g |
| Agar | 15.0g |
| Vitamin assay casamino acids | 2.5g |
| KH$_2$PO$_4$ | 1.8g |
| MgSO$_4$·7H$_2$O | 0.1g |
| Inositol | 50.0mg |

pH 6.8 ± 0.2 at 25°C

**Source:** This medium is available as a premixed powder from BD Diagnostic Systems.

**Preparation of Medium:** Add components to distilled/deionized water and bring volume to 1.0L. Mix thoroughly. Gently heat and bring to boiling. Distribute into tubes. Autoclave for 15 min at 15 psi pressure–121°C. Allow tubes to cool in a slanted position.

**Storage/Shelf Life:** Store dehydrated media in the dark in a sealed container below 30°C. Prepared media should be stored under refrigeration (2-8°C). Media should be used within 60 days of preparation. Media should not be used if there are any signs of deterioration (shrinking, cracking, or discoloration) or contamination, or if the expiration date supplied by the manufacturer has passed.

**Use:** For the differentiation of the *Trichophyton* species.

## *Trichophyton* **Agar 3**

**Composition** per liter:

| | |
|---|---|
| Glucose | 40.0g |
| Agar | 15.0g |
| Vitamin assay casamino acids | 2.5g |
| KH$_2$PO$_4$ | 1.8g |
| MgSO$_4$·7H$_2$O | 0.1g |
| Inositol | 0.05g |
| Thiamine·HCl | 0.2mg |

pH 6.8 ± 0.2 at 25°C

**Source:** This medium is available as a premixed powder from BD Diagnostic Systems.

**Preparation of Medium:** Add components to distilled/deionized water and bring volume to 1.0L. Mix thoroughly. Gently heat and bring to boiling. Distribute into tubes. Autoclave for 15 min at 15 psi pressure–121°C. Allow tubes to cool in a slanted position.

**Storage/Shelf Life:** Store dehydrated media in the dark in a sealed container below 30°C. Prepared media should be stored under refrigeration (2-8°C). Media should be used within 60 days of preparation.

Media should not be used if there are any signs of deterioration (shrinking, cracking, or discoloration) or contamination, or if the expiration date supplied by the manufacturer has passed.

**Use:** For the differentiation of the *Trichophyton* species.

## *Trichophyton* **Agar 4**

**Composition** per liter:

| | |
|---|---|
| Glucose | 40.0g |
| Agar | 15.0g |
| Vitamin assay casamino acids | 2.5g |
| KH$_2$PO$_4$ | 1.8g |
| MgSO$_4$·7H$_2$O | 0.1g |
| Thiamine·HCl USP | 200.0µg |

pH 6.8 ± 0.2 at 25°C

**Source:** This medium is available as a premixed powder from BD Diagnostic Systems.

**Preparation of Medium:** Add components to distilled/deionized water and bring volume to 1.0L. Mix thoroughly. Gently heat and bring to boiling. Distribute into tubes. Autoclave for 15 min at 15 psi pressure–121°C. Allow tubes to cool in a slanted position.

**Storage/Shelf Life:** Store dehydrated media in the dark in a sealed container below 30°C. Prepared media should be stored under refrigeration (2-8°C). Media should be used within 60 days of preparation. Media should not be used if there are any signs of deterioration (shrinking, cracking, or discoloration) or contamination, or if the expiration date supplied by the manufacturer has passed.

**Use:** For the differentiation of the *Trichophyton* species.

## *Trichophyton* **Agar 5**

**Composition** per liter:

| | |
|---|---|
| Glucose | 40.0g |
| Agar | 15.0g |
| Vitamin assay casamino acids | 2.5g |
| KH$_2$PO$_4$ | 1.8g |
| MgSO$_4$·7H$_2$O | 0.1g |
| Nicotinic acid | 2.0mg |

pH 6.8 ± 0.2 at 25°C

**Source:** This medium is available as a premixed powder from BD Diagnostic Systems.

**Preparation of Medium:** Add components to distilled/deionized water and bring volume to 1.0L. Mix thoroughly. Gently heat and bring to boiling. Distribute into tubes. Autoclave for 15 min at 15 psi pressure–121°C. Allow tubes to cool in a slanted position.

**Storage/Shelf Life:** Store dehydrated media in the dark in a sealed container below 30°C. Prepared media should be stored under refrigeration (2-8°C). Media should be used within 60 days of preparation. Media should not be used if there are any signs of deterioration (shrinking, cracking, or discoloration) or contamination, or if the expiration date supplied by the manufacturer has passed.

**Use:** For the differentiation of the *Trichophyton* species.

## *Trichophyton* **Agar 6**

**Composition** per liter:

| | |
|---|---|
| Glucose | 40.0g |
| Agar | 15.0g |
| KH$_2$PO$_4$ | 1.8g |

NH$_4$NO$_3$ .......................................................................1.5g

MgSO$_4$·7H$_2$O ...............................................................0.1g

pH 6.8 ± 0.2 at 25°C

**Source:** This medium is available as a premixed powder from BD Diagnostic Systems.

**Preparation of Medium:** Add components to distilled/deionized water and bring volume to 1.0L. Mix thoroughly. Gently heat and bring to boiling. Distribute into tubes. Autoclave for 15 min at 15 psi pressure–121°C. Allow tubes to cool in a slanted position.

**Storage/Shelf Life:** Store dehydrated media in the dark in a sealed container below 30°C. Prepared media should be stored under refrigeration (2-8°C). Media should be used within 60 days of preparation. Media should not be used if there are any signs of deterioration (shrinking, cracking, or discoloration) or contamination, or if the expiration date supplied by the manufacturer has passed.

**Use:** For the differentiation of the *Trichophyton* species.

## *Trichophyton* Agar 7

**Composition** per liter:

Glucose ...........................................................................40.0g

Agar .................................................................................15.0g

KH$_2$PO$_4$.............................................................................1.8g

Ammonium nitrate...........................................................1.5g

MgSO$_4$·7H$_2$O ...............................................................0.1g

Histidine·HCl ................................................................0.03g

pH 6.8 ± 0.2 at 25°C

**Source:** This medium is available as a premixed powder from BD Diagnostic Systems.

**Preparation of Medium:** Add components to distilled/deionized water and bring volume to 1.0L. Mix thoroughly. Gently heat and bring to boiling. Distribute into tubes. Autoclave for 15 min at 15 psi pressure–121°C. Allow tubes to cool in a slanted position.

**Storage/Shelf Life:** Store dehydrated media in the dark in a sealed container below 30°C. Prepared media should be stored under refrigeration (2-8°C). Media should be used within 60 days of preparation. Media should not be used if there are any signs of deterioration (shrinking, cracking, or discoloration) or contamination, or if the expiration date supplied by the manufacturer has passed.

**Use:** For the differentiation of the *Trichophyton* species.

## *Trichophyton* HiVeg Agar 1

**Composition** per liter:

Glucose ...........................................................................40.0g

Agar .................................................................................15.0g

Vitamin-free casein enzymic hydrolysate......................2.5g

KH$_2$PO$_4$.............................................................................1.8g

MgSO$_4$ .............................................................................0.1g

pH 6.8 ± 0.2 at 25°C

**Source:** This medium is available as a premixed powder from Hi-Media.

**Preparation of Medium:** Add components to distilled/deionized water and bring volume to 1.0L. Mix thoroughly. Gently heat and bring to boiling. Distribute into tubes. Autoclave for 15 min at 15 psi pressure–121°C. Allow tubes to cool in a slanted position.

**Storage/Shelf Life:** Store dehydrated media in the dark in a sealed container below 30°C. Prepared media should be stored under refrigeration (2-8°C). Media should be used within 60 days of preparation.

Media should not be used if there are any signs of deterioration (shrinking, cracking, or discoloration) or contamination, or if the expiration date supplied by the manufacturer has passed.

**Use:** For the differentiation of the *Trichophyton* species.

## *Trichophyton* HiVeg Agar 2

**Composition** per liter:

Glucose ...........................................................................40.0g

Agar .................................................................................15.0g

Vitamin-free plant hydrolysate ......................................2.5g

KH$_2$PO$_4$.............................................................................1.8g

MgSO$_4$ .............................................................................0.1g

Inositol ..........................................................................5.0mg

pH 6.8 ± 0.2 at 25°C

**Source:** This medium is available as a premixed powder from Hi-Media.

**Preparation of Medium:** Add components to distilled/deionized water and bring volume to 1.0L. Mix thoroughly. Gently heat and bring to boiling. Distribute into tubes. Autoclave for 15 min at 15 psi pressure–121°C. Allow tubes to cool in a slanted position.

**Storage/Shelf Life:** Store dehydrated media in the dark in a sealed container below 30°C. Prepared media should be stored under refrigeration (2-8°C). Media should be used within 60 days of preparation. Media should not be used if there are any signs of deterioration (shrinking, cracking, or discoloration) or contamination, or if the expiration date supplied by the manufacturer has passed.

**Use:** For the differentiation of the *Trichophyton* species.

## *Trichophyton* HiVeg Agar 3

**Composition** per liter:

Glucose ...........................................................................40.0g

Agar .................................................................................15.0g

Vitamin-free plant hydrolysate ......................................2.5g

KH$_2$PO$_4$.............................................................................1.8g

MgSO$_4$ .............................................................................0.1g

Inositol ..........................................................................5.0mg

Thiamine .......................................................................5.0mg

pH 6.8 ± 0.2 at 25°C

**Source:** This medium is available as a premixed powder from Hi-Media.

**Preparation of Medium:** Add components to distilled/deionized water and bring volume to 1.0L. Mix thoroughly. Gently heat and bring to boiling. Distribute into tubes. Autoclave for 15 min at 15 psi pressure–121°C. Allow tubes to cool in a slanted position.

**Storage/Shelf Life:** Store dehydrated media in the dark in a sealed container below 30°C. Prepared media should be stored under refrigeration (2-8°C). Media should be used within 60 days of preparation. Media should not be used if there are any signs of deterioration (shrinking, cracking, or discoloration) or contamination, or if the expiration date supplied by the manufacturer has passed.

**Use:** For the differentiation of the *Trichophyton* species.

## *Trichophyton* HiVeg Agar 4

**Composition** per liter:

Glucose ...........................................................................40.0g

Agar .................................................................................15.0g

KH$_2$PO$_4$.............................................................................1.8g

MgSO$_4$ .............................................................................0.1g

Vitamin-free plant hydrolysate .......................................................2.5g
Thiamine hydrochloride...............................................................0.2mg
pH 6.8 ± 0.2 at 25°C

**Source:** This medium is available as a premixed powder from Hi-Media.

**Preparation of Medium:** Add components to distilled/deionized water and bring volume to 1.0L. Mix thoroughly. Gently heat and bring to boiling. Distribute into tubes. Autoclave for 15 min at 15 psi pressure–121°C. Allow tubes to cool in a slanted position.

**Storage/Shelf Life:** Store dehydrated media in the dark in a sealed container below 30°C. Prepared media should be stored under refrigeration (2-8°C). Media should be used within 60 days of preparation. Media should not be used if there are any signs of deterioration (shrinking, cracking, or discoloration) or contamination, or if the expiration date supplied by the manufacturer has passed.

**Use:** For the differentiation of the *Trichophyton* species.

### *Trichophyton* HiVeg Agar 5
**Composition** per liter:
Glucose ............................................................................40.0g
Agar ..................................................................................15.0g
Vitamin-free plant hydrolysate ...........................................2.5g
KH$_2$PO$_4$..........................................................................1.8g
MgSO$_4$ ............................................................................0.1g
Nicotinic acid .................................................................0.02g
pH 6.8 ± 0.2 at 25°C

**Source:** This medium is available as a premixed powder from BD Diagnostic Systems.

**Preparation of Medium:** Add components to distilled/deionized water and bring volume to 1.0L. Mix thoroughly. Gently heat and bring to boiling. Distribute into tubes. Autoclave for 15 min at 15 psi pressure–121°C. Allow tubes to cool in a slanted position.

**Storage/Shelf Life:** Store dehydrated media in the dark in a sealed container below 30°C. Prepared media should be stored under refrigeration (2-8°C). Media should be used within 60 days of preparation. Media should not be used if there are any signs of deterioration (shrinking, cracking, or discoloration) or contamination, or if the expiration date supplied by the manufacturer has passed.

**Use:** For the differentiation of the *Trichophyton* species.

### Tween™ 60-Esculin Agar
**Composition** per liter:
Agar ..................................................................................15.0g
Peptone.............................................................................10.0g
Glucose .............................................................................10.0g
Yeast extract.......................................................................2.0g
Esculin ...............................................................................1.0g
Ferric ammonium citrate....................................................0.05g
Tween™ 60 ......................................................................5.0mL
pH 6.0 ± 0.2 at 25°C

**Preparation of Medium:** Add components to distilled/deionized water and bring volume to 1.0L. Mix thoroughly. Gently heat and bring to boiling. Autoclave for 15 min at 15 psi pressure–121°C. Pour into sterile Petri dishes.

**Storage/Shelf Life:** Store dehydrated media in the dark in a sealed container below 30°C. Prepared media should be stored under refrigeration (2-8°C). Media should be used within 60 days of preparation.

Media should not be used if there are any signs of deterioration (shrinking, cracking, or discoloration) or contamination, or if the expiration date supplied by the manufacturer has passed.

**Use:** For the cultivation of *Malassezia* species.

### V-8 Agar
**Composition** per liter:
Agar ..................................................................................15.0g
CaCO$_3$.............................................................................2.0g
V-8 canned vegetable juice .............................................200.0mL

**Preparation of Medium:** Add components to distilled/deionized water and bring volume to 1.0L. Mix thoroughly. Gently heat and bring to boiling. Distribute into tubes or flasks. Autoclave for 15 min at 15 psi pressure–121°C. Pour into sterile Petri dishes or leave in tubes.

**Storage/Shelf Life:** Store dehydrated media in the dark in a sealed container below 30°C. Prepared media should be stored under refrigeration (2-8°C). Media should be used within 60 days of preparation. Media should not be used if there are any signs of deterioration (shrinking, cracking, or discoloration) or contamination, or if the expiration date supplied by the manufacturer has passed.

**Use:** For the cultivation of numerous yeasts and filamentous fungi.

### Water Agar
### (Tap Water Agar)
**Composition** per liter:
Agar ..................................................................................15.0g
Tap water ............................................................................1.0L

**Preparation of Medium:** Add agar to 1.0L of tap water. Mix thoroughly. Gently heat and bring to boiling. Autoclave for 15 min at 15 psi pressure–121°C. Pour into sterile Petri dishes.

**Storage/Shelf Life:** Store dehydrated media in the dark in a sealed container below 30°C. Prepared media should be stored under refrigeration (2-8°C). Media should be used within 60 days of preparation. Media should not be used if there are any signs of deterioration (shrinking, cracking, or discoloration) or contamination, or if the expiration date supplied by the manufacturer has passed.

**Use:** For the cultivation and differentiation of fungi and aerobic actinomycetes based on filament and aerial hyphae morphology.

### Yeast Carbon Base, 10X
### (Wickerham Carbon Base Broth)
### (Assimilation Broth for Yeasts--Nitrogen)
**Composition** per liter:
Glucose .............................................................................10.0g
KH$_2$PO$_4$..........................................................................1.0g
MgSO$_4$·7H$_2$O...................................................................0.5g
NaCl....................................................................................0.1g
CaCl$_2$·2H$_2$O.....................................................................0.1g
DL-Methionine....................................................................0.02g
DL-Tryptophan...................................................................0.02g
L-Histidine·HCl...................................................................0.01g
Inositol ...............................................................................2.0mg
H$_3$BO$_3$.............................................................................0.5mg
ZnSO$_4$·7H$_2$O....................................................................0.4mg
MnSO$_4$·4H$_2$O....................................................................0.4mg
Thiamine·HCl .....................................................................0.4mg
Pyridoxine..........................................................................0.4mg

Niacin................................................................0.4mg

Calcium pantothenate ........................................0.4mg

*p*-Aminobenzoic acid.........................................0.2mg

Riboflavin...........................................................0.2mg

$FeCl_3$................................................................0.2mg

$Na_2MoO_4 \cdot 4H_2O$..............................................0.2mg

KI.......................................................................0.1mg

$CuSO_4 \cdot 5H_2O$ ..................................................0.04mg

Folic Acid .......................................................... 2.0µg

Biotin ................................................................. 2.0µg

pH 5.5 ± 0.2 at 25°C

**Source:** This medium is available as a premixed powder from BD Diagnostic Systems.

**Preparation of Medium:** Add components to distilled/deionized water and bring volume to 1.0L. Mix thoroughly. Filter sterilize.

**Storage/Shelf Life:** Store dehydrated media in the dark in a sealed container below 30°C. Prepared media should be stored under refrigeration (2-8°C). Media should be used within 60 days of preparation. Media should not be used if there are any signs of deterioration (discoloration) or contamination, or if the expiration date supplied by the manufacturer has passed.

**Use:** Used as a base to which different nitrogen sources may be added. For the cultivation and differentiation of bacteria based on their ability to utilize diverse added nitrogen sources.

## Yeast Extract Agar

**Composition** per liter:

Agar ...................................................................20.0g

Yeast extract.......................................................1.0g

Buffer solution ................................................. 2.0mL

pH 6.0 ± 0.2 at 25°C

**Buffer Solution:**

**Composition** per 400.0mL:

$KH_2PO_4$...........................................................60.0g

$Na_2HPO_4$.........................................................40.0g

**Preparation of Buffer Solution:** Add 40.0g of $Na_2HPO_4$ to 300.0mL of distilled/deionized water. Mix thoroughly. Add 60.0g of $KH_2PO_4$. Mix thoroughly. Adjust pH to 6.0.

**Preparation of Medium:** Add components to distilled/deionized water and bring volume to 1.0L. Mix thoroughly. Autoclave for 15 min at 15 psi pressure–121°C. Pour into sterile Petri dishes.

**Storage/Shelf Life:** Store dehydrated media in the dark in a sealed container below 30°C. Prepared media should be stored under refrigeration (2-8°C). Media should be used within 60 days of preparation. Media should not be used if there are any signs of deterioration (shrinking, cracking, or discoloration) or contamination, or if the expiration date supplied by the manufacturer has passed.

**Use:** For the identification of *Histoplasma capsulatum*, *Blastomyces dermatitidis*, and *Coccidioides immitis*.

## Yeast Extract Phosphate Agar
### (YEP Agar)

**Composition** per liter:

Agar ...................................................................20.0g

Yeast extract.......................................................1.0g

$KH_2PO_4$.............................................................0.3g

$Na_2HPO_4$...........................................................0.2g

Phenol Red..........................................................1.0mg

**Source:** This medium is available as a premixed powder from BD Diagnostic Systems.

**Preparation of Medium:** Add components to distilled/deionized water and bring volume to 1.0L. Mix thoroughly. Gently heat and bring to boiling. Distribute into tubes or flasks. Autoclave for 15 min at 15 psi pressure–121°C. Pour into sterile Petri dishes or leave in tubes.

**Storage/Shelf Life:** Store dehydrated media in the dark in a sealed container below 30°C. Prepared media should be stored under refrigeration (2-8°C). Media should be used within 60 days of preparation. Media should not be used if there are any signs of deterioration (shrinking, cracking, or discoloration) or contamination, or if the expiration date supplied by the manufacturer has passed.

**Use:** For the isolation of dimorphic pathogenic fungi from clinical specimens.

## Yeast Extract Phosphate Agar with Ammonia
### (YEP Agar with Ammonia)
### (Smith's Agar)

**Composition** per liter:

Agar ...................................................................20.0g

Yeast extract.......................................................1.0g

$KH_2PO_4$.............................................................0.3g

$Na_2HPO_4$...........................................................0.2g

Phenol Red..........................................................1.0mg

$NH_4OH$ (58%) ...........................................1 drop per plate

**Source:** This medium is available as a premixed powder from BD Diagnostic Systems.

**Preparation of Medium:** Add components to distilled/deionized water and bring volume to 1.0L. Mix thoroughly. Gently heat and bring to boiling. Distribute into tubes or flasks. Autoclave for 15 min at 15 psi pressure–121°C. Pour into sterile Petri dishes or leave in tubes. Add 1 drop of ammonium hydroxide to each plate.

**Storage/Shelf Life:** Store dehydrated media in the dark in a sealed container below 30°C. Prepared media should be stored under refrigeration (2-8°C). Media should be used within 60 days of preparation. Media should not be used if there are any signs of deterioration (shrinking, cracking, or discoloration) or contamination, or if the expiration date supplied by the manufacturer has passed.

**Use:** For the isolation of dimorphic pathogenic fungi from clinical specimens. For the primary recovery of *Blastomyces dermatitidis* and *Histoplasma capsulatum* from contaminated specimens. The medium is designed to be used with ammonium hydroxide, a selective agent that improves the recovery of dimorphic pathogens by inhibiting bacteria, yeasts, and saprophytic fungi.

## Balamuth Medium

**Composition** per 200.0mL:

| | |
|---|---|
| Dehydrated egg yolk | 36.0g |
| Dried liver concentrate | 1.0g |
| Rice starch | 0.2g |
| Potassium phosphate buffer, pH 7.5 | 125.0mL |
| NaCl solution | 125.0mL |

pH 7.3 ± 0.2 at 25°C

### NaCl Solution

**Composition** per 200.0mL:

| | |
|---|---|
| NaCl | 1.6g |

**Preparation of NaCl Solution:** Add NaCl to distilled/deionized water and bring volume to 200.0mL. Mix thoroughly.

### Potassium Phosphate Buffer, 0.067*M*

**Composition** per 200.0mL:

| | |
|---|---|
| $K_2HPO_4$ (1*M* solution) | 8.6mL |
| $KH_2PO_4$ (1*M* solution) | 4.66mL |

**Preparation of Potassium Phosphate Buffer:** Combine the $K_2HPO_4$ and $KH_2PO_4$ solutions. Bring volume to 200.0mL with distilled/deionized water. Adjust pH to 7.5.

**Preparation of Medium:** Add dehydrated egg yolk to 36.0mL of distilled/deionized water. Add 125.0mL of 0.8% NaCl. Mix thoroughly in a blender. Heat in a covered double boiler until infusion reaches 80°C and maintain at this temperature for 20 min. Add 20.0mL of distilled/deionized $H_2O$. Filter through a layer of cheesecloth. To 90–100.0mL of filtrate add 0.8% NaCl solution to bring volume to 125.0mL. Autoclave for 20 min at 15 psi pressure–121°C. Cool to 4°C. Filter. To filtrate, add an equal volume of 0.067*M* potassium phosphate buffer, pH 7.5. Add 1.0g of dried liver concentrate. Mix thoroughly. Distribute into tubes or flasks in 10.0mL volumes. Autoclave for 20 min at 15 psi pressure–121°C. Prior to inoculation, add 0.01g of rice starch to each tube.

**Storage/Shelf Life:** Store dehydrated media in the dark in a sealed container below 30°C. Prepared media should be stored under refrigeration (2-8°C). Media should be used within 60 days of preparation. Media should not be used if there are any signs of deterioration (discoloration) or contamination, or if the expiration date supplied by the manufacturer has passed.

**Use:** For the cultivation and maintenance of *Entamoeba histolytica* and other intestinal protozoa.

## CTLM Medium

**Composition** per 1100.0mL:

| | |
|---|---|
| Beef liver, infusion from | 125.0g |
| Tryptose | 25.0g |
| Proteose peptone | 2.5g |
| L-Cysteine·HCl | 1.75g |
| Maltose | 1.25g |
| NaCl | 1.25g |
| Agar | 1.15g |
| L-Ascorbic acid | 0.25g |
| $NaHCO_3$ | 0.075g |
| Horse serum, heat inactivated | 100.0mL |
| Ringer's salt solution, 10× | 75.0mL |

pH 6.0 ± 0.2 at 25°C

### Ringer's Salt Solution, 10×:

**Composition** per 100.0mL:

| | |
|---|---|
| NaCl | 9.0g |
| KCl | 0.42g |
| $CaCl_2$ | 0.24g |

**Preparation of Ringer's Salt Solution, 10×:** Add components to distilled/deionized water and bring volume to 100.0mL. Mix thoroughly.

**Preparation of Medium:** Add components, except horse serum, to distilled/deionized water and bring volume to 1.0L. Mix thoroughly. Adjust pH to 6.0. Gently heat and bring to boiling. Autoclave for 25 min at 15 psi pressure–121°C. Cool to 25°C. Aseptically add 100.0mL of sterile, heat-inactivated horse serum. Mix thoroughly. Aseptically distribute into sterile, screw-capped tubes or flasks.

**Storage/Shelf Life:** Store dehydrated media in the dark in a sealed container below 30°C. Prepared media should be stored under refrigeration (2-8°C). Media should be used within 60 days of preparation. Media should not be used if there are any signs of deterioration (discoloration) or contamination, or if the expiration date supplied by the manufacturer has passed.

**Use:** For the cultivation of *Trichomonas vaginalis*.

## Kupferberg *Trichomonas* Base

**Composition** per liter:

| | |
|---|---|
| Pancreatic digest of casein | 20.0g |
| L-Cysteine·HCl·$H_2O$ | 1.5g |
| Agar | 1.0g |
| Maltose | 1.0g |
| Methylene Blue | 3.0mg |
| Bovine serum | 50.0mL |

pH 6.0 ± 0.2 at 25°C

**Source:** This medium is available as a premixed powder from BD Diagnostic Systems.

**Preparation of Medium:** Add components, except bovine serum, to distilled/deionized water and bring volume to 950.0mL. Mix thoroughly. Gently heat and bring to boiling. Autoclave for 15 min at 15 psi pressure–121°C. Cool to 45°–50°C. Aseptically add 50.0mL of bovine serum. If desired, additional selectivity can be obtained by aseptically adding 250,000U of penicillin and 1.0g of streptomycin or 1.0g of chloramphenicol. Mix thoroughly. Pour into sterile Petri dishes or distribute into sterile tubes.

**Storage/Shelf Life:** Store dehydrated media in the dark in a sealed container below 30°C. Prepared media should be stored under refrigeration (2-8°C). Media should be used within 60 days of preparation. Media should not be used if there are any signs of deterioration (discoloration) or contamination, or if the expiration date supplied by the manufacturer has passed.

**Use:** For the cultivation of the *Trichomonas* species from clinical specimens.

## Kupferberg *Trichomonas* Broth

**Composition** per liter:

| | |
|---|---|
| Enzymatic digest of protein | 20.0g |
| L-Cysteine·HCl·$H_2O$ | 1.5g |
| Agar | 1.0g |
| Maltose | 1.0g |
| Chloramphenicol | 0.1g |

Methylene Blue .......................................................................... 3.0mg
Bovine serum .......................................................................... 50.0mL

pH 6.0 ± 0.2 at 25°C

**Source:** This medium is available as a premixed powder from BD Diagnostic Systems.

**Preparation of Medium:** Add components, except bovine serum, to distilled/deionized water and bring volume to 950.0mL. Mix thoroughly. Gently heat and bring to boiling. Autoclave for 15 min at 15 psi pressure–121°C. Cool to 45°–50°C. Aseptically add bovine serum. If desired, additional selectivity can be obtained by aseptically adding 250,000U penicillin and 1.0g streptomycin or 1.0g chloramphenicol. Mix thoroughly. Pour into sterile Petri dishes or distribute into sterile tubes.

**Storage/Shelf Life:** Store dehydrated media in the dark in a sealed container below 30°C. Prepared media should be stored under refrigeration (2-8°C). Media should be used within 60 days of preparation. Media should not be used if there are any signs of deterioration (discoloration) or contamination, or if the expiration date supplied by the manufacturer has passed.

**Use:** For the cultivation of the *Trichomonas* species from clinical specimens.

### Kupferberg *Trichomonas* HiVeg Broth Base with Serum and Selective Supplement
### (*Trichomonas* HiVeg Broth Base, Kupferberg)

**Composition** per liter:
Plant hydrolysate ......................................................................... 20.0g
Agar ........................................................................................ 1.0g
L-Cysteine·HCl ........................................................................... 1.5g
Maltose ..................................................................................... 1.0g
Methylene Blue ........................................................................... 3.0mg
Bovine serum .......................................................................... 50.0mL
Selective supplement ............................................................... 10.0mL

pH 6.0 ± 0.2 at 25°C

**Source:** This medium, without bovine serum and selective supplement, is available as a premixed powder from HiMedia.

**Selective Supplement Solution:**
**Composition** per 10.0mL:
Penicillin ............................................................................. 250,000U

**Preparation of Selective Supplement Solution:** Add penicilliln to distilled/deionized water and bring volume to 10.0mL. Mix thoroughly. Filter sterilize.

**Preparation of Medium:** Add components, except bovine serum and selective supplement, to distilled/deionized water and bring volume to 950.0mL. Mix thoroughly. Gently heat and bring to boiling. Autoclave for 15 min at 15 psi pressure–121°C. Cool to 45°–50°C. Aseptically add 50.0mL of bovine serum and 10.0mL selective supplement.

**Storage/Shelf Life:** Store dehydrated media in the dark in a sealed container below 30°C. Prepared media should be stored under refrigeration (2-8°C). Media should be used within 60 days of preparation. Media should not be used if there are any signs of deterioration (discoloration) or contamination, or if the expiration date supplied by the manufacturer has passed.

**Use:** For the cultivation of *Trichomonas* species from clinical specimens.

### Kupferberg *Trichomonas* HiVeg Broth Base with Serum and Selective Supplement
### (*Trichomonas* HiVeg Broth Base, Kupferberg)

**Composition** per liter:
Plant hydrolysate ........................................................................ 20.0g
Agar ......................................................................................... 1.0g
L-Cysteine·HCl ............................................................................ 1.5g
Maltose ..................................................................................... 1.0g
Methylene Blue ........................................................................... 3.0mg
Bovine serum .......................................................................... 50.0mL
Selective supplement ............................................................... 10.0mL

pH 6.0 ± 0.2 at 25°C

**Source:** This medium, without bovine serum and selective supplement, is available as a premixed powder from HiMedia.

**Selective Supplement Solution:**
**Composition** per 10.0mL:
Streptomycin .............................................................................. 1.0g

**Preparation of Selective Supplement Solution:** Add streptomycin to distilled/deionized water and bring volume to 10.0mL. Mix thoroughly. Filter sterilize.

**Preparation of Medium:** Add components, except bovine serum and selective supplement, to distilled/deionized water and bring volume to 950.0mL. Mix thoroughly. Gently heat and bring to boiling. Autoclave for 15 min at 15 psi pressure–121°C. Cool to 45°–50°C. Aseptically add 50.0mL of bovine serum and 10.0mL selective supplement.

**Storage/Shelf Life:** Store dehydrated media in the dark in a sealed container below 30°C. Prepared media should be stored under refrigeration (2-8°C). Media should be used within 60 days of preparation. Media should not be used if there are any signs of deterioration (discoloration) or contamination, or if the expiration date supplied by the manufacturer has passed.

**Use:** For the cultivation of *Trichomonas* species from clinical specimens.

### Kupferberg *Trichomonas* HiVeg Broth Base with Serum and Selective Supplement
### (*Trichomonas* HiVeg Broth Base, Kupferberg)

**Composition** per liter:
Plant hydrolysate ........................................................................ 20.0g
Agar ......................................................................................... 1.0g
L-Cysteine·HCl ............................................................................ 1.5g
Maltose ..................................................................................... 1.0g
Methylene Blue ........................................................................... 3.0mg
Bovine serum .......................................................................... 50.0mL
Selective supplement ............................................................... 10.0mL

pH 6.0 ± 0.2 at 25°C

**Source:** This medium, without bovine serum and selective supplement, is available as a premixed powder from HiMedia.

**Selective Supplement Solution:**
**Composition** per 10.0mL:
Chloramphenicol .......................................................................... 1.0g

**Preparation of Selective Supplement Solution:** Add chloramphenicol to distilled/deionized water and bring volume to 10.0mL. Mix thoroughly. Filter sterilize.

**Preparation of Medium:** Add components, except bovine serum and selective supplement, to distilled/deionized water and bring volume to 950.0mL. Mix thoroughly. Gently heat and bring to boiling. Autoclave for 15 min at 15 psi pressure–121°C. Cool to 45°–50°C. Aseptically add 50.0mL of bovine serum and 10.0mL selective supplement.

**Storage/Shelf Life:** Store dehydrated media in the dark in a sealed container below 30°C. Prepared media should be stored under refrigeration (2-8°C). Media should be used within 60 days of preparation. Media should not be used if there are any signs of deterioration (discoloration) or contamination, or if the expiration date supplied by the manufacturer has passed.

**Use:** For the cultivation of *Trichomonas* species from clinical specimens.

## Nonnutrient Agar Plates

**Composition** per liter:

Agar ....................................................................................15.0g
Page's amoeba saline ........................................................ 1.0L

### Page's Amoeba Saline:
**Composition** per liter:

$Na_2HPO_4$ ...........................................................................0.142g
$KH_2PO_4$ ...........................................................................0.136g
NaCl ....................................................................................0.12g
$MgSO_4 \cdot 7H_2O$ ...............................................................4.0mg
$CaCl_2 \cdot 2H_2O$ ....................................................4.0mg

**Preparation of Page's Amoeba Saline:** Add components to distilled/deionized water and bring volume to 1.0mL. Mix thoroughly.

**Preparation of Medium:** Add agar to 1.0L of Page's amoeba saline. Mix thoroughly. Gently heat and bring to boiling. Autoclave for 15 min at 15 psi pressure–121°C. Cool to 60°C. Pour into sterile Petri dishes in 20.0mL volumes. Store at 4°C for up to 3 months.

**Storage/Shelf Life:** Store dehydrated media in the dark in a sealed container below 30°C. Prepared media should be stored under refrigeration (2-8°C). Media should be used within 60 days of preparation. Media should not be used if there are any signs of deterioration (shrinking, cracking, or discoloration) or contamination, or if the expiration date supplied by the manufacturer has passed.

**Use:** For the isolation and cultivation of pathogenic free-living amoebae.

## *Toxoplasma* Medium

**Composition** per liter:

NaCl ....................................................................................6.8g
$NaHCO_3$ .............................................................................2.2g
Glucose ...............................................................................1.0g
KCl ......................................................................................0.4g
$CaCl_2$ .................................................................................0.2g
$NaH_2PO_4 \cdot H_2O$ ...........................................................0.125g
Arginine .............................................................................0.105g
$MgSO_4$ ..............................................................................0.1g
L-Cystine ...........................................................................0.024g
Glutamine ..........................................................................0.292g
Histidine .............................................................................0.031g
Lysine .................................................................................0.058g
Isoleucine ...........................................................................0.052g
Leucine ...............................................................................0.052g
Phenol Red .........................................................................0.050g
Threonine ...........................................................................0.048g
Valine .................................................................................0.046g

Tyrosine ..............................................................................0.036g
Phenylalanine .....................................................................0.032g
Methionine ..........................................................................0.015g
Tryptophan ..........................................................................0.010g
Inositol ................................................................................2.0mg
Choline ................................................................................1.0mg
Folic acid ............................................................................1.0mg
Nicotinamide .......................................................................1.0mg
Pantothenic acid ..................................................................1.0mg
Pyridoxal·HCl .....................................................................1.0mg
Thiamine·HCl ......................................................................1.0mg
Riboflavin ...........................................................................0.1mg
Fetal bovine serum, heat inactivated ..................... 100.0mL
pH 7.2–7.4 at 25°C

**Preparation of Medium:** Add components, except fetal bovine serum, to distilled/deionized water and bring volume to 905.0mL. Mix thoroughly. Adjust pH to 7.2–7.4. Autoclave for 15 min at 15 psi pressure–121°C. Aseptically add 100.0mL of sterile, heat-inactivated fetal bovine serum. Mix thoroughly. Aseptically distribute into sterile tubes or flasks.

**Storage/Shelf Life:** Store dehydrated media in the dark in a sealed container below 30°C. Prepared media should be stored under refrigeration (2-8°C). Media should be used within 60 days of preparation. Media should not be used if there are any signs of deterioration (discoloration) or contamination, or if the expiration date supplied by the manufacturer has passed.

**Use:** For the cultivation of *Toxoplasma gondii*.

## *Trichomonas* HiVeg Agar Base with Serum

**Composition** per liter:

Plant extract No. 2 .......................................................25.0g
NaCl ....................................................................................6.5g
Glucose ...............................................................................5.0g
Agar ....................................................................................1.0g
Horse serum .................................................................. 80.0mL
pH 6.4 ± 0.2 at 25°C

**Source:** This medium, without horse serum, is available as a premixed powder from HiMedia.

### Horse Serum:
**Composition** per 80.0mL:

Horse serum .................................................................. 80.0mL

**Preparation of Horse Serum:** Gently heat sterile horse serum to 56°C for 30 min. Aseptically adjust pH to 6.0 with 0.1*N* HCl. Use immediately.

**Preparation of Medium:** Add components, except horse serum, to distilled/deionized water and bring volume to 920.0mL. Mix thoroughly. Gently heat and bring to boiling. Autoclave for 15 min at 15 psi pressure–121°C. Cool to 45°–50°C. Aseptically add 80.0mL of freshly prepared sterile horse serum. Mix thoroughly. Aseptically distribute into sterile tubes or flasks.

**Storage/Shelf Life:** Store dehydrated media in the dark in a sealed container below 30°C. Prepared media should be stored under refrigeration (2-8°C). Media should be used within 60 days of preparation. Media should not be used if there are any signs of deterioration (shrinking, cracking, or discoloration) or contamination, or if the expiration date supplied by the manufacturer has passed.

**Use:** For the cultivation of *Trichomonas vaginalis*.

## *Trichomonas* HiVeg Agar Base
## with Serum and Selective Supplement

**Composition** per liter:
Plant extract No. 2 ........................................................25.0g
NaCl ...........................................................................6.5g
Glucose ......................................................................5.0g
Agar ...........................................................................1.0g
Horse serum ............................................................80.0mL
Selective supplement ..............................................10.0mL

pH 6.4 ± 0.2 at 25°C

**Source:** This medium, without horse serum or selective supplement, is available as a premixed powder from HiMedia.

**Horse Serum:**
**Composition** per 80.0mL:
Horse serum ..............................................................80.0mL

**Preparation of Horse Serum:** Gently heat sterile horse serum to 56°C for 30 min. Aseptically adjust pH to 6.0 with 0.1*N* HCl. Use immediately.

**Selective Supplement:**
**Composition** per 10.0mL:
Streptomycin ................................................................0.5g
Penicllin ...........................................................1,000,000U

**Preparation of Selective Supplement:** Add components to distilled/deionized water and bring volume to 10.0mL. Mix thoroughly. Filter sterilize.

**Preparation of Medium:** Add components, except horse serum and selective supplement, to distilled/deionized water and bring volume to 910.0mL. Mix thoroughly. Gently heat and bring to boiling. Autoclave for 15 min at 15 psi pressure–121°C. Cool to 45°–50°C. Aseptically add 80.0mL of freshly prepared sterile horse serum and 10.0mL sterile selective supplement. Mix thoroughly. Aseptically distribute into sterile tubes or flasks.

**Storage/Shelf Life:** Store dehydrated media in the dark in a sealed container below 30°C. Prepared media should be stored under refrigeration (2-8°C). Media should be used within 60 days of preparation. Media should not be used if there are any signs of deterioration (shrinking, cracking, or discoloration) or contamination, or if the expiration date supplied by the manufacturer has passed.

**Use:** For the cultivation of *Trichomonas vaginalis*.

## *Trichomonas* Medium

**Composition** per liter:
Liver digest ...............................................................25.0g
NaCl ...........................................................................6.5g
Glucose ......................................................................5.0g
Agar ...........................................................................1.0g
Horse serum ............................................................80.0mL

pH 6.4 ± 0.2 at 25°C

**Source:** This medium is available as a premixed powder from Oxoid Unipath.

**Horse Serum:**
**Composition** per 80.0mL:
Horse serum ..............................................................80.0mL

**Preparation of Horse Serum:** Gently heat sterile horse serum to 56°C for 30 min. Aseptically adjust pH to 6.0 with 0.1*N* HCl. Use immediately.

**Preparation of Medium:** Add components, except horse serum, to distilled/deionized water and bring volume to 920.0mL. Mix thoroughly. Gently heat and bring to boiling. Autoclave for 15 min at 15 psi pressure–121°C. Cool to 45°–50°C. Aseptically add 80.0mL of freshly prepared sterile horse serum. Mix thoroughly. Aseptically distribute into sterile tubes or flasks.

**Storage/Shelf Life:** Store dehydrated media in the dark in a sealed container below 30°C. Prepared media should be stored under refrigeration (2-8°C). Media should be used within 60 days of preparation. Media should not be used if there are any signs of deterioration (discoloration) or contamination, or if the expiration date supplied by the manufacturer has passed.

**Use:** For the cultivation of *Trichomonas vaginalis*.

## *Trichomonas* Medium No. 2

**Composition** per liter:
Glucose .....................................................................22.5g
Liver digest ...............................................................18.0g
Pancreatic digest of casein ........................................17.0g
NaCl ...........................................................................5.0g
Pancreatic digest of soybean meal ..............................3.0g
K$_2$HPO$_4$ .............................................................2.5g
Chloramphenicol .......................................................0.125g
Horse serum ............................................................250.0mL
Calcium pantothenate (0.5% solution) .......................1.0mL

pH 6.2 ± 0.2 at 25°C

**Source:** This medium is available as a prepared medium from Oxoid Unipath.

**Preparation of Medium:** Add components, except horse serum, to distilled/deionized water and bring volume to 750.0mL. Mix thoroughly. Autoclave for 15 min at 5 psi pressure–108°C. Cool to 45°–50°C. Aseptically add 250.0mL of sterile horse serum. Mix thoroughly. Aseptically distribute into sterile tubes or flasks.

**Storage/Shelf Life:** Store dehydrated media in the dark in a sealed container below 30°C. Prepared media should be stored under refrigeration (2-8°C). Media should be used within 60 days of preparation. Media should not be used if there are any signs of deterioration (discoloration) or contamination, or if the expiration date supplied by the manufacturer has passed.

**Use:** For the isolation of *Trichomonas vaginalis*.

## *Trichomonas* Selective Medium

**Composition** per liter:
Liver digest ...............................................................25.0g
NaCl ...........................................................................6.5g
Glucose ......................................................................5.0g
Agar ...........................................................................1.0g
Horse serum ............................................................80.0mL
Antibiotic inhibitor ..................................................10.0mL

pH 6.4 ± 0.2 at 25°C

**Source:** This medium is available as a premixed powder from Oxoid Unipath.

**Horse Serum:**
**Composition** per 80.0mL:
Horse serum ..............................................................80.0mL

**Preparation of Horse Serum:** Gently heat sterile horse serum to 56°C for 30 min. Aseptically adjust pH to 6.0 with 0.1*N* HCl. Use immediately.

**Antibiotic Inhibitor:**
**Composition** per 10.0mL:

Streptomycin ............................................................................500.0mg
Penicillin G ........................................................... 1,000,000U

**Preparation of Antibiotic Inhibitor:** Add components to distilled/deionized water and bring volume to 10.0mL. Mix thoroughly. Filter sterilize.

**Preparation of Medium:** Add components, except horse serum and antibiotic inhibitor, to distilled/deionized water and bring volume to 910.0mL. Mix thoroughly. Gently heat and bring to boiling. Autoclave for 15 min at 15 psi pressure–121°C. Cool to 45°–50°C. Aseptically add 80.0mL of freshly prepared sterile horse serum and 10.0mL of sterile antibiotic inhibitor. Mix thoroughly. Aseptically distribute into sterile tubes or flasks.

**Storage/Shelf Life:** Store dehydrated media in the dark in a sealed container below 30°C. Prepared media should be stored under refrigeration (2-8°C). Media should be used within 60 days of preparation. Media should not be used if there are any signs of deterioration (discoloration) or contamination, or if the expiration date supplied by the manufacturer has passed.

**Use:** For the cultivation of *Trichomonas vaginalis* from specimens with a mixed bacterial flora.

### *Trichomonas* Selective Medium

**Composition** per liter:

Liver digest ...........................................................................25.0g
NaCl .........................................................................................6.5g
Glucose ...................................................................................5.0g
Agar .........................................................................................1.0g
Horse serum ..................................................................... 80.0mL
Antibiotic inhibitor ......................................................... 10.0mL

pH 6.4 ± 0.2 at 25°C

**Horse Serum:**
**Composition** per 80.0mL:

Horse serum ..................................................................... 80.0mL

**Preparation of Horse Serum:** Gently heat sterile horse serum to 56°C for 30 min. Aseptically adjust pH to 6.0 with 0.1*N* HCl. Use immediately.

**Antibiotic Inhibitor:**
**Composition** per 10.0mL:

Chloramphenicol.....................................................................100.0mg

**Preparation of Antibiotic Inhibitor:** Add chloramphenicol to distilled/deionized water and bring volume to 10.0mL. Mix thoroughly. Filter sterilize.

**Preparation of Medium:** Add components, except horse serum and antibiotic inhibitor, to distilled/deionized water and bring volume to 910.0mL. Mix thoroughly. Gently heat and bring to boiling. Autoclave for 15 min at 15 psi pressure–121°C. Cool to 45°–50°C. Aseptically add 80.0mL of freshly prepared sterile horse serum and 10.0mL of sterile antibiotic inhibitor. Mix thoroughly. Aseptically distribute into sterile tubes or flasks.

**Storage/Shelf Life:** Store dehydrated media in the dark in a sealed container below 30°C. Prepared media should be stored under refrigeration (2-8°C). Media should be used within 60 days of preparation. Media should not be used if there are any signs of deterioration (discoloration) or contamination, or if the expiration date supplied by the manufacturer has passed.

**Use:** For the cultivation of *Trichomonas vaginalis* from specimens with a mixed bacterial flora.

### Trichosel™ Broth, Modified

**Composition** per liter:

Pancreatic digest of casein.............................................12.0g
Yeast extract........................................................................5.0g
Liver extract.........................................................................2.0g
Maltose ................................................................................2.0g
L-Cysteine·HCl....................................................................1.0g
Agar .....................................................................................1.0g
Chloramphenicol.................................................................0.1g
Methylene Blue....................................................................3.0mg
Horse serum .................................................................. 50.0mL

pH 6.0 ± 0.2 at 25°C

**Source:** This medium is available as a premixed powder from BD Diagnostic Systems.

**Preparation of Medium:** Add components, except horse serum, to distilled/deionized water and bring volume to 950.0mL. Mix thoroughly. Gently heat while stirring and bring to boiling. Autoclave for 15 min at 13 psi pressure–118°C. Cool to 45°–50°C. Aseptically add 50.0mL of sterile horse serum. Mix thoroughly. Aseptically distribute into sterile tubes or flasks.

**Storage/Shelf Life:** Store dehydrated media in the dark in a sealed container below 30°C. Prepared media should be stored under refrigeration (2-8°C). Media should be used within 60 days of preparation. Media should not be used if there are any signs of deterioration (discoloration) or contamination, or if the expiration date supplied by the manufacturer has passed.

**Use:** For the isolation and cultivation of *Trichomonas* species.

## Eagle Medium

**Composition** per 99.1mL:

| | |
|---|---|
| Eagle MEM in Hanks BSS | 87.0mL |
| Fetal bovine serum | 10.0mL |
| NaHCO$_3$ (7.5% solution) | 1.0mL |
| Penicillin-streptomycin solution | 1.0mL |
| Amphotericin B solution | 0.1mL |

pH 7.2–7.4 at 25°C

### Eagle MEM in Hanks BSS:

**Composition** per liter:

| | |
|---|---|
| NaCl | 8.0g |
| Glucose | 1.0g |
| KCl | 0.4g |
| CaCl$_2$·2H$_2$O | 0.14g |
| MgSO$_4$·7H$_2$O | 0.1g |
| KH$_2$PO$_4$ | 0.06g |
| Na$_2$HPO$_4$ | 0.05g |
| L-Isoleucine | 0.026g |
| L-Leucine | 0.026g |
| L-Lysine | 0.026g |
| L-Threonine | 0.024g |
| L-Valine | 0.0235g |
| L-Tyrosine | 0.018g |
| L-Arginine | 0.0174g |
| L-Phenylalanine | 0.0165g |
| L-Cystine | 0.012g |
| L-Histidine | 8.0mg |
| L-Methionine | 7.5mg |
| Phenol Red | 5.0mg |
| L-Tryptophan | 4.0mg |
| Inositol | 1.8mg |
| Biotin | 1.0mg |
| Folic acid | 1.0mg |
| Calcium pantothenate | 1.0mg |
| Choline chloride | 1.0mg |
| Nicotinamide | 1.0mg |
| Pyridoxal·HCl | 1.0mg |
| Thiamine·HCl | 1.0mg |
| Riboflavin | 0.1mg |

**Preparation of Eagle MEM in Hanks BSS:** Add components to distilled/deionized water and bring volume to 1.0L. Mix thoroughly.

### Penicillin-Streptomycin Solution:

**Composition** per 1.0mL:

| | |
|---|---|
| Streptomycin | 0.01g |
| Penicillin | 10,000U |

**Preparation of Penicillin-Streptomycin Solution:** Add components to distilled/deionized water and bring volume to 1.0mL. Mix thoroughly.

### Amphotericin B Solution:

**Composition** per 1.0mL:

| | |
|---|---|
| Amphotericin B | 1.0mg |

**Preparation of Amphotericin B Solution:** Add amphotericin B to distilled/deionized water and bring volume to 1.0mL. Mix thoroughly.

**Preparation of Medium:** Combine components. Mix thoroughly. Filter sterilize.

**Storage/Shelf Life:** Store dehydrated media in the dark in a sealed container below 30°C. Prepared media should be stored under refrigeration (2-8°C). Media should be used within 60 days of preparation.

Media should not be used if there are any signs of deterioration (discoloration) or contamination, or if the expiration date supplied by the manufacturer has passed.

**Use:** For the cultivation of animal tissue culture cell lines.

## Eagle Medium

**Composition** per liter:

| | |
|---|---|
| Hanks balanced salt solution (10X) | 100.0mL |
| Calf serum | 50.0mL |
| NaHCO$_3$ (7.5% solution) | 29.6mL |
| Tissue culture amino acids (50X) | 20.0mL |
| Tissue culture vitamins (100X) | 10.0mL |
| Glutamine solution | 10.0mL |
| Phenol Red (0.5% solution) | 4.0mL |
| Penicillin solution | 1.0mL |
| Streptomycin solution | 0.4mL |

pH 7.0 ± 0.2 at 25°C

### Hanks Balanced Salt Solution (10X):

**Composition** per 100.0mL:

| | |
|---|---|
| NaCl | 8.0g |
| Glucose | 1.0g |
| KCl | 0.4g |
| NaHCO$_3$ | 0.35g |
| CaCl$_2$·2H$_2$O | 0.14g |
| MgCl$_2$·6H$_2$O | 0.1g |
| MgSO$_4$·7H$_2$O | 0.1g |
| Na$_2$HPO$_4$ | 0.06g |
| KH$_2$PO$_4$ | 0.06g |
| Phenol Red | 0.02g |

**Preparation of Hanks Balanced Salt Solution (10X):** Add components to distilled/deionized water and bring volume to 100.0mL. Mix thoroughly.

### Tissue Culture Amino Acids (50X):

**Composition** per liter:

| | |
|---|---|
| L-Arginine | 0.1g |
| L-Lysine | 0.058g |
| L-Isoleucine | 0.052g |
| L-Leucine | 0.052g |
| L-Threonine | 0.048g |
| L-Valine | 0.046g |
| L-Tyrosine | 0.036g |
| L-Phenylalanine | 0.032g |
| L-Histidine | 0.031g |
| L-Cystine | 0.024g |
| L-Methionine | 0.015g |
| L-Tryptophan | 0.01g |

**Preparation of Tissue Culture Amino Acids (50X):** Add components to distilled/deionized water and bring volume to 1.0L. Mix thoroughly.

### Tissue Culture Vitamins (100X):

**Composition** per liter:

| | |
|---|---|
| Inositol | 2.0mg |
| Calcium pantothenate | 1.0mg |
| Choline chloride | 1.0mg |
| Folic acid | 1.0mg |
| Nicotinamide | 1.0mg |
| Pyridoxal | 1.0mg |
| Thiamine·HCl | 1.0mg |
| Riboflavin | 0.1mg |

**Preparation of Tissue Culture Vitamins (100X):** Add components to distilled/deionized water and bring volume to 1.0L. Mix thoroughly.

**Glutamine Solution:**
**Composition** per 100.0mL:
L-Glutamine ........................................................................2.9g

**Preparation of Glutamine Solution:** Add glutamine to distilled/deionized water and bring volume to 100.0mL. Mix thoroughly.

**Penicillin Solution:**
**Composition** per 1.0mL:
Penicillin ..........................................................................200,000U

**Preparation of Penicillin Solution:** Add penicillin to distilled/deionized water and bring volume to 1.0mL. Mix thoroughly.

**Streptomycin Solution:**
**Composition** per 1.0mL:
Streptomycin ............................................................................0.5g

**Preparation of Streptomycin Solution:** Add streptomycin to distilled/deionized water and bring volume to 1.0mL. Mix thoroughly.

**Preparation of Medium:** Combine components. Mix thoroughly. Adjust pH to 7.0 with 1*N* NaOH. Filter sterilize.

**Storage/Shelf Life:** Store dehydrated media in the dark in a sealed container below 30°C. Prepared media should be stored under refrigeration (2-8°C). Media should be used within 60 days of preparation. Media should not be used if there are any signs of deterioration (discoloration) or contamination, or if the expiration date supplied by the manufacturer has passed.

**Use:** For the cultivation of animal tissue culture cell lines, especially for use with rhinoviruses.

## Eagle Medium

**Composition** per 100.1mL:
Eagle MEM in Earle BSS ........................................... 94.0mL
NaHCO₃ (7.5% solution) ............................................. 3.0mL
Fetal bovine serum, inactivated ................................... 2.0mL
Penicillin-streptomycin solution ................................. 1.0mL
Amphotericin B solution ............................................. 0.1mL
pH 7.2–7.4 at 25°C

**Eagle MEM in Earle BSS:**
**Composition** per liter:
NaCl .........................................................................6.8g
Glucose ....................................................................1.0g
KCl............................................................................0.4g
CaCl₂·2H₂O...............................................................0.2g
MgCl₂·6H₂O..............................................................0.2g
NaH₂PO₄ ..................................................................0.15g
L-Arginine ...............................................................0.1g
L-Lysine ..................................................................0.06g
L-Isoleucine ............................................................0.05g
L-Leucine.................................................................0.05g
L-Threonine .............................................................0.05g
L-Valine ..................................................................0.05g
L-Tyrosine ...............................................................0.04g
L-Phenylalanine ......................................................0.03g
L-Histidine ..............................................................0.03g
L-Cystine ................................................................0.02g
L-Methionine...........................................................0.02g
L-Tryptophan ..........................................................0.01g

*i*-Inositol................................................................2.0mg
Calcium pantothenate ...............................................1.0mg
Choline chloride........................................................1.0mg
Folic acid .................................................................1.0mg
Nicotinamide............................................................1.0mg
Pyridoxal..................................................................1.0mg
Thiamine·HCl............................................................1.0mg
Riboflavin.................................................................0.1mg

**Preparation of Eagle MEM in Earle BSS:** Add components to distilled/deionized water and bring volume to 1.0L. Mix thoroughly.

**Penicillin-Streptomycin Solution:**
**Composition** per 1.0mL:
Streptomycin ............................................................0.01g
Penicillin ................................................................. 10,000U

**Preparation of Penicillin-Streptomycin Solution:** Add components to distilled/deionized water and bring volume to 1.0mL. Mix thoroughly.

**Amphotericin B Solution:**
**Composition** per 1.0mL:
Amphotericin B .............................................1.0mg

**Preparation of Amphotericin B Solution:** Add amphotericin B to distilled/deionized water and bring volume to 1.0mL. Mix thoroughly.

**Preparation of Medium:** Combine components. Mix thoroughly. Filter sterilize.

**Storage/Shelf Life:** Store dehydrated media in the dark in a sealed container below 30°C. Prepared media should be stored under refrigeration (2-8°C). Media should be used within 60 days of preparation. Media should not be used if there are any signs of deterioration (discoloration) or contamination, or if the expiration date supplied by the manufacturer has passed.

**Use:** For the cultivation of animal tissue culture cell lines.

## Eagle Medium, Modified

**Composition** per liter:
Eagle MEM (10X)........................................................ 100.0mL
Fetal bovine serum....................................................... 100.0mL
Glucose solution .......................................................... 20.0mL
HEPES (*N*-2-hydroxyethyl
   piperazine-*N′*-2-ethanesulfonic acid)
   buffer, 1*M*, pH 7.2 ............................................... 20.0mL
Glutamine solution........................................................ 10.0mL
NaHCO₃ (7.5% solution) ............................................... 7.5mL
Gentamicin sulfate solution ........................................... 0.2mL
pH 7.2 ± 0.2 at 25°C

**Eagle MEM (10X):**
**Composition** per 100.0mL:
Sterile salt solution ...................................................... 97.0mL
TC amino acids, minimal Eagle 50X.............................. 2.0mL
TC vitamins, minimal Eagle 100X ................................ 1.0mL

**Preparation of Eagle MEM (10X):** Combine components. Mix thoroughly. Filter sterilize.

**Sterile Salt Solution:**
**Composition** per 100.0mL:
NaCl.........................................................................6.8g
Glucose ....................................................................1.0g
KCl............................................................................0.4g
CaCl₂........................................................................0.2g

MgCl₂ ..................................................................................0.2g

NaH₂PO₄ ..........................................................................0.15g

**Preparation of Sterile Salt Solution:** Add components to distilled/deionized water and bring volume to 100.0mL. Mix thoroughly. Filter sterilize.

## TC Amino Acids (50X):

**Composition** per liter:

L-Arginine ........................................................................0.1g

L-Lysine ..........................................................................0.06g

L-Isoleucine ....................................................................0.05g

L-Leucine ........................................................................0.05g

L-Threonine .....................................................................0.05g

L-Valine ..........................................................................0.05g

L-Tyrosine .......................................................................0.04g

L-Phenylalanine ...............................................................0.03g

L-Histidine ......................................................................0.03g

L-Cystine .........................................................................0.02g

L-Methionine ...................................................................0.02g

L-Tryptophan ...................................................................0.01g

**Preparation of TC Amino Acids (50X):** Add components to distilled/deionized water and bring volume to 1.0L. Mix thoroughly. Adjust pH to 7.2–7.4. Filter sterilize.

## TC Vitamins, Minimal Eagle 100X:

**Composition** per liter:

Inositol ...........................................................................2.0mg

Calcium pantothenate .......................................................1.0mg

Choline chloride ..............................................................1.0mg

Folic acid ........................................................................1.0mg

Nicotinamide ...................................................................1.0mg

Pyridoxal ........................................................................1.0mg

Thiamine·HCl ..................................................................1.0mg

Riboflavin .......................................................................0.1mg

**Preparation of TC Vitamins, Minimal Eagle 100X:** Add components to distilled/deionized water and bring volume to 1.0L. Mix thoroughly. Filter sterilize.

## Glucose Solution:

**Composition** per 100.0mL:

Glucose ...........................................................................27.0g

**Preparation of Glucose Solution:** Add glucose to distilled/deionized water and bring volume to 100.0mL. Mix thoroughly. Filter sterilize.

## Glutamine Solution:

**Composition** per 10.0mL:

L-Glutamine ......................................................................5.0g

**Preparation of Glutamine Solution:** Add glutamine to distilled/deionized water and bring volume to 10.0mL. Mix thoroughly. Filter sterilize.

## Gentamicin Solution:

**Composition** per 1.0mL:

Gentamicin sulfate ...........................................................0.05g

**Preparation of Gentamicin Solution:** Add gentamicin sulfate to distilled/deionized water and bring volume to 1.0mL. Mix thoroughly. Filter sterilize.

**Preparation of Medium:** Combine components. Mix thoroughly. Filter sterilize.

**Storage/Shelf Life:** Store dehydrated media in the dark in a sealed container below 30°C. Prepared media should be stored under refrigeration (2-8°C). Media should be used within 60 days of preparation. Media should not be used if there are any signs of deterioration (discoloration) or contamination, or if the expiration date supplied by the manufacturer has passed.

**Use:** For the cultivation of animal tissue culture cell lines, especially for McCoy cells.

# Eagle's Minimal Essential Medium with Earle's Salts and Nonessential Amino Acids (MEM with Earle's Salts and Nonessential Amino Acids) (BAM M46)

**Composition** per liter:

NaCl ................................................................................6.8g

NaHCO₃ ...........................................................................2.2g

Glucose ...........................................................................1.0g

KCl ..................................................................................0.4g

CaCl₂·2H₂O ...................................................................0.265g

MgSO₄·7H₂O ...................................................................0.2g

L-Arginine·H₂O ..............................................................0.15g

NaH₂PO₄·H₂O .................................................................0.14g

L-Arginine·HCl ..............................................................0.126g

L-Lysine·HCl .................................................................72.5mg

L-Tyrosine, disodium salt ..............................................52.1mg

L-Leucine ......................................................................52.0mg

L-Threonine ...................................................................48.0mg

L-Valine ........................................................................46.0mg

L-Histidine·HCl·H₂O ......................................................42.0mg

D-Phenylalanine .............................................................32.0mg

L-Cysteine·2HCl ...........................................................31.29mg

L-Methionine .................................................................15.0mg

L-Glutamic acid .............................................................14.7mg

L-Aspartic acid ..............................................................13.3mg

L-Proline .......................................................................11.5mg

L-Serine ........................................................................10.5mg

L-Tryptophan .................................................................10.0mg

L-Alanine ........................................................................8.9mg

Phenol Red ....................................................................10.0mg

L-Glycine ........................................................................7.5mg

*i*-Inositol .......................................................................2.0mg

D-Calcium pantothenate ..................................................1.0mg

Choline chloride ..............................................................1.0mg

Folic acid ........................................................................1.0mg

Nicotinamide ...................................................................1.0mg

Pyridoxal·HCl .................................................................1.0mg

Thiamine·HCl ..................................................................1.0mg

Riboflavin .......................................................................0.1mg

pH 7.2 ± 0.2 at 25°C

**Preparation of Medium:** Add components to 1.0L of distilled/deionized water. Mix thoroughly. Filter sterilize.

**Use**: For the cultivation of animal cells in tissue culture, for example, cells for viral detection and identification by characteristic cytopathic effects.

## Earle's Balanced Salts, Phenol Red-Free
**Composition** per liter:
NaCl ................................................................................6.8g
NaHCO$_3$ .........................................................................2.2g
Glucose ............................................................................1.0g
KCl ..................................................................................0.4g
CaCl$_2$·2H$_2$O ...............................................................0.265g
MgSO$_4$·7H$_2$O .................................................................0.2g
NaH$_2$PO$_4$·H$_2$O .............................................................0.14g

pH 7.2 ± 0.2 at 25°C

**Preparation of Medium:** Add components to distilled/deionized water and bring volume to 1.0L. Mix thoroughly. Filter sterilize.

**Storage/Shelf Life:** Store dehydrated media in the dark in a sealed container below 30°C. Prepared media should be stored under refrigeration (2-8°C). Media should be used within 60 days of preparation. Media should not be used if there are any signs of deterioration (discoloration) or contamination, or if the expiration date supplied by the manufacturer has passed.

**Use:** For the preparation of tissue culture media where Phenol Red is not desired.

## Tissue Culture Amino Acids, HeLa 100X
## (TC Amino Acids, HeLa 100X)
**Composition** per liter:
L-Lysine ...........................................................................0.029g
L-Isoleucine ......................................................................0.026g
L-Leucine ..........................................................................0.026g
L-Threonine ......................................................................0.023g
L-Valine ............................................................................0.023g
L-Tyrosine .........................................................................0.018g
L-Arginine ........................................................................0.017g
L-Phenylalanine ................................................................0.016g
L-Cystine ..........................................................................0.012g
L-Histidine ........................................................................7.8mg
L-Methionine .....................................................................7.5mg
L-Tryptophan .....................................................................4.1mg

pH 7.2–7.4 at 25°C

**Preparation of Tissue Culture Amino Acids, HeLa 100X:** Add components to distilled/deionized water and bring volume to 1.0L. Mix thoroughly. Adjust pH to 7.2–7.4. Filter sterilize.

**Storage/Shelf Life:** Store dehydrated media in the dark in a sealed container below 30°C. Prepared media should be stored under refrigeration (2-8°C). Media should be used within 60 days of preparation. Media should not be used if there are any signs of deterioration (discoloration) or contamination, or if the expiration date supplied by the manufacturer has passed.

**Use:** For the preparation of Eagle HeLa medium for tissue culture procedures and virus studies.

## Tissue Culture Amino Acids, Minimal Eagle 50X
## (TC Amino Acids, Minimal Eagle 50X)
**Composition** per liter:
L-Arginine ..........................................................................0.1g
L-Lysine ...........................................................................0.058g
L-Isoleucine ......................................................................0.052g
L-Leucine ..........................................................................0.052g
L-Threonine ......................................................................0.048g

L-Valine ............................................................................0.046g
L-Tyrosine .........................................................................0.036g
L-Phenylalanine ................................................................0.032g
L-Histidine ........................................................................0.031g
L-Cystine ..........................................................................0.024g
L-Methionine .....................................................................0.015g
L-Tryptophan .....................................................................0.01g

pH 7.2–7.4 at 25°C

**Preparation of Tissue Culture Amino Acids, Minimal Eagle 50X:** Add components to distilled/deionized water and bring volume to 1.0L. Mix thoroughly. Adjust pH to 7.2–7.4. Filter sterilize.

**Storage/Shelf Life:** Store dehydrated media in the dark in a sealed container below 30°C. Prepared media should be stored under refrigeration (2-8°C). Media should be used within 60 days of preparation. Media should not be used if there are any signs of deterioration (discoloration) or contamination, or if the expiration date supplied by the manufacturer has passed.

**Use:** For the preparation of TC minimal medium Eagle for tissue culture procedures and virus studies.

## Tissue Culture Dulbecco Solution
## (TC Dulbecco Solution)
**Composition** per liter:
NaCl ................................................................................8.0g
Na$_2$HPO$_4$ .......................................................................1.15g
KH$_2$PO$_4$ ...........................................................................0.2g
KCl ..................................................................................0.2g
CaCl$_2$·2H$_2$O ....................................................................0.1g
MgCl$_2$·6H$_2$O ....................................................................0.1g

pH 7.2–7.4 at 25°C

**Preparation of Tissue Culture Dulbecco Solution:** Add components to distilled/deionized water and bring volume to 1.0L. Mix thoroughly. Adjust pH to 7.2–7.4. Filter sterilize.

**Storage/Shelf Life:** Store dehydrated media in the dark in a sealed container below 30°C. Prepared media should be stored under refrigeration (2-8°C). Media should be used within 60 days of preparation. Media should not be used if there are any signs of deterioration (discoloration) or contamination, or if the expiration date supplied by the manufacturer has passed.

**Use:** For use in tissue culture and virus preparations.

## Tissue Culture Earle Solution
## (TC Earle Solution)
**Composition** per 1002.0mL:
NaCl ................................................................................6.8g
NaHCO$_3$ .........................................................................2.2g
Glucose ............................................................................1.0g
KCl ..................................................................................0.4g
CaCl$_2$·2H$_2$O ....................................................................0.2g
NaH$_2$PO$_4$ .......................................................................0.125g
MgSO$_4$·7H$_2$O ...................................................................0.1g
Phenol Red (1% solution) ................................................2.0mL

pH 7.2–7.4 at 25°C

**Preparation of Tissue Culture Earle Solution:** Add components, except Phenol Red, to distilled/deionized water and bring volume to 1.0L. Mix thoroughly. Add 2.0mL of Phenol Red solution. Adjust pH to 7.2–7.4. Filter sterilize.

**Storage/Shelf Life:** Store dehydrated media in the dark in a sealed container below 30°C. Prepared media should be stored under refrigeration (2-8°C). Media should be used within 60 days of preparation. Media should not be used if there are any signs of deterioration (discoloration) or contamination, or if the expiration date supplied by the manufacturer has passed.

**Use:** For use in tissue culture and virus preparations.

## Tissue Culture Hanks Solution
## (TC Hanks Solution)

**Composition** per liter:

| | |
|---|---|
| NaCl | 8.0g |
| Glucose | 1.0g |
| KCl | 0.4g |
| NaHCO$_3$ | 0.35g |
| CaCl$_2$·2H$_2$O | 0.14g |
| MgCl$_2$·6H$_2$O | 0.1g |
| MgSO$_4$·7H$_2$O | 0.1g |
| Na$_2$HPO$_4$ | 0.06g |
| KH$_2$PO$_4$ | 0.06g |
| Phenol Red | 0.02g |

pH 7.2–7.4 at 25°C

**Source:** This medium is available as a premixed solution from BD Diagnostic Systems.

**Preparation of Tissue Culture Hanks Solution:** Add components to distilled/deionized water and bring volume to 1.0L. Mix thoroughly. Adjust pH to 7.2–7.4. Filter sterilize.

**Storage/Shelf Life:** Store dehydrated media in the dark in a sealed container below 30°C. Prepared media should be stored under refrigeration (2-8°C). Media should be used within 60 days of preparation. Media should not be used if there are any signs of deterioration (discoloration) or contamination, or if the expiration date supplied by the manufacturer has passed.

**Use:** For use in tissue culture procedures.

## Tissue Culture Medium 199
## (TC Medium 199)

**Composition** per 1050.0mL:

| | |
|---|---|
| NaCl | 8.0g |
| Glucose | 1.0g |
| KCl | 0.4g |
| NaHCO$_3$ | 0.35g |
| DL-Glutamic acid | 0.15g |
| CaCl$_2$·2H$_2$O | 0.14g |
| DL-Leucine | 0.12g |
| L-Glutamine | 0.1g |
| MgSO$_4$·7H$_2$O | 0.1g |
| L-Arginine | 0.07g |
| L-Lysine | 0.07g |
| DL-Aspartic acid | 0.06g |
| Na$_2$HPO$_4$ | 0.06g |
| KH$_2$PO$_4$ | 0.06g |
| DL-Threonine | 0.06g |
| DL-Alanine | 0.05g |
| Glycine | 0.05g |
| DL-Phenylalanine | 0.05g |
| DL-Serine | 0.05g |
| Sodium acetate | 0.05g |
| DL-Valine | 0.05g |
| DL-Isoleucine | 0.04g |
| L-Proline | 0.04g |
| L-Tyrosine | 0.04g |
| DL-Methionine | 0.03g |
| L-Cystine | 0.02g |
| L-Histidine | 0.02g |
| Phenol Red | 0.02g |
| DL-Tryptophan | 0.02g |
| Adenine | 0.01g |
| L-Hydroxyproline | 0.01g |
| Tween™ 80 | 5.0mg |
| Adenosine triphosphate | 1.0mg |
| Choline | 0.5mg |
| Deoxyribose | 0.5mg |
| Ribose | 0.5mg |
| Guanine | 0.3mg |
| Hypoxanthine | 0.3mg |
| Thymine | 0.3mg |
| Uracil | 0.3mg |
| Xanthine | 0.3mg |
| Adenylic acid | 0.2mg |
| Cholesterol | 0.2mg |
| Calciferol | 0.1mg |
| Fe(NO$_3$)$_3$·9H$_2$O | 0.1mg |
| L-Cysteine | 0.1mg |
| Vitamin A | 0.1mg |
| α-Tocopherol phosphate | 0.01mg |
| Biotin | 0.01mg |
| Calcium pantothenate | 0.01mg |
| Folic acid | 0.01mg |
| Menadione | 0.01mg |
| Riboflavin | 0.01mg |
| Thiamine·HCl | 0.01mg |
| *p*-Aminobenzoic acid | 0.05mg |
| Ascorbic acid | 0.05mg |
| Glutathione | 0.05mg |
| Inositol | 0.05mg |
| Niacin | 0.025mg |
| Niacinamide | 0.025mg |
| Pyridoxine·HCl | 0.025mg |
| Pyridoxal·HCl | 0.025mg |
| Serum | 50.0–100.0mL |

pH 7.2–7.4 at 25°C

**Preparation of Medium:** Add components, except serum, to distilled/deionized water and bring volume to 1.0L. Mix thoroughly. Adjust pH to 7.2–7.4 with 10% Na$_2$CO$_3$ solution. Filter sterilize. Aseptically add 50.0–100.0mL of sterile serum. Human serum, bovine serum, horse serum, or fetal calf serum may be used. Mix thoroughly. If desired, antibacterial inhibitors may be added. Aseptically add 500,000U of penicillin and 0.5g of streptomycin to 1050.0mL of the complete medium to increase selectivity.

**Storage/Shelf Life:** Store dehydrated media in the dark in a sealed container below 30°C. Prepared media should be stored under refrigeration (2-8°C). Media should be used within 60 days of preparation. Media should not be used if there are any signs of deterioration (discoloration) or contamination, or if the expiration date supplied by the manufacturer has passed.

**Use:** For the cultivation of a wide variety of cell lines in tissue culture. It is especially useful for the detection, titering, and identification of viruses in tissue culture cells.

## Tissue Culture Medium Eagle with Earle Balanced Salt Solution (TC Medium Eagle with Earle BSS)

**Composition** per 1056.0mL:

| | |
|---|---|
| NaCl | 6.8g |
| NaHCO$_3$ | 2.2g |
| Glucose | 1.0g |
| KCl | 0.4g |
| CaCl$_2$·2H$_2$O | 0.2g |
| NaH$_2$PO$_4$ | 0.125g |
| MgSO$_4$·7H$_2$O | 0.1g |
| L-Isoleucine | 0.026g |
| L-Leucine | 0.026g |
| L-Lysine | 0.026g |
| L-Threonine | 0.024g |
| L-Valine | 0.0235g |
| L-Tyrosine | 0.018g |
| L-Arginine | 0.0174g |
| L-Phenylalanine | 0.0165g |
| L-Cystine | 0.012g |
| L-Histidine | 8.0mg |
| L-Methionine | 7.5mg |
| Phenol Red | 5.0mg |
| L-Tryptophan | 4.0mg |
| Inositol | 1.8mg |
| Biotin | 1.0mg |
| Calcium pantothenate | 1.0mg |
| Choline chloride | 1.0mg |
| Folic acid | 1.0mg |
| Nicotinamide | 1.0mg |
| Pyridoxal·HCl | 1.0mg |
| Thiamine·HCl | 1.0mg |
| Riboflavin | 0.1mg |
| Serum | 50.0–100.0mL |
| Glutamine solution | 6.0mL |

pH 7.2–7.4 at 25°C

**Glutamine Solution:**

**Composition** per 100.0mL:

| | |
|---|---|
| L-Glutamine | 5.0g |
| NaCl (0.85% solution) | 100.0mL |

**Preparation of Glutamine Solution:** Add the glutamine to the 0.85% NaCl solution. Mix thoroughly. Filter sterilize.

**Preparation of Medium:** Add components, except glutamine and serum, to distilled/deionized water and bring volume to 1.0L. Mix thoroughly. Adjust pH to 7.2–7.4. Filter sterilize. Aseptically add 6.0mL of sterile glutamine solution and 50.0–100.0mL of sterile serum. Human serum, bovine serum, horse serum, or fetal calf serum may be used. Mix thoroughly.

**Storage/Shelf Life:** Store dehydrated media in the dark in a sealed container below 30°C. Prepared media should be stored under refrigeration (2-8°C). Media should be used within 60 days of preparation. Media should not be used if there are any signs of deterioration (discoloration) or contamination, or if the expiration date supplied by the manufacturer has passed.

**Use:** For the cultivation of HeLa, KB, and other tissue culture cell lines.

## Tissue Culture Medium Eagle with Hanks Balanced Salt Solution (TC Medium Eagle with Hanks BSS)

**Composition** per 1056.0mL:

| | |
|---|---|
| NaCl | 8.0g |
| Glucose | 1.0g |
| KCl | 0.4g |
| CaCl$_2$·2H$_2$O | 0.14g |
| MgSO$_4$·7H$_2$O | 0.1g |
| KH$_2$PO$_4$ | 0.06g |
| Na$_2$HPO$_4$ | 0.05g |
| L-Isoleucine | 0.026g |
| L-Leucine | 0.026g |
| L-Lysine | 0.026g |
| L-Threonine | 0.024g |
| L-Valine | 0.0235g |
| L-Tyrosine | 0.018g |
| L-Arginine | 0.0174g |
| L-Phenylalanine | 0.0165g |
| L-Cystine | 0.012g |
| L-Histidine | 8.0mg |
| L-Methionine | 7.5mg |
| Phenol Red | 5.0mg |
| L-Tryptophan | 4.0mg |
| Inositol | 1.8mg |
| Biotin | 1.0mg |
| Folic acid | 1.0mg |
| Calcium pantothenate | 1.0mg |
| Choline chloride | 1.0mg |
| Nicotinamide | 1.0mg |
| Pyridoxal·HCl | 1.0mg |
| Thiamine·HCl | 1.0mg |
| Riboflavin | 0.1mg |
| Serum | 50.0–100.0mL |
| Glutamine solution | 6.0mL |

pH 7.2–7.4 at 25°C

**Glutamine Solution:**

**Composition** per 100.0mL:

| | |
|---|---|
| L-Glutamine | 5.0g |
| NaCl (0.85% solution) | 100.0mL |

**Preparation of Glutamine Solution:** Add the glutamine to the 0.85% NaCl solution. Mix thoroughly. Filter sterilize.

**Preparation of Medium:** Add components, except glutamine and serum, to distilled/deionized water and bring volume to 1.0L. Mix thoroughly. Adjust pH to 7.2–7.4. Filter sterilize. Aseptically add 6.0mL of sterile glutamine solution and 50.0–100.0mL of sterile serum. Human serum, bovine serum, horse serum, or fetal calf serum may be used. Mix thoroughly.

**Storage/Shelf Life:** Store dehydrated media in the dark in a sealed container below 30°C. Prepared media should be stored under refrigeration (2-8°C). Media should be used within 60 days of preparation. Media should not be used if there are any signs of deterioration (discoloration) or contamination, or if the expiration date supplied by the manufacturer has passed.

**Use:** For use as a base in the preparation of liquid media used for the cultivation of tissue culture cell lines.

# Tissue Culture Medium Eagle, HeLa
## (TC Medium Eagle, HeLa)

**Composition** per 1056.0mL:

| | |
|---|---|
| NaCl | 5.85g |
| NaHCO₃ | 1.68g |
| Glucose | 0.9g |
| KCl | 0.373g |
| NaH₂PO₄ | 0.138g |
| MgCl₂·6H₂O | 0.12g |
| CaCl₂·2H₂O | 0.11g |
| L-Lysine | 0.0269g |
| L-Isoleucine | 0.0262g |
| L-Leucine | 0.0262g |
| L-Threonine | 0.0238g |
| L-Valine | 0.0234g |
| L-Tyrosine | 0.0181g |
| L-Arginine | 0.0174g |
| L-Phenylalanine | 0.0165g |
| L-Cystine | 0.012g |
| L-Histidine | 7.8mg |
| L-Methionine | 7.5mg |
| Phenol Red | 5.0mg |
| L-Tryptophan | 4.1mg |
| Folic acid | 0.44mg |
| Thiamine·HCl | 0.34mg |
| Biotin | 0.24mg |
| Pantothenic acid | 0.22mg |
| Pyridoxal·HCl | 0.2mg |
| Choline chloride | 0.14mg |
| Nicotinamide | 0.12mg |
| Riboflavin | 0.04mg |
| Serum | 50.0mL–100.0mL |
| Glutamine solution | 6.0mL |

pH 7.2–7.4 at 25°C

**Glutamine Solution:**

**Composition** per 100.0mL:

| | |
|---|---|
| L-Glutamine | 5.0g |
| NaCl (0.85% solution) | 100.0mL |

**Preparation of Glutamine Solution:** Add the glutamine to the 0.85% NaCl solution. Mix thoroughly. Filter sterilize.

**Preparation of Medium:** Add components, except glutamine and serum, to distilled/deionized water and bring volume to 1.0L. Mix thoroughly. Adjust pH to 7.2–7.4. Filter sterilize. Aseptically add 6.0mL of sterile glutamine solution and 50.0–100.0mL of sterile serum. Mix thoroughly.

**Storage/Shelf Life:** Store dehydrated media in the dark in a sealed container below 30°C. Prepared media should be stored under refrigeration (2-8°C). Media should be used within 60 days of preparation. Media should not be used if there are any signs of deterioration (discoloration) or contamination, or if the expiration date supplied by the manufacturer has passed.

**Use:** For the cultivation and maintenance of HeLa and other cell lines in tissue culture, and for studying the cytopathogenicity of viral agents.

# Tissue Culture Medium Ham F10
## (TC Medium Ham F10)

**Composition** per 1050.0mL:

| | |
|---|---|
| NaCl | 7.4g |
| Glucose | 1.1g |
| Na₂HPO₄ | 0.29g |
| KCl | 0.285g |
| L-Arginine | 0.211g |
| MgSO₄·7H₂O | 0.153g |
| L-Glutamine | 0.1462g |
| Sodium pyruvate | 0.11g |
| KH₂PO₄ | 0.083g |
| CaCl₂·2H₂O | 0.044g |
| L-Cystine | 0.0315g |
| L-Lysine | 0.0293g |
| L-Histidine | 0.021g |
| L-Asparagine | 0.015g |
| L-Glutamic acid | 0.0147g |
| L-Aspartic acid | 0.0133g |
| L-Leucine | 0.0131g |
| L-Proline | 0.0115g |
| L-Serine | 0.0105g |
| L-Alanine | 8.91mg |
| Glycine | 7.51mg |
| L-Phenylalanine | 4.96mg |
| L-Methionine | 4.48mg |
| Hypoxanthine | 4.0mg |
| L-Threonine | 3.57mg |
| L-Valine | 3.5mg |
| L-Isoleucine | 2.6mg |
| L-Tyrosine | 1.81mg |
| Cyanocobalamin | 1.3mg |
| Folic acid | 1.3mg |
| Phenol Red | 1.2mg |
| Thiamine·HCl | 1.0mg |
| FeSO₄·7H₂O | 0.83mg |
| Calcium pantothenate | 0.7mg |
| Thymidine | 0.7mg |
| Choline chloride | 0.69mg |
| Niacinamide | 0.6mg |
| L-Tryptophan | 0.6mg |
| *i*-Inositol | 0.54mg |
| Riboflavin | 0.37mg |
| Lipoic acid | 0.2mg |
| Pyridoxine·HCl | 0.2mg |
| ZnSO₄·7H₂O | 0.028mg |
| Biotin | 0.024mg |
| CuSO₄·5H₂O | 2.5µg |
| Fetal calf serum | 50.0–100.0mL |

pH 7.2–7.4 at 25°C

**Preparation of Medium:** Add components, except fetal calf serum, to distilled/deionized water and bring volume to 1.0L. Mix thoroughly. Adjust pH to 7.2–7.4 with 10% Na₂CO₃ solution. Filter sterilize. Aseptically add 50.0–100.0mL of sterile fetal calf serum. Mix thoroughly.

**Storage/Shelf Life:** Store dehydrated media in the dark in a sealed container below 30°C. Prepared media should be stored under refrigeration (2-8°C). Media should be used within 60 days of preparation. Media should not be used if there are any signs of deterioration (discoloration) or contamination, or if the expiration date supplied by the manufacturer has passed.

**Use:** For the cultivation of a wide variety of cell lines in tissue culture.

## Tissue Culture Medium NCTC 109
## (TC Medium NCTC 109)

**Composition** per 1050.0mL:

| | |
|---|---|
| NaCl | 6.8g |
| NaHCO$_3$ | 2.2g |
| Glucose | 1.0g |
| KCl | 0.4g |
| L-Cysteine | 0.26g |
| CaCl$_2$·2H$_2$O | 0.2g |
| NaH$_2$PO$_4$ | 0.14g |
| L-Glutamine | 0.14g |
| MgSO$_4$·7H$_2$O | 0.1g |
| Sodium acetate | 0.05g |
| Ascorbic acid | 0.05g |
| L-Alanine | 0.03g |
| L-Lysine | 0.03g |
| L-Arginine | 0.026g |
| L-Valine | 0.025g |
| L-Leucine | 0.02g |
| Phenol Red | 0.02g |
| L-Histidine | 0.019g |
| L-Threonine | 0.019g |
| L-Isoleucine | 0.018g |
| L-Tryptophan | 0.017.g |
| L-Phenylalanine | 0.017g |
| L-Tyrosine | 0.016g |
| Glycine | 0.014g |
| Tween™ 80 | 0.012g |
| L-Serine | 0.011g |
| L-Cystine | 0.01g |
| Glutathione | 0.01g |
| Cyanocobalamin | 0.01g |
| Deoxycytidine | 0.01g |
| Deoxyguanosine | 0.01g |
| Deoxyadenosine | 0.01g |
| Thymidine | 0.01g |
| L-Aspartic acid | 9.91mg |
| L-Glutamic acid | 8.26mg |
| L-Arginine | 8.09mg |
| L-Ornithine | 7.38mg |
| Nicotinamide adenine dinucleotide | 7.0mg |
| L-Proline | 6.13mg |
| L-α-N-butyric acid | 5.51mg |
| L-Methionine | 4.44mg |
| L-Taurine | 4.18mg |
| L-Hydroxyproline | 4.09mg |
| D-Glucosamine | 3.2mg |
| Coenzyme A | 2.5mg |
| Glucuronolactone | 1.8mg |
| Sodium glucuronate | 1.8mg |
| Choline chloride | 1.25mg |
| Cocarboxylase | 1.0mg |
| Flavin adenine dinucleotide | 1.0mg |
| Uridine triphosphate | 1.0mg |
| Nicotinamide adenine dinucleotide phosphate | 1.0mg |
| Vitamin A | 0.25mg |
| Calciferol | 0.25mg |
| *i*-Inositol | 0.125mg |
| *p*-Aminobenzoic acid | 0.125mg |

| | |
|---|---|
| 5-Methylcytosine | 0.1mg |
| Pyridoxine·HCl | 0.0625mg |
| Pyridoxal·HCl | 0.0625mg |
| Niacin | 0.0625mg |
| Niacinamide | 0.0625mg |
| Biotin | 0.025mg |
| Folic acid | 0.025mg |
| Menadione | 0.025mg |
| Pantothenate | 0.025mg |
| Riboflavin | 0.025mg |
| Thiamine·HCl | 0.025mg |
| α-Tocopherol phosphate | 0.025mg |
| Serum | 50.0–100.0mL |

pH 7.2–7.4 at 25°C

**Preparation of Medium:** Add components, except serum, to distilled/deionized water and bring volume to 1.0L. Mix thoroughly. Adjust pH to 7.2–7.4 with 10% Na$_2$CO$_3$ solution. Filter sterilize. Aseptically add 50.0–100.0mL of sterile serum. Human serum, bovine serum, horse serum, or fetal calf serum may be used. Mix thoroughly.

**Storage/Shelf Life:** Store dehydrated media in the dark in a sealed container below 30°C. Prepared media should be stored under refrigeration (2-8°C). Media should be used within 60 days of preparation. Media should not be used if there are any signs of deterioration (discoloration) or contamination, or if the expiration date supplied by the manufacturer has passed.

**Use:** For the cultivation of a wide variety of cell lines in tissue culture.

## Tissue Culture Medium RPMI No. 1640
## (TC Medium RPMI #1640)

**Composition** per liter:

| | |
|---|---|
| NaCl | 6.46g |
| Glucose | 2.0g |
| NaHCO$_3$ | 2.0g |
| NaH$_2$PO$_4$ | 1.512g |
| KCl | 0.4g |
| L-Glutamine | 0.3g |
| L-Arginine | 0.2g |
| Calcium nitrate | 0.1g |
| MgSO$_4$·7H$_2$O | 0.1g |
| L-Asparagine | 0.05g |
| L-Cystine | 0.05g |
| L-Isoleucine | 0.05g |
| L-Leucine | 0.05g |
| L-Lysine·HCl | 0.04g |
| Inositol | 0.035g |
| L-Serine | 0.03g |
| Hydroxy-L-proline | 0.02g |
| L-Aspartic acid | 0.02g |
| L-Glutamic acid | 0.02g |
| L-Proline | 0.02g |
| L-Threonine | 0.02g |
| L-Tyrosine | 0.02g |
| L-Valine | 0.02g |
| L-Histidine | 0.015g |
| L-Methionine | 0.015g |
| L-Phenylalanine | 0.015g |
| Glycine | 0.01g |
| L-Tryptophan | 5.0mg |
| Phenol Red | 5.0mg |
| Choline chloride | 3.0mg |

| | |
|---|---|
| *p*-Aminobenzoic acid | 1.0mg |
| Folic acid | 1.0mg |
| Glutathione | 1.0mg |
| Nicotinamide | 1.0mg |
| Pyridoxine·HCl | 1.0mg |
| Thiamine·HCl | 1.0mg |
| Calcium pantothenate | 0.25mg |
| Biotin | 0.2mg |
| Riboflavin | 0.2mg |
| Vitamin B$_{12}$ | 5.0mg |
| Serum | 50.0–100.0mL |

pH 7.2–7.4 at 25°C

**Source:** This medium is available as a premixed powder and solution from BD Diagnostic Systems.

**Preparation of Medium:** Add components, except serum, to distilled/deionized water and bring volume to 1.0L. Mix thoroughly. Adjust pH to 7.2–7.4 with 10% Na$_2$CO$_3$ solution. Filter sterilize. Aseptically add 50.0–100.0mL of sterile serum. Human serum, bovine serum, horse serum, or fetal calf serum may be used. Mix thoroughly.

**Storage/Shelf Life:** Store dehydrated media in the dark in a sealed container below 30°C. Prepared media should be stored under refrigeration (2-8°C). Media should be used within 60 days of preparation. Media should not be used if there are any signs of deterioration (discoloration) or contamination, or if the expiration date supplied by the manufacturer has passed.

**Use:** For the cultivation of a wide variety of cell lines in tissue culture.

## Tissue Culture Minimal Medium Eagle

**Composition** per liter:

| | |
|---|---|
| Sterile salt solution | 944.0mL |
| TC amino acids, minimal Eagle 50X | 20.0mL |
| TC NaHCO$_3$, 10% | 20.0mL |
| TC vitamins, minimal Eagle 100X | 10.0mL |
| TC glutamine, 5% | 6.0mL |

pH 7.2–7.4 at 25°C

**Sterile Salt Solution:**

**Composition** per 944.0mL:

| | |
|---|---|
| NaCl | 6.8g |
| Glucose | 1.0g |
| KCl | 0.4g |
| CaCl$_2$ | 0.2g |
| MgCl$_2$ | 0.2g |
| NaH$_2$PO$_4$ | 0.15g |

**Preparation of Sterile Salt Solution:** Add components to distilled/deionized water and bring volume to 944.0mL. Mix thoroughly. Filter sterilize.

**TC Amino Acids, Minimal Eagle 50X:**

**Composition** per liter:

| | |
|---|---|
| L-Arginine | 0.1g |
| L-Lysine | 0.06g |
| L-Isoleucine | 0.05g |
| L-Leucine | 0.05g |
| L-Threonine | 0.05g |
| L-Valine | 0.05g |
| L-Tyrosine | 0.04g |
| L-Phenylalanine | 0.03g |
| L-Histidine | 0.03g |
| L-Cystine | 0.02g |

| | |
|---|---|
| L-Methionine | 0.02g |
| L-Tryptophan | 0.01g |

**Preparation of TC Amino Acids, Minimal Eagle 50X:** Add components to distilled/deionized water and bring volume to 1.0L. Mix thoroughly. Adjust pH to 7.2–7.4. Filter sterilize.

**TC NaHCO$_3$, 10%:**

**Composition** per 100.0mL:

| | |
|---|---|
| NaHCO$_3$ | 10.0g |

**Preparation of TC NaHCO$_3$, 10%:** Add NaHCO$_3$ to distilled/deionized water and bring volume to 100.0mL. Mix thoroughly. Filter sterilize.

**TC Vitamins, Minimal Eagle 100X:**

**Composition** per liter:

| | |
|---|---|
| Inositol | 2.0mg |
| Calcium pantothenate | 1.0mg |
| Choline chloride | 1.0mg |
| Folic acid | 1.0mg |
| Nicotinamide | 1.0mg |
| Pyridoxal | 1.0mg |
| Thiamine·HCl | 1.0mg |
| Riboflavin | 0.1mg |

**Preparation of TC Vitamins, Minimal Eagle 100X:** Add components to distilled/deionized water and bring volume to 1.0L. Mix thoroughly. Filter sterilize.

**TC Glutamine, 5%:**

**Composition** per 100.0mL:

| | |
|---|---|
| L-Glutamine | 5.0g |
| NaCl (0.85% solution) | 100.0mL |

**Preparation of TC Glutamine, 5%:** Add the glutamine to the 0.85% NaCl solution. Mix thoroughly. Filter sterilize.

**Preparation of Medium:** Aseptically combine 944.0mL of sterile salt solution, 20.0mL of sterile TC amino acids, minimal Eagle 50X, 20.0mL of sterile TC NaHCO$_3$, 10%, 10.0mL of sterile TC vitamins, minimal Eagle 100X, and 6.0mL of sterile TC glutamine, 5%. Mix thoroughly. Adjust pH to 7.2–7.4 if necessary.

**Storage/Shelf Life:** Store dehydrated media in the dark in a sealed container below 30°C. Prepared media should be stored under refrigeration (2-8°C). Media should be used within 60 days of preparation. Media should not be used if there are any signs of deterioration (discoloration) or contamination, or if the expiration date supplied by the manufacturer has passed.

**Use:** For the cultivation of mammalian cells in monolayer or suspension for tissue culture procedures and virus preparation.

## Tissue Culture Minimal Medium Eagle with Earle Balanced Salts Solution (TC Minimal Medium Eagle with Earle BSS)

**Composition** per 1056.0mL:

| | |
|---|---|
| NaCl | 6.8g |
| Glucose | 1.0g |
| KCl | 0.4g |
| CaCl$_2$·2H$_2$O | 0.2g |
| MgCl$_2$·6H$_2$O | 0.2g |
| NaH$_2$PO$_4$ | 0.15g |
| L-Arginine | 0.1g |
| L-Lysine | 0.06g |
| L-Isoleucine | 0.05g |

L-Leucine..................................................................0.05g
L-Threonine.............................................................0.05g
L-Valine...................................................................0.05g
L-Tyrosine................................................................0.04g
L-Phenylalanine.......................................................0.03g
L-Histidine..............................................................0.03g
L-Cystine.................................................................0.02g
L-Methionine...........................................................0.02g
L-Tryptophan...........................................................0.01g
*i*-Inositol..............................................................2.0mg
Calcium pantothenate..............................................1.0mg
Choline chloride.......................................................1.0mg
Folic acid................................................................1.0mg
Nicotinamide............................................................1.0mg
Pyridoxal.................................................................1.0mg
Thiamine·HCl...........................................................1.0mg
Riboflavin ..............................................................0.1mg
Serum ........................................................ 50.0–100.0mL
Glutamine solution..................................................6.0mL
CaCl$_2$·2H$_2$O solution (optional)........................2.0mL
pH 7.2–7.4 at 25°C

**Glutamine Solution:**
**Composition** per 100.0mL:
L-Glutamine...............................................................5.0g
NaCl (0.85% solution) .......................................... 100.0mL

**Preparation of Glutamine Solution:** Add the glutamine to the 0.85% NaCl solution. Mix thoroughly. Filter sterilize.

**Preparation of Medium:** Add components, except glutamine and serum, to distilled/deionized water and bring volume to 1.0L. Mix thoroughly. Adjust pH to 7.2–7.4 with 10% Na$_2$CO$_3$ solution. Filter sterilize. Aseptically add 6.0mL of sterile glutamine solution and 50.0–100.0mL of sterile serum. Human serum, bovine serum, horse serum, or fetal calf serum may be used. Mix thoroughly. To grow cells in a monolayer, aseptically add 2.0mL of a sterile 10% CaCl$_2$·2H$_2$O solution. To grow cells in suspension, omit the CaCl$_2$·2H$_2$O solution.

**Storage/Shelf Life:** Store dehydrated media in the dark in a sealed container below 30°C. Prepared media should be stored under refrigeration (2-8°C). Media should be used within 60 days of preparation. Media should not be used if there are any signs of deterioration (discoloration) or contamination, or if the expiration date supplied by the manufacturer has passed.

**Use:** For preparation of Eagle's minimal medium for the cultivation of cells in monolayer or suspension in tissue culture.

## Tissue Culture Minimal Medium Eagle Spinner Modified

### (TC Minimal Medium Eagle Spinner Modified MEM-S)
**Composition** per 1056.0mL:
NaH$_2$PO$_4$...........................................................1.35g
NaCl.........................................................................6.8g
NaHCO$_3$................................................................2.2g
Glucose ....................................................................1.0g
KCl...........................................................................0.4g
CaCl$_2$·2H$_2$O.......................................................0.2g
NaH$_2$PO$_4$..........................................................0.125g
MgSO$_4$·7H$_2$O......................................................0.1g
L-Isoleucine............................................................0.026g
L-Leucine................................................................0.026g
L-Lysine..................................................................0.026g

L-Threonine...........................................................0.024g
L-Valine.................................................................0.0235g
L-Tyrosine .............................................................0.018g
L-Arginine..............................................................0.0174g
L-Phenylalanine.....................................................0.0165g
L-Cystine................................................................0.012g
L-Histidine...............................................................8.0mg
L-Methionine............................................................7.5mg
Phenol Red...............................................................5.0mg
L-Tryptophan............................................................4.0mg
Inositol.....................................................................1.8mg
Biotin.......................................................................1.0mg
Calcium pantothenate ..............................................1.0mg
Choline chloride........................................................1.0mg
Folic acid ................................................................1.0mg
Nicotinamide............................................................1.0mg
Pyridoxal·HCl...........................................................1.0mg
Thiamine·HCl ..........................................................1.0mg
Riboflavin ...............................................................0.1mg
Serum..........................................................50.0mL–100.0mL
Glutamine solution...................................................6.0mL
pH 7.2–7.4 at 25°C

**Glutamine Solution:**
**Composition** per 100.0mL:
L-Glutamine .............................................................5.0g
NaCl (0.85% solution)........................................... 100.0mL

**Preparation of Glutamine Solution:** Add the glutamine to the 0.85% NaCl solution. Mix thoroughly. Filter sterilize.

**Preparation of Medium:** Add components, except glutamine and serum, to distilled/deionized water and bring volume to 1.0L. Mix thoroughly. Adjust pH to 7.2–7.4 with 10% Na$_2$CO$_3$ solution. Filter sterilize. Aseptically add 6.0mL of sterile glutamine solution and 50.0–100.0mL of sterile serum. Human serum, bovine serum, horse serum, or fetal calf serum may be used. Mix thoroughly.

**Storage/Shelf Life:** Store dehydrated media in the dark in a sealed container below 30°C. Prepared media should be stored under refrigeration (2-8°C). Media should be used within 60 days of preparation. Media should not be used if there are any signs of deterioration (discoloration) or contamination, or if the expiration date supplied by the manufacturer has passed.

**Use:** For the cultivation of mammalian cells in suspension.

## Tissue Culture Tyrode Solution (TC Tyrode Solution)

**Composition** per 1002.0mL:
NaCl.........................................................................8.0g
Glucose ....................................................................1.0g
NaHCO$_3$................................................................1.0g
CaCl$_2$·2H$_2$O.......................................................0.2g
KCl...........................................................................0.2g
MgCl$_2$·6H$_2$O.......................................................0.1g
NaH$_2$PO$_4$...........................................................0.05g
Phenol Red (1% solution)........................................ 2.0mL
pH 7.2–7.4 at 25°C

**Preparation of Tissue Culture Tyrode Solution:** Add components, except Phenol Red, to distilled/deionized water and bring volume to 1.0L. Mix thoroughly. Add 2.0mL of Phenol Red solution. Adjust pH to 7.2–7.4. Filter sterilize.

**Storage/Shelf Life:** Store dehydrated media in the dark in a sealed container below 30°C. Prepared media should be stored under refrigeration (2-8°C). Media should be used within 60 days of preparation. Media should not be used if there are any signs of deterioration (discoloration) or contamination, or if the expiration date supplied by the manufacturer has passed.

**Use:** For use in tissue culture procedures.

## Tissue Culture Vitamins Minimal Eagle, 100X
## (TC Vitamins Minimal Eagle, 100X)

**Composition** per liter:

| | |
|---|---|
| Inositol | 2.0mg |
| Calcium pantothenate | 1.0mg |
| Choline chloride | 1.0mg |
| Folic acid | 1.0mg |
| Nicotinamide | 1.0mg |
| Pyridoxal | 1.0mg |
| Thiamine·HCl | 1.0mg |
| Riboflavin | 0.1mg |

pH 7.2–7.4 at 25°C

**Preparation of TC Vitamins, Minimal Eagle 100X:** Add components to distilled/deionized water and bring volume to 1.0L. Mix thoroughly. Filter sterilize.

**Storage/Shelf Life:** Store dehydrated media in the dark in a sealed container below 30°C. Prepared media should be stored under refrigeration (2-8°C). Media should be used within 60 days of preparation. Media should not be used if there are any signs of deterioration (discoloration) or contamination, or if the expiration date supplied by the manufacturer has passed.

**Use:** For the preparation of Tissue Culture minimal medium Eagle used in tissue culture procedures.

## Viral Transport Medium
## (VTM)

**Composition** per 104.1mL:

| | |
|---|---|
| Bovine serum albumin | 0.5g |
| Veal infusion broth | 100.0mL |
| Phenol Red | 0.4mL |
| Amphotericin B solution | 2.0mL |
| Gentamicin solution | 1.0mL |
| Vancomycin solution | 0.2mL |

pH 7.4 ± 0.2 at 25°C

**Veal Infusion Broth:**
**Composition** per liter:

| | |
|---|---|
| Veal, infusion from | 500.0g |
| NaCl | 5.0g |
| Pancreatic digest of casein | 5.0g |
| Peptic digest of animal tissue | 5.0g |

**Preparation of Veal Infusion Broth:** Add components to distilled/deionized water and bring volume to 1.0L. Mix thoroughly. Distribute into tubes or flasks. Autoclave for 15 min at 15 psi pressure–121°C. Use freshly prepared solution.

**Amphotericin B Solution:**
**Composition** per 10.0mL:

| | |
|---|---|
| Amphotericin B | 2.5g |

**Preparation of Amphotericin B Solution:** Add amphotericin B to distilled/deionized water and bring volume to 10.0mL. Mix thoroughly. Filter sterilize.

**Gentamicin Solution:**
**Composition** per 10.0mL:

| | |
|---|---|
| Gentamicin | 0.5g |

**Preparation of Gentamicin Solution:** Add gentamicin to distilled/deionized water and bring volume to 10.0mL. Mix thoroughly. Filter sterilize.

**Vancomycin Solution:**
**Composition** per 10.0mL:

| | |
|---|---|
| Vancomycin | 0.5g |

**Preparation of Vancomycin Solution:** Add vancomycin to distilled/deionized water and bring volume to 10.0mL. Mix thoroughly. Filter sterilize.

**Preparation of Medium:** To 100.0mL of sterile veal infusion broth, aseptically add bovine serum albumin, Phenol Red, amphotericin B solution, gentamicin solution, and vancomycin solution. Mix thoroughly. Dispense 2.0mL of medium into serum vials. Store at 4°C and use for up to 2 months.

**Storage/Shelf Life:** Store dehydrated media in the dark in a sealed container below 30°C. Prepared media should be stored under refrigeration (2-8°C). Media should be used within 60 days of preparation. Media should not be used if there are any signs of deterioration (discoloration) or contamination, or if the expiration date supplied by the manufacturer has passed.

**Use:** For the maintenance and transport of specimens suspected of being virally infected.

## Agars

Below are some agars used as solidifying agents in various media.

Agar Bacteriological (Agar No. 1)

An agar with low calcium and magnesium. Available from Oxoid Unipath.

Agar, Bacto

A purified agar with reduced pigmented compounds, salts, and extraneous matter. Available from Difco, BD Diagnostic Systems.

Agar, BiTek™

Agar prepared as a special technical grade. Available from Difco, BD Diagnostic Systems.

Agar, Flake

A technical grade agar. Available from Difco, BD Diagnostic Systems.

Agar, Grade A

A select grade agar containing minerals. Available from BD Diagnostic Systems.

Agar, Granulated

A high grade granulated agar that has been filtered, decolorized, and purified. Available from BD Diagnostic Systems.

Agarose

A low sulfate neutral gelling fraction of agar that is a complex galactose polysaccharide of near neutral charge.

Agar, Purified

A very high grade agar that has been filtered, decolorized, and purified by washing and extraction of refined agars. It has reduced mineral content. Available from BD Diagnostic Systems.

Agar Technical (Agar No. 3)

A technical grade agar. Available from Difco, BD Diagnostic Systems, and Oxoid Unipath.

Ionagar

A purified agar. Available from Oxoid Unipath.

Noble Agar

An agar that has been extensively washed and is essentially free of impurities. Available from Difco, BD Diagnostic Systems.

Purified Agar

An agar that has been extensively washed and extracted with water and organic solvent. Available from Difco, BD Diagnostic Systems, and Oxoid Unipath.

## Peptones

Below is a list of some of the peptones that are used as ingredients in various media.

Acidase™ Peptone

A hydrochloric acid hydrolysate of casein. It has a nitrogen content of 8% and is deficient in cystine and tryptophan. Available from BD Diagnostic Systems.

Bacto Casitone

A pancreatic digest of casein. Available from Difco, BD Diagnostic Systems.

Bacto Peptamin

A peptic digest of animal tissues. Available from Difco, BD Diagnostic Systems.

Bacto Peptone

An enzymatic digest of animal tissues. It has a high concentration of low molecular weight peptones and amino acids. Available from Difco, BD Diagnostic Systems.

Bacto Proteose Peptone

An enzymatic digest of animal tissues. It has a high concentration of high molecular weight peptones. Available from Difco, BD Diagnostic Systems.

Bacto Soytone

A enzymatic hydrolysate of soybean meal. Available from Difco, BD Diagnostic Systems.

Bacto Tryptone

A pancreatic digest of casein. Available from Difco, BD Diagnostic Systems.

Bacto Tryptose

An enzymatic hydrolysate containing numerous peptides including those of higher molecular weights. Available from Difco, BD Diagnostic Systems.

Biosate™ Peptone

A hydrolysate of plant and animal proteins. Available from BD Diagnostic Systems.

Casein Hydrolysate

A hydrolysate of casein prepared with hydrochloric acid digestion under pressure and neutralized with sodium hydroxide. It contains total nitrogen of 7.6% and NaCl of 28.3%. Available from Oxoid Unipath.

Gelatone

A pancreatic digest of gelatin. Available from Difco, BD Diagnostic Systems.

Gelysate™ Peptone

A pancreatic digest of gelatin deficient in cystine and tryptophan and which has a low carbohydrate content. Available from Oxoid Unipath.

Lactoalbumin Hydrolysate

A pancreatic digest of lactoalbumin, a milk whey protein. It has high levels of amino acids. It contains total nitrogen of 11.9% and NaCl of 1.4%. Available from Difco, BD Diagnostic Systems, and Oxoid Unipath.

Liver Digest Neutralized

A papaic digest of liver that contains total nitrogen of 11.0% and NaCl of 1.6%. Available from Oxoid Unipath.

Mycological Peptone

A peptone that contains total nitrogen of 9.5% and NaCl of 1.1%. Available from Oxoid Unipath.

Myosate™ Peptone

A pancreatic digest of heart muscle. Available from BD Diagnostic Systems.

Neopeptone

An enzymatic digest of protein. Available from Difco, BD Diagnostic Systems.

Peptone Bacteriological Neutralized

A mixed pancreatic and papaic digest of animal tissues. It contains total nitrogen of 14.0% and NaCl of 1.6%. Available from Difco, BD Diagnostic Systems, and Oxoid Unipath.

Peptone P

A peptic digest of fresh meat that has a high sulfur content and contains total nitrogen of 11.12% and NaCl of 9.3%. Available from Difco, BD Diagnostic Systems, and Oxoid Unipath.

Peptonized Milk

A pancreatic digest of high grade skim milk powder. It has a high carbohydrate and calcium concentration. It contains total nitrogen of 5.3% and NaCl of 1.6%. Available from Oxoid Unipath.

Phytone™ Peptone

A papaic digest of soybean meal. It has a high vitamin and a high carbohydrate content. Available from BD Diagnostic Systems.

Polypeptone™ Peptone

A mixture of peptones composed of equal parts of pancreatic digest of casein and peptic digest of animal tissue. Available from BD Diagnostic Systems.

Proteose Peptone

A specialized peptone prepared from a mixture of peptones that contains a wide variety of high molecular weight peptides. It contains total nitrogen of 12.7% and NaCl of 8.0%. Available from Difco, BD Diagnostic Systems, and Oxoid Unipath.

Proteose Peptone No. 2

An enzymatic digest of animal tissues with a high concentration of high molecular weight peptones. Available from Difco, BD Diagnostic Systems.

Proteose Peptone No. 3

An enzymatic digest of animal tissues. It has a high concentration of high molecular weight peptones. Available from Difco, BD Diagnostic Systems.

Soya Peptone

A papaic digest of soybean meal with a high carbohydrate concentration. It contains total nitrogen of 8.7% and NaCl of 0.4%. Available from Oxoid Unipath.

Soytone

A papaic digest of soybean meal. Available from Difco, BD Diagnostic Systems, and Oxoid Unipath.

Special Peptone

A mixture of peptones, including meat, plant and yeast digests. It contains a wide variety of peptides, nucleotides, and minerals. It contains total nitrogen of 11.7% and NaCl of 3.5%. Available from Oxoid Unipath.

Thiotone™ E Peptone

An enzymatic digest of animal tissue. Available from BD Diagnostic Systems.

Trypticase™ Peptone

A pancreatic digest of casein. It has a very low carbohydrate content and a relatively high tryptophan content. Available from BD Diagnostic Systems.

Tryptone

A pancreatic digest of casein. It contains total nitrogen of 12.7% and NaCl of 0.4%. Available from Oxoid Unipath.

Tryptone T

A pancreatic digest of casein with lower levels of calcium, magnesium, and iron than tryptone. It contains total nitrogen of 11.7% and NaCl of 4.9%. Available from Difco, BD Diagnostic Systems, and Oxoid Unipath.

Tryptose

An enzymatic hydrolysate containing high molecular weight peptides. It contains total nitrogen of 12.2% and NaCl of 5.7%. Available from Difco, BD Diagnostic Systems, and Oxoid Unipath.

## Meat and Plant Extracts

Below is a list of some of the meat and plant extracts that are used as ingredients in various media.

Bacto Beef

A desiccated powder of lean beef. Available from Difco, BD Diagnostic Systems.

Bacto Beef Extract

An extract of beef (paste). Available from Difco, BD Diagnostic Systems.

Bacto Beef Extract Desiccated

An extract of desiccated beef. Available from Difco, BD Diagnostic Systems.

Bacto Beef Heart for Infusion

A desiccated powder of beef heart. Available from Difco, BD Diagnostic Systems.

Bacto Liver

A desiccated powder of beef liver. Available from Difco, BD Diagnostic Systems.

Lab-Lemco

A meat extract powder. Available from Oxoid Unipath.

Liver Desiccated

Dehydrated ox livers. Available from Oxoid Unipath.

Malt Extract

A water soluble extract from germinated grain dried by low temperature evaporation. It has a high carbohydrate content. It contains total nitrogen of 1.1% and NaCl of 0.1%.

## Growth Factors

Many microorganisms have specific growth factor requirements that must be included in media for their successful cultivation. Vitamins, amino acids, fatty acids, trace metals, and blood components often must be added to media. Most often mixtures of growth factors are used in microbiological media. Acid hydrolysates of casein commonly are used as sources of amino acids. Extracts of yeast cells also are employed as sources of amino acids and vitamins for the cultivation of microorganisms. Many media, particularly those employed in the clinical laboratory, contain blood or blood components that serve as essential nutrients for fastidious microorganisms. X factor (heme) and V factor (nicotinamide adenine dinucleotide) often are supplied by adding hemoglobin (BBL and Difco, BD Diagnostic Systems), Iso-VitaleX (BD Diagnostic Systems), and Supplement VX (Difco, BD Diagnostic Systems).

Below is a list of some of the growth factors that are used as ingredients in microbiological media.

Bacto Casamino Acids

A mixture of amino acids formed by acid hydrolysis of casein. Available from Difco, BD Diagnostic Systems.

Bacto Vitamin Free Casamino Acids

A mixture of amino acids formed by acid hydrolysis of casein that is free of vitamins. Available from Difco, BD Diagnostic Systems.

Bovine Albumin

Bovine albumin fraction V 0.2% in 0.85% saline solution. Available from BD Diagnostic Systems.

Bovine Blood, Citrated

Calf blood washed and treated with sodium citrate as an anticoagulant. Available from BD Diagnostic Systems.

Bovine Blood, Defibrinated

Calf blood treated to denature fibrinogen without causing cell lysis. Available from BD Diagnostic Systems.

*Campylobacter* Growth Supplement

Sodium pyruvate, sodium metabisulfite, and $FeSO_4$.

Castenholtz Salts

Agar, $NaNO_3$, $Na_2HPO_4$, $KNO_3$, nitrilotriacetic acid, $MgSO_4 \cdot 7H_2O$, $CaSO_4 \cdot 2H_2O$, NaCl, $FeCl_3$, $MnSO_4$, $H_3BO_3$, $ZnSO_4$, $CoCl_2 \cdot 6H_2O$, $Na_2MoO_4$, $CuSO_4$, and $H_2SO_4$.

CVA Enrichment

Glucose, L-cysteine·HCl·$H_2O$, vitamin $B_{12}$, L-glutamine, L-cystine·2HCl, adenine, nicotinamide adenine dinucleotide, cocarboxylase, guanine·HCl, $Fe(NO_3)_3$, *p*-aminobenzoic acid, and thiamine·HCl.

Cysteine Sulfide Reducing Agent

L-Cysteine·HCl·$H_2O$ and $Na_2S \cdot 9H_2O$.

Dubos Medium Albumin

Albumin fraction V, glucose, and saline solution. Available from Difco, BD Diagnostic Systems.

Dubos Oleic Albumin Complex

Alkalinized oleic acid, albumin fraction V, and saline solution. Available from Difco, BD Diagnostic Systems.

Egg Yolk Emulsion

Chicken egg yolks and whole chicken egg. Available from Difco, BD Diagnostic Systems, and Oxoid Unipath.

Egg Yolk Emulsion, 50%

Chicken egg yolks, whole chicken egg, and saline solution. Available from Difco, BD Diagnostic Systems.

EY Tellurite Enrichment

Egg yolk suspension with potassium tellurite. Available from Difco, BD Diagnostic Systems, and Oxoid Unipath.

Fresh Yeast Extract Solution

Live, pressed, starch-free, hydrolyzed Baker's yeast.

Fildes Enrichment

A peptic digest of sheep or horse blood that is a rich source of growth factors including hemin and nicotinamide adenine dinucleotide. Available from BD Diagnostic Systems and Oxoid Unipath.

Hemin Solution

Hemin and NaOH.

Hemoglobin

Dried bovine hemoglobin. Used to provide hemin required by many fastidious microorganisms. Available from BD Diagnostic Systems.

Hemoglobin Solution 2%

Provides hemin required by many fastidious microorganisms. Available from BD Diagnostic Systems.

Hoagland Trace Element Solution, Modified

$H_3BO_3$, $MnCl_2 \cdot 4H_2O$, $AlCl_3$, $CoCl_2$, $CuCl_2$, KI, $NiCl_2$, $ZnCl_2$, $BaCl_2$, $Na_2MoO_4$, $SeCl_4$, $SnCl_2 \cdot 2H_2O$, $NaVO_3 \cdot H_2O$, KBr, and LiCl.

Horse Blood, Citrated

Horse blood washed and treated with sodium citrate used as an anticoagulant. Available from BD Diagnostic Systems.

Horse Blood, Defibrinated

Horse blood treated to denature fibrinogen without causing cell lysis. Available from BD Diagnostic Systems and Oxoid Unipath.

Horse Blood, Hemolysed

Horse blood treated to lyse cells. Available from Oxoid Unipath.

Horse Blood, Oxalated

Horse blood treated with potassium oxalate as an anticoagulant. Available from Oxoid Unipath.

Horse Serum

Horse blood is allowed to clot at 2°–8°C so that the serum separates; the serum is filter sterilized. Serum usually is inactivated by heating to 56°C for 30 min to eliminate lipases that would cause degradation of lipids and inactivation of complement. Available from Difco, BD Diagnostic Systems, and Oxoid Unipath.

IsoVitaleX® Enrichment

Glucose, L-cysteine·HCl, L-glutamine, L-cystine, adenine, nicotinamide adenine dinucleotide, vitamin $B_{12}$, thiamine pyrophosphate, guanine·HCl, $Fe(NO_3)_3 \cdot 6H_2O$, *p*-aminobenzoic acid, and thiamine·HCl. Available from BD Diagnostic Systems.

*Legionella* Agar Enrichment

L-Cysteine and ferric pyrophosphate. Available from Difco, BD Diagnostic Systems.

*Legionella* BCYE Growth Supplement

ACES buffer/KOH, ferric pyrophosphate, L-cysteine-HCl, and α-ketoglutarate. For the enrichment of *Legionella* species. Available from Oxoid Unipath.

*Leptospira* Enrichment

Lyophilized pooled rabbit serum containing hemoglobin that provides long chain fatty acids and B vitamins for growth of *Leptospira* species. Available from Diagnostic Systems.

Middlebrook ADC Enrichment

NaCl, bovine albumin fraction V, glucose, and catalase. The albumin binds free fatty acids that may be toxic to mycobacteria. Available from BD Diagnostic Systems.

Middlebrook OADC Enrichment

NaCl, bovine albumin, glucose, oleic acid, and catalase. The albumin binds free fatty acids that may be toxic to mycobacteria; the enrichment provides oleic acid used by *Mycobacterium tuberculosis* for growth. Available from BD Diagnostic Systems.

*Mycoplasma* Enrichment without Penicillin

Horse serum, fresh autolysate of yeast—yeast extract, and thallium acetate. Provides cholesterol and nucleic acids for growth of *Mycoplasma* species. The thallium selectively inhibits other microorganisms. Available from BD Diagnostic Systems.

*Mycoplasma* Supplement

Yeast extract and horse serum. Available from Difco, BD Diagnostic Systems.

Nitsch's Trace Elements

$MnSO_4$, $H_3BO_3$, $ZnSO_4$, $Na_2MoO_4$, $CuSO_4$, $CoCl_2 \cdot 6H_2O$, and $H_2SO_4$.

Oleic Albumin Complex

NaCl, bovine albumin fraction V, and oleic acid. The albumin binds free fatty acids that may be toxic to mycobacteria and the enrichment provides oleic acid that is used by *Mycobacte-*

*rium tuberculosis* for growth. Available from BD Diagnostic Systems.

PPLO Serum Fraction

Serum fraction A. Available from Difco, BD Diagnostic Systems.

Rabbit Blood, Citrated

Rabbit blood washed and treated with sodium citrate as an anticoagulant. Available from BD Diagnostic Systems.

Rabbit Blood, Defibrinated

Rabbit blood treated to denature fibrinogen without causing cell lysis. Available from BD Diagnostic Systems.

RPF Supplement

Fibrinogen, rabbit plasma, trypsin inhibitor, and potassium tellurite. For the selection and nutrient supplementation of *Staphylococcus aureus*. Available from Oxoid Unipath.

Sheep Blood, Citrated

Sheep blood washed and treated with sodium citrate as an anticoagulant. Available from BD Diagnostic Systems.

Sheep Blood, Defibrinated

Sheep blood treated to denature fibrinogen without causing cell lysis. Available from Oxoid Unipath.

SLA Trace Elements

$FeCl_2 \cdot 4H_2O$, $H_3BO_3$, $CoCl_2 \cdot 6H_2O$, $ZnCl_2$, $MnCl_2 \cdot 4H_2O$, $NiCl_2 \cdot 6H_2O$, $CuCl_2 \cdot 2H_2O$, $Na_2MoO_4 \cdot 2H_2O$, and $Na_2SeO_3 \cdot 5H_2O$.

Soil Extract

African Violet soil and $Na_2CO_3$.

Supplement A

Yeast concentrate with Crystal Violet. Available from Difco, BD Diagnostic Systems.

Supplement B

Yeast concentrate, glutamine, coenzyme, cocarboxylase, hematin, and growth factors. Available from Difco, BD Diagnostic Systems.

Supplement C

Yeast concentrate. Available from Difco, BD Diagnostic Systems.

Supplement VX

Essential growth factors V and X. Available from Difco, BD Diagnostic Systems.

VA Vitamin Solution

Nicotinamide, thiamine·HCl, *p*-aminobenzoic acid, biotin, calcium pantothenate, pyridoxine·2HCl, and cyanocobalamin.

Vitamin $K_1$ Solution

Vitamin $K_1$ and ethanol.

Vitox Supplement

Glucose, L-cysteine·HCl, L-glutamine, L-cystine, adenine sulfate, nicotinamide adenine dinucleotide, cocarboxylase, guanine·HCl, $Fe(NO_3)_3 \cdot 6H_2O$, *p*-aminobenzoic acid, vitamin $B_{12}$, and thiamine·HCl. Available from Oxoid Unipath.

Yeast Autolysate Growth Supplement

Yeast autolysate fractions, glucose, and $NaHCO_3$. Available from Oxoid Unipath.

Yeast Extract

A water soluble extract of autolyzed yeast cells. Available from BD Diagnostic Systems, and Oxoid Unipath.

Yeastolate

A water soluble fraction of autolyzed yeast cells rich in vitamin B complex. Available from Difco, BD Diagnostic Systems.

## Selective Components

Below is a list of selective components that are used to inhibit the growth of nontarget microorganisms and favor the growth of specific organisms. Selective media are especially useful in the isolation of specific microorganisms from mixed populations.

Ampicillin Selective Supplement

Ampicillin. Used in media for the selection of *Aeromonas hydrophila*. Available from Oxoid Unipath.

Anaerobe Selective Supplement GN

Hemin, menadione, sodium succinate, nalidixic acid, and vancomycin. For the selection of Gram-negative anaerobes. Available from Oxoid Unipath.

Anaerobe Selective Supplement NS

Hemin, menadione, sodium pyruvate, and nalidixic acid. For the selection of nonsporulating anaerobes. Available from Oxoid Unipath.

*Bacillus cereus* Selective Supplement

Polymyxin B. For the selection of *Bacillus cereus*. Available from Oxoid Unipath.

*Bordetella* Selective Supplement

Cephalexin. For the selection of *Bordetella* species. Available from Oxoid Unipath.

*Brucella* Selective Supplement

Polymyxin B, bacitracin, cycloheximide, nalidixic acid, nystatin, and vancomycin. For the selection of *Brucella* species. Available from Oxoid Unipath.

*Campylobacter* Selective Supplement Blaser-Wang

Vancomycin, polymyxin B, trimethoprim, amphotericin B, and cephalothin. For the selection of *Campylobacter* species. Available from Oxoid Unipath.

*Campylobacter* Selective Supplement Butzler

Bacitracin, cycloheximide, colistin sulfate, sodium cephazolin, and novobiocin. For the selection of *Campylobacter* species. Available from Oxoid Unipath.

*Campylobacter* Selective Supplement Preston

Polymyxin B, rifampicin, trimethoprim, and cycloheximide. For the selection of *Campylobacter* species. Available from Oxoid Unipath.

*Campylobacter* Selective Supplement Skirrow

Vancomycin, trimethoprim, and polymyxin B. For the selection of *Campylobacter* species. Available from Oxoid Unipath.

CCDA Selective Supplement

Cefoperazone and amphotericin B. For the selection of *Campylobacter* species. Available from Oxoid Unipath.

Cefoperazone Selective Supplement

Cefoperazone. For the selection of *Campylobacter* species. Available from Oxoid Unipath.

CFC Selective Supplement

Cetrimide, fucidin, and cephaloridine. For the selection of pseudomonads. Available from Oxoid Unipath.

Chapman Tellurite Solution

Potassium tellurite 1% solution. Available from Difco, BD Diagnostic Systems.

Chloramphenicol Selective Supplement

Chloramphenicol. For the selection of yeasts and filamentous fungi. Available from Oxoid Unipath.

*Clostridium difficile* Selective Supplement

D-Cycloserine and cefoxitin. For the selection of *Clostridium difficile*. Available from Difco, BD Diagnostic Systems, and Oxoid Unipath.

CN Inhibitor

Cesulodin and novobiocin. It inhibits enteric Gram-negative microorganisms. Available from BD Diagnostic Systems.

CNV Antimicrobic

Colistin sulfate, nystatin, and vancomycin. Available from Difco, BD Diagnostic Systems.

CNVT Antimicrobic

Colistin sulfate, nystatin, vancomycin, and trimethoprim lactate. Available from Difco, BD Diagnostic Systems.

Colbeck's Egg Broth

Egg emulsion and saline solution. Formerly available from Difco, BD Diagnostic Systems—replaced with egg emulsion.

Fraser Supplement

Ferric ammonium sulfate, nalidixic acid, and Acriflavin hydrochloride. For the selection of *Listeria* species. Available from Oxoid Unipath.

*Gardnerella vaginalis* Selective Supplement

Gentamicin sulfate, nalidixic acid, and amphotericin B. For the selection of *Gardnerella vaginalis*. Available from Oxoid Unipath.

GC Selective Supplement

Yeast autolysate, glucose, $Na_2HCO_3$, vancomycin, colistin methane sulfonate, nystatin, and trimethoprim. For the selection of *Neisseria* species. Available from Oxoid Unipath.

*Helicobacter pylori* Selective Supplement Dent

Vancomycin, trimethoprim, cefulodin, and amphotericin B. For the selection of *Helicobacter pylori*.

Kanamycin Sulfate Selective Supplement

Kanamycin sulfate. For the selection of enterococci. Available from Oxoid Unipath.

LCAT Selective Supplement

Lincomycin, colistin sulfate, amphotericin B, and trimethoprim. For the selection of *Neisseria* species. Available from Oxoid Unipath.

*Legionella* BMPA Selective Supplement

Cefamandole, polymyxin B, and anisomycin. For the selection of *Legionella* species. Available from Oxoid Unipath.

*Legionella* GVPC Selective Supplement

Glycine, vancomycin hydrochloride, polymixin B sulfate, and cycloheximide. For the selection of *Legionella* species. Available from Oxoid Unipath.

*Legionella* MWY Selective Supplement

Glycine, polymyxin B, anisomycin, vancomycin, Bromthymol B, and Bromcresol Purple. For the selection of *Legionella* species. Available from Oxoid Unipath.

*Listeria* Primary Selective Enrichment Supplement

Nalidixic acid and acriflavin. For the selection of *Listeria* species. Available from Oxoid Unipath.

*Listeria* Selective Enrichment Supplement

Nalidixic acid, cycloheximide, and acriflavin. For the selection of *Listeria* species. Available from Oxoid Unipath.

*Listeria* Selective Supplement MOX

Colistin and moxalactam. For the selection of *Listeria monocytogenes*. Available from Oxoid Unipath.

*Listeria* Selective Supplement Oxford

Cycloheximide, colistin sulfate, acriflavin, cefotetan, and fosfomycin. For the selection of *Listeria* species. Available from Oxoid Unipath.

Modified Oxford Antimicrobic Supplement

Moxalactam and colistin sulfate. Available from Difco, BD Diagnostic Systems.

MSRV Selective Supplement

Novobiocin. For the selection of *Salmonella*. Available from Oxoid Unipath.

*Mycoplasma* Supplement G

Horse serum, yeast extract, thallous acetate, and penicillin. For the selection of *Mycoplasma* species. Available from Oxoid Unipath.

*Mycoplasma* Supplement P

Horse serum, yeast extract, thallous acetate, glucose, Phenol Red, Methylene Blue, penicillin, and *Mycoplasma* broth base. For the selection of *Mycoplasma* species. Available from Oxoid Unipath.

*Mycoplasma* Supplement S

Yeast extract, horse serum, thallium acetate, and penicillin. Available from Difco, BD Diagnostic Systems.

Oxford Antimicrobic Supplement

Cycloheximide, colistin sulfate, acriflavin, cefotetan, and fosfomycin. Available from Difco, BD Diagnostic Systems.

Oxgall

Dehydrated fresh bile. For the selection of bile tolerant bacteria. Available from Difco, BD Diagnostic Systems.

Oxytetracycline GYE Supplement

Oxytetracycline in a buffer. For the selection of yeasts and filamentous fungi. Available from Oxoid Unipath.

PALCAM Selective Supplement

Polymyxin B, acriflavin hydrochloride, and ceftazidime. For the selection of *Listeria monocytogenes*. Available from Oxoid Unipath.

Perfringens OPSP Selective Supplement A

Sodium sulfadiazine. For the selection of *Clostridium perfringens*. Available from Oxoid Unipath.

Perfringens SFP Selective Supplement A

Kanamycin sulfate and polymyxin B. For the selection of *Clostridium perfringens*. Available from Oxoid Unipath.

Perfringens TSC Selective Supplement A

D-Cycloserine. For the selection of *Clostridium perfringens*. Available from Oxoid Unipath.

Sodium Desoxycholate

Sodium salt of desoxycholic acid. Available from Difco, BD Diagnostic Systems.

Sodium Taurocholate

Sodium salt of conjugated bile acid—75% sodium taurocholate and 25% bile salts. For the selection of bile tolerant bacteria. Available from Difco, BD Diagnostic Systems.

STAA Selective Supplement

Streptomycin sulfate, cycloheximide, and thallous acetate. For the selection of *Brochothrix thermosphacta*. Available from Oxoid Unipath.

Staph/Strep Selective Supplement

Nalidixic acid and colistin sulfate. For the selection of *Staphylococcus* species and *Streptococcus* species. Available from Oxoid Unipath.

*Streptococcus* Selective Supplement COA

Colistin sulfate and oxolinic acid. For the selection of *Streptococcus* species. Available from Oxoid Unipath.

Sulfamandelate Supplement

Sodium sulfacetamide and sodium mandelate. For the selection of *Salmonella* species. Available from Oxoid Unipath.

Tellurite Solution

A solution containing potassium tellurite. Inhibits Gram-negative and most Gram-positive microorganisms. It is used for the isolation of *Corynebacterium* species, *Streptococcus* species, *Listeria* species, and *Candida albicans*. Available from BD Diagnostic Systems.

Tinsdale Supplement

Serum, potassium tellurite, and sodium thiosulfate. For the selection of *Corynebacterium diphtheriae*. Available from Oxoid Unipath.

V C A Inhibitor

Vancomycin, colistin, anisomycin, and trimethoprim. Inhibits most Gram-negative and Gram-positive bacteria and yeasts. It is used for the isolation of *Neisseria* species. Available from BD Diagnostic Systems.

V C A T Inhibitor

Vancomycin, colistin, anisomycin, and trimethoprim lactate. Inhibits most Gram-negative and Gram-positive bacteria and yeasts. It is used for the isolation of *Neisseria* species. Available from BD Diagnostic Systems and Oxoid Unipath.

V C N Inhibitor

Colistin, vancomycin, and nystatin. Inhibits most Gram-negative and Gram-positive bacteria and yeasts. It is used for the isolation of *Neisseria* species. Available from BD Diagnostic Systems and Oxoid Unipath.

V C N T Inhibitor

Colistin, vancomycin, nystatin, and trimethoprim lactate. Inhibits most Gram-negative and Gram-positive bacteria and yeasts. It is used for the isolation of *Neisseria* species. Available from BD Diagnostic Systems and Oxoid Unipath.

*Yersinia* Selective Supplement

Cefsulodin, irgasan, and novobiocin. For the selection of *Yersinia enterocolitica*. Available from Oxoid Unipath.

## pH Buffers

The pH adjusted using phosphate buffers is established by using varying volumes of equimolar concentrations of $Na_2HPO_4$ and $NaH_2PO_4$.

| pH | $Na_2HPO_4$ (mL) | $NaH_2PO_4$ (mL) |
|---|---|---|
| 5.4 | 3.0 | 97.0 |
| 5.6 | 5.0 | 95.0 |
| 5.8 | 7.8 | 92.2 |
| 6.0 | 12.0 | 88.0 |
| 6.2 | 18.5 | 81.5 |
| 6.4 | 26.5 | 73.5 |
| 6.6 | 37.5 | 62.5 |
| 6.8 | 50.0 | 50.0 |
| 7.0 | 61.1 | 38.9 |
| 7.2 | 71.5 | 28.5 |

| pH | $Na_2HPO_4$ (mL) | $NaH_2PO_4$ (mL) |
|---|---|---|
| 7.4 | 80.4 | 19.6 |
| 7.6 | 86.8 | 13.2 |
| 7.8 | 91.4 | 8.6 |
| 8.0 | 94.5 | 5.5 |

## Differential Components

Below is a list of some commonly used pH indicators and their color reactions.

| pH Indicator | pH Range | Acid Color | Alkaline Color |
|---|---|---|---|
| *m*-Cresol Purple | 0.5–2.5 | Red | Yellow |
| Thymol Blue | 1.2–2.8 | Red | Yellow |
| Bromphenol Blue | 3.0–4.6 | Yellow | Blue |
| Bromcresol Green | 3.8–5.4 | Yellow | Blue |
| Chlorcresol Green | 4.0–5.6 | Yellow | Blue |
| Methyl Red | 4.2–6.3 | Red | Yellow |
| Chlorphenol Red | 5.0–6.6 | Yellow | Red |
| Bromcresol Purple | 5.2–6.8 | Yellow | Purple |
| Bromthymol Blue | 6.0–7.6 | Yellow | Blue |
| Phenol Red | 6.8–8.4 | Yellow | Red |
| Cresol Red | 7.2–8.8 | Yellow | Red |
| *m*-Cresol Purple | 7.4–9.0 | Yellow | Purple |
| Thymol Blue | 8.0–9.6 | Yellow | Blue |
| Cresolphthalein | 8.2–9.8 | Colorless | Red |
| Phenolphthalein | 8.3–10.0 | Colorless | Red |

## Preparation of Media

The ingredients in a medium are usually dissolved and the medium is then sterilized. When agar is used as a solidifying agent the medium must be heated gently, usually to boiling, to dissolve the agar. In some cases where interactions of components, such as metals, would cause precipitates, solutions must be prepared and occasionally sterilized separately before mixing the various solutions to prepare the complete medium. The pH often is adjusted prior to sterilization, but in some cases sterile acid or base is used to adjust the pH of the medium following sterilization. Many media are sterilized by exposure to elevated temperatures. The most common method is to autoclave the medium. Different sterilization procedures are employed when heat-labile compounds are included in the formulation of the medium.

## Autoclaving

Autoclaving uses exposure to steam, generally under pressure, to kill microorganisms. Exposure for 15 min to steam at 15 psi—121°C is most commonly used. Such exposure kills vegetative bacterial cells and bacterial endospores. However, some substances do not tolerate such exposures and lower temperatures and different exposure times are sometimes employed. Media containing carbohydrates often are sterilized at 116°–118°C in order to prevent the decomposition of the carbohydrate and the formation of toxic compounds that would inhibit microbial growth. Below is a list of pressure–temperature relationships.

| Pressure—psi | Temperature—°C |
|---|---|
| 0 | 100 |
| 1 | 101.9 |
| 2 | 103.6 |
| 3 | 105.3 |

| Pressure—psi | Temperature—°C |
|---|---|
| 4 | 106.9 |
| 5 | 108.4 |
| 6 | 109.8 |
| 7 | 111.3 |
| 8 | 112.6 |
| 9 | 113.9 |
| 10 | 115.2 |
| 11 | 116.4 |
| 12 | 117.6 |
| 13 | 118.8 |
| 14 | 119.9 |
| 15 | 121.0 |
| 16 | 122.0 |
| 17 | 123.0 |
| 18 | 124.0 |
| 19 | 125.0 |
| 20 | 126.0 |
| 21 | 126.9 |
| 22 | 127.8 |
| 23 | 128.7 |
| 24 | 129.6 |
| 25 | 130.4 |

## Tyndallization

Exposure to steam at 100°C for 30 min will kill vegetative bacterial cells but not endospores. Such exposure can be achieved using flowing steam in an Arnold sterilizer. By allowing the medium to cool and incubate under conditions where endospore germination will occur and by repeating the 100°C–30 min exposure on three successive days the medium can be sterilized because all the endospores will have germinated and the heat exposure will have killed all the vegetative cells. This process of repetitive exposure to 100°C is called tyndallization, after its discoverer John Tyndall.

## Inspissation

Inspissation is a heat exposure method that is employed with high protein materials, such as egg-containing media, that cannot withstand the high temperatures used in autoclaving. This process causes coagulation of the protein without greatly altering its chemical properties. Several different protocols can be followed for inspissation. Using an Arnold sterilizer or a specialized inspissator, the medium is exposed to 75°–80°C for 2 h on each of three successive days. Inspissation using an autoclave employs exposure to 85°–90°C for 10 min achieved by having a mixture of air and steam in the chamber, followed by 15 min exposure during which the temperature is raised to 121°C using only steam under pressure in the chamber; the temperature then is slowly lowered to less than 60°C.

## Filtration

Filtration is commonly used to sterilize media containing heat-labile compounds. Liquid media are passed through sintered glass or membranes, typically made of cellulose acetate or nitrocellulose, with small pore sizes. A membrane with a pore size of 0.2mm will trap bacterial cells and, therefore, sometimes is called a bacteriological filter. By preventing the passage of microorganisms, filtration renders fluids free of bacteria and eukaryotic microorganisms, that is, free of living organisms, and hence sterile. Many carbohydrate solutions, antibiotic solutions, and vitamin solutions are filter sterilized and added to media that have been cooled to temperatures below 50°C.

## Caution about Hazardous Components

Some media contain components that are toxic or carcinogenic. Appropriate safety precautions must be taken when using media with such components. Basic fuchsin and acid fuchsin are carcinogens and caution must be used in handling media with these compounds to avoid dangerous exposure that could lead to the development of malignancies. Thallium salts, sodium azide, sodium biselenite, and cyanide are among the toxic components found in some media. These compounds are poisonous and steps must be taken to avoid ingestion, inhalation, and skin contact. Azides also react with many metals, especially copper, to form explosive metal azides. The disposal of azides must avoid contact with copper or achieve sufficient dilution to avoid the formation of such hazardous explosive compounds. Media with sulfur-containing compounds may result in the formation of hydrogen sulfide which is a toxic gas. Care must be used to ensure proper ventilation. Media with human blood or human blood components must be handled with great caution to avoid exposure to human immunodeficiency virus and other pathogens that contaminate some blood supplies. Proper handling and disposal procedures must be followed with blood-containing as well as other media that are used to cultivate microorganisms.

T - #0248 - 071024 - C0 - 279/216/31 - PB - 9780367379315 - Gloss Lamination